Engineering Geology Practice in Southern California

Edited by

Bernard W. Pipkin and Richard J. Proctor

Special Publication No. 4
ASSOCIATION OF ENGINEERING GEOLOGISTS
Southern California Section

Star
PUBLISHING COMPANY

This Special Publication was prepared by the Southern California Section of the Association of Engineering Geologists and reference to any commercial product, process, or service by trade name, trademark, manufacturer, or otherwise does not imply endorsement, recommendation or favoring by the Association of Engineering Geologists.

These articles are reprinted from material submitted by the contributing authors, with editorial review by Bernard Pipkin and Richard Proctor for SC-AEG. Photographs, illustrations, charts, tables and maps are reproduced from material submitted by the contributing authors.

Special Publication No. 4
ASSOCIATION OF ENGINEERING GEOLOGISTS
Southern California Section

Published and distributed by:

Star Publishing Company
940 Emmett Avenue
Belmont, California 94002
U.S.A.

Printed in the United States of America

99 98 97 96 95 94 93

ISBN 0-89863-171-8

0 9 8 7 6 5 4 3 2 1

CONTENTS

EDITORS PREFACE

This book by the Southern California Section of the Association of Engineering Geologists is a collection of papers illustrating the state-of-the-practice for engineering geology in the geographic region that contains more engineering geologists than any place in the world.

We offer this volume in conjunction with the 1992 AEG Annual Meeting in Long Beach. It is mainly composed of original papers, but also contains several reprints of classic case histories and geotechnical papers from less circulated sources. A review of the Table of Contents will reveal the diverse and comprehensive coverage. We have emphasized case histories because they illustrate what actually happened and how geologists, engineers, politicians, and planners have coped with each situation. For example, the 1928 St. Francis Dam disaster, usually cited as the real beginning of engineering geology, is discussed by J. David Rogers in a critical re-examination of its mode of failure, including previously unpublished drawings.

We have grouped the 60 papers into ten chapters. The chapter entitled Environmental Issues includes papers in hazardous waste and hydrocarbon contamination and cleanup (A. Hatheway; S. Testa)—new specialties in engineering geology. Review or overview articles open each chapter, such as Geologic Hazards (E. Gath), Seismicity (Hauksson), Groundwater (R. Bean & G. Brown), Foundation Problems (R. Hollingsworth and D. Grover), Landslides (L. Cann and E. Steiner), Beach Erosion (B. Pipkin, H. Robertson and R. Mills), and Dams and Tunnels (R. Proctor).

Some papers update the reader on old problems, such as the Portuguese Bend landslide (P. Ehlig) and sea-water intrusion (M. Johnson), whereas others provide state-of-the-practice techniques (J. Birman; R. Larson and J. Slosson; R. Ramirez; R. Hollingsworth and D. Grover). Two papers focus on the all-important legal pitfalls of the consulting profession (J. Byer; J. Patton). One paper describes the largest legal settlement for a landslide in history—$92 million (D. Evans and R. Lownes). There are several new papers on seismicity of the L. A. Basin and on the activity of several well-known faults in southern California, including two papers on the San Andreas fault (K. Sieh; D. Padgett). Geotechnical aspects of two of the largest recent tunnel projects are described—the Metro Rail and the NORS sewer tunnel beneath Los Angeles International Airport.

We thank the contributing authors, only through the generous sharing of their knowledge have we been able to assemble this special publication.

Bernard W. Pipkin
Co-editor
Palos Verdes, CA

Richard J. Proctor
Co-editor
Arcadia, CA

"Every work of man is built on, in, or with the earth, except those things that fly, float, or fall down, and these last three must start or end with some earth contact."

— P. S. Meallie, Geotechnical Board, National Research Council

CONTRIBUTORS

Company, Agency or Independent Consultant (IC)

	Chapter
Applied Environmental Services (Testa), Laguna Hills	2
Bakhtar Associates (Bakhtar), Newport Beach	10
Robert T. Bean, (IC), La Crescenta	5
J. Byer Group (Byer), Glendale	4
California Department of Water Resources (Stankov), Palmdale	9
California GeoSystems (Ramirez), Glendale	8
California Institute of Technology (Hauksson, Hutton, Douglass, Sieh), Pasadena	3
California State Polytechnic University (Herber), Pomona	8
California State University (Foster), Fullerton	3
California State University (Ehlig, Stout), Los Angeles	8
Converse Consultants West (Padgett, Santo, Schluter), Pasadena	3
Richard Crook, Jr. (IC), Sierra Madre	3
Dames & Moore (Stirbys), Los Angeles	10
Paul Davis (IC), Huntington Beach	3
Earth Logics Consultants (Audell), San Juan Capistrano	4
Earth Sciences Associates (D. Hamilton), Palo Alto	2
Earth Technology (Guptill), Irvine	3
Emcon (Farrell), Burbank	10
GeoTech Imagery (Higdon), Oceanside	8
Geotechnical Consultants (Hood), Glendale	6
Grover-Hollingsworth and Associates (Grover, Hollingsworth), Westlake Village	6
GSI/Water (Birman, Hibner), South Pasadena	5,8
Hayward Baker (Gularte), Ventura	6
Edward G. Heath (IC), Newport Beach	3
Lawrence B. James (IC), retired, Folsom	9
Daniel C. Kalin (IC), Yorba Linda	10
Donald L. Lamar (IC), Eugene, Oregon	3
Law/Crandall (Escandon, Latiolait, Hammock), Los Angeles	10
Law Environmental (G. Brown), Burbank	2,5
Leighton and Associates (Cann, Steiner), Irvine	8
Leighton and Associates (Drumm), Woodland Hills	3
Leighton and Associates (Gath), Diamond Bar	1
Los Angeles City Department of Building & Safety (Cobarrubias)	2
Los Angeles City Department of Water & Power (Mayeda, Wilson)	9,10
Los Angeles County Dept. of Public Works (Larson), Alhambra	3,8
Los Angeles County Dept. of Public Works (Johnson), Alhambra	5,9
Lownes Geologic Services (Lownes), Pasadena	8
Richard L. Meehan (IC), Palo Alto	2
Paul M. Merifield (IC), Los Angeles	3,8
Metropolitan Water District (Proctor, Payne, Kalin), Los Angeles	9,10
Moore and Tabor (Eagan, Tims, Nonamaker), Anaheim	6
Parsons-Brinkerhoff (Daugherty), Dallas, Texas	10
City of Pasadena (Peugh)	9

James H. Patton, Jr., Attorney, Agoura Hills	4
C. Marshall Payne (IC), Dana Point	10
Bernard W. Pipkin (IC), Palos Verdes	7
Richard J. Proctor, Inc. (IC), Arcadia	3,9,10
City of Rancho Palos Verdes (Ehlig)	8
Robertson Geotechnical (Robertson, Mills), Westlake Village	7
Rogers/Pacific (Rogers, Scullin), Pleasant Hill	4,8,9
Shepardson Engineering (Evans), Carlsbad	8
Roy J. Shlemon, Inc. (IC), Newport Beach	3
Slosson and Associates (Slosson), Van Nuys	3
Southland Geotechnical (Custenborder), El Cajon	6
Southern California Edison Company (P. Hamilton), Rosemead	9,10
Steve Stankov (IC), Laguna Niguel	9
Thomas F. Thompson (IC), deceased	10
U. S. Bureau of Reclamation (Trefzger), Sacramento	10
U. S. Geological Survey (L. Jones), Pasadena	3
University of California (Brillinger), Berkeley	3
University of California (Merifield), Los Angeles	3,8
University of Missouri (Hatheway), Rolla	2
University of Southern California (Pipkin, Osborne, Yeh), Los Angeles	7
University of Washington (Stuiver), Seattle	3
Karl VonderLinden (IC), Palo Alto	3
Woodward-Clyde Consultants (Freeman, Waggoner), Santa Ana	3

Chapter 1

Overview

Photo of numerous shallow debris flows along Llano Drive in Woodland Hills following heavy rains in 1980. The record-setting rains here on February 10, 1991 (8 inches in 6 hours) caused similar problems.

Overview

Geologic Hazards and Hazard Mitigation in the Los Angeles Region
Eldon Gath

Editors Note: The opening chapter identifies and discusses the various geologic hazards in southern California and methods for their mitigation. Throughout Eldon's paper we have inserted the names of other contributors within this book who discuss in more depth specific problems or case histories.

GEOLOGIC HAZARDS AND HAZARD MITIGATION IN THE LOS ANGELES REGION

Eldon Gath
Principal Geologist
Leighton and Associates, Inc.
1470 S. Valley Vista Drive, Suite 150
Diamond Bar, California 91765

CONTENTS

"It is the duty of investigators - seismologists, geologists, and scientific engineers - develop the theory of local danger spots, to discover the foci of recurrent shocks, to develop the theory of earthquake-proof construction. It is the duty of engineers and architects so to adjust construction to the sound character of the ground that safety shall be secured. It should be the policy of communities in the earthquake district to recognize the danger and make provisions against it."

– G.K. Gilbert (1909)

INTRODUCTION

The presence of geologic hazards is an accepted condition of living in the greater Los Angeles region. During the 1990s, the International Decade of Natural Disaster Reduction, the population of about 14 million is expected to grow an additional million. Most of that new growth will be into areas of increasingly severe geologic hazards. Los Angeles is, therefore, an interesting and challenging place to be a practicing engineering geologist. Engineering geologists play a crucial role in geologic hazard identification and mitigation, and will continue to play an increasingly necessary role in public awareness and education. The purpose of this paper is to introduce the diversity of the geologic hazards present, and introduce some of the public safety practices being implemented. The paper is modified from the report *"Hazard Reduction in Los Angeles County"*, the technical appendix to the Los Angeles County Safety Element (Leighton and Associates, 1990a & b).

The Los Angeles region lies at the juncture of two very active tectonic systems (Fig. 1): the San Andreas system of right-lateral strike slip faults, including the Whittier, Newport-Inglewood, Palos Verdes Hills, and several offshore borderland faults, and the east-west system of faults forming the Transverse Ranges complex of compressive folding and thrust faults, including blind thrusts, a newly emerging tectonic concept south of the Transverse Ranges (Wright, 1991; Davis, et al., 1989; Hauksson, this volume). Approximately 50 onshore active and potentially active faults are located in the Los Angeles region. Some 21 of these have evidence to suggest they are active Holocene faults (Ziony and Yerkes, 1985; Ziony and Jones, 1989; and papers in Chapter 3 in this book).

Between 1800 and mid-1991, active faults in the Los Angeles region were responsible for approximately 57 damaging earthquakes; thus the region as a whole experiences about one earthquake every four years. Since 1933, four damaging earthquakes (Long Beach, San Fernando, Whittier Narrows, and Sierra Madre) have caused 182 deaths, several thousands of injuries, and significant damage in the greater Los Angeles metropolitan area. Strong shaking, fault rupture, and liquefaction have collapsed buildings and structures on the average of once every 20 years. Earthquake recurrence cannot be reliably determined for the majority of faults in the region due to the short historic seismic record, and the few geologic in-

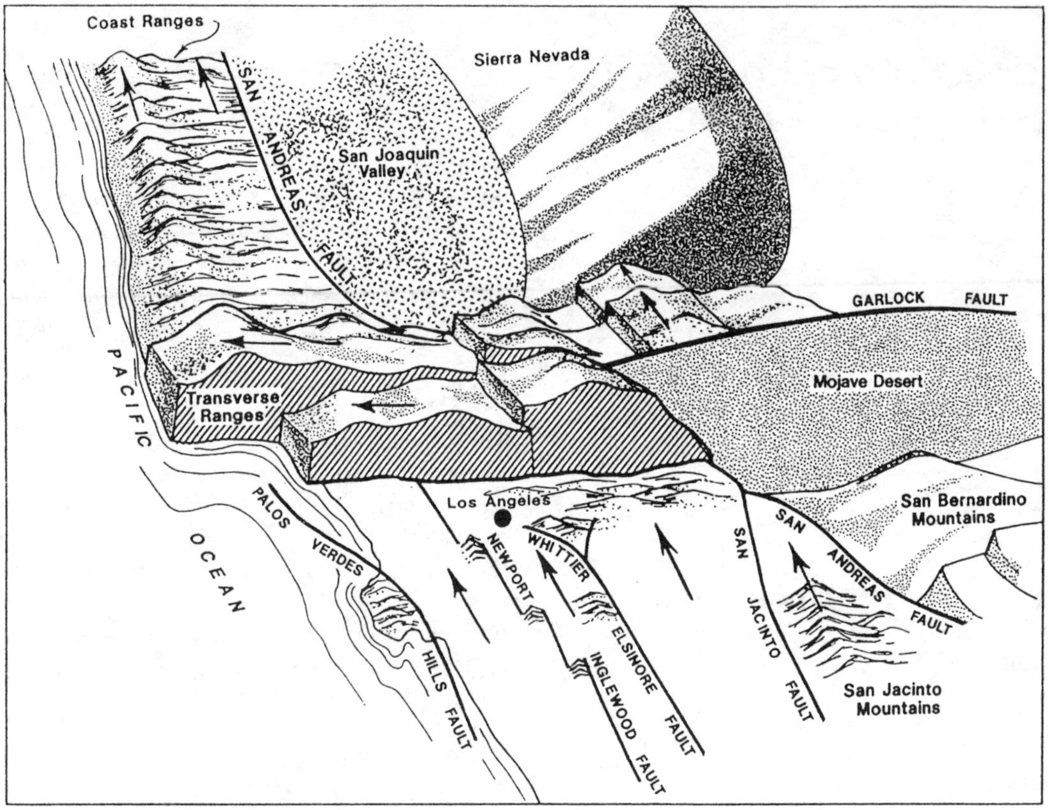

Figure 1. Illustrative block diagram of the major tectonic features and geomorphic provinces of Los Angeles County.

vestigations designed to recover paleoseismic information (Wesnousky, 1986; Ziony and Yerkes, 1985).

Liquefaction has the potential to cause enormous damage to local areas. Some of the most expensive developments in the region lie within areas potentially susceptible to liquefaction, and, particularly for low rise structures, this hazard has not been consistently addressed in preconstruction design. Areas of the Harbor, Marina and Playa del Rey, as well as southern San Fernando Valley are all susceptible to liquefaction. Other areas may also be seasonally susceptible, but our knowledge of the shallow groundwater localities in the LA Basin is incomplete. It is estimated that about 11 percent of the total area is liquefiable.

Eight thousand old masonry structures are at risk from ground shaking in the greater Los Angeles area. But, the pattern of destructive shaking cannot be predicted in detail in advance. It is most dependent on the characteristics of the earth materials through which earthquake waves pass (Tinsley and Fumal, 1985). The most cited earthquake forecast in the region is for a Magnitude 7.5-8.0 earthquake on the San Andreas fault (Working Group on California Earthquake Probabilities, 1988). The risk is estimated to be about two percent per year, or a 30 percent probability in the next 30 years (Fig. 2). A more damaging scenario is a M 7.0 on the northern segment of the Newport- Inglewood fault (Toppozada et al., 1988). According to the scenarios, the San Andreas event could cause 12,000 to 55,000 casualties requiring hospital care and economic losses near $20 billion dollars (Davis et al., 1982). The Newport-Inglewood event could cause 16,000 to 80,000 casualties requiring hospital care, and up to $35 billion dollars in economic losses. A Federal Emergency Management Agency (FEMA) study forecasts 22,000 persons permanently and 37,000 temporarily homeless in the aftermath of an 8+ Magnitude earthquake on the San Andreas fault (Spangle, 1987).

Half of the region is at risk of slope failures. Many areas have been developed, such as the Portuguese Bend and Big Rock Mesa slides (see papers in chapter 8). Landslide damage generates millions of dollars in litigation claims each year, but it is a hazard which early and thorough geologic investigations can reduce.

Next to earthquakes, flood hazards probably pose the greatest potential for a major disaster or loss of life in southern California. The winter storms of March 1938 flooded vast areas of the LA Basin, and generated thousands of mud and debris flows in the Santa Monica and San Gabriel Mountains, and the Puente Hills. There have been two catastrophic dam failures and one near failure in Los Angeles County. The county's excellent flood control system is aging, and increased urbanization is causing peak runoffs greater than the system was designed to accommodate. Mud and debris flows killed 18 persons in the winter storms of 1980 (Chin et al.,

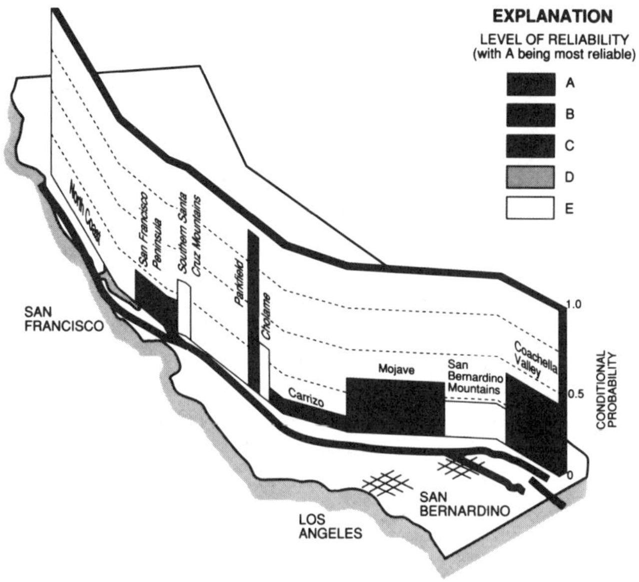

Figure 2. Conditional probability for a major earthquake occurring on the identified segments of the San Andreas fault, in a 30-year period 1988-2018. (Working Group on California Earthquake Probabilities, 1988)

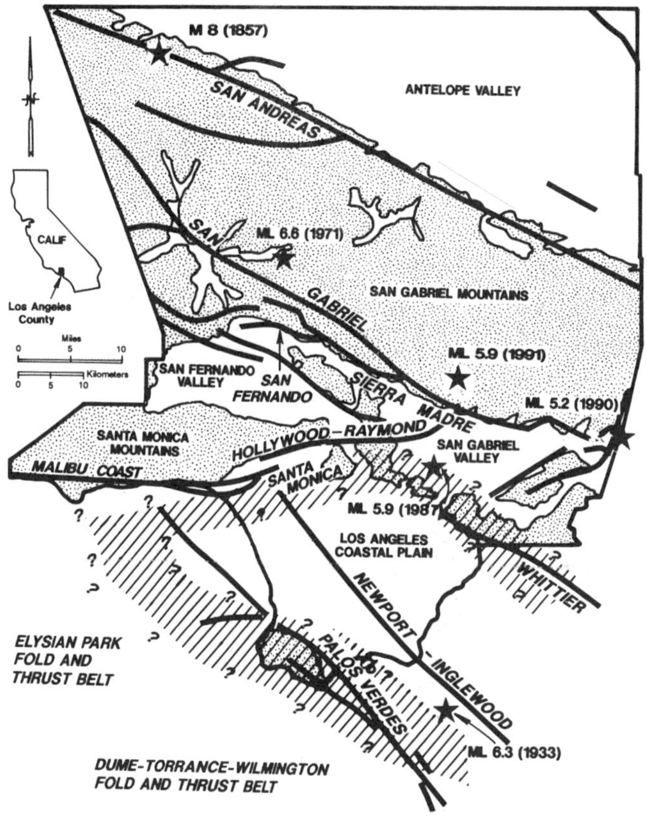

Figure 3. Location map of Los Angeles County showing physiographic provinces, selected faults and significant historic earthquakes. Fold and thrust belts shown from Hauksson (1990) represent potentially significant "blind" seismogenic sources. Figure modified from Sanchez et al. (1991).

1991). Debris retention basins are accumulating sediments resulting in skyrocketing maintenance costs.

Hazard reduction is everyone's concern. A municipal Safety Element provides a road map for long range hazard mitigation planning, but there are many other pressing demands for resources. Without a public-private cooperation and understanding, and an upwelling of public support for the financial commitment needed, future safety planning will continue to fall victim to more immediate needs. Unfortunately, disasters have a habit of creating their own immediate needs.

GEOLOGIC SETTING OF THE LOS ANGELES REGION

The geology and geomorphology of the Los Angeles region have been controlled by the evolution of the tectonic boundary between the Pacific and North American plates (Atwater and Molnar, 1973). About 20 million years ago, the plate boundary evolved from one of oblique convergence and subduction west of modern Los Angeles, to transform (lateral) motion along a series of subparallel strike slip faults of the San Andreas fault system. The oblique nature of the convergence is accommodated by north-south crustal shortening across a complex set of fold and thrust belts, resulting in the development of physiographic, and geologic "subprovinces" (Fig. 3).

Southern California is still evolving as the tectonic slip is partitioned between the northwest-trending dextral slip faulting, the east-west trending fold and thrust fault terrain, including a conjugate set of sinestral slip faults. As shown in Fig. 3, the mountain highland areas of the Transverse Ranges; the Santa Monica, Santa Susana, and San Gabriel Mountains, transect the middle of the County, effectively separating the urbanized coastal and valley lowland areas from the Antelope Valley, an elevated portion of the Mojave Desert geomorphic province. The geologic evolution of this region is discussed in this context of physiographic provinces.

Coastal and Valley Lowland Areas

The lowland areas south of the San Gabriel Mountains are part of the Los Angeles structural basin. The onset of the present tectonic setting deposited as much as 14,000 feet of marine sediments in the newly formed northwest-trending basin, with the majority of deposition occurring between 5 and 10 million years ago (Wright, 1991; Yerkes et al., 1965). Commencing about 2 to 3 million years ago, north south compression began in the Basin, likely due to the large bend in the San Andreas fault. The Santa Monica, Santa Susana, and San Gabriel Mountains, and, within the last million years, the Elysian Park, Repetto, Puente, and San Jose Hills all formed due to southward migration of the east-west trending compressive front, and the accompanying structural warping occurring below the Los Angeles Basin. The physiographically distinct mountain and valley topography partition the present day landscape into the geographic subregions of the San Gabriel Valley, the San Fernando Valley, and the Los Angeles Coastal Plain (Fig. 3).

As the folding and uplift migrates south of the Transverse Ranges and integrates with the strike slip faults within the Basin, the structural interaction becomes complex. Farther to the south are the Coyote Hills, where uplift appears to have initiated only in the last 200,000 to 300,000 years in response to partitioned slip from the Whittier fault, and the southward growth of the compressive front. The Baldwin, Dominguez, and Signal Hills began to form as recently as 100,000 years ago, in response to movement along the Newport-Inglewood fault (Barrows, 1974).

The young sedimentary rocks forming the hills are extremely susceptible to landsliding, especially because the anticlinal nature of uplift produces out-of-slope bedding attitudes on all sides of the structure. Highly expansive soils are produced by the weathering of the marine shales and silts. Mud flow scars were common following the heavy rains of 1938. Occasional volcanic ash beds within the marine sediments cause severe, but localized bedrock expansion upon wetting and oxidation.

Numerous ground water basins are present within the valley lowlands. Topographically isolated by the uplifted hills, the basins are also frequently subdivided by groundwater barriers caused by youthful faulting through the valleys. Many of the potentially active faults in the Los Angeles Basin are known (or inferred) only from their differences in groundwater elevations across them.

Mountain Highlands

The Santa Monica, Santa Susana, and San Gabriel Mountains are part of the Transverse Ranges uplift which are bounded on the south by active north-dipping thrust faults, and are cut internally by other faults of various orientations and histories of movement (Fig. 3). Debris flows accompanying heavy seasonal rains are a particular hazard of the San Gabriel Mountains due to the extensive fracturing and deep erosion of the crystalline rocks. Seismically induced rockfalls are also common, and there are a few large landslides within the crystalline core rocks of the range (Rogers et al., Chapter 8 herein; Morton and Streitz, 1969). However, landslides are most common in the sedimentary rock sequences where large bedding plane translational slides occur.

Antelope Valley

Northeast of the San Gabriel Mountains lies the Antelope Valley, which is an elevated desert between 2,300 and 3,500 feet in altitude. The valley is part of the Mojave Desert structural block bounded by the San Andreas and Garlock faults, Antelope Valley contains as much as 4,000 feet of Plio-Pleistocene alluvial fill accumulated in response to the uplift and erosion of the San Gabriel Mountains. It is characterized by broad, gently sloping alluvial plains and fans, with remnant buttes and ridges many miles from the mountain front. Flooding is a constant threat in the Antelope Valley, and collapsible soils are common in the fan and lake deposits. Ground fissuring, whether by groundwater drawdown-induced subsidence, desiccation of clays, tectonic stresses, karst- like solution, or other causes, is a newly observed geologic threat to the Antelope Valley, and one that may pose significant constraints to future development in this rapidly growing area.

Channel Islands and Offshore Area

Southwest of the mainland, the offshore continental borderland is crossed by numerous northwest-trending strike-slip faults. Two of the Channel Islands, Santa Catalina and San Clemente Islands, comprise basement rocks that have been uplifted alongside deep sedimentary basins. The basement rocks consist of pervasively sheared low and high-grade metamorphic rocks of volcanic and marine origin, and are prone to landsliding.

SEISMIC HAZARDS

The Los Angeles region occupies a junction between two major zones of intersecting strike-slip and thrust faults. Tectonic stresses are governed by regional north-south compression and northwest trending right-lateral strike slip faulting (Lettis and Hanson, 1991). The main trace of the San Andreas fault accommodates over half the strain produced by this motion, with the remaining strain variously distributed across nearly one hundred regional faults.

Major northwest-trending right strike-slip faults, such as the Whittier, Palos Verdes Hills, and Newport-Inglewood faults appear to lose their structural identity as they approach within 5 miles of the frontal thrust faults of the Santa Monica and San Gabriel Mountains (Fig. 3) (Hauksson, 1987 and this book). Crustal shortening is manifested by an imbricate system of thrust fault segments, coupled with a major system of left-lateral strike-slip faults across the front of the range. While several faults are exposed as frontal thrust faults along the southern boundaries of the Santa Monica, Santa Susana, and San Gabriel Mountains, other thrusts probably extend beneath the Los Angeles Basin as blind structures (Davis et al., 1989; Hauksson, 1990; Wright, 1991). Two points requiring emphasis when discussing tectonic deformation in the Los Angeles area are: "the deformation caused by these fault structures is as active now as it has ever been and is expected to continue at the present rate" (Atwater and Molnar, 1973); and "the risk of a damaging earthquake to the Los Angeles Basin is approximately five percent per year" (Spangle, 1987).

Fault Rupture Hazard

A number of active faults cross the Los Angeles region. The designation of a fault's activity is dependent on the classification system used, and the purpose of the definition. An active fault considered in siting and designing critical facilities such as nuclear plants, dams, hospitals or other critical facilities, may not be as seriously considered in building conventional residential and commercial developments. Following the severe damage sustained during the 1971 San Fernando Earthquake, the California legislature adopted the 1973 Alquist-Priolo Act within which a definition of an active fault was prescribed by law (Hart, 1990).

The definition of an active fault within the Alquist-Priolo Act uses geologic evidence to prove or disprove prior surface displacement within Holocene time (about the last 11,000 years). If the fault has had Holocene surface rupture, it is active; if not, it is inactive. If a fault has Quaternary displacement, but Holocene activity cannot be confirmed then the fault is considered "potentially active". The Alquist-Priolo Act definition applies only to structures intended for human occupancy, and then only to subdivision developments of four residences or more.

A list of faults, either suspected or proven to have experienced surface rupture during Holocene time, is shown in Table 1 along with other potentially active faults. Active and potentially active faults in Los Angeles region are shown on Figure 4, with specific details of the Fault Rupture Map shown in Figure 5.

The County of Los Angeles has adopted the Alquist-Priolo definitions for an active fault, but has supplemented them with more stringent geologic interpretation and investigation requirements (Leighton and Associates, 1990a). Further, the county has mandated that certain critical and high occupancy facilities be held to a higher standard than merely identifying lack of Holocene offset (Leighton and Associates, 1990b). The California Division of Mines and Geology has identified only six faults as being "sufficiently active and well defined" such that special geologic study zones were established for the faults, as prescribed under the Alquist- Priolo Act (Hart, 1990). Fifteen additional faults or fault segments are considered as active by the county, and 23 are considered potentially active (Table 1).

Given the state of our knowledge on fault characteristics, interpreting the relative risk that each fault poses is a qualitative judgment. Several factors to consider when interpreting the fault rupture hazard for a particular fault are: type of fault; reliability of the fault location; evidence for age of faulting; regulatory requirements.

To be classified active in the County of Los Angeles, one of the following criteria must be met (Leighton and Associates, 1990a):

- The fault is within an area zoned by the State as an Alquist-Priolo Special Studies Zone (APSSZ).
- The fault is suspected or reported to displace Holocene age alluvium, as noted in published geologic reports.

FAULT RUPTURE HAZARDS

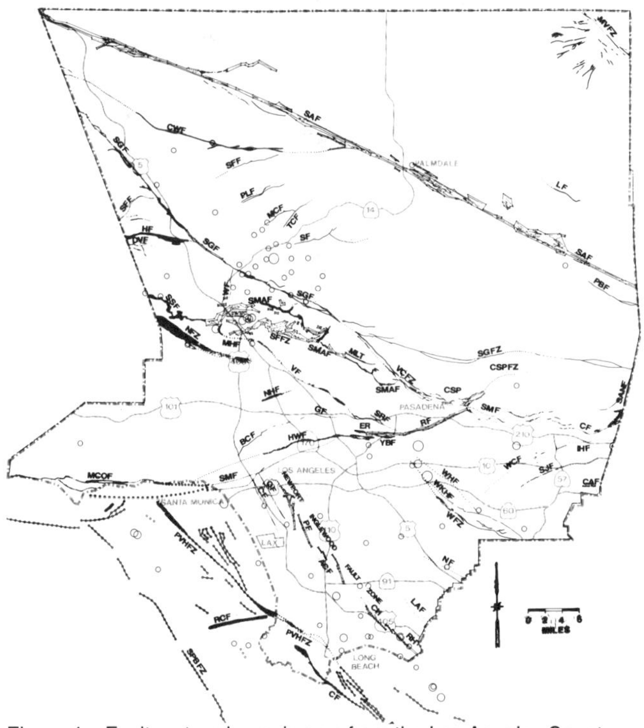

Figure 4. Fault rupture hazards map from the Los Angeles County Safety Element (Leighton and Associates, 1990a), showing locations of faults identified as active and potentially active. Small circles represent historic earthquake epicenters greater than magnitude 4. Fault name abbreviations used are listed on the next page and on Table 1 at end of article.

- A fault segment lies between two APSSZ faults, adjoins and appears to extend an APSSZ fault, or a distinct fault segment includes an APSSZ fault.
- The fault parallels an active fault, or the fault intersects an existing or proposed special study area, and the fault displays the same tectonic character (i.e., geometry, slip mechanism, etc.)
- The fault is not well studied, but is covered by one of the above criteria.

Blind Seismogenic Structures

As illustrated by the well-dispersed pattern of earthquakes (Hauksson, 1990, and this book), essentially the entire Los Angeles area is underlain by smaller seismogenic structures. Large damaging earthquakes have been concentrated near major known Quaternary faults mapped at the surface. Nevertheless, the thrust fault mechanism identified as causing the 1987 Whittier Narrows earthquake suggests that many other unknown faults lie below the surface. These may be accommodating up to 13.5 mm/year of north-south compressional shortening between the offshore continental borderland and the San Andreas fault (Davis et al., 1989). Consistent with thrust fault mechanics, upward verging blind thrust tips produce fault propagation folds manifested at the surface as an anticlinal fold belt (Fig. 6). Shear along the thrust ramps appears to die out into ductile deformation in the overlying anticlinal structures.

ABREVIATIONS USED ON FIGURE 4

ACF	AVALON - COMPTON FAULT
BCF	BENEDICT CANYON FAULT
CF	CABRILLO FAULT
CAF	CENTRAL AVENUE FAULT
CF	CHARNOCK FAULT
CHF	CHERRY HILL FAULT
CSPFZ	CLAMSHELL SAW PIT FAULT ZONE
CWF	CLEARWATER FAULT
DVF	DEL VALLE FAULT
ERF	EAGLE ROCK FAULT
GF	GRIFFITH FAULT
HF	HOLSER FAULT
HWF	HOLLYWOOD FAULT
IHF	INDIAN HILLS FAULT
LF	LIANO FAULT
LAF	LOS ALAMITOS FAULT
MCOF	MALIBU COAST FAULT
MCF	MINT CANYON FAULT
MVFZ	MIRAGE VALLEY FAULT ZONE
MHF	MISSION HILLS FAULT
MLF	MOUNT LUKENS FAULT
NIFZ	NEWPORT INGLEWOOD FAULT ZONE
NFZ	NORTH HOLLYWOOD FAULT
NF	NORTHRIDGE FAULT ZONE
NF	NORWALK FAULT
OV	OVERLAND FAULT
PVHFZ	PALOS VERDES HILLS FAULT ZONE
PLF	PELONA FAULT
PF	POTRERO FAULT
PBF	PUNCH BOWL FAULT
RF	RAYMOND FAULT
RCF	REDONDO CANYON FAULT
RH	RESERVOIR HILL
SAF	SAN ANDREAS FAULT
SANF	SAN ANTONIO FAULT
SFFZ	SAN FERNANDO FAULT ZONE
SFF	SAN FRANCISQUITO FAULT
SGF	SAN GABRIEL FAULT
SJF	SAN JOSE FAULT
SPBFZ	SAN PEDRO BASIN FAULT ZONE
SRF	SAN RAFAEL FAULT
SFF	SANTA FELICIA FAULT
SSF	SANTA SUSANA FAULT
SMF	SANTA MONICA FAULT
SMAF	SIERRA MADRE FAULT
SF	SOLEDAD FAULT
TCF	TICK CANYON FAULT
VCFZ	VASQUEZ CREEK FAULT ZONE
VF	VERDUGO FAULT
WCF	WALNUT CREEK FAULT
WFZ	WHITTIER FAULT ZONE
WHF	WHITTIER HEIGHTS FAULT
WF	WHITNEY FAULT
YBF	YORK BOULEVARD FAULT

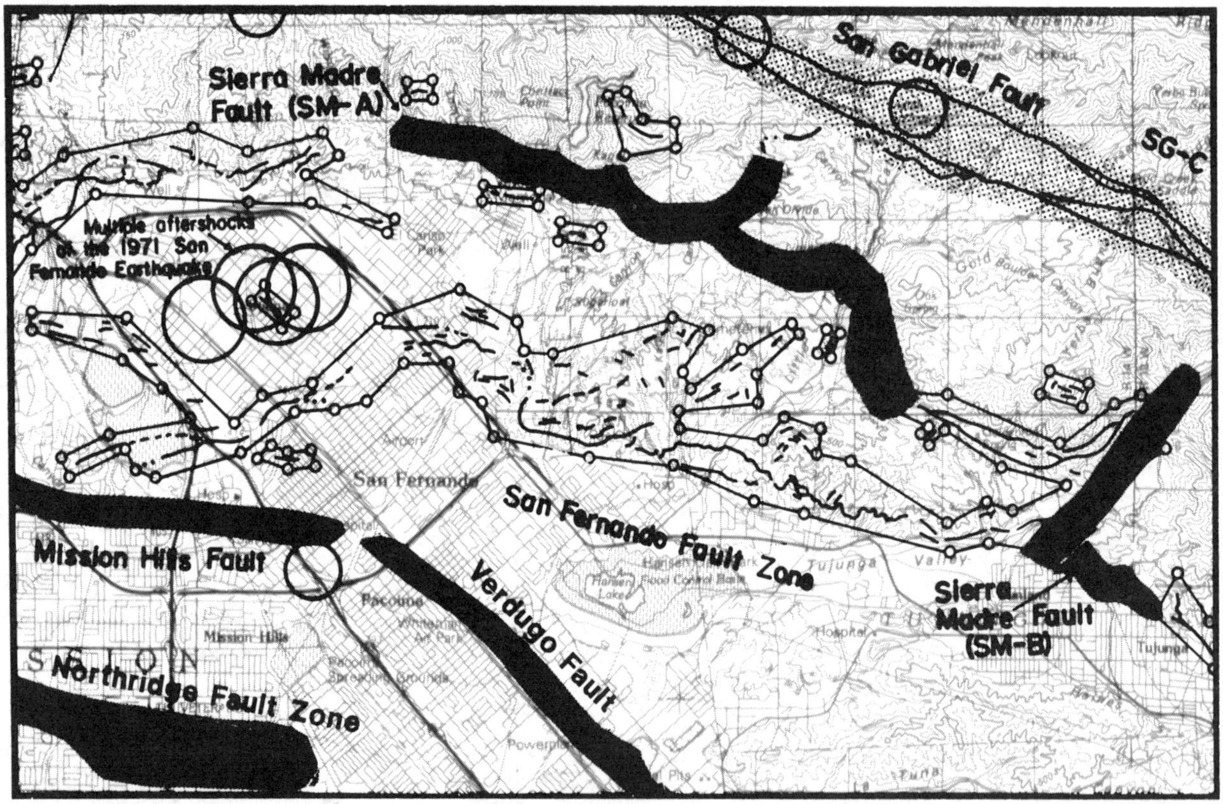

Figure 5. San Fernando Valley portion of Figure 4, shown at actual published size of 1-inch = 2 miles in the County Safety Element (Leighton and Associates, 1990a). Polygonal boxes with small circles on the end-points represent the APSSZ boundaries for selected faults (Hart, 1990). Darkly shaded faults are those additional structures considered as active and requiring special geologic studies for all new development by Los Angeles County policy. Lightly shaded faults are considered as potentially active by County policy, and require special study only for critical facilities.

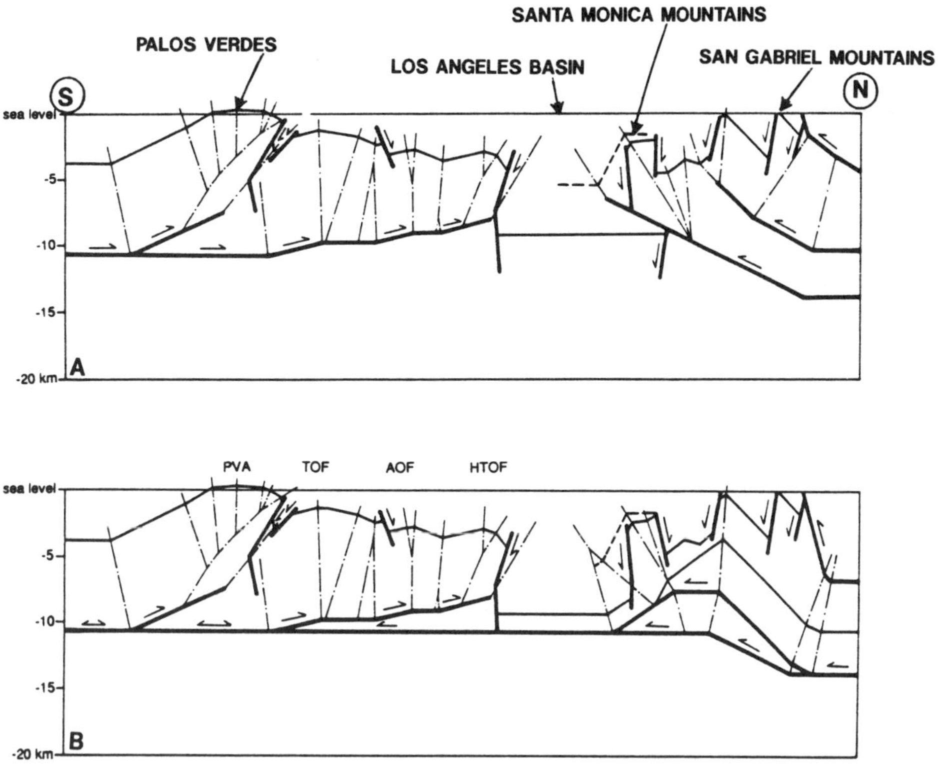

Figure 6. Two of several alternative structural models across the western Los Angeles basin from Palos Verdes to the San Gabriel Mountains, illustrating the fold and thrust belts shown on Figure 3, as well as the complexity of subsurface structures underlying the Los Angeles region. Figure A models the basin with no through-going detachment surface under the central basin, and Figure B presents a model with a detachment surface. Figure modified from Davis et al. (1989).

Buried thrust faults add a new dimension to the earthquake threat in metropolitan Los Angeles. Relatively large earthquake sources may underlie vast areas of the coastal and valley lowlands. The potential size or recurrence of earthquakes originating on seismogenic structures associated with buried thrust faults is still the subject of intense research (Davis et al., 1989). The seismic risk of these buried structures, in terms of recurrence and maximum potential magnitude, is not yet well established. Consequently, the potential for an earthquake on these surface-verging splays at magnitudes larger than 6.0, the approximate magnitude of the 1987 Whittier Narrows earthquake, cannot be precluded (Hauksson, this volume; Whitney and Gath, 1991).

Historic Seismicity

Catalogs of earthquake activity for southern California extend as far back as the earliest "felt" event, recorded in 1769 by the first Spanish exploration party, the Portola expedition, as they traveled between the Santa Ana and Los Angeles Rivers. The earliest chronicles of earthquake activity are generally skewed to larger events and their documented effects are related only through subjective personal accounts. The earliest events, locations, and descriptions are found in Toppozada, et al. (1980). The accuracy and completeness of the seismic record has increased over the years, with a major boost in 1926 with the establishment of the first Southern California Seismic Network. Systematic calculations of earthquake epicenter locations, and the compilation of ongoing earthquake activity began in 1932 with the establishment of the U.S. Geological Survey's seismographic instrumentation program at the California Institute of Technology in Pasadena.

Figure 7 is a map showing 200 years of earthquake epicenters reported in the Los Angeles area showing both the regional dissemination of small epicenters, and the close spatial relationship with some of the known surface faults. Of particular concern to earthquake hazards preparedness is the apparent presence of seismic gaps on many of the major structures. The two most striking aseismic gaps are along the Sierra Madre fault in the central San Gabriel Mountains, and the Mojave segment of the San Andreas fault across the northeast margin of the San Gabriels. Smaller gaps that are also apparent include the Whittier fault along the southern Puente Hills, the entire San Fernando Valley area south of the 1971 events, and the San Gabriel fault north of the Santa Clarita area.

Table 2 shows a list of 57 damaging earthquakes, generally Modified Mercalli (MM) > VII, that have occurred in southern California from 1800 through 1991, and recorded or felt in the Los Angeles region yielding a return rate of one damaging earthquake every 3.5 years. Implicit in this observation is the extreme risk of the region to future earthquakes.

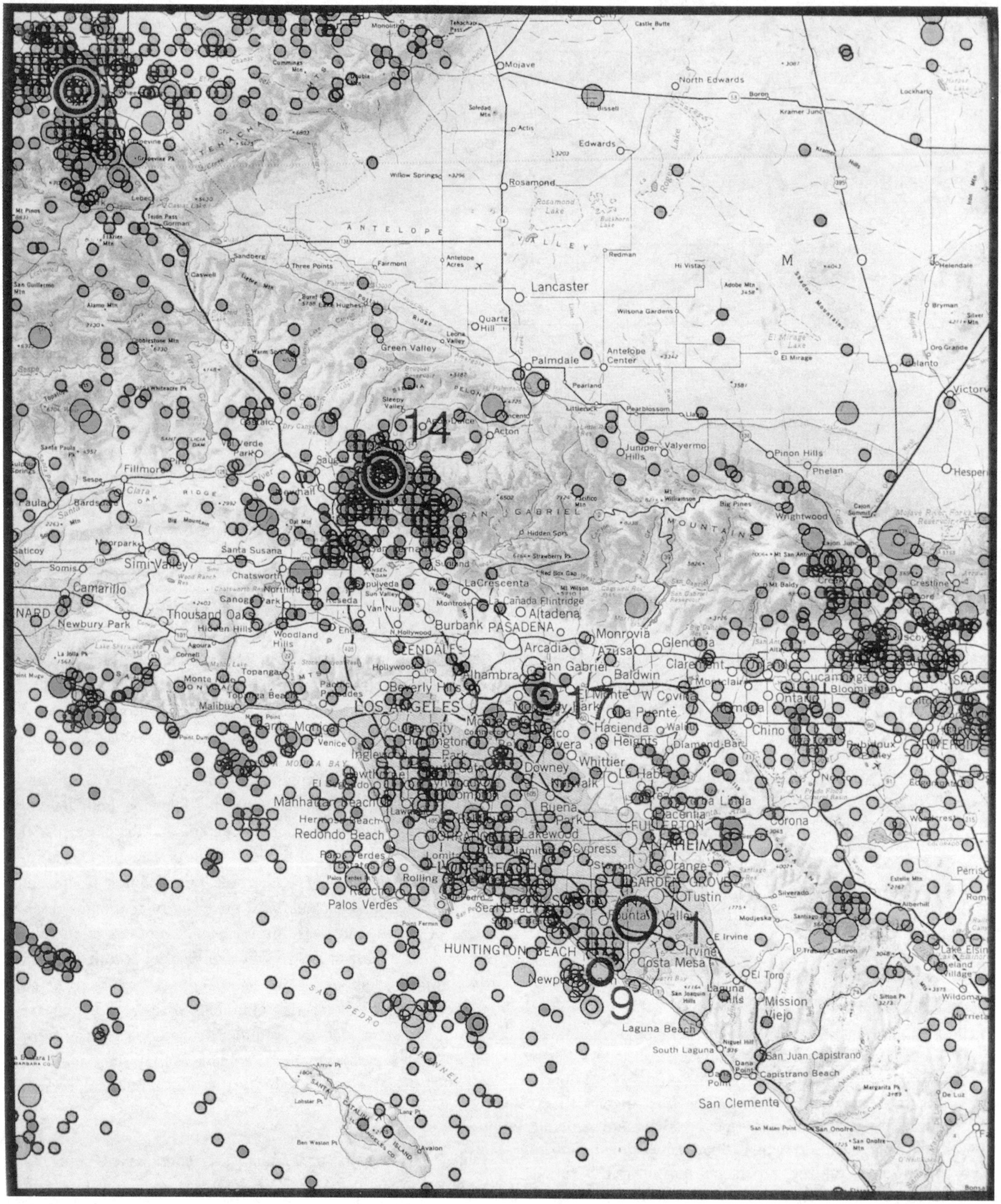

Figure 7. Map of seismicity from 1808-1987 in the greater L.A.area (Goter, 1988).

Several of the larger earthquakes listed in Table 2 have resulted in changes in the regulatory codes and public perception of the earthquake hazard. The following earthquakes, have had an impact on safety planning in California. Earthquakes from 1933 to the present are stressed because some fundamental lessons can be learned from a review of their damage patterns.

Larger Earthquakes in Southern California

1855 July 11 - The earthquake was felt locally as a strong shock with intensities as high as VIII. A number of buildings were damaged, bells were thrown down at the San Gabriel Mission and an adobe dwelling was destroyed in an area now occupied by the Los Angeles Arboretum in Arcadia. A total of 26 buildings were damaged. The earthquake was probably located on one of the surface faults bordering the Los Angeles Basin, such as the Hollywood, Raymond or Whittier faults, or it may have been related to the subsurface rupture of a buried thrust fault, as was the case in the 1987 Whittier Narrows earthquake.

1857 January 9 *"Fort Tejon Earthquake"* Just seven years after California gained statehood, southern California experienced the great Fort Tejon Earthquake (ML = 7.9). Surface rupture along the south-central portion of the San Andreas fault extended southeast from the Carrizo Plain in San Louis Obispo County to Ratted, California and perhaps as far as Burro Flat in San Bernardino County. In Los Angeles, the actual damage reported was less than that reported in the 1855 earthquake, although there was a very strong, long duration swaying motion, and water from the Los Angeles River was thrown from its bed. Houses were reported knocked down in the San Fernando Valley. The comparatively small loss of life is attributed to the sparsely settled character of the most heavily shaken areas.

1920 June 22 - This earthquake was attributed to the Newport-Inglewood fault near the City of Inglewood (Toppozada and Parke, 1982). Considerable damage occurred but there was no loss of life. The earthquake exposed the weakness of the local types of construction, especially brick masonry, which had been common in California.

1933 March 10 *"Long Beach Earthquake"* - One of the most destructive earthquakes in the history of southern California occurred on the southern segment of the Newport-Inglewood fault. The Cal Tech seismic network determined that the hypocenters occurred at a depth of 10 kilometers (six miles) and that the rupture propagated 25 kilometers to the northwest from its epicentral location off Newport Beach, into the Long Beach area. The Isoseismal Map shown in Figure 8 shows the heavily damaged area as an elliptical pattern; the long axis extends from Laguna Beach to Marina del Rey. Considerable damage still occurred outside this area, but it was sporadically distributed. Lower damage totals in Orange

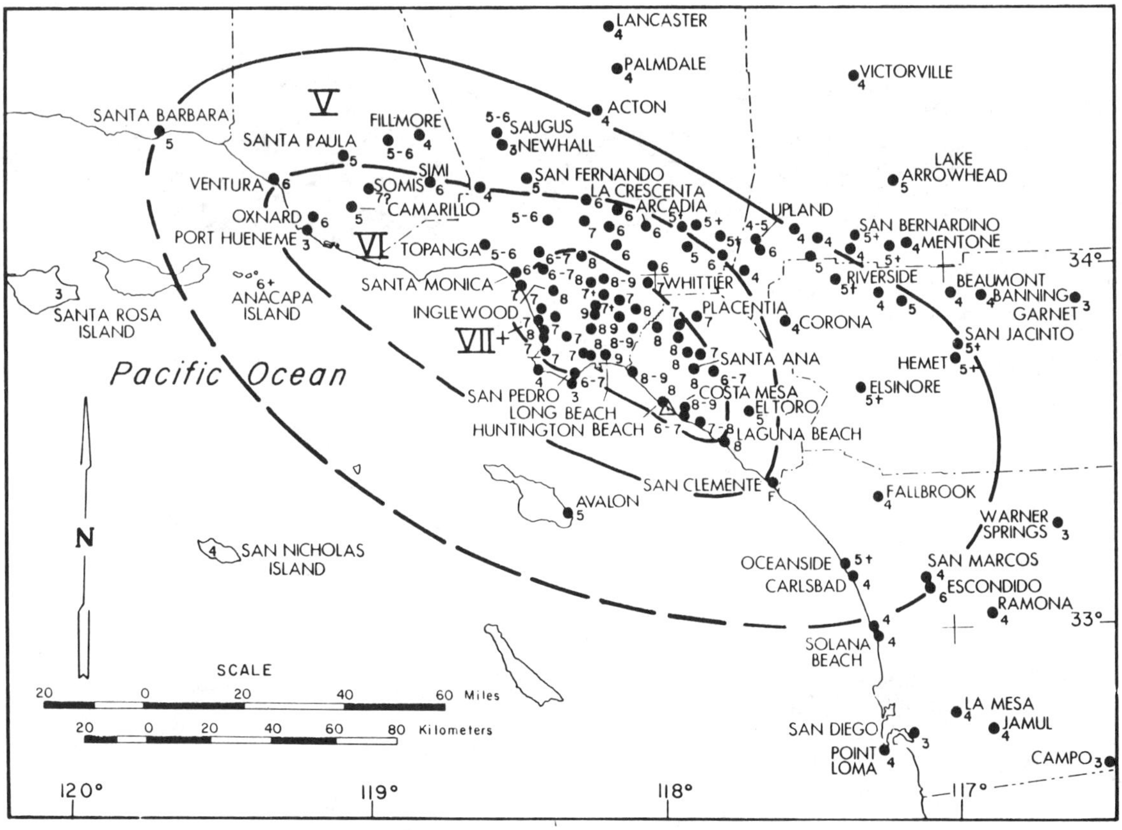

Figure 8. Isoseismal map of the 1933 Long Beach earthquake showing Modified Mercalli intensities in Arabic numerals. Modified from Toppozada and Parke (1982)

Photo 1. True split-level house in Sylmar after 1971 San Fernando earthquake. Heavier bedroom collapsed onto 2-car garage during strong ground shaking.

Photo 2. At time of 1971 San Fernando earthquake, overpass abutment sank, creating a ramp just as unknown vehicle slammed on brakes, causing vehicle to become airborne for about 15 feet.

County, closest to the epicenter, was probably the result of the sparsely populated nature of Orange County in the 1930's.

Wood (1933) found that "conspicuous damage is greatest in areas underlain by man-made fill or loose alluvium, especially where saturated at shallow depths with ground water". Consolidated terrace deposits, compacted man-made fill, hilly ground such as Signal Hill, and even areas underlain by sedimentary rock escaped with considerably less damage even though located in high MMI shaking intensity zones. Well-constructed buildings in even the most vigorously shaken areas, suffered relatively little damage, especially when sited on "well chosen or well-prepared foundations" (Wood, 1933).

The 1933 Long Beach event exposed the vulnerability of the alluviated Los Angeles basin area to liquefaction and earthquake-induced ground settlement. As noted by Wood, (1933): "along the shore between Long Beach and Newport Beach, and in a few localities a short distance inland, road fills across marshy land, and similar earth construction resting in wet sand or mud, settled, shook apart, or moved laterally, causing considerable damage to concrete highway surfaces and to approaches to highway bridges". Analogous phenomena were observed where piers and landings adjoined the shore in the harbor area. A summary of specific damage is compiled in Table 3.

1971 February 9 *"San Fernando Earthquake"* - In the years prior to 1971, the San Fernando Area was characterized by low to moderate seismicity, not unlike many other areas in the Los Angeles region. The area experienced about ten earthquakes of Magnitude 3.0 between 1934 and 1971. The only other earthquake attributed to the area prior to the San Fernando Earthquake was a small felt event in 1893 (Weber, 1975). The 1971 San Fernando Earthquake measured 6.6 and caught many earth scientists unprepared because the area had not shown any anomalous activity prior to the event. The main shock and the two very strong aftershocks resulted in one of the most damaging earthquakes that had yet occurred in the United States.

The San Fernando Earthquake provided the first test of the newly developed and heavily populated southern California region to the damaging effects of a moderate size earthquake. Most of the economic loss from this earthquake resulted from a combination of intense ground shaking and extensive and severe ground rupture. Principal fault rupture was distributed along the San Fernando and western Santa Susana fault zones over an area as wide as 1.8 miles. Also contributing to the heavy property losses were the widespread ground lurching, liquefaction, lateral spreading, and differential settlement of non-engineered fills and loose alluvium (Weber, 1975). A summary of selected damages is provided in Table 3. As a result of this earthquake, previously accepted Uniform Building Code Standards and construction techniques common to southern California were reassessed.

1987 October 1 *"Whittier Narrows Earthquake"* - The Whittier Narrows Earthquake occurred on an unsuspected seismogenic feature, a buried, "blind" thrust fault underlying the Elysian Park-Montebello Hills area (Davis et al., 1989; Hauksson et al., 1988, Hauksson and Jones, 1989). The blind thrust system had been known, but not quantified in terms of seismic risk analysis. Consequently, the earthquake potential of these faults caught planners, and geologists by surprise. The implications of these buried thrust faults range from increasing previously accepted annual probabilities for damaging earthquakes to the most disconcerting notion, that such faults may be capable of a Magnitude 7.5 earthquake at depths as shallow as 6 Km (four miles) below an urbanized area. In any case, it is clear that the earthquake threat is increased for this region.

No evidence of fault rupture or ground failure was revealed during this earthquake but structural and nonstructural property damage was extensive. The damage caused by the earthquake provided much information on the survivability of retrofitted freeway interchanges and certain buildings (unreinforced masonry and concrete tilt-up). In addition, the lessons learned from the Whittier Narrows Earthquake were also significant, because it tested the Federal, State, and local earthquake mitigation, emergency response, and preparedness programs that had evolved since the 1971 San Fernando Earthquake. A summary of selected damage, losses, and lessons learned from the 1987 Whittier Narrows Earthquake is presented in Table 3.

1991 June 28 *"Sierra Madre Earthquake"* - This early morning earthquake, 7 miles beneath the San Gabriel Mountains, was significant in that it confirmed the seismogenic potential of the central reach of the Sierra Madre fault zone (Crook et al., 1987). The earthquake caused landsliding in the San Gabriel Mountains, and damaged numerous older homes and structures in the foothills communities. Only one death was attributed to the earthquake, but the implications of a seismically active frontal fault pose severe implications of much greater losses in the future.

Strong Ground Motion

"Near field" ground motion is generally composed of high frequency waves that are more efficient in causing low-rise buildings and private residences to vibrate. Longer period wave forms, usually associated with surface waves, are manifested at greater distances from the earthquake source, and are more likely to cause high-rise buildings and buildings with large floor areas to vibrate more vigorously. When the frequency composition of an earthquake, the site, and the building period coincide, resonance can occur. Buildings or structures which experience this type of response can suffer severe damage or total structural collapse.

The size and type of fault mechanism (e.g., strike-slip or thrust), the rupture orientation, the basin geometry and topography, the subsurface geology and near-site conditions are all important variables affecting the spatial distribution and character of strong ground motion. Accelerograph records of the M_L 5.9, 1987 Whittier Narrows event have confirmed that ground acceleration, frequency composition, and duration of both the horizontal and vertical components of ground vibration can vary widely in a given earthquake in the Los Angeles area. The variance can be a function of the earthquake rupture mechanism or the influence of site conditions. Alluvial basin areas are especially vulnerable to site amplification where higher earthquake accelerations and magnification of wave forms and frequencies can induce resonance effects on structures.

For these reasons strong motion records and post-earthquake damage investigations consistently show that peak acceleration alone can be an unreliable parameter for both seismic design and predictive ground motion mapping (Sharpe, 1982). Dynamic response spectra and time history analysis are the seismic design methods that are recommended for determining the earthquake-resistant design of critical facilities, and are the focus of future design and predictive mapping research (Joyner and Fumal, 1985; Rogers et al. 1985).

Predictive strong ground motion models are made to fit specific applications. Regional ground motion analysis is usually conducted with more generalized information. The most commonly used method for predicting regional ground motion uses seismic intensity, an empirical method that evaluates shaking in terms of structural damage and human response (Evernden and Thomson, 1985; Davis et al., 1982; Gulliver, 1986; Toppozada et al., 1988). Many versions of this method have been used by State and Federal government agencies in insurance, damage, and emergency response studies.

The effect of saturated soil on ground motion is controversial and may not be directly correlated with higher seismic intensities (Rogers et al., 1985). However, the potential for liquefaction in these areas may indicate that more damage may be sustained which is consistent with the definition of seismic intensity. Strong contrasts in shear wave velocity between the near surface alluvium and underlying rock can cause dramatic increases in the amplitude and frequency composition of the seismic vibration, and prolong the duration of shaking when alluvial thicknesses are as little as 30 to 100 feet (Hays, 1980). Lastly, all earthquake sources must be accounted for in a predictive ground motion model.

LIQUEFACTION

Liquefaction and lateral spreading failures during earthquake shaking can result in severe disruption of near-surface structural and foundation distress. Structures may settle several feet below grade while buried tanks or other buoyant structures may rise to the surface. Especially at risk to liquefaction are retaining walls and levees at port and harbor facilities. Damaged roadways can prevent access for emergency response vehicles and complicate evacuation operations. Severed water distribution lines can impair fire fighting capability or, if associated with waste waterline breaks, contaminate the drinking water supply. Natural gas pipelines may rupture causing explosions or fires.

The 1933 Long Beach and 1971 San Fernando Earthquakes were accompanied by costly damage from earthquake-induced ground failure. Liquefaction was a factor in the failure of the hydraulic fill embankment on the Lower San Fernando Dam. The potential impact of earthquake-induced ground failure makes a regional assessment of areas prone to such phenomena useful to disaster response planners. Because damaging liquefaction can be induced with seismic intensities as low as MM VII, and predictive ground motion maps indicate that such intensities can occur relatively frequently, the Los Angeles region is extremely vulnerable to future property damage from earthquake-induced ground failure.

Groundwater Conditions

Since the early 1900's, the agricultural development of groundwater in the coastal lowland and valleys of Los Angeles has caused regional lowering of ground water levels. In a general way, that depletion of groundwater can be credited with reducing the liquefaction hazard. However, the reduction in agricultural land use in the Basin, coupled with high residential landscape watering, has resulted in the reestablishment of some of the previous historic high groundwater levels.

High groundwater conditions exist quite extensively in parts of the Antelope Valley, the flood plains of the Los Angeles and San Gabriel Rivers, and the Santa Clarita Valley. More localized high groundwater conditions also persist, partly due to the discontinuous nature of the sand and clay lenses associated with perched water bodies, and partly from human activity involved in recharging water into shallow water aquifers. Tinsley et al., (1985) used an extensive compilation of 1944 to 1981 data on water level measurements from selected shallow aquifer water wells and engineering boreholes to delineate areas of shallow groundwater. Their work provides the best data available for metropolitan Los Angeles.

There are important limitations to the groundwater compilation. Future changes in groundwater management could elevate regional groundwater and perched water to pre-1944 levels. Seasonal fluctuations and basin recharge could be significant for localized areas.

Based on the conclusions of Tinsley et al., (1985) and Youd et al., (1978), the depths to groundwater chosen to represent relative susceptibility to liquefaction are:

Depth Below the Surface (feet)

0 - 10:	high
10 - 30:	moderate
30 - 50:	low
50 and below:	very low

In areas along the coast, along youthful stream drainages, and above the perched water body in the Antelope Valley, where detailed stratigraphic studies have not yet been performed, liquefaction susceptibility is based solely on judgment. The map should be used with a sufficient understanding of the limitations of the compilation.

Liquefaction Susceptibility

In the Los Angeles region, where seismic shaking opportunities are everywhere, liquefaction susceptibility is reduced to two component variables: 1) the distribution of sediments prone to liquefaction, represented by youthful, cohesionless, granular surficial and/or subsurface materials, and 2) the distribution of shallow or perched ground water.

Appropriate for regional interpretations of material susceptibility is the influence of sediment age and the mode of deposition (Lajoie and Helley, 1975). Sediment generally becomes more resistant to ground failure with increasing age as natural processes tend to cement, compact, and gradually prestress and pre-strain the material. Youd and Perkins (1978) link the recency of deposition with susceptibility, such that materials deposited in the past few hundred years are the most susceptible.

The most comprehensive liquefaction studies conducted in Los Angeles County (Tinsley et al., 1985), utilized soil profile development, and patterns of historical flooding and flood deposits as mapping criteria to delineate Quaternary units. Cone penetrometer data from selected boreholes provided textural control for these mappable geomorphic units. Their study delineated the relative susceptibility of three units, defining sediments deposited within the past 1,000 years as the most susceptible, sediments deposited during the last 10,000 years as highly to moderately susceptible, depending upon sorting, and Pleistocene deposits as unlikely to liquefy.

Tinsley and Fumal (1985) utilized measurements of shear wave velocity at 84 sites in the Los Angeles area to verify the correlation between velocity and the age and textural characteristics (grain-size, sorting, relative density) of alluvial sediments. The age and textural characteristics of the surficial geologic materials as defined by Tinsley and Fumal (1985) were used in conjunction with the liquefaction units defined by Tinsley et al., (1985) to classify the relative susceptibility to liquefaction in the Los Angeles County Safety Element (Leighton and Associates, 1990a). Figure 9 is a detail from the Engineering Geologic Materials Map to illustrate the differentiation of geologic formational units on the basis of their typical engineering performance. Further differentiation was also possible for the highest Pleistocene terraces in the Antelope Valley.

The liquefaction susceptibility map (Fig. 10 and 11) illustrates the extremely large areas of Los Angeles that are susceptible to liquefaction. Perhaps the most unique area of susceptibility to ground failure is the ancient Rosamond Dry Lake (pluvial) deposits in the northern Antelope Valley. The interbedded sequence of clays and unconsolidated sediments form a perched water body, an important factor conducive to saturated ground failures and lateral spreads. The map illustrates a relative susceptibility based first upon the presence of liquefiable sediments, and then upon the availability of groundwater. The liquefaction map is constrained most by the existing groundwater data base and the regional nature of the mapping, making it unsuitable for site specific applications. For that reason, Holocene sediments at highest risk (fine-medium grained silts and sands) were mapped as liquefiable even though shallow groundwater had not been reported.

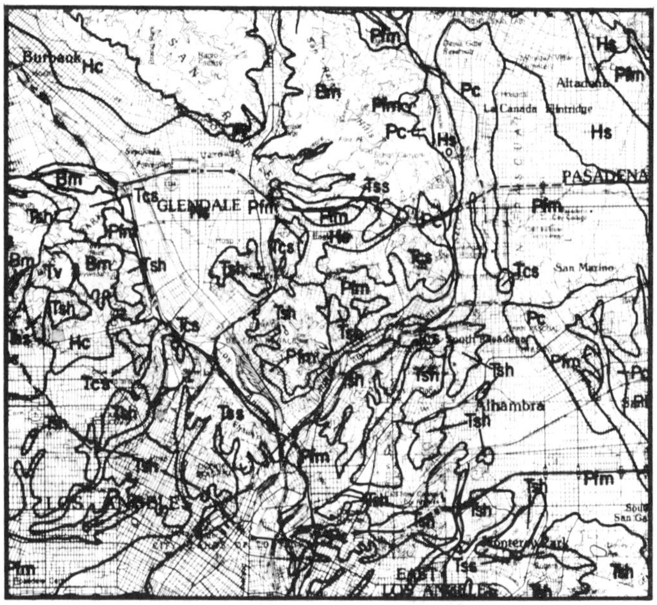

Figure 9. Glendale-Pasadena-Los Angeles portion of the Engineering Geologic Materials map of the Los Angeles County Safety Element, shown at the published scale of 1-inch = 4-miles (Leighton and Associates, 1990a). This map compilation was used for both liquefaction susceptibility analysis, and for the slope stability compilation map, Figure 12 below. Unit designations are intended to be descriptive of the anticipated engineering performance of the unit, and are only geologically approximate. Abbreviations used are: Time stratigraphic [H-Holocene, P-Pleistocene, T-Tertiary]; textural [c-coarse, fm-fine to medium]; behavior [s-susceptible to liquefaction]; and lithologic [v-volcanic, sh-shale, ss-sandstone, cs-conglomeratic sandstone, and Bm-crystalline].

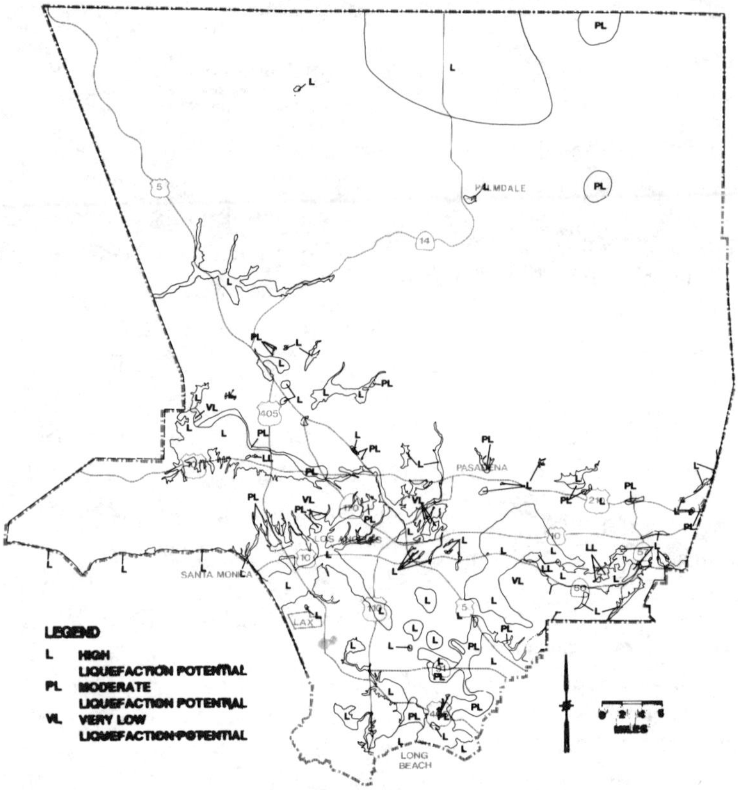

Figure 10. Liquefaction susceptibility map of Los Angeles modified from the Los Angeles County Safety Element (Leighton and Associates, 1990a).

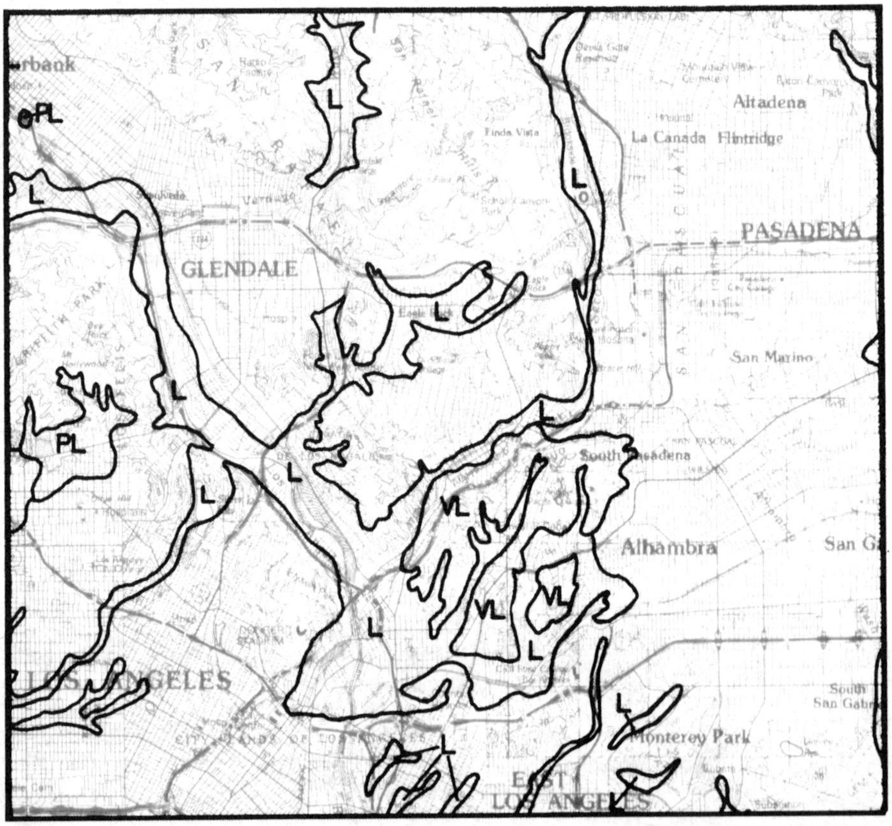

Figure 11. Glendale-Pasadena-Los Angeles portion of the Liquefaction Susceptibility map of the Los Angeles County Safety Element, shown at the published scale of 1-inch = 3-miles (Leighton and Associates, 1990a).

SLOPE INSTABILITY

Consistently rising real estate costs in southern California highlight concerns for increased economic losses from this type of ground failure. Slopes may become unstable due to construction activities, but most unstable slopes can also be stabilized by accepted geotechnical methods. Landslide damage alone in the western part of southern California amounted to several hundred million dollars as a result of the intense rainstorms of 1978 and 1980 (Slosson and Krohn, 1982). The losses were attributed to landslides, debris flows, mudflows and artificial fill slope failures.

The debris and mudflows associated with the intense rainstorms of 1969, 1978, 1980, and 1992 caused nearly 30 fatalities and extensive damage to residential development, small businesses, and agriculture. Public protection from the life-threatening effects of mud and debris flows needs to become a key area of focus in the coming years. Development is crowding into the mountain foothills, and as it does, the historic effects of debris flows need to be considered.

Landslide Susceptibility

Geologic information related to slope stability is commonly only gained by site inspections using air photo interpretation and large scale mapping. Even then, subsequent subsurface investigations often reveal important geologic conditions not observable from surface inspection alone. Landslide hazard and slope stability maps at 1-inch = 2,000-feet exist for some areas of the region, published by the state-funded Landslide Hazard Identification Program (LHIP) of the California Division of Mines and Geology (CDMG, 1987). LHIP used a more comprehensive approach for designating relative slope stability by using adequately scaled geologic and topographic maps. The program produced detailed slope stability maps suitable for land use planning (e.g. Treiman, 1987). Importantly, the program also evaluated the relative susceptibility of some terrain to debris flows and other particularly dangerous forms of shallow, rapid slope failures.

For the Los Angeles County Safety Element, the production of a relative slope susceptibility map for Los Angeles County was considered inappropriate at the scale of geologic resolution existing for most of the County. Until the influence of bedrock lithology, structure, slope angle and vegetation can be extended to areas outside of the limited mapping by the LHIP, and at scales of at least 1 inch = 2000 feet, a susceptibility map should not be attempted. The landslide inventory presented in the Safety Element when used in conjunction with the Engineering Geologic Map (Fig. 9), provide an initial step in the future production of a complete Slope Failure Susceptibility Map by identifying areas where information is lacking and by keying failure styles to geologic rock types. The emphasis has been on using relative slope stability maps for regional planning. However, adequate geologic investigation is necessary regardless of the assigned relative slope stability, and should be emphasized at the site specific scale where terrain modifications are also occurring.

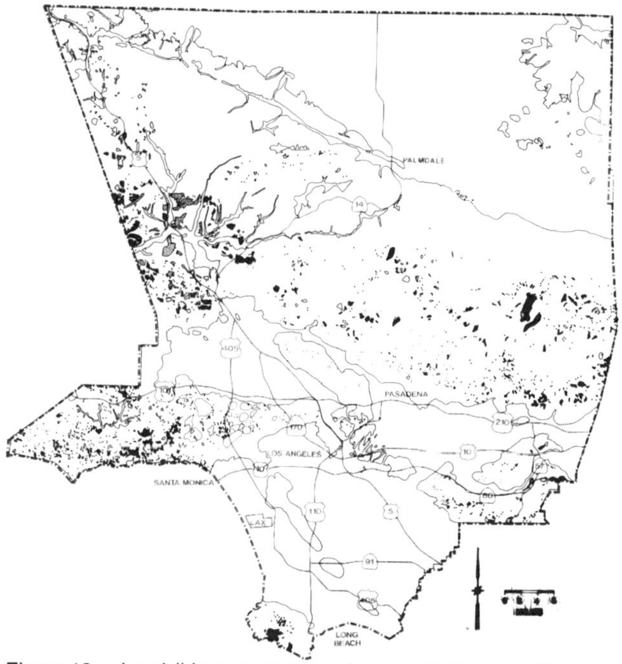

Figure 12. Landslide occurrence and susceptibility map of Los Angeles County, modified from Leighton and Associates (1990a). Landslides are shown as dots [generally not visible at this reduced scale] where they are between 5-100 acres in extent, and in their mapped configuration where they exceed 100 acres. The absence of landslides in susceptible geologic units may be as much related to the lack of geomorphic mapping, as to the absence of failures.

Earthquake-induced Landslides

Because of Southern California's mountainous terrain and seismic activity, it is at high risk from earthquake-induced landslide hazards (Wilson and Keefer, 1985). The 1971 San Fernando Earthquake triggered hundreds of small slumps, debris slides, and rock falls on the slopes of the San Gabriel Mountains. A larger earthquake could cause thousands of landslides within just a few minutes, exposing the population to a far greater risk. Although extremely large slides were not observed, the smaller slides showed the adverse consequences that occur from widespread road blockages. Landsliding is capable of incapacitating vital evacuation or emergency response activities.

Earthquake-induced slope failures in the Los Angeles region closely correlate with geologic units. Many failures consisted of rockfalls that were concentrated in highly jointed crystalline basement and in massive, cliff-forming conglomeratic units (Clark et al., 1979). There were limited occurrences of reactivated, large deep-seated landslides in some well bedded Tertiary units, but they represented only a small percentage of the total slope failures (Weber, 1982). Shallow, surficial failures in elevated Plio-Pleistocene deposits were common, however, they were small in total volume.

The 1991 Sierra Madre earthquake closed most roads in the San Gabriel Mountains due to the pervasive number of rockfalls. The pattern of failures exhibited during this earthquake is consistent with earthquake- induced failures observed from worldwide historical records of earthquakes of similar magnitude (Keefer, 1984). Rockfalls and rockslides are more likely to occur on slopes greater than 45 degrees, whereas, surficial soil failures tend to be concentrated on intermediate slopes (35 to 40 degrees) of weathered colluvial material (Yen et al., 1978). Rock failures will be abundant in the San Gabriel Mountains, where large outcrops of crystalline rocks and massive sandstone occur on steep slopes throughout the entire area. Rockfalls and slides will occur in well-jointed intact rocks, while rock avalanches will result from failure of the steep, unstable talus slope. Harp and Keefer (1985) raise concern that the massive conglomerates on Cahuenga Peak and Mount Hollywood may be subject to large, catastrophic rock avalanches.

Rainfall-induced Landslides and Debris Flows

The winter rains of March, 1938 generated thousands of shallow debris flows in the Puente and Chino Hills, the scars clearly visible on the air photos of the period. In 1969, flow failures were responsible for 24 deaths throughout southern California, many of them locally (Weber, 1982). Flows became associated with burnt watershed areas, and costly damage was incurred on developments situated below the recently burned areas of the San Gabriel Mountains. However, Campbell (1975), notes that all of the fatalities resulted from flows originating on unburnt slopes. Damage was countywide, but the largest concentration of slides and flows was located in the Santa Monica and San Gabriel Mountains, and the Santa Clarita Valley area. A low estimate of $6 million in property damage was recorded, prompting State and Federal disaster agencies to emphasize accurate documentation of landslide related financial losses (Campbell, 1985).

In 1978, two major rainstorms triggered destructive debris flows from burnt hillside areas in the San Gabriel Mountains. Also, there were debris flows originating from unburnt slopes in the Santa Monica Mountains and hillside areas in the coastal lowlands and valleys, particularly in Monterey Park and the Baldwin Hills. Areas underlain predominately by units more susceptible to rotational and block glide failures (Topanga, Monterey, and Puente Formations), although also experiencing debris flows, received most of their damage through reactivation of deep-seated bedrock failures.

Pervasive and costly failures occurred on both graded and natural slopes in the Repetto Hills (including Monterey Park), in the form of shallow slumps, earth flows, and debris flows. Storm-triggered landslide damage in the City of Los Angeles alone, to both private and public property, was estimated at $100 million (Slosson and Krohn, 1982). The effect of damage on small businesses due to the loss of productivity, on agricultural activities because of damaged crops, or other subtle consequences were not factored into the loss estimation.

In 1980, another series of six rainstorms hit southern California between February 13 and February 21, causing major landsliding. Several people were missing and presumed killed in the San Gabriel Mountains and one woman was killed in the Santa Monica Mountains by debris flows (Davis, 1982). Fifty people were evacuated from an apartment in Montebello when a large slope behind the complex came close to failing. In the Santa Monica Mountains, sewer mains broke as result of landslide movement, releasing 80 million gallons of raw sewage. Throughout the region, but especially in the Santa Monica Mountains, huge boulders, and masses of earth from roadcuts, partially severed roads traversing steep terrain. During that 8-1/2 day period, 111 homes were destroyed and more than 1,350 were damaged (Weber, 1980). Landslide damage from the storms was estimated at $500 million dollars in southern California. Loss estimations began to include direct costs from public and private repairs and indirect costs from socioeconomic effects (Campbell, 1985). More than $130 million dollars were requested in the form of Federal disaster relief. The Small Business Administration made nearly $255 million dollars in low interest loans to 2,965 businesses and 12,851 homeowners (Weber, 1980); $95 million dollars were paid directly from the Federal Emergency Management Agency National Flood Insurance Program.

Debris flows are independent of the underlying rock type; flow failures originate from soil slip features in colluvial soil (slope wash) and ravine fill. Their distribution depends on an adequate volume of debris in the source area, the necessary slope angles, and a prolonged period of saturation. Flows of various viscosities can attain velocities of ten feet per second as they scour through ravines downslope and impact and inundate areas below. Rapid debris flows are generated when perched water causes a pore water pressure buildup in the soil (Reneau and Dietrich, 1987). Such conditions often occur on the bedrock-colluvium contact. Geomorphic features may be employed as a method of identifying debris flow sources and hence debris flow susceptibility (Reneau and Dietrich, 1987). Hollows and bowl-shaped geomorphic scars located at the top of swales serve as a source for debris flows. They constitute an important mappable feature in hillside areas that could help to better guide development away from these narrow debris chutes. A consistent methodology to quantify debris flow volumes from discrete swales is not yet being applied in the Los Angeles region.

FLOODING

The Los Angeles region is vulnerable to a wide range of flood hazards, including intense storms, and failure of manmade structures. Several local dams have failed (see Proctor, chapter 9). The failure of the St. Francis Dam in 1928 (see Rogers, chapter 9) resulted in the death of over 400 people in the sparsely populated Santa Clarita Valley downstream of the dam. The failure today of just one of the larger reservoirs during peak storage could devastate a large urban area.

Intense storms present a flood hazard for developments located in narrow canyons in mountainous areas and on alluvial fans. Because of the difficulty in modeling flooding on alluvial fans, certain 500 year floods may also pose an equal or only slightly greater risk than 100 year floods in these areas. A particularly dangerous combination, intense storms coupled with burnt watersheds, place developments located at the base of steep mountainous terrain at extreme risk to mud floods and mudflows.

Tsunami Hazards

Tsunamis are impulsively generated, long-period sea waves produced by displacement of the sea floor during earthquakes, volcanic eruptions, or large submarine landslides. Once generated, tsunami can travel thousands of miles at speeds as high as 400 miles per hour. The coastline geomorphology and sea bottom topography may accentuate the wave, by focusing tsunami energy into narrow inlets, harbors, embayments, or other low- lying areas along the coast (URR, 1988). Tsunami approach is typically preceded by near-shore water withdrawal followed by a series of wave surges.

Particularly relevant to coastal southern California are the potential secondary hazards caused by tsunamis. Debris dislodged by the initial wave impact can contribute to damage by impacting other structures and hampering search and rescue operations. Fires caused by the collision of dislodged fuel tanks and fuel trucks with electrical facilities have accompanied a number of historical tsunami incidents. Water surges have been known to spread fire rapidly and ignite other flammable structures (URR, 1988). The coastline is subject to tsunami risk from distant, Pacific-wide sources and from local near field sources. The Pacific wide areas of concern to southern California appear to be the Colombia-Ecuador Region, the Alaska/Aleutian Trench area and the Peru-Chile Trench area (Houston, 1980; Houston and Garcia, 1974). The 1960 M 8.6 Chile and 1964 M 8.4 Alaska earthquakes produced the largest tsunamis recorded in southern California. The Chile event resulted in a 5-foot wave height, and caused little structural damage in the harbor, but did destroy over 300 small pleasure boats and produced concern over fire from spilled gasoline. The 1964 tsunami produced wave heights of less than four feet in tide gage records in San Pedro Bay, Los Angeles Harbor and Long Beach Harbor, causing about $500,000 damage to several small boat docks, pilings, and the Union Oil Company fuel dock (Lander and Lockridge, 1989). Most of the damage was the result of swift currents at the node of the seiche, of up to 12 knots in the Cerritos Channel and wave oscillation (seiching) of up to five feet in the inner harbor (McCulloch, 1985).

Although the local impact from the 1964 Alaskan Earthquake tsunami was relatively minor, it must be recognized that faults off the coast of Los Angeles could generate large local tsunamis. The concern for tsunami impact is heightened mainly because the short historical record does not adequately characterize the long-term tsunami risk. In addition, the potential impact on the number one port facility in the United States from a large tsunami could seriously impact the economy of the entire country.

Little data exist to evaluate the potential for a near field, or local tsunami generated off the coast of southern California. Historically, the southern California area has been affected by only two locally generated tsunamis. The 1812 Santa Barbara Earthquake reportedly produced run-up heights of approximately 10 to 12 feet at Gaviota (McCulloch, 1985), although Lander and Lockridge (1989) suggest a wave height closer to 7 feet. The better documented, 1927 Point Arguello Earthquake (M_L = 7.3), produced wave heights as much as 6 feet (Lander and Lockridge, 1989).

The lower magnitude threshold for tsunami development is considered to be approximately M_L 6.5. Faults offshore Los Angeles, including the San Pedro Basin, Anacapa, Palos Verdes, and Santa Monica faults, all appear capable of producing this size earthquake and the necessary vertical offset making it conceivable that a local earthquake-induced tsunami could impact the coast in the future (McCulloch, 1985; Griggs and Savoy, 1985). Adding to the risk is the much more likely possibility of a seismically triggered submarine landslide into one of the several submarine basins, which would be large enough to generate a damaging tsunami with little or no warning. Like earthquakes, geologic evidence is needed to estimate the actual long-term tsunami threat (Lander, 1987). Tsunami vulnerability is evaluated using numerical simulation techniques involving modeling the source, run-up, and velocities of the incoming surges (URR, 1988; Shuto, 1987; Goto and Shuto, 1983; and Liu and Erickson, 1983). This method is critical in southern California because there is a lack of historical precedent to show the location of particularly vulnerable areas. Vulnerability is then determined by estimating the impacts of direct and secondary tsunami effects on land uses and building types in the inundation zone. There are no numerical simulations of a tsunami from a near field source in the Los Angeles area (Camfield, 1987).

The primary source of information used in constructing the tsunami component of the Flood and Inundation Hazard Map (Fig. 13; Leighton and Associates, 1990a) is a FEMA study (Lee et al., 1982), which is basically an updated and more sophisticated version of the studies put forth by Houston and Garcia (1974) and Houston (1980). Their study demonstrates the probability composite effect of two distant tsunamis originating on the Alaskan-Aleutian and the Peru-Chile Trench. For planning purposes, tsunami risk for southern California should be based on a larger recurrence interval (500-years), because of the general agreement between historical run-up heights in southern California (6 to 12 feet maximum), and the 2 to 14 foot run-up heights predicted in Lee's study for a 500-year event. Tsunami run-up heights are based on the height of the wave above mean low-low water level at the shoreline.

Dam Failure Hazards

The issue of dam safety can be traced to three major events within the County of Los Angeles. The sudden failure of the St. Francis Dam (see Rogers, chapter 9) led to the 1929 enactment of legislation providing for supervision over non-Federal dams in the State. Following the 1963 failure of the Baldwin Hills Dam, the State water code was amended to require a more stringent inspection and review of existing and proposed off-stream storage facilities and entire reservoir basins. After the near failure of the Lower San Fernando Dam (Van Norman Reservoir) following the 1971 San Fernando earthquake, an extensive reevaluation of acceptable risk and seismic safety for all hydraulic fill dams was initiated (see Proctor, chapter 9). The principles which were almost universally accepted for establishing seismic safety of dams prior to the 1971 San Fernando earthquake were shown to be inadequate.

Because of the nearly disastrous consequences of the 1971 San Fernando Dam failure, Dam Failure Inundation Zones became a required components of public Safety Elements. Figure 13 shows the multitude of overlapping inundation areas from water storage and flood and debris control structures in the southern portion of Los Angeles County. Many of the dams shown retain water only during the most intense storm events, nevertheless, the inundation zones are based on the dam failing during high water.

There are more than 90 dams and debris basins under State and Federal jurisdiction within the County of Los Angeles (California Department of Water Resources, 1988), Reservoirs serve several functions including water supply, flood control, and recreation. Many dams are designed for temporary detention of flood waters to prevent downstream flooding. During flash flood events, these dams can fill rapidly, as dozens of motorists discovered by being trapped behind the Sepulveda Dam during the intense storms in February 1992.

Structural failure of water facilities is not limited to dams and debris basins. Other structures capable of causing floods upon failure are dikes in the waterfront areas, flood control dikes (levees) along channelized streams, and large ground-level storage tanks. Levees and water tanks are exempt from jurisdiction by the State, yet they are vulnerable to ground failure, ground rupture, and earthquake shaking. Dikes are vulnerable to the same style of failure criteria that dams are exposed to during an earthquake, except that simultaneous high river flow is a prerequisite during the earthquake. Extremely rapid flash floods beyond the design capacity of the

FLOOD INUNDATION HAZARDS

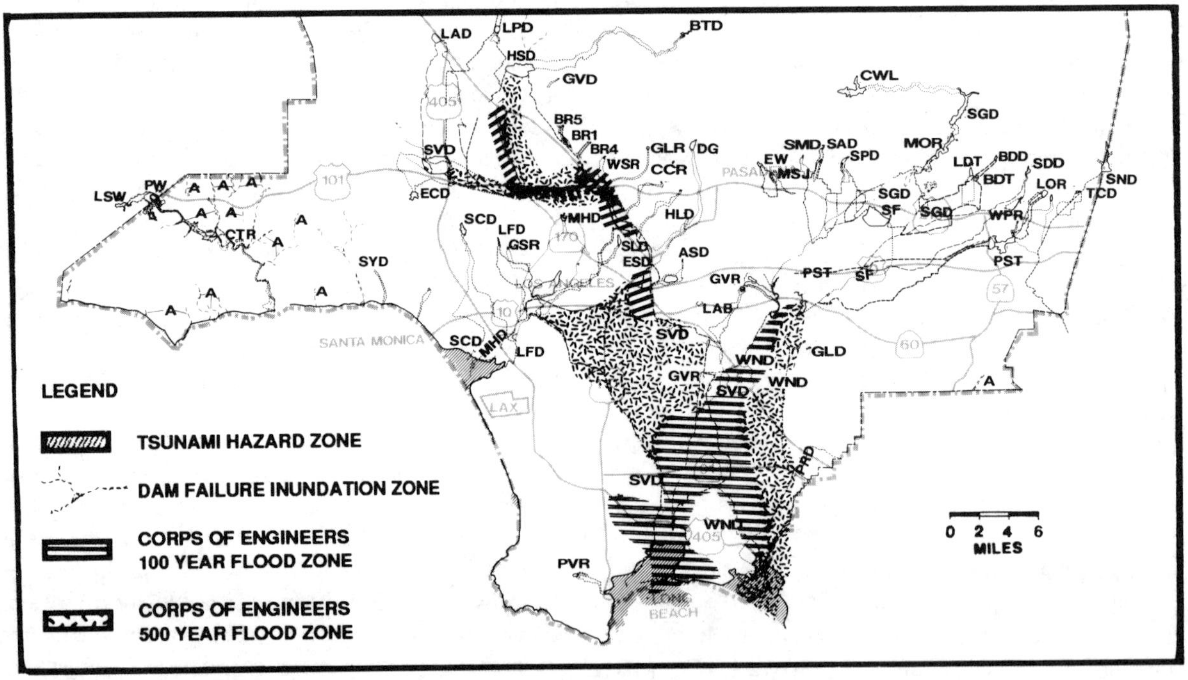

Figure 13. Flood inundation hazards map, modified from the Los Angeles County Safety Element (Leighton and Associates, 1990a). The figure shows the potential tsunami run-up areas along the western and southern coasts, the extent of postulated flooding for a 100- and 500-year flood in the southern basin, and the downstream hazard zones in case of a catastrophic dam failure at high water levels. Abbreviations are for names of the dams and debris basins. Not shown at this reduced scale are all of the 100 year flood zones for the smaller mountain drainages or the northern, unregulated portion of the County.

system can also cause failure of these flood control devices, as occurred in many parts of western Los Angeles and Ventura counties during the February 1992 storms. During those storms, over 14 inches of rain fell in a three day period, often at rates of one inch per hour.

Seiche Hazards

Seiches are oscillating waves on a body of water. Pronounced "saash", they can be triggered by long period earthquake ground motion from distant earthquakes, or by large displacements of the ground surface beneath a body of water. In reservoirs, seiches can generate overtopping and inundation of surrounding or downstream areas. In addition, earthquake-induced landsliding can also result in short-term overtopping. Seiches have not caused significant historical damage in the Los Angeles area and there is not, at this time, a readily-applicable method for predicting their occurrence or their area of inundation. In addition to reservoirs, seiching can be a significant damage factor in port and harbor facilities, particularly dock structures. A tsunami surge or simply the shock wave from even a moderately-sized near field earthquake can cause boats and ships docked in an embayment to rock as the seiche builds in amplitude. Most boat moorings are designed to accommodate the vertical motion of sea swells. Damage from a seiche is caused by strong lateral currents induced across the node of oscillation.

Storm Flooding

During the history of Southern California, the mountains, hillside areas, the coastal lowlands, and alluvial valleys have all proven vulnerable to various degrees of flooding from intense winter storms. In the 1940s, the U.S. Army Corps of Engineers designed the major flood control channels along the Los Angeles, San Gabriel, and Rio Hondo Rivers. These projects were built mainly in the 1950s. With increased urbanization, the standards and conditions under which these projects were built have changed, however.

Today, the Los Angeles County Department of Public Works, assisted by the Corps of Engineers, provides protection to life and property from flood hazards in most areas of the County, including the San Fernando and San Gabriel Valleys, the Los Angeles Basin, and some drainages in foothill and mountain areas. The structural flood protection system consists of roughly 2,000 miles of concrete-lined channels, 20 reservoirs, over 130 debris basins, as well as an extensive system of dams, levees, and stabilization structures (California Department of Water Resources, 1980 & 1988; Department of Public Works, 1986). The system is designed for the Corps of Engineers "standard project floods" which vary between 1-in-100 year to 1-in-250 year flood events depending on the drainage (California Department of Water Resources, 1980). New flood channels, if maintained by the Department of Public Works, have to be designed for a "capital flood", which is based on a rainfall with a probability of occurring once in 25 to 50 years, depending on the development project.

Despite localized failures, in general, the main channelized flood control system has performed well over the years. Tests occurred during the intense storms of 1978, and 1980. Statewide, these flood events caused a total of 56 deaths and an estimated 500 million dollars in property damage (Garza and Peterson, 1982; Chin et al., 1991).

Some problems and failures with the County structural system have occurred over the years, yet the cost-benefit ratio and performance of the system has proved that flood control is a necessary component of the urban infrastructure that should be adequately maintained (Evelyn, 1982). The severity of the storms was well within the flood level used in designing the flood control system, but was much less than that which could occur during the operating life of the project. These concerns were borne out in the storms of February 1992. The entire Los Angeles area received about 7 to 10 inches of rainfall, but the storms were most concentrated in the western portion of the Los Angeles region where over 16 inches was recorded. The flood control system did not necessarily fail in those areas, it was simply not designed to accommodate the rate at which precipitation occurred in a very localized area, nor was the rainfall prediction accurate enough to have initiated contingency plans.

Despite the high level of protection that the system provides against flood hazards in the Basin, some projects may still be at risk. After the intense storms of 1978 and 1980 in southern California, Pryde (1982) concluded that higher runoff resulting from increased urbanization may require upgrading of the major downstream flood control channels, facilities which were designed and built to older Corps of Engineer's standards. After the 1980 storms, the Corps developed new "flood frequency standards" to identify those areas that could be damaged by severe floods because of such deficiencies; preliminary 1-in-100 and very conservative 1-in-500 year flood limits have been prepared for several main drainages, including the San Gabriel, Rio Hondo, and Los Angeles Rivers (Figure 13). These maps pose a much more severe flood hazard than the flood limits used by the National Flood Insurance Program (NFIP), which are based on 1-in-100 year flood standards developed in 1974.

To avoid future deficiencies related to the adequacy of runoff parameters, the Department of Public Works has modified its flood design methodology. New Department of Public Work's projects must conform to current "level of flood and drainage protection standards", design standards that are based on rainfall flow rates. Unlike runoff rates, rainfall flow rates do not increase with increasing urbanization.

Northern portions of the county are not within the Los Angeles Flood County Control District because the area lacks a large enough tax base to install the necessary, but expensive, structural flood control facilities that would be required. Unregulated drainages are without structural flood control protection, such as concrete-lined channels or debris control basins. Unregulated drainages include those upstream of debris basins in mountainous areas which are also prone to flood problems. The rapid development occurring in the Santa Monica, Santa Susana, and San Gabriel Mountains makes this flood hazard an issue because new development can expedite the flow of flood waters from previously harmless drainages.

The guidelines currently used by the Federal Emergency Management Agency assume that the location of the stream channel on a fan is essentially random, with an equal probability of flooding occurring anywhere across the fan. Hence, all development on the fan is exposed to some probability of flooding (Dawdy et al., 1988). Significantly, this probability may not correlate with normal assumptions regarding 100 year and 500 year flood plains used for normal stream channels. Therefore, although Federal Emergency Management Agency Maps used to administer the National Flood Insurance Program (NFIP) show 100 year flood plains and 500 year flood plains in the Antelope Valley, the essentially random nature of flooding on alluvial fans emphasizes that some 500 year flood plains may be at higher risk than designated (Dawdy et al., 1988).

Structures such as levees, highway embankments, and lined channels, if capable of withstanding major floods, can confine probabilistically determined flow paths within specific portions of a fan. Detention or debris basins constructed above the fan apex can also regulate or reduce the flood water and sediment flow on to the fan, but should be viewed as a temporary solution. Maintenance of debris basins in active fan beds would be a very intensive operation, requiring almost annual debris removal.

The Flood and Inundation Hazard Map in the Safety Element portrays flood magnitude and frequency, showing those areas that would be affected by 1-in-100 and 1-in-500 year floods. The boundaries of flood hazard zones were determined from flood insurance rate maps issued by FEMA, and augmented with information supplied by the Los Angeles County Department of Public Works. Selected additional drainages in need of future flood plain management or additional flood protection, as identified by Public Works, are also included on the map.

SUBSIDENCE

Regional subsidence has affected large portions of the Los Angeles region. Broad areal subsidence associated with the withdrawal of underground fluids caused areas to subside between 2 and 5 feet through the mid-1960s throughout the Los Angeles Basin. A world record subsidence of 29 feet is due to petroleum extraction in the Long Beach area (Wentworth et al., 1970). Subsidence due to oil extraction in the Baldwin Hills is suspected of being a factor in the failure of the Baldwin Hills Dam in 1963 (see Proctor in chapter 9). To avoid future disasters similar to these justifies mitigation of regional subsidence through artificial recharge of water into oil reservoirs. The rapid expansion of development onto distal alluvial fan sediments in the Antelope Valley area, including sites near Lancaster, Palmdale, and Leona Valley, necessitates a concern for potential damage from collapsible soils. Restrictive landscaping and water use are difficult to regulate after completion and sale of homes. By not adequately considering collapsible soils mitigation in project design, and construction, widespread areas of a development can be damaged by differential subsidence.

Oil Field Subsidence

Subsidence related to oil extraction in the Los Angeles area became an issue in the 1950's and 60's. Of the 15 main oil fields in the Los Angeles Basin (Figure 14), the Beverly Hills/Cheviot Hills, Santa Fe Springs, Wilmington, and Inglewood oil fields displayed significant subsidence during that time period (Wentworth et al., 1970). In the Wilmington oil field, subsidence caused damage to industrial facilities, buildings, utilities, and transportation facilities. Some areas of Long Beach sank below sea level, and the threat of inundation prompted oil companies to recharge the reservoir rock to counteract the subsidence. This had the effect of repressurizing the oil reservoir and immediately began slowing subsidence rates. By the early 1960's, water-injection operations slowed subsidence in the other rapidly subsiding oil fields in the Los Angeles Basin (Castle and Yerkes, 1976). The State Division of Oil and Gas presently administer a post-oil production plan to ensure minimum subsidence. Similar post-fluid production plans, which include repressurization of underground aquifers and oil reservoirs, are advocated in other subsidence areas in the Los Angeles Basin (Castle and Yerkes, 1976).

Ground Water Withdrawal

Subsidence caused by ground water withdrawal is documented at many locations in southern California, such as the San Joaquin, Santa Clara, Elsinore, and San Jacinto Valleys. There are several areas in Los Angeles County that are also potentially sensitive to withdrawal of ground water. Depending on the depth and mechanical properties of the aquifer and overlying sediment, alluvial areas in the Antelope and Santa Clarita Valleys can subside if ground water resources are not managed properly. (See also Shlemon et al. in chapter 3.)

Considering the ongoing rapid development in the Antelope Valley area, subsidence is a critical issue to be thoroughly investigated. Severe damage to urban development could result if ground fissures develop in response to subsidence, and such ground rupture has been documented in the Antelope Valley. Holzer and Clark (1981), in an unpublished report for the Soil Conservation Service, described an arcuate earth fissure approximately 2000 feet long and up to 8 feet deep about 7 miles east-northeast of Lancaster. Additional ground fissuring north of

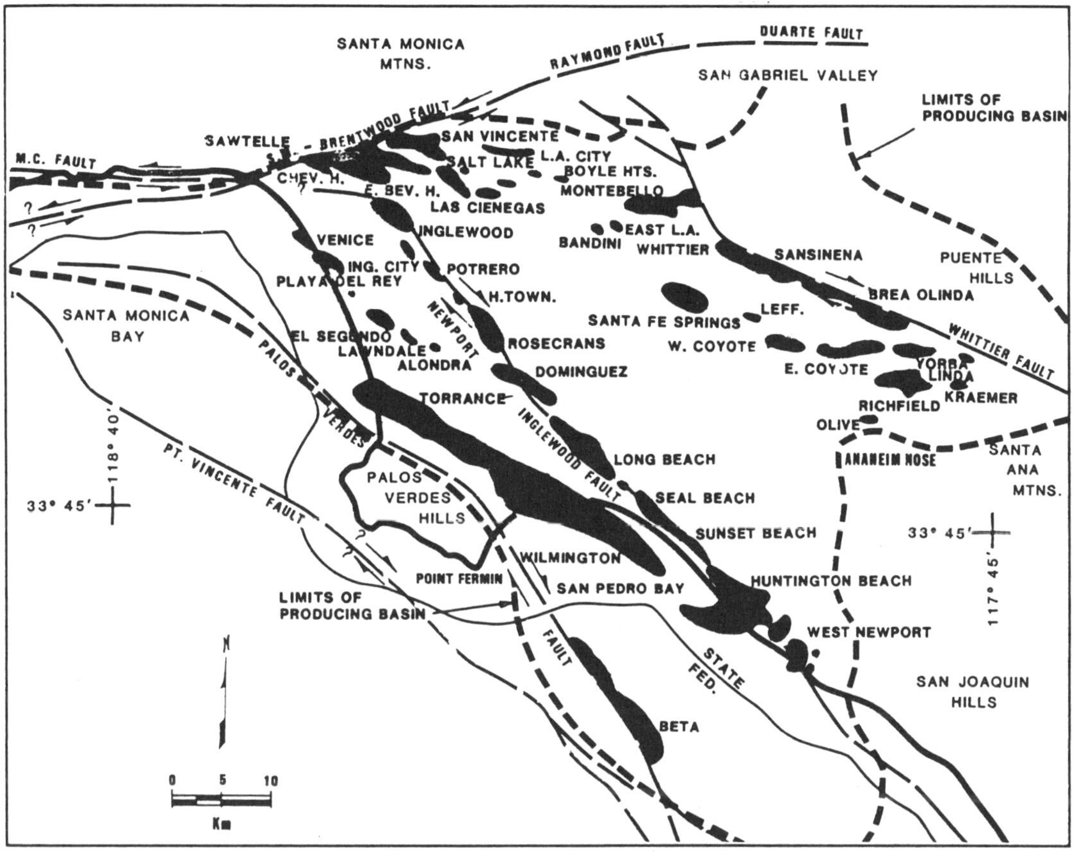

Figure 14. Map showing locations of producing oil fields in the Los Angeles basin.

Quartz Hill and in other localized areas is also reported but no regional studies have been conducted to determine the extent of the potential problem. When several jurisdictions are utilizing a continuous ground water aquifer resource such as in the Antelope Valley, subsidence management requires regional cooperation among the participating agencies.

Collapsible and Expansive Soils

Collapsible soils are low density, silty to very fine-grained, predominantly granular soils with high void ratios. (See Hollingworth and Grover in chapter 6.) They generally result from rapid deposition close to the source of the sediment, in either debris flow or aeolian deposits. When saturated, these soils undergo a rearrangement of their grains resulting in substantial and rapid settlement under relatively low loads. A rise in the ground water table or an increase in surface water infiltration, combined with the weight of a new building or structural fill, can initiate rapid settlement and cause foundations and walls to crack.

Expansive soils are widely distributed throughout the greater Los Angeles area. Although most likely to occur in those areas underlain by finer- grained sedimentary rocks, they could be found locally nearly anywhere in the region. Proper site investigation, soil testing, foundation design, and enforcement of construction grading practices are the most efficient tools for locating, assessing the severity of, and mitigating expansive soils in a proposed development. Presaturation (pre-expansion) and/or post-tensioned foundation slabs are the most prevalent methods used for mitigating expansive soils.

EXPLOSIVE METHANE

Several local areas have experienced episodes of methane accumulations. The best known is the Fairfax District in Los Angeles where, in March 1985, a methane explosion damaged a Ross Dress for Less store, injuring 21 persons, and damaging other nearby buildings (see papers in chapter 2, this book). As a result of this tragedy, the City of Los Angeles enacted an ordinance requiring businesses in the area to install gas meters and monitor them daily. The methane occurrences are associated with oil and gas, as the Fairfax District lies directly over the Salt Lake oil field (Fig. 14). New development in abandoned oil fields must consider the lessons of the Fairfax District. Methane is a common occurrence wherever old oil well casings have corroded and lost their seal integrity. New construction can be safely designed despite the methane threat through the use of vents, floor slab vapor barriers, automatic sensing alarms, and community awareness.

HAZARD MITIGATION POLICIES

Safety Element's are required components of a city or county's General Plan under Sections 65302(g) and 8875 of the California Government Code. Established as a requirement after the 1971 San Fernando earthquake, the Safety Element provides the long range planning guidelines and technical documentation for hazard identification and mitigation in the public sector. In February, 1991, the Los Angeles County Board of Supervisors adopted a totally revised and up to date Safety Element (Leighton and Associates, 1990b). The objective of this new Safety Element is to reduce the consequences of a natural disaster to the citizens of Los Angeles County. Through its goals and policy statements, the Los Angeles County Safety Element states the philosophy and direction taken by the County to serve its citizens. and also provides the geologic, cultural, and hydrologic data needed for hazard reduction awareness and planning.

The Safety Element incorporates natural and urban hazard awareness into the framework of Los Angeles County through policy, regulation, ordinances, and zoning. Both new and existing developments are addressed, and action programs are provided. Existing emergency response, preparedness, and recovery options are an integral part of the County's daily departmental operations.

Much of the present and future development in the Los Angeles region is and will be into areas of increasingly greater geologic complexity, and hazard identification and mitigation will continue to increase in importance for most projects (Sanchez et al., 1991). The County already mandates site specific investigations for most new projects, and provides a technical review of all submitted project consulting reports. Adoption of the Safety Element Hazard Maps will facilitate both the consultant and the agency reviewer's efforts by providing a common medium for focusing the investigation efforts. Future developments are regulated under the California Environmental Quality Act (CEQA) and several provisions of existing County codes, including: Title 21: Subdivision Code, Title 22: Zoning Code, Title 26: Building Code, and Title 32: Fire Code.

Minimum investigation standards are mandated for most new projects. The incorporation of a set of hazard maps into the Zoning Code as a Hazard Overlay Zone was initially proposed, but considered unnecessary due to the strict requirements for detailed investigations at the site specific level. Nevertheless, the Safety Element's ready availability of county-wide hazards maps will facilitate the public dissemination of hazard information for siting decisions. Existing development located in hazard zones is a more difficult problem. Severe financial loss could be incurred by property owners, many of whom do not have the resources necessary to bring a property up to current code conformance. Despite this concern, the hazards are real, and will not go away by ignoring them.

The implementation of the Safety Element is the responsibility of many County departments. One of the major contributions is to increase public awareness and support of safety conscious planning. There are numerous other demands on resources and time, but decision makers are responsive to public priorities (Sanchez et al., 1991). Public agencies alone cannot achieve or afford to eliminate natural hazards. Most of the structures in the Los Angeles region are private property, and unless or until the private property owner embraces the concept of a safer environment, disaster will continue to occur to both the property owner, and the occupants of the property.

ACKNOWLEDGMENTS

This paper was revised and extensively abstracted from the Technical Appendix to the Safety Element of the Los Angeles County General Plan, prepared by Leighton & Associates (1990a), with Paul Sanchez as the principal investigator. His contribution to that effort is gratefully acknowledged, as is that of Jene McKnight, the Los Angeles County Department of Regional Planning's project manager. Acknowledgement and appreciation is also made to Bruce Clark for his review and to Dorothy Bergman for preparation of the graphics, both of Leighton and Associates. Thanks are due to the editors of this volume for their review, guidance, and patience.

Author's Note: Copies of the Los Angeles County Safety Element will be available by the time this paper is published, although neither the price nor the sales location have been finalized. For purchase information, contact the Los Angeles County Department of Regional Planning, 13th Floor of the County Administration Building, 320 W. Temple Street, Los Angeles, CA 90012.

REFERENCES

Atwater, T., and Molnar, P., 1973, Relative motion of the Pacific and North American plates deduced from sea-floor spreading in the Atlantic, Indian, and South Pacific oceans: in Kovach, R.L., and Nur, A., (eds.), Proceedings of the conference on tectonic problems of the San Andreas fault system: Stanford University Publications, Geological Sciences, v. 13, p. 136-148.

Barrows, A.G, 1974, A review of the geology and earthquake history of the Newport-Inglewood structural zone, southern California: California Division of Mines and Geology Special Report 114, 115 p.

California Department of Water Resources, 1980, California flood management: an evaluation of flood damage prevention programs: Department of Water Resources Bulletin 199, 277 p.

———, 1988, Dams within jurisdiction of the State of Californi:; Department of Water Resources Bulletin 17-88, 121 p.

California Division of Mines and Geology (CDMG), 1987, Landslide hazard identification program: Annual report 1987; 8 p.

Camfield, P.E., 1987, Insufficient data effects on tsunami flood level predictions summar: in Bernard E.N., (ed.), Proceedings of the international tsunami symposium; International Union of Geodesy and Geophysics, National Technical Information Service, U.S. Department of Commerce, Springfield, VA, p. 247-251.

Campbell, R.H., 1975, Soil slips, debris flows, and rainstorms in the Santa Monica Mountains and vicinity, Southern California: U.S. Geological Survey Professional Paper 851, 50 p.

———, 1985, Methods and costs for the delineation of susceptibility to mudflows and other landslide:; in Campbell, R.H., (ed.), Feasibility of a nationwide program for identification and delineation of hazards from mudflows and others; U.S. Geological Survey Open File Report 85-276, 109 p.

Castle, R.O., and Yerkes, R.F., 1976, Recent surface movements in the Baldwin Hills, Los Angeles County, California: U.S. Geological Survey Professional Paper 882, 125 p.

Chin, E.H., Aldridge, B.N., and Longfield, R.J., 1991, Floods of February 1980 in southern California and central Arizona: U.S. Geological Survey Professional Paper 1494, 126 p.

Clark, B.R., Leighton, F.B., and Cann, L.R., 1979, Surficial landslides triggered by seismic shaking, San Fernando earthquake of 1971; U.S. Geological Survey Contract No. 14-08-001-16810, 42 p.

Clark, S.H., Jr., Greene, H.G., and Kennedy, M.P., 1985, Identifying potentially active faults and unstable slopes offshore; in Ziony, J.I, (ed.), Evaluating earthquake hazards in the Los Angeles region - an earth- science perspective; U.S. Geological Survey Professional Paper 1360, p. 347- 374.

Clements, T., 1982, St. Francis dam failure of 1928; in Fife, D.L, and Minch, J.A., (eds.), Geology and mineral wealth of the California Transverse Ranges; South Coast Geological Society, Santa Ana, CA, p. 657-660

Cobarrubias, J.W., (Chairman), 1985, Methane gas task force report; Unpublished technical report to the Los Angeles City Council on the Fairfax District explosion and fire, dated June 10, 1985.

Cobarrubias, J.W., (Chairman), 1990, Task force report II on the methane gas incursion, Fairfax area, City of Los Angeles; Unpublished technical report to the Los Angeles City Council, Council file 88-1826, dated January 5, 1990.

Crook, R. Jr., Allen, C. R., Kamb, B., Payne, C. M., and Proctor, R. J., 1987. Quaternary geology and seismic hazard of the Sierra Madre and associated faults, western San Gabriel Mountains, California; U. S. Geological Survey Professional Paper 1339, p. 27-63, and 1978 California Institute of Technology, Contribution no. 3191.

Davis, J.M., 1982, Rare and unusual post-fire flood events experienced in Los Angeles County during 1978 and 1980; in National Research Council, Committee on Natural Disasters, Storms, floods, and debris flows in southern California and Arizona 1978 and 1980; Proceedings of a Symposium, National Academy Press, p. 243-256.

Davis, J., Bennett, J., Borchardt, G., Kahle, J., Rice, S., and Silva, M., 1982, Earthquake planning scenario for a magnitude 8.3 earthquake on the San Andreas fault in southern California; California Division of Mines and Geology Special Publication 60, 128 p.

Davis, T.L., Namson, J., and Yerkes, R.F., 1989, A cross section of the Los Angeles area: seismically active fold and thrust belt, the 1987 Whittier Narrows earthquake, and earthquake hazard; J. of Geophysical Research, v. 94, n. B7, p. 9644-9664.

Dawdy, D.R., Wang, W., Hill, J., 1988, The delineation of flood hazard areas on alluvial fans for the Flood Insurance Program; in Natural Hazard Reduction, Contribution from Geology, A Professional Conference (October 13- 14), p. 1-9.

Department of Public Works, 1986, Level of Flood Protection, County of Los Angeles, File No. 2-15.321.

Evernden, J.F., and Thomson, J.M., 1985, Predicting seismic intensities; in Ziony, J.I., (ed.), Evaluating earthquake hazards in the Los Angeles region - an earth-science perspective; U.S. Geological Survey Professional Paper 1360, p. 151-202.

Evelyn, J.B., 1982, Operation and performance of Corps of Engineers flood control projects in southern California and Arizona during 1978 and 1980; in National Research Council, Committee on Natural Disasters, Storms, floods, and debris flows in southern California and Arizona 1978 and 1980; Proceedings of a Symposium, National Academy Press, p. 131-150.

Garza, C., and Peterson, C., 1982, Damage producing winter storms of 1978 and 1980 in southern California: a synoptic view; in National Research Council, Committee on Natural Disasters, Storms, floods, and debris flows in southern California and Arizona 1978 and 1980; Proceedings of a Symposium, National Academy Press, p. 43-53.

Gilbert, G.K., 1909, Earthquake forecasts; Science N.S., v. 29, n. 734, p. 121-136.

Goter, S.K., 1988, Seismicity of California, 1808-1987; National Earthquake Information Center Open File Report 88-286.

Goto, C., and Shuto, N., 1983, Numerical simulation of tsunami propagations and run-up; in Iida, K., and Iwasaki, T., (eds.), Tsunamis - their science and engineering; Terra Scientific Publishing Company, Tokyo, Japan, p. 439- 451.

Griggs, G., and Savoy, L., 1983, Living with the California coast; Duke University Press, Durham, NC, 395 p.

Gulliver, R.M., 1986, Some insights on the use of shaking intensities in earthquake loss studies; in Brown, W.M., III, et. al., (eds.), Future directions in evaluating earthquake hazards in southern California, Proceedings of Conference XXXII; U.S. Geological Survey Open File Report 86- 401, p. 158-172.

Harp, E.H., and Keefer, D.K., 1985, Landsliding: predicted geologic and seismological effects of a postulated magnitude 6.5 earthquake along the northern part of the Newport-Inglewood zone; in Ziony, J.I., (ed.), Evaluating earthquake hazards in the Los Angeles region - an earth-science perspective; U.S. Geological Survey Professional Paper 1360, p. 434-436.

Hart, E.W., 1990, Fault rupture hazard zones in California; California Division of Mines and Geology Special Publication 42, 25 p.

Hauksson, E., 1987, Seismotectonics of the Newport-Inglewood fault zone in the Los Angeles basin, southern California; Bulletin of the Seismological Society of America, v. 77, n. 2, p. 539-561.

Hauksson, E., 1990, Earthquakes, faulting, and stress in the Los Angeles basin; Journal of Geophysical Research, v. 95, n. B10, p. 15,365-15,394.

Hauksson, E., Jones, L.M., Davis, T.L., Hutton, L.K., Brady, A.G., Reasenberg, P.A., Michael, A.J., Yerkes, R.F., Williams, P., Reagor, G., Stover, C.W., Bent, A.L., Shakal, A.K., Etheredge, E., Porcella, R.L., Bufe, C.G., Johnston, M.J.S., and Cranswick, E., 1988, The 1987 Whittier Narrows earthquake in the Los Angeles metropolitan area, California; Science, v. 239, p.1409-1412.

Hauksson, E. and Jones, L.M., 1989, The 1987 Whittier Narrows earthquake sequence in Los Angeles, southern California: seismological and tectonic analysis: Journal of Geophysical Research, v. 94, n. B7, p. 9569-9589.

Holzer, T.L., and Clark, M.M., 1981, Earth fissuring in the Antelope Valley; Unpublished U.S. Geological Survey report for the Soil Conservation Service.

Houston, J.R., 1980, Type 10 flood insurance study: tsunami predictions for southern California; U.S. Army Corps of Engineers, Waterways Experiment Station Technical Report HL-80-18, 172 p.

Houston, J.R., and Garcia, A.W., 1974, Type 16 flood insurance study: tsunami predictions for Pacific coastal communities; U.S. Army Corps of Engineers, Waterways Experiment Station Technical Report H-74-3.

James, L.B., 1968, Failure of the Baldwin Hills reservoir, Los Angeles, California; in Kiersch, G.A., (ed.), Engineering geology case histories, number 6; Geological Society of America, p. 1-12.

Joyner, W.B., and Fumal, T.E., 1985, Predictive mapping of earthquake ground motion; in Ziony, J.I., (ed.), Evaluating earthquake hazards in the Los Angeles region - an earth-science perspective; U.S. Geological Survey Professional Paper 1360, p. 203-217.

Keefer, D.K., 1984, Landslides caused by earthquakes; Geological Society of America Bulletin, v. 95, p. 406-421.

Lajoie, K.R., and Helley, E.J., 1975, Classification and mapping of Quaternary sediments for purposes of seismic zonation; in Borchердt, R.D., (ed.), Studies for seismic zonation of the San Francisco Bay region; U.S. Geological Survey Professional Paper 941-A, p. A39-A51.

Lander, J.F., 1987, Impact of tsunamis on the U.S. and associated territories; in Bernard E.N., (ed.), Proceedings of the international tsunami symposium, International Union of Geodesy and Geophysics; National Technical Information Service, U.S. Department of Commerce, Springfield, VA, p. 49-58.

Lander, J.F, and Lockridge, P.A., 1989, United States tsunamis (including United States territories) 1690-1988; U.S. Department of Commerce Publication 41-2, 265 p.

Lee, Y.K., Wu, F.H., Wier, W., Parnicky, P., and DeMers, D., 1982, Methodology for computing coastal flood statistics in southern California; Federal Emergency Management Agency Report TC-3205, Tetra Tech, Inc., Pasadena, California, 75 p.

Leighton and Associates, 1990a, Hazard reduction in Los Angeles County: Technical Appendix of the Safety Element, Los Angeles County General Plan; Department of Regional Planning, 458 p., 8 plates.

Leighton and Associates, 1990b, Safety Element of the General Plan, County of Los Angeles; Department of Regional Planning, 48 p.

Lettis, W.R. and Hanson, K.L., 1991, Crustal strain partitioning: implications for seismic hazard assessment in western California; Geology, v. 19, p. 559-562.

Liu, P.L., and Erickson, J., 1983, A numerical model for tsunami generation and propagation; in Iida, K., and Iwasaki, T., (eds.), Tsunamis - their science and engineering; Terra Scientific Publishing Company, Tokyo, Japan, p. 227-240.

McCulloch, D.S., 1985, Evaluating tsunami potential; in Ziony, J.I., (ed.), Evaluating earthquake hazards in the Los Angeles region - an earth-science perspective; U.S. Geological Survey Professional Paper 1360, p. 375-413.

Morton, D.M., and Kennedy, M.P., 1979, Landsliding and mudflows at Wrightwood, San Bernardino County, California; California Division of Mines and Geology Special Report 136, 21 p.

Morton, D.M., and Streitz, R., 1969, Preliminary reconnaissance map of major landslides, San Gabriel Mountains, California; California Division of Mines and Geology MS-15, Scale 1:62,500.

National Research Council, 1982, Selecting a methodology for delineating mudslide hazard areas for the National Flood Insurance Program; Advisory Board on the Built Environment, National Academy Press, Washington, D.C., 35 p.

Outland, C.F., 1963, Man-made disaster: the story of the St. Francis Dam; Arthur H. Clarke Co., Glendale, CA., 249 p.

Ponti, D.J., 1985, The Quaternary alluvial sequence of the Antelope Valley, California; in Weide, D.L., and Faber, M.L., (eds.), Soils and Quaternary geology of the southwestern United States; Geological Society of America Special Paper 203, p. 79-96.

———, 1989, Aminostratigraphy and chronostratigraphy of Pleistocene marine sediments, southwestern Los Angeles basin, California; unpublished PhD dissertation, University of Colorado at Boulder, 409 p.

Pryde, P.R., 1982, Comment on peak floodflows in San Diego's Mission Valley; in National Research Council, Committee on Natural Disasters, Storms, floods, and debris flows in southern California and Arizona 1978 and 1980; Proceedings of a Symposium, National Academy Press, p. 205-206.

Redin, T., 1991, Oil and gas production from submarine fans of the Los Angeles basin; in Biddle, K.T., (ed.), Active margin basins; American Association of Petroleum Geologists Memoir 52, p. 239-259.

Reneau, S.L., and Dietrich, W.E., 1987, The importance of hollows in debris flow studies: examples from Marin County, California; in National Hazard Reduction, Contribution from Geology; Geological Society of America Reviews in Engineering Geology, v. VII, p. 165-180.

Rogers, A.M., Tinsley, J.C., and Borcherdt, R.D., 1985, Predicting relative ground response; in Ziony, J.I., (ed.), Evaluating earthquake hazards in the Los Angeles region - an earth-science perspective; U.S. Geological Survey Professional Paper 1360, p. 221-248.

Sanchez, P.E., Ziony, J.I., McKnight, J.S., Clark, B.R., and Gath, E.M., 1991, Seismic zonation of the Los Angeles region; Proceedings of the Fourth International Conference on Seismic Zonation, Earthquake Engineering Research Institute, Oakland, CA, p. 797-844.

Sharpe, R., 1982, An investigation of the correlation between earthquake ground motion and building performance; prepared for the Applied Technology Council, U.S. Geological Survey, Contract No. 14-08-0001-19892, 113 p.

Shuto, N., 1987, Spread of oil and fire due to tsunami; in Bernard E.N., (ed.), Proceedings of the international tsunami symposium; International Union of Geodesy and Geophysics, National Technical Information Service, U.S. Department of Commerce, Springfield, VA, p. 247-251.

Slosson, J.E., and Krohn, J.P., 1982, Southern California landslides of 1978 and 1980; in National Research Council, Committee on Natural Disasters, Storms, floods, and debris flows in southern California and Arizona 1978 and 1980; Proceedings of a Symposium, National Academy Press, p. 291-319.

Spangle, W.E., 1987, Pre-earthquake planning for post-earthquake rebuilding (PEPPER); Report to City of Los Angeles, 96 p.

Tinsley, J.C. and Fumal T.E., 1985, Mapping Quaternary sedimentary deposits for areal variations in shaking response; in Ziony, J.I., (ed.), Evaluating earthquake hazards in the Los Angeles region - an earth-science perspective; U.S. Geological Survey Professional Paper 1360, p. 101-125.

Tinsley, J.C., Youd, T.L., Perkins, D.M., and Chen A.T.F., 1985, Evaluating liquefaction potential; in Ziony, J.I., (ed.), Evaluating earthquake hazards in the Los Angeles region - an earth-science perspective; U.S. Geological Survey Professional Paper 1360, p. 263-315.

Toppozada, T.R., Bennet, J.H., Borchardt, G., Saul, R., and Davis, J.F., 1988, Planning scenario for a major earthquake on the Newport-Inglewood fault zone; California Division of Mines and Geology Special Publication 99, 193 p.

Toppozada, T.R, and Parke, D.L., 1982, Areas damaged by California earthquakes, 1900-1949; California Division of Mines and Geology Open File Report 82-17 SAC, 65 p.

Toppozada, T.R, Real, C.R., and Parke, D.L, 1980, Preparation of isoseismal maps and summaries of reported effects for pre-1900 California earthquakes; California Division of Mines and Geology Open File Report 81-11 SAC, 182 p.

Treiman, J.A., 1987, Landslide hazards in the east half of the Newhall quadrangle, Los Angeles County, California; California Division of Mines and Geology Landslide Hazard Identification Map No. 7., Open File Report 86-16 LA.

Urban Regional Research (URR), 1988, Planning for risk: comprehensive planning for tsunami hazard areas; National Science Foundation Grant CEE 8408237, 246 p.

Weber, F.H., 1975, Surface effects and related geology of the San Fernando earthquake in the Sylmar area; in Oakeshott, G.B., (ed.), San Fernando, California, earthquake of 9 February, 1971; California Division of Mines and Geology Bulletin 196, p. 71-96.

———, 1980, Landsliding and flooding in southern California during the winter of 1979-1980; California Division of Mines and Geology Open File Report 80-3, 67 p.

———, 1982, Landsliding and flooding in southern California during the winter of 1979-80; in National Research Council, Committee on Natural Disasters, Storms, floods, and debris flows in southern California and Arizona 1978 and 1980; Proceedings of a Symposium, National Academy Press, p. 321-332.

———, 1984, Geology of the Calabasas-Agoura eastern Thousand Oaks area, Los Angeles and Ventura counties, California; California Division of Mines and Geology Open File Report 84-1LA, 191 p.

Weber, F.H., Hsu, E.Y., Saul, D., Tan, S.S., and Treiman, J.A., 1982, Slope stability and geology of the Baldwin Hills, Los Angeles County, California: California Division of Mines and Geology Special Report 152, 93 p.

Wentworth, C.M., Ziony, J.I., Buchanan, J.M., 1970, Preliminary geologic environmental map of the greater Los Angeles areas, California; U.S. Geological Survey TID-25363, 41 p.

Wesnousky, S.G., 1986, Earthquakes, Quaternary faults, and seismic hazard in California; Journal of Geophysical Research, v. 91, n. B12, p. 12,587- 12,631.

Whitney, R.A, and Gath, E.M., 1991, Structure, tectonics, and surface rupture hazard at the Las Posas anticline, Ventura County, California; in Blake, T.F., and Larson, R.A., (eds.), Engineering geology along the Simi- Santa Rosa fault system and adjacent areas, Simi Valley to Camarillo, Ventura County, California; Association of Engineering Geologists, Southern California section field trip guidebook, v. 2, p. 164-182.

Wilson, R.C, and Keefer, D.K, 1985, Predicting areal limits of earthquake- induced landsliding; in Ziony, J.I., (ed.), Evaluating earthquake hazards in the Los Angeles region - an earth-science perspective; U.S. Geological Survey Professional Paper 1360, p. 317-338.

Wood, H.O., 1933, Preliminary report on the Long Beach earthquake; Bulletin of the Seismological Society of America, v. 23, p. 43-56

Working Group on California Earthquake Probabilities, 1988, Probabilities of large earthquakes occurring in California on the San Andreas fault; U.S. Geological Survey Open File Report 88-398, 62 p.

Wright, T.L., 1991, Structural geology and tectonic evolution of the Los Angeles basin, California; in Biddle, K.T., (ed.), Active margin basins; American Association of Petroleum Geologists Memoir 52, p. 35-134.

Yen, B.C., Wang, W.L., and McNeill, R.L., 1978, Seismically induced shallow hillside failures; U.S. Geological Survey Technical Report, Contract 14-08-0001-G-512, 19 p.

Yerkes, R.F., McCulloh, T.H., Schoellhamer, J.E., and Vedder, J.B., 1965, Geology of the Los Angeles basin, California - an introduction; U.S. Geological Survey Professional Paper 420-A, 57 p.

Youd, T.L., Tinsley, J.C., Perkins, D.M., King, E.J., Preston, R.F., 1978, Liquefaction map for San Fernando Valley, California; Second international conference on microzonation for safer construction - research and applications; Proceedings, San Francisco, California, v. 1, p. 267-278.

Youd, T.L., and Perkins, D.M., 1978, Mapping liquefaction-induced ground failure potential; Proceedings of the American Society of Civil Engineers, Journal of Geotechnical Engineering Division, v. 104, n. GT4, p. 433-446.

Ziony, J.I. and Jones, L., 1989, Map showing late Quaternary faults and 1978 through 1984 seismicity of the Los Angeles region, California; U.S. Geological Survey MF 1964, 1:250,000.

Ziony, J.I., and Yerkes, R.F, 1985, Evaluating earthquake and surface faulting potential; in Ziony, J.I., (ed.), Evaluating earthquake hazards in the Los Angeles region - an earth-science perspective; U.S. Geological Survey Professional Paper 1360, p. 43-92.

Photo 3. Landslide in Los Feliz district of Los Angeles resulted in rock mass pushing house into street.

Photo 4. Sudden debris flood of February 1978 out of Shields Canyon in La Crescenta flowed over roof of house. Note boulders inside car. (Photos by Richard Proctor)

TABLE 1: LATE QUATERNARY FAULTS IN THE LOS ANGELES REGION

ABR. FIG. 4	FAULT NAME	ACTIVITY STATUS	SEGMENT LENGTH	MAXIMUM CREDIBLE EQ	SLIP RATE
✖	Anacapa-Dume	Potentially Active*	45 km	7.0	Not Reported
CF	Cabrillo	Active**	18 km	6.2	Not Reported
CAF	Central Avenue	Potentially Active	8 km	6.1	Not Reported
CF	Charnock	Potentially Active	>10 km	6.2	Not Reported
✖	Chino	Active	28 km	?	<0.2 mm/yr
CSPFZ	Clamshell-Sawpit	Active	13 km	6.4	Not Reported
CWF	Clearwater	Potentially Active	33 km	6.9	Not Reported
✖	Cucamonga	Active	20 to 25 km	6.6	2.9 to 6.4 mm/yr
DVF	Del Valle	Potentially Active	4 km	5.8	Not Reported
ERF	Eagle Rock	Potentially Active	5 km	6.3	Not Reported
HWF	Hollywood	Active	17 km	6.4	0.3 to 0.8 mm/yr
HF	Holser	Active	19 km	6.6	Not Reported
IHF	Indian Hill	Potentially Active	9 km	6.2	Not Reported
LF	Llano	Active	>6 km	6.1	Not Reported
LAF	Los Alamitos	Potentially Active	12 km	?	Not Reported
MCOF	Malibu Coast	Active	>27 km	6.9	0.04 to 0.09 mm/yr
MVFZ	Mirage Valley	Potentially Active	30 km	6.8	Not Reported
MHF	Mission Hills	Active	10 km	6.3	Not Reported
MLF	Mount Lukens	Potentially Active	7 km	?	Not Reported
NIFZ	Newport-Inglewood	Active	>43 km	6.9	0.1 to 6.0 mm/yr
NFZ	North Hollywood	Active	2 km	?	Not Reported
NF	Northridge	Active	15 to 21 km	6.6	Not Reported
NF	Norwalk	Potentially Active	>14 km	?	Not Reported
OV	Overland	Potentially Active	>10 km	6.2	Not Reported
PVHFZ	Palos Verdes Hills	Active	>77 km	6.7	0.02 to 0.7 mm/yr
RF	Raymond	Active	22 km	6.7	0.1 to 0.2 mm/yr
RCF	Redondo Canyon	Active	13 km	6.4	Not Reported
SAF	San Andreas	Active	>120 km	8.0	30.0 mm/yr
SANF	San Antonio	Active	18 km	6.2	Not Reported
SFFZ	San Fernando	Active	17 km	6.5	Not Reported
SGF	San Gabriel	Active-Pot. Active	130 km	>7.0	5.0 mm/yr
SJF	San Jose	Active	14 km	6.7	Not Reported
SPBFZ	San Pedro Basin	Potentially Active	70 km	>7.0	Not Reported
SRF	San Rafael	Potentially Active	6 km	6.3	Not Reported
✖	Santa Catalina Ridge	Potentially Active	>120 km	>7.0	Not Reported
✖	Santa Cruz Island	Active	68 km	?	0.2 to 0.9 mm/yr
SFF	Santa Felicia	Potentially Active	8 km	6.1	Not Reported
SMF	Santa Monica	Active	>40 km	6.7	0.1 to 1.0 mm/yr
✖	Santa Rosa Island	Potentially Active	72 km	?	Not Reported
SSF	Santa Susana	Active	28-38 km	6.9	3 to 4.5 mm/yr
SMAF	Sierra Madre	Active	62 km	>7.0	<0.4 mm/yr
VF	Verdugo	Active	21 km	6.7	Not Reported
WCF	Walnut Creek	Potentially Active	17 km	6.5	Not Reported
WFZ	Whittier	Active	>40 km	>7.0	1.7 to 5.5 mm/yr

* "Potentially Active" is used here to classify those faults for which there is evidence for mid-Pleistocene age offsets, but evidence for or against Holocene activity has not (yet) been developed.

** "Active" is used here to indicate a fault for which there is geologic or geomorphic evidence to infer Holocene offset, or if there is current seismologic activity on the fault structure.

✖ Faults not shown on Figure 4.

Compiled of multiple sources from Appendix 2 of Hazard Reduction in Los Angeles County (Leighton and Associates, 1990a). Updated based on preliminary research findings by the Southern California Research Center and other researchers.

Table 2: Damaging Earthquakes in Southern California 1800-1991

Event No.-	Year	Month	Day	Latitude North	Longitude West	Mag. Richter	MMI Mercalli	Fault or Structure
1--	1800	Nov.	22	33.0	117.3	6.5	VII	Rose Canyon (?)
2--	1812	Dec.	8	33.7	118.1	6.9	VII	Newport-Inglewood or San Andreas
3--	1812	Dec.	21	34.2	119.9	7.1	VIII	Santa Barbara Channel
4--	1855	July	11	34.1	118.1	6.0	VIII	Raymond (?)
5--	1857	Jan.	9	35.3	119.8	7.9	IX	San Andreas
6--	1858	Dec.	16	34.0	117.3	6.5	IX	San Andreas or San Jacinto
7--	1892	Feb.	24	32.6	115.7	6.8	IX	Laguna Salada (Elsinore)
8--	1893	Apr.	4	34.3	118.7	5.5	VIII	Santa Susana or Simi (?)
9--	1899	July	22	34.3	117.5	6.7	VIII	San Andreas (?)
10--	1899	Dec.	25	33.8	117.0	6.8	IX	San Jacinto
11--	1902	July	28	34.7	120.4	5.4	VIII	Los Alamitos
12--	1907	Sept.	20	34.2	117.4	6.0	VII	——(?)
13--	1910	May	15	33.7	117.4	6.0	VII	San Jacinto
14--	1912	Dec.	14	34.0	119.0	5.0	VII	White Wolf
15--	1915	Jan.	12	34.73	120.23	5.8	VIII	Newport-Inglewood
16--	1916	Oct.	23	34.9	118.9	5.6	VII	San Jacinto
17--	1918	Apr.	21	33.75	117.00	6.8	IX	Santa Barbara Channel
18--	1919	Feb.	16	35.0	119.0	5.0	VII	Santa Ynez (?)
19--	1920	June	22	33.97	118.37	4.7	VIII	Offshore
20--	1923	July	23	34.0	117.25	6.0	VII	Norwalk
21--	1925	June	29	34.3	119.8	6.8	IX	Santa Barbara Channel
22--	1926	June	29	34.5	119.7	5.5	VII	Anacapa
23--	1927	Nov.	4	34.6	120.9	7.3	IX	Newport-Inglewood
24--	1929	July	8	33.9	118.04	4.7	VIII	San Jacinto
25--	1930	Aug.	5	34.3	119.5	5.0	VII	Santa Barbara Channel
26--	1930	Aug.	31	33.95	118.68	5.2	VII	Anacapa
27--	1933	Mar.	11	33.62	117.97	6.3	IX	Newport-Inglewood
28--	1937	Mar.	25	33.41	116.26	6.0	VII	San Jacinto
29--	1941	July	1	34.37	119.58	6.0	VIII	Santa Barbara Channel
30--	1941	Oct.	22	33.8	118.2	4.9	VII	Newport-Inglewood (?)
31--	1941	Nov.	14	33.78	118.25	5.4	VII	Newport-Inglewood (?)
32--	1942	Oct.	21	32.97	116.0	6.6	VII	San Jacinto
33--	1947	Apr.	10	34.98	116.55	6.5	VII	Unknown (prob. not Mannix)
34--	1948	Dec.	4	33.93	116.38	6.0	VII	Mission Creek
35--	1952	July	21	35.0	119.02	7.5	XI	White Wolf
36--	1954	Jan.	12	35.0	119.02	5.9	VII	White Wolf
37--	1968	Apr.	9	33.19	116.13	6.5	VII	Coyote Creek (San Jacinto)
38--	1969	Apr.	28	33.34	116.35	5.8	VII	San Jacinto
39--	1970	Sept.	12	34.27	117.54	5.4	VII	No. San Jacinto
40--	1971	Feb.	9	34.41	118.40	6.6	IX	San Fernando
41--	1971	Mar.	31	34.29	118.52	5.1	VII	San Fernando
42--	1973	Feb.	21	34.07	119.04	5.3	VII	Anacapa
43--	1973	Aug.	6	33.97	119.45	5.0	—	Santa Cruz Island
44--	1975	May	31	34.52	116.49	5.2	—	Galway Lake
45--	1978	Aug.	13	34.40	119.68	5.1	VI	Santa Barbara Channel
46--	1979	June	30	34.23	116.89	4.9	—	San Bernardino Mountains
47--	1980	Feb.	25	33.52	116.55	5.6	—	San Jacinto
48--	1981	Sept.	4	33.68	119.10	5.3	—	No. of Santa Barbara Island
49--	1985	Oct.	2	34.030	117.25	4.7	—	——(?)
50--	1987	Oct.	1	34.06	118.08	5.9	VIII	Elysian Park
51--	1987	Oct.	4	34.07	118.10	5.3	VII	Elysian Park - aftershock
52--	1988	June	10	34.94	118.74	5.2	VII	——(?)
53--	1988	June	26	34.14	117.71	4.6	—	San Jose/Upland
54--	1988	Dec.	3	34.142	118.16	5.0	VI	Raymond
55--	1989	Jan.	18	33.42	118.63	5.0	VI	Santa Monica Bay
56--	1990	Feb.	28	34.14	117.7	5.2	VI	San Jose/Upland
57--	1991	June	28	34.26	117.99	5.8	VIII	Sierra Madre

Compiled from Toppozada, et. al. (1980) and Toppozada and Parke (1982); and the earthquake catalog of the Southern California Seismographic Network, U.S. Geological Survey, California Institute of Technology, Pasadena, CA.

TABLE 3: DAMAGE PATTERNS OF THE 1933 LONG BEACH, 1971 SAN FERNANDO, AND 1987 WHITTIER NARROWS EARTHQUAKES

1933 LONG BEACH EARTHQUAKE

PROPERTY AND LIFE LOSS

- 120 casualties; escape and panic are the main causes of injuries.
- $40,000,000 dollar losses (1933).
- Isolated fires in residential areas. No fires occur in mercantile/industrial districts or where water pipes are broken. A lack of wind prevents higher losses from fire.
- Oil facilities suffer damaged internal piping/storage tanks.

LIFELINES

- Water supply pumping stations rendered inoperable; underground distribution mains damaged in areas of ground failure.
- Catastrophic failure of ground level water tank releases approximately 6,000,000 gallons in violent "explosion".
- Broken gas services and devices cause 7 to 19 fires.
- of 25 elevated storage tanks collapse; others damaged.
- Value of post-earthquake communication for emergency response is recognized.

GENERAL CONSTRUCTION

- Unreinforced masonry suffers severe damage; walls collapse, catastrophic collapse of school buildings; thousands of lives spared due to late time of day (5:54 p.m.).
- Extensive structural and nonstructural damage to steel frame fireproof and reinforced concrete frame buildings.
- Simple architectural design is recognized to inhibit severity of damages.
- Well-built frame houses survive with minimal damage; no wholesale destruction of residential buildings.

1971 SAN FERNANDO EARTHQUAKE

PROPERTY AND LIFE LOSS

- 58 dead and 2,543 injured; escape, panic, and collapse of structures are the main causes of injuries.
- $511,000,000 direct dollar loss (1971).
- Approximately 20,000 houses damaged, nearly 830 destroyed; others collapsed or razed later; fires destroy some.
- Oil facilities suffer damaged internal piping/storage tanks.
- Four major hospitals severely damaged and evacuated; Olive View Hospital built under inadequate Building Code Standards.
- Collapse of structural and nonstructural building components traps emergency vehicles.
- Freeway overpasses collapse.

LIFELINES

- Sylmar converter station heavily damaged.
- Ground level water supply tanks buckle and again prove vulnerable to ground shaking.
- Gas distribution lines show propensity for explosion and fire.
- Underground utilities (gas, water, liquid) again prove vulnerable to "buckling" and "pull-apart" through ground failure and fault rupture.
- Embankment failure in lower San Fernando Dam alarms emergency officials; evacuation declared.
- Modern freeway overpasses prove vulnerable to collapse from intense ground shaking.
- Roads damaged from ground rupture and/or failure –many blocked by shallow rockfalls, rock avalanches, slope/colluvial failures.

GENERAL CONSTRUCTION

- Precast tilt-up, prestressed concrete construction used in light industrial buildings prove vulnerable to severe structural damage, especially roof failure caused by collapse of walls.
- Unreinforced concrete and improperly designed reinforced concrete buildings suffer moderate to severe damage.
- Structural defects and poor construction quality cited as main cause of damage.
- "Soft-story" multi-story buildings such as Olive View Hospital suffer partial or total collapse.
- Unanchored/unbraced nonstructural equipment in industrial facilities experience severe damages (electrical transformers, coolers, boilers, machinery, equipment).
- Nonstructural damages include: collapse of unsupported interior walls, stairways, and lights. Office equipment also cause many injuries.
- Mobile homes and dilapidated buildings perform poorly.
- 1-story homes perform better than 2-story and much better than 1 and 2-story homes with irregular configuration.

1987 WHITTIER NARROWS

PROPERTY AND LIFE LOSS

- Main shock: three dead, 932 injured; non- structural hazards and panic are the main causes of injuries; ethnicity becomes a factor in emergency response activity.
- $358,000,000 direct dollar loss (1987).
- 123 single-family homes destroyed (Whittier). 134 destroyed apartment units (City of Los Angeles).
- 508 medical rescue situations.
- Interjurisdictional earthquake exercise, "Response 87 and ongoing emergency response and preparedness credited with reducing overall impact of earthquake.
- Only five earthquake-caused fires; three were linked to natural gas leaks.
- 22 hazardous material situations.
- Large chlorine tank collapses and forces evacuations in Santa Fe Springs; isolated incidents involve spilled chemicals and released asbestos.
- Warning of large chlorine gas release is impeded by telephone/emergency communication.
- I-5/I-605 (built mid-1960's) freeway overpass comes perilously close to collapse during major aftershock. Caltrans initiates retrofit of single-column design. Phase 1 retrofit found vulnerable.

LIFELINES

- Burned down or wrapped overhead transmission lines cause brief service disruption; damaged substations cause wider outages, but overall system performance good because damages were dispersed.
- Overhead distribution lines may cause fire, posing danger of electric shock and blockage of major thoroughfares.
- Mainframe computer system at Southern California Edison down for 32 hours. Widespread belief that computer systems are vulnerable to damage.
- Limited areas without fire fighting water; low water pressure caused by cracks in cast-iron water pipelines. Damage concentrated along welds and corroded segments in 4- to 24-inch diameter water mains.
- Water pipelines, water storage tanks damaged at piping connections, and other types of liquid storage tanks are damaged.
- 1,400 earthquake-caused natural gas leaks. 75 percent of leaks associated with appliance water heater connections; the remainder occur in corroded cast-iron distribution pipes, service connections, and internal residence system.
- 81 seismically-activated natural gas shut-off valves (houses, hotels, fire stations, etc.).
- Los Angeles County Earthquake Response: Responsible agencies lose one hour because regular telephone system is saturated so that all access was limited; emergency radio frequencies are overloaded.

GENERAL CONSTRUCTION

- Tilt-up construction is again identified as seismically vulnerable. Out of 120 tilt-ups surveyed in areas experiencing seismic intensity VIII, many pre-1971 tilt-ups partially collapse. Post-1971 buildings also sustained significant damages.
- Partial collapse of pre-1971 tilt-up commercial buildings due to same structural problems experienced during the 1971 San Fernando Earthquake. Some retrofitting techniques perform well.
- Recent trends in design and construction practice of tilt-up buildings reveal a new seismic vulnerability not previously recognized. Building Codes have been improved since 1971, but many 2- and 3-story tilt-ups with irregular configuration minimize code improvements.
- Interior and exterior nonstructural hazards: falling bookcases, collapse of lighting fixtures and desktop computers, unsupported parking lot walls.
- Strengthened unreinforced masonry buildings sustain no major collapse. The absence of total collapse of unstrengthened URM buildings suggests the magnitude and duration or the earthquake did not provide a true test of seismic capacity.
- Much incipient damage to strengthened and unstrengthened URM buildings: shear cracking of URM piers, shear failure and fallen bricks.

TABLE 4: CHRONOLOGY OF POST 1933 EARTHQUAKE RELATED LEGISLATION

Year	Event	National Level	State/Local Level
1933	Long Beach Earthquake		Widespread adoption of Uniform Earthquake Building Code Provisions. (UBC) Field Act: Public school construction becomes a State responsibility.
1939			Riley Act: State regulation of buildings larger than two-family dwellings.
1959			*SEAOC develops program for improvement of seismic codes in UBC.
1969			Garrison Act: First Hazard Abatement effort.
1971	San Fernando Earthquake		Inadequacy of building codes: Earthquake provisions in UBC are reevaluated. Hospital Seismic Safety Act: Hospital construction a State responsibility. Safety and Seismic Safety Elements required in General Plan.
1974			California Seismic Safety Commission established:
1977		Earthquake Hazards Reduction Act passed: National program for earthquake hazard reduction.	
1978		Applied Technology Council (ATC) established to strengthen seismic codes.	
1979		Federal Emergency Managment Agency (FEMA) established: Lead agency to implement the National EHRA program.	
1980			*SCEPP established: Regional Agency to coordinate local and State Hazard Reduction programs.
1984			*BAREPP established.
1986			Recovery and Reconstruction Act; California Earthquake Hazard Reduction Act; Building (URM) Earthquake Safety Act established.
1987	Whittier Narrows Earthquake		Los Angeles County Safety Element update initiated.
1989	Loma Prieta Earthquake		Governor's Board of Inquiry into Caltrans bridge failures; Temporary sales tax increase for earthquake recovery.
1990			California Seismic Hazards Mapping Act; California Residential Earthquake Recovery Fund established.
1991		Southern California Earthquake Center established.	

* SEAOC - Structural Engineers Association of California
* SCEPP - Southern California Earthquake Preparedness Project
* BAREPP - Bay Area Earthquake Preparedness Project

Chapter **2**

Environmental Issues

Aerial photo of oil refineries in Carson-Wilmington area adjacent to residential and commercial buildings.

Environmental Issues

Editors Note: Environmental degradation of soils and groundwater has led to the new wave of employment of engineering geologists who must identify, monitor and remediate specific sites. After an in-depth history of waste management, several case histories of specific problem sites and their mitigation are presented. The litigation over leakage of toxicants from the Stringfellow Acid Pits is continuing. The controversy over the cause of the methane seeps and resulting explosions in the Fairfax District is discussed in two papers. Woody Higdon explains the use of a neglected tool in the use of investigative aerial photography.

PRE-RCRA HISTORY OF INDUSTRIAL WASTE MANAGEMENT IN SOUTHERN CALIFORNIA

Allen W. Hatheway
Professor of Geological Engineering
129 McNutt Hall
University of Missouri
Rolla, MO 65401-0249

CONTENTS

ABSTRACT

Historical aspects of industrial waste management in southern California are complex and interwoven with the then-developing national awareness over the threats of hazardous waste management. The first attention to industrial waste management in California came with the Dickey Water Pollution Control Act of 1949, only a year after the first Federal legislation (Clean Water Act of 1948). The Federal act was inconsequential to the manner in which industrial waste was disposed to the land and the California experience began with the assumed right that waste "dischargers" had as much right to make use of the waters of the State, in waste management, as did any other user. Connections between health risk ("contamination") and synthetic organic chemicals were known in some sectors but poorly appreciated in terms of fate and transport in the natural environment. "Pollution" was regarded as an incremental degradation of water quality, short of the chemical toxicity and strength of concentration bringing sickness to humans.

Into this arena were thrust the seven-person Regional Water Pollution Control Boards, each assigned to a major California watershed and governed by politically-appointed citizens of local prominence but nearly always without technical qualification. Until 1972, after a major reorganization dictated by the Porter-Cologne Water Pollution Control Act (successor to the Dickey Act), Regional Boards were without significant guidance from the overseeing State Board. For 22 years (1950-1972), the Regional Boards struggled with small civil-servant staffs and reliance upon the Division of Water Resources for technical support. In this climate Regional Boards were charged with reviewing the characteristics of individual sites and with developing "Waste Discharge Requirements" that would minimize ensuing pollution.

Would-be operators of waste disposal sites, commercial or plant-site, had only to receive county approval for operation, through the process of planning and zoning waivers. Regional Water Pollution Control Boards had no real power to deny an applicant the right to discharge all manner of wastes, except to heed the its Discharge Requirements. Technical information to support its decisions relating to formulation of the Discharge Requirements came from the local District office of the DWR. Engineering geologists and civil engineers of DWR made limited site visits as the basis for their advisory reports to the Regional Boards.

In all cases it was the responsibility of the applicants to design, construct and operate facilities so that environmental impairment would not occur. State law did not anticipate the need to enforce proper siting, design, construction and operation. Applicable Federal laws for the management of hazardous waste did not appear until 1976 and the implementing regulations began to appear only in late 1980. This was as much as 25 years after California saw fit to attempt to control discharge of industrial waste to the ground and waters of the State.

The State and Federal courts are now trying landmark cases determining the financial responsibility for remediating real damages to the land and waters of the State and alleged detrimental health effects to the citizens of the State. This paper has as its goal, the presentation and assessment of several geological-technical issues and a number of jurisdictional questions relating to the activities and implied responsibilities of State and County agencies in the oversight of industrial waste management in the period before passage of the Federal Resource Conservation and Recovery Act of 1976.

1. INTRODUCTION

Knowledgeable Californians may rightly presume that the bulk of today's technologies and regulations for the management of hazardous waste have their origin in California law and practice of civil engineering and applied geology.

Today we view hazardous waste from a rather sophisticated viewpoint. This residue of our modern society is potentially harmful to our lives and to our environment for many reasons. This realization has not always been the case, for we, as a society, were quite reluctant to come to believe that the very same chemicals that have done so much to improve our standard of living could be so dangerous. The purpose of this paper is not to detail the history of awareness. It is to provide a short introductory list of the major factors surrounding what was originally termed "industrial waste", now more aptly and legally known as "hazardous waste".

As of this writing, the State of California is besieged by litigation initiated by industries charged as *Potential Responsible Parties* (PRPs) or legally declared or voluntarily ascribed as *Responsible Parties.* The claims are being lodged on the presumption that the State is responsiblie for creating as many as 11 failed commercial Class I (hazardous waste) disposal facilities. For the Stringfellow Acid Pits alone the potential liability to the State is more than $750,000,000. A Federal jury, in 1989, acquitted the State of three charges of responsibility associated with siting, design and construction of the Stringfellow facility and guilty of three charges related to pollution associated with the site after its abandonment by the Stringfellow interests. On review of the jury findings, the judge invalidated the first three findings and held the State entirely responsible for all associated damages. The case is in appeal as of this writing (early 1992).

2. HISTORIC PRECEDENCE OF AN ACTIVIST GOVERNMENT

California has had two major dilemmas in dealing with waste management. The first dilemma was due to expanding population and significant transportation distances, leaving the long and mountain-divided land needing of independent or "home" rule. County government came to California on 18 February 1850, on the heels of statehood, when Los Angeles was established as one of the first 27 counties. By 1857 the need for mining water had produced a water supply industry with 1120 km (700 mi) of diversion ditches in Nevada County alone. In 1861 the legislature established a Board of Reclamation Commissioners to oversee plans and construction of levees of the Sacramento Valley This was in the wake of the legendary "Noachian" floods between November 1861 and January 1862, caused by a nearly two-month rainfall throughout the State, and the wettest winter on record.

Health Departments have traditionally been the first source of regulation of industrial waste. The California State Health Department was created in 1870, the second in the nation after Massachusetts. California's first real waste producing industry was James Walsh's gas works in Los Angeles, franchised by the City on 5 March 1866, and cited later on the grounds that "sewage from this plant was on August 19, 1872, found to be a public nuisance and ordered not to be emptied into the public sewer." On 24 February, 1873, Los Angeles established a garbage and dead-animal plot, requiring the town marshall to supervise burial at three feet below the ground surface.

The Drainage Act of 1880 brought about not only a means of publically financing an integrated system of levees to respond to the need to channelize the Sacramento River and to assist in removal of mining-induced sediment for deposition in San Francisco Bay. Such sedimentation was so profound that U.S. Circuit Judge Lorenzo Sawyer's decision of 1884 created a perpetual injunction against discharge of hydraulic mining debris into the rivers of California. The final chapter in control of mining debris was written by the Federal Congress on 1 March 1893, with creation of the California Debris Commission, a board of Army officers working to clear the rivers of mining debris and to restore the navigable channel of the Sacramento River from the city to San Francisco Bay (Hagwood, 1981).

Between 1886 and 1888, the State Engineer, William Hamilton Hall, published two of a projected, but never completed, three-volume series on *Irrigation Development,* a grand first attempt at cataloging the State's surface water resource. Creation of irrigation districts was insured by the Wright- Bridgeford Act of 1887, providing the means for a minimum of 50 property owners to band together to mutually create a means of distributing agricultural waters. The law was later shown to be a failure and repealed in 1897, which was followed by enabling acts of 1909 and 1911.

With provisions for the first flood control and irrigation construction in place, the State Legislature passed, on 31 March 1891, the Act for Formation of Sanitary Districts at the County level (Crouch and McHenry, 1949, p. 243) for managing sewage. The means of large-scale construction was brought by 1894 to the State, through establishment of the Commission of Public Works. It had been envisioned in the

1880s by then State Engineer Hall. Counties were given the right, in 1897, through the County Government Act, to initiate drainage protection as well as irrigation and sanitation districts.

The first major evidence of concern for surface water quality came in 1898 with a California Superior Court suit (Folwell, 1903, p. 20) brought against the City of Santa Rosa for discharge of "impure effluent" from sewage irrigation into a local creek. San Diego, on 3 March 1930, adopted an *Ordinance to Prevent the Contamination and Pollution of The Water Supply of This City* (1937 Ord., p. 277).

Following the examples of Cornell University and the Massachusetts Institute of Technology, the University of California established, in 1899, a curriculum in Sanitary Engineering, headed by Professor Charles Gilman Hyde, who was to figure prominently in California consulting engineering until the 1950s. By 1900, it has been estimated that 40 percent of U.S. industrial pollution was being discharged into rivers (notably highest in the eastern States and doubled by 1968). In that same year, the bulk of municipal wastes were nonrecyclable food wastes and yard-incineration ashes. Food came in bulk, sacks were used to start winter fires, soft food scraps went to yard animals,and ragmen carried away the recyclables. The U.S. Public Health Service, which was created in 1798 as the U.S. Marine Hospital Service, was just beginning to organize to provide for widespread assistance to the States in combating the major causes of death from infectious diseases (pneumonia, influenza and tuberculosis).

Anaheim opened the first recorded municipal disposal site in 1901 and this remained active until its closure in the early 1930s (Refuse Rem. Jour., 1962). Sewage treatment plants were in place in 1902 in seven California cities (Folwell, 1903), with populations as shown:

Bakersfield:	4,836
Los Angeles:	102,474
Pasadena :	9,117
Pomona :	5,526
Redlands:	6,673
Santa Rosa:	6,673
Visalia:	3,985

California's first major investment in public works came after the 1902 amendment to the State Constitution which permitted construction of State highways and subsidized the counties to undertake such construction. A Department of Engineering was created and an $18,000,000 bond issue authorized (Crouch and McHenry, 1945, p. 325). Equally important was the January 1905 inauguration of the Department of Public Works (DPW) of the City of Los Angeles (Los Angeles City Council, 1905, p. 62), in which, among other duties, included "disposal of garbage, sewage and street refuse."

On the heels of public construction of highways came the determination of the City of San Francisco to gain a water supply to increase the thin margin of the reservoirs along the San Mateo peninsula, south of the city. This led to the controversial 1908 award (Harding, 1960, p. 119) for use of the Hetch Hetchy Reservoir site, in Yosemite Valley, by the Secretary of Interior. Actual granting of water rights did not come until passage of the Federal Raker Act of 1913. In that same year the Los Angeles Department of Public Service initiated the Owens River Aqueduct, completed 1913, as a means of providing water supply beyond the dwindling amounts held within reach in the alluvial basins of Southern California.

By 1909 engineers of the U.S. Geological Survey (Stabler and Pratt, 1909) were aware of the "effects of special industrial wastes on streams, their persistence, their relation to health, their damage to natural water resources" and the need to identify the "best methods of disposing of them, purifying them, or recovering valuable materials contained in them." The results are found in the Water Supply Papers on the era. The next substantial boost in awareness of industrial waste was the initiation (1918) of a series of U.S. Public Health Service reviews of existing state-of-the- art industrial waste treatment methods, beginning with the strawboard and tannery industries.

"Progressive" political thinking brought about by former President Theodore Roosevelt is attributed to the overwhelming vote of citizens for County Home Rule in 1911 (Crouch and McHenry, 1949, p. 197), in which Los Angeles County became first to adopt the new charter (1912). County government became more direct, specific and encompassing, with the ability to adopt charters for services to citizens. One of the first applicable legislative results was the 1913 provision for formation of county water works. Keeping the onus of development of public services in the south, the disastrous rains and floods of the winter of 1913-1914 led to the passage of the Los Angeles County Flood Control Act of 1915 (Crouch & McHenry, 1945, p. 332), creating the first such special public service district, Los Angeles Co. Flood Control District (LACFCD). A 1915 Act further provided for formation of municipal planning agencies, though the modern system that we know today was not added until the amendments of 1937.

3. AWARENESS OF INDUSTRIAL POLLUTION

Awareness of potential health threats has been the single most damaging factor in the worldwide struggle to manage industrial wastes. Table 1 lists the general time frames of presence or appearance of general classes of hazardous substances. Aside from the useful works of the U.S. Geological Survey and the U.S. Public Health Service, it was not until 1920 that the first industrial recognition of such threats began to appear. In that year, the American Gas Association (AGA, 1920) reported on the unpleasant "presence and origin of "idoform" tastes in chlorinated drinking water", recognized today as "polyaromatic hydrocarbons", wastes typical of the production of manufactured gas.

As late as 1959 the official *Manual of Engineering Practice* of the American Society of Civil Engineers (ASCE) did not address land disposal of either industrial waste or hazardous waste.

The Southern California Section of the Association of Engineering Geologists published in 1973 its second-volume sequel to an ongoing series (this present volume is the third) summaries of engineering geology in southern California. Known as *Geology, Seismicity, and Environmental Impact*, the book contains surprisingly little reference to the waste management situation. Arthur Keene (1973, p. 359) noted the existence of "contamination of shallow groundwater resources through negligence or inadequate consideration of geologic structures underlying sanitary dumps.".

Chronologic Awareness of Industrial Pollution

Worldwide industrialization and the periodic occurrence of wars have brought about advances related mainly to industrial chemistry and metallurgy. Associated contaminants have appeared at regular intervals, as the result of those advances (Table 1).

While Congress was bickering over potential water pollution control legislation, Turneaure and Russell published (1940) the 4th edition of the 1901 technological standard of the day *Public Water Supply*, containing an extensive discussion of public-health aspects of water supply, but without mention of organic or inorganic chemicals as sources of concern, or at least beyond an awareness of the potential for lead poisoning. Water quality was judged mainly on the basis of their classification of hardness. The authors recite the basis for the old and untrue axiom (ibid, p. 191) "That flowing streams carrying soil, sewage, or other added wastes undergo natural purification there can be no question. Dilution only spreads pollution so, in reality, aside from making a wide sphere of action for agencies of purification, it has no effect."

By 1947 a general classification of industrial wastes was in use: phenols, hydrocarbon wastes, miscellaneous chemical wastes, wastes from biological materials and food processing wastes, considered capable of producing taste and odor in surface waters. Types of industries were also generally considered linked with production of each class of waste.

In 1948, *Public Health Engineering* (by Phelps, Pratt, and Velz) recommended five classes of chemical analyses as "quantitative determination of any poisonous, injurious, or deleterious chemical constituent, organic or inorganic", and defined "pure water" as "a water which contains no harmful or deleterious substances with respect to human physiology".

The third major advance in development and distribution of toxic chemicals came in the late 1940s, with introduction of the organochlorine-type pesticides. These products were made to be effective at low rates of application.

California State Code of Regulations (1949; Div. 7, Sec. 13005) defined "Industrial waste" as "any/all liquid or solid waste substances ... from ... producing, manufacturing or processing operation ...." The first record of use of "hazardous waste" appears in the act created the South Carolina Water Pollution Control Authority in 1950, which became part of the State Board of Health.

California government began using the term "hazardous waste" on or about 14 January 1974, as evidenced by an internal memorandum of the State Department of Health (H.F. Collins to D.L. Storm) entitled *Hazardous Waste Interactions* and making specific reference to the Otay Disposal Site, in San Diego County.

Awareness on the Part of Industry

We should all be aware that industry is managed by people, and that people in general are not prone to go out of their way to anticipate trouble. "Trouble" in the sense of this paper is recognition of the environmental threats of industrial waste. The aircraft industry is a leading example of southern California industry which began to flourish prior to World War II. In it, one can find industrial managers beginning to think about dealing with the problem of industrial chemicals and solvents and safety in dealing with industrial chemicals and solvents During the War, one prominent aircraft manufacturer, the Autoneticcs Division of North American Aviation, had a program of personal safety regarding the use of thinners, solvents and paints. Once consumed or spent and contaminated, the waste solvents were no more than an unwanted burden about the plant. A reclaimer or waste hauler was called and the problem was driven out the gate. On 4 November 1968 Douglas Aircraft Corporation, Long Beach plant, issued a *"Standard Practice Bulletin, Precautionary Labeling of Hazardous Substances"* and Stauffer Chemical Company, in preparing a 1 January 1970 blanket order for waste chemical disposal at the Stringfellow Acid Pits, included the caveat "These chemical waste materials may be contaminated with toxic, explosive, or otherwise dangerous foreign matter." On 29 July 1970, ALCAN Corp, a generator of wastes disposed at the Stringfellow Acid Pits, proposed (internal memo Gill to Nichols; Stringfellow case document) a volume reduction program, and to "provide for eventual recycling of waste chemical concentrates." In the interim, company officials were to use "existing disposal sites and take advantage of reduced volumes."

Further evidence of the growing awareness and sensitivity toward hazardous waste in southern California industry comes from the 22 August 1972 McDonnell-Douglas Corporation adoption of *Industrial Waste Disposal Practice*, as signed by Chairman and CEO, C.R. Abel, in which there is a disallowance of disposal of such plant wastes to drains and sewers.

On 4 April 1973, Rocketdyne Division, Rockwell International Corporation, issued its *Operating Procedure on Hazardous Substances* in which it defines "Hazardous Substance" as having the characteristics of flammability, toxicity, corrosivity, oxidation capacity, and pyrophoricity.

Impact of Federal Clean Water Legislation

The Congress had begun to seriously consider water pollution control legislation in 1934, when Senator Augustine Lonergan (Connecticut) called a working conference in Washington, D.C., on the general subject of water pollution. In late 1947 the Congress had passed the first of its environmental acts, the Fungicide and Rodenticide Act (PL 80-104), but yet this body is remembered for the years prior to 1948, with the fact that it "took almost no responsibility for control of water pollution" (NRC, 1966).

On 30 June, 1948, Federal legislation relating to clean water was enacted, when Senator Lonergan's dream come true with passage of the Federal Water Pollution Control Act (PL 80-845). Lonergan had introduced legislation on four occasions between 1935 and 1940). This was during Senator Dickey's California hearings, and only in the year before California's landmark Dickey Water Pollution Control Act.

Known as the first *Clean Water Act*, this legislation assumed what traditionally had been held in the American democracy as a State or local responsibility. The Act called for development of a comprehensive nationwide pollution control plan and created limited Federal assistance to the States for its implementation through construction grants for sewage treatment works. Responsibility for the program was assigned to the Surgeon General of the United States and administered by the Public Health Service. Limited Federal assistance for such works had been forthcoming during the depression years, but the program was managed more for creation of jobs that for water pollution abatement (using the popular terminology of the day). Research to support the effort was established in Cincinnati, Ohio, at the PHS station, which today is the Taft and Breidenbach Centers of the Environmental Protection Agency.

On 30 June, 1956, the first Water Pollution Control Act (1948, amended 1952) expired and President Eisenhower signed its new amendments on 9 July. With this legislation, the Congress began the first funding of municipal sewage treatment plants and additional consulting sanitary engineering firms began to form.

4. U.S. PUBLIC HEALTH SERVICE

Created in 1795 as the Marine Hospital Service, the Public Health Service received its current name through an act of 1910. Its mission was limited to advice on communicable diseases. In 1914, the Service issued the first drinking water quality criteria, dealing with bacterial count and these applied only to waters used in interstate commerce (such as on rail passenger and dining cars). The first revision of the *Standards for Drinking Water* came in 1924, with the addition of (U.S.P.H.S., 1925) the first non-biological criteria; copper, lead, and zinc. The Standards included "methods employed for protecting the water supply against pollution by sewage treatment, waste disposal and the like," and a warning to be aware of the "Proximity of sources of character of industrial wastes."

Beginning about 1915, the PHS began to provide a limited number of sanitary engineers as advisors to States. The number grew substantially after the lessons of World War I, though the total number were still few and the attention of these people was placed mainly on sewage collection and treatment.

Nationwide, the PHS began the first *Inventories of Municipal Waste Facilities* in 1940, followed in 1945, and at subsequent dates to the present time. Separate industrial facilities were not included in the compendium. By 1942, PHS had enough concern for the basic toxicity of industrial wastes in the environment to recommend methods of reduction of pollution through changes within plants, reuse ("recycling" of the day), changes in plant processes, or greater care in operations. Pretreatment was recognized as often being required in order to prevent damage to equipment or processes at POTWs. Treatment at the plants was recommended.

5. LOS ANGELES COUNTY AND ORANGE COUNTY SANITATION DISTRICTS

Enactment of the California Joint Powers Act (1921; Crouch and Dinerman, 1963, p. 112) enabled units of municipal or county government to share responsibilities for joint operations, such as a number of County Sanitation Districts. Out of this legislation grew the County Sanitation Districts Act (Orange County, 1947) which provided the opportunity to consolidate and/or initiate sewage districts through bond debt for construction, and annual operating costs through direct property tax to each District. In that year, Los Angeles County already had a population of 1,300,000 (versus a population of 500,000 in 1910) of which 775,000 people resided in the City.

Los Angeles County Sanitation Districts

Quick on the heels of the 1923 County Sanitation Districts Act, Los Angeles became the first county to create a Sanitation Districts organization, consisting of eight separate Districts, two of which were later dissolved. A.K. Warren was appointed Chief Engineer and General Manager and the operation was underway by May 1925. The dual appointment of Warren established a strong engineer's control of the Districts and a model for other California entities to follow. At the time of creation, there were more than 500 water supply firms and agencies in the County and groundwater levels were rapidly declining at the same time that significant groundwater pollution was recognized in the unsewered areas. Construction of the LACSD Joint Disposal Plant No. 1 (Bixby), at Long Beach, was begun in 1927 and completed in February 1928. This first modern sewage treatment plant in California utilized three activated sludge units of single capacity at 3 million gallons per day (mgd). The system could be stretched to 14 mgd (Rawn, 1965, p. 53).

By 1932 the Bixby Plant was being impacted by frequent biological upsets due to illicit industrial waste discharges to the sewers. In July 1932 a major fuel oil discharge from Columbia Steel Mills, in Torrance (Rawn, 1965, p. 57) required "weeks of work to get the plant and equipment into good operating order again." Columbia Steel Mills paid the cost of cleaning. At the time, industrial waste waters were accepted on a negotiated basis for treatment. Treatment was only possible at the time because of the general scarcity of the new and more toxic chemicals being introduced in the 1930s. Oilfield brines, originally ruled by District officials not to constitute industrial wastes, were not routinely discharged into the sewers. Beginning in 1927, separate agreements were negotiated at $35, plus pumping charges, per million gallons. Acceptance of oilfield brines continued until 1950.

Prior to World War II, LASDs additionally were operating three small inland POTWs, with the Pomona plant discharging treated effluent directly to San Jose Creek. Areas not served by the Bixby Plant were collected and discharged directly to the ocean (Pomeroy, 1989, p. 15-17). "A M" (unpunctuated) Rawn, originally appointed as Deputy Chief Engineer in 1924, became Chief Engineer in 1940.

In May 1955, the Los Angeles County Joint Trash Disposal Committee received a report from W.J. Fox, Consulting Engineer, on Program Report on Refuse Practices in Los Angeles County, 1948-1955. In the same year LACSDs completed its first Master Plan for County Solid Waste Management, believed herein to be the first such document anywhere. Six new landfills were scheduled for activation, beginning with Palos Verdes in 1957 and terminating with Scholl Canyon in 1961 (Table 2).

That POTWs are not designed as industrial wastewater treatment facilities was as true in the 1955-1966 period as it is today. Former State Board Executive Officer Paul Bonderson, in speaking of dealing with Rawn noted in 1980 (p. 25) that "L.A. County Sanitation District (sic) was the biggest polluter in the State of California. In fact, it was so bad, I deliberately never went out to his plant."

LACSDs began operation of two sanitary landfills for placement of refuse in 1957, New Palos Verdes, in May, and Spadra, in Pomona, in July. Neither facility employed open burning as the basis of their standard operating procedures prior to the 1 October 1957 announcement by the Los Angeles Board of Supervisors. The Board of Supervisors also terminated back-yard incineration and burning at all dumps (landfills). At the time, there were more than 150 dumps operating in the county (USEPA, 1971, SW-17c). The same ban went into effect in the six-county San Francisco Bay Air Pollution Control District. By February 1960, LACSDs were heavily into the disposal business, when its fifth landfill, at Calabasas, was activated.

Orange County Sanitation Districts

Orange County, just beginning to feel the reality of urban growth, contracted (on 25 June 1946) for a *Comprehensive Report on Sewage and Industrial Waste Treatment,* to be prepared by AM Rawn, Charles Gilman Hyde, and F. Thomas. The consultants employed as staff four young engineers who were to become major players in California waste management. Harvey F. Ludwig, Vinton W. Bacon, Gordon W. Ludwig and John D. Parkhurst. The team operated out of a temporary office in a Santa Ana residence and issued its report on 30 June 1947, estimating that 4.1 mgd (15,518,000 l) of industrial wastes were placed in Orange County sewers, containing large concentrations of organic materials (such as sugar beet residues) and that this content should constitute a basic design factor for the necessary combined, joint- operation sewage works, to serve ten constituent Sanitation Districts. As a result of the report, the Orange County Sanitation Districts (OCSDS) were formed in 1946-1947, though operation did not begin until 1954.

A second 1947 measure, this by the Sate Legislature authorized Sanitation Districts to collect and dispose of refuse, including industrial waste. At this point in time a convergence began in waste management between the traditional city responsibility for refuse and the Sanitation District responsibility for waste waters.

In 1949 the California Sanitation District Act was amended to allow Districts to "enter the field of refuse collection and disposal." On 13 October of that year, the LACSDs instructed its Chief Engineer to report on the feasibility of coordinated refuse collection and disposal. On 10 February 1954, the Districts took a long-range measure that was to have significant, but unknown, secondary impacts on the management of hazardous wastes in southern California. Ordinance No. 1, its very first , *Rules and Regulations for the Admission of Sewage from the Sanitation Districts*, defined industrial wastes as "... the wastes from industrial processes, as distinct from sanitary sewage." The delayed impact was an increasing prohibition against discharge of concentrated industrial waste liquids into the sanitary sewer system, a common industrial practice of the time."

Article IV :

"No district shall discharge, or cause to be discharged,any sewage, waters, vapors or industrial wastes thathas any one of the following characteristics": 1) Gasoline, benzene,

naptha, fuel oil, or other flammable or explosive liquid, solid or gas. 2) Any sewage or industrial wastes containing toxic or poisonous substance in sufficientquantity to injure or interfere with any sewage treatment process. 3) Liquid or vapor at temperatures $85^{o}F$. *4) Fat, oil, or grease at concentrations 200 ppm (weight). 5) Dissolved sulphides at > 0.5 ppm . 6) pH , 6.5 or > 9.0, or having corrosive property damage or hazard to structures, equipment, and personnel of the sewage works. 7) Suspended solids of such a character and quantity that unusual attention or expense is required to handle such materials in trunk sewers or at the sewage treatment plant."*

Apparently the 10 February 1954 Ordinance was not enforced until mid-August 1956, when the Districts were instructed not to receive liquid industrial waste discharges in nonconformance with the above specifications. This "Orange County Sanitation District crisis" occured just weeks before the Stringfellow Acid Pits dump was rushed to completion and opened on 31 August 1956. The crisis is documented in Stringfellow Federal case interrogatory of the Rainbow Canyon Manufacturing Company (23 Sept. 1983).

6. SOUTHERN CALIFORNIA WASTE MANAGEMENT

Great Britain is alleged to have developed the first trial of sanitary landfilling in 1916. By the 1920s this was known as the "controlled tipping" of solid waste and the pioneer location was Bradford, where layers of less than two feet of refuse were covered daily and control was in place for windblown debris.

Situation Prior to 1949

With the rising amounts of non-yard-incinerable municipal wastes, San Diego, in 1919 (City Ordinances of 1937, p. 68-69) adopted an *Ordinance Providing for the Collection and Disposal in a Sanitary Manner of City Refuse and Other Waste Matter*, in which the choice of disposal was given to the City Manager, who was obligated to remove garbage on a weekly basis through Collection Districts. Garbage Disposal Districts became enabled by a California Legislative act of 1927 (Crouch and McHenry, 1949, p. 237).

California rightly claims the first modern sanitary landfill in the United States, as Fresno County sanitary engineer Jean Vincenz introduced the cut- and-cover method of landfilling under a daily cover. Vincenz later went to Washington, D.C. as the Army's chief civilian sanitary engineer, to effect an integrated program for waste disposal during World War II. The example espoused by Vincenz became the subject of a landmark 1940 paper by Carpenter and Setter, entitled "Some notes on sanitary landfills", establishing the need for engineered waste disposal facilities.

Municipal waste disposal in southern California has been unremarkable. Prior to 1941, the City of Los Angeles Department of Public Works provided refuse collection, which was subsequently transferred (Crawford, 1955, p. 60) to the City Bureau of Maintenance and Sanitation later to become the Department of Sanitation). The national post-War urge to make civic improvements prompted the City of Los Angeles to pass an ordinance in 1946 for construction of a municipal incinerator in Vernon, to be leased to the City for nine years (Crawford, 1955, p. 161). The City of Vernon reneged on the contract and the site was changed to Avenue 21 and Lacy Street, in Los Angeles, and reapproved in 1946. In 1947, on the heels of the City having instituted incineration, the County Air Pollution District banned burning of waste disposal "dumps." This was probably the first such measure in the nation, possibly in the world.

By 1948 Los Angeles Harbor had become seriously polluted by ocean-going ship bilge and leakage and by numerous inflows of polluted water from heavy industry which ringed the harbor. Early in 1948 the County appointed the Advisory Committee on Pollution in Los Angeles Harbor, first chaired by Deputy District Attorney Fred Wichello, to study the state of pollution in the harbor.[1] In August 1948 the Committee's Technical Subcommittee issued its second interim report on the harbor with several surprising findings:[2]

1) Harbor pollution came largely from upstream discharges intoDominguez Channel;

2) The offending discharges were all under Industrial Waste Permits issued by the City of Los Angeles Bureau of Sanitation;

3) Rudimentary discharge criteria were in place, including limits on settleable solids, oil, grease, fats, waxes, garbage, sewage, pH,"toxic ingredients (cyanides, phenols less than toxic concentrations)", sulfides, and temperatures in excess of 14^{0}F, and;

4) "Specific tests are not made for toxic substances".

Situation from 1949 to 1972

Just as the California Department of Water Resources (DWR) had struggled to characterize California's surface and ground water, Los Angeles County Sanitation Districts began to study the *Collection and Disposal of Refuse in the County Sanitation District of Los Angeles County* in 1950. While this study was being conducted, and southern California industry and urban growth were exploding.

[1] Preoccupation with harbor pollution continued and led to an early action of the Los Angeles RWPCB, taken on 11 January 1951, with sponsorship of the Los Angeles-Long Beach Harbor Pollution Control Committee, itself independently reorganized on 2 January 1951 from the 1948 County committee. The invitation was accepted.

[2] On 5 February 1945 the Committee determined to issue an annual report, and later did so, with First Annual Report, Los Angeles-Long Beach Harbor Water Pollution Committee; Report - January 1, 1954 to December 31, 1954."

Hazardous waste was burgeoning in variety and quantities and there was no legal classification for the waste other than the generalization of "industrial waste". Most of what is now classified as hazardous waste was therefore placed in sanitary landfills. Completion of the 1950 survey provided the first status of refuse collection and disposal in Los Angeles County (Table 2). The County was not in the formal business of collection and disposal; rather it saw that such happened. In 1950 there were 40 municipal and commercial waste collection and/or disposal systems in place in the County. Only three of these were considered by LACSDs to represent "well located" sites, though the record does not stipulate the criteria for that judgement.

Based on the 1950 findings, LACSDs issued a 1955 Master Plan for County Solid Waste Management. This was the first such plan in California and probably the first in the nation. During this time a crisis in industrial waste management was slowly developing in California, especially in the more industrialized southern half.

Sanitary engineers were without substantial published guidance on sanitary landfilling until the 1959 publication of the American Society of Civil Engineers *Manual of Engineering for Sanitary Landfill Practice*. The manual defines "sanitary landfill" as "a method of disposal of refuse on land, without creating nuisances or hazards to public safety, by utilizing the principles of engineering to confine the refuse to the smallest practical area to reduce it to the smallest practical volume and to cover it with a layer of earth at conclusion of each day's operation or at more frequent intervals as may be necessary." The manual does not address the management of industrial waste or hazardous waste but does emphasize the need to eliminate possibility of ground or surface-water contact with the refuse.

Industrial Waste Disposal Crisis in Southern California (1956)

California population increased by 3,500,000 in the decade of the 1940s, increasing from 6,900,000 to 10,600,000. There was serious concern at the State Water Pollution Control Board (SWPCB, 1966) and the Legislature introduced several bills in 1963 to strengthen Board functions.

Population growth was running in tandem with an explosion of industrialization. The aircraft industry, which had carefully recycled (mainly by distillation) scarce solvents during World War II, now disposed of such directly to waste haulers. Of the solvents, TCE, yet had some residual value and was often sold for reclaiming; much of it went to the main reclamation facility, in Azusa (Lambie, 1984, p. 200), along with some solvents uncontaminated with wastewaters and at solvent contents of 66 percent. Other loads of solvents went directly to Class I disposal facilities, unless non-specified as to destination, in which case, some drivers managed to unload the wastes at reclamation plants, for a second income.

While sanitary engineers and public health specialists and officials were becoming aware, in the mid-1950s, of the threat of industrial contamination, other informed sectors were yet unaware. There was a full-blown roadside- dumping crisis underway in southern California by 1956, triggered by expansion of industrial processes and a gradual cessation of acceptance of hazardous waste liquids by the OCSDs. Engineers at the State and Regional Board were becoming aware, as were professionals at DWR. Nevertheless, a Statewide Conference on "The California Ground Water Situation" was convened at the University of California, Berkeley in 1956 (Todd and Glendenen, 1956), and not one of sixty-five papers presented, not one of which recognized what was going on at the roadside or at the POTWs.

Retired DWR civil engineer Robert Clawson recalls (1989, p. 42) that the mid-1950's were a time of common knowledge "that toxic wastes were being generated", such as ". ..had caused groundwater pollution", and further, that "I don't know anybody that worked in the field that didn't recognize that there was a need". Clawson further notes (ibid, p. 25) that San Bernardino was "one of the few counties that had an industrial waste ordinance". DWR engineering geologist Robert C. Fox, in 1983 depositions taken for the Stringfellow case (1983, v. VIII, p. 398), noted that he had witnessed "dirt road" dumping of industrial wastes in Ventura and Orange Counties, mainly oilfield sludges. Max Bookman, DWR Southern District Chief Engineer recalled (1989, p. 88) that dumping of industrial wastes on "roadsides and vacant land" was widespread.

The "crisis" erupted in August 1956, when the OCSDs began to enforce the 10 February 1954 limitation on discharges of industrial waste liquids to sanitary sewers. This was the very month that the Stringfellow Company was completing preparations to take such liquids. That site opened for discharge on 31 August, 1956. By October of 1956, San Diego County was experiencing uncontrolled roadside- and-field dumping, as "There was really no control over where industrial wastes were being disposed of in this area (O'Leary, 1989, p. 125)." Additionally, POTWs were being "poisoned and not operating properly." Regional Water Pollution Control Board (RWPCB) Executive Officer Dennis Leary took the problem to the San Diego Inter-Governmental Relations Committee, which returned with a selection of what became the Otay Landfill.

LACSDs followed suit on 1 June 1957 with a prohibition of dumping of sulfuric and chromic acids to POTWs. According to the Santa Ana RWPCB weekly summary for 21 December, the Stringfellow Company was aware of this turn of events and stepped up work to accept diverted wastes.

An important waste-discharge condition was discovered in 1964 in northern California. Extensive groundwater monitoring had been imposed in the permitting process for Del Valle Dam, built by DWR and operated jointly by Alameda County and the Pleasanton Township County Flood Control and Water Conservation District, to capture waters of a tributary to Alameda Creek. In its Excessive nitrate (200 ppm) was detected in the eastern portion of Livermore Valley and traced to on-site injection-well disposal by the West Coast Manufacturing Corporation, a munitions manufacturer. The investigator, Alvin L. Franks, engineering geologist of the State Water Rights Board, found that the company had stopped shipping nitrate waste waters from the plant four years previously and had begun, at that time, to inject the waters in an abandoned on-site water well. The facts were reported before the San Francisco Bay Water Pollution Control Board but the matter was put aside and the practice was allowed to continue due to a shortcoming in the Dickey Water Pollution Control Act; that sickness or death literally had to be proven in order to stop the discharge. This was one of the cases presented to the Legislature as the basis for including ground water protection in the Porter-Cologne Act of 1967.

The Situation from 1968 to Implementation of RCRA

After 18 years of water pollution control the Dickey Act was replaced on 1 January 1968 by the Porter-Cologne Act, which more specifically addressed waste management. The Act called for combining the existing State Water Rights Board, the State Water Quality Control Board and the existing Regional Board sytem, into one regulatory agency. It was no longer necessary to show death or sickness; a discharge could be stopped on a threat of pollution. At that time, the engineering geology dueies of Dr. Alvin L. Franks were expanded from the Water Rights function to include water quality policy, regulation and enforcement, under the new Division of Water Quality.

Franks found that there was little or no control of liquid industrial waste from its generation to disposal or reclamation. It was estimated that at least 70 percent of the waste was unaccounted for and was being dumped along roadways or into waterways. A review of all technical regulatory requirements for land disposal facility siting showed that no real means of protecting ground water were in place. Franks also collected and reviewed all the available data on the state-of-the-art of industrial waste management, the literature and any written requirements that had been developed by the Regional Boards, Public Health Service (after 1 January 1970, the EPA) Laboratories, County and City agencies.

Based on these studies, a bill was drafted in 1969 and submitted to the Legislature to place control on liquid wastes, from generation to disposal and/or reclamation, and to establish a classification system for all wastes and land disposal sites. The *Liquid Waste Haulers Act* of 1969, as passed, did not contain the tracking system, as proposed, but maintained the required licensing of Liquid Waste Haulers and most importantly, the mandate that the SWRCB establish a classification system for wastes and facilities for all disposal to the land (excluding sewage and its treatment sludge). Formulation of the resulting specific waste disposal facility guidance is discussed in a later section.

Passage of the Federal Resource Recovery Act (AKA Solid Waste Act of 1970; 26 Oct 1970) included the first statutory definition of "hazardous waste" to appear in Federal law, along with a requirement that each State include management of hazardous wastes as part of the *State Hazardous Waste Management Plans* that had been a requirement of the 1965 Federal Solid Waste Act.

Recognizing the potentially huge task of managing solid wastes in California, the Legislature passed, on 13 July 1972, the California *Solid Waste Management & Resource Recovery Act* in which each city and county must have formulated a solid waste management plan. Both types of documents were to be submitted by 1 January 1976 to a newly-formed *State Solid Waste Management Board* (SSWMB, 7 members), located under the Resources Agency. As an assist the the cities and counties, the SSWMB was required to develop and have in place (not later than 1 January 1975) a *State Solid Waste Management Policy* and a *State Solid Waste Resource Recovery Program.*

In 1977, the California Department of Health held the first nationwide EPA-sponsored conference on hazardous waste management, as one of the grants associated with previous financial support under the Federal Resource Recovery Act (a broad-scale amendment of the Solid Waste Management Act of 1965). The proceedings are a valuable source of information on the status of industrial waste management in California at the time. As of 1977, there were about 50 Class II-1 commercial landfills accepting what later became known as "special wastes". This is inert material of a predictable nature, such as non-toxic industrial process wastes, construction debris and the growing volume of petroleum-contaminated soils encountered in site grading for redevelopment of former industrial properties.

By this time, the problem of disposal of hazardous wastes had grown acute in California. The wastes had been so defined under terms of RCRA (Federal Resource Conservation and Recovery Act), in 1976, but State and Federal regulations for their management and landburial were not in place, and would not be for another three years. The BKK Class I facility in West Covina was taking in about 70,000,000 gallons per year (gpy) of hazardous waste liquids under its Unlimited Class I permit. Unlike the situation in northern California, local air pollution control districts had shut down the evaporation facilities and the liquids were being routinely co-disposed as admixtures with solid wastes on a 1:2 (liquids to garbage) ratio.[3]

[3] This means of "co-disposal" is still (as of this writing) practiced in the UK and certain former Commonwealth countries, especially South Africa. "Sorption" is the grand hope of this technology.

Two means of co-disposal were employed at the BKK facility; 1) direct-pour to garbage- bermed sorption areas of recently disposed refuse; and 2) gravity discharge into uncased infiltration wells 5-10 ft diameter drilled into older refuse to depths of 100-175 ft below the landfill surface. As an air-pollution control measure, 15-ft diameter containment domes were placed over the infiltration wells. BKK Corporation maintained that lateral migration of the liquid wastes was limited to 60 ft.

By late 1981, California's hazardous waste landburial capacity had decreased to six operating Class I facilities, while the Class II operations had increased to about 65 operating sites. On 13 October of that year, Governor Jerry Brown published an Executive Order to phase out landburial of many wastes by 1 June, 1983. The order also specified that both Class I and Class II-1; disposal facilities must be "designed so that there is no possibility of pollutant discharge to usable waters...."

By 1983 the State and various waste management organizations had a good inventory of types and volumes. *The Southern California Hazardous Waste Management Plan,* released in July, 1983, presented the following statistics:

1) Regional waste generation: (tons/year)

North L.A. area	524,000 tpy
Upper coastal L.A. area	336,000
L.A. County total	1,490,000
Orange County	486,000
San Diego County	316,000
Long Beach area	315,000

2) Largest quantities of HW sent off-site were in the following categories:

   Pesticides and pesticide wastes
   Aqueous solutions @ 10% heavy metals
   Aqueous solutions with reactive anions
   Mixtures of soil, solvents and water
   Asbestos wastes

3) Land disposal at region Class I sites had reached 500,000 tons/year.

4) Solvents and used oil were the only hazardous wastes being recycled, and there were 12 companies in the solvent reclamation business and;

5) An estimated 233,000 tons/year hazardous wastes wereamenable to treatment and subsequent disposal at RCRAHW landburial facilities.

7. ROLE OF DIVISION OF WATER RESOURCES BEFORE 1949 (Pre-Dickey Act)

In an official sense the public works activities of the State of California have never played any direct role in the management of wastes. This was a serious item of contention during the Stringfellow Acid Pits federal trial (1989-1990) and needs clarificcation. Attorneys for federally-charged responsible parties strove to establish responsibility on the part of the State of California for remediation of the world's first legally-recognized hazardous waste disposal site (see Hatheway and Brown, in this volume). It was charged (and rejected by the jury, though immediately reversed by the presiding judge) that the State of California had taken charge of siting, design and construction of this commercial facility. Nothing could be further from the truth.

In 1927 the Department of Public Works was formed as a means of integrating State works agencies, under a general reorganization plan for state government (Crouch and McHenry, 1945, p. 323). The California Division of Water Resources was also created under this plan, in 1929, in a merger of the Division on Engineering and Irrigation and the Division of Water Rights. The new agency was headed by the State Engineer. Its purpose was to develop and understand the State's water resources and to design, construct and operate public conveyance systems to place agricultural and drinking water supplies in the hands of local jurisdictions. Geologic failure of the St. Francis Dam on 12 March 1928 led to enactment of the California Dam Safety Act.

Division of Water Resources (DWR) engineers had been working since the early 1920s on classifying and measuring surface waters of the State. Concern for groundwater supplies was not a major issue until the Division began to appreciate the finite supply of water and long-term nature of the need for this resource. Rollin C. Eckis, an undergraduate geology student of A.O.Woodford's at Pomona College, had completed his doctoral research at UCLA on southern Californian regional geology with support from DWR, through the good offices of Harold Conkling, DWR civil engineer, on Professor Woodford's recommendation. Eckis was offered one of the State's first jobs in Engineering Geology, at the influence of incumbent Chester Marliave, the State's first engineering geologist. The year was 1932 and Eckis was hired as the first "Assistant Engineer-Geologist".

Eckis' task was to organize the regional geology of all of southern California and to identify the modes and classes of water-bearing materials. Eckis' product, now a rare book, was released in 1934, as DWR Bulletin No. 45, *Geology and Ground Water Storage Capacity of Valley Fill in South Coastal Basin Investigation.* The product and philosophy of this "bible" of southern California ground water has had a profound effect on the management of industrial wastes in California and elsewhere. Eckis was faced with a situation in which the published literature on areal geology was scant, large areas were poorly or incompletely mapped, and basin-wide age relationships were not yet well established. Eckis observed that ground water was found in so-called "pervious" or "Water Bearing" formations, and that the remaining units were, therefore, the "Nonwater-Bearing Series", lumped into the category of the "Basement Complex," something very familiar to California-educated geologists over the present age of 40. Water-Bearing units were further subdivided into "Older Alluvium" and "Recent Alluvium." The original reference contains several word-descriptions of the general character of each of the series. Suffice to say that "The Non-water-bearing series, ... is made up of the relatively impervious formations ....", which were felt to store comparatively little water, which was yielded to wells very slowly, and in most places, the sedimentary and crystalline rocks were found to be so "tight" that they yielded 'very little water to wells' and, consequently, were of 'no value except for domestic or stock watering' purposes (p. 76)."

8. DICKEY WATER POLLUTION CONTROL ACT (1949)

Since 1934 Connecticut U.S. Senator Lonergan had been the congressional champion for Federal "clean water" legislation. Lonergan's hope for comprehensive Federal legislation were not realized until the first Clean Water Act (Water Pollution Control Act), was signed into law on 30 June 1948. California followed with the first comprehensive state water quality protection laws in the nation. Beginning in 1947, the California Assembly passed House Resolution No. 27, fostered by Assemblyman Randall F. Dickey, to establish the Interim Fact-Finding Committee on Water Pollution. Hearings were held in 1948 in major California cities.

Dickey's Committee came to the conclusion that the waters of California were useful for three major purposes; 1) for drinking supplies; 2) for agriculture, and; 3) for industrial purposes. It was clear at the time that industrial wastes were routinely discharged into City and County sanitary sewers and storm sewers. The two major preliminary findings of the Committee (19 January 1949) were

1) Recognition "...that existing laws and regulations do not provide for effective and equitable control of water pollution under the present conditions of growth of population in California."

2) That "...maximum usefulness of the water resources of the State isimpaired by the effects of unreasonable practices in the disposalof sewage and industrial waste; at the same time economic development of the State is being jeopardized by undue restriction on the reasonable use of land and water for necessary waste disposal purposes."

This single act of the California Legislature probably is the most important environmental act of any State or the Federal Government. It was enacted at a time when no other state had a comprehensive water pollution control and waste management act and at a time when the Congress was highly reflective of states' rights and the constitutional rights and obligations of the states to manage their own affairs. The Act (SWPCB, 1957) was forged of 11 separate bills and created a management organization unlike anything in place in the nation at the time. DWR had determined that California's surface watersheds and groundwater basins had distinctively geologic and hydrologic conditions between the seven separate regions. These water regions became the operational basis for water pollution control. Seven autonomous Regional Water Pollution Control Boards were to representatively undertake the task. On review of the proposed act, Governor Goodwin Knight convinced the legislative authors to create semi-autonomous Boards acting under the purview of a State Water Pollution Control Board. The Act passed with the Governor's recommendation and the general organization commenced activity early in 1950.

Regional Boards operated with significant freedom of action and were minimally staffed with one Executive Director (generally a civil engineer, but not so specified) and a clerical secretary. Although there was a provision for a "technical staff" at the Regional Board (Bonderson, 1980, p.4), there is no evidence that such persons were actually authorized or employed in the early years (pre-1960). The Boards were made up of appointed citizens from various backgrounds. Boards functioned without standardized rules or procedures. In his Bancroft Library oral interview, State Board second executive director Paul Bonderson notes (ibid, p. 6) that some States "... were beginning to move into the field of classifying streams; California didn't," meaning that the approach was one of "prevention", rather than classification and "abatement" , which constitute the "corrective approach".

Critical definitions contained in the Dickey Act are:

Contamination: Impairment of water quality as sewage/industrial waste to such a degree as creates a public health hazard, resulting from disposal activities;

Pollution: Impairment of water quality by sewage or industrial waste, short of creating an actual public health hazard.

Regional Water Pollution Control Boards were charged with:

1) Prescribing "requirements" relative to any particular condition of pollution or nuisance;
2) Accepting "reports " of proposed discharges;
3) Prescribing "requirements" " ...with relation to conditions existing in the disposal area or receiving waters," and;
4) Utilizing authority to order corrections of non-complying facilities.

In reality, the Boards generally were constituted of five non-compensated citizens meeting together to hear and review "reports" of proposed waste dischargers, and served by the Board's one Executive Secretary and one clerical secretary. The Board members generally were each lacking appropriate technical expertise and were highly autonomous. The Boards initially operated without standard procedures, and were forced, by dint of legislative assignment, to employ a non-existing state-of-the-art for management of the wastes under their purview.

Actions taken by the Santa Ana RWPCB, a central factor in today's litigation over the Stringfellow Acid Pits, were carried out by Executive Officer Paul G. Brown, a civil engineer.

Implications of the Dickey Act

Nowhere in the act or in documents relating to its implementation does one find the terms "permit "or "license" used as nouns representing forms of environmental regulatory control. The activities of the SWPCB or the Regional Board were not then (nor till implementation of the 1972 SWRCB *Disposal Site Design and Operational Information*) similar to the waste management permit systems that we are today familiar with. Operators of California waste disposal facilities were termed "dischargers" and that term was carefully chosen by the Legislature to carry the meaning that citizens of the State were allowed the right to discharge wastes to the land and to utilize the waters of the State for the purpose of managing wastes. In terms of today's environmental awareness, this concept may be disturbing to many readers, but the intent of the drafters of the Dicky Act (Table 3) was to allow the constructive use of California waters to enhance the standard of living and economy of the State. responsibility to approve or disapprove the right to open and manage a disposal site, county health officials had the responsibility to oversee the facilities in terms of control of "contamination", and county building and safety officials were responsible for oversight of disposal facility construction.

The intent of the Legislature, in the 1949 Act, was to give State Government the responsibility to protect water quality, not to manage the disposal of wastes as such. The Board found itself responsible for not only the oversight of management of sewage and industrial waste, but of water quality in general. There is, however, a subtle but important fact in this situation. The Board was indeed responsible for developing the conditions of discharge ("discharge requirements"), in recognition of maintaining its "regional water quality objectives", but it was not responsible for the actual management of the wastes. County planning and zoning commissions had the basic responsibility and authority to allow such facilities to be built and to operate. County planning and zoning commissions would have to issue variances for the operation, building and safety departments would issue building permits and inspect quality and completeness, as with any other form of construction, and the health department would naturally monitor the operation.

1951 Dickey Act Amendments

In 1951 the Legislature amended the 1949 act, increasing the number of Regional Boards from the original seven to the present nine. The concept of "self-monitoring" was introduced and the Boards could require disposal facility operators to provide "technical reports" on any topic or instance of importance discovered by the Boards. This move had two implications of significance: 1) to recognize the responsibility of the operator to manage the performance of sites with respect to impacts on water quality; and 2) that there was now a distinct field of opportunity for California consulting engineers in the fields of waste management (sanitary engineering) and soil mechanics. California had, at that time, one of the largest concentrations of engineering consulting firms in the nation. Geologists, on the other hand, were generally precluded from open consulting practice because of the 1929 Civil Engineer's Act, which created the first professional registration of any engineering discipline in California.

1959 Dickey Act Amendments

The legislature again amended the Dickey act in 1959, recognizing that the State Board did not have the duty of power to establish general statewide policies for Regional Boards. The changes appeared in Division 7 of the California Water Code. All boards were remanded to refer conditions of contamination to the State Department of Public Health for resolution. Boards were given the authority to revise discharge requirements or to specify non-discharge conditions for waste disposal sites. The power was also given boards to issue immediate cease-and-desist orders and to direct compliance with discharge requirements.

9. STATE WATER POLLUTION CONTROL BOARD (AND SUCCESSORS)

In January 1950 (Bonderson, 1980, p.2), the State Water Pollution Control Board (SWPCB) was activated, with Frank Stead as its Acting Executive Officer. The 14 Board members were appointed and convened in January to select the permanent officer from assembled candidates. Retired Army Engineer General Warren T. Hannum was appointed Board Chairman and Vinton W. Bacon, AM Rawn's working associate from the Orange County study (1947) was chosen as Executive Officer. On 3 April 1950 the California Legislature passed Assembly Joint Resolution No. 4, asking the Governor to request that the Federal Water Pollution Control Administration (created by the Water Pollution Control Act of 1948) "...recognize the State Board as the State water pollution control agency in connection with all matters relating to the Water Pollution Control Act ...." In April, 1950, the State Board held its first working meeting in San Francisco.

General Hannum established the parameters and goals of his Board. On 16 October 1950 he wrote Attorney General Fred N. Howser to ask if the Boards had the authority to raise funds to make loans to local authorities under the provision of Dickey Act Chapter 1551. Howser returned a negative opinion: "This statute possess several basic legal defects of a very acute nature, which render its administration extremely difficult, if not impossible." The Attorney General's summary proved to be all too accurate, and such is the very crux of the dilemma that the Federal Court placed on the people of California in its May 1990 decision regarding remediation of the Stringfellow Acid Pits.

In responding to one of the requirements of its enacting legislation, the State Board, on 17 October, 1952, issued a Statement of *Long-Range Plans/Policies for Water Pollution Control.* In this document it is clear that the Board was relying on the fact that "...determination of compliance with all requirements will be based primarily on information required to be furnished by the "waste discharger" with the right reserved for any pubic agency to make any useful investigations."

The fact that the Regional Boards were nearly autonomous should be acknowledged in assessing the roles played by various parties in waste management in the Pre-RCRA (Federal Resource Conservation and Recovery Act; pre-1976) period. Paul Bonderson became Bacon's retirement successor at the SWPCB in 1956. Bonderson notes that for the period 1950-1956, while he was Executive Officer of the Central Coastal Board, that contact between the Regional Board and the SWPCB was "somewhat limited" and that Regional Executive Officers would "occasionally attend state board meetings." In that same period, Bonderson remembers (ibid, p. 10-12) that most Region III Board activity (Region III;) was in dealt with discharges from municipal wastewater treatment plants, rather than with solid or industrial waste disposal.

In 1959 the California Assembly reported on its investigation of *Bay and Water Pollution in* the State. The findings were:

1) There was confusion over working relationships between State and Regional Boards;
2) Enforcement procedures for control of water pollution lacked clarity and effectiveness;
3) Boards have taken the position that they have no right to question a discharge, but must merely set requirements for it;
4) Regional Boards should have the power to prohibit a waste dischargealtogether;
5) DWR investigations were a proper exercise of the Regional Board's statutory mandate; and,
6) During 1950-1959, approximately 335 technical investigations were made by "other agencies" under contract with Regional Boards; more than 100 additional investigations were conducted by board staffs.

Shortly after reviewing the Assembly findings, the Legislature, in July, 1959, made substantive changes in Division 7 of the Water Code as amendments to the Dickey Act, establishing, in particular, that:

1) Conditions of contamination would be referred to SDH
2) The State Board did not have the power to establish general Statewide policies for local or Regional Boards[4];
3) Regional Boards could revise discharge requirements or specify non-discharge conditions; and,
4) Regional Boards had the power to issue immediate cease-and-desist orders and to direct compliance.

On 11 January 1961 the State Board adopted its Article 2, *Principles and Factors to be Considered in Formulating Discharge Requirements*, instructing the Regional Boards to "secure that degree of care in planning and operations of the works for the treatment and disposal of sewage and industrial wastes as will adequately protect the public health and the beneficial uses of the waters of the State" (1963 edition of Water Quality Criteria).

Enforcement

In reality the entire Dickey Act proved to be virtually unenforceable. Recognized failure of the Stringfellow site, in 1972, after nearly 17 years of operation, left the Santa Ana Regional Board essentially powerless to affect any sort of abatement actions. And, the Stringfellow site was only the tip of the iceberg.

Regional Boards could require a discharger to submit "technical reports" on any subject for which the members wished additional clarification[5]. Retired State Board Executive Officer Paul Bonderson recalled (1980, p. 17) that "The state and regional boards had limited access to legal counsel from the attorney general's office.

The enforcement mechanism that was provided for in the original Dickey Act was the administrative hearing process, where an administrative hearing officer from general services or somebody, would sit as judge, and carry on for them." "We not being legal people, and having limited access to legal counsel, were afraid of the complexities of this type of approach. Consequently, very limited enforcement was taken,..." "Today, circumstances are such that the Dickey enforcement provisions would be totally inadequate."

[4] This establishes the fact that there were no standards for siting or design of waste disposal facilities in 1959; a fact; that did not change until 1972.

[5] Bonderson (1980, p. 13) notes that "After we had been in operation for about a year, Cecil Geraghty, Executive Secretary of the Dickey committee, informally asked the regional executive officers if our experience showed a need for amending the act. I felt that not having the authority to require a discharger to file technical reports was a major weakness and suggested that this power be given to the Boards."

Board Sponsored Research

Incorporated into the Dickey Act[6] (Sec. 13024, Water Code) was a program to initiate research on water pollution control, with the intent to develop a sound technological basis for its management. Grants were issued to leading academic researchers at each of the major universities in the State. The research was practical and provided some of the first underpinnings for modern waste management. Reports of findings were issued through the State Board in its "Publications" series (Table 4)

Report No. 2 (Merz, 1952) was entitled *Investigation of Leaching of Ash Dumps*, a very apt title. The "ash" was that of the ubiquitous backyard incinerators; the concept of "leaching" was hitherto unknown in the literature, and the ugly term "dump" was in vogue. This report is evidence of the beginning awareness that disposal sites could produce and release substances of a chemical nature. The authors concluded that rainfall could penetrate the average "ash dump" at a vertical rate of about 0.20 ft/day and recommended that such facilities be sited where affected only by incident precipitation, not groundwater intrusion.

The State Board, in 1952, as Publication No. 3, issued the world's first water quality reference book, compiled under the direction of Jack E. McKee, at the California Institute of Technology. The book was revised twice, most recently in 1963, having been superceeded in the long run by the first of the U.S. Public Health Service Water Quality books in 1962.

In its Publication No. 10, *Final Report on the Investigation of Leaching of a Sanitary Landfill*, the Board concluded that "a sanitary landfill, if so located as to be in intermittent or continuous contact with ground water, will cause the ground water in the immediate vicinity to become grossly polluted." Yet, the only technical guidance available for practicing sanitary engineers was that of Carpenter and Setter (1940).

CalTech Professor Jack McKee, and co-worker W.E.Wolfe, produced under State Board sponsorship, in 1952, the world's first water quality reference book. The volume always has been a huge success and a revision was forthcoming in 1963.

California Water Resources Control Board

A Legislative amendment to the Dickey Act required a change in State and Regional Boards to "Water Quality Control" on 1 January 1963. Late in 1967, Assembly Bill 163 was passed, creating the State Water Resources Control Board (SWRCB), through a 1 January 1968 merger of the SWPCB, the Water Rights Board and the Water Quality Board.

The reorganization act included specified definitions for two terms that have become embedded in water quality and waste management terminology, later appearing in the Federal Safe Drinking Water Acts:

Standard: "...a plan that is established by governmental authority as a program for water pollution prevention and abatement", and;

Criteria: " ..a scientific requirement on which a decision or judgement may be based concerning the suitability of water quality to support a designated use."

Only five members made up the new Board. Presumably the use of increased technical staff gave the organization the ability to function in its increased role. An Enforcement Program was initiated but its overall effect has not yet become apparent to the author.

In November of 1972 the SWRCB issued its first Statewide siting and design guidance for waste disposal sites, as previously noted. These elements of Subchapter 15 of the California Administrative Code are herein considered the first such comprehensive guidance worldwide. This was the result of over two years' compilation effort by Alvin Franks. The new guidance was entitled *Waste Disposal to Land, Disposal Site Design and Operation Information* and reaffirmed the original responsibility of the applicant in that all necessary information to be utilized by the Regional Boards in their deliberations was to be provided by the proposed discharger. "Sufficient information to be provided by the discharger ... to enable evaluation of the disposal operation in relation to conditions in the disposal area, such as local geohydrology and the surface water hydrology ...." By January 1973 Dr. Franks reported that State Board engineering geologists were providing critical reviews of reports on landfill siting and expansion and that such reports were now being compiled by consultants to the applicants.

Another form of waste-related geologic investigation standard had appeared in June 1975 with issuance of the State Division of Mines & Geology Guidelines for *Geologic /Seismic Consideration in Environmental Impact Reports*, which had been prepared in cooperation with the State Board to include such "waste disposal problems" as; 1) change in groundwater level; 2) disposal of excavated materials, and; 3) percolation of waste materials.

California Division of Mines & Geology

This State Agency, except for its technical assistance to the Division of Forestry in land use planning for logging, has devoted most of its talent over the decades to classical geologic and mineral resources studies. One exception, however, was the 1973 report on *Potential Disposal Sites for Class I Waste in Southern California*.

[6] Sec. 13052(e), Water Code, specified that "Studies and research are urgently needed to determine the region-wide overall safe and reasonable rate of discharge, through sewage and industrial waste, of the various soluble deleterious materials, particularly with respect to their effect on underground water."

10. REGIONAL WATER POLLUTION CONTROL BOARDS

Regional Boards initially were appointed through the Governor's Office and had to meet as a first order of business to select their own Executive Officers. Most Boards, according to Paul Bonderson (1980, p. 3) were cautious about staffing and did not reach their Legislature-authorized employee strengths until about 1962-1963.

In searching and reviewing many documents issued by the Regional Boards, it appears that the Stringfellow Acid Pits represents not only the first commercial hazardous waste disposal site to have completed construction and opened for operation under board discharge requirements, but also the site represents the first regulatory-influenced (note that the "regulated" is not herein chosen as that wording would be incorrect to the situation) commercial hazardous waste disposal site in the world.

DWR did issue three reports to the Los Angeles Board that pre-date the 2 June 1955 site report for the Stringfellow site. These were an 18 April 1951 report for *Investigation of Oil Wastes in Hall Canyon, Ventura County* , an 8 April, 1954 report for *Proposed Industrial Waste Discharge - Elkins Ranch Company, Filmore,* 14 October 1954 on *Gillard Petroleum Service, Inc., Pierpont Boulevard Disposal Site, Ventura County* and a 28 September 1954 report for the Santa Ana Regional Board, entitled *Proposed Industrial Waste Disposal Site - Parker Brothers, Orange County.* It is believed that these sites either were not established or that they did not initiate operations before the Stringfellow Acid Pits (31 August 1956).

From 1957 to 1959 a California Legislature Interim Committee reviewed the role and performance of the Regional Boards. Three findings are of interest (California Legislature, 1960):

1) The Boards took a position of not having the right to question a discharge; even into open-water bodies such as streams, lakes and reservoirs; they merely set objectives (if local jurisdictions granted permits for such activities);

2) The Boards sometimes utilized the "cooperative approach", such as that of the Central Valley RWPCB which assisted oilfield operators in providing disposal facilities through the proper exercise of stautory mandate (p. 37-38), and

;3) Regional Boards had requested about 345 technical investigations, submitted by support agencies (mainly DWR) as 320 technical reports. It was typical for Boards to request DWR to performspecial investig-ations (p. 38).

Regional Boards were never involved in the actual "siting" process of identifying suitable locations for disposal facilities. Furthermore, in the words of Dennis O'Leary (BSCE), former Executive Director of Regional Board IX (San Diego; in his deposition of 1989, p. 147) the Regional Board "does not approve or disapprove of sites for location." O'Leary further stated that "The Board is concerned with the waste disposal operations insofar as they may adversely or unreasonably affect the ground or surface waters of the State."

The Los Angeles Regional Board, in 1961, undertook the first known investigation of hazardous waste pollution of southern California ground water, by asking DWR Southern District to make a field investigation of *Groundwater-Quality Impairment, near Live Oak Ave. and Peck Road,* in the now recognized and widely polluted San Gabriel Basin. Southern California Water Company had requested Board assistance on the basis of taste and odors first detected in its distribution system from the Mayflower Well, in February 1958. The State Board allocated $15,290 for this study. The source was concluded to have been carbon dioxide and methane from a small refuse dump operated by Valley Transfer Station, Inc., about 1000 ft from the Mayflower water supply well. This case constituted the first real recognition by DWR of the existence of groundwater pollution.

11. DWR IN POST-DICKEY ACT YEARS

With passage of the Dickey Act, the entire organization and mission of the Division of Water Resources changed immensely. The Central Valley Plan was underway as a giant study and engineering geologists were acquired with such rapidity that the graduating classes of all of California's geology departments were quickly absorbed by the traditional petroleum company employers and by the Division. Most California engineering geologists in their sixties today were of the original DWR entry "class" of 1950. California university campuses (their being really only four at the time producing geologists) seemed just able to supply the numbers. Civil engineers, however, were being hired in greater numbers and many of the young engineers were imported from all over the country.

DWR began at once to support the Regional Water Pollution Control Boards, though the volume of this work never touched the traditional water quality and quantity studies that were underway to discover and quantify all of California's water, for the purpose of protecting and distributing it to the people. Geologists and engineers reviewed published literature, studied well logs and conducted confirmatory field work.

When the calls for technical assistance came from the Boards, they were received at the assigned DWR District office, there was no formal instruction in evaluating proposed waste disposal sites, as recalled by former Junior Engineer Robert F. Clawson (Stringfellow case deposition, 1989, p. 9).

Among the first engineering geologists taken into DWR Class of 1950 were Association of Engineering Geologists (AEG) senior members Robert C. Fox, Robert T. Bean, Glenn A. Brown, Robert G. Thomas, and Don McCann. In the Southern California District, for example, support for geologic and engineering review of proposed waste disposal sites was performed from the Water Quality Section. There were no specifications or guidelines for these office and field reviews. Clawson recalls (1989, p. 27-28) and emphasizes the limited capacity of the three-person Board staffs and the fact that the Santa Ana Board did not always ask DWR for supporting information before developing its *Waste Discharge Requirements*, especially if the site was not thought to be "complex."

In its *Water Quality Report No. 1* (Dec. 1952, p.11), DWR defined "industrial waste" as "Any and all liquid or solid substance not sewage, from any producing, manufacturing or processing operation of whatever nature". Renamed by the Legislature, DWR became the Department of Water Resources on 2 February 1957 and the Water Rights Section was split off and its staff was reassigned to the new State Water Rights Board.

Waste-Site Classification

As a means of establishing a systematic means of maintaining water quality, DWR, in 1950, began to compile *Water Quality Objectives* for each major watershed, sometimes for County-wide areas. Once established, Objectives were meant to be modified and revised, especially in consideration of differing geologic conditions in and between WPC Regions and the autonomous nature of the Boards, which were the prime users of the studies. Searching for the means of broadly classifying water quality, the State Board came upon the works of Dr. D.L. Doneen, a University of California, Davis, agronomist. DWR's Bulletin No. 1 (1951) became *Water Resources of California* and the Doneen scheme of classification of waters, based on their quality, was adopted:

Class 1	Excellent to Good
Class 2	Good to Injurious
Class 3	Injurious to Unsatisfactory.

Throughout the waste management profession, many persons refer to regulatory-controlled industrial waste disposal sites as "Class I Facilities." This terminology is found through the United States and in some overseas areas. It refers to a grass-roots classification developed among engineering geologists at the Southern California District of DWR in the time frame 1952-1953. As Glenn Brown recalls (personal communication, 1988), the new system was the result of an informal group effort, mainly as coffee-break talks in Los Angeles. Jack Scheliga codified the new system and reported upon it at the 1956 Cordilleran Section Meeting of The Geological Society of America, in San Diego. It appears that the earliest printed use of the system was in the December 1953 *Investigation of Safe Dump Sites in Coastal Basin Within Los Angeles Basin.* Hydrological engineer David B. Willets recalled in Stringfellow deposition (1984, p. 16) that "the classification of wastes and dump sites which we made has been adopted throughout California and I believe throughout a good part of the nation. And to my knowledge, no one had ever done that before." Robert Clawson noted (1989, p. 58) that Region 4 got the first DWR report utilizing the new classification system.

To go along with the site classification scheme, three groups of wastes were identified:

Group 1: Industrial Wastes

Group 2: Municipal Wastes

Group 3: Inert Materials

This duality exhibits an awareness of the need to consider site integrity on the basis of both geologic conditions and nature of waste. This classification system was unique at the time, and it has served since 1953 as the basis of most regulation of waste disposal to landfills.

Within an unspecified period of time Southern District personnel come to conclude that persons proposing discharges had the option of modifying sites, on an engineered basis, so as to "make sure that nothing was getting out at a rate that would exceed the capacity of the adjacent water system" (Clawson, ibid, p. 68), at sites founded on materials that "restricted hydraulic continuity."

Official use of the DWR classification system is recorded in a letter of 12 June 1958, from Max Bookman, then Chief Engineer of the Southern District, to Dennis O'Leary, then Executive Officer of the San Diego Regional Board. Here Bookman lays out the meaning of each Class of waste disposal sites:

Class 1 Sites: Industrial Wastes: "Dump sites located on rocks through which no appreciable seepage to usable water can occur or underlain by isolated bodies of unusable groundwater, but which are protected from surface runoff and where surface drainage can be restricted to the site or discharged safely."

Class 2 Sites: Sanitary Landfills: "Dump sites underlain by confined or free ground water where the minimum elevation of the dump can be maintained above the record and/or anticipated high ground water elevation, and which are or can be protected from surface runoff, and where surface drainage can be restricted to the site or discharged to a suitable waste-way."

Class 3 Sites: Inert Materials: "Dump sites so located as to afford little or no protection to usable waters of the State."

Standard of Care for Review of Proposed Disposal Sites

Persons reflecting on the manner in which DWR personnel conducted their review of proposed disposal sites are often confused as to the standard of care exercised by Division geological and engineering personnel. Coordination between the State Board and DWR management had previously arranged yearly funding for technical support to the Regional Boards, at their request. Nominal funding was to be provided by the Regional Boards for each request for assistance. Evidence indicates that the funding never matched the level of effort expended by DWR, generally to the extent that the Boards received more DWR work product than would normally be purchased by their expenditures to DWR.

DWR offices began working for the Boards almost immediately after their formation in the spring of 1950. The earliest identifiable report (courtesy of Glenn A. Brown) seems to be the April 1951 *Investigation of Oil Wastes in Hall Canyon, Ventura County*, a five-page document with geologic map and geologic cross section. Subsequent reports followed the format. At the time, there was no significant body of literature to describe the nature and occurrence of ground water in California. The Southern District had completed Eckis's study in 1934 as a basis for its water supply role, but individual groundwater basins were yet uncharacterized. The U.S. Geological Survey had studies underway but results were not yet issued. At the same time, the Boards were not yet receiving reports of proposed industrial waste disposal sites. Coupled with this situation is the fact that until 1952-1953 there was no technical definition or differentiation between sanitary landfills and what was required for proper disposal of industrial wastes.

DWR and SWPCB determined that studies should be made of individual watersheds, beginning in those areas apparently destined to receive proposals for disposal sites and therein to determine how surface water and ground water occurred and to describe locations of waste disposal so encountered. In describing ground water, DWR relied directly on the precedence set by the 1934 Eckis study. There was no definition of what constituted water-bearing rock, no quantitative standard (as barely exists today), no standard among hydrologists and no quantitative amount of water available for extraction (Clawson, 1989, p. 120-121).

In reality, the unwritten rule of the day, a rule-of-thumb that still stands the test in dealing with water supply studies (i.e. those not concerned with contaminant transport or pollution control), is the "two-gallons-per-minute" value. It has been long regarded appropriate that a well that does not produce a volume of water sufficient for one residence (2 gpm) is not worth the expense associated with its drilling and installation, hence "impermeability" came to be associated with earth materials yielding less than 2 gpm.

DWR introduced a quantification of permeability of the "nonwater-bearing formation" in its March 1966 *Planned Utilization of Ground Water Basins; San Gabriel Basin, Appendix A, Geohydrology* stipulated a yield of 5-15 gpm from a properly-constructed well as indicative of a nonwater-bearing formation. Water-bearing formations were referenced as yielding in the range of 100- 4600 gpm.

Regional Studies

Regional Boards needed summaries of the geology and occurrence of surface water, groundwater and existing waste disposal facilities ("dumps") within their areas of jurisdiction. DWR Southern District got this effort underway in 1952, for each of its Regional Boards. The first of a series of reports completed was the 1953 *Investigation of Safe Dumping Sites in South Coastal Basin within Los Angeles Region*

These regional surveys were made over about a year's period of time, relying mainly on existing geologic literature, particularly graduate-school theses. Photogeology was used on a non-standard basis and when the regional geologic maps were completed small teams of geologists and engineers (usually one engineering geologist, one civil engineer and one civil engineer-hydologist) would take to the field on a "windshield" survey to verify their geologic assumptions.

Regional surveys provided DWR personnel with a total-reference basis for undertaking review of specific sites proposed for discharge. Regional surveys were not only to provide a common basis for development of Regional Board discharge requirements but were also to serve as a one-stop reference for the geologist and engineers assigned to review a particular site. The introduction to the 10 May 1954 *Groundwater Occurrence and Quality, Colorado River Basin Region*, work supervisor Elmer C. Marliave (assisted by Robert T. Bean, Robert C. Fox, and Glenn A. Brown) asserted that their purpose was "To present under one cover a comprehensive summary of available geologic, hydrologic, and water quality data concerning ground water in the Colorado River Basin ." Eckis' 1934 concept was embraced in explaining (Table 2) that the "Metamorphic Crystalline Complex" included "almost 20 percent of the nonwater-bearing rocks".

In June, 1954 Southern District completed its *"Region-wide Municipal and Industrial Waste Site Studies, Santa Ana Region*, including parts of Riverside, San Bernardino and Orange Counties. The study was completed in one year at a SARWPCB cost-contribution of $5,000 and identifying 37 candidate areas for potential waste-facility siting; 28 of which were eliminated in the course of the associated field "reconnaissance ." Nine of the sites received a "detailed survey" during January and March of 1955. Some might question the motive of the "detailed surveys" and the general identification of candidate or prospective sites. The absolute motive behind identifying specific sites was never stated, but it is apparent that this effort was made to portray to the Boards examples of the most suitable types of ground within their region. There is no evidence that the Santa Ana Regional Board, for instance, was proactively selecting specific or individual sites for suggestion to potential dischargers. In any event, the actual site selected by Mr. James Stringfellow, Sr., was not one of those locations identified and reviewed by DWR prior to his actions leading to establishment of the Acid Pits.

Within the scope of the Santa Ana Regional study was the following level of attention toward specific sites:

1) Identify existing dumps
2) Identify possible future dump sites
3) Study possible pollution aspects of existing dumps

DWR had an important mind-set involving what we would today refer to as the "hydraulic conductivity" of crystalline rocks. Based on the Eckis work (Bull. No. 45) bodies of igneous plutonic and metamorphic rock in general were considered nonwater-bearing unless shown otherwise. Ideally water supply wells were utilized as evidence in the vicinity of a particular site under "investigation" for a Regional Board. DWR was without funding, equipment or authorization to conduct field exploration for the benefit private-sector applicants for discharge requirements. There were no guidelines in place in any form, anywhere, as to what constituted "impermeability." However, sites on nonwater-bearing rock were generally considered safe for disposal of all types of waste. The Santa Ana report of 1954 notes that sites were de-selected when "Rock or surface material appeared permeable or deeply fractured...." (ibid, p. 24;) [7]

Class I Site-Specific Reviews

Clawson (1989, p. 68) recalls the typical menu for DWR Southern District field activities at sites which were submitted by owners as proposed Class I site. Subsurface exploration was never undertaken, though hand augers were employed (p. 117) for the purpose of determining soil types. The District had no "heavy drilling equipment" and funds were not available for subcontracting such drilling, which was the obvious responsibility of the applicant. Clawson recalls that some auger drilling was occasionally utilized in the period 1950-1960 at some sites where the "quality of water or degree of isolation of site were in question" (1989, p. 75).

There has been serious contention over the intent and standard of care represented by the DWR technical review of proposed Class I sites. *"Detailed inspection"* is the term generally encountered in official transcripts and reports. Clawson (1989, p. 97) uses "reconnaissance" but had no exact definition of *detailed inspection* . He believed it to be a site visit and walk-over to "look at it in some detail."

[7] In May 1956 DWR California Water Project Authority classified nonwater-bearing rock in the following terms: 1) "In relation to the occurrence of ground water and to water-bearing characteristics, the lithologic units can be divided into two groups"; 2) "The second group consists of deposits of minor importance as a source of ground water such as the consolidated pre-Cretaceous basement complex...."; and 3) "The units in this second group are designated as the nonwater-bearing series and will be only briefly described in this report."

Much has been made during litigation over use of "detailed inspection" to represent the level of effort expended on the site-specific reviews. The State of California literally has been charged (Stringfellow case) with "locating ("siting"), designing, and supervision of construction" of the Stringfellow Acid Pits. No evidence has been found by this author to indicate that such a role was the intention of the Legislature nor was such activity then or now allowed of DWR or under taken by that agency. One-third of the Stringfellow case decision lay on the meaning of "detailed inspection" to the jury, and the fact that the three-person party from DWR Southern District spent about 90 minutes each in their field visit to the proposed site. Therefore, it was charged, the State should be held responsible on the basis of not meeting a superior level of care in reviewing site conditions, ascribed by the plaintiff's attorneys as represented by the term "detailed inspection."

In reality, use of "detailed inspection" comes from the fact that the site- specific reviews were a second-level or second-phase of DWR participation; the next phase, as it were, beyond the Regional Studies. On page 40 of the December 1953 South Coastal Basin study, DWR states that "Reliable and positive classification of potential dump sites can be made only after thorough and detailed study of all pertinent factors pertaining to each site." It is clear here that the main purpose of the "detailed study" was to give DWR personnel the basis for classifying individual sites on the basis of the new system.

Former Southern District Chief Engineer Max Bookman (1984, p. 77) testified that DWR did not undertake subsurface investigations that "drilling of holes is not considered that degree of knowledge necessary. And it would be only when plans were made to go ahead with the site that consideration would be given to exploration, if necessary" (implying, of course, that such was not the responsibility of DWR or the Regional Board). In fact, there is no evidence that such programs were ever contemplated, funded or actually carried out. On page 95 of the same deposition Bookman noted that drilling for "test holes" would not be called for. According to Robert Clawson (1989, p. 40) in the mid-1950's DWR had "no money or allocations for using any sophisticated equipment" and that "modern geophysical apparatus weren't available in those days anyway."

On 2 June 1955, Santa Ana RWPCB received a five-page site-review report on the proposed Stringfellow Acid Pits site, with a few provisional suggestions for the Board to consider in developing its Discharge Requirements for the Stringfellow Company. Among the suggestions from DWR was the provision that a "ground water monitoring well" be installed and observed by the operator. This is the first such well to have been specified in the history of waste management.

Replacement by State Board Geologic Staff

With implementation of Subchapter 15, on 7 March 1972, by the State Board, along with the siting and design guidelines, there was a shift away from involvement of the Department of Water Resources as the technical support group, especially in terms of engineering geologic assistance. Alvin Franks had been given the authority to build an internal engineering geologic staff and was prepared to provide direct assistance to the Regional Boards.

12. CONSULTING ENGINEERS AND GEOLOGISTS

As noted previously, the era of accelerated industrial waste disposal opened with a considerable array of talented consulting professionals. The engineers were situated, in terms of the legal right to practice in the field,and California's Universities were among the leaders in sanitary engineering. Geologists, on the other hand, had virtually no opportunity to work independently in the area of waste management for they had no legal professional status. Sanitary engineering consulting firms, when aware of the need for geologic input, turned to the very few established soil mechanics firms. When the soils engineers felt so disposed, they would consult with geologists known to them, but never held on the payroll.

Robert C. Fox, then employed in the Water Quality Section of DWR Southern District, from June, 1950, recalls (1983, p. 383) that private parties frequently came to the Division to seek help in finding sites for industrial waste discharge. Assistance was provided in general terms, indicating general areas that might have siting potential. Fox maintains that DWR personnel did not comment on factors relating to the manner of discharge.

A significant turn of events came about on 26 August, 1952, when the City of Los Angeles Building & Safety Department received City Council approval of its new Grading Ordinance, a worldwide first (McGill, 1954). The ordinance called for input by Engineering Geologists but there was no established means of judging competency and many of the new grading reports submitted by developers were signed by unqualified geologists. The ordinance, however, was a sound piece of work and Scullin reports (1983, p. 14) that 16 California counties had used it as their pattern for their guidelines by 1954.

By 1955, with assignments coming along with Federal and local funds, sanitary engineering consulting was well in place in southern California, including the following consulting engineers and consulting engineering firms which were active an and about Los Angeles (G.A. Brown, pers. comm., 1991); Harold Barnett, Engineering Sciences, Holmes & Narver, Koebig & Koebig, Leeds Hill & Jewett, Pafford Pomeroy Johnson & Bailey, Psomas, and Kenneth Volk. San Bernardino and Riverside Counties were served by the area of influence of C. Davidson, King & Associates, and Albert Webb.

The first evidence of the use of a consultant by an applicant for industrial waste discharge was for the non-commercial, plant-site operation of the Pacific Coast Borax Company (PCBC) at its Boron, Kern County plant. PCBC engaged Dr. John Mann to develop its site investigation and such is noted as a source of background data in the DWR report of 19 September 1955, to the Los Angeles Regional Water Pollution Control Board. In 1958 the San Diego Board issued discharge requirements to the Omar Rendering Company for its Otay Valley facility near San Diego. The applicant had retained Engineering- Science, Inc., of Pasadena, as its consultant. This is the first surviving DWR site report that indicates utilization of a consulting engineering firm by the applicant. The Board required installation of a subsurface drain system and monitoring sampling of leachate at a collecting sump along the drain (O'Leary, 1989, p. 116).

Faced with enormous construction and hazards abatement problems relating to the stability of hillsides, the City of Los Angeles was the first unit of government anywhere to recognize the need to certify engineering geologists and did so as part of its new building code. An Engineering Geologists Qualification Board was established in 1958 by the City to screen persons competent to perform and sign pre-construction grading permit reports for the City Building and Safety Department (Scullin, 1983; Neel, 1988). The County of Los Angeles followed suit in 1960, and Orange and Ventura Counties, in 1962 and 1967, respectively.

With respect to consulting assignments for Water Pollution Board applicants, Engineering-Science, Inc., Pasadena, completed a *Report on Industrial Waste Disposal - Omar Rendering Company - Otay Valley, San Diego* in October 1958, giving clear evidence of the presence of qualified engineering firms and of the recognition of need by an applicant to utilize such services.

13. AMENDMENTS TO CALIFORNIA ADMINISTRATIVE CODE (1972)

On 2 March 1972, the State Water Quality Control Board (SWQCB) adopted the substance of Subchapter 15, Title 23 of the California Administrative Code, to become effective 6 April of that year. Sweeping changes were made, as based on two years of legislative hearings. Inherent in this move was a program of uniform Statewide requirements for the siting and design of waste disposal facilities. Still not known as a "permitting" program, the "discharger" was advised formally that he/she was responsible to furnish all necessary information; DWR was removed from the loop of review and the Board would rely on its own technical staff to review the submittals of the applicant, both for breadth and depth of content and for technical accuracy and quality.

Relying on the original DWR waste and site classification system, SWQCB employed a system utilizing Groups and Classes, and Arabic and Roman numerals; for instance:

Group 1 Wastes (industrial) "consist of or contain toxic substances which could impair the quality of usable waters"

Class I Disposal sites require complete protection for all time.

For example, Group 1 wastes would be placed in Class I disposal sites.

Alvin Franks had been selected on 1 March 1968 to take on additional duties with the new State Board as its only engineering geologist. The concept of "permitting" of the waste management facilities in California had arrived and to Franks had fallen the task of compiling a new and comprehensive set of standards for the permitting of waste management facilities. Franks found the SWQCB to be without any form of rules governing Class I sites (1988, p. 130), and that "very few of the Regional Boards required much of anything" from proposed dischargers (Franks, ibid, p. 131). Such a condition did not, of course, relieve the proposed discharger from the inherent responsibility to obtain whatever data or information are necessary to responsibly manage risks associated with the intended disposal operation.

Dr. Franks drafted Subchapter 15 as an addition to Chapter 3, Title 23 of the California Administrative Code. After about two years of public hearings, the revision was adopted with minor modifications, on 2 March 1972 and became effective statewide. During this period, Franks worked on guidelines for siting, permitting and construction practices that would be required for implementation by the Regional Boards, and as a basis for understanding by local governments, as well as by industrial applicants. These siting and design requirements, "Disposal Site Design and Operational Information; Waste Discharge Requirements for Regulating Waste Disposal to Land", first issued on 2 March, 1972, constituted the first codification nationally, and, therefore, on a worldwide basis. Franks was assisted with some of the engineering features by Larry Birch, a civil engineer staff member of the Board.

These guidelines brought about a new order of business for industrial waste management in California. Operators of all Class I, II, and III waste disposal facilities had a new and specific set of site characterization and design criteria to meet. This was yet four years ahead of the passage of the Federal Resource Conservation and Recovery Act of 1976, the regulations for which did not begin to appear until 1980.

New Waste Disposal Facility Rules (1972)

For the first time ever, applicants for the right to establish and operate waste disposal sites were given a Statewide set of criteria to meet (Table 5). As before, the responsibility to gather and present such information to the Regional Board remained with the applicant, but now that responsibility was clear and unmistakable.

14. SUMMARY

In one way or another, California has seen and experienced everything in hazardous waste management, generally before they have been encountered elsewhere. Along with its other perceived forms of magic, the Golden State has been the microcosm of the industrialized planet; it is a young place, formed from wilderness populated with energetic people who have made progress and waste, and all at the same time, yet usually with a mindful thought and voice that environmental care must be exercised along with freedom.

The end result and concern have created a means of industrial waste management that has not been duplicated anywhere for its detail. It has produced a wealth of experiment and substance that has been adopted in parts, first nationally, then, in round about way, by most other States. The progress that came from the California experiment is lodged everywhere in successful environmental management; the shortcomings inevitably linked with first passage, are yet the battleground of the courts.

This paper deals with a period of time during which the Federal government had virtually no regulation or guidance on the management of industrial waste; no means of assistance to the States, and nothing to give the States for a task that had not yet acquired national attention. It may seem that the author portrays southern California as being somehow ahead not only of the rest of the nation, but a step out in front of even its northern half. If that seems to the case, it is because that is the case. The largest part of the California experiment was concluded before the Congress drafted the Resource Conservation and Recovery Act (RCRA) in 1976. Still, in subsequent years, California has maintained a technically-oriented lead in development of regulation and regulatory programs to manage what came to be known as hazardous waste. In random order, here are what emerge from the California experiment in industrial waste management:

1) Industrial waste was managed to a greater degree in California than elsewhere in the nation or world;
2) All industrial waste, of that time, was managed under provisions of the Dickey Water Pollution Control Act and its successor Porter-Cologne Act; both of which were unequaled in positive terms of environmental protection, anywhere;
3) The Dickey Act was driven by dual concerns: the use and protection of the waters of the State;
4) Use of water to manage waste was the over-riding theme;
5) Waste management sites were judged on their subsurface water- bearing characteristics, as based on an existing water-supply experience, in which crystalline rocks were regarded as being impermeable;
6) "Pollution" and "contamination" have separate and distinct meanings; "pollution" was linked to degradation of

water quality, and "contamination" was linked to the issue of public health;

7) Responsibility to detect contamination lay in the chain of action from local health official, through the County, to the State Department of Health;

8) The Water Pollution Control Boards, State and Regional, were never invested with a practical or workable enforcement authority by the State Legislature;

9) Waste disposal sites created under the provisions of the Dickey Water Pollution Control Act were never subject to a system of regulatory permitting that is familiar today under the provisions of Federal law;

10) The Boards never had the right to disallow the initiation of waste disposal; that right was vested in Counties, through the same planning and zoning acts that governed all other manner of private-sector developments;

11) Board power was limited only to issuance of "discharge requirements" formulated to lessen the chances of water-pollution;

12) Boards were not created with technical staff and were to rely on the Division (later Department) of Water Resources for technical expertise in "development of discharge requirements" other technical questions relating to water pollution control;

13) In assisting the Boards, DWR personnel were never charged with providing information for use of and by the "proposed discharger";

14) The fact that DWR utilized the terminology of "detailed investigation" to describe its site-specific reviews of proposed disposal sites in no way indicates a standard of care greater than a limited site visit for the purpose of verifying general geologic and hydrologic conditions already studied (at that time) in regional site assessments completed for the benefits of the Regional Boards;

15) There was no call for, nor any way to justify, a level of effort on the part of DWR that would provide a greater base of information than the aforementioned verification of regional geologic and hydrologic information, or to provide unique subsurface geologic information for a proposed site; such was, and still is, the inherent responsibility of the person, organization, or interests wishing to construct and operate the facility;

16) "Proposed dischargers" inherently bore all responsibility-for any form of risks associated with siting, construction and operation of their facilities;

17) In the time covered by this paper, California had the largest population of competent sanitary engineering consultants, individual and by firms, of any State in the Union; such resources were easily available to any "proposed discharger";

18) There is no evidence that there was an intent of the State Legislature to create a role for DWR to assist the privatesector in any aspect of the siting, construction or operation of any waste disposal facility that received discharge requirements from a Regional Board;

19) "Self-monitoring" of disposal site performance was always the intent of the Legislature; inclusion of the first groundwater monitoring well, anywhere, in the discharge requirements for the Stringfellow Acid Pits (1955) is an extension of that policy,and;

20) An inspection role for the Regional Boards was not indicated by the Legislature or implemented by the State Board; such was within the normal activities of the Counties, especially from the standpoint of public health, which relates to contamination.

Those who are familiar with the sequential development of Federal waste management legislation, from the first such law, the Solid Waste Disposal Act of 1965 (PL 89-272, 20 October), to the Resource Recovery Act of 1970 (PL 91-512, 26 October), and finally, the currently-effective Resource Conservation and Recovery Act of 1976 (PL 94-580, 21 October) are aware of the strong elemental relationship between the California experiment and the substance of the Federal acts. California was the bed of development and the proving ground for most of this Federal legislation. It was not until 19 May 1980 that the first EPA Subititle C regulations (hazardous waste management) appeared, more than 30 years after California took action in form of the Dickey Act.

ACKNOWLEDGEMENTS

The author is grateful to the Office of The Attorney General, State of California, for its interest and in supplying some of the historic facts of waste management in California. In particular, I wish to thank Donald A. Robinson, Esq. and William L. Carter, Esq., Deputy Attorneys General, and Brian A. Hembacher, Esq., Assistant Attorney General; all of the Los Angeles Office. To Glenn A. Brown, for providing access to his remarkable memory and outstanding personal library on southern California geology and its practice, along with numerous discussions and review of this paper. To Diane R. Anderson, my research colleague, for ferreting out some of the more difficult-to-locate facts, and for closely reviewing and editing the manuscript. To Alvin L. Franks, who reviewed the entire paper for correctness, increased its accuracy by at least several percent, and provided three obsucre-but-valuable references that I had not known of. To volume editors Richard J. Proctor and Bernard Pipkin for their in-depth reviews. To James Vitale, AIA, Wagner, Hohns & Inglis, Pasadena, CA, thanks for his bibliophilic help and ability to find books that no one else has, along with providing me with a base for research during the late summer of 1991. At the law firm of Breidenbach, Swainston, Crispo, and Way (Los Angeles), thanks go to Howard L. Halm, Esq. and Dulany Harms, Esq., for their encouragement, interest, and comment, and to their documents manager Susan Andrews, who manages huge volume of Stringfellow case literature, keeping such under accessible control. For much of the base of literature that is never very well duplicated between individual libraries, I thank the following universities: My alma mater, UCLA, my present academic home, University of Missouri-Rolla, along with libraries at UC-Berkeley, UC-Davis, California Institute of Technology, Stanford University, California Polytechnic State University, University of Colorado, Colorado School of Mines, University of Missouri-Columbia, University of New Mexico, University of Montana, University of Utah, University of Oregon, Oregon State University, Washington University, St. Louis, Missouri, and the U.S. Geological Survey libraries at Reston, Denver, and Menlo Park, the California Resources Agency Library, Sacramento, The U.S. Army Engineer Center Library and historical collection, Fort Leonard Wood, Missouri, Linda Hall Library of Technology, Kansas City, Missouri, the USEPA National Enforcements Investigation Center, Denver, Colorado (special thanks to Dottie Biggs and Barbara Greenman) and the National Institutes of Health, all of whom I have visited and worked at during travels since 1988.

REFERENCES CITED

American Gas Association, 1920, Report of Committee on Disposal of Water from Gas Plants: unknown location and pagination.

American Society of Civil Engineers, 1959, Sanitary Landfill: The Society, NYC, NY, Manual of Engr. Practice, No. 39, 61 p.

Bonderson, P.R. (1919-1983) former Executive Director, CA SWPCB and SWQRCB; 1956-1966), 1980, Interview by Malca Chall, Univ. California Program in Regional Oral History: Bancroft Library, Univ. Calif., Berkeley; in, California Water Issues, 1950-1960 (covers dates 1950-1966; Mr. Bonderson died in 1983; see 1966 entry of Wm. Attwater, CA SWRCB, in AWH Chronology), 81 p., Bancroft Library call no. 82/32c

Bookman, M., 1984, Deposition, U.S.A., et al, Plaintiffs, vs. J.B. Stringfellow, Jr., et al, Defendants: Held at Los Angeles, CA, on 12 Jan, recorded by Traci A. Dokich, Inc., 145 p.

Brown, G.A., 1991, Letter designating consulting engineering firm that were active in southern California in 1955: to Allen W.Hatheway, dated 14 May, 2 p. California Division of Mines & Geology:

_____, 1973, Potential Disposal Sites for Class I Waste in Southern California: The Division, Sacramento, Spec. Publ. 44, Sep, by J. L. Burnett and G.C. Taylor, 12 p.

_____, 1975 (rev. 1982), Guidelines for Geologic/Seismic Consideration in Environmental Impact Reports: CDMG Notes, Number 46, Sacramento, 2 p. (by P.Y. Amimoto, and J.E. Slosson, State Geologist). California Division of Water Resources:

_____, 1934, Geology and Ground Water Storage Capacity of Valley Fill; South Coastal Basin Investigation: Sacramento, Bull. 45, by R.C. Eckis, 279 p., with 6 map plates.

_____, 1951, Investigation of Oil Wastes in Hall Canyon, Ventura County: Report to Los Angeles Regional Water Pollution Control Board, 18 Apr, 5 p, map, and geologic section.

_____, 1953-1972, Unpublished "Compilation of Dump Site Classifications" developed and utilized by State Division/Department of Water Resources (pre-1972) and State Water Resources Control Board (1972 and later): Law environmental, Inc., Office of Glenn A. Brown, Burbank, CA, var. p.

_____, 1953, Investigation of Safe Dump Sites in South Coastal Basin Within Los Angeles Region: Los Angeles, Report to Los Angeles Regional Water Pollution Control Board, Dec., 40 p., plus 2 app. (Premier report of a series on surveys, work upon which began in 1952).

_____, 1954a, Proposed Industrial Waste Discharge - Elkins Ranch Company, Filmore: 8 Apr, 4 p, plus table.

_____, 1954b, Gilliard Petroleum Service,Inc., Pierpont Boulevard Disposal Site, Ventura County: 14 Oct, 3 p, table, and map.

_____, 1954c, Proposed Industrial Waste Disposal Site - Parker Brothers, Orange County, 28 Sep, 3p, plus table and map.

_____, 1954d, Region-Wide Municipal and Industrial Dump Site Study, Santa Ana Region: Los Angeles, Report to Santa Ana Regional Water Pollution Control Board, 41 p., plus 2 app.

California Legislature, 1960, Interim Committee Report (for 1957-1959) Period: Sacramento, unk. p.

California State Water Pollution Control Board, 1952, Report on the Investigation of Leaching of Ash Dumps: Sacramento, Publ. 2, Jun, by Dr. R.C. Merz, Univ. Southern California, 91 p.

_____, 1954, Report on the Investigation of Leaching of a Sanitary Landfill: Sacramento, Univ. So. California, Publ. 10, Aug., 92 p.

_____, 1957, Water Pollution Control Act (includes 1956) [1st. Exec. Sess.] and 1957 Legislative Ammendments): Sacramento, Nov, 14 p.

_____, 1959. Water Pollution Control Act; Div. 7, Calif. Water Code (Including 1956 [1st. Exec. Sess.], 1957 and 1959 Legislative Amendments): Sacramento, Jul, 19 p.

California State Water Quality Control Board, 1966, History of the State Water Resources Control Board; A Compendium: The Board, Law Library, compiled by Wm. Attwater, Esq., Sacramento, Jul, unpaginated.

_____, 1969, Recommended Changes in Water Quality Control, Final Report of the Study Panel; Study Project - Water Quality Control Program: Sacramento, prepared for the California Legislature, Mar, 119 p., plus 5 variously paginated appendices.

California State Water Resources Control Board, 1972, Disposal Site Design and Operational Information; Waste Discharge Requirements for Regulating Waste Disposal to Land: Sacramento, Nov., 61 p. (authored by Dr. A.L. Franks).

_____, 1979, Inventory, Inspection, and Condition of Class II-1 Disposal Sites and Class II Disposal Sites Receiving High-Moisture Content Waste: Sacramento, Hydrogeologic/Geotechnical Section, Div. of Planning and Research, Dec, under the participation and direction of A.L. Franks, 71 p.

_____, 1980, Disposal Site Design, Operation, and Closure Information; Waste Discharge Requirements for Newsewerable Waste Disposal to Land: Sacramento, Jul., 53 p., plus 40-page Appendix (authored by Dr. A.L. Franks).

_____, 1980, The Porter-Cologne Water Quality Control Act; Addenda and 1979 Amendments: Sacramento, Jun, 113 p.

Carpenter, L.V., and Setter, L.R., 1940, Some notes on sanitary landfills: American Jour. Publ. Health, v. 30, Apr., p. 385-393.

Clawson, R.F., 1989, Deposition, U.S.A, et al, Plaintiffs, vs. J. B. Stringfellow, Jr., et al, Defendants: Held at Redding, CA, on 19 Jan, Office of Challe and Fisher, Deposition Reporters, 125 p.

Crawford, F.G., 1955, Organizational and Administrative Development of the Government of the City of Los Angeles; During the Thirty-year Period July 1, 1925 to September 30, 1955: Univ. of Southern Calif., Sch. of Publ. Admin., Sep, 281 p.

Crouch, W.W., and Dinerman, B., 1963, Southern California; A Study in Development of Government for a Metropolitan Area: Univ. Calif. Press, Berkeley and Los Angeles, 443 p.

Crouch, W.W., and McHenry, D.E., 1945, California Government; Politics and Administration, rev. ed.: Univ. Calif. Press, Berkeley and Los Angeles, 344 p.

_____, 1949, California Government; Politics and Administration, rev. ed.: Univ. Calif. Press, Berkeley and Los Angeles, 406 p.

Eckis, R.C., 1934, Geology and Ground Water Storage Capacity of Valley Fill; South Coastal Basin Investigation: Sacramento, Bull. 45, 279 p., with 6 map plates (authored by R.C. Eckis; CIT Library has two of the few remaining copies).

Folwell, A.P., 1903, Sewerage; The Designing, Construction, and Maintenance of Sewerage Systems, 5th ed.: John Wiley & Sons, New York City, NY, 455 p.

Fox, R.C., 1983a, Deposition, U.S.A., et al., Plaintiffs, vs. J.B. Stringfellow, Jr., et al., Defendants: Held at Newport Beach, CA, on 29 Jun, Vol. I, Office of Latham and Watkins, CSR, 204 p.

_____, 1983b, Deposition, U.S.A., et al., Plaintiffs, vs. J.B. Stringfellow, Jr., et al., Defendants: Held at Newport Beach, CA, on 29 Jun, Vol. II, Office of Latham and Watkins, CSR, p. 205-351.

_____, 1983c, Deposition, U.S.A., et al., Plaintiffs, vs. J.B. Stringfellow, Jr., et al., Defendants: Held at Newport Beach, CA, on 6 Jul, Vol. III, Office of Latham and Watkins, CSR, p. 352-451.

Franks, A. L., 1988a, Deposition, U.S.A., et al, Plaintiffs, vs. J. B. Stringfellow, Jr., et al, Defendants: Held at Sacramento, CA, on 22 Dec, recorded by Dokich, Murray & Assoc., Vol. I, 180 p.

_____, 1988b, Deposition, U.S.A., et al, Plaintiffs, vs. J. B. Stringfellow, Jr., et al, Defendants: Held at Sacramento, CA, on 23 Dec, recorded by Dokich, Murray & Assoc., Vol. II, p. 181-328.

_____, 1989a, Deposition, U.S.A., et al, Plaintiffs, vs. J. B. Stringfellow, Jr., et al, Defendants: Held at Sacramento, CA, on 29 Dec, recorded by Dokich, Murray & Assoc., Vol. III, p. 329-468 (plus exhibits).

_____, 1989b, Deposition, U.S.A., et al, Plaintiffs, vs. J. B. Stringfellow , Jr., et al, Defendants: Held at Sacramento, CA, on 18 Jan, recorded by Capitol Reporters, Vol. IV, p. 469-581.

_____, 1989c, Deposition, U.S.A., et al, Plaintiffs, vs. J. B. Stringfellow, Jr., et al, Defendants: Held at Sacramento, CA, Exhibits 1690 through 1713, assembled by Dokich, Murray & Assoc., Vol. I.

_____, 1989d, Deposition, U.S.A., et al, Plaintiffs, vs. J. B. Stringfellow, Jr., et al, Defendants: Held at Sacramento, CA, Exhibits 1714 through 1749, assembled by Dokich, Murray & Assoc., Vol. II.

Geological Society of America:

_____, 1986, Hydrogeology of Southern California: Volume and Guidebook, 82nd Ann. Mtg., Cordilleran Sec., Los Angeles, CA, 25-28 Mar, compiled by P. K. Saint, 148 p.

Part Two, Landfills and Their Hydrology, p. 77-136.

Campbell, A.M., The sanitary landfill; A bird's eye view, p. 81.

Carry, C.W., and Dowdell, C.A., Geology and operations of Puente Hills Landfill,p. 82-83.

Ray, M.E., BKK Landfill hydrogeological considerations, p. 85-104.

Gross, W., Summary of remedial activities, geology, and hydrolgeology at the Stringfellow hazardous wastge site, Riverside County, California, p. 105-114.

Audinis, J., Leachate production in relation to climate at Riverside Landfill, p. 115-123.

Raede, D.L., Bertine, K.K., and Huntley, D., Major ion and trace metal conamination of groundwater at the Riverside Sanitary Landfill, p. 125-126.

Reilly, J.F., Organic contaminants in theOrange County Basin, p. 127-136.

Hagwood, J.J., Jr., 1981, The California Debris Commission; A History of the Hydraulic Mining Industry in the Western Sierra Nevada of California, and of the Governmental Agency Charged with its Regulation: U.S. Army Engineer District, Sacramento, CA, 102 p.

Harding, S.T., 1960, Water in California: N-P Publications, Palo Alto, CA, 231 p.

Lambie, J., 1984, Deposition, U.S.A, et al, Plaintiffs, vs. J. B. Stringellow, Jr., et al, Defendants: Held at Los Angeles, CA, on 24 Feb, Vol. II, by Dokich, Murray & Associates, Court Reporters, San Clemente, CA, v. I, 122 p. (Mr. Lambie was a truck driver for BKK-Chancellor & Ogden, Inc.).

Los Angeles, City of:

_____, 1905, Charter of the City of Los Angeles, as Adopted January 1889, Amended January, 1903, and Amended January, 1905: The City Council, publ. by Out West Co., 115 S. Broadway, Los Angeles, 117 p.

_____, 1952, An Ordinance Amending Article I of Chapter 9 of the Los Angeles Municipal Code: Ordinance no. 100,347, City Council, 26 August. Los Angeles County Sanitation Districts:

_____, 1950, Report Upon the Collection and Disposal of Refuse in the County Sanitation Districts of Los Angeles County, California: Los Angeles, prepared under direction of AM Rawn (note: "Mr. Rawn" as he preferred, had the first name of "AM", no periods), Chief Engineer and Genl. Manager, Oct, 102 p., + tables and charts.

McGill, J.T., 1954, Residential building-site problems in Los Angeles, California, in, Jahns, R.H., ed., Geology of Southern California: California Div. Mines & Geology, Ch. X, Engineering Aspects of Geology, p. 11-18.

National Research Council, 1966, Waste Management and Control: National Academy of Sciences, Committee on Pollution, Publ. 1400, Rept. to Federal Council for Science and Technology, 257 p.

O'Leary, D.A., 1989, Deposition, U.S.A, et al, Plaintiffs, vs. J. B. Stringellow, Jr., et al, Defendants: Held at San Diego, CA, on 19 Jan, Vol. I, by Dokich, Murray & Associates, Court Reporters, 81 p.

Orange, County of, 1947, Report Upon the Collection, Treatment, and Disposal of Sewage and Industrial Wastes of Orange County, California; Orange County Sewage Survey, 1946-1947: Board of Consulting Engineers, AM Rawn, Chairman, Chas. Gilman Hyde, and Franklin Thomas, 30 Jun,. Santa Ana, CA, 470 p.

Orange County Sanitation Districts, 1990, Annual Report, 1990; Source Control: Fountain Valley, CA, v. 2, var. p.

Phelps, E.B., H.D. Pratt, and C.J. Velz, 1948, Public Health Engineering; A Textbook of the Principles of Environmental Sanitation, John Wiley & Sons, Inc., New York, NY, 655 p.

Refuse Removal Journal, 1962, Play center for 132,000 from landfill site: v. 6, no. 6, Jun, unknown pagination.

Ruchhoft, C.C., and Ettinger, M.B., 1947, Tastes and odors in water resulting from industrial wastes: Proc., 3rd Industrial Waste Conference: Lafayette, IN, 21-22 May, Ext. Ser. No. 64, p. 321-350.

San Diego, City of, 1937, Ordinances of the City of San Diego, California: Publ. by Authority of the Council of the City of San Diego, Jul, 616 p.

Schleiga, J. T., 1961, The application of geology to the selection of waste disposal facilities: Reprint of paper before Cordilleran Section, Geological Society of America, San Diego State College, 28 Mar 1961, 5. (author was with Calif. DWR, Southern California Distr.).

Scullin, C.M., 1983, Excavation and Grading Code Administration, Inspection, and Enforcement: Prentice-Hall, Englewood Cliffs, NJ, 405 p

Todd, D.K., and Clendenen, F.B., eds., 1956, Proceedings, Conference on California's Ground Water Situation: University Committee on Research in Water Resources, Berkeley, 3-4 Dec, 212 p.

Turneaure, F.E., and Russell, H.L., 1940, Public Water Supplies; Requirements, Resources, and the Construction of Works, 4th. ed.: John Wiley & Sons, Inc., New York City, NY, 702 p.

U.S. Public Health Service:

_____, 1925, Drinking Water Standards: Rept., 10 Apr, p. 693-721.

_____, 1942, Ohio River Pollution Survey; Industrial Waste Guides: Cincinnati, Stream Pollution Investigations Stn., 14 p., NTIS PB 214-894.

_____, 1946, Drinking Water Standards: Public Health Rept. no. 61, Mar, v. 6, p. 371-384

_____, 1962, Drinking Water Standards: Public Health Rept. 61, Publ. no. 956, Washington, DC, 70 p. (The "Blue Book").

University of California, Davis, 1966, Comprehensive Studies of Solid Waste Management: Sanitary Engr. Lab., SERL Report 72-3, unk. p.

Willets, D.B. 1984, Deposition, U.S.A., et al, Plantiffs, vs. J.B. Stringfellow, Jr., et al, Defendants: Held at San Diego, CA, on 24 Jan, recorded by Traci A. Dokich, 83 p.

TABLE 1

CHRONOLOGIC APPEARANCE OF HAZARDOUS SUBSTANCES

Date	Group Name	Implications
18th Cent.	Phenols/Polyaromatic hydrocarbons	Carcinogens
1807	Copper sulfate becomes fungicide	
1854	Carbon disulfide used as fumigant	
1860s	*Paris green* (copper acetoarsenite)	Applied against Colorado potato beetle
1900	Arsenicals, sulfur, petroleum oils, pyrethrum	Common pesticides
1917	Isoproponol, other C3 derivitives	First produced by Standard Oil Co.
1917-18	Hydrocyanic gas and pyrethrum	New pesticides for vermin control
1919	Ethylene produced	Union Carbide, Tonawanda, NY
1920	Isopropyl alcohol (rubbing) alcohol	First petrochemical production, by EXXON, Linden, NJ refinery
1920	Phenol, cresol hydrocarbons, anthrazene or napthalene	Found to produce disagreeable "idoform" tastes in chlorinated drinking water
1925	Ethylene, ethylene oxide, ethylene glycol	Union Carbide, Charleston, WV
1927	PCBs introduced commercially	Heat transfer for electrical transformers, motors, appliances
1930	Polystyrene introduced commercially	
1930	First synthetic pesticides; fluorine compounds Dinitro- and Thiocyanate	
1932	Butylenes refined to produce MEK, other halogenated solvents	Shell Chemical Co.; first highly toxic groundwater contaminants
1930's	Polyethelene discovered	Imperial Chemical Industries (ICI), UK
1930-32	Phenol-formaldehyde alkydes, amines, epoxy, phenolics, polyesters, polyurethanes, vinyl emulsions	First synthetic resins; paint base
1931	Polyvinyl chloride available	

TABLE 1 (Cont.)

Date	Group Name	Implications
1934	Union Carbide makes 50 derivitives	
1937	Butyl synthetic rubber developed	Exxon; first commercial producer
1939	Synthetic chemicals	Vastly outstrip coal-tar derivatives
1939	Polyethylene and nylon available	
1940	Dow produces styrene and polystyrene	From ethylene and benzene
1941	Urethanes introduced commercially	
1942	GR-S rubber and fiberglass available	
WW II	DDT in Switzerland; BHC in UK, France;	
	US, napthalene feedstock for toluene	Used as aviation fuel additive
1946	More-toxic chlorinated hydrocarbon	Chlordane, toxaphene, dieldrin, aldrin insecticides
	Germans produce organo-phosphorous compounds	Including TEPP and parathion
1948	Acrylic fibres, ABS, and latex paints	
1949	Organochlorine pesticides	1) Highly effective, low rates of application 2) Persistent for years 3) Toxic to animal life 4) Incremental increases in the environment 5) Bioaccumulative in humans 6) Low solubility in water 7) Sorption to colloids 8) Low biodegradation
1949	Polyester fibres	
1953	Polyethylene terephthalate film	
1955	Toxicity of halogenated hydrocarbons	Discovered by Van Oettingen, USPHS
1956	High-density polyethylene (HDPE)	
1957	Polypropylene	
1959	DBCP (dibromochloropropane)	California nematode killer; banned in 1977

TABLE 1 (Cont.)

Date	Group Name	Implications
1960	Polycarbonates	Beverage containers
1965	Hydrocarbon resins (hot-melt adhesives)	EXXON
1968	Phase-out of white lead paint pigment	
1970	PCB manufacture ceases	Monsanto Corp., "Aroclor"

TABLE 2

MAJOR WASTE DISPOSAL FACILITIES IN SOUTHERN CALIFORNIA: 1950-PRESENT

Dates Operated	Name	County	Remarks
ca. 1928	LACSDs POTW Plant No. 1 (Bixby; Long Beach)	L.A.	Direct discharge to outfall, beyond sewage treatment plant
ca. 1950-85	Baldwin Hills Landfill	L.A.	Fill & cover @ $0.30-0.60/ton; Total costs @ $0.40-1.00/ton
ca. 1950-75	City Terrace Landfill	L.A.	Cut & Cover added $0.20/ton
pre-1950-54	Old Palos Verdes Landfill	L.A.	Expansion of former disposal site?
1954+ Jun88	Mission Bay Landfill	S.D.	Sea World Drive, San Diego; Class II landfill CA SWAT Rept.alleges drum burial below gw surface; gw contamination indicated
1954+	National City Landfill	S.D.	AKA Standard Bay Landfill
pre-1954-Present	Fountain Valley WWTP (OCSDs beginning 1954)	OR	Plant No. 1, Ellis Ave. & Verano St.
1954-1957	Huntington Beach WWTP Plant No. 2	OR	Hwy 101 at Santa Ana River; Acids, cyanides, caustics
Aug 1956 Nov72 Feb74 Sep75 1982	Stringfellow Acid Pits (Stringfellow Quarry Co.)	RIV	Pyrite Rd., Glen Avon; all manner of liquid hazardous wastes to evap. ponds Closed by owner Riverside CO. revokes permit Owner abandons site IRM temporary cover, gw pumping
Oct56-Nov80 Sep89	Otay Class I Landfill (SD CO Refuse Disp. Div.)	S.D.	City's candidate, 2 mi E. of Otay Liquids, sludges (no acids), solid wastes CA SWAT finds gw contamination
May 1957+ Oct80 Jun90	New Palos Verdes* (LASCD's first facility	L.A.	Class II landfill; 26301 Crenshaw Blvd., Rolling Hills; Liquids, sludges, solid wastes; alleged solvents (Lambie, 1984); Open burning ceased May57 Acids w/o hvy metals dischgd to treat ment. system Acids w/ metals dischgd to outfall ($0.02/gal); Chgd. $6/ton for HW in 1980 CA SWAT finds leachate transport; DHS conducting RI/FS
Jul 1957 Sep90	Spadra Landfill, Pomona	L.A.	Class II landfill, 4125 W. Valley Blvd., Walnut; open burning ceased Jul57 CA SWAT finds elev. VOCs in gw
May-Jul57	Independent dumps operating	L.A.	More than 150 separate landfills; Nominally solid wastes
Pre-1957	TCL	L.A.	Industrial wastes; replaced by Industry Hills facility
Jan 1958	South Gate Transfer Station	L.A.	
Jun 1958	Mission Canyon Landfill (LACSDs)	L.A.	2201 N. Sepulveda Blvd., W.L.A; Solid wastes
1958 Sep89	Omar Rendering Co. Facility*	S.D.	1.75 mi E of Otay; commercial disposal Liquids, slurries only (no oily wastes) CA SWAT finds gw contamination
Feb60-Aug80	Calabasas Landfill* (LACSDs)	L.A.	6919 W,. Vent. Fwy, Agoura; Liquids, sludges, solid wastes; 5th such facility
1957	Industry Hills Landfill (AKA Azusa St. Dumps Nos. 1&2)	L.A.	137 Azusa St., City of Industry; Industrial wastes (AKA Industry City Dump) City of Industry municipal landfill
Mar 1961	Scholl Canyon Landfill (LACSDs)	L.A.	LACSDs; 7721 N. Figueroa St., L.A.
Post-1954 1987	Elkins Ranch, Filmore*	VEN	Mainly oilfield brine wastes; took some IW diverted from Stringfellow Acid Pits during Mar 69 rains SWAT-classed in 1987 as consisting of 10 brine sumps

(cont.)

TABLE 2 (Cont.)

Dates Operated	Name	County	Remarks
1960s	Operating Industries, Inc.	L.A.	1000 Potrero Grande Dr. Monterey Pk; IW disposal began 1981 Had Class II unit by 1981 Now a CERCLA NPL site
1960s-1972	Pacific Ocean Disposal*	L.A.	Wilmington; Industrial wastes to ponds and into two shallow injection wells
1960s	Capri, Wilmington	L.A.	Industrial wastes
1960s	BKK, Wilmington*	L.A.	Ind. wastes; @ $0.03/gal in 1969
1960s-1972	Basin Byproducts (Chancellor & Ogden)	L.A.	Wilmington; shallow inject. borings; no tanks; discharge directly to wells
1960s-May80	ANCON Landfill, Wilmington	L.A.	Acids placed in landfill
1960s	ANCON Landfill, Huntington Beach	L.A.	Industrial waste; on Golden West Ave.
Unk. to Jul80	Unidentified Landfill, Carson	L.A.	Co-disposal of SW/HW
1962-1974	West Covina Landfill* (Home Savings & Loan Assoc.)	L.A.	2210 S. Azusa Ave., W. Covina; 130 ac County Unclass. Use Permit No. 71
1963	(LARWPCB Ord. No. 63-16)		Household and commercial refuse
1963	(LARWPCB Ord. No. 63-58		HW disposal only in E. portion of site; requires constr. of impermeable ("Initial") bari alluvium
1968	(County Unconditional Use Permit)		Additional 120 ac at rear of site for refuse
1971			Approval of expansion to 583 ac and 30-year operational life
1972	(BKK Corp.; Benjamin K. Kasarian)		HW disposed to landfill; "Initial Barrier" completed; liquid HW co-disposed with SW, p drum disposal
1974	(LARWQCB Ord. No. 74-380)		Concept. approval of 40-ac Cl I site downstream, westerly into adjacent 100-ac area required construction of Barriers 1 and 2; gwm and ext. wells
1975			Liquid HW waste injection wells, continued co-disposal of liq. HW into refuse
ca. 1976	Liquid Waste Disposal Co. (Chemical Waste Mgmt. Co.)	KERN	Kettleman Hills; vic. Bakersfield; 210 ac. Evaporation ponds, PCBs
Pre-Oct81			170 ac for landfarming non-halogenated solvents @ $\leq$ 10% or < 2% TOH
Jul83			Feasibility study, construction of CN destruction plant underway
1988			Proposed treatment facility
1973-Present	Casmalia Resources Landfill*	S.BAR	Nr. Santa Maria; Ponding, landfarming, 10% drum burial @ $\leq$ 10% organics PCBs, oi control
1981			Only EPA-approved PCB landfill in CA
Jul83			Planned startup of wet-air oxidation facility
Pre-1981	MP Oil Co., Round Mountain	KERN	Class II (oilfield wastes)
Pre-1981	Environmental Protection Co. (EPC)	KERN	Round Mountain Rd., vic. Bakersfield
ca 1983			Class II (oilfield wastes); taking CL I wastes at evaporation impoundments
ca. 1987			Closed
Jan91			RCRA Closure Rept. pending
Pre-1981	Getty Oil Co. (Catch Basin No. 3)	VEN	Class II, Oilfield wastes
Pre-1981	Fellows site	KERN	Class II, Oilfield wastes
Pre-Jul83	Imperial Valley Class I Facility (IT Corp.)	IMP	New HW landburial facility
Early 1983	BKK Otay Aqueous Treatment Fac.	S.D.	1st commercial tmt facility in So. CA
Late 1983	BKK Wilmington Aqueous Tmt Fac.	L.A.	Planned startup
July 1983	Triple-J Pacification Corp., Vernon	L.A.	Proposed unidentified tmt. facility

(cont.)

TABLE 2 (Cont.)

Dates Operated	Name	County	Remarks
	West Covina Landfill (cont.)		
1975	(LARWQCB Ord. No. 75-145)		Increased Class I to final 100 ac.; needed data on effectiveness of Barrier 1
1976	(County Unconditional Use Permit)		SW landfill for add. 30 yrs, to 2006
1978	(Tinnan, in CADOH, 1978)		All forms HW; 3.1 E5 tpy liq., sludges, solid wastes (=4 E6 tpd).; Liquids poured to CL II as sponge on 2 tons SW to 1 ton liquid; Drum burial
1979	(LARWQCB Order No. 78-140)		No HW disposal upgradient from proposed Barrier 2; separate area for drum disposal
1980			Use of injection wells discontinued; Odor complaints from adj. residents
Oct81			Only remaining CL I site in So.Calif.
1982			HW disp = $20/ton + $10.50 State fee; Trans. cost = $5-10/ton;
1983			HW disposed at 4.6 E5 tpy
Pre-Jul 83			Receiving 80% of all So. CA HW
09Oct83			Subsurface investigation of site starts
1984	(EPA CERCLA Sec. 106 order)		Reduces amount of HW liq. disposal
Jul84			Methane gas leakage off-site detected
24Sep84			BKK withdraws RCRA Pt. B Permit application
1985			BKK ceases HW disposal at site; files draft Closure Plan for Cl I site
1986			BKK submits Final Closure Plan; constructs final cover at Cl I site
	McColl Disposal Site	OR	SW of Rosecrans & Sunnyridge, Fullerton
1987			Abandoned; oil ref. sludge ponds; CA DHS site investigation complete
Aug88			CA SWAT finds gw contamination
Late 1972	Hunter Disposal Site (K.R. Hunter, Jr.)	S.BAR	7 mi S of Guadalupe; Liquids, sludges, pesticides
Pre-1973	Simi Valley Landfill* (Ventura CO PWD)	VEN	Nr. Int. Brea Canyon Rd and Hwy 118; Group 1 liquids, sludges, SW
1981			Simi No. 2 operating
1987	(AKA Waste Management, Inc.)		Union Oil Co. landowner
Pre-1988	Petroleum Waste, Inc., Taft	KERN	Unknown
Pre-1988	Carney & Sons*	UNK	Unknown

*Denotes facility ascribed by A.L. Franks (1989, p. 344) as having evidenced contaminant leakage during his tenure as engineering geologist, California State Water Resources Control Bo period 1968-1980.; Names reviewed against SWRCB SWAT list for 17 Dec 1987.

TABLE 3

KEY IMPLICATIONS OF DICKEY WATER POLLUTION CONTROL ACT (1949)

Chapter	Provision	Implication
N/A	No Board "could specify the design, location,	Clearly disassociates State from such activities; type of construction, or particular manner in places such responsibility entirely on the which an operation causing or threatening to operator cause pollution or nuisance was to be corrected."
1550	Amends Health & Safety Code: Any Public Health inspection-Officer, on finding a condition of contamination to order situation abated and to furnish the Regional Board with report	Assigns primary responsibility for detection to County health officials
H&S Code Sec. 2141 Sec. 5460 Kaufman 1979	"If contamination is involved, a peremptory order can be issued by either the Department of Health or locoal Health Officer requiring the abatement of the cotamination, and an injunction suit can be filed by either at once or upon the failure to comply."	Places responsibility for monitoring performance on County; also the requirement to initiate any neccessary abatement action
California CSR Div.7, Sec. 13005	Defines "Industrial Waste a"s "any/all liquid or solid waste substances ... from ... producing, manufacturing or processing operation ...."	Includes most of what today is considered to be hazardous waste
	Duties of Regional Boards:	
	1) Obtain coordinated action through formal or informal meetings of persons involved persons involved	o Role of facilitator in assisting persons wishing to operate waste management facilities
	2) Encourage/assist in self-policing waste (disposal programs for industry	o Operator must self-inspect; Board does not have primary responsibility
	3) On receipt of application, advise on conditions to be maintained at disposal area or receiving waters	o Does not resemble today's waste management permitting process in any way
	4) Require assistance of any State agency to inspect and report on any technical factors involved in water pollution	o In actuality, Division of Water Resources was engaged by State Board to provide regional Boards with this service
	5) Formulate and adopt long-range plans	
Sec. 13053	"Each regional board shall prescribe requirements relative to any particular conditions of pollution or nuisance, existing or threatened, in the region."	Broad powers
Sec. action; 13054	Specifies "*Report of Proposed Sewage Discharge*"	Applicant submits this as the basis for Board applicant had the basic right to undertake waste disposal as long as applicable County planning zoning and building and safety ordinances are met.
	1) Content of report not specified 2) Board to hold hearing, "then prescribes require-ments as to nature of the discharge" 3) Requirements may be revised from time to time 4) After receiving such notice the person so notified shall provide adequate facilities to meet any such requirements with respect t the discharge of sewage and industrial waste	Implies some sort of ongoing review of site performance; but not the primary regulatory responsibility, since self-policing is written into the Act.

TABLE 4

KEY PUBLICATIONS OF THE STATE BOARD OF WATER POLLUTION CONTROL

As California State Water Pollution Control Board (1950-1966)

Publ. No. 1, 1951, Digest of Sewerage Enabling Acts of the State of California: Sacramento, by R.E. Ramseier, Berkeley, CA, unk. p. (revised 1954, 1955).

Publ. No. 2, 1952, Report on the Investigation of Leaching of Ash Dumps: Sacramento, Jun, by Dr. R.C. Merz, Univ. Southern California, 91 p.

Publ. No. 3, 1952, Water Quality Criteria: Sacramento, Publ. 3, Dr. J.E. McKee (CALTECH), Chairman, 512p. (famous as the first water quality reference volume, worldwide).

Publ. No. 4, 1952, Preliminary Statement of Objective and Policy and Report on Water-Quality Evaluation: Sacramento, 8p.

Publ. No. 10, 1954, Report on the Investigation of Leaching of a Sanitary Landfill: Sacramento, Univ. So. California, Aug., 92 p.

Publ. 11, 1954, Report on the Investigation of Travel of Pollution: Sacramento, 31 Dec, by P.H. McFarland and R.B. Krone, Univ. California, Berkeley, 218 p.

Publ. No. 1, rev, 1954, Digest of Sewerage Enabling Acts of the State of California: Sacramento, 2nd. rev. ed., unk. p.

Publ. No. 3, rev., 1954, Water Quality Criteria: Sacramento, Publ. 3, Dr. Jack.E. McKee, Chairman, Addendum to Criteria of 1952, unk p.

Publ. No. 9, 1954, Studies of Waste Water Reclamation and Utilization: Sacramento, Publ. 9, unk. p.

Publ. No. 1, rev., 1955, Digest of Sewerage Enabling Acts of the State of California: Sacramento, 3rd. rev. ed., 68 p.

Publ. No. 23, 1961, Travel Studies and Pollutional Analyses of Estuaries: Sacramento, by Univ. Calif. Berkeley, San. Engr. Res. Lab., 139 p.

Publ. No. 24, 1961, Effect of Refuse Dumps on Ground Water Quality: Sacramento, 107 p.

Publ. No. __, 1962, Status Report on Water Pollution and Water Quality in California: Sacramento, unk. p.

Publ. No. __, 1962, In-Situ Investigation of Movements; of Gases Produced from Decomposing Refuse: Rept. by Engineering-Science, Inc., Arcadia and Oakland, 1 Oct, 36 p.

Publ. No. 3-A, 1963, Water Quality Criteria: Sacramento, Publ. 3-A, revision of 1952 and 1954 eds., by Dr. J.E. McKee and H.W. Wolfe; 548 p..

Publ. No. _, 1964, Report on Collation, Evaluation, and Presentation of Scientific and Technical Data Relative to the Marine Disposal of Liquid Wastes: Sacramento, by Engineering-Science, Inc., Arcadia, CA., 105 p. + numerous unpag. appendices.

As California State Water Resources Control Board (1967-1972)

Publ. No. 36, 1967, Problems of Setting Standards and of Surveillance of Water Quality Control: Sacramento, unk. p.

Publ. No. 37, 1967, Useful Waters for California: Sacramento, 72 p.

TABLE 5

CLASS I HAZARDOUS WASTE DISPOSAL FACILITY CRITERIA

(California State Water Quality Control Board [Franks], 1972)

---

1. Geological conditions naturally capable of preventing vertical hydraulic continuity between liquids and gases emanating from the waste in the site and usable surface or groundwaters.
2. Geological conditions naturally capable of preventing lateral continuity between liquids and gases emanating from wastes in the site and usable surface or groundwaters, or the disposal facility has been modified to achieve such capability.
3. Underlying geological formations which contain rock fractures or fissures of questionable permeability must be permanently sealed to provide a competent barrier to the movement of liquids or gases from the disposal site.
4. Inundation of disposal areas shall not occur until the site is closed in accordance with the requirements of the regional board.
5. Disposal areas shall not be subject to washout.
6. Leachate and subsurface flow into the disposal area shall be contained within the site unless other disposition is made in accordance with requirements of the regional board.
7. Sites shall not be located over zones of active faulting or where other forms of geologic change would impair the competency of natural features or artificial barriers which prevent continuity with usable waters.
8. Sites made suitable for use by manmade physical barriers shall not be located where improper operation or maintenance of such structures could permit the waste leachate, or gases, to contact usable ground or surface water.
9. Sites which comply with 1,2,3,5,6,7, and 8 (above), but would be subject to inundation by a tide or a flood of greater than 100-year frequency, may be considered by the Regional Board as a limited Class I disposal site.

GROUNDWATER REMEDIATION AT PETROLEUM-HANDLING FACILITIES, LOS ANGELES COASTAL PLAIN

Stephen M. Testa
Engineering Geologist
Applied Environmental Services, Inc.
23113 Plaza Pointe Drive, Suite 100
Laguna Hills, California 92653

ABSTRACT

California has an enriched history of oil exploration and production dating back to 1876. The Los Angeles coastal plain was a prime location for the construction of numerous refineries, tank farms, and major pipeline corridors in the early 1920s, reflecting the close geographic proximity to both major production zones and the Los Angeles harbor; consequently, the coastal plain is currently occupied by a large number of petroleum-handling facilities and underground storage tanks. Pools of light nonaqueous phase liquid (LNAPL) hydrocarbon product are widespread in the subsurface beneath many of these facilities, occurring as both isolated perched layers of limited extent, as well as large pools overlying the water table. Some of these LNAPL pools encompass tens to hundreds of acres in extent, with a minimum estimated total volume range of all sites of 1.5 to 7.5 million barrels. Remediation of LNAPL was initiated during the late 1970's at some sites, and accelerated by regulatory order in the mid-1980's. Several of these remediation programs will continue for tens of years for both LNAPL and dissolved hydrocarbon constituents in groundwater. The development of short- and long-term remediation strategies will require a perspective and understanding on a regional scale. An understanding of the regional and local hydrogeologic settings serves as an important foundation in the designation of aquifer(s) for beneficial use, determination of level of remediation warranted, implementation of remediation strategies, prioritization of limited manpower and funds, and general long-term management of remediation programs.

INTRODUCTION

California has an enriched history of oil and gas exploration and development dating back to 1876, the first year of commercial production (Rintoul, 1990). California is comprised of 56 counties, of which 30 produce oil and gas. Eighteen of the 30 counties produce chiefly oil, while the remaining 12 counties predominantly produce natural gas. Hydrocarbon occurrence is predominantly oil south of 37° North Latitude, and the major production in Southern California occurs within four sedimentary basins: the Los Angeles Basin, Ventura Basin, San Joaquin Basin, and Santa Maria Basin. The Los Angeles Basin and the Los Angeles coastal plain, are known to include more than 10 giant oil fields (ultimate recovery potential of 100 million barrels or more). Environmental issues associated with oil fields and their subsequent redevelopment are discussed by Dezfulian (1988), Garcia and Henry (1989), Lovegreen (1989), and Testa (1991a). Prior to the turn of the century, the Los Angeles coastal plain was comprised of large Spanish land grants used primarily for agricultural purposes (Grenier and Gillingham, 1987).

With the discovery of oil, numerous refineries and associated above-ground bulk liquid-storage tank farms were constructed during the early 1920's in close proximity to both petroleum production areas and the nearby shipping facilities of the Los Angeles Harbor. The majority of these facilities continued to expand their operations and areal extent through the late 1940s, while many others were initially isolated or located in moderate to heavily industrialized areas. However, with the encroachment of urban development, many are now in close proximity to densely populated light- manufacturing, commercial and residential-zoned areas.

Sixteen major refineries and 33 above-ground bulk-liquid tank farms are situated on the Los Angeles coastal plain (Table 1). The locations of these refineries and tank farms, and associated major pipeline corridors, are shown in Figure 1. This paper presents discussions of regulatory framework, regional and site specific hydrogeologic setting, and subsurface hydrocarbon presence. Upon definition of subsurface conditions, those factors which play an important role in the approach to both short- and long-term remediation strategies, as well as overall efficiency and effectiveness are discussed. Current and planned strategies are also discussed.

REGULATORY FRAMEWORK

Several refineries, tank farms, and other petroleum-handling facilities nationwide are included on the United States Environmental Protection Agency (EPA) National Priorities List (EPA, 1986), or are regulated under the Resource Conservation and Recovery Act (RCRA). Such facilities present several potential subsurface environmental concerns, reflecting approximately 80 years of continued operation. These concerns include:

- Hydrocarbon-affected soil which could potentially be characterized as a hazardous material;
- Accumulation of hydrocarbon vapors posing a potential fire or explosion hazard;
- Dissolved hydrocarbons in groundwater which may adversely affect beneficial-use water-bearing zones or drinking water wells; and

- Light Nonaqueous Phase Liquid (LNAPL) hydrocarbon pools occurring as perched zones and generally overlying the water table which serves as a continued source of both soil and groundwater contamination;

In early 1985, oil droplets were evident on beach sands near high-priced beachfront real estate located just west of the Chevron refinery in El Segundo (Figure 1). The immiscible hydrocarbons evidently leaked from the nearby refinery, migrated downward to the shallow water table, and then moved downward with the regional groundwater flow direction toward the ocean and beachfront properties. Elevated concentrations of hydrocarbon vapors presenting a potential explosion hazard were also detected at the bottom of a construction pit under excavation near the refinery and several adjacent homes. A significant pollution problem was recognized by the environmental regulatory community. At that time, an estimated (although exaggerated) 6,000,000 barrels of LNAPL hydrocarbon product were thought to exist beneath the majority of the 1,000-acre, 80-year-old refinery.

The highly visible presence of petroleum and potential hazard to public health, safety and welfare subsequently prompted a minimum of 16 oil refineries and tank farms to be designated as health hazards by the California Department of Health Services (DOHS). This designation reflected both potential and documented occurrences of leaked hydrocarbon product derived from such facilities during 70 years of operations, which migrated through subsurface soils and accumulated as LNAPL hydrocarbon pools overlying the water table. Although several of these refineries were listed as hazardous waste sites, with assessment and remediation required under RCRA, the majority of such facilities fell under the jurisdiction of the Los Angeles Region California Regional Water Quality Control Board's (CRWQCB) Order No. 85-17 which was adopted in February 1985, reflecting the potential regional adverse impact on overall groundwater quality.

Fifteen refineries were included under CRWQCB Order 85-17. By December 1985, this order was expanded to include above-ground bulk-liquid storage facilities, notably the GATX Carson Terminal, Paramount Petroleum Lakewood Terminal, and ARCO's Four Corners Pipeline (FCPL) Terminal, among others. Order 85-17 was the first regulatory mandate nationwide to address large-scale regional subsurface environmental impacts by the petroleum-refining industry. This order required, in part, the assessment of the subsurface presence of hydrocarbons and other associated pollutants which could affect subsurface soils and groundwater beneath and adjacent to such facilities. Specifically, the following items required addressing:

- Characterization of subsurface geologic and hydrogeologic conditions;

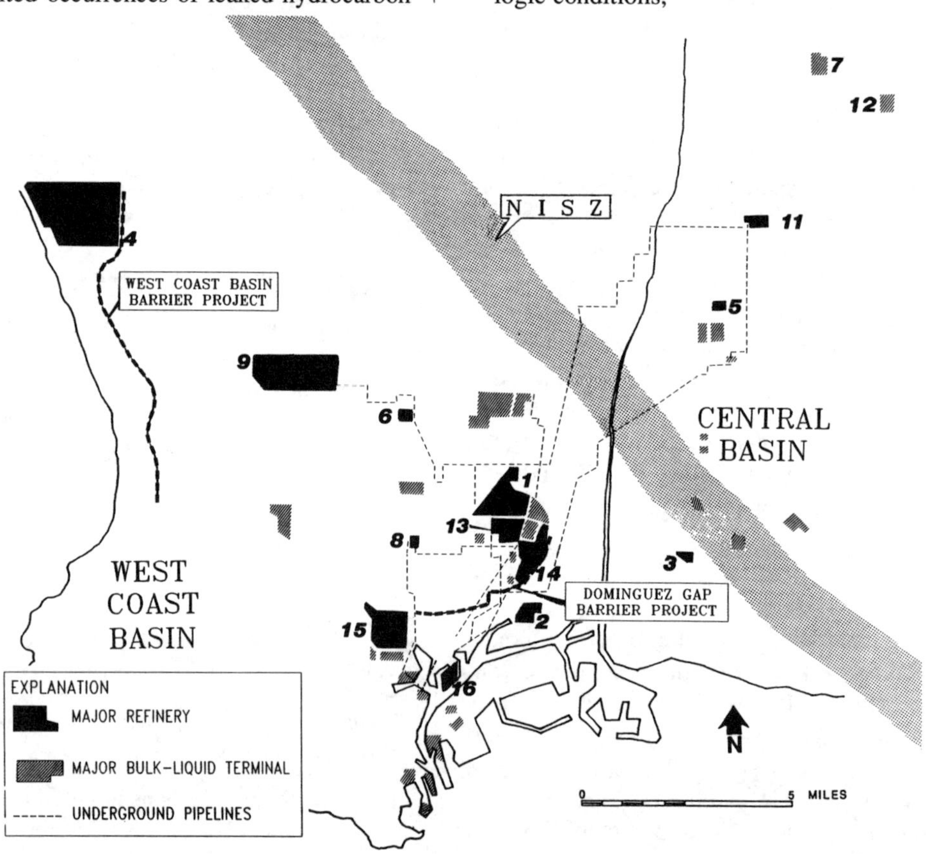

Figure 1. Location of major refineries, bulk-liquid terminals, and pipeline corridors (modified after Testa and Winegardner, 1990, and Central and West Basin Water Replenishment District, 1989).

- Delineation of LNAPL pools including volume and chemical characterization;
- Implementation of LNAPL recovery;
- Overall aquifer restoration of dissolved hydrocarbons and associated contaminants; and
- Eventual remediation of residual hydrocarbons in soils.

At the time this order was issued, approximately six refineries, including the ARCO Los Angeles and Chevron El Segundo refineries (site nos. 1 and 4, respectively; Figure 1), had already commenced LNAPL hydrocarbon recovery programs, although such programs were limited in scope. In these cases, the large volumes of LNAPL present and initial ease of hydrocarbon recovery proved economically favorable, since the recovered LNAPL product could be easily recycled and sold. The overall impact on groundwater resources at the major refineries and bulk-liquid terminals is summarized in Table 1.

HYDROGEOLOGIC SETTING

The Los Angeles coastal plain extends approximately 50 miles in a northwest-southeast direction, and roughly 15 to 20 miles in width between the Pacific Ocean and the base of the Puente Hills and Santa Ana Mountains. Encompassing approximately 775 square miles, the coastal plain is characterized by low relief and gentle surface gradients seaward. The low relief is interrupted by the Newport-Inglewood structural zone (NISZ), characterized by a northwesterly trending line of gentle topography extending roughly 40 miles in length (Barrows, 1974). The NISZ is the major structural feature in the area and consists of en-echelon faults, anticlines, and domes, and is marked by a series of low hills and coastal mesas which are broken by six topographic gaps or low areas.

The Los Angeles coastal plain is divided into four principal groundwater basins (CDWR, 1961). The Santa Monica Basin and West Coast Basin are located southwest of the NISZ; the Central Basin and Hollywood Basin are situated northeast of the NISZ. The NISZ separates the West Coast Basin from the Central Basin (Figure 2). This discussion will focus upon the West Coast Basin and thc western portion of the Central Basin, where the majority of the petroleum refineries and bulk-liquid storage terminals are situated.

Initial studies concerning groundwater beneath the coastal plain were reported by Mendenhall (1905a, 1905b, 1905c, and 1909). Since these early studies, the most comprehensive information on aquifers beneath the coastal plain was presented by Piper et al. (1953), Poland et al. (1948), Poland et al. (1956), Poland et al. (1959), and Zeilbauer et al. (1962). The Silverado aquifer, which occurs within the Lower Pleistocene San Pedro Formation, is of very good qu-

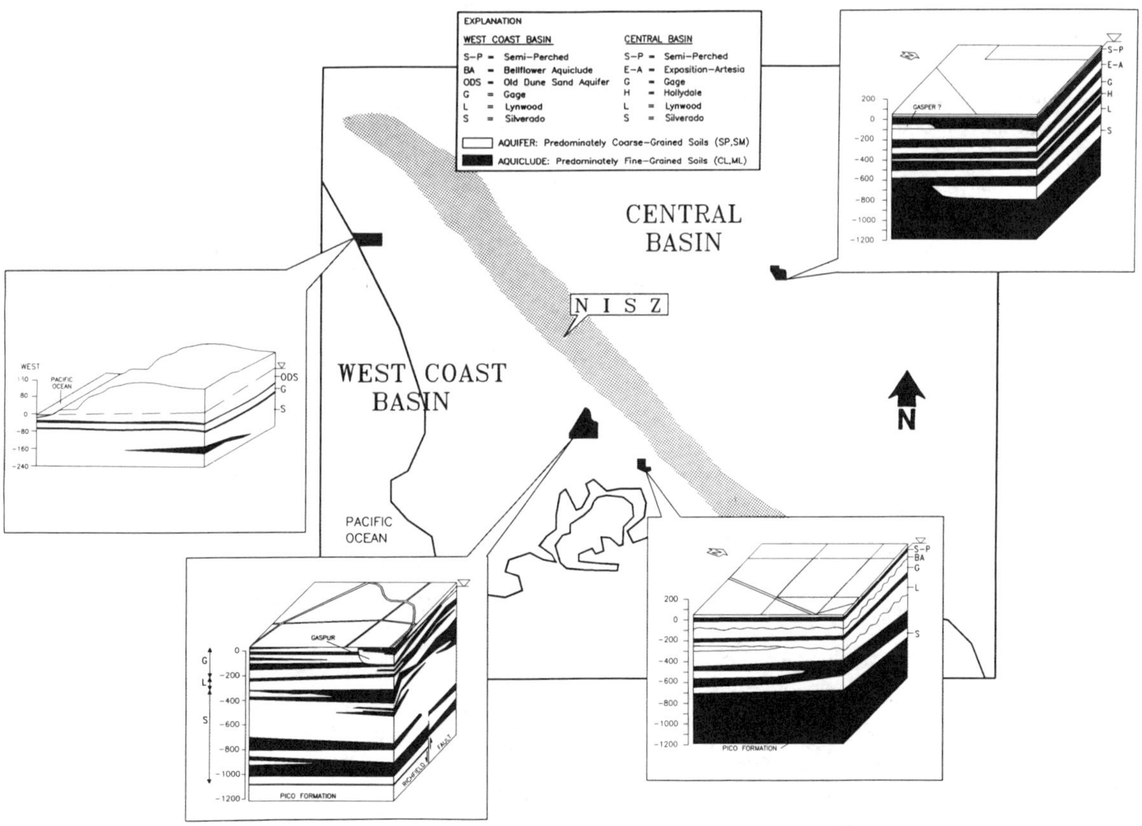

Figure 2. Subsurface hydrogeological setting beneath major petroleum-handling facilities situated within the Central and West Coast Basin.

ality and is the primary source of beneficial use groundwater from the West Coast and Central basins.

Hydrostratigraphic units underlying these portions of the West Coast and Central basins where major refineries and bulk-liquid terminals occur are shown in Figure 2. Salient hydrostratigraphic units beneath the area encompassing site nos. 1, 13 and 14 (Figure 1) in the West Coast Basin are, in descending stratigraphic position, the Gaspur, Gage ("200-foot sand"), Lynwood ("400-foot sand") and Silverado aquifers. Beds of relatively low-permeability soils which act as aquicludes or confining layers separate these aquifers in some but not all places. In particular, angular unconformities developed along the flanks of folds within the NISZ form contact zones between younger aquifers and one or more older saturated zones. Pertinent to the occurrence of LNAPL is the Gage aquifer, which is encountered under water table conditions (Testa et al., 1989). In this portion of the coastal plain, the hydrogeologic regime is influenced by numerous factors, all of which play a role in remediation strategy (Testa, 1989). These factors include regulated pumpage from the Silverado aquifer as well as continued artificial recharge into the Gaspur, Gage and Lynwood aquifers through injection wells associated with the Dominguez Gap Barrier Project and seawater intrusion (Zielbauer et al., 1962; Central and West Basin Water Replenishment District, 1986). The NISZ also has a significant, multifaceted impact on groundwater occurrence, quality, and usage. The impact of the NISZ and hydrogeologic effects of pumpage and reinjection are discussed by Testa et al. (1988) and Testa (1991b).

The area encompassed by site no. 4 (Figure 1) is situated near the western margin of the West Coast Basin. This portion of the coastal plain is underlain by sand dune deposits of Holocene age, and marine and continental deposits of Pleistocene age. In descending stratigraphic order, the key hydrostratigraphic units include the Old Dune Sand aquifer, the Manhattan Beach aquiclude, the Gage aquifer, the El Segundo aquiclude, and the Silverado aquifer (Figure 2). Pertinent to the occurrence of LNAPL hydrocarbon is the Old Dune Sand aquifer, which is encountered under water table conditions. This portion of the West Coast Basin is also influenced by artificial recharge as part of the West Coast Barrier project.

Site no. 3, situated in the southern portion of the West Coast Basin (Figure 1), is underlain by, in descending stratigraphic position, the Semi-perched, Gage, Lynwood, and Silverado aquifers (Figure 2). Most pertinent to LNAPL hydrocarbon occurrence is the Semi-perched aquifer, which is encountered under water-table conditions.

Site nos. 5, 7, 11 and 12 are situated in the Central Basin northeast of the NISZ. This area is generally underlain by shallow perched and semi-perched aquifers, and nine distinct regional aquifers (CDWR, 1961). In descending stratigraphic position, these aquifers are the Semi-perched, Gaspur, Ex-

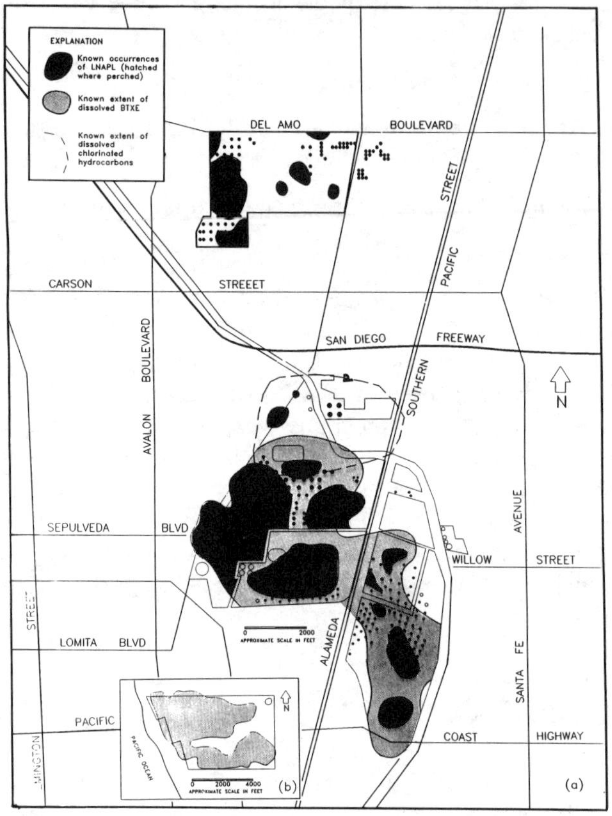

Figure 3. Spatial distribution of LNAPL beneath portions of the West Coast Basin

position-Artesia, Gardena, Gage, Hollydale, Jefferson, Lynwood, and Silverado (Figure 2). The Exposition-Artesia and deeper aquifers are the primary water-bearing zones tapped for municipal and industrial uses. The shallow aquifers, primarily the Semi-perched but also to a large extent the Gaspur, in general are polluted with brines and industrial wastes, rendering them unsuitable for domestic use throughout the region. LNAPL in the Central Basin occurs chiefly within shallow, perched zones of limited lateral extent and within the Semi-perched aquifer, which is encountered under water table conditions. The deeper aquifers are generally of good water quality, although some saltwater intrusion and localized occurrences of contaminated groundwater has been reported within the Exposition-Artesia aquifer.

FREE HYDROCARBON OCCURRENCE

The ubiquitous occurrence of hydrocarbon-affected soil and groundwater at petroleum-handling facilities in the Los Angeles basin is documented by Testa and Winegardner (1990) and Testa (1991c). The occurrence, areal extent, and estimated volume of major LNAPL hydrocarbon pools beneath these facilities situated on the Los Angeles coastal plain are summarized in Table 2. For purposes of this discussion, a pool is defined as an areally continuous accumulation of LNAPL. Two or more pools which have distinct differences in their respective physical and chemical properties are referred to as coalesced pools. Individual accumulations of relatively uniform product are referred to as subpools, since it is inferred that they have coalesced to form areally continuous occurrences. The occurrence of several pools and subpools at a particular site reflect releases from multiple sources at various times. The combined areal extent of these LNAPL pools is on the order of 1,500 acres. The estimated cumulative minimum volume is on the order of approximately 1.5 million barrels; an estimated cumulative maximum volume is on the order of 7.5 million barrels. The discrepancy in LNAPL hydrocarbon volume reflects varying methodologies involved in the estimates (Testa and Winegardner, 1990).

The largest known subsurface accumulation of LNAPL hydrocarbon in the coastal plain area occurs beneath an active refinery (refer to site no. 4; Figures 1 and 3). Groundwater beneath this site occurs under unconfined water table conditions within the Old Dune Sand aquifer. The LNAPL hydrocarbon pool is elongated toward the northwest and encompasses approximately 690 acres in areal extent (Figure 3). Apparent LNAPL thickness in this pool ranges up to approximately 12 feet. The LNAPL pool is comprised of a variety of refined petroleum product types, including light oil, diesel fuel, gasoline, jet fuel, and reformat.

One of the more complex LNAPL-impacted areas (refer to site no. 4), is in the vicinity of sites no. 1, 13 and 14. Several major refineries, terminals, and pipeline corridors are located in this area. Both groundwater and LNAPL beneath this area occurs within the Gage aquifer under both perched and water table conditions, at depths of 40 to 50 and 30 to 60 feet below ground surface, respectively. Some facilities are underlain by several pools, which in turn represent two or more coalescent pools of diverse product types (Testa et al., 1989). The release of bulk-liquid hydrocarbon product in this area has occurred over time via several pathways, including surface spillage, breached pipelines, corroded storage tank bottoms, and failed reservoir bottoms. Many of these occurrences have resulted from very slow releases over long periods of time.

Source identification of dissolved hydrocarbons in groundwater and refined LNAPL is difficult due to several factors. These factors include numerous microbiological, chemical and physical processes, the complex historical industrial development of the area, numerous property ownership transfers, and the close proximity of several crude and petroleum-handling facilities including clusterings of refineries, bulk storage tank farms, underground pipelines in industrialized areas, and numerous underground storage tanks associated with gasoline service stations in less industrialized areas. Several methodologies have been used to identify not only the crude and/or refined product type, but also the brand, grade, and in some instances, the source crude (Testa and Winegardner, 1990). These methods have included routine determination of API gravity, development of distillation curves, and analysis for trace metals (notably, organic lead and sulfur). More sophisticated methods have included gas chromatography, statistical comparisons of the distribution of paraffinic or n-alkane compounds of specific molecular weight, and determination of isotopic ratios of carbon and hydrogen ($^{13}C/^{12}C$ and $^{2}H/^{1}H$ or D/H, respectively) for lighter gasoline-range fractions, and $^{15}N/^{14}N$ and $^{34}S/^{32}S$ ratios for the heavier petroleum fractions (Kaplan, 1987).

LNAPL HYDROCARBON RECOVERY

Most major petroleum-handling facilities have implemented aquifer restoration programs, with some facilities further ahead in their respective programs relative to others. Effective recovery of LNAPL hydrocarbon can be accomplished by several means, as discussed by Testa and Winegardner (1990). Where perched or water table conditions are deep (greater than 30 feet below ground surface), the approaches to LNAPL hydrocarbon recovery have conventionally included both single- and two-submersible pump type systems. Under shallow perched water table conditions, as commonly experienced at port facilities or shallow perched conditions, passive skimmer-type systems are also utilized. In low yielding formations with minimal LNAPL hydrocarbon presence, simply bailing the LNAPL periodically is also being performed. The more conventional one- and two-pump submersible systems are briefly discussed below. The advantages and disadvantages of the one- and two-pump systems are summarized in Table 3.

One-pump LNAPL recovery systems typically utilize a 6-inch diameter or larger Schedule 40 PVC well casing with slotted screen. Four-inch diameter electrical submersible stainless-steel pumps are used for production of both LNAPL and coproduced water. Pneumatic pumps are also used in aquifers of low transmissivity, but are less common. The produced fluids are piped via overhead pipelines to an oil-water separator. The LNAPL is subsequently drained off and recycled, while the water is collected and treated on-site. Since most refineries and large bulk-storage terminals contain a wastewater treatment plant, the coproduced water is thus treated and reused if possible before ultimately being discharged into a local county sanitation district industrial sewer system. A typical one-pump LNAPL hydrocarbon recovery system is schematically illustrated in Figure 4.

A two-pump system typically utilizes a 16-inch diameter Schedule 40 PVC well casing. The well is designed to accommodate two independently-operated pumps placed at different levels within the well. The two pumps are controlled by a series of electrodes that are positioned at predetermined levels within the well. The lower (water) recovery pump utilizes a power interrupter probe to direct LNAPL hydrocarbon. This probe is positioned above the water intake and is adjusted to automatically turn off the pump when the LNAPL-water interface approaches the pump intake, preventing the lower pump from accidently pumping LNAPL.

The upper (LNAPL hydrocarbon) recovery pump is switched on and off by a probe which measures the resistivity of fluids adjacent to the intake. When the probe senses relatively high resistivity, as in the case with LNAPL, the pump is switched on and accumulated LNAPL is removed. As soon as the water reaches the probe, the probe automatically shuts down the LNAPL recovery pump until the probe again senses that a sufficient quantity of LNAPL has accumulated in the well. The well continues to cycle in this manner as long as LNAPL is entering the well. A typical two-pump LNAPL hydrocarbon recovery system is schematically illustrated in Figure 5.

Recovered LNAPL is routinely recycled at refineries. Coproduced groundwater is typically reused and subsequently treated utilizing the facility's existing wastewater treatment, and finally discharged into a sewer system. In two known cases, sites no. 1 and 4, coproduced groundwater is reinjected without treatment, for purposes of plume control and enhanced LNAPL recovery. In these two cases, coproduced water is reinjected back into the same stratigraphic zone from which the groundwater was originally withdrawn. In these cases, reinjection wells are utilized, commonly comprised of 6-inch diameter Schedule 40 PVC casing. Proper screen selection is more important with injection wells, since screen type affects injection well performance more than screen diameter. Thus, for long service lives, wells with high screen open area and injection rates are desirable. Unlike typical water wells, the consequences of water-chemistry, air entainment, thermal interference, and sand pumping are also considerably more serious and common for injection wells. Clogging of screens is the most serious problem in injection well operation, notably bacterial action where bacterial growth is promoted by the change of temperature during injection when warmer water is added to a "cool" aquifer. A typical injection well design is schematically illustrated in Figure 6.

Injection well design for unconsolidated sediments often includes screening the well through some portion of the unsaturated zone. With injection into unsaturated deposits, lower injection efficiency results, reflecting 1) a loss in pressure head (i.e. injected water leaves the injection tubing at atmospheric pressure), and 2) capillary and surface tension forces in the void spaces, which provide resistance to water movement. Injection into only the saturated zone allows maintenance of a positive head in the water until it exits the well screen, limiting chemical reactions that might occur in the well bore. In addition, since the void spaces are saturated, there are minimal capillary or surface tension forces to overcome. When injection into a LNAPL hydrocarbon pool is deemed necessary, a difference exists in the manner in which water and LNAPL will mix. In one instance, water is injected directly into the LNAPL hydrocarbon. In the other, the water mound created will try to displace and thus mix with the LNAPL hydrocarbon.

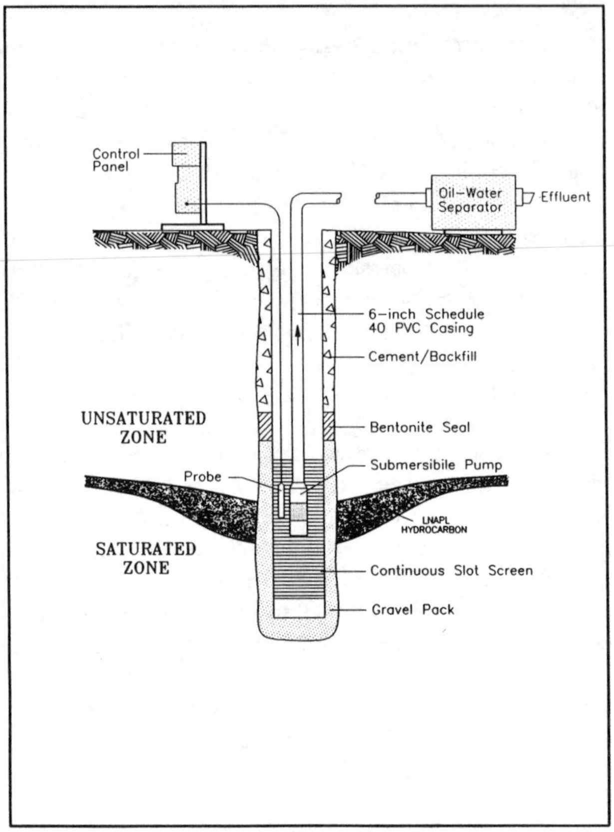

Figure 4. Schematic showing typical one-pump LNAPL hydrocarbon recovery system configuration.

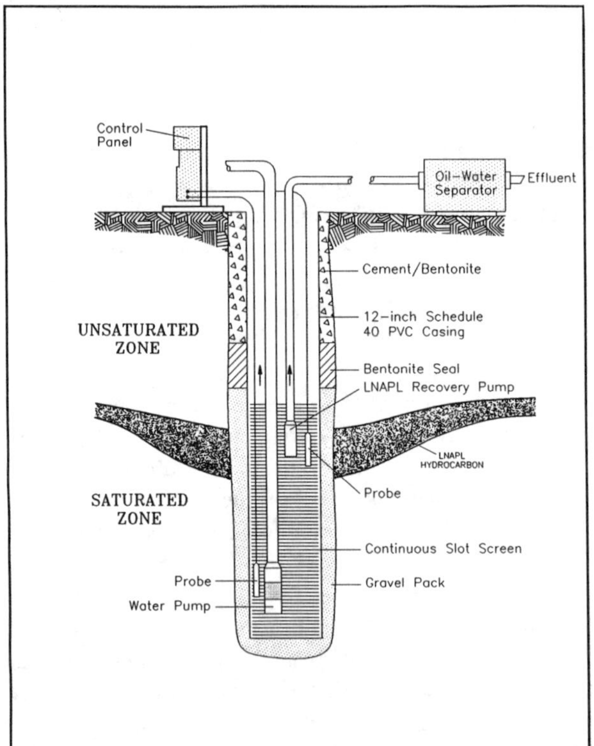

Figure 5. Schematic showing typical two-pump LNAPL hydrocarbon recovery system configuration.

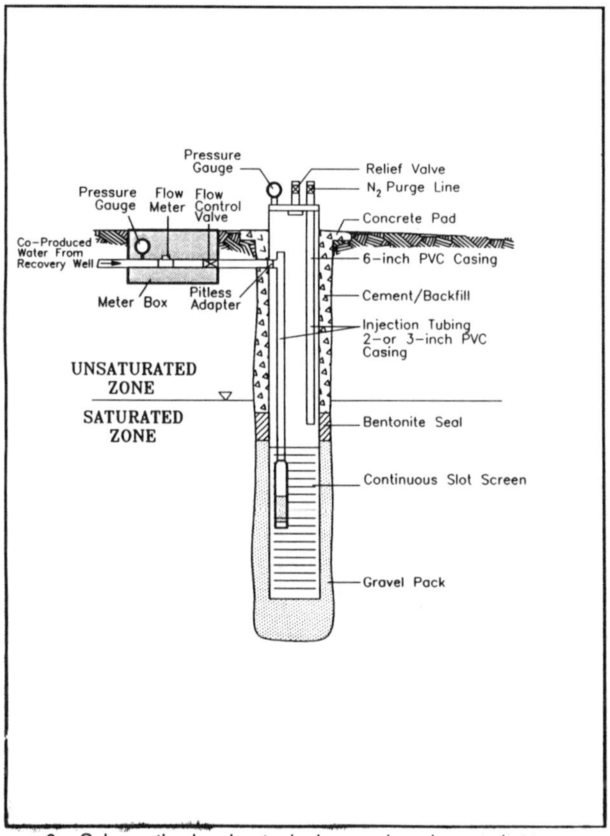

Figure 6. Schematic showing typical coproduced groundwater injection well configuration.

Overall efficiency and effectiveness of the several LNAPL hydrocarbon recovery programs in operation in the Los Angeles coastal plain are limited by numerous factors. The most important is the inability to handle and treat coproduced waters which, depending upon the size of the facility and the scale of the recovery program, can potentially exceed 1,000 gallons per minute. This reflects antiquated wastewater treatment systems (or lack thereof) at the facilities, in conjunction with existing systems functioning at capacity, cutbacks in the volume of hydrocarbon-affected groundwater which the Los Angeles County Sanitation District is willing to accept, and increased regulatory pressure against injection without treatment.

The overall efficiency and effectiveness of LNAPL hydrocarbon recovery programs can also be impacted by other factors. These factors include LNAPL recovery from low yielding formations, inability to gain access to optimal recovery and offsite locations, and economic constraints.

REGIONAL LONG-TERM REMEDIATION STRATEGY

The large extent of petroleum-affected groundwater, complexities of the regional hydrogeological setting, and the numerous factors involved in developing short- and long-term remediation strategy have resulted in an avoidance of a site specific approach to setting regional objectives in the Los Angeles area groundwater basins, and prioritizing these objectives. The primary objective is the protection of the Silverado aquifer and other regional beneficial use aquifers.

Regulatory emphasis has recently focused on the following issues:

- Complete delineation of the areal extent of LNAPL, and the horizontal and vertical extent of its respective dissolved constituents;
- Enhance efficiency and effectiveness of existing LNAPL recovery programs;
- Development of analytical and numerical groundwater models to 1) predict the fate and transport of LNAPL and its dissolved constituents; 2) provide more reliable LNAPL volume determinations; and 3) enhance design for optimal groundwater clean-up strategies;
- Identify the extent and source(s) of other organic compounds (e.g. chlorinated solvents, pesticides, etc.);
- Develop a source elimination program to detect leakage from above-ground tanks and underground piping in the early stages of LNAPL release; and
- Review current technologies and develop a soil clean-up strategy consistent with the depth and quantity of contaminants present.

In response to the issues outlined above, horizontal delineation of LNAPL is near completion, with off-site delineation required at a few localities. The horizontal and notably the vertical delineation of dissolved hydrocarbon constituents has yet to be addressed for most of the affected sites. Despite the large volumes of LNAPL released over time, the lateral extent of dissolved hydrocarbon constituents, notably benzene, toluene, ethylbenzene and total xylenes (BTEX) have been limited to distances of about 200 feet or less hydraulically downgradient of most of the larger LNAPL pools at those sites where plume delineation has been addressed. The vertical extent of these constituents has not been fully ascertained. Delineation programs to address the vertical extent of dissolved hydrocarbon constituents in groundwater is currently being implemented at some sites. Enhancing the efficiency and effectiveness of existing LNAPL recovery programs is more difficult in nature, notably for the larger LNAPL recovery programs. The primary factor limiting most recovery programs as discussed above, is the inability of the facilities to handle the generated coproduced water.

Certainly if protection of the Silverado aquifer is the primary objective, and the water table aquifers (i.e., Gage, Old Dune Sand, or Semi-perched aquifers) are essentially a write-off, then the intervening aquifer(s), although not used for water supply, may serve the purpose of "guardian" aquifers.

Two of the larger LNAPL hydrocarbon occurrences, site nos. 1 and 4 (Figure 1), reinject coproduced groundwater into generally the same hydrostratigraphic zone from which it is withdrawn; site no. 1 reinjects without treatment into the Gage aquifer, whereas site no. 4 reinjects into the Old Dune Sand aquifer. Because of the presence of dissolved hydrocarbons, notably benzene, in the coproduced water that is typically returned to the aquifer during LNAPL recovery operations, immediate application of EPA's toxicity characteristic rule (EPA, 1990) may result in classification of the reinjected water as disposal of a hazardous waste. This in turn would terminate use of UIC Class V wells (which many of these operations currently use). These wells would then be automatically reclassified as UIC Class IV wells which would be prohibited under RCRA. A two-year extension (EPA, 1991) has recently been given to allow oil companies time to comply with this rule so that LNAPL recovery efforts involving the return of benzene-affected groundwater via reinjection wells will not be readily interrupted. Although this relieves certain facilities from immediately complying with this ruling, many of these LNAPL hydrocarbon recovery programs will continue for tens of years, thus, despite the extension, the effects of future regulatory requirements and restrictions are uncertain.

A conceptual program proposed by the author, and currently under consideration, is the feasibility of using coproduced groundwater associated with the LNAPL hydrocarbon recovery program at site no. 1 for offshore injection. Coproduced groundwater would essentially be transported via an existing underground pipeline to the THUMS transfer station. The coproduced groundwater would then be used by THUMS to inject into the subsurface for subsidence control and enhanced oil recovery purposes, in association with offshore oil production-related activities.

The feasibility of reuse of onsite coproduced groundwater for offshore reinjection is dependent upon certain primary technical issues. These issues are currently being addressed and are outlined below:

- THUMS requirements for groundwater quality reinjection suitability;
- Volume requirements;
- Pipeline availability; and
- Overall regulatory perspective.

Overall, such water reclamation-type projects, if feasible, are generally encouraged and afford a substantial savings of fresh water currently used for this purpose.

Only rudimentary modeling has been performed to date to predict the fate and transport of dissolved hydrocarbons in groundwater over time. In addition, much uncertainty exists concerning actual volume present. Other organic compounds, notably chlorinated hydrocarbons, are of significant concern due to their high mobility in groundwater systems. However, these constituents are not typically derived from petroleum-handling facilities and are inferred to be derived from old landfill cells and chemical-related industry-type sources.

Petroleum-affected soil and groundwater primarily reflect historic releases although present contributions also exist at some sites. Source elimination programs are thus currently being developed on a site-specific basis. The objectives of this program are to assure that the majority of releases are historic in nature, and to identify current releases. This program in general will include:

- Periodic piping testing and replacement, as appropriate;
- Eventual conversion of underground piping to above-ground piping;
- Periodic inspection of surface and/or above-ground storage tanks. Tanks will be periodically taken out of service for inspection and repaired as needed. Double bottoms are also being installed, as appropriate; and
- Testing, removal and replacement, if appropriate, of underground storage tanks.

Clean-up strategies for hydrocarbon-affected soil will most likely be the last issue addressed and certainly the most difficult technically to address. This difficulty reflects the large deep-seated volumes of residual hydrocarbon presence, and the current lack of efficient, cost-effective methodologies for in-situ remediation of residual hydrocarbons in low permeability, fine-grained soils.

CONCLUSION

Within the NISZ, structures such as folds and faults are critical with respect to the effectiveness of the zone to act as a barrier to the inland movement of salt water. A nearly continuous set of faults are aligned along the general crest of the NISZ, notably within the central reach from the Dominguez Gap to the Santa Ana Gap. The position, character and continuity of these faults are fundamental to the discussion of groundwater occurrence, regime, quality, and usage. In addition, delineation and definition of aquifer inter-relationships with a high degree of confidence is essential. The multi-faceted impact of the NISZ is another aspect of the level of understanding required prior to addressing certain regional groundwater issues. A primary issue is which aquifers are potentially capable of being of beneficial use versus those that have undergone historic degradation. Those faults which do act as barriers with respect to groundwater flow, may, in fact, be one of several factors used in assigning a part of one aquifer to beneficial use status as opposed to another. A second issue, based on beneficial use status, is the level of aquifer rehabilitation and restoration to be required in association with the numerous LNAPL recovery and aquifer remediation programs being conducted within the Los Angeles coastal plain.

Overall understanding of the regional hydrogeologic conditions, including proper delineation of the relationship between the various aquifers, is essential to implementing both short- and long-term regional aquifer restoration and rehabilitation programs (Testa, 1989). The Silverado aquifer is recognized presently as the primary source for municipal supplies in the area. The NISZ, where many of the oil fields beneath the coastal plain are situated, has a significant multifaceted impact on groundwater occurrence, quality, usage, and remediation strategy (Testa et al., 1988). Present groundwater management programs include fresh water injection as part of the seawater intrusion barrier project which tends to raise the water levels (or pressure head) in the major water-bearing zones, and regulated pumpage which is limited by court order which tends to lower the water levels within the Silverado aquifer (and Lynwood aquifer) as documented by Testa (1991b). Future land use and site-specific concerns such as wastewater handling will all play a role in the approach to remediation, thus, remediation strategies need to be based on a case-by-case basis as well as the regional situation.

ACKNOWLEDGEMENTS

Appreciation is expressed to my wife, Lydia, for her patience in editorial assistance and word processing, to Regina Moore and Ellen Patton for word processing assistance, to Tony Carmouche for preparation of the technical illustrations, and to Eric Hendrix for technical review and editorial assistance.

REFERENCES

Barrows, A.G., 1974, A review of the geology and earthquake history of the Newport-Inglewood structural zone, southern California: California Division of Mines and Geology Special Report 114, 115 p.

California Department of Water Resources (CDWR), 1961, Planned utilization of the ground water basins of the coastal plain of Los Angeles County, Appendix A: Ground Water Geology, 181 p.

Central and West Basin Water Replenishment District, 1986, Evaluation of Saline Water Inland of West Coast Basin Barrier: Bookman-Edmonston Engineering, Inc., Glendale, California, 55 p.

Central and West Basin Water Replenishment District, 1989, 30th Annual Survey Report on Groundwater Replenishment: Bookman-Edmonston Engineering, Inc., Glendale, California, 63 p.

Dezfulian, H., 1988, Site of an oil-producing property: In Proceedings of the Second International Conference on Case Histories in Geotechnical Engineering, vol. 1, p. 43-49.

Garcia, D.H. and E.C. Henry, 1989, Environmental considerations for real estate development of oil well drilling in California. In Testa, S.M. (ed.), Environmental Concerns in the Petroleum Industry: Pacific Section of the American Association of Petroleum Geologists, Palm Springs, CA, Symposium Volume, p. 117-127.

Grenier and Gillingham, R.C., 1987, California legacy: the Watson family: The Watson Land Company, 520 p.

Henry, M.J., 1987, Los Angeles basin - an overview: In Clarke, D. (ed.), Geologic Field Guide to the Long Beach Area, Pacific Section of the American Association of Petroleum Geologists, Los Angeles, p. 1-29.

Lovegreen, J.R., 1989, Environmental concerns in oil-field areas during property transfer:. In Testa, S.M. (ed.), Environmental Concerns in the Petroleum Industry, Pacific Section of the American Association of Petroleum Geologists, Palm Springs, CA, Symposium Volume, p. 129-158.

Mendenhall, W.C., 1905a, Development of underground waters in the eastern coastal plain region of southern California, U.S. Geol. Survey Water Supply Paper No. 137, 140 p.

__________, 1905b, Development of underground waters in the central coastal plain region of southern California: U.S. Geol. Survey Water Supply Paper No. 138, 162 p.

__________, 1905c, Development of underground waters in the western coastal plain region of southern California: U. S. Geological Survey Water Supply Paper No. 139, 103 p.

_________, 1909, A phase of ground water problems in the West: Economic Geology, vol. IV, no. 1, p. 35-45.

Piper, A.M., Garrett, A.A., and others, 1953, Native and contaminated groundwaters in the Long Beach-Santa Ana area, California: U.S. Geol. Survey Water Supply Paper No. 1136, 320 p.

Poland, J.F., 1959, Hydrology of the Long Beach-Santa Ana area, California: U.S. Geol. Survey Water Supply Paper No. 1471, 257 p.

Poland, J.F., Garrett, A.A. and Sinnot, A., 1948, Geology, hydrology and chemical character of ground waters in the Torrance - Santa Monica area, Los Angeles County, California: U.S. Geol. Survey Open-file Report.

______, A.M. Piper, and others, 1956, Ground water geology of the coastal zone, Long Beach - Santa Ana area, California: U.S. Geol. Survey Water Supply Paper No. 1109, 162 p.

Richter, R.C., 1950, Geology of the West Basin, Los Angeles County, California: California State Department of Public Works, Div. of Water Resources, unpub. prelim. report.

Rintoul, W., 1990, Drilling through time: California Department of Oil and Gas, Sacramento, 178 p.

Testa, S.M., Baker, D.M., and Avery, P.L., 1989, Field studies on occurrence, recoverability and mitigation strategy for free phase liquid hydrocarbon: In Testa, S.M. (ed.), Environmental Concerns in the Petroleum Industry: Pacific Section of the American Association of Petroleum Geologists, Palm Springs, CA, Symposium volume, p. 57-81.

Testa, S.M. and Winegardner, D.L., 1990, Restoration of petroleum-contaminated aquifers: Lewis Publishers, Chelsea, Michigan, 269 p.

Testa, S.M., 1989, Regional hydrogeologic setting and its role in developing aquifer remediation strategies: In Proceedings, Geological Society of America 1989 Annual Meeting Abstracts with Programs, vol. 21, no. 6, p. A96.

Testa, S.M., Henry, E.C. and Hayes, D., 1988, Impact of the Newport-Inglewood structural zone on hydrogeologic mitigation efforts - Los Angeles Basin, California: In Proceedings of the National Water Well Association of Ground Water Scientists and Engineers FOCUS Conference on Southwestern Groundwater Issues, p. 181-203.

Testa, S.M., 1991a, Environmental aspects relating to redevelopment of oil field properties within the California regulatory framework: In Proceedings, Hazardous Materials Control Research Institute National Research and Development Conference on the Control of Hazardous Materials, Anaheim, CA, p. 4-10.

Testa, S.M., 1991b, Elevation changes associated with groundwater withdrawal and reinjection in the Wilmington area, Los Angeles coastal plain: In A.I. Johnson, (ed.), Proceedings of the Fourth International Association of Hydrological Sciences Conference on Subsidence, Houston, TX, May, Publ. No. 200, p. 485-502.

Testa, S.M., 1991c, Light non-aqueous phase liquid hydrocarbon occurrence and remediation strategy, Los Angeles coastal plain, California: In proceedings of the International Association of Hydrogeologists Canadian National Chapter Correspondence on Subsurface Contamination by Immiscible Fluids, April 18-20, 1990, Calgary, Alberta, Canada, in press.

United States Environmental Protection Agency, 1986, Amendments to national oil and hazardous substances contingency plan national priorities list, final rule and proposed rule: Federal Register, vol. 51, no. 111, June 10, p. 21053-21112.

_____, 1990, Hazardous waste management system; identification and listing of hazardous waste; toxicity characteristics revisions, final rule: Federal Register, vol. 55, no. 61, March 29.

_____, 1991, Hazardous waste management system; identification and listing of hazardous waste; toxicity characteristics; hydrocarbon recovery operations: Federal Register, vol. 56, no. 63, April 2, p. 13406-13411.

Yerkes, R.F., McCulloh, T.H., Schoellhamer, J.E. and Vedder, J.G., 1965, Geology of the Los Angeles Basin, California - an introduction: U.S. Geol. Survey Professional Paper 420-A, 57 p.

Zielbauer, E.J., and others, 1962, Dominguez Gap Barrier Project, geologic investigation: Los Angeles County Flood Control District.

Table 1. Summary of Subsurface Hydrocarbon Impact from Major Refineries and Bulk-Liquid Storage Terminals

Facility No.	Facility Name	Location	LNAPL Presence	Dissolved Hydrocarbons In Groundwater (BTXE)(a)	Dissolved Chlorinated Hydrocarbons
REFINERIES					
1	ARCO	Carson	Yes	Yes	Yes
2	Champlin	Wilmington	Yes	Yes	Yes
3	Chemoil	Signal Hill	Yes	Yes	Yes
4	Chevron U.S.A.	El Segundo	Yes	Yes	Unknown
5	Edgington Oil Company	Long Beach	Yes	Yes	Yes
6	Golden Eagle Refining Company	Carson	No	Yes	Yes
7	Golden West Refining Company	Santa Fe Springs	Yes	Yes	Yes
8	Fletcher Oil Company	Carson	Yes	Yes	Yes
9	Mobil Oil	Torrance	Yes	Yes	Yes
10	Newhall Refining Company(b)	Newhall	No	No	No
11	Paramount Petroleum	Paramount	Yes	Yes	Yes
12	Powerine Oil Company	Santa Fe Springs	Yes	Yes	Unknown
13	Shell Oil Company	Wilmington	Yes	Yes	Yes
14	Texaco	Wilmington	Yes	Yes	Unknown
15	Union Oil Company	San Pedro	Trace	Yes	Unknown
16	Western Fuel Oil Company	Long Beach	Yes	Yes	Unknown
TERMINALS/ TANK FARMS					
17	ARCO	Vinvalle	Yes	Yes	Yes
18	ARCO FCPL	Paramount	Yes	Yes	Unknown
19	British Petroleum	San Pedro	Unknown	Unknown	Unknown
20	Chemoil	Carson	No	Yes	Unknown
21	Chevron USA	San Pedro (Berths 97–102)	Unknown	Unknown	Unknown
22	Chevron Chemicals	San Pedro (Berths 163–164)	Unknown	Unknown	Unknown
23	Department of Water & Power	San Pedro (Berth 181)	Unknown	Unknown	Unknown
24	Edgington	Long Beach	Unknown	Unknown	Unknown
25	ECO	Long Beach	Unknown	Unknown	Unknown

Page 1 of 2

Table 1. Summary of Subsurface Hydrocarbon Impact from Major Refineries and Bulk-Liquid Storage Terminals

Facility No.	Facility Name	Location	LNAPL Presence	Dissolved Hydrocarbons In Groundwater (BTXE)(a)	Dissolved Chlorinated Hydrocarbons
26	GATX	Carson	Yes	Yes	Unknown
27	GATX	San Pedro (Berths 70–71)	Unknown	Unknown	Unknown
28	GATX	San Pedro (Berths 118–121)	Yes	Yes	Unknown
29	GATX	San Pedro (Berths 171–173)	Unknown	Unknown	Unknown
30	Golden Eagle Refining Company	San Pedro (Berths 163–164)	Unknown	Unknown	Unknown
31	L. A. Terminal	San Pedro (Berths 148–151)	Unknown	Unknown	Unknown
32	Mobil Oil	Terminal Island	Unknown	Unknown	Unknown
33	Mobil Oil	San Pedro (Berths 238–240)	Unknown	Unknown	Unknown
34	Pacific Northern Oil	San Pedro (Berth 215)	Unknown	Unknown	Unknown
35	Paramount	Lakewood	Yes	Yes	Unknown
36	Penzoil	Lakewood (Berth 70–71)	Unknown	Unknown	Unknown
37	Petrolane	Lakewood (Berth 120)	Unknown	Unknown	Unknown
38	Santa Fe Pacific Pipleine	Carson	Unknown	Unknown	Unknown
39	Shell Oil	Dominquez	Yes	Unknown	Unknown
40	Shell Oil	Signal Hill	No	No	No
41	Shell Oil	San Pedro (Berth 167–169)	Unknown	Unknown	Unknown
42	Supertanker Terminal	San Pedro (Berth 45–47)	Unknown	Unknown	Unknown
43	Refiners Marketing	Terminal Island	Unknown	Unknown	Unknown
44	Union Oil 22nd Street	San Pedro (Berth 148–151)	Unknown	Unknown	Unknown
45	Union Oil	San Pedro (Berth 37–40)	Unknown	Unknown	Unknown
46	Unocal	Torrance	No	Yes	Unknown
47	U. S. Navy Fuel Depot	San Pedro (Berth 37–40)	Unknown	Unknown	Unknown
48	Western Fuel Oil	San Pedro (Berth 120–121)	Unknown	Unknown	Unknown
49	Wilmington Liquid Bulk Terminal	San Pedro (Berth 187–190)	Unknown	Unknown	Unknown

(a) BTXE = Benzene, toluene, total xylenes and ethylbenzene
(b) Not shown in Figure 1
2113–1(21)

Table 2. Estimated Extent and Volume of LNAPL for Major Refineries and Bulk-Liquid Storage Terminals(a)

Facility Reference No.	Facility Name or Operator	Hydrostratographic Unit Affected	Pool Occurrence No.	Maximum Apparent Thickness (Feet)	Areal Extent (Acres)	Minimum Estimated Volume (b) (Barrels)	Maximum Estimated Volume (Barrels)
Refineries							
1	ARCO	Gage	1	26.0	86.5	174,469.7	436,174.3
			2	24.0	83.6	155,706.1	389,265.3
			3	12.0	38.5	35,873.5	89,683.7
			4	0.6	1.8	85.9	214.7
			5	7.5	14.8	8,587.0	21,467.4
			1-P(c)	2.0	43.0	667.9	1,669.6
			2-P	15.5	29.9	35,984.8	89,962.0
			3-P	28.0	37.7	81,923.8	204,809.5
2	Champlin	Gage(?)	1	1.92	0.8	116.2	290.6
3	Chemoil	Semi-Perched	1	1.04	.4	463	ND(e)
			2	0.01	.1	2.7	ND
			3	0.01	.1	21	ND
4	Chevron U.S.A.	Old Dune Sand	1	12.0	690	2,000,000	6,000,000
5	Edgington Oil Company	Semi-Perched	1	2.59	ND	ND	ND
6	Golden Eagle Refining Company	Gardena	ND	NA(d)	NA	NA	NA
7	Golden West Refining Company	Artesia	1	7.16	80.6	ND	ND
		Semi-Perched	2	1.64	14.5	ND	ND
			1-P	11.02	ND	ND	ND
8	Fletcher Oil Company	Gage	1	>6.00	ND	ND	ND
			1-P	ND	ND	ND	ND
							ND
9	Mobil Oil	Gardena	1	ND	ND	ND	ND
		perched	1-P	8.70	ND	ND	ND
10	Newhall Refining Company	ND	NP	NA	NA	NA	NA
11	Paramount Petroleum	Semi-Perched	1	8.81	3.8	6,098	10,854
12	Powerine Oil Company	Artesia	ND	ND	ND	ND	ND
13	Shell Oil Company	Gage	1	9.6	141.5	105,416.0	263,540.1
			2	4.8	9.6	3,565.6	8,914.0
			3	0.4	0.6	19.4	48.4
14	Texaco	Gage	1	16.42	24.2	30,809.8	77,024.6
			2	3.87	19.4	5,815.4	14,538.4
			3	3.03	3.0	698.6	1,746.6
			4	0.95	0.5	37.7	94.4
15	Union Oil Company	Gage	NP	NA	NA	NA	NA
16	Western Fuel Oil Company	perched	ND	ND	ND	ND	ND

Page 1 of 2

Table 2. Estimated Extent and Volume of LNAPL for Major Refineries and Bulk-Liquid Storage Terminals(a)

Facility Reference No.	Facility Name or Operator	Hydrostratographic Unit Affected	Pool Occurrence No.	Maximum Apparent Thickness (Feet)	Areal Extent (Acres)	Minimum Estimated Volume (b) (Barrels)	Maximum Estimated Volume (Barrels)
Terminals/ Tank Farms							
17	ARCO	ND	1	ND	ND	ND	ND
18	ARCO FCPL	Semi-Perched	ND	ND	ND	ND	ND
28	GATX	perched	1	12.72	12	11846.1	29615.2
			2	2.25	4.7	1220.8	2051.7
			1-P	12.4	0.8	734.8	1837.1
			2-P	14.92	0.3	310.4	776
35	Paramount	Semi-Perched	1	724	2.8	11,450	ND
39	Shell	ND	1	7.2	35.8	19,979.2	49,948.1

(a) Based on review of information available at the CRWQB files as of December, 1989.
(b) One barrel = 42 gallons
(c) NP = Not Present
(d) NA = Not Applicable
(e) ND = Not Determined
2113-2(21)

Table 3. Comparison of one-and two-pump LNAPl Recovery Systems

System Type	Advantages	Disadvantages
One-Pump	Minimizes the amount of residual hydrocarbon-affected soil Appropriate for large and small volumes of LNAPL Appropriate for low to high yielding formations Fully automated	Above ground separation for LNAPL and water is required Emulsification
Two-pump	LNAPL is separated in the well, thus,no need for above ground separator Minimal mixing of LNAPL and water in well, thus, solid components being added to the pumped water is minimized Can handle large quantities of LNAPL Fully Automated	Increase amount of residual hydrocarbon-affected soil Routine maintenance essential Limited to high yielding formations More costly

STRINGFELLOW ACID PITS; WORLD'S FIRST LEGALLY-CONSTITUTED HAZARDOUS WASTE DISPOSAL SITE

Allen W. Hatheway
Professor of Geological Engineering
129 V.H. McNutt Hall
University of Missouri-Rolla
Rolla, Missouri 65401

Glenn A. Brown
Sr. Vice President and
Senior Consultant in Engineering Geology
Law/Crandell, Inc.
200 Citadel Drive
Los Angeles, California 90040

CONTENTS

INTRODUCTION

The Stringfellow Acid Pits occupies several niches of extreme importance in the history and technology of industrial waste management and in the more recent history of the State of California. There is hardly enough that can be written and substantiated about this "first" regulated hazardous waste disposal site on the planet. Exact details of its failure are poorly known; the nature of its waste is yet to be fully characterized, the history of its origination is poorly understood by most Californians, including geologists and engineers who practice in hazardous waste management and remediation; it is the whipping boy of the press; the tug-of-war of the century in California's courts, and it cannot be substantiated that it is the major threat to human health that many hold it to be.

The site was created without the benefit of any form of engineering and without any statement as to the means and purpose of disposal of liquid hazardous wastes. In retrospect it may be safe to assume that the largest hope of the operator was that evaporation would remove the disposed wastes, making room for more. No tests, computations or consultations on the feasibility and extent of evaporation were ever made. In reality, significant amounts of the liquid industrial wastes seeped downward to the Younger Alluvium and escaped downgradient in that material, through the portion of the valley cross-section that was to have seen placement and maintenance of an "impermeable" barrier constructed from the ground surface to sound bedrock. Only 60 feet (18.2 m) of a total of 250 feet (76.2 m) set forth in the Waste Discharge Requirements, adopted by the Regional Water Pollution Control Board, were ever constructed. That fact was concealed from visiting personnel of the oversight agencies of the County of Riverside as well as personnel of the Regional Board.

Operating for 17 years, the facility ceased operation on 29 November, 1972, due to the occurrence of visible indications of subsurface leakage of wastes and from mismanagement of surface water flooding by the operator, who chose not to respond to directives by the Santa Ana Water Quality Control Board. The facility lost its only legal permit, its zoning variance, as rescinded by the County of Riverside Department of Building and Safety (DB&S) on 24 February 1974.

From cessation of operations on 29 November 1972 until the Spring of 1975, James Stringfellow Jr., operated leachate transfer pumps to remove surface collection of hazardous liquids and contaminated waters to some of the on-site evaporation ponds that he had previously utilized to dispose of site wastes. The Santa Ana Board, which on 10 January 1975, issued its Cleanup and Abatement Order No. 75-38, for Stringfellow to produce a plan for closure of the site and abatement of the existing environmental threat.

With abandonment of the County-permitted site by Stringfellow, the Regional Water Quality Control Board sought abatement of the obvious condition of pollution. On 14 February 1975 the Board found the Acid Pits to constitute a "non-operating industry" and three days later formally requested Riverside County take necessary abatement actions in accordance with California Administrative Code. On 11 March 1975 the

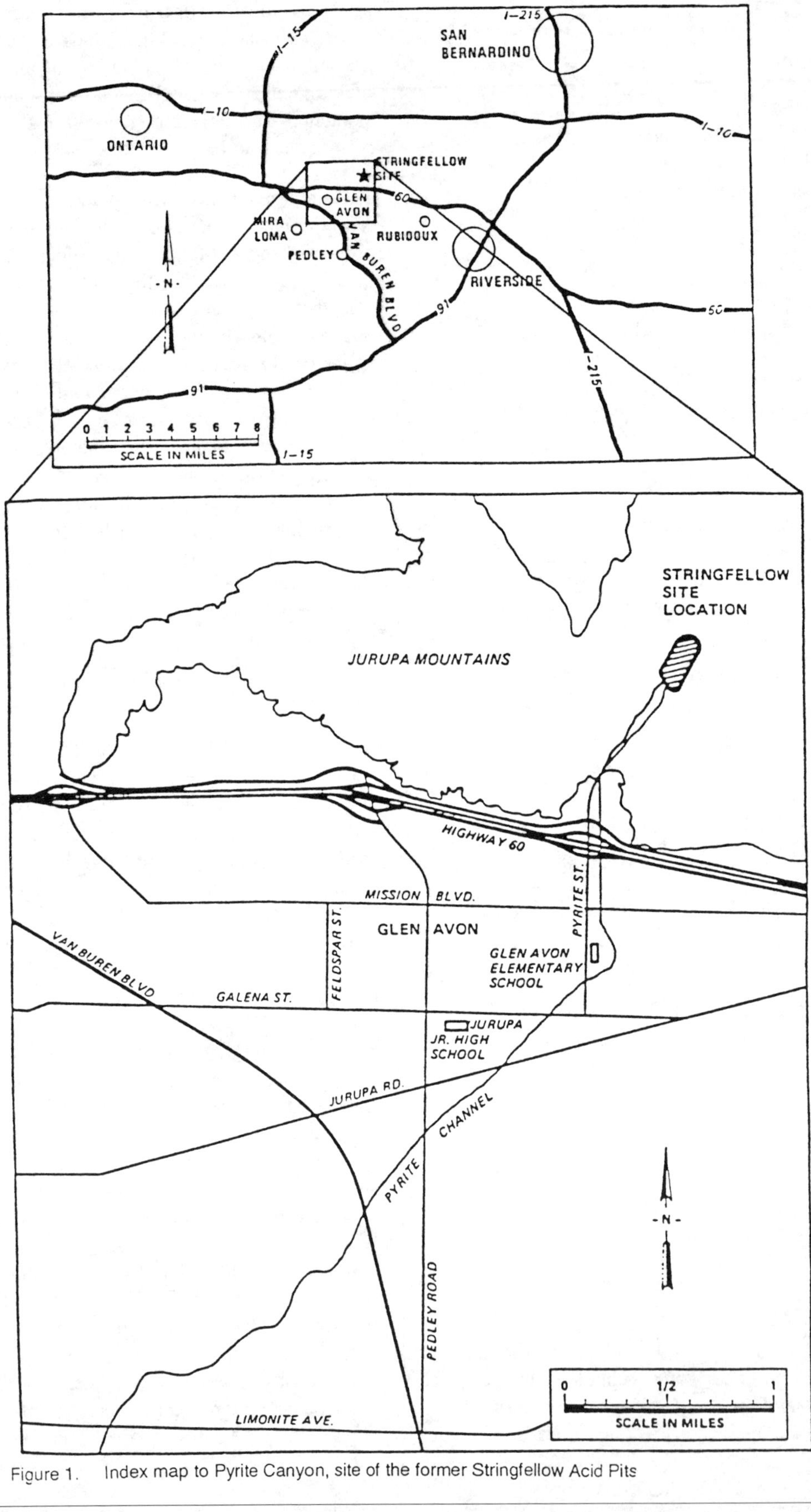

Figure 1. Index map to Pyrite Canyon, site of the former Stringfellow Acid Pits

County Board of Supervisors voted unanimously to decline the request to undertake abatement of their formerly-permitted disposal site. Later, on 13 November 1975 the Board of Supervisors held a hearing and revoked, for the second time, the Stringfellow dumpsite land use variance, which had originally been revoked by the DB&S on 24 February 1974.

J.B. Stringfellow Jr. had continued to operate leachate transfer pumps during 1974 and early 1975 and entered into an agreement on 21 May 1975 to receive $13,000 for purchase and maintenance of equipment to be used for site maintenance efforts. On 12 June 1975, the State Board of Water Resources Control was forced by circumstances to begin activities to remediate the site, in the face of abandonment by both Stringfellow and the County of Riverside.

At the time of abandonment there was no formal legal pathway or means for ordering remediation of a failed hazardous waste disposal site in California. For that matter none of the other 49 states had such provisions. In fact, not one single hazardous waste disposal facility anywhere on Earth had been subjected to a legally-declared site remediation (Hatheway, 1992a). Appropriate Federal law was nonexistent and there were no protocols for undertaking the string of necessary investigations and engineering deliberations now required to remediate and close such sites.

In acting responsibly, the State became entangled in a web of events which have served as the basis of a formidable countersuit by Responsible Parties (RPs) in which the State has been accused of post-abandonment operation of the Acid Pits and thereby creating certain additional environmental threats. In the same suit, tried in 1990 and 1991, the RPs charged that the State of California located the Acid Pits site, provided its design, and supervised its construction. To address these issues, the presiding judge devised six interrogatory questions to be answered by a jury, then retracted one of the questions and found for the RPs. Of the remaining five questions, the jury found for the State in three and against the State in two. In an early review of the jury's findings, the presiding judge overturned the three State-favorable jury findings and gave the entire determination to the RPs, thus declaring that the State of California was wholly responsible for environmental degradation associated with the Stringfellow Acid Pits. This paper presents the facts for consideration of the reader.

1. BACKGROUND TO ESTABLISHMENT OF THE ACID PITS

What is this infamous "dump" site? We have found this site to be of prime importance since it is a prototype of what industrial waste management has been in California and also a precursor of what was to be encountered in later years across the nation.

To understand what has happened in and around the origination, operation, and failure of the Stringfellow Acid Pits, one must to know that the State of California has been the birthplace for the vast majority of waste management laws and regulations in the United States. The dilemmas, debates, decisions, mistakes, corrections and arguments that have contributed to development of today's hazardous waste laws and regulations, statewide and nationally, have direct correlation with what has happened at the Stringfellow site and what has been brought about by actions of various parties. We have done our best to reconstruct and weave together the threads that make up the tapestry of the Acid Pits. A companion paper in this volume, attempts to identify a pattern that emerges for waste management in southern California.

About 1904 the Bly Brothers, of Los Angeles, suppliers of dimension stone, monuments and riprap, were pulling Woodson Peak Granodiorite out of a quarry they had established in Pyrite Canyon, site area of the future Acid Pits (Fig. 1 - Site Location Map). By 1916, the State Mineralogist reported that the firm of Bly Brothers and McGilliard had 20 men working at the quarry, believed to have been located several hundred meters ENE of the present site-remediation wastewater treatment plant.

Impetus for creation of the Stringfellow facility incorporates several events important nationally and to the State of California. As explained in a companion paper (Hatheway, 1992b) in this volume, the rising tenor of southern Californian industrialism had followed the huge increases generated for the war effort of 1941-1945.

2. GEOLOGIC CONDITIONS IN PYRITE VALLEY

Geologic mapping (1:12,000) of the Jurupa Hills was completed and published in February, 1951, by McKevett, accurately classifying the general geology of the range, including igneous (primarily Woodson Peak Granodiorite) and metamorphic rock types and Younger Alluvium, and Recent Alluvium. McKevett noted two dominant joint sets (E-W primarily, and N-S, secondarily), both spaced at about 8-10 ft. No evidence was found of "large continuous faults" in the Jurupa Mountains, joints and foliation were prevalent, and foliation "was abundantly developed." Such description did not (then or now) indicate that these discontinuities were open. Quartz diorite (tonalite) was ascribed to the subdued, grass- covered, non-bouldery hillsides, while the granodiorite was noted to represent the most weather-resistant rock of the hills.

Over the years since the State of California became involved in the County of Riverside's permitting process for the Stringfellow Acid Pits (1955), the Pyrite Canyon watershed, site of the Acid Pits, has been geologically mapped either fully or partially on several occasions. Geologic details of the site are not the subject of this paper, but the authors and

Figure 2. Stringfellow Acid Pits co-owner Lawrence Earl Nutt standing against the first and only segment (about 60 ft; 18 m) of the subsurface barrier wall undertaken by James Stringfellow Jr. and partners to meet one of the Discharge Requirements of the Santa Ana Regional Water Pollution Control Board. View is to the west (unknown photographer; original photo print machine-dated "Nov 1961"). Mr. Nutt (now deceased) is reputed to have been six ft (1.8 m) tall.

other colleagues (see Acknowledgements) did independently compile a geologic map of the entire watershed.

3. PRE-STRINGFELLOW EVENTS

By the early 1950s, sanitary engineers practicing in southern California were generally aware of a narrowing margin between the physio-biological treatment capacities of their wastewater plants and the increasing volume and concentration of industrial wastes being discharged directly into sewers. The stage was set for a chain of events that would produce an economic climate favorable to establishment of commercial industrial liquid waste evaporation impoundment facilities (Hatheway, 1992b, this volume).

Activities by Santa Ana Regional Water Pollution Control Board

On 17 October 1952, the State Water Quality Control issued a Statement of Long-Range Plans and Policies for itself and its Regional Boards, including a "determination of compliance with primarily on information required to be furnished by the waste discharger with the right reserved for any public agency to make additional useful investigations. As technical backup for its service to the Regional WPC Boards, DWR Southern District released (Dec 1953) the first of a series of basinal surveys as *Investigation of Safe Dump Sites in Coastal Basin Within Los Angeles Region.*

In 1952 a series of meetings were held in Riverside and San Bernardino Counties, called by RWPCB for officials and representatives of major industries, to teach them methods of dealing with improper discharge of wastes.

Orange County Sanitation Districts (OCSDs) passed, on 4 August 1954, its Ordinance No. 1, as the basis for *Rules and Regulations for the Admission of Sewage from the Sanitation Districts,* which applied to its eleven constituent Districts. In particular, gasoline, benzene, naptha, fuel oil or other flammable or explosive liquid, solid or gas were not to be accepted by any of the Districts for discharge to the two wastewater treatment plants operated jointly by the authority. Furthermore, OCSD Article IV stipulated that, "No District shall discharge, or cause to be discharged, any sewage, waters, vapors or industrial wastes that has any one of the following characteristics," of which there were seven (Hatheway,1992b, this volume). There is no direct evidence of how widely this declaration was known at the time (but it is not unreasonable to presume that southern California industry and the Regional Boards were aware of the situation, as well as persons who were interested in offering or participating in operating waste disposal sites).

Division of Water Resources Studies the Santa Ana Region

As the means for its long-range planning, the RWPCB had contracted DWR to produce a "Region-wide Municipal and Industrial Waste Site Studies, Santa Ana Region," which it completed in June, 1954. Of the 37 sites reviewed during this $5000 (budgeted) study, 28 were eliminated and a "reconnaissance" was made of the remaining nine sites, none of which were to be what was later proposed by the Stringfellow Company as the Acid Pits facility. What should be remembered was that the study was undertaken primarily to provide a background for Class II (garbage) and Class III (inert wastes, typically construction debris), not Class I (industrial) wastes. The scope of the study was therefore to:

1) identify existing dumps;
2) identify possible future dump sites, and;
3) study possible pollutional aspects of existing dumps.

Terminology associated with this 1954 study has haunted the determination of liability ever since. The entire study was a "reconnaissance" and the nine individual sites selected for further work were given "detailed studies" (p. 2) in the context of that level of effort. Furthermore, "Each individual site" was to be "carefully investigated in the field before a final decision can be made as to its suitability for unrestricted disposal of wastes." Sites were de-selected from further consideration when "Rock or surface material appeared permeable or deeply fractured" (p. 24).

Southern District DWR personnel who made the Santa Ana Regional survey were employing R.C. Eckis's 1934 (Bulletin 45) compendium of regional geology from the standpoint of water supply (see Hatheway, 1992b, this volume). Geologic aspects of the Santa Ana Regional survey were observed and compiled by Robert C. Fox, who, in his 1983 deposition (p. 30) recalled that the terminology "Basement Complex" was borrowed from Eckis' work as one of the "nonwater-bearing groups of rocks."

As a prelude to its role in oversight of water quality, the RWPCB passed its Resolution No. 54-5 on 5 November 1954, adopting a general policy of rudimentary groundwater protection through siting conditions for all types of "dumps." Included is a statement to the effect that sites located on ground "through which no appreciable seepage to usable waters can occur" would generally be amenable to disposal of Group 1 wastes. "Nonwater-bearing rock" was not actually defined by DWR until 1966, in its Bulletin No. 104-2, on the San Gabriel Valley, in which quantities less than the range of 5-15 gpm were considered diagnostic (p.25) of being equivalent to the terminology of 1954; "impermeable" and having "no appreciable seepage."

Retired DWR civil engineer Robert Clawson recalled in Stringfellow deposition (1989, p. 24) that the first Class I site review by the Southern District, for the Santa Ana Region was possibly for the existing Culligan Corporation zeolite recharging plant on-site disposal facility. This was in the "mid-1950s," in the upper Santa Ana Basin, near San Bernardino, where "we took a look at geology and hydrology and recommended location for monitoring wells." Clawson took care (ibid, p.29) to point out that the Culligan application was for a "disposal" site as a place of waste discharge at the owner's property, whereas a "dump" site was a facility that took in wastes belonging to others, as a service.

Board of Trade and Orange County Sanitation District Events of 1954-1955

By 3 February 1955, the Riverside County Board of Trade had met in Palm Springs to hear the report of Mr. Dennis Flannagan (one of its employees) report "...on and industrial inquiry which was sightly out of the ordinary in that the prospect is looking for irregular land with gullies which can be filled in with industrial waste. As yet, the Board has no idea what the product is, or the name of the company, with which they are dealing." At its next monthly meeting (7 March), jointly with the San Bernardino Board of Trade, at Riverside, the minutes reflect:

> "Industrial Waste Disposal: Due to increased need for industrial waste disposal by manufacturing companies presently located in the area or potential locaters, a great need for a waste disposal site in accord with the Santa Ana Water Pollution Control Board standards for water, is evidenced.
>
> A qualifying site seems available:
>
> (1) It is located on the Stringfellow Quarry property in a canyon about a mile north of Highway 60;
>
> (2) Mr. Paul Brown, of the Santa (sic) Water Pollution Control Board asked that his engineers be requested to make a geological survey of the area; and
>
> (3) If found satisfactory, it is planned to allow development of the area by the Stringfellow Company as a private enterprise. This matter is to be taken up at the joint meeting of the San Bernardino-Riverside Boards of Trade."
>
> "BI-COUNTY WASTE DISPOSAL PROGRAM: Dennis Flannagan introduced the subject of a bi-county waste disposal program. He pointed out that, he together with Arthur Groos, San Bernardino Industrial Director, and Paul Brown, Manager of the Santa Ana Water Pollution Control Board, had visited the Stringfellow Quarry. Brown's feeling was that the site would be most feasible for establishment of a bi-county dump. The owners, it was reported, appeared favorable to the proposals for establishment of such a dump. They pointed out that they would prefer to maintain the roads, of which about two to three miles are required in each county, rather than request the counties to do such work. They are desirous, however, that proper zoning be accomplished for protection of the area as an industrial dump. Following discussion, it was moved by William Mulligan of the San Bernardino Board of Trade, seconded by William Watson of the Riverside Board of Trade, and unanimously passed that "the two Boards approve a request be made to Paul Brown, Manager of the Santa Ana Water Pollution Control Board, asking that the district, at its own expense, conduct a survey of the feasibility of establishment of a waste disposal dump at the Stringfellow Quarry." In additional discussion the question was raised as to cost and fees involved, and it was generally agreed that, under private business, competitive rates would control the use of this dump site."

With all of this activity in mind, it is interesting to find the record of 8 March 1955 in which the two Boards of Trade, jointly request RWPCB Executive Officer Paul Brown to conduct a siting survey, at its expense, for feasibility of waste disposal site at Stringfellow Quarry.

DWR is Requested to Review Stringfellow Site Conditions

Board Executive Officer Brown replied via letter on 16 March, to the Boards of Trade, noting that they would need the "owner of the parcel of land (Stringfellow) to direct letter to show intent to open a "dump." On the same day, Brown requested DWR to undertake an "investigation," with findings and recommendations for operation of site to minimize threat of water pollution. Mr. Brown specified a budget of $300 for the study, though retired San Diego Regional Board Executive Officer Dennis O'Leary (1989, p. 139) recalled that DWR put in some of their funds too, so this $300 tends to be misleading. "They might have spent a thousand dollars. Our money was used as seed money and they added funds to carry out investigations for us." Thus, with these early site characterization efforts there started a complex eleven-phased site history (Table 1):

TABLE 1

ELEVEN GENERAL PHASES OF SITE HISTORY

1	Pre -1954	Pre-Siting
2	1954-1955	Siting, Pre-permitting Inspection, and County Permitting
3	1956	Construction
4	1956-1972	Operational Phase
5	1973-1974	Stringfellow Corporation Maintains Inactive Site
6	1974	Stringfellow Abandons Site
7	1974-1975	RWQCB Attempts to Initiate County Abatement Actions
8	1975	Closure Studies
9	1976-1980	Pre-CERCLA Maintenance and Remediation
10	1981-1982	CWA Section 311 (k) Removals
11	1982-Present	CERCLA Site Remediation

A close understanding of the events associated with the Stringfellow site also calls for familiarity of the persons most closely associated with the origination, construction, operation, oversight, mitigation, and closure activities associated with the Acid Pits (Table 2).

On 25 March 1955, Lawrence Earl Nutt, of Stringfellow Company, wrote to the RWPCB requesting that it investigate a site owned by the Company, indicating that "if permission could be obtained they intended to establish an industrial waste dump." Brown responded on 28 March by forwarding Nutt's letter to DWR, and in April, Chief Engineer Max Bookman assigned Robert C. Fox to investigate the Stringfellow site (Fox, 1983, p. 140).

DWR Conducts the Site Review

Fox assembled the usual three-person investigation team, adding Donald Phillips (hydraulic engineer) and Claude Hewitt (civil engineer) and the three went to visit the site to confirm the generalities of the Jurupa Hills, as determined during the 1954 report on regional waste disposal conditions. It is clear, then that the team went forward with a clear bias toward expecting site conditions not to be variable from those of the Jurupa Hills in general.

Fox noted in his 1983 deposition (p. 53) that he expected the granitic bedrock of the area to be "dense and of good potential" for siting. Once in the site area, the team stopped by the former Shannahan Quarry (not operated by Jim Stringfellow, Jr., and his proposed dump partners Earnest ("Moe") McCook and Earl Nutt). Fox recalls (ibid, p. 288) that he "extrapolated" his observations from the quarry to the proposed dump site (the distance being as little as 1400 ft; 425 m), assuming that the bedrock at the dump site was similar to what was exposed in the quarry.

Seeing no evidence for major fracture or fissures or apparent faults" (p. 406), Fox became "excited" about the site on the basis of the volume of waste that it could potentially hold (p. 402). Once the confirmation of rock type and general structural geologic assessment was made, Fox turned to assessing the site geomorphic characteristics and came to the conclusion that the site needed a barrier to hold the waste above the dense bedrock (p. 406), the greater the volume, the higher the barrier (p. 424).

In reviewing proposed sites for the Regional Boards, DWR operated on the legacy of Eckis' 1934 geologic classification and the developing site classification that was later (1956) reported by Scheliga. That is, if a waste disposal site was to be considered for placement on bedrock, then the bedrock must be in characteristically impermeable terrane and not to be underlain by useable ground water (Fox, 1983, p. 84). David Willets recalled (1984, p. 26) that "... the underlying rock was granitic and not susceptible to any percolation."

Detail of the DWR Site Review

Fox and colleagues made one visit to the site, for about an hour's duration (ibid, p. 73) during which time they concluded that the site was "relatively good" in their experience. Fox noted a "cemented promontory" (p. 73) which he termed as a "conglomerate" and that "the area is particularly well adapted for this use because of a natural barrier..." (p. 272).

In their 1989 plaintiffs presentation for the Stringfellow Federal case, attorneys for the RPs (Responsible Parties) claimed that Fox should have undertaken a proper site investigation on behalf of Mr. Stringfellow. This assertion is patently absurd, for DWR never has provided such siting assistance, or assistance in design or construction inspection, of private-sector projects. In his 1984 deposition, former DWR Southern District Chief Engineer Max Bookman (p. 92) made that fact clear, and that design was the responsibility of the applicant (Stringfellow)

Fox could not recall (1983 deposition) if he had made use of aerial photographs in preparing for the site visit. He was subsequently criticized in the hindsight of court proceedings for not having done so. The fact is that most of the southern California groundwater basins were covered by some form of black/white aerial photographs and that coverage of the Stringfellow site probably could have been had by referring to files. There has been a good deal of comment over this matter. Fox did apparently utilize such photos for the 1954 regional study, but the nature of his visit of 25 May 1955 was only to confirm the general geologic conditions. What ever could have been interpreted from the aerial photographs would have been related to the actual facility design, which was and remains the responsibility of the applicant. In actuality, the authors and colleagues have geologically mapped (1991) the entire Pyrite Canyon watershed and have utilized aerial photographs in the process. We know of no particular structural geologic feature that could have been detected by photointerpretation in 1955 that has been proven to have significant impact on the fate and transport of wastes from the Acid Pits.

The standard geologic reference for the site area was McKevett's 1951 report *Geology of the Jurupa Mountains*, which Fox recalls he consulted (ibid, p. 90). Additionally, Fox obtained, from the Stringfellow Company, the driller's log of well 25/6W-1Q1 (USGS system), which is alleged to have been the fifth percussion-drilled water supply that Mr. Stringfellow, Sr. commissioned in order to obtain one well "with water in it," for his quarrying operations (in a discussion with Alvin L. Franks; Franks, 1990, p. 320).The well is located about 2900 ft (884 m) downgradient from the location Fox recommended to the RWPCB for installation of the waste containment barrier. Its driller's log is as follows:

TABLE 2

PERSONALITIES ASSOCIATED WITH PRE-CERCLA STRINGFELLOW ACID PITS ACTIVITIES

Time Frame	Person	Role
1950-1956	Paul Brown, CE	1st Executive Officer of RWPCB; Retired late in 1956
1955	Max Bookman	DWR Southern District Engineer
	Robert Edmonston	DWR SD Assistant District Engineer
	Laurence B. James	DWR SD District Geologist
	David B. Willets	DWR SD Supervising Engineer
	Robert C. Fox	DWR SD Engineering Geologist
1955-1972	J.B. Stringfellow, Sr.	Owner of parent corporation and property (deceased about 1965)
	J.B. Stringfellow, Jr.	Partner in Acid Pits
	Lawrence E. Nutt	Partner in Acid Pits (deceased pre -1972)
	E. Moe McCook	Partner in Acid Pits (bought out late 1950s)
1957-1963	Raymond Stone, MSCE	2nd Executive Officer of RWPCB Founder, Neste, Brudin & Stone, 1963
1959-1983	John Zasadinski, BSCE	1st Staff Engineer at RWPCB
1 Jan 1968	Name Changed to Regional Water Quality Control Board	
1963-1972	Richard Bueermann, CE	3rd Executive Officer of SARWQCB
08Dec1972	Joins Neste, Brudin & Stone (NBS)	
1 Jan 1970	Names Changed:	State Water Resources Control Board Regional Water Quality Control Board
1968-1973	James Anderson, CE, JD	Assistant Executive Officer of SARWQCB
1 Jan 1973-1984		Executive Officer of SARWQCB
1972-1975	Neste, Brudin & Stone	Consulting Engineers to Stringfellow
	Robert C. Fox,	Consultant to NBS
1972	James Doody	DWR SD Chief Engineer
	Jack Coe	DWR SD Asst. Chief Engineer
	David B. Willets	DWR SD Supervising Engineer
	Sanford Werner	DWR SD Engineering Geologist
08 Jun 1976	J.M. Montgomery Engrs.	Contracted for Closure Study
	Dr. F.A. Grant	Project Manager
	Karl Wiebe	1st Project Geologist (1976-1978)
	Ronald Barto	2nd Project Geologist (1978-1980)
	George Curtin	3rd Project Geologist (assigned 1980)
	John Vasconcellos	Engineer
01 May 1968 - 21Oct1980	Alvin L. Franks	State Board Engineering Geologist
23 May 1973		Named Chief, Hydrologic Geotechnical Sec.
21 Sep 1975		Named Case Manager for State Board
01 Aug 1980		Leaves involvement with case
21 Oct 1980		Retires from State service
	Robert Ford	Replaced Franks on retirement
	Gil Torres	Engineering Geologist (hired 1970)
21 Oct 1980	James Parsons	State Board Engineering Geologist assigned to Stringfellow case (hired 1971)
05-12Mar 1980	Robert M. Mandel	Federal OSC for Emergency Removal
12 Mar 1980	Robert Clawson	Federal OSC for Emergency Removal
08 Aug 1980	Gerald Klug	Federal OSC for Emergency Removal
08 Aug 1980	LCDR J. O'Brien	USCG Pacific Strike Team Leader
08 Aug 1980	Vince Marci	Ecology & Environment TAT representative

0-12 ft:	"sandy material"
12-51 ft:	"decomposed" granite
51-56 ft:	"broken rock"
56-81 ft:	"decomposed" granite
81-88 ft:	"broken rock"
88-98 ft:	"broken rock"

Fox considered the position of this well and then recommended that the Board consider specifying (as a Discharge Requirement) that Stringfellow "monitor" the water quality in this well. In so doing, the authors believe that Fox specified the first groundwater monitoring well in history. Franks (ibid, p.89) recalls that Stringfellow, Sr. noted that the well had later been deepened, to 200 ft (61 m) but that the intake screen remained at 70-93 ft (21.3-28.3 m).

Once in receipt of the 2 June "Bookman" report on the proposed Stringfellow site, the RWPCB staff (then just Executive Officer Brown and one clerical secretary) prepared a draft Resolution No. 55-11, containing a set of "discharge requirements" for consideration of the Board at its next meeting. The DWR report (5 pages in length, including 3 pages of text, one geologic map and a table of analyzed water quality parameters from Stringfellow's well below the proposed site) was incorporated. Board members were unpaid volunteers of questionable professional or technical expertise.

DWR's technical report of findings of the brief site reconnaissance was completed on 2 June 1955 and signed by Max Bookman, District Engineer. In terminology of the day, this "detailed inspection" consumed about 45 minutes or an hour on site for each of the three technical authors, Robert Fox, Don Phillips and Claude Hewitt. The authors believe that this report was the first such report ever issued for a proposed commercial hazardous waste disposal site anywhere.

The report, contains only three pages of text of this "detailed inspection" of the proposed site. The term "detailed inspection" took on huge proportions in the Federal trial, as the RPs portrayed the nature of the DWR confirmation visit to fall far short of what they offered as a scope of work for "detail." In fact, in today's terminology, such a site visit would not likely be termed as "detailed." But, on the other hand, neither did Fox and his colleagues use the term "detailed investigation," which would more represent the level of work that the RPs were trying to portray to the jury.

Geologic detail utilized by Fox in the site review was all that was required to confirm the general findings of the 1954 regional study. Based on the Eckis philosophy of the occurrence of ground water in southern California, Fox utilized only three geologic units in his site assessment: 1) the Basement Complex, which grouped all of the "impermeable" crystalline rocks (and minor meta-limestones in the vicinity); 2) the Older Alluvium; and 3) the Recent Alluvium. Fox also accurately interpreted movement of ground water in the valley alluvium to move to the southwest and reported the presence of a "natural dike or barrier ....". This referred to the "cemented conglomerate," now believed by the authors and colleagues to represent a post-depositionally-cemented prehistoric debris flow.

Fox and his colleagues concluded that the site was "...underlain by impermeable rock" but that there was a need to consider an artificial groundwater barrier below the "dump" to "...confine liquid wastes." The presence of the "cemented conglomerate" led to the advice to the Board to consider a 250-ft (83 m) wide barrier connecting the prominent ridge of cemented conglomerate (to the east) with sound bedrock at the west side of the canyon. The DWR report of 2 June 1955 clearly does not offer any design-related specifics and this report was the whole product of the DWR site review.

Santa Ana Board Actions Leading to Discharge Requirements

Santa Ana Board Executive Officer Paul Brown wrote DWR Chief Engineer Bookman on 6 June 1955 to compliment them on the quality of the work and to say that "The subject report will be used as a basis for formulating tentative requirements for the proposed waste discharge ...."

Brown wrote, on 7 June, to Lawrence Nutt, of the Stringfellow interests, enclosing a copy of the 2 June DWR report, noting the presence of the "natural barrier," and noting the general comments of DWR This letter appears to represent a courtesy advisory notification of the intent of RWPCB to issue Discharge Requirements.

Stringfellow, Sr. responded by letter of 9 June 1955 confirming his intention to proceed with the project under the conditions already released in the DWR report of 2 June. Brown noted to Stringfellow, by letter of 14 June, that the RWPCB "staff" had conducted a "field investigation" and that Discharge Requirements for the Class I site would include avoidance of obnoxious odors and confinement of wastes to the facility. Brown further specified that Water-Quality Objectives for the Chino Ground Water Basin, when adopted by the Board, would apply, and that his own well (1Q1) would be the facility monitoring well, to be self-monitored by the applicant. Further, Stringfellow might be required to submit reports on the quality and quantity of waste discharge.

On 22 July 1955 the RWPCB met and discussed establishment of the dump, thereon passing its Resolution 55-11, the Discharge Requirements to be levied on the dump and to be forwarded to the County Planning Commission for incorporation into the terms of any variance granted for the facility.

RWPCB Resolution No. 55-11 was composed of 11 parts: five paragraphs of "stipulations," six "resolutions," and three "discharge requirements," "In order to prevent unreasonable impairment in quality of the ground water in the Chino Ground-Water Basin for the aforesaid beneficial uses and of the quality of surface or underground water ...."

The Discharge Requirements:

1) "The wastes shall be handled and controlled in a manner which will prevent nuisance."

2) "The wastes shall be confined to the above desig-nated property, above the concrete barrier controlled by Stringfellow Quarry Company and will not be allowed to flow on other private or public property."

3) "Storm water drainage shall be effectively excluded from the disposal site."

The five "specifications" were for technical reports that the Board might see fit to request of Stringfellow, to provide additional information necessary to its oversight role. The Discharge Requirements were amended (Res. 58-4) on 29 September 1961 to introduce the *Chino Ground Water Basin Water Quality Objectives* (DWR, 1957) of 28 March 1958, to list 25 industrial wastes not requiring specific clearance, for initial dumping from specific waste haulers and a requirement that Stringfellow routinely sample well 1Q1 and and submit such for water quality analysis.

Role of Riverside County Planning Commission

All manner of constructed facilities in California were, by 1955, subject to land use zoning. In the case of the proposed Stringfellow Acid Pits the only regulatory permit that was required to go into operation was a land use variance to be obtained on petition to the Riverside County Planning Commission. Stringfellow's request was to have been heard before the Commission (as recorded in RWPCB log) on 9 July 1955. The variance decision was postponed until September, pending receipt of conditions to be levied by the Regional Board, and the County Health Department and Flood Control District.

On 9 August 1955 the County Planning Commission met, heard testimony by Earl Nutt, partner of the ownership, and future manager of the "dump." This was Variance Case No. 66. Commission representatives met at the site on 1 September and "discussed and agreed on conditions of use variance." At its regular meeting of 13 September the Commission approved the zoning variance request, subject to construction of stormwater diversion channels and "other requirements." Stringfellow's variance application was approved by the Planning Commission on 26 September and the site had the only "permit" needed to begin construction. Approval of the variance was based on various conditions to be met by Stringfellow; 1) to be located as presented; 2) to employ a "dam" constructed at the SW boundary of the site, extended to bedrock and sealed so "as to preclude the passage of liquids." All in all, there were 13 conditions to be met by Stringfellow. At the time there were no commercial Class I dumps being operated in San Bernardino, Orange or Riverside Counties (Willets, 1984, p. 71).

4. CONSTRUCTION OF THE STRINGFELLOW ACID PITS

Construction of the dump began sometime after notification of variance approval by the Planning Commission and was observed from time to time by Executive Officer Brown, of the RWPCB, although the legal authority and responsibility for inspection of construction lay with the County Building & Safety Department. Brown visited the site for purpose of informing the Board of progress, since construction inspection was the basic responsibility of the County Department of Building and Safety. On 20 January 1956, Brown observed that the site had suffered drainage damage during a violent two-day rainstorm that had wrought extensive hillside damage in Los Angeles. Brown estimated that flow in two 36-in (0.9 m) culverts below the dump office was at 20-30 cfs (566-850 lps). On 23 May 1956, Brown was at the site and when Mr. Nutt brought up the subject of construction of the dam, Brown "asked him to confer with RCFCD (Riverside County Flood Control District) and DWR, re dam construction." Since DWR has never served the private sector, the reference to that agency is taken in the sense that DWR would provide information to the public relating to technical matters, but did not partake in design or assumption of any other form of private-sector risk assumption.

Regional Board files originally contained six small black and white snapshot-type photographs of construction of the barrier "dam" at the site, dated in printing "Nov 56," and mounted by transparent tape on legal note paper. Location of these original prints is presently unknown, having been lent, on legal demand, to responsible party (RP) attorneys by the Attorney General's office in 1989 and subsequently reported lost by the borrowers. Fortunately, copies of the photos were made via hand-held camera, in window light before the transmittal. It is not known who made the photographs, but it is generally presumed by the State that the photographer was a member of the County Department of Building & Safety (DB&S). These photos show the first 60 feet (18 m) of the dam as formed for concrete casting in plywood and lying in a bulldozer trench excavated in the Younger Alluvium of the east side of Pyrite Canyon. No records of construction or County inspection of the "dam" exist. In reality, Stringfellow stopped construction of the "dam" at this length, leaving the remaining 190 ft (58 m) of the distance untouched in natural alluvium. Later, under oath in the Federal case, Stringfellow was to testify that he completed the entire required distance of the "dam," and that (1990, p. 42) the barrier key-way was excavated four feet into the "blue granite" of the site bedrock. (See Fig. 2, p. 84).

Non-presence of the "dam" was proven in the 1980 emergency abatement actions undertaken by the State, during which time the natural casts or impressions of the alluvium abutting the west end of the 60-ft. (20 m) dam were observed and photographed. On 9 August 1956 RWPCB made a "reinspection" (Board Log) of the site and reported that dumping had not yet begun.

5. OPERATION OF THE STRINGFELLOW ACID PITS (1956-1972)

While Stringfellow was readying the dump, the Orange County Sanitation District crisis occurred. On 10 February 1954, just before activating its wastewater management operation, Orange County Sanitation Districts (OCSDs) had issued its Ordinance No. 1, which stipulated limitations for discharge of liquid wastes by the member Sanitation Districts. Apparently the Ordinance was not enforced until August 1956, at which time (17 August) Earl Nutt telephoned RWPCB to state that an emergency existed and that potential clients of his were needing access to the dump. That there was indeed such an "emergency" related to liquid industrial waste disposal was supported in a Federal case interrogatory (23 Sep 1983) provided by RP Rainbow Canyon Manufacturing Co., but citing only the month of August 1956.

RWPCB Executive Officer Paul Brown visited the site on 23 August and reported that the dam was eight feet high and that Stringfellow would set aside an isolated area for cyanides. Brown made no other comments as to the length of the dam or percentage of its length completed at that time. Robert Fox recalled (1983, p. 219) that DWR personnel were on site during the period 23 to 29 August 1956 and noted that "roads and dam were under construction" and that "Haulers being advised, can use dump 8-27." It is not known under what circumstances the DWR personnel were present. Fox further recalled (ibid, p. 220) that the dam was 8 ft high, as measured from sound bedrock.

On 31 August 1956 the Stringfellow Acid Pits began operation, with the initial load being a batch of missile-refurbishing wastes from Norton Air Force Base, San Bernardino (RWPCB log). The Board "reinspects" (using its logbook terminology) the site on 4 September during which visit the Hunter-Douglas Company dumped wastes. A Board representative (presumably Paul Brown) again visited the site on 4 October, to "reinspect" dam construction, during which time photographs were taken. These photographs could be those bearing the November 1956 imprint, if they had been developed within the following month. Brown returned on 30 October, with L.C.Parsons, Arthur Pickett and Harold Simmons, all of the Los Angeles County Engineer's office. Riverside County Planning Commission (RCPC) inspector R.C. Smith was on site that day and reported (memo of 7 Nov 1956) that "All conditions have been complied with except Condition No. 2 regarding a dam. At present the dam is under construction." Smith also reported that "Mr. Stringfellow advised that upon completion of the dam he would notify the Riverside County Planning Commission so an inspection can be made." Neither Stringfellow, Sr., or Jr., apparently notified the Planning Commission and the final, confirmatory inspection was never apparently made by Riverside County DB&S. On 13 November RWPCB again visited the site for an "Inspection and conversation with Nutt." Presumably the incomplete "dam" was by then covered up and presented by Stringfellow as having been completed.

In May 1957, DWR published its *Ground Water Quality Objectives, Chino Basin* as the operational technical reference for RWPCB administration of its "discharge requirements" for the Stringfellow Acid Pits. Robert Fox (1983, p. 368) assisted in this compilation and noted that no field work was performed for this evaluation and that ground water south of the Jurupa Mountains was of poor quality (very hard) but acceptable according to the Federal Drinking Water Standards of the day.

Stringfellow's business again picked up on 1 June 1957 when Los Angeles County Sanitation Districts (LACSDs) followed the lead of OCSDs and prohibited the dumping of sulfuric and chromic acids to the sanitary sewer system, for fear of damage to bacteria at the publically-owned treatment works (POTWs). Stringfellow was aware of this turn of events (RWPCB summary, 21 Dec 1972) and prepared to accept the diverted wastes.

Complaints Against the Stringfellow Disposal Operation

During the County zoning variance process, including hearings, there seems to be no record of public opposition to the dump operation. After the dump had been in operation for nearly 11 months, on 16 July 1957 RWPCB received the first citizen complaints lodged against the Stringfellow operation. Mrs. Durette, of Pyrite Street, in neighboring Glen Avon, called the Board and complained of "odors and water pollution." This was followed by a call from Mrs. Ruth Kirkby (founder of "Parents of Jurupa" in that year, beginning a struggle that has gone on ever since). A Board representative (probably the new Executive Officer Ray Stone) and residential water samples were taken for analysis. On 17 July the Board notified Mr. Nutt who responded that there was no dumping going on at unapproved locations. On 18 July, Stringfellow began self-monitoring well 1Q1, with laboratory analyses being conducted by E.S. Babcock & Sons, of Riverside. On 8 August Mr. Gene Derring, Chairman of a "citizen's committee, called the Board for a "long discussion on Stringfellow and the relative schools" of Glen Avon.

A Board representative visited the dump on 27 September 1957 and was confronted by local citizens protesting a sanitary landfill for which Stringfellow was proposing to seek a zoning variance.

On 2 December 1960 Board Staff Engineer Joh
visited citizens Stark and Durette in Glen Av

response to a complaint call. Stark's toilet was inspected for reported oil-in-water and a small quantity of flesh and hair samples were taken from Durette's dog. On 12 December Zasadinski was back at the site and Glen Avon to report Babcock Lab's findings that the dog had been burned by caustic material.

Case records are devoid of evidence of any action taken by concerned citizen's groups until a 15 August 1972 letter from the "Jurupa Steering Committee" (curiously employing terminology favored by environmental lawyers....), directed to the Riverside County Board of Supervisors. The Board Directed that the letter be answered individually by the Planning Commission, Air Pollution Control District and the Flood Control District.

Mrs. Kirkby was again in contact with the Regional Board, seeking assistance, and asking of the disposal quantities involved. Mr. Bueermann replied by letter on 3 November 1972 that 3,583,428 gallons (13,563,275 l) of liquid waste had been deposited in the three-month period from 1 January through 31 March of 1972 and that an average of 278,895 gallons (1,055,618 l) per month were being discharged at the site.

Stringfellow's Legal Position

Federal Case documents originally contained two rather condemning letters of legal advice, of 1956 and 1957. These two letters were later removed (1991) on court order by way of appeal by Stringfellow's defense attorneys. On 20 November 1956 Stringfellow's second legal counsel, Lonergan & Jordan, San Bernardino, advised that "... we believe that every precaution should be taken to insure the partnership against liability," and that the "... most effective devise is the formation of a corporation to own and operate the dump...." The 1 August 1957 letter from Lonergan & Jordan, to L.E. Nutt, discussing the liability of a "land owner who permits dangerous material to be placed on his property." This letter advises that the dump should be operated with the "utmost care" and "strictly in accordance with laws, rules, and orders of public bodies having jurisdiction and continue to follow their every suggestion."

James Stringfellow, Sr. died in 1965 and the Stringfellow Quarry Co. (a partnership) transferred title of the dump site on 26 October to Stringfellow Quarry Co. (a limited partnership of J.B. Stringfellow, Jr., Lawrence E. Nutt and E. Moe McCook). On 26 November the title interests of deceased Mr. Stringfellow were transferred to the Quarry Company. Federal case documents indicate that thereafter numerous ownership transfers were made (Table 3), sometimes at several per year between various Stringfellow interests.

Early Dump Operation (1957-1962)

On 25 September 1957 a Board representative visited the 'ump to note that site grading operations had cut a swath through the "uppermost" pit by dozer, exposing color-evidence of waste liquid percolation to a depth of three feet (1.0 m). The color at three feet was "faint" and the report was that the percolation probably stopped by a depth of four feet.

Waste arrived at the Acid Pits via tank trucks. The driver would produce a manifolded *Waste Hauler's Report*, signed by the generator. Dumping was allowed on a 24-hour basis. If the office was closed, the driver made the discharge and brought the waste haulers report back on the next daytime trip. Night-time dumping was frequent and access was gained via a key given to select drivers after about 1959. Each pit had a three-by- three-foot (1 m x 1 m) sign indicating which acids would be accepted. A "caustics" pond was also so labeled, located ".... all the way up the hill, way back from the acid." Filled ponds were signed with "Closed" or "Do Not Dump," and some "hot acids" were dumped at temperatures of 180-240° F (82-116° C).

TABLE 3

SOME OF THE LEGAL TRANSFERS OF OWNERSHIP OF THE ACID PITS FACILITY

Date	Transferor	Transferee
26Oct65	James Stringfellow, Sr. and SF Quarry Co. (partnership)	Stringfellow Quarry Co. (limited partnership) Stringfellow, Jr., Nutt, McCook)
26Nov65	Title interests of J.B. Stringfellow, Sr.(deceased)	To Stringfellow Quarry Co.
19Jun72	Stringfellow Quarry Co, (ltd. partnership)	Stringfellow Quarry Co. (CA Corp)
10Apr75	Stringfellow Quarry Co., Inc.	Stringfellow Quarry Co. (limited partnership).; J.B. Stringfellow, Jr. and Ruthella Nutt [widow of Earl])
10Apr75	Stringfellow Quarry Co. (limited partnership.)	Stringfellow Quarry Co. (A Calif. Corp. (estate of J.B. Stringfellow)
31Jan77	Stringfellow Quarry Co. (limited partnership)(CA Corp;	Stringfellow Quarry Co., Inc. Paul and Lucille Hubbs, San Bernardino, CA)
07Feb77	Stringfellow Quarry Co., Inc. Quitclaim	Stringfellow Quarry Co. (limited partnership)
08Oct79	Stringfellow Quarry Co. (Ltd. partnership)	Stringfellow Quarry Co., Inc.
01Jul80	Stringfellow Acid Pits property soldfor non-payment of taxes since 1974	Sold to State of California for $878.62
04Mar81	Transfer of site ownership From State of California to California Board of Public Works	

At least two of the generators were given specially designated exclusive ponds for disposal by haulers. On an inspection of 3 January 1958, Riverside County Health Department noted that the dump was not yet completely fenced and reported such to the Planning Commission for action (RWQCB, 21 Dec 1972 Summary).

In inspecting the site on 11 March 1960 Board staff engineer Zasadinski noted that one pond containing dried bottom-cake of 12-in (30 cm) of "black material," a fact that he again saw at a dried caustic pit on 3 November 1960, which was scheduled for cleanout excavation by Stringfellow. There is no direct evidence of the fate of the bottom-cake, but site foreman "Hoot" Gibson later testified at the Federal trial that such material was used to construct berms for new ponds. As ponds wer[e] inactivated for drying they were often reconstructed at variable locations. At the Federal trial, James Stringfellow Jr. testified that he had lined the ponds with "clay" from the site, thereby insuring that bottom leakage did not occur. Thorough geologic mapping of the site area indicates, as would be geologically expected from such terrane, that clay soil does not exist.

The entire nature of the Acid Pits operation suggests that Stringfellow was relying on evaporation to reduce the volume of the 34,000,000 gallons (128,690,000 l) of liquid wastes eventually disposed of at the site, aided, of course, by the significant amounts of seepage through semi-permeable pond bottoms.

Disposal charges at the Acid Pits were $0.015/gal ($0.0040/l), with quantity price reduction to $0.005 per gallon ($0.0013/l). Stringfellow had computed his available volume (letter Nutt to San Bernardino Board of Trade, 19 April 1962) at 5,639,782 gallons (21,346,575 l).

Dump Operation (1963-1968)

The end of routine operation of the Acid Pits came on 23 August 1963, when fumes from hot nitric acid caught fire in the morning and were extinguished in that afternoon.

In its later aerial photogeologic interpretation of site imagery, USEPA consultant JRB Associates (ca. 1982) noted that aerial photographs of the site, taken in late 1965 to early 1966, showed impounded wastes in contact with the body of "cemented conglomerate." This was thought possibly to have initiated acid-dissolution of the natural "dike" on the east side of the site. Field work by the authors and colleagues, in 1990, indicate that the conglomerate is actually a siliceously-cemented ancient debris flow, and therefore, not highly susceptible to acid dissolution.

ALCAN Aluminum Corporation, an Acid Pits generator with a plant in Riverside, was notified by the City (early 1966) of enforced specified limits on chemical constituents of its wastewater. An ALCAN internal memo of 5 January 1966 (Federal case exhibit) stated that hexavalent chromium above 0.1 ppm (1000 ppb) might represent a potential problem to the company as they were then "dumping regenerant brines in a quarry along with spent acid solutions" (the Stringfellow site was not mentioned by name) and that there could be "dangers of subterranean leaks and overflow into Santa Ana underground water." This comment was proven to be very accurate.

Sometime prior to April 1968 Stringfellow constructed a concrete retention dam at the ground surface at the southernmost tip of the dump site. The dam was a surface retention structure, apparently part of the drainage system, placed there in an attempt to keep pond wastes on site.

The first recorded indication of failure of the Stringfellow facility occurred on 10 April 1968, when Regional Board Executive Officer Bueermann's log notes that he "Discussed leak through dam with Sectry" (sic), probably "Marge," the facility office clerk. The "dam" apparently was the surface retention structure.

Declining Years of Dump Operation (1969-1972)

On the eve of its thirteenth year of operation the dump began to experience a string of negative events which brought the undetected subsurface leakage to the attention of the County, the Regional Board and to those local citizens who had maintained smoldering opposition to the operation. The facility had escaped damage from the heavy rains of mid-January 1960, but was to suffer greatly from the cloudbursts of mid-March 1969, and even more disastrously from the three months of rain in early 1978, after Stringfellow, Jr. had abandoned the site. Another pond fire broke out on 4 February 1971 and was reported by County Building & Safety inspector S.C.Taylor (noted in B&S letter of 17 Aug 1972, dealing with a followup inspection visit).

On 18 February 1972, the day after the visit of S.C. Taylor of the County DB&S, Stringfellow ceased leaving the dump-site open after normal business hours and instituted a policy that the gate was to be locked at all times facility personnel were not present and that certain "responsible haulers had access to a key."

Torrential Rains of March 1969

One of those rare, by-the-decade, occurrences of torrential cloudburst rainfall struck the Jurupa Hills on 17 March 1969. Floodwaters overcame the discharge and routing capacity of Stringfellow's drainage system and hazardous waste was carried out of the impoundments. By all accounts the rainfall was of historic proportions. There is no evidence that the Stringfellow interests had ever engaged professional engineering assistance in any aspect of the facility, especially in regards to operator responsibility and the assumption of risks inherent in operation of the dump. This was a serious reason for the permitting and oversight agencies to be concerned.

Stringfellow Junior's Federal case testimony (Defendant's Memo on Points, 1983, p. 135, and Declaration, 1983, p. 43) was that there was no discharge of wastes from the site. RCFCD in a 17 March 1969 visit (SARWQCB Summary, 21 Dec 1972), reported that wastes were being carried out of the dump and across US Highway 60 (meaning into the north edge of the community of Glen Avon).

A Regional Board representative visited the site on 18 March and collected water samples at the surface dam and in the drainage ditch at the NW corner of Pyrite and Mission Boulevards. A photograph was made and captioned "Chemical Laden Brown Water Slowing South about 750 ft Downstream from Dumpsite." The samples were analyzed as being acidic.

Surface seepage at south end of dump (1969-1972)

Sometime in 1969 Stringfellow began to note seepage of site waste and contaminated water from below the concrete retention dam that he had constructed at the south tip of the dump. He began to collect these wastes in an unlined surface pit and to pump it back to site ponds. This action continued from 1969 into 1972 (Fox, 1983, p. 181). By 1 January 1970 there were two such retention dams in place (NBS report Jan 1973) and leachate was appearing only at the toe of Dam No. 1. NBS reported that the seepage sump between dams 1 and 2 had dried up by late 1970.

By August 1971 (NBS, 19 Jan 1973) Stringfellow had constructed two more concrete retention dams and had activated a sump known as "Well No. 3" and collected leachate was being pumped to the dumpsite behind Dam No. 4. On 10 January 1973, Bueermann, acting in his new capacity as NBS consultant had a telephone discussion with Stringfellow relating to the possible use of bentonitic clay as a seepage sealing agent and suggested use of a 55-gal (208 l) drum as a test model.

In November 1972, Stringfellow was still operating his leachate recovery sumps ("Wells" Nos. 1 and 2), installed upstream from Dams No. 1 and 2, respectively, and pumping recovered leachate back to the dump site upgradient of Dam No. 4 (NBS Rept. 19 Jan 1973).

Degradation of groundwater quality in 1Q1 (1969-1972)

In early 1969 laboratory analyses on water from monitoring well 1Q1 were returned to Stringfellow showing new and marked shifts in quality from a Ca-Mg bicarbonate chemistry to that of Ca SO_4-Cl. It's increased hardness and its sulfate and chloride content gave it a character of influence of site wastes. By 22 February the Regional Board had judged Well 1Q1 chemistry to be high in salinity, and hexavalent chromium was detected at 0.05 ppm.

On 18 May 1972 the Glen Avon School Board had its Well No. 2 sampled and chemically analyzed. A trace of hexavalent chromium (Cr^{+6}; 0.013 ppm) was present, along with and a slight increase in salinity since the last water-quality test in 1957. Bueermann was called by Stringfellow and on 19 May sampled the school well and a well on the road into the Stringfellow Quarry.

Sampling of monitoring well 1Q1 on 1 October 1972 revealed high salinity, very high nitrate and 0.05 ppm Cr^{+6}. Followup sampling and analysis of the Glen Avon School well on 18 November 1972 indicated Cr^{+6} at 0.0013 ppm, equivalent to the value noted on 18 May of that year; salinity remained "slight."

In one of his last official actions, just before resigning to join the consulting firm of NBS, Bueermann wrote Stringfellow (14 November 1972) to note the increase in salinity as an indication of an apparent violation of Discharge Requirements and to ask that Stringfellow provide a technical report on probable causes and proposed remedial measures. Again, on 4 December 1972 the Board sampled monitoring well 1Q1 to find that Cr^{+6} had risen in concentration to 0.07 ppm and that the "high" salinity condition was yet present.

Stringfellow Closes the Site (29 November 1972)

In consideration of the ongoing surface seepage at the southern end of the site and the ongoing indications of contamination in the site groundwater monitoring well (1Q1), along with possible contamination of the Glen Avon School well, the Santa Ana Board (Bueermann) had requested Stringfellow to submit a technical report on 14 November 1972, as noted above. Stringfellow replied, by letter, on 28 November that he had visually inspected the "face of the concrete dam" and had found no visible leaks and that he had engaged engineering and geological consultants, presumably to file the requested technical report. On the following day Stringfellow telegraphed the Board to note that the dump was closed at 2200 hrs. that date.

6. ACTIONS BY REGULATORY AGENCIES DURING OPERATION (1957-1972)

Without a specific program of industrial waste management existing or in force during the operating years of the Stringfellow Acid Pits, worldwide, California State, County and local authorities continued to allow direct disposal of liquid industrial wastes to the land. This was, then, as now, environmentally more preferable than to allow for widespread dumping on raw land, agricultural land and into storm sewers, as was already the case in southern California. In fact, during this period, there existed an informal co-regulation of the disposal, based mainly on the right of "home rule", initiated in California on 23 January, 1889 (Hunter,

1933, p. 71), which granted most discretionary regulatory actions and responsibilities to the Counties, and primarily enforced by Building Superintendents appointed to issue building permits and to oversee the erection of structures, to include construction such as the Acid Pits.

Santa Ana Regional Board (1957-1972)
Once the Stringfellow facility had been reviewed and Discharge Requirements formulated for use by the County of Riverside in its permitting program, the Regional Board took on a non-permit-related oversight role.

1957 through 1968 (Normal operation)

Persuant to the issuance of the DWR in May 1957 of its *Ground Water Quality Objectives, Chino Basin*, the RWPCB hearing of 28 March 1958 approved its Resolution No., 58-4, *Prescribing Water Quality Objectives for the Ground Waters of the Chino Basin Within Region No. 8.* Three groundwater "indicator wells" are specified for each disposal site receiving Discharge Requirements.

Stringfellow (usually in the person of Earl Nutt) presumably called for permission to allow discharge from each new waste source. The notifications are also found in Board logs and in its files as confirmatory followup letters.

In 1959, probably late in the year, John Zasadinski was hired as the first staff engineer of the Board, becoming the third staff member after the Executive Director and the clerical secretary. The Board log shows that Zasadinski visited the site on about a quarterly basis beginning on 18 January 1960. Thereafter, it appears that the Acid Pits became one of Engineer Zasadinski's case load.

Zasadinski, in an effort to reduce the paperwork load of responses to Stringfellow, authorizing specific chemical discharges from individual generators, on 20 December 1963, discussed with County Health Director Stone, "formulation" of a standard list of materials that would be deemed satisfactory for dumping at the site. Dr. Stone accepted the proposition and was to check with the Planning Department for approval on their part. This action was not completed until after the March 1969 rains and flooding at the site. On 10 July 1969 the Board approved Exhibit B to Resolution No.55-11 (as amended Sep 1961) as a generic list of wastes not requiring Board and County Health Department approval prior to acceptance for dumping.

From 1969 to 1972 (Declining performance)

A Board representative was on site on the second day of the torrential rains of March 1969 and X.O. Bueermann provided Nutt with a letter on 19 March noting the observed overflow leakage from the impoundments and requesting action in reconstruction of the storm-drainage works. Nutt replied by letter on the 20th and confirmed that they had stopped "receiving all wastes and started repairs to the dams and channels which were damaged in the recent floods." Bueermann inspected the site in Stringfellow's company on 29 April and found that drainage was still a problem along the east and north sides and that the dump was receiving a "limited" amount of waste.

On 1 January 1970, as a result of enactment of the Porter-Cologne Act (successor to the Dickey Water Pollution Control act of 1949), the governing unit became the State Water Resources Control Board and the Regional Water Quality Control Boards took on its new name.

Throughout 1972 the Board took a special interest in attempting to correct the deficient status of site drainage, finally advising DB&S of this deficiency on 24 October (see below).

State Water Resources Control Board
As long as the Stringfellow Acid Pits were apparently operating in general conformance with County and Regional Board guidelines there was no need for participation in oversight by the State Board. A major event in Board affairs occurred, however, on 2 March 1972, with adoption of SubChapter 15, Title 23 of the California Administrative Code. These were the new Statewide requirements for siting and design of waste disposal facilities. These regulations clearly stated the previously-implied responsibilities of the disposal facility applicant. All information necessary for siting and design were to be furnished by the proposed discharger.

Engineering geologist Alvin L. Franks had worked with the task force on the Porter-Cologne Water Quality Act in early 1967 and took over water quality responsibilities when the Water Rights Board and Water Pollution Control Board were combined to form the SWRCB on 1 May 1968. After establishment of the new State Board, he set about studying all of the existing Regional Board materials pertaining to siting of disposal facilities and included all of the other available references he could find on design of such facilities. The resulting two-and-one-half year study provided input for the 1970 Waste Haulers Act and formulation of the 1972 *Waste Discharge Requirements for Waste Disposal to Land, Disposal Site Design and Operation Information.* In their early years both the regulations appearing in the California Administrative Code and the Waste Discharge guidelines were revised frequently (1974, 1976, 1978, 1979, 1980 and 1981) as Porter-Cologne amendments were passed by the Legislature.

Santa Ana Air Pollution Control District
The first participation in air pollution oversight of the Acid Pits appears to be the 10 May 1972 order of the Santa Ana Air Pollution Control District ordering termination of discharge of paint thinner and paint stripping solvents in the ponds. Stringfellow, then nearing the end of his willingness to comply with oversight orders, carried out the order. The

District, in the 10 May 1972 letter, stated that its representative would make visits to the dump periodically to insure compliance with Rule 71 of its Rules and Regulations. Such a visit occurred in September of the same year when the District representative reported compliance

Riverside County Agencies

There is no doubt that the various agencies of Riverside County were the legally-constituted regulatory body of primary influence in the siting, design, construction and operation of the Stringfellow Acid Pits. Without County approval, in the form of "facility permits" there would have been no way for the facility to have legally opened its gates or to have remained open for more than 17 years. Permitted activities in California Counties are under the direction of the Board of Supervisors who delegate their regulatory authority to several agencies that provide forms of inspection and permitting.

Board of Supervisors

Even though it is the supreme County Agency, the Board of Supervisors does not appear to have been directly involved in the Stringfellow facility until receipt of a letter of concern from the "Jurupa Steering Committee," a citizens' group, on 15 August 1972. As noted elsewhere, the letter was remanded to the Planning Commission, Air Pollution Control District and Flood Control District for individual attention.

Planning & Zoning Commission and Planning Department

The County Planning Department, the public agency arm of the Planning and Zoning Commission, remained involved in the regulatory oversight of the Acid Pits. In addition to the 1963 involvement noted above, on 2 November 1966 the Stringfellow facility called the Regional Board for permission to accept Edington Refinery "contaminated road mix oil." The Board told Stringfellow employee Marge that it was in agreement but that the company must seek approval from the County Planning Commission (CPC) "re their permit." This plainly indicates that the Regional Board regarded the County as the permitting agency and that the County was closely associated with the oversight.

Strangely, in the midst of complaints by the Regional Board and by the County Flood Control District (FCD) of marginal stormwater routing and pond retention capacities, the Planning Commission met on 25 October 1972 to find that the Stringfellow facility was in compliance with all conditions of Riverside County Zoning Variance No. 66. Copies of the letter were to be provided to all appropriate county agencies.

County Building & Safety Department

In addition to its original participation in the consideration and issuance of the County zoning variance, DB&S inspectors occasionally visited the site. One such visit, on 4 February,1971, placed inspector S.C. Taylor as witness to a fire on one of the "sumps" (an undetermined facility component). On 17 August 1972 an internal letter from inspector S.C. Taylor to DB&S Director R.C. Smith noted that Taylor had visited and toured the facility on that day and that he noted no violations, keeping in mind the fire of 18 months earlier.

The real and substantial role of the County Building & Safety Department has remained somewhat obscured during Stringfellow legal battles to date. In actuality, the DB&S was the action arm of the county agencies, making contact with land use variance permittees and other owners of structures requiring building licenses and all activities influencing public safety, any conditions deemed necessary for correction. On 31 August 1972 DB&S Director Smith was informed by letter from FCD Chief Engineer John W. Bryant that "nominal attempts ... have been made to keep storm runoff from entering the (disposal) basins ...." and that "The measures, however, taken (by Stringfellow) are considered to be inadequate." Bryant warned that the dumpsite was subject to retention failure from erosion and that the site was needing structural flood-control measures.

Regional Board X.O. Bueermann advised DB&S on 24 October 1972 that the Stringfellow Quarry Company was in violation of Waste Discharge Requirements "by not providing adequate surface drainage."

County Health Department

Under California law of the day, the County Health Departments were the de facto action agencies in maintaining proper operational conditions at waste disposal sites. On 3 July 1958, Riverside County Health Department (CHD) advised the CPC that the "...operation of this industrial waste site does not constitute a threat of pollution." In a letter of 22 October 1958 County Health Officer Stone wrote to the County Board of Supervisors noting that "... the entire operation is observed from time to time by engineers from the Department of Water Resources, the Water Pollution Control Board and from this Department." The statement is in error with respect to DWR participation as there is no such evidence in the record.

On the basis of its ongoing inspection program, CHD found it necessary to request addition of two more Discharge Requirements and had requested such to the California Department of Health (CDH), which concurred and made official request to the Regional Board on 30 August 1961. These were: 1) Breeding of insects shall be controlled, and; 2) Discharge shall be confined to the designated areas and shall be controlled by the discharger.

In response to the torrential rains and associated site flooding of 1969, County Health Officer Erickson advised Stringfellow (letter of 19 March) to "discontinue the drainage from the chemical disposal ponds," confirming his verbal directive of the previous day. Nutt responded by letter (20 March) that the "precautions" specified in the Health Department letter of 19 March were being taken.

County Flood Control District

The role of the Riverside County Flood Control District (FCD) was secondary to those of the other County agencies. The Planning and Zoning Commission would have reviewed the District's maps and plans for inclusion in its original land use variance deliberations. On 23 May 1956, in response to questions from Earl Nutt regarding drainage provisions, the Regional Board suggested contact by Nutt with the RCFCD.

In a letter of 9 August 1972, to the Regional Board, FCD Chief Engineer Bryant included a revision of the District's 1965 topographic map showing the acid pits (as of 1972) and a recent aerial photograph, apparently as a courtesy. The District Chief Engineer was present on 16 January 1973 to inspect site drainage, which NBS (Jan 1973 report) declared he had "approved."

7. THE CONSULTANTS

Stringfellow had the implied obligation to use whatever resources were available to design, construct and operate a disposal facility that would not pollute surface and ground waters or result in contamination that would pose a risk to the nearby populace. Competent consulting civil, sanitary and soils engineers and engineering geologists were plentiful in southern California at the time of his original permit application to the Riverside County. After the initial site failure of 1969, it was nearly another four years (late 1972) before Stringfellow sought consulting engineering assistance.

Neste, Brudin & Stone, Consulting Engineers
On the event of his 1963 retirement from the position of Santa Ana Regional Water Pollution Control Board Executive Officer, Ray Stone became a founding partner of Neste, Brudin & Stone, Consulting Engineers. His replacement as Board Executive Office, Richard Bueermann, left the Board on 8 December 1972 and joined Stone's firm.

Robert C. Fox, Consulting Engineering Geologist
As a result of the general shortage of qualified southern California engineering geologists in the early 1950s, Fox began to engage in an individual private practice outside of his State DWR employment. In his testimony Fox was not specific about the timing of this additional work experience. Prior to 1969, the year of professional registration for Geologists and Engineering Geologists began in California, most consulting engineering geology was performed by individual consultants or very small associations (say, less than three to five geologists), and almost always under the supervision of registered civil engineers. After receiving professional registration, Fox left State service in 1969.

J.M. Montgomery Engineers (for SWQCB)
Pioneer Pasadena sanitary engineering consulting firm, J.M. Montgomery Engineers (JMM) was engaged by the State Board for the purpose of characterizing the site and waste to the degree sufficient to bring the Acid Pits to closure. Selection of JMM was based on a competitive announcement through a Request for Proposals (RFP) on 12 June 1975.

JMM was finally placed under contract to conduct an "abatement study" on 1 July 1976. It is apparent that the long time lag representing contracting was a result of this being the first such abatement work contracted by the State. The JMM contract was to develop an optimal plan for either safely opening or safely closing the Acid Pits, providing that abatement was achieved.

Acting without any formal guidance for such work (again, there being no such body of knowledge worldwide), JMM and the State Board devised a protocol for exploring the site, assessing the probable environmental threat, comparing options for abatement, and advising the State on the optimal measures. By January 1977 the JMM abatement study was complete, having identified 11 "alternatives" (which is the very terminology adopted five years later in the Federal CERCLA National Contingency Plan; NCP) for opening the site and 14 alternatives for closing the Acid Pits.

Before substantial site remediation work could be undertaken under the direction of JMM, the disastrous storms and rainfall of March 1978 struck, leaving the site in such disruption that additional funding was required prior to initiation of any substantial remediation. As a result of flooding, in March 1980, Region IX was able to discuss possible funding under the National Oil & Hazardous Substances Pollution Contingency Plan developed in accordance with Section 311(k) of the Clean Water Act of 1977. The JMM report of 1977 was the basis for deliberations and it was chosen to move toward closure of the site. (Reiter and Mandel, 1980).

JRB Associates (For U.S. EPA, Region IX)
It is not entirely clear just why Region IX chose to undertake a separate remedial investigation of the site

Ecology & Environment, Inc. (For U.S. EPA, Region IX)
As noted below, Ecology & Environment, Inc., of Buffalo, NY, became the EPA regional representative, through the Technical Assistance Team (TAT) national contract to oversee its funded activities, on 8 August 1980.

Woodward-Clyde Consultants (For SWRCB, 1980-1981)

Woodward-Clyde Consultants (WCC; Orange, CA) provided geological and geotechnical field support to Montgomery Engineers. In late 1981, WCC produced an undated report entitled *Draft Final Report; Geologic and Geophysical Investigation of the Stringfellow Class I Disposal Site.*

Assistant Professor Donald Stierman (Univ. Calif. Riverside, 1982-1983)

In 1981 Donald Stierman, an Assistant Professor of Geophysics at UCR became interested in the site and moved to undertake site studies that he judged were needed but which had not yet been accomplished. His case study was completed on 31 March 1982 and was provided without cost to DHS. Stierman searched for the presence of chromium, believing that such contamination lay to depths of 85 ft (26 m) below the collar of well 1Q1 and perhaps to 40-50 ft (12-15 m) below the site. He analyzed the February 1981 WCC rock core and found evidence of rock-quality degradation, due to natural rock alteration, to depths of 199 ft (61 m) below the site. This alteration, he advanced, was evidence of the previous circulation of ground water during geologic time.

8. STRINGFELLOW MAINTAINS THE INACTIVE SITE (1973-1974)

On hearing of Stringfellow's closure declaration the Regional Board stepped up its pace of contact with Stringfellow, a task which grew to consuming proportions for the Board Executive Officer.

Cooperative Efforts to Characterize the Site (early 1973)

On 14 December 1972, and for the first time in perhaps 16 years, DWR sent a representative to the site on the basis of technical assistance to the Regional Board. On 19 December NBS principal Ray Stone recorded in a memo to file that Stringfellow, Jr. had uncovered two "cracks in granite" below the concrete dam, along which wastes were flowing. Sumps were in place to intercept this leachate (that term was not used at the time of the memo) and Stringfellow had apparently sprayed gunite as an apron over an area below the "cut-off dam" (Fox, 1983, p. 249) and Stringfellow was talking bankruptcy.

Bueermann made his last visit to the site as a State employee on 20 December, bringing with him James Anderson, slated to become his replacement as of 1 January 1973. Bueermann's log entry shows that two streams of leachate were leaking from beneath the dam, at about two gpm. The Board agreed to sample wells in and around Pyrite Street and Highway 60. In his last official action, Bueermann wrote to DWR Southern District Engineer J.J. Doody to ask for assistance in developing an evaluation of the site and to determine if water quality problems were present.

NBS engaged Robert C. Fox as geological consultant for its assistance to Stringfellow, and Fox returned to the site for the first time since operation began, probably in December 1972. NBS asked Fox to identify the nature of the problem, to coordinate with the State and Regional Boards and to arrive at a mutual solution to the problem (Fox, 1983, pp. 95, 140, 143).

Stone soon asked Stringfellow to exhume the original barrier dam (Fox, 1983, p. 147) as part of the NBS field investigation. Fox's 1983 deposition details the general character of the barrier dam but does not indicate that such was ever excavated. Case documents indicate that total exhumation of the original barrier dam did not occur until the site explorations of 1980, supervised by J.M. Montgomery Engineers.

Stringfellow placed some borings in the area, apparently short (unrecorded depth) wagon-drill percussion holes as used for blasting at the quarry. Fox did not log the uncored borings, noting that cuttings were "lying around the holes" (ibid, p. 157 & 258). Some of the drill holes were blown out with compressed air and Fox recalls (ibid, p. 259) that contaminated water was found in one drill hole located east near the east abutment. Fox also observed the original dam, as partially excavated on its downstream site, where only "fine cracks" existed in the bedrock. Leachate at a "couple" of gpm was moving through the "dam" (ibid, p. 247).

NBS recognized that leachate was moving out of the site and proposed a concept of "interference" wells as a "barrier" to movement (ibid, p. 166). Known as "interception wells" today, this remedy was to be placed so as to capture leachate and to recirculate such to the impoundments. The State Board had sent Alvin Franks to the site and he and Werner were in concurrence that this initial remedy should be initiated and that the interference well system and guniting or barrier grouting would be in order (pp. 172 & 193).

Stringfellow's reaction to the NBS remediation proposal was (Fox, 1983, p. 170) that he could not manage to undertake remediation if the site was not to remain open, and that he felt that the site would then become the responsibility of the State.

DWR's Doody furnished a memorandum (5 January 1973) to the Regional Board containing recommendations for a monitoring program to evaluate water quality (Fox, 1983, p. 266). NBS presented its exploration findings at a meeting at the Regional Board offices on 8 January 1973. Nine drill holes had been placed perpendicular to the stream channel and about 60 ft (18 m) south of the earthen retention dams. Eight more drill holes were made on an EW axis just south of the (supposed) location of Stringfellow's original barrier "dam"; the most shallow was at 30 feet (10 m) and the deepest extended to a depth of 80 ft (24.4 m). Degraded groundwater was founded in drill holes located "east of the low-flow channel." NBS offered three potential contaminant-migration pathways; 1) dam-bedrock contact; 2) infiltration from the temporary leachate collection sump, and; 3) infiltration from the ground surface. On 14 January NBS issued its findings as a draft report.

Franks was back on site on 18 January 1973 to make geologic observations and to ascertain if prior reports were accurate and reasonable (Franks, 1988, p. 26). Franks saw some of the original concrete "dam" exposed by excavation to the granitic bedrock and noted "... to east of dam, you could walk across a cemented conglomerate"; and there was a good exposure of the cemented conglomerate down in a stream bed off to the east." Other of Franks "preliminary conclusions were that waste was leaking through weathered granite, fractures in the granite and the conglomerate (ibid, p. 62). Visible rates of leakage were about one to two gpm (3.8 to 7.6 lpm; ibid, p. 398). Franks detailed his preliminary findings in a memo of 13 February 1973.

NBS issued its *Report to Stringfellow Quarry Company on Industrial Waste Disposal* on 19 January 1973. In the report, the original barrier "dam" is characterized as being made of reinforced concrete, key-blasted into "impermeable bedrock" for two to three feet (60-90 cm), two feet (60 cm) thick at the base and 14 inches (36 cm) wide at the crest, rising nine feet (2.7 m) above bedrock and backfilled with "imported" clayey soil. That this was based on Stringfellow's own account is evidenced by the fact that NBS noted that there was no evidence of the "dam" and the ground surface. In its cover letter to Stringfellow, NBS recommended that a second monitoring well be activated (at well 12B2) closer to Glen Avon, in the Older Alluvium and that all observable bedrock fractures be sealed downgradient of the existing "dam." To gain appropriate site geologic knowledge NBS proposed that nine boreholes, ranging in depth from 20 to 80 feet (6.1 to 24.4 m), be placed within 60-80 feet downstream of the "barrier." Ground water was noted at depths of 5, 16 and 23 feet (1.5, 4.9 and 7.0 m) below the ground surface and water quality analyses below the "barrier" indicated the following:

SO_4:	22000 - 46000 ppm
NO_3:	1670 - 9240 ppm
Cl:	550 - 2100 ppm

On 22 January 1973 Stringfellow wrote a letter to the Regional Board proposing to reopen the site. Permission was never granted. Franks reported to the Board, on 13 February 1973, that the reports of chromium in the Glen Avon School well were proven false and he then recommended a "hydraulic barrier" (1988, p. 45) downgradient of the site to "pick up any fluid that had already leaked from the site and recycle it back to the site." Franks further recommended that "The discharger shall submit ... for approval, a final as-built plans and specifications describing the corrective measures ...." (ibid, p. 50).

Stringfellow then contracted with the Haliburton Company to drill and inject sodium silicate grout into the bedrock and the "dam" Franks reported (ibid, p. 550) that the grout holes were placed on a single row, at two-three feet (60-90 cm) of depth, and not as a three-line system as he had previously recommended. The visible leakage was reduced to about one gpm (3.8 lpm) and Franks recommended that additional grouting be undertaken in weathered rock and possibly under the cemented conglomerate.

Disagreements Over Site Findings (1973)

Werner took issue with the observations of the other professionals on the case and made his position known through a letter from DWR Southern District Engineer J. Haley, on 6 April, 1973. Werner's findings were:

1) Site may have leaked through rock joints;
2) Gel grout injection program needed to be engineered;
3) Contaminants at well 1Q1 came from the site subsurface;
4) Abatement of groundwater contamination was questionable, and;
5) Presence of an extensive SW-trending vertical fracture system exposed in Pyrite Canyon downgradient from the earth dam.

The three principal geologists dealing with the site, Franks, Werner and Fox, met on site on 13 April 1973, in the company of Regional Board staff Anderson and Zasadinski, to discuss possible Discharge Requirements if the site were allowed to reopen. This was immediately before the combined State and Regional Boards meeting of that night. Franks (1988, p. 80) recalls that the meeting was 'controversial" and that he had presented a concept that sodium silicate gel injection grouting could possibly solve the leakage problem. Franks felt that joints in the metamorphic rock below the "dam" were spaced at about five feet, whereas downgradient of the dam the rock quality was apparently poorer and joints there might be only "inches apart."

The three geologists met again at the site on 8 June 1973, in the presence of Regional Board X.O. Anderson. Fox recalls (1983, p. 255) a spirited discussion about the presence of bedrock fractures. Franks recalls (198), p. 343) discussions related to leakage through the "dam" or "overtopping of dam system around the drainage" and through the cemented conglomerate and weathered bedrock. At that time Franks recalls the first indication of a slug of contamination moving downgradient at least 1200 feet south of the "dam." Franks noted (ibid, p. 343) that "in all of the Class I sited he had explored, he found lateral leakage."

Werner disagreed to the extent that "there is continuous fractured and jointed bedrock under the disposal site ponds," and noted such in a DWR memo of 12 June 1973.

State and Regional Boards Consider Reopening of Site

On 13 April 1973, the State Board met in Orange to consider new Discharge Requirements if Stringfellow were allowed to reopen the dumpsite. During that meeting the Regional Board issued its Order No. 73-20 rescinding the amended (Sep 1961) Discharge Requirements for the facility, as based on a site evaluation report of 22 January 1973. The new Requirements were stipulated as:

1) Positive hydraulic barrier with sump;
2) Treatment/disposal of wastes shall not create a nuisance;
3) Discharge of wastes to surface or ground water is prohibited;
4) Disposal area is to be protected from washout/ erosion;
5) Exterior faces of barrier shall be protected;
6) No disposal of organic solvents, and;
7) Remove/relocate any wastes discharged in violation of these requirements.

On 15 June 1973 the Regional Board issued its Order 73-20 *Monitoring and Reporting Program Number 73-20, General Provisions for Sampling and Analyses*. This order prescribed new Waste Discharge Requirements and allowed for re-opening of the site. The provisions were as follows:

1) Construction of positive hydraulic barrier and sump;
2) Recycling of collected leachate to site ponds;
3) Injection of "sealant material" beneath the original barrier "dam";
4) One new monitoring well to be placed immediately downstream of the barrier dam, but outside of the natural drainage channel;
5) Sampling of well 1Q1; assessment for complete water quality parameters plus electrical conductivity and total chromium, and;
6) Annual report, by registered civil engineer, as to capacity to divert runoff.

Fox completed the additional borings, probably by air rotary drilling, as he said (NBS letter to RWQCB, 12 June 1973) that he did not log the borings. These borings were placed at distances down Pyrite Creek channel at 300, 600, 900 and 1200 feet (91, 182, 273, 364 m) south of the site, and taken to depths of 20 to 80 feet (6.1 to 24.4 m) and encountered bedrock at only inches of depth. The maximum inflow of water to the borings was at 13.9 gpm (52.6 lpm; Fox, 1983, p. 270). Stringfellow reported that he displaced a total of 4000 gallons (15,140 l) of water in 48 hours, from six boreholes. Findings were submitted in the NBS report of 22 June 1973.

As noted below, the County of Riverside revoked Stringfellow's legal right to operate the dump on 24 February 1974. State Board Executive Officer Bill Dendy announced, on 6 March, a public hearing on the facility, to be held in Riverside on 13 March. According to Fox (1983, p. 226) four issues were on the agenda for discussion:

1) Does Order No. 73-20 comply with requirements ofthe State Water Code, Section 13263?
2) Does the Stringfellow facility, with modifications-specified by the Regional Board, meet with criteriaof the California Administrative Code, Title 23,Sub-Chapter 15, for a Class I site?
3) Is the site subject to inundation by a 100-year flood?If so, what will be the effect?
4) Are monitoring provisions adequate?

Fox gave testimony at the hearing, in support of Stringfellow's remaining wish to reopen the facility.

By 20 June 1974 the State Board had apparently accepted the *fait acompli* of the County's revocation, leaving Stringfellow without a legal means to operate. The State Board accordingly then issued its Order No. WQ 74-10 ordering Stringfellow to effect site closure and then held a public hearing on the order, in Riverside, on 3 March. It was this event, claim the Acid Pits generator RPs, that constituted the act of the State of California "taking control" of the site. On 8 July 1974 the State Board held another public hearing in Riverside, disclosing some modifications the the Discharge Requirements, which would still be required of Stringfellow in closing the site.

The Regional Board completed its administration actions by way of a 13 November 1974 request of Stringfellow to provide a preliminary report on the method of closure, along with time schedule, in accordance with the California Administrative Code, Title 23, Chapter 15, Section 2535, by 4 December 1974. That day came and went without the anticipated Stringfellow closure plan, and Stringfellow's attorney had written on that day to say that the 1973 NBS report would have to suffice. The Board, on 10 January 1975, issued its Cleanup and Abatement Order No. 75-38, for Stringfellow to produce the plan.

Citizen Complaints

Parents of Jurupa, following developments at the site, reactivated their efforts petitioning the Regional Board, on 16 July 1973, that Order 73-20 was "inappropriate and improper." By October Congressman George Brown had been brought into the issue and had written the Regional Board. Apparently the USEPA, Region IX, San Francisco, was also notified by Congressman Brown, constituting the first record of its participation. Region IX returned a letter to Brown staffer Mr. Lynch on 17 October.

A Flurry of Activity by Riverside County (1974)

Sometime in this interval (probably late 1973) the County Planning Commission called Stringfellow to appear and to report on the proposed groundwater recovery system. Fox, Stone and Bueermann appeared on Stringfellow's behalf (Fox, 1983, p. 177). The County Health Department met with Stringfellow and NBS in the same time frame (ibid, p. 188) voicing its concern for public health. The County Board of Supervisors requested (15 Jan 1974) the Health Department to sample water downstream from the site, noting that the Stringfellow Quarry Company was in violation of Variance Case No. 66. The Supervisors took the important and telling position of asking the Department of Building & Safety to issue orders effecting permanent closure of the site, with steps to be taken to remove existing waste. A month later (14 February) the Health Department reported that five wells had been sampled, and these samples, along with six soil samples, had been analyzed by Babcock & Sons, of Riverside, showing no indication of contamination of soil or ground water.

County Department of Building & Safety officials concluded on 26 February 1974 that the site was detrimental and constituted a public nuisance and had not been used for one year (the degree of activity prescribed in its permit), and County Variance No. 66 was revoked. Stringfellow, for the first time in nearly 18 years, was finally left without a permit to operate the dump.

In response to Stringfellow's 12 November 1974 declaration of cessation of operations and his actual shutdown of the recycling pumps on the threatened day, 21 November, the County Health Department advised Stringfellow, on 4 December, of his continuing responsibility to maintain the site.

Stringfellow's Continuing Legal Position

By March of 1974 Stringfellow had lost his permit to operate the dump, yet still had the advantage of the State and Regional Boards being willing to allow him that chance, along with the obligation to remediate, if his permit could be re-established with the County. His attorney (Don C. Brown, Esq.) wrote to the County DB&S (24 March) to advise that it was Stringfellow's position that he had been in continuous operation for the period 9 December 1972 to 22 March 1974, implying that the County's basis for revocation of his operating permit was flawed. The letter further states that Stringfellow intended to cease operations at the site (meaning mitigation activities) unless the order of revocation was rescinded immediately. On 18 April 1974 there occurred an appeal hearing before the Planning Commission. Fox (1983, p. 215) gave testimony in advocacy of Stringfellow, giving a version of the original permitting of the site and claiming that the contamination noted in monitoring well 1Q1 had been due to floodwaters of the 1969 torrential rains (presumably by way of travel down the outer casing interface).

In the wake of the Regional Board's 13 November request for a preliminary report of closure method and its schedule, Stringfellow's attorney (Thompson & Colgate, by Brown, Esq.) responded to the County on 19 November that the leachate recycling pumps would cease operation on 21 November 1974 and that responsibility for further operation "will lie with the County of Riverside."

Stringfellow attorney Brown, now juggling both the Regional Board and County interfaces, wrote the Board, on the day the closure plan was to be due, that Stringfellow considered the 19 January 1973 NBS report to be sufficient for the purpose and that there was now insufficient funding for the closure.

9. REGIONAL BOARD ATTEMPTS FOR CLOSURE OF THE ACID PITS (1975)

The new year of 1975 brought about a standoff between the permittee (Stringfellow), the permitting agency (County) and the oversight Agency (Regional Board). Stringfellow maintained that he no longer had a source of income to carry on the closure and remediation and both the County and the Board had exhausted their administrative options.

The Board chose to act responsibly and in doing so issued its *Cleanup and Abatement Order No. 75-38* that the site be "closed and abated as soon as possible." The Board knew that any abatement by a public agency would bear costs responsible to the owner and, if unpaid, would constitute a lien on the land title. A hearing was scheduled for 14 February 1975. The Board further found that the site constituted a "non-operating industry" and that it be posted as a pollution or nuisance, as required by California Administrative Code (CAC), Section 13305.

On 10 February the Board held the public hearing and requested Riverside County to abate the site via Order No. 75-39. A second hearing was held on 14 February regarding the actions specified in Order No. 75-39, and there were no protests. On 18 February the Board formally requested that Riverside County take necessary abatement actions in accordance with CAC, Sec. 13305(e).

Riverside County Declares Itself Disassociated with the Acid Pits

In response to the action orders and requests by the Regional Board, the County Board of Supervisors met on 11 March 1975 to vote unanimously to decline the request to undertake abatement of their formerly-permitted disposal site (Book 92, p. 209, Supervisor's minutes). On 13 November 1975 the Board of Supervisors held a hearing and revoked, for the second time, the Stringfellow dumpsite land use variance, which had originally been revoked by the DB&S on 24 February 1974.

Stringfellow's Response to the Regional Board Abatement Orders

Stringfellow responded to the two-pronged closure orders by two more property title transfers, both on 10 April; 1) Stringfellow Quarry Co., Inc. transferring title to Stringfellow Quarry Co. (ltd. partnership; J.B. Stringfellow, Jr., and Ruthella Nutt [widow of L.E. Nutt); and 2) a subsequent transfer to Stringfellow Quarry Co. (a California Corp., J.B. Stringfellow, Sr. [estate of the deceased elder Stringfellow]). The site thus became owned by the estate of a dead man.

On 11 August 1975 Stringfellow attorneys Thompson & Colgate had written to the State Board (unknown content) and received a 19 August response from the Riverside County Office of Counsel in which the County advised Stringfellow to present his situation to the County Department of Public Health. In response to this letter, Stringfellow's counsel replied to Riverside County that counsel had instructed J.B. Stringfellow, Jr. to terminate pumping of leachate on 15 September, as Stringfellow had loaned personal funds to the company in order to continue the pumping beyond his stipulated termination date for this maintenance of 21 November of the previous year. It has not been established how much of this pumping was actually undertaken after 21 November 1974. Much of the pumping was apparently linked to an effort to spray the leachate on return, thus enhancing evaporation. Franks noted (1989, p. 297) that the spray nozzles plugged rapidly and the resulting spray pattern spread only two or three feet (60 or 90 cm) into the air from the standpipes.

On 21 May the Regional Board made an individual agreement with J.B. Stringfellow, Jr. to pay him the sum of \$13,000 for purchase and maintenance of equipment to be used for the following site maintenance efforts:

1) Removal of trash and silt;
2) Cleaning of the ports of the recycling system;
3) Removal of weeds from the flood control channels;
4) Recycling waste materials, and;
5) Repair of ponds.

Early 1977 brought another round of title transfer within and outside of the Stringfellow holdings (see Table 4). The earliest (31 Jan 1977) placed the site or some of the adjacent ground in the hands of Paul and Lucille Hubbs, now the present operators of the rock quarry and persons who have contracted in the meantime for specific site remedial actions.

State Board Moves Ahead in the Vacuum Created by County Retraction

Moving responsibly ahead, while searching for alternatives for abatement management and funding, the Regional Board assessed its options. These were stipulated by Section 13005 of the California Administrative Code, as enacted with the Porter-Cologne Act of 1969:

(a) "Upon determining that a condition of pollution or nuisance exists which has resulted from a nonoperating industrial or business location within its region, a regional board may cause notice of such condition to be posted upon the property in question. The notice shall state that such condition constitutes either a condition of pollution or nuisance which must be abated by correction of such condition, otherwise it will be corrected by the city, county, other public agency or regional board at the property owner's expense...."

(e) After final action is taken by the board on the disposition of any protests or objections, or in case no protests or objections are received, the board shall request the city, county, or other public agency in which the conditions of pollution or the nuisance exists to abate it. In the event that such city, county, or other public agency does not abate such condition within a reasonable time the board shall cause the condition to be abated. It may proceed by force account contract or other agreement or any other method deemed most expedient by the board, and shall apply to the state board for the necessary funds.

The State Board was asked for assistance in estimating the required costs. Franks issued his *Cleanup and Abatement and Closure of the Stringfellow Class I Disposal Site on 17 March 1975*. These were the major cost elements:

1) Remove 1,000,000 gal. liquid wastes (\$0.05- \$0.07/gal; (\$0.013-\$0.018/l; \$5000-\$7000);
2) Install 8-in. recovery well to 100 ft depth (30 m), 1400 ft (427 m) downgradient (\$8000);
3) Level and fill 15 ac. (6 ha) of old ponds (\$0.60/cy; \$0.78/m3; \$30,000), and;
4) Install clay soil interim cover (\$2.00/cy; \$2.60/m3; \$30,000).

Internal review of the cost itemization placed a total cost at $600,000 as an upper limit.

Loosing no time, the Regional Board issued, on 12 June 1975, a *Request for Proposal, Cleanup and Abatement of Stringfellow Class I Dump Site.* For some years the State Board had accumulated limited funds in a Cleanup and Abatement Fund, monies that had apparently never been requested or allocated. The decision was to utilize these funds while action could be taken to secure emergency legislation if the County persisted in not meeting its responsibilities in taking the lead in abatement. Since the Regional Board had never disallowed Stringfellow from continuing his operations with accommodation of the new Discharge Requirements, the RFP was written to provide either for safely opening or closing the site. Copies of the RFP were issued to consulting engineering firms. Proposals were received in July and J.M. Montgomery Engineers, of Pasadena, was selected to undertake the study.

After prosecuting the contract to initiate closure studies, Anderson remained in contact with the site, visiting it again on 23 September, when he met with Stringfellow, and photographing the lower portion of the site, a practice that he was to maintain through the long ordeal. Apparently still in contact with the County, his log for 8 October notes that the County Counsel had again advised the Supervisors "not to get involved" in the abatement. On that same day Stringfellow had noted that the lower soil dike was saturating and might break out soon. Rains followed on 11 October and Anderson visited the site on the 14th, to note seepage at the south corner of the side and at the east side of the "last" (probably lowest, or most down-canyon) earth barrier.

On 5 November the State Board held a workshop dealing with Order 75-233, apparently to consider how the abatement issues forced on the Regional Board were carried out to date. The Regional Board, again on 13 November 1975, requested Stringfellow to file a Report of Closure. The title of this request was innovative for its time, and was later utilized as a feature of the USEPA regulations written to promulgate the Federal Resource Conservation and Recovery Act (RCRA) of 1976, but generally introduced after 1980, in order to deal with hazardous waste-handling facilities that wished to cease operations rather than to comply with the new permitting actions.

Alvin Franks was appointed project director for State Board involvement, on 21 September, and another abatement planning meeting was held on the following day.

Franks became the State Board's in-house geological consultant for site and waste characterization. Observing (1988, pp. 244-268, and 569) the exhumed conditions for a portion of the original barrier "dam" he estimated 1-2 gpm (3.8-7.6 lpm) of leachate "coming out of the east abutment" (not actually the abutment contact with the "cemented conglomerate") and felt that this leakage was likely along blast-induced fractures stemming from Stringfellow's excavation of the minimal keyway into which he poured the concrete. Some of the leachate, noted Franks, could have been near-surface ground water being liberated from the surrounding hillslopes, through the "weathered and fractured portion of the upper part of the granitics ...."

Sanford Werner, the sole DWR engineering geologist yet remaining assigned to the case, provided a memo of 26 January 1977 in which he put forth the premise that leachate was also leaving the site via rock fractures at a greater depth than that proposed by Franks. Werner wanted determination of inflow gradient of water toward the site, presumably from the hillsides along both sides and at the north.

J.M. Montgomery Engineers Conducts the Closure Study

Regional Board Resolution No. 75-233, of 12 September 1975 dealt with an agreement with Montgomery Engineers for "engineering studies for closure and maintenance of site to prevent pollution or nuisance." Funds were to come from the State Board and ensuing costs were to be charged to the Stringfellow Quarry Company for recovery as a lien on its property.

Procurement of the services of Montgomery Engineers proceeded slowly, possibly because of the fact that this was a first contract of its kind and time for the State Board. The State Board passed its version of the Regional Board resolution on 9 April 1976 and Montgomery Engineers presented its scope of work and budget on 12 May. In the meantime the Regional Board contracted for carriers to pump and remove the bulk of liquid wastes for discharge at the LACSD ocean outfall during April 1976. Twenty truck loads were discharged at the Chino Basin Municipal Water District White's Point ocean outfall and an undetermined amount was taken to he BKK site in West Covina. Hauling was accomplished by Chancellor & Ogden Truck Co.

Authority to enter into the contract with Montgomery Engineers came to the Board on 16 May. A planning meeting was held on that same day and the firm was placed under contract on 1 July 1976.

Montgomery Engineers completed its *Engineering Study of Stringfellow Class I Disposal Site* in January 1977. The study concluded:

1) Stringfellow's original Variance-required concrete barrier "dam" was probably 120 ft (36.6 m) long, 24 in (61 cm) wide at base, 14 in (36 cm) wide at crest, and 9 ft (3.1 m) high;

2) About 300,000 gal (1,125,000 l) of liquid wastes remained on site;

3) About 333,500 cubic yards (254,960 m3) of contaminated soil required treatment, and;

4) Groundwater recovery of about 2,600,000 gpy (9,841,000 lpy) of contaminated ground water, with subsequent evaporation on site or haulage from the site.

the County Board of Supervisors wrote the Senator concurring with the content of his SB 1131. On 25 August, the Presley Bill ($296,000) passed in the Assembly and on 30 August cleared the Senate and was signed by Governor Regan on 13 September 1977. $370,000 was allowed for the abatement and to prevent recurrence of threatening conditions.

The Regional Board held a public hearing on 12 January 1977 to openly discuss and choose the appropriate cleanup remedy among the alternatives developed by JMM.

Regional Board Efforts to Raise the Funds for Abatement

With refusal by the County to manage the abatement and realizing that recovery of abatement costs was appropriate and that legal means would be required, the State Board, acting on behalf of the Regional Board, adopted its Resolution No. 75-105 on 20 November 1975, requesting the California Attorney General (AG) to take appropriate actions and to transfer the matter of cost recovery to the AG. That transfer was completed on 10 December. The lien against the Stringfellow property was filed on 1 February 1976 for $15,205.44, equivalent to Anderson's computed cost of response actions, to the State.

Efforts to remove the ponded acidic wastes were initiated on 1 December, anticipating the forthcoming Springtime rains and to provide a temporary means of controlling seepage. Franks returned a review of the 14 January 1973 NBS report to Stringfellow, from the standpoint of closure, with the determination that the document was inadequate for that purpose. On 6 January 1976, Anderson reported to the Board that the calculated 2,500,000 gallons(9,462,500 l) of waste liquids could be removed at a cost of $50,000.

Bad news regarding legal solutions arrived from Attorney D.R. Goldman of the AG's office on 1 March, giving the appraisal that there was no basis for suing officers of Stringfellow Quarry Company, no basis for suing the County of Riverside, and only a poor basis for lodging suits against Stringfellow's insurers. The only "viable solution" opined Goldman was to seek legislative funding.

Regional Board X.O. Anderson determined that his first request for funding should be at $370,000 and requested such from SWRCB, this was sometime in late January or early February 1977. On 14 February, Anderson wrote to County Supervisor Donald Schroeder noting that he could raise but $60,000 in Cleanup and Abatement funds (presumably that was all that could be released by SWRCB) and he again requested the County to assist in the funding.

Sometime between February and July, State Senator Robert Presley was introduced to the case. Presumably Anderson, of the Regional Board, had contacted him for support in seeking passage of special legislation to fund abatement. On 26 July,

State Department of Health Services Becomes Involved (March 1977)

The Department of Health Services (DHS) charged with dealing with "contamination" completed its own draft study in March 1977, *Stringfellow Industrial Waste Disposal Site; A Technical Assessment of Environmental Impact.*

For the first time, a suite of impoundment liquid contaminants emerged; Pb, Cr, Fe, Mn, Cu, Ni Zn, Cd, along with sulfates, nitrates, chlorides, and fluorides; all could have been predicted from the general nature of wastes that Stringfellow had accepted. Most of the inorganic content was made up of metals solubilized in the range of 100-500 ppm and pockets of pesticides (DDT and degradation products) in the NW corner of the site. Highly-concentrated leachate was found as far as 900 ft (300 m) downgradient, along the natural Pyrite Canyon flood channel ("wash") and lower concentrations were believed to have been transported further toward the community of Glen Avon.

DHS made the following recommendations (without apparent resort to the feasibility-selection process employed by JMM):

1) Closure employing a minimum of a) pH barrier between the accumulation of acids and a clay-soil cover; alternatively, raise the pH from 2-3 to the range of 6-7;

2) Conduct a surface cleanup to include contaminated soil in Pyrite Channel at least as far south as the beginning of Hubb's quarry property;

3) A cleanup of surficial contaminated soil to a radial distance of at least 75 ft (25 m) outward from the peripheral site road;

4) Investigate the extent to which containers may have been disposed at the site;

5) Delimit the extent and chemical character of contamination plumes (requiring deep, cored boreholes);

6) Determine the furthest southward extent of contamination from the site, and;

7) Remove (to depth of 6 in; 15 cm) contaminated soil that may exist in the area of flood ponding at Jurupa Rd. and Agate St.

10. PRE-CERCLA MAINTENANCE AND REMEDIATION (1978-1981)

Actions toward actively remediating the site began in early 1978. A draft agreement between the State Board and the State Department of Finance and the County of Riverside was completed on 24 February. The State Water Pollution Fund was to finance the abatement with $74,000 of its funds and would be reimbursed for $148,000 from the County, to replace a like disbursement.

Torrential Rains and Site Flooding (March 1978)

The "once-in-a-decade" storm and associated flooding began in January 1978 and continued, on and off, for three months. About 30 in (76 cm) of rain fell in this period, about 300 percent of normal. In March the former dumpsite exceeded its capacity to retain the combined pond waste and on-site precipitation. As can be appreciated from the account that follows, the State Board was obliged to undertake a planned release of about 850,000 (3,217,250 l) of liquid wastes and contaminated rainwater, in order to prevent a catastrophic failure and total release of all impounded wastes and contaminated rainwater.

On 1 March it had become apparent to James Anderson that the abandoned dumpsite was in danger of total destruction from flooding. Anderson kept the events recorded in his daily log at the Board. County Flood Control District employee Pierce had inspected the site and noted that thee was a three-foot (0.91 m) freeboard. On 2 March, Anderson contacted the County Road Department for assistance and Mr. A.E. Newcomb arrived on site a day later with equipment to build a new retention pond and a pump to begin displacing waste liquids to the new pond. On 4 March an additional pump was added (1700 hrs.) and one pond began overflowing at 2400 hrs. The fifth brought California National Guard troops to provide sandbagging at 0230 hours. Pumping was still underway at 2400 when the last retention pond within the site complex began to overflow. On the seventh, DWR geologist Werner and engineer J. Walker visited and predicted that the earthen retention dam would fail soon.

Anderson spent 7 April in efforts to gain assistance. His daily log entry records that he met with no less than 12 County officials, in their offices, and was again advised that the County would provide no more assistance. The County suggested contact with the State Office of Emergency Services (OES).

On 8 March, Anderson made an emergency telephone call to the Governor's office and was told that efforts to locate funding would be made. On that day Anderson learned the the only viable alternative for off-site shipment, BKK, West Covina, was closed due to the same ongoing rains. Anderson turned to the OES only to be told by its representative Vickery that the "County has primary responsibility." OES said that it would not pay for repair to the dams (berms) but this it might pay for hauling "as relief of public threat" and advised Anderson to contact DHS for declaration of "Public Threat and Emergency," to notify DWR's Division of Dam Safety for certification of the failing situation and to call the Governor's office. DHS (E. Margistan) promised to visit the site on the 9th.

On 9 March, Dam Safety personnel declined to assist, Los Angeles County FCDs offered a loan of funding, and the IT Corporation would provide tank trucks if a disposal site could be located. DWR Southern District provided lighting equipment for night-time operations, and both BKK and IT sent in tank trucks. In search for any willing expertise, Anderson contacted the Los Angeles District, Army Corps of Engineers, Dam Safety Branch and was told (Mr. Baumley) that "Dam Safety does not want to get involved." Anderson, continuing his telephone crusade, contacted Attorney Atwater at the Attorney General's Office (Sacramento) for advice in dealing with DWR. E. Orme, of the Federal Disaster Assistance Administration (predecessor to the Federal Emergency Management Administration; FEMA), called with word that the Board would be eligible for disaster funds and that Anderson could "proceed with corrective work.", and it representative A. Haun told him that "you've got authority to expend funds. We can reimburse under our program." Margistan, of DHS, assisted in having the BKK facility opened, Len Tinnen (BKK/Chancellor & Ogden Corp.) and Mike Donohue (IT Corp.) would send trucks on the 10th.

Before all of the assistance could arrive the major retention structure began to fail at 0200 hours on 10 March. Failure was underway at 0400 and trucks were pumping at 0500. County DPH (Boling) and Disaster Office (Horrigan) were notified by Anderson and hauling continued all day and into the night until 2130.

On 11 March the release of site wastes was stopped and pumping to tank trucks continued. On 13 March, Hubbs Construction Co. personnel arrived for repair of the major retention structure, actions which were complete on the 14th. Off-site hauling of waste liquids was completed on the 17th. Hauling was reinstated on 3 April and John ("Hoot") Gibson, long-time Stringfellow employee was installed as caretaker of the site.

Anderson was back in Sacramento on the 23rd, meeting with Senators Schroeder and Presley. On 29 March, County Supervisor Donald Schroeder requested Congressman Brown and Senator Cranston to author bills to provide assistance to the Regional Board in abatement. Congressman Brown contacted HQ, USEPA, in Washington and the files are devoid of a response from Senator Cranston.

In a surprise move, the County Office of Community Development authorized, on 1 June, $30,000 in support of the Board's abatement efforts. This support was not forthcoming, according to a letter of 6 June, by which time the County had withdrawn its support.

State Senator Presley held a public hearing of his Subcommittee, in Riverside, and the Board did as well, on 27 December, with Montgomery Engineers making a presentation of findings and recommendations to date.

Montgomery Engineers Conducts Site Work (1978)

By August, Montgomery Engineers were excavating the site and uncovered evidence of subsurface leakage west of the last-known westward extension of Stringfellow's original barrier dam, constructed in accordance with the Board's Discharge Requirements of 1956. Groundwater monitoring wells were planned at an all-party site meeting on 11 October 1978, for a total number of 10 to 14, to be allocated on the basis of the water quality measures collected from the first three or four installations.

The U.S.G.S. released D.M. Morton's *Geologic Map of Fontana 7-1/2' Quadrangle* in 1978, at 1:62,500 scale. For the first time, schists are shown projecting under the north half of the site, a fact not noted by previous workers. In March 1979 DHS completed its *Technical Assessment* of the site, as a final draft of the study released the year before and bearing essentially the same findings.

The County (DPH), in responding to an SDH Hazardous Waste Section Report of 3 April, issued a health advisory for five parcels of Stringfellow site land. On the 13th, OES provided $34,317 in State funds, raising the available funding to $131,704. On 25 May the Regional Board contracted with A&A Equipment & Construction Co., Riverside, to remove 4,500 (3440 m3) of contaminated soil piled near Pyrite Street and Agate Road; the soils were to be placed on the dumpsite. DHS was in the area sampling in June, presumably to verify removal of contaminated soil.

An additional 18 in (46 cm) of rainfall fell in the Fall of 1979 and the ponds again became filled, short of release, but with continued visible leachate seepage to the collection sump immediately south of the dumpsite.

Montgomery Engineers had continued site exploration work and by 6 November 1979 drilling and construction specifications for interceptor wells were complete. These were to serve as the first groundwater monitoring wells and were to be located 1800 ft (550 m) downstream of the site and 300 ft (91 m) apart, as spaced east to west. On the 15th a plan to inject sodium silicate gel grout across the 350-ft (107 m) width of the alluvium-filled and colluvium-covered valley was submitted to the State Board. This grouting program was to provide the underbase of a planned clay soil dam, to be constructed in an excavated trench across the southern tip of the site.

In November, 1979, DHS issued an update on waste characterization with its *The Stringfellow Industrial Waste Disposal Site; Follow-up Sampling and Analysis*, and Montgomery Engineers completed (Dec 1979) its *Report on Phase I Work for Closure of Stringfellow Class I Hazardous Waste Disposal Site.*

By 9 January 1980 Montgomery Engineers were advising the Regional Board that bedrock encountered in the drilling and grouting program was more highly weathered and of greater permeability than originally considered. The grout holes were completed on 5 February, just before a series of heavy rainstorms of 5-10 March 1980 brought 24 inches (61 cm) of rainfall to the site, all of which provoked subsurface leakage, observed in the grout holes. Montgomery recognized four critical problems at that time:

1) a need to immobilize heavy metals;

2) a need to eliminate leachate seepage by both the grout curtain and the "impermeable" clay-soil subsurface barrier, as proposed by Franks;

3) that contaminated ground water may well be downstream of the site, and;

4) a need to intercept or divert potential subsurface flows into the site.

On 5 July the Board received a letter from UCR Assistant Professor of Geology Donald J. Stierman, a geophysicist believing that bedrock joints below the site were capable of transporting contamination.

Return of Organized Opposition to the Dumpsite (1979-80)

In December 1979, Governor Jerry Brown requested political activist Tom Hayden to investigate the Stringfellow case. In that month, Concerned Neighbors in Action (CNA) was organized by the California Campaign for Economic Democracy (CED), also led by Mr. Hayden. Penny Newman became the local leader and Jeffrey Robinson, a CED organizer, went door-to-door with the first members of the CNA, to gain support and membership amongst local citizens.

CED rallied to demonstrate before a special meeting of RWQCB on 2 July to discuss the six possible "Abatement Alternatives" and held a public meeting of its own on 15 July to feature Tom Hayden, Ms. Newman, and Lois Gibbs, the well-known citizen-spokeswoman of Love Canal (Niagara County, New York).

Participation by the State Board

Engineering geologist Franks provided the Regional Board with his abatement recommendations on 4 January 1980:

1) Stringfellow's original "dam" had been excavated and found to be only 60 ft (18.3 m) long instead of the required 250 f (76.2 m).

2) There existed a buried stream channel, located to the west end of the original "dam," and such had been a leachate transport pathway since 1956.

3) Franks was recommending that the area of the original "dam" be excavated and stripped to bedrock, with subsequent grouting and installation of a clay-core subsuface barrier (designed to function at 1 X 10^{-8} cm/sec hydraulic conductivity; K) and to be bottom-faced on the upgradient side by a gravel leachate collection system containing a perforated collection pipe.

State Board Chairman Carla Bard (an appointee of Governor Jerry Brown) declared, in a letter of 7 January, to EPA/IX Regional Administrator Paul deFalco, that the State Board was preparing to "solve ... problems presented by the Stringfellow site." Ms. Bard noted that $2,500,000 to $3,000,000 would probably be required and that the State would probably be seeking the emergency funding in accordance with Section 311 (k) of the Clean Water Act [2] through his Regional office.

Continued Participation by Riverside County

The County continued to sample surface water in Pyrite Creek, down stream of the site, its latest sampling round during the Closure Study period being on 30 January 1980. On 3 March, DPH analyzed the "Main Pond" waters from two locations (1B and 2B), just before the 24-inch rainfall, to find the following;

	1B (ppm)	2B (ppm)
TDS:	11,280	52,400
Cr^{+3}:	30	563
Cr^{+6}:	0.00	0.00
Sulfate:	914	563

Following the termination of the Federal removal action, DPH was back on 5 June, 1980, to conduct soil sampling on the site, reporting concentrations of phalates (nil to 8,128 ppb) and VOCs and pesticides (to hundreds of ppb).

Th last official act of the County in divesting itself of the responsibility for abatement was the 21 November 1980 approval of County Legal Counsel for purchase of the Stringfellow site property by the State Controller, for the Board of Public Works, via payment of $878 in taxes gone delinquent since 1974.

[2] The Water Quality Improvement Act of 3 April 1970 (Clean Water Act of 1970, or Federal Water Pollution Control Act of 1970) resulted in required issuance, by USEPA, of the *National Multiagency Oil and Hazardous Materials Pollution Contingency Plan (NCP)*, a major revision of the first NCP, issued in 1968, and which set the procedure to respond to spills and for emergency funding for response actions, in accordance with its Section 311 (k), to be administered by the U.S. Coast Guard.

EPA Region IX Conducts Emergency Removal (Spring 1980)

Apparently Ms. Bard's letter of 7 January 1980 to Regional Administrator deFalco carried weight. On 3 March a Federal Fact Finding Group was formed, and met at Riverside, to determine if the Stringfellow site met Section 311 (k) criteria for funding under the Clean Water Act of 1977 (Reiter and Mandel, 1989), in accordance with the NCP. A recommendation was made that funds might be requested for use as a Phase I Response. The Regional Response Team (RRT), required to make the official consideration, was assembled at Long Beach on the 4th and voted to activate 3111(k) revolving funds to initiate emergency action. A Federal On-Scene Coordinator (OSC) was appointed of the Coast Guard representative and began initiating actions on 6 March. Through 4 August the EPA sponsored an $825,000 "removal action" of 10,000,000 gal (37,850,000 l) of liquid wastes by tank truck, reinforced surface retention features and repaired an on-site truck loading area. Removal began on 6 March. Since most of the liquids were now just contaminated rain water (total phenols at 950 ppb; analyzed at the time of screening for the USEPA-designated 129 Priority Pollutants), and the object was to effect an "Interim Remedial Measure" in temporarily closing the site, the hauling was made to the Santa Ana Regional Interceptor (SARI). 721 truck loads were discharged to the SARI line, and only 3 loads were discharged at the BKK Class I facility in West Covina. The OSC kept the usual daily log and the initial operation was carried out in the face of heavy rain falling from 6-17 March, during which time the liquid levels in three of the ponds was within 6 in (15) cm of overflowing. Surface seepage appearing in the vicinity of Stringfellow's old seepage pit just downgradient of the site was about 2.5 gpm (9.7 lpm). By employing 24-hour operations 3,500,000 gal (13,563,000 l) of hazardous waste liquids were removed by 10 March.

On 10 March the Regional Board formally requested additional 311(k) assistance from EPA/IX on the advent of expenditure of the $370,000 in state emergency funding, which had been consumed in the first of the removal actions of March. The RRT met again, in Riverside, on 12 March, to plan termination of 311 (k) support on 18 March.

The State Board again resumed responsibility for site cleanup on 18 March and on 20 March. Montgomery Engineers requested the Regional Board to approve funding for Woodward-Clyde Consultants to conduct additional site exploratory drilling, a seismic refraction survey, and to locate interceptor wells downgradient of the site.

The Regional Board secured a visit from the Federal Disaster Assistance Administration (Dept. HUD) on 24 March and a claim was completed on the spot for $99,654 in reimbursal to the Regional Board. DHS issued a 30 March draft *Development and Initial Screening of Alternatives* report.

The Board was back at EPA/IX on 22 April attempting to secure site closure funds and on that day the Congress held a hearing on the site in Washington, D.C. In April, Montgomery Engineers filed its *Phase 1 Work for Closure of Stringfellow Class I Hazardous Waste Disposal Site, Addendum No. 1* in which the following was reported:

1) The original barrier "dam" was found not to be keyed into solid granitic bedrock;

2) 107 sodium-silicate gel injection holes had been drilled on 3-ft (91 cm) centers to an average depth of 20 ft (6.1 m) ;

3) "there is subsurface flow which surfaces within the contaminated disposal area....", and;

4) The refraction profile was recommended, to extend exploration coverage from the east end of the new clay-core subsurface barrier , to the east side of the valley alluvium.

FEMA notified the Regional Board, on 26 June, that $112,589 had been approved for reimbursement of emergency removal costs, though the allocation required Congressional approval of a FEMA supplemental request.

The RWQCB held a special meeting on 2 July to discuss the six possible "Abatement Alternatives." Mr. Anderson presented the concepts of Best Available Technology (BAT) and Best Practicable Technology (BPT), recommending a complete removal of waste from the site, at an estimated cost of $14,000,000. CED organized a small demonstration prior to the meeting. On the same day the Legislature allocated $297,556, in accordance with the Presley Bill.

Governor Brown's Toxic Substances Coordinating Council, on 8 July, requested a summary status report from the Board. On the 14th, the DHS brought in a team of physicians, to Glen Avon, for medical screening of citizens; 30 appointments had been made by residents, of which only 22 were honored.

After completion of EPA's Spring 1980 Emergency Removal, Region IX was induced by the State Board to consider funding additional site remediation efforts. On 8 August 1980, in concurrence with the 20 August 1971 *National Multiagency Oil and Hazardous Materials Pollution Contingency Plan (NCP; Clean Water Act amendments of 3 Apr 1970)*, RRT allocated $608,000 for FY 1980 and appointed Ecology and Environment, Inc., as its on-site representative.

Rains were continuing when TAT arrived on site and Anderson submitted the RWRCB Abatement Plan to implement Phase II elements such as continued waste hauling to lower the pond levels, to complete the gel grouting program initiated by Montgomery Engineers and to complete construction of the clay-soil subsurface barrier and leachate collection drain. As a part of its responsibility, the Regional Board contracted for a seismic refraction survey which was undertaken by Woodward-Clyde Consultants (WCC, Orange, CA) as the basis for determining depth of required excavation. WCC also undertook logging and installation of extraction wells and monitoring wells, under the direction of Montgomery Engineers.

Coast Guardsmen set up an on-site command post, held an explanatory meeting with CNA and began contracting with BKK for off-site haulage on a 5-1/2 day-a-week basis, to be completed on 18 September. Costs of hauling and disposal were $0.099/gal ($0.026/1) for weekdays and $0.11/gal ($0.029/1) for weekends. On 13 August the budget for EPA Phase I/II participation was raised from $500,000 to $1,000,000.

By the end of August, Montgomery Engineers had begun work on the subsurface barrier design, working with WCC's preliminary results, indicating that depth to sound bedrock was deeper than anticipated. TAT reported the following schedule for abatement:

Phase I:	1975-1981	Establish security of site; prevent washouts of waste
Phase II:	1976-1981	Engineering studies and evaluation of potential abatement alternatives
Phase III:	1981-1982	Construction of initial abatement ($2,722,333 anticipated)
Phase IV:	1982-1984	Final construction, operation an maintenance (extraction wells and waste water treatment plant)

In October 1980, Montgomery Engineers discovered evidence of a lens of waste-saturated alluvium at 40-60 ft (12-18 m) raising the possibility that site wastes had migrated past the partial concrete barrier constructed by Stringfellow. In the same month the State Board completed a contract negotiation with the Regional Board for $4,035,000 of SWRCB funds, which were not released until July 1981. Also in October, SWRCB released its *Stringfellow Hazardous Waste Disposal Site; Interagency Status Report*, in which the DHS assessment of on-site contamination was adopted and the Board concluded that:

1) Significant differences in the nature of contamination listed in the Sep 1979 Montgomery Engineers Report and the Mar 1980 EPA report were due to an incomplete waste characterization;

2) The major portion of on-site contamination now lay in the pond-bottom sludges now that 10,000,000 gal (37, 850,000 l) of contaminated water had been hauled from the site;

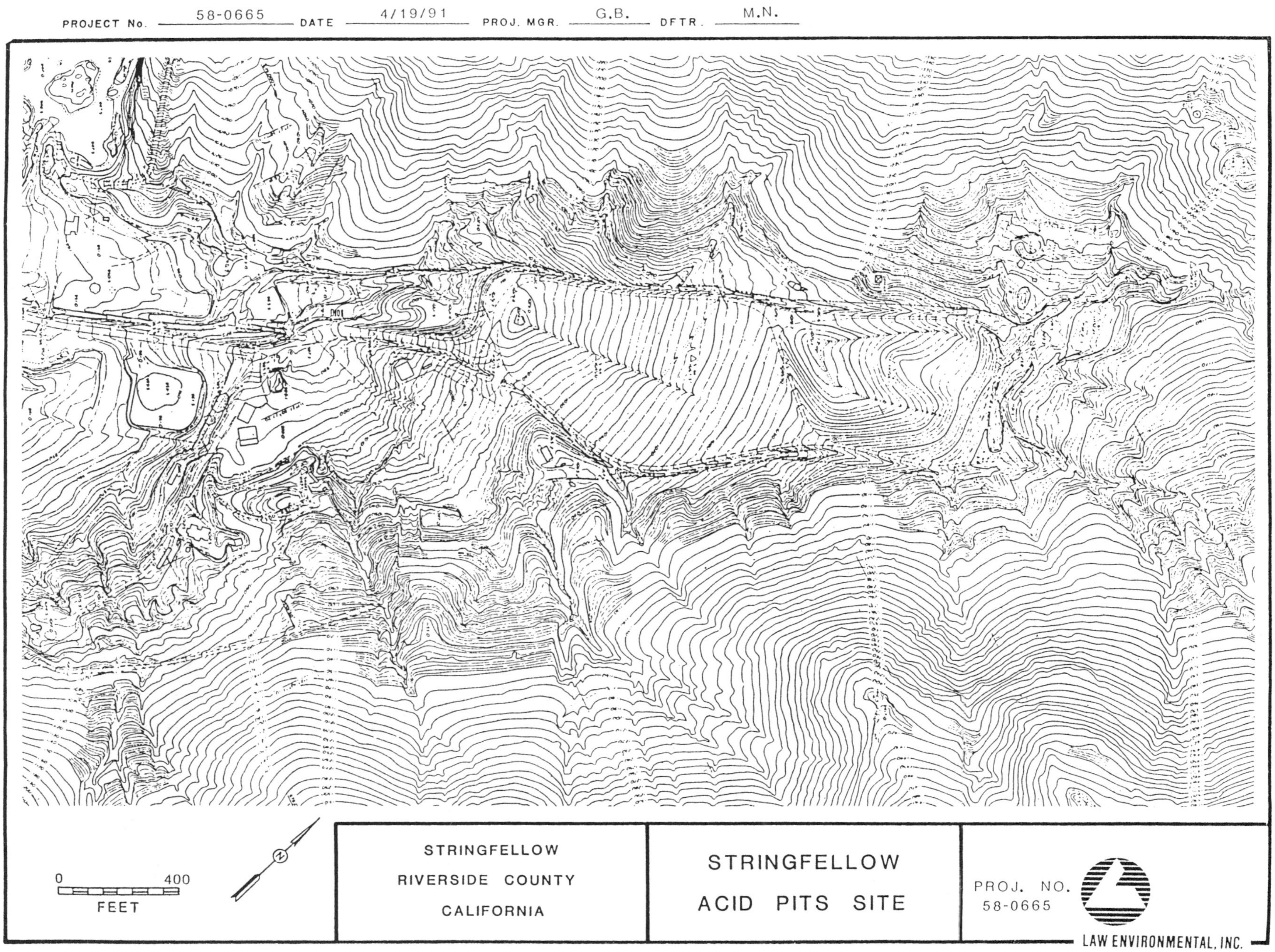

Figure 3. Topographic map of the abated former Stringfellow Acid Pits, showing the interim abatement program began in 1980, as it appears at the time of this paper (Photogrammetric topographic base from Riverside County Flood Control District).

Figure 4. Present (1991) condition of the former Shannahan, then Stringfellow, now Hubbs quarry in the Woodson Peak Granodiorite, about 1/4-mile (400 m) southwest of the former Stringfellow Acid Pits (Photo by A.W. Hatheway).

Figure 5. Character of the interim abatement program (started in 1980) for the northern half of the Stringfellow Acid Pits, as it appeared in 1991. View is to the NE (Photo by A.W. Hatheway).

3) Groundwater contamination had moved some 1500 ft (457 m) downgradient of the site;

4) There was no indication that contamination had migrated further than monitoring well 1Q1;

5) The proposed extraction well system should prevent further downgradient migration of contamination;

6) Sediment of Pyrite Creek was contaminated with Cr and Cd in excess of the Federal Interim Primary Drinking Water Standards, and were the result of releases from the site during the 1969 floods, and;

7) Contamination released during the 1978 floods was to be found immediately downgradient of the site.

Fourteen monitoring wells were in place by 14 November and by 5 December the three interception wells had been installed, two located 1800 ft (548 m) and one at 800 ft (244 m) downgradient of the new subsurface barrier. Two 8000 gal (31,000 l) storage tanks were in place and daily loads of recovered ground water were removed to a Class I disposal site.

Remediation planning went forward and on 12 December the Regional Board met and voted to employ pH neutralization of contaminated soil, classifying such as an "Interim Remedial Measures" using terminology that was to be adopted in the forthcoming CERCLA NCP. The Board approved the plan at its December meeting (Resolution 80-219), in the following plan "Interim Abatement Program"

1) removal of 1000 cy (765 m^3) of waste;

2) neutralization of about 75 percent of all contaminated soils;

3) install concrete surface-drainage channels;

4) grade the site for runoff control;

5) install interim cover made up of 12 in (30 cm) of kiln dust and overlying 2 ft (60 cm) clay soil;

6)install 14 groundwater monitoring wells upstream of the new subsurface barrier and 3 intercepter wells downstream of the barrier;

7) gel-injection grouting beneath the new subsurface barrier;

8) construct a pretreatment plant for extracted and intercepted contaminated ground water, and;

9) construct 1.0 mile (1.6 km) pipeline to Santa Ana Regional Interceptor.

E&E, Inc., acting as the EPA TAT, completed on 29 December 1980 its *Cost Estimate for Alternatives for Land Burial*, in which 330,000 cy (252,285 m^3) were slated for removal by December 1981 and placement at a Class I landburial facility. Depending on the specific landfill, these costs ranged from $16,000,000 to $41,000,000. The Coast Guard received a 21 April 1981 authorization of $6,742,875 to conduct site remediation. 21 May brought $950,000 in RCRA funds to the State Board for payment of significant portions of the Interim Abatement Program.

Construction of the Interim Abatement Program was well underway in October 1981 with the grouting; 150 holes, 20-100 ft (6.1 - 30.4 m) deep, on 6-ft (2 m) centers, for an 800-ft (244 m) width, with secondary holes spaced at 3 ft (1.1 m). In two "highly fractured areas" tertiary grouting was undertaken.

State abatement activities were chronicled by a periodic *Abatement Report*, issued in numbered sequence. The new subsurface barrier was completed in November 1981 (Rept. no. 10), at a cross-valley length of 800 ft (244 m), 15 ft (4.6 m) width at its base, 8 ft (2.4 m) at its top, and 25-90 ft (7.6-27.4 m) to reach sound bedrock. Its upgradient face was faced with a 3-ft (1.1 m) French drain leachate collector system. Compaction specifications for the clay-soil barrier were 95 percent maximum dry density yield correlative hydraulic conductivity of 1 X 1—5 cm/sec. Site extraction wells and downstream interception wells were placed into operation.

SWRCB issued its report of evaluation of remedial alternatives on 11 December, as based on a similar Montgomery Engineers report of 24 November. The State Board added operational considerations and selected groundwater recovery and treatment, with on-site storage to accumulate off-site haul loads until an onsite wastewater treatment plant could be designed and built. On-site solidification and stabilization of contaminated soils encountered in excavation and grading utilized a 1:10 mix ratio of low-Ca) cement kiln dust . State Abatement Report no. 15 judged that about 75 percent of contaminated soils on the site had been encountered and treated. Late in 1981, the State Board granted an additional $4,035.000 to the Regional Board, in an effort to sustain site remediation until receipt of CERCLA funding.

Beginning in January 1982, the State again hauled contaminated water from the site, the surface of which had not yet been given the final cover and runoff measures. Ninety-four truck loads of this contaminated water, including recovered contaminated ground water, were taken to an unidentified Class I disposal facility (probably BKK, West Covina) in Los Angeles County, at a charge of $0.11/ gal ($ 0.03/ l) hauled and disposed. At the end of calendar year 1981 (period Jan 1975 through Dec 1981), the State had expended a total of something more than $2,000,000 in a composite of State Board and State General Fund supplements, CWA (Sec. 311

(k)) funds and augmenting RCRA grants from EPA/IX, and Federal disaster funds.

Montgomery Engineers reported on 28 January that neutralization of contaminated soil would be necessary only where pH was observed to be 6.0 and 7.5, in which case the interim cover to be shortly installed might forego the underlayer of kiln dust. Neutralization of exhumed and graded on-site soil was completed as of 1 February, to be followed at mid-month by completion of the leachate-control system (combination of the new subsurface barrier and the its upgradient French drain). Groundwater interception was terminated and now reinstituted until the Summer of 1984 (Anderson, 1989, p. 1250). As of 31 March, site contamination was observed to have migrated as far downgradient as 1800 ft (549 m).

11. INITIATION OF SUPERFUND CLEANUP ACTIONS (1980-1982)

The SUPERFUND Law (Comprehensive Environmental Response, Compensation and Liability Act; PL 96-510) was signed on 11 December 1980. Its provisions were to be carried on under the procedures outlined in a *National Contingency Plan* (NCP) modeled after the original 1968 oil spill NCP. The nation was disappointed over the progress of release of EPA funding as the 1980 appropriations did not materialize into allocations. Finally, after nearly seven months, on 5 June 1981, the Congress passed the *Supplemental Appropriations & Recission Act* (PL 97-129) appropriating $68,000,000 from the SUPERFUND to EPA. Even with this word in hand the Congress did not actually authorize expenditures against the fund until 14 August. EPA's Comptroller determined that the actual charge date for incurrence of site-related work was 8 June.

On 16 July 1982, nearly 18 months after passage of the SUPERFUND Act, EPA published the first CERCLA NCP in the Federal Register. The manner of application for Federal funding and for the entire site characterization and remediation sequence was finally established.

USEPA Support Ramping Up to CERCLA Funding (1980-1981)

The State of California continued to struggle with securing funding for ongoing Stringfellow site work in the face of the year-long debacle in release of CERCLA funds by the Congress. EPA Region IX was apparently solidly behind the State Board in its assistance and began to search for the means of continuing the abatement. In June 1981, $1,000,000 in RCRA funds were granted by EPA/IX toward the Montgomery Engineers design work, removal of 1000 cy (765 m^3) DDT wastes by Chemical Waste Management, Inc. (CWMI), to the Kettleman Hills (Kern Co.) Class I facility, and off-site haulage of additional contaminated surface water that had accumulated in site ponds.

Also in June 1981, a formal application for CERCLA support had been completed, although EPA had not yet provided application guidelines. On 2 July, RWQCB held a public hearing to review remediation alternatives. Bids were opened in July for remedial actions and such work was awarded in August to adjacent land owner Paul Hubbs for $1,974,946.

State Actions to Receive CERCLA Funding (1981-1982)

On 14 August 1981 the dealings in the Congress were complete and Presidential Executive Order No. 12315 delegated CERCLA authority to the USEPA. With this action, the EPA Administrator received authority to expend CERCLA funds authorized by the Congress eight months earlier. On 28 August Governor Jerry Brown designated DHS as the CERCLA lead agency for the State. EPA/IX denied the State's "2nd. Cooperative Agreement" draft for release of CERCLA funds to Stringfellow site remediation as the State must then nominate the site as its top-priority site on the National Priority List (NPL) of uncontrolled hazardous waste sites (UHWSS) authorized to receive CERCLA funding3. By 1 October EPA had obligated $40,143,028 in CERCLA funds, but these contracts, grants and interagency agreements placed 84 % of the monies in the hands of EPA Regional Administrators, yet only New Hampshire had received funds ($2,305,600) for a State Lead site remediation.

Heading the CERCLA requirement that State Lead for remediation of UHWSs would require a ten-percent cost contribution, the California Legislature passed the State SUPERFUND Law in December 1981, to ready the funding when allocation-approval was received from EPA/IX.

On 9 April 1982, SWRCB passed its Resolution no. 82-137 choosing an interim remedy of containment ("encapsulation" in terminology of the day) and groundwater recovery and treatment. DHS was requested to proceed with its application to EPA/IX for SUPERFUND assistance.

By 9 June, DHS has communicated with EPA/IX with the opinion that the RWQCB Interim Abatement Plan was the most cost effective solution for site mitigation, thus clearing the way for release of SUPERFUND monies. On 2 July, a cooperative agreement was signed, between the State (DHS) and Region IX. A 19 July HQUSEPA memo (R.H. Weyer to Wm. Hedeman) noted that the State had spent $3,499,210 and had completed the following remedial actions:

[3] The first 100 NPL sites were reserved to nomination by Governors of the 50 States without recourse to the point-score tabulation and minimal 25.0- (later 28.5) point ranking for submittal to the list. Subsequent NPL sites have required Hazard Ranking System scores 28.5 points. The first NPL contained 115 sites and was issued in October 1981.

- constructed a barrier dam integrated with a silica-gel grout curtain;
- leachate extraction wells placed upgradient of the new barrier dam;
- conducted a Feasibility Study for treatment/disposal of recovered leachate;
- neutralized certain acidic sludges;
- constructed a clay-soil interim cover, and;
- implemented runoff-abatement measures.

Operable Unit no. 1 was declared to be the remedial support measures required to manage the contaminated water being held within the source area. A budget of $2,180,640 was being allocated to complete this work, of which $709,000 was to be used by the Army Corps of Engineers to construct an on-site wastewater treatment plant (began operation in Dec 1985). At the same time, Operable Unit No. 2 was designated as management of off-site groundwater contamination, funded at $456,200. The State was to make its required ten-percent funding contribution as a credit from its previous pre-CERCLA contribution (1 Jan 1978 to 11 December 1980, the date of passage of the Federal law).

CERCLA remediation of the Stringfellow site began on 27 July 1982 as EPA Administrator Anne M. Burford-Gorsuch announced a Federal grant of $6,100,000 for cleanup action, to be forthcoming as a cooperative agreement would be signed within weeks. The grant was to be made up of two equal halves, one for reimbursal of previous State expenditures and the other for construction of additional containment and treatment facilities at the site.

Discovery of Major Unknown Site Features (1980-1981)

Two major, unknown site features were uncovered during the 1980-1981 site remediation efforts.

At the completion of a Coast Guard removal action aimed at transporting impounded liquids, JMM issued (April 1980) its *Phase I Work for Closure of Stringfellow Class I Hazardous Waste Disposal Site, Addendum No. 1*, in which 107 gel-injection grout holes would be drilled on 3-ft (1 m) centers just above the presumed location of Stringfellow's original subsurface barrier. It was at this time that Stringfellow's original barrier was found to be only 60 ft (18 m) in length, rather than the specified 240 ft (73 m). Where present, the barrier was a reinforced concrete wall 8 ft (2.4 m) high and without the keying into "solid granitic bedrock", as claimed then, and later in testimony, by Stringfellow, Jr.

Additionally, JMM was to supervise construction of a new remedial barrier trench, to lie above the gel grout curtain, with construction of a new subsurface barrier with upstream drainage system from which to recover leachate from the body of alluvium and weathered crystalline rock below the source area (original ponds). Woodward-Clyde Consultants were engaged to undertake exploratory drilling, geologic mapping and seismic reflection profiling as an assist to designing the new leachate collection trench.

On 5 November 1981 the Riverside Enterprise reported that "Workmen find giant underground leak in Stringfellow Acid Pits. In the course of excavation for the new subsurface barrier wall, subcontractors managed by Montgomery Engineers had unearthed an "ancient streambed" (Riverside Enterprise) which has since been characterized as a buried alluvial channel resting on bedrock and overlain by the ancient and now cemented debris flow that had been recommended by Fox for consideration as the east abutment of the impermeable barrier that Stringfellow never completed. It is the opinion of the authors that this channelway was of a lesser influence in release of site contaminants than was the 180-ft (55-m) gap left by Stringfellow in the RWPCB's Discharge Requirement for a subsurface barrier.

12. ACTIONS BY EPA REGION IX (1982)

USEPA Region IX, on 16 February 1982, compiled a list of 200 generators of hazardous waste for which evidence led to their disposal of such wastes at the Stringfellow site. On 24 August 1982, the Region informed more than 200 "Potentially Responsible Parties" (PRPs) of the fact that evidence at hand substantiated that they were, in some way, responsible for the condition of environmental threat at the Stringfellow site. In September, EPA/IX referred the case to the US Department of Justice for potential litigation under CERCLA Sections 104 and 107. HQUSEPA, on 5 November, promulgated *Ownership Status of Candidate Sites for the National Priority List.*

By late 1982 EPA had taken actions in preparation of legal suites against Potential Responsible Parties (PRPs; organizations that had generated or transported wastes to the site). By November, the EPA National Enforcement Investigations Center, Denver, Colorado, had generated a computer database of 11,000 shipments of more than 23,000,000 gallons (86, 9000,000 l) of liquid waste to the Acid Pit.

PRP actions culminated in creation of the Stringfellow Steering Committee on 15 December 1982, incorporating legal representation from the firms charged with complicity in discharge of wastes at the Acid Pits. The Steering Committee immediately (16 December) began negotiations with the Federal Government. On 21 April, 1983 the US Department of Justice filed suit against 88 generator responsible parties (RPs), and on the following day a joint US Congressional and State Legislature hearing was held in Glen Avon.

13. CALIFORNIA DEPARTMENT OF HEALTH SERVICS (1982-PRESENT)

DHS had undertaken a significant activity on site by mid 1982. Its first participation had begun in March, 1977 in its role of characterizing the site waste and its threat to the population of Glen Avon. With passage of CERCLA and Governor Brown's designation (28 Aug 1981) of DHS as the CERCLA action agency for California, it was clear that SUPERFUND monies would be granted to DHS. On 1 September, while EPA Region IX was referring the case to USDOJ for litigation, SWRCB Report no. 18 noted completion of three on-site recovery wells and the fact that DHS was to take charge of remediation on 1 October, 1982. The planned wastewater treatment plant was to be constructed under the auspices of DHS.

In January 1983 DHS began "extensive" monitoring of selected public and private ground water sells in the Glen Avon area. TDS content of well water south of US Highway 60 were in the range of 426-1630 ppm. DHS sampling was extended to surface water and soils in Glen Avon on 26 May. Hastening to move toward remediation, DHS issued a request for proposals (RFP) for a combined Remedial Investigation and Feasibility (RI/FS) under the new CERCLA National Contingency Plan, and awarded the project to JRB Associates, a Washington, DC-area consulting firm (and subsidiary of Science Applications International Corp., La Jolla). On 10 June, DHS amended its application to EPA for a Cooperative Agreement, raising the requested funding to $12,000,000, to include $7,700,000 for the planned RI/FS and indicated remedial actions to be selected as a result of FS deliberations. Of the requested funding, EPA/IX, on 28 July, provided $2,800,000 of a total award of $9,780,000 or 81.5 percent of the amount requested.

JRB Associates delivered its RI/FS report on 18 November 1984 and the Acid Pits entered its current phase of remediation, a subject that is beyond the scope of this paper.

14. ACTIONS BY RESPONSIBLE PARTIES (1983-PRESENT)

Steering Committee activities led to legal opposition to the Federal Motion for Summary Judgement and the PRPs filed a Memo of Points and Authorities on 18 June 1983. In this document, the PRPs began their continued remonstration that the State of California has been responsible (by taking a "uniquely dominant and controlling roll...." beginning in 1975, for conditions of environmental degradation at the Acid Pits and that the State had been negligent in those actions. The Steering Committee charged that the very actions of the California Legislature in voting four statutes for funding of abatement actions, between 1977 and 1982 had imposed an obligation on the State for cleanup.

15. SUMMARY

As of this writing (1992) the Acid Pits have not yet received their final remediation and entered into the post-closure period. In 1988 the Federal Government completed an initial settlement with 183 of the PRPs, for $5,600,000 in interim remedial measures (IRMs) for the actions to contain and recover contaminated ground water in the lower canyon area, and to construct surface-water diversion channels. Remedial actions have been hampered by suits and countersuits regarding ultimate responsibility for costs incurred to date and for responsibility and costs for ultimate closure and out-year monitoring. An additional suit is underway by Glen Avon residents, against government and PRPs alike, the Stringfellow interests (and insurers) having reached a settlement with the Glen Avon residents in the Fall of 1991.

It is the opinion of the authors that no better site than an untouched Pyrite Canyon can be identified today, in Southern California, for a high-technology hazardous waste management facility designed, constructed and operated within the context of current California State and USEPA regulations.

ACKNOWLEDGEMENTS

Work on this paper began in 1988 when the senior author was assigned as expert consultant to the Stringfellow Acid Pits case by the California Attorney General, for the areas of site characterization, regulatory and permitting history, and remedial engineering. Later, the junior author was brought into the case as an historic witness in recognition of a long career of involvement with engineering geology, hydrogeology, how these bodies of technical knowledge have been applied, and commentary on the inner workings of the State Division/Department of Water Resources and the Water Pollution Control Boards and its successor bodies. We were encouraged by Deputy Attorney General Donald A. Robinson, of the Los Angeles office to bring together this history of the Acid Pits so that all interested parties would have a single reference to materials otherwise found in literally hundreds of unpublished documents.

During our service and writing we had the pleasure of working with William L. Carter, Esq. and Brian Hembacher, Esq. of Mr. Robinson's staff and with two supporting legal firms. First, at Gray, Cary, Ames & Frye, of San Diego, we were assisted in many ways by Browning E. Marean, Esq. and Roger Britt, Senior Science and Engineering Analyst. Later, with Breidenbach, Swainston, Crispo & Way, of Los Angeles, our efforts were extended by case manager Howard L. Halm, Esq., Dulany Harms, Esq. (also a graduate civil engineer), and by paralegal assistant and documents manager Susan Andrews. Dr. Alvin L. Franks, former SWRCB engineering geologist (1968-1980) and case manager for the State Board for the period December 1972 - October 1980, reviewed a nearly-complete earlier version of the manuscript and made significantly helpful remarks and comments.

In the Spring of 1991 the senior author completed geologic mapping of the Pyrite Creek watershed, from crest to below the former Acid Pits. This effort was conducted at 1:2400 scale on a composite site-area topographic map photogrammetrically-compiled by the Riverside County Flood Control District (with the smaller San Bernardino County area enlarged by us from the USGS Fontana topographic quadrangle), and quality reviewed by the junior author. The following persons assisted in the field mapping; Richard J. Proctor, Kathleen Considine (DHS), Robin Kim (Law/Crandall, Inc.), Richard Bailey (DHS), Dr. Fletcher G. Driscoll (Geraghty & Miller), and Dr. S. Kerry Grant, Department of Geology & Geophysics, University of Missouri-Rolla, who completed petrographic evaluations of all rock types (as reviewed by the senior author).

REFERENCES

Bookman, M., 1984, Deposition, U.S.A., et al, Plaintiffs, vs. J.B. Stringfellow, Jr., et al, Defendants: Held at Los Angeles, CA, on 12 Jan, recorded by Traci A. Dokich, Inc., 145 p.

Brown, Glenn A., 1989a, Deposition, U.S.A., et al, Plaintiffs, vs. J.B. Stringfellow, Jr., et al, Defendants: Held at Los Angeles, CA, on 17 Feb, recorded by Dokich, Murray & Assoc., p. 1-160, plus exhibits.

_____, 1989b, Deposition, U.S.A., et al, Plaintiffs, vs. J. B. Stringfellow, Jr., et al, Defendants: Held at Los Angeles, CA, on 15 Feb, recorded by Dokich, Murray & Assoc., p. 161-318, plus exhibits.

_____, 1989c, Deposition, U.S.A., et al, Plaintiffs, vs. J. B. Stringfellow, Jr., et al, Defendants: Held at Los Angeles, CA, on 17 Feb, recorded by Dokich, Murray & Assoc., p. 319-446.

_____, 1989d, Deposition, U.S.A., et al, Plaintiffs, vs. J. B. Stringfellow, Jr., et al, Defendants: Held at Los Angeles, CA, on 21 Feb, recorded by Dokich, Murray & Assoc., p. 447-550, plus exhibits.

_____, 1989d, Deposition, U.S.A., et al, Plaintiffs, vs. J. B. Stringfellow, Jr., et al, Defendants: Held at Los Angeles, CA, on 21 Feb, recorded by Dokich, Murray & Assoc., p. 551-775, plus exhibits.

California Auditor General, 1983, Review of Selected Contracts for Cleanup of the Stringfellow Toxic Waste Disposal Site: Sacramento, Rept. P-244 October, 39 p.

California Department of Health Services, 1979a, The Stringfellow Industrial Waste Disposal Site; A Technical Assessment of Environmental Impact: Hazardous Materials Lab., Berkeley, by H.K. Hatayama, B.P. Simross, and R.D. Stephens, March, 104 p.

_____, 1979b, The Stringfellow Industrial Waste Disposal Site; Follow-up Sampling and Analysis: Hazardous Materials Lab., Berkeley, by H.K. Hatayama, B.P. Simross, and R.D. Stephens, Nov, 15 p.

California Division of Water Resources, 1934, Geology and Ground Water Storage Capacity of Valley Fill; South Coastal Basin Investigation: Sacramento, Bull. 45, 279 p., with 6 map plates (authored by R.C. Eckis).

_____, 1953, Investigation of Safe Dump Sites in South Coastal Basin Within Los Angeles Region: Los Angeles, Report to Los Angeles Regional Water Pollution Control Board, Dec., 40 p., plus 2 app. (Premier report of a series on surveys, work upon which began in 1952).

_____, 1954a, Region-Wide Municipal and Industrial Dump Site Study, Santa Ana Region: Los Angeles, Report to Santa Ana Regional Water Pollution Control Board, 41 p., plus 2 ap.

_____, 1954-1957, Various waste disposal site inspections: Southern District: Report to Santa Ana Regional Water Pollution Control Board, dates and pages as shown.

_____, 1955a, Proposed Dump Site at Little Hill Quarry, Jurupa Avenue and Maple Avenue, Crestmore Village, San Bernardino County, 21 Mar, 4 p, plus table, map, and geologic section.

_____, 1955b, Proposed Industrial Dump Site, Jurupa Mountains, Riverside Co.; Los Angeles, Report to Santa Ana Regional Water Pollution Control Board, 2 June, 3 p., plus appendices.

_____, 1956, Study on Planned Utilization of Ground Water Basins; San Gabriel Basin: Bull.

104-2, Appendix A, Geology, Sacramento, 230 p., plus plates.

_____, 1957, Ground Water Quality Objectives, Chino Basin: Los Angeles, Report to Santa Ana Regional Water Pollution Control Board, Mar., 70 p., maps.

California Regional Water Quality Control Board, 1977, Public Hearing; Stringfellow Disposal Site: Transcript of Proceedings of 14 Jan, by Macauley & Manning, Court Reporters, Santa Ana, CA, held at Riverside, CA, 79 p.

_____, 1980-1982, Stringfellow Reports: Santa Ana, CA, Selection of 18 On-Scene Coordination Reports of on-site activities, var. pages.

_____, 1980, Abatement Alternatives for the Stringfellow Class I Disposal Site: Santa Ana, CA, Summary of Special Board Meeting, 2 Jul., 9p.

_____, 1980, Abatement Alternatives for the Stringfellow Class I Disposal Site: Santa Ana, CA, Minutes of Special Board Meeting, 2 Jul., 160 p.

_____, 1980, Stringfellow Hazardous Waste Disposal Site; Interagency Report: Sacramento, with Cooperation of State Water Resources Control Board, Dept. Health Services, Air Resources Board, and Dept. of Food and Agriculture, 40 p. plus 3 appendices.

California State Water Resources Control Board, 1969, Recommended Changes in Water Quality Control, Final Report of the Study Panel; Study Project - Water Quality Control Program: Sacramento, prepared for the California Legislature, Mar, 119 p., plus 5 variously-paginated appendices.

_____, 1971, Laboratory and Manpower Needs for Surveillance and Enforcement of Waste Discharge Requirements: Sacramento, Water Quality and Planning (Unit?), Nov, 89 p.

_____, 1972, Notice, Adoption of Regulations, Order Adopting and Repealing Regulations

of the State Water Resources Control Board (Adoption of SubChapter 5, Waste Disposal to Land): Sacramento, CA, 2 Mar, 9 p.

_____, 1972, Disposal Site Design and Operational Information; Waste Discharge Requirements for Regulating Waste Disposal to Land: Sacramento, Nov., 61 p. (authored by Dr. A.L. Franks).

_____, 1973, Evolution of State Board (Water Quality Control) Membership: Memo from Gavin M. Craig to Executive Officer Bill Dendy, Sacramento, 31 Aug, unpaginated.

_____, 1974, Disposal Site Design and Operational Information; Waste Discharge Requirements for Waste Disposal to Land: Sacramento, Oct., 63 p. (authored by Dr. A.L. Franks).

_____, 1976, Water Quality Criteria: Sacramento, Publ. 3-A, Jun, (J.E. McKee, Chairman) 2nd printing, 2nd ed, 1 Jun, 548 p. (1954, 1957, 2nd reprinting with addendum,1963, 2nd ed.)

_____, 1976, Disposal Site Design and Operational Information; Waste Discharge Requirements for Nonsewerable Waste Disposal to Land: Sacramento, Dec., 60 p. (authored by Dr. A.L. Franks).

_____, 1977, Water Quality Research Needs in California: Problems, Research Needs,Implementation: Publ. no. 59, Aug., 83 p., plus 4 variously-paginated appendices.

_____, 1978, Disposal Site Design and Operational Information; Waste Discharge Requirements for Nonsewerable Waste Disposal to Land: Sacramento, Nov., 61 p. (authored by Dr. A.L. Franks).

_____, 1979, Disposal Site Design and Operational Information; Waste Discharge Requirements for Nonsewerable Waste Disposal to Land: Sacramento, Oct., __ p. (authored by Dr. A.L. Franks).

_____, 1979, Inventory, Inspection, and Condition of Class II-1 Disposal Sites and Class II Disposal Sites Receiving High-Moisture Content Waste: Sacramento,Hydrogeologic/ Geotechnical Section, Div. of Planning and Research, Dec, under the participation and direction of A.L. Franks, 71 p.

_____, 1980, Disposal Site Design, Operation, and Closure Information; Waste Discharge Requirements for Nonsewerable Waste Disposal to Land: Sacramento, Jul., 53 p., plus 40-page Appendix (authored by Dr. A.L. Franks).

_____, 1980, The Porter-Cologne Water Quality Control Act; Addenda and 1979 Amendments: Sacramento, Jun, 113 p.

_____, 1980, Effluent Charges for Water Pollution: Sacramento, Rept. by W.D. Carson, May, 68 p.

_____, 1981, Disposal Site Design and Operational Information; Waste Discharge Requirements for Nonsewerable Waste Disposal to Land: Sacramento, Jul., 93 p. (authored by Dr. A.L. Franks).

_____, 1980, 1980 Stringfellow Hazardous Waste Disposal Site; Interagency Status Report: Sacramento, 40 p., plus references.

*Carter, Bruce, 1979-1980, Photographs of Stringfellow site response: Stringfellow Federal Case Exhibit no. 3146.

Clawson, R.F., 1989, Deposition, U.S.A, et al, Plaintiffs, vs. J. B. Stringfellow, Jr., et al, Defendants: Held at Redding, CA, on 19 Jan, Office of Challe and Fisher, Deposition Reporters, 125 p.

Ecology and Environment, Inc., 1983, Draft Monitoring Well Completion Report, Stringfellow Hazardous Waste Site, Glen Avon Heights, CA: San Francisco, Rept. to USEPA, Region IX, Toxic Waste Mgmt. Div., San Francisco, CA, 21 Dec., var. pages.

Fox, R.C., 1983a, Deposition, U.S.A., et al., Plaintiffs, vs. J.B. Stringfellow, Jr., et al., Defendants: Held at Newport Beach, CA, on 29 Jun, Vol. I, Office of Latham and Watkins, CSR, 204 p.

_____, 1983b, Deposition, U.S.A., et al., Plaintiffs, vs. J.B. Stringfellow, Jr., et al., Defendants: Held at Newport Beach, CA, on 29 Jun, Vol. II, Office of Latham and Watkins, CSR, p. 205-351.

_____, 1983c, Deposition, U.S.A., et al., Plaintiffs, vs. J.B. Stringfellow, Jr., et al., Defendants: Held at Newport Beach, CA, on 6 Jul, Vol. III, Office of Latham and Watkins, CSR, p. 352-451.

Franks, A. L., 1988a, Deposition, U.S.A., et al, Plaintiffs, vs. J. B. Stringfellow, Jr., et al, Defendants: Held at Sacramento, CA, on 22 Dec, recorded by Dokich, Murray & Assoc., Vol. I, 180 p.

_____, 1988b, Deposition, U.S.A., et al, Plaintiffs, vs. J. B. Stringfellow, Jr., et al, Defendants: Held at Sacramento, CA, on 23 Dec, recorded by Dokich, Murray & Assoc., Vol. II, p. 181-328.

_____, 1989a, Deposition, U.S.A., et al, Plaintiffs, vs. J. B. Stringfellow, Jr., et al, Defendants: Held at Sacramento, CA, on 29 Dec, recorded by Dokich, Murray & Assoc., Vol. III, p. 329-468 (plus exhibits).

_____, 1989b, Deposition, U.S.A., et al, Plaintiffs, vs. J. B. Stringfellow , Jr., et al, Defendants: Held at Sacramento, CA, on 18 Jan, recorded by Capitol Reporters, Vol. IV, p. 469-581.

_____, 1989c, Deposition, U.S.A., et al, Plaintiffs, vs. J. B. Stringfellow, Jr., et al, Defendants: Held at Sacramento, CA, Exhibits 1690 through 1713, assembled by Dokich, Murray & Assoc., Vol. I.

_____, 1989d, Deposition, U.S.A., et al, Plaintiffs, vs. J. B. Stringfellow, Jr., et al, Defendants: Held at Sacramento, CA, Exhibits 1714 through 1749, assembled by Dokich, Murray & Assoc., Vol. II.

Hatheway, A.W., 1989a, Deposition, U.S.A., et al, Plaintiffs, vs. J.B. Stringfellow, Jr., et al, Defendants: Held at Los Angeles, CA, on 13 Feb, recorded by Dokich, Murray & Assoc., Vol. I, 172 p.

_____, 1989b, Deposition, U.S.A., et al, Plaintiffs, vs. J. B. Stringfellow, Jr., et al, Defendants: Held at San Francisco, CA, 24 Feb, recorded by Dokich, Murray & Assoc., Vol. II, p. 173-366.

_____, 1989c, Deposition, U.S.A., et al, Plaintiffs, vs. J. B. Stringfellow, Jr., et al, Defendants: Held at San Francisco, CA, 24 Feb, Exhibits 1951 through 1981, recorded by Dokich, Murray & Assoc.,

_____, 1989d, Deposition, U.S.A., et al, Plaintiffs, vs. J. B. Stringfellow, Jr., et al, Defendants: Held at San Francisco, CA, 25 Feb, Exhibits 1982 through 1999 and 3081 through 3499, and 3648 through 3652, recorded by Dokich, Murray & Assoc., Vol. III, p. 367-499.

_____, 1989d, Deposition, U.S.A., et al, Plaintiffs, vs. J. B. Stringfellow, Jr., et al, Defendants: Held at San Francisco, CA, 22 Apr, Exhibits 3850 through 3853, recorded by Dokich, Murray & Assoc., Vol. IV, p. 500-643.

_____, 1989f, Trial Testimony, U.S.A., et al, Plaintiffs, vs. J. B. Stringfellow, Jr., et al, Defendants: Held in United States District Court, Central District of California, Los Angeles, 22 May, recorded by National Network Reporting Co., Los Angeles, ___ p..

_____, 1991a Penny Newman, et al, Plaintiffs, vs. J.B. Stringfellow, Jr., et al, Defendants: Held at Los Angeles, CA, on 19 Apr, v. I, recorded by John F. Biehl, 173 p.

_____, 1991b, Penny Newman, et al, Plaintiffs, vs. J.B. Stringfellow, Jr., et al, Defendants: Held at Los Angeles, CA, on 20 Apr, v. II, recorded by John F. Biehl, p. 174-363.

_____, 1992a, Chronologic History of Industrial and Hazardous Waste Management, Worldwide:

Unpublished Report, Department of Geological and Petroleum Engineering, School of Mines & Metallurgy, University of Missouri-Rolla, 367 p. (with bibliography).

_____, 1992b, Pre-RCRA history of industrial waste management in southern California, in,

Hunter, B.L., 1933, The Evolution of Municipal Organization and Administrative Practice in the City of Los Angeles: Parker, Stone & Baird Co., Publishers, Los Angeles, 283 p.

JRB Associates (later a Division of SAIC), 1983, A Proposal for The Stringfellow Facility Remedial Investigation/ Feasibility Study: La Jolla, CA, prepared for California Dept. Health Services,Sacramento,CA, 19 Jul, var. pages.

McKevett, E.M., 1951, Geology of the Jurupa Mountains, San Bernardino and Riverside Counties, California: Calif. Div. Mines & Geology, Spec. Rept. No. 5, 14 p., plus map at 1:12,000 scale.

Montgomery, J.M., Consulting Engineers, Inc., 1977, Engineering Study of Stringfellow Class 1 Disposal Site: Pasadena, CA, 19 Dec., Report to Santa Ana Regional Water Quality Control Board, 13 p.

_____, 1979, Report on Phase I Work for Closure of Stringfellow Class 1 Hazardous Waste Disposal Site: Pasadena, CA, 19 Dec., Report to Santa Ana Regional Water Quality Control Board, var. pages, appendices A through F.

_____, 1980a, Report on Phase 1 Work for Closure of Stringfellow Class 1 Hazardous Waste Disposal Site; Addendum No. 1: Pasadena, CA, Apr., Report to Santa Ana Regional Water Quality Control Board, var. pages.

_____, 1980b, Engineering and Economic Evaluation of Transferring Waste From the Stringfellow Class I Waste Disposal Site: Pasadena, CA, 1 Dec, Report to Santa Ana Regional Water Quality Control Board, 20 p., plus 4 exhibits.

_____, 1981a, Report on Phase 2 Work for Closure of Stringfellow Class 1 Hazardous Waste Disposal Site; Hydrogeologic Evaluation: Pasadena, CA, 19 May., Report to Santa Ana Regional Water Quality Control Board, var. pages.

_____, 1981b, Bidding Documents and Specifications for Construction of Closure Facilities and the Stringfellow Class I Disposal Site: Pasadena, CA, Jun., Report to Santa Ana Regional Water Quality Control Board, var. pages.

_____, 1981c, An Evaluation of Ground Water Disposal Alternatives and Post-Closure Monitoring for Stringfellow Class I Disposal Site: Pasadena, CA, 2 Oct, Report to Santa Ana Regional Water Quality Control Board, var. pages.

_____, 1982a, The Gel-Injection Phase for Final Closure of Stringfellow Class I Disposal Site: Pasadena, CA, 19 Feb., Report to Santa Ana Regional Water Quality Control Board, 17 p., plus drawings and tables.

_____, 1982b, Bench-Scale Treatability Studies for Liquid Wastes at the Stringfellow Site: Pasadena, CA, Jul., Report to Santa Ana Regional Regional Water Quality Control Board, var. pages.

_____, 1984, Proposal to Provide Engineering Services for Design and Construction Management, Interim Program for Treatment and Disposal of Extracted Groundwater from the Stringfellow Hazardous Waste Site: Presented to the Santa Ana Watershed Project Authority, Pasadena, Aug, var. p.

Morton, D.M., 1978, Geologic Map of the Fontana Quadrangle, San Bernardino and Riverside Counties, California: U.S. Geol. Survey Open File Report 78-19, one sheet, with explanation, 1:24,000 scale.

Neste, Brudin, and Stone, Consulting Engineers, 1973, Report to Stringfellow Quarry Company on Industrial Waste Disposal: Jan, unsigned, 21 p., with appendices.

O'Leary, D.A., 1989a, Deposition, U.S.A, et al, Plaintiffs, vs. J. B. Stringfellow, Jr., et al, Defendants: Held at San Diego, CA, on 19 Jan, Vol. I, by Dokich, Murray & Associates, Court Reporters, 81 p.

Orange County (California) Sanitation Districts, 1954, Ordinance No. 1, Rules and Regulations for Admission of Sewage from the Sanitary Districts: The Districts, Santa Ana, CA, 10 Feb, 7 p. (same document passed on this date by all member Districts having wastewater collection responsibilities; Nos 1,2,3,5,6,7, & 11).

_____, 1958, Ordinance No. 101, Rules and Regulations Relating to Connections to District Trunk Sewers: The Districts, Santa Ana, CA, 5 Feb, 6 p. (follows Ordinance No. 1, of 10 Feb 54).

Phillips, D.F., 1984, Deposition, U.S.A., et al, Plaintiffs, vs. J.B. Stringfellow, Jr., et al, Defendants: Held at Los Angeles, CA, on 26 Jan, recorded by Traci A. Dokich, Inc., 63 p.

Reiter, G.A., and Mandel, R.M., 1980, Use of 311(k) Pollution Contingency Fund at Stringfellow Class I hazardous waste site, Riverside, California, 3-17 Mar 1980: Paper delivered before Conf. on Management of Uncontrolled Haz. Waste Sites, Washington, DC, (unpubl. ms in case files), 10 p.

Santa Ana Regional Water Pollution Control Board, 1954a Resolution 54-5 (Pertaining to Stringfellow Site): Santa Ana, 5 Nov, 4 p.

_____, 1954b, Resolution 54-5, (General Policy Relating to Control of "Dump" Sites): Sacramento, 2 p

_____, 1955, Letter Relating to Proposed Industrial Dump Site: from Paul G. Brown, Executive Director, to California Division of Water Resources, Los Angeles, 15 March, Attention, Max Bookman, Engineer-in-Charge, 1 p.

_____, 1980a, Minutes, Special Board Meeting, Abatement Alternatives for the Stringfellow Class I Disposal Site: Held 02 Jul, Santa Ana, CA, 7 p., 2 tables.

_____, 1980b, Transcript of Minutes, Special Board Meeting, Abatement Alternatives for the Stringfellow Class I Disposal Site: Held 02 Jul, Santa Ana, CA, recorded by Arthur Spring, Orange, CA, 160 p.

_____, 1982, Compendium of On-Site Emergency Response Reports for the Stringfellow Acid Pits: Santa Ana, 18 reports, from 15 Aug 1980 through Aug 1982, var. pages.

Scheliga, J.T., Jr., 1961, The Application of Geology to the Selection of Waste Disposal Facilities: Preprint of paper presented before the Cordilleran Section, Geological Society of America, San Diego State College, 28 Mar, 5 p., typewritten.

Science Applications International Corp., 1986a, Draft Data Report, Stringfellow Facility Remedial Investigation/Feasibility Study; Routine Groundwater Monitoring Survey; Annual Report for 1985: La Jolla, 11 Jun, 46 p., prepared for California Department of Health, Toxic Substances Control Div., Sacramento

Stierman, D.J., 1984a, Deposition, U.S.A, et al, Plaintiffs, vs. J. B. Stringfellow, Jr., et al, Defendants: Held at Riverside, CA, 7 Mar, Vol. I, by Traci A. Dokich, Court Reporter, 204 p.

_____, 1984b, Deposition, U.S.A, et al, Plaintiffs, vs. J. B. Stringfellow, Jr., et al, Defendants: Held at Riverside, CA, on 8 Mar, Vol. II, by Traci A. Dokich, Court Reporter, p. 205-292.

_____, 1984c, Deposition, U.S.A, et al, Plaintiffs, vs. J. B. Stringfellow, Jr., et al, Defendants: Held at Riverside, CA, on 5 Apr, Vol. III, by Traci A. Dokich, Court Reporter, p. 293-433.

_____, 1984d, Deposition, U.S.A, et al, Plaintiffs, vs. J. B. Stringfellow, Jr., et al, Defendants: Held at Riverside, CA, on 6 Apr, Vol. IV, by Traci A. Dokich, Court Reporter, p. 434-582.

_____, 1984e, Deposition, U.S.A, et al, Plaintiffs, vs. J. B. Stringfellow, Jr., et al, Defendants: Held at Riverside, CA, on 10 May, Vol. V, by Traci A. Dokich, Court Reporter, p. 583-740.

_____, 1984f, Deposition, U.S.A, et al, Plaintiffs, vs. J. B. Stringfellow, Jr., et al, Defendants: Held at Riverside, CA, on 21 May, Vol. VI, by Traci A. Dokich, Court Reporter, p. 741-931.

Stone, R.V., 1984a, Deposition, U.S.A, et al, Plaintiffs, vs. J. B. Stringfellow, Jr., et al, Defendants: Held at San Diego, CA, on 9 Jul, Vol. I, by Traci A. Dokich, Court Reporter, 189 p.

_____, 1984b, Deposition, U.S.A, et al, Plaintiffs, vs. J. B. Stringfellow, Jr., et al, Defendants: Held at San Diego, CA, on 10 Jul, Vol. II, by Traci A. Dokich, Court Reporter, p. 190-354.

_____, 1984c, Deposition, U.S.A, et al, Plaintiffs, vs. J. B. Stringfellow, Jr., et al, Defendants: Held at San Diego, CA, on 11 Jul, Vol. III, by Traci A. Dokich, Court Reporter, p. 355-537.

_____, 1984d, Deposition, U.S.A, et al, Plaintiffs, vs. J. B. Stringfellow, Jr., et al, Defendants: Held at San Diego, CA, on 12 Jul, Vol. IV, by Traci A. Dokich, Court Reporter, p. 538-716.

U.S. Congress, 1983, Hazardous Waste and the Stringfellow Acid Pits: Hearing before the Subcomm. on Natural Resources, Agriculture, Research, and Environment, of the Comm. on Science and Technology, House of Representatives, 98th Congress, 1st. Session, Hearing no. 136, 23 Apr, 234 p.

U.S. Geological Survey, 1973, Ground-Water Quality in the Upper Santa Ana River Basin, Southern California: Water-Resources Investigations Rept. 79-113, 51 p., NTIS PB 80-161888/LL.

_____, 1980, Distribution of Nitrate in the Unsaturated Zone, Highland- East Highlands Area, San Bernardino County, California: Water-Resources Investigations Rept. 80-48, 70 p., NTIS, PB 81-117004/LL.

Vasconcelos, J.J., Barto, R.L., Anderton, R.L., and Grant, F.A., 1980, Progress on the Containment of Hazardous Waste at the Stringfellow Class I Disposal Site: Preprint of paper presented at 3rd Ann. Madison Conf. of Applied Research and Practice on Municipal and Industrial Waste, Madison, WI, Sep, unk. p.

Werner, S.L., 1983a, Deposition, U.S.A., et al, Plaintiffs, vs. J.B. Stringfellow, Jr., et al, Defendants: Held at Los Angeles, CA, on 20 Sep, recorded by Traci A. Dokich, Inc., 108 p.

_____, 1983b, Deposition, U.S.A., et al, Plaintiffs, vs. J.B. Stringfellow, Jr., et al, Defendants: Held at Los Angeles, CA, on 21 Sep, recorded by Traci A. Dokich, Inc., p. 109-253.

_____, 1983c, Deposition, U.S.A., et al, Plaintiffs, vs. J.B. Stringfellow, Jr., et al, Defendants: Held at Los Angeles, CA, on 28 Sep, recorded by Traci A. Dokich, Inc., p. 254-380.

_____, 1983d, Deposition, U.S.A., et al, Plaintiffs, vs. J.B. Stringfellow, Jr., et al, Defendants: Held at Los Angeles, CA, on 29 Sep, recorded by Traci A. Dokich, Inc., p. 381-506.

_____, 1983e, Deposition, U.S.A., et al, Plaintiffs, vs. J.B. Stringfellow, Jr., et al, Defendants: Held at Los Angeles, CA, on 17 Nov, recorded by Traci A. Dokich, Inc., p. 507-613.

_____, 1983f, Deposition, U.S.A., et al, Plaintiffs, vs. J.B. Stringfellow, Jr., et al, Defendants: Held at Los Angeles, CA, on 18 Nov, recorded by Traci A. Dokich, Inc., p. 614-763.

_____, 1991, Penny Newman, et al, Plaintiffs, vs. J.B. Stringfellow, Jr., et al, Defendants: Held at Woodland Hills, CA, on 28 Jan, v. I, recorded by Laurie Held-Biehl, 159 p.

Willets, D.B. 1984, Deposition, U.S.A., et al, Plaintiffs, vs. J.B. Stringfellow, Jr., et al, Defendants: Held at San Diego, CA, on 24 Jan, recorded by Traci A. Dokich, 83 p.

Woodward-Clyde Consultants, 1981, Draft Final Report, Geologic and Geophysical Investigation of the Stringfellow Class I Hazardous Disposal Site, Riverside: Orange, California, Report to Santa Ana Regional Water Quality Control Board, 15 Jun, var. pages.

_____, 1981, Seismic Refraction Investigation; Stringfellow Quarry, Class I Hazardous Waste Disposal Site, Riverside County, California: Santa Ana, California, June var. p.

HYDROGEOLOGIC CHARACTERIZATION AND MONITORING SYSTEM DESIGN CASMALIA RESOURCES FACILITY

Stephen M. Testa
Engineering Geologist
Applied Environmental Services, Inc.
23113 Plaza Point Drive, Suite 100
Laguna Hills, California 92653

ABSTRACT

The Casmalia Resources Facility is a major hazardous waste management facility located in the northwestern pportion of Santa Barbara County, California. The Facility maintains numerous favorable environmental characteristics which makes it suitable for the siting of a hazardous waste disposal facility. These characteristics include abundant low permeability soil and consolidated rock, relatively slow groundwater flow rates, underlying non-beneficial use groundwater, high evapotranspiration rates which exceed precipitation rates, waste management unit waste type and anticipated life span, existing downgradient cut-off trenches and extraction systems, and unit specific engineering design elements. The overall detection monitoring program incorporates a surface water and detection groundwater monitoring program, and a planned vadose zone monitoring program. Although beneficial use or potential beneficial use groundwater occurs within 1/2 mile downgradient of the Facility, the potential for adverse impact on this groundwater resource is considered very low.

INTRODUCTION

The Casmalia Resources Facility is a hazardous waste management facility located in the northwest portion of Santa Barbara County, California (Figure 1). The Facility provides services for the treatment and disposal of hazardous wastes that are generated off site. The Facility started receiving wastes for disposal in 1972. A wide spectrum of chemical and industrial wastes has been received, a significant portion by volume of which was generated by oil and gas exploration and development activities and are considered non-hazardous waste under the Resource, Conservation and Recovery Act (RCRA). The remainder consists of solvents, pesticides, alkalines, acids, cyanides, resins, phenols, laboratory wastes, heavy metals, and PCB-contaminated equipment.

Extensive investigations have been conducted at and in the vicinity of the Facility since the early 1980's to characterize geologic and hydrogeologic conditions, and develop a groundwater monitoring program including the installation of a groundwater monitoring system designed to detect the immediate release of constituents from the waste management

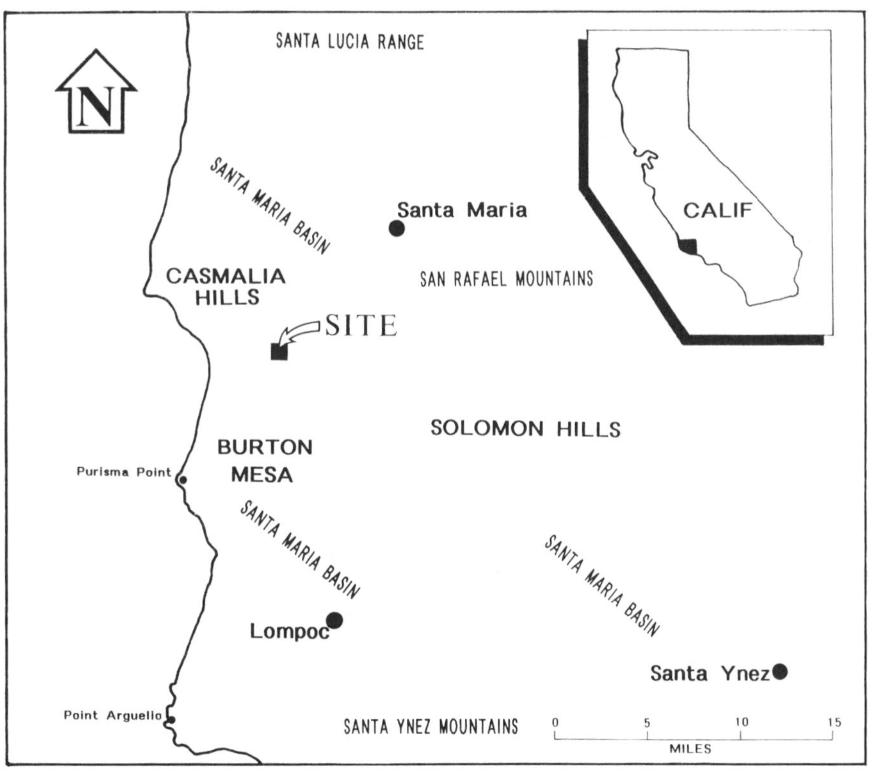

Figure 1. Site location map

units (Canonie Environmental, 1988; Woodward-Clyde Consultants, 1988). Presented in this paper is discussion of site description and subsurface geologic and hydrogeologic conditions. In addition, the relationship between subsurface conditions to site suitability and design of a detection groundwater monitoring system is also discussed.

SITE DESCRIPTION

The Casmalia Resource Facility is situated in the Casmalia Hills, a region of low, rolling hills which lie between the Santa Maria Valley and the Pacific Coast. The Facility is located about 1.5 miles north-northwest of the town of Casmalia, and about 5 miles southwest of the city of Santa Maria. Generally situated with south-facing exposures, ground surface elevations generally decrease from north to south ranging from about 835 to 405 feet above mean sea level (MSL), respectively. The northern portion of the Facility is situated between northwest- and northeast-trending ridges; the southern portion is situated between two low, adjacent knolls. Casmalia Creek and Casmalia Canyon, and an unnamed drainage are located west and northeast of the Facility, respectively. Both drainages are relatively broad and eventually merge with Shuman Creek and Shuman Canyon about 2 miles south of the Facility, and about one mile west of the town of Casmalia. Toqnuzini Creek and Toqnuzini Canyon also merge with Shuman Creek south of the Facility.

The overall climate is generally semi-arid to arid resulting in sparse vegetation and intermittent streams. The vast majority of the surrounding area is open grassland with shrubs, live oak, and scattered willow and eucalyptus trees found along some of the intermittent drainages along Casmalia and Shuman Creek. Some land to the southwest of the Facility is used for dry-farming, notably, garbanzo beans and wheat. Overall land use within one mile of the Facility is developed as large parcel agricultural consisting mostly of livestock grazing and associated ranching operations. Oil and gas developments and a few residences are located southeast of the Facility along Black Road.

FACILITY LAYOUT AND OPERATION

At present, the Facility has four operating landfill units, two operating process treatment systems, and approximately 43 ponds and 15 evaporation pads that are currently or will be undergoing closure in the near future. The ponds and pads, which comprise the liquid waste management facilities, are presently being closed. Substantial modification of the Facility is anticipated to occur within the near future to maximize the use for solid waste management. A Facility layout map showing these features is presented in Figure 2. An overview of the existing Facility and the modification plans are discussed below.

Existing Landfills

The four operating landfill units are identified by the types of wastes they receive (i.e. acids, caustics, metals, and solvents/pesticides). Two additional landfills, the PCB Landfill and the RCRA Solid Waste Landfill, exist at the Facility; however, they are currently not operated and will be closed in the near future. Approximate existing landfill unit volumes (including both waste and interim cover material) as well as calculated remaining capacities (including waste, interim and final cover material) as of March, 1988 are presented in Table 1.

The landfill units will be operated until their capacities are reached, at which time they will be closed. The polychlorinated biphenyls (PCB) and RCRA Solid Waste Landfills which are characterized as existing waste disposal units are permitted by the United States Environmental Protection Agency (EPA) and the California Regional Water Quality Control Board - Central Coast Region (CRWQCB) for the disposal of waste; nevertheless, pursuant to the 1986 Conditional Use Permit (CUP) Agreement with the County of Santa Barbara, operation of these landfills was suspended in 1986, and anticipated to be closed by the early 1990's.

Existing Surface Impoundments

Surface impoundments at the Facility have comprised the site's liquid waste management system. The existing RCRA

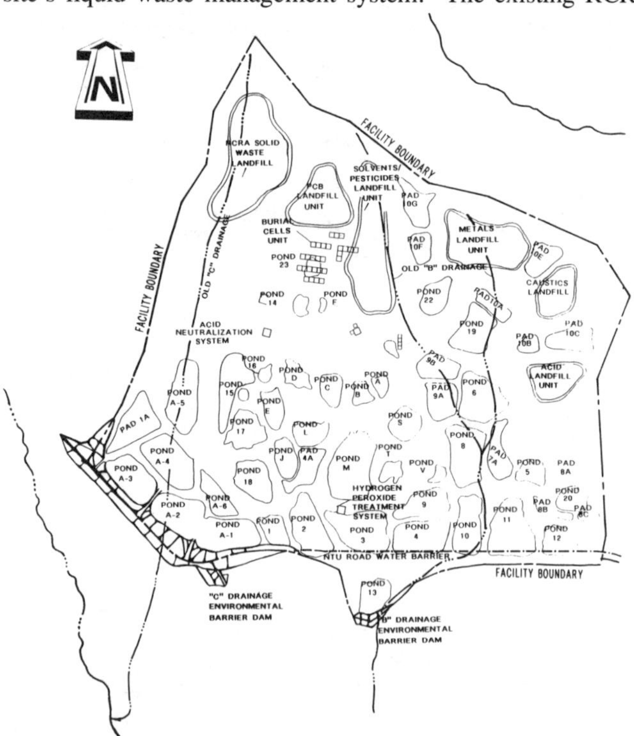

Figure 2. Facility layout map showing the approximate location of major ponds, evaporation pads, landfills and associated structures

surface impoundments are presently undergoing closure activities (i.e., dewatering, excavating, sampling, and sample analysis). Closure of all existing surface impoundments is anticipated to be complete by the early 1990's. This system has historically consisted of 23 surface impoundment units, containing a total of 40 individual ponds and 15 evaporation pads which have been used for temporary storage, treatment, and evaporative disposal of liquid waste and on-site surface runoff. Each surface impoundment unit has consisted of a combination of ponds and evaporation pads, a series of ponds only, or an individual pond. Construction of the ponds occurred over a period of several years, beginning in 1972 with Ponds 13, A, B, and C. In general, all ponds are unlined and used excavated materials (typically weathered claystone) to form dikes. Liquids that have been managed in the surface impoundments include acid and alkaline wastes, oil field wastes, various bulk industrial wastewaters, and contaminated waters from State and Federal cleanup activities.

Three ponds, presently classified as non-RCRA ponds, are also present on site: Sludges 1, Sludges 2, and the Catch Basin. Sludges 1 and 2 have been used for disposal of non-RCRA waste, which includes sewage sludges and putrescible wastes, such as restaurant grease-trap wastes. The Catch Basin is used solely for holding rainfall runoff.

Liquids have historically been transferred from pond to pond through the site's piping and culvert network. Many ponds and all pads have overflow culverts. Ponds that do not have overflow culverts are equipped with either a valved siphon drain, a valved underdrain, or are located near a riser and can be connected to the site's piping network. All ponds and pads can be made to drain (by gravity or by pumping) into one another except Pond 13. Pond 13 is located at the southern edge of the Facility and has been and is reserved for emergency storage of excessive rainfall and runoff.

Enhanced evaporation through pond water spraying along berms was practiced up to 1985. Graded evaporation ponds were constructed adjacent to many ponds in 1985 and are presently operated through drip nozzle discharge, rather than spraying.

Liquid wastes are no longer being accepted for treatment or disposal in the ponds. At present, Pond 18 receives the treated effluent from the Neutralization System. All other ponds now function as precipitation and runoff containment ponds. Evaporation pads have been used for enhanced evaporation of pond water.

Existing Process Treatment Systems

The existing process treatment systems at the Facility include the Neutralization System (NS) and the Hydrogen Peroxide Injection/Aeration (HPIS) System. Another existing system, the Zimpro Wet Air Oxidation (WAO) System, is being retrofitted to be used as a proposed Bulk Liquid Solid Solidification (BLS) plant.

The NS, located in the north-central portion of the Facility, treats acid and alkaline liquid waste streams by neutralization and removes heavy metals from acid, alkaline, and neutral liquid waste streams. The HPIS, located between Ponds 2 and M in the southcentral portion of the Facility, treats odorous pond waters through the injection of hydrogen peroxide and subsequent aeration.

Environmental Barrier Dams and Leachate Collection Facilities

Clay barriers (or containment structures) have been installed in six locations at the Facility. These structures were designed and installed to control the migration of potential leachate both within and from the Facility. The clay material used in constructing these dams was obtained on site or adjacent to the Facility.

Four of the six clay barriers have been installed with drainage collection galleries that have pump-out facilities. The largest of these three is called "C" Environmental Barrier Dam and is located south of Ponds A-1, A-2, and A-3 along the southern perimeter of the Facility. The "B" Environmental Barrier Dam is located at the southern edge of Pond 13, along the southern perimeter of the Facility. The remaining two clay barriers with drainage collection galleries are located at the foot of the Solvents/Pesticides Landfill Unit and at the southern edge of the Catch Basin in West

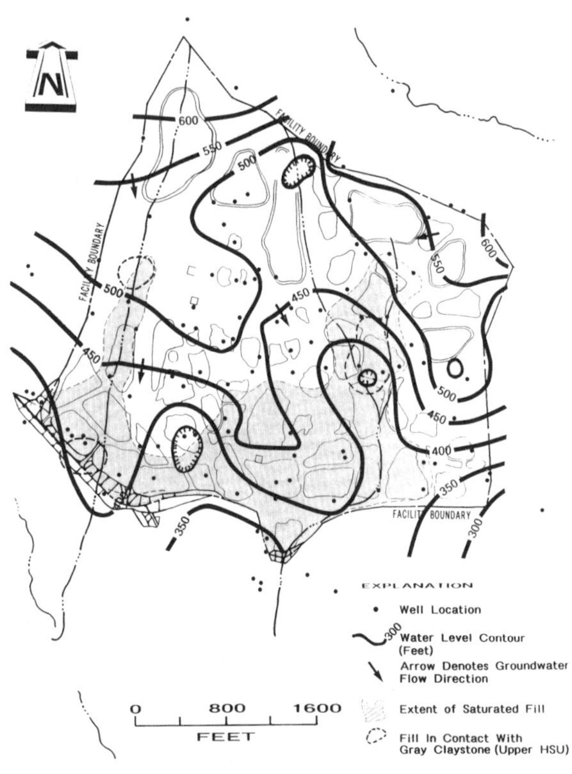

Figure 3. Geology map of area within a one-mile radius of the Facility

Canyon. Two other clay barriers (without drainage collection galleries) are located at the southwestern edge of the PCB Landfill Unit and around the southern and eastern perimeter of Pond 20.

NTU Road Barrier

In September, 1979, as-built drawings were completed for the reconstruction of NTU Road. Drawings indicated that the old NTU Road was removed, a barrier subsequently constructed of fill material compacted to 95 percent along the line of the old road, and the new NTU Road then constructed on top of the barrier. In 1978, the NTU Road was being reconstructed to attain an additional 12-foot elevation that would act as a containment barrier in the event of a 100-year flood.

Early Waste Disposal Practices

During the early development years at the Casmalia Resources Facility, four additional areas were used for waste disposal. These areas include the Burial Cells Unit, Shallow Disposal Wells Unit, RCRA Solid Waste Landfill Unit, and the Burial Drums Unit.

Solid wastes were disposed of in the Burial Cells Unit, a 4.5-acre area north of Ponds 14 and P. These wastes were placed in individually excavated cells, or trenches, that were aligned in rows. Totalling approximately 40,000 tons (300 cy), the wastes include acids, caustics, solvents, pesticides, and metals. Separate cells were constructed for each type of waste material. The cells were excavated with a backhoe to approximately 40 feet by 40 feet and 20 to 30 feet deep. According to the site personnel, when a cell reached capacity a new cell was constructed in the adjacent position in the row, and the clean soil removed from the new cell was used as cover for the completed cell. Each row or trench was numbered and each cell within a row was identified with a letter (e.g., Cell 7E). Cell numbers and letters were recorded for each load.

Eleven shallow disposal wells were used for disposal of hazardous liquid wastes from approximately December, 1977 through September, 1982. These disposal wells, located adjacent to the Burial Cells Unit, were not deep well injection facilities. The wells consisted of shallow borings about 4 feet in diameter and about 40 feet in depth. The wells were filled adjacent to shallow disposal cells of approximately the same depth. The borings were made as close to the sidewall of the cells as possible. The intent was that liquids discharged into these shallow borings would be absorbed into the cells by the buried waste materials, which would act as a sponge. Each well was fitted with a metal cover which contained a vent pipe and discharge line. Each well was used for the disposal of a different waste type (i.e. solvents, pesticides, acids, gas/diesel, etc.).

In April, 1984, a small amount of waste was placed in the RCRA Solid Waste Landfill. As part of the construction of Pond 23, wastes that were previously placed in the Burial Cells Unit were excavated. These wastes were placed, over a period of several days, in the area designated as the RCRA Solid Waste Landfill.

According to site personnel, disposal of drums occurred on an experimental basis in an area now occupied by Pond 19, designated as the Buried Drums Unit. Drums containing a small quantity of acidic waste were placed in this area from December, 1979 to approximately June, 1980 prior to the operation of the Acid Landfill Unit in July, 1980. The drums were then covered during the construction of the base of the current Pond 19.

Future Site Modifications

Future modification plans for the Facility include the construction and operation of new landfills, surface impoundments, and process systems. It is presently anticipated that the existing ponds will all be dewatered and closed by the early 1990's. After this process, the Facility would become primarily a solids residual repository facility.

GEOLOGIC SETTING

The Facility is situated within the Casmalia Hills which is characterized as a structurally uplifted region situated between the Santa Maria Valley and the Pacific Coast. Regional geologic conditions have been discussed by Woodring and Bramlette (1950), Dibblee (1950), Worts (1951), and Sylvester and Darrow (1989). Regionally, late Tertiary marine sedimentary rocks overlie Mesozoic Franciscan basement rocks. The late Tertiary sedimentary formation within a several mile radius of the Facility generally appears to become progressively younger from southwest to northeast. In order of decreasing age, the sedimentary formations include the Miocene Series, Miocene-Pliocene Series, Pliocene Series, Pliocene and Pleistocene Series and the Holocene Series. Geologic conditions at and within a one-mile radius of the site is shown in Figure 3. An east-west and north-south trending hydrogeologic cross-section is presented in Figure 4.

The stratigraphic units characterized at the Facility include the Miocene-Pliocene Todos Santos Claystone Member of the Sisquoc Formation, Pleistocene terrace deposits, and Quaternary alluvium and colluvium. Artificial fill was also characterized to a limited extent.

Todos Santos Claystone Member of the Sisquoc Formation

The entire Facility resides on thin surficial deposits and weathered portions of the Todos Santos Claystone Member of the Sisquoc Formation (hereinafter Todos Santos). Based on site specific studies the Todos Santos is divided into two in-

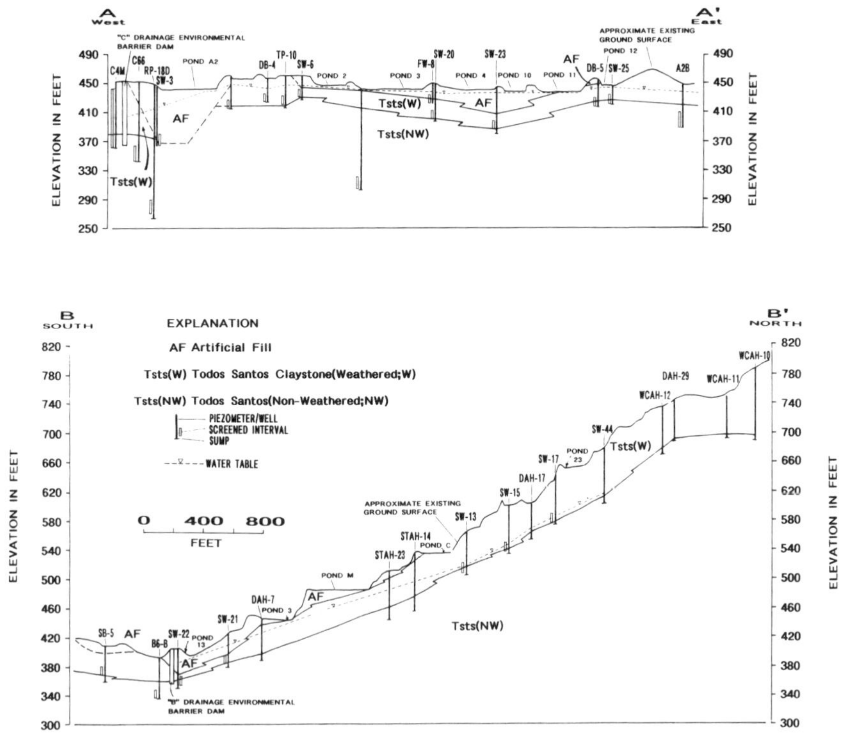

Figure 4. Hydrogeologic cross-section of east-west trending A-A' and north-south trending B-B'

formal units: the lower, unweathered claystone and the upper, weathered claystone. This division does not reflect two distinct faces with stratigraphic and structural relations within the Todos Santos, but instead the two units reflect a pervasive weathering profile.

The Todos Santos forms low, smoothly rounded hills which are covered by thin, predominantly clayish soil that locally supports a variety of grasses, isolated shrubs, and small groves of live oak, eucalyptus, and willow trees. The Todos Santos is exposed along fairly extensive cut-slopes and cut-areas within the Facility boundary.

Unweathered Claystone

Because unweathered claystone comprises less than 10 percent of the exposures on site, most of the lithologic descriptions are from core samples. Based on visual examination of core samples and surface exposures, the unweathered unit typically consists of olive-black to gray olive-green (wet), and medium bluish-gray to greenish-gray (dry) claystone to silty claystone. It is generally massive with some thin laminations and fissile zones.

Estimates of strength, based on gross field tests, show the unweathered claystone as firm to strong. This unit is less fractured than the weathered claystone and contains a small percentage of diatoms.

Weathered Claystone

Within the Facility, the weathered claystone accounts for most of the exposed, mappable bedrock. The weathered claystone is present over most of the Facility with the exception of areas beneath the environmental barrier dams in the B- and C-drainages, portions of West Canyon, and possibly beneath many of the solid waste disposal units.

Because the division between the weathered and unweathered claystone is based on the presence of weathering rather than depositional relationships, the petrography of the weathered claystone is essentially the same as that of the unweathered claystone. The most obvious differences between the weathered and unweathered units is a distinctly different color, the presence of prevalent discontinuities (fractures and joints), and secondary mineralization within the weathered unit.

Terrace Deposits

Woodring and Bramlette (1950) mapped erosional remnants of Pleistocene terrace deposits which cap hills located just south of the Facility. Lithologic data obtained during drilling suggest that the terrace deposits at locations shown by Woodring and Bramlette (1950) generally consist of poorly sorted, gravelly sand that may be up to 12 feet thick. Review of pre-construction aerial photographs and the geologic mapping conducted by Woodring and Bramlette (1950), indicate that terrace deposits are not present within the boundary of the Facility.

Alluvium and Colluvium

Quaternary alluvium does not occur within the Facility boundary, but is exposed within the lower portions of Casmalia Creek Canyon, along and at the mouths of minor intermittent tributaries to Casmalia Creek, and along the unnamed drainage northeast of the Facility boundary. Alluvial deposits along Casmalia Creek west of the Facility reach an apparent maximum thickness of at least 50 feet according to boring logs prepared for shallow water supply wells installed by Hoover and Associates (1982).

When exposed in arroyos and as encountered in various monitoring wells located outside of the Facility boundary, the alluvium consists of a light brown to gray, clayey sandy silt and silty clay with scattered gravel clasts. Within the Facility boundary, clay-rich soil and thin colluvium occur along natural slopes undisturbed by Facility activities. Outside the Facility boundary, colluvium and clay-rich soil occurs along hillsides, covering weathered claystone and in some cases overlying alluvial deposits. Colluvium in the vicinity of the Facility is generally 5 to 10 feet in thickness with oriented tubular or elongate claystone and siltstone clasts in a gray silty clay matrix.

Artificial Fill

Artificial fill occurs on site and is composed of silty clay with some sand and pebble to cobble size fragments of claystone and silty claystone. Generally, the fill is more extensive along the southern Facility boundary, becoming less extensive northward.

Structural Setting

The Casmalia Hills are generally characterized by broad, northwest-trending folds. The San Antonio Syncline and the Casmalia Anticline are structural elements identified from surface geologic mapping and oil field explorations at oil producing areas located east and southeast of the Facility. To the northwest, well data are more widely scattered, and the locations and extent of the San Antonio Syncline and the Casmalia Anticline are less precise.

Based on the mapping of Woodring and Bramlette (1950), the axis of the San Antonio Syncline is mapped roughly parallel to the southwest boundary and the Facility, whereas the axis of the Casmalia Anticline, also based on the mapping of Woodring and Bramlette (1950), appears to trend towards and through the Facility. However, the entire Facility appears to lie in an area characterized by low-dipping to essentially flat bedding inclinations. This suggests that tight folds which significantly influence bedding attitudes on a large scale are not present in the near-surface portion of the Sisquoc Formation in the vicinity of the Facility. It also appears that the entire Facility is situated within what could be considered the transitional area between adjacent, subparallel flanks of the San Antonio Syncline and the Casmalia Anticline. Based on surface geologic mapping, this transitional area appears to be largely characterized by a gradual, broad, low amplitude flexure.

Folds

The Casmalia Anticline appears to trend northwestward and, based on low bedding inclinations, appears to diminish in surface expression as it approaches the Facility. Bedding attitudes measured on site generally do not support the axial trend mapped by Woodring and Bramlette (1950). Along the southeastern Facility boundary, bedding attitudes strike generally east-west and dip 5 to 15 degrees southward. Towards the northeast corner of the Facility, the dip angles decrease to 3 to 5 degrees but maintain their southerly direction before gradually dipping to the north. In the central portion of the Facility, near pond 22 and along the western Facility boundary, bedding generally strikes within 20 degrees east or west of north, and dips from 3 to 20 degrees eastward. Based on these local bedding attitudes, which are generally low and somewhat variable across the Facility, it appears that it is not possible to confidently establish the location of the axial trace of either the San Antonio Syncline or the Casmalia Anticline in the vicinity of the Facility. It appears likely that surface expression of folding in the Facility vicinity is generally weak. Structural expression of the Casmalia Anticline also appears to be weak or absent in the near surface and the upper portion of the Sisquoc Formation.

The Facility is not within 200 feet of a known Holocene fault or in an area of rapid geologic change. The nearest known faults of Holocene age are the Lion's Head Fault located about one mile to the south, and the Orcutt Frontal Fault located about two miles to the north.

HYDROGEOLOGIC SETTING

The Santa Maria Basin lies to the north and east of the Facility. The western edge of the basin is situated approximately 2.5 miles from the Facility boundary. There are two main alluvial plains in the basin: one in the Santa Maria Valley and one in the Sisquoc Valley. Surface terraces occur between the alluvial plains and the bordering hills. The near-surface geology is composed primarily of unconsolidated water-bearing formations such as Quaternary alluvium, the Orcutt Sand, the Paso Robles Formation, and the Careaga Sand. These consist primarily of lenticular bodies of gravels, sands, silts, and clays. Deeper consolidated formations such as the Sisquoc and Monterey are generally considered incapable of yielding significant amounts of groundwater to wells and thus from the bottom and sides of the basin (Worts, 1951).

The San Antonio Creek Valley lies to the south of the Casmalia Hills (Muir, 1964). The northern margin of this basin is situated more than three miles south of the Facility boundary. Unconsolidated water-bearing formations lie near the surface and consolidated non-water-bearing formations lie deeper from the borders and bottom of the basin (Muir, 1964). The consolidated rocks that make up the Casmalia Hills form the southern and western boundaries of the Santa Maria Valley groundwater basin (Worts, 1951) and the northern boundary of the San Antonio Creek Valley groundwater basin (Muir, 1964). The Facility is located in the Casmalia Hills which are not situated within either of these groundwater basins.

The Santa Maria and San Antonio groundwater basins are connected by thin, overlapping, unconsolidated deposits that bend over the anticlinal structures that separate the two basins. Based on a review of published data (Hutchinson, 1980) there are two zones located four or more miles east of the Casmalia Resources Facility where unconsolidated materials occur between the two valleys. It is in these areas that the San Antonio Basin is hydrologically connected with the Santa Maria Basin (Hutchinson, 1980). Based on the data provided by Hutchinson (1980), this area of basin interconnection does not appear to be hydraulically downgradient of the Facility. While unconsolidated deposits in these groundwater basins are the primary water-bearing formations for the region (Muir, 1964; Worts, 1951), the potential for shifting of the groundwater divide between these two basins due to excessive pumpage in either basin is considered to be very slight due to the intervening anticlinal structures.

The primary direction of groundwater flow in the unconsolidated deposits in both basins is westward towards the Pacific Ocean (Muir, 1964; Worts, 1951). In the Santa Maria Basin, in addition to the overall westward movement of the groundwater, there is also a discernible flow component away from the northeastern slope of the Casmalia Hills (Worts, 1951). Correspondingly, in the San Antonio Creek Basin, groundwater divides coincide with the topographic divides (Hutchinson, 1980), suggesting that groundwater recharge to the basin is derived almost entirely from local rainfall.

Groundwater quality is of generally poor quality with background total dissolved solids values ranging from about 1,500 to 4,500 milligrams per liter. In addition, naturally occurring sulfate and chloride levels in groundwater exceeds EPA recommended drinking water standards.

In summary, the regional hydrogeologic setting appears to enhance the isolation of the Facility from the areas of groundwater usage in the Santa Maria and San Antonio Groundwater Basins. The town of Casmalia receives water from the Casmite well located approximately three to four miles to the northeast of the Facility in the Santa Maria Valley. Based on the available hydraulic head near the Facility and regional groundwater level data presented in reports by previous investigators (Muir, 1964; Worts, 1951), groundwater within the local flow system around the Facility does not appear to flow towards the Santa Maria Groundwater Basin. The regional configuration of the groundwater table and its similarity to regional topography and directions of stream gradients are the bases for inferring that, on a regional basis, shallow groundwater flows generally southward from the Facility vicinity toward Shuman Canyon and then probably westward towards the ocean. Furthermore, the low mean permeability of the consolidated rocks lying beneath the Facility, and those consolidated rocks lying between the Facility and the Santa Maria Basin, would restrict groundwater flow toward the Santa Maria Groundwater basin. Groundwater in the local flow system near the Facility is also not likely to flow into the San Antonio Groundwater Basin because of the position of the groundwater flow divides that appear to surround that basin, and the relatively low mean permeability of the consolidated rocks which occur between this basin and the Facility. The Facility is not within a 100-year flood plain or within an area subject to tidal waves.

Well Inventory Within Three Miles of the Facility

Water supply wells and oil and gas wells within three miles of the Facility have been located and inventoried. No geophysical exploration wells or geothermal wells are located within the survey area. Based on information obtained from the well survey and on local land usage, current groundwater use within one mile of the Facility is limited to the on-site non-portable water supply. One agricultural water supply well (Righetti Water Well No. 1) has been located within one mile of the Facility, but it is presently inactive. Current land use information obtained from the Santa Maria Zoning Department in Santa Maria, California, indicates that most of the land within one mile of the Facility is classified as large parcel agricultural. Cattle is presently grazed in this area. In the areas located between about one and three miles from the Facility, there are an additional 46 water wells and numerous oil and gas wells. The current usages of water wells in this area based on available data are summarized in Table 2.

Groundwater Occurrence

Groundwater occurs under water table conditions at depths of about 150 feet below ground surface in the northern portion of the Facility, to about 40 feet below most impoundments, to surface springs within the Todos Santos Member of the Sisquoc Formation. The Todos Santos is characterized as the most extensive, vertically and laterally, hydrostratigraphic unit (HSU) underlying and in the vicinity of the Facility. For the purposes of hydrogeologic characterization, an HSU is defined as a body of rock having considerable lateral extent and composing a geologic framework for a reasonably distinct hydrologic system. The presence of a pervasive weathering profile within the upper portion of this unit allows the Todos Santos to be subdivided into distinct units. Hereinafter, the upper portion of the Todos Santos hydrostratigraphic unit is referred to as the Upper HSU, while the lower portion is referred to as the Lower HSU. A water table contour map is presented in Figure 5.

The Upper HSU is composed of the weathered claystone portion of the Todos Santos. The transition zone of the weathering profile between the weathered and unweathered claystone observed in piezometers and core borings is relatively thin, ranging from about two to five feet in thickness. Since this zone exhibits some weathering characteristics similar to those of the weathered claystone, it is included within the Upper HSU. Near-surface geologic deposits (i.e., surficial clay-rich soil and colluvium) are thought to be of minor importance to groundwater flow at the Facility, as compared to the weathered claystone, due to their relatively limited extent beneath and in the immediate vicinity of the Facility. In addition, the waste material itself is not considered to be part of the Upper HSU due primarily to its discontinuous subsurface extent between waste management units within the Facility. The Lower HSU consists entirely of the unweathered portion of the Todos Santos.

The classification of the Todos Santos HSU into two units reflects the following:

- Stratigraphic data allowing identifi-cation and verification of the presence of a laterally-extensive weathering profile within the claystone;
- Structural data including seismic refraction surface geophysical surveys, borehole geophysical logs, and the substantial decrease in fracture density below the weathering profile observed during detailed on-site geologic mapping and in the recovered cores;
- Contrasting hydraulic conductivity values measured between the Upper and Lower HSU from short- and long-term hydraulic tests. Although permeabilities vary widely, the geometric means of the hydraulic conductivity values based on all available borehole hydraulic tests of the Lower HSU (1.5 X 10-6 cm/sec) is approximately 45 times lower than that of the Upper HSU (6.8 X 10-5 cm/sec). This contrast in mean hydraulic conductivity is expected to be sufficient to retard groundwater flow on a large scale between the Upper and Lower HSU as compared to the flow rates within the Upper HSU; and
- Hydrochemical data from adjacent clustered piezometers completed to different depths showing distinct differences in water quality between the Upper and Lower HSU. Both the natural geochemistry of the groundwater as well as the levels of site-related constituents detected are dissimilar for each unit.

Occurrence of Perched Water

Perched water is defined as unconfined groundwater that is separated from an underlying main zone of groundwater by an unsaturated zone. Perched water conditions occur when vertically migrating water from the ground surface encount-

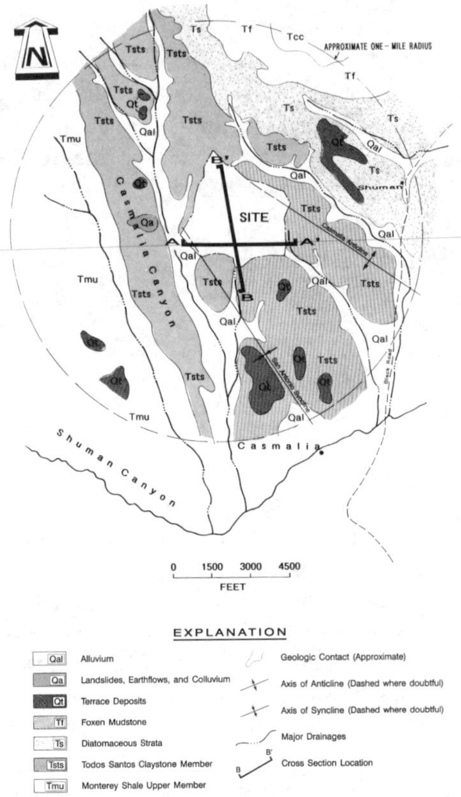

Figure 5. Contour map showing groundwater occurrence under water table and perched conditions

ers zones of relatively lower permeability which retard its migration. The water then accumulates above the low permeability material, causing the matrix above to become saturated. Such perched conditions have been observed at some locations within the northern portions of the Facility at depth intervals near the contact between the Upper and Lower HSU.

Minor zones of seepage have been observed and described in the east-side excavation of the Caustics Landfill Unit. These seeps, and similar ones in West Canyon, occur near the weathered/unweathered claystone contact and are laterally discontinuous over distances of several feet to several tens of feet. These seeps are also vertically discontinuous over distances of less than an inch to several feet. Zones of moisture and minor seepage that occur along bedding planes tend to be associated with intersections of geologic features which create contrasting hydraulic conductivities. Bedding in the claystone is typically not well developed and bedding plane partings, where they are discernible, are often laterally discontinuous and obscured by subhorizontal fractures related to erosional unloading. However, in some localized zones, bedding may influence flow pathways over short distances.

An excavation in the claystone on the east side of the Caustics Landfill Unit exposed a subhorizontal alignment of minor seeps. These seeps occur along a very thin discontinuous zone (in places approximately one-half inch thick) that occurs near the base of the excavation. Although bedding is not clearly discernible in this excavation, these seeps may coincide with a subhorizontally oriented, bedding-related zone or discontinuity. Some of the seeps appear to occur where steeply-dipping fractures intersect this subhorizontal discontinuity. In the northern portion of the excavation, the apparent line of seeps is approximately 10 to 12 feet below the contact between the weathered and unweathered claystone. However, in the southern portion of the excavation, the seepage line appears to coincide with the weathered/unweathered claystone contact. This alignment of seeps represents a minor zone of perched water that is formed by groundwater that has been restricted in its downward movement by a subhorizontal zone of relatively low permeability. These seeps are probably related to the position of the contact between the weathered and unweathered claystone, the possible presence of a bedding plane discontinuity, and intersecting fractures.

Other than the weathered/unweathered contact zone, it is unclear which stratigraphic zones, beds, or horizons within the weathered or unweathered claystone could predictably enable water to be perched. While it also may be possible that a certain geometry or set of fractures could enable some groundwater to be isolated or essentially perched at any depth interval above the water table, perched groundwater, if it were to occur in the claystone, is mostly like to occur near the contact between the Upper and Lower HSU.

GROUNDWATER MONITORING SYSTEM DESIGN

Although geologic materials underlying the Facility do not meet prescriptive permeability criteria for Class I units as defined under CCR (Subchapter 15, Section 2531(b)(1)), provisions for alternative standards consistent with the performance goal affords equivalent protection of water quality. The groundwater monitoring system was thus designed to monitor off-site migration of groundwater from the entire site rather than from specific waste management units, therefore, individual monitoring wells have not been used to identify vertical or lateral migration of constituents from specific waste management units within the Facility boundary.

The detection monitoring program incorporates surface water, groundwater and vadose zone soil pore liquids monitoring. The system itself is comprised of approximately 24 surface water sampling locations, most of which are situated offsite along Casmalia Creek and its tributaries. These locations are situated both upstream and downstream and constitute the points of compliance for surface water. The groundwater detection monitoring program includes approximately 18 background monitoring wells and approximately 59 downgradient monitoring wells which constitute the points of compliance with respect to groundwater. These wells are screened at various intervals including opposite the water table within weathered claystone (Upper HSU), fill and alluvium; at the contact between the weathered and unweathered claystone (Upper and Lower HSU, respectively), and within the unweathered claystone (Lower HSU) at depths of about 30 to 40 feet and 100 feet below ground surface. The vadose zone monitoring program which is in preparation is also planned for the Facility. Approximate location of surface water and groundwater monitoring wells which comprise the system is shown in Figure 6.

CONCLUSIONS

Relative to the overall operation of the Facility, the detection monitoring program design is consistent with the established zero discharge site boundary. The environmental barriers and collection galleries were installed to improve upon the favorable hydrogeologic characteristics of the site. In that regard, the design of the detection groundwater monitoring system in regards to selection of points of compliance is not significantly dependent upon horizontal to vertical hydraulic conductivity values within the uppermost aquifer (i.e., Upper or Lower HSU) as illustrated by Testa (1988, 1991). Instead, most of the wells comprising the groundwater system are downgradient of the Facility and located in respect to base flow toward and within the four drainages that have been identified as being the primary potential groundwater flow paths. These monitoring wells thus intersect the paths of potential fluid migration and allow detection of off-site migration of site-specific chemicals.

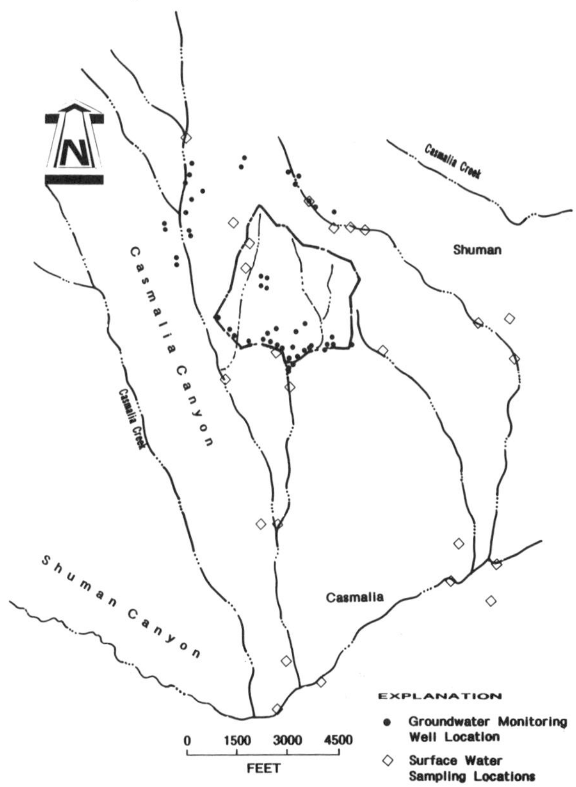

Figure 6. Approximate location of surface water sampling locations and detection monitoring wells which comprises the detection monitoring program

Overall, the Facility maintains numerous favorable environmental characteristics which make it suitable for the siting and operation of a hazardous waste disposal facility. These characteristics are:

- Abundant low permeability soil which underlie impoundments;
- Relatively slow groundwater flow rates on the order of five to 30 feet per year;
- Natural groundwater quality beneath the Facility exceeds EPA drinking water standards for several constituents, thus, limiting potential groundwater uses;
- Average evapotranspiration rates exceed average rainfall rates by approximately 40 inches per year. This condition in conjunction with the water needed for on-site reuse will minimize the amount of water stored in the impoundments, thus, lowering the possibility of the impoundments impacting State's water;
- Liner systems will exceed prescribed standards and be retrofitted with new containment systems every 10 years, if required;
- Liquids in impoundments will be limited to less than one percent halogenated organic compounds, pH between 5.0 and 9.0, and total dissolved solids on the order of 10,000 part per million;
- Existing cut-off trenches and extraction systems which capture and retain groundwater underlying the Facility;
- Underdrain systems installed under each new impoundment will maintain a five-foot separation between waste and highest anticipated groundwater, and serve as an additional collection system in the unlikely event of a failure of the primary, secondary and tertiary containment system; and
- Establishment of a closure fund.

Although beneficial use or potential beneficial use groundwater occurs within 1/2 mile downgradient of the Facility, the potential for adverse impact on this groundwater resource is considered very low.

ACKNOWLEDGEMENTS

I would like to gratefully acknowledge the word processing assistance of Brenda Saltiban and my wife Lydia, the figure preparation skills of Tony Carmouche, and the helpful assistance of Mr. Michael Patton of Applied Environmental Services, Inc., and representatives of the County of Santa Barbara Health Care Services, and the City of Santa Maria Environmental Management Office.

REFERENCES

Canonie Environmental, 1988, Hydrogeologic assessment report, Casmalia Resources Class I Hazardous Waste Management Facility: unpublished report prepared for Casmalia Resources, Santa Barbara, California, May 11.

Dibblee, T.W., Jr., 1989, Geologic map of the Casmalia and Orcutt Quadrangles, Santa Barbara County, California: Dibblee Geolgic Foundation, P.O. Box 60560, Santa Barbara, CA 93160.

Hoover and Associates, 1982, Water well completion report: unpublished report prepared for Casmalia Resources, Santa Barbara, California, August.

Hutchinson, C.B., 1980, Appraisal of ground-water resources in the San Antonio Creek Valley, Santa Barbara County, California: USGS Open-File Report No. 809-0750, 52 p.

Miller, G.A. and Evenson, R.E., 1966, Utilization of groundwater in the Santa Maria Valley area: USGS Water Supply Paper 1819-A, p. A1-A24.

Muir, K.S., 1964, Geology and ground water of San Antonio Creek Valley, Santa Barbara County, California: USGS Water Supply Paper 1664, 53 p.

Sylvester, A.G. and Darrow, A.C., 1979, Structure and neotectonics of the western Santa Ynez Fault System in Southern California: Tectonophysics, vol. 52, p. 389-406.

Testa, S.M., 1988, Hazardous waste disposal site hydrogeologic characterization: Proceedings of the Second International Conference on Casee Histories in Geotechnical Engineering, June 1-5, St. Louis, MO, Paper No. 1.26, p. 97-106.

Testa, S.M., 1991, Site characterization and monitoring well network design, Columbia Plateau Physiographic Province, Arlington, Oregon: Geol. Soc. Amer. Abstracts with programs, vol. 23, no. 5, p. A325.

Woodring, W.P. and Bramlette, M.N., 1950, Geology and paleontology of the Santa Maria District, California: USGS Professional Paper 222, 185 p.

Woodward-Clyde Consultants, 1988, Hydrogeologic site characterization and evaluation report, Casmalia Resources Hazardous Waste Management Facility: unpublished report prepared for Casmalia Resources, Santa Barbara, California, May 11.

Worts, G.F., Jr., 1951, Geology and ground-water resources of the Santa Maria Valley area, California: USGS Water Supply Paper 1000, 169 p.

TABLE 1

MATERIAL VOLUME AND CAPACITY

Material	Existing Volume of waste and cover (cubic yards)	Remaining Capacity of waste and cover (cubic yards)
. Acids	153,200	138,200
. Caustics	221,300	140,000
. Metals	243,800	433,000
. Solvents/ Pesticides	660,100	690,600
. PCB	390,400	

TABLE 2

SUMMARY OF WATER WELL USE

Number of Wells	Use
10	Domestic
1	Public supply
2	Industrial
4	Stock supply
2	Irrigation
13	Unused
3	Abandoned
11	Unknown use

METHANE GAS HAZARD WITHIN THE FAIRFAX DISTRICT, LOS ANGELES

Joseph W. Cobarrubias
Department of Building and Safety
Los Angeles, CA 90012

ABSTRACT

The March 24, 1985 incursion of gas into a retail clothing store within the Fairfax District caused an explosion and fire resulting in injury and extensive damage to the building. The District is situated over the old Salt Lake Oil Field where approximately 528 improperly abandoned oil wells are located. In response to citizens concerns, the City Council acted to create a Task Force to investigate the cause of the explosion, evaluate the risk, and provide recommendations that would ameliorate this hazard. The Department of Building and Safety was designated to chair the Task Force and to report our findings back to Council within 60 days. No funds were allocated for the study. The investigation was conducted with the aid of various technical consultants, each with specified information on local geologic conditions and methane gas behavior. In general, the area is underlain by up to 35 meters of unconsolidated alluvial deposits which overlie early Pleistocene San Pedro Formation, a faulted and folded sequence of sandstones and shales of the Pliocene Repetto and Pico Formations, and Miocene Puente Formation. Bedrock has been folded into an east-west trending anticlinal structure bounded by fault sets on the north and south. The southerly fault set consists of the Sixth and Third Street Faults.

The geologic conditions within the District which have a strong influence on methane behavior were found to be numerous and complex. A rising ground water table, discontinuous braided channels of high-permeability sands, decomposing peat bogs, the unknown locations of oil wells, faulting, and recent deformation of the area were conditions to evaluate. The possibility of a regional methane venting program was considered and rejected on the basis of cost. Instead, the Task Force recommended an approach which involves the identification of potential and high potential risk zones and methane gas surveillance by the City within these zones. New construction within the designated zones requires shielding of the building, methane detection systems, and perimeter venting of paved parking areas. Existing buildings require installation of methane detection systems to give early warning of possible methane incursion before an explosive concentration is reached.

INTRODUCTION

The Metropolitan Los Angeles area extends over an area of 755 square kilometers and lies predominantly on a lowland basin surrounded by youthful mountains and the emergent coastline of the Pacific Ocean (Figure 1).

In 1781, the Pueblo de los Angeles was established along the high ground of the Porciuncula (later called Los Angeles) River as a place to provide a food source for Spanish troops. Today, the pueblo has grown into an urban center for three million people with much of the city development occurring without regard for the geological processes active in the basin and hillside areas. Pavement of the natural soil and recontouring of the hills were unregulated developments that encountered floods and slope instabilities with large scale property damage and loss of life as the result. The City responded to these hazards by flood control measures and the country's first comprehensive grading ordinance in 1952.

Another type of geological hazard went undetected for 60 years, until the 1985 explosion and fire occurred unexpectedly within the Fairfax District. Here, dense commercial development covered almost all of the ground surface. The community hazard was created by extensive paving which blocks the natural seepage of methane gas from an abandoned oil field known as the Salt Lake Field. In fact, it was the gas and tar seeps that occur at the southerly end of the field that led to its discovery in 1902.

The number of oil wells in the Salt Lake Field expanded to 547 and by 1937, urbanization was well underway while most of the cable-tool wells were abandoned without regulation. It was the intrusion of urban sprawl into the field that caused the early demise of the Salt Lake Field in that it discouraged deepening wells or expanding the field to increase production. The special focus of this paper is on the conditions inherent with construction over an abandoned oil field and the response the City initiated to provide the necessary protection to people and property in the District.

THE 1985 METHANE EVENT

On March 24, 1985, an undetected incursion of methane gas occurred in the Ross Dress for Less located near the intersection of 3rd Street and Ogden Drive within the Fairfax District of Los Angeles. The smell of burnt coffee or sewage had been noticed in two buildings in the area for much of the day, but the fact that methane gas was seeping into the building was not recognized until after the explosion and fire occurred in the clothing store. Approximately 75 people were in the building at the time of the explosion, with bodily injury and extensive building damage limited to the office portion where the source of ignition originated.

The responding Fire Department units arrived to observe flames in exterior planter boxes, street storm-drain vaults and pavement joints. To reduce the risk of further explosions, the fires were permitted to burn in the hope that gas pressures would diminish. The force of the explosion left the office portion of the building without a roof and with the front exterior wall tilted outward. Analysis of the gas indicated that the explosion was caused by almost pure natural methane gas rather than pipeline gas. In an attempt to relieve the gas pressures a relief well (Anthony 1) was located and drilled. Gas pressures in excess of 25 psi were encountered between a clay bed and underlying perched ground water at 42 feet, which dissipated within 48 hours.

The Response

In response, the Los Angeles City Council acted to create a City Task Force to investigate the event, determine the cause of the explosion, why it took place at this location and what measures could be adopted to avoid such incidents occurring in the future. The Task Force was directed to investigate and report back within 60 days, with no allocation of funds. The course of action taken by the Task Force in the investigation of the area was made possible by the availability of professional expertise in the City.

THE INVESTIGATION

Information available to the Task Force was limited to the following:

- The location of the explosion was within the old Salt Lake Oil Field which was abandoned in the 1920's without modern controls for well abandonment.
- Oil seeps and low-pressure methane gas were known to have occurred in the area for the past 60 years.
- The methane seepage found at pressures in excess of 25 psi did not appear directly connected to a large volume of gas.
- The gas appeared confined to a stratigraphic zone 42 feet below the ground level, between a thin clay layer and the underlying ground water.
- Geologic and soil engineering data had been obtained for the area by project consultants on building excavations and the proposed Metrorail project.

Two separate technical committees were formed, composed of engineering geologists and petroleum geologists on one committee and landfill engineers on the other. The data and experience that was brought to the committee was augmented by available geologic and oil field literature and consultation with leading geophysicists, hydrogeologists, and geochemists.

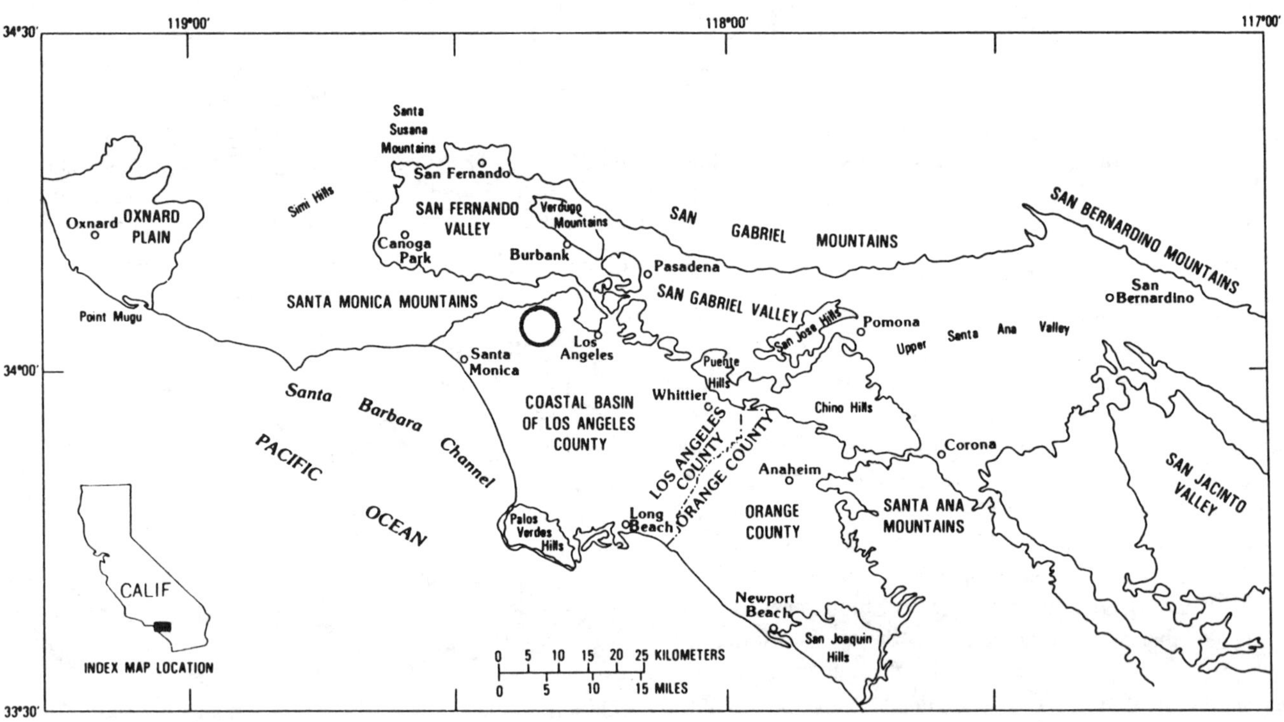

Figure 1. Location map showing the Los Angeles area, the Fairfax District lies between Los Angeles and Santa Monica, within circle.

GEOLOGIC SETTING

Physiography

The area under investigation is a portion of Los Angeles called the Fairfax District. The District is situated on a broad alluviated plain known as the La Brea Plain. It extends south-westerly from the Santa Monica Mountains onto the Los Angeles basin.

Stratigraphy

Wells drilled in the Salt Lake Field encountered braided channel deposits 50 to 100 feet deep consisting of interbedded clays, course sands, and gravels which were deposited by streams before they reached the ocean. The deposits are Mid-Pleistocene to Recent in age and may, in fact, represent sediments deposited from the Los Angeles River which flowed through the Cahuenga Pass area during Pleistocene and Post-Pleistocene time.

Underlying the alluvium is a 30,000 feet thick section of marine sediments in the central down dropped block of the Los Angeles Basin. Of primary interest to this investigation is the oil-bearing strata of the Repetto and Pico Formations of Miocene-Pliocene age (Figure 2).

Oil Field Structure and Faults Matrix

Within the Fairfax District the stream channel deposits conceal bedrock structure. Subsurface information from test hole borings, along with oil well logs from the Salt Lake Field, indicate that the bedrock is folded and faulted along a general east-west trend. This type of deformation has created traps in which oil and gas have accumulated in four distinct zones, separated by permeable strata. Almost all production of oil and gas was taken from the third oil bearing zone lying about 3,400 feet below the surface. As these folds extend easterly from the Salt Lake Field, they may connect with similar flexures noted within the Los Angeles City Oil Field. The surface occurrence of oil and gas at the Rancho La Brea Tar Pits lies at the southern extent of the Salt Lake Field and may be derived from the uppermost oil bearing Zone A.

Faulting has been observed within the subsurface of the Salt Lake Field. Fault sets occupy the northern and southern limits of the field underlying recent sediments. The southerly set composed of the Sixth and Third Streets Faults is known to breach the folded sediments creating structural traps for gas and oil within the oil-bearing zones. However, this entrapment more than likely is caused by the juxtaposition of oil-bearing sediments against impermeable strata, thereby restricting the flow of gas and oil beyond the fault plane (Figure 3). The Santa Monica Fault transects the westerly and northerly portions of the Field. This fault is

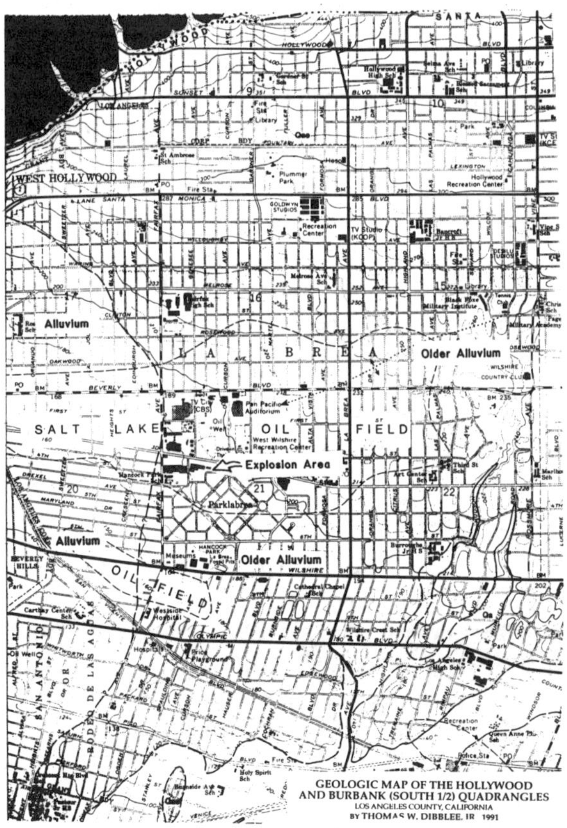

Figure 2. Geologic map of Fairfax - Wishire area

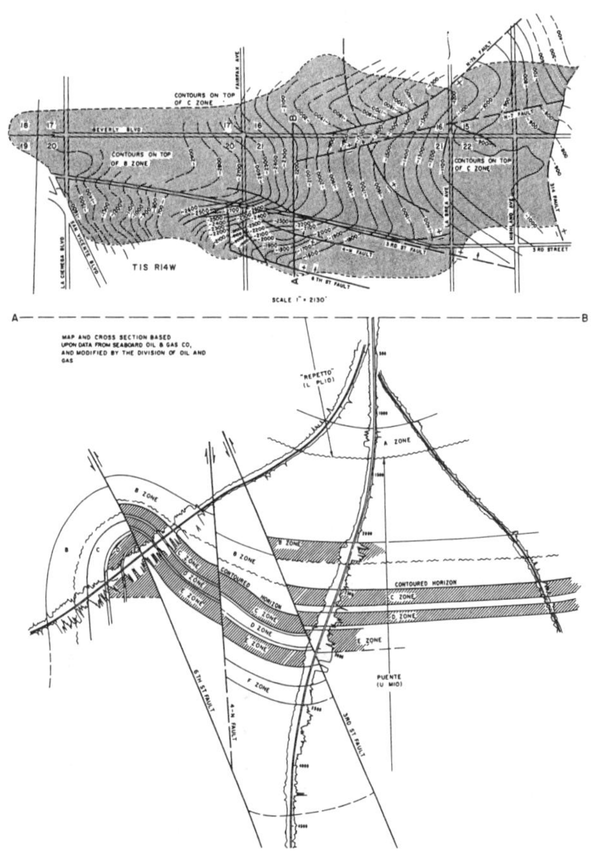

Figure 3. Salt Lake oil field

classified by the State of California as being potentially active.

A known fact about the Fairfax District is the occurrence of swamps, oil seepage, and gas in the area between the Sixth Street and Third Street Faults. Virtually all of the fossil finds at Rancho La Brea are located northerly of the Sixth Street Fault, thereby, indicating the location of swamps and tar pits of the past (Figure 4).

Today, occurrences of oil and gas are found in abundance northerly of the Sixth Street Fault. The other paralleling fault, the Third Street Fault, is located such that the area of explosion is located just northerly of that fault. The fault locations within the Fairfax District appear to have an effect on the occurrence of oil and gas at the surface.

SEISMIC ACTIVITY

According to Dr. Leon Teng of the University of Southern California, seismic activity has been monitored in this portion of the Los Angeles Basin. Seismic activity levels within the District are measurably less than those found to the northwest. On March 24, 1985, the day of the Ross Store explosion and fire, no measurable earthquake was recorded. The depths of minor earthquakes are reported to be 3 miles below the lowest production zone of the Salt Lake Field.

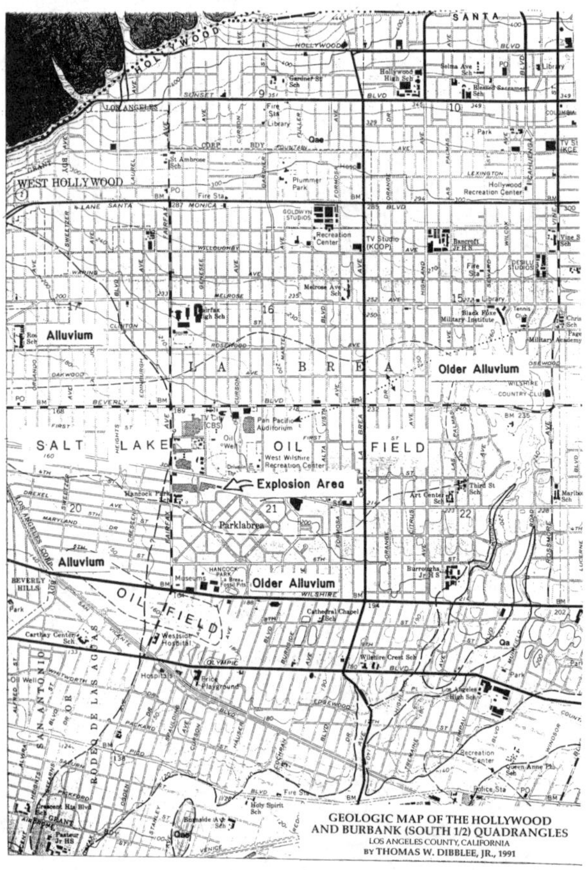

Figure 4. Fault location map showing area of explosion

There seems to be a correlation between the location of the Sixth and Third Street faults and oil and gas seeps, however, fault displacement within the recent sediments is not evidenced in the literature. The chemical analysis of surface methane gas has detected traces of free hydrogen for which there is no obvious source. Free hydrogen is not known to exist in the deep production zones or in the ground water regime. Its presence may have significance as studies in China have identified free hydrogen with other precursors of future seismic activity. Obviously, additional instrumentation is necessary to clearly define seismic activity within this portion of the Los Angeles Basin.

GROUND WATER

Monitoring of local water wells by the County of Los Angeles indicates that since 1978 ground water levels within the Fairfax District have been rising. This may have an important effect on the oil and gas migration within the oil field.

The total amount of ground water flowing beneath the Fairfax District is determined by the amount of inflow, outflow, and the consumption. The City of Beverly Hills drew upon this water supply for its municipal use. Between 1975 and 1976, however, the quality of the ground water was deemed unhealthful and pumping was terminated. Since that time, the City of Los Angeles has experienced a period of record rainfalls which have increased runoff and infiltration into the recharge areas of this portion of the basin. These occurrences are believed to be the main reason for increased inflow. The rising perched water table generally lies from 12 to 19 feet below ground level in the central part of the District, but intersects the ground near Wilshire Boulevard. The main ground water table lies at an approximate depth of 190 feet. Normal field pressures recorded for the ground water ranges between 4 and 7 psi.

OIL AND GAS OCCURRENCES

Oil and gas seeps have been active in this area for thousands of years. Evidence of swamps north of the Sixth Street fault are clearly seen in the geological record and still exist as the Rancho La Brea Tar Pits. As the hydrocarbons are transported to the surface, gas is expelled resulting in an accumulated layer of heavy tar or asphaltum at the bottom of the swamps.

The fact that surface gas and tar seeps occur at the southerly end of the field led to the discovery of the oil-bearing formations of the Salt Lake Field in 1902. As the Salt Lake Field developed, oil well operations expanded to approximately 574 wells over an area of 1,200 acres. The oil zones, from 50 to 166 meters thick, occurs at four separate and distinct oil levels ranging from 216 to 1,600 meters in depth. The largest yield of oil was extracted from the third zone. Production oil was described as of good quality, ranging from 9°-22° API and with considerable methane gas. The Salt

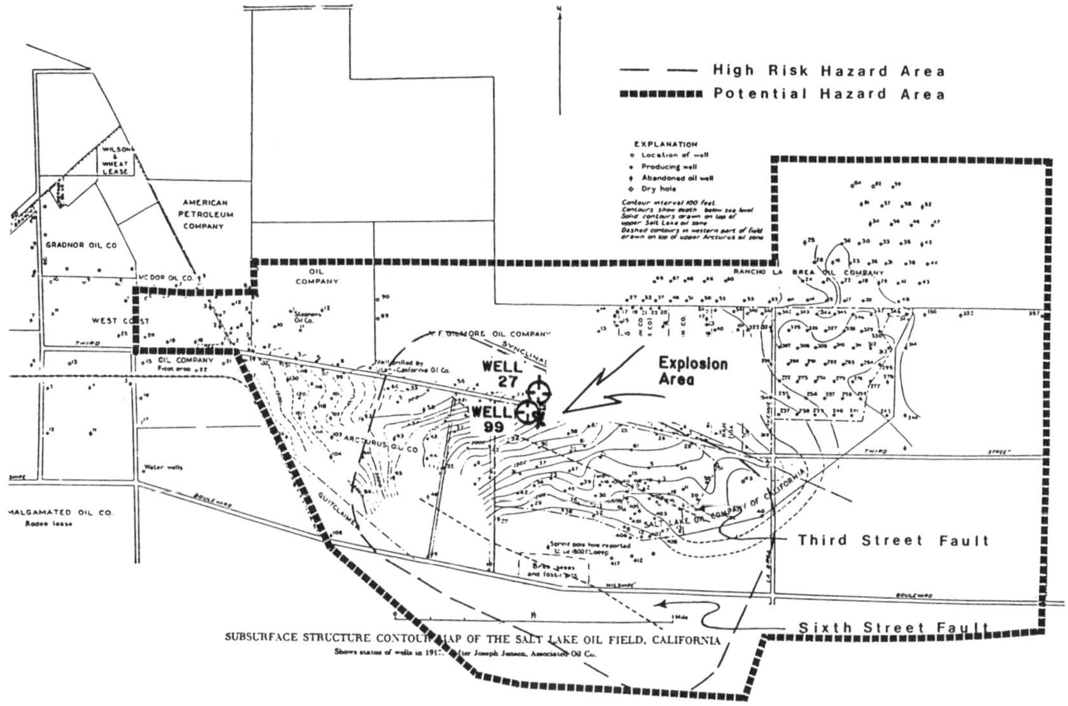

Figure 5. Map of the Salt Lake FIeld showing explosion area and oil well locations. (Modified from Plate 33, U.S. Geological Survey Professional Paper 165).

Lake Field was at one time, the highest producing field in California.

By 1937, almost all of the old cable-tool wells had ceased operation. Abandonment procedures were completed virtually without regulation and in most cases, the wells were severed 6-8 feet below grade level and plugged with wooden posts or backfilled with waste debris. Some of these wells encountered during later construction contained oil, water, and bubbles of methane gas. Figure 5 illustrates the relationship between the wells and explosion area.

METHANE PROBLEMS

Historically, oil and methane have been known to exist at the Rancho La Brea Tar Pits, in numerous major structure basement excavations, oil wells, and in the several oil sumps installed by the City between 3rd and 6th Streets. In these occurrences, the gas produced has always been low-volume seepage which dissipates quickly and has not been the cause of a building fire. Events of 1985 and 1989 are the first known high-volume seepage of methane known in the City. The initial identification of the gas as natural occurring methane caused great public concern. The numerous abandoned oil wells were considered to be the most probable cause of the problem. The obvious course of action therefore, was to obtain gas samples and test them for chemical composition.

The attention to chemical analysis or chromatography of the gas became an important consideration of the investigation. This focused attention became even more intense after the 1989 event, when isotopic analyses by two individual studies became available to the Task Force.

Gas Chemistry

Methane is a simple hydro-carbon formed as the by-product of organic decomposition. It is colorless and odorless when free of impurities. Due to the gaseous nature of methane and a lighter-than-air density, methane has the ability to diffuse rapidly upward through rock fractures and permeable materials. The upward diffusion, however, may be terminated against confining layers of impermeable clays or impeded by saturated sands or ground water. To some extent, the environment of the travel path can alter the chemical composition of the gas and, therefore, yield clues to the place of origin and some insight into the directness of the migration route. The ability to interpret these data is directly proportional to knowledge of these influences and the amount of data collected.

Within the Fairfax District the deep-production zone gas (C-D and E Zones shown in Figure 3) is high in heavy hydrocarbons and carbon dioxide, but contains little hydrogen sulfide and no nitrogen, oxygen, helium or hydrogen. However, the chemical constituents of the gas samples taken at the relief well at the clothing store was found to be almost pure methane, with small amounts of heavy hydrocarbons and traces of nitrogen, oxygen, carbon dioxide and hydrogen. No hydrogen sulfide was detected in the seepage gas. On the basis of gas chromatography, it could be hypothesized that the gas at the retail clothing store was not deep oil-field gas, but youthful methane formed at higher elevations where less heat and pressures would be encountered. This is termed biogenic methane such as forms in shallow, buried peat bogs less than about 100 feet deep. This origin would reasonably exclude the abandoned oil wells as a direct supplier of gas to the surface. It would also exempt current oil field operations, such as fluid injection, as a causative agent.

However, with the 1989 event, new technical data in the form of isotopic analysis came to the attention of the Task Force and Committee. This analysis, based on carbon-isotope ratios, yields a more sophisticated approach to gas origin and a controversy developed from its interpretation.

GAS ORIGIN CONTROVERSY

New information pertaining to gas origin was generated by Global Geochemistry under a grant from the Gas Research Institute and a GeoScience Analytical Report funded under Senate Bill 1458 authored by Senator David A. Roberti. Both studies included the more sophisticated isotopic analysis of the methane gas which was not available at the time of the 1985 Task Force Report. This analysis permitted the carbon-isotope ratio of samples taken from the retail clothing store location to be compared with samples taken at the Rancho La Brea Tar Pits and production field gas of the Salt Lake Field.

Thermogenic Methane Theory (Oil Field Gas Origin)

The isotopic ratio analysis is considered more credible than chromatographic (chemical) analyses, because chemical stripping can occur within the field gas as gas migrates to the surface, whereas carbon isotopes remain virtually unaffected. Many subtle characteristics are employed to discern probability of origin.

Based upon this testing information, Global Geo-chemistry concluded that the methane gas seepage at the Third and Ogden site is similar to the ratio of gas isotopes found in samples taken at Rancho La Brea Tar Pits and oil field gas. The study concludes that the gas incursions at the 3rd and Ogden site most probably have their origins in the thermogenic A-Zone oil field gas reservoirs. Methane derived from the very high pressures of the production zone would more clearly explain the high-pressures occurring within the Third and Ogden area than would the biogenic origin theory. The deep-field origin theory would best account for the rapid accumulation of high-pressures noted within the episodal zone than a slowly rising ground water system.

Biogenic Methane Theory (Shallow Origin)

The report of GeoScience Analytical, Incorporated was funded by the State through the Division of Oil and Gas. Based upon isotopic analyses of samples taken from 157 sites within the Salt Lake Field, the report concludes that the origin of methane in the Fairfax area is biogenic. This conclusion recognizes that isotope ratios characteristic-ally found in deep gases can be obtained in the shallow biogenic gases through microbial alteration of a thermogenic material. The resultant gas with biodepleted lighter hydrocarbons would appear to be of thermogenic origin as concluded by Global Geotechnical. In other words, the methane appears to be a product of aerobic microbes feeding on oil sump wastes within the oxide zone underlying the Third and Ogden area. If microbes produce methane from consumption of deep crude oil, the shallow biogenic methane will have isotopic ratios similar to thermogenic gas derived from the oil from which it is produced. It is also stated that the gas contains significant amounts of biogenic carbon dioxide, the source of which is not explained in the Global Report. GeoScience Analytical has photo information that locates five large oil sumps in the Third and Ogden area (circa 1927). The crude oil contamination, they believe, could be the source of the heavy hydrocarbons found in surface methane.

Conclusion on Origin

Obviously, the origin of the methane gas is complex and not clearly understood. Two reputable firms with qualifications and experience seem to have opposing opinions on the origin of the gas. But, if we intend to shield buildings or vent areas of probable methane gas seepage, the origin of the gas really does not matter. The most probable origin of the gas may be a combination of thermogenic and shallow biogenic sources with the high-pressure events from thermogenic (deep) sources and the low-pressure background gas from a combination of biogenic and thermogenic sources.

So, in reality, the origin of the methane has implications in the total understanding of the problem, but not in remediation or control. Methane gas is explosive whether produced within deep fields or in shallow marshes. Both types are capable of creating explosive conditions and both can be safely vented to the atmosphere.

Recurrence Potential

Prior to the 1985 event, it was believed that high-pressure methane events had not occurred in the past 60 years and would not likely occur in the future. The similarity of measured gas pressures of 27$\pm$ psi in 1985 and 20$\pm$ psi in 1989 may indicate that a high-pressure reservoir exists at depth and, perhaps more importantly, an established route for migration to the Third and Ogden area exists.

Theories have been suggested to explain surface migration into the Third and Ogden seepage area and the physical processes involved in these episodal events. Converse Consultants working for the Metrorail Project, noted that the area immediately north of Third Street to the area south of Lindenhurst Avenue is underlain by some of the more permeable sediments within the Fairfax area. This may explain the apparent ease of gas emplacement within the impact area, but does little to define migration routes between the venting area and place of origin.

Abandoned Wells as a Source of Methane

The potential hazard imposed by abandoned wells was not considered in the 1985 report because the gas was believed to have formed within the shallow biogenic areas above the production zone. However, the extent of chemical stripping of gas constituents and microbial alterations during migration probably could not occur within gas bubbles rising upward in a vertical shaft filled more or less with water or brine. These wells could be the source of low-volume methane that mixes at the surface with ambient gas at low pressures. All available chemical and isotopic information exonerates these abandoned wells as the source of the higher-pressure methane events.

The locations of various key wells, such as numbers 27 and 99 are shown on Figures 5 and 13, but these are not precise locations. According to information from the Division of Oil and Gas, Well #27 was abandoned by installing cement plugs and removing the casing down to a depth of 60 feet. A metal detector survey would not locate casing at that depth and no accurate survey information is available. The details of the abandonment of Well #99 were not known to the Task Force.

These wells should be considered in the future or as they are uncovered by excavation. Reabandonment to modern standards would be required under regulations of the California State Division of Oil and Gas.

Pressure Injection Operations

Within the Salt Lake Field, about 42 wells are currently producing from the "C" and "D" Zones. According to production records of the Division of Oil and Gas, more fluids are being withdrawn than are returned to the production zones. The type of injection used by McFarland Energy is a water disposal program (flooding) that does not cause field repressurization. Current oilfield operations within the Fairfax area have been recently evaluated by the Division of Oil and Gas and they believe the operations do not have an effect on the area methane seepage.

THE FINDINGS

The Task Force reports of 1985 and 1989 provided conclusions and recommendations which were adopted by the City Council as Division 15 of the Building Code. Basically, the Code regulations are directed towards identification of the

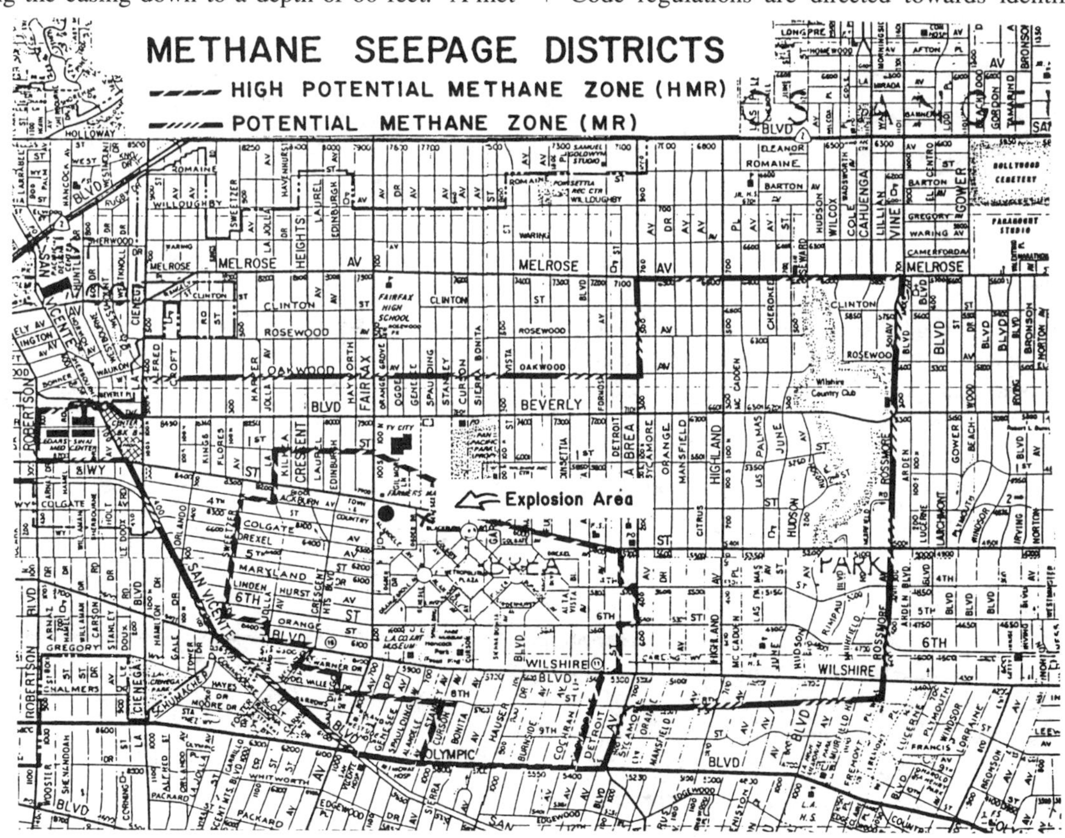

Figure 6. Methane seepage districts

potential areas of methane migration and either venting or shielding of buildings within the zoned areas.

The Solution

There is no simple solution to the Fairfax-Wilshire methane problem. Gas bubbles will, from time to time, rise to the surface. The best protection relies upon methane surveillance detection and precautionary measures to either remove the gas or to prevent the incursion of the gas into the building. The first phase of the solution is to identify existing or future structures located within the potential methane hazard areas. Then to prescribe the method of protection depending upon the type of structure, use, and degree of risk ascribed to the building location. Paved parking lots located within the high potential risk areas should be vented to retard methane accumulation beneath the pavement (Tables 1 through 3).

Areas of Risk

Methane Potential Hazard Zones were created based upon the most probable location for methane gas seepage and the established outline of the Salt Lake Oil Field. The zones are termed, in order of increasing risk: Zone of Potential Methane and Zone of High Potential Methane (Fig. 6). All permanent structures within the Methane Zones are classified into one of the following categories:

- Existing or Proposed Construction
- Commercial Buildings
- Major Structures or High-Rise Buildings
- Multiple Residential Buildings
- Single-Family Dwellings

Methane Gas Detectors

Continuous-monitoring methane detectors consist of a remote gas sensor connected to a central control box. In practice, the gas sensor is placed in a location of most probable hazard or near the building's ventilation system thereby being exposed to a maximum circulation and air volume. These devices are commercially available and do require technical expertise in their installation and calibration.

The detectors provide continuous monitoring of methane gas and sounds an alarm for building evacuation prior to the accumulation of explosive levels of gas. The system does not rely on knowledge of methane origin or migration paths. In buildings equipped with security systems coupled with an off-site alarm-monitoring center, the methane gas hazard could be monitored on a 24-hour basis. Figure 7 depicts the standard installation of gas control devices within a building.

Shielding or Building Barrier Techniques

Shielding of the building under construction will probably offer the best protection for occupants and structure. This procedure involves the placement of layers of PVC sheeting such as hypolon or its oil resistant equivalent below the concrete slab-on-grade. A blanket of sand with slotted venting pipes below the PVC sheeting will provide for a methane monitoring system and protection for the shielding material. It is of utmost importance that attention be given to the air-tight sealing of sheet overlaps and at contact points with foundation footings and conduits. Prior to pouring the concrete, the space blocks used to raise the steel rebars or wire-mesh fabric must not puncture and damage the sheeting fabric. An inspection check for imperfect seals should be performed prior to the pouring of concrete.

Perimeter Venting

Venting the perimeter of an existing or proposed building with various types of vent pipes or trenches serves to provide the gas with a negative pressure zone that represents the path of least resistance. Once the gas enters the vent system, passive or dynamic venting to an approved air outlet, automatic flaring device, or methane holding tank can be provided. These systems can vary on a building-to-building basis, but examples are as follows:

a. Foundation vents are trenches excavated against the exterior walls of buildings with a slotted 4-inch PVC pipe and

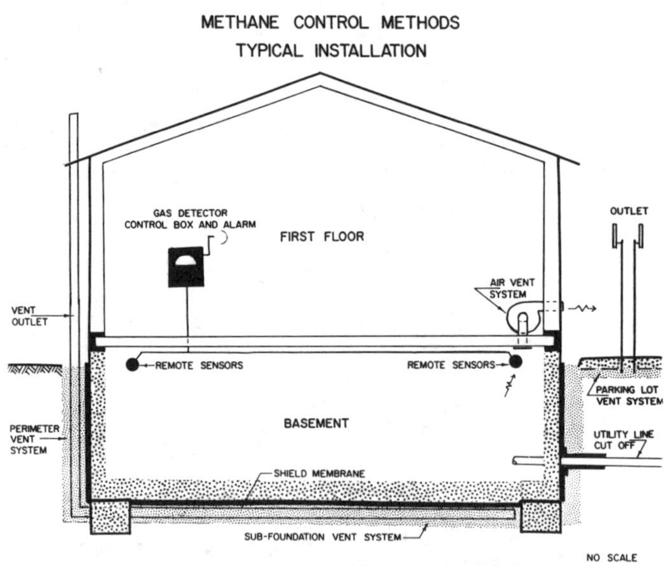

Figure 7. Drawing provides illustration of standard methane gas control devices

coarse-gravel backfill placed on and against the foundation footings.

b. Vertical perimeter vent pipes horizontally spaced along the outside footings can be either "driven" perforated metal pipes or PVC slotted pipes placed in predrilled small diameter borings of 10 feet or more in depth below foundation footings.

Utility Trench Cut-Off Walls

Subsurface conduits entering through slabs or foundation walls can provide open access routes for gas to enter into buildings. This hazard can be eliminated in existing and new construction by the two available design methods; the sleeved entry and the flexible entry. This procedure is also applicable to the shielding or barrier techniques where conduits are required to pass through the PVC sheeting.

Reabandonment of Oil Wells at the Site

In general, the Salt Lake Field wells are cable-tool wells with loose-joint casing. As a standard procedure used at the time in the abandonment process, the well casing was severed 6 to 8 feet below surface grade and the well covered with redwood planks and waste rubble. As these wells were never accurately located, the positions shown on the oil field maps are approximately located and no suggested method of locating them, such as use of metal detectors, seems effective.

Parking Lot Venting

One of the major obstacles of natural gas diffusion to the atmosphere appears to be pavement. Major concentrations of methane are reported below parking lot pavement and, where the pavement abuts a building, concentrations of methane and possible incursions can occur. The venting of significant paved parking areas appears prudent and feasible.

Many varieties of venting systems can be employed, but basically, the Task Force recommended a shallow trench at intervals determined by the porosity of the soil which is backfilled with a 4-inch slotted PVC pipe and gravel. In the event the pavement is to be restored over the top of the trench, passive stand-pipe vent outlets would be appropriate (Fig. 8).

Automatic Methane Ventilation System

A standard "Economizer" air conditioning system can be modified to provide the minimum required 4 air changes per hour ventilation within the building. When interior methane gas concentrates reach 25 percent of the lower explosive limit (L.E.L.), the building's "Economizer" can be activated by the detector system to provide 100 percent outside air intake and room exhaust. All existing and proposed commercial and institutional buildings within the High Potential Methane Zone or area of persistent methane seepage should be equipped to provide this automatic ventilation.

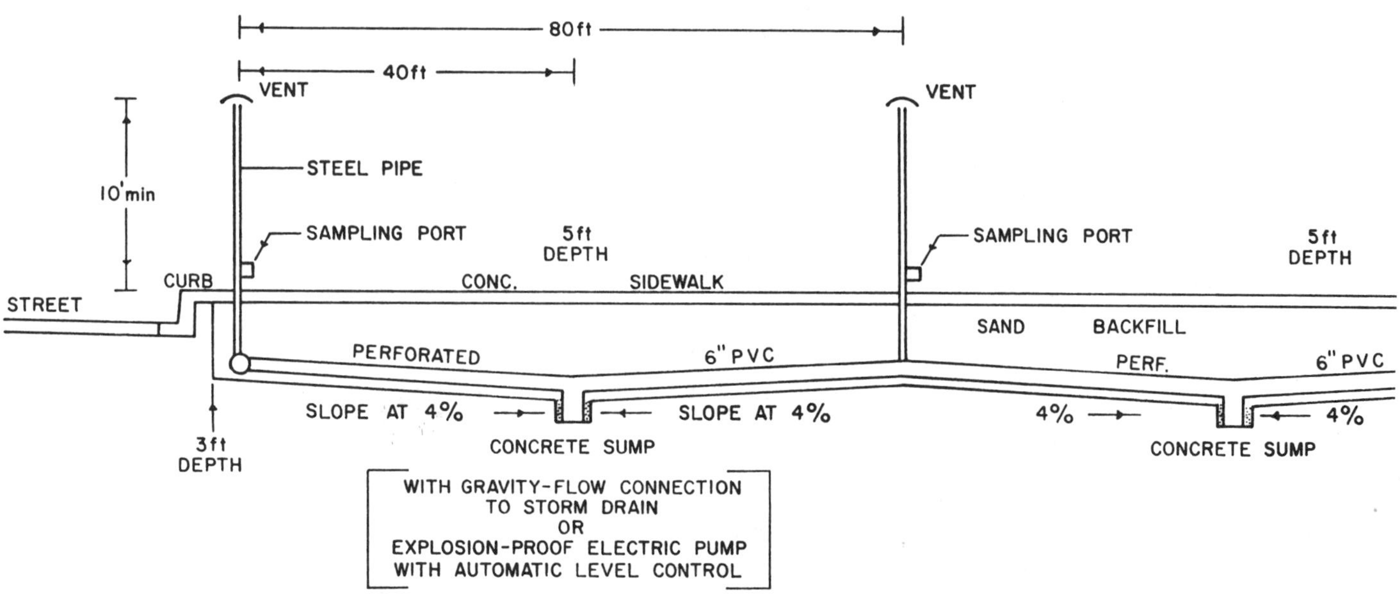

Figure 8. Suggested design for subpavement vent system, including separation of ground water and methane gas feature and samping port on vent riser

Existing Street Sumps

The City's oil collection sump program currently in operation appears effective and should be continued under the direction of the Bureau of Engineering. This involves the periodic removal of oil and water from the sumps.

THE CITY'S 3-POINT PROGRAM

The City adopted a regional plan which incorporates three approaches to the methane problem;

- Venting by interceptor relief wells and collector trenches (Fig. 9),
- An area-wide monitoring survey team,
- City inspection for Code compliance.

It is the intention of the 3-Point Program to reduce the threat, not necessarily eliminate it. What this program provides is an initiation of a course of action by both City and private property owners that should lead to an advanced warning system, gas collection, and venting intended to reduce or minimize bodily injury and property losses resulting from future methane gas incursions. The advantages of the Program extend beyond ameliorations of the Third and Ogden hazard as it can also provide useful information that might lead to a better understanding of the episodal high-pressure methane and the ambient low-pressure methane.

Venting

Methane Relief Wells

The various property owners are to install at least three wells at the locations indicated on the location map. The design of the relief well shown on Figure 10 is a multiple vent type consisting of a main vent to a depth of 52 feet and a surface vent pipe sealed just below the surface pavement. Instrumentation and design will facilitate sampling of gas and determination of gas pressures at either the subpavement depth or below the 42-foot deep clay bed. The existing relief well (Anthony 1) located in the parking lot in front of the retail clothing store should be periodically inspected and maintained by the owner.

Collector Trenches

Collector trenches may be designed according to available access and location of substructures, but the location must be within the areas of persistent methane seepage. The suggested trench design is intended to vent trapped gas below the pavement to the atmosphere. The sloping design of the collector pipes to a sump is to eliminate ground water from the gas venting system. In the event that the sumps cannot be gravity-drained to the storm drain, then explosion-proof pumps will be required. The Department of Public Works should construct or coordinate the trench construction because they will require excavations within the City street

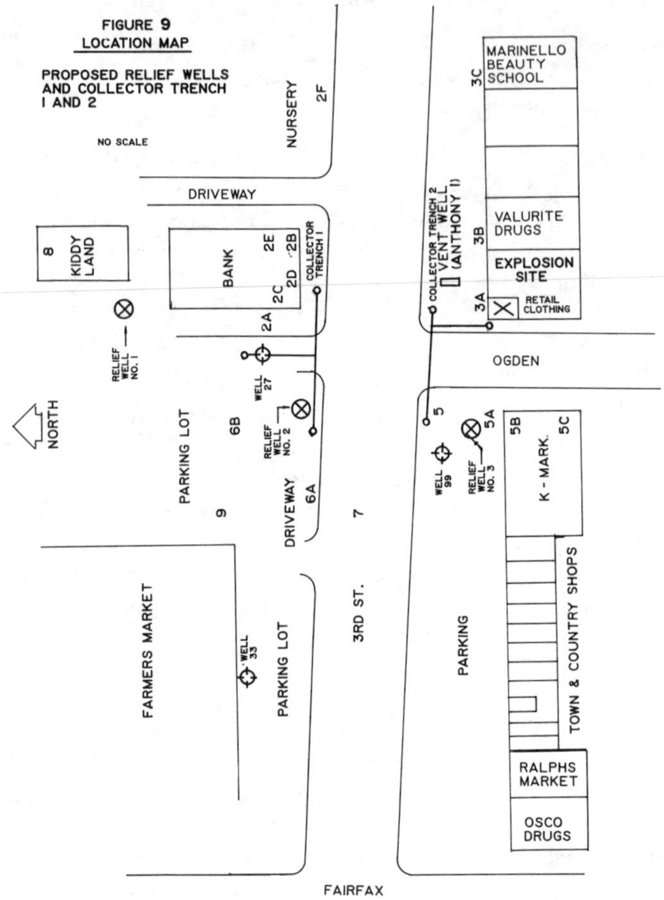

Figure 9. Location map - proposed relief wells and collector trench 1 and 2

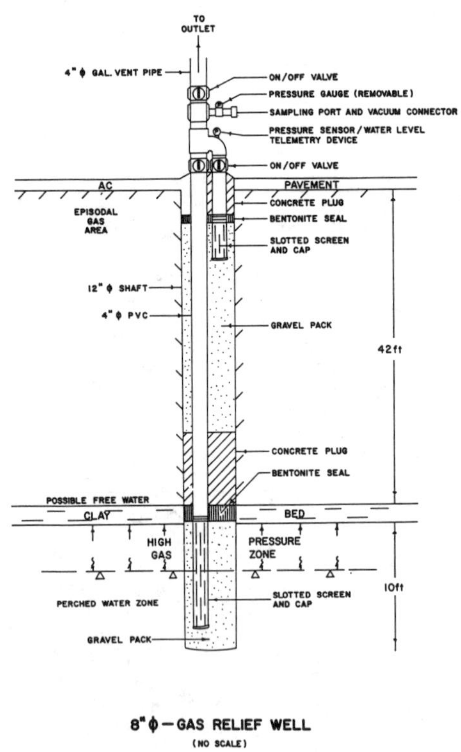

Figure 10. Suggested methane gas relief well featuring double well design and shallow/deep sampling capabilities.

easements in areas where underground utilities underlie the pavement.

Baseline and Ground Monitoring Surveys

Baseline Mobile Survey

A one-time baseline survey is to be performed by a specially equipped mobile vehicle to establish the ambient methane seepage patterns on a single time-line data base. The type of survey vehicle needed for a rapid regional survey is available in the private sector and is generally used in pipe-line surveys. Due to the nature of the work and equipment needed to perform the mobile survey, the work is to be contracted out to a private contractor.

Network Surveillance

The "Network Surveillance" Program was adopted as a one to two year program for the City to monitor areas of persistent methane seepage. Once identified, areas shall be considered for installation of a permanent monitoring device to provide early warning and response capabilities. Due to the special equipment and type of work to be performed, it appears most practical to contract the survey work to a private contractor who should report survey findings to the City. The Methane Task Force should be reconvened to analyze the survey data and to report its findings back to the City Council. At that time, the feasibility and value of installing vent systems, permanent monitoring stations, or a full-time perpetual ground surveillance survey of the area could be considered.

CONCLUSION

The incursion of high-volume methane gas in 1985 was unprecedented in the history of the City. The occurrence of low-volume methane seepage has been experienced in association with oil and water in basements and subterranean excavations, but not within buildings at explosive levels.

The findings and recommendations of the Task Force were derived from existing accumulated knowledge of the selected consultants and public agencies each with various expertise relating in some way to the methane gas problem. When the multiple expertise of the technical committees was brought together, a relatively complete understanding of the methane problem seemed to emerge. The one area of controversy was, and principally remains, the question of methane origin. However, the Task Force chose the defensive tactics consisting of detection, shielding, and venting to protect people and property - not locating and intercepting the gas at its source.

The re-occurrence of high-volume gas in 1989 was significant as it served to justify regional venting and team surveillance in the impact zone that had been established. One additional technique implemented into the Building Code was the energy savings "economizer" heating and ventilation system. This system provides normal room heating and air-conditioning with the option for automatic switching from conditioned air to 100% outside air intake and exhaust. This innovation uses the gas detectors to signal the "economizer" to switch to 100% outside air intake and exhaust to provide building ventilation in the event of methane gas presence. The desired ventilation rate is 4 air changes per hour.

Some techniques were discussed but not adopted. These include building pressurization to prevent gas entry and wick drains. The wick drains have potential in large areas where a regular pattern of venting may be desired. They consist of a woven fabric which is punched into the ground like thread in a sewing machine. The wicks draw fluids upward by capillary action to the surface. Similarly, it is possible that they may also provide an unobstructed path for methane gas to easily reach the surface. Major drawbacks of the wick system are equipment setup costs and possible damages to underground utilities. Contaminated water and oil drawn up by the wicks would require treatment before disposal into the storm drain system. The experimental nature of the wicks was the prime reason for not recommending this system.

It is emphasized, however, that the methods adopted by the Task Force are not intended to eliminate methane incursion, but to provide detection and a more convenient escape route for the methane than venting through existing buildings. Based upon early detection during the 1989 reoccurrence and prevention of an explosion, it appears that these recommendations are effective. However, the inspection of buildings by City inspectors for compliance to the methane regulations and other City action plans must be considered as important to the success of the program as the recommendations themselves.

ACKNOWLEDGEMENTS

The author wishes to recognize the efforts of numerous geologists, geophysicists, gas control specialists, and engineers from both private and public sectors who gave the Task Force their time and expertise. A very special thanks is given to Ms. Sharon Estrin, a fine geologist who assisted me with the Technical Committees and 1985 Task Force Report.

REFERENCES

Anderson, R.P. and Vogh, J.W., 1989, Various tracers identify injected storage gas: Oil and Gas Journal, March, p. 52-56.

Converse Consultants, Inc., 1984, Interim Geotechnical Report-Metrorail Project Design Unit A 250: Unpublished Report to Southern California Rapid Transit District (SCRTD).

Engineering-Science, Inc., 1984, Report of Subsurface Gas Investigation-Southern California Rapid Transit District Metrorail Project-Wilshire Corridor Alignment, Volume II: Unpublished Report.

GeoScience Analytical, Inc. 1986, A study of abandoned oil and gas wells and methane and other hazardous gas accumulations: Division of Oil and Gas, Limited Printing.

Hedberg, H.D., 1980 Methane generation and petroleum migration: American Association Petroleum Geologists, p. 179-206.

Jerden, H.D., 1985, Analysis of gases in the Earth's crust: Gas Research Institute, Contract 5081-360-0533.

Jerden, H.D., 1986, The origin and characteristics of seep gases in the Fairfax District, City of Los Angeles: Global GeoChemistry, Unpublished Report.

Methane Gas Task Force, 1985, Report on the March 24, 1985 Methane Gas Explosion and Fire in Fairfax Area: Unpublished Report, City of Los Angeles.

Methane Gas Task Force, 1989, Report on the February 7, 1989 Methane Gas Incursion: Unpublished Report.

Rice, Dudley D., 1975, Origin of the conditions for shallow accumulations of natural gas: Twenty-Seventh Annual Field Conference, Wyoming Geological Association Guidebook.

TABLE 1

METHANE CONTROL SYSTEMS

EXISTING CONSTRUCTION

TYPE OF STRUCTURE	HIGH POTENTIAL METHANE ZONE: REQUIRED SYSTEMS	HIGH POTENTIAL METHANE ZONE: PLANS AND PERMITS REQUIRED (See footnotes 1 & 2)	POTENTIAL METHANE ZONE: REQUIRED SYSTEMS	POTENTIAL METHANE ZONE: PLANS AND PERMITS REQUIRED (See footnotes 1 & 2)
Commercial, Industrial, and Institutional Buildings	1. Ventilation system activated by a gas detection system. *	Mechanical Electrical Fire Dept.	1a. Gas detection system or 1b. Ventilation system	Mechanical Electrical Fire Dept.
	2. Paved area vent system. Paved areas over 5,000 sq. ft. and within 15 feet of the exterior wall of the building	Plumbing	2. Paved area vent system. Paved areas over 5,000 sq. ft., and within 15 feet of the exterior wall of the building when methane is detected on the site	Plumbing
	3. Qualified Engineer to design system in all commericial, industrial, and institutional buildings *	Grading Division to review and approve system design report	3. Qualified Engineer. Not required	
	4. Emergency procedures required for all commercial, industrial, and institutional buildings	Fire Dept. approval	4. Emergency procedures for buildings with a methane detection system or ventilation system	Fire Dept. approval
Multiple Residential Buildings	1a. Gas detection system or 1b. Ventilation system	Mechanical Electrical Fire Dept.	1a. Gas detection system or 1b. Ventilation system	Mechanical Electrical Fire Dept.
	2. Paved area vent system. Paved areas over 5,000 sq. ft. and within 15 feet of the exterior wall of the building	Plumbing	2. Paved area vent system. Paved areas over 5,000 sq. ft., and within 15 feet of the exterior wall of the building when methane is detected on the site	Plumbing
	3. Emergency procedures for buildings with a methane detection system or ventilation system	Fire Dept. approval	3. Emergency procedures for buildings with a methane detection system or ventilation system	Fire Dept. approval
Single Family Dwellings with basement	1. Gas detection system	Electrical Fire Dept.	1. Gas detection system	Electrical Fire Dept.

1. Fire Department permits are to be obtained from the Engineering Unit, 9th Floor, City Hall East.
2. In addition to Mechanical, Electrical, Plumbing, and Fire Department Permits, a Building Permit is required for all work required by Division 15 of the L.A.B.C. except for single family dwellings.

* Pending change in Section 91.1504(a) of the L.A.B.C.

TABLE 2

METHANE CONTROL SYSTEMS

NEW CONSTRUCTION

TYPE OF STRUCTURE	HIGH POTENTIAL METHANE ZONE REQUIRED SYSTEMS	PLANS AND PERMITS REQUIRED (See footnotes 1 & 2)	POTENTIAL METHANE ZONE REQUIRED SYSTEMS	PLANS AND PERMITS REQUIRED (See footnotes 1 & 2)
Commercial, Industrial, and Institutional Buildings	1. Ventilation system activated by a gas detection system *	Mechanical Electrical Fire Dept.	1a. Gas detection system or 1b. Ventilation system	Electrical Mechanical Fire Dept.
	2. Membrane shield under the building	Building	2. Membrane shield under the building	Building
	3. Foundation vent system. Buildings over 50 feet in width	Plumbing Fire Dept.	3. Foundation vent system. Buildings over 50 feet in width when methane is detected on the site	Plumbing Fire Dept.
	4. Paved area vent system. Paved areas over 5,000 sq. ft. and within 15 feet of the exterior wall of the building	Plumbing	4. Paved area vent system. Paved areas over 5,000 sq. ft. and within 15 feet of the exterior wall of the building, when methane is detected on the site	Plumbing
	5. Qualified Engineer to design system in all commercial, industrial, and institutional buildings *	Grading Division to review and approve system design report	5. Qualified Engineer to design system in buildings covering over 50,000 sq. ft. of lot area or with more than one level of basement	Grading Division to review and approve system design report
	6. Emergency procedures required for all commercial, industrial, and institutional buildings	Fire Dept. approval	6. Emergency procedures for buildings with a methane detection system or ventilation system	Fire Dept. approval
Multiple Residential Buildings	1a. Gas detection system or 1b. Ventilation system	Electrical Mechanical Fire Dept.	1a. Gas detection system or 1b. Ventilation system	Electrical Mechanical Fire Dept.
	2. Membrane shield under the building	Building	2. Membrane shield under the building	Building
	3. Foundation vent system. Buildings over 50 feet in width	Plumbing Fire Dept.	3. Foundation vent system. Buildings over 50 feet in width when methane is detected on the site	Plumbing Fire Dept.
	4. Paved area vent system. Paved areas over 5,000 sq. ft. and within 15 feet of the exterior wall of the building	Plumbing	4. Paved area vent system. Paved areas over 5,000 sq. ft. and within 15 feet of the exterior wall of the building, when methane is detected on tne site	Plumbing
	5. Qualified Engineer to design system in buildings covering over 50,000 sq. ft. of lot area or with more than one level of basement	Grading Division to review and approve system design report	5. Qualified Engineer to design system in buildings covering over 50,000 sq. ft. of lot area or with more than one level of basement	Grading Division to review and approve system design report
	6. Emergency procedures for buildings with a methane detection system or ventilation system.	Fire Dept. approval	6. Emergency procedures for buildings with a methane detection system or ventilation system	Fire Dept. approval

1. Fire Department permits are to be obtained from the Engineering Unit, 9th Floor, City Hall East.
2. In addition to Mechanical, Electrical, Plumbing, and Fire Department Permits, a Building Permit is required for all work required by Division 15 of the L.A.B.C. except for single family dwellings.

* Pending change in Section 91.1505(a) of the L.A.B.C.

TABLE 3

METHANE CONTROL SYSTEMS

NEW CONSTRUCTION

TYPE OF STRUCTURE	HIGH POTENTIAL METHANE ZONE REQUIRED SYSTEMS	PLANS AND PERMITS REQUIRED (See footnote 1 & 2)	POTENTIAL METHANE ZONE REQUIRED SYSTEMS	PLANS AND PERMITS REQUIRED (See footnotes 1 & 2)
Single Family Dwelling or Duplex with concrete slab floor	1. Membrane shield under the building	Building	No requirements	
	2. Foundation vent system	Plumbing	No requirement	
Single Family Dwelling or Duplex with basement	1. Gas detection system	Electrical Fire Dept.	1. Gas detection system	Electrical Fire Dept.
Single Family Dwelling or Duplex with raised floor	No requirement		No requirement	
Swimming pool	Engineering geologic study	Grading Division	No requirement	

1. Fire Department permits are to be obtained from the Engineering Unit, 9th Floor, City Hall East.
2. In addition to Mechanical, Electrical, Plumbing, and Fire Department Permits, a Building Permit is required for all work required by Division 15 of the L.A.B.C. except for single family dwellings.

CAUSE OF THE 1985 ROSS STORE EXPLOSION AND OTHER GAS VENTINGS, FAIRFAX DISTRICT, LOS ANGELES

Douglas H. Hamilton
Principal Geologist
Earth Sciences Associates, Inc.
701 Welch Road
Palo Alto, California 94304

Richard L. Meehan
Consulting Civil Engineer
701 Welch Road
Palo Alto, California 94304

INTRODUCTION

Late in the afternoon of March 24, 1985, methane gas that had been accumulating ignited in an auxiliary room of the Ross Dress-For-Less Department Store located on Third Street, in the Wilshire-Fairfax District of Los Angeles. The resulting explosion blew out the windows and partially collapsed the roof of the structure, reduced the store interior to a heap of twisted metal and resulted in injuries requiring hospital treatment of twenty-three people. Police closed off four blocks around an eerie scene of spouting gas flames that continued through the night.

In the following days, a drill rig brought to the site was used to test for possible gas accumulations in the alluvial soil beneath the store. A "pocket" of pressurized gas was encountered at a depth of 42 feet beneath the parking lot between the store building and Third Street. Gas was also encountered in several other borings at the site in smaller quantities and at lower pressures. Pressure gauges, control valves, and, on the hole where the high pressure pocket was encountered, a valved flare pipe, were installed. Following a brief period during which gas was flared and bled off into the air, the anomalous gas condition at the Ross Store site gradually declined to the normal gas concentrations characteristic of the local area. In 1989 another venting incident occurred, this time at several sites on the north side of Third Street. This second venting fortunately was detected in time, and did not ignite. In this case, water and silt were ejected from outdoor vents along with the gas, in addition to accumulation of dangerous levels of gas in several buildings. A blow-out crater several feet deep, from which dirt and small stones were ejected several feet into the air was formed during this episode which lasted about 24 hours.

The setting of the accident — an old-world Levantine market place a few miles from Hollywood; the famed tarry graveyard of the sabre-toothed tigers; pillars of fire dancing in the darkened streets — these biblical images attracted attention of the press, the bar, and local politicians. And yet, three months later when a hastily convened panel of experts announced that the event was caused by digestive rumblings of an ancient and invisible swamp the whole thing had been mostly forgotten, the explanation accepted as yet another production of Los Angeles' quirky environment. Outside of a lawsuit that was settled quietly in 1990, the possibility that the accident was caused by the knowing agency of Los Angeles' lesser known industry or that the official report of the experts, rather than being a serious statement of the scientific community, was a heavily edited script with a happily blameless ending, was not made known to the public, as we shall proceed to do here.

THE SETTING — TAR PITS AND OIL FIELDS IN THE URBAN ENVIRONMENT

The vicinity of the gas ventings lies within a sloping plain between the easternmost Santa Monica Mountains and the isolated topographic high of the Baldwin Hills (Fig. 1). The area drains via Ballona Creek, which cuts through Ballona Gap to the coastal plain fringing Santa Monica Bay. The Ross Store is in a neighborhood that has been attracting attention since the mid-Pleistocene on account of its hydrocar-

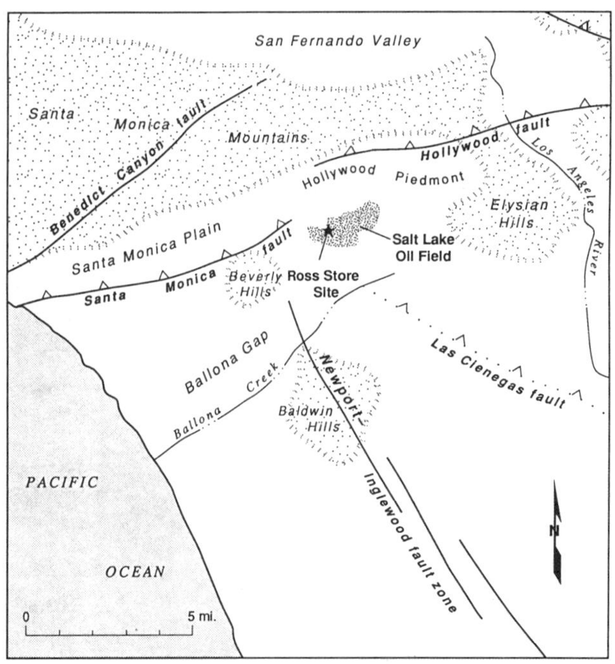

Note: Fault locations from Wright, 1991.

Figure 1. Location and Regional Setting

bon wealth; the La Brea tar pits are about two blocks to the south.

The old Salt Lake oil field, once the biggest producing field in California, directly underlies and extends west and east of the immediate neighborhood (Fig. 2). Methane gas has been a ubiquitous feature of this area so that urban engineering for both public and private facilities has required consideration of the hazards of flammable gas present in the soil. Examples include the excavation for the L.A. County Flood Control Pan Pacific storm runoff basin and park, the storm drain along Third Street, and the (formerly planned but since relocated) L.A. Metro subway along Fairfax Avenue.

Before the Ross Store explosion in 1985, there was general acceptance of the view that the hydrocarbon presence in the surface and shallow subsurface environment had its source in the underlying reservoir of the Salt Lake oil field. Upward migration of oil from parts of this reservoir was thought to be the source of the "breas" or tar accumulations that had trapped many of the Pleistocene residents of the area (Shaw and Quinn, 1986). Gas known to be present at shallow depth within the alluvial soil was likewise considered to have escaped from the reservoir and migrated toward the surface (Converse et al., 1981). The surface accumulation of the residual bitumin "breas" (Spanish for pitch) led to the discovery of the once-prolific Salt Lake oil field in 1902. As the field developed, mainly before 1917, further communication between the confined oil, gas, and water in the reservoir

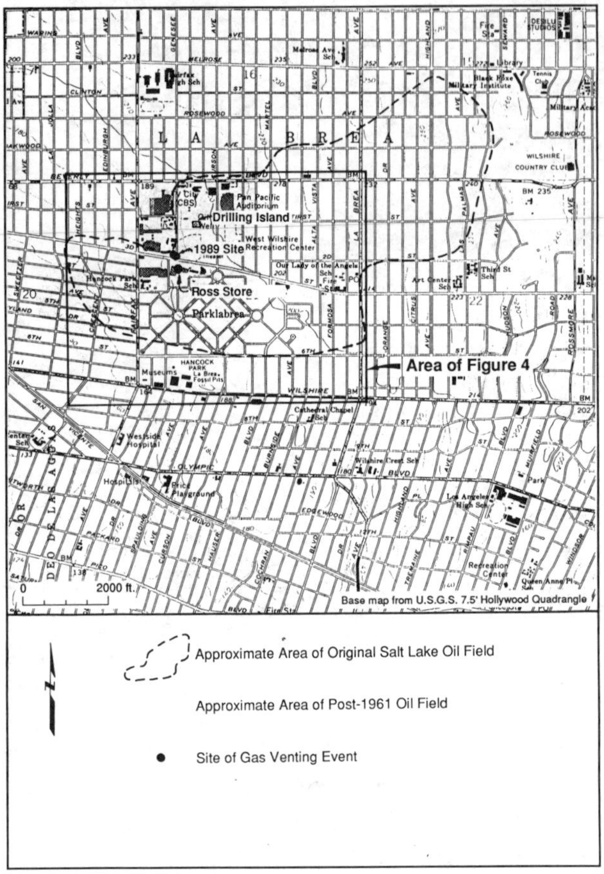

Figure 2. Wilshire-Fairfax area and the Salt Lake Oil Field

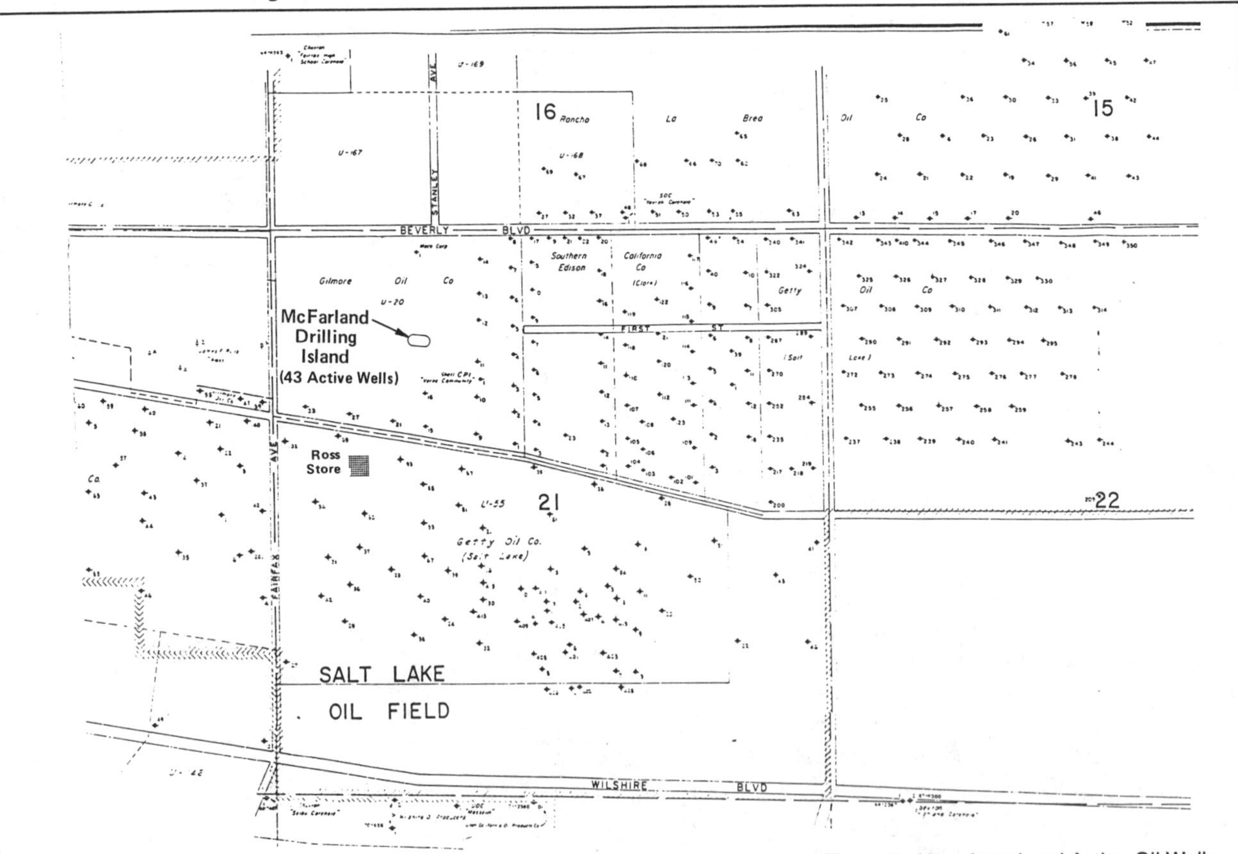

Figure 3 Abandoned and Active Oil Wells in the Salt Lake Oil Field

Source: C.D.O.G. Map 118

and the surface was established by drilling more than 450 wells (Fig. 3). Nearly 50 million barrels of oil were brought to the surface through these wells. Most was conveyed off-site to markets, initially in tanks and thereafter mostly by pipeline, but some was spilled and left in sump basins that were then a standard feature of the production system. By the 1930s, the urban development value of the land overlying the Salt Lake oil field exceeded its value as a production site for a declining oil field, and the field was largely abandoned. The legacies of a thirty-year oil boom included several large family and corporate fortunes, a few square miles of land where the natural upward hydrocarbon migration was complicated by the presence of hundreds of abandoned wells plus numerous filled sumps and covered-over spills, all underlain by a partially depleted hydrocarbon reservoir. Following the abandonment of most oil production, the area was rapidly covered by the homes, apartments, businesses and industrial facilities that exist today (Fig. 2). The remaining and most notable hydrocarbon presence is manifest in that popular and scientifically important exhibit of the "La Brea Fossil Pits" and Page Museum in Hancock Park on Wilshire Boulevard near Fairfax Avenue.

Meanwhile, beneath this urban environment, the underlying Salt Lake oil field was again tapped, in 1961, through skillful application of modern oil drilling techniques and production technology. This time, however, the entire surface installation was contained in an unobtrusive, screened complex covering only one acre. This compact facility included the heads of 43 wells which were slant-drilled to exploit several lease block holdings over an area of about a square mile (Figs. 3, 4), together with necessary equipment for separation of the well-head crude oil into oil, gas, and water components. The gas was disposed of by injection back into the Gilmore block of the field between 1961 and 1971. Salt water was initially disposed of by discharge into a local storm drain, but for environmental reasons was required to be injected back into the field, also in the Gilmore block, since 1964 (Crowder and Johnson, 1963; California State Division of Oil and Gas [C.D.O.G.], 1974).

AFTERMATH OF THE 1985 ROSS STORE EXPLOSION

After the flames subsided and the post-explosion exploration and gas-control drilling was completed, a special Task Force was convened at the direction of the Los Angeles City Council. The Council wanted to know what had happened and how future accidents could be prevented. The Task Force was made up of representatives from both City and State agencies and from industry, and it had available a Geologic Technical Committee which included experts in engineering geology and other relevant disciplines. Concurrent with the deliberations of the Task Force, the Senate of the California State Legislature enacted SB1458, the Roberti Bill, which required a study of the location of all abandoned oil and gas wells within the State of California.

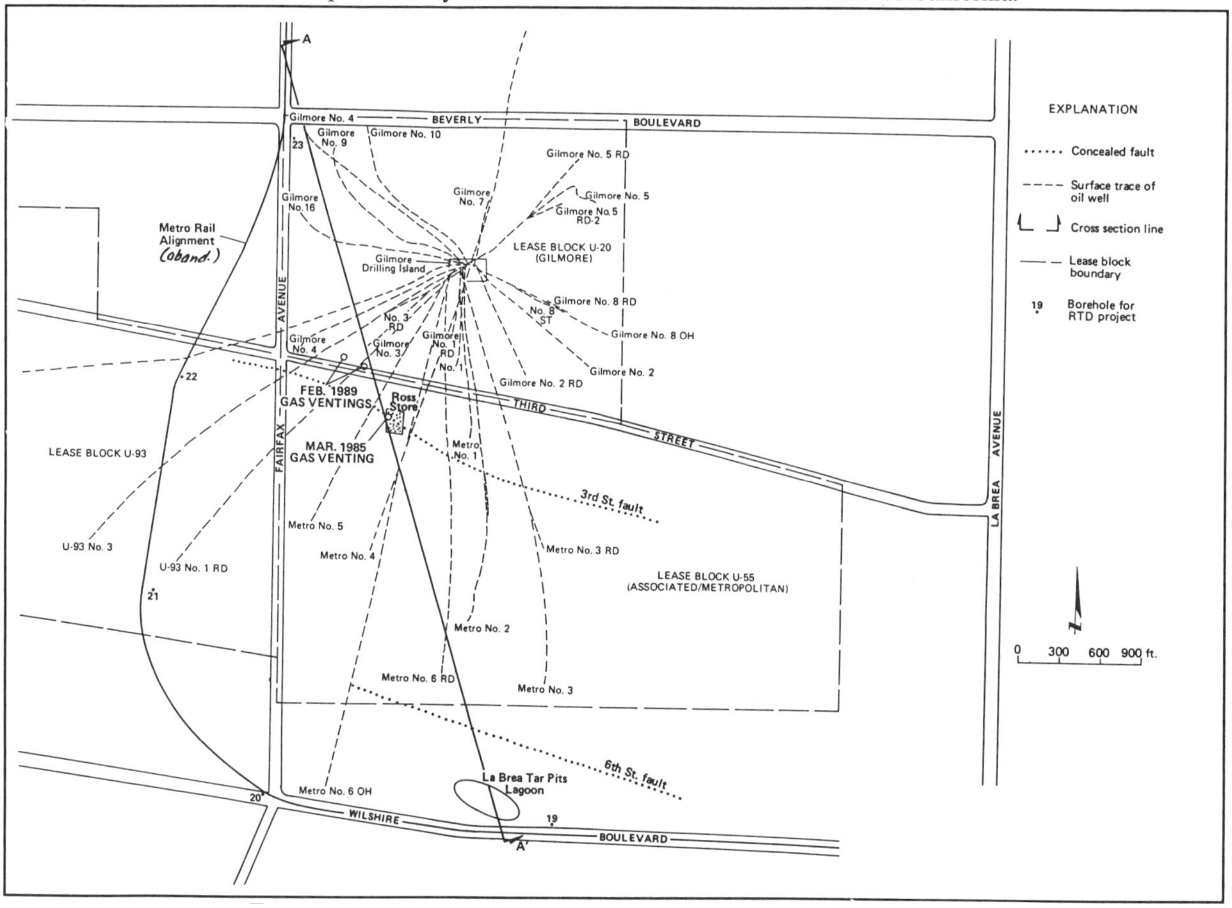

Figure 4. Surface plan of wells, faults and gas ventings, vicinity of Ross Store Site

Despite its own initial interest in abandoned wells as a potential source of the gas and the mandated purpose of the Roberti Bill to study such wells, the Task Force's surprising conclusion was that neither the existence of dozens of such abandoned wells, nor even the presence of the underlying oil and gas field which had been venting to the surface for a hundred millennia, had anything to do with the gas venting accidents. In fact, the oil field operation was not even mentioned as a possible agent in the Task Force report. Instead, a chemical analysis of a sample of the venting gas led the Task Force to conclude that the gas was of "biogenic" origin, presumably derived from near surface decaying of organic matter in the alluvial soil, rather than being of petrogenic or thermogenic origin, i.e., derived from rock-source hydrocarbon (oil field type) accumulations. The Task Force then went on to present a scenario of shallow "biogenic" methane being displaced and pressurized by a rising water table in the perched fresh-water aquifer beneath the Ross Store. It was an imaginative explanation: what school child is not aware of that primordial scene of mastodons thrashing about in a gassy swamp (as is artfully represented at full scale in the Hancock Park exhibit, only two blocks from the Ross Store). Troublesome legal issues were eliminated by this conclusion, the implication being that the methane hazard could exist virtually anywhere, so no human agency was at fault for its workings. The presence of the abandoned wells, the filled sumps, or anything else having to do with the past and current exploitation of the Salt Lake oil field evidently was considered no more than coincidence.

About a year after the issuance of the Task Force report, the results of two separate studies of the composition and indicated origin of near surface gas samples both at the Ross Store and elsewhere in the Los Angeles Basin, became available. One report was prepared under contract to the C.D.O.G., in response to the Roberti Bill (Geoscience Analytical, Inc., 1986). It presented the results of a study of eight selected urban areas sited over oil and gas fields where there were wells abandoned prior to 1930, and where there was a history of oil and/or gas seepage. This report concluded that with few exceptions, the methane present in these areas was, as at the site of the Ross Store building, of biogenic origin and thus unrelated to the underlying oil and gas accumulations. In their opinion it was essentially swamp gas. This remarkable conclusion was all the more interesting since it turned out that most of the areas sampled are local topographic highs. In any case, the results of this massive study assured the public the hazardous conditions identified by the Task Force as existing at the Ross Store, as well as "the vast majority of the high concentration gas samples" taken from the surface areas overlying oil and gas deposits throughout the Los Angeles basin, were also of purely natural swamp gas origin unrelated to human activities or oil field presence.

The other study, based on isotopic analysis of samples of gas at the Ross Store building escaping on the day following the explosion, concluded that the gas was of thermogenic origin, hence, came from an oil reservoir source (Global Geochemistry, 1986). These contradictory studies, one compatible with the Task Force's "rising ground-water displacing shallow biogenic gas" and the other clearly not, failed to elicit any notice outside of the groups of attorneys and technical consultants who were by then engaged in the inevitable legal maneuvering among the injured explosion survivors and the possible responsible parties.

Official disinterest continued into early 1989, when a second notable gas excursion occurred in the Ross Store neighborhood. This event, referred to officially as "Incident #113", is described in the subsequent "Task Force Report II on the Methane Gas Incursion" (1989) as follows:

"At or before 6 a.m. on February 7, 1989, an incursion of methane gas occurred within the K-Mart Store located at Third Street and Ogden Drive. The gas was detected on a portable unit which had detected near-explosive levels of methane within the store when employees entered the building. The Fire Department was notified and the employees evacuated the building. The responding Fire Department units observed a fountain of gas, water, and mud at the southwest corner of the Gilmore bank and additional emergency units were dispatched to the scene. Within 2 hours, explosive levels of methane were measured within the Gilmore Bank, K-Mart store, and in subsurface locations outside of these buildings including the Ross Dress-For-Less Store. The response team soon recognized similarities to the 1985 event in location and that the gas incursion was a higher-pressure type event. The area was then cordoned off to prevent ignition and explosion of the gas."

Fortunately, no explosion occurred this time and, as it had in 1985, the gas "incursion" dissipated within a few days time. The venting event did, however, trigger a reconstitution of the Task Force that had been convened following the 1985 explosion, and this body generated the "Task Force Report II" noted above. The 1989 Task Force had to grapple with the reality that there was no way the "rising ground water" component of its original (1985) position could account for the fountain of gas, water, and mud that occurred in 1989, while steadfastly maintaining that past and contemporary drilling and production in and from the Salt Lake oil field had nothing to do with the venting occurrences. Because the arguments and discussion in the 1989 Task Force Report II are distinct from those in the 1985 Task Force Report, those two reports are discussed separately in the following sections.

TASK FORCE REPORTS - DISCUSSION

The facts that established the point of departure for the 1985 Task Force's investigation were the following:

1) The explosion and ensuing fire were caused by ignition of methane gas venting from the ground beneath the site and vicinity, and

2) The methane gas did not originate within the utility distribution system. The latter point, as noted in the Task Force Report, was confirmed both by chemical analysis of the gas which showed it to be free of the tracer chemicals always added to commercial gas, and by the observation that the gas continued to vent after all pressure mains into the area were shut off.

Proceeding from these facts, the conceivable scenarios were the Task Force Report conclusion of gas of shallow biogenic origin displaced and pressurized by rising ground water (or a variation of this not given credence in the report), in which the gas originates in residual surface accumulations of crude left over from the early days of the Salt Lake oil field exploitation and subsequently buried during the course of urban development. The other scenarios addressed but rejected by the Roberti Bill Study were gas escaping from the underlying oil field via an inadequately abandoned well, to various possibilities of gas escaping from the underlying field, either as a continuation or resumption of the natural leakage that had previously characterized the field, or as a totally new process, related in some way to the current operation of the field. At the time the Task Force proposed it, there was no evidence to disprove the "shallow biogenic gas displaced by a rising water table" theory. Its plausibility, however, was another matter, especially in regard to the ability of a rising water table to pressurize shallow gas to the 28 psi found at 40 feet depth in the relief well drilled in the Ross Store parking lot the day after the explosion. The absence of this condition as a hazard in urban areas underlain by organic material and subject to variable water table levels raises some question as to whether it is a common condition, especially in semi-arid areas like Los Angeles, hardly known for extensive shallow organic soils.

The Task Force's biogenic gas theory would appear to have several requirements, as follows:

- The methane gas should be uniquely of shallow biogenic origin, (hence have no possibility of having originated in the underlying oil and gas reservoir). This conclusion relied on the results and interpretation of the chemical test on a gas sample obtained following the explosion. The results of the Roberti Bill study in which application of the same chemical criterion to gas sampled from the soil overlying known gas-producing fields throughout the Los Angeles Basin yielded the same conclusion, i.e., that in "the vast majority" of cases the gas was also of shallow biogenic origin. This might be taken as overwhelming support for the validity of the Task Force Scenario. Alternatively, these results can be viewed as being so geologically implausible as to cast serious doubt on the validity of the chemical criterion to prove the origin of methane sampled from near-surface soil. In any case, the results of a different study employing isotopic analysis released by Global Geochemistry in June 1986, indicated that the Ross Stores gas was in fact of "thermogenic" (oil field) origin.
- The surficial deposits in the area should locally contain organic matter capable of generating methane gas upon decomposition. With regard to this issue, the data available to us are scanty, but offer no support for the biogenic gas hypothesis. Borings for the RTD line along Fairfax (Converse et al, 1981) yielded plenty of gas from soil samples but no organic matter. The soil encountered to a depth of 100 feet in these borings appeared to be of alluvial origin, and was of "rust brown" color indicating generally oxidizing conditions of deposition. Similarly, borings made for the Pan Pacific flood control basin by the L.A. County Flood Control staff (1977) encountered gassy soil, but no organic deposits. Turning again to the Roberti Bill study, one notes that many of the areas sampled were topographic highs, unlikely sites for geologically recent organic swamp deposits. Thus we find no plausible natural source for shallow biogenic gas in the Ross Store area.
- The mechanism for mobilizing and pressurizing the gas was identified as a rising water table, occasioned by cessation of pumping extraction of ground water from wells in the vicinity. This consideration can be examined from two standpoints: 1) whether such a rise in the water table of the perched zone actually occurred; and 2) whether the rise, either actual or postulated, could plausibly mobilize and pressurize shallow methane gas.

With regard to whether a rise in water table did in fact occur, as opposed to "should have occurred because extraction stopped in 1976," data from several wells located in the general area show no significant rise in the water table during the period 1976 to 1985. As with the issue of organic matter in the shallow subsurface, there may be other data that indicate some rise in the local water table did in fact occur between 1976 and 1985, but such information would be inconsistent with the pattern established by the data known to us.

With regard to the plausibility of the rising water table mechanism, we are unaware of a model which demonstrates how a rising water table displaces gas from natural traps beneath clay (or asphalt) caps in the soil section. Methane gas generated by decomposition of organic matter in a confined system can reach pressures above atmospheric. It is not so clear, however, how a rising water table could affect a pocket of gas that was confined effectively enough to maintain this excess pressure. The pressure effect of the water can be no more than hydrostatic unless the water itself is confined and pressurized. This does not occur under water table conditions.

The foregoing discussion simply addresses the position adopted by the Task Force in its' June 1985 report. As discussed below, the character of the 1989 venting incident led the subsequently reconstituted Task Force to abandon the "rising water table" mechanism, but to otherwise leave as open questions where the methane actually comes from and the means by which it is forcefully vented to the surface.

TASK FORCE REPORT II — DISCUSSION

The Methane Gas Task Force was reestablished by the Los Angeles City Council three days after the second major gas venting in the Fairfax District, which occurred on February 7, 1989. Its charge was again to report on the recent incident and "present recommendations necessary to protect health and safety within the area". The revived Task Force began its work with less urgency than had existed in 1985, since no actual damage had occurred in connection with the 1989 venting, but with a larger and more confusing data set. Some of the data that had accumulated since 1985 were simply contradictory, mainly the two different gas-origin studies. One indicated a shallow-source biogenic origin for the gas, the other indicated an oil field origin for the Fairfax gas, but the forceful ejection of gas, water, and mud during the 1989 venting clearly required a driving mechanism with more energy than a "rising water table". Faced with these contradictions and uncertainties, the 1989 Task Force elected to, in effect, throw up its hands with the following paragraph.

"*CONCLUSION ON ORIGIN — Obviously, the origin of the methane gas is complex and not clearly understood. Two reputable firms with qualifications and experience seem to have opposing opinions on the origin of the gas. But, if we intend to shield buildings or vent areas of probable methane gas seepage, the origin of the gas really does not matter. The most probable origin of the gas 'may' be a 'combination of' thermogenic and shallow biogenic sources with the high-pressure events from thermogenic (deep) sources and the low-pressure background gas from a combination of biogenic and thermogenic sources. So, in reality, the origin of the methane has its implications in the total understanding of the problem, but not in remediation or control. Methane gas produced within deep fields or in shallow marshes will not physically function nor react in the same manner. Both are capable of explosive conditions and both can be safely vented to the atmosphere.*"

Before going on to its recommendations for dealing with the problem, by a 3-point program comprising venting by relief wells, use of an area-wide monitoring survey team, and City inspection for Code compliance, the 1989 Task Force paused to provide assurance that neither past nor present activities related to exploitation of the underlying Salt Lake oil field had anything to do with the 1985 or 1989 gas venting incidents. With regard to the possibility that the gas might have escaped via one of the 400 plus abandoned wells that formerly tapped that field, the Task Force stated the following:

"*ABANDONED WELLS AS A SOURCE OF METHANE — The potential hazard imposed by old abandoned wells was not considered in the 1985 report because the gas was believed to have formed within the shallow biogenic areas above the production zone. However, the extent of chemical stripping of gas constituents and microbial alternations during migration probably could not occur within a gas bubble rising upward in a vertical shaft filled more or less with water or brine. These could, however, be the source of low-volume methane that mixes at the surface with ambient gas at low pressures. However all available chemical and isotopic information exonerates these abandoned wells as the source of the higher-pressure methane events.* (Underlining added for emphasis).

These wells are an entity to be considered in the future or as they are uncovered by area construction. But, even if the wells could be located, reabandonment to modern standards would be an expensive program."

The writers do not understand how the isotopic information reported by Global Geochemistry (1986), which indicated the Ross Store gas to be thermogenic, hence of oil field origin, in any way exonerates a potential conduit from the oil field to the surface such as an abandoned well.

With regard to the possibility that the problem might be related to disposal by pressure injection back into the subsurface of salt water stripped from the crude oil currently being extracted from the field, the Task Force stated the following:

"*PRESSURE INJECTION OPERATIONS — Within the Salt Lake field about 42 wells are currently producing from the "C" and "D" Production Zones. According to production records of the Division of Oil and Gas, more fluids are being withdrawn than are returned to the production zones. The type of injection used by McFarland Energy is a water disposal program that does not cause field repressurization. Current oil field operations within the Fairfax area, therefore, do not appear to have an adverse affect on the area methane seepage.*" (Underlining added for emphasis).

We address this latter conclusion in the discussions presented in following sections.

Although Task Force Report II explicitly "exonerated" past and present oil field operations from any role in the 1985 and 1989 gas ventings, it did, unlike the 1985 report, at least mention them. But, even this was evidently regarded as stepping outside of some kind of bounds by the Task Force member representing the C.D.O.G., since that member wrote a letter providing comments regarding the second draft of the Task Force Report II which seemed to generally take his fellow members to task for even mentioning oil field operations (Baker, 1989). This letter was provided as an attachment to Task Force Report II.

At the same time, a contradictory view was set forth in another letter, this from Dr. James E. Slosson, former California State Geologist and Consultant to the Task Force in both 1985 and 1989. In this letter, Slosson (1989) details his disagreement with virtually every point made in the Task Force Report II that dissociates the 1985 and 1989 gas venting incidents from the Salt Lake oil field. Unlike the letter from the C.D.O.G. member, Slosson's letter was not attached to or referenced in Task Force Report II, even though it was dated six weeks earlier.

GEOLOGIC SETTING OF THE FAIRFAX AREA

The geology of the Ross Store area has been studied and described in terms variously of surficial geology, emphasizing the formation of the La Brea tar pits (Shaw and Quinn, 1986), hydrogeology (DWR, 1965), neotectonics (Richards, 1973; Hill and others, 1979), engineering geology (Converse, 1981; L.A. County Flood Control, 1977) and petroleum geology (e.g., Hoots, 1931; Soper, 1943; Crowder and Johnson, 1963; Wright, 1991). To summarize, the area is situated near the northern edge of the Los Angeles Basin, and the central axis of the Salt Lake oil field nearly parallels and lies about 2 miles south of the Santa Monica fault. This fault forms the local boundary between the Western Transverse Ranges and the Los Angeles Basin. The Santa Monica fault is generally represented as a steeply north-dipping reverse fault, which has been active at least as recently as during late Quaternary time (Hill and others, 1979; Crook and Proctor in this volume). Hill and others (1979) based on data provided them by the City of Los Angeles Bureau of Engineering, and following Grant and Sheppard (1939), studied the surface subsidence in the region of the Santa Monica fault, including the Salt Lake oil field area. Although Hill and others (1979) viewed the results of analysis of recent micro-earthquake activity as suggesting that subsurface faults in the Santa Monica-Raymond fault zone are undergoing tectonic strain accumulation, the pattern of surface subsidence they mapped within a 5-mile wide zone along the inferred trace of this fault could be interpreted as indicating surface relaxation rather than compression (as well as indicating response to consolidation in the subsurface resulting from prior extraction of ground water). So it is not clear that the Santa Monica fault is contributing to any near-surface tectonic deformation in the area. Within the ground south of the Santa Monica fault, the folds and faults that form the structural trap of the Salt Lake oil field are the principal local geologic structures. This area is underlain by about a 200-foot thickness of Quaternary shallow marine and alluvial deposits, which overlie late Miocene and Pliocene age Repetto and Pico Formations (undivided) and late Miocene Puente Formation. The Tertiary strata are both tightly folded and displaced by steeply inclined reverse faults, but the overlying Quaternary strata are apparently undeformed (Fig. 5). Therefore, the tectonism that formed the compressional structure of the Salt Lake oil field therefore died out or has been quiescent since approximately Plio-Pleistocene or at least early Pleistocene time.

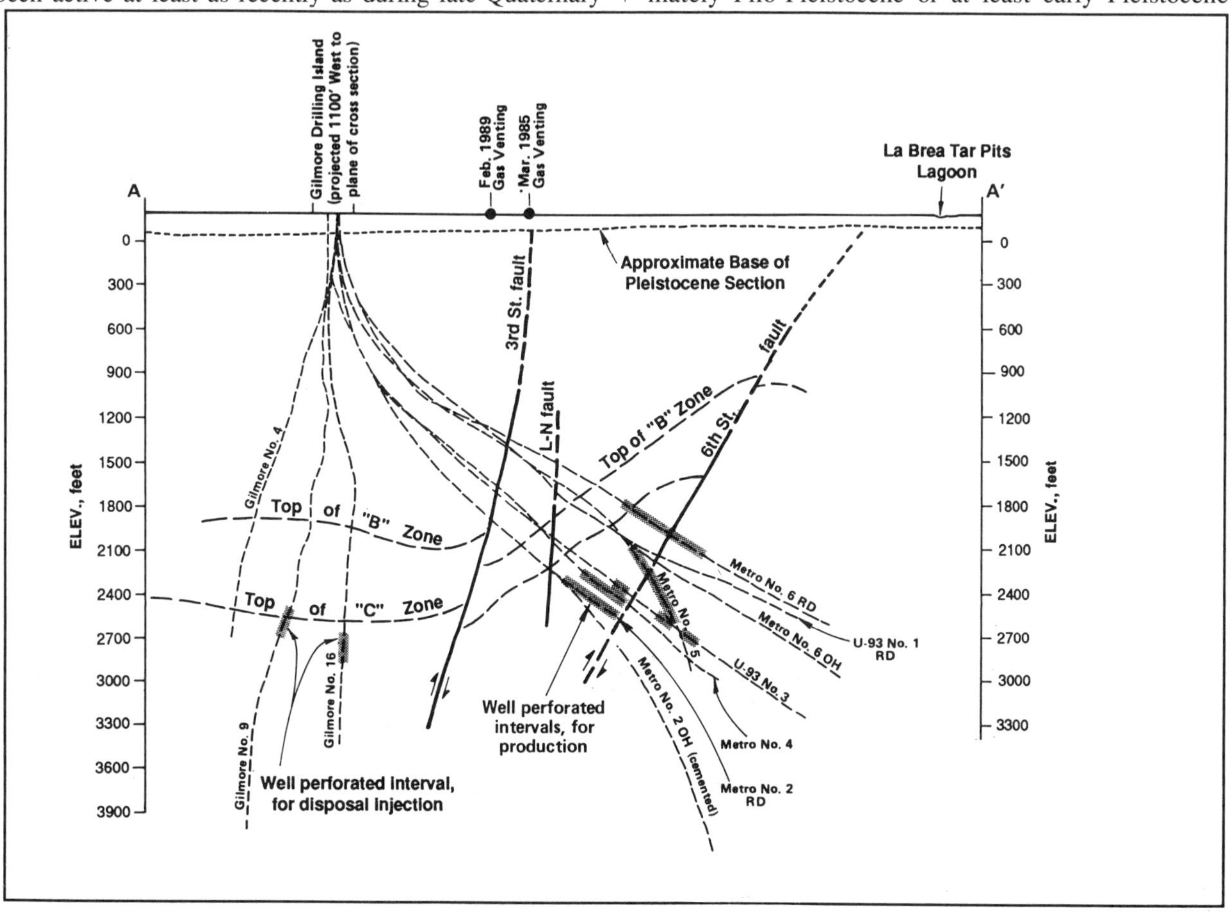

Figure 5. Geologic cross section A-A' vicinity of Ross Store Site

The pulse of tectonism which probably corresponds to the "Pasadenan Orogeny" (Stille, 1936 Wright, 1991), resulted in substantial north-south shortening, partly accommodated within anticlinal folds and partly along the several generally east-west aligned reverse faults. One of these, known as the Third Street fault owing to its location beneath the part of Third Street that extends east from Fairfax, locally divides the Salt Lake oil field into separate blocks with different production and pressure histories. Other faults also form boundaries of more or less isolated compartments within the field. Some of these breaks, notably the Sixth Street fault, may well have provided cross-cutting surfaces along which upward migration of oil and gas occurred from the reservoir to the overlying Quaternary capping deposits and ultimately to the surface environment. The Sixth Street fault breaches the crest of the tight anticline in the block south of the Third Street fault and is the closest recognized fault to the La Brea tar pits (Fig. 5).

The Salt Lake oil field comprises a succession of at least 6 producing horizons, referred to in order from shallowest to deepest as the A through F zones (Crowder and Johnson, 1963; Table A). The earliest oil production was from the A horizon but by the time the original development of the field reached its maximum in 1908, production was largely from the B and C horizons. The renewed production that was initiated in 1961 by wells slant-drilled from the "Gilmore Drilling Island", is altogether from the intermediate and deeper horizons. We have not had access to formation pressure data from the original period of exploitation of the Salt Lake oil field but Hoots (1930) mentions that some wells initially produced at a fairly high rate (up to 1000 barrels per day) and also that considerable gas was produced. From this we infer that the field started with a significant gas or solution-gas drive.

When the new development of parts of the Salt Lake oil field was started in 1961, a few wells flowed for a short time but all eventually required pumping; average initial tubing pressures were about 50 psi; casing pressures were approximately 100 psi (Crowder and Johnson, 1963). As of mid-1963, average daily production of the re-developed field, from an average of 13 production wells, was 650 bbl oil, 150 bbl water, and 270 MCF gas. In 1973, average daily production was 970 bbl oil, 1760 bbl water, and 686 MCF gas. Waste (i.e., salt water) disposal was discharged into the L.A. City sewer system (C.D.O.G., 1974); however, this was converted to subsurface injection back into the field starting in late 1973, through well Gilmore No. 16, completed in the "C" zone at about 2,900 feet in depth. By 1980, average daily production had declined to about 466 bbl oil and 200 bbl water. Oil production slipped gradually through the 80's reaching a daily level of about 340 bbl during our latest year of record, 1988, while water production rose, at least through 1984. Evidently, oil production was being limited by the available waste disposal options.

Production statistics for oil and water for the period 1980-1988 on record with C.D.O.G. are plotted in Fig. 6. This

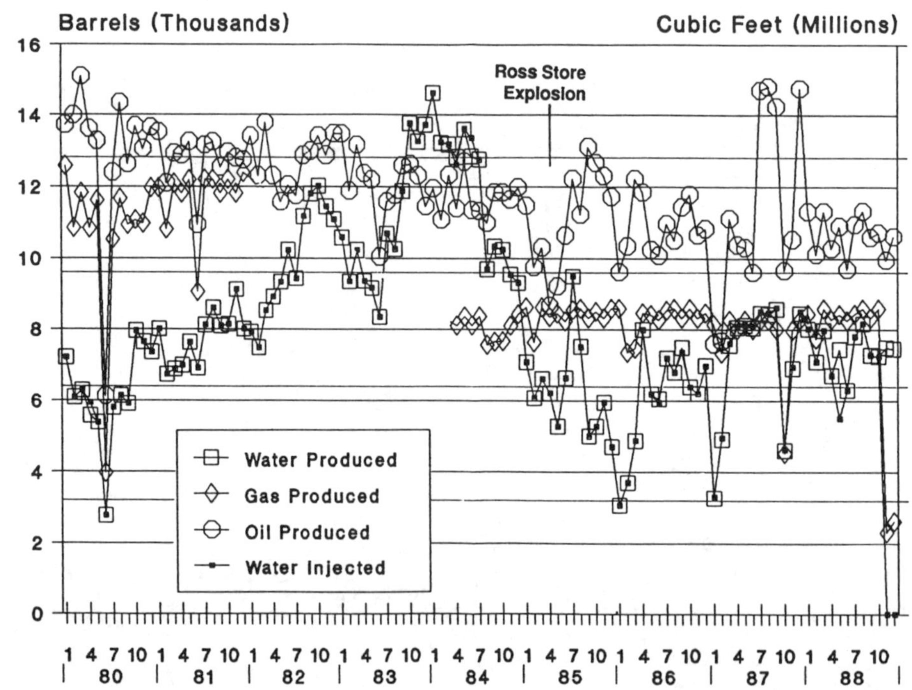

Figure 6. Reported monthly production/Injection: all wells

plot shows the gradual decline in oil production noted above. Of particular interest in the context of the present discussion, however, is the history the production (and disposal) of water, brought to the surface along with the crude oil and gas hydrocarbon fluids. This is plotted separately in Fig. 7, showing monthly production and injection back into the field of the salt water byproduct of oil production. The indicated pattern is one of steadily increasing water production as oil production decreased, Fig. 6, from the beginning of 1980 through mid-1984. Between mid-1984 and about mid-1985, however, reported water production declined precipitously from around 14,000 to around 6,000 and less bbl/mo, the low being about 3,000 bbl in January 1986. A gradual upward trend in water production was re-established in 1985, so that average water production had increased to about 8,000 bbl/mo. by late 1988.

Taken at face value, the water production data shown in Figs. 6 and 7 would seem to indicate either a radical change in the oil-water ratio of the field's production, or a remarkably successful exercise in selective control of water production (for example, by shutting in certain wells or grouting off certain producing zones). In fact, however, well records on file with the C.D.O.G. show no evidence that anything like this occurred. What did happen, however, was that a second well, Gilmore No. 9, was reconditioned and converted from a production to an injection disposal well, in 1984. Meanwhile the operator, McFarland Energy, Inc., sought and received permission from C.D.O.G. to raise well head injection pressures fourfold from 200 psi to 770 psi. The workover records show that considerable effort was expended in the No. 9 conversion, including acid treatment with 5,000 gallons of hydrochloric acid and 2,500 gallons of hydrofluoric acid, in January, 1985. The well was then tested by injecting back into the C zone of the field at 475-500 psi. Thereafter, the C.D.O.G.-tabulated records of production for the Salt Lake oil field show Gilmore No. 9 as a disposal well, but no injection disposal is recorded for it prior to March, 1989. However the dramatic decrease in the quantity of water disposed of by injection into the original disposal well, Gilmore No. 16, coincided with the time Gilmore No. 9 was converted to a disposal well. Since the production of oil did not suffer a corresponding decrease, but instead actually increased by 3,000 to 4,000 bbl/mo. from early 1985, when the decrease in water disposal was most dramatic, to early 1986, the explanation for the pattern of water disposal must lie elsewhere. Given the generally constant (or at least only gradually changing) ratio of oil to water for production from this field, we suggest that the explanation may involve omission of data from the record rather than implausibly effective control of water production. According to this scenario, by mid-1984 water production would have exceeded the injection capacity of the only disposal well in the field, Gilmore No. 16, so work was begun to convert a second well, Gilmore No. 9, completed in the same lease and zone as No. 16, to function as a second disposal well; this was initially accomplished by late 1984, and an increasing percentage of the field's water production was diverted into the new disposal well through

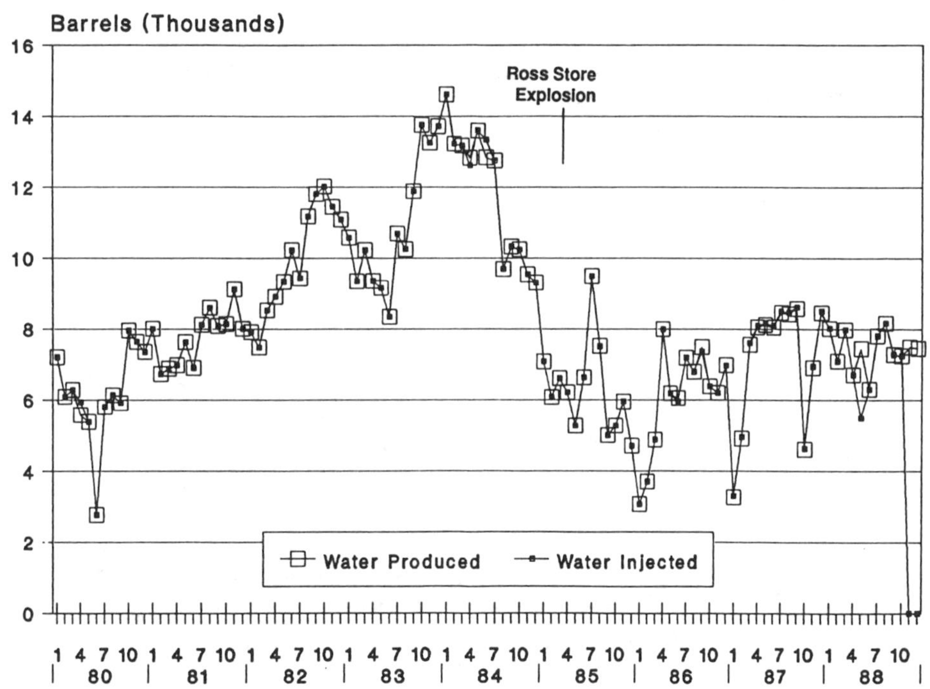

Figure 7. Reported monthly production/injection of water

mid-1985. Thereafter, we suggest that the water production may have been disposed of in both wells, probably in about equal quantities between them. For some reason, however, none of the (inferred) injection into Gilmore No. 9 was entered into the C.D.O.G. statistics, so C.D.O.G. records show only the production-injection volumes plotted in Figs. 6 and 7.

It is of interest to note that the gas venting that caused the Ross Store explosion (on March 24) occurred about two thirds of the way down the sharp decline in Well No. 16's recorded disposal history (Fig. 7). Declining injection into Well No. 16 seemingly would tend to decrease a tendency toward formation destabilization, if the factors of injection pressure and interference from recently established injection through Well No. 9 are neglected. However, it seems possible that one or both of these factors may have caused or contributed to an excursion or incremental buildup of formation pressure in the Gilmore Block "C" zone adjacent to the Third Street fault. If the injection performance of Well No. 16 is considered alone, the prior history of that well showed that the operator attempted to compensate for declining injection (due to clogging) by increasing the well head injection pressure as early as 1980. This would result in a corresponding increase in formation pressure, even if the actual amount of injection take did not increase (or continued to decline). The effect of transferring some of the injection to Well No. 9, located on the opposite side of Well No. 16 relative to the Third Street fault (Figs. 4, 5), would be to create a new pressure bulb which would form a barrier to the radial dissipation of pressure around Well No. 16, and would result in further concentration of pressure in the zone between Well No. 16 and the fault.

ALTERNATIVE SCENARIOS FOR EPISODIC GAS VENTINGS

There appear to be two basic scenarios that have been proposed, and one we propose, to explain the 1985 and 1989 gas ventings in the vicinity of the Ross Store near Third and Fairfax:

Scenario 1:	The 1985 Task Force biogenic gas/rising ground-water scenario;
Scenario 2:	The Roberti Bill oil field gas/abandoned well conduit scenario; and
Scenario 3:	The oil field gas/intermittent fracture conduit scenario.

Scenario 1 would be favored by the interpretation that the chemical composition of gas samples obtained following the March 24th explosion indicate the methane gas could not have originated under petrogenic or thermogenic (oil field) conditions, hence could not have had its source in the underlying oil field. This interpretation, of course, is contrary to the indication from the isotopic analyses that the gas in fact did originate under thermogenic conditions. Neglecting this, we believe that beyond this particular coincidence of a gas blowout occurring directly over a pressurized oil field but being completely unrelated to that field, the dubious conclusion of the Roberti Investigation, that the "vast majority" of gas above oil-gas fields in the Los Angeles Basin is unrelated to the underlying fields, effectively destroys the credibility of these "biogenic source" interpretations. Note that the 1989 Task Force abandoned the rising water table as a driving mechanism, while preserving the option that the gas might still be of shallow biogenic origin.

Scenario 2. The Roberti Bill "leaking abandoned well" scenario, corresponds to a recognized phenomenon, wherein pressurized gas and/or oil, water and entrained sediment escape via an abandoned well (or other conduit, including natural fractures and fault zones), either after the source is repressured, after the blockage or seal in the well fails, or after the fracture conduit is opened by some means. This scenario was dismissed by the 1985 Task Force on the basis of the "biogenic gas" analysis, since gas escaping from the underlying oil field would presumably be of "petrogenic" or "thermogenic" character. The 1989 Task force explicitly "exonerated" leaking abandoned wells as the source of either the 1985 of 1989 venting incidents even though the 1986 Global Geochemistry report had at least cast doubt on, if not, in the view of the Task Force, altogether disproved, a shallow biogenic origin for the gas. Endres and others (1991) on the other hand, cite leakage from naturally repressurized upper zones in the underlying oil field, most probably via old wells or well bores, as the probable source of the vented gas. We consider there to be three objections to the argument that abandoned wells are the cause, or at least the exclusive cause, of gas migration during the 1985 and 1989 venting events at Fairfax:

a) There is no known abandoned well within about 300 feet of the Ross Store building (nor within 100 feet of either of the 1989 venting sites), according to C.D.O.G. Map 118 (C.D.O.G., 1984);

b) The ventings were of brief duration — 24 hours or less, whereas an abandoned well blowout or fracture venting usually lasts longer and may continue indefinitely if not contained (a point also recognized in the Task Force Report II); and

c) The Ross Store site overlies Lease Block U-55 of the Salt Lake oil field. This block is isolated in the subsurface from the main part of the field to the north by the Third Street fault, but is undergoing active pumping extraction via wells slant drilled from the Gilmore Drilling Island. The abandoned original vertical wells in this lease block therefore bottom in a part of the field where oil, water and dissolved gas are being pumped out and pressure is thereby maintained below hydrostatic. Even in case the seal in one of these abandoned wells should fail, the ambient reservoir fluids should not have enough pressure to vent to the surface.

Endres (pers. comm., 1991) suggests that gravity drainage of oil down the steep north flank of the anticline in the Metropolitan lease block would leave a gas cap at the crest of the structure, and this together with a process of natural repressuring by access to high pressures from deeper within the structure means reservoir sands beneath the Ross Store would form a preferred source of gas. We believe, however, that in case repressuring (or some relict overpressuring) of the oil and gas-bearing zone beneath the site existed, the most likely route of fluid (especially gas) escape would be via the Sixth Street fault. This fault breaches the top of the tight anticlinal fold in the Metropolitan Lease Block (south of the Third Street fault - Fig. 5). Another route for gas would be via abandoned wells with inadequate or failing seals located in the area over the crest of this anticline. The crestal part of this structure, however, trends parallel to Third Street and lies about a thousand feet south of the Ross Stores site (Fig. 5). There would be required, therefore, for considerable lateral migration of pressurized gas (and water) for escaping oil field fluids to reach the Ross Store and Gilmore Bank venting sites.

Scenario 3. We propose a third alternative scenario, which we call the oil field gas/intermittent fracture conduit scenario. This scenario, not mentioned in the Task Force report nor set forth in the Roberti Bill charge, envisions a mechanism of intermittent releases of pressurized gas from the Salt Lake oil field, via passages temporarily opened along the Third Street fault, through the process of hydraulic fracturing. The fracturing would be triggered by buildup of fluid pressure caused by injection disposal of salt water in the Lease Block U-20 fault-isolated compartment of the field. This mechanism is similar to the fault activation process that occurred in the Inglewood Oil Field in response to pressure injection for secondary oil recovery, which in that case was instrumental in the breaching and ultimate failure of Baldwin Hills Dam in 1963 (Hamilton and Meehan, 1971). In that case, there was both escape of injected fluid to the surface along one or more fault conduits, and differential movement across faults. In the Ross Store explosion scenario proposed here, the Third Street fault fractured enough to allow upwardly migrating pulses of gas-pressurized water and free gas discharging into and through the overlying alluvial section, without detectible differential movement of the fault itself. With migration of the water upward, gas would be released either into the soil, or could vent directly to the surface.

THE OIL FIELD GAS/INTERMITTENT FRACTURE CONDUIT SCENARIO

The actual mechanism of fracturing and migration is indicated by the following calculations, in which the process is modelled by considering a fault which acts as a membrane separating adjoining fluid bearing zones within Production Zone C. We place our calculations at 3000-feet depth, roughly representative of the injection depth. This condition

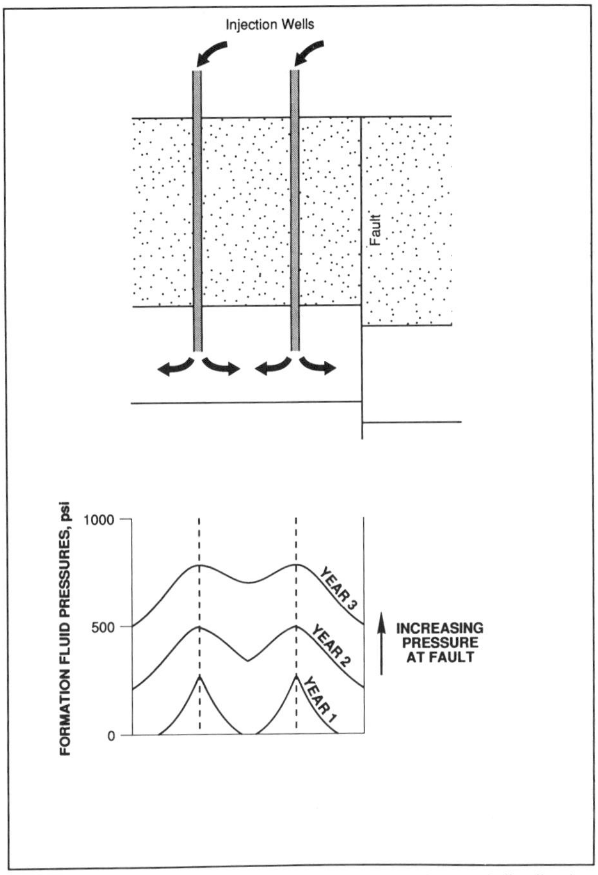

Figure 8. Diagram showing hypothetical evolution and distribution of pressure from injection wells

is shown on Fig. 8. We assume that the producing Zone C has the following history: fluid pressure initially hydrostatic, or 3000 x 0.45 = 1350 psi, with resulting partial buoyancy of sediments. Reduction of fluid pressure to about 200 psi, following extended periods of production causes compression (consolidation) of sediments with increases in both vertical and horizontal effective intergranular stresses. Assuming during this process zero horizontal strain, (the so-called "at rest" or Ko condition, in which the ratio Ko of horizontal to vertical intergranular stress is about 0.5), and a unit weight of sediments corresponding to 1 psi per foot depth (144 pcf), results in horizontal total pressure of (3000-200) x 0.5 + 200 = 1600 psi. Under this condition, any pressure, Px, from the injection side of the block acting on the fault membrane that exceeds the 1600 psi may cause incipient displacement or parting of the fault.

Now, with injection at X, fluid pressure will increase. Restoration of hydrostatic pressure in the X side of the producing zone will create original fluid pressure of 3000 x .45 = 1350 psi. This is 250 psi (at ground level) short of the fluid pressure that would equal the 1600 psi necessary to displace the fault membrane slightly. This incipient displacement condition would require an injector well head pressure of at least 150 psi, but in the short-term it would need more than 150 psi because of pressure loss in the well bore and time-dependent pressure loss between the well and the fault. This pressure decay process including both geometric and temporal decay factors can be simulated by modelling, as is routinely done by petroleum engineers. Even without doing so, however, it appears likely, based in part on other studies of the field, that the permitted 770 psi wellhead injection pressure would suffice to raise the formation pressure enough to displace the fault membrane. (Note that fracturing occurs at the fault discontinuity at some significant time after the well is pressurized, not at the well, therefore short-term hydrofracture tests of the well, which are sometimes required to check for hydrofracture hazards, would not indicate a potential for later hydrofracture at nearby faults).

Displacement of the fault membrane would permit opening of a parting lenticular fracture transmitting a mobile pulse of fluid which would migrate rapidly and upward (Fig. 9). Escape of gas from the mobile and increasingly less confined fluid into the alluvium (or into abandoned wells) would follow. The fracture opening, however, would collapse as the fluid pressure was released, and would remain closed until an adequate pressure gradient again built up in consequence of continuing injection into the formation. This would account for the episodic and brief character of the surface venting events.

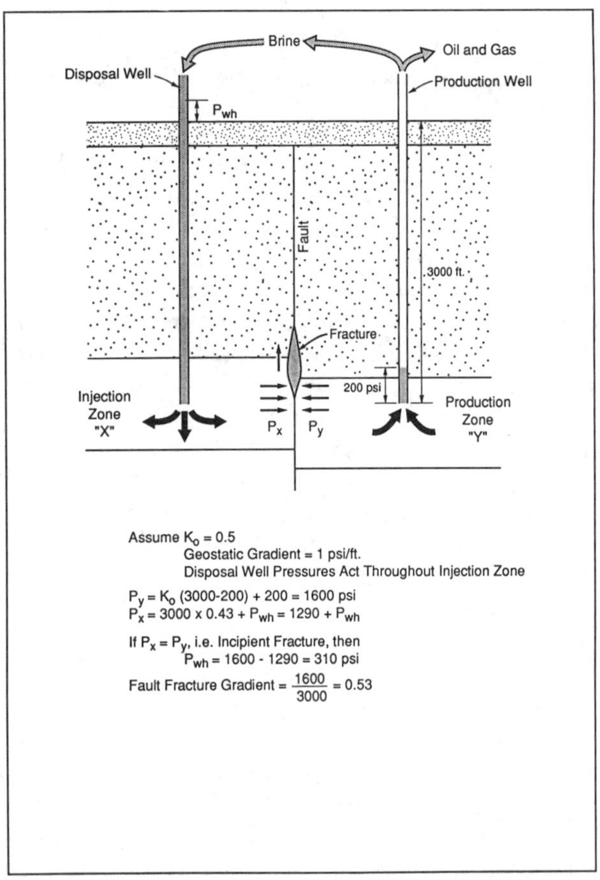

Figure 9. Fault injection pressure for case of long term injection

CONCLUSION

Our analysis indicates that the raise of well head pressures at Gilmore #16, sanctioned by C.D.O.G., from 200 to 770 psi, was sufficient to fracture the formation at the nearby fault. This possibility is not recognized in current C.D.O.G. review of oil field operations, though the analytical techniques necessary for its consideration are certainly available in the oil industry. Moreover, there is evidence suggestive that additional injection, unreported in C.D.O.G. records, was taking place at the time of the disaster. It seems to us that future studies of the Los Angeles gas problem can hardly be considered complete without full investigation of these issues. Finally, our analysis suggests that the criteria currently employed by C.D.O.G. and industry may not adequately protect against formation fracturing and escape of formational or injected fluids and gas to the surface environment, resulting in disastrous consequences comparable to the 1985 Ross Store explosion.

ACKNOWLEDGEMENTS

The writers benefitted from assistance in analysis of data and formulation of geologic interpretations provided by Philip A. Frame and John Wakabayashi. A petroleum geologist who wishes to remain anonymous provided helpful insights into several aspects of the new development and exploitation of the Salt Lake oil field. Bernard Endres, equally at home in the law and the sciences, shared his views on the processes involved in the methane venting in 1985, and also provided reports and data not available directly from the public files of the C.D.O.G. The Long Beach office of the C.D.O.G. helpfully responded to our request for waterflood testing information. Dr. James E. Slosson provided a copy of his letter commenting on the 1989 Task Force Report II.

Each of the authors was involved in technical analyses of the causes of the methane gas ventings in the Fairfax District, in connection with the disputes that arose following the 1985 explosion. The views and interpretation presented in this article, however, are purely those of the writers and do not necessarily correspond to the views and interpretations of any party to these disputes, or to any other entity.

REFERENCES

Baker, 1989, letter provided as attachment to Task Force Report II.

California Division of Oil and Gas, 1974, California Oil and Gas Fields, Volume II.

____________________, 1984, Las Cienegas, Lake, W. Portion, Los Angeles, City, Beverly Hills East Area, San Vicinte E. Portion, South Salt Lake, Map 118.

California Department of Water Resources, 1965, Water well standards, Central, Hollywood, Santa Monica basins, Los Angeles County, California Department of Water Resources Bulletin 74-4.

Converse Consultants, Earth Sciences Associates, and GeoResource Consultants, 1981, Metro Rail Project Geotechnical Investigations Report, Volume I: Report submitted to the Southern California Rapid Transit District.

Crowder, R.E., and Johnson, R.A., 1963, Recent developments in Jade-Buttram area of the SAlt Lake Oil Field: California Division of Oil and Gas Summary of Operations, California Oil Fields, Annual Report, v. 49, p. 53-59.

Endres, B., Chilingarian, G.V., and Yen, T.F., 1991, Environmental hazard of urban oil field operations: Jour. Petroleum Science and Engineering, vol. 6 (in press).

Geoscience Analytical, Inc., 1986, A study of abandoned oil and gas wells and methane and other hazardous gas accumulations: Report to California Department of Conservation, Division of Oil and Gas.

Global Geochemistry, 1986, Unpublished report referenced in Task Force Report II.

Grant, U.S., and Sheppard, W.E., 1939, Some recent changes of elevation in the Los Angeles basin of southern California and their possible significance: Bulletin — Seismological Society of America, vol. 29, p. 299-326.

Hamilton, D.H., and Meehan, R.L., 1971, Ground rupture in the Baldwin Hills: Science, vol. 172, p. 333-344.

Hill, R.L., Sprotte, E.C., Bennett, Real, C.R., and Slade, R.C., 1979, Earthquake hazards associated with faults in the greater Los Angeles metropolitan area, Los Angeles County, California, including faults in the Santa Monica-Raymond, Verdugo-Eagle Rock, and Benedict Canyon fault zones: U.S. Geological Survey Open File Report 79-16.

Hoots, 1930, Geology of the eastern part of the Santa Monica Mountains, Los Angeles County, California: U.S. Geological Survey Professional Paper 165-C, p. 83-134.

Los Angeles County Flood Control District, 1977, Proposed Pan Pacific Retention Basin, Report of Investigation, Summary of Geotechnical Investigation.

Richards, C.A., 1973, Engineering geology aspects of petroleum in the urban environment: in Moran, D.E., Slosson, J.E., Stone, R.O., and Yelverton, C.A., (eds.), Geology, Seismicity and Environmental Impact: Association of Engineering Geologists Special Publication, Los Angeles, p. 391-399.

Shaw, C.A., and Quinn, J.P., 1986, Rancho La Brea: a look at coastal southern California's past: California Geology, vol. 39, p. 123-133.

Slosson, J.E., 1989, letter regarding conclusions of Task Force Report II.

Stille, H., 1936, Present tectonic state of the earth, Am. Assoc. Petro. Geol. Bull. 20, p. 849-880.

Soper, 1943, Salt Lake Oil Field in Jenkins, O.P., (ed.), Geologic formations and economic development of the oil and gas fields of California: California Division of Mines and Geology, Bulletin 118, p. 284-286.

Task Force Report, 1985, Task Force Report on the March 24, 1985 methane gas explosion and fire in the Fairfax area: City of Los Angeles.

Task Force Report, 1989, Task Force Report II on the methane gas incursion: City of Los Angeles.

Wright, T.L., 1987, Geological Setting of the Rancho LaBrea Tar Pits in T. Wright and R. Heck, (eds.) Petroleum Geology of Coastal Southern California: Pacific Section AAPG Guidebook 60, p. 47-91.

Wright, T.L., 1991, Structural Geology and Tectonic Evolution of the Los Angeles Basin, California, in K.T. Biddle, (ed.), Active Margin Basins; Am. Assoc. Petro. Geol. Memoir 52, p. 35-134

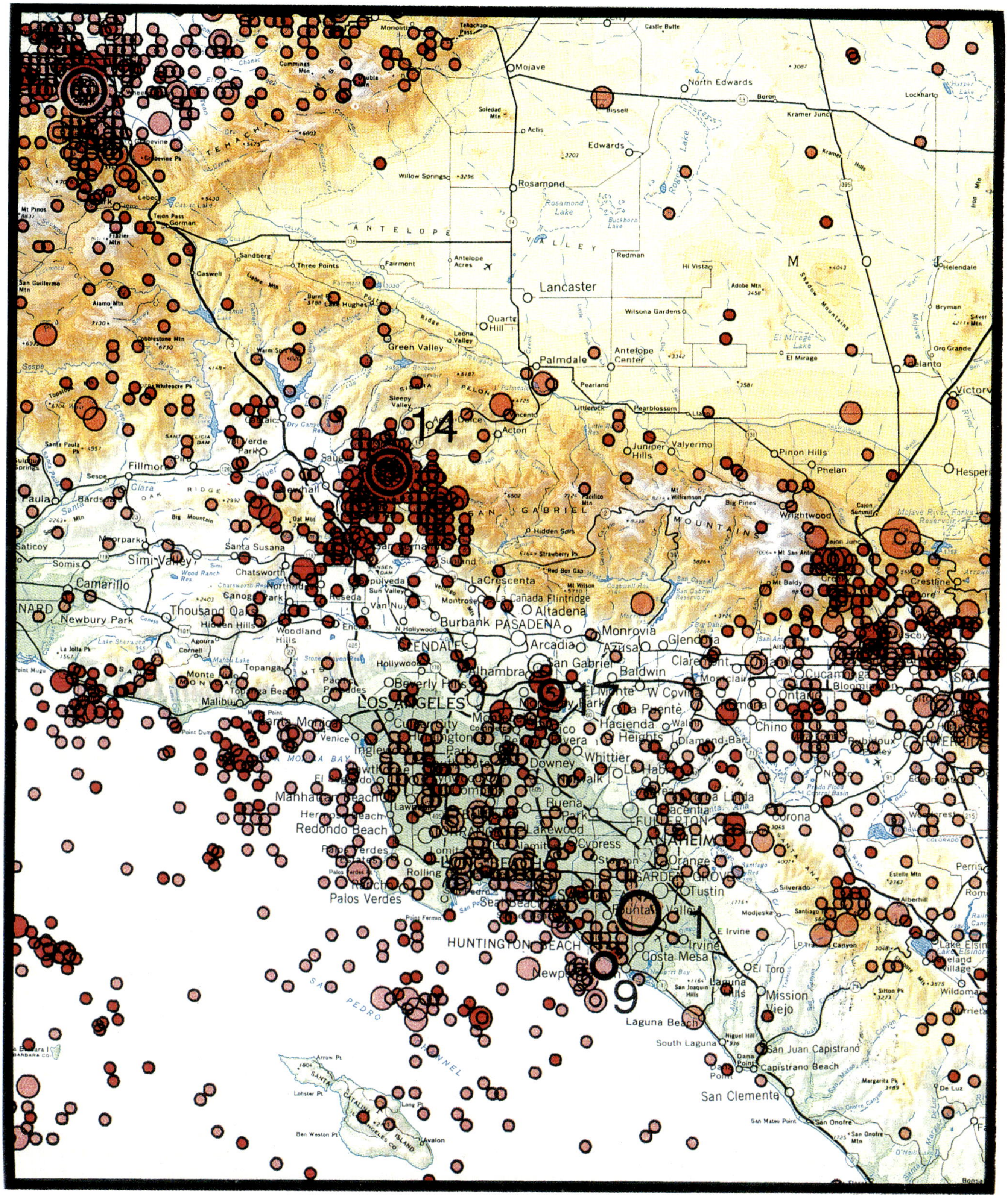

Map of seismicity from 1808-1987 in the greater L.A. area (Goter, 1988); from page 10.

Aerial Photo 1 from page 162

Aerial Photo 2 from page 162

Southern fissure area ground cracks trending across graded pads and toward newly-contstructed homes in background; from page 277.

Representative 1987 ground cracks in street pavement and passing through the large industrial building in background, north-central fissure area; from page 287.

Camp Pendleton Harbor and Oceanside Marina looking north. San Luis Rey river in lower right and Santa Margarita River in left; from page 475.

Tempory steel pile wall was constructed at the toe of slope after failure of the lower freeze wall. This was later removed; from page 521.

Severe scour in January, 1988 stripped the sand cover from Malibu beaches exposing piles to wave attack scour, and battering by flotsam; from page 478.

Near-vertical aerial view of the Seaside Landslide repair. The beach and Pacific Ocean are at the bottom of the picture. The conveyer-belt system utilized to remove slide debris from the excavation is shown on the left side of the picture. The buttress key was approximately 75 percent completed at this stage; from page 517.

Development along the Malibu coast has produced a variety of designs. Homes with shallow foundations directly on the beach surface were severely damaged by storms in January, 1988. Homes on deep-pile foundations that were elevated above anticipated wave runup faired much better; from page 479.

GEOTECHNICAL APPLICATIONS OF INFRARED OBLIQUE STEREO PHOTOGRAPHY

Woodrow L. Higdon, President
Geo-Tech Imagery
P.O. Box 2179
Oceanside, California 92051

INTRODUCTION

Two general areas of aerial photography will be discussed. The first is what affects the appearance of photographic images, and how we can control those aspects to improve information quality. The second area will cover some of the basic interpretations that can be made from this improved image quality with particular emphasis on oblique stereo infrared photography. Oblique stereo photographic images, in particular infrared images, have the capability of being one of the more significant investigative support tools available to the geologist, and yet they are the least used and most misunderstood form of visual information in use today. Generally, the term infrared photography produces a response from most geologists that heat variations are not significant in their investigation, or oblique images are useful for office walls, but not as an investigative tool. The fact is that infrared photography has nothing to do with heat and oblique aerial photography has as many capabilities and uses as the common vertical image. To put this statement in context it is necessary to define some terms and limits for infrared compared to standard color photography, and oblique stereo imaging.

WAVELENGTHS & FILM

The zone of electromagnetic radiation usually referred to as the visible light spectrum, has wavelengths ranging from approximately 400 to 700 nanometers (Fig. 1). The short end of this zone consist of wavelengths that our eyes perceive as the color blue. As one proceeds up the wavelength scale one passes through green, yellow, orange, and red, which is the longest wavelength the human eye can perceive. Beyond the short end of the visible spectrum is ultraviolet. The effects of these wavelengths on photographic images are most apparent as a hazy skyline in standard color photographs. This effect is produced from ultraviolet and blue light scattering by airborne particles or moisture. On the opposite end of the visible spectrum is the infrared zone starting at approximately 700 nanometers wavelength. These wavelengths do not produce sensible heat until they attain a wavelength of approximately 1350 nanometers. Imagery that is generated as a function of heat radiation from the long wavelength part of the infrared spectrum is generally produced by a computer. This type of imagery is referred to as scan imagery. The computer has a series of scanners, each of which is sensitive to a specific wavelength or group of wavelengths. The computer receives information from the scanners, and then builds an image based on preprogrammed criteria. This type of imagery has many uses, but it is not infrared photography.

Infrared photographic images are recorded directly on a film base with specific wavelengths of reflected visible and near infrared electromagnetic radiation. Near Infrared is a zone just outside of the long end of the visible spectrum, and the lower end of the infrared range. This zone has wavelengths ranging from approximately 700 to 900 nanometers. This section of the infrared range is not sensible heat, but it can be thought of as an invisible extension of the visible light spectrum. This is why infrared images should not always be equated with heat. If infrared films were heat sensitive they could not be handled for camera loading, as body heat would fog the film. Some type of refrigerated and insulated camera would be needed to shoot the film. So, what actually exists is a film that is not heat sensitive and that has been specially designed to respond to only the near infrared part of the infrared spectrum, and all shorter visible wavelengths. The film was developed during the Second World War as the military wanted a film that would differentiate between natural vegetation and camouflage in the field. It was determined that normal healthy chlorophyll in plants reflect the near infrared wavelengths. This reflectivity provides some significant interpretative parameters that will be discussed later. As

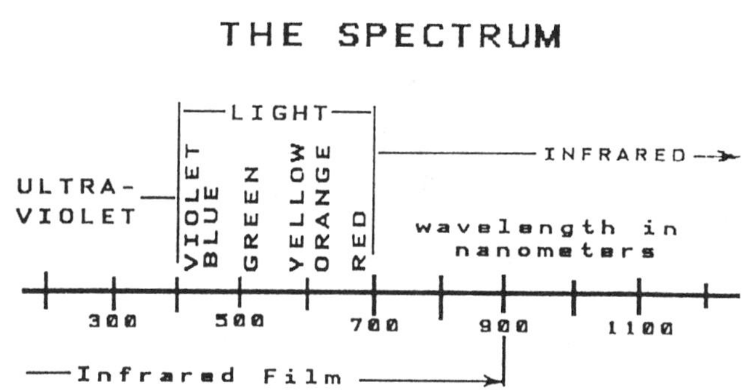

Figure 1. The Spectrum

the film is also sensitive to all of the shorter wavelengths, including ultraviolet, it is recommended a number 12 yellow filter be used. This filter restricts the passage of blue light and shorter wavelengths that tend to degrade the image resolution, and shift colors. It is important to use the number 12 yellow specified by the film instructions.

There are several types of color infrared film available, and consideration should be given to which one to use for best results. The only color 35mm infrared used in the U.S. today is Ektachrome Infrared, which is an E-4 process film. Even though most geologists have access to a 35mm camera, and the processing is not impossible to find, this format is not recommended because 35mm format is just too small for good aerial photos. The format limitations are compounded by the fact that the E-4 process is an older process that is in little use today, which means the film will have to be shipped to some major processing center with E-4 capabilities. The turn around times varies from four to six weeks. The commonly used infrared films are designated as aerial infrared and are available in negative and positive (Color Slide) film types. The smallest film size available is 70mm roll film, followed by 4 X 5 inch format films and larger. Even though a medium format camera with 70mm back capabilities is not the average geological camera, it is still the best all around system for low elevation aerials, ease of handling, transport, versatility, and image economics.

Negative-type films tend to be less expensive than positive-type films. and the prints generated from negatives are also generally less expensive than prints from positives. However, positive (color slide) films are superior for geologic applications. Positives do not require prints to view and evaluate images. This capability eliminates the need, as well as the cost, of printing every single frame. It also allows direct evaluation of original images. These originals are superior in resolution and image quality compared to prints produced from any type of film. There are no photographic papers on the general market capable of reproducing all of the detail and color variation that exist in an original film. This positive infrared film is called Aerochrome Infrared 2443 and is available from authorized Kodak professional film dealers.

Aerial infrared films do not have standard ASA ratings that can be used to set camera light meters for appropriate exposure. In fact, there is no standard exposure for all conditions, and it will vary widely depending upon a number of factors. If a qualified professional experienced with infrared films is not being hired for your project shooting, then start with an ASA 200 and experiment to establish norms for various conditions including the following:

1. Type of light metering system being used.
2. Percent of ground vegetation cover.
3. Percent and type of lithologies exposed.
4. Sun angle/Time of day.
5. Percent of horizon or skyline included in picture (Obliques).
6. Filtration used over lens.
7. Light transmission characteristics of optics.

There are also a number of secondary image enhancements associated with positive color infrared films. Positive films tend to be higher contrast than negative films. Many times there are ground features with subtle light reflection differences that are not recorded by lower contrast negative type films or they become lost in reproduction. This pumped up contrast also affects color difference separation which can add additional interpretative capabilities.

If positive films are so much better for airphoto interpretation, why don't the aerial companies use positive films instead of negative films as their standard? The answers are money, ease of use, and the primary application was not planned for use by geologists. As indicated earlier, negative films are easier to shoot, and color prints must be generated for analysis. The ease of use factor is that negative films have a very wide exposure latitude compared to positive films. This means that best exposure can be missed by several F-Stops with negative films, and you will still be able to generate a good color print from the negative. If you miss best exposure on a positive film by more than one half F-Stop, there will be substantial loss in image quality. If an aerial company has to re-fly a project due to bad exposure, this costs them money, and the chance of getting poorly exposed film on negatives is substantially less compared to using positive films. The primary planned use of airphotos for more than fifty years has been cartographic, or map making applications. The map making industry and the equipment they use has been designed to use negative type films and their associated prints. Aerial photographic images seldom, if ever, are preplanned specifically for geotechnical interpretation.

Botany might be considered to be a little out of place here, however it is extremely important. Geologist have known for years that many formations or their contacts could be mapped by the types of vegetation growing on them. The chlorophyll structure is one of the wonders of nature that allows plants to generate the nutrients, in this case glucose, that feeds the plant. The plant extracts carbon dioxide (CO_2) from the atmosphere and water (H_2O) from the ground or air. The chlorophyll structure combines the carbon and hydrogen from the two molecules to produce a form of sugar or glucose. Let's extend the thought processes one step further keeping in mind the responses of near infrared light to chlorophyll, the function of chlorophyll, and structural geology.

1. The more chlorophyll a plant has, the more water it needs to function.

2. Plants tend to seek out ground and groundwater conditions that are most conducive to its long term growth and development.

3. Broad leaf varieties of plants have abundant chlorophyll and require abundant and long term water supplies to survive, as compared to desert varieties that have little to no leaf structures and only bloom for short periods each year.

4. Chlorophyll is the most sensitive indicator of a plants health. With adequate water and basic nutrients the chlorophyll will have high activity and therefore high reflectivity in the near infrared.

5. Geologic structure controls groundwater, and water controls vegetation.

APPLICATIONS

With theses concepts in mind let's find a landslide and a fault that cannot be seen in the field, and that do not show up in standard color or black & white photographic images. Landslides produce ground rupture along their boundaries and normally produce various types of perched groundwater or springs. Even though the visible surface expression will change or even be obliterated with time, the subsurface conditions remain. Head and lateral grabens will infill with colluvium, and many times produce perched water conditions that provide a reservoir for months, or years. These areas of abundant near surface water will attract appropriate types of vegetation. The vegetation in the areas of good water supply will be healthier and therefore have a higher near infrared reflectance than nearby plants of a similar type with less water, or plants of other varieties with smaller leaf structures. These structural groundwater controls of vegetation can be made visible to the human eye with infrared films. These trends, lineaments, and curvatures can then be interpreted.

Aerial photograph number 1 is a frame taken from a series of oblique stereo images in southern Orange County, California. The area is mostly Miocene Capistrano Formation siltstone capped by terrace deposits. The creek channel is marked by vivid red vegetation due to the abundant water supply stored in alluvial materials. Just upslope and parallel to the channel, a well defined lineament can be seen. Upslope of the lineament the vegetation is thicker and larger with prominent bushes showing vivid red coloration. Below the lineament, and downslope, the vegetation is smaller with little to minimal red vegetation reflections. The lineament is a fault that is generating a ground water barrier that restricts the flow of water to the downslope side of the fault. The abundant water backing up on the upslope side of the fault attracts vegetation forms that are broad leaf types with abundant chlorophyll, and therefore high infrared reflection. The right visual limit of the fault terminates at the boundary of an ancient landslide. The lateral boundaries and the head graben are marked by large vivid red vegetation. A bulge in the channel marks the toe of the feature. Neither the fault or the landslide were detected during an earlier geotechnical investigation.

The basic interpretations can be extended in a number of different directions and over a variety of ground conditions. The result is that "invisible" ground features can be detected and delineated. An extensive knowledge of plants and their environments is not a necessity, however, the more knowledge you have the better your interpretations will become.

There is no one time of the year that is best for all conditions. The number of detectable features can vary weekly, monthly, or yearly. Some features drain slowly with large reservoir capacity and control vegetation year round. However, this same feature may be obscured in the rainy season in a high rainfall climate as all surrounding vegetation is also well watered and healthy. This feature may only be visible during the middle or late part of the dry season. Other features, in the same area, may drain quickly and leave no trace in middle or late summer. In a somewhat dryer climate many features may be seen best during the middle or late rainy season. For these reason and others, flight time considerations are important. The best possible alternative, based on economics, is to fly the site at least two times during the year with a six month separation between flights. If time and economics permit, make additional flights. Each flight should be preplanned and areas of priority detected in previous flights targeted for special shooting.

Potential and old debris flows can also be detected and delineated in much the same way. These features tend to build up in old channels or pockets on hillsides with a variety of slope angles and sizes. These alluvial or colluvial structures are normally loose and porous compared to surrounding bedrock materials and they transmit and trap water. These perched water conditions also control vegetation and are often found along the boundaries or within old landslides that obscure old slides on standard photographic images. Other potential debris flows, like deep weathering areas with subsurface water, will show many of the same earmarks described above.

Aerial photograph number 2 shows a series of these potential debris flow structures. The right hand wall of the canyon is marked by a number of linear and pocket plumes of high reflection (red) vegetation. The life of these reflections, like landslides, will vary with size, storage capacity, water resupply, and climate. Anytime steep natural hillsides are to be left adjacent to developed areas, they should be investigated for potential debris flows. There is also an unrepaired reactivated ancient landslide visible in the head of the canyon. This landslide, like many landslides occurring in developed areas today, are not new landslides. They are reactivated ancient landslides that were missed by standard investigation techniques.

Photo 1. (See text and color inserts)

Photo 2. (See text and color inserts)

GROUND WATER AND LITHOLOGY

All of the objects we see as color are a combination of both reflection and absorption of light. Near Infrared is totally absorbed by water. This characteristic means that areas with near surface groundwater, adequate to produce a moisture variation at surface, will photograph a little darker on infrared film, than the same ground condition nearby with no surface moisture. This is an additional interpretative parameter to add to the above infrared capabilities. Often high or perched groundwater conditions will not only produce the deep red vegetations reflections, but also darker areas on the ground indicating the near surface limits of the water body. In some areas there may be no vegetation and only the slightly darker areas marking groundwater springs or seepage.

The combined effects of vegetation reflections, increased contrast, and infrared absorption by water, also makes infrared films an excellent tool for delineating and mapping land/water interfaces in heavily vegetated areas. The deep red color of vegetation and the dark blue to black of water is a substantial contrast improvement over conventional color photos that show green and dirty green or brown.

High infrared contrast rock types can add an additional interpretative parameter. Rock types containing quartz, like sandstones, reflect more near infrared than siltstone or claystone. Limestones produce a high reflection like sandstones. These variations in rock reflections may produce lineaments on ground surfaces that are not visible to the eye or standard color or black and white imagery forms, but can be seen in color infrared. These lithologic variations may be produced as the slide block displaces bedding structures, or as the graben infills with colluvium that may contrast with adjacent slide debris or bedrock. This type of variation is very useful in finding faults as the lineaments become longer and easier to see. For landslides it may provide that one additional piece of information necessary to make an interpretation.

USE OF OBLIQUE STEREO PHOTOGRAPHS

It has long been the opinion of most professionals that vertical aerial photos are the prime investigation image form, and that oblique aerials are mostly useful as office displays. The fact is that for every piece of information obtained from a set of verticals, two to three pieces of information can be obtained from a set of planned stereo obliques. The same ratio applies in available applications. This does not mean that verticals should be replaced by obliques. It is obvious that there are important applications in verticals that cannot be replaced by oblique images. Oblique aerial imagery is just one more tool that must be added to the geologic tool box. Oblique aerial images can be generated in stereo, just like verticals. The biggest single advantage of an oblique geotechnical flight over a vertical flight is perspective. Perspective, in this case, is a position of view. Vertical images have only one perspective relative to the ground, no matter how many frames are generated. Obliques have an infinite number of possible perspectives. Many ground features do not show up in vertical images simply because the angle of view is wrong to see the shape or texture of the feature. This capability has been utilized, in a limited way, by geologists in the form of low sun angle photography. In this case lighting is allowed to move to a very different position from normal, and in some cases oblique photography was even used. This perspective benefit associated with oblique images increases significantly as terrain steepens, and is optimized with stereo viewing. As terrain steepness increases, vertical exaggeration increases and many features located in those areas become obscured in vertical images. Oblique stereos are the best image form to use on project that have vertical or near vertical faces to investigate. These types of projects include coastal bluff studies, dam and reservoir sites, and any kind of steep high relief terrain. For every feature detected in a vertical flight, at least one more feature will be found in a planned oblique stereo flight. This translates to at least 100% improvement in information yield. If any of those oblique detected features are major hazards, that would have gone undetected until grading or development, substantial amounts of time and money can be saved. Therefore the economic benefits to clients far outweighs the small cost of generating such imagery. Other secondary benefits associated with oblique image flights are the ease of varying scale in photos without substantial cost increases, and improved communications and documentation. Vertical aerial companies charge for a new flight each time they change scales. Technical photographers shooting from helicopters seldom charge extra for elevation changes, and a helicopter is the best method of shooting oblique stereo photos. Non-professionals do not visualize vertical images like most geologists. Obliques are superior when communicating with lay groups or documenting vertical features. It is a rare project that cannot benefit from planned geotechnical oblique stereo photos.

The aspects of relatively expensive cameras, special shooting techniques, and films that are somewhat difficult to use should not be allowed to deter a geologist from attempting these capabilities. The difference between using a regular commercial photographer and placing these photographic and interpretative skills in the hands of a geologist, is literally like night and day. The geologists ability to preplan imagery based upon geotechnical skills, and then evaluate site conditions during flight and make appropriate adjustments is critical for best results.

Chapter 3

Faults and Seismicity

Photo of John Muir School in Long Beach after the 1933 earthquake. Typical unreinforced masonry construction.

Faults and Seismicity

Editors Note: "California is earthquake country," we've all heard—and with good reason. Egill has prepared a synopsis of the local seismic environment of the second most populous region in the U.S. He then joins his colleagues at Caltech, along with his wife Lucy of the USGS, to create an Atlas of local earthquakes. For the remaining papers we were fortunate to have experts prepare papers on several of the most active, and in some cases controversial, faults in the southland.

SEISMICITY, FAULTS, AND EARTHQUAKE POTENTIAL IN LOS ANGELES, SOUTHERN CALIFORNIA

Egill Hauksson
Seismological Laboratory
California Institute of Technology
Pasadena, CA 91125

INTRODUCTION

This paper focuses on the earthquake potential of both surficial and concealed seismogenic faults in the Los Angeles area. Lessons from significant earthquakes along these faults are discussed. What is known about slip rates, expected magnitudes and return periods is summarized. The results indicate that the earthquake potential of the concealed faults is more than twice the potential of the surficial faults. During the last 200 years the seismic moment release in the Los Angeles basin has lagged significantly behind the average seismic moment release in California. Only one large earthquake in the M7-7.75 range, however, is needed to erase this discrepancy.

Moderate-sized and large damaging earthquakes have occurred in the past and will occur in the future on late Quaternary faults in the greater Los Angeles area (Fig. 1). These earthquakes reflect the tectonics of both the Peninsular Ranges and the Transverse Ranges as these two tectonic provinces intersect in the Los Angeles basin. This paper focuses on evaluating the earthquake potential of the faults that have generated these earthquakes based on both seismological and geological data. "Earthquake potential" is a descriptive term that is used to evaluate the likelihood of a fault having a damaging event in the future (Ziony and Yerkes, 1985). Because geological data collected since the turn of the century and instrumental seismological data since 1932 are available, the earthquake potential of these faults can be estimated to first order. In this study the Los Angeles basin is defined as the Los Angeles basin proper together with the San Gabriel Valley and the adjacent offshore Santa Monica and San Pedro Bays.

The earthquake sequences that have occurred since 1930 have demonstrated that both surficial and concealed late Quaternary faults represent significant seismic hazards. The surficial faults show right-lateral or left-lateral strike-slip and reverse faulting, whereas the concealed faults exhibit thrust faulting on gently dipping crustal ramps. The right-lateral faults are a part of the northwest trending tectonics of the Peninsular Ranges, which is dominated by the relative strike-slip motion between the Pacific and North American Plates. The left-lateral strike-slip and thrust faults are associated

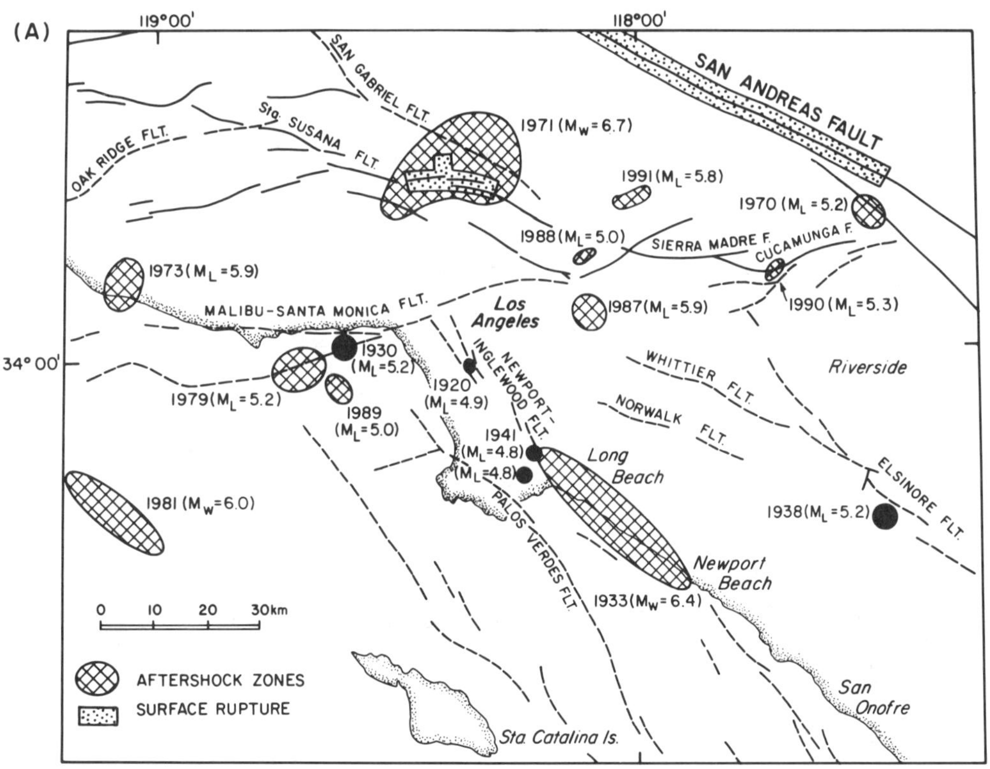

Figure 1. Significant earthquakes of M4.8 that have occurred in the greater Los Angeles basin area since 1920. Aftershock zones are shaded with cross hatching. Dotted areas indicate surface rupture, including the rupture of the 1857 earthquake along the San Andreas fault.

with the tectonics of the east-west trending Transverse Ranges, which are often attributed to the "big bend" in the San Andreas fault (Morton and Matti, 1987). Oblique faulting that could be inferred by the merging of strike-slip and compressional tectonics does not occur in the basin. Instead, the coexistence of zones of thrusting and large strike-slip faults in the basin suggests that the thrust and strike-slip movements are mostly decoupled from one another.

In addition to moderate-sized or large earthquakes, small earthquakes occur in the Los Angeles basin and the adjacent offshore areas at a rate similar to the average rate for southern California (Allen et al., 1965). In Figure 2, all of the earthquake locations in the Los Angeles basin in the California Institute of Technology/US Geological Survey (CIT/USGS) catalog from January 1978 through December 1991 are plotted. Since 1973 the dense CIT/USGS and University of Southern California (USC) networks of seismograph stations have been in operation in and around the Los Angeles basin, providing excellent coverage of seismic activity. The densest clusters of events in Figure 2 are mainshock-aftershock sequences such as 1979 and 1989 (M_L=5.2 and 5.0) Malibu (Hauksson and Saldivar, 1986), 1987 (M_L=5.9) Whittier Narrows (Hauksson and Jones, 1989), 1990 (M_L=5.3) Upland, and 1991 (M_L=5.8) Sierra Madre sequences.

While the moderate-sized to large earthquakes show good correlation with late Quaternary faults, the epicenters of small earthquakes do not show a simple correlation. As pointed out by Hauksson (1987), a broad zone of activity around the Newport-Inglewood fault can be seen in Figure 2. The absence of epicenters to the north of the southernmost north-dipping reverse faults (e.g. the Santa Monica-Hollywood-Sierra Madre fault zones) suggests that the small earthquakes are not associated with these faults. The almost continuous activity of small earthquakes to the south of these faults, along the zone west of the Whittier fault, and in the partly offshore zone to the west of the Newport-Inglewood fault, suggests that the deformation is concentrated along the flanks of the basin (Hauksson, 1990). The focal mechanisms of M>2.5 earthquakes in the Los Angeles basin are more systematic than the distribution of epicenters, and correlate well with the late Quaternary faults and folds (Hauksson, 1990). These were used by Hauksson (1990) to map the concealed faults in the Los Angeles basin. Because nearly all of the strain release is accommodated by the moderate-sized and large earthquakes, these events are the main focus of this paper.

SIGNIFICANT EARTHQUAKES 1920-1991

Since 1920, sixteen moderate-sized (M4.8-6.6) mainshock-aftershock sequences have occurred in the greater Los Angeles area (Figure 1 and Table 1). These earthquakes are associated with many low slip-rate, late Quaternary faults that are distributed over a large area (Ziony and Yerkes, 1985;

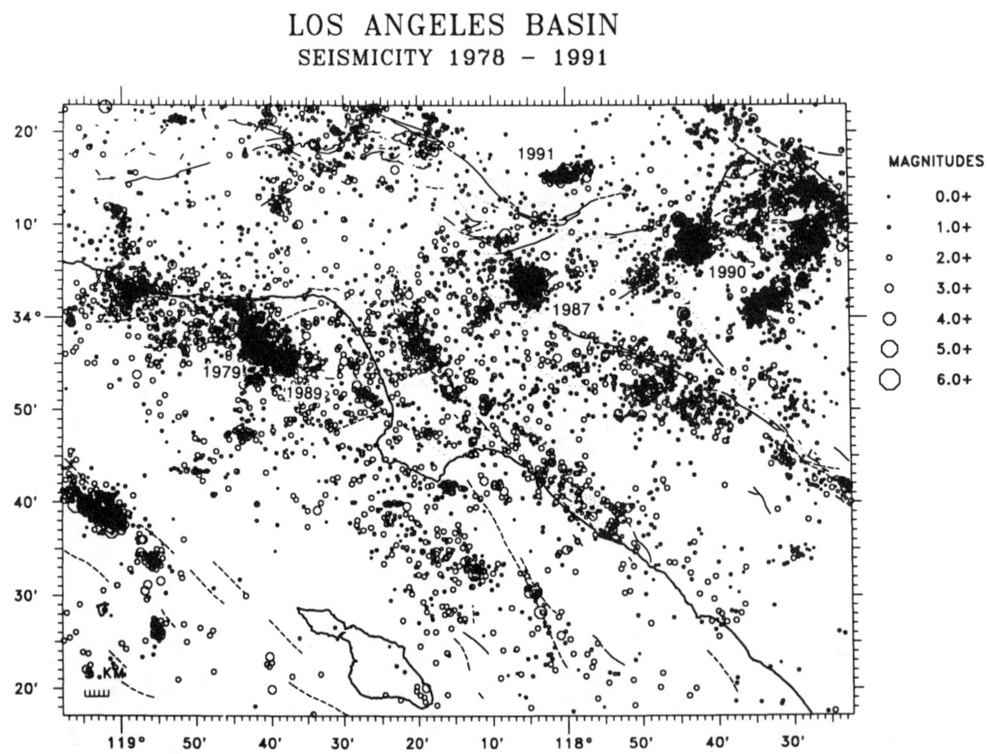

Figure 2. Seismicity in the Los Angeles basin from 1978 to 1991 recorded by the CIT/USGS Southern California Seismic Network and the USC Los Angeles Basin Seismic Network. The following sequences are labeled on the map with the year: 1979 and 1989 Santa Monica Bay; 1987 Whittier Narrows ; 1990 Upland sequence; and 1991 Sierra Madre sequence.

Ziony and Jones, 1989). Because surface rupture has only occurred once since the 1920's, in the 1971 San Fernando earthquake, the association between a mainshock hypocenter and a nearby fault is inferred from the mainshock focal mechanism and the distribution of aftershocks.

Some of these earthquakes, such as the 1933 (M_W6.4) Long Beach, 1971 (M_W6.7) San Fernando, and 1987 (M_W5.9) Whittier Narrows, have caused significant damage. Most recently the 1991 (M_L5.8) Sierra Madre earthquake caused $40 million worth of damage in western San Gabriel Valley. The damage patterns depend on the magnitude of the event, on local ground conditions, and the location of the epicenter relative to major population concentrations.

Style of Faulting

In general, the style of faulting determined from focal mechanisms of significant earthquakes is in excellent agreement with the surficially mapped style of faulting on late Quaternary faults. From a data set of eleven (M_L>4.8) events only two focal mechanisms shown in Figure 3, exhibit right-lateral strike-slip, two events show left-lateral strike-slip movement, and seven events show mostly reverse or thrust faulting.

Before instrumentally determined epicenters became available, the location of an earthquake was inferred from damage patterns or seismic intensities. For instance, some of the first research on Los Angeles basin earthquakes was done by Taber (1920), who was commissioned by the Seismological Society of America to study the 1920 Inglewood (M_L=4.9) earthquake. He attributed the earthquake to slip on the Newport-Inglewood fault just west of the town of Inglewood, because the event caused concentrated heavy damage in the town.

The 1933 Long Beach earthquake occurred a few years after the installation of the Caltech seismographic network, which at that time consisted of seven stations. The Caltech network and seismographic stations around the world provided sufficient data to show that the event was caused by right-lateral strike-slip movement along the southern segment of the Newport-Inglewood fault, from Newport Beach to the southern edge of the City of Long Beach (Wood, 1933; Richter, 1958; Hauksson and Gross, 1991).

The 1971 San Fernando earthquake is the only historic earthquake in the Los Angeles basin that is known to have caused surface rupture. The focal mechanism showed reverse faulting with a component of left-lateral strike slip motion (Whitcomb et al., 1973). The mainshock caused surface faulting along the 19 km long San Fernando member of the Sierra Madre fault zone (Proctor et al., 1972; Heaton and Helmberger, 1979, Heaton, 1982).

The 1987 Whittier Narrows earthquake, which occurred on a previously unrecognized concealed thrust fault, showed pure thrust motion on a gently north-dipping plane (Hauksson and Jones, 1989). The 1988 and 1990 Upland earthquakes were caused by left-lateral slip on the San Jose fault, which splays

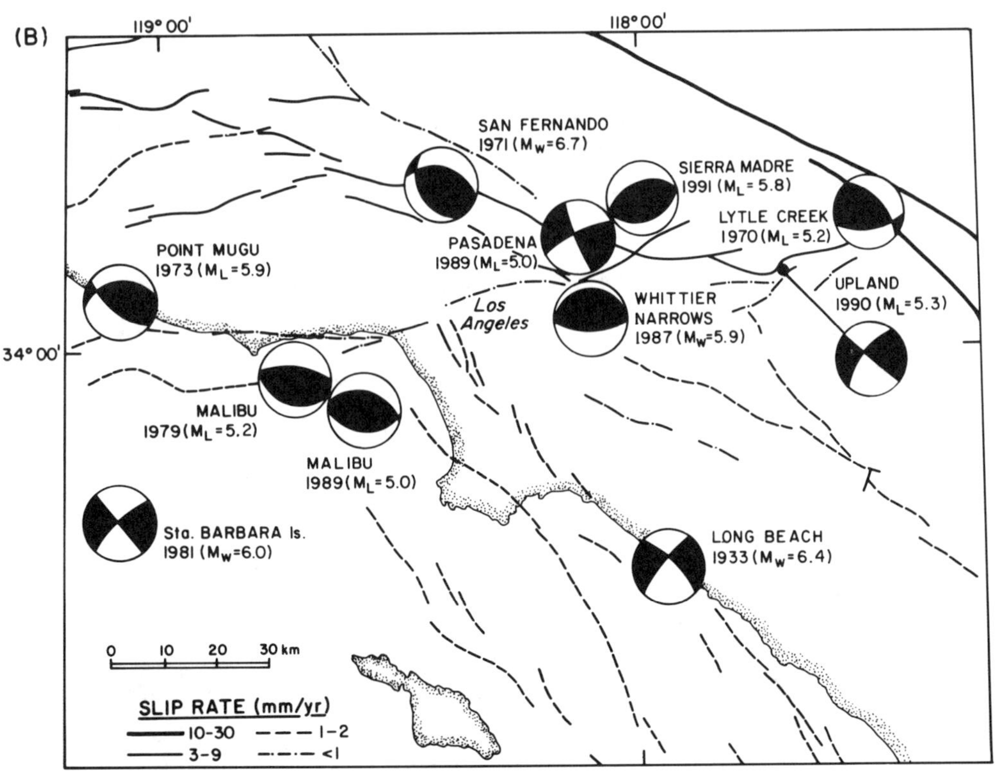

Figure 3. Single-event lower hemisphere focal mechanisms of the M4.8 events in the greater Los Angeles basin. Focal mechanisms are from Hauksson and Gross (1991), Jones (1984), Whitcomb *et al.* (1973), Stierman and Ellsworth (1976), Corbett (1984), Hauksson and Saldivar (1986), Hauksson and Jones (1989), Jones *et al.* (1990), Hauksson and Jones (1991), and Hauksson (1992).

southwest from the main frontal fault of the central Transverse Ranges (Hauksson and Jones, 1991). These left-lateral faults allow fragments of crustal blocks from the Peninsular Ranges block to move westward around the Transverse Ranges.

The 1991 Sierra Madre earthquake was the second moderate-sized event caused by thrust faulting along the Sierra Madre fault zone (Hauksson, 1992). Both surficial mapping of the Sierra Madre fault zone and the occurrence of the San Fernando and Sierra Madre earthquakes demonstrates that the 110 km long fault zone is seismically active and may cause larger damaging earthquakes (Crook et al., 1987).

Temporal Distribution

The temporal distribution of M>4.8 earthquakes in the Los Angeles basin is illustrated in Figure 4. Two periods of activity can be recognized. The first period began in 1920 and ended in 1941. The second period of activity began in 1970 and is still continuing in 1992. These two periods of activity are separated by a 30 year-long period of seismic quiescence when there were no earthquakes of M>4.8 in the greater Los Angeles basin.

Nearly all of the M>4.8 earthquakes that occurred during the first period of activity from 1920 to 1941 were located along the western edge or in the southwestern part of the Los Angeles basin. The 1933 Long Beach $M_w6.4$ earthquake was the largest member of the earlier sequence. Two events that occurred in 1941 were also located in the South Bay area. One $M_L5.2$ event in 1930 was located in Santa Monica bay and another $M_L5.2$ in 1938 one near the Elsinore fault. This activity was terminated with two events in 1941 in the South Bay.

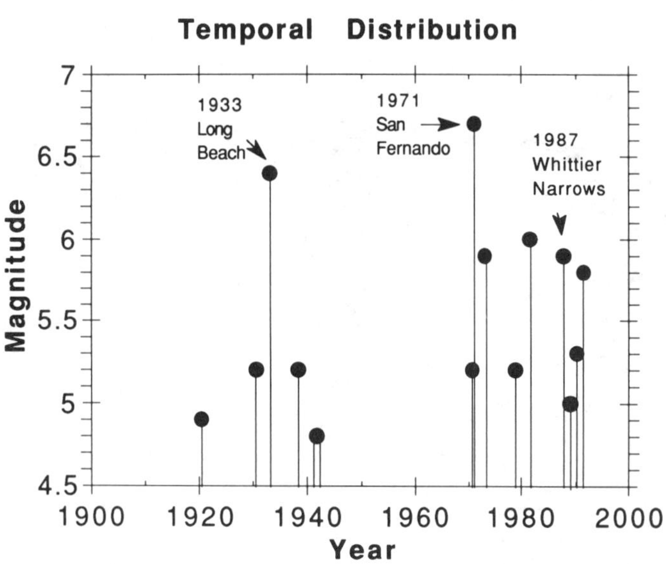

Figure 4. The temporal distribution of (M4.8) earthquakes in the Los Angeles basin since 1900. Magnitudes are listed in Table 1.

Since the early seventies the earthquake activity has mostly remained along the northern edge of the Los Angeles basin. The largest earthquake to occur in the region so far this century is the 1971 San Fernando earthquake. The spatial relationship between regional tectonic structures and the most recent moderate-sized earthquakes, the 1987 Whittier Narrows, 1988 Pasadena, 1988 and 1990 Upland, and 1991 Sierra Madre earthquakes, in San Gabriel Valley, suggest a pattern of northward migration.

The rate of M>4.8 earthquakes during the first period of activity was fairly uniform with one to two events per decade. During the second period of activity the number of earthquakes has increased with time. In the last five years (1987-1991) five moderate-sized earthquakes have occurred in the Los Angeles area. This is the highest rate of occurrence of moderate-sized earthquakes in the Los Angeles basin reported since the beginning of the CIT/USGS earthquake catalog.

Magnitude and Seismic Moment

The most often reported magnitude for an earthquake in the Los Angeles area is the local magnitude (M_L), which is determined from the photographic Wood-Anderson seismograms (Richter, 1958; Hutton and Boore, 1987; Hutton and Jones, 1992). For larger events, data from low gain (amplification of 100 times) instruments are also used. In the case of earthquakes of magnitude 6.0-6.5 this magnitude scale, however, may saturate.

"Seismic moment" is a different measure of earthquake size, which is often used for moderate-sized or large earthquakes. A seismic moment is a measure of the size of an earthquake based on the area of fault rupture, the average amount of slip, and the shear modulus of the rocks offset by faulting. Seismic moment is usually determined as part of the inversion of waveforms for the parameters of the earthquake source (Aki and Richards, 1980).

The seismic moment (M_o) of an earthquake is:

$$M_o = \mu A u$$

where μ is shear modulus, A is fault area undergoing slip, u is the relative slip along the ruptured fault segment.

An empirical relationship has been found between magnitude and seismic moment (M_o) (Hanks and Kanamori, 1979):

$$M = 2/3(\log_{10})M_o - 10.7$$

From this relationship, moment magnitude (M_w) has been defined. With this more accurate estimate of an earthquake size, seismic moment can be used to estimate magnitude.

In Table 1, three different magnitudes are provided if these are available. The first is the M_L (old) listed in the CIT/USGS earthquake catalog for southern California. The second is the M_L (new) which are redetermined magnitudes by Hutton and Jones (1992). They reread all the amplitudes and redetermined the magnitudes in a consistent manner for M4.8 earthquakes in southern California since 1932. The third is the M_w from seismic moments that have been determined in various studies. The fourth magnitude column (M) indicates the current best available estimates of magnitudes for M>4.8 earthquakes in the Los Angeles basin since 1920.

The local magnitude (M_L old) and the preferred magnitude (M) from Table 1 are plotted versus seismic moment in Figure 5. All of the data follow a logarithmic relationship except possibly the 1981 Santa Barbara Island earthquake which appears to have an underestimated magnitude (M_L old). The saturation of the M_L scale can be seen for the 1933 Long Beach and 1971 San Fernando earthquakes, the two largest events, because the moment magnitudes are systematically higher.

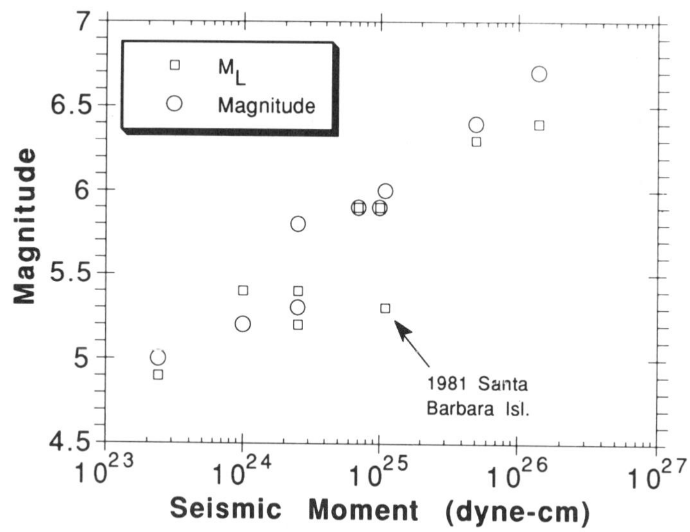

Figure 5. Preferred magnitude and local Richter magnitude (M_L) plotted as a function of seismic moment for earthquakes listed in Table 1.

Depth of Faulting

The overall depth of faulting is important because the maximum size of an earthquake that a fault can produce depends on the fault width. In addition, the distribution of damage depends on whether the fault rupture is shallow or deep. In Table 1 the focal depth of the mainshocks are listed with best estimates of uncertainty in the depth. In Figure 6 the hypocenter and depth distribution of aftershocks is plotted for eight events with well-known aftershock distributions. The three deepest events are the 1973 Point Mugu, 1987 Whittier Narrows and the 1988 Pasadena earthquake. The only earthquake that ruptured the surface is the 1971 San Fernando earthquake which had surface offsets of 1.5-2.3 m (Proctor et al., 1972). The Whittier Narrows earthquake is the only event that is known to have occurred on a concealed fault. The other events that did not rupture to the surface can all be associated with surficial faults.

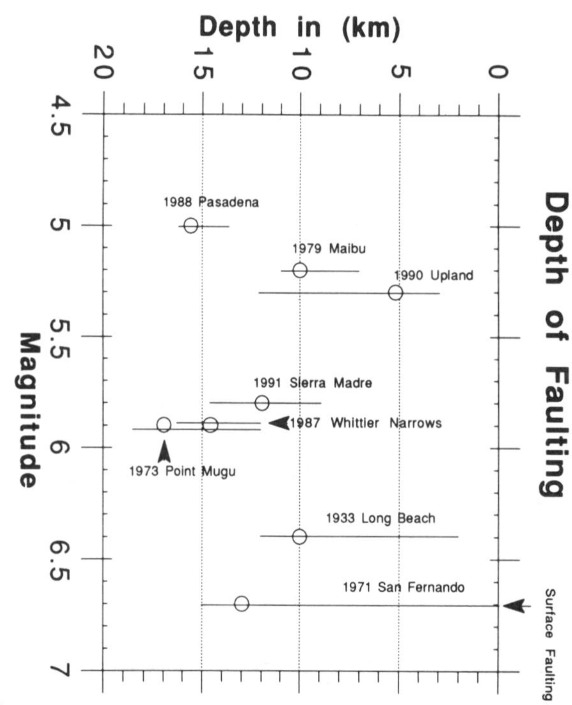

Figure 6. Depth of faulting determined from aftershock distribution for moderate-sized earthquakes in the Los Angeles basin. The hypo- center of each mainshock is indicated with an open circle. The vert- ical line through each circle represents the depth distribution of after- shocks. Only the 1971 San Fernando earthquake caused surface faulting.

SURFICIAL AND CONCEALED FAULTS

Some of the parameters of seismogenic faults in the Los Angeles area are summarized in Table 2. Most of the information in Table 2 is from Ziony and Yerkes (1985), Wesnousky (1986), and Davis et al. (1989).

The purpose of Table 2 is to compare the slip rates, expected magnitudes, and return periods for the relevant faults. The numbers in Table 2 are in some cases approximate averages or best guesses (flagged with a question mark) and the reader is referred to the above references for further details. Each fault can be divided into number of segments, as is shown in Table 2. The total return rate refers to the cumulative return rate for all the segments along that particular fault

Newport-Inglewood Fault

The Newport-Inglewood fault that is located in the western part of the Los Angeles basin is well studied because several major oil fields are situated adjacent to it (Wright, 1991). The Newport-Inglewood fault can be divided into two segments, the south segment that broke in 1933 and the north segment that extends from Signal Hill to the Santa Monica fault. Both segments are 25-30 km long and may rupture in similar-sized earthquakes. The state of stress along the north segment is more akin to that resulting in thrust faulting than the stress along the south segment (Hauksson, 1987). The possibility of both segments breaking in one large earthquake is small but cannot be excluded. A vertical slip rate of 0.12-0.6 mm/yr is derived from estimates of vertical offsets along the northern segment (Ziony and Yerkes, 1985). A vertical slip rate of 0.1-1.2 mm/yr is derived from offsets of rock formations along the southern segment (Ziony and Yerkes, 1985). Guptill and Heath (1981) stated that the strike-slip or horizontal component of motion along the fault was 0.5 mm/yr since late Miocene and early Pliocene (see Freeman et al., this chapter). In this paper the vertical and horizontal slip rates are assumed to be similar. Further studies are needed to better constrain the slip rate along the Newport-Inglewood fault because even a moderate-sized earthquake along this fault could cause significant damage in the metropolitan area.

To the south of Newport Beach, the Newport-Inglewood fault projects seaward into a system of faults having similar trends that can be traced more than 250 km south into Baja California. This part of the Newport-Inglewood fault could cause seismic shaking in the Los Angeles area as well as along the coastal area extending south to San Diego.

Palos Verdes Fault

The Palos Verdes fault, which coincides with the Torrance-Wilmington fold and thrust belt along most of its length, is interpreted to be either a right-lateral strike-slip fault or a steeply dipping reverse fault with oblique displacement (e.g., Ziony and Yerkes, 1985; Davis et al., 1989). Although numerous efforts have been made to find horizontal surficial offsets on the Palos Verdes fault, no clear evidence for horizontal offsets has been found (Darrow and Fisher, 1983). Small earthquakes with strike-slip focal mechanisms have occurred near the Palos Verdes fault as it heads offshore, both in San Pedro Bay and Santa Monica Bay (Hauksson, 1990).

Vertical offsets have been documented from uplifted marine terraces, which show a vertical uplift rate of 0.1-0.8 mm/yr (Patterson and Freeman, 1990). Because almost no evidence for strike-slip motion has been found and the vertical motion appears to be small, the Palos Verdes fault appears to accommodate only a small fraction of the total convergence being taken up by the southwest flank of the basin. Alternatively, moderate-sized or large earthquakes on the Palos Verdes fault may rarely rupture up to the surface.

Sierra Madre Fault Zone

The 110 km-long Sierra Madre fault zone follows the base of the San Gabriel mountains from Cajon Pass to San Fernando in a series of seven arcuate, 13-20 km-long fault segments (Proctor et al., 1972; Crook et al., 1987). The distribution of aftershocks from the 1971 San Fernando and 1991 Sierra Madre earthquakes show that the downdip width of these segments is about 23-28 km (Hauksson, 1992). The repeat time for earthquakes has been estimated along the San Fernando segment to be 200 years (Bonilla, 1973). Similarly, along the Cucamonga segment the repeat time was estimated to be 700 years by Matti et al. (1982). Based on the results obtained by Bonilla (1973) and Matti et al. (1982), Ziony and Yerkes (1985) assigned a maximum M6.5-7.0 for non-critical structures near the Sierra Madre fault zone. Similarly, Wesnousky (1986) proposed a maximum earthquake magnitude of 6.4-6.6 along each segment by comparison with the 1971 San Fernando earthquake.

A future M7.7 earthquake that could rupture all seven segments or the whole length of the Sierra Madre fault cannot be excluded. The 19 km-long surface rupture of the 1971 San Fernando earthquake (Proctor et al., 1972) and the two separate 4 km-long aftershock distributions of the 1991 Sierra Madre earthquake are seismological evidence for the existence of the surficially mapped segmentation of the fault zone at depth (Hauksson, 1992). Furthermore, the San Fernando and Cucamonga segments, which have relatively short repeat times, appear to behave at least in some cases as independent segments (Bonilla, 1973; Matti et al., 1982). This seismological and geological evidence for segmentation along the fault thus suggests that 6.4-6.6 earthquakes occur more frequently and that the M7.7 earthquake may be a very rare event.

Whittier Fault

The Whittier fault is a right-lateral strike-slip fault with slip rates of 2-3 mm/yr (Gath et al., 1992). It is 74 km long and extends from the north end of the Elsinore fault into the eastern Los Angeles basin (Wesnousky, 1986). The Whittier fault has had up to 30 km of right-lateral slip since 4-9 Ma (Lamar, 1990, and this chapter). The vertical offset on it is more difficult to quantify but is much less and averages about 1.0-1.5 km (Wright, 1991). Although the Whittier fault has not had a large damaging earthquake during historic times, it could generate a M7-7.3 earthquake (Wesnousky, 1986).

Santa Monica and Hollywood Faults

The Santa Monica and Hollywood faults form the north edge of the Los Angeles basin (Crook and Proctor, this chapter). They appear to have low slip rates, and almost no background seismicity is associated with these faults. The low slip rate of 0.04-0.4 mm/yr (Ziony and Yerkes, 1985) leads to long return rates. Because these faults can be traced for a distance of more than 100 km, they are thought to be important in hazard evaluations for the Los Angeles Basin (Ziony and Yerkes, 1985). The expected magnitude estimate is beset with large uncertainties. These faults have also been interpreted to be old overturned normal faults that have been reactivated as thrust faults (Davis et al., 1989). Thus some of the old geological offsets may not be representative of the current geological slip rate.

Elysian Park Fold and Thrust Belt

The importance of the Elysian Park fold and thrust belt as a deep-seated seismogenic structure was not fully recognized until the time of the 1987 Whittier Narrows earthquake. It extends along the east and north flanks of the Los Angeles basin for a distance of approximately 100 km (Fig. 7). The location of the thrust and fold belt was determined by mapping of fold axes and thrust-faulting earthquakes (Davis et al., 1989; Hauksson, 1990). The thrust or reverse focal mechanisms are mostly located to the south of reverse faults such as Santa Monica and Hollywood faults and form broad clusters along the flanks of the basin. Most of the folding of the basin sediments also occurs along the flanks of the basin. This spatial coincidence of the folding and thrust faulting can be explained with fault-bend or fault-propagation folding models (e.g. Suppe 1985), where the thrust-faulting earthquakes at depth are associated with folding of the cover sequence. The 1987 Whittier Narrows earthquake, which showed thrust faulting and caused coseismic uplift in the Whittier Narrows region, demonstrated how the folding and faulting are causally related (Hauksson and Jones, 1989; Lin and Stein, 1989; Davis et al., 1989).

To the south-southeast of Whittier Narrows the thrust-faulting focal mechanisms delineate a zone of thrust faulting dipping 25-35° north. It is not possible to tell, based on hypocentral distribution and focal mechanisms alone, if this zone consists of one continuous fault or many small abutting or overlapping faults. This zone of thrust faults is closest to the surface at 4-6 km depth, near Yorba Linda. The southernmost mapped thrust fault along the eastern flank of the Los Angeles basin, the Peralta Hills thrust fault is located in this area, 5 km south of Yorba Linda (Bryant and Fife, 1982).

The Elysian Park fold and thrust belt also extends to the west of Whittier Narrows, beneath downtown Los Angeles, and continues into Santa Monica Bay, where it is found at a depth of 10-15 km. It may be offset or segmented by the northern end of the Newport-Inglewood fault, because the maximum focal depths to the east of the intersection of the two faults are 3-4 km deeper than to the west.

The Elysian Park thrust zone has moved 5.5-21 km over the last 4.0-2.2 Ma (Davis et al., 1989). The slip rate of the

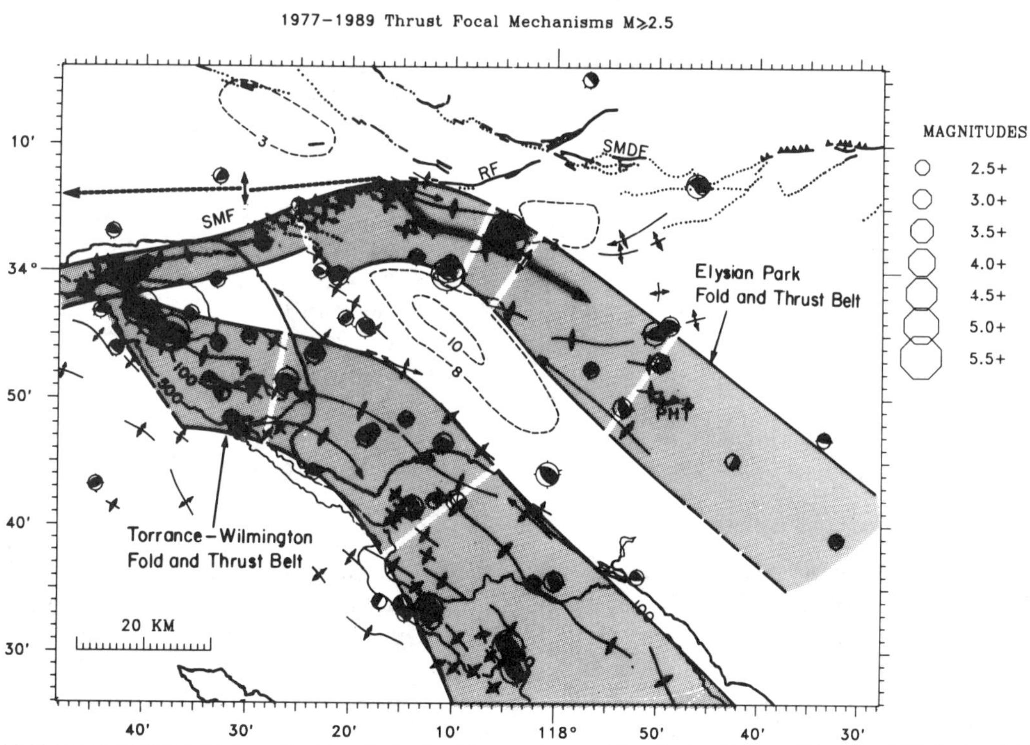

Figure 7. Map of fold axes and thrust focal mechanisms in the Los Angeles basin. The Elysian Park and Torrance-Wilmington fold and thrust belts are shaded. Breaks in the shading are at segment boundaries. RF, Raymond fault; SMDF, Sierra Madre fault; SMF, Santa Monica fault; and PHT, Peralta Hills thrust (from Hauksson, 1990).

major faults within the fold and thrust belt is derived from balanced cross sections by Davis et al. (1989). The segmentation of this belt is poorly understood and needs further study to adequately constrain the expected magnitude of earthquakes within it. In addition, more geological studies of Quaternary deformation and geodetic surveys are needed to confirm the fault slip rates as indicated by the retrodeformable cross section technique.

Torrance-Wilmington Fold and Thrust Belt

The Torrance-Wilmington fold and thrust belt, which is approximately 60 km long, is outlined by fold axes and thrust focal mechanisms extending from offshore Newport Beach, across the Palos Verdes Peninsula, and into Santa Monica Bay along the southwest flank of the Los Angeles basin (Fig. 7). The depth distribution of thrust-fault events within this fold and thrust belt is more complex than within the Elysian Park fold and thrust belt. In Santa Monica Bay, a zone of thrust faults dips north 30-40°, while, farther to the southwest (onshore) the hypocenters show a steeply dipping zone beneath the Palos Verdes fault trace. A similar zone of thrusting is observed in San Pedro Bay, dipping southwest at 20-30°.

In Santa Monica Bay the two thrust belts merge into a north-dipping imbricate thrust zone. The Elysian Park thrust zone is located on the north side, beneath the Malibu shelf. On the south side the north-dipping thrust zone of the Torrance-Wilmington belt changes strike from northwest to west as it merges with the Elysian Park zone. The Shelf Projection anticline in Santa Monica Bay is probably a surface expression of this imbricate thrust zone (Hauksson and Saldivar, 1989).

Because most faults in the study area are either north or northeast dipping, the south-dipping thrust in San Pedro Bay is unexpected and requires a special explanation. One possible explanation for this zone of thrusting, which has two M_L=4.0 and 4.5 events near its downdip end, could be a back thrust, which commonly appears at the later stages of thrust fault formation (J. Suppe, personal communication, 1989). Usually a set of thrust faults is configured such that, on a regional scale, the upper block overrides the lower block, while often a conjugate set of thrust faults, called back thrusts, may develop. In San Pedro Bay, this implies a deeper east- or northeast-dipping thrust fault, which would be the primary member of the conjugate set.

Davis et al. (1989) called the primary concealed fault beneath the Palos Verdes Peninsula fault A. Using the retrodeformable cross section technique, they argued that fault A has a slip rate of 1.9-3.45 mm/yr, which is a significantly higher slip rate than those estimated for either the Palos Verdes or Newport-Inglewood faults. The Torrance-Wilmington fold and thrust belt needs further studies of the long-term slip rates and segmentation of the concealed faults to make it possible to determine the expected magnitude and return rate more accurately.

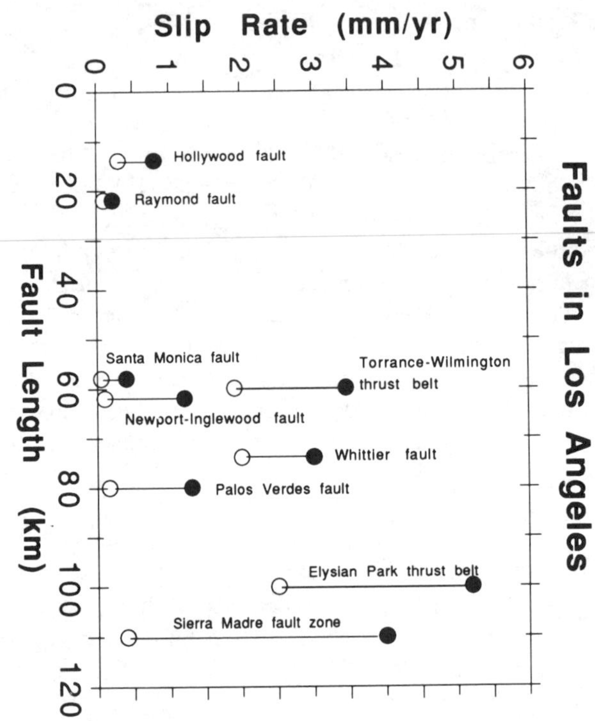

Figure 8. Slip rates versus fault length for faults in the Los Angeles area. Both low (open circle) and high (closed circle) estimates are provided. The slip rates are also listed in Table 2.

Fault Slip Rates and Lengths

The range of slip rates and lengths of faults in the Los Angeles basin are listed in Table 2 and plotted in Figure 8. All of the faults have the slip rates limited to a range of values, often reflecting that these values are only known to an order of magnitude. Faults with slip rates less than 0.01 mm/yr are not included in the estimates of the earthquake potential. In general the slip rate appears to increase with fault length. Although most of the faults are surficial faults, two systems of buried faults, the Torrance-Wilmington and the Elysian Park thrusts are also included.

EARTHQUAKE POTENTIAL

Seismic moment, seismicity and slip rate along faults or rates of relative plate motions have been used to compare the historical seismicity with the long-term seismicity predicted by geological slip rates (e.g. Brune, 1968; Anderson, 1979; Molnar, 1979; Doser and Smith, 1982; and Ekstrom and England, 1989). Below, I use a similar approach to these previous studies to evaluate how much long-term seismic moment has accumulated and how much seismic moment has been relieved in earthquakes on faults in the Los Angeles basin (i.e. their earthquake potential).

We can determine how much seismic moment is stored on each fault (here referred to as cumulative seismic moment) as long as we know the time interval, geological slip rate, and fault length and width. We can also determine from the recorded seismograms of each earthquake how much seismic moment was released in each event. The difference between accumulated and released seismic moment indicates how much seismic moment is stored and available to generate the next earthquake.

To get an average accumulated seismic moment for the Los Angeles area we have divided the faults in the region into surficial and concealed. This is in part done to illustrate which group of faults makes a larger contribution to the earthquake hazards in the Los Angeles basin. There are two concealed fault systems, the Elysian Park and Torrance-Wilmington fold and thrust belts (Davis et al., 1989; Hauksson, 1990). All other faults listed in Table 2 are presumed to be surficial. The accumulated seismic moment (CSM_O) from geological data on faulting can be determined from (Brune, 1968):

$$CSM_O = \mu Ast$$

where s is geologic slip rate and t is time interval. To determine the fault area the average width of faulting is assumed to be 10 km, in part based on the data shown in Figure 6. This is a minimum estimate because in large earthquakes on a dipping fault, the width could be as much as 25-30 km (Hauksson, 1992). The fault length is determined by measuring the length of the surface traces from fault maps such as Jennings (1975), or provided by Ziony and Yerkes (1985) and Wesnousky (1986), or for the concealed faults from Hauksson (1990).

Surficial Faults

The minimum and maximum estimates of the cumulative seismic moment for surficial faults in the Los Angeles basin are shown in Figure 9. A somewhat arbitrary start time of 1800 was chosen because the historic seismicity is known fairly well back to that time (Toppozada et al., 1981). In Figure 9 it is also shown how the two largest earthquakes in the Los Angeles area, 1933 Long Beach and 1971 San Fernando earthquakes, have released most of the seismic moment on surficial faults. Thus in the case of the surficial faults, the discrepancy between the minimum cumulative seismic moment since 1800 and the released moment is small. A future earthquake of M6.6, similar in size to the 1971 San Fernando earthquake, would release most of the maximum seismic moment accumulated on surficial faults since 1800.

Concealed Faults

The minimum and maximum seismic moments accumulated on concealed faults since 1800 are shown in Figure 9. The cumulative seismic moments indicate that seismic moment is accumulating faster on the concealed faults than on the surficial faults. Only the 1987 $M_W5.8$ Whittier Narrows earthquake is known to have occurred on the concealed faults. A second M6 event in 1855 that caused similar damage to the San Gabriel Mission is assumed to have occurred on the Elysian Park thrust (Ellsworth, 1990). Both of these earthquakes are relatively small and released only small amounts of seismic moment. A large discrepancy thus exists between the cumulative seismic moment and the seismic moment that has been released in earthquakes on concealed faults since 1800.

Segmentation of fold and thrust belts can limit the size of earthquake rupture and is thus important for evaluating the earthquake potential of these concealed faults. The available data can not resolve whether the fold and thrust belts are underlain by contiguous or segmented faults (Hauksson, 1990). Furthermore, the seismicity from 1977 to 1989 is a snap shot in time and may not illuminate all of the relevant concealed fault structures. Several lines of evidence, however, suggest that the thrust belts may be underlain by segmented faults.

The Elysian Park fold and thrust belt is approximately 100 km-long, and based on limited evidence can be divided into at least five large segments (Fig. 7). The geological and earthquake data alone cannot resolve if each of these large segments are underlain by one continuous thrust fault or

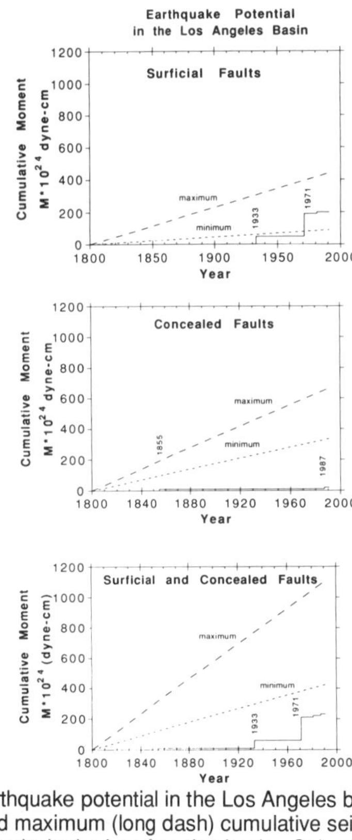

Figure 9. Earthquake potential in the Los Angeles basin. Minimum (short dash) and maximum (long dash) cumulative seismic moment since 1800 on faults in the Los Angeles basin. Seismic moment released in earthquakes is indicated by the solid line. The earthquakes are indicated with the year. (a) surficial faults only; (b) concealed faults only; (c) both surficial and concealed faults.

many small thrust faults. The segment boundaries are identified from (1) the eastern and western limits of the aftershock zone of the 1987 Whittier Narrows earthquake; (2) the Yorba Linda seismicity trend; and (3) the intersection with the Newport-Inglewood fault where the depth of the hypocenters appears to be offset with greater focal depths found to the east.

The Torrance-Wilmington fold and thrust belt is approximately 60 km long and can be divided into three large segments mostly based on the boundaries of oil fields within the belt. Like the Elysian Park thrust zone, whether each of the large segments consists of one large or many small thrust faults cannot be resolved. One of the segmentation boundaries is in San Pedro Bay where the Wilmington oil field is offset from the Beta oil field (Wright, 1987). The other is offshore from Torrance, at the western end of the Torrance oil field. This segmentation boundary is supported by some strike-slip focal mechanisms near the north offshore extension of the Palos Verdes fault. The Torrance-Wilmington and Elysian Park fold and thrust belts appear to merge in Santa Monica Bay and continue to the west beyond Point Dume as one fold and thrust belt (see also Davis et al., 1989).

Any of these eight large segments could rupture in a moderate-sized (M>5) or large (M>6.7) earthquake. If several segments could rupture simultaneously, a large (M>6.7) or great (M>7.8) earthquake could result. The compressional tectonics of the Los Angeles basin are similar to the tectonics of the southern San Joaquin Valley with concealed thrust faults beneath thick folded sediments. The 1952 (M_S7.7) Kern County earthquake ruptured a 75-km-long concealed fault with three distinct segments near the southern end of the San Joaquin Valley (Stein and Thatcher, 1981). Although the concealed faults in the basin appear to be segmented, a large earthquake that would break multiple fault segments beneath the Los Angeles basin cannot be precluded.

Surficial and Concealed Faults

The minimum and maximum accumulative seismic moments for both surficial and concealed faults are shown in Figure 9. Because more than half the seismic moment is accumulating on the concealed faults and no large earthquakes have occurred on these, the discrepancy between accumulated and released moment is more prominent. A future large (M7-7.5) earthquake on one of the two concealed faults is needed to bring the cumulative and released seismic moments into closer agreements.

DISCUSSION

The Los Angeles basin is not a passive depositional sedimentary basin, but rather a region of complex tectonics (e.g., Yerkes et al., 1965; Davis et al., 1989; Wright, 1991). This complexity has made it difficult to interpret the ongoing tectonic deformation and the associated earthquake activity, which, in turn, has resulted in "surprise" earthquakes causing significant property damage (e.g., Hauksson et al., 1988). Hauksson (1990) contributed toward understanding the complex tectonics through mapping of fold and thrust belts using geologic data and earthquake focal mechanisms and suggested that the coexistence of strike-slip and thrust faults is the result of decoupling of oblique fault motion.

Decoupling of Fault Slip

The Los Angeles basin forms a transition zone between the strike-slip tectonics of the Peninsular Ranges and the convergent tectonics of the Transverse Ranges. Both strike-slip and compressional tectonic structures exist within this transition zone. Traditionally, the compressive structures such as folds have been interpreted as secondary structures caused by movements on the primary late Quaternary strike-slip faults. This interpretation is called the wrench tectonics model (e.g., Wilcox et al., 1973). The wrench faulting models assume layers of soft sediments that deform in response to motion on a strike-slip fault at depth below them. The Newport-Inglewood fault on the southwest flank of the basin is often presented as the classic example of wrench faulting (Wilcox et al., 1973). Recently, Wright (1987 and 1991) has pointed out the strong disagreement between the actual distribution and orientation of fold axes along the Newport-Inglewood and Whittier faults and those predicted by the wrench faulting model. Similarly, the folding along the north flank of the basin cannot be associated with a known strike-slip fault. Thus the folding along the flanks of the basin cannot easily be explained as secondary effects from strike-slip faulting.

If both the strike-slip and compressional tectonics in the basin are of similar importance, their coexistence could suggest that the faulting in the basin should be mostly oblique. Alternatively, the merging of these two tectonic styles could take place as decoupled strike-slip and thrust faulting. Such decoupling occurred on a smaller scale in the 1987 Whittier Narrows sequence where the mainshock ruptured a gently dipping thrust fault and the largest aftershock ruptured a steeply dipping strike-slip fault (Hauksson and Jones, 1989).

Decoupling of fault motion, or slip partitioning, implies that two sets of faults with distinctly different fault motions coexist and together accommodate oblique fault motion in the same region. Decoupling of fault motion has not been thought important in the past because short faults commonly exhibit oblique slip and focal mechanisms of small earthquakes showing oblique faulting are also common. However, unlike the small earthquakes in the Los Angeles basin, the moderate and large earthquakes rarely exhibit oblique faulting (Fig. 3). Recent studies of the San Andreas fault in central California infer that slip is decoupled near the fault such that strike-slip movement takes place on the main trace of the fault and thrust or normal movements are accommodated on smaller subsidiary faults (Zoback et al., 1987; Mount and Suppe, 1987; Mount, 1989). Similarly, along the

San Andreas fault in southern California, Jones (1988) found a good correspondence between off-fault earthquakes, late Quaternary tectonic deformation, and state of stress. This suggests that the decoupling of strike-slip from normal and thrust motion in some cases is an important part of faulting.

Earthquake Hazards

When earthquake hazards are evaluated, the most important data input to the evaluation are the determination of which faults and how much each fault contributes to the hazard. For instance, Ziony and Yerkes (1985) identified 95 late Quaternary faults in southern California that could contribute to earthquake hazards in the Los Angeles area. Their study did not include the more recently recognized concealed faults in the Los Angeles basin. When the concealed faults were recognized in 1987 following the Whittier Narrows earthquake, the question was immediately raised as to whether the surficial faults or the concealed faults made a greater contribution to earthquake hazards in the Los Angeles area. This study has provided a qualitative answer to that question, showing that the concealed faults more than double the earthquake hazards in the Los Angeles area.

The high earthquake potential of the concealed faults is the result of their combined high slip rate and long length. Because I assume that the downdip width of the concealed thrust faults is only 10 km, the calculated cumulative seismic moment is probably a lower bound. Furthermore, no large earthquakes are known to have occurred on these faults over the last 200 years. This suggests either that (1) significant slip is accommodated aseismically by fault creep, which seems possible in view of the depth of the main detachment surface although crustal ramps may extend up into the brittle part of the crust, or that (2) the slip is accommodated by infrequent large (M7) earthquakes, none of which happens to have occurred within the historic record. Currently this issue is unresolved.

Ekstrom and England (1989) did a study similar to the one presented here for the whole plate boundary in California. They compared the cumulative seismic moment along the plate boundary with the seismic moment released in earthquakes. During 1977-1987 the released seismic moment can only account for less than 10% of the relative motion between the Pacific and the North America plates. During 1812-1987 they found that 45-60% of the cumulative seismic moment was released mostly in four large earthquakes, 1857 Fort Tejon, 1872 Owens Valley, 1906 San Francisco, and 1952 Kern County. Their results, combined with the findings of this study, show that over the last 200 years the Los Angeles basin, with only 20-55% of the seismic moment released, has lagged behind the average seismic moment release in California. Only one large earthquake in the M7-7.75 range, however, is needed to erase this discrepancy.

Anderson (1979) used a similar approach to estimate cumulative seismic moment in southern California and showed that there was good agreement between cumulative and released seismic moment between 1857 and 1978. He also pointed out that for smaller subregions of southern California where the repeat time of large earthquakes is much larger than 200 years, the cumulative seismic moment was often much larger than the released seismic moment. Thus, the results of this study are consistent with his findings because only a fraction of the cumulative seismic moment has been released and the average repeat times of earthquakes in the Los Angeles basin are much larger than 200 years.

CONCLUSIONS

The merging of the strike-slip faulting of the Peninsular Ranges and the thrust faulting of the Transverse Ranges does not occur through oblique faulting but rather through slip partitioning, or decoupled strike-slip and thrust faulting. Moderate-sized damaging earthquakes have occurred on both the strike-slip and the thrust faults. Because the majority of historic damaging earthquakes in southern California have occurred on faults with surface evidence of geologically recent displacements, significant progress has been made in estimating the associated hazards. It is more difficult to estimate seismic hazards associated with concealed thrust faults, because they are not accessible and the understanding of their mechanics is only gradually evolving. Because the concealed faults more than double the earthquake hazards in the Los Angeles area, improved understanding of their nature and behavior through more detailed geological studies, GPS measurements, earthquake monitoring and onshore and offshore seismic-reflection profiling is needed.

ACKNOWLEDGMENTS

I thank L. Jones, J. Dolan, and the editors of this book for critical review of the manuscript. This work was supported by USGS Grant 14-08-0001-G1761. Division of Geological and Planetary Sciences, California Institute of Technology, contribution 5138.

REFRENCES

Aki, K., and P. G. Richards, 1980, Quantitative Seismology: Theory and Methods: W. H. Freeman and Co., NY, 932 p.

Allen, C. R., P. St. Amand, C. F. Richter, and J. M. Nordquist, 1965, Relationship between seismicity and geologic structure in the southern California region: Bull. Seismol. Soc. Amer., vol. 55, p. 753-797.

Anderson, J. G., 1979, Estimating the seismicity from geological structure for seismic-risk studies, Bull. Seismol. Soc. Am., vol. 69, p135-158.

Bent, A. L., 1990, Source characteristics of recent and historic earthquakes in central and southern California: Results from forward modeling, Ph.D Thesis, 149 p., California Institute of Technology.

Bent, A. L., and D. V. Helmberger, 1991, Seismic characteristics of earthquakes along the offshore extension of the Western Transverse Ranges: Seismol. Soc. Am. Bull., vol. 81, p. 399-422.

Bonilla, M. G., 1973, Trench exposures across surface fault ruptures associated with San Fernando earthquake in San Fernando, California, earthquake of February 9, 1971: vol. 3, U. S. Department of Commerce, Washington DC, p. 173-182.

Brune, J. N., 1968, Seismic moment, seismicity, and rate of slip along major fault zones: J. Geophys. Res., vol. 73, p. 777-784.

Bryant, M. E., and D. L. Fife, 1982, The Peralta Hills fault, a Transverse Ranges structure in the northern Peninsular Ranges, southern California: in Geology and mineral wealth of the California Transverse Ranges, edited by D. L. Fife and J. A. Minch, South Coastal Geological Society, Santa Ana, Calif., p. 403-409.

Corbett, E. J., 1984, Seismicity and crustal structure studies of southern California: Tectonic implications from improved earthquake locations: Ph. D. Thesis, 231 p., California Institute of Technology.

Crook, R., Jr., C. R. Allen, B. Kamb, C. M. Payne, and R. J. Proctor, 1987, Quaternary geology and seismic hazard of the Sierra Madre and associated faults, western San Gabriel Mountains, in Recent Reverse Faulting in the Transverse Ranges, California: U. S. Geol. Surv. Prof. Paper 1339

Darrow, A. C., and P. J. Fisher, 1983, Activity and earthquake potential of the Palos Verdes fault: final technical report, U. S. Geol. Survey, 90p.

Davis, T. L., J. Namson, and R. F. Yerkes, 1989, A cross section of the Los Angeles area: Seismically active fold and thrust belt, the 1987 Whittier Narrows earthquake and earthquake hazard: Jour. Geophys. Res., p. 9644-9664.

Doser, D. I., and R. B. Smith, 1982, Seismic moment rates in the Utah Region: Bull. Seismol. Soc. Am., vol. 72, p. 52-552.

Dreger, D., and D. Helmberger, 1991, Source parameters of the Sierra Madre earthquake from regional and local body waves: Geophys. Res. Lett., vol. 18, p. 2015-2018.

Dreger, D., and D. V. Helmberger, 1991, Complex faulting deduced from broadband modeling of the Feb. 28, 1990 Upland earthquake (ML=5.2): Bull. Seismol. Soc. Amer., vol. 81, p. 1129-1144.

Ekstrom, G., and P. England, 1989, Seismic strain rates in regions of distributed continental deformation: Jour.. Geophys. Res., vol. 94, p. 10,231-10,257.

Ellsworth, W. L., 1990, Earthquake history, 1769-1989: in The San Andreas Fault System, California, edited by R. Wallace, U. S. Government Printing Office, Washington DC, p. 153-188.

Gath, E. M., T. Gonzalez, and T. K. Rockwell, 1992, Slip rate of the Whittier Fault based on 3-D trenching at Brea, southern California: GSA Cordilleran Section Meeting, abstract.

Guptill, P. D., and E. G. Heath, 1981, Surface faulting along the Newport-Inglewood zone of deformation: California Geology, vol. 34, p. 142-148.

Gutenberg, B., C. F. Richter, and H. O. Wood, 1932, The earthquake in Santa Monica Bay, California on August 30, 1930, Bull. Seism. Soc. Am., vol. 22, p. 138-154.

Hanks, T., and H. Kanamori, 1979, A moment magnitude scale, Jour. Geophys. Res., vol. 84, p. 2348-2350.

Hauksson, 1992, The 1991 Sierra Madre Earthquake Sequence in Southern California: Seismological and Tectonic Analysis, submitted, J. Geophys. Res.,

Hauksson, E., 1987, Seismotectonics of the Newport-Inglewood fault zone in Los Angeles basin, southern California: Bull. Seismol. Soc. Amer., vol. 77, p. 539-561.

Hauksson, E., 1990, Earthquakes, faulting and stress in the Los Angeles Basin: Jour. Geophys. Res., vol. 95, p. 5,365-15,394.

Hauksson, E., and S. Gross, 1991, Source parameters of the 1933 Long Beach earthquake, Seismol. Soc. Amer. Bull., vol. 81, p. 81-98.

Hauksson, E., and L. M. Jones, 1989, The 1987 Whittier Narrows earthquake sequence in Los Angeles, southern California: Seismological and tectonic analysis, Jour. Geophys. Res., vol. 94, p. 9569-9590.

Hauksson, E., and L. M. Jones, 1991, The 1988 and 1990 Upland earthquakes: Left-lateral faulting adjacent to the Central Transverse Ranges: Jour. Geophys. Res., vol. 96, p. 8143-8165.

Hauksson, E., L. M. Jones, T. L. Davis, L. K. Hutton, A. G. Brady, P. A. Reasenberg, A. J. Michael, R. F. Yerkes, P. Williams, G. Reagor, C. W. Stover, A. L. Bent, A. K. Shakal, E. Etheredge, R. L. Porcella, C. G. Bufe, M. J. S. Johnston, and E. Cranswick, 1988, The 1987 Whittier Narrows earthquake in the Los Angeles metropolitan area, California: Science, vol. 239, p. 1409-1412.

Hauksson, E., and G. V. Saldivar, 1986, The 1930 Santa Monica and the 1979 Malibu, California, earthquakes: Bull. Seismol. Soc. Amer., vol. 76, p. 1542-1559.

Hauksson, E., and G. V. Saldivar, 1989, Seismicity and active compressional tectonics in the Santa Monica Bay, southern California: Jour. Geophys. Res., vol. 94, p. 9591-9606.

Heaton, T. H., 1982, The 1971 San Fernando earthquake: A double event?: Bull. Seismol. Soc. Amer., vol. 72, p. 2037-2062.

Heaton, T. H., and D. V. Helmberger, 1979, Generalized ray models of the San Fernando earthquake: Bull. Seismol. Soc. Am., vol. 69, p. 1311-1341.

Hileman, J. A., C. R. Allen, and J. M. Nordquist, 1973, Seismicity of the southern California region, 1 January 1932 to 31 December 1972, p., Seismological Laboratory, California Institute of Technology, Pasadena, California.

Hutton, L. K., and D. M. Boore, 1987, The ML Scale in Southern California: Bull. Seismol. Soc. Amer., vol. 77, p. 2074-2093.

Hutton, L. K., and L. M. Jones, 1992, submitted, Local magnitudes and apparent variations in seismicity rates in southern California: Bull. Seismol. Soc. Am.,

Jennings, P., 1975, Fault map of California with volcanoes, thermal springs and thermal wells, 1:750,000 scale, geol. data map 1: Calif. Div. Mines and Geol., Sacramento, CA.

Jones, L. M., 1984, Foreshocks (1966-1980) in the San Andreas System, California: Bull. Seismol. Soc. Amer., vol. 74, p. 1361-1380.

Jones, L. M., 1988, Focal mechanisms and the state of stress on the San Andreas fault in southern California: Jour. Geophys. Res., vol. 93, p. 8869-8891.

Jones, L. M., K. E. Sieh, E. Hauksson, and L. K. Hutton, 1990, The 3 December 1988 Pasadena, California earthquake: Evidence for strike-slip motion on the Raymond fault: Bull. Seismol. Soc. Amer., vol. 80, p. 474-482.

Kanamori, H., J. Mori, and T. H. Heaton, 1990, The December 3, 1988 Pasadena earthquake (ML=4.9) recorded with the very broadband system in Pasadena: Seismol. Soc. Amer. Bull., vol. 80, p. 483-487.

Lamar, D. L., 1990, Right-slip on Whittier fault system, southern California, in Field Guide of Whittier Fault, edited by E. Gath, Southern California Section, Associate of Engineering Geologists, Los Angeles, California.

Lin, J., and R. Stein, 1989, Coseismic folding, earthquake recurrence, and the 1987 source mechanism at Whittier Narrows, Los Angeles basin, California: Jour. Geophys. Res., p. 9614-9632.

Matti, J. C., and D. M. Morton, 1982, Geologic history of the Banning fault zone, southern California: Geol. Soc. Am. Abstr. Programs, vol. 14, 184 p.

Molnar, P., 1979, Earthquake recurrence intervals and plate tectonics: Bull. Seim. Soc. Am., vol. 69, p. 115-134.

Morton, D. M., and J. C. Matti, 1987, The Cucamonga fault zone: Geological setting and Quaternary history, in Recent Reverse Faulting in the Transverse Ranges, California, U. S. Geol. Surv. Prof. Paper 1339, edited by D. M. Morton and R. F. Yerkes, U. S. Govt. Print. Office, Washington DC, p. 179-203.

Mount, V. S., 1989, State of stress in California and a seismic structural analysis of the Perdido fold belt northwest Gulf of Mexico, Ph.D. Thesis, pp., Princeton Univ., Princeton, NJ.

Mount, V. S., and J. Suppe, 1987, State of stress near the San Andreas fault: Implications for wrench tectonics: Geology, vol. 15, p. 1143-1146.

Patterson, R. H., and S. T. Freeman, 1990, Characterization of the Palos Verdes fault-onshore: POLA Seismic Workshop, Proceedings, p. 1-21.

Proctor, R. J., R. Crook, M. H. McKeown, and R. L. Morseco, 1972, Relation of known faults to surface ruptures, 1971 San Fernando earthquake, southern California: Geo. Soc. Amer. Bull., vol. 83, p. 1601-1618.

Richter, C. F., 1958, Elementary Seismology, W. H. Freeman and Co., San Francisco, Calif., p. 768

Sharp, R. V., 1981, Variable rates of late Quaternary strike-slip on the San Jacinto fault zone, southern California: Jour. Geophys. Res., vol. 86, p. 1754-1762.

Stein, R. S., and W. Thatcher, 1981, Seismic and aseismic deformation associated with the 1952 Kern County, California, earthquake and relationship to the Quaternary history of the White Wolf fault: Jour. Geophys. Res., vol. 86, p. 4913-4928.

Stierman, D. J., and W. L. Ellsworth, 1976, Aftershocks of the February 21, 1973, Point Magu, California earthquake: Bull. Seismol. Soc. Amer., vol. 66, p. 1931-1952.

Suppe, J., 1985, Principles of Structural Geology, Prentice-Hall, Englewood Cliffs, N. J., p. 537

Taber, S., 1920, The Inglewood earthquake in southern California, June 21, 1920: Bull. Seismol. Soc. Am., vol. 10, p. 129-145.

Thatcher, W., J. A. Hileman, and T. C. Hanks, 1975, Seismic slip distribution along the San Jacinto fault zone, southern California, and its implications: Geol. Soc. Am. Bull., vol. 86, p. 1140-1146.

Toppozada, T. R., C. R. Real, and D. L. Parke, 1981, Preparation of isoseismal maps and summaries of reported effects for pre-1900 California earthquakes: California Div. of Mines and Geol. Open-File Report 81-11 SAC, 182p.

Wesnousky, S. G., 1986, Earthquakes, Quaternary faults and seismic hazard in California: Jour. Geophys. Res., vol. 91, p. 12,587-12,632.

Whitcomb, J. H., C. R. Allen, J. D. Garmany, and J. A. Hileman, 1973, San Fernando earthquakes series, 1971: Focal mechanisms and tectonics, Rev. Geophys., vol. 11, p. 693-730.

Wilcox, R. E., T. P. Harding, and D. R. Seely, 1973, Basic wrench tectonics: AAPG Bull., vol. 57, p. 74-96.

Wood, H. O., 1933, Preliminary report on the Long Beach earthquake: Bull. Seism. Soc. Amer., vol. 23, p. 43-56.

Wright, T. L., 1987, Geologic summary of the Los Angeles basin, in Petroleum Geology of Coastal Southern California, Guidebook, Pacific Section, American Assoc. of Petroleum Geologists, Los Angeles, Calif., p. 21-31.

Wright, T. L., 1991, Structural geology and tectonic evolution of the Los Angeles basin, in Active Margin Basins, edited by K. T. Biddle, AAPG Memoir, p. 35-134.

Yerkes, R. F., T. H. McCulloh, J. E. Schoellhamer, and J. G. Vedder, 1965, Geology of the Los Angeles basin California-An introduction: U. S. Geol. Surv. Professional Paper, 420A, p. A1-A57.

Ziony, J. I., and L. M. Jones, 1989, Map showing late Quaternary faults and 1978-1984 seismicity of the Los Angeles region, California, U. S. Geological Survey Misc. Series Map MF-1964.

Ziony, J. I., and R. F. Yerkes, 1985, Evaluating earthquake and surface faulting potential, in Evaluating Earthquake Hazards in the Los Angeles Region - *An Earth-Science Perspective, U. S. Geological Survey Professional Paper 1360*, edited by J. I. Ziony, U. S. Government Printing Office, p. 43-91.

Zoback, M. D., M. L. Zoback, V. S. Mount, J. Suppe, J. P. Eaton, J. H. Healy, D. Oppenheimer, P. Reseanberg, L. M. Jones, C. B. Raleigh, I. G. Wong, O. Scotti, and C. Wentworth, 1987, New evidence on the state of stress of the San Andreas fault system, Science, vol. 238, p. 1105-1111.

TABLE 2

SURFICIAL AND CONCEALED FAULTS IN THE GREATER LOS ANGELES AREA

Fault	Slip Rate (mm/yr)	Expected Mag.	# of Segm.	Total Return Rate (years)	Length of Fault [km]
Newport-Inglewood	0.1-1.2V?(1)	7	2	(1500)?(8)	62(8)
Palos Verdes	0.16-1.3?(5)	7?	?	(3000)?(8)	80
Sierra Madre	0.4-4.0?(6)	6.6	7	(50-500)?(9)	110(6)
Raymond	0.1-0.22(6)	6.5	1	3000(6)	22(8)
Whittier	2.0-3.0(3)	7.3	4	730(8)	74(8)
Santa Monica	0.05-0.4V?(1)	7?	?	(5000)?	58(8)
Hollywood	0.3-0.8?(8)	6?	1	(1600)?(8)	14(8)
Elysian Park (Concealed)	2.5-5.2?(4)	7-7.5?	6?	(1000)?	100(7)
Torrance-Wilmington (Concealed)	1.9-3.45?(4)	7-7.5?	3?	(1000)?	60(7)

1) Ziony and Yerkes (1985); V-vertical component of slip rate; 2) Sharp (1981); 3) Gath et al. (1992); 4) Davis et al. (1989); 5) Patterson and Freeman (1990). 6) Crook et al. (1987); 7) Hauksson (1990); 8) Wesnousky (1986); 9)Hauksson (1992).

TABLE 1

SIGNIFICANT EARTHQUAKES IN THE GREATER LOS ANGELES BASIN AND ADJACENT OFFSHORE REGIONS

DATE	EARTHQUAKE	Mo *10^{25} (dyne-cm)	M_L old	M_L new1)	M_w	M	depth (km)	FM2)	REFERENCE
1920 June 21	Inglewood		4.9			4.9			Taber (1920)
1930 Aug. 30	Santa Monica Bay		5.2			5.2			Gutenberg et al. (1932)
1933 Mar. 10	Long Beach	5	6.33)		6.4	6.4	10±2	r-ss	Wood (1933); Hauksson & Gross (1991)
1938 May 31	Elsinore Fault		5.5	5.2		5.2			Hileman et al., (1973)
1941 Oct. 22	Gardena		4.9	4.8		4.8			Richter (1958)
1941 Nov. 14	Torrance-Gardena		5.4	4.8		4.8			Richter (1958)
1970 Sept. 12	Lytle Creek	0.1	5.4	5.2		5.2	9	th	Jones (1984), Thatcher et al. (1975)
1971 Feb. 9	San Fernando	14	6.43)		6.7	6.7	13±3	th	Whitcomb et al. (1973); Heaton (1982)
1973 Feb. 21	Point Mugu	0.7	5.93)		5.9	5.9	17±5	th	Stierman & Ellsworth (1976); Bent & Helmberger (1991)
1979 Jan. 1	Malibu		5.0	5.2		5.2	10±4	th	Hauksson & Saldivar (1986)
1981 Sept. 4	Santa Barbara Isl.	1.1	5.3	5.5	6.0	6.0	11±5	r-ss	Corbett (1984); Bent (1990)
1987 Oct. 1	Whittier Narrows	1	5.9	5.9	5.9	5.9	14.6±1	th	Hauksson et al. (1988); Hauksson and Jones (1989)
1988 Dec. 3	Pasadena	0.024	4.9	5.0		5.0	15.6±1	l-ss	Jones et. al. (1990); Kanamori et al. (1990)
1989 Jan. 19	Malibu		5.0	5.0		5.0	14±3	th	Hauksson (1990)
1990 Feb. 28	Upland	0.25	5.2	5.3		5.3	5.2±0.5	l-ss	Hauksson & Jones (1991); Dreger & Helmberger (1991a)
1991 June 28	Sierra Madre	0.25	5.4	5.84)		5.8	12±1	th	Hauksson (1992); Dreger & Helmberger (1991b)

1) Hutton and Jones, (1992); 2) FM: focal mechanism, r-ss: right-lateral strike-slip, l-ss: left-lateral strike-slip, th: thrust faulting;
3) No Wood-Anderson records on scale in southern California; 4) from TERRAscope data, H. Kanamori written communication, 1991.

EARTHQUAKE ATLAS OF SOUTHERN CALIFORNIA 1978 - 1990

Egill Hauksson, Kate Hutton, Katrin Douglass
Seismological Laboratory
California Institute of Technology
Pasadena, CA 91125

Lucile M. Jones
U. S. Geological Survey at
California Institute of Technology
Pasadena, CA 91125

INTRODUCTION

The Southern California Seismographic Network (SCSN) (Fig. 1) has recorded more than 139,000 earthquakes in southern California since 1978. This earthquake atlas presents maps to describe this ongoing earthquake activity and to facilitate a comparison of the activity with mapped tectonic structures such as late Quaternary faults. These maps also show what earthquake data are available from the Southern California Seismographic Network. Furthermore, when new significant activity occurs, these maps will facilitate comparison with past activity.

The relationship between the seismicity and tectonic structures in southern California has been the subject of broad regional studies such as Wood (1947) and Allen et al. (1965). Wood (1947) analyzed a data set of about 1200 earthquakes while Allen et al. (1965) analyzed 10,126 earthquakes. Allen et al. (1965) recognized that events of magnitude 6.0 or larger were usually associated with major late Quaternary faults, but that the smaller events are scattered and in most cases cannot be associated with mapped surficial faults. They also pointed out the lack of small earthquakes or background seismicity along major late Quaternary faults such as the San Andreas fault. Since the Allen et al. (1965) study was completed, the number of seismographic stations in southern California has grown from 18 to 220 in 1990. Most of the new stations were deployed in the 1970s and, with the advent of digital recording and data processing in 1977, make the data collected and analyzed since 1978 of significantly higher quality. This atlas thus allows a comparison between the new and old data and provides an opportunity to check if some of the earlier conclusions about seismicity, such as the scattering of the small events, are real or artifacts caused by insufficient data.

The seismicity maps in this atlas cover two time windows, 1978-1990 and 1990. Because the 13-year maps 1978-1990 show all the recorded activity, they are dominated by the smaller earthquakes. These illustrate the relationship between late Quaternary faulting and seismicity. The 1990 maps include lower-hemisphere single-event focal mecha-

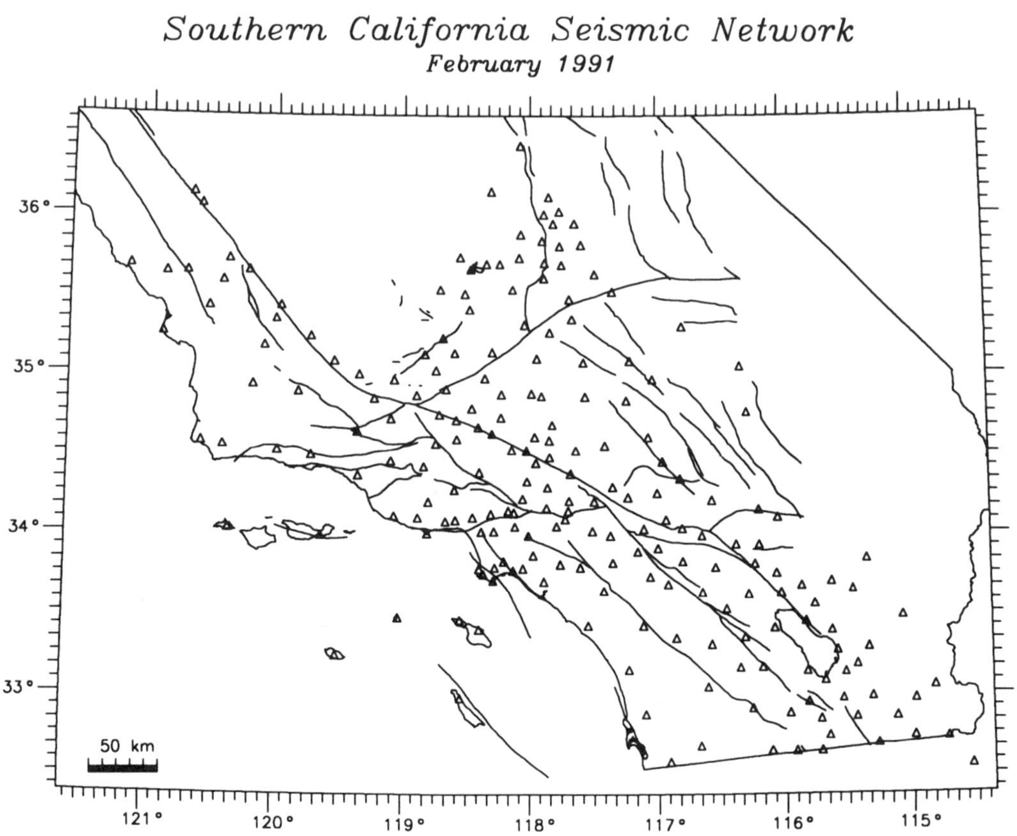

Figure 1. The Southern California Seismographic Network, triangles indicate locations of stations.

nisms of the larger earthquakes, which reflect the local tectonics. Comparison of the two maps illustrates how many years of monitoring can give a much more complete picture of the earthquake activity than a short time interval, such as a year.

TECHNICAL DISCUSSION

An overview of southern California's recent seismic activity is shown in Figures 2a and 2b. The data were recorded by the Southern California Seismographic Network, operated jointly by the California Institute of Technology and the U.S. Geological Survey. During the 13-year period, 1978 through 1990, 139,207 earthquakes were recorded, of which 26,539 were magnitude 2.0 or larger. The level of completeness of the catalog (where all earthquakes of this magnitude or larger are recorded within the primary coverage region) is at least 2.0 for this whole time period (Hutton et al., 1991). During 1990, 10,410 earthquakes were recorded with 3,160 above the level of completeness of magnitude 1.8. The most active regions of southern California (Figure 2) are: 1) the major strike-slip faults of the Peninsular Ranges including the San Jacinto and southern Elsinore faults, 2) the Imperial Valley including the Imperial fault and the Brawley seismic zone, 3) the southern Mojave desert, the San Bernardino and Little San Bernardino Mountains, including the knot of seismicity associated with the split of the San Andreas fault into the Banning and Mission Creek segments, 4) a band just south of

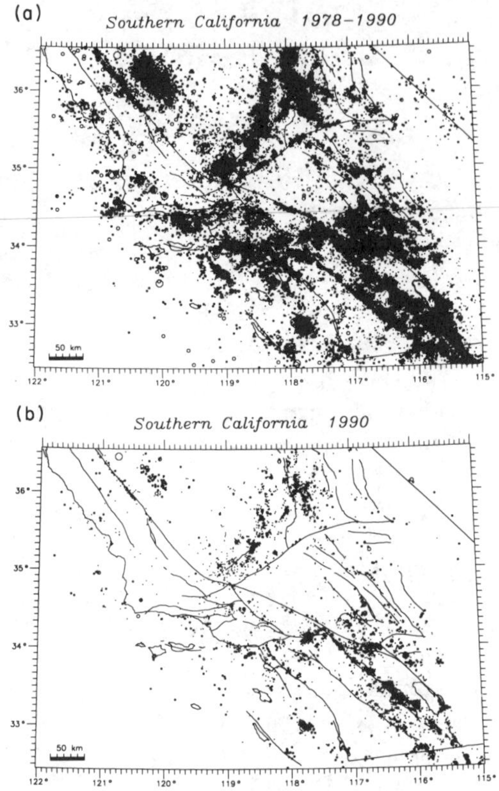

Figure 2. (a) An overview of southern California seismicity, showing all earthquakes recorded in the time period from 1978 - 1990. See later figures for the names of the major faults. (b) Seismicity during 1990 in southern California.

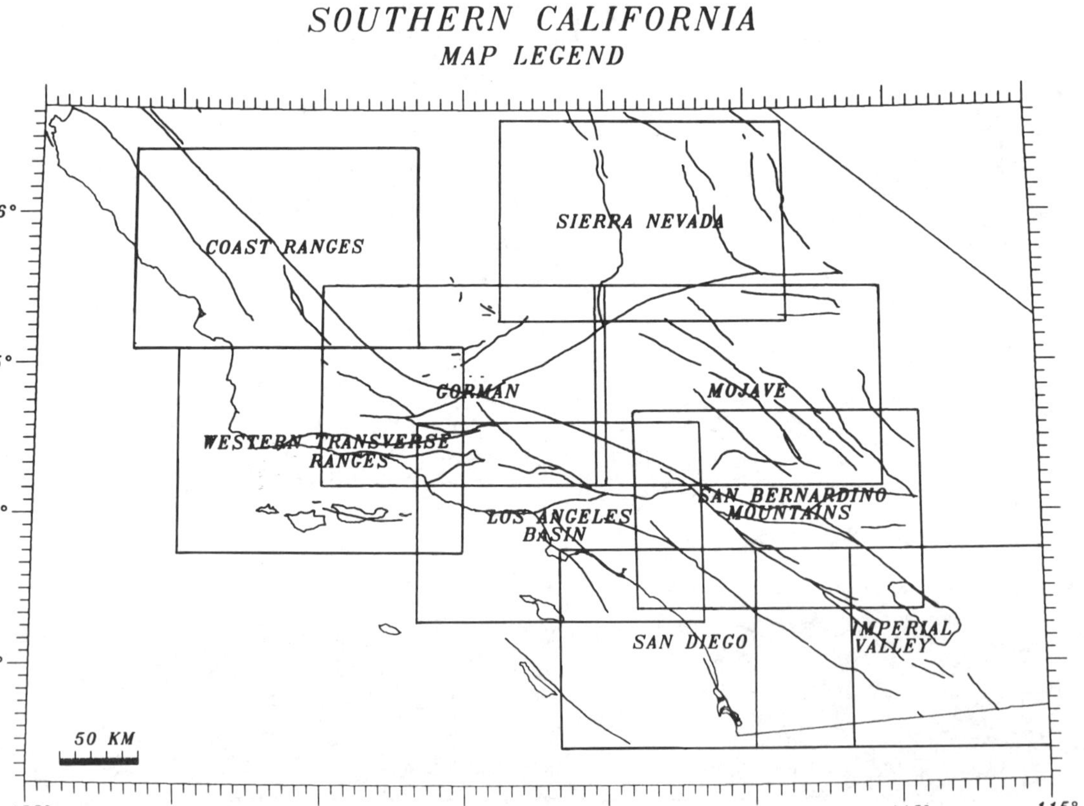

Figure 3. Key to subregions examined in the following figures.

the intersection of the Peninsular Ranges with the western Transverse Ranges including the Los Angeles and Ventura basins and the Santa Barbara channel and 5) the southern Sierra Nevada, from the Kern County aftershock zone at the southern end of the Great Valley northeastward through the Coso Range in Owen's Valley. In addition to these areas that are constant sources of microseismicity, individual mainshock-aftershock sequences can be clearly recognized in the seismicity maps, including the 1983 Coalinga sequence (M_L 6.2) (36° 10'N, 120° 10'W), the 1986 Oceanside sequence (M_L 5.3) (33° 0'N, 117° 50'W), and the 1987 Whittier Narrows sequence (M_L 5.9) (34° 3'N, 118° 4'W).

Some earthquakes cluster onto mapped faults, like the San Jacinto fault, the Elsinore fault, or the Garlock fault, similar to the clustering observed along faults in central California (Hill et al., 1990). Other earthquakes are scattered across the region. Some major faults, such as the San Andreas fault, the likely site of most of southern California's great earthquakes, are relatively quiet as far as microearthquakes are concerned (Allen et al., 1965; Jones, 1988) whereas some easily seen lineations of seismicity, like the one extending from north to south in the Sierra Nevada, cannot be associated with any mapped faults. Some of these will be discussed as we focus on subregions of southern California.

To facilitate more detailed examination of the seismicity, Figures 4 through 13 examine particular subregions of interest. As in Figure 2, each one has a 13 year (1978 - 1990) and a 1 year map (1990). Figure 3 is as an index to the subregions.

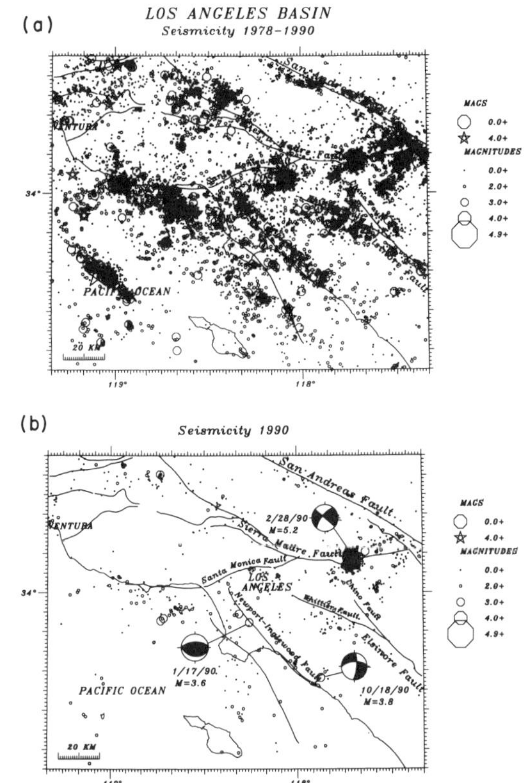

Figure 4. (a) Seismicity of the Los Angeles basin subregion 1978-1990. Symbol size is keyed to the magnitude and earthquakes of magnitude greater than 4.0 are plotted as stars. (b) Seismicity of the Los Angeles basin subregion 1990; lower-hemisphere focal mechanisms for the larger earthquakes are also included.

Figure 4: Los Angeles Basin

Allen et al. (1965) reported that the rate of seismicity in the greater Los Angeles area from 1934 to 1963 was similar to the average rate of seismicity in southern California. From 1978 to 1990, the Los Angeles basin has been one of the more active regions of southern California (Figure 4a). The aftershock sequences of numerous moderate earthquakes dominate Figure 4a including the 1979 and 1989 Malibu earthquakes (both M_L 5.0, 33° 55'N, 118° 40'W) (Hauksson and Saldivar, 1986), the 1981 Santa Barbara Island earthquake (M_L 5.3, 33° 38'N, 119° 5'W) (Corbett and Johnson, 1982; Corbett,1984), the 1987 Whittier Narrows earthquake (M_L 5.9, 34° 3'N, 118° 4'W) (Hauksson et al., 1988; Hauksson and Jones, 1989), and the 1990 Upland earthquake (M_L 5.2, 34° 8'N, 117° 42'W) (Hauksson and Jones, 1991). Continued aftershock activity from the 1971 San Fernando mainshock (M_L 6.4, 34° 25'W, 118° 25'W) has also been recorded (Whitcomb et al., 1973). The 1988 Pasadena earthquake (M_L 4.9, 34° 8'N, 118° 8'W) (Jones et al., 1990), which had few aftershocks, was most likely associated with the Raymond fault in Pasadena.

Between March of 1986 and early 1989, the rate of seismicity in the general Los Angeles area doubled (Jones and Reasenberg, 1989). The 1987 Whittier Narrows and 1989 Malibu events were part of this trend, but many smaller earthquakes also contributed. The rate of microseismic activity appears to have returned to its previous level in 1990. The most significant event of 1990 was the February 28 Upland earthquake (M_L 5.2), which was associated with the San Jose fault (Figure 4b). In addition, two events of (M_L 3.6) occurred near the Newport-Inglewood fault in 1990.

Trends of seismicity follow the Whittier fault and the Newport-Inglewood fault as far south as Newport Beach (Hauksson, 1987), but most of the rest of the activity is amorphous, or at least not associated with surficially mapped faults. A number of the events with thrust focal mechanisms, including both Malibu earthquakes and the Whittier Narrows earthquake, are probably associated with the concealed Elysian Park and Torrance-Wilmington Fold and Thrust Belts (Hauksson, 1990). Several northeasterly striking left-lateral seismicity trends, such as the Fontana trend, are also observed in the region (e.g. Hauksson, 1990).

A comparison of Figure 4a showing the 1978-1990 seismicity and Figure 7 in Allen et al. (1965), showing the 1934-1963 seismicity, shows that the larger and higher quality 1978-1990 data set indeed reveals patterns in the seismicity that could not be seen in the older data. Some of these can be correlated with mapped tectonic structures; in particular, the two largest strike-slip faults of the region, the Newport-Ingle-

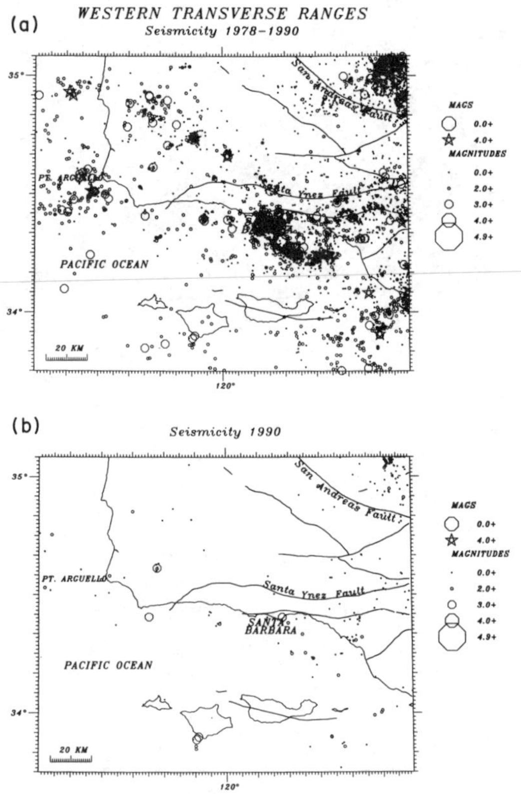

Figure 5. (a) Seismicity of the western Transverse Ranges subregion 1978-1990. (b) Seismicity of the western Transverse Ranges subregion 1990.

wood and the Elsinore faults, have concentrated bands of seismicity associated with them. However, much of the seismicity is not associated with surficial late Quaternary faults. Hauksson (1990) has argued that the complexity of the microseismicity patterns of the Los Angeles basin and the variety of strike-slip, thrust and normal faulting focal mechanisms recorded there are a product of the transition between the strike-slip tectonics of the Peninsular Ranges and the thrust faulting tectonics on the Transverse Ranges.

Figure 5: Western Transverse Ranges

The Western Transverse Ranges are dominated by west striking faults with abundant evidence of late Quaternary offsets. The spatial distribution of seismicity in Figures 5a and b shows a scattered pattern of epicenters, in part because most of the earthquakes are associated with dipping and geometrically complex structures. The rate of seismicity in this subregion is noticeably lower than the rate in the Los Angeles basin. It also exhibits episodic behavior, and 1990 is in one of the quiet periods. In 1978 the seismicity rate was significantly higher and culminated in the M_L 5.1 Santa Barbara earthquake (34° 17'N, 119° 42'W) in August, which occurred on a shallow thrust under the Santa Barbara channel (Corbett and Johnson, 1982; Yeats and Olson, 1984).

Numerous earthquakes of M_L greater than 3.0 have occurred beneath the Santa Maria basin. Small swarms of earthquakes are also fairly common in the Santa Barbara Channel offshore of Carpenteria. Although the Ventura basin is closing at a rapid rate (Yeats, 1988), it has a relatively low level of seismicity with a few magnitude 3 and one magnitude 4.0 events.

Figure 6: Coast Ranges

This part of the San Andreas fault system is well known for its slip partitioning, where the oblique plate motion is partitioned into right-lateral shear on the San Andreas fault and a decoupled compressional component that is accommodated on north-northwest striking thrust faults (Fig. 6) (Zoback et al., 1987). This causes folding and thrust faulting earthquakes adjacent to the San Andreas fault while earthquakes that occur on the fault show strike-slip movement.

The San Andreas fault is clearly outlined by background seismicity in the Parkfield region (35° 50'N, 120° 30'W) and further northwest on the creeping section of the fault (Fig. 6). However, south of the rupture zone of the 1966 Parkfield earthquake in the Carrizo plain, the San Andreas fault becomes almost completely aseismic. The quietest segment of the San Andreas fault is between 35°N and 35° 25'N (Hill et al., 1990) which is the section of the fault with the greatest offsets (10 m) during the 1857 Fort Tejon earthquake (Sieh, 1978).

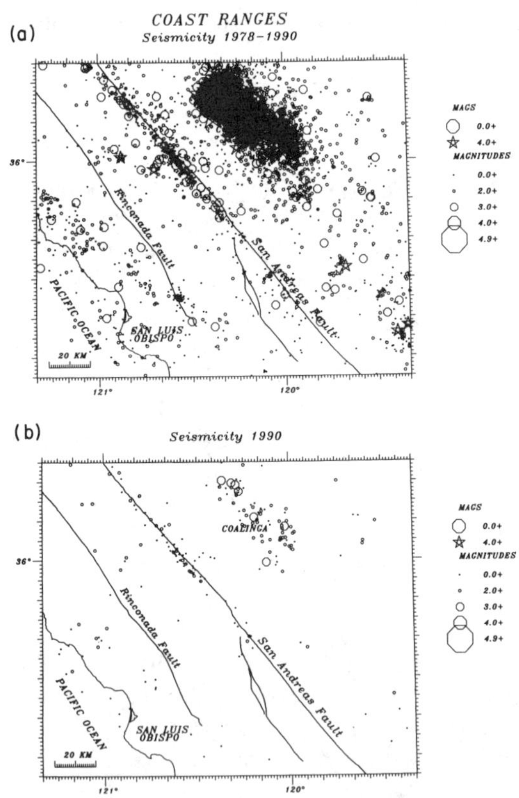

Figure 6. (a) Seismicity of the Coast Ranges subregion 1978-1990. (b) Seismicity of the Coast Ranges subregion 1990.

The prominent cluster of seismicity in the Franciscan terrane east of the San Andreas fault is the Coalinga earthquake sequence which consisted of at least three mainshocks: 1) 1983 New Idria (M_W 5.4), 2) 1983 (M_W 6.7) Coalinga, and 3) 1985 (M_W 6.1) Kettleman Hills earthquake. All of these mainshocks that were followed by prolific aftershock sequences occurred on buried thrust faults (Hill et al., 1990). The granitic Salinian block west of the fault is relatively aseismic. Some of the seismicity that is recorded along the coast, to the north of San Luis Obispo, may be associated with offshore faults such as the Hosgri fault.

Figure 7: Sierra Nevada

Four major structures can be recognized in the seismicity of the southern Sierra Nevada (Fig. 7). These are: 1) a northerly striking seismicity lineation 30 km west of the Sierra Nevada frontal fault that does not follow a mapped fault structure; 2) an isolated northeasterly lineation within the Sierran block; 3) numerous clusters of seismicity in the Coso geothermal area; and 3) some seismicity aligned with the Garlock fault .

The north-striking seismicity lineation is parallel to, but 15 km east of the Kern River fault and is sometimes called the southern Sierra seismic lineation. Most of the earthquakes within the southern Sierra lineation occur at shallow depth (less than 6 km) and normal faulting on north striking planes

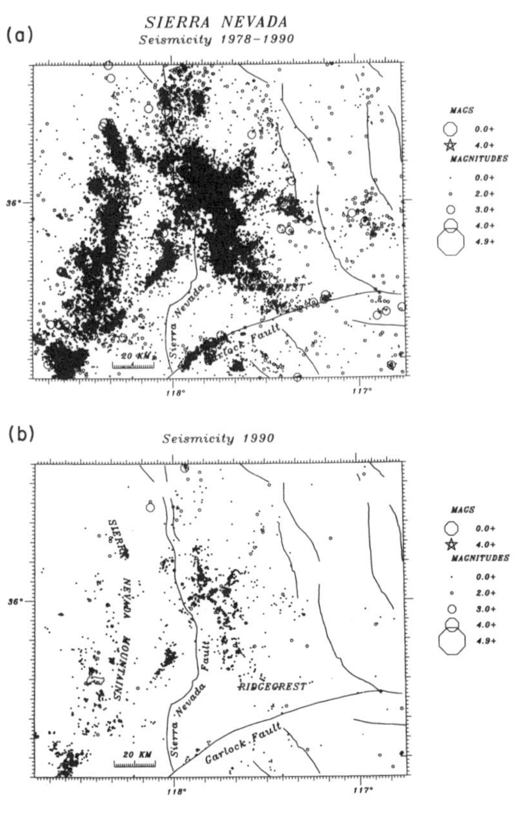

Figure 7. (a) Seismicity of the Sierra Nevada subregion 1978-1990. (b) Seismicity of the Sierra Nevada subregion 1990.

is common in the northern part of the lineation (Jones and Dollar, 1986). Jones and Dollar (1986) interpreted this lineation as the initiation of a basin and range normal fault within the Sierra Nevada block. The southern Sierra lineation extends to the south through Lake Isabella, where strike-slip earthquakes have been recorded, and merges with the aftershock zone of the 1952 M_S 7.7 Kern County earthquake. The northeasterly lineation merges with the northerly lineation south of Lake Isabella (35° 25'N, 118° 30'W). The northern end of this lineation at 35° 45'N, 118° 0'W is near the epicenter of the 1946 Walker Pass earthquake (M_L 6.3).

The series of seismicity clusters located east of the Sierra Nevada fault are in the Coso Range, a site of Quaternary volcanism. Many of these events exhibit strike-slip faulting and occur at normal depths for southern California earthquakes (5-10 km) and thus do not seem to be volcanic events (Walck and Clayton, 1987).

The Garlock fault extends from the San Andreas fault 265 km to the east. Late Quaternary left-lateral strike-slip offsets and continued background seismicity have been associated with the Garlock fault. The level of seismicity associated with the western half of the Garlock fault (Fig. 8) is somewhat less than seen on much of the eastern part (Astiz and Allen, 1983). Because the Garlock fault is an active strike-slip fault and has not caused an event of M_L4.3 since 1932, Astiz and Allen (1983) argued that it is a temporal seismic gap with potential for large earthquakes.

In 1990 the seismicity in this subregion consisted of swarm activity both in the Sierra Nevada and in the Coso Range. The 1990 seismicity showed similar spatial patterns as the 1978-1990 activity.

Figure 8: Gorman

The Gorman subregion includes the "Big Bend" of the San Andreas fault, the relatively quiescent western Mojave desert, and the San Emigdio Mountains at the southern end of the Great Valley (Fig. 8). The seismicity along the "Big Bend" is low except in a small area where the Garlock fault intersects the San Andreas fault. The persistent low level of seismicity along this part of the San Andreas fault is an example of the quiescence of the San Andreas fault during interseismic periods. The seismicity in the western Mojave desert is very low. This lack of seismic activity and the absence of any significant late Quaternary faults in the area suggest that the western Mojave is either a decoupled crustal block or most of the deformation is occurring aseismically.

In the San Emigdio Mountains the seismicity is clustered at the ends of the Pleito-White Wolf fault system. These features have persisted since the 1952 Kern County earthquake (M_S 7.7) and are considered aftershocks to that large mainshock. The more northeasterly of the two clusters merges into the southern Sierra lineation in the Sierra Nevada block, mentioned above. The largest earthquake that occurred in

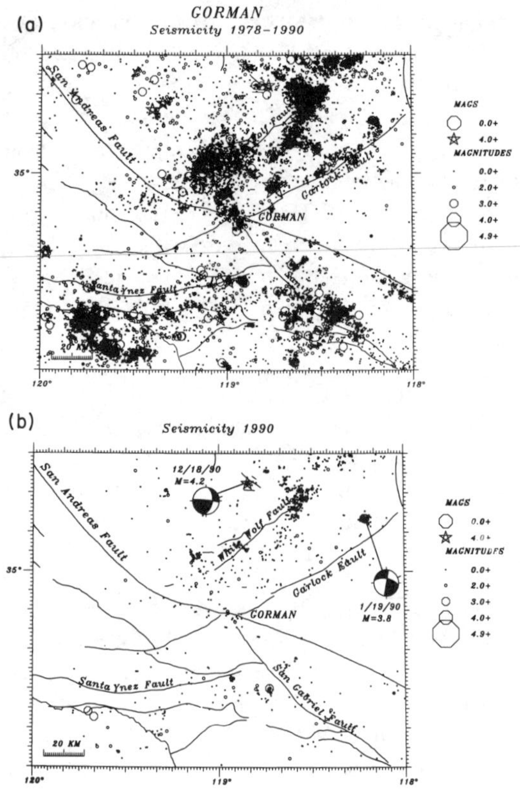

Figure 8. (a) Seismicity of the Gorman subregion 1978-1990. (b) Seismicity of the Gorman subregion 1990; lower-hemisphere focal mechanisms for the larger earthquakes are also included.

this area 1978-1990 was the June 1988 (M_L 5.4) Tejon Ranch earthquake. It was located to the north of the Garlock fault and exhibited oblique reverse faulting.

In 1990, two earthquakes of magnitude greater than 3.5 occurred in the region. The first event occurred just north of the Garlock fault and showed strike-slip faulting. The second event occurred in Sierra Nevada foothills and showed oblique strike-slip faulting.

Figure 9: Mojave

In contrast to the western Mojave, the eastern Mojave desert is cut by several late Quaternary faults and is much more seismically active (Fig. 9). At least seven northwest striking right-lateral, strike-slip faults have been mapped in the eastern Mojave desert. In addition, a few west striking faults have been mapped on the southern edge of the Mojave block, near the San Andreas fault. The seismicity in the area is diffuse and does not show a simple relationship with the surface faulting. Seismicity lineaments within individual clusters can show northwest, northeast or easterly trends. The prominent east-west trend that intersects the Calico fault at 35°N (Fig. 9a) includes the epicenter of the 1947 Manix earthquake (Richter, 1958; Doser, 1990). Most of the seismicity occurs in the southern Mojave, near the San Bernardino Mountains. This activity is continuous with the abundant seismicity of the San Gorgonio Pass area of the San Andreas fault zone. The cross shaped-feature in the lower right-hand corner of Figure 9a is the Homestead Valley earthquake sequence (largest M_L 5.2) in 1979 (Hutton et al., 1980; Stein and Liwoski, 1983).

Some of the clusters in the Mojave subregion are actually quarry blasts. When quarries are recognized, the blasts are marked as such in the catalog and not used for these seismicity maps. However some of the blasts from the Boron (35° 02'N, 117° 41'W), Oro Grande (34° 37'N, 117° 18'W), Victorville (34° 38'N, 117° 7'W), and Hector (34° 43'N,116° 25'W) quarries have not been recognized, as seen by the clusters at the sites of these quarries. The first three are the tight clusters along the Helendale fault and east of it. The Hector blasts form a less populous cluster in the middle right-hand portion of Figure 9a. During finalization of the catalog, events are compared to known quarry locations to catch this sort of problem but finalization is not yet complete for the 1977 through 1983 section of the Caltech/USGS earthquake catalog.

The 1990 seismicity in the Mojave subregion consisted of scattered background activity and three earthquakes of magnitude greater than 3.5. All three events exhibited strike-slip faulting on north to northwesterly striking planes.

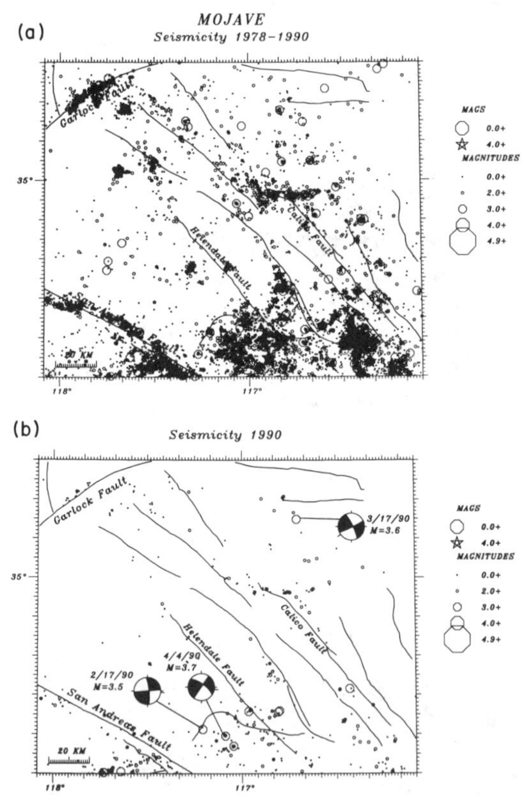

Figure 9. (a) Seismicity of the Mojave subregion 1978-1990. (b) Seismicity of the Mojave subregion 1990; lower-hemisphere focal mechanisms for the larger earthquakes are also included.

Figure 10: San Bernardino Mountains

This and the next subregion (Figs. 10 and 11) are by far the most complicated, since they cover the active areas around the multiple strands of the San Andreas fault in the eastern Transverse Ranges and the San Jacinto fault (Webb and Kanamori, 1985; Nicholson et al., 1986; Jones, 1988). This area has experienced several moderate earthquakes and a high level of background seismicity. The largest earthquakes since 1979 have been the 1986 North Palm Springs earthquake (M_L 5.6, 34° 0'N, 116° 40'W) (Jones et al., 1986), the 1979 Homestead Valley sequence (M_L 5.2, 34° 20'N, 116° 27'W), and the San Bernardino Mountains earthquake in June of 1979 (M_L 4.9, 34° 15'N, 116° 55'W).

The microseismicity in the San Bernardino and Little San Bernardino Mountains is clustered without any specific directional trends. Each cluster is either a mainshock-aftershock sequence or an earthquake swarm. In the Little San Bernardinos east of the San Andreas fault, however, the clusters form east-west trends, terminating at the San Andreas fault.

The San Jacinto fault is the most seismically active fault in southern California and is characterized by microseismically active and quiet segments. The northernmost section in San Bernardino and Redlands, north of the intersection with the Banning fault, is very active at the micro-earthquake level. From the intersection with the Banning fault south to about

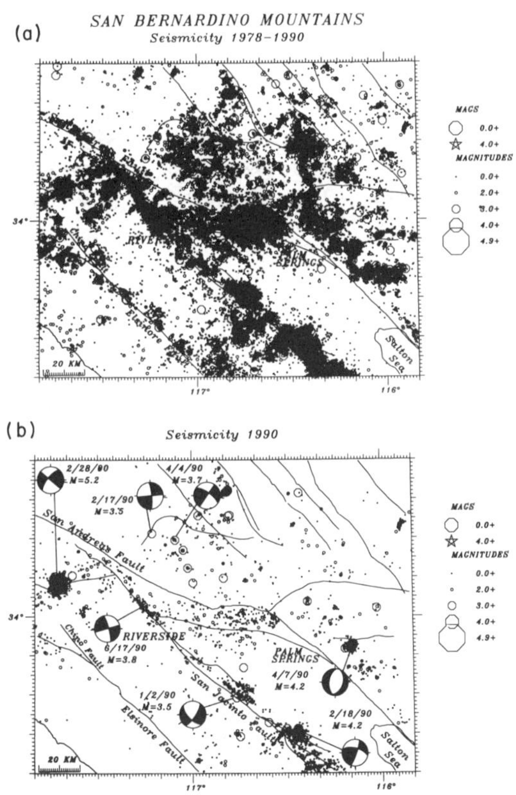

Figure 10. (a) Seismicity of the San Bernardino Mountains subregion 1978-1990. (b) Seismicity of the San Bernardino Mountains subregion 1990; lower-hemisphere focal mechanisms for the larger earthquakes are also included.

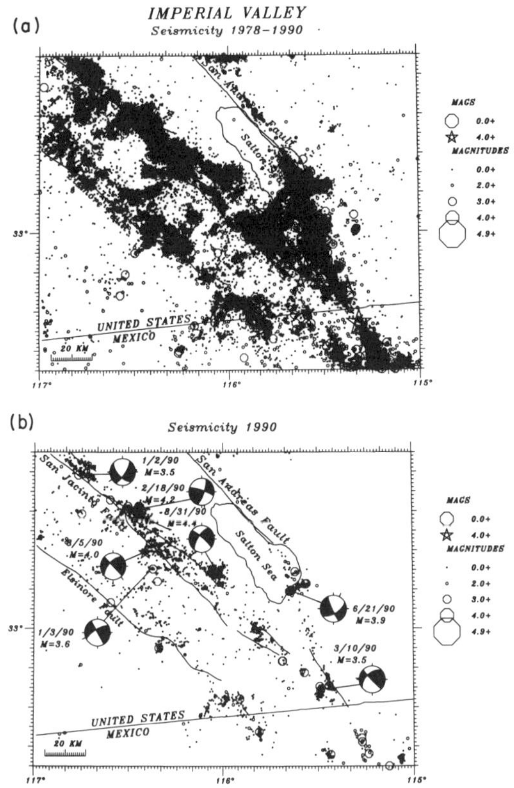

Figure 11. (a) Seismicity of the Imperial Valley subregion 1978-1990. (b) Seismicity of the Imperial Valley subregion 1990; lower-hemisphere focal mechanisms for the larger earthquakes are also included.

Hemet (33° 45'N), the fault is relatively quiet, but not entirely so. From Hemet to Lake Hemet (33° 40'N) the fault is moderately active, but south of Lake Hemet to Anza (33° 33'N) is the quiet zone known as the Anza gap (Sanders and Kanamori, 1984). South of the Anza area where the San Jacinto forks into three parallel branches, the earthquake activity increases again. Depths of earthquakes on the San Jacinto fault correlate well with heat flow being deeper in the colder regions (Doser and Kanamori, 1986).

Seismicity outlines the Elsinore fault as well, although not as prominently as the San Jacinto. The block between the Elsinore and San Jacinto faults has a number of spots which consistently produce small earthquakes, whereas the block between the San Jacinto and San Andreas faults, which includes the Coachella Valley is very quiet.

In 1990 seven earthquakes of M_L3.5 were recorded in this area. Three right-lateral strike-slip events occurred along the San Jacinto fault zone. The largest event (M_L 4.2) in an earthquake swarm that occurred east of Palm Springs, exhibited pure normal faulting.

Figure 11: Imperial Valley

Three moderate to large earthquake sequences occurred within the Imperial Valley between 1978 and 1990 (Fig. 11), the 1979 Imperial Valley earthquake (M_L 6.6, 32° 37'N, 115° 19'W), the 1981 Westmoreland earthquakes (M_L 5.6, 33° 6'N

115° 35'W), and the 1987 Elmore Ranch and Superstition Hills earthquakes (M_S 6.2, 33° 11'N, 115° 47'W; M_S 6.6, 33° 3'N, 115° 55'W). The Elmore Ranch earthquake was a foreshock to the Superstition Hills earthquake, preceding it by 12 hours. Two of the events occurred on northwest-striking faults, the Imperial Valley earthquake on the Imperial fault (Johnson and Hutton, 1982) and the Superstition Hills earthquake on the Superstition Hills fault (Magistrale et al., 1989), while the two smaller events occurred on northeast-striking structures within the Brawley seismic zone.

The large star just south of the Mexico/U.S. border represents the 1979 Imperial Valley earthquake. The band that stretches from there to the southern end of the San Andreas fault is the Brawley Seismic Zone (Johnson and Hill, 1982), an area of numerous earthquake swarms. At least two and possibly three northeasterly trends intersect the Brawley Seismic Zone. The most northerly of these is formed by the aftershocks of the 1987 Elmore Ranch earthquake. The next one to the south developed during the 1981 Westmorland sequence. The most southerly is barely visible on Figure 11a but formed during the aftershock sequence of the 1979 Imperial Valley earthquake (Heaton et al., 1983). The seismicity in the Brawley Seismic Zone itself consists mainly of earthquake swarms, whereas the cross trends result from aftershocks to large earthquakes and do not persist.

The seismic characteristics of the San Jacinto and Elsinore faults described in the previous section can also be seen in this region to the south (Fig. 11). Several of the individual traces of the San Jacinto fault are clearly delineated by the seismicity. In 1989 and 1990, a continuing series of earthquakes up to M_L 4.5 have occurred near Obsidian Butte in the northern Imperial Valley. These events have been confined to the Brawley Seismic Zone. The focal mechanisms of the seven earthquakes of M_L3.5 in the Imperial Valley in 1990 all show right-lateral strike-slip faulting.

Figure 12: San Diego

This subregion includes the part of the Peninsular Ranges west of the Elsinore fault and the southeastern part of the Continental Borderland (Fig. 12). The two major earthquake sequences that occurred in this subregion are the 1986 Oceanside earthquake sequence (Hauksson and Jones, 1988), and a series of small quakes near San Diego (Heaton, 1987).

The July 13, 1986 Oceanside event only had a magnitude of M_L 5.3, but, for its size, was followed by the most prolific aftershock sequence in the entire CIT/USGS southern California earthquake catalog. It probably was associated with the San Diego Trough-Bahia Soledad fault zone and the focal mechanism of the mainshock showed thrust faulting on an east-southeast-striking plane. The Oceanside aftershock activity continued through 1990 (Fig. 12b).

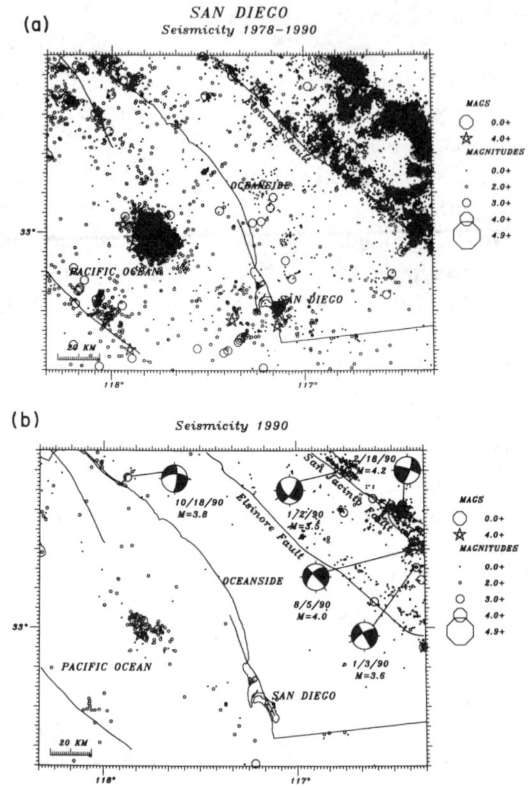

Figure 12. (a) Seismicity of the San Diego subregion 1978-1990. (b) Seismicity of the San Diego subregion 1990; lower-hemisphere focal mechanisms for the larger earthquakes are also included

Most of the scattered San Diego events occurred in 1986, and the largest was M_L 4.1. Many of these events appear to be associated with the southernmost onshore extension of the Rose Canyon fault zone. This fault is the southern extension of the Newport-Inglewood fault. A few of the 1986 events that occurred offshore 10-20 km west of San Diego may be associated with offshore faults in the Continental Borderland.

CONCLUSIONS

The seismicity recorded in southern California during the last 13 years reflects the distributed deformation around the Pacific-North America plate boundary. The main plate boundary, the San Andreas fault, has been mostly seismically quiescent during this time period and appears to cause only infrequent large earthquakes. Other faults such as the San Jacinto and the Garlock fault are seismically active at both small and large magnitudes and clearly demonstrate the relationship between the seismicity and the late Quaternary faulting.

The improved seismic network instrumentation since the late 1970s has made it possible to associate nearly all earthquakes of M_L5.0 to late Quaternary faults. Most of these events were followed by significant aftershock sequences. Several north or northeast striking seismicity lineations (for instance, the southern Sierra lineation and the Fontana trend) that are not associated with mapped late Quaternary faults or fold axes have been identified. High levels of seismicity are recorded in the areas adjacent to the two bends in the strike of the San Andreas fault, near Gorman and Banning Pass. The Los Angeles basin has been an area of increased seismicity over the last five years. Analysis of earthquake focal mechanisms from the Los Angeles basin has revealed two concealed fault systems beneath the basin sediments, the Elysian Park and Torrance-Wilmington fold and thrust belts. Similar fold and thrust belts exist in the western Transverse Ranges and to the east of the San Andreas fault in central California. Some of the other more seismically active areas are the Imperial Valley and the Coso region in Owens Valley, which are both associated with extensional tectonics and late Quaternary volcanisms.

ACKNOWLEDGMENTS

I thank the editors for critical reviews of the manuscript. This research was partially supported by USGS grant 14-08-0001-G1761 and USGS cooperative agreement 14-08-0001-A0613 and Southern California Earthquake Center under NSF cooperative agreement No. EAR-8920136 and USGS cooperative agreement 14-08-0001-A0899. Contribution No. 5058, Division of Geological and Planetary Sciences, California Institute of Technology, Pasadena, CA 91125.

REFERENCES

Allen, C. R., P. St. Amand, C. F. Richter, and J. M. Nordquist, 1965. Relationship between Seismicity and Geologic Structure in the southern California Region, Bull. Seismol. Soc. Amer., 55, 753-797.

Aztiz, L. and C. R. Allen, 1983. Seismicity of the Garlock fault, California, Bull. Seismol. Soc. Amer., 73, 1721-1734.

Corbett, E. J., 1984. Seismicity and crustal structure studies of southern California: Tectonic implications from improved earthquake locations, Thesis, 231 pp., California Institute of Technology.

Corbett, E. J., and C. E. Johnson, 1982. The Santa Barbara, California earthquake of 13 August 1978, Bull. Seismol. Soc. Amer., 72, 2201-2226.

Doser, D. I., 1990. A re-examination of the 1947 Manix, California earthquake sequence and comparison to other earthquake sequences within the Mojave block, Bull. Seismol. Soc. Amer., 80, 267-277.

Doser, D. I., and H. Kanamori, 1986. Depth of seismicity in the Imperial Valley region (1977-1983) and its relationship to heat flow, crustal structure, and the October 15, 1979, earthquake, J. Geophys. Res., 91, 675-688.

Hauksson, E., 1987. Seismotectonics of the Newport-Inglewood fault zone in Los Angeles Basin, southern California, Bull. Seismol. Soc. Amer., 77, 539-561.

Hauksson, E., 1990. Earthquakes, faulting and stress in the Los Angeles Basin, J.Geophys. Res., 95, 15,365 - 15,394.

Hauksson, E., and L. M. Jones, 1988. The July 1986 Oceanside (ML 5.3) earthquake sequence in the Continental Borderland, southern California,, Bull. Seismol. Soc. Amer., 78, 1885 - 1906.

Hauksson, E., and L. M. Jones, 1989. The 1987 Whittier Narrows earthquake sequence in Los Angeles, southern California: Seismological and tectonic analysis, J. Geopys. Res., 94, 9569-9590.

Hauksson, E., and L. M. Jones, 1991. The 1988 and 1990 Upland earthquakes: Left-lateral faulting adjacent to the central Transverse Ranges, J. Geophys. Res., 96, 8143-65.

Hauksson, E., L. M. Jones, T. L. Davis, L. K. Hutton, A. G. Brady, P. A. Reasenberg, A. J. Michael, R. F. Yerkes, P. Williams, G. Reagor, C. W. Stover, A. L. Bent, A. K. Shakal, E. Etheridge, R. L. Porcella, C. G. Bufe, M. J. S. Johnston, and E. Cranswick, 1988. The Whittier Narrows earthquake in the Los Angeles metropolitan area, California, Science, 239, 1409-1412.

Hauksson, E., and G. V. Saldivar, 1986. The 1930 Santa Monica and the 1979 Malibu, California, earthquakes, Bull. Seismol. Soc. Amer., 76, 1542-1559.

Heaton, T. H., 1987. Anomalous seismicity in the San Diego coastal region,in Proceedings of Workshop XXXVII, Physical and Observational Basis for Intermediate-term Earthquake Prediction, U.S. Geol. Surv. Open File Report 87-591.

Heaton, T. H., J. G. Anderson, and P. T. German, 1983. Ground failure along the New River caused by the October 1979 Imperial Valley earthquake sequence, Bull. Seismol. Soc. Amer., 73, 1161-1171.

Hill, D. P., J. P. Eaton, and L. M. Jones, 1990. Seismicity, 1980-86, in The San Andreas Fault System, California, Prof. Paper 1515, U. S. Geological Survey.

Hutton, L. K., C. E. Johnson, J. C. Pechman, J. E. Ebel, D. W. Given, D. M. Cole, and P. T. German, 1980. Epicentral locations for the Homestead Valley earthquake sequence, March 15, 1979, Calif. Geol., 33, 110-114.

Hutton, L. K., L. M. Jones, E. Hauksson, and D. D. Given, 1991. Seismotectonics of southern California, Geological Society of America, in press.

Johnson, C. E., and D. P. Hill, 1982. Seismicity of the Imperial Valley,in The Imperial Valley, California Earthquake of October 15, 1979, U.S. Geological Survey Professional Paper 1254.

Johnson, C. E., and L. K. Hutton, 1982. Aftershocks and pre-earthquake seismicity, in The Imperial Valley, California Earthquake of October 15, 1979, U.S. Geological Survey Professional Paper 1254.

Jones, L. M., L. K. Hutton, D. D. Given,and C. R. Allen, 1986. The North Palm Springs, California earthquake sequence of July 1986, Bull. Seismol. Soc. Amer., 76, 1830-1837.

Jones, L. M., and R. S. Dollar, 1986. Evidence for basin-and-range extensional tectonics in the Sierra Nevada: the Durrwood Meadows swarm, Tulare County, California (1983-1984), Bull. Seismol. Soc. Amer., 76, 439-461.

Jones, L. M., 1988. Focal mechanisms and the state of stress on the San Andreas fault in southern California, J. Geophys. Res., 93, 8869-8891.

Jones, L. M., and P. A. Reasenberg, 1989. A preliminary assessment of the recent increase in earthquake activity in the Los Angeles region, U.S. Geological Survey Open-File Report 89-162.

Jones, L. M., K. E. Sieh, E. Hauksson, and L. K. Hutton, 1990. The 3 December 1988 Pasadena, California earthquake: Evidence for strike-slip motion on the Raymond fault, Bull. Seismol. Soc. Amer., 80, 474-482.

Magistrale, H., L. Jones, and H. Kanamori, 1989. The Superstition Hills, California, earthquakes of 24 November 1987, Bull. Seismol. Soc. Amer, 79, 239 - 251.

Nicholson, C., L. Seeber, P. Williams, and L. R. Sykes, 1986. Seismic evidence for conjugate slip and block rotations within the San Andreas fault system, southern California, Tectonics, 5, 629-648.

Richter, C. F., 1958. Elementary Seismology, 768 pp., W. H. Freeman and Co., San Francisco.

Sanders, C. O., and H. Kanamori, 1984. A seismotectonic analysis of the Anza seismic gap, San Jacinto fault, southern California, J. Geophys. Res., 89, 5873-5890.

Sieh, K. E., 1978. Slip along the San Andreas fault associated with the great 1857 earthquake, Bull. Seismol. Soc. Amer., 68, 1421-1448.

Stein, R. S. and M. Liwoski, 1983. The 1979 Homestead Valley earthquake sequence, California: control of aftershocks and postseismic deformation,J. Geophys. Res., 88, 6477-6490.

Walck, M. C., and R. W. Clayton, 1987. P wave velocity variations in the Coso Region, California, derived from local earthquake travel times, J. Geophys. Res., 92, 393-405.

Webb, T. H., and H. Kanamori, 1985. Earthquake focal mechanisms in the eastern Transverse Ranges and San Emigdio Mountains, southern California and evidence for a regional decollement, Seismol. Soc. Amer. Bull., 75, 737-757.

Whitcomb, J. H., C. R. Allen, J. D. Garmany, and J. A. Hileman, 1973. San Fernando earthquakes series, 1971: Focal mechanisms and tectonics, Rev. of Geophys., 11, 693-730.

Wood, H. O., 1947. Earthquakes in southern California With Geologic Relations, (Part One), Bull. Seismol. Soc. Amer., 37, 107-258.

Yeats, R. S., and D. J. Olson, 1984. Alternate fault model for the Santa Barbara, California, earthquake of 13 August 1978, Bull. Seismol. Soc. Amer., 74, 1545 - 1553.

Yeats, R. S., 1988. Late Quaternary Slip Rate on the Oak Ridge Fault, Transverse Ranges, California: Implications for Seismic Risk, J. Geophys. Res., 93, 12,137-12,149.

Ziony, J. I., and L. M. Jones, 1988. Map showing late Quaternary faults and 1978-84 seismicity of the Los Angeles region, California, U. S. Geological Survey Misc. Field Studies Map MF-1964.

Zoback, M. D., M. L. Zoback, V. S. Mount, J. Suppe, J. P. Eaton, J. H. Healy, D. Oppenheimer, P. Reasenberg, L. Jones, C. B. Raleigh, I. G. Wong, O. Scotti, and C. Wentworth, 1987. New evidence on the state of stress of the San Andreas fault system, Science, 238, 1105-1111.

THE ROLE OF SEISMIC HAZARD EVALUATION IN ENGINEERING GEOLOGY REPORTS

Robert A. Larson
District Engineering Geologist
Los Angeles County Department of Public Works
900 S. Fremont Ave., Alhambra, CA 91803

James E. Slosson
Chief Engineering Geologist
Slosson and Associates
15500 Erwin St., Suite 1123, Van Nuys, CA 91411

ABSTRACT

Eventually, all sites in southern California will be subject to recognizable seismic hazards, thus seismic hazard evaluation should be included in all engineering geology reports as a basis for providing recommendations for the reduction of the hazard. While government regulatory agencies may require reports for some sites, per Seismic Hazards Mapping Act of 1990 (AB 3897), these requirements will not be uniformly enforced across the state, nor will all projects be included. Geologists should take it upon themselves to provide the data and analysis, such that a complete report is provided to the client, regardless of regulatory requirements. When the data are presented design, civil, geotechnical, and structural engineers involved in the project can utilize the data to develop mitigative design specifications within the purview of their specialties. When the data are not presented, code values are usually assumed regardless of the validity. A review of the general methodology for evaluation of the hazard is provided in this paper as a prelude to a discussion of recommendations that can be provided to reduce the hazard to people and structures. Education of the public should begin with consulting engineering geology reports and not be delegated by default to the evening news program.

INTRODUCTION

In this Decade of Natural Disaster Reduction it is appropriate for engineering geologists to ask themselves what they can do to achieve the goals envisioned by Frank Press, President of the National Academy of Sciences, at the Eighth World Conference on Earthquake Engineering. We submit that one of the fields that engineering geologists can have the greatest effect is in seismic hazards. Considering the 6 billion dollars damage and considerable loss of life due to the Loma Prieta earthquake (Ward and Page, 1991), which occurred with no surface fault-rupture, improvement in our profession's analysis, predictions, and recommendations is warranted. In this regard, seismic hazard evaluation should be included in all engineering geology reports in order to identify and assess potential hazards and provide recommendations for the reduction or elimination of hazards. The purpose of this paper is to discuss seismic hazard evaluation and associated recommendations that should be provided in engineering geology reports for structures and that may be beyond that required in proposed State of California guidelines.

EVALUATION OF SEISMIC HAZARD

The evaluation of seismic hazard includes three major steps. These are: 1) a review of historical seismicity in the vicinity of the site that may have affected the site in the past; 2) determination of seismic parameters that the site may experience in the future; and 3) based on the seismic parameters determined in step 2, an evaluation of primary and secondary seismic hazards that may affect the site. The analysis of historical data in conjunction with reasonable estimates from various published charts of seismic parameters allows sites to be characterized seismically.

The goal is to establish the realistic horizontal and vertical ground acceleration that the site may experience within the lifespan of the project. To do this, a "design earthquake" is determined for the site (Leeds, 1973). From this a horizontal and vertical acceleration value may be calculated. The design earthquake in the vicinity of known faults (i.e. southern California) is the maximum probable earthquake or, if this earthquake cannot be determined due to insufficient scientific data, the maximum credible earthquake that will generate the highest ground acceleration at the site. The "design acceleration" is then the maximum horizontal and vertical ground acceleration from the maximum probable earthquake or the repeatable high ground acceleration from the maximum credible earthquake.

The review of historical seismicity allows estimation (usually from charts) of the ground acceleration that the site may have experienced in the past and could experience in the future. In a few cases, ground acceleration values may be available from seismic recorders situated on nearby properties. Historical seismicity is available from many sources including the Caltech Seismological Laboratory, the California Division of Mines and Geology (CDMG), as well as published reports such as Real and others (1978), Toppozada and others (1978, 1981), and Ellsworth (1990). All major earthquakes of magnitude 5 or greater within 60 miles of the site should be reviewed. In addition, patterns of smaller earthquakes that could indicate higher seismic potential at the site should be reviewed. Utilizing published seismic attenuation charts and the measured horizontal distance between the site and the epicenters or the nearest trace of surface fault-rupture, it is possible to estimate the ground acceleration that the site has experienced within historic time. This calculated value then becomes the lowest possible value for the "design acceleration".

In order to assess future seismicity at a site, all active and potentially active (Quaternary-age) faults within 60 miles of the site should be evaluated for their potential for generating earthquakes. A list should be made from data in Hart (1990), Jennings (1975, 1985), Ziony and Yerkes (1985), Mualchin and Jones (1992), and other pertinent reports. The list should include: 1) the name of the fault, 2) the type of fault, 3) the activity on the fault (historically active, active, or potentially active), 4) reasonable estimates of the anticipated maximum credible earthquake magnitude and, if appropriate, the maximum probable earthquake magnitude, 5) the horizontal distance between the site and the closest part of the surface trace of the fault, as well as 6) the sources of data and methodology used. Added to this list and based on the previous data it is then easy to calculate reasonable estimates of the maximum horizontal and vertical acceleration, repeatable high ground acceleration, and the duration of strong ground motion. Once a list of all active and potentially active faults and their associated potential earthquake magnitudes is established, it is possible to calculate by hand within 30 minutes the distance to the faults, acceleration, and duration values for the site. Relatively inexpensive computer programs can generate the data in much less time.

Maximum credible earthquakes for southern California faults are available from several sources such as Mualchin and Jones (1992), Evernden and Thomson (1988), Wesnousky (1986), and Greensfelder (1978). These publications list slightly different values for different faults, but all are fairly close. In order to be diligent and prudent, simply choose the highest value from any publication for each fault, with deference to your own professional knowledge and opinions. It should be noted that publications often give anticipated earthquake values to the nearest 0.1 magnitude, even though this accuracy is not available in determining the magnitude of a recorded event. We advocate estimating the magnitude of future earthquakes to the nearest half magnitude, such that unrealistic accuracy is not built into the analysis, nor the mindframe of the user.

Maximum ground accelerations can be estimated from a plethora of attenuation charts for various ground conditions. The charts that appear to be most used in southern California are Schnabel and Seed (1973), Joyner and Boore (1981), Seed and Idriss (1982), Joyner and Fumal (1985), and Campbell (1987, 1989). However, acceleration values derived from charts should not be utilized if the calculated values are below those of Mualchin and Jones (1987, 1992), or Greensfelder (1978) which should take precedence as minimum values. Repeatable high ground acceleration (RHGA) from maximum credible earthquakes can be used as a design value for potential earthquakes greater than magnitude 5.5 within 20 miles of the site (Ploessel and Slosson, 1974). However, the RHGA should not be used with the maximum probable earthquake to derive a design value. Care must be taken in using attenuation charts because seismic energy may be focused along fault zones and at the top of ridges resulting in higher acceleration.

The duration of strong ground shaking can be estimated from Bolt (1973). Presently the duration of an event is only used as a design criteria for major structures, however it may be utilized for residential structures in the future. Qualitatively, reporting the duration of shaking is probably more important than acceleration because people relate better to time than fractions of gravity. The discussion of duration should include an estimate of total duration of shaking from aftershocks over the few days after the event, since many structures are more sensitive to prolonged shaking than peak ground acceleration.

MITIGATION OF SEISMIC HAZARDS

The mitigative design of seismic hazards is provided by design civil, geotechnical, and structural engineers. The role of the engineering geologist is to recognize that geologic/seismic hazards affect the site, and to provide appropriate recommendations, seismic parameters, the type or mode of anticipated failures, and the engineering geologic characteristics of site materials. Geologists should recommend in their reports that the seismic data be utilized by design civil, geotechnical, and structural engineers and architects. Another recommendation that should be included is that all future property owners be provided with a copy of the report and that they should be responsible for reading it. The information should be provided in a format such that it is understandable by and useful to the engineer and homeowner. Engineering geologists can also provide recommendations (or suggestions) to the developer and property owner to reduce or eliminate the potential for personal injury and structural/cosmetic damage, such as:

1) Utilize non-brittle roofing materials on structures in proximity to a major fault (e.g. no brick tiles) to preclude expensive roof repairs. If these materials must be used, such as in areas of high fire hazard, special provisions should be followed for proper attachment.

2) Have a structural engineer review the design specifications for nonuniform structures (e.g. cathedral ceilings, split-levels, irregular-shaped two-story, etc.) in all areas where anticipated ground motion is greater than UBC Seismic Zonation requirements (i.e. 0.4g in Seismic Zone 4).

3) Use flexible utility connections and shut off valves/switches as necessary.

4) Provide furniture, especially refrigerators, with blocks to prevent sliding across a room.

5) Anchor bookshelves, pictures, and other objects that could come loose during an earthquake, including water heaters.

We should also question the wisdom of constructing major roadways over faults that may rupture the surface and that contain underground utilities without shut-off valves and switches . If this type of alignment in unavoidable, the utilities should be designed with automatic shut-off valves on each side of the potential rupture zone and appropriate access for rerouting along the surface until complete repairs can be made. Homeowners should also be informed that street access to their homes may be limited until repairs can be completed.

DISCUSSION

Geologists should take it upon themselves to provide the data and analysis, such that a complete report is provided to the client, regardless of regulatory requirements. Government regulatory agencies may require reports for some sites, per Seismic Hazards Mapping Act of 1990 (Brown, AB 3897), in order to reduce hazards from strong ground motion and related secondary effects. However, the requirements will not be uniformly enforced across the state, nor will all projects be included. Guidelines for compliance with AB 3897 may be available by the time this paper is published and should be followed. Until these guidelines are available, existing CDMG guidelines should be used such as "Guidelines to Geologic/Seismic Reports", "Recommended Guidelines for Determining the Maximum Credible and the Maximum Probable Earthquakes", and "Recommended Guidelines for Preparing Engineering Geologic Reports".

When an evaluation determines that the design acceleration is greater than UBC standards (e.g. 0.4g in Seismic Zone 4) realistic data and recommendations are necessary to inform the structural engineer, geotechnical engineer, design civil engineer, and property owner. The structural engineer must be provided with correct seismic data, such that it may be included in the professional design of the structure. The design civil and geotechnical engineers must have the correct data in order to assess the potential for ground failure and any mitigative design required. The property owner, if properly informed of anticipated seismic effects and the cost-effectiveness of mitigation, can request structural reinforcement, require alterations in the pad location or grading design of the property, or buy/build a home in another locality. The majority of property owners are not aware that structures, built to code, are designed to remain standing after an earthquake in order for occupants to safely evacuate, but may then have to be demolished. Single-family residences in the 300-600 thousand dollar range can be properly designed and built to withstand an earthquake without structural damage for a 1-2% increase in price of 3-6 thousand dollars (Cota and Slosson, 1992). Many home buyers and builders would opt for this increase in price and safety if they were aware of the option. Considering the cost of earthquake insurance and the related deductible it appears to be very cost effective to build the structures for anticipated strong ground motion, rather than rely on insurance to pay for major repairs or demolition and rebuilding.

The recommendation currently given by geologists in many reports is to follow UBC Seismic Zonation requirements or to simply state that these requirements are sufficient for the design, even though published data or their own analysis indicate otherwise. UBC standards are only valid for simple, wood-framed, box-like structures on firm materials at least a few miles from an earthquake-generating fault. It must be recognized that the UBC represents a compromise of many interested parties and specific provisions are not valid for many sites. Thus, analysis by engineering geologists is required to establish site conditions. We believe that reports should have more data, not less, and that they must contain accurate information.

The review of many reports indicates that some consultants choose questionable and inapplicable methods for determining seismic parameters in order to report low values that may reduce construction costs, but increase the hazard. An increasing common but invalid practice is utilizing a vertical distance from the site to a hypothetical hypocenter (e.g. 10 miles), rather than horizontal distance (e.g. 1 mile) to the fault, to determine ground acceleration from attenuation charts for sites situated very close to active faults. This method may be appropriate for calculating attenuation of ground motion from earthquakes at subduction zone depths, but not the relatively shallow hypocenters characteristic of California earthquakes. Some consultants also misinterpret the classification of faults by the California Division of Mines and Geology and thus ignore many potential seismic sources. Many faults that are active are not included within Alquist-Priolo Special Studies Zones because they do not meet the criteria that was established for surface fault-rupture hazard. An example was provided by the 1991 Sierra Madre earthquake that occurred at depth on the part of the Sierra Madre fault zone that is not presently zoned. Careful review of Hart (1990) and Jennings (1975, 1985) is required for understanding the various fault classifications utilized by CDMG.

The public is sufficiently aware of seismic hazards to be further educated by reading engineering geology reports, if the hazards are discussed in the report. The publicity in southern California since the 1971 San Fernando earthquake has increased public awareness of the hazard and earthquake processes. Public knowledge and participation is required to enforce recommendations for reduction of minor hazards such as movement of furniture, severing of utility connections, breakage of roofing tiles, destruction of chimneys, etc. The education of the public should begin with engineering geology reports.

CONCLUSIONS

Engineering geologists can help effectively eliminate or greatly reduce seismic hazards if practitioners choose to participate in seismic hazard reduction. If the profession does not actively participate, other professions will usurp the role that we can best fulfill. Engineering geology reports must include a realistic evaluation of seismic hazards and appropriate recommendations for mitigation.

ACKNOWLEDGEMENTS

We thank Jeffrey A. Johnson and Charles G. Sudduth for reviewing an early draft of this paper: their critique is appreciated. This paper does not represent the opinions or policies of, nor endorsement by, the Los Angeles County Department of Public Works.

REFERENCES

Bolt, B.A., 1973, Duration of strong ground motion: Proceedings of the Fifth World Conference on Earthquake Engineering, Paper no. 292, Rome.

Campbell, K.W., 1987, Predicting strong ground motion in Utah: in Hays, W.W., and Gori, P.L., Eds., Assessment of Regional Earthquake Hazards and Risk along the Wasatch Front, Utah, U.S. Geol. Survey, Open-file Report 87-585, vol. 2, p. L1-L90.

Campbell, K.W., 1989, Empirical prediction of near-source ground motion for the Diablo Canyon Power Plant site, San Luis Obispo County, California: U. S. Geol. Survey, Open-file Report 89-484, 115 p.

Cota, J.A., and Slosson, T.L., 1992, Seismic hazard analysis of the Mission Wells fault zone, Los Angeles County, California, utilizing evaluation guidelines of Assembly Bill 3897: Proceedings of the Annual Meeting of the Association of Engineering Geologists, Long Beach, California, (in preparation).

Ellsworth, W.L., 1990, Earthquake history, 1769-1989: in Wallace, R.E., Ed., The San Andreas Fault System, California, U.S. Geol. Survey Professional Paper 1515, p. 153-181.

Evernden, J.F., and Thomson, J.M., 1985, Predicting seismic intensities: in Ziony, J.I. (ed.), Evaluating earthquake hazards in the Los Angeles region - an earth science perspective: U. S. Geol. Survey Professional Paper 1360, p. 151-220.

Greensfelder, R.W., 1974, Maximum credible rock acceleration from earthquakes in California: California Div. of Mines and Geology, Map Sheet 23.

Hart, E.W., 1990, Fault-rupture Hazard Zones in California: California Div. of Mines and Geology, Special Publication 42, 26 pgs.

Jennings, C.W., 1975, Fault map of California with location of volcanoes, thermal springs, and thermal wells: California Div. of Mines and Geology, Geologic Data Map #1.

Jennings, C.W., 1985, An explanatory text to accompany the 1:750,000 scale fault and geologic maps of California: California Div. of Mines and Geology, Bull. 201.

Joyner, W.B., and Boore, D.M., 1981, Peak horizontal acceleration and velocity from strong-motion records including records from the 1979 Imperial Valley, California, earthquake: Bull. of the Seismology Society of America, vol. 71, pgs. 2011-2038.

Joyner, W.B., and Fumal, T.E., 1985, Predictive Mapping of Earthquake Ground Motion; in Ziony, J.I., ed., Evaluating earthquake hazards in the Los Angeles region - an earth science perspective: U. S. Geol. Survey Professional Paper 1360, p. 203-220.

Leeds, D.J., 1973, The design earthquake: in Geology, Seismicity, and Environmental Impact, Association of Engineering Geologists, Special Publication, Los Angeles, p. 337-347 (out of print).

Mualchin, L., and Jones, A.L., 1987, Peak acceleration from maximum credible earthquakes in California: California Div. of Mines and Geology, Map Sheet 45.

Mualchin, L., and Jones, A.L., 1992, Peak acceleration from maximum credible earthquakes in California: Department of Conservation, California Div. of Mines and Geology, scale 1:1,000,000.

Ploessel, M.R., and Slosson, J.E., 1974, Repeatable high ground accelerations from earthquakes - an important design criteria: California Geology, vol. 27, no. 9, p. 195-199.

Real, C.R., Toppozada, T.R., and Parke, D.L., 1978, Earthquake Epicenter Map of California: California Div. of Mines and Geology, Map Sheet 39.

Schnabel, P.B., and Seed, H.B., 1973, Accelerations in rock for earthquakes in the western United States: Bull. of the Seismological Society of America, vol. 63, no. 2, p. 501-516.

Seed, H.B., and Idriss, I.M., 1982, Ground motions and soil liquefaction during earthquakes: Earthquake Engineering Research Institute, Monograph no. 5, 134 p.

Toppozada, T.R., Parke, D.L., and Higgins, C.T., 1978, Seismicity of California 1900-1931: California Div. of Mines and Geology, Special Report 135, 39 p.

Toppozada, T.R., Real, C.R., and Parke, D.L., 1981, Preparation of isoseismal maps and summaries of reported effects for pre-1900 California earthquakes: California Div. of Mines and Geology, Open-file Report 81-11, 182 p.

Ward, P.L., and Page, R.A., The Loma Prieta earthquake of October 17, 1989; in Baldwin, J.E., and Sitar, N., eds., Loma Prieta Earthquake: Engineering Geologic Perspectives, Association of Engineering Geologists, Special Publication no. 1, p. 5-15.

Wesnousky, S.G., 1986, Earthquakes, Quaternary faults, and Seismic Hazards in California: Jour. of Geophysical Research, vol. 91, no. B12, p. 587-613.

Ziony, J.I., and Yerkes, R.F., 1985, Evaluating Earthquake and Surface-faulting Potential: in Zioney, J.I., Ed., Evaluating earthquake hazards in the Los Angeles region - an earth science perspective: U. S. Geol. Survey Professional Paper 1360, p. 43-91.

IMPLICATIONS OF THE PRECISE CHRONOLOGY OF EARTHQUAKES PRODUCED BY THE SAN ANDREAS FAULT AT PALLETT CREEK

Kerry Sieh
Seismological Laboratory
Division of Geological and Planetary Sciences
California Institute of Technology, Pasadena

Minze Stuiver
Department of Geological Sciences and
Quaternary Research Center
University of Washington, Seattle

David Brillinger
Department of Statistics
University of California, Berkeley

This article is a synopsis of our longer article that was published in 1989 in the Journal of Geophysical Research (vol. 94, p. 603-623). Here we present several figures and part of the discussion found in that paper. Improved methods of radiocarbon analysis have enabled us to date more precisely the earthquake ruptures of the San Andreas fault that are recorded in the sediments at Pallett Creek, 35 miles northeast of downtown Los Angeles. Previous dates of these events had 95% confidence errors of 50-100 calendar years. New error limits are less than 23 calendar years for all but two of the dated events. This greater precision is due to larger sample size, longer counting time, lower background noise levels, more precise conversion of radiocarbon ages to calendric dates, and better stratigraphic constraints and statistical techniques. The new date ranges, with one exception, fall within the broader ranges estimated previously, but our estimate of the average interval between the latest 10 episodes of faulting is now about 132 years.

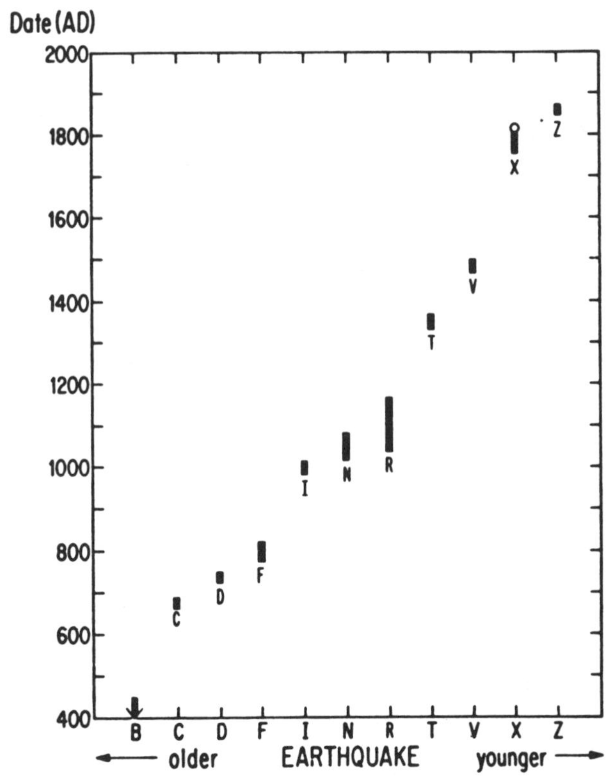

Figure 1. Dates of past earthquakes at Pallett Creek

Figure 1 shows our new estimates of the dates for earthquakes recorded in the sediments at Pallett Creek. The vertical bars give 95% confidence intervals. The open circle on the top of the bar of event X indicates our preferred date for that earthquake of A.D. 1812.

Variability about the mean interval is much greater than was suspected previously. Five of the nine intervals are shorter than a century; three of the remaining four intervals are about two to three centuries long. Despite the wide range of these intervals, a pattern in the occurrence of large earthquakes at Pallett Creek is apparent in the new data. Figure 1 shows that the past 10 earthquakes occur in four clusters, each of which consists of two or three events. Earthquakes within the clusters are separated by periods of several decades, but the clusters are separated by dormant periods of two to three centuries. This pattern may reflect important mechanical aspects of the fault's behavior. If this pattern continues into the future, the current period of dormancy will probably be greater than two centuries. This would mean that the section of the fault represented by the Pallett Creek site is currently in the middle of one of its longer periods of repose between clusters, and sections of the fault farther to the southeast are much more likely to produce the next great earthquake in California.

The precision of the new Pallett Creek dates encourages us to attempt correlations with earthquakes recorded at other paleoseismic sites along the southern half of the San Andreas fault. If we could accurately correlate events at Pallett Creek with those identified at other sites along the San Andreas fault, we would be able to establish more firmly the lengths and locations of individual fault ruptures. Data of this sort would provide important clues about the mechanical behavior of the fault and about its future behavior.

Figure 2 is a graphical comparison of earthquake dates derived from the Pallett Creek site and those estimated from the Wallace Creek site, nearly 300 kilometers to the northwest along the fault. The date ranges determined from Pallett Creek for the ten earthquakes are shown as vertical bands beneath the diagonal line. The diagonal line represents the long-term slip rate of the fault at Wallace Creek, 34 mm/yr. The date ranges above the line are derived by dividing the amount of offset during a past earthquake at Wallace Creek by the long-term slip rate there. In this part of the figure, the length of the earthquake cycle is assumed to be proportional to the amount of offset during the earthquake at the beginning of the cycle. This, then, assumes a time-predictable behavior for the fault along the segment of the San Andreas fault represented by Wallace Creek. The figure suggests that the second or third earthquake in each cluster at Pallett Creek might involve rupture at Wallace Creek and that the first event in each cluster does not.

The possibility that slip events at Wallace Creek correlate with the last or middle event in each cluster at Pallett Creek is intriguing. Perhaps the last large event in each cluster is an event like the great 1857 event; that is, an event that involves rupture of the Cholame, Carrizo and Mojave segments of the fault. (Fig. 3 shows the locations of these hypothetical segments.) Perhaps the earlier one or two large events in each cluster represent events which involved rupture of the Mojave segment alone or the Mojave segment in concert with the San Bernardino and Indio segments to the southeast.

The new dates at Pallett Creek are precise enough to enable a weak test of this hypothesis. Using data from all of the paleoseismic sites along the southern half of the fault, which are labeled in Figure 3, we attempt to correlate dated events. The uncertainties in radiocarbon dating of earthquakes at each site precludes us from being certain of any of the correlations, of course.

In Figures 3 and 4, we give those ruptures documented and dated at Pallett Creek the greatest lengths allowed by the data from the other sites. This seems to be the most reasonable assumption, given that both the great 1906 and 1857 earthquakes were generated by fault ruptures that spanned several hundred kilometers of the fault . Figure 3 displays six proposed large earthquake ruptures. The oldest event, R, is in the lower right box, and the youngest event, Z, is in the upper left box. We justify this version of history more fully in our 1989 paper.

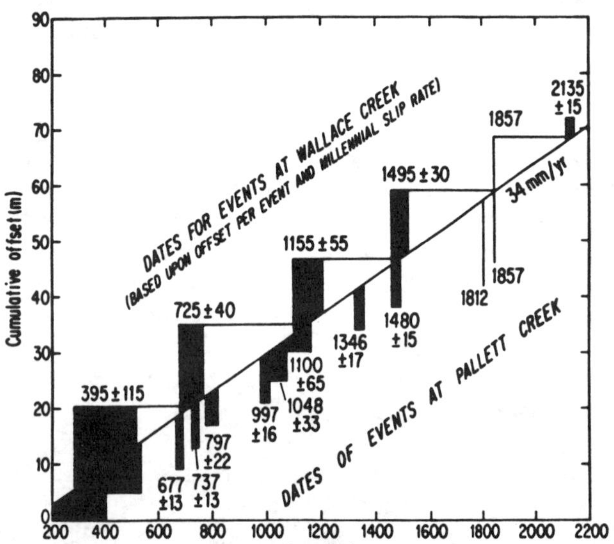

Figure 2. Graphical comparison of earthquake dates

The correlation of events proposed and illustrated in Figure 3 is speculative and is intended only to provoke discussion and further paleoseismic studies and theoretical modeling. If the historical scenario of Figure 3 is roughly correct, the past three large earthquakes along the southern half of the fault progressed from southeast to northwest during a period of about 180 years. Two of the three previous large events in this case ruptured the entire southern half of the fault.

Figure 4 illustrates a less likely, but plausible alternative earthquake pattern. In this case, we assume event X occurred in the late 17th century and that evidence of the 1812 earthquake at Pallett Creek is hidden in the massive unit between events X and Z. We also correlate the A.D. 1450 event at Indio with event T at Pallett Creek. This scenario presents a more regular pattern of earthquake occurrence; in it three northwestward progressions of large earthquakes occur: one between A.D. 1000 and 1100, another between A.D. 1300 and 1480, and another between A.D. 1680 and 1857. Each of these progressions involves overlapping ruptures along the Mojave segment of the fault and corresponds to one of the periods of earthquake clustering in the record at Pallett Creek.

Other viable earthquake scenarios, consistent with the sparse data now available, can be constructed. If, for example, one abandons the attempt to make the earthquake ruptures as large as the paleoseismic data allow, the possibilities are legion. We believe that more precise dating of slip events at paleoseismic sites other than Pallett Creek will lead to more certain correlation of earthquake ruptures between sites. This would enable more reliable estimates of magnitude for prehistoric events and recognition of spatial and temporal patterns of large earthquake occurrence.

REFERENCES

Agnew, D., and K. Sieh, 1978, A documentary study of the felt effects of the great California earthquake of 1857: Bull. Seismol. Soc. Am., vol. 68, p. 1717-1729.

Brillinger, D., 1982, Seismic risk assessment: some statistical aspects: Earthquake Predic. Res., vol. 1, p. 183-195.

Davis, T., 1983, Late Cenosoic structure and tectonic history of the western "Big Bend" of the San Andreas fault and adjacent San Emigdio Mountains: Ph.D. dissertation, Univ. of Calif., Santa Barbara, 580 p.

Jacoby, G., P. Sheppard, and K. Sieh, 1988, Irregular recurrence of large earthquakes along the San Andreas fault-Evidence from trees: Science, vol. 241, p. 196-199.

Lindh, A., 1983, Preliminary assessment of long-term probabilities for large earthquakes along selected segments of the San Andreas fault system in California: U. S. Geol. Surv. Open File Rept. 83-63, p. 1-5.

Sieh, K., 1978, Prehistoric large earthquakes produced by slip on the San Andreas fault at Pallett Creek, California: J. Geophys. Res., vol. 83, p. 3907-3939.

Sieh, K., 1984, Lateral offsets and revised dates of large earthquakes at Pallett Creek, California: J. Geophys. Res., vol. 89, p. 7641-7670.

Sieh, K., 1986, Slip rate across the San Andreas fault and prehistoric earthquakes at Indio, California: Eos Trans. AGU, vol. 67, p. 1200.

Sieh, K. , and R. H. Jahns, 1984, Holocene activity of the San Andreas fault at Wallace Creek, California: Geol. Soc. Am. Bull., vol. 95, p. 883-896.

Stuiver, M., and B. Becker, 1986, High precision decadal calibration of the radiocarbon timescale, AD 1950-2500 BC: Radiocarbon, vol. 28 (2B), p. 863-910.

Sykes, L., and S. Nishenko, 1984, Probabilities of occurrence of large plate rupturing earthquakes for the San Andreas, San Jacinto, and Imperial faults, California, 1983-2003: J. Geophys. Res., vol. 89, p. 5905-5927.

Sykes, L., and L. Seeber, 1985, Great earthquakes and great asperities, San Andreas fault, southern California: Geology, vol. 13, p. 835-838.

Weldon, R., and K. Sieh, 1985, Holocene rate of slip and tentative recurrence interval for large earthquakes on the San Andreas fault, Cajon Pass, southern California: Geol. Soc. Am. Bull, vol. 96, p. 793-812.

Wesnousky, S., 1986, Earthquakes, Quaternary faults, and seismic hazard in California, J. Geophys. Res., vol. 91, p. 12,587-12,631.

Working Group on California Earthquake Probabilities, 1988, Probabilities of large earthquakes occurring in California on the Andreas fault: U. S. Geol. Surv. Open File Rept. 88-398.

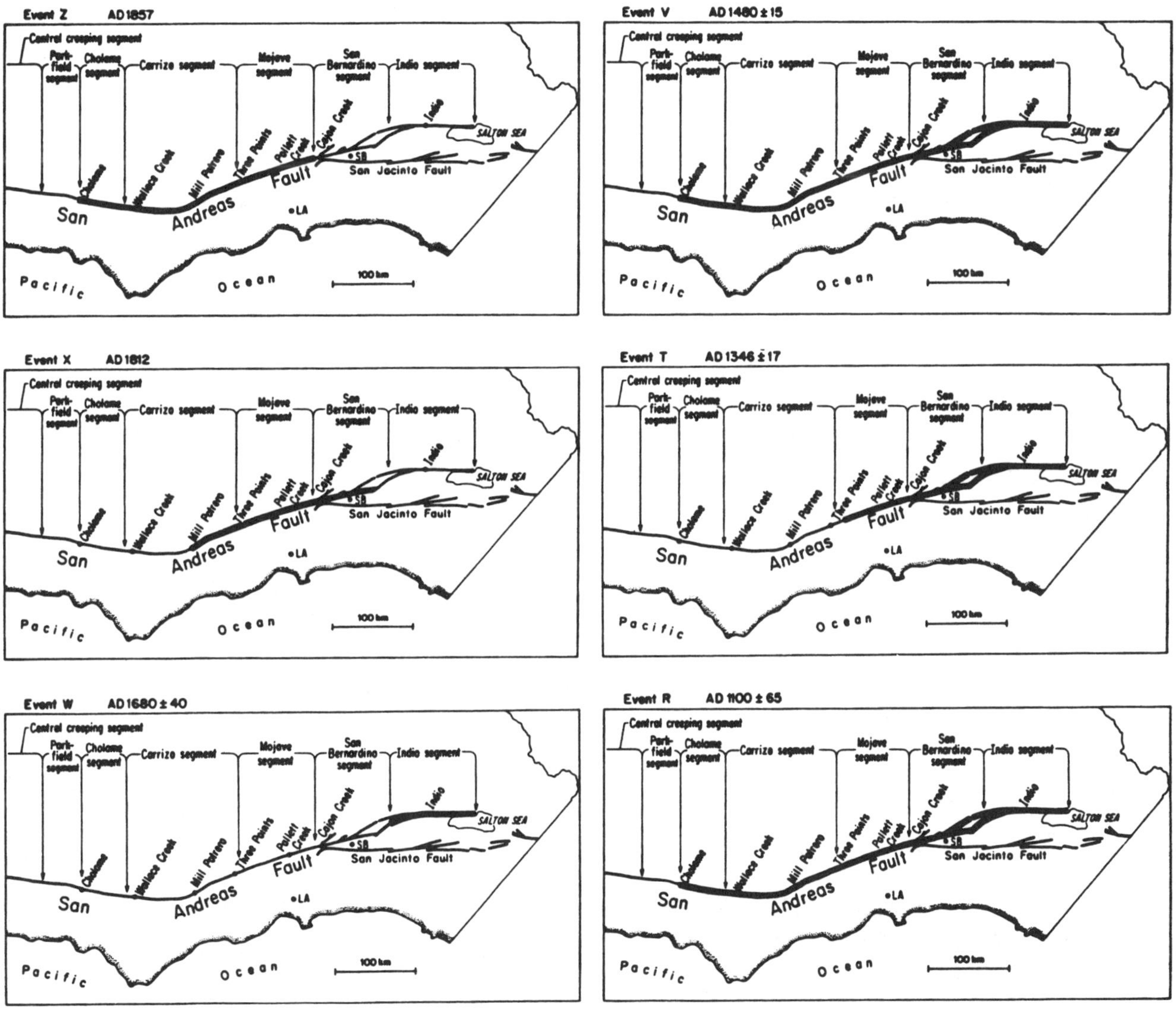

Figure 3. Segment location

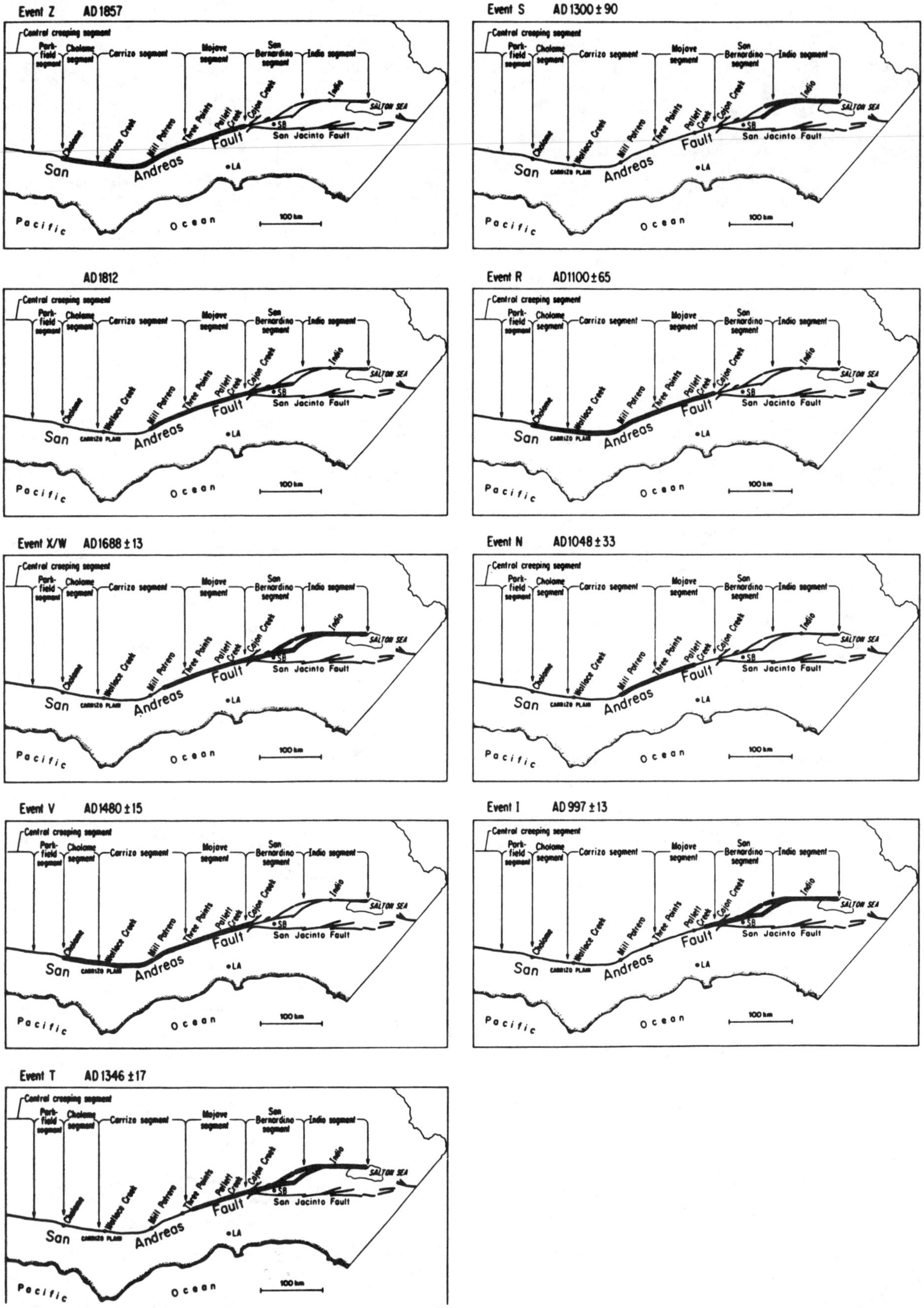

Figure 4. Alternative earthquake pattern

GEOTECHNICAL HAZARDS FACING A RAPIDLY URBANIZING CITY, PALMDALE, CALIFORNIA

Deems C. Padgett
Converse Consultants West
3393 E. Foothill Boulevard
Pasadena, CA 91107

ABSTRACT

The city of Palmdale is at the margin between the Mojave Desert and Transverse Ranges geomorphic provinces. The San Andreas fault transects the city and forms the structural boundary between these two provinces. Owing to its location in an active tectonic setting, Palmdale presents a challenge for providing thorough geotechnical reviews while supporting the city's desire for a rapid urban growth rate. The main geotechnical hazards in Palmdale are faulting and related seismic hazards. Other geotechnical concerns are subsidence, slope stability, collapsible soils, expansive soils, flooding, and rising water.

INTRODUCTION

Palmdale is about 40 miles north of Los Angeles. Palmdale was the fastest growing city in the United States during the 1980s. It grew 229% between 1980 and 1988, with the city population increasing from 12,277 to 40,419 (Los Angeles County Dept. of Regional Planning, pers. comm., 1989). As real estate values skyrocketed in the Los Angeles basin many residents sought refuge in outlying communities like Palmdale where homes were still affordable to the first-time home buyer. Because of this radial migration away from the Los Angeles basin, many portions of southern Palmdale along the foothills came into demand for development. These areas were not previously considered developable from an economic standpoint because of overwhelming geotechnical hazards that required relatively expensive geologic investigations. Additionally, setback areas from active faults would reduce buildable areas and thereby detrimentally impact the economics of the development.

Converse Consultants West (CCW) was retained by the city of Palmdale to act as geotechnical reviewer between April, 1988 and June, 1990. In order to promote consistency in geotechnical reports they developed a series of guidelines. Preliminary drafts of these guidelines were circulated to the City and the principal geologists of several firms consulting in the area for their critical review and comment. Based on their comments a final set of geotechnical guidelines was produced. The intent of this document is to provide an objective outline of geotechnical issues to be addressed and to

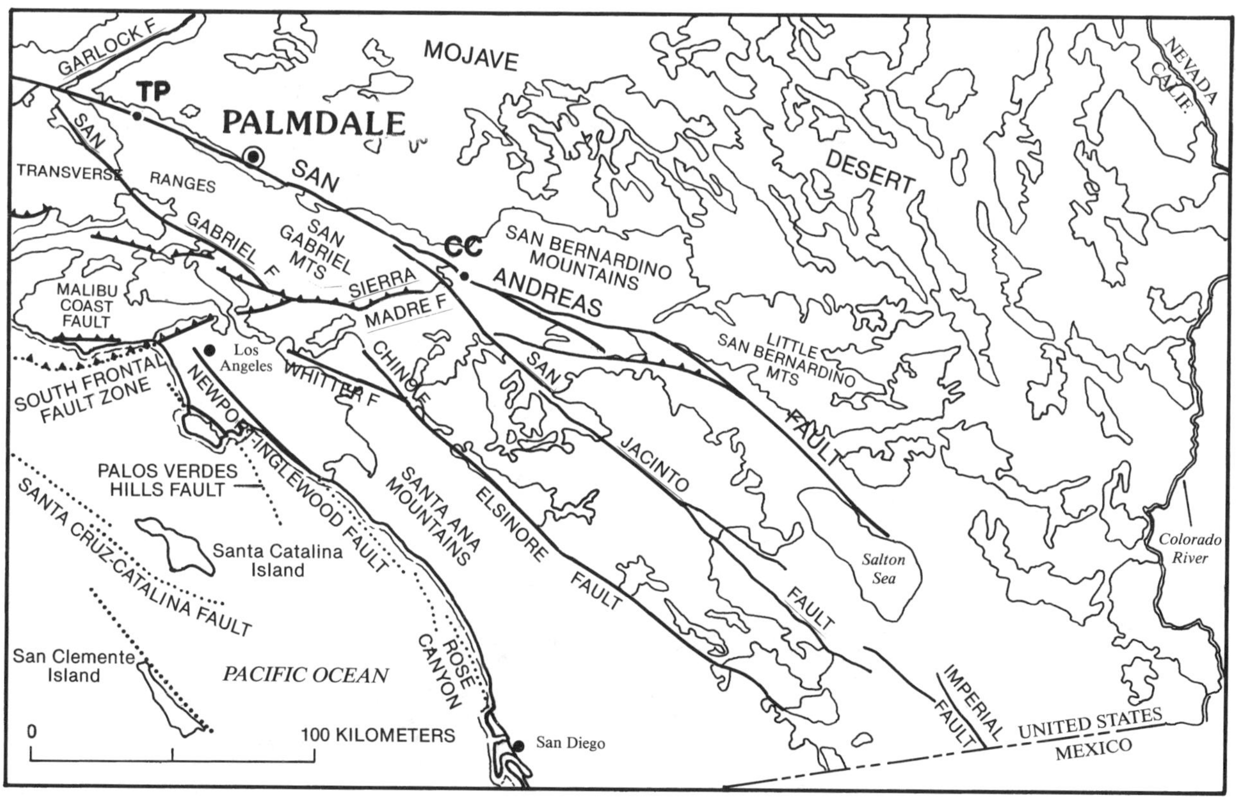

Figure 1. Regional physiographic setting of southern California. Northwest and southeast end of Mojave segment of the San Andreas fault indicated by Three Points (TP) and Cajon Creek (CC) (After Brown, 1990).

present geotechnical review guidelines developed by Converse (which are available on request).

During the course of geotechnical review, all tracts and associated geotechnical hazards were documented on a 1" = 1,000' legal map encompassing the majority of Palmdale. This included documentation of collapsible soils, active faults, flood prone areas and related geotechnical concerns. Geotechnical hazard index maps were maintained and updated throughout the review period. This geotechnical hazard index map was available for review by private consultants and this document assisted both consultants and geotechnical reviewers in determining geotechnical site conditions for proposed developments.

GEOMORPHIC AND CLIMATIC SETTING

Palmdale is situated along the margin between the Mojave Desert and the Transverse Range geomorphic provinces (Figure 1). The Mojave Desert geomorphic province is a triangular shaped area characterized by a series of low-relief desert basins separated by low mountain ranges. This province is bounded approximately by the Garlock fault on the northwest and the San Andreas fault on the southwest. The eastern boundary of this province is arbitrarily defined by the Nevada state line and the Colorado River (Norris and Webb, 1990).

The Transverse Ranges are a series of essentially west-trending ranges and valleys which contrast with the predominant north-northwest structural grain of most of California. The San Gabriel Mountains are the local transverse range in the Palmdale area. The rugged relief of this range can be attributed to its rapid uplift over the last 2 million years. Uplift has been accommodated through a combination of reverse faults along the southern margin and arching along the northern margin (Ehlig, 1975).

Rapid incision and alluvial deposition has accompanied this uplift and produced a number of distinctive Quaternary alluvial units. Subsequent to deposition these alluvial units were repeatedly offset by faults; these deposits are generally narrow and elongated relative to the northwest trending faults in the area. Correlating these alluvial units is complicated by successive base level changes and movement on the San Andreas and subsidiary faults. Nonetheless, a number of these units can be traced throughout the Palmdale area; several are helpful in constraining the timing of fault activity due to their degree of weathering and soil profile development.

Palmdale is within the Antelope Valley. The Antelope Valley is a closed topographic and groundwater basin located within the western Mojave Desert. This basin contains a thick section of alluvial and lacustrine deposits. The alluvial deposits were shed from adjacent mountain ranges surrounding the basin including the Tehachapi Mountains to the north and the

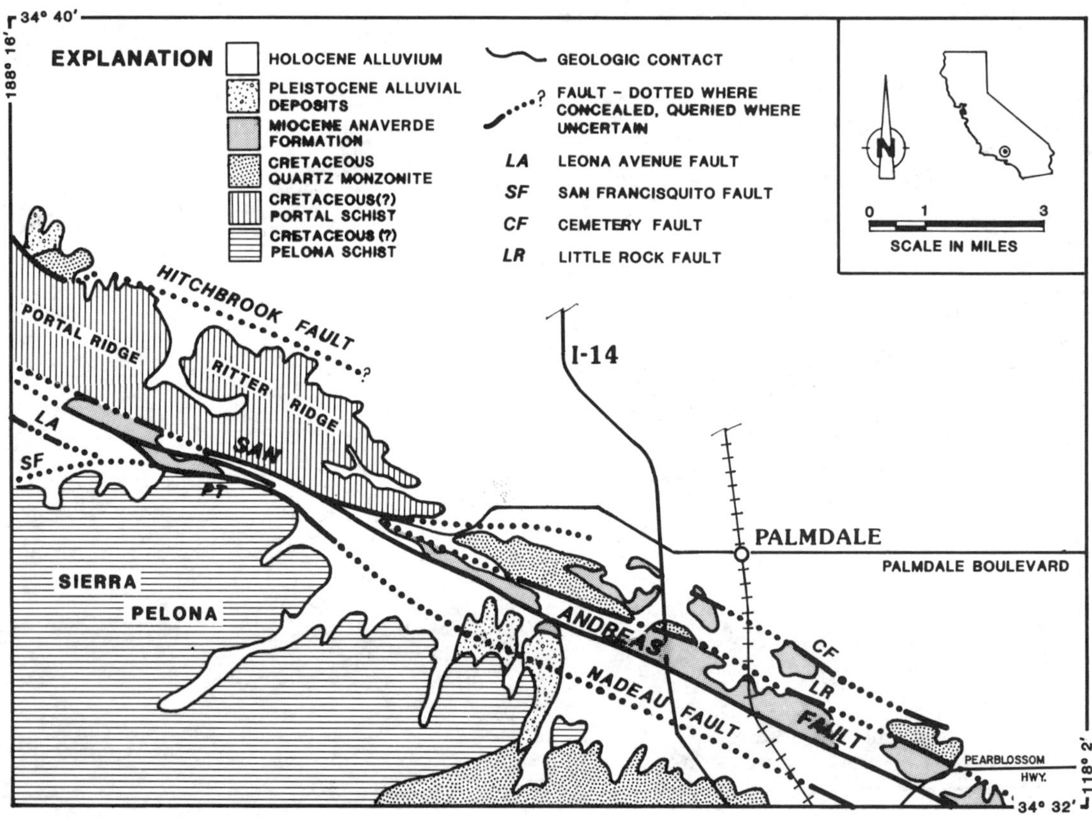

Figure 2. Generalized geologic map of Palmdale and vicinity (After Dibblee, 1967).

San Gabriel Mountains to the south. The relatively flat-lying valley floor is locally punctuated by small bedrock ridges.

The arid climate of the Palmdale area reflects its physiographic setting in a rain shadow caused by the orographic effect of the San Gabriel Mountains to the south and the Tehachapi Mountains to the north. As a result, annual average rainfall in Palmdale is only about 4 to 6 in/yr (USDA, 1970). The foothill regions south of Palmdale receive up to about 20 in/year rainfall. The combination of Palmdale's arid climatic setting and juxtaposition to active faults puts special demands on land development.

The general geologic setting of the Palmdale area is presented on Figure 2. For more detailed information the reader is referred to Barrows et al. (1985).

TECTONIC SETTING

The city of Palmdale is in an active tectonic setting. The majority of the city's inhabitants are within one to two miles of the Mojave segment of the San Andreas fault as defined by the Working Group on California Earthquake Probabilities (1988). The Mojave segment extends about 60 miles from near Three Points to the vicinity of Cajon Creek (Figure 1). The Mojave segment of the San Andreas fault sustained ground rupture during the 1857 magnitude 8.0 Fort Tejon earthquake. Surface rupture extended for about 190 miles from north of Cholame to south of Wrightwood (Sieh, 1978b).

This segment has been subject to extensive examination and analyses. A late Holocene seismicity record has been determined based on detailed logging and analysis of faulted Holocene sediments at Pallett Creek (Sieh, 1978a, 1984, 1989, and this chapter). Based on radiocarbon dating of displaced peat beds, Sieh calculated a recurrence interval of about 131 years for this segment of the fault. The documented earthquake events tend to occur in clusters with recurrence intervals of about 50 to 130 years separated by aseismic intervals of 200 to 300 years in duration (Figure 3). Based on this Holocene record and other evidence, a slip rate of 30 ±5 mm/yr has been assigned by the Working Group (1988).

Based on the 131 year recurrence interval provided by Sieh, the Mojave segment is past due for an earthquake. It is therefore possible that the Mojave segment San Andreas fault is currently within an extended aseismic interval. The Working Group (1988) discusses some of the inconsistencies in earthquake data for this segment of the fault. Based on their analyses they assigned a conditional probability of 0.3 for a 7.0M earthquake on the Mojave Segment for the 30-year period extending from 1988 to 2018 (Figure 4).

EVENT	DATE
Z	9 Jan 1857
X	1785±32
V	1480±15
T	1346±17
R	1100±65
N	1048±33
I	997±16
F	797±22
D	737±13
C	677±13
B	>529

EVENTS	INTERVAL
Z/	131 years
X/Z	44
V/X	332±15
T/V	134±23
R/T	246±67
N/R	52±73
I/N	52±37
F/I	200±27
D/F	60±26
C/D	60±18
B/C	≥148±31

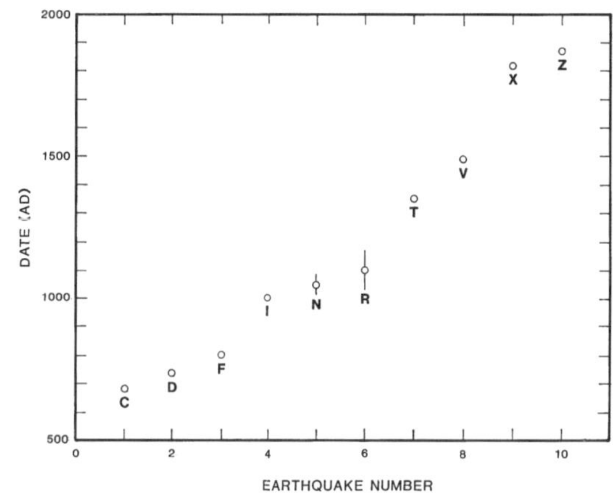

Figure 3. Dated historical and prehistoric slip events at Pallett Creek on the Mojave segment of the San Andreas fault (After Sieh et al., 1989).

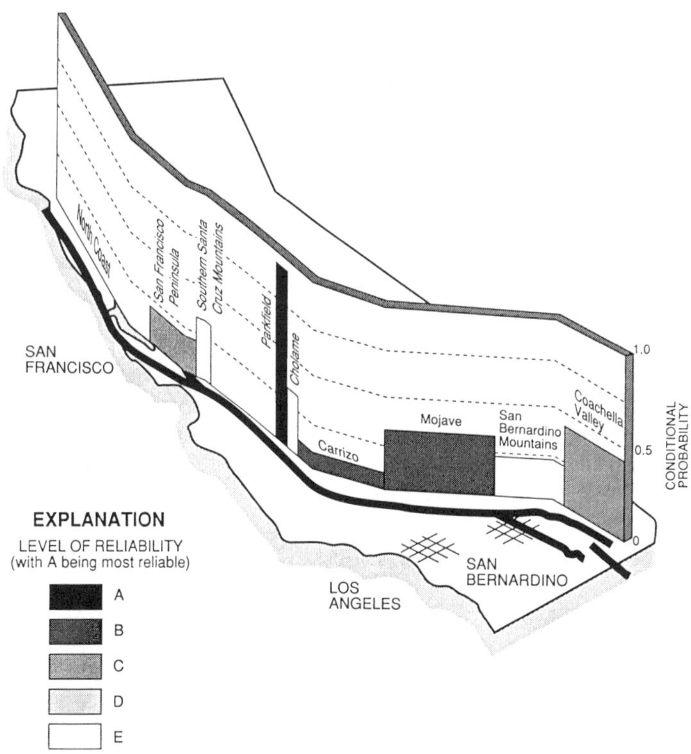

Figure 4. Conditional probability for the occurrence of major earthquakes along the San Andreas fault in the 30-year interval from 1988 to 2018 (After USGS Working Group, 1988).

Alquist-Priolo Special Study Zones

Alquist-Priolo Special Studies Zones encompass a number of active and potentially active faults in the Palmdale area. These are zones within which faults must be evaluated by a geologist prior to development (Hart, 1988). An active fault is defined by the California Division of Mines and Geology as one that has had surface displacement during Holocene time (about the last 11,000 years). Potentially active faults are faults which exhibit evidence of displacement within the Quaternary (last 2 million years).

As the state began developing the Alquist-Priolo Special Studies Zones they realized that there were so many potentially active faults in California that it would be meaningless to zone all of them. In 1975 the State Geologist decided that only "sufficiently active" and "well-defined" potentially active faults would be zoned. A fault is considered sufficiently active if there is evidence of Holocene surface displacement along one or more of its segments or branches. Well-defined faults are faults that can be clearly detected as a physical feature at or just below the ground surface by a trained geologist (Hart, 1988). The active and potentially active faults in the Palmdale area are discussed below.

San Andreas Fault Zone

The San Andreas fault is perhaps the most well known fault in the world. This can be attributed to its well defined trace, its historic activity and its location with respect to major metropolitan cities (e.g. Los Angeles and San Francisco). The San Andreas fault extends continuously for 620 miles from Point Arena in northern California to the eastern side of Salton Sea where it is concealed by alluvium. The San Andreas fault is a vertical right-lateral strike-slip fault which has sustained a cumulative offset of 190 miles since the middle Miocene (Crowell, 1975).

The San Andreas fault zone comprises a number of subsidiary fault segments. The zone is about 1 to 2 miles wide throughout the Palmdale area. The main trace is a remarkably straight segment that trends about N60W to N65W. The trace of the fault is clearly delineated by aligned topography, compression ridges, shutter ridges, offset gullies and linear vegetation. The most spectacular compression ridge exposure is in the I-14 (Antelope Freeway) roadcut through the San Andreas fault in southern Palmdale. Anaverde Formation clay shale and sandstone exposed in the roadcut has been complexly folded and faulted and bears testimony to the long history of movement on the San Andreas fault (Figure 5).

The main trace of the San Andreas fault is the most likely candidate for fault rupture in the future. This is supported by the following facts: 1) almost all historic large earthquakes (exceeding magnitude 6.0) have occurred along preexisting recognizable fault traces (Allen et al., 1965), and 2) the main trace has the most abundant geomorphic evidence suggesting recency of activity. The subsidiary faults to the San Andreas fault zone are generally limited in extent and are not considered capable of generating large magnitude earthquakes. Most of these segments are old main traces of the San Andreas fault zone which have been abandoned. These segments do show evidence of moving sympathetically in response to movement on the San Andreas fault and based on field evidence some of these subsidiary faults have sustained Holocene rupture. The main subsidiary and related faults in the Palmdale area are the Nadeau, Little Rock, Cemetery, San Francisquito, Leona Valley, Powerline Thrust and Hitchbrook faults (Figure 2). The subsidiary faults are discussed below to provide background for the geotechnical hazards section.

Nadeau Fault

The Nadeau fault is a series of short discontinuous strands that are subparallel and south of the San Andreas fault. In southeast Palmdale this fault commonly forms uphill facing scarps with attendant deranged drainages in old alluvial fans. In contrast, to the west the Nadeau fault trace is obscure. This has been attributed to rapid alluviation and burial of the fault within Anaverde Valley and the Lake Palmdale depression (Barrows et al., 1985). The fault is currently zoned as active by the State of California. Although the Nadeau fault clearly cuts Pleistocene age alluvial fans it does not clearly displace Holocene age sediments. However, there is geomorphic evidence suggesting recency of activity; this includes the youthful appearance of fault scarps and closed depressions along its trace.

Little Rock Fault

The Little Rock fault is a prominent subsidiary fault of the San Andreas fault zone. It is located about 1,500 to 2,000 feet north of the San Andreas fault, generally trending subparallel to the main fault trace. The Little Rock fault forms prominent scarps through south central and southeast Palmdale. The Little Rock fault is classified as active by the State of California. The style of faulting is variable ranging from single south-dipping or vertical traces to a series of branching en-echelon traces. The Little Rock fault appears to split into two segments just west of I-14. The northern segment is a north dipping thrust fault; the south trace continues as a nearly vertical, linear to slightly curving strand which extends nearly to Elizabeth Lake Road (Barrows et al., 1985).

Cemetery Fault

The Cemetery fault is subparallel and approximately 1 mile north of the San Andreas fault. The Cemetery fault is classified as active by the State of California. The Cemetery fault has been observed to clearly cut modern soil profiles in fault trench exposures in south central Palmdale. This fault is generally a nearly vertical strike-slip fault, but some component of dip-slip may be present (Barrows et al., 1985). The Cemetery fault forms prominent scarps in old alluvium along its trace. The Alquist-Priolo Special Studies Zone for this fault terminates to the southeast at 37th Street East where the trace appears to break up into a series of discontinuous strands.

San Francisquito Fault

The San Francisquito fault is located near the northern front of the Sierra Pelona range about 0.5 to 1.0 mile south of the San Andreas fault (Figure 6). This fault trends northeast through the Sierra Pelona and curves to the east near the front of the Sierra Pelona. This fault is zoned as active by the State of California. The fault appears to be primarily a vertical, strike-slip fault. To the east it apparently accommodates displacement through reverse dip-slip movement along the Powerline Thrust fault (discussed below). This fault trace forms the contact between Pelona Schist on the south and granitic rocks on the north. Locally old alluvium is thrust over granitics. Based on review of a number of fault trenches across its trace there is no clear evidence for Holocene activity (Buena, 1989).

Powerline Thrust

The Powerline thrust fault is a nearly continuous thrust fault located along the base of the Sierra Pelona (Figure 2). Along its trace, the Pelona Schist is thrust over the Ritter Formation. The Powerline Thrust fault is classified as active by the State of California. A series of fault trenches were excavated across the trace of this fault to assess its geometry, location and relative age (Buena Engineers, 1989). The Powerline thrust fault was observed to be displaced repeatedly by landslides within the trench exposures. The landslide masses themselves, of presumed Pleistocene age based on morphology, do not appear to be faulted. This suggests that Holocene rupture has not occurred along this fault.

Leona Avenue Fault

The Leona Avenue fault forms a series of prominent south facing normal fault scarps in old alluvium along its trace. This fault is located about 0.5 miles south of the San Andreas fault and is currently enclosed by the Alquist-Priolo Special Studies Zones for the San Andreas fault zone (Figure 6). The fault forms a groundwater barrier commonly separating the Ritter formation on the south from the Harold formation on the north. Although the trace is well developed, no clear evidence of Holocene rupture was observed in a series of trenches placed across its trace (Buena Engineers, 1989).

Hitchbrook Fault

The Hitchbrook fault is a northwest trending slightly curved fault. It appears to be an ancient inactive splay of the San Andreas fault. It terminates against the San Andreas fault to the northwest in the vicinity of Elizabeth Lake. Its trace fol-

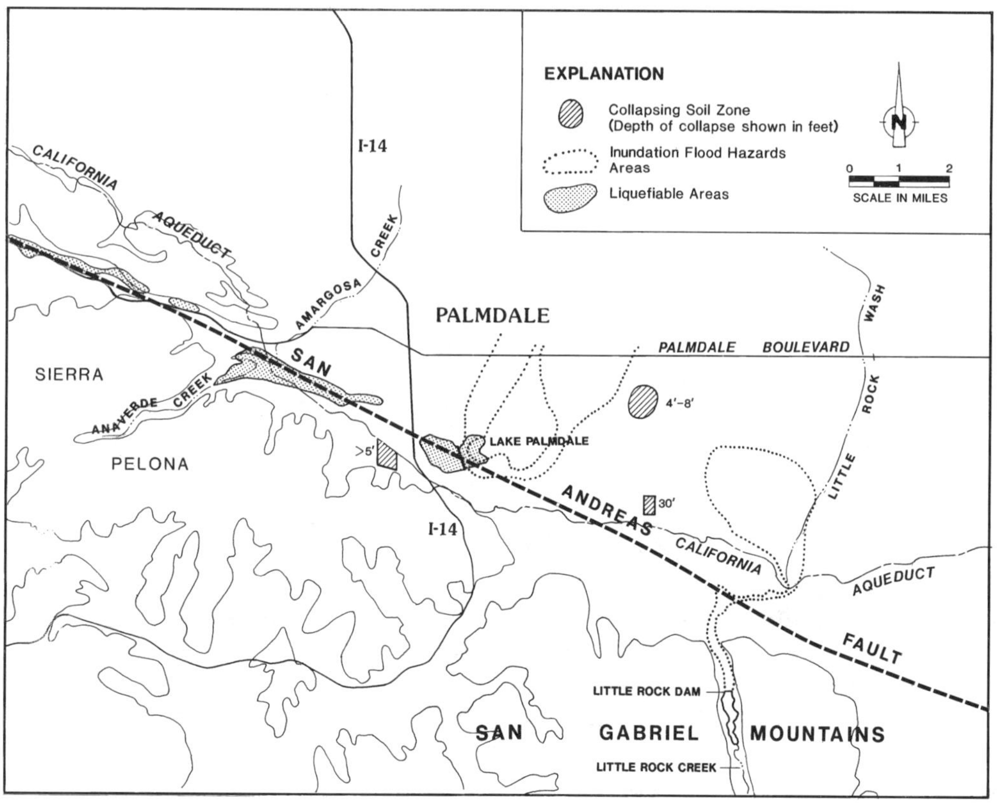

Figure 7. Geotechnical hazard map of the Palmdale area. Hazard areas are based on a variety of sources as specified in text.

Figure 5. Compression ridge in Antelope Valley Freeway road cut through the San Andreas fault. Road cut is about 90 feet high and exposes complexly folded sandstone and shale beds of the Anaverde formation.

Figure 6. View southwest of San Francisquito fault (SF) and Leona Avenue fault (LA). Bouquet Canyon road follows valley along trace of San Francisquito fault. (Photo by Jim Hickey).

Figure 8. Little Rock Dam with north flowing Little Rock Creek in foreground. Lake Palmdale (LP) and California Aqueduct shown in background.

Figure 9. Northwest facing large landslides in Pelona Schist. Photo taken near south central portion of Sierra Pelona. South draining Hauser Canyon in foreground.

lows the front of Portal Ridge and Ritter Ridge (Figure 2) separating Portal Schist from gneissic granodiorite; the trace is commonly concealed by old alluvium. Based on seismic refraction studies across its trace, it is buried by a thick section of continuous, unbroken old alluvial deposits (Smith, 1979). It is not zoned by the State of California as active.

GEOTECHNICAL HAZARDS

Because the San Andreas fault transects Palmdale, seismic issues are the primary geotechnical hazard. These include ground rupture and secondary seismic hazards such as liquefaction, flooding and seismically induced landslides. Other potential geotechnical hazards include collapsible soils, slope stability concerns and expansive bedrock.

Geotechnical hazards are discussed below commencing with seismic hazards. The reader is referred to the Geotechnical Hazard Map presented as Figure 7. This map is based on a variety of sources including: 1) Technical Appendix to the Safety Element of the Los Angeles County General Plan, Hazard Reduction in Los Angeles County, 2) unpublished data from Los Angeles County Geology Section and 3) CCW documentation during the course of geotechnical review for the city. This map is intended only as a general overview of geotechnical hazard prone areas and should not be considered an exhaustive representation of all hazards in the Palmdale area. Site investigations should always be conducted on a case-by-case basis and should stand alone based on their own merit.

SEISMIC HAZARDS

Ground Rupture

Trenching is performed to determine the location and activity of faults within proposed developments. Once faults have been located and analyzed, they should be accurately located by a licensed land surveyor on a suitable topographic base map. Civil engineers can then begin preliminary tract design based on fault setbacks provided by the geologist.

Fault rupture is mitigated by providing structural setbacks from the fault trace. Typically a 50-foot or less setback from the trace of the active fault for all habitable structures is employed based on the "resolution" of available geologic data. The geometry of the fault and the general nature of the earth materials around the fault dictates any modifications to the standard 50-foot setback. Thrust faults may require a wider setback to accommodate the larger area subject to ground shaking and rupture in the event of an earthquake. The presence of alluvium or other unconsolidated deposits should also be considered since faults typically splay and diverge when they rupture through unconsolidated earth material at the earths surface. A soil stratigrapher should be consulted if relative age dating of soils becomes critical in constraining fault activity.

If all these issues are taken into account and a thorough fault investigation has been completed, fault rupture can generally be mitigated by providing the appropriate structural setback. In some cases where faults are questionably active (no clear evidence of Holocene movement) consultants have proposed combining 25-foot setbacks with an additional 25-foot special foundation zone. Special foundations have also been used in areas adjacent to faults where the rock is highly fractured and broken. Typically the special foundations are post-tensioned slabs.

Ground Shaking

Ground shaking is the greatest seismic hazard facing Palmdale. It has been well documented that the majority of earthquake distress is related to strong ground motion effects and not from ground rupture (Borcherdt, 1985). The majority of the densely developed portion of the city is within about 1 to 2 miles of the San Andreas fault. Consequently, the city as a whole is subject to strong to severe ground shaking.

Prediction of the location, magnitudes and ground accelerations of future earthquakes is subjective. Geotechnical consultants provide ground motion response estimates in terms of peak ground acceleration data. Peak ground acceleration values are based on the maximum credible (MCE) and maximum probable earthquake (MPE). Given Palmdale's location with respect to the active San Andreas fault, the MCE and MPE are equivalent. The maximum probable earthquake on this segment of the fault is generally considered to be about magnitude 7.5 to 8.0 (Working Group, 1988).

A number of empirical engineering formulas are available to determine site ground motion characteristics. Based on empirical curves (Seed and Idriss, 1982), earthquakes of $M \geq 6.5$ will generate peak ground accelerations of 0.7g for bedrock areas and about 0.5g for deep cohesionless soils sites within 1 mile of the causative fault. Inhabitants of Palmdale (within 1 to 2 miles of the San Andreas fault) are subject to about 33 to 37 seconds bracketed duration ground shaking (Bolt, 1977) for a magnitude 8.0 earthquake on the San Andreas fault. The typical residential structure in Palmdale is wood-frame stucco construction. These structures are relatively ductile and generally perform well during earthquakes, if of modern design.

Liquefaction

Liquefaction is the transformation of a soil into a liquid-like mass due to loss of shear strength. It develops in response to pore pressure buildup due to ground shaking associated with an earthquake. Liquefaction is most likely to occur in low density sands and silty sands which are saturated to within 30 to 50 feet of the surface (Youd, 1975). In general, most of the soils around Palmdale are not subject to liquefaction because:

1) the soils are generally too dense,

2) the soils are generally too fine-grained and

3) groundwater is generally too deep.

In the Palmdale area, liquefaction susceptible areas are generally confined to recent alluvial soils with high groundwater conditions. Areas of historic high groundwater such as Barrel Springs Road should be carefully evaluated for liquefaction potential. Liquefiable areas are shown on the enclosed Geotechnical Hazard Map (Figure 7).

Lurching

Lurching is the cracking and fracturing of earth materials in response to strong ground shaking. This phenomenon is common in active tectonic areas like Palmdale. Lurching is denoted by the presence of closely spaced angular fractures which are infilled with modern soil. They are sometimes difficult to distinguish from desiccation cracks associated with soil drying and wetting processes. One way to distinguish them is by mapping their orientation and location and comparing them to the location of recognized faults in the area. Additionally, assuming the same earth material on either side of the causative fault, lurching features commonly decrease in frequency radially away from the fault.

Lurching can be mitigated several ways. One method is to remove and replace the fractured and disturbed ground and replace it with an engineered fill. The fill blanket can then act as a "cushion" to foundation distress in the event of future earthquakes. The other common means of mitigation is through structural setbacks, special foundations or a combination of both. Special attention should be given to ridgetops which commonly focus seismic waves during earthquakes.

Seiches

Seiches are seismically induced floods due to overtopping of reservoirs, lakes or other isolated water bodies in response to large rolling water waves generated from ground shaking. In the Palmdale area there are three potential seiche sources: 1) Lake Palmdale, 2) Little Rock Reservoir and 3) the California Aqueduct (Figure 8). Flood inundation zones in the event of failure of the Palmdale or Little Rock reservoirs are shown on Figure 7. These zones approximate areas potentially impacted by seiches. Developments directly adjacent to downstream areas of the reservoirs should be avoided. Areas directly adjacent to the aqueduct or further down gradient along the flood inundation zones can be mitigated by elevating finished grade or by providing engineered flood diversion devices.

Seismically Induced Landslides

Seismically induced landslide evaluation has not been standardized to state-of-the practice yet. Most predictive models available are not designed to determine the potential for seismically induced landslides at a specific site location. It therefore appears that the most practical approach to identifying areas of potential seismically induced landslides is through the preparation of landslide hazard maps. The landslide hazard maps can be used to target areas requiring additional study and/or field exploration. At the time of field exploration, the following parameters should be quantified:

- Geometry of subsurface earth materials,
- Shear strength of earth materials,
- Groundwater table, and
- Proximity of active faults.

Pseudostatic analyses can be performed to evaluate the stability of a given slope in the event of an earthquake. This is usually done by specifying a .15g horizontal earthquake coefficient and using peak soil shear strengths. Higher ground accelerations have been used on projects closer to active faults. The problem with pseudostatic analyses is that its a simplified approach to a very complex problem and does not model the "real world" very well. For example, it is possible to obtain higher safety factors for pseudostatic versus static conditions depending on soil shear strengths employed. For these reasons it appears that the best approach to evaluate areas of potential seismically induced landslides is through accurate geologic mapping and subsurface exploration coupled with sound geotechnical engineering recommendations to mitigate slope stability. Effective control of drainage is critical to the stability of these areas. Portions of the Sierra Pelona (Figure 7) are considered susceptible to seismically induced landslides due to the presence of weak micaceous foliation and the proximity of the range to active faults.

Collapsing Soils

The upper sediments in the Palmdale area are primarily alluvial sands, silty sands, and clayey sands with localized gravel beds. These soils are commonly collapsible, but only to a depth of 3 to 5 feet. As such, most collapsing soils can be mitigated readily by removal and reprocessing of the upper 3 to 5 feet of soil cover. Holocene alluvial deposits are the main candidate for collapsible soils. These soils are generally low density and poorly cemented. Infilled depressions along the traces of faults contain soils which are also subject to hydrocollapse. Relatively deep collapsible soils have been identified locally and are shown on Figure 7. Areas of deep collapsing soils are commonly mitigated by pre-soaking site soils to promote collapse of soil structure.

Expansive Soils and Bedrock

Expansive soils and/or bedrock are not common in the Palmdale area. Most of the soils are derived from crystalline granitic-type rocks which generally produce sandy soils. There are three general areas which potentially contain expansive earth materials:

1) Soils derived from the Pelona schist are commonly clay rich due to the weathering and breakdown of chlorite and other micas into clay minerals.

2) The Anaverde formation contains expansive clay shale members. The Anaverde formation is generally confined to pressure ridges between the San Andreas and Little Rock fault (Figure 2). Expansive clay shale bedrock of the Anaverde formation has reportedly contributed to cracking and distress of residential homes in the Palmdale area.

3) The Juniper Hills formation in southeast Palmdale and vicinity contains clay shale members which are potentially expansive.

Subsidence

Palmdale lies within the Antelope Valley, a closed topographic basin underlain by up to 5,000 feet of alluvial and lacustrine deposits (Durbin, 1978). All drainage ultimately feeds to Rogers Lake or Rosamond Lake playas. The basin is underlain by two aquifers which are separated by lacustrine beds. The upper aquifer has been providing a major source of groundwater for Palmdale and the vicinity since development of this area was initiated. In 1990, 42% of the consumptive water for Palmdale was provided by the State Aqueduct and the remainder was provided by groundwater pumping (Palmdale Water District, personal communication, 1991). During about the last 60 years pumping of groundwater from this aquifer has exceeded natural recharge. This has created an overdraft condition which has led to subsidence in the basin.

Initially, subsidence was generally confined to the dry lake areas where there was a greater concentration of groundwater wells. This has created problems for Edwards Air Force Base which utilizes the playas as landing fields for the shuttle and other aircraft. The playas are underlain by fine-grained silts and clays which are very sensitive to changes in moisture. Overdraft conditions have lowered the water table in this area by as much as 200 feet by about 1978 (Durbin, 1978).

There was a large increase in the rate of withdrawal from the upper aquifer due to rapid development in the Antelope Valley over the last decade. As a result of this prolonged overdraft condition, subsidence-related ground cracking has radiated outward from the dry lakes and has recently impacted areas within Lancaster.

This is a common problem for rapidly developing communities in the southwestern United States and it will continue to spread as long as the overdraft conditions are maintained. Mitigation of this condition depends on effective groundwater management of the basin. This requires reduction of the pumping of the aquifer and recharge of the aquifer through percolation basins, deep well injection, or some other means.

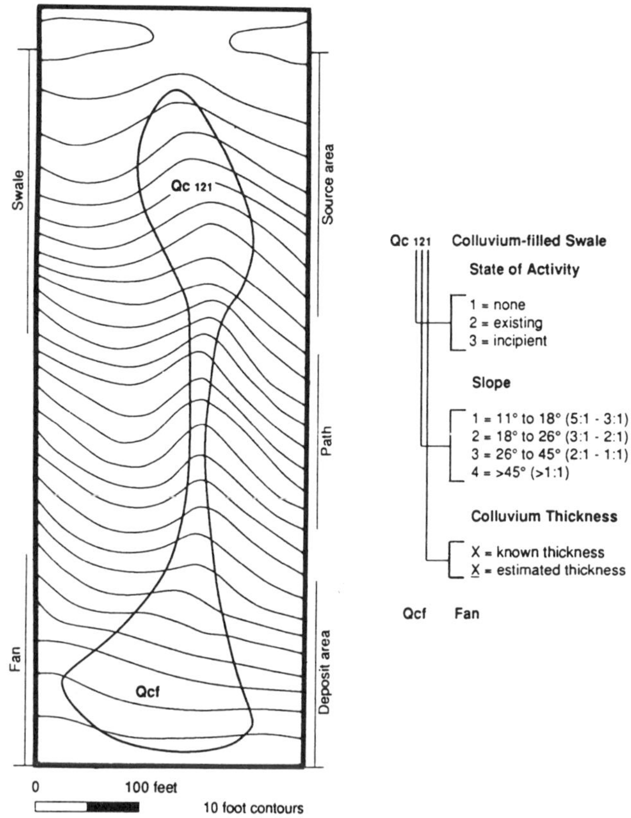

Figure 10. Geomorphic classification system for colluvial filled swale and downslope debris fan (After Montgomery et al., 1991).

Landslides and Debris Flows

Most developed areas within Palmdale have gentle slopes not subject to landslides and related hazards. However, as development has spread to the foothills in southern Palmdale, a number of slope stability problems have appeared. This is especially true for Ritter Ridge and the Sierra Pelona which are areas underlain by foliated schistose rocks. Pelona schist in particular commonly contains weak micaceous and chloritic units which are susceptible to failure.

Large nested sequences of landslides, as much as 100-acres in size, are present in the Sierra Pelona (Figure 9). The slide masses appear to be mostly block-glide features based on their morphology. It is most likely that these slides are controlled by the presence of weak units and developed in response to wetter climates during the Pleistocene. An alternative possibility is that they were seismically induced possibly in combination with wetter climatic conditions. These landslide prone areas should be carefully evaluated for proposed developments in these areas.

Debris flows are yet another potential hazard in the foothill regions. In past years usually only the main drainage basins were considered susceptible to debris flow activity. However, the heavy rainstorms in the Bay area in 1982 advanced our understanding of the potential debris flow threat posed by hollows or colluvial filled swales. Hollows are subtle depressions located along the flanks of ridges. The hollows are generally subparallel to one another and are oriented parallel to ridge slopes (Baldwin et al., 1987). Many of these hollows were incised during wetter climatic conditions during the Pleistocene. During the Holocene these hollows have been infilled with colluvium. This infilling processes masks their presence making them difficult to recognize in the field. During periods of heavy rainfall these unconsolidated colluvial materials become saturated and can mobilize as a fluid like mass composed of rock, soil and vegetation.

These features can be recognized on high quality large-scale topographic maps. Standardized mapping of these features facilitates review and analyses of debris flow threat to proposed developments. With this in mind, a geomorphic classification scheme for colluvial filled swales and associated debris fans is presented as Figure 10 (Montgomery et al., 1991). Recognition and mitigation of these features should be taken into account in the foothill regions of Palmdale. Typical mitigation procedures include:

- removal and reprocessing of loose colluvial sediments
- construction of debris flow impact walls and
- development of non-erosive engineered drainage devices.

Flooding

Flood prone areas within the Palmdale area are presented on FEMA and FIRM maps. Generally, flood-prone areas are confined to major drainages and associated floodplains along Little Rock Creek, Amargosa Creek and Anaverde Creek (Figure 7). Local sheet wash areas are also identified on these maps. Failure of Little Rock reservoir or Lake Palmdale are also potential sources of flooding as discussed previously under seiches. In general, the localized flood-prone areas can be accommodated by setbacks from flood-prone areas, elevating grades, or providing flood water diversion devices. Lining of stream channels may be prudent to reduce the potential for erosion and undermining of channel banks.

Rising Water

Areas of rising water have been identified in the Palmdale area in recent years. The rising water problems are manifested as daylighting water on graded slopes and in front yards. The rising water problems have been attributed to impermeable groundwater flow conditions created by fault juxtaposition of dissimilar rock types. There are a number of major faults that locally act as groundwater barriers. The San Andreas fault is a major groundwater barrier that blocks the flow of subsurface flow and contributes to swamp like conditions along its trace.

Potential groundwater barriers to subsurface flow should be identified in geotechnical investigations. Drainage related problems are one of the most common sources of residential distress. Rising water problems can be mitigated through 1) engineered surface and subsurface drainage facilities and 2) proper grading techniques to reduce the potential for permeability contrast of earth materials in graded slopes.

Nuisance Water

Nuisance water is the runoff created from landscape watering, washing of cars, and other similar activities that contribute to gutter flow. Palmdale does not have an established storm drain system to conduct this water to a disposal site. Consequently all nuisance water is disposed via dry wells. Dry wells are essentially seepage pits which are designed to intercept gutter flow throughout residential and commercial developments. Construction and use of the dry wells has both geotechnical and environmental implications.

Palmdale has established guidelines for dry well design. Nuisance water dry wells are required for all developments 1-acre or larger in size. Each residential household is considered capable of generating 40 gpd of nuisance water. A safety factor of 4 is applied to generate 160 gpd of nuisance water per household. For commercial developments, each acre of development is considered capable of generating 640 gpd of nuisance water.

The dry well capacity is determined based on percolation testing performed on the soil interval designated to receive water. Based on the percolation test results, the number and depths of dry wells are calculated based on the size of the proposed development. Consolidation or SPT tests are generally conducted to determine the maximum depth of collapsible soils. The dry well is designed so that nuisance water is discharged below the soil interval susceptible to hydrocollapse.

In several tracts in the Palmdale area nuisance water has to be disposed of off-site due to the presence of impermeable or low permeability earth materials. This can create substantial additional costs and time to locate a suitable area for disposal and to install piping to convey nuisance water offsite. Feasibility of nuisance water disposal should be determined early on in proposed developments as it may severely impact its economic viability.

In general, most of the nuisance water is percolating into the soil about 15 to 30 feet below adjacent grade. This water locally contains grease, oil and other residues generated from washing of cars and spraying down industrial equipment and the like. Over the long term these contaminants may percolate deep into the subsurface and affect the potability of the groundwater aquifer.

Palmdale's master drainage plan calls for construction of a storm drain system and a series of basins to intercept both nuisance water and rainfall runoff. Once this master drainage plan is in effect, dry wells will be phased out (City of Palmdale Engineering Department, Pers. Comm., 1991).

ACKNOWLEDGEMENTS

Early drafts of this document were reviewed by Dr. Perry Ehlig of California State University, Los Angeles and by Dr. Roy J. Shlemon. Their critical review and comments greatly improved the technical quality of this document. Appreciation is extended to Converse Consultants West for sponsoring this paper and providing assistance with drafting and word processing.

CONCLUSION

There are several geotechnical hazards that affect developments in the vicinity of Palmdale. These hazards include fault rupture and related seismic hazards, flooding, slope stability issues, expansive earth materials, rising water, and collapsing soils. These hazards can be attributed to the location of Palmdale in an arid environment adjacent to an active fault zone. Future development of the area should take into account the interdependent effects of climate, geology and physiography. In this manner, the area can be developed more safely and in concert with its climatic setting.

REFERENCES

Allen, C. R., 1975, Geological Criteria for Evaluating Seismicity: Geol. Soc. America Bulletin, Vol. 86, p. 1041-1057.

Baldwin, J.E. Ii, Donley, H.F., and Howard, T. R., 1987, On debris flow/avalanche mitigation and control, San Francisco Bay area, California. In Costa, J.E. and Wieczorek, G.F., (ed.), Debris Flows/Avalanches: Process, Recognition, and Mitigation: Geol. Soc. America Reviews in Engineering Geology, Vol. 7, p. 223-236.

Baldwin, J.E., Foster, J.H., Lewis, W.L., and Hardy, J. K. (editors), 1989, San Andreas Fault Cajon Pass to Wallace Creek: South Coast Geological Society, Inc., Guidebook Number 17, Vol. 1 and 2, 688 p.

Barrows, A. G., Kahle, J.E., and Beeby, D. J., 1985, Earthquake Hazards and Tectonic History of the San Andreas Fault Zone, Los Angeles County, California: California Division of Mines and Geology Open-File Report 85-10 LA, 139 p.

Bolt, B.a., Horn W.L., Mac Donald, G.A., and Scott, R.F., 1977, Geological Hazards: Second Ed., Springer-Verlag, 330 p.

BorchedT, R.D., 1985, Predicting Earth Ground Motion: An Introduction. In Ziony, J.I., (ed.), 1985, Evaluating Earthquake Hazards in the Los Angeles Region - An Earth-Science Perspective: U.S. Geological Survey Professional Paper 1360, p. 93-100.

Brown, R.D., Jr., 1990, Quaternary Deformation. In The San Andreas Fault System, California: U.S. Geological Survey Professional Paper 1515, p. 83-114.

Buena Engineers, Inc., 1989 Preliminary Geotechnical Engineering Report, Ritter Ranch Property, Elizabeth Lake Road, Palmdale, Los Angeles, County, California: Project No. B-3345-LO1.

Dibblee, T. W., Jr., 1967, Areal Geology of the Western Mojave Desert California: U. S. Geological Survey Professional Paper 522, 153 p.

Durbin, T. J., 1978, Calibration of a Mathematical Model of the Antelope Valley Ground-Water Basin, California: U. S. Geological Survey Water-Supply Paper 2046, 51 p.

Ehlig, P. L., 1975, Geologic Framework of the San Gabriel Mountains In Oakeshott, G.B. (editor), San Fernando, California, Earthquake of 9 February 1971: California Division of Mines and Geology Bulletin 196, p. 7-18.

Ehlert, K. W. and Schell, B. A., 1989, Geologic hazards and engineering geology in the Palmdale area: in San Andreas Fault, South Coast Geological Society, Guidebook No. 17, P.O. Box 10244, Santa Ana, CA 92711, p555-567

Hart, E.W., 1988, Fault-Rupture Hazard Zones in California: California Division of Mines and Geology Special Publication 47, 24 p.

Leighton & Associates, 1990, Technical Appendix to the Safety Element of the Los Angeles County General Plan, Hazard Reduction in Los Angeles County, Vol. 1 and 2.

Montgomery, D.R., Wright, R.H., And Booth, T., 1991, Debris Flow Hazard Mitigation for Colluvium-Filled Swales: Bulletin of the Association of Engineering Geologists, Vol. 28, No. 3, p. 303-323.

Morton, D. M. and Sadler, P. M., 1989, The failings of the Pelona Schist: landslides and sackungen in the Lone Pine Canyon and Wrightwood areas of the eastern San Gabriel Mountains, southern California. In Sadler, P.M. and Morton, D.M. (editors), Landslides in a Semi-Arid Environment, Inland Geological Society, Vol. 2, p. 301-322.

Norris, R.M., and Webb, R.W., 1990, Geology of California: Second Edition, John Wiley & Sons, Inc., New York, 541 p.

Ponti, D. J., 1985, The Quaternary alluvial sequence of the Antelope Valley, California. In Weide, D.L. (editor) Soils and Quaternary Geology of the Southwestern United States: Geological Society of America Special Paper 203, p. 79-96.

Romani, F. and Hick, B. A., 1989, Collapsible Soils in the Antelope Valley - California. In Kulhany, F.H. (editor), Foundation Engineering: Current Principles and Practices, Vol. 1, p. 135-142.

Seed, H.B. and Idriss, I.M., 1982, Ground Motions and Soil Liquefaction During Earthquakes: Earthquake Engineering Research Institute Monograph.

Sieh, K.E., 1978a, Prehistoric large earthquakes produced by slip on the San Andreas fault at Pallett Creek, California: Journal of Geophysical Research, Vol. 83, p. 3907-3939.

Sieh, K. E., 1978b, Slip along the San Andreas fault associated with the great 1857 earthquake: Seismological Society of America Bulletin, Vol. 68, p. 1421-1448.

Sieh, K. E., 1984, Lateral offsets and revised dates of large prehistoric earthquakes at Pallett Creek, southern California: Journal of Geophysical Research, Vol. 89, p. 7641-7670.

Sieh, K. E, Stuiver, M., and Brillinger, D., 1989, A more precise chronology of earthquakes produced by the San Andreas fault in southern California: Journal of Geophysical Research, Vol. 94, p. 603-623.

Smith, M.V., 1979, Engineering Geology Study, Rancho Vista, Palmdale, California: Michael V. Smith Engineering Geology, Project No. 7027-1.

The Working Group On California Earthquake Probabilities, 1988, Probabilities of Large Earthquakes Occurring in California on the San Andreas fault: U. S. Geological Survey Open-File Report 88-398.

Youd, T.L., Nichols, D.R., Helley, E.J., and Lajoie, K.R., 1975, Liquefaction Potential. In Studies for Seismic Zonation of the San Francisco Bay Region: U.S. Geological Survey Professional Paper 941-A, p. A68-A74.

SEISMIC HAZARD ASSESSMENT, NEWPORT-INGLEWOOD FAULT ZONE

S. Thomas Freeman
Principal Geologist
Woodward-Clyde Consultants
2020 E. First Street, Suite 400
Santa Ana, California 92705

Edward G. Heath
Consulting Engineering Geologist
123 Apolena Avenue
Newport Beach, California 92662

Paul D. Guptill
Engineering Geologist
The Earth Technology Corporation
13900 Alton Parkway
Irvine, California 92718

John T. Waggoner
Associate Geologist
Woodware-Clyde Consultants
2020 First Street, Suite 400
Santa Ana, California 92705

ABSTRACT

The active Newport-Inglewood Fault Zone (NIFZ) is one of the most significant seismic hazards in southern California. Yet, the fault is one of the least understood seismic sources in terms of its potential for future hazards. Rupture along the southern portion of the fault caused the destructive 1933 (M_S6.3) Long Beach earthquake, but no surface displacements were reported along its trace. It is unclear whether the 1933 event represents a characteristic earthquake or whether a larger event can be expected and whether such an event would be associated with surface fault displacements.

Oil field data from oil-bearing Miocene and Pliocene strata delineate a complex fault deep in the subsurface. The surface along the fault is typified by elevated mesas and hills that are underlain by Pleistocene sediments and intervening lowlands that are floored by Holocene sediments. The characteristics of the NIFZ in these near-surface deposits is not well documented. Old aerial photographs show discontinuous lineaments, particularly across the elevated hills and mesas and more faintly in some local lowland areas. In part based on the presence of these lineaments, the California Division of Mines and Geology has assigned Alquist-Priolo Special Study Zones along most of the NIFZ's trace so that the potential for surface fault rupture hazards will be addressed prior to building new habitable structures. This has resulted in increased efforts by many to delineate the active surface traces of the fault. In spite of these efforts, many of the late Quaternary seismic characteristics of the fault remain poorly constrained.

Dense urban development along the NIFZ severely limits the number of feasible sites where data regarding the late Quaternary seismic characteristics of the fault can be collected. In particular, shallow groundwater in many of the lowland areas greatly restricts the use of conventional geomorphic and geologic trenching methods for uncovering the most recent evidence of past seismic behavior.

The authors have been working on a number of projects to collect data and improve our understanding of the late Quaternary seismic characteristics of the fault. From our experiences, we have selected three examples to illustrate the approaches we have utilized to overcome some of the limitations imposed on the use of conventional geomorphic and geological trenching methods. Each of these examples focuses on our approach toward resolving one or more of the fault's seismic characteristics. Our multiple layered approach describes stacking oil field data, high resolution reflection geophysics, closely spaced cone penetrometer tests (CPTs), and borings to locate near surface fault traces in Holocene (lowland) deposits and to estimate Holocene recurrence rates. In the last example we discuss the use of sea level changes and marine terraces to delineate the timing of Quaternary deformation along the zone.

In this paper, we convey our realization that the results from isolated investigations, which utilize one or a limited number of investigation methods, can be misleading. Just to resolve potential surface fault rupture hazards, much less understand the seismic characteristics of the NIFZ, requires costly and time-consuming efforts to collect data from a number of different investigative methods. Only through these efforts and the cumulative evidence of the combined data can reasonable conclusions be drawn about the seismic characteristics of the NIFZ and its future hazard potential.

INTRODUCTION

The Newport-Inglewood Fault Zone (NIFZ) extends across the western margin of the Los Angeles Basin from offshore Newport Beach to Beverly Hills — along the way, it crosses beneath one of California's most populated areas. The destructive 1933 (M_S6.3) Long Beach Earthquake, the 1920 M_S 4.9 Inglewood Earthquake and the numerous other small events that have occurred on the fault illustrate its potential for generating future earthquakes (Barrows, 1974). This combination of conditions makes the NIFZ one of the most significant seismic hazards in southern California.

In evaluating the seismic hazard potential of a fault, the typical procedure is to characterize the fault in terms of its geometry and level of late Quaternary activity. The typical parameters used include fault location, length and width, style of movement or sense of slip, displacement per event, and slip rate. If these variables are known, or can be reasonably estimated, there are a number of different empirical and analytical approaches that can be used to estimate the maximum earthquake magnitude and earthquake recurrence for the fault. The conventional methods of investigating faults for their seismic parameters involves the use of aerial photos to identify geomorphic evidence of the fault's surface displacement history and the use of trenching to uncover evidence of the faults displacements in late Pleistocene and Holocene deposits.

The NIFZ does not readily lend itself to conventional seismic hazard analyses. While many geological studies have been completed along the NIFZ, good data is still scarce on some of its key late Quaternary seismic parameters. There are few accessible locations where late Pleistocene and Holocene seismic characteristics can be successfully studied. This problem results partly from the focus of the previous geologic studies of the fault. Unfortunately, at a time when much of the surface over the fault was unobstructed, most of the earlier investigation efforts were directed toward the faults influence on oil or groundwater. However, these previous studies provide invaluable data to form the framework for todays seismic hazard studies.

The character of the fault and the terrain it crosses provides significant challenges to answering todays geoseismic questions. In the subsurface, the fault zone is characterized by en echelon main fault traces with conjugate and splay faults, which in section appear to form flower-like patterns dissecting a series of anticlines. Along the fault, Pleistocene sediments occur beneath elevated mesas and hills that often lack an appreciable Holocene cover to estimate the timing of latest fault displacements. Most Holocene sediments are found in intervening lowlands where shallow groundwater and deep soft sediments often preclude effective trenching. Finally, along most of its length, dense urban development severely restricts access to locations to study the seismic characteristics of the fault.

The authors of this paper have investigated the fault, utilizing a variety of different methods at a number of different locations. In this paper we present three examples of our work and how we attempted to overcome some of the limitations of conducting seismic hazard investigations of the NIFZ. First, we discuss a study that was completed to characterize the NIFZ and define its maximum earthquake. This work examined many of the fault's characteristics, including its slip rate, which was estimated through an analysis of facies and structural data from several oil fields. The slip rate was then used to help estimate the fault's maximum earthquake through comparisons with historical events on similar faults in the southern California region and worldwide. Secondly, we discuss studies that focused on delineating potential surface fault rupture hazards. These studies also provided insight into Holocene recurrence and the use of Quaternary sea level changes to assess the timing of past earthquake activity.

As a result of these studies, we believe we have gained some valuable insight on how to approach investigations of the fault's seismic characteristics. We have prepared this paper to share our experiences using different investigation techniques and to stimulate the interest of other investigators in using and improving upon these techniques to investigate the NIFZ or other similar faults. Further, we believe that the data we have collected on the fault's seismic characteristics, much of which have not previously been widely available, will be useful in evaluating the potential future seismic hazards imposed by the NIFZ.

The authors of this paper are listed alphabetically, although each has contributed about equally to the paper. We have all worked together on this project at one time or another either as employees of Woodward-Clyde Consultants or separately. In the first case discussed, Ed Heath was the principal investigator with primary support from Paul Guptill and limited support from Tom Freeman and John Waggoner. In the second example, the majority of the investigations at Huntington Beach and Long Beach were completed by Tom Freeman and John Waggoner with additional support from Ed Heath and Paul Guptill. Paul Guptill was the principal investigator of the West Newport Mesa study, with support from Ed Heath and a limited effort by Tom Freeman. Other key workers that deserve credit for their efforts during these studies include Ray Sugiura, Tom Rockwell, Roy Shlemon, Clarence Allen, Casey Armstrong, and the owners of the properties, Signal Landmark/The Koll Company, The Bixby Ranch Company and the West Newport Oil Company.

EVALUATING THE MAXIMUM EARTHQUAKE MAGNITUDE

In the late 1970s the authors were involved on a project where the primary goal was to define the maximum earthquake for the NIFZ. This was not a trivial assignment, given the limited data and analytical tools, which were available at that time. Typical approaches to evaluating the maximum earthquake magnitude use empirical relationships between earthquake magnitude and the length of surface fault rupture or amount of surface displacement per event. However, because the NIFZ is not a continuous through-going fault near the surface, information on fault length for use in maximum magnitude estimates was incomplete and difficult to verify. Displacement-per-event relationships could not be readily applied because surface traces or escarpments, needed to evaluate displacement histories, were limited. Even with todays improved understanding of seismic geology, such an assessment of the NIFZ can be a difficult exercise because the database on the late Pleistocene and Holocene seismic characteristics of the fault remain limited. While our late 1970s studies of the fault might now be considered somewhat dated, some of its aspects remain unique, and merit wider exposure. In fact, the results of the study still represent some of the few data on slip rate of the NIFZ.

The approach used in our study of the assessment of the fault's seismic potential was to compare the NIFZ's characteristics and degree of activity with those of better known faults in southern California as well as those located in similar tectonic settings worldwide. By comparing geoseismic parameters of each fault in Table 1, the NIFZ can be placed in relation to other faults and its relative seismic risk assessed.

At the outset of this study, not all the fault characteristics and parameters needed for a complete fault-ranking analysis were readily available or well known. Therefore, it was necessary to investigate these further. In addition to our review of the regional and local tectonic setting, we focused on establishing the NIFZ'S degree of activity, total displacement, slip rate, displacement per event, and estimates of its maximum earthquake. The results, and conclusions of this overall fault-ranking methodology are presented in the following sections.

Regional Tectonics

The San Andreas Fault System dominates the regional tectonics of southern California. This fault system consists of the long northwest-trending San Andreas and San Jacinto fault zones and other parallel fault zones to the west: Whittier-Elsinore Fault Zone, NIFZ, Palos Verdes Fault Zone, and several offshore faults. Like the San Andreas Fault Zone, all the northwest-trending faults in the western part of southern California appear to be predominantly right-lateral, strike-slip, crustal discontinuities. The faults appear to show an apparent general westward decrease in total displacement, continuity of surface trace, and apparent amount of seismic activity, see Table 1. In order to compare the NIFZ with these other faults it is first helpful to review some of its characteristics.

The NIFZ extends for approximately 70 kilometers as a complex series of discontinuous faults and folds, all of which present strong evidence for dominant right-lateral slip. To the north, the NIFZ merges with and/or is overridden by the Santa Monica Mountains along the Santa Monica Fault; to the south, it branches into several splay faults that appear to die out against the San Joaquin Hills structural high and its offshore extension.

Two distinctly different basement rock types have been recognized on either side of the NIFZ: a "continental facies" on the east and an oceanic facies on the west (Woodford et al., 1954; Hill, 1971). Overlying these basement rocks is a thick sequence of marine sedimentary rocks of Miocene through late Pleistocene age, with minor local volcanic rocks. In the vicinity of the NIFZ these units are 10,000 to 14,000 feet thick. The middle Miocene rocks are a varied succession of marine clastic sediments, volcanic rocks, and schist breccia. The upper Miocene and Pliocene rocks are a repetitiously interbedded sequence of fine to coarse, clastic, marine sediments. Overlying the Pliocene rocks are Pleistocene marine sandstones which today are exposed in the Mesas and other highlands along the fault. Pleistocene to Holocene down-cutting through these Pleistocene deposits, and subsequent backfilling have left a series of Holocene filled gaps through the Pleistocene deposits which today are expressed as lowlands.

There are 24 separate faults found in the subsurface along the NIFZ (Barrows, 1974) that have significantly displaced the sequence of sedimentary rocks. Where they are associated with major folds, second- and third-order normal, reverse, and left-lateral faults are also found. The northwest-trending en echelon fault segments and northwest- to west-trending folds associated with or lying between the en echelon faults suggest predominant wrench faulting (Harding, 1973). Some of these faults reach the surface where they form topographic escarpments in Pleistocene alluvial materials; however, observed Holocene displacements are very limited.

Timing and Rate of Deformation Along the NIFZ

The latest period of deformation along the NIFZ appears to have begun soon after the Los Angeles sedimentary basin formed and the middle Miocene marine sediments were deposited and continued through Holocene time. The observations leading to this conclusion are: (1) the Miocene units are displaced laterally a greater distance than the overlying units, showing that displacement was occurring at the same time as deposition of the younger units; and (2) the Miocene units also show the greatest structural relief on the major anticlinal structures, indicating that they were growing structures through the Plio-Pleistocene. The initiation of late Miocene deformation along the NIFZ is believed to coincide with the opening of the Gulf of California when the San Andreas Fault started dominating southern California tectonics.

Historic Seismicity

Documented historical earthquake activity on the NIFZ is dominated by the occurrence of the magnitude 6.3 (M_S) Long Beach earthquake of March 11, 1933. The source parameters for this earthquake are the only such data from large earthquakes generated by the NIFZ.

The style of faulting of the 1933 earthquake was evaluated by waveform analysis of teleseismic bodywave records of the earthquake (Woodward-Clyde Consultants, 1979; Hauksson and Gross, 1991). The best-fit focal mechanism indicates that the displacement during the 1933 event was right slip on a N45^o west-trending fault, with a near vertical dip of 80° northeast. This matches the geologic observation of orientation of the NIFZ and also agrees with the aftershock pattern. The waveform study indicates that the focal depth was 10 to 13 kilometers and the seismic moment was about 5 x 10^{25} dyne-cm (Woodward-Clyde Consultants, 1979; Hauksson and

Gross, 1991). Woodward-Clyde (1979) calculated the average displacement at 31 to 46 centimeters; Hauksson and Gross (1991) calculated a slip of 85 to 120 centimeters.

SLIP RATES

Slip Rates Estimated from Displaced Stratigraphic Facies

In comparing the faults geoseismic parameters as shown in Table 1 especially in analyzing the degree-of-activity of the faults, it became apparent that the geologic slip rates on the faults was one of the most pertinent parameters; therefore, it was necessary to obtain more definitive data on geologic slip rates along the NIFZ. Hill (1954 and 1971) presented a methodology for determining lateral displacements across the NIFZ by matching sedimentary rock facies and stratigraphic thickness changes across the fault. By comparing electric logs (E-logs) from adjacent oil wells, one can determine what facies changes have occurred between the wells for any specific horizon. Thus a series, or line of E-logged wells can show graphically what facies changes are occurring. If two lines of wells are picked parallel to and on each side of a fault it is often possible to match these facies changes within each horizon across the fault. Using these methods, Hill (1971) demonstrated approximately 3 kilometers of right-lateral slip on the NIFZ. With the large number of oil fields associated with the NIFZ and the availability of oil-well E-logs, it was possible to obtain and develop abundant structural and stratigraphic control on either side of the fault. Therefore, it was decided to use Hill's methodology to estimate fault displacements of facies changes, and then by relating these displacements to the age of the sediments, calculate the long-term geologic slip rate at several locations along the NIFZ.

Hill's methods of facies correlations can produce unique matches with a high degree of certainty; and can provide reliable estimates of the displacement of these facies changes across the fault. The accuracy of determining the exact displacements, however is limited by several factors: (1) spacing between wells; (2) distance from wells to the fault; and (3) the possibility that the facies changes can be occurring at an oblique angle to the fault. However, when considering displacements ranging from 1 to 3 kilometers, such as along some portions of the NIFZ, these error limits can generally be held down to 200 to 300 meters and thus represent less than 20% and generally no more than 10% of the calculated displacement at any one locality. Furthermore, combining numerous measurements from several locations along the fault can lead to offsetting errors, thus, improving the reliability of the overall displacement estimates.

The initial study (Woodward-Clyde Consultants, 1979) was completed in two phases. The first phase consisted of reviewing the available published literature about oil field structures and their stratigraphy in order to develop initial estimates of displacements, both horizontal and vertical, along the NIFZ. In addition, locations were identified that would provide the best opportunities to use the E-log correlation approach.

Criteria used in the site selection process included: (1) the fault zone in the study area needed to be narrow and well defined, with a high degree of assurance that all or most of the zone of faulting was being spanned by the data set; (2) subsurface structure and stratigraphy needed to be relatively simple and well understood; and (3) there needed to be a sufficient number of oil wells from which data would be available.

The literature review provided data about horizontal and vertical displacement and showed that displacement along the NIFZ is predominantly right-lateral strike-slip. The data also suggested that the Long Beach, Seal Beach, and Huntington Beach oil fields were the most promising locations for further analysis.

The second phase of the initial study involved the evaluation of horizontal displacements in the three oil fields selected for detailed analysis. At the selected sites, specific oil well E-logs were collected, correlations where developed, and displacements of facies changes were evaluated. To support the displacements developed using the E-log correlations, we also evaluated interpreted cross sections and stratigraphic information.

In addition to the initial Woodward-Clyde (1979) study, investigations were later completed to obtain supplemental displacement data to confirm the geologic slip rates developed in the initial study. To accomplish this objective, oil well data were used to evaluate displacements of higher and lower stratigraphic horizons than those identified in the Long Beach and Seal Beach oil fields and in one additional area, the Inglewood oil field.

Figures 1 and 2 present an example of correlations of E-log data from the Long Beach field. This example shows the correlation of Well Dodge 3 between Wells B-18 and B-38. Examination of the characteristics of the well logs (Figure 2) shows a distinctive change from a sandy facies to a more silty facies toward the southeast (right). Most notable on the Well Dodge-3 log is the blocky shape of the signature of the uppermost sandy unit and the presence of the three lower sandy units with signatures like serrated funnels. This log signature is most similar to the Well B-18 log, differs somewhat from the Well B-38 log, but differs distinctly from wells farther down the section.

The E-log from one well seldom provides an exact match with a log from a well located on the other side of the fault. However, the well often can be placed at an intermediate point between the two wells across the fault. The horizontal displacement can then be estimated by the distance between the present location of the correlation well and its relative correlation position with respect to the reference wells across

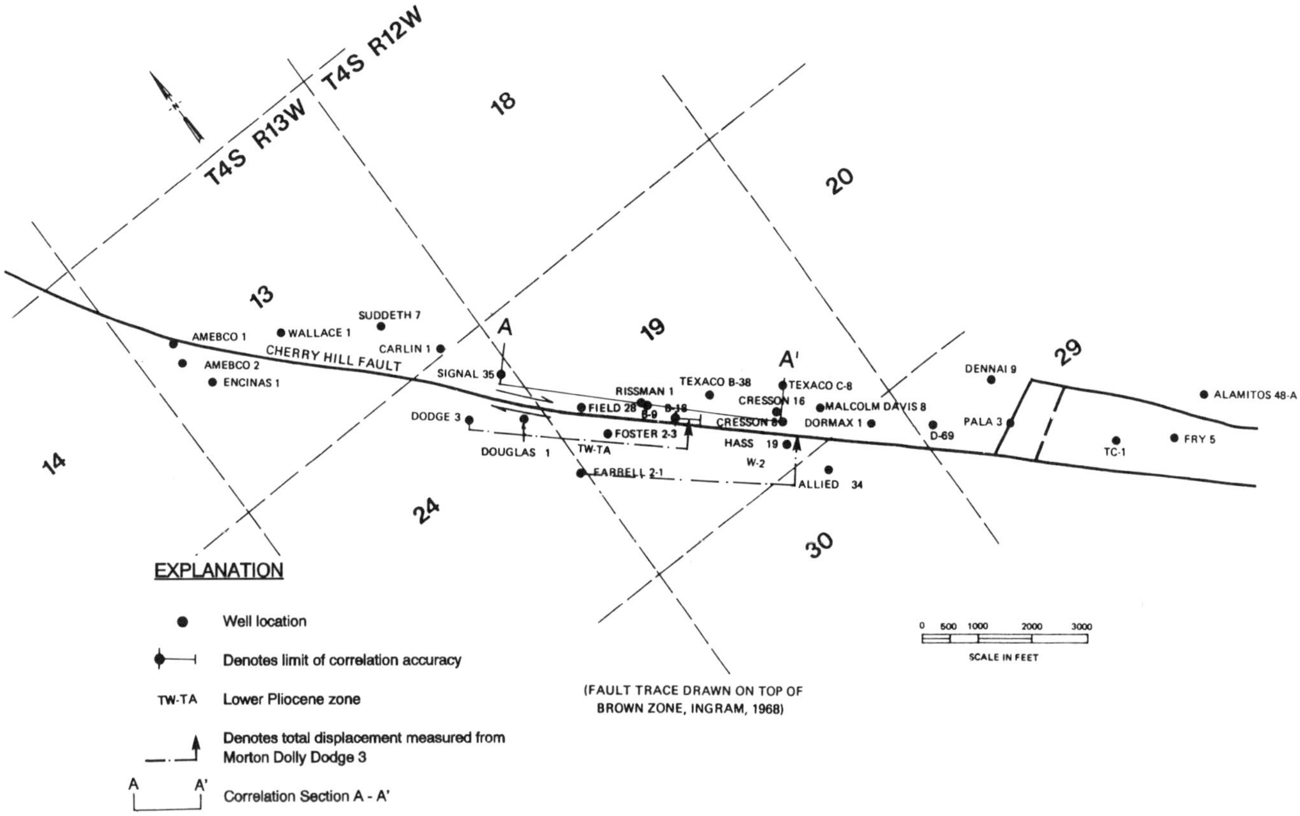

Figure 1. Map location of reference and correlation wells Long Beach oil field.

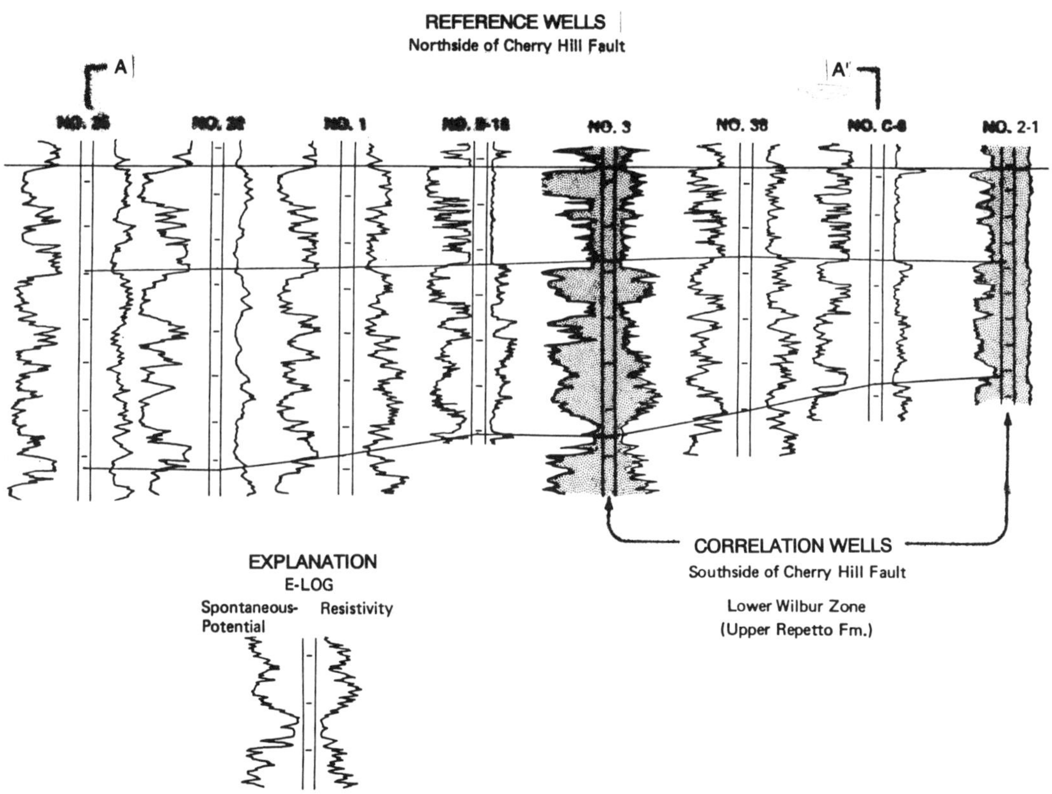

Figure 2. Section A-A', Long Beach oil field Example E - Logs

the fault. The horizontal displacements, so measured, can be assigned plus and minus values that represent the distance between the nearest reference well location and the selected correlation position. Similar estimates of horizontal displacements, on a well-by-well basis, were made at several stratigraphic intervals for the other selected oil fields.

Data from the California Division of Oil and Gas and the American Association of Petroleum Geologists (Knapp et al., 1962) made it possible to relate the correlated intervals on the E-logs to various stratigraphic horizons. The ages of the stratigraphic horizons were estimated based on the Upper Cenozoic Stratigraphic Column of the Western Los Angeles Basin (Nardin and Henyey, 1978). The range of possible geologic age for each E-log correlation interval was estimated, based on the relative position of the interval within a known stratigraphic unit.

The horizontal displacement measurements were plotted with respect to the estimated age of the units and are presented graphically in Figures 3 and 4; the boxes surrounding each data point reflect the possible range of uncertainty associated with each age and displacement estimate. These figures indicate that displacement along the NIFZ appears to have been rather uniform and that in all areas studied, the long-term horizontal slip rate has been approximately 0.5 mm/yr from as early as the deposition of late Miocene sediments.

Slip Rates Estimated from Displaced Structural Features

We also estimated the displacement of structural features along the NIFZ utilizing structural data and interpretations in the literature for the oil fields. The structural data was used to develop geologic slip rates that could be compared to the slip rate estimates developed from the facies displacement studies.

Although 16 oil fields are located along the NIFZ, only a few of them have structural relationships that can be used for analysis of horizontal displacement. Estimates of horizontal displacement were based on the measurement of displaced structural features interpreted from published structural contour or isopach maps of selected stratigraphic horizons. The geologic ages of the horizons defining the displaced structures were interpreted from the published stratigraphic columns accompanying the structural maps and the literature (Knapp et al., 1962). The timing of growth of folds and development of faults was interpreted from the structures in each oil field and their effects on the geologic horizons of various ages. Age, in years for the various horizons, was estimated from a published stratigraphic column (Nardin and Henyey, 1978).

Wright et al. (1973) provided a detailed structural analysis of the Inglewood oil field; it will be used here to illustrate the methodology. Evaluation of structural maps for the Inglewood oil field presented by Wright et al. (1973),and the re-

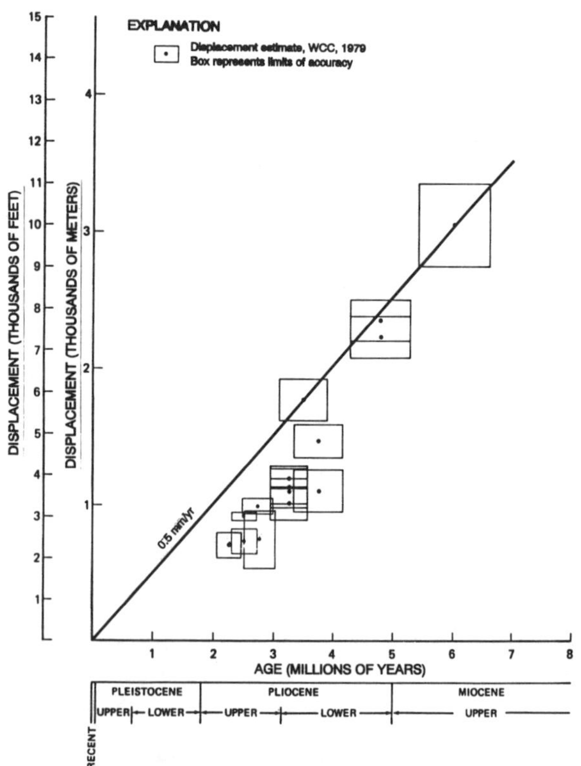

Figure 3. Horizontal geologic slip rate - E-Log Data, Long Beach oil field, NIZD

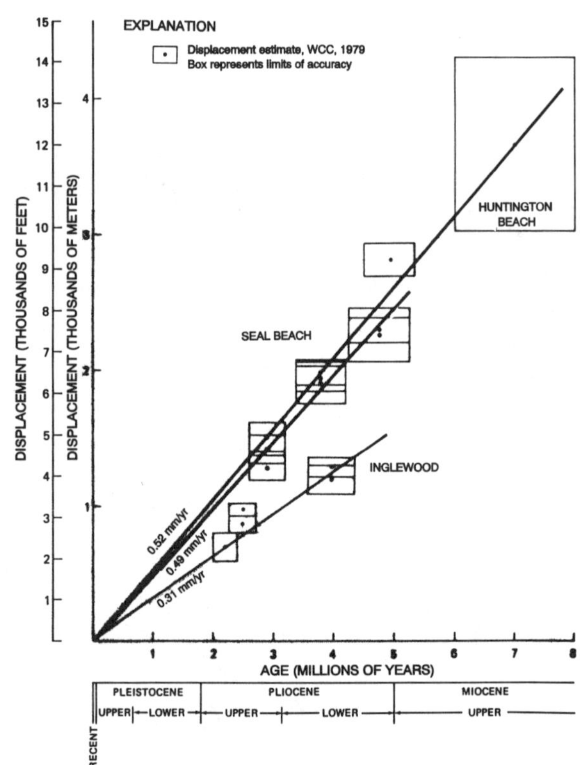

Figure 4. Horizontal geologic slip rate - E-Log Data, Seal Beach and Huntington Beach oil field, NIZD

cords of the California Division of Oil and Gas confirms a right-lateral horizontal displacement on the NIFZ. The estimated horizontal displacement, measured on the displaced anticlinal axis, ranges from 3,000 to 4,000 feet since latest Pliocene time. These displacements suggest a long-term average horizontal slip rate of 0.48 mm/yr for the Inglewood oil field segment of the NIFZ since that time (Fig. 5).

Uplift of the Inglewood anticline appears to have begun in late Miocene time, probably as a result of deep-seated dextral movement along a fault in the basement rocks underlying the thick sedimentary cover (Wright et al., 1973). However, the change in slope in the slip-rate curve of Figure 5 indicates that strike-slip faulting on the Inglewood fault in the Inglewood oil field did not start until late Pliocene time.

This conclusion agrees with Wright et al. (1973) and Wright (1991) and indicates that displacements, measured in the late Miocene to late Pliocene sediments, show greater offsets along the southern portion of the NIFZ than in the northern portion; also, that faulting was initiated earlier in the south and slowly propagated northerly along the fault zone.

Examination of the structural features across the fault in Long Bach and Seal Beach oil fields indicates that the measurable apparent vertical separation is largely a function of the effects of horizontally offset plunging anticlines.

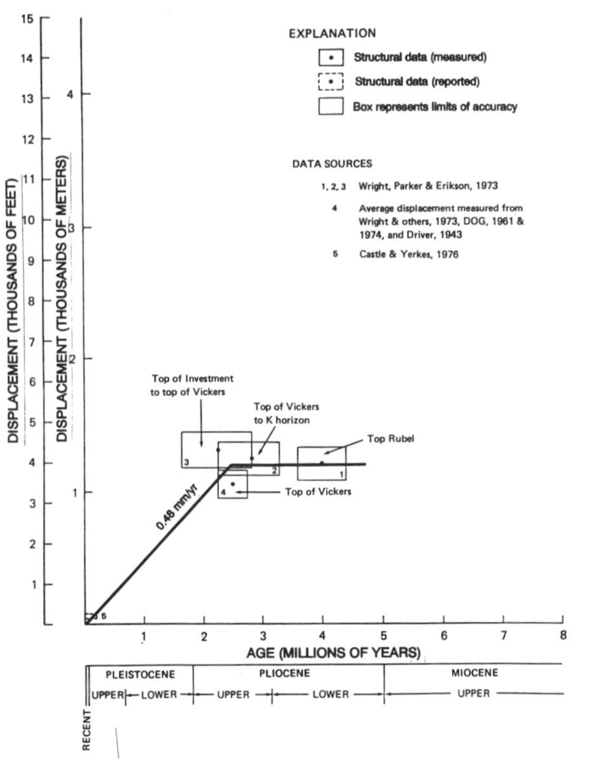

Figure 5. Horizontal geologic slip rate - E-Log data, structural data, Inglewood oil field, NIZD

A review of the structural contour map of the top of the Brown Zone (Lower Pliocene) at the Long Beach oil field showed that apparent vertical separation across the NIFZ decreases from the crest of the fold to each nose of the doubly plunging anticline. At the crest of the anticline, the apparent vertical separation for the top of the Brown Zone is about 800 feet. The amount of apparent vertical separation decreases to the northwest and southeast along the fault trend to the extremities where the fold is least developed. This decreasing apparent vertical separation away from the crest demonstrates that a sizeable part of the apparent vertical separation across the fault in the Long Beach oil field results from horizontal offset of the plunging anticlines. At the northwest and southeast edges of the fold, the apparent vertical separation across the fault is approximately 100 to 200 feet within the Brown Zone (Early Pleistocene). This is considered the best estimate for the maximum vertical tectonic displacement across the fault within the lower Pliocene in the Long Beach oil field. It is on the same order of magnitude as the vertical separation observed in the Seal Beach oil field where folding is much more gentle.

Thus, the conclusions of the study are that the displacement of the NIFZ is dominantly right-lateral strike-slip in character with a long-term slip-rate of approximately 0.5 mm/yr. Further, the vertical to horizontal ratio (V:H) of tectonic displacement is highly variable and is estimated to be about 1:20, where least affected by local folding.

Fault Ranking by Slip Rate

In reviewing known methods of estimating maximum magnitude, Chinnery (1978) suggested that Smith's (1976) use of slip rate was one of the most reasonable. Smith (1976) used the geologic slip rate to obtain the average rate of seismic activity and to limit the maximum earthquake that can be expected to occur along the fault. Cluff (1978) and Packer and Cluff (1977) described differences in relative degree-of-activity in terms of slip rate, recurrence interval, and slip per event. Thus, in order to further assess the degree-of-activity of the NIFZ in comparison to other strike-slip faults, we compared the geologic slip rates for the NIFZ with the slip rates of other strike-slip faults in southern California.

A review of the slip-rate data developed for southern California strike-slip faults (Table 1) indicates that both geologic slip rate and the magnitude of the largest historical earthquakes decrease to the west — beginning with the largest value on the San Andreas Fault Zone and decreasing with each successive, major strike-slip fault. This analyses of the relationships between geologic and seismologic characteristics of faults and earthquakes suggested that there may be a substantial basis for establishing a relationship between a fault's geologic slip rate and its maximum earthquake.

To pursue this relationship, it was necessary to expand the data set from the major strike-slip faults in southern California. Therefore, fault displacement, geologic slip rates, and magnitudes of historical earthquakes were gathered for other strike-slip faults in California, Alaska, and other areas of the world with tectonic regimes similar to that of southern California.

Empirically, Slemmons (1977 and 1982) and theoretically, Kanamori and Anderson (1975), have found that earthquake magnitude (M_S) is proportional to the logarithm of the associated displacement of the fault on which the earthquake occurred. The slip rate on a particular fault can be measured by the sum of the fault displacements associated with all the earthquake activity on the fault as a function of time (Brune, 1968; Molnar, 1979). In general, the total amount of displacement on a fault is dominated by the displacements associated with the single or several largest characteristic earthquakes to occur during the selected time period. Thus, these observations on theoretical and empirical relationships suggest that the logarithmic value of the slip rate may scale linearly with the size of the maximum earthquakes associated with the fault.

In order to test these observations, the assembled geologic slip-rate data and the magnitude of the larger events on these faults were plotted in a semi-log format of slip rate versus magnitude (Figure 6). The pattern of historical earthquake data presented on Figure 6 suggests a trend of increasing maximum earthquakes with increasing geologic slip rates. This observation together with an evaluation of the slip rate and magnitude parameters suggests that an upper-bound limit may exist to the right of the data points. Such an outer bounding line may then be useful in estimating the maximum characteristic earthquake magnitude for a fault within this data set or for other faults in similar tectonic settings.

A conservative approach in evaluating the slip rate versus magnitude data set can be taken by considering the ranges of slip rate and magnitude data obtained from published and unpublished sources. To account for these uncertainties, a range can be assigned to each earthquake magnitude along with the possible ranges of slip rates for each of the faults. A line that bounds the lowest slip-rate/maximum-magnitude ranges of all the data points then would provide a significant level of conservatism (Figure 7). This line does not mean that each of these faults is capable of the earthquake estimated from the line; only that the line may represent an upper bound for future earthquakes.

Figures 6 and 7 suggest that the largest characteristic earthquake for each fault represented would fall to the left of the bounding line. Using this relationship to further assess the earthquake potential of the Newport-Inglewood Fault and using the 0.5 mm/yr slip rate calculated earlier in Figure 7 suggests that the characteristic maximum earthquake would fall in the range of M6.0 to M6.5 and would probably not exceed M7.0.

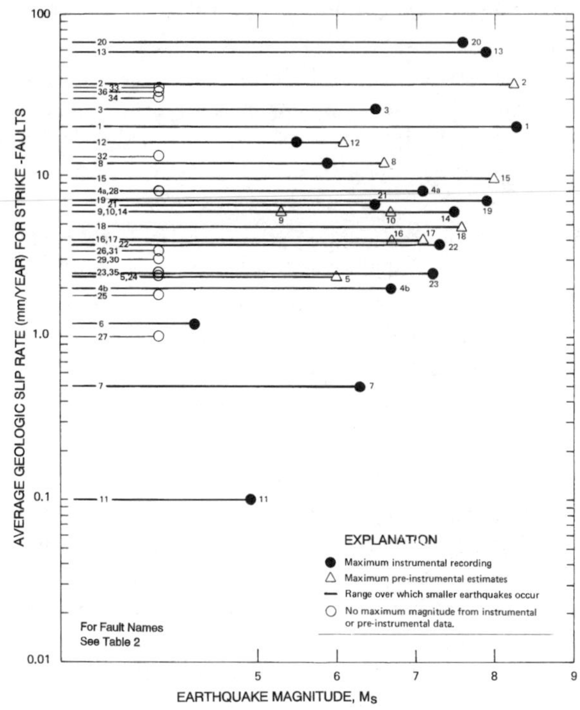

Figure 6. Empirical plot, geologic slip rates vs historical magnitude for strike-slip faults

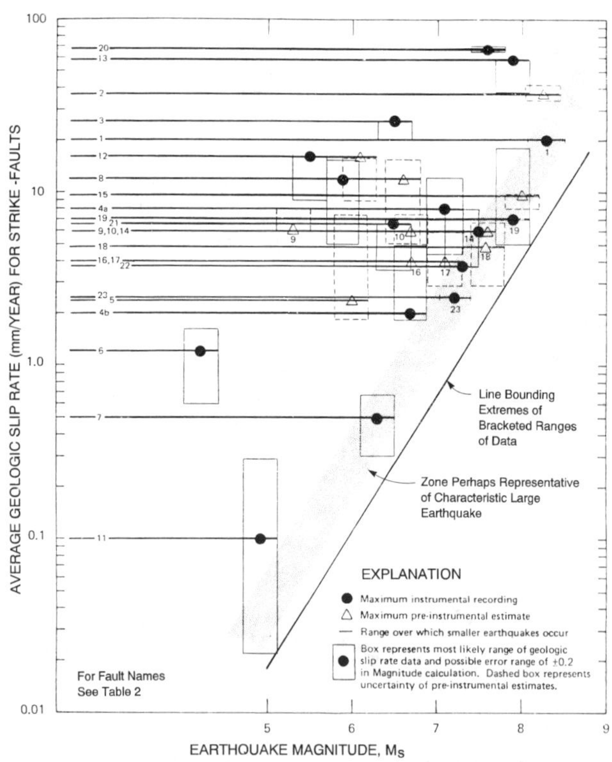

Figure 7. Maximum earthquake limit geologic slip rate vs historical magnitude for strike-slip faults

In discussing the initial presentation of this slip rate vs. magnitude relation (Woodward-Clyde Consultants, 1979), dePolo and Slemmons (1990) state that this relation gives reasonable earthquake size estimates for high-slip-rate faults, but believe it may not be representative for low-slip-rate faults due to the low number of low-slip-rate faults used in the data set. It is pointed out that the data set shown on Figures 6 and 7 was not purposely biased. By their nature, the high-slip-rate faults experience large earthquakes more often than low-slip-rate faults. To resolve this question, additional data on the occurrence of large earthquakes on low-slip-rate faults is needed. Unfortunately, this solution may be problematic because a low slip rate dictates a very long recurrence time for large earthquakes. On the other hand, this problem may be offset by the fact that there are many more low-slip-rate faults than high slip-rate faults.

Also, in comparing fault characteristics it is apparent that high slip rate faults are characterized by long fault segments, and low slip rate faults are characterized by much shorter fault segments. Thus, the established relationship of fault rupture length to earthquake magnitude further suggests a close relationship to the geologic slip rate.

It must be pointed out that Figures 6 and 7 are based on a limited data set of strike-slip faults and probably pertains only to strike-slip faults in similar tectonic settings. These figures are not expected to apply to other styles or types of faults such as normal faults, high angle reverse faults, blind thrust faults, or other faults dissimilar to those used in the data set. Further, the use of slip rates for comparing the state of activity of faults and for possibly gauging the maximum earthquake potential is based on only one fault parameter, and thus, should not stand alone, it should be used in conjunction with other parameters and fault characteristics as discussed earlier.

"LAYERED" MULTIPLE GEOSEISMIC STUDIES

In the early 1980s, the authors began series of investigations of the Newport-Inglewood fault, where it crosses a large undeveloped parcel of land near Huntington Beach. Studies at a similar site, in Long Beach, began in 1990. These sites are both relatively large, comprising about 1,500 and 100 acres respectively, and are two of the few remaining large undeveloped areas along the fault. As a result, they provided somewhat unique opportunities to gain additional knowledge on the location and character of the NIFZ. The impetus for completing studies at these sites resulted from the owners' desire to develop the property. Although development plans have varied, concepts have primarily focussed on residential and commercial improvements. As would be expected, and in accordance with California state law (the Alquist-Priolo Special Studies Zones Act of 1972), definition of the location and activity of faulting across the sites was needed for development planning and permitting.

The Huntington Beach site is only a few hundred feet from the Pacific Ocean, and topographically consists of two mesas and an intervening tidal lowland. Elevations within the lowland area are generally within a few feet of mean sea level (MSL), whereas the mesas rise to elevations of about +30 to +50 feet MSL. This configuration resulted from the regional uplift of Pleistocene deposits (forming the mesas), which were breached by a broad late Pleistocene fluvial channel (the ancient Santa Ana River) during periods of lowered sea level. In response to subsequent sea level rise, a sequence of Holocene fluvial and estuarine deposits filled the ancient channel. The base of this channel is about 7,000 feet wide, relatively flat, and ranges from about 60 to 80 feet beneath MSL. The NIFZ cuts across both mesas and the intervening tidal lowland.

The Long Beach site occupies a portion of the floodplain of the San Gabriel River, and lies about a mile from the ocean. It is characterized by flat topography, generally ranging in elevation from about +5 MSL to a few feet below MSL. A tidal slough is located along the northwest edge of the site. Subsurface geology at the Long Beach site includes Pleistocene deposits, which were planed off and deeply incised by a late Pleistocene fluvial channel (the ancient San Gabriel River). The ancient channel and adjacent terrain was subsequently backfilled with a series of Holocene fluvial deposits in response to sea level rise. According to Zielbauer (1962) and Lundeen and Roth (1976), the ancient channel is relatively narrow, and its center, which lies near the east side of the site, is about 50 to 60 feet deep. Across much of the Long Beach site, the base of the Holocene deposits is estimated to be about 30 to 40 feet deep. The NIFZ crosses the central portion of the site.

Both sites overlie faulted anticlines which have been, and continue to be, active oil fields. Because of their similar characteristics, these two sites were investigated using a similar approach. This approach used a variety of investigation techniques, which were combined in a "layered" analysis to evaluate the location and age of fault traces.

Investigative Techniques

When planning our investigation of faulting on the mesas, for example at the Huntington Beach site, trenching was the obvious choice. It was recognized however, that for the lowland area at the Huntington Beach site and for the entire Long Beach site, the deep Holocene deposits and shallow depth (a few feet) to groundwater made trenching impractical. If Holocene faulting was to be assessed in these areas, exploration methods other than the "tried and true" trenching method had to be employed.

In evaluating alternative investigation methods, it was recognized that most methods had inherent resolution limitations, and none were considered suitable for use by themselves. As a result, it was decided that our approach had to use several investigation methods, each of which would provide a data set that could be combined in our overall interpretation. By using this approach, the various methods could be used to supplement the weak points of each other, and locally provide corroborative evidence for faulting.

The investigative methods consisted of analysis of oil well E-logs, marine and terrestrial reflection surveys and closely spaced cone penetration tests and borings. Each of these methods provided subsurface data from various depths. As a result, our approach to investigating the two sites might be referred to as "layered" since the data from each method could be stacked over the deeper layer and the combined stack of data helped overcome the limitations of each method. The following paragraphs provide a summary of the methods used in a "bottom-to-top" sequence.

Analysis of Oil Well Electric Logs - Starting at depth, we first examined the location and character of the NIFZ by collecting and reviewing available E-logs from the onsite oil wells. Depending on the site, this included about 100 to nearly 250 E-logs. Using these E-logs, cross sections were developed across the fault zone to identify the subsurface location of its individual traces and their possible vertical extent. This review and analysis focussed on:

1. Refining the previous subsurface structural interpretation, which was completed for oil field purposes.
2. Providing better control on the subsurface location and width of faulting beneath the sites by analyzing the E-log data for depths less than about 2,000 feet (typically, the previous structural interpretations had focussed on deeper depths).
3. Evaluating the possible existence of near-surface faults other than the main trace.
4. Correlating fault locations and characteristics with the findings from the other investigation methods.

Terrestrial and Marine Geophysics - A series of terrestrial and marine geophysical reflection profiles were completed at both sites. Terrestrial reflection profiling was generally completed along traverses that paralleled the cross sections developed from the oil well E-log analyses. Reflecting horizons apparent on the records generally ranged from about 400 to 600 feet.

Another somewhat unique aspect of the two sites was that there were shallow-water channels, either on or adjacent to the sites, where marine geophysical reflection surveys were completed. At the Long Beach site, the Cerritos Channel provided a means to collect these data. At the Huntington Beach site the marine data were collected within tidal and flood control channels, where water depths were only about 2 to 3 feet. In these shallow-water-depth cases, small-scale equipment was used that was adjusted to provide optimal signal return. Reflecting horizons apparent in the marine survey data were generally in the depth range of about 100 to 250 feet.

Cone Penetration Tests (CPTs) and Continuous Core Borings - To provide shallow data (in the upper 100 feet), a series of CPTs and borings were completed at each site. These were completed along linear arrays, which crossed roughly perpendicular to the suspected fault traces. The CPTs were completed first, at a spacing of about 100 feet. Additional CPTs were subsequently added between the original locations to reduce the spacing as needed to allow correlation of subsurface units and to refine the locations of fault traces. Having completed most or all of the CPT program, several continuous-cored borings were advanced at each site to provide physical samples that could be correlated with the signatures obtained from the CPT records. In addition, in many cases, the borings provided samples of organic material for radiocarbon dating.

Aerial Photo Review and Trenching - Faulting at the surface at both sites was investigated through the examination of aerial photographs, and by trenching on the mesas at the Huntington Beach site.

Investigation Results

Analysis of Oil Well Electric Logs - Through our analysis of E-logs, we were able to identify marker beds, which were used for lithologic correlation and the development of a series of interpretive cross sections across the fault at each site. Gaps or elevation discordance in these marker beds on the cross sections were interpreted to be caused by faults. In addition, faults were interpreted to have been penetrated by wells, as recognized either by a repeat in lithologic section, or by missing section. Where more than one lithologic correlation appeared to cross over a suspect fault trace unaffected, that fault was inferred to not project to the surface. Figure 8 provides an example of one of these cross sections.

By examining a series of these sections along the fault, maps were developed showing the geometry of the fault zone at various depths. The results of this analysis provided a greatly improved understanding of the fault zone's geometry. For example, it was noted that at the Huntington Beach site, the fault tended to be a relatively narrow, single trace across the mesas; whereas beneath the lowland area the fault tended to be more complex. In fact, through the E-log analysis, we identified a right step of the main fault trace and several associated splay and antithetic faults (Fig. 9). The E-log correlations also provided data useful in targeting the other investigation efforts of the layered approach.

Terrestrial and Marine Geophysics - The terrestrial and marine geophysical surveys provided data for the zone from about 100 to 600 feet deep, above the zone covered by the E-log correlation. Typical examples of the marine and terrestrial geophysical data are provided in Figures 10 and 11, respectively.

Of the two methods, the marine technique generally provided better definition of subsurface reflecting horizons. However, with both, marker horizons could be identified that extended for several hundreds of feet across the site, apparently unfaulted within the resolution of the data.

At several locations, the geophysical data indicated faulting that coincided well with the projection of faults from below (interpreted from the E-log data). An example of this correlation is shown at the top of the E-log profile on Figure 8. The geophysical data also indicated several "dead zones" where return signals were absent. Several of these dead zones aligned with the projection of faults from depth, and presumably resulted from the shattered character of the sediments along the fault. However, such dead zones are not always attributable to faulting. One such dead zone coincided precisely with the location of a buried erosional channel, which was evident on old aerial photographs.

A significant observation made from the terrestrial data included a characteristic warp (anticlinal limb) in the Pleistocene units north of the main fault trace. South of the fault zone, bedding within the Pleistocene units appeared to be flatter. As shown on Figure 10, the marine data suggested similar warping on both sides of the main fault.

Cone Penetration Tests (CPTs) and Continuous Core Borings - Like E-log data, the CPTs provided a characteristic signature that corresponded to the properties of the subsurface soils. These data were collected along linear arrays so that interpretive cross sections could be developed across areas selected from the results of the oil well E-log correlations and the geophysical surveys. An example of CPT cross sections prepared for the Huntington Beach and Long Beach sites are provided in Figures 11 and 12, respectively. Continuous core borings were used to check the actual subsurface materials against the CPT signatures, as well as to provide the opportunity to confirm the age of sediments through the collection of samples for radiocarbon age determinations.

Core borings are often required for making stratigraphic correlations (and locating faults) because subtle changes in texture or color of the sediments across a fault may not be detected by CPT. In this regard, it is interesting to note that in several instances the CPT data provided characteristic signatures, which could be correlated for significant distances, but for which no cause was visually apparent in core samples retrieved nearby.

Significant findings shown on the interpretive profiles (Figures 12 and 13) include:

- The contact between Holocene and Pleistocene units is often very distinctive in the CPT data due to the lithologic contrast between the Holocene (sand and gravel) aquifer

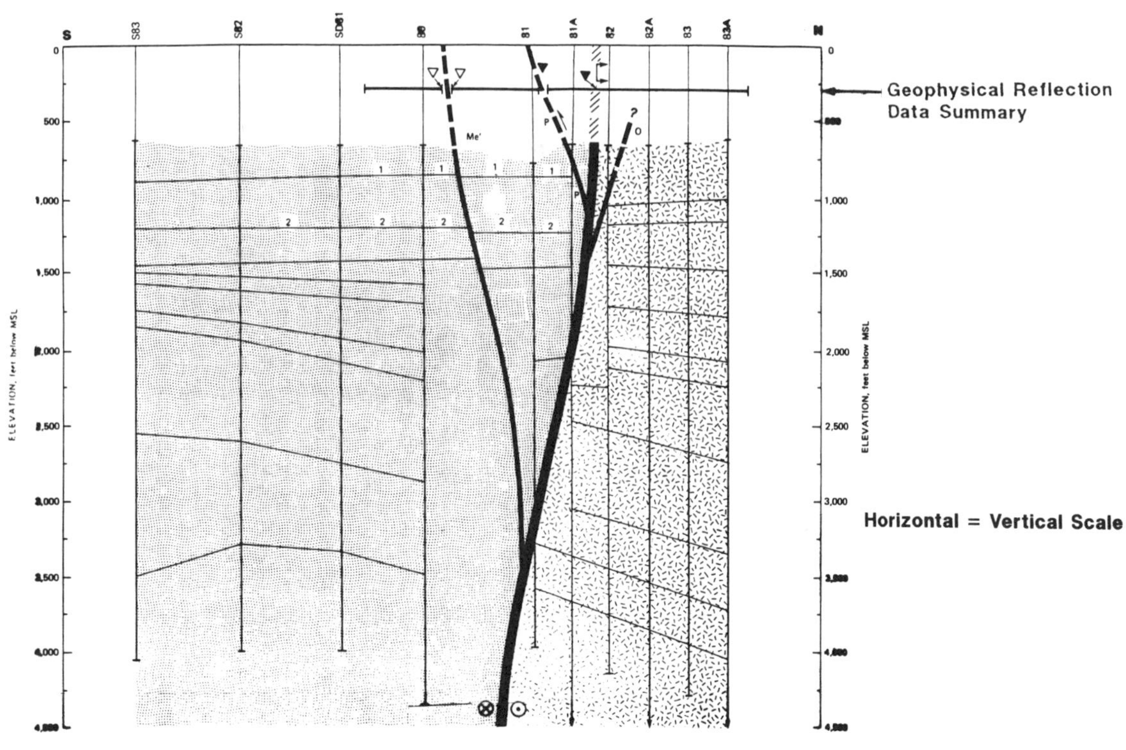

Figure 8. Typical profile showing E-log interpretation and geophysical results

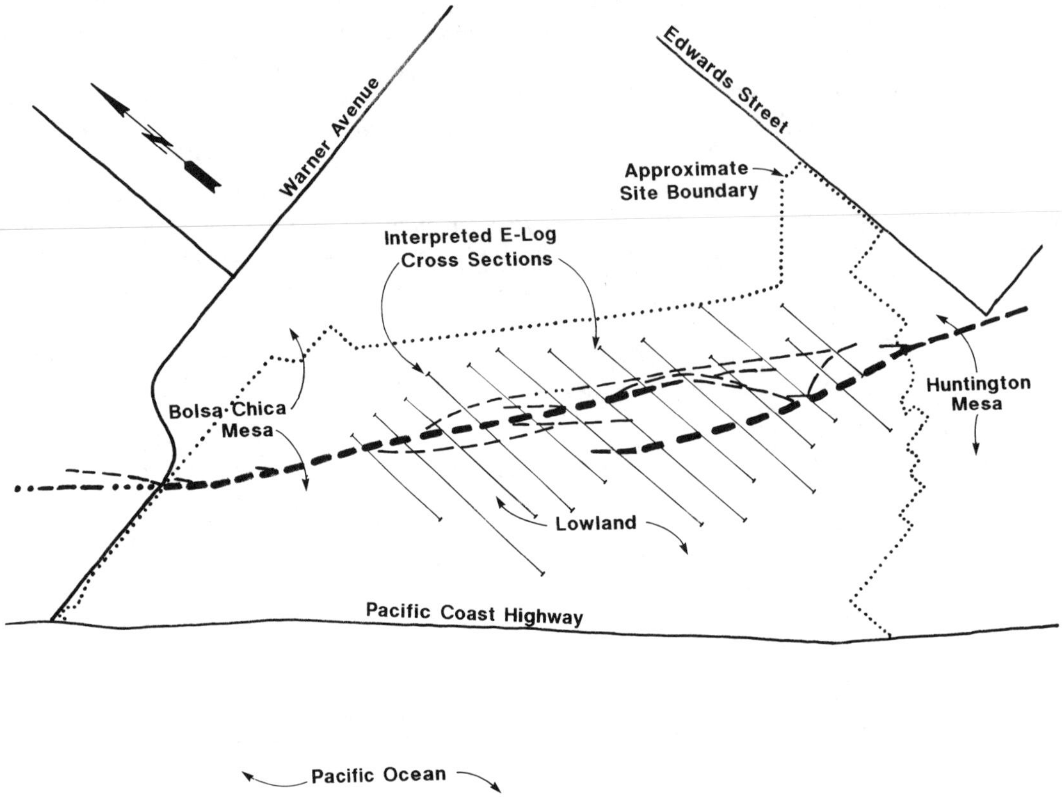

Figure 9. Fault traces across the Huntington Beach site

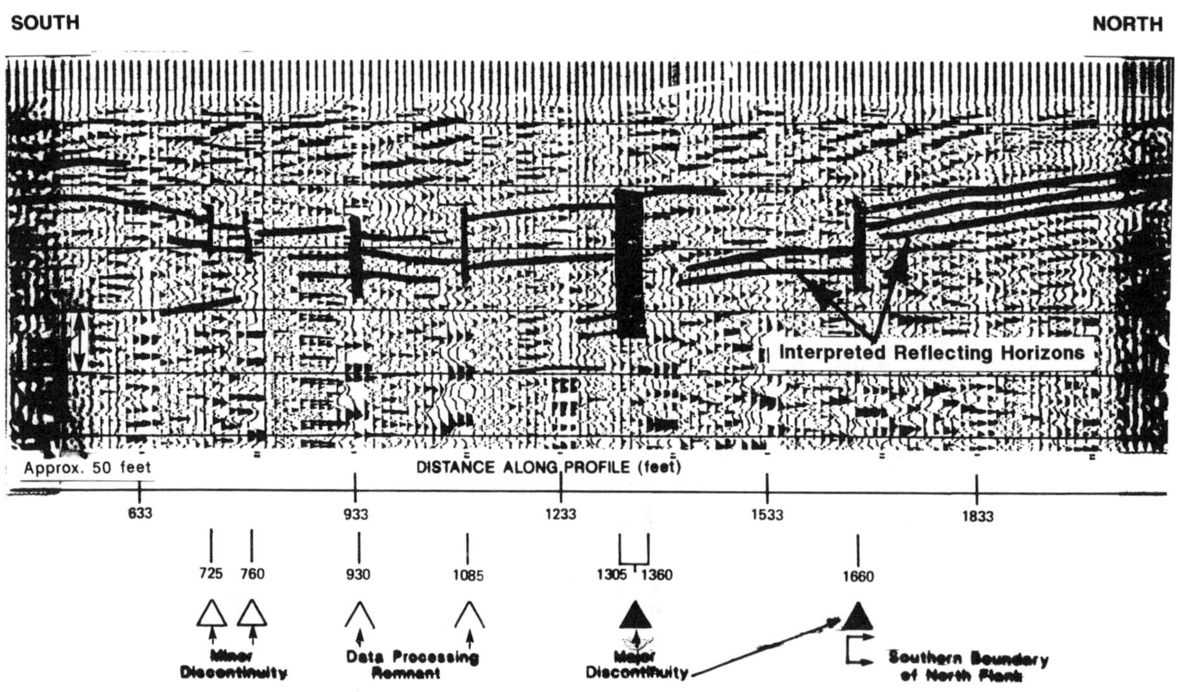

Figure 10. Typical terrestrial geophysical reflection profile

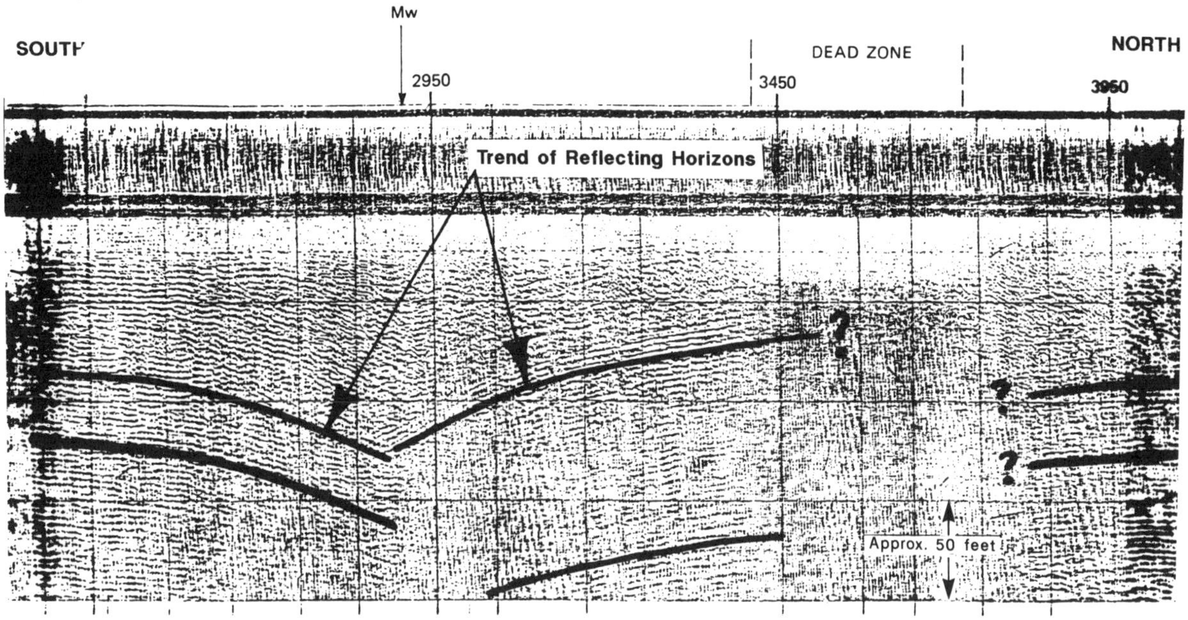

Figure 11. Typical marine geophysical reflection profile

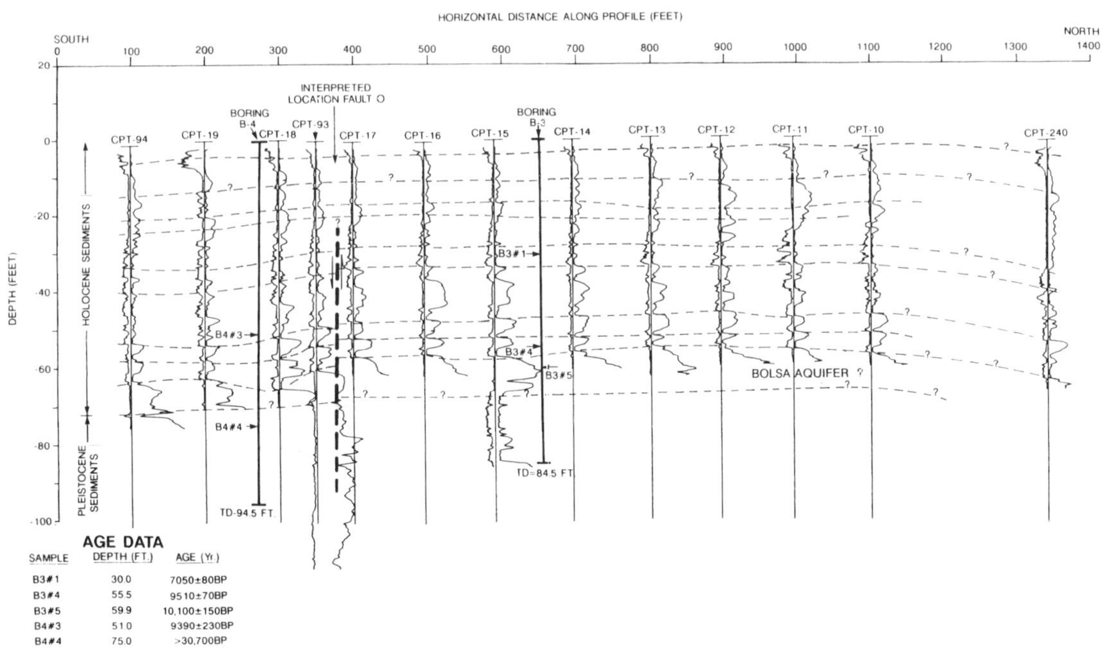

Figure 12. Typical profile of cone penetration data - Huntington Beach site

- and the underlying fine sands and silty sands of the Pleistocene units (as observed on the north side of the section on Figure 13, and at the deeper-penetrating CPTs on Figure 12).

- The broad anticlinal warp in the Pleistocene units observed in the geophysical data on the northeast side of the fault, is also locally evident in the CPT and shallow boring data (see Figure 13).

- A splay fault located north of the main trace of the North Branch fault (at the Huntington Beach site) was detected using the oil well data. Apparent offsets in stratigraphy across the splay fault suggest that its dominant sense of offset is normal. Detailed analysis of large-scale CPT profiles (1 inch = 10 feet; normal scale) across this area indicates normal offset of Holocene units across this splay fault and suggests an increase in vertical separation of units with depth. Based on apparent offsets and radiocarbon dates, it appears that there may be evidence for at least 3 large earthquake events (during which surface rupture may have occurred) during the period from about 4,000 to 10,000 years bp. As discussed below, trench data suggests another large event occurred during the last 3,000 to 4,000 years bp. CPT evidence for this event is not currently apparent.

- CPT profiles from the Long Beach site provide clear evidence of offset Holocene units, possibly to within about 10 feet of the ground surface.

It is recognized that lateral offsets across strike slip faults may be difficult to discern when analyzing this type of data. Fortunately, at least at these sites, the lateral offsets produced discernable vertical separations, either due to facies changes or dipping beds, or both. In fact, the fluvial environment (with its cut/fill structures) at the Long Beach site (Fig. 13) proved to be quite illustrative of the type of offsets that are typical across strike slip faults.

Aerial Photo Review and Trenching - Subtle tonal lineaments, identified on aerial photographs, across portions of the lowlands of both the Huntington Beach and Long Beach sites, coincided with fault traces that were identified in the subsurface. A trench excavated across one of the mesas in the Huntington Beach area generally penetrated a gray sandy loam "A" soil horizon about 1 to 2 feet in thickness, and a strong red brown argillaceous "B" soil horizon about 3 to 5 feet thick, overlying fine sandy Pleistocene deposits ("C" horizon materials). The "A" soil horizon in this region is noted for its locally rich content of shells and cultural materials, which were dropped by prehistoric habitants of the area. Based on archaeological work by Scientific Resource Surveys, it is our understanding that shells contained in the "A" soil horizon, in the vicinity of our trench, generally yield radiocarbon dates ranging from about 3,000 to 4,000 years bp.

The trench revealed a single main fault trace, with a few associated splay faults. Separations across the splay faults

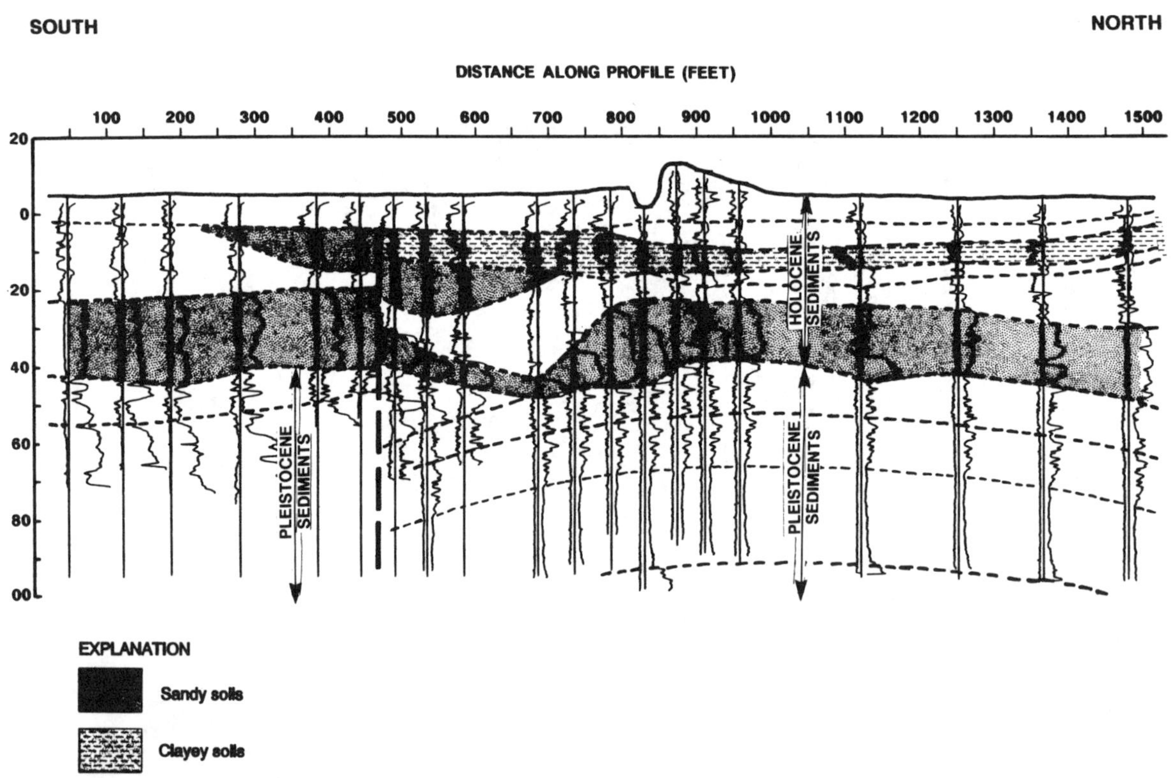

Figure 13. Profile of cone penetration test data - Long Beach site

were typically minor, locally difficult to observe in the "C" horizon, and usually could not be traced very far upward into the "B" horizon. By contrast, the trace of the main fault was very distinct, with a clear lithologic mis-match at the floor of the trench ("C" horizon), and an easily discernable trace upward through the "B" horizon and into the "A" horizon. Vertical separations of about 1.5 and 5 feet were observed on the bases of the "A" and "B" horizons, respectively. Also, soils from the "A" horizon were observed to occur as infilling of a half-foot-wide fissure in the upper "B" horizon. While this is not believed to be evidence of the 1933 Long Beach event, it does confirm that late Holocene surface rupture has occurred along this portion of the fault.

NEWPORT BEACH STUDY SITE

Western Newport Beach is one of the few locations along the NIFZ where offsets in Pleistocene sediments can be observed in outcrop. Investigations of fault exposures were conducted by Guptill and Heath (1981) with hand tools to expose areas of potential faulting. The exposures revealed apparent en echelon faults with offset soils and possible offset man-made fill. These studies resulted in conclusions that the faulting may be wrench style with historical offset however, more detailed investigations were required to confirm the findings. Those additional studies were conducted later but did not confirm the 1981 conclusions.

In 1986, follow-up investigations consisted of a combination of methods including aerial photographic analysis, oil well E-log interpretation, pedogenic analysis, and logging of outcrops and trench exposures in Pleistocene age marine and nonmarine sediments. Two distinct zones of faulting were found on West Newport mesa. However, neither of these structural trends correspond to the main branches of the NIFZ, which based on oil well data lies southwest of the property at the Santa Ana River mouth. Evidence discovered during the 1986 study indicate the observed faulting is secondary normal faulting in response to localized extension adjacent to a change in trend of the NIFZ.

Ages of latest faulting were established utilizing pedogenic soil stratigraphy, rates of soil development, and correlation of morphostratigraphic units to world-wide marine oxygen-isotope stages. At least three stages (Stages 5, 7, and 9) of high sea level are represented at the site. Older sediments are offset generally more than the younger sediments and soils. At no location do faults penetrate the entire column of soil, which is Pleistocene in age. No evidence of Holocene activity is present in the soils exposed in a trench across the same fault trace investigated in 1981. The more detailed methods of study and greater amount of data refute postulated evidence of historical activity reported in earlier work.

Site Description

The West Newport Oil Field, located just north of Pacific Coast Highway between the Santa Ana River to the west and Superior Avenue to the east (Figure 14) encompasses approximately 500 acres. Subsurface faults that cut the oil producing "B" horizon are shown on Figure 14. The oil field consists of a lowland area and an elevated mesa area.

This study focused on the southwestern mesa portion of the oil field. This mesa is a terrace surface (predominately marine) that rises abruptly above the surrounding lowland and the coastline. The bluff face separating the lowland and the mesa surface is a steep erosional feature that provides numerous exposures where stratigraphic and structural relationships trending beneath the mesa surface can be observed.

Methods of Study

Guptill and Heath (1981) investigated the artificial and natural exposures of the Pleistocene sediments and reported several minor offsets of marine sediments as well as the overlying soils. At one location, one of the fault traces coincided with an apparent offset in fill overlying the soil horizon.

The more detailed investigations conducted in 1986 utilized conventional office and field methods consisting of aerial photograph analyses and reconnaissance mapping of the mesa to locate faults and develop an understanding of the site stratigraphy. Trench sites were selected and trenches were excavated by backhoe. Soil units logged in detail using U.S. Soil Conservation Service Soil Taxonomy guidelines where useful for evaluating ages of soils with respect to faults. Twenty-one exposures consisting of 3400 feet of trenches and cut slopes were logged providing nearly continuous exposures across the width of the central and southern portions of the mesa. The steep bluff face that forms the northern, western, and southern edges of the mesa also provide excellent continuous exposures across most of the property. With guidance by Dr. Roy Shlemon, ages of soils (pedogenic units) were based on the relative ages of the geomorphic surfaces, soil stratigraphy, and relative degree of pedogenesis (soil development).

Geomorphology

Geomorphology of the mesa was evaluated by stereoscopic analysis of aerial photographs and field reconnaissance of the site. The aerial photographs, obtained mainly from the Whittier College collection, were taken in the years 1927, 1928, 1953, 1969, and 1980. Two distinct geomorphic surfaces occur on the mesa within the limits of the site. The highest elevation, and older geomorphic surface, occupies the northeast portion of the site. This relatively flat surface ranges in elevation from 75 to 105 feet above mean sea level rising to the east a approximately 20 feet per 1000 feet. This geomor-

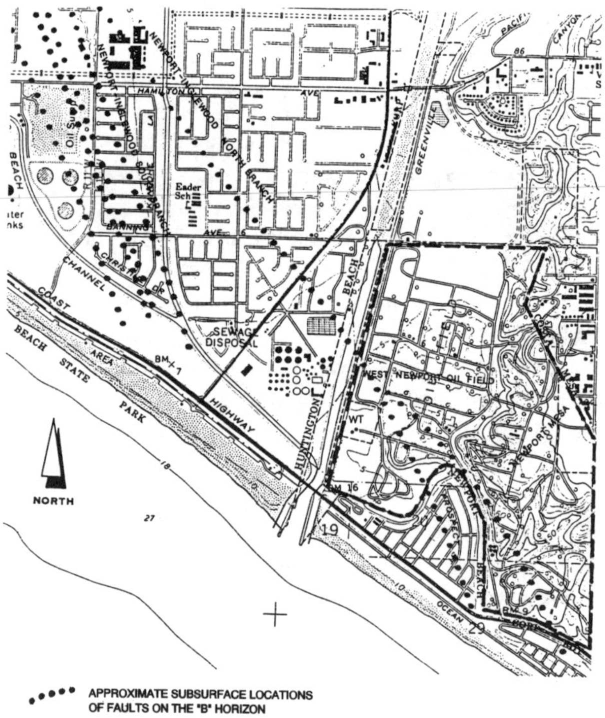

Figure 14. Location map

phic surface is tentatively correlated with marine oxygen isotope Stage 7 as described later.

A lower elevation, and hence younger, geomorphic surface forms the mesa surface in the southwestern portion of the mesa. The surface is relatively flat and ranges in elevation from 50 to 70 feet above mean sea level, rising to the northeast at approximately 10 feet per 1000 feet. The lower geomorphic surface is tentatively correlated with marine oxygen-isotope Stage 5. A gentle, yet distinct, south-facing escarpment with several feet of relief forms the topographic break between the higher and lower geomorphic surfaces on the meas (Figure 14). Because this escarpment superficially could appear fault controlled, an important task for this study was to place trenches across the feature to evaluate its origin. A trench excavation across the escarpment revealed an erosional shoreline angle. Although surface faulting at the site had been postulated previously (Guptill and Heath, 1981), no geomorphic evidence of young displacement (such as scarps) was observed associated with faulting exposed in trenches. The two geomorphic surfaces of the mesa appear to be undisturbed by faulting since the mesa was formed and uplifted.

General Stratigraphy

To evaluate the latest age of fault movement on the site, it was necessary to develop an understanding of both the local and regional stratigraphy. Characterization of the stratigraphy is based on analyses of the sediments exposed along the mesa escarpment, in trench excavation and in cut-slope exposures. These analyses included discussions in the field with Dr. Roy Shlemon, who assisted with the correlations of the sediments, soils, and geomorphic surfaces (marine terraces) to estimate their relative ages. these were estimated mainly by correlating stratigraphic and geomorphic relationships on the meas with regional marine terraces and the world-wide, marine, oxygen-isotope stages. Regional marine terraces have been correlated and assigned relative ages in accordance with the marine isotope stages of Shackleton and Opdyke (1973).

The isotope stages defined by Shackleton and Opdyke (1973) represent fluctuations in sea level caused by major global climatic changes corresponding to glacial and interglacial periods. The present day or Holocene interglacial period is Stage 1 and the past glacial period is Stage 2, etc., such that odd-numbered stages represent high sea level stands and even-numbered stages represent low sea level stands. In coastal Orange County, California, the lowest elevation marine terrace usually correlates with the Stage 5 high stand of sea level, which occurred approximately 80,000 to 120,000 years before present (Wehmiller and others, 1977). Stage 3 never reached present day sea level and thus, if preserved, is submerged offshore.

The lowest elevation platform at West Newport is tentatively assigned to the Stage 5 interglacial period. Older marine sediments upon which the Stage 5 platform is cut at Newport Mesa are assigned progressively earlier interglacial periods, namely Stage 7 (approximately 200,000 years b.p.) and Stage 9 (approximately 300,000 years b.p.), when sea levels approached present day elevations. We refer to the sediments associated with the various stages as Stage 5 sediments, Stage 7 sediments, etc., and, the terrace surfaces as the Stage-5 terrace surface and the Stage-7 terrace surface. Although these terrace surfaces are most likely younger than the stages assigned because they are deposited on the wave-cut platforms. Figure 15 depicts schematic relationships of the stratigraphic units at Newport Mesa.

Stratigraphic relationships observed in cut slopes along the east side of the site indicate that at least one fluvial channel incised Stage-5 deposits and others incised Stage-7. The sediments in the channel are either contemporaneous or younger than the respective age of incised sediments (i.e., Stages 5 or 7).

Soil Horizons

Cumulic soil profiles up to 12 feet thick and consisting of more than one horizon have developed on both the Stage-5 and Stage-7 terrace surfaces. The physical character of the soils is primarily a function of the nature of the underlying parent material as well as the stability of the surface they formed upon combined with climatic changes. Because the

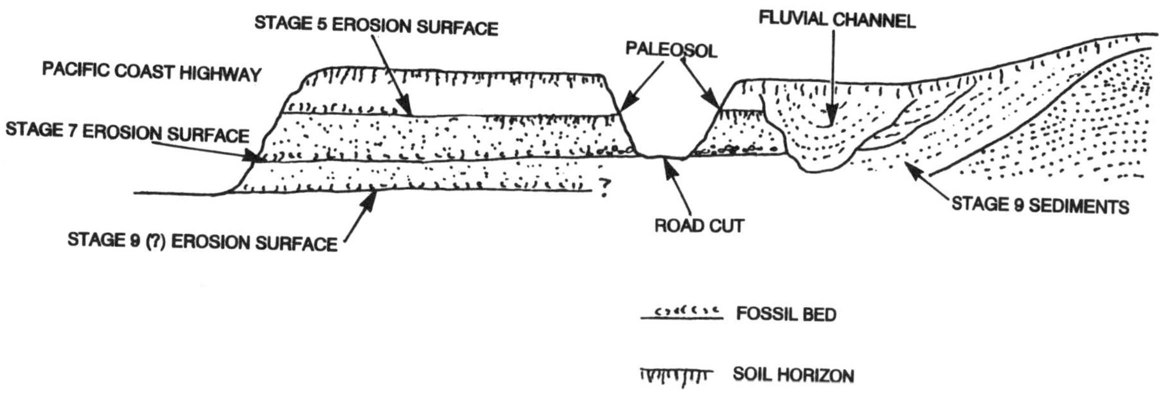

Figure 15. Schematic cross section West Newport mesa

sediments at the site are primarily fine-grained, and are exposed to salt air, soils have formed relatively rapidly and are generally well developed where not eroded. Most of the soils logged were developed on the Stage-5 sediments, and are younger than Stage-5 (or less than 80,000 to 120,000 years old) in that they developed after the high sea levels of Stage 5.

Within the cumulic soils, textural changes and parent material differences indicate episodic deposition, suggesting that the soils have been subject to several climatic changes and thus are pre-Holocene age. Except for local channels, most of the mesa surface is undissected. Accordingly, most soils probably began to form after cessation of deposition during late Stage-5 time. Thus, the soils on the lower mesa reflect approximately 80,000 years or more of weathering affected by several climatic cycles, and probably were most strongly influenced by the climate during Stage 3, about 40,000 years ago.

Local Structural Setting

The local geologic structure is best understood by interpreting the great amount of subsurface data now available for the West Newport Oil field. By correlating distinct electric log horizons, and interpreting oil-water interfaces, the subsurface structure at the oil field is found to consist of two primary faults and numerous associated smaller faults such as the Banning fault. Locations of offset Pleistocene deposits iden-

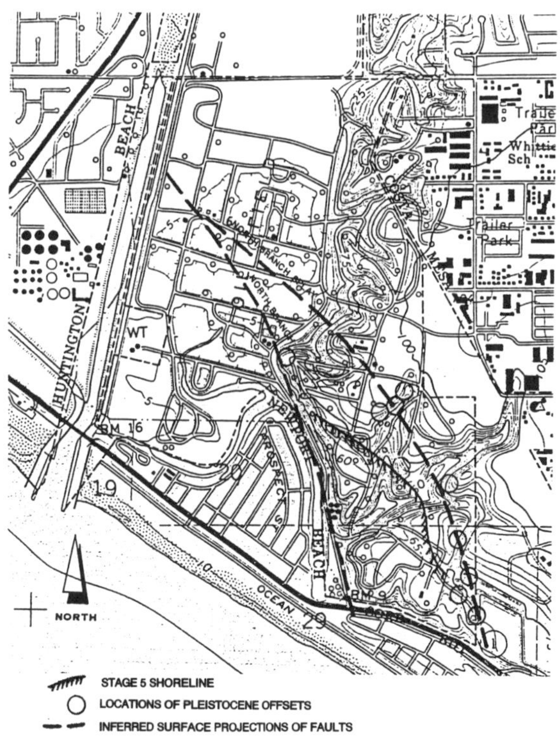

Figure 16. Inferred fault locations West Newport mesa

tified during this study are shown as fault trends or branches (Fig. 16).

The North Branch fault offsets subsurface Miocene-age horizons by 400 to 450 feet of apparent dip-slip separation (down to the west). The North Branch Splay fault appears to splay from the North Branch near the west side of the oil field and offsets Miocene horizons by only about 70 to 75 feet of dip-slip separation. Surface projections of both faults are illustrated on Figure 16 to make it possible to compare associations of faulting identified in trench excavations with the subsurface structure. The faulting exposed on the mesa approximately correlates with the surface projections of the North Branch fault and the North Branch Splay fault as shown on Figure 16. Two general zones of faulting are exposed on the mesa, one trending north northwest from the southeastern corner of the property and the second trending north northwest but intersecting only a small northwestern corner of the mesa.

As shown on Figure 14, the trend of the South Branch is more southerly (N25W) in the vicinity of Newport Mesa. Along a northwest-trending dextral strike-slip fault, the southerly curvature results in an area of tension across the curvature (or right step). That geometry should give rise to local normal dip-slip movement, comparable to that which is observed both on the North Branch and North Branch Splay faults. We believe that the North Branch and North Branch Splay faults are secondary, predominantly normal dip-slip faults that developed in response to the local change in strike of the South Branch Fault (Figure 14). The general tensional environment across the mesa is reflected in the near surface apparent dip-slip separation observed in trenches and in the bluff exposures.

Character and Age of Faulting

The older sediments (Stage 7 to Stage 9) are offset generally more than Stage-5 sediments, which are offset more than soils suggesting periodic displacement on the faults through time. The largest offset observed in Stage-7 to Stage-9 sediments was approximately 12 inches. In some areas offsets of the base of soil horizons are on the order of a few inches. At no location do faults penetrate through the entire column of soil, which is Pleistocene in age.

We conclude that the faults on site have not moved in Holocene time based on the field inspections and trench logging. In order to test this conclusion the fault trace identified by Guptill and Heath (1981) as having potential evidence of historic slip was trenched a short distance away along strike. The magnitude of offsets of Pleistocene-age cumulic soil horizons diminished from the base of the trench to 2 feet below the top. The faulting terminated below a horizon within the Pleistocene soil, which does not support the notion of historical surface faulting a few feet away.

CONCLUSIONS

Characterizing the NIFZ in terms of its geoseismic parameters to facilitate seismic hazards assessments is fraught with difficulties and uncertainties. In the subsurface, the en echelon stepovers, splay faults, and close associations with anticline structures complicate identifying its active traces. This can lead to unresolved questions about its potential future rupture lengths, and makes it difficult to determine its slip rate for estimation of future seismic displacements. On the surface, mesas and hills, which are underlain by Pleistocene sediments, offer the opportunity to observe traces of the fault in outcrop. But, a paucity of Holocene sediments limits the ability to study the geologic record for its late quaternary displacement history. Although thick sections of Holocene deposits are found crossing the fault in paleo-channels of the Los Angeles, San Gabriel, and Santa Ana Rivers, trenching for geoseismic data in these deposits is restricted by their loose, water-saturated character. Compounding these natural limitations, dense urban development over the majority of the NIFZ's trace limits available sites for investigations. In spite of these limitations, we can estimate the NIFZ's geoseismic characteristics by utilizing a variety of indirect investigation methods such as comprehensive geoseismic analysis of the E-log database (particularly at higher elevations), high resolution marine and terrestrial geophysics, closely spaced CPT and borings, and where possible, trenches. Each of the methods has inherent limitations but the weight of the combined data at any given location can overcome these limitations. Further, the combination of detailed investigations utilizing a broad database provides a more complete understanding of the origin, age, and characteristics of fault. These techniques were successful in delineating the near surface trace of the fault through Holocene sediments, and that evidence strongly suggests the occurrence of past earthquakes large enough to cause surface displacement. Using these approaches, evidence strongly supports a long-term slip rate of 0.5 mm/yr, beginning in Miocene time. Data from more recent sediments suggest a return period of about 2,000 to 3,000 years for large events, which is consistent with a relatively low rate of slip. At the Newport Mesa site, the combination of geomorphology, soil stratigraphy, correlation of sediments to world-wide isostatic sea levels, and subsurface structural interpretation had to be interpreted together to develop a viable understanding of the significance of observed faulting.

REFERENCES

Barrows, A.G., 1974, A review of the geology and earthquake history of the Newport-Inglewood Structural Zone, Southern California: California Div. of Mines and Geology, Special Report 114, 115 p.

Brune, J.N., 1968, Seismic moment, seismicity and rate of slip along major fault zones: Jour. of Geophysical Research, vol. 73, no. 2, p. 777-784.

Butzer, K.W., 1974, Geological and Ecological Perspectives on the Middle Pleistocene: Quaternary Research, vol. 4, p. 136-148.

Chinnery, M., 1978. Investigations of the seismological input to the safety design of nuclear power reactors in New England: U.S. Nuclear Regulatory Commission, NUREG/CR-0563.

Cluff, L.S., 1978. Geologic considerations for seismic microzonation: in Proceedings of the second international conference on microzonation: vol. I, San Francisco, Nov. 26 - Dec. 1, 1978, p. 135-152.

dePolo G.M., and D.B. Slemmons, 1990. Estimation of earthquake size for seismic hazards in neotectonics in earthquake evaluation, Krinityski, E.L. and Slemmons, D.B. (eds.), GSA Reviews in Engineering Geology, vol. VIII, p.1-28.

Guptill, P.D., and Heath, E.G., 1981, Surface Faulting Along the Newport-Inglewood Zone of Deformation: California Geology, p. 142-148.

Harding, T.P., 1973, Newport-Inglewood trend, California - an example of wrench style deformation: American Association of Petroleum Geologists Bulletin, vol. 57, no. 1, p. 97-116.

Harding, T.P., 1974, Petroleum Traps Associated with Wrench Faults: American Association of Petroleum Geologists Bulletin, vol. 58, no. 7, p. 1290-1304.

Harrison, J.C., R.E. Von Huene and C.E. Corbato, 1966. Bouguer gravity anomalies and magnetic anomalies off the coast of southern California: Jour. of Geophysical Research, vol. 71, p. 4921-4941.

Hauksson, E. and S. Gross, 1991, Source Parameters of the 1933 Long Beach Earthquake: Bull. Seismological Society of America, vol. 81, no. 1.

Heath, E.G., 1954, Geology along the Whittier fault north of Horseshoe Bend, Santa Ana Canyon, California: Master's thesis, Claremont Graduate School, 84 p.

Heath, E.G., D.E. Jensen and O.W. Lusesh, 1982, Style and age of deformation on the Chino fault: Neotectonics in southern California, Field Trip Nos. 3, 4 and 14.

Hill, M.L., 1954, Tectonics of faulting in southern California: in Jahns, R.H., (ed.), Geology of southern California: California Div. of Mines and Geology Bulletin 170, p. 5-13.

Hill, L., 1971, Newport-Inglewood zone and Mesozoic subduction, California: Geological Society of America Bulletin, vol. 82, p. 2957-2962.

Hunter, A.L., and Allen, D.R., 1957. Recent developments in West Newport Oil Field: Summary of Operations, California Oil Fields, California Division of Oil and Gas, vol. 42, no. 2, p. 11-18.

Lundeen, E.W. and N.N. Roth, 1976, Engineering Geology Report, Alamitos Barrier Project: Los Angeles County Flood Control District, Materials Engineering Division, 23 p.

Kanamori, H. and D.L. Anderson, 1975, Theoretical basis of some empirical relations in seismology: Bull. of the Seismology Society of America, vol. 65, p. 1073-1095.

Knapp, R.R., J.D. Traxler, T.J. Newbill, D.J. Laughlin, R.D. Stewart, E.G. Heath, H.E. Stark, S.G. Wissler and W.H. Holman, 1962, Cenozoic correlation section across Los Angeles basin from Beverly Hills to Newport, California: American Association of Petroleum Geologists, Pacific Section.

Matsuda, T., 1975, Magnitude and recurrence interval of earthquake from a fault (in Japanese): Earthquake, Series 2, vol. 28, p. 269-283.

Matsuda, T., 1977, Estimation of future destructive earthquakes from active faults on land in Japan: Jour. of Geophysics of the Earth, vol. 25, supplement, p. S251-S260.

Molnar, P., 1979. Earthquake recurrence intervals and plate tectonics: Seismological Society of America Bull., vol. 69, no. 1, p. 115-133.

Nardin, T.R., and T.L. Henyey, 1978, Pleistocene-Pliocene Diastrophism of Santa Monica and San Pedro Shelves, California Continental Borderland: American Association of Petroleum Geologists Bulletin, vol. 62, no. 2.

Packer, D.R. and L.S. Cluff, 1978, Comparison of activity of late Cenozoic faults int eh western Sierra foothills, California, with other active faults (abs.): Earthquake Notes, vol. 49, no. 1, p. 89-90.

Shackleton, N.J., and Opdyke, N.D., 1973. Oxygen Isotope and paleomagetic stratigraphy of Equatorial Pacific Core V-28-238: Oxygen Isotope Temperatures and Ice Volumes on a 10 and 100 Year Scale, Quaternary Research, vol. 3, no. 1, p. 39-55.

Slemmons, D.B., 1977. State-of-the-Art for assessing earthquake hazards in the United States, Report 6, Faults and Earthquake Magnitude: U.S. Army Corps of Engineers, Waterways Experiment Station, Vicksburg, MI, Miscellaneous Paper S-73-1, 129 p.

Slemmons, D.B., 1982, Determination of design earthquake magnitudes for microzonation: 3rd International Earthquake Microzonations Conference Proceedings, 28 June - 1 July, Seattle, WA, vol. 1, p. 119-130.

Smith, W.S., 1976, Determination of maximum earthquake magnitude: Geophysical Research Letters, vol. 3, p. 351-354.

Wehmiller, J.F., Lajoie, K.R., Kvenvolden, K.A., Peterson, E., Belknap, D.F., Kennedy, G.L., Addicott, W.O., Vedder, J.G., and Wright, R.W., 1977, Correlation and chronology of pacific coast marine terrace deposits of continental United States by fossil and amino acid stereochemistry-technique, evaluation, relative ages, kinetic model ages and geologic implications: U.S. Geological Survey, Open File Report 77-680, Figure 1B.

Woodford, A.O., J.E. Schoellhamer, J.G. Vedder and R.F. Yerkes, 1954. Geology of the Los Angeles basin: Geology of southern California, California Division of Mines Bulletin 170, Chap. II, p. 65-81.

Woodward-Clyde Consultants, 1979, Report of the evaluation of maximum earthquake and site ground motion parameters associated with the offshore zone of deformation, San Onofre Nuclear Generating Station: prepared for Southern California Edison.

Wright, T.L, S. Parker and R.C. Erickson, 1973. Stratigraphic evidence for the timing and nature of late Cenozoic deformation in the Los Angeles Basin: American Association of Petroleum Geologists, Anaheim, California.

Wright, T.L., 1991, Structural geology and tectonics evaluation of the Los Angeles Basin, California, in Biddle, Kautnti (ed.) Active Margin Basins, Am. Assoc. Petrol. Geologists, Memoir 52, p. 35-134.

Yeats, R.S., 1973, Newport-Inglewood Fault Zone, Los Angeles Basin: Bull. American Association of Petroleum Geologists, vol. 57, p. 117-135.

Yerkes, R.F., T.H. McCulloh, J.E. Schoellhamer and J.G. Vedder, 1965. Geology of the Los Angeles Basin - an introduction: U.S. Geological Survey Professional Paper 420-A, 57 p.

Zielbauer, E.J., H.A. Kues, W.L. Burnham, and A.G. Keene, 1962. Dominguez Gap Barrier Project Geologic Investigation: Coastal Basins Barrier and Replenishment Investigation: Los Angeles County Flood Control District, no. 33

TABLE 1

SOUTHERN CALIFORNIA STRIKE SLIP FAULT ZONES
CHARACTERISTICS AND RANKING CRITERIA

FAULT ZONE CHARACTERISTICS	SAN ANDREAS	SAN JACINTO	WHITTIER-ELSINORE	NEWPORT-INGLEWOOD
Dimensions and Segmentation	Total Length - 1300 km Imperial-Cerro Prieto segment: 180 km (Imperial Valley to Gulf of California) Southern Segment: 225 km (Cajon Pass to Imperial Valley) Central Segment: 330 km (Parkfield to Cajon Pass) Creep Segment: 135 km (Hollister to Parkfield) Northern Segment: 435 km (Cape Mendocino to Hollister)	Total Length - 260 km Loma Linda-Claremont: 97 km Casa Loma-Clark: 126 km Coyote Creek: 60 km Superstition Mountain: 50 km Superstition Hills: 53 km	Total Length - 260 km Whittier: 42 km Chino: 32 km Eagle-Glen Ivy: 43 km Wildomar-Elsinore: 160 km	Total Length - 70 km Various segment lengths: 6.5 to 36 km
Total Displacement	300 km (Miocene to Cretaceous)	24 km (Pliocene)	8 to 13 km (Tertiary)	3 to 3.5 km (Upper Miocene)
Distance From San Andreas Fault (Plate Boundary)	0 km	0 to 48 km	40 to 80 km	62 to 70 km
Historic Rupture Length	435 km (Northern Segment)	33 km (Coyote Creek)	N/A	30 km (Southern Segment)
Historic Displacement	6.1 m	0.38 m (Coyote Creek)	N/A	85 to 120 cm (Hauksson and Gross, 1991) 31 to 46 cm (Woodward-Clyde, 1979)
Continuity and Geomorphic Features	Great Continuity Long linear surface scarps, numerous traces; traces suggest great continuity; sag ponds offset streams and topography	En echelon Segments Strong linear trends in young alluvium; water barriers; sag depressions; offset streams and topography; linearity and continuity not as pronounced as San Andreas	En echelon segments Linear scarps; offset alluvial fans and streams fault trace vanishes frequently in younger sediments; sag depressions	Discontinuous en echelon segments En echelon large folds and local linear fault scarps.
Historic Seismicity	Very high	Very high	Moderate	Moderate
Maximum Historic Magnitude, M_s	8.2 (1957)	6.7 (1968 Coyote Creek) 7.1 (1940 Imperial)	5.5 to 6 (1910)	6.3 (1933)
Geologic Slip Rate	24 to 35 mm/yr	8 to 14 mm/yr	1.2 to 5 mm/yr	0.5 mm/yr (Woodward-Clyde, 1979)

File: nif-1.tbl

TABLE 2

LIST OF FAULTS SHOWN ON FIGURES 6 AND 7

Reference Number	Fault Name
1	San Andreas (northern)
2	San Andreas (central)
3	San Andreas (southern)
4a	San Jacinto fault zone
4b	San Jacinto fault - Coyote Creek segment
5	Elsinore
6	Whittier
7	Newport-Inglewood
8	Calaveras-Paicines (south of Hayward branch)
9	Calaveras-Sunol (north of Hayward branch)
10	Hayward
11	Antioch (and Vaca)
12	San Gregorio
13	Fairweather
14	Motagua, Guatemala
15	Bocono, Venezuela
16	Hope, New Zealand
17	Awatere, New Zealand
18	West Wairarapa, New Zealand
19	North Anatolian, Turkey
20	Sumatra
21	Jordan-Dead Sea
22	Kopet-Dagh, Iran-USSR
23	Dasht-E-Bayaz

THE SANTA MONICA AND HOLLYWOOD FAULTS AND THE SOUTHERN BOUNDARY OF THE TRANSVERSE RANGES PROVINCE

Richard Crook, Jr.
Consulting Engineering Geologist
93 N. Sunnyside Avenue
Sierra Madre, CA 91024

Richard J. Proctor
Consulting Engineering Geologist
327 Fairview Avenue
Arcadia, CA 91007

SUMMARY

Six locations in western Los Angeles and Beverly Hills were investigated by trenching in 1982 to attempt to determine the location, characteristics and most recent time of seismic activity along the Hollywood and Santa Monica faults. In addition, recent subsurface observations by others at seven other localities are included here to present the latest information on these faults.

The Hollywood fault has been located at 8 subsurface points, in addition to the several scarps that define parts of its surface trace. These localities include two trenches, four boring locations, and two building excavations (Fig. 1).

- Wattles Park trench at Franklin and Sierra Bonita Avenue exhibited several thin, mainly southerly-dipping shears with clay gouge in diorite that are believed to be displaced and reoriented by local landsliding in the hanging wall of the fault.
- An excavation at a roadcut at the east side of Greystone Park displayed steep northerly-dipping shears where sandstone juxtaposes slate at a brecciated clay-rich zone.
- 1981-83 borings for the L. A. Metro Rail subway on Cahuenga Boulevard just north of Hollywood Boulevard revealed brecciated Miocene sedimentary rock over old alluvium (Fig. 2).
- 1991-92 borings for the subway at the north end of Camino Palmero showed diorite over gouge and old alluvium.
- At the north end of La Cienega Boulevard a boring on the south side of Sunset Boulevard was more than 200 feet deep in old alluvium, but diorite crops out on the north side of Sunset (Glenn Brown, pers. comm., 1992).
- A 30-foot difference in water levels exists in adjacent borings for a building foundation just south of Franklin Avenue at Las Palmas Avenue (Richard Slade, pers.

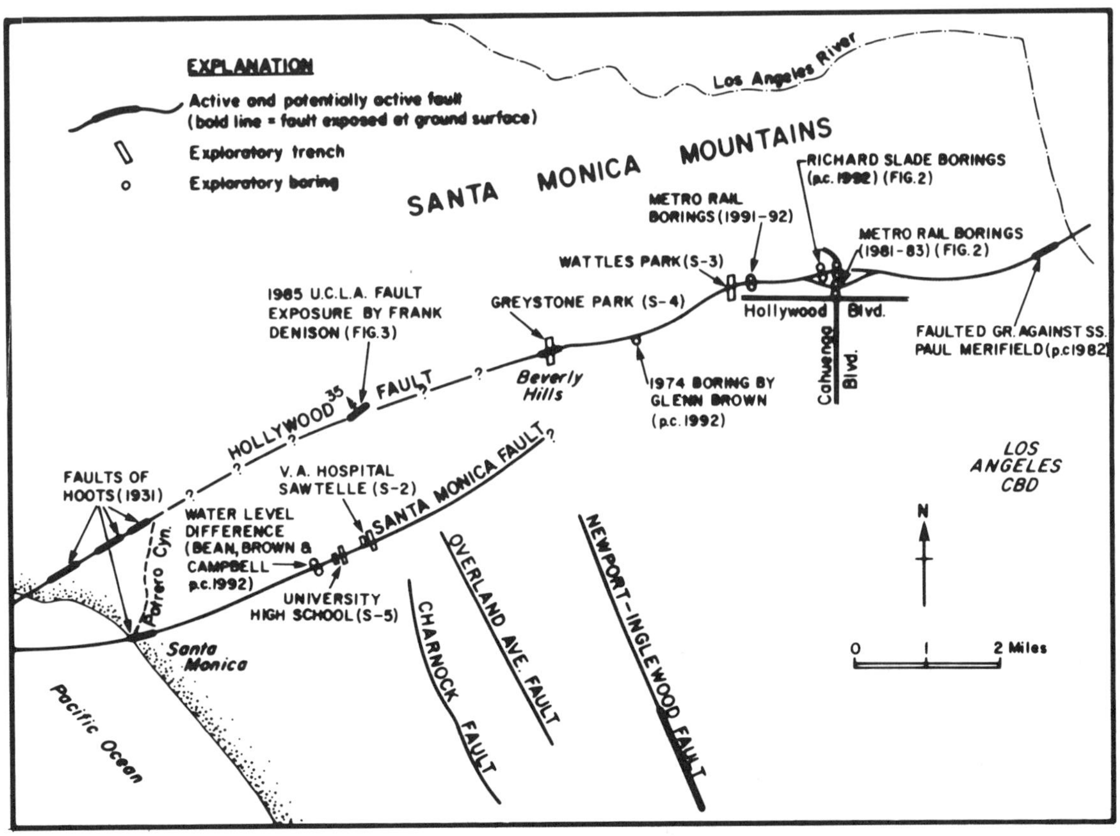

Figure 1. Location map, showing trench and boring locations, and active and potentially active faults.

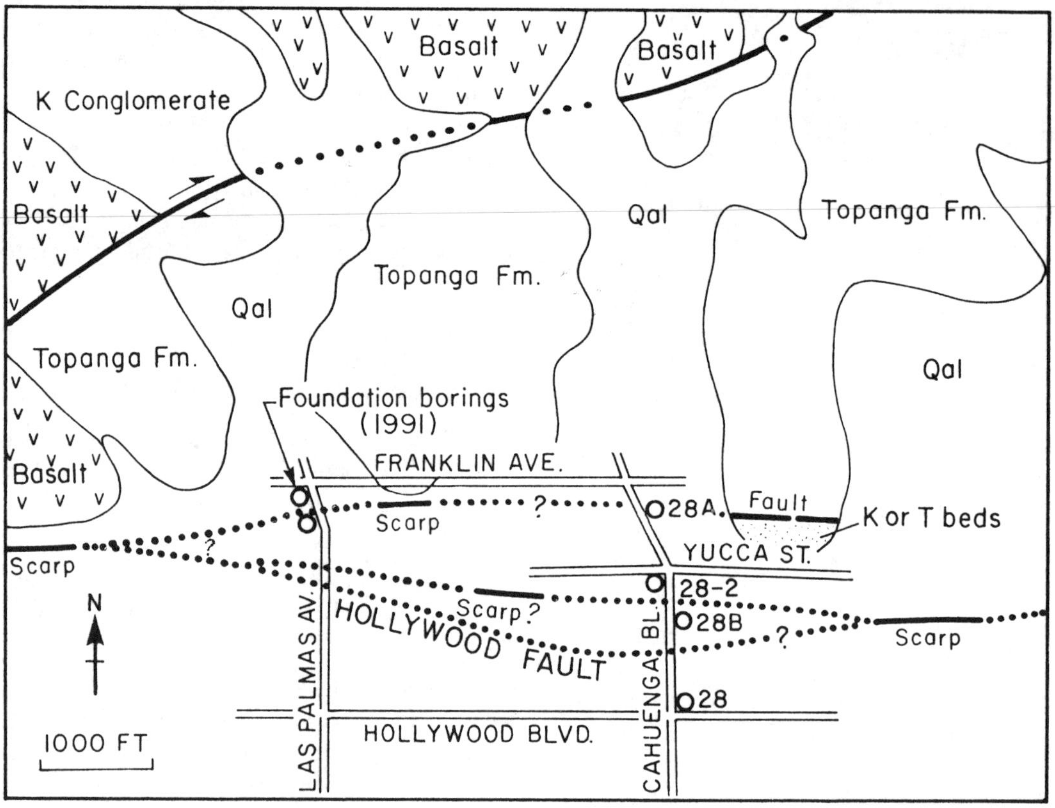

Figure 2a. Geologic map showing location of three branches of the Hollywood fault crossing Cahuenga Boulevard (former Metro subway alignment) based on borings. The borings along Cahuenga are from Converse and others (1981) and the 2 borings near Las Palmas Avenue showed a difference in water levels of 30 feet in borings about 150 feet apart in alluvium (Richard Slade, pers. comm., 1992). Surface geology from Hoots (1931), Weber and others (1980), and Dibblee (1991).

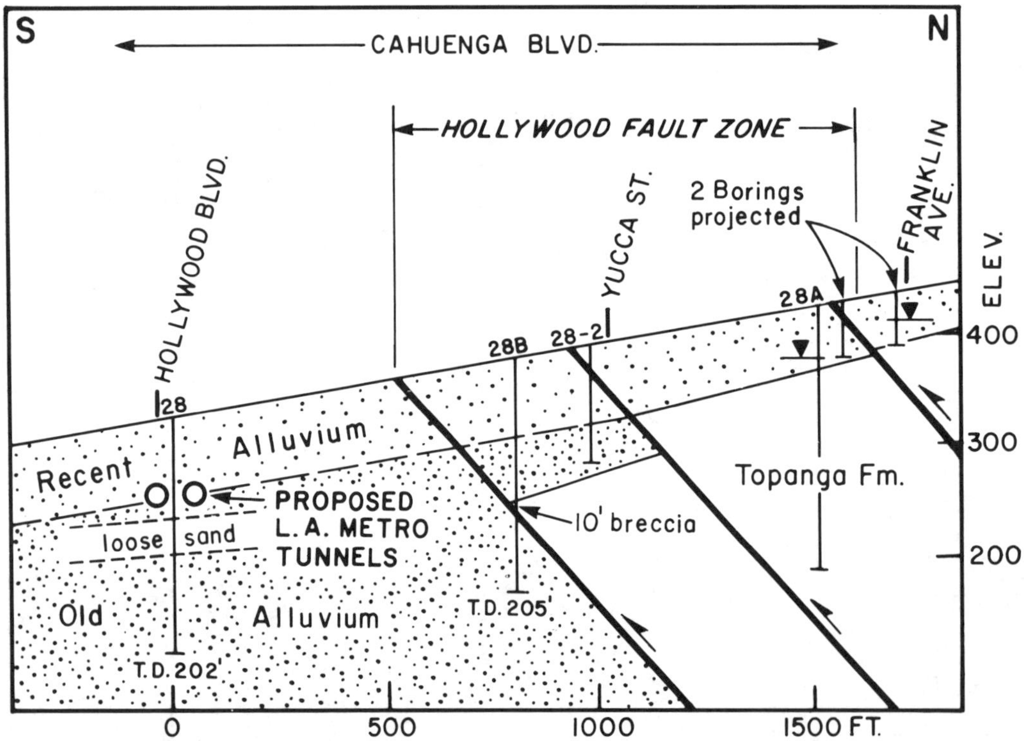

Figure 2b. Geologic section along Cahuenga Boulevard showing two buried Hollywood fault locations from Metro Rail borings, and a third northern branch fault from water level differences in borings projected from near Las Palmas Avenue.

comm., 1992); the northern boring reportedly encountered water at 15 feet and the southern boring at 45 feet (Fig. 3).

- Granite was seen faulted against Puente Formation sandstone in a Los Feliz district residential excavation (Paul Merifield, pers. comm., 1982).
- Frank Denison (pers. comm., 1985) mapped Modelo Formation shale faulted over Pleistocene old alluvium, with the fault striking N60E and dipping an average of 35 degrees north, in a library building excavation at the west part of the UCLA campus north of Gayley Avenue and east of Veteran Avenue (Fig. 3). This exposure is the only hard evidence that a potentially active fault exists in this reach. However, his northeast strike appears to align with a small fault mapped by Hoots (1931, Plate 16), rather than the projected westward trace of the Hollywood fault. We made a long south-trending trench at the north part of the Brentwood Hospital Veterans property, 1000 feet west of UCLA, across a stepped surface of an old fan, but found no fault.

All these localities, except at UCLA, show an 8-mile Hollywood fault segment extending northeasterly from Beverly Hills to the Los Angeles River, defining the south base of the Santa Monica Mountains. West of the west boundary of Beverly Hills, at the mouth of Benedict Canyon, the topography and geology changes to dissected large old alluvial fans, and the fault has not been proven to exist in this 6-mile segment extending to the Pacific Ocean. It should be noted that Hoots (1931) shows faults in three canyons near the coast that seem to line up with a westward extension of the Hollywood fault (Fig. 1).

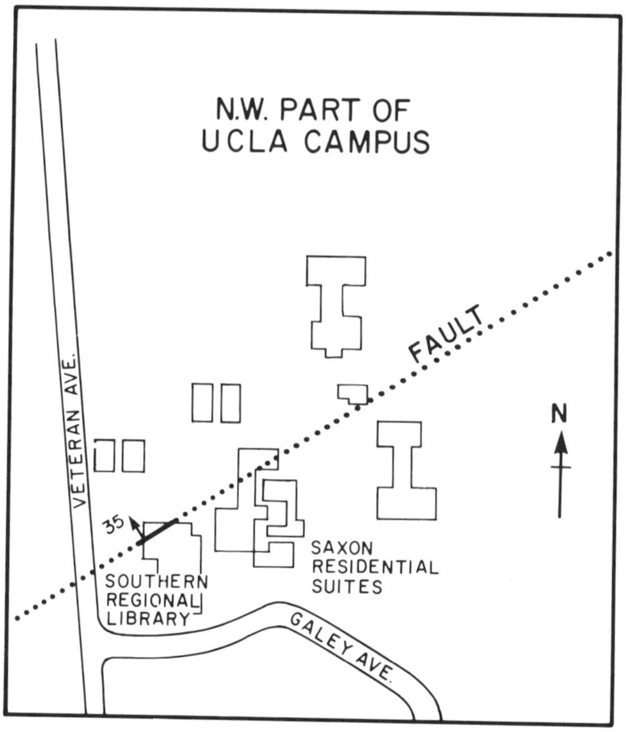

Figure 3a. Location map of western UCLA campus showing (for the first time) the location of the thrust fault exposure in the excavation for the library building mapped by Frank Denison in 1985 (personal communication). Modelo Formation shale overrides old alluvium, with two parallel faults at N60E, 35N. The location of faults in the foundations of adjacent buildings has not been reported.

Recency of displacement on the Hollywood fault could not be absolutely determined, although geomorphic expression (Weber and others, 1980; Dolan and Sieh, 1991) suggests late Quaternary movement.

The Santa Monica fault has several scarps extending 6 miles on-shore from Santa Monica Bay, and is exposed at the mouth of Potrero Canyon north of the City of Santa Monica. It was also exposed in a foundation excavation and by our trenching at two locations:

- A foundation excavation and pumping test at Wilshire Boulevard and Bundy Drive defines a segment of the fault (Robert Bean, Glenn Brown and Alice Campbell, pers. comm., 1992).
- Our trenches at the southwest corner of the U. S. Veterans Administration Sawtelle property exhibited near-vertical to south-dipping normal shear zones with clay gouge trending roughly east-west.
- Trenches at University High School displayed two well-defined steep southerly-dipping normal faults, consisting of 2 to 12 inches of sheared clay, with apparent lateral displacements. No datable materials were found in any displaced units.

Proposed trenching on both the Santa Monica and Hollywood faults by Caltech (James Dolan and Kerry Sieh, pers. comm., 1992) may be completed by the time this paper is printed and may reveal new evidence on recency of faulting. Hauksson and Saldivar (1986) believe the 1930 M5.2 earthquake in Santa Monica Bay was on the Santa Monica fault. West of here, recent work by Rzonca and others (1991) and Drumm (in this chapter) indicates that the Malibu Coast fault moved in Holocene time.

OVERVIEW AND CONJECTURE ON THE SOUTHERN BOUNDARY OF THE TRANSVERSE RANGES PROVINCE

The Santa Monica fault zone is a portion of the much longer Anacapa-Malibu Coast-Santa Monica-Hollywood-Raymond-Sierra Madre-Cucamonga fault system that stretches more than 100 miles from offshore west of Ventura County to Cajon Pass (Fig. 4). This system of faults roughly defines the southern structural boundary between the Transverse Ranges (mainly left-lateral and thrust faults) and the Peninsular Ranges (mainly steep right-lateral faults) provinces of

southern California. At Cajon Pass, this boundary is right-laterally offset 15 miles along the San Jacinto fault, and resumes eastward as the Banning thrust fault. (See Fig. 5.) Although this fault system may have been a through-going active entity at some time in the past, Yerkes and Lee (1979a, b), Chapman and Chase (1979) and Weber and others (1980) suggest that since late Pleistocene time this is no longer so. Our work at Caltech in 1975-78 (Crook and others, 1987), led us to believe that the Sierra Madre fault zone and the Raymond fault behave as several discrete segments that act independently to local tectonic stresses and that they therefore should be treated as individual segments with regard to future seismic activity. This conclusion is also extended to the Anacapa-Malibu Coast-Santa Monica-Hollywood fault system. Figure 3 shows thc interrupted nature of the faults, particularly the nearby Sierra Madre fault zone that is actually a connected series of thrust faults that curve northeastward to become steep left-slip faults. We believe each thrust segment moves independently to cause typically moderate-sized earthquakes, such as the M 6.5 San Fernando earthquake in 1971.

In our opinion, therefore, a scenario of a long reach of the fault system moving to cause an earthquake with a magnitude approaching 7.5 is unlikely. Bonilla and others (1984) indicate such a large event would rupture the ground surface for almost 50 miles; a long rupture length does not appear to fit the direct arcuate thrust fault traces, nor the local tectonics (e.g. Allen, 1975; Hauksson, 1990, and Hauksson this volume).

Until approximately 1987 the boundary between the Transverse Ranges and the Peninsular Ranges provinces was considered to be the obvious geomorphic southern limit of the Transverse Ranges. In other words, the southernmost steep parts of the Santa Monica, San Gabriel and San Bernardino mountain ranges. However, work by Davis (1987) suggested that the actual boundary might lie south of these areas as a series of north-dipping blind thrusts. This model appears to have been confirmed by the October 1987 Whittier Narrows earthquake. Geologic and seismic studies of this earthquake by Davis and others (1989) and Hauksson and Jones (1989), and of other subsequent earthquakes (Hauksson, 1990; Hauksson and Jones, 1991), have suggested that the boundary between the two provinces lies south of the Santa Monica-Hollywood-Raymond-Sierra Madre-Cucamonga fault system. The boundary exists as a zone of exposed and buried northeast-trending, left lateral strike-slip faults (San Jose, Indian Hill, Walnut Creek and Red Hill) east of the Los Angeles Basin and a series of northwest- to west-trending blind thrusts (Davis and others, 1989) in the central part of the Los Angeles basin.

At the far western end of the Los Angeles basin this boundary is less defined at present. Here the presently active bounding fault appears to be the Santa Monica fault, although on-shore Holocene movement has yet to be proven.

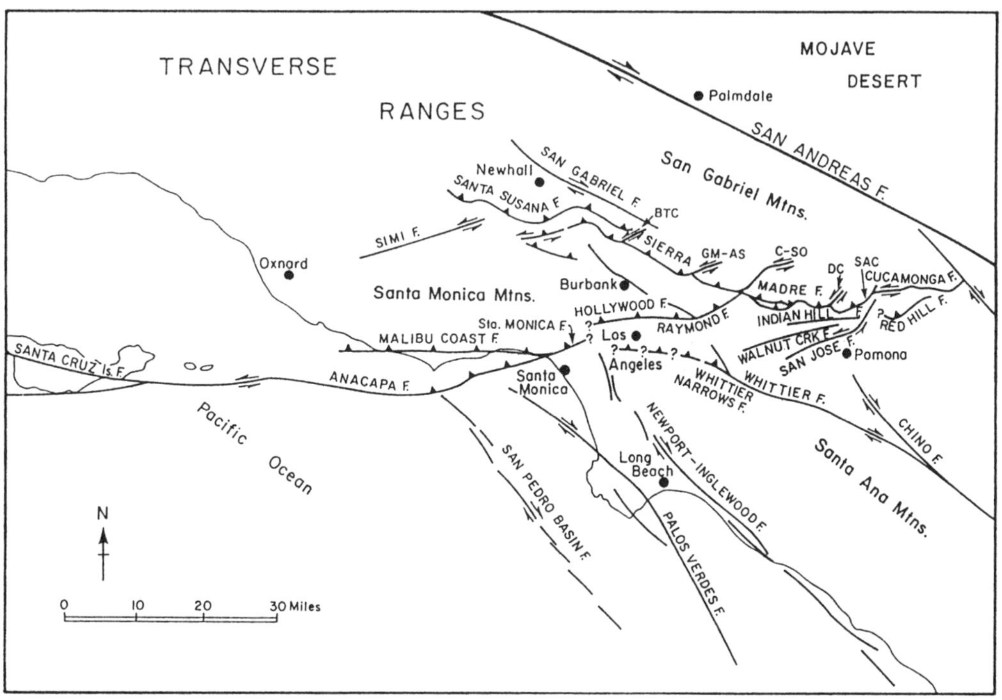

Figure 4. Regional fault map, showing uncertainty of south boundary of Transverse Ranges Province in this part of southern California. Queries indicate doubts as to connection between E-W faults. See text for discussion. The Sierra Madre thrust fault forms distinct segments as it swings NE into canyons and becomes steep left-lateral faults: BTC = Big Tujunga Canyon; GM-AS = Gould Mesa-Arroyo Seco; C-SC = Clamshell-Sawpit Canyon; DC = Dalton Canyon; SAC = San Antonio Canyon.

Work by the authors indicated late Pleistocene movement at two locations (University High School and U. S. Veterans Administration, Sawtelle). More recent field work by Dolan and Sieh (1991) suggests that Holocene movement has indeed occurred on this fault, which they hope to substantiate by their ongoing subsurface investigation.

Curiously, the subsurface evidence for faulting revealed by our 1982 work on the Santa Monica and the Hollywood faults all indicated nearly vertical to south-dipping faults with apparent normal (dip-slip) displacements. Such faulting could be expected within the hanging wall of a deeper blind thrust (Avouac and others, 1992) and might also explain a portion of what Wright (1991) has concluded to be monoclinal folding due to recent extension in this area (Fig. 6a-c).

HISTORICAL REVIEW

Onshore segments of the Santa Monica and Hollywood faults were first shown on the map prepared by Hoots (1931). His map and text indicate that the only surface trace observed by him on the Santa Monica fault is at the mouth of Potrero Canyon. Hill (1979) and Johnson (1932) reported that two traces are exposed here; one is vertical and the other dips to the north at approximately 45 degrees. The latter trace can be seen to displace terrace deposits thought to be late Pleistocene in age, possibly 125,000 years old (Hill, 1979).

Hill and others (1979), in a review of all studies of these zones up to that time, state that except for an exposure seen by John McGill in Rustic Canyon (quoted in Hill, 1979) no additional site-specific information was available regarding the surficial traces of these two faults. Analyses of oil well and exploratory well data by Lang and Dreesen (1975), Knapp and others (1962), Jacobson and Lindblom (1977), and Wright (1991) indicate that movement on the Santa Monica fault has been negligible since the Pliocene. However, Hill and others (1979), on the basis of their investigation conclude that:

1. The near-surface trace of the Santa Monica fault in the Beverly Hills-Hollywood area is defined by a zone of differential subsidence;
2. The fault has been active during at least part of the Pleistocene time;
3. Holocene movement cannot be precluded on the basis of current knowledge;
4. Subsurface fault traces within the Santa Monica-Raymond fault zone in the eastern part of the Beverly Hills-Hollywood area are actively undergoing tectonic strain accumulation and release.

This last conclusion was partly based on the results of a study of seismicity in this area by Buika and Teng (1978).

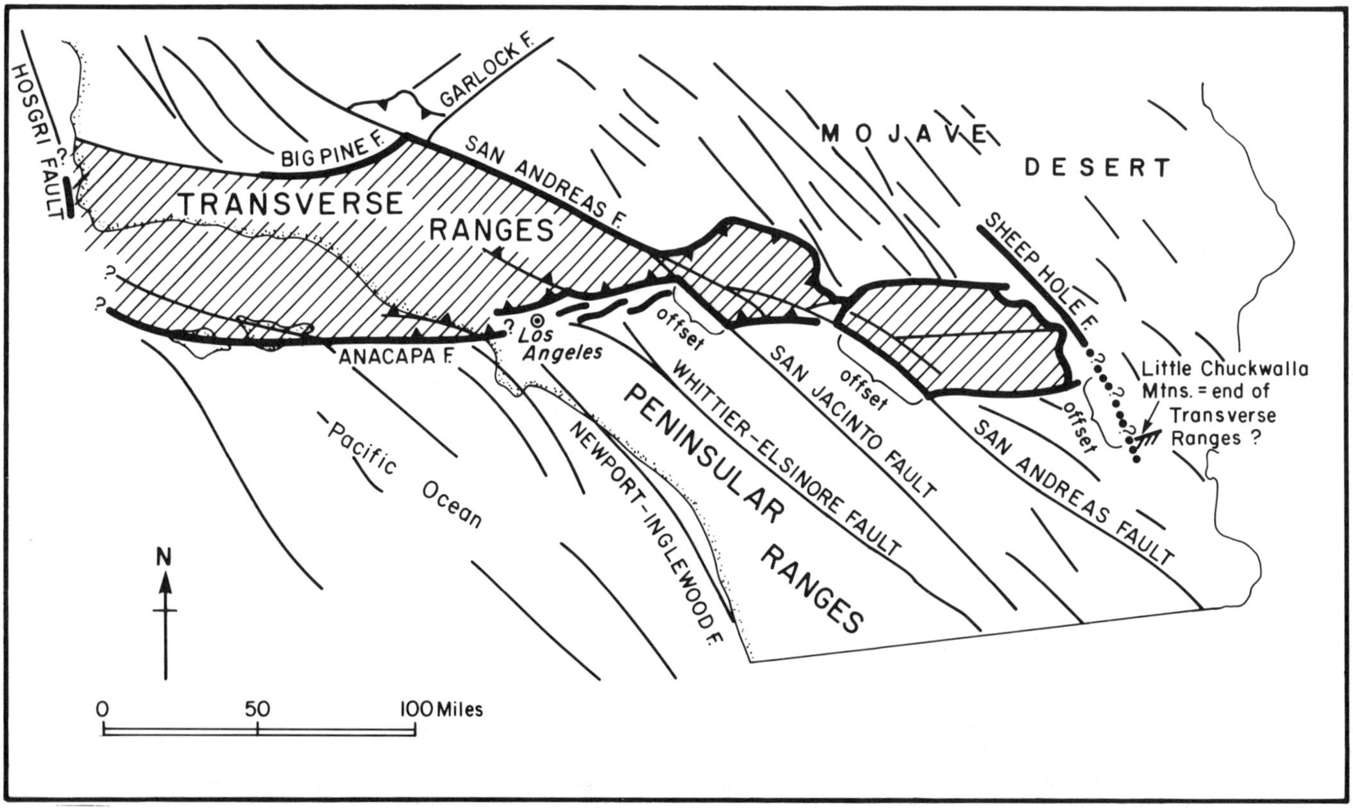

Figure 5. The Transverse Ranges province. See text for comment on apparent right-lateral offsets of southern boundary. Queries indicate unknown end of faults and unknown east and west ends of Transverse Ranges province.

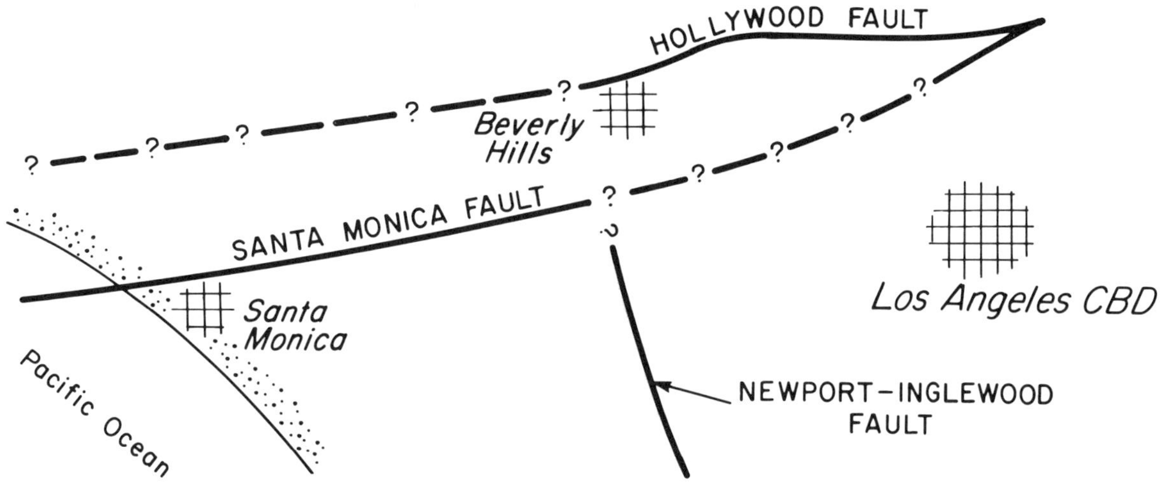

Figure 6a. Fault interpretation shown on many maps to 1980s. Does the Hollywood fault extend westward? does the Santa Monica fault extend eastward?

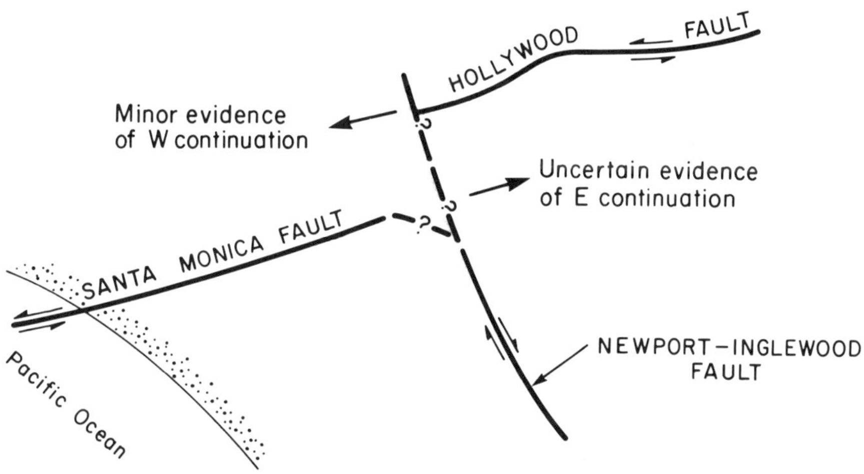

Figure 6b. Possibility of left-stepping separation of 1.5 miles by the right-lateral Newport-Inglewood fault (wrong sense of historic movement); or possibility of merging of Santa Monica and Newport-Inglewood faults as shown by Ziony and Jones (1989), but causing conflicting sense of historic movement on each fault.

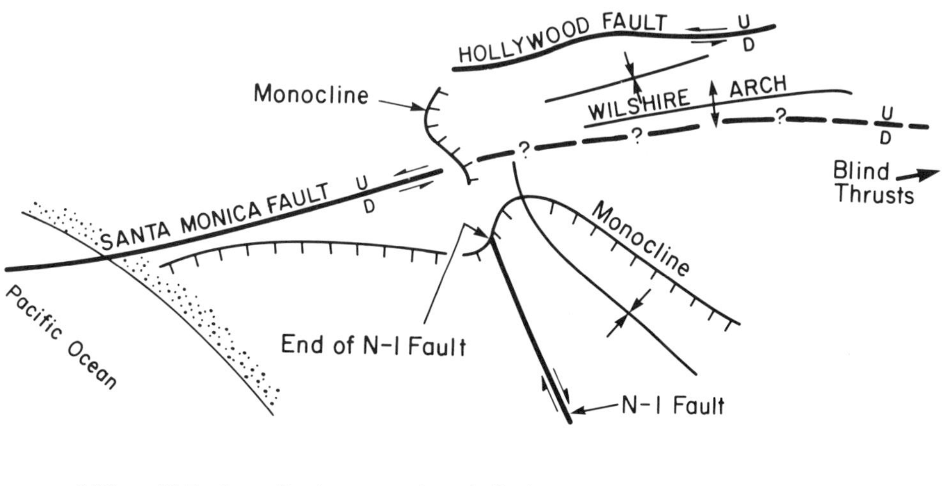

Figure 6c. Two new possibilities: (1) Faults ending in east- and south-dipping monoclines (Wright, 1992, Fig. 16c). (2) Our conjecture of as yet undiscovered buried eastward extension of the Santa Monica fault, possibly associated with the newly discovered Wilshire Arch of Robert S. Yeats (pers. comm., 1992; Hummon and others, 1992), extending eastward to connect to known E-W blind thrusts toward Whittier Narrows.

Subsequently, Weber and others (1980) report that the part of the Hollywood fault zone between Hollywood and the Los Angeles River is primarily expressed by geomorphic features—"scarp-like features in older and younger alluvial deposits and steeply inclined spurs at the south ends of spurs". No exposures of the fault are reported.

Foundation excavations and borings near the corner of Wilshire Boulevard and Bundy Drive in west Los Angeles yielded evidence of a ground water barrier in late Pleistocene deposits (Glenn Brown and Alice Campbell, pers. comm., 1982). They report that ground water was encountered at a depth of approximately 15 feet under the lot at the northwest corner of the intersection and at depths of up to 70 feet under the lot at the southwest corner. A pump test performed on the latter lot (Robert Bean, pers. comm., 1992) indicated a nearby ground water barrier about 350 feet south of Wilshire.

This location coincides with a geomorphic feature—possibly a scarp on a branch of the Santa Monica fault—as seen on 1927 Fairchild aerial photographs and old topographic maps. It also aligns with a projection of similar geomorphic features at University High School and the U. S. Veterans Administration property within a mile eastward. Current geomorphology studies by Dolan and Sieh (1991, and pers. comm., 1992) are discovering subtle new scarps in this area.

THE 1982 TRENCHING

Two sets of older vertical and a series of oblique aerial photographs of the project area were studied to assist in picking fault scarps and prospective trenching sites. The verticals were 1927 and 1928 Fairchild photographs; the obliques were Spence photographs with dates ranging from 1921 to 1931. (These rare photos are preserved at the Whittier College Geology Department.) We located possible fault scarps, then plotted these onto Thomas Guide maps, to help find vacant sites in a densely urbanized area.

Hollywood Fault Trenches

Wattles Park, Los Angeles, site S-3, Plate 1. Two trenches were excavated at this site. The first trench was 166 feet long, and exposed massive- to locally crudely-bedded, silty sand colluvium with scattered pebbles and cobbles. No evidence for faulting was encountered in this trench.

The second trench was excavated upslope of the first trench. This trench was started in granodiorite and extended downhill toward Trench No. 1. It was short, by about 6 feet, from overlapping trench No. 1 owing to the existence of a paved driveway. In the southernmost 20 feet of trench several thin gouge layers were exposed that dipped northward into the hill at 10 to 15 degrees. As these layers approached the depositional rock-colluvium contact, they were displaced downward to the south along several south-dipping shears that probably are small landslide planes. A mass of

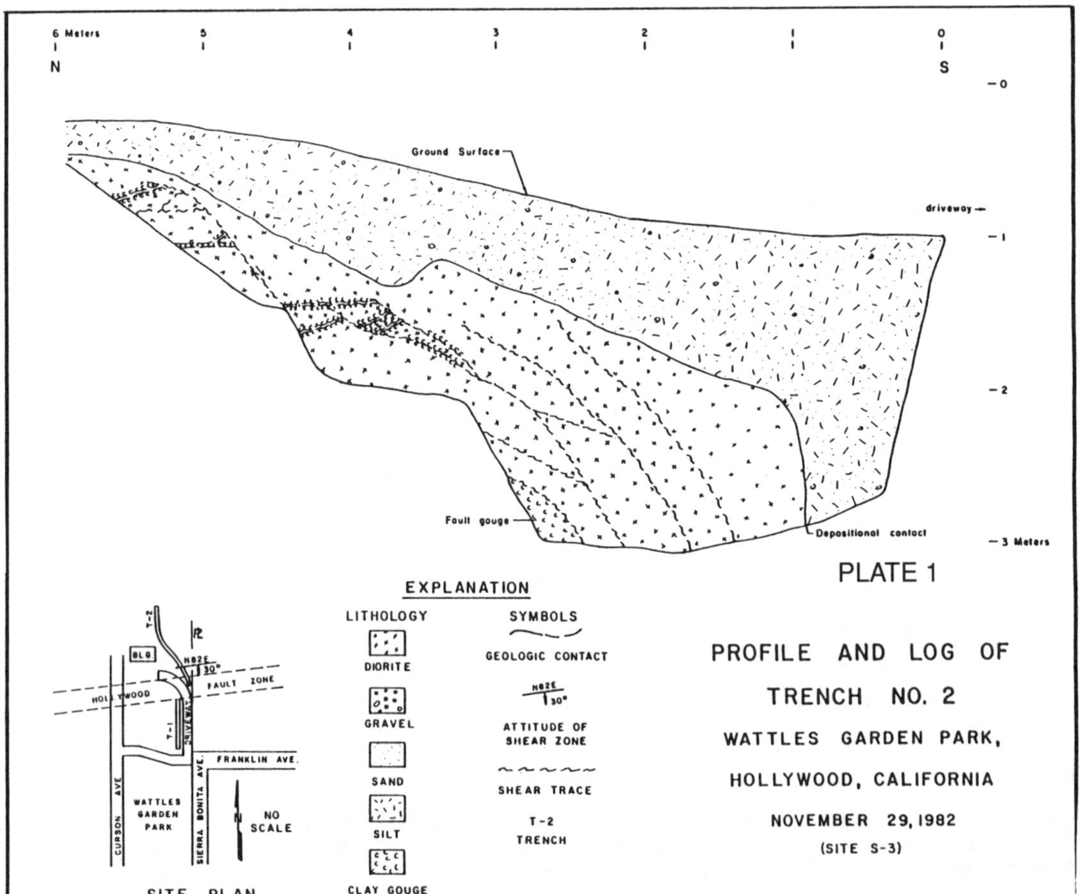

PLATE 1

PROFILE AND LOG OF TRENCH NO. 2 WATTLES GARDEN PARK, HOLLYWOOD, CALIFORNIA

NOVEMBER 29, 1982

(SITE S-3)

bluish-green clay gouge at least 2 feet thick was exposed at the bottom of the trench beneath rock displaced along one of the slide planes. It is probable that this gouge is part of the main zone of the Hollywood fault. No datable materials were exposed.

Greystone Park, Beverly Hills, site S-4, Figure 7. Portions of a road cut on Loma Vista Drive north of Doheny Drive were excavated by hard. Materials encountered were the Santa Monica slate and a sedimentary unit consisting of yellow-tan, massive arkosic sandstone with abundant pebble-size fragments of slate. The relationship between the two units was difficult to ascertain. Both units are extremely weathered and clay-rich and generally have a sheared appearance. The contact has the general appearance of being depositional; however, at two locations sheared planar inclusions of sandstone are within the sheared slate, dipping 72° and 75° north. The age of the sedimentary unit is not known and may be as old as Tertiary. This is based on the highly weathered and clayey state of the material, as well as the well-developed soil at its surface. We also found that on the park grounds there are several oil and water seeps that appear to align with the road cut exposure. It thus appears that the Hollywood fault passes through the Greystone Park property.

Santa Monica Fault Trenches

Two sites on the Santa Monica fault were investigated by trenching, these being the U. S. Veterans Administration property (site S-2) and the University High School property (site S-5).

U. S. Veterans Administration, Sawtelle, site S-2, (Plates 2 and 3). Trench No. 2 at this site was 141 feet long and 12 to 18 feet deep with two approximately 12-foot long gaps due to caving ground and a concrete substructure. Materials encountered in this trench were massive- to crudely-bedded fluvial and sheet-wash fan deposits varying from firm silty clay to loose, coarse, silty and sandy gravel. Contacts were generally gradational and only locally were sharp and well defined. The sharper contacts generally were between deposits with large grain-size difference.

The southern 100 feet of trench consisted primarily of the crudely-bedded fluvial silt, sand and gravel deposits overlying a south-sloping, locally stripped, older alluvial fan surface. No evidence of faulting was observed in this section of trench nor were there any features that might be attributable to liquefaction.

In the northern 30 feet of trench (Plate 2) the "older" fan surface is within 3 feet of the ground surface and most of the exposed deposits were less well-bedded and sorted and much more fine grained and cohesive than those to the south. These materials contained numerous features suggestive of faulting. Some of the features consist of planar zones of gray clay and clay-rich material that coincide with discontinuities in the crudely-bedded material. Two of the zones

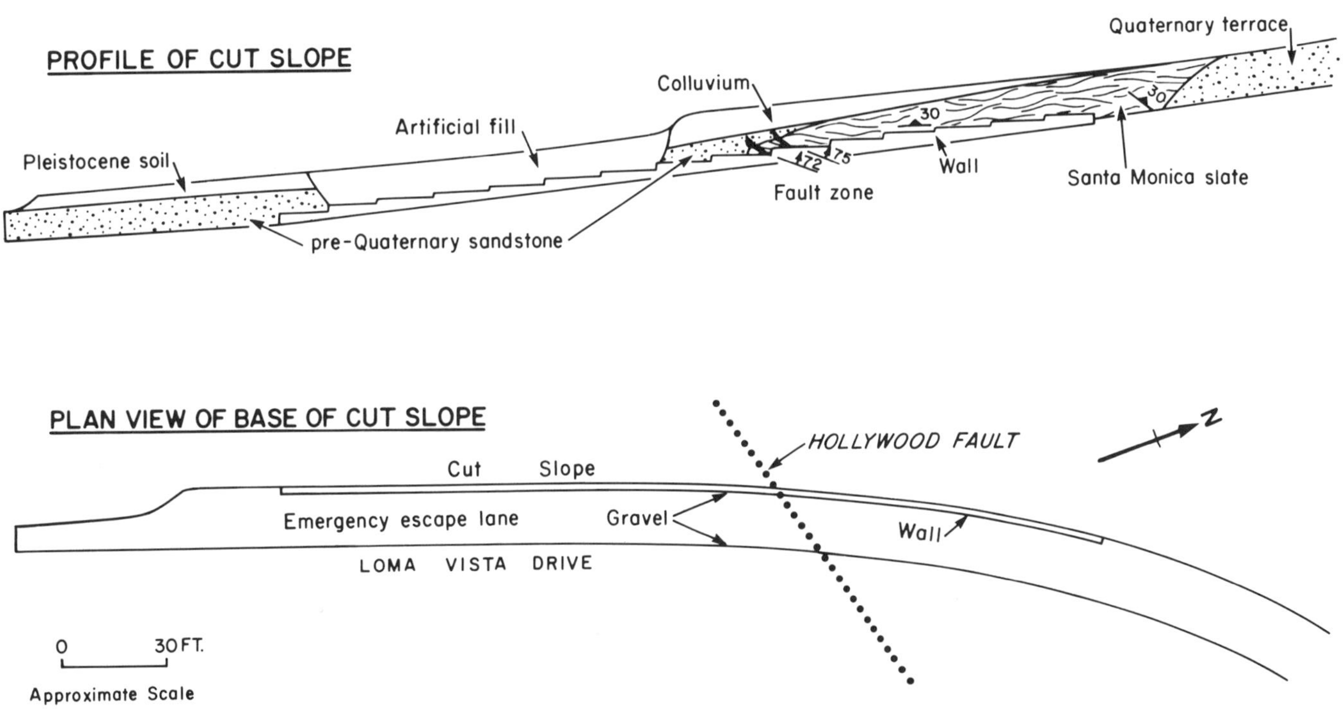

Figure 7. Sketch map of roadcut excavation on Loma Vista Drive (Trench Site S-4) adjacent to Greystone Park, Beverly Hills. The faults in the Santa Monica slate contain sheared pre-Quaternary sandstone.

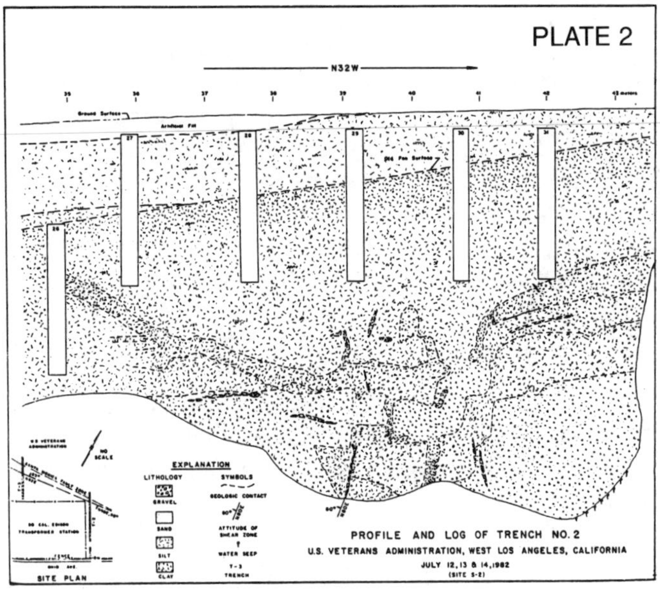

near the bottom of the trench could be found on both trench walls and they defined a strike of N 80 E. Other features include the apparent stepped configuration of some crude bedding and slight warping.

It should be noted that the above described features could suggest normal movement, down to the south, on south-dipping planes as well as north-dipping reverse-slip features.

The above described features are in the "older" alluvial fan whose buried surface appears to become coincident with the ground surface a short distance north of the trench. This fan surface is underlain by an argillic B horizon approximately 2.5-feet thick. Evidence for faulting could not be found within 6 feet of this surface although the massive, heterogenous nature of the materials within this section might make such evidence difficult to recognize. The fan surface did not appear to be displaced but this could not be confirmed as the surface north of Shoring #27 looks to have been stripped prior to deposition of the overlying deposits. Additionally, the thickness of the argillic horizon did not appear to change above the faulted material.

Trench No. 3 (Plate 3) at this site was excavated approximately 145 feet west of Trench No. 2. The north end of the trench was excavated in the same "older" fan unit seen in Trench No. 2. These materials were crudely-bedded with gradational contacts and consisted of massive heterogeneous units of slightly clayey sand, silt, and silty, sandy gravel.

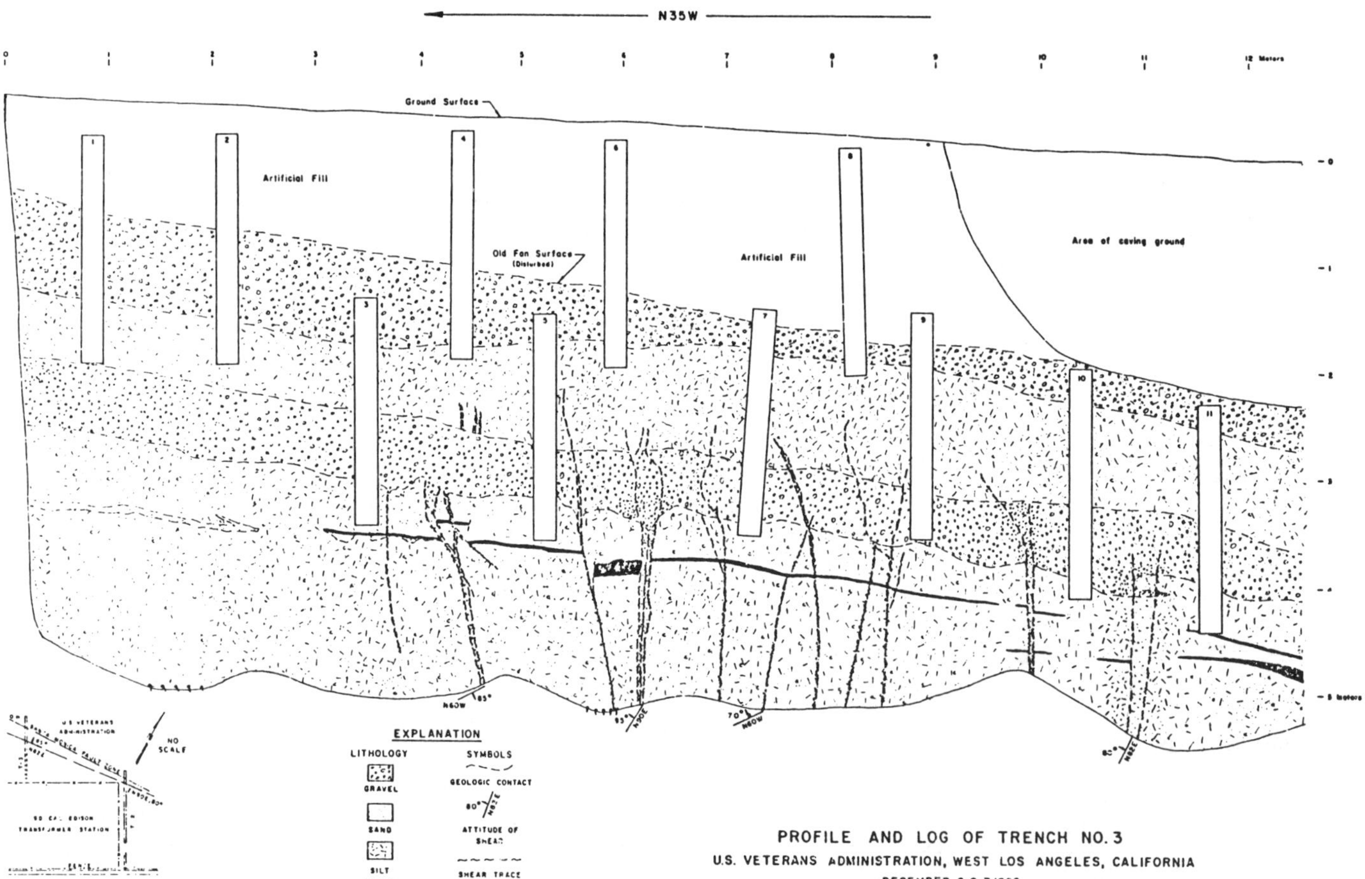

PROFILE AND LOG OF TRENCH NO. 3
U.S. VETERANS ADMINISTRATION, WEST LOS ANGELES, CALIFORNIA
DECEMBER 6 & 7, 1982
(SITE S-2)

PLATE 3

The lower silt unit contained what appeared to be a burned paleo-fan surface with a peculiar reddish color not attributable to pedogenic processes. The "older" fan surface in this area is overlain by 3 to 6 feet of artificial fill.

Numerous features were exposed in this trench that are attributed to faulting although displacements did not appear to be substantial and were questionable in many instances. All of these features consisted of planar zones of gray, silty, sandy clay in contrast to the brown to yellow-brown silt and gravel units containing them. The thicker zones appeared to have been sheared but the thinner zones appeared to differ from the surrounding material only by color and grain-size distribution. All of these features have steep dips that vary from 80° north through vertical to 80° south.

Only the zones beneath Shoring #4 and #6, and possibly the one north of Shoring #10, showed a relatively convincing component of vertical separation. As in Trench No. 2, these features appeared to stop short of the "older" fan surface, in this case by at least 2 feet, and maybe more, depending upon how much of the surface has been removed by erosion.

Although we are confident that the Santa Monica fault was exposed in both Trench No. 2 and No. 3, the geologic features on which we base this conclusion are less than ideal. There was certainly no clear evidence, for example, comparable to that seen in our 1976-78 trenches (Crook and others, 1987) on the Sierra Madre and Raymond faults. In fact, our strongest evidence is that the features attributed to faulting in both trenches lie on a N 80 to 85 E trend that coincides with the geomorphic expression of the scarp-like features through this area.

We propose that sharp, well-defined shears and offsets are lacking because the faulting recorded here occurred at a time when the deposits were far less consolidated than at present. Movement under these conditions would result in more of a mixing process along wide indistinct zones rather than shearing along narrow discreet planes. Similar features were noted in unconsolidated deposits exposed in trenches across some 1971 San Fernando earthquake breaks (Chapter 11 in Oakeshott, 1975; Proctor and others, 1972).

University High School, site S-5, Plate 4. A total of five trenches were excavated at this site and only Trench No. 1 exposed faulted material (Plate 4). This trench was 102 feet long, 9 to 12 feet deep and was excavated on the scarp-like feature crossing the school property.

Materials exposed in this trench varied considerably in grain-size distribution, and hence depositional environment, in the downslope direction. The upslope materials consisted of crudely-bedded to massive and heterogeneous sandy silt, silty clay and slightly clayey, silty, sandy gravel. These materials have the appearance of distal alluvial fan deposits.

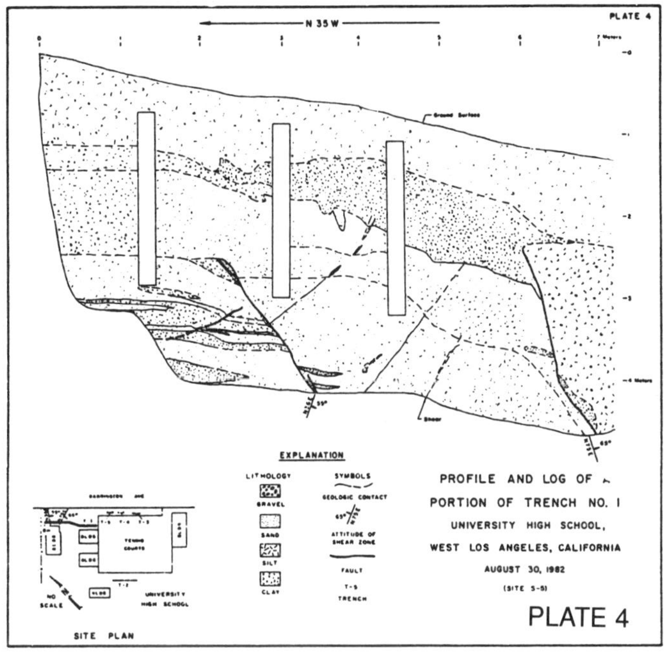

These deposits are bound in the downslope direction by a fault that has juxtaposed against them a dark olive-gray to black, massive, heterogeneous, slightly sandy silty clay with scattered pebbles and cobbles. Depositional environment of this material is unknown but judging from the high clay content and color it may have been a sag pond.

Further downslope the dark clay unit grades into a fairly uniform brown, massive, clayey, sandy silt with scattered pebbles. This material has the appearance of colluvium that probably was derived from erosion of the fault scarp upslope. South of this trench this unit is overlain by relatively young, bedded fluvial deposits consisting of unconsolidated silt, sand and gravel.

Two well-defined south-dipping normal faults were exposed in the northernmost 22 feet of this trench, within and bounding the alluvial fan deposits (Plate 4). The northernmost fault is the most sharply defined and consists of a 1/4-inch thick clay layer that truncates several thin gravel and silt layers. It also appears to offset a small north-dipping fault and exhibits an apparent dip-slip component of 13 inches, down to the south. This fault appears to die out upward in a massive silty, sandy clay unit approximately 5 feet below the present ground surface.

The southernmost normal fault is less clearly defined and consists of an approximately 12-inch wide zone of sheared clay and gravel. The materials on either side of this zone are significantly different suggesting considerably more displacement than on the northernmost fault. The apparent vertical component of displacement is down to the south as suggested by the sense of drag exhibited by depositional contacts in the footwall. Significant lateral slip is suggested by the total lack of similarity of units on either side of the fault. This fault also appears to die out upward approximately 4 feet below the present ground surface and within the central portion of a massive gravel bed; this was seen on both walls of the trench.

Three sub-parallel north-dipping features were exposed in the trench between the two south-dipping faults. Two of these features consist of gray colored streaks with no evidence of shearing or displacement. The third feature consists of a 1/8-inch thick clay layer that contains two thin pods of sand. At its upper end this layer widened to 1/2- to 3/4-inch and extends approximately 2.5 inches into the base of the gravel deposit that is faulted by the southernmost normal fault. No displacement was evident along this feature. Another sub-parallel feature, north of the northernmost normal fault, exhibited a minor amount of reverse displacement in the form of a deformed silt bed.

A peculiar occurrence noted in a portion of the trench is that of abundant calcium carbonate in the form of caliche. These deposits were found only filling pores in the matrix of the uppermost gravel deposit and fractures in the upper portion of the clay unit on the south side of the southernmost normal fault. Caliche was found nowhere else in this trench nor in the other trenches at University High School.

Charcoal was collected from a silty sand layer at depths of 3.5 and 5 feet in Trench No. 2 (log not included herein). Dates were determined by Beta Analytic, Inc., Coral Gables, Florida. The samples yielded radiocarbon dates of 630 ±70 (Beta-6110) years B.P. and 380 ±130 (Beta-6111) years B.P. respectively and are in conflict.

1981-83 BORINGS FOR L.A. METRO RAIL

Until 1990, the Southern California Rapid Transit District was the public agency designated to design and build a subway system for Los Angeles. During the period 1979-1984, the District convened an eight-person Board of Geotechnical Consultants to prepare an exploration program and to advise on anticipated geologic conditions for subway construction. The geologists on the Board were Ronald Heuer, Richard Jahns, Eric Lindvall and Richard Proctor. Several of the recommended exploratory borings for a proposed subway route along Cahuenga Boulevard have bracketed traces of the Hollywood fault north of Hollywood Boulevard (see Fig. 2). Boring #28 at the south encountered only Quaternary fluvial deposits to a total depth of 202 feet and boring #28A, 1,000 feet north, encountered Tertiary rock at a depth of 63 feet (Converse and others, 1981). The Hollywood fault has to lie between these two borings. The Board recommended that additional borings be drilled to try to better define the fault location. In February 1983 boring #28B was drilled which encountered the fault as 10 feet of brecciated sandstone, alluvium and siltstone at a depth of 122 feet in a hole otherwise entirely in alluvium to a total depth of 205 feet. From the geologist's log, the breccia is described, from one 8-inch sample, as follows: "Mixed 1" to 2" masses, densely packed at skewed angles. 1. Sandstone fragments, angular, mottled light brown/dark yellowish-orange, fine to medium, cemented, massive. 2. Gravelly sand (1" gravel), light to mod. brown, well-graded, unconsolidated, very dense, with irregular dark reddish-brown stained masses. 3. Siltstone, grayish orange, at near-vertical angle, minor".

PHYSIOGRAPHY AND AGE OF FAULTING

It appears that the Hollywood fault between Coldwater Canyon on the west and at least to Western Avenue on the east, lies at the break in slope at the south edge of the Santa Monica Mountains. We were unable to ascertain the recency of faulting. It does appear, however, from the steepness of the alluvial fan surfaces in this reach and the lack of significant drainage entrenchment on these surfaces that late Quaternary movement has occurred on this fault.

West of Coldwater Canyon the fan surfaces are older, highly dissected and entrenched, suggesting that the Hollywood fault in this area has not been so recently active, or as Hoots (1931) suggested, the deformation has been by folding rather than faulting. This geomorphic boundary between areas of differing geomorphology also coincides with a northwest projection of the Newport-Inglewood fault zone (Dolan and Sieh, 1991).

The Santa Monica fault through west Los Angeles is fairly well known at depth from oil well data (especially Wright, 1991). The surface expression is best defined by a 2.5-mile long stretch of scarp-like features between Stanford Street and Washington Avenue on the west, and Manning Avenue and Santa Monica Boulevard on the east. These features have the appearance of being relatively old, as compared to the Raymond fault scarp in Arcadia, San Marino and South Pasadena (Crook and others, 1987). Nevertheless, these features were the primary reason for trenching at both University High School and the U. S. Veterans Administration property. We were successful in finding faults at these sites. None of the Holocene fluvial deposits exposed in these trenches exhibited any faults or features attributable to seismic shaking, such as infilled lurch cracks or liquefaction structures.

One surprising feature of the Santa Monica fault is that no evidence was found that would confirm that it is a north-dipping reverse fault. This may be due to either the fact that we did not expose the main fault in any of our trenches and that the features we saw are antithetic faults above a thrust fault (e.g., Avouac and others, 1992), or that the most recent movement has been predominantly strike-slip along high angle faults.

It might be argued that the scarp on the fan surface east of Potrero Canyon (Hoots, 1931; Hill, 1979) indicates Holocene movement on the Santa Monica fault. The age of the fan surface in that area, however, almost certainly is pre-Holocene.

Obviously a recurrence interval for the Santa Monica fault cannot be determined from our data. Our work also cannot prove Holocene movement on the on-shore portion of this fault, even though the geomorphic evidence suggests it.

REFERENCES

Allen, C. R. 1975, Geological criteria for evaluating seismicity: Geological Society of America Bulletin, vol. 86, p. 1041-1057.

Avouac, P. P., Meyer, B., and Tapponnier, P., 1992, On the growth of normal faults and the existence of flats and ramps along the El Asnam active fold and thrust system: Tectonics, vol. 11, p. 1-11.

Bonilla, M. G., Mark, R. K. and Lienkaemper, J. J., 1984, Statistical relations among earthquake magnitude, surface rupture length, and surface fault displacement: Bull. Seismological Society of America, vol. 74, no. 6, p. 2379-2441.

Buika, J. A., and Teng, Ta-liang, 1978, Recent seismicity and fault-plane solutions of Santa Monica Mountains, Santa Monica Bay and northern Los Angeles basin, in Geologic Guide and Engineering Geology Case Histories Los Angeles Metropolitan Area, D. L. Lamar, Chairman; Association of Engineering Geologists First Annual California Sections Conference, p. 64-72.

Chapman, R. H., and Chase, G. W., 1979, Geophysical investigation of the Santa Monica-Raymond fault zone, Los Angeles County, California, in Earthquake hazards associated with faults in the greater Los Angeles Metropolitan Area, Los Angeles County, California, including faults in the Santa Monica-Raymond, Verdugo-Eagle Rock, and Benedict Canyon fault zones: California Div. of Mines and Geology, OFR 76-16 LA, Chapter E, p. E-1 to E-30.

Converse Consultants, Earth Sciences Associates and Geo/Resources Consultants, 1981, Metro Rail Project, Geotechnical Investigation: Report for Southern California Rapid Transit District, Los Angeles, California: Volumes I and II.

Crook, R. Jr., Allen, C. R., Kamb, B., Payne, C. M., and Proctor, R. J., 1987, Quaternary geology and seismic hazard of the Sierra Madre and associated faults, western San Gabriel Mountains, California: U. S. Geological Survey Professional Paper 1339, p. 27-63, and 1978 California Institute of Technology, Contribution No. 3191.

Davis, T. L., 1987, Subsurface study of the late Cenozoic structural geology of the Los Angeles Basin: USGS Open-File Report 87-374, p. 143-147.

Davis, T. L., Namson, J., and Yerkes, R. F., 1989, A cross-section of the Los Angeles area: seismically active fold and thrust belt, the 1987 Whittier Narrows earthquake, and earthquake hazard: Geophys. Res. vol. 94, p. 9644-9664.

Dibblee, T. W., Jr., 1991, Geologic Map of the Beverly Hills and Van Nuys South Half Quadrangles: Dibblee Geologic Foundation Map DF-31, P. O. Box 60560, Santa Barbara, CA 93160 ($10.00).

Dolan, J. F. and Sieh, K. E., 1991, Structural style and geomorphology of the Santa Monica-Hollywood fault system: Abstract in EOS (Transactions of Amer. Geophys. Union), vol. 72, p. 319-320.

Durrell, Cordell, 1954, Geology of the Santa Monica Mountains, Los Angeles and Ventura Counties, in Geology of southern California: California Division of Mines Bulletin 170, Map Sheet 8.

Elliott, J. L., 1951, Geology of the eastern Santa Monica Mountains between Laurel Canyon and Beverly Glen Boulevards, Los Angeles County, California: Unpublished Master's thesis, University of California, Los Angeles, 57 p.

Ellsworth, W. L., Campbell, R. H., Hill, D. P., and others, 1973, Point Mugu, California, earthquake of 21 February 1973, and its aftershocks: Science, vol. 182, p. 1127-1129.

Erickson, R. C., and Spaulding, A. O., 1975, Urban oil production and subsidence control — a case history, Beverly Hills (east) oil field, California: Society of Petroleum Engineers of the American Institute of Mining, metallurgical and Petroleum Engineers Paper No. SPE 5603, 13 p.

Hauksson, E., 1990, Earthquakes, faulting and stress in the Los Angeles basin: J. of Geophys. Res., vol. 95, p. 15,365-15,394.

Hauksson, E., and Jones, L. M., 1989, The 1987 Whittier Narrows earthquake sequence in Los Angeles, southern California: seismological and tectonic analysis: J. of Geophys. Res. vol. 94, p. 9569-9589.

Hauksson, E., and Jones, L. M., 1991, The 1988 and 1990 Upland earthquakes: left-lateral faulting adjacent to the central Transverse Ranges: J. of Geophys. Res. vol. 96, p. 8143-8165.

Hauksson, E., and Saldivar, G. V., 1986, The 1930 Santa Monica and the 1979 Malibu earthquakes: Bull. Seismological Society of America, vol. 76, no. 6.

Hill, R. L., 1979, Potrero Canyon fault and University High School escarpment, in Field Guide to Selected Engineering Geologic Features, Santa Monica Mountains, J. R. Keaton, Chairman: Association of Engineering Geologists, Southern California Section, Annual Field Trip, Map 19, p. 83-103.

Hill, R. L., Sprotte, E. C., Chapman, R. W., and 6 others, 1979, Earthquake hazards associated with faults in the greater Los Angeles metropolitan area, Los Angeles County, California, including faults in the Santa Monica-Raymond, Verdugo-Eagle Rock, and Benedict Canyon fault zones: California Division of Mines and Geology Open File Report 79-16LA, 201 p., 2 plates.

Hill, R. L., Slade, R. C., Sprotte, E. C., and Bennett, J. H., 1978, Location and activity assessment of the Santa Monica fault, Beverly Hills-Hollywood area, California: in Geologic Guide and Engineering Geology Case Histories Los Angeles area, D. L. Lamar, Chairman: AEG First Annual California Sections Conference, Los Angeles, p. 167-174.

Hoots, H. W., 1931, Geology of the eastern part of the Santa Monica Mountains, Los Angeles County, California: U. S. Geological Survey Professional Paper 165-C, p. 83-134.

Hummon, C., Schneider, C. L. and Yeats, R. S., 1992, late Quaternary deformation of late Pleistocene marine gravels in the northern Los Angeles basin: Abstract, Cordilleran Section Mtg., Geol. Soc. Am., Eugene, OR.

Jacobson, J. B., and Lindblom, R. G., 1977, Timing - geological and governmental - in urban oil development, Los Angeles, California: Oil and Gas Journal, vol. 75, no. 6, p. 160-176.

Johnson, H. R., 1932, Folio of plates to accompany geologic report, Quelinda Estate: Unpublished report containing 25 plates including annotated photographs, resistivity survey data, geologic maps and palinspastic cross-sections illustrating the geologic history of the Pacific Palisades area.

Knapp, R. R., Traxler, J. D., Newhill, T. J., Laughlin, D. J., Stewart, R. D., Heath, E. G., Stark, H. E., Wissler, S. G., and Holman, W. H., 1962, Cenozoic correlation sections across the Los Angeles basin from Beverly Hills to Newport, California: American Association of Petroleum Geologists, Pacific Section, scale 1 inch to 1,000 feet vertical, and 1 inch to 5,000 horizontal.

Lamar, D. L., 1975, Relationship between Hollywood and Raymond Hill faults: Association of Engineering Geologists, Field trip guidebook, John W. Byer, Chairman, Los Angeles, p. 44-45.

Long, H. R., and Dreesen, R. S., 1975, Subsurface structure of the northwestern Los Angeles basin: California Division of Oil and Gas Technical Papers Report No. TP01, p. 15-21.

Oakeshott, G., (ed.) 1975, San Fernando, California, earthquake of 9 February 1971: Bulletin 196, California Division of Mines and Geology, 463 p.

Proctor, R. J., Crook, R. Jr., McKeown, M. H., and Moresco, R. L., 1972, Relation of known faults to surface ruptures, 1971 San Fernando earthquake, southern California: Geological Society of America Bulletin, vol. 83, p. 1601-1618.

Rzonca, G. F., H. A. Spellman, E. W. Fall and R. J. Shelmon, 1991, Holocene displacement of the Malibu Coast fault zone, Winter Mesa, Malibu, CA: engineering geologic implications: *Bull. Assoc. Engineering Geologists*, vol. 28, no. 2, p. 147-158.

Weber, F. H., Bennett, J. H., Chapman, R. H., and others, 1980, Earthquake hazards associated with the Verdugo-Eagle Rock and Benedict Canyon fault zones, Los Angeles County, California: California Division of Mines and Geology Open File Report 80-1OLA, 166 p., 4 plates.

Wright, T. L., 1991, Structural geology and tectonic evolution of the Los Angeles Basin: *in* Biddle, K. T. (ed.) Active Margin Basins, American Assoc. Petroleum Geologists Memoir 52, Chapter 3.

Yerkes, R. F., and Lee, W. H. K, 1979a, Late Quaternary deformation in the western Transverse Ranges of California: U. S. Geological Survey Circular 799-B, p. 27-37.

Yerkes, R. F., and Lee, W. H. K, 1979b, Faults, fault activity, epicenters, focal depths, focal mechanisms, 1970-1975 earthquakes, western Transverse Ranges, California: U. S. Geological Survey, Map MF 1032.

Ziony, J. I. (ed.), 1985, Evaluating Earthquake Hazards in the Los Angeles Region: U. S. Geological Survey Professional Paper 1360, 505 p.

Ziony, J. I., and L. M. Jones, 1989, Map showing late Quaternary faults and 1978-84 seismicity of the Los Angeles region, California: U. S. Geological Survey Map MF-1964.

HOLOCENE DISPLACEMENT OF THE CENTRAL SPLAY OF THE MALIBU COAST FAULT ZONE, LATIGO CANYON, MALIBU

Patrick L. Drumm
Geologist
Leighton & Associates
6430 Independence
Woodland Hills, CA 91367

ABSTRACT

The Malibu Coast Fault (MCF), located along the southern flank of the central Santa Monica Mountains, is an east-west deformation zone consisting of several splays south of the main trace; it is a north dipping thrust with a left-lateral sense of displacement.

A geotechnical investigation was performed for residential development along the central splay of the MCF between Latigo and Corral Canyons adjacent to Pacific Coast Highway in the City of Malibu. The fault was exposed in trenches across a high saddle which was marked by an anomalous 25 feet thick colluvial-filled swale. The swale is part of an ancient basin that parallels the fault and was created by periodic uplift of the northerly block. A cumulic soil profile within the swale records several episodes of fault-induced colluviation.

Detrital charcoal from a colluvial unit 8 feet below the surface yielded a radiocarbon date of 17,810 ± 240 years b.p. and suggests an average sedimentation rate of about 0.5 foot/1000 years. The uppermost colluvial layers are therefore Holocene in age. The swale records episodic tectonic activity from late Pleistocene into Holocene time for this particular sediment of the MCF. Estimated fault rupture recurrence is about 4,000 to 5,000 years.

INTRODUCTION

It has long been argued whether or not the Malibu Coast Fault (MCF) is active within Holocene time. Recent trench exposures southeast of Pepperdine University (Rzonca and others, 1991) indicate that at least one splay of the MCF is active. Still, Leighton and Associates (1989) suggests that the MCF "main trace" is not active within Holocene time owing to analysis of trench excavations along that particular segment. This study describes another locality that proves Holocene activity.

Information and interpretations presented here stem from trench exposures across the central splay of the MCF in the Latigo Canyon area. The site is located adjacent to Pacific Coast Highway east of Point Dume (Fig. 1). This splay has

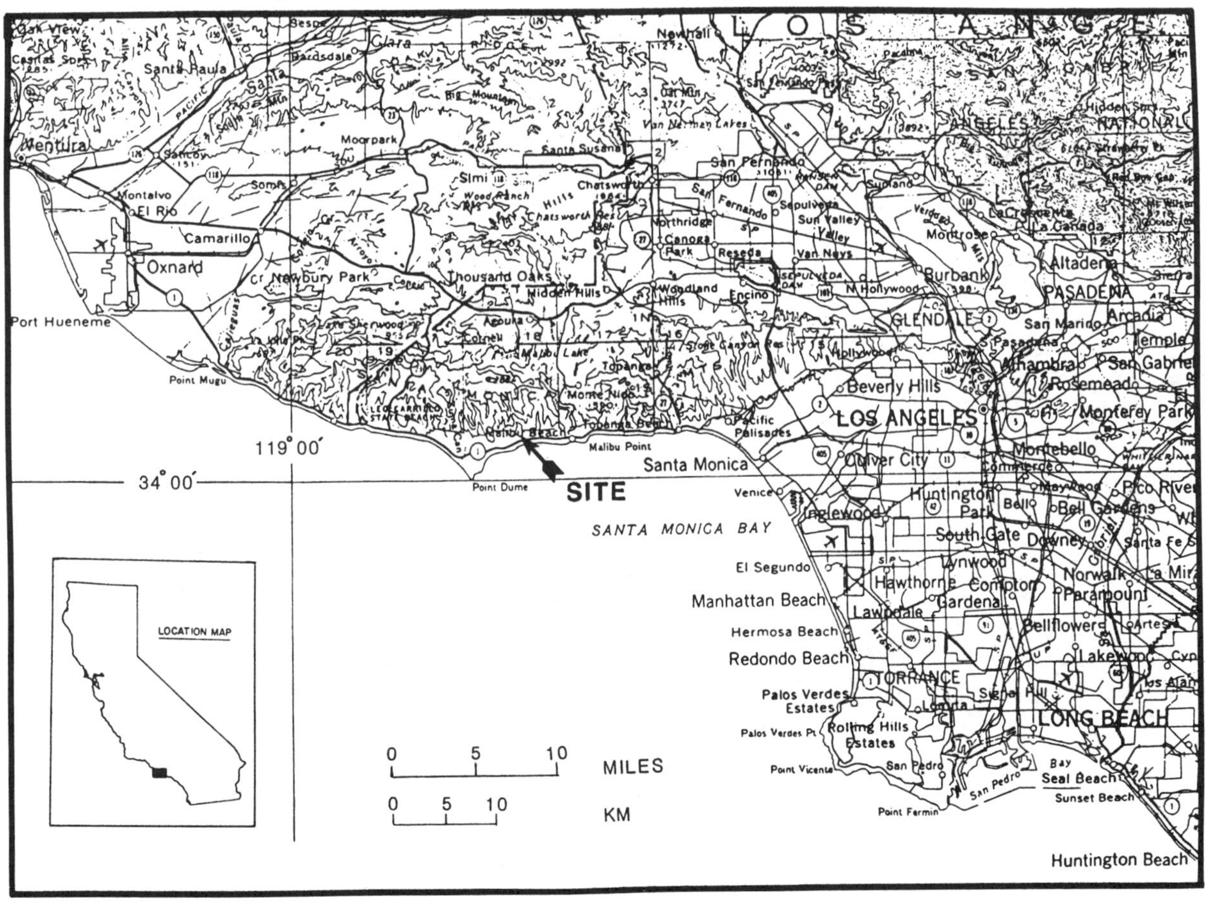

Figure 1. Location map

undergone repeated displacement in latest Pleistocene and Holocene time and therefore should be regarded as "active" for engineering design purposes.

GEOLOGICAL SETTING

The west trending Santa Monica Mountains make up the southwesterly border of the Traverse Ranges province. In the Malibu area, the Santa Monica Mountains are generally bound on the south by the MCF. The fault trends east-west and is responsible for the straightness of the Malibu coastline. It includes several splay faults south of the main trace (Campbell et al., 1970a; 1970b; Yerkes and Campbell, 1988). The MCF consists of a 1,000 - 2,000 feet wide zone of asymmetric faulted blocks that have been intensely folded and sheared. The zone can be mapped for 15 miles onshore. The splay faults are sub-parallel to the main trace and the entire zone is inferred to be high angle north dipping thrust faults (Yerkes and Wentworth, 1965).

Eastward, the Malibu Coast Fault may extend to or become the Santa Monica Fault (the Santa Monica Fault is discussed in this book). To the west the MCF probably connects to the Anacapa - Santa Cruz Island Fault System offshore (Fig. 2).

The Malibu Coast Fault demonstrably juxtaposes late Cretaceous to middle Miocene sedimentary and middle Miocene Volcanics and intrusive igneous bedrock on the north against middle to upper Miocene sedimentary and middle Miocene Volcanics bedrock on the south. The displacement of bedrock units across the fault zone is such that neither strike-slip movement of less than 25 miles nor high angle dip-slip movement can explain the geometry. Further, the MCF is assumed to be the result of a large magnitude, north blocks over south, crustal shortening (Yerkes and Wentworth, 1965).

PREVIOUS SITE INVESTIGATIONS

Several local engineering geological studies have been carried out in the Latigo Canyon area. One pertinent study was a regional investigation by the U.S. Geological Survey for the U.S. Atomic Energy Commission for potential siting of a nuclear power plant near the mouth of the Corral Canyon as shown on Figure 3 (Yerkes & Wentworth, 1965). This investigation focused on the seismicity and recurrence interval of the MCF and related faults. At that time, the last fault displacement was determined to be within late Pleistocene time, approximately 35,000 years ago. That age was based on topographic features, offset Pleistocene deposits, and the fact that Holocene non-marine deposits did not appear to be displaced along fault projections. Late Miocene is the youngest bedrock unit faulted within the proposed reactor site. The abandoned Corral Canyon nuclear site is located about one half mile east of the preset trench site as shown by Campbell et al. (1970a and b) and Figure 3a.

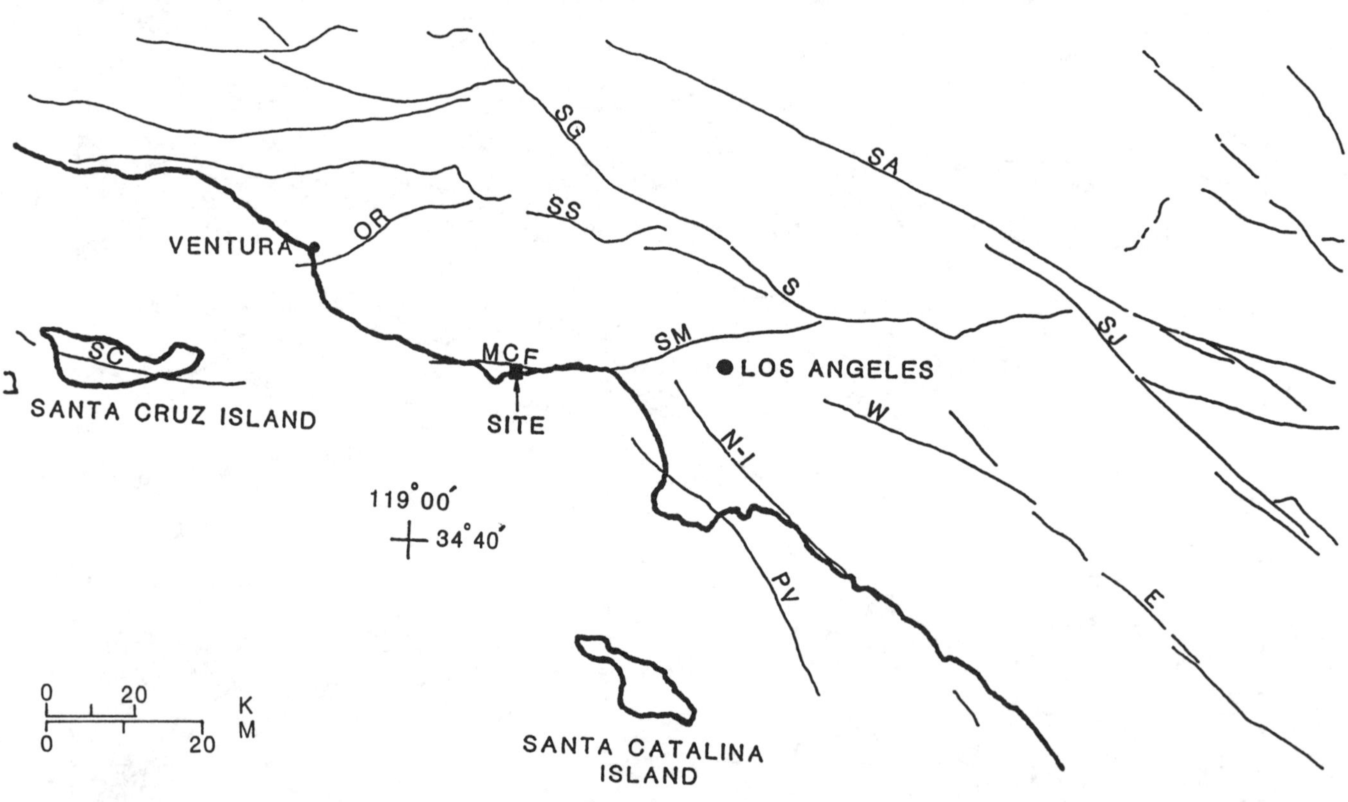

Figure 2. Fault map of Southern California

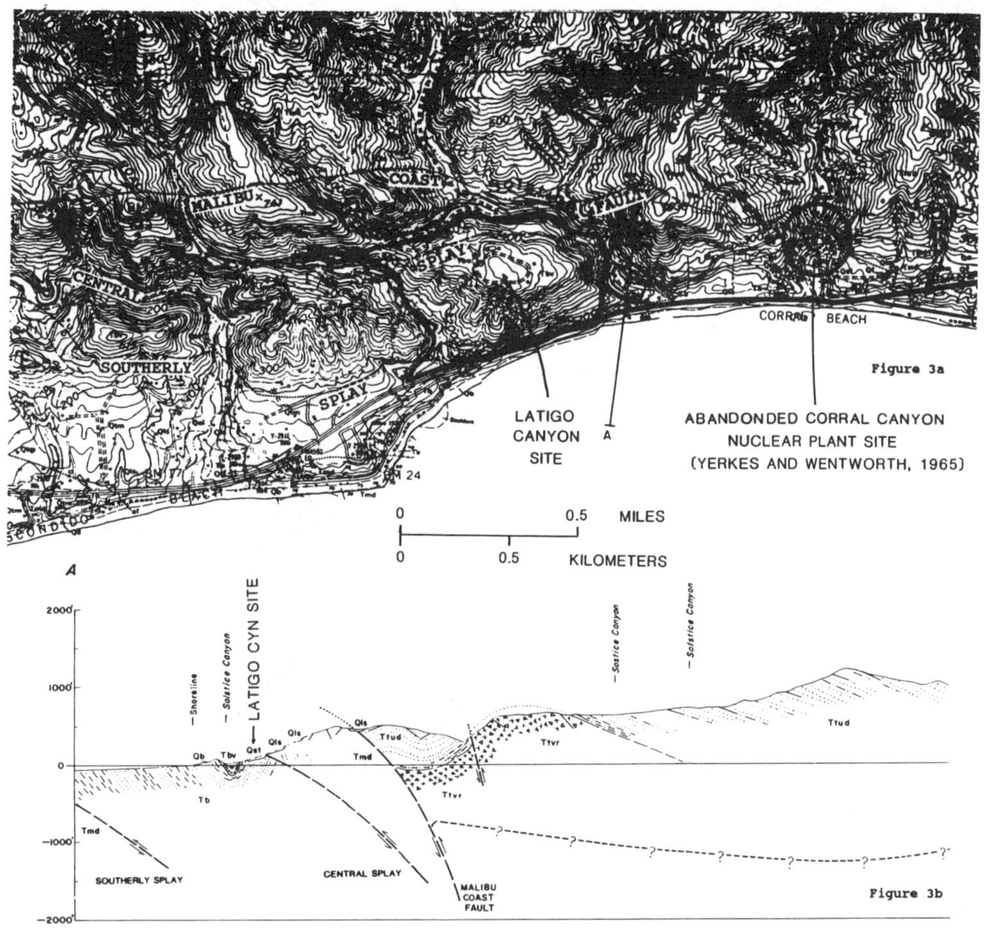

Figure 3. Project site

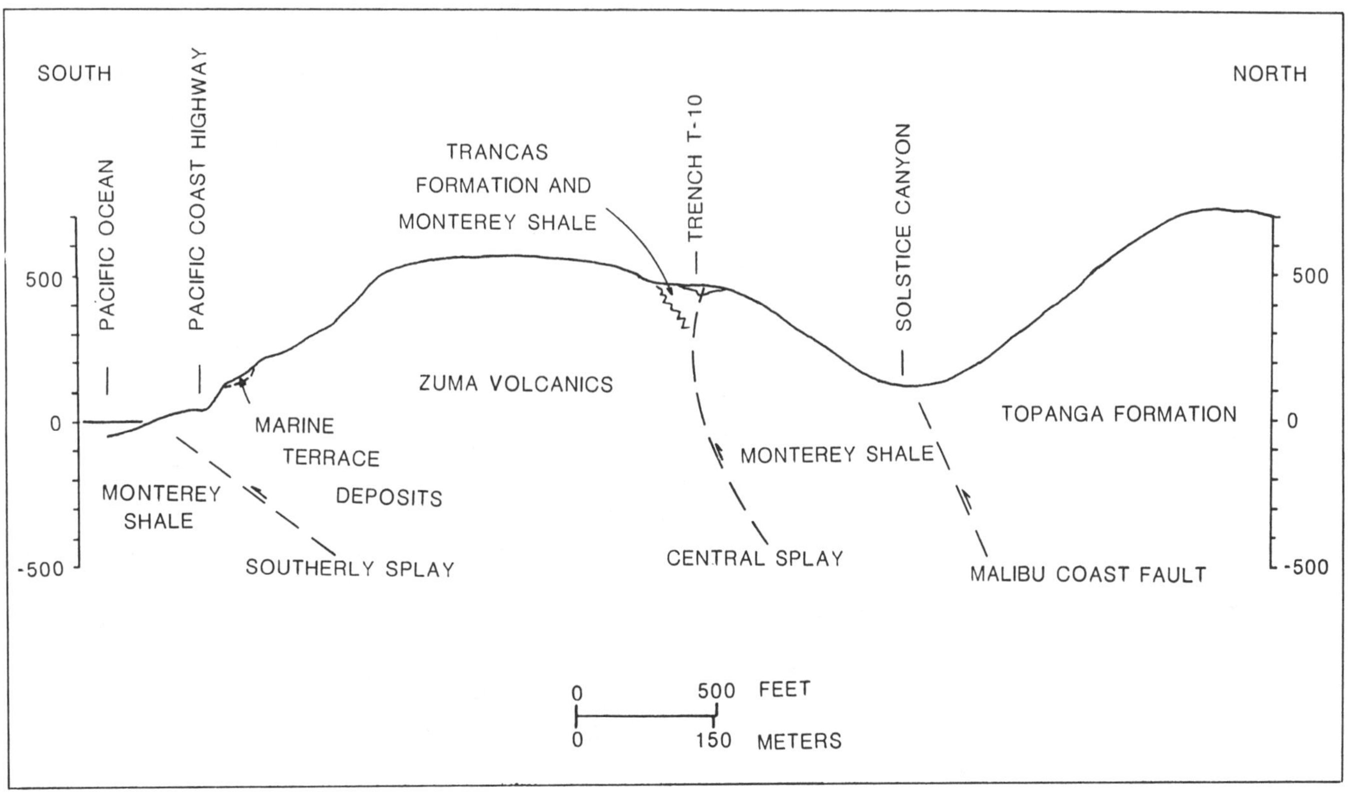

Figure 4. North-South section through site

THE LATIGO CANYON INVESTIGATION

In 1990, we performed a geotechnical investigation for single family housing near the central splay if the MCF (RSA, 1990). The project site is located along the north side of Pacific Coast Highway between Latigo and Solstice Canyons (Fig. 3a). The 122-acre parcel is undeveloped except for two single family homes. A trench site was located on a relatively-level, east-west ridge at about the 500-foot elevation (Fig. 4). The south flank becomes almost vertical along Pacific Coast Highway. In the central portion of the ridge the northerly flank slopes gently for approximately 200 feet forming a high-level saddle. The slope breaks sharply and steeply descends to Solstice Canyon. Solstice Canyon is about 100 feet above sea level at this point and drains easterly-southeasterly around the property before reaching the Pacific Ocean near the southwest property corner. Previous mapping showed the central splay of the MCF to cross the gentle northerly flank of this ridge (Campbell et al., 1970 a and b). The main trace crosses the northerly adjacent ridge approximately 250 feet away and the southerly splay is mapped along Pacific Coast Highway near the southerly property boundary. The central splay of the MCF trends east-west across the northern 1/3 of the site (Fig. 5). Sedimentary bedrock north of the fault is thrust up over volcanic bedrock on the south. Landslides occur along and north of the fault trace owing to fracturing of sedimentary bedrock on the up-thrown block. The site was once used as a military outpost during World War II and some grading of the site presumably occurred at that time. However, grading along the northerly slope was probably less than a few feet.

Figure 5. Photograph of site looking southeasterly

North of the Fault Zone

The Monterey Formation occurs north of the fault zone. The Monterey crops out northwest of the site where the beds are nearly vertical. These are mainly reddish-brown to light gray siliceous shales and claystones, highly fractured and pulverized to form a brecciated zone. Cherty zones are also common along with abundant gypsum infillings along fractures and alteration zones of green epidote. The epidote zones range from 2 inches to 2 feet thick and parallel the fault. The entire unit appears to have undergone weak metamorphism and may in fact be the Monterey Shale "deformed unit" described by Yerkes and Campbell (1979), mapped separately from the parent Monterey Formation shale.

Bedding north of the fault is indistinguishable, for it is almost completely masked by tight folding and fracturing. Bedding dips usually to the northwest northeast and east and is shallow to steep. At least three landslides exist along the fault trace within the Monterey shale.

South of the Fault Zone

The Trancas Formation of Middle Miocene age overlies and is interbedded with the Monterey Formation of middle and upper Miocene age (Fig. 3b). The Trancas Formation is usually interbedded with Zuma Volcanics and the two formations have been mapped together as "Unit B" in previous literature (Campbell, et al., 1970 a and b). Zuma Volcanics make up most of the ridge and surrounding southerly slopes from Latigo to Corral Canyons. Trancas bedrock is limited to a thin, narrow strip along the south side of the fault along the central portion of the ridge.

The Monterey shale consists of light gray massive claystones and siltstones and white to pale brown bentonitic shales. Claystones and siltstones are highly fractured. The bentonitic shales are highly expansive and waxy in appearance. Units are thick-bedded.

The Trancas Formation is strongly deformed and consists of laminated shales and mudstones with few sandstone lenses and pods. Shales and mudstones are dark brown, gray and blue, slickensided, and plastic. Units are laminated to thinly bedded, poorly consolidated and highly fractured. Sandstones are light brown, medium grained, poorly cemented, friable and occur as occasional pods and lenses paralleling bedding. This alignment of sandstone pods probably represents pre-existing sandstone beds that have been segmented during deformation within the fault zone (Yerkes and Wentworth, 1965). The Trancas Formation exposed in trenches is much contorted and deformed.

Interbedded with the Trancas Formation is the Zuma Volcanics. The volcanics consist predominately of dark reddish-brown flow breccias with yellowish-red volcanic sands occurring along portions of the southerly slopes along Pacific Coast Highway. The breccia is very resistent to erosion and extremely hard as evident by the bold outcrops along the ridge and the existence of the broad ridge itself. Units are moderately to highly fractured with fracture spacing of 1 to 2 feet.

Bedding south of the fault strikes approximately east-west and dips southerly at moderate angels. Flow foliation within the interbedded volcanics are not well developed, but generally dip westerly at shallow angles. Joint orientations within the volcanics are random and predominately dip steeply to vertical.

Colluvial-Filled Swale

A thick colluvial-filled swale occurs adjacent to the south side of the fault. Colluvium thickness is about 25 feet. The deposit extends about 50 to 60 feet upslope from the fault before gradually blending into the modern slopewash/colluvium. The normal colluvium layer is about 3 to 5 feet thick and overlies most of the sedimentary bedrock throughout the site. The lower part of the colluvium dips steeply as it nears the fault forming a deep pocket that rounds back up and is truncated by the fault (Fig. 6a).

The colluvium generally consists of black (Munsell 2.5Y 2/0-5Y 2.5/1) organic-rich sandy to silty clay. Close examination showed several discrete colluvial layers separated by basal (volcanic) stone lines, columnar to prismatic peds, manganese "sheen", and the presence of local detrital charcoal fragments. The individual colluvial wedges that fill the swale lack either cambic or argillic horizon and thus were not likely exposed to long periods of surface weathering before burial by the next epoch of colluviation (Birkeland, 1984). Carbonate coatings along ped faces and shears are common, especially in the upper colluvial units. Paleosoils were absent from the entire colluvial-filled swale. At least ten separate colluvial wedges are present ranging in thickness from 1 inch to 7 1/2 feet. These layers are identified as "A" through "J" on the western exposure of the T-10 trench (Fig. 6a and 7). The geometry of individual layers generally parallels that of the entire colluvial unit with noticeable upward progression toward the fault.

Colluvial unit "F" about eight feet below the existing ground surface and approximately 6 inches thick, contains small amounts of detrital charcoal. This unit was sampled and shipped to Beta Analytic, Inc., of Coral Gales, Florida for radiocarbon dating. The sample yielded a date of 17,810 + 240 years b.p. (Beta-35836). The age and depth relation suggests a crude average colluviation rate of about 0.5 foot per 1000 years taking into account the probable removal of a few feet of sediment during previous grading. Although the colluviation rate is

FAULT TRENCH LOG T-10 (Westerly Exposure)

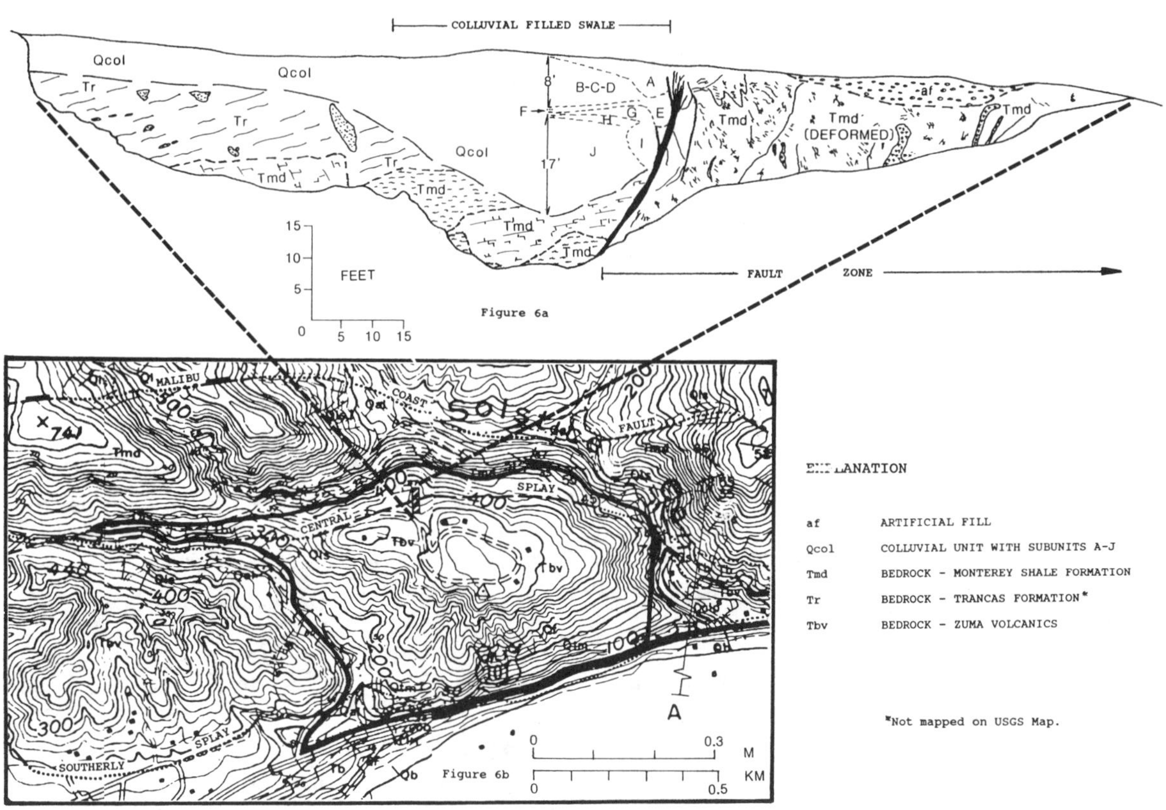

Figure 6. Fault trench

Figure 7. Photo of trench T-10

uncertain, it implies that at least the upper 5 feet of colluvium within Trench T-10 is probable Holocene in age.

Trench T-10

Seven borings were drilled to a maximum depth of 100 feet and trenches were dug. In order to penetrate the tick colluvium, a deep slot was excavated across the saddle and labelled trench T-10 (Fig. 6). The maximum depth of T-10 was 33 feet with a length of 170 feet. The trench exposed a major fault, presumably the central splay of the MCF (Figure 4b).

Trench T-10 exposed a thick wedge of black colluvium south of the fault and reddish-yellow to light gray breccia north of the fault. The fault is one foot wide, green clayey gouge that is plastic and slickensided. The brecciated fault zone is more than 60 feet wide north of the gouge. Exploration was not feasible beyond this point owing to the steepness of the descending slope. Secondary splay faults, shears, and associated mineralized zones sub-parallel the main fault. Carbonate coated shears occur within the colluvial wedge and at least three shears extend into the uppermost colluvial soils (Fig. 7).

The fault strikes approximately N68E to N81E and dips 45 to 74 degrees south, a direction different from that of the generally northerly dip. Slickensides in the fault gouge indicate multiple directions of slip. For example, slickensides plunge S39E at 60 degrees indicating dip-slip movement, and plunge N75E at 12 degrees indicating strike-slip movement with the rake 22 degrees east.

DISCUSSION AND CONCLUSIONS

The presence of a high topographic saddle is anomalous and originated by repeated movement by the central splay of the Malibu Coast Fault. Apparently, each vertical movement would create a scarp. Each scarp would then be slowly buried by discrete colluvial wedges derived from higher terrain to the south. Thus, it appears that most, if not all, of the ten colluvial wedges within the high-level swale resulted from repeated displacements of the central splay of the MCF (Fig. 8).

From a soil stratigraphic standpoint alone, fault recurrence is likely on the order of several hundred to a few thousand years because neither cambic nor argillic horizons cap the discrete colluvial wedges. Additionally, the rate of colluviation was slow as indicated by the cambic organic horizons. In particular, vegetation growth kept pace with sediment accretion.

The approximate 0.5 foot/1000 years average colluviation rate calculated by the radiocarbon dating of charcoal suggests that the entire 25 foot thick colluvial swale exposed in Trench T-10 may well record the last 40,000 to 50,000 years of displacement along the central splay of the MCF in the Latigo Canyon area. Assuming that the ten discrete colluvial

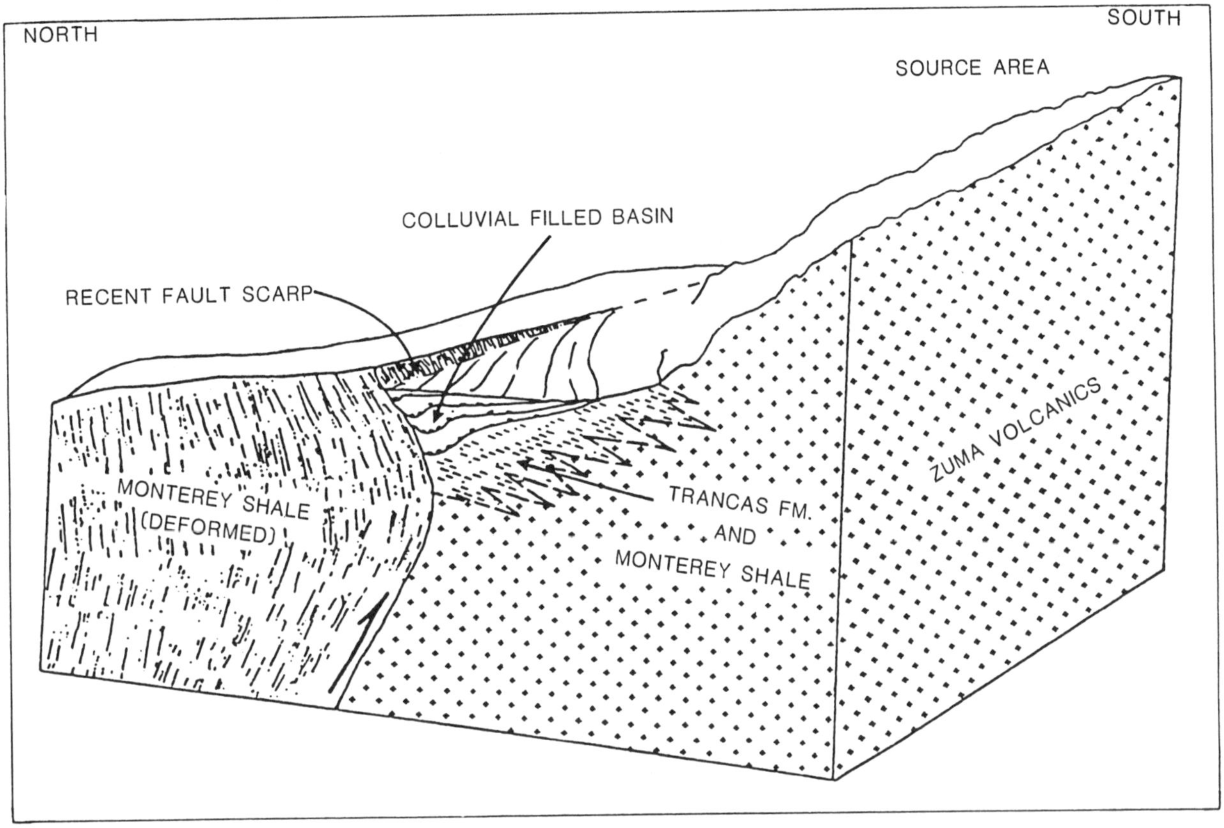

Figure 8.

wedges in T-10 are all fault-derived, recurrence is on the order of 4,000 to 5,000 years.

In addition to dip-slip movement, there is also evidence suggesting that left-lateral strike-slip movement has taken place. For example, Escondido and Latigo Canyons to the west and Solstice Canyon to the east of Trench T-10 appear to be deflected to the east (Fig. 9) where they cross the fault zone (Campbell and others, 1970a and b). The slickenside orientations in Trench T-10 substantiate this hypothesis.

These Latigo Canyon studies suggest that at least this segment of the central splay of the MCF is active according to State of California criteria (Hart, 1988). Whether all MCF splays are active remains unknown and awaits site specific investigations.

ACKNOWLEDGMENTS

I thank Roy Shelmon and Michael Scullin for their contributions to the manuscript, and acknowledge the assistance of Eric Smith and Christopher Koepke during the project as well as the staff of RSA Associates, Inc.

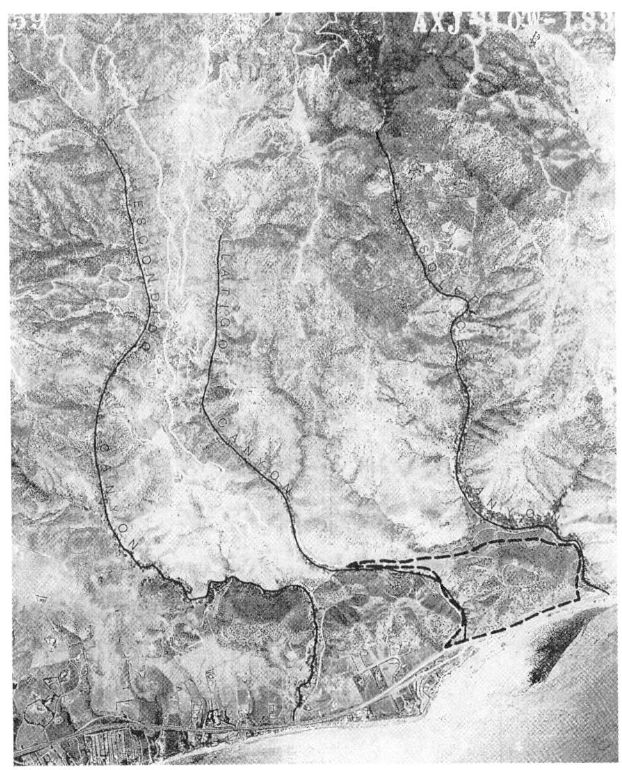

Figure 9.

REFERENCES

Birkeland, P.W., 1984, Soils and Geomorphology: Oxford University Press, New York, NY, p. 372.

Campbell, R.H., Blackerby, B.A., Yerkes. C.F., Schoellhamer, D.E., Birkeland, P.W. and Wentworth, C.M., 1970a, Preliminary Geological Map of the Malibu Beach Quadrangle, Los Angeles County, California: U.S. Geological Survey Open-file Map 70-53, scale 1:24,000.

Campbell, R.H., Blackerby, B.A,; Yerkes. C.F., Schoellhamer, D.E., Birkeland, P.W. and Wentworth, C.M., 1970b, Preliminary Geological Map of the Point Dume Quadrangle, Los Angeles County, California: U.S. Geological Survey Open-file Map 70-53, scale 1:24,000.

Fall, E.W., Rzonca, G.F., Spellman, H.A. and Shlemon, R.J., 1987, Late Quaternary Faulting, Malibu Coast Fault Zone, Malibu, California: Association Engineering Geologists, Abstract and Program, 30th Annual Meeting, Atlanta, GA, p. 38.

Hart, E.W., 1988, Fault Rupture Hazard Zones in California: California Divisions of Mines and Geology Special Publication 42, 24 p.

Leighton and Associates, Inc., 1989, Report of Geotechnical Investigation, Rancho Malibu Mesa Project, Pacific Coast Highway at Malibu Canyon Road, Malibu, California: Technical Report Project No. 3831025-04, August 4, 1989.

RSA Associates, Inc., 1989, Supplemental Geotechnical Investigation and Review Report, Residential Development on 122 Acre Parcel, Tentative Tract 48519, Along Northerly Portion of Pacific Coast Highway Between Latigo Canyon Road and Corral Canyon Road, Malibu, California: Consultants Technical Report, Job No. 4510-01 Log No. 16780, March 1, 1990.

Rzonca, G.F., Spellman, H.A., Fall, E.W. and Shlemon, R.J., 1991, Holocene Displacement if the Malibu Coast Fault Zone, Winter Mesa, Malibu, California: Engineering Geologic Implications: Bulletin of the Association of Engineering Geologists, Vol. XXVII, p. 147-158.

Yerkes, R.F. and Campbell, R.H., 1979, Stratigraphic Nomenclature of the Central Santa Monica Mountains, Los Angeles County, California: U. S. Geological Survey Bulletin 1457-E.

Yerkes, R.F. and Campbell, R.H., 1980, Geologic Map of the east-central Santa Monica Mountains, Los Angeles, County, California: U. S. Geological Survey Map I-1146

Yerkes, R.F. and Wentworth, C.M., 1965, Structure, Quaternary History, and General Geology of the Corral Canyon Area, Los Angeles County, California: U. S. Geological Survey Open-file Report, p.65-179.

Ziony, J.I. and Jones, L.M., 1989, Map Showing Late Quaternary Faults and 1978-84 Seismicity of the Los Angeles Region, California: U. S. Geological Survey MF-1964.

WHITTIER-ELSINORE FAULT SYSTEM, SOUTHERN CALIFORNIA

Donald L. Lamar
Engineering Geologist
P.O. Box 10784
Eugene, OR 97440

ABSTRACT

The Whittier-Elsinore fault system extends from the Mexican border to the northern margin of the San Fernando Valley. The distribution of thicknesses and facies in Paleocene sediments, middle Miocene volcanic rocks and upper Miocene sediments demonstrates 19 mi (30 km) of post upper Miocene right-slip. Variations in Upper Cretaceous and Eocene sediments across the Elsinore fault zone suggest minimums of 8 mi (13 km) and 6 mi (10 km) of right-slip. Folds along the Whittier fault system are secondarily related to strike-slip. An unconformity at the base of the Pliocene Fernando Formation resting on folds and a strand of the Whittier fault system in late Miocene sediments indicates that a portion of the right-slip was pre-Pliocene. Along the Whittier fault in the Puente Hills, Pliocene rocks are right slipped 3 mi (5 km) and stream channels across the fault are deflected to the right as much as 8800 ft (2600 m). Displaced late Quaternary sediments in the walls of test trenches reveal slip rates of 2.0 to 2.5 mm/yr on the Whittier fault zone and 3 to 9.3 mm/yr on the Elsinore fault zone. These unconformable relationships, offset Cretaceous through Pliocene rocks and projection of late Quaternary slip rates into the past support upper Miocene initiation of right-slip. The principal aftershock of the October 1988 Whittier Narrows earthquake was the result of right-slip on a northwest trending strand of the Whittier fault system between the Puente and Elysian Park Hills.

INTRODUCTION

The Whittier-Elsinore fault system comprises the Whittier fault system and the Elsinore fault zone extending from the San Fernando Valley to the Mexican border (Figs. 1, 2 and 3). Similar orientation and Quaternary right-slip relate the Whittier-Elsinore fault system to the San Andreas and San Jacinto major right-slip faults to the east (Fig. 1) (Crowell, 1962). Displacement on the Whittier-Elsinore fault system is enigmatic because relationships commonly are obscured by Quaternary deposits and irregular, branching and segmented en echelon traces (Mann, 1955; Engle, 1959; Kennedy, 1977; Anderson and others, 1989).

Faults splay from the northwest and southeast ends of the Whittier-Elsinore fault system. On the southeast the Agua Caliente and related faults split from the Elsinore fault zone (Fig. 1). To the northwest the Whittier fault system forms a

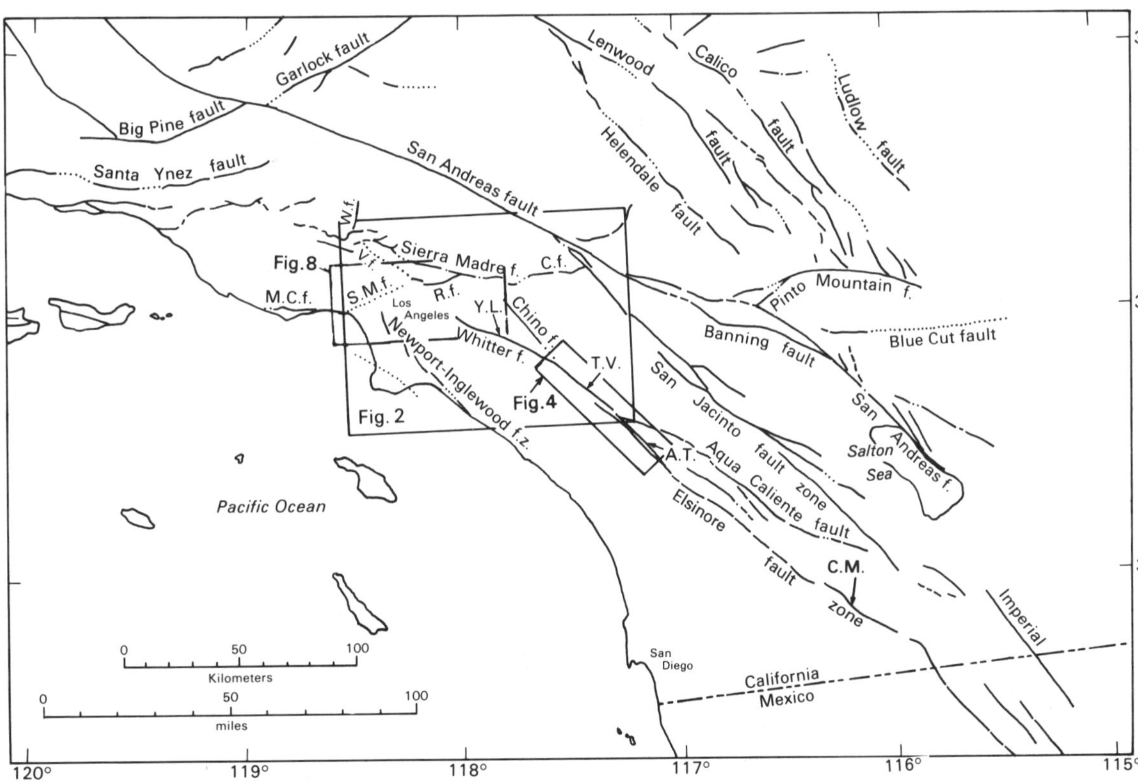

Figure 1. Map of southern California showing principal faults, modified from Proctor (1973), and area covered by Figures 2, 4 and 8. Abbreviations: A.T.: Agua Tibia Mountains; C.f.: Cucamonga fault; C.M.: Coyote Mountains; M.C.f.: Malibu Coast fault; R.f.: Raymond fault; S.M.f.: Santa Monica fault; T.V.: Temescal Valley; Y.L.: Yorba Linda; V.f.: Verdugo fault; W.f.: Whitney fault.

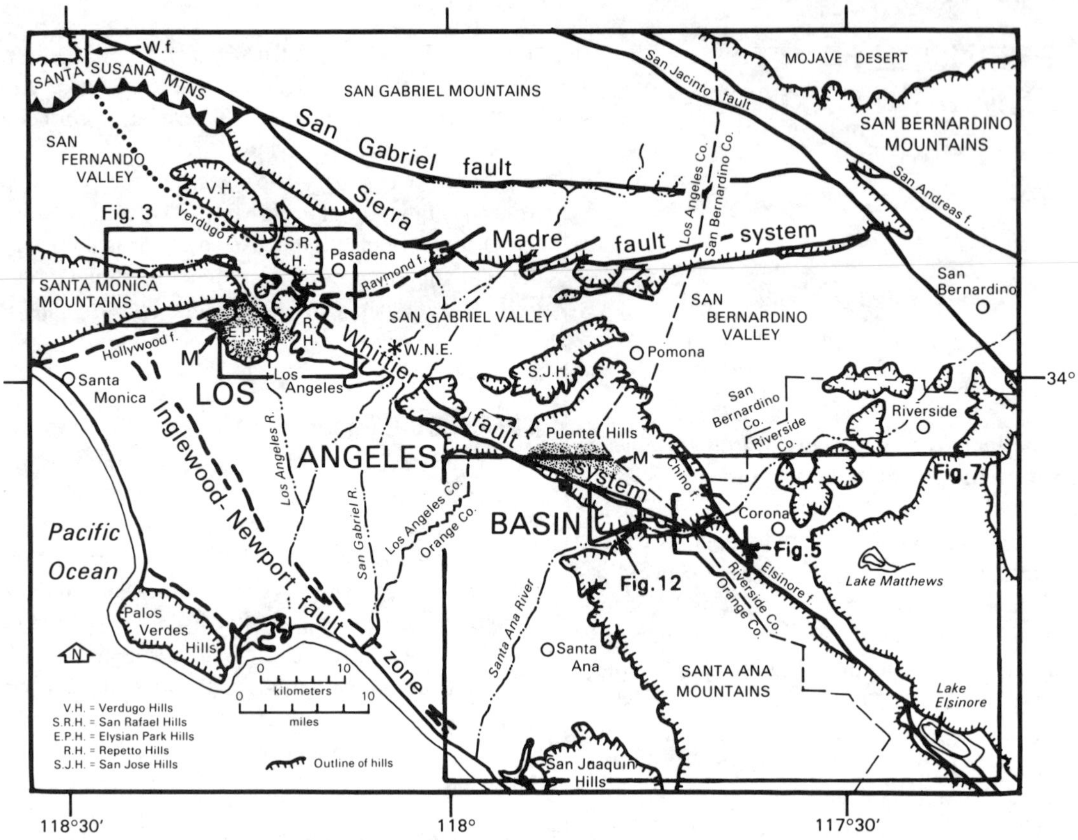

Figure 2. Map of Los Angeles basin area, modified from Kundert (1955) and Yerkes and others (1965), showing principal faults and areas covered by Figures 3, 5, 7 and 12. Areas of maximum thickness of the Puente Formation in the Elysian Park Hills and Puente Hills are indicated by a pattern of dots and M. Abbreviation: E.P.H.: Elysian Park Hills; S.R.H.: San Rafael Hills; R.H: Repetto Hills; V.H.: Verdugo Hills; W.f.: Whitney fault; W.N.E.: Whittier Narrows earthquake epicenter. Location of this figure is shown on Figure 1.

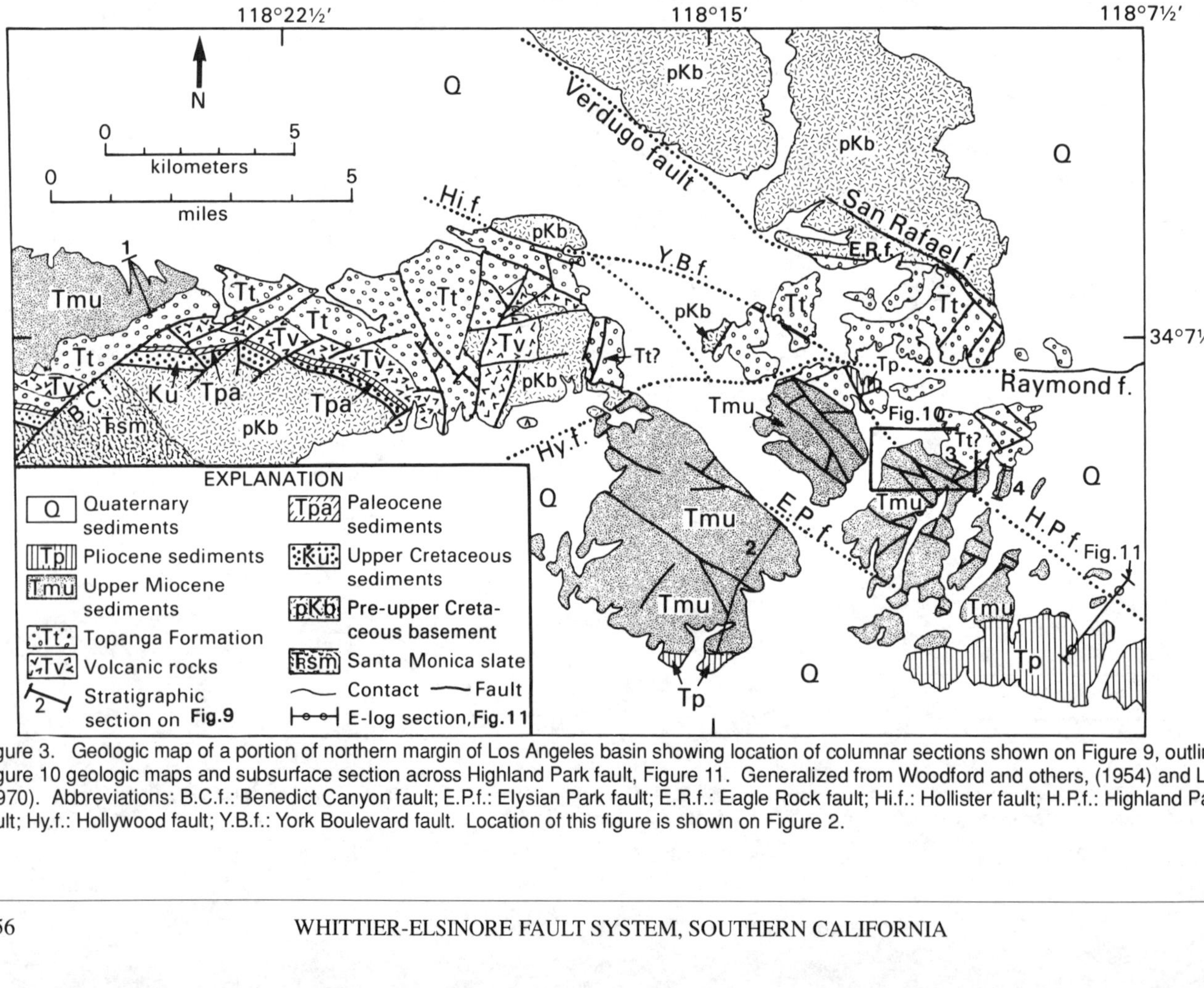

Figure 3. Geologic map of a portion of northern margin of Los Angeles basin showing location of columnar sections shown on Figure 9, outline of Figure 10 geologic maps and subsurface section across Highland Park fault, Figure 11. Generalized from Woodford and others, (1954) and Lamar (1970). Abbreviations: B.C.f.: Benedict Canyon fault; E.P.f.: Elysian Park fault; E.R.f.: Eagle Rock fault; Hi.f.: Hollister fault; H.P.f.: Highland Park fault; Hy.f.: Hollywood fault; Y.B.f.: York Boulevard fault. Location of this figure is shown on Figure 2.

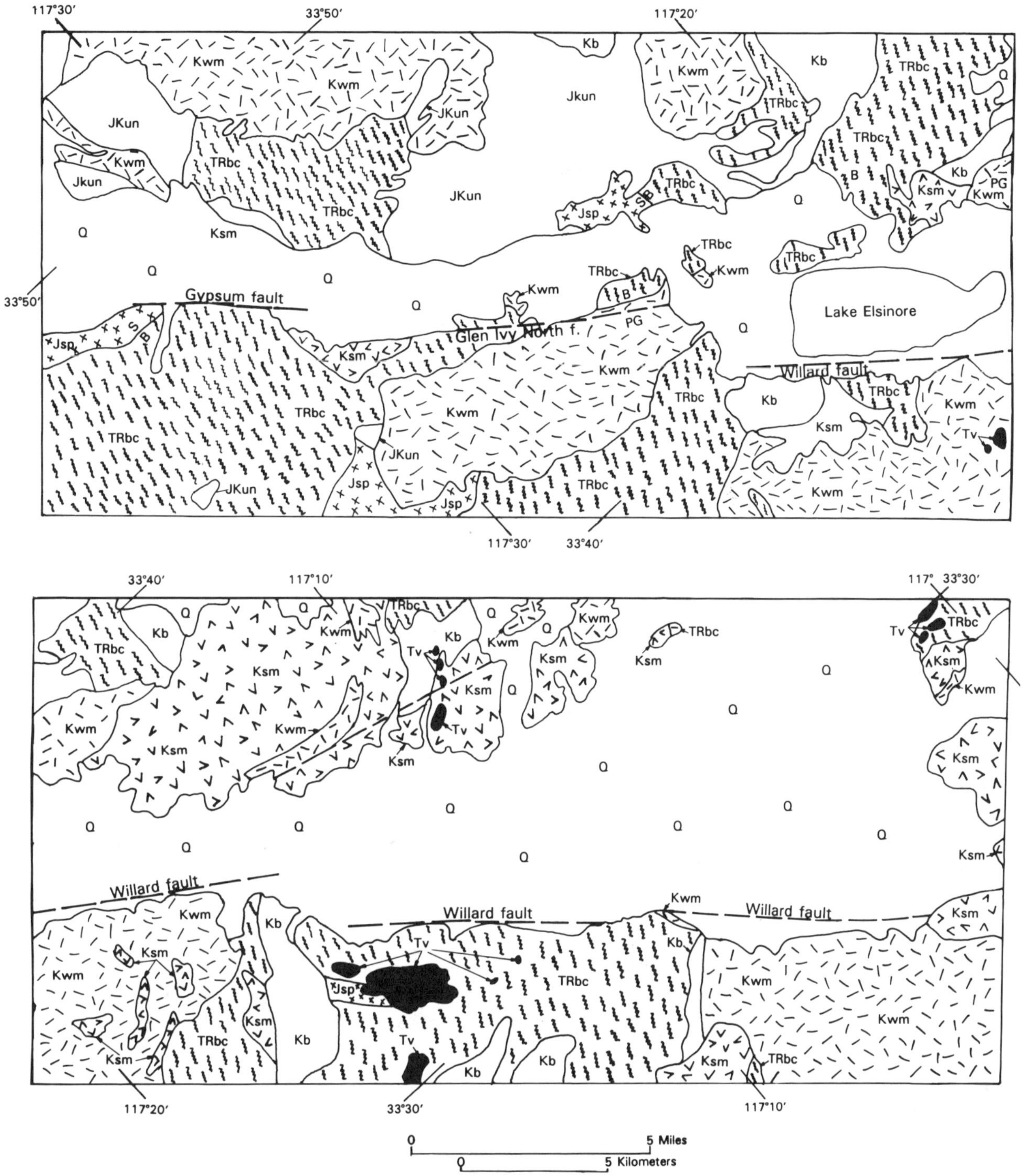

Figure 4. Geologic map of Elsinore trough area generalized from Larsen (1951). Abbreviations: Q: Quaternary sediments; Tv: Tertiary volcanic rocks; Kb: Bonsall Tonalite; Kwm: Woodson Mountain Granodiorite; Ksm: San Marcos Gabbro; JKun: undifferentiated Jurassic and Cretaceous rocks; Jsp: Santiago Peak Volcanics; TRbc: Bedford Canyon Formation. Abbreviations of features recognized by Weber (1977): PG: pegmatite dikes and diorite-gabbro in Woodson Mountain granodiorite; B: Bedford Canyon Formation terrane; S/B: contact between Santiago Peak Volcanics (S) and Bedford Canyon Formation (B). Right side of upper map joins left side of lower map. Location of this figure is shown on Figure 1.

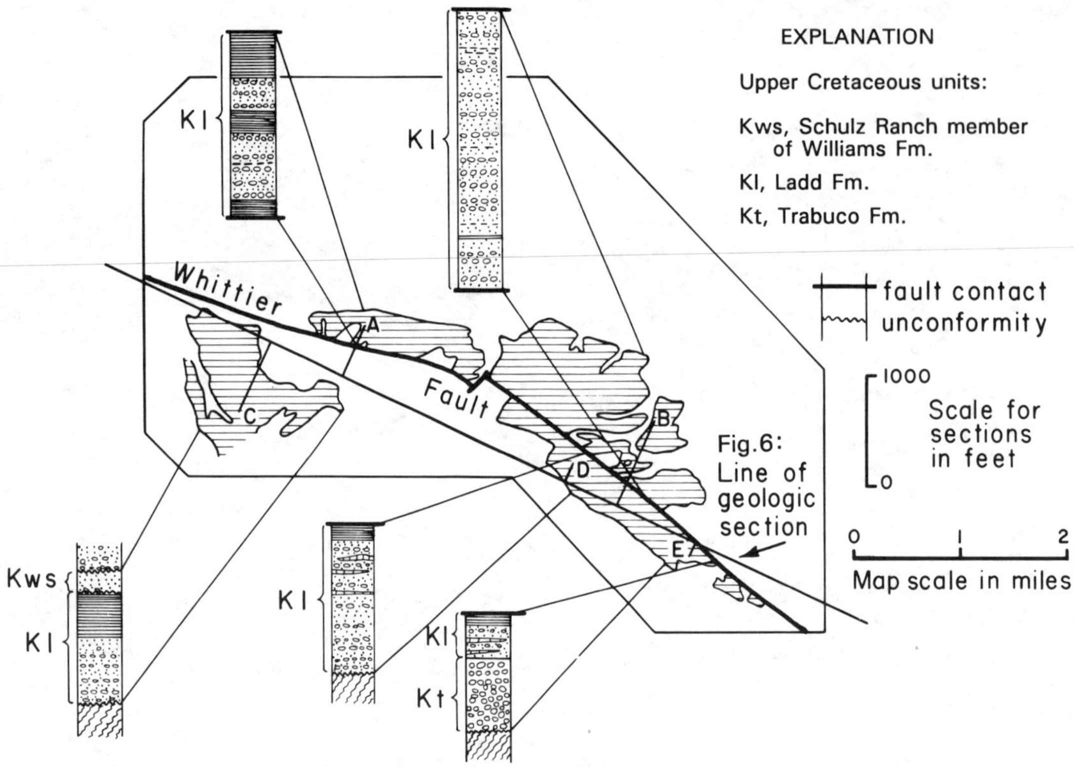

Figure 5. Present distribution of Upper Cretaceous rocks in the northeast Santa Ana Mountains and location of Figure 6 line of section. Location of this figure is shown on Figure 2.

5 mi (8 km) wide zone of faults in the Elysian Park-Repetto Hills area (Lamar, 1970). The Verdugo and Whitney faults in the northern San Fernando Valley are northwest continuing strands (Proctor and others, 1972). These branching faults accommodated some of the displacement on the Whittier-Elsinore fault system so that total displacement may be obscure (Figs. 1, 2 and 3). Stratigraphic separation commonly is so large that similar aged rocks across a single fault trace rarely are available to ascertain displacement.

PRE-UPPER CRETACEOUS BASEMENT ROCKS

Weber (1977) has proposed that Woodson Mountain Granodiorite (PG, Fig. 4) and Bedford Canyon Formation (B, Fig. 4) along the Elsinore fault zone in the Lake Elsinore area have been right-slipped 6 to 6.5 mi (9 to 11 km) (Fig. 4). These rocks are widely distributed so that left-slip, no-slip and right-slip larger than proposed by Weber (1977) also seem possible. Weber (1977) suggests that the Bedford Canyon-Santiago Peak Volcanics contact (S/B, Fig. 4) is displaced. The proposed displaced terranes are different: a large area of Bedford Canyon Formation is exposed southwest of the fault, whereas to the northeast the Bedford Canyon Formation exposure is relatively small in an area of intrusive rocks (Fig. 4).

According to Morton and Miller (1987) contours of biotite ages in basement rock suggest 6 mi (10 km) of right-slip on the Elsinore fault zone. Oxygen isotope contours of age determined from biotite in basement rocks preclude right-slip on either the Elsinore or San Jacinto faults (Taylor and Silver, 1978).

UPPER CRETACEOUS SEDIMENTS

Upper Cretaceous sediments in the northeast Santa Ana Mountains (Fig. 5; Lamar, 1959) consist of: Ladd Formation (Popenoe, 1942, 1954) composed of interbedded marine conglomerate, sandstone and siltstone resting on Trabuco Formation non-marine red conglomerate (Packard, 1916). These sediments unconformably overlie pre-Upper Cretaceous basement.

Scaphites, a small thin-shelled ammonite, in the Ladd Formation northeast of the Whittier fault (B, Fig. 5) indicates deposition in relatively deep water of the infraneritic zone (120 to 600 ft (36 to 180 m) depth, Scott, 1940). Opposite these deep water sediments, southwest of the Whittier-Elsinore fault (D and E, Fig. 5) thick shelled pelecypods and gastropods evidence shallow water deposition in the epineritic zone (120 ft (36 m) or less in depth, Scott, 1940) or shallow to moderate depths (Sundberg, 1982; Saul, 1982). The Ladd Formation northeast of the Whittier fault (A and B, Fig. 5) and in the Coal Canyon area (C, Fig. 5) include graded conglomerate and sandstone beds typical of deep water turbidity current deposition; Ladd Formation sediments

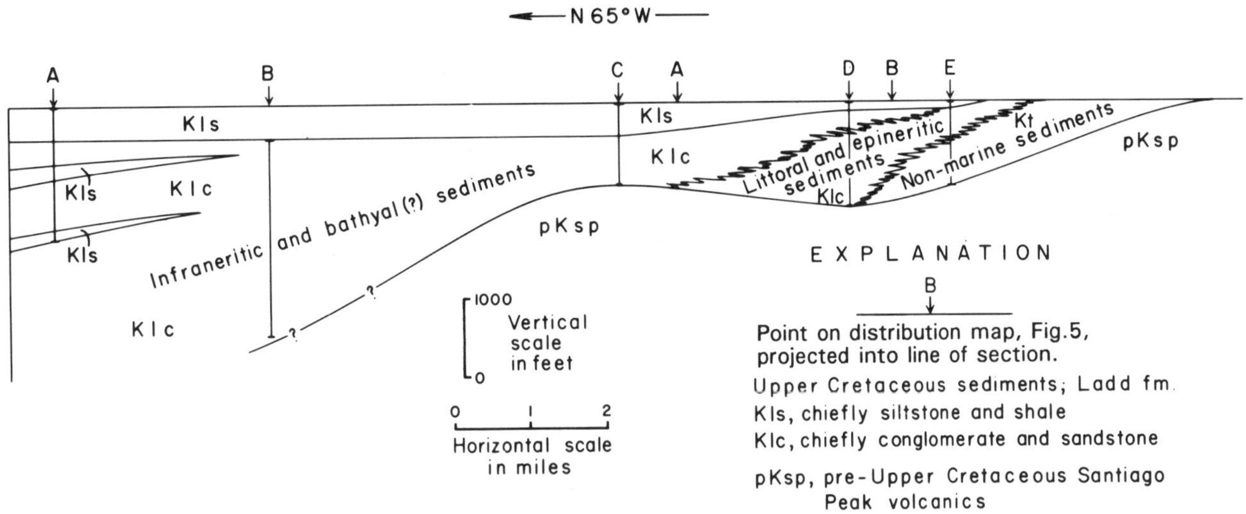

Figure 6. Hypothetical geologic section through Upper Cretaceous rocks in the northeast Santa Ana Mountains prior to assumed right-slip of 8 mi (13 km) on Whittier fault. Points A through E correspond to sections indicated on Figure 5.

southwest of the Whittier fault (D and E, Fig. 5) are massive with no graded beds, implying shallow water deposition.

The deep water facies Ladd Formation northeast of the Whittier-Elsinore fault zone (A and B, Fig. 5) and the shallow water facies Ladd Formation southwest of the Whittier-Elsinore fault (D and E, Fig. 5) were juxtaposed by right-slip on the Whittier-Elsinore fault system. A reconstructed cross section through the basin of deposition (Fig. 6) illustrates a minimum of 8 mi (13 km) of post-Ladd Formation right-slip on the Whittier-Elsinore fault (Lamar, 1959). The magnetostratigraphy of Upper Cretaceous strata from the Santa Ana Mountains indicate 1200 mi (1900 km) of right-slip relative to the Great Valley (Fry and others, 1985). This displacement could have been distributed on several faults, such as the San Andreas, San Jacinto and Whittier-Elsinore.

PALEOCENE SEDIMENTS

Paleocene Silverado Formation sediments are exposed in the Santa Ana Mountains (Woodring and Popenoe, 1945) and were penetrated in the Texaco Inc. A-13-1 well (Fig. 7). The Silverado Formation rests unconformably on Upper Cretaceous sediments in the Santa Ana Mountains and on pre-Upper Cretaceous basement rocks northeast of the Elsinore fault (Fig. 7).

The lower, non-marine member of the Silverado Formation is composed of a basal conglomerate overlain by coarse grained sandstone interbedded with clay and siltstone (Lamar, 1959). The upper marine member in Coal Canyon (CC, Fig. 7) consists of interbedded conglomerate and sandstone, resting on a basal sandstone containing up to 75% angular shell fragments. Northeast of the Whittier-Elsinore fault, in the Fresno Canyon area (FC, Fig. 7), the upper member consists of a basal calcareous conglomeratic sandstone containing scattered shell fragments; the overlying sediments are predominately interbedded shale and siltstone with subordinate sandstone and conglomerate. Graded beds are common. In the Tin Mine Canyon area (TM, Fig. 7) the upper member consists of a basal sandstone containing 75% shell fragments overlain by interbedded 80% sandstone containing shell fragments, and 20% siltstone and shale (Lamar, 1959).

The relatively thick section of fine grained, probable deep water deposits of the upper member northeast of the Whittier-Elsinore fault zone moved opposite the thin section of shallow water deposits southwest of the fault by 19 mi (30 km) of right-slip on the Whittier-Elsinore fault zone (Fig. 7) (Lamar, 1961). Similar right-slip of Paleocene sediments by the Whittier-Elsinore fault system have been proposed: Yerkes and Campbell (1971), 15 mi (24 km); Sage (1973), 25 mi (40 km).

Eckis (1989) has suggested that the Peninsular Range Divide has been right-slipped 25 mi (40 km) by the Elsinore fault zone. Although the age is not given, comparison with the

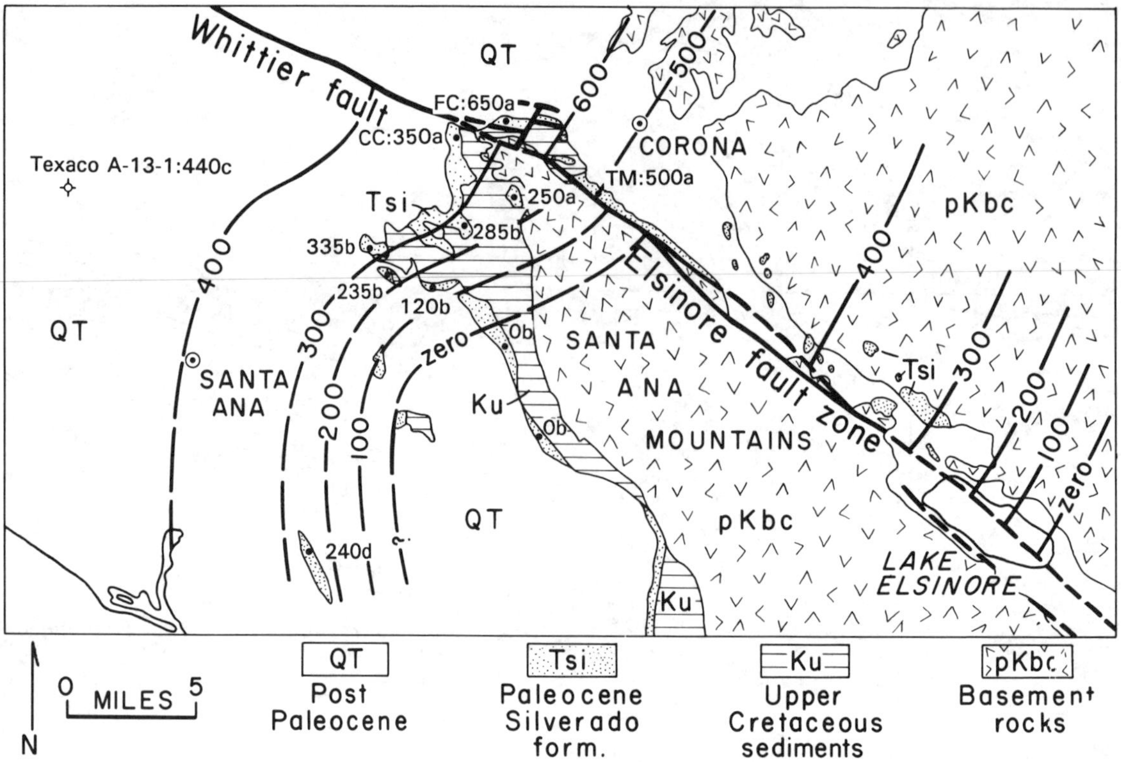

Figure 7. Isopach map of upper (marine) member of the Paleocene Silverado Formation in the Santa Ana Mountains area. Numbers indicate thickness in feet. Sources of thickness information: a: Lamar, 1961; b: Woodring and Popenoe (1945); c: well data; d: estimated from geologic map (Vedder, 1957). Abbreviations: CC: Coal Canyon; FC: Fresno Canyon; TM: Tin Mine Canyon. Location of this figure is shown on Figure 2.

above displacements suggest that the right-slip was Paleocene to Upper Cretaceous.

EOCENE SANTIAGO FORMATION

Fossiliferous middle Eocene Santiago Formation crops out in the Santa Ana Mountains (Woodring and Popenoe, 1945). Northeast of the Whittier-Elsinore fault (FC, Fig. 7), the Santiago Formation consists of interbedded conglomerate, sandstone and siltstone containing middle Eocene fossils (Lamar, 1959). Opposite the Whittier-Elsinore fault in the Coal Canyon area (CC, Fig. 7), the Santiago Formation interval consists of unfossiliferous, coarse grained conglomeratic sandstone identical to the underlying non-marine Sespe Formation (Woodring and Popenoe, 1945; Lamar, 1959). From Tin Mine Canyon (TM, Fig. 7) southeast, fossiliferous Santiago Formation is absent; non-marine Sespe Formation, including sediments in the Santiago Formation interval, rests on Paleocene Silverado Formation (Lamar, 1959; Gray, 1961) similar to the Coal Canyon area (CC, Fig. 7). Present locations of non-fossiliferous Santiago Formation in the Tin Mine Canyon area (TM, Fig. 7) and Coal Canyon (CC, Fig. 7) relative to fossiliferous Santiago Formation northeast of the Whittier-Elsinore fault zone (FC, Fig. 7) indicates at least 6 mi (10 km) of right-slip.

TERTIARY VOLCANIC ROCKS

Middle Miocene volcanic rocks in the Santa Monica Mountains and San Gabriel Valley area (Fig. 8) are similar in composition and perhaps similar in age: Santa Monica Mountains, Relizian stage (Eaton, 1958) or Luisian and Relizian stages (Hardey, 1958); San Gabriel Valley, Luisian and Relizian stages (Eaton, 1958). Middle Miocene volcanic rocks thin to the north in the San Gabriel Valley subsurface and Santa Monica Mountains (Eaton, 1958) (Fig. 8). Similar compositions, ages and thickness variations support continuity of the volcanic rocks prior to 19 mi (30 km) of post-Paleocene right-slip on the Whittier-Elsinore fault system. Such displacement explains the absence of middle Miocene volcanic rocks in the Eagle Rock-Highland Park area (Figs. 3 and 8).

Kennedy (1977) has suggested that the distribution of Tertiary volcanic rocks (Tv, Fig. 4) capping mesas and isolated high areas on opposite sides of the Elsinore fault zone southeast of Lake Elsinore provides a measure of right-slip; estimated to be 9-10 mi (15-16 km) (Fig. 4). Morton and Morton (1979) have torpedoed this reconstruction by reporting that volcanic rocks southwest of the fault are 7 to 8 m.y. old, whereas volcanic rocks northeast of the fault are 10.6 m.y. old.

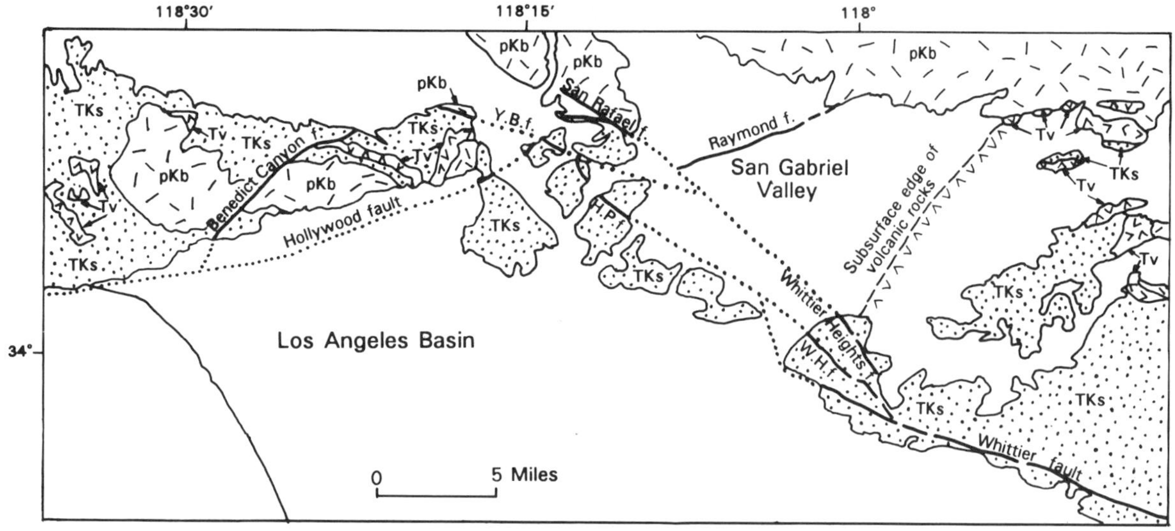

Figure 8. Geologic map showing distribution of Middle Miocene volcanic rocks along the northern margin of the Los Angeles Basin and around the San Gabriel Valley. Generalized from Durrell (1954), Eaton (1958) and Woodford and others (1954). Abbreviations: blank: Quaternary sediments; TKs: undifferentiated Tertiary and Cretaceous sediments; Tv: middle Miocene volcanic rocks; pKbc: pre-Cretaceous basement rocks; H.P.f.: Highland Park fault; W.H.f.: Workman Hill fault; Y.B.f.: York Boulevard fault. Location of this figure is shown on Figure 1.

UPPER MIOCENE SEDIMENTS

The upper Miocene Puente Formation consists of interbedded conglomerate, sandstone, siltstone and shale containing microfossils of the Delmontian and Mohnian Stages (Natland and Rothwell, 1954). The Puente Formation is exposed in the Elysian Park-Repetto Hills area (Lamar, 1970) (Figs. 3 and 9) and the Puente Hills (Durham and Yerkes, 1964; Yerkes, 1972). The Modelo Formation of similar age and lithology (Hoots, 1931; Sullwold, 1960) is exposed in the Santa Monica Mountains (Fig. 3). The Puente-Modelo Formations rest unconformably on the Topanga Formation and older rocks. The Topanga Formation contains fossils of the Relizian-Luisian stages (Natland and Rothwell, 1954) and consists of shale, sandstone and conglomerate and in the Santa Monica Mountains and in the San Gabriel Valley area (Fig. 8) includes volcanic and shallow intrusive rocks.

The maximum thickness of the Mohnian lower Puente Formation north of the Whittier fault is located in the central Puente Hills (Woodford and others, 1944, 1954; Yeats and Beall, 1991). Prior to 19 mi (30 km) of right-slip along the Whittier fault system, this area of maximum thickness would have been directly northeast of the area of maximum thickness of similar aged Mohnian rocks in the Elysian Park area (Fig. 2).

The Highland Park fault strand of the Whittier fault system is the contact between Fernando Formation on the east against Topanga Formation on the west (Lamar, 1970, plate 1) directly south of York Boulevard. Continuation of the fault to the southeast is evidenced by: 1. truncation of sandstone beds, 2. discordance in attitudes, 3. truncated "drag" folds (Fig. 10, upper; Lamar, 1970, plate 1), 4. an abrupt change in depth to basement rock and differences in upper Miocene sections in wells on opposite sides of the fault. Well data indicate a complete, 1300 feet (370 meters) thick section of the Mohnian portion of the Puente Formation in the Continental Oil Company well located northeast of the Highland Park fault (Fig. 11). The Texaco Inc., well, southwest of the fault, penetrated a 2100 feet (600 meters) thick, incomplete section of Mohnian sediments with different electric log characteristics than similar aged sediments in the Continental Oil Company well located directly to the northeast (Fig. 11). Neither Weber (1979) nor Dibblee (1989) (Fig. 10, lower) recognized the Highland Park fault; Stark and Thompson (1970) show the fault in a somewhat different location.

Northeast of the Highland Park fault the Puente Formation consists of diatomaceous shale, shale and siltstone overlying a basal conglomeratic sandstone containing siliceous shale and sandstone fragments derived from the underlying Topanga Formation (Fig. 9). Interbedded siltstone, sandstone and shale beneath the Puente Formation are assigned to the Topanga Formation (Fig. 9) (Lamar, 1970), on the basis of predominantly Luisian age determinations (chart 2, Lamar, 1961) and an unconformable contact evidenced by the basal Puente conglomeratic sandstone and differences in rocks, microfauna and attitudes above and below. Stark and Thompson (1970) recognized the unconformity, but report lower Mohnian microfossils from sediments beneath the unconformity. Apparently because the Mohnian/Luisian Stage boundary at many locations corresponds to the Modelo-Puente/Topanga Formations contact, Stark and Thompson (1970) assign sediments beneath the unconformity to the Puente Formation.

The unconformity in the Highland Park area is the base of the Puente Formation, rather than the Mohnian/Luisian boundary, because formation boundaries must not be drawn on the basis of time horizons, however determined, but on physical characteristics (The North American Commission on Stratigraphic Nomenclature, 1983); an unconformity is con-

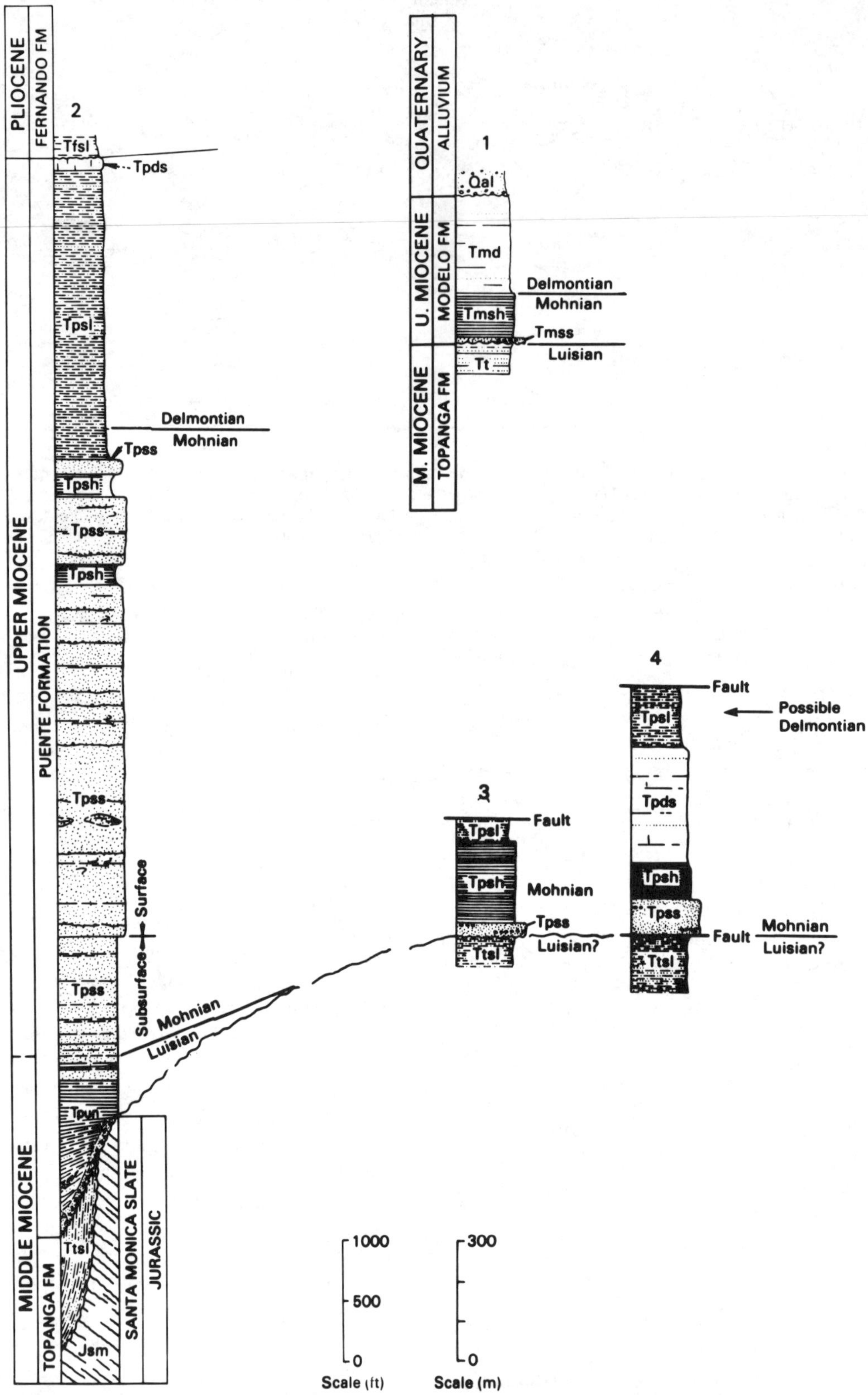

Figure 9. Columnar sections of Modelo Formation in Santa Monica Mountains from Elliott (1951), Section 1 and Puente Formation in Elysian Park-Repetto Hills area from Lamar (1970, p. 25), Sections 2, 3 and 4. Abbreviations: Topanga Formation: Ttsl: siltstone; Tt: interbedded sandstone and siltstone; Puente Formation: Tpun: undifferentiated shale, siltstone, sandstone and conglomerate; Tpss: sandstone and conglomerate; Tpsh: shale; Tpsl: siltstone; Tpds: diatomaceous shale; Modelo Formation: Tmss: sandstone and conglomerate; Tmsh: shale; Tmd: diatomaceous shale; Fernando Formation: Tfsl: siltstone; Alluvium: Qal. Locations of sections shown on Figure 3.

sidered an ideal physical boundary and formations should never straddle regional unconformities.

The Puente Formation directly northeast of the Highland Park fault (Figs. 3 and 9, sections 3 and 4) is similar to the lower portion of the Modelo Formation exposed in the eastern Santa Monica Mountains (Elliott, 1951; Harding, 1952; Hardey, 1958; Sullwold, 1960) (Fig. 9, section 1). In both areas the section consists of approximately 2000 ft (600 m) composed of a basal conglomeratic sandstone overlain by siltstone, shale and diatomaceous shale containing Delmontian and Mohnian microfossils resting unconformably on the Topanga Formation. These similarities in the Puente-Modelo Formations across the Highland Park fault are explained by 6 to 9 mi (10 to 15 km) of right-slip. The remaining 9 to 12 mi (15 to 20 km) of the 19 mi (30 km) of right-slip on the Whittier-Elsinore fault system is distributed on other northwest trending faults in the Elysian Park-Repetto Hills area such as the Elysian Park, Eagle Rock-Verdugo, San Rafael and York Boulevard faults (Fig. 3). Lamar (1970, 1975), Hill and others (1976) and Dibblee (1989) debate "the York Boulevard fault".

UPPER MIOCENE STRUCTURES

Right-slip on the Whittier fault system is further evidenced by the wrench tectonic (Harding and Lowell, 1979) pattern and orientation of folds along individual northwest trending faults in the Elysian Park-Repetto Hills area (Lamar, 1970, plate 1) and along the Whittier fault in the Puente Hills (Kundert, 1952; Durham and Yerkes, 1964; Fig. 12). Along the Whittier fault in the Kraemer area (Fig. 12) overturned beds and folds are deciphered from graded beds and other sedimentary structures (Lamar, 1961). Durham and Yerkes (1964) interpret the area of Figure 12 differently. Figure 12 and Tan and others (1984, plate 2) show east-west trending, overturned folds which die out or become upright away from the Whittier fault. Similar relationships are seen in the Elysian Park-Repetto Hills area where overturned folds die out or become upright at a distance from individual northwest trending strands of the Whittier fault system (Lamar, 1970).

In the Kraemer area (Fig. 12) Pliocene Fernando Formation rests unconformably on folded Puente Formation and in the Repetto Hills area (Lamar, 1970, plate 1) the Fernando Formation rests unconformably across the Elysian Park fault illustrating pre-Pliocene displacement on the Whittier fault system. The Elysian Park fault lines up with the Whittier fault at the northwest end of the Puente Hills (Lamar, 1970, Fig. 24) and is a pre-Pliocene strand of the Whittier fault system.

Folds in the uppermost Puente Formation in the Kraemer area die out up section (sect. A-A', Fig. 12). This illustrates fold growth during deposition secondary to right-slip on the Whittier fault, similar to relationships along the San Andreas fault in the San Joaquin Valley (Harding, 1976). Uncon-

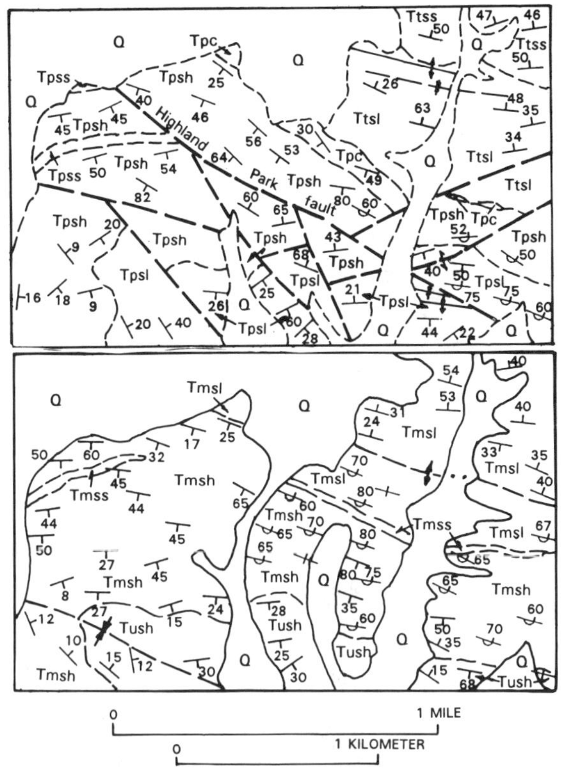

Figure 10. Geologic maps of same area in hills southeast of Highland Park. Upper map from Lamar (1970, plate 1), lower map from Dibblee (1989). Abbreviations, upper map: Q: Quaternary sediments; Tertiary Puente Formation: Tpsl: siltstone; Tpsh: shale; Tpss: sandstone; Tpc: conglomeratic sandstone; Tertiary Topanga Formation: Ttsl: siltstone; Ttss: sandstone, lower map: Q: Quaternary sediments; Tush: unnamed shale; Tertiary Monterey Formation: Tmsh: shale; Tmsl: shale and siltstone; Tmss: sandstone. Location of this figure is shown on Figure 3.

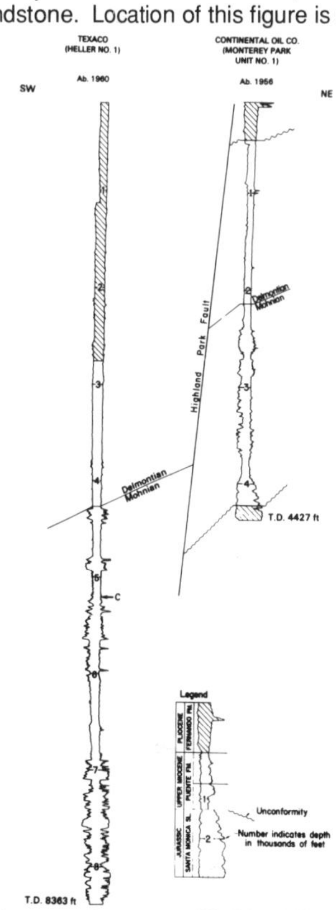

Figure 11. Subsurface section across Highland Park fault showing electric log characteristics and paleontologic data on rocks penetrated by wells on opposite sides of the fault in the Repetto Hills. Location of section shown on Figure 3.

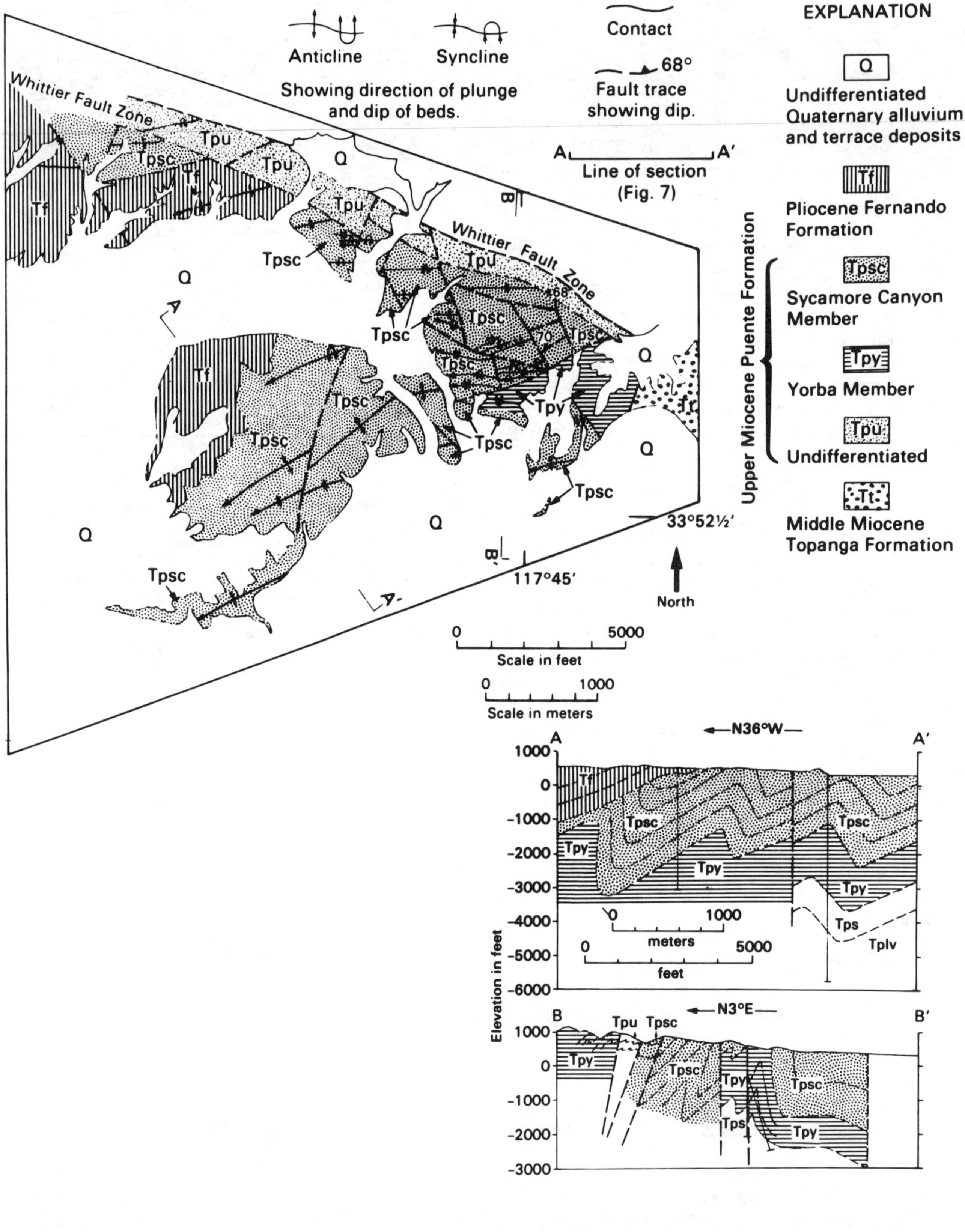

Figure 12. Geologic map and structure sections of the Kraemer area, generalized from Lamar (1961, plate III). Explanation of symbols for rocks in subsurface, not shown above: Puente Formation: Tps: Soquel Member; Tplv: La Vida Member. Location of this figure shown on Figure 2.

formities and the distribution of coarser sediments in upper Miocene rocks in the Sansinena Oil Field adjacent to the Whittier fault also suggest faulting or folding contemporaneous with deposition (Woodward, 1958).

OTHER OPINIONS

According to Woodford and others (1971) large post-Cretaceous right-slip on the Whittier-Elsinore fault system is precluded by the distribution of distinctive Upper Cretaceous and Paleocene rocks on both sides of the fault system; the size of "large" or any supporting rock descriptions or references are not given. Woodford and Gander (1977), are correct when they state: "The displacements on the Whittier-Elsinore fault zone . . . have been misunderstood by some geologists", they may not be correct in stating that Miocene or later strike-slip has been near zero; left-slip of 20 mi (32 km) of the Santiago Peak Volcanics along the Whittier-Elsinore fault system is discussed. Gourley (1971) found similar thicknesses of Puente Formation units on opposite sides of the Whittier fault in the northwest Puente Hills and on this basis rejected significant post-Puente Formation horizontal slip. According to Wright (1991) the eastern limits of the middle Puente sandstones show right-separation of about 5 mi (8 km) by the eastern trace of the Whittier fault zone which may be the result of variations in the shape of the sand body rather than right-slip. Yeats and Beall (1991) also question significant post-Mohnian right-slip on the Whittier fault. Gray (1961) limits right-slip on the Whittier-Elsinore fault system to 1.5 mi (2.4 km).

POST-MIOCENE DISPLACEMENT

Pliocene and Miocene displacements on the Whittier fault are indicated by landslide debris composed of upper Miocene sediments in Pliocene sediments (Woodward, 1958; Kundert, 1952), steep dips in and an unconformity at the base of the Pleistocene La Habra Formation (Kundert, 1952; Yerkes, 1972). Yerkes' (1972) estimate of 15,000 feet (4580 meters) of right-oblique slip on the Whittier fault zone in the Puente Hills is similar to the 3 mi (5 km) offset of Conrey's (1967) lower Pliocene marine fans proposed by Lamar (1961).

According to Kennedy (1977) southeast of Lake Elsinore an unnamed Pleistocene sandstone and conglomerate unit is right-slipped more than 3 mi (5 km) along the Wildomar strand of the Elsinore fault zone. Late Quaternary displacement is indicated by right-slip of stream canyons in the Puente Hills of up to 8800 ft (2600 m) (Durham and Yerkes, 1964) Wright (1991) argues that stream canyons may erode along the fault zone and give a false picture of displacement. Stream canyons along the San Rafael fault in the San Rafael Hills (Fig. 3) show 1000 feet (300 meters) of right-slip (Lamar, 1970). Although much of the Elsinore fault zone underlies straight canyons that obscure offset minor stream courses, right-slip of stream courses along the Elsinore fault zone have been reported from the north end of the Santa Ana Mountains to the Mexican Border (Lamar and Rockwell, 1986).

Study of trenches across the Whittier fault in the Puente Hills yields a late Quaternary slip rate of 2.0 to 2.5 mm/yr (Gath, in prep.). Slip rates on the Elsinore fault zone are 3.0 to 9.3 mm/yr (Lamar and Rockwell, 1986). The higher slip rates along the Elsinore fault compared to the Whittier fault could be distributed on the Whittier and Chino faults (Fig. 2) and folding in the Puente Hills (Rockwell and others, 1988). But right-slip on the Chino fault is no more than 0.06 mm/yr (Heath and others, 1982) so the required folding seems excessive.

The main shock of the Whittier Narrows earthquake was on an east-west trending north dipping thrust fault (Weber, 1987) interpreted by Hauksson and others (1988) to be a major thrust beneath the Elysian Park anticline. This thrust fault has the proper orientation to be secondarily related to right-slip (Hancock and Atiya, 1979) on a strand of the Whittier fault system and need not be the regional feature visualized by Hauksson and others (1988). The major aftershock on the Whittier Narrows earthquake was caused by right-slip along a steeply dipping, north-northwest striking fault which truncates the thrust fault causative of the main shock (Hauksson and others, 1988). The epicenter lies between the Puente and Elysian Park-Repetto Hills (Fig. 2). Thus data on the Whittier Narrows earthquake substantiate a wrench or strike-slip regime and connections between right-slip faults in the Puente and Elysian Park-Repetto Hills (Fig. 2).

DISCUSSION

Division of the late Quaternary 2 to 2.5 mm/yr slip rate on the Whittier fault into the proposed 19 mi (30 km) of right-slip yields 12 to 15 million years, Mohnian to Luisian; 3 to 9.3 mm/yr on the Elsinore fault system yields 10 to 3 million years, late Mohnian to late Pliocene (Berggren, 1972; Berggren and others, 1985). According to Atwater (1970), at the latitude of Los Angeles, strike slip on the San Andreas and related faults, such as the Whittier-Elsinore fault system, began 20 million years ago. Assuming late Quaternary slip rates in the past, all rocks of Mohnian to Luisian age and older along the Whittier fault and late Mohnian to middle Pliocene age and older along the Elsinore fault should be offset the same 19 mi (30 km) and younger rocks and features should show offset that diminishes with decreasing age.

Although projection of late Quaternary slip rates along the Whittier fault system into the past fits the geologic data, Woodford and others (1971), Woodford and Gander (1977), Gourley (1971), Gray (1961), Wright (1991) and Yeats and Beall (1991) reject 19 mi (30 km) of post Mohnian right-slip on the Whittier-Elsinore fault system. The discrepancy could be explained if past slip rates were different from late Quaternary rates; slip rates have varied along the San Jacinto fault (Sharp, 1981). Another possibility is that the previous studies of Mohnian sediments across the Whittier-Elsinore fault system do not cover enough area to detect 19 mi (30 km) of right-slip or do not consider slip on faults that splay off of the Whittier-Elsinore fault system.

Although Woodford and others (1971) and Woodford and Gander (1977) also question post Paleocene right-slip, others agree on 15 to 25 mi (24 to 40 km) of right-slip of Paleocene sediments along the Whittier-Elsinore fault system (Lamar, 1961; Yerkes and Campbell, 1971; Sage, 1973). If this displacement occurred, then the suggested right-slip of Upper Cretaceous and Eocene sediments also occurred. Although Upper Cretaceous sediments and fossils on the northeast slope of the Santa Ana Mountains have long been known (Packard, 1916; Lamar, 1959; Gray, 1961; Schoellhamer and others, 1981), regional studies of Upper Cretaceous sediments in the Santa Ana Mountains have not ventured into this area (Lang, 1978; Saul, 1982; Sundberg and Cooper, 1982) and leave the suggestion of Upper Cretaceous facies displacement along the Whittier-Elsinore fault system (Lamar, 1959) untested.

Although right-slip of basement rocks must be at least as great as the Paleocene offset, Weber (1977) and Morton and Miller (1987) indicate that basement is right-slipped 6 mi (10 km) and Taylor and Silver (1978) dispute any offset. The displaced terranes proposed by Weber (1977) seem too dissimilar to support the suggested right-slip (Fig. 4). Morton and Miller (1987) cannot be evaluated because their map has not been published. The oxygen isotope contours of Taylor and Silver (1978) showing no right-slip and geologic data indicating 19 mi (30 km) of right-slip on the Whittier-Elsinore fault system cannot be reconciled.

ACKNOWLEDGMENTS

Jeannine Lamar, the late Mason Hill, Tom Dibblee and Robert Yerkes provided the major impetus for this effort. Howard Stark was generous with micropaleo data on the sediments below the Highland Park unconformity.

REFERENCES

Anderson, J. F., T. Rockwell and D. C. Agnew, 1989, Seismic hazard of San Diego: Earthquake Spectra.

Atwater, T., 1970, Implications of plate tectonics for the Cenozoic tectonic evolution of western North America: Geol. Soc. Am. Bull., vol. 81, p. 3513-3536.

Berggren, W. A., 1972, A Cenozoic time scale - some implications for regional geology and paleobiogeography: Lethaia, vol. 5, p. 195-215.

Berggren W. A., D. V. Kent, J. J. Flynn and J. A. Van Couvering, 1985, Cenozoic Geochronology: Geol. Soc. Am. Bull., vol. 96, p. 1407-1418.

Conrey, B. L., 1967, Early Pliocene sedimentary history of the Los Angeles basin, California: Calif. Div. Mines and Geology Spec. Rpt. 93, 63 p.

Crowell, J. C., 1962, Displacement along the San Andreas fault, California: Geol. Soc. America Spec. Paper 71, 61 p.

Dibblee, T.W., Jr., 1989, Geologic map of the Los Angeles Quadrangle, Los Angeles County, California: Dibblee Geologic Foundation, Santa Barbara, CA.

Durham D. L. and R. F. Yerkes, 1964, Geology and oil resources of the eastern Puente Hills area, southern California: U. S. Geol. Survey, Prof. Paper 420-B, 62 p.

Eaton, G. P., 1958, Miocene volcanic activity in the Los Angeles basin: in J. W. Higgins, (ed.), A guide to the geology and oil fields of the Los Angeles and Ventura regions, Pacific Section of the Assoc. of Petroleum Geologists, p. 55-58.

Eckis, R., 1989, Geomorphic evidence for 80 Km of right-slip on the Elsinore and San Jacinto fault zones, California: Abstracts with Program, Geol. Soc. of America, Cordilleran/Rocky Mountain Section, p. 75.

Elliott, J. L., 1951, Geology of the eastern Santa Monica Mountains between Laurel Canyon and Beverly Glen Boulevards, Los Angeles County, California: unpublished Master's thesis, University of California, Los Angeles, 57 p.

Engle, R., 1959, Geology and mineral deposits of the Lake Elsinore quadrangle, California: Calif. Div. Mines, Bull. 146, 154 p.

Fry, J. G., D. J. Bottjer and S. P. Lund, 1985, Magnetostratigraphy of displaced Upper Cretaceous strata in southern California: Geology, vol. 13, p. 648-651.

Gath, Eldon, in preparation, Tectonic geomorphology and neotectonics of the Whittier fault in southern California: Masters thesis, California State University, Los Angeles.

Gourley, J. W., III, 1971, Stratigraphy and sedimentation patterns upper Miocene Puente Formation (Mohnian - Delmontian) northwestern Puente Hills, Los Angeles County, California: unpublished Master's thesis, University of Southern California.

Gray, C. H., Jr., 1961, Geology and mineral resources of the Corona South Quadrangle: Calif. Div. Mines and Geol., Bull, 178, 120 p.

Hancock, P. L. and M. S. Atiya, 1979, Tectonic significance of mesofracture systems associated with the Lebanese segment of the Dead Sea transform fault: Jour. Structural Geology, vol. 1, p. 143-153.

Hardey, G. W., 1958, Stratigraphy of the Topanga Formation in the eastern Santa Monica Mountains: unpublished Master's thesis on file at University of California, Los Angeles, 85 p.

Harding, T. P., 1952, Geology of the eastern Santa Monica Mountains between Dry Canyon and Franklin Canyon: unpublished Master's thesis on file at University of California, Los Angeles. 54 p.

Harding, T. P., 1976, Tectonic significance and hydrocarbon trapping consequences of sequential folding synchronous with San Andreas faulting, San Joaquin Valley, California: Am. Assoc. Petrol. Geol., Bull., vol. 60, p. 356-378.

Harding, T. P. and J. D. Lowell, 1979, Structural styles, their plate-tectonic habitats, and hydrocarbon traps in petroleum provinces: Am. Assoc. Petrol. Geol., Bull. vol. 63, p. 1016-1058.

Hauksson, E., L. M. Jones, T. L. Davis, L. K. Hutton, A. G. Brady, P. A. Reasenberg, A. J. Michael, R. F. Yerkes, P. Williams, G. Reagor, C. W. Stover, A. L. Bent, A. K. Shakal, E. Etheredge, R. L. Porcella, C. G. Bufe, M. J. S. Johnson and E. Cranswick, 1988, The 1987 Whittier Narrows earthquake in the Los Angeles Metropolitan area, California: Science, vol. 239, p. 1409-1412.

Heath, E. G., D. E. Jensen and D. W. Lukesh, 1982, Style and age of deformation on the Chino fault: in Neotectonics in southern California, guidebook for 78th Annual Meeting, Cordilleran Section, Geol. Soc. Amer., p. 123-134.

Hill, R. L., E. C. Sprotte, J. H. Bennett, R. H. Chapman, G. W. Chase, C. R. Real and R. W. Sherburne, 1976, Santa Monica-Raymond Hill fault zone, Los Angeles County, California: Annual Technical Report for Fiscal Year 1976, U. S. Geol. Survey Contract No. 14-08-0001-G263, 117 p.

Hoots, H. W., 1931, Geology of the eastern part of the Santa Monica Mountains, Los Angeles County, California: U. S. Geol. Survey Prof. Paper 165-c, 134 p.

Kennedy, M. P., 1977, Recency and character of faulting along the Elsinore fault zone in southern Riverside County, California: Calif. Div. Mines and Geol. Spec. Report 131, 12 p.

Kundert, C. J., 1952, Geology of the Whittier-La Habra area, Los Angeles County, California: Calif. Div. Mines, Spec. Report 18, 22 p.

Lamar, D. L., 1959, Geology of the Corona area, Orange, Riverside, and San Bernardino Counties, California: M.A. thesis, Univ. Calif., Los Angeles, 95 p.

Lamar, D. L., 1961, Structural evolution of the northern margin of the Los Angeles basin: Ph.D. thesis, Univ. Calif., Los Angeles, 142 p.

Lamar, D. L., 1970, Geology of the Elysian Park-Repetto Hills area, Los Angeles County, California: Calif. Div. Mines and Geol., Spec. Report 101, 45 p.

Lamar, D. L., 1975, Correction to CDMG Special Report 101: Calif. Geology, vol. 28, p. 170.

Lamar, D. L. and T. K. Rockwell, 1986, An overview of the tectonics of the Elsinore fault: in Neotectonics and Faulting in Southern California, Geol. Soc. Amer., Cordilleran Sec., Guidebook and Volume, p. 149-157.

Lang, H. R., 1978, Late Cretaceous biostratigraphy of the southeastern Los Angeles Basin: Calif Div. Oil and Gas Report TR20, 15 p.

Larsen, E. S., Jr., 1951, Crystalline rocks of the Corona, Elsinore and San Luis Rey Quadrangles, southern California: Calif. Div. Mines and Geology Bull. 159, p. 7-50.

Mann, J. F., Jr., 1955, Geology of a portion of the Elsinore fault zone, California: Calif Div. Mines and Geology Spec. Report 43, 22 p.

Morton, D. M. and F. K. Miller, 1987, K/AR apparent ages of plutonic rocks from the northern part of the Peninsular Ranges batholith, southern California: Abstracts with Program, Cordilleran Section, Geol. Soc. of America, p. 435.

Morton, J. L. and D. M. Morton, 1979, K-AR ages of Cenozoic volcanic rocks along the Elsinore fault zone, southwestern Riverside County, California: Cordilleran Section, Geol. Soc. Am., abstracts with programs, p. 119.

Natland, M. L. and W. T. Rothwell, Jr., 1954, Fossil foraminfera of the Los Angeles and Ventura regions, California: Calif. Div. Mines and Geology Bull. 170, chapt III. p. 33-42.

Packard, E. L., 1916, Faunal studies in the Cretaceous of the Santa Ana Mountains of southern California: Univ. Calif. Publ., Bull. of the Dept. of Geology, vol. 9, p. 137-159.

Popenoe, W. P., 1942, Upper Cretaceous formations and faunas of southern California: Amer. Assoc. Petrol. Geol., Bull., v. 26, p. 162-187.

Popenoe, W. P., 1954, Mesozoic formations and faunas, southern California and northern Baja California: Calif. Div. Mines and Geol., Bull. 170, Chapt. III, p. 15-21.

Proctor, R. J., R. Crook, Jr., M. H. McKeown, and R. L. Moresco, 1972, Relation of known faults to surface ruptures, 1971 San Fernando earthquake, southern California: Geol. Soc. Am. Bull., vol. 83, p. 1601-1618.

Proctor, R. J., 1973, Map showing recently active faults in the southern California region: in Geology, Seismicity and Environmental Impact, Assoc. Eng. Geol. Special Publication, Los angeles.

Rockwell, Thomas, Eldon Gath and Doug Cook, 1988, Soil chronology and tectonic geomorphology of the Whittier fault zone, Yorba Linda, California, a preliminary estimate of the late Quaternary lateral slip rate: in Gath, Eldon and Larry Cann, (ed.), Neotectonics of the Whittier fault, temporary field trip guidebook, Assoc. Engineering Geologists, Los Angeles.

Sage, O. G., Jr., 1973, Paleocene geography of the Los Angeles region: in Proc. of Conf. on Tectonic Problems of San Andreas Fault System, Stanford Univ. Publ., Geol. Sci., vol. XIII, p. 348-357.

Saul, L. R., 1982, Water depth indicators from Late Cretaceous mollusks, Santa Ana Mountains, California: in Bottjer, D. J., I. P. Colburn and J. D. Cooper, (ed.), Late Cretaceous depositional environments and paleogeography, Santa Ana Mountains, southern California: Pacific Section Society of Economic Paleontologists and Mineralogists, p. 69-88.

Schoellhamer, J. E., J. G. Vedder, R. F. Yerkes, and D. M. Kinney, 1981, Geology of the Northern Santa Ana Mountains, California: U. S. Geol. Survey, Prof. Paper. 420-D, 109 p.

Scott, G., 1940, Paleoecological factors controlling the distribution and mode of life of Cretaceous ammonoids in the Texas area: Jour. of Paleontology, vol. 14, p. 299-323.

Sharp, R. V., 1981, Variable rates of late Quaternary strike slip on the San Jacinto fault zone, southern California: Jour. Geophy. Research, vol. 86, p. 1754-1762.

Stark, H. E. and J. H. Thompson, 1970, Geology of the western San Gabriel Valley: unpublished petroleum consultant report.

Sullwold, H. H., Jr., 1960, Tarzana fan, deep submarine fan of late Miocene age, Los Angeles County, California: Am. Assoc. Petrol. Geol. Bull., vol. 44, p. 433-457.

Sundberg, F. A., 1982, Late Cretaceous paleoenvironments and paleoecology, Santa Ana Mountains, Orange County, California: in Bottjer, D. J., I. P. Colburn and J. D. Cooper, (eds.), Late Cretaceous depositional environments and paleogeography, Santa Ana Mountains, southern California, Pacific Section Society of Economic Paleontologists and Mineralogists, p. 59-68.

Sundberg, F. A. and J. D. Cooper, 1982, Late Cretaceous (Campanian) paleogeography of southern California: in Bottjer, D. J., I. P. Colburn and J. D. Cooper, (eds.), Late Cretaceous depositional environments and paleogeography, Santa Ana Mountains, southern California, Pacific Section Society of Economic Paleontologists and Mineralogists, p. 115-121.

Tan, S. S., R. V. Miller and J. R. Evans, 1984, Environmental geology of parts of the La Habra, Yorba Linda and Prado Dam quadrangles, Orange County, California: Calif. Div. Mines and Geol., Open-File Report 84-24 LA, 113 p.

Taylor, H. P. and L. T. Silver, 1978, Oxygen isotope relations in plutonic igneous rocks of the Peninsular Ranges batholith, southern and Baja California: in R. E. Zartman, editor, short papers of the fourth International conference, geochronology, cosmochronology, isotope geology, U. S. Geol. Surv. Open-File Report 78-701, p. 423-426.

Weber, F. H., Jr., 1977, Seismic hazards related to geologic factors, Elsinore and Chino fault zones, northwestern Riverside County, California: Calif. Div. Mines and Geol., Open-file Report 77-4 LA, 96 p.

Weber, F. H., Jr., 1979, Geology and surface features related to character and recency of movement along faults, north-central Los Angeles area, Los Angeles County, California - a preliminary report: in Earthquake hazards associated with faults in the greater Los Angeles Metropolitan area, Los Angeles County, California, including faults in the Santa Monica-Raymond, Verdugo-Eagle Rock, and Benedict Canyon fault zones, Calif. Div. Mines and Geology Open-File Report 79-16, p. D1-D56.

Weber, F. H., 1987, Whittier Narrows earthquakes: Calif. Geol., December 1987, p. 275-281.

Woodford, A. O. and C. Gander, 1977, Los Angeles erosion surface of Middle Cretaceous age: Am. Assoc. Petrol. Geol. Bull., vol. 61, p. 1979-1990.

Woodford, A. O., J. E. Schoellhamer, J. G. Vedder, and R. F. Yerkes, 1954, Geology of the Los Angeles Basin: Calif. Div. Mines and Geology, Bull. 170, chapt. II, p. 65-81.

Woodford, A. O., J. S. Shelton, D. O. Doehring and R. K. Morton, 1971, Pliocene-Pleistocene history of Perris block, southern California: Geol. Soc. Am. Bull. vol. 82, p. 3421-3448.

Woodford, A. O., J. S. Shelton and T. G. Moran, 1944, Geology and oil possibilities of Puente and San Jose Hills, California: U. S. Geol. Survey, Oil and Gas Inv. Prelim. Map 23.

Woodring, W. P. and W. P. Popenoe, 1945, Paleocene and Eocene stratigraphy of northern Santa Ana Mountains, Orange County, California: U. S. Geological Survey Oil and Gas Inv. Prelim. Chart 12.

Woodward, A. F., 1958, Sansinena Oil Field: in Higgins, J.W. (ed.), A guide to the geology and oil fields of the Los Angeles and Ventura regions: Pacific Sec., Amer. Assoc. Petrol Geol., p. 109-118.

Wright, T. L., 1991, Structural geology and tectonic evolution of the Los Angeles basin, California: in Biddle, K. T. (ed.), Active margin basins, Amer. Assoc. Petrol. Geol., Memoir 52, p. 35-134.

Yeats, R. S. and J. M. Beall, 1991, Stratigraphic controls of oil fields in the Los Angeles basin, California: in Biddle, K. T., (ed.), Active margin basins, Amer. Assoc. Petrol. Geol. Memoir 52, p. 221-235.

Yerkes, R. F., 1972, Geology and oil resources of the western Puente Hills area, southern California: U.S. Geol. Survey, Prof. Paper 420-C, 63 p.

Yerkes, R. F. and R. H. Campbell, 1971, Cenozoic evolution of the Santa Monica Mountains-Los Angeles basin area: U. S. Geol. Survey, Open-file report.

THE SANTA YNEZ FAULT AT JUNCAL DAM, SANTA BARBARA COUNTY

Karl VonderLinden
Consulting Geologist
5 Stowe Lane, Menlo Park, CA 94025

ABSTRACT

Juncal Dam is a major water storage structure on the Santa Ynez River in southeastern Santa Barbara County. The Santa Ynez Fault, regarded as an active fault by many geologists, is a major structural feature that essentially parallels the Santa Ynez River in the vicinity of Juncal Dam. Here the fault comprises two roughly parallel, east-west traces, the more northerly of which passes beneath part of the dam structure. Fortunately, the weight of geologic evidence indicates that the most recent movement on the Santa Ynez Fault in this area has occurred on the more southerly trace that passes well south of the dam, and that the trace that does pass beneath the dam has not moved for several tens of thousands of years.

THE SANTA YNEZ FAULT

The Santa Ynez Fault is a major structural feature in southern California. It extends from Sespe Canyon, in eastern Ventura County, approximately 90 miles westward to western Santa Barbara County. The fault is the boundary between the western part of the Transverse Ranges geomorphic province and the southern end of the Coast Ranges geomorphic province. Its trace is, in a gross sense, continuously visible throughout most of its length, either by breaks in the topography or steep, relatively straight scarps that define borders of mountain ranges. Typical examples are the northern boundaries of the Santa Ynez Mountains and the Topatopa Mountains.

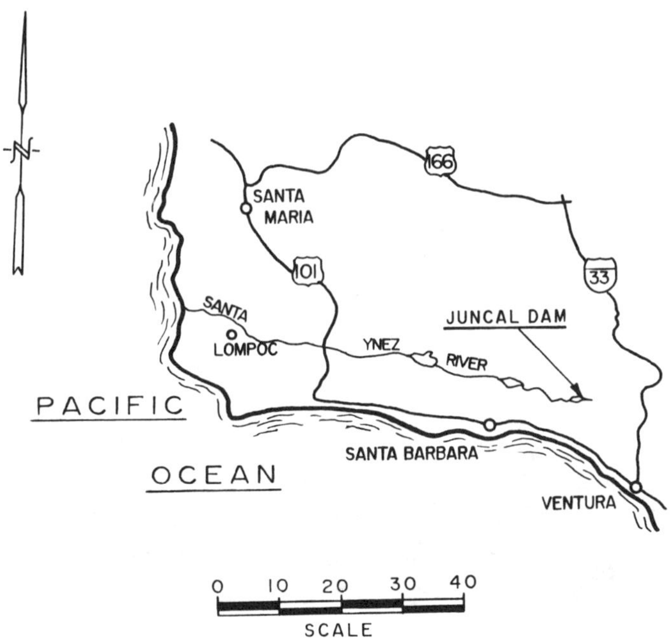

Figure 1. Location map showing major cities, highways, the Santa Ynez River, and Juncal Dam.

Movement along the fault has been oblique. Vertical offset appears to be greater than lateral offset, and is of the order of several miles (Dibblee, 1982). The sense of vertical displacement is characterized by ground on the southern side of the fault being elevated relative to that on the northern side. The horizontal component of movement is left-lateral; i.e., the southern block is moving eastward relative to the northern block. Magnitude of horizontal displacement as indicated by stream offset along the fault, is at least three miles (Dibblee, 1966).

Clear exposures of the Santa Ynez Fault are few because of widespread landslide, fanglomerate, and alluvial deposits along its trace. Dibblee, who has mapped essentially the entire length of the fault, describes the larger exposures as gouge zones ranging in width from several tens of feet to as much as a thousand feet (1966). In several areas the fault consists of two or more closely spaced and roughly parallel traces.

Dip along the fault is not uniform (Dibblee, 1982). The western part dips generally to the south while the central section is vertical or near-vertical. Along its eastern stretch, in Ventura County, the fault dips to the north.

The Santa Ynez Fault is considered active by the U.S. Geological Survey, the California Division of Mines and Geology, and many other individual geologists. No historic seismic events can be related directly to the fault; hence, the active classification is based on circumstantial evidence. Foremost is this category is the sharp topography associated with the fault. With respect to bedrock offsets, the fault clearly has offset mid-Tertiary units a few tens of thousands of feet. These are the youngest indurated units mapped along the fault. The fault trace, or traces, do pass under Quaternary deposits at numerous locales, but invariably these deposits are loose and are largely coarse rocks. Because of these characteristics, it is essentially impossible to recognize any fault-related offset, if indeed, any exists.

JUNCAL DAM

Juncal Dam is a major water storage structure on the Santa Ynez River in southeastern Santa Barbara County (Fig. 1). The impounded body of water, Jameson Reservoir, is the major source of water for the coastal community of Montecito, California. Water from Jameson Reservoir reaches Montecito via gravity flow, with much of the route passing through the Doulton Tunnel, a two-mile long passage driven through the Santa Ynez Mountains.

Structurally, Juncal Dam is a composite feature. the main structure is a concrete arch dam that spans a deep, narrow canyon (Fig. 2). Two other structures together form an auxiliary dam that caps the lower part of the ridge south of the concrete arch section (Fig. 2). Together, the concrete arch, the

Figure 2. View to northeast showing Juncal Dam complex along ridge in near distance. Multiple arch section on extreme right and Wegmann gravity section to its immediate left. Concrete arch section in extreme left part of view.

Figure 4. Aerial view of Santa Ynez River Valley showing Juncal Dam, Jameson Reservoir, the Santa Ynez River, and Alder Creek. View is to east. Note narrow but continuous ridge between Alder Creek and Santa Ynez River (Photo courtesy of George Ford).

Figure 3. Aerial view of Juncal Dam and Jameson Reservoir, viewed to northeast. Alder Creek in lower right portion of view (Photography courtesy of George Ford).

Figure 5. Westerly view down Alder Creek. Santa Ynez Mountains to left and Jameson Reservoir to right. Note steep, sharp character of Alder Creek, low ridge between Alder Creek and reservoir, and abrupt changes in slope, due to triangular facets, of ridges aligned to left of Alder Creek.

Figure 6. Juncal Dam and Jameson Reservoir, view slightly east of north. North slope of Santa Ynez Mountains in foreground. Alder Creek flows northward down steep valley in lower right of view then abruptly turns and flows westward at base of mountains. Note low but continuous ridge between Alder Creek and Jameson Reservoir.

bedrock ridge and the auxiliary dam form an impoundment barrier approximately 1500 feet in length (Figs. 2 and 3).

The concrete arch dam is a variable radius structure. Arc length along the crest is 430 feet and the crest is 160 feet above the canyon bottom. The abutment elevation of the arch section is 2230 feet above sea level. The central part of the arch is designed as an overpour spillway, with an elevation of 2224 feet.

The auxiliary dam at the southern end of the ridge consists of two integral structures of different design. The more southerly structure is an inclined multiple arch section (Figs. 2 and 3). The northern part of the auxiliary structure is a Wegmann gravity dam. The multiple arch section is 475 feet long and the Wegmann gravity section is 420 feet in length. Maximum height of the auxiliary dam, near the center of the multiple arch section, is 50 feet.

Construction of Juncal Dam began in the late 1920's and was completed in 1931. During excavation for the inclined multiple arch section, the project geologist, Bailey Willis, reported the exposure of a fault, trending approximately east-west and dipping 72 degrees to the north (Willis, 1929).

Review of the field reports of existing dams in California by the state Division of Safety of Dams following the Teton Dam failure led to a concern by this organization that the fault trace described by Willis could be an active trace of the Santa Ynez Fault. Hence, an engineering geology investigation and seismic stability analysis were undertaken by the firm of Lindvall, Richter and Associates to determine if further movement along the Santa Ynez Fault was likely, and if so, what effect it would have on Juncal Dam. The late Richard H. Jahns and the author, acting as an independent consultant, supervised the geologic and geophysical studies related to the Juncal Dam investigation. This paper is a summary of the geologic findings of that investigation.

Juncal Dam is on private property within a fairly remote section of Los Padres National Forest. The dam can be viewed from a distance from a Forest Service trail that runs south of the dam and reservoir; however, one must walk about two and one-half miles from the locked gate at Juncal Campground. Anyone wishing to visit the dam for a closer look should contact the Montecito Water District.

GEOLOGY AT JUNCAL DAM

Physiography

The Juncal dam site spans the upper part of the Santa Ynez River approximately 14 miles northeast of the City of Santa Barbara. Here, the Santa Ynez River follows a topographically sharp valley that trends generally east to west (Fig. 4). The Santa Ynez Mountains lie south of the Santa Ynez River and the San Rafael Mountains lie to the north. Elevations within the area range from about 1750 feet along the river valley to about 4500 feet at points along the crest of the Santa Ynez Mountains. The area surrounding Juncal Dam features rugged and locally steep topography. Many areas have extremely dense vegetative cover. This combination of steep terrain and dense vegetation renders some areas inaccessible.

Bedrock

Bedrock in the vicinity of Juncal Dam comprises Mesozoic and Tertiary units. The Mesozoic rocks are Franciscan and they form the local basement. Identified Franciscan exposures are predominantly greenish-blue, pervasively sheared and slickensided shale. Other Franciscan lithologies identified are grayish-green graywacke and both red and green chert. The graywacke and chert occur as thin, isolated lenses and layers within the shale. Due to its sheared nature, the Franciscan shale is not particularly resistant to erosion.

The Tertiary rocks comprise three thick Eocene sedimentary units: the Juncal Formation, the Matilija Sandstone, and the Cozy Dell Shale. The Juncal is the oldest of the three. Lithologically, the Juncal primarily is a highly fissile, dark-gray shale. Thin to medium beds of harder sandstone are scattered stratigraphically throughout. Total thickness of the Juncal Formation is of the order of 3500 to 5000 feet. The Juncal is moderately resistant to erosion, with the sandstone beds standing out in higher relief than shaley intervals.

The Matilija Sandstone rests conformably on the Juncal Formation. The Matilija is a hard, light-tan, arkosic sandstone, with a few shale lenses. It is thick-bedded to massive and is highly resistant to erosion, as evidenced by its presence along the crest of the Santa Ynez Mountains. Total thickness is about 1200 feet.

The Cozy Dell Shale rests conformably on the Matilija Sandstone. The Cozy Dell consists mainly of soft-gray micaceous clay shale that is prominently layered and moderately fissile. A few thin intervals of sandstone have been reported in the Cozy Dell (Dibblee, 1966). The Cozy Dell is relatively soft and erodes more readily than the other Eocene rocks within the subject area. Total thickness is of the range of 1600 to 1800 feet

Surficial Units

Many of the less-steep slopes around Juncal Dam are mantled by fan deposits and landslide deposits. The presence of these rather extensive units makes comprehensive bedrock mapping impossible. The Santa Ynez River Valley features relatively deep deposits of coarse, bouldery alluvium.

Structural Framework

As noted earlier, the Santa Ynez Fault separates the Transverse Ranges from the Coastal Ranges. The Santa Ynez Mountains, the imposing range overlooking the City of Santa Barbara, are the westernmost part of the Transverse Ranges. They are characterized by a prevailing east-west strike that reflects the structure of a major anticline with bedding dipping essentially parallel to its northern and southern flanks. The Santa Ynez Mountains have been displaced by geologically recent (Post-Eocene) uplift and left-lateral slip along the Santa Ynez Fault.

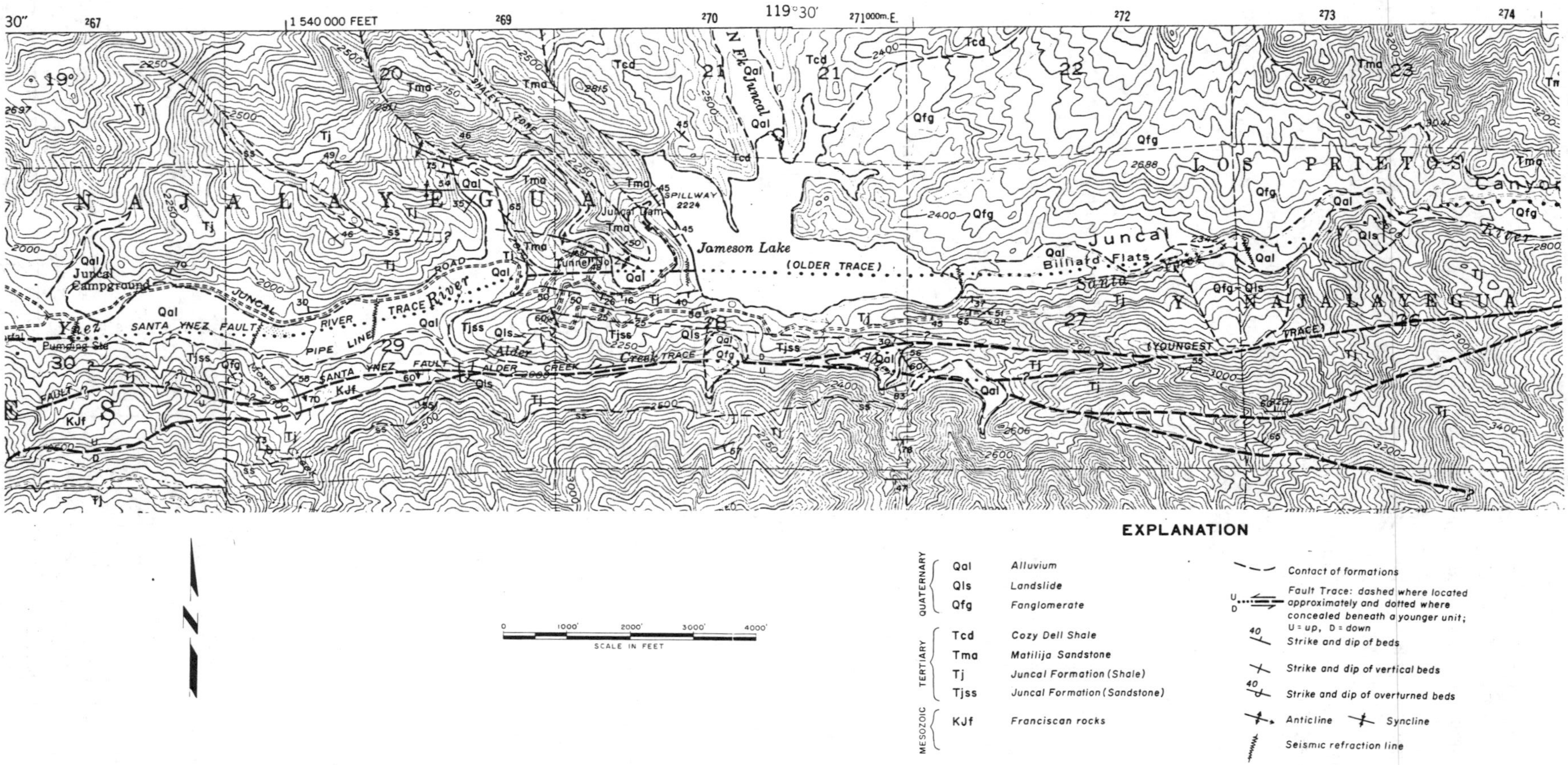

Figure 7. Geologic Map of Santa Ynez Fault in the vicinity of Juncal Dam. Geology from Dibblee (1966), Lindvall, Richter & Associates (1982) and Ford (unpublished).

The San Rafael Range is the southernmost Coast Range in the vicinity of Juncal Dam. This range is characterized by a more northwesterly orientation of fold axes than that of the Santa Ynez Mountains. The lateral component of fault movement in the San Rafaels is largely right-lateral and fault traces also are oriented to the northwest.

The overall structural pattern strongly suggests that the principal tectonic stress in the region has been a north-south compressional force that has elevated the Santa Ynez Range with respect to the Coast Ranges to the north. Left-lateral displacement along the primary Santa Ynez Fault and right-lateral slip along the more northwesterly trending faults in the San Rafael Range indicate a relatively strong northeasterly compression also has been exerted in the region.

FAULTING AT JUNCAL DAM

As part of this geologic investigation, a stretch of approximately 4 1/2 miles of the Santa Ynez Fault, and marginal areas north and south of the fault, were mapped by George Ford, then an M.S. candidate in Geology at Stanford University. This stretch of the fault is essentially centered on Juncal Dam, extending about two miles west of the dam to Juncal Campground and approximately two and one-half miles east of the dam to Upper Santa Ynez Campground. Mr. Ford's mapping of this area, as yet unpublished, was vital in interpreting fault movement at and adjacent to the dam and reservoir.

The Santa Ynez Fault between Juncal Campground and Upper Santa Ynez Campground consists of at least two separate traces (Fig. 7). The more northerly trace, herein identified as the River Valley Trace, generally follows the path of the Santa Ynez River. The path of the river is essentially linear except for a pair of incised meanders at and directly west of the dam (Fig. 3). Here the River Valley Trace crosses two bedrock ridges, one of which is the foundation ridge of the dam and the other is directly west of the dam. At the foundation ridge, the River Valley Trace passes under the multiple-arch section of the dam. In all likelihood, this is the location of the exposure viewed by Willis during construction of the dam.

The River Valley Trace is poorly exposed within the interval mapped. West of the two ridges it is covered by alluvium and east of the dam it is covered by the reservoir and by several styles of surficial units. However, the River Valley Trace was identified upstream of the reservoir and downstream from the ridges by seismic refraction surveys (Lindvall, Richter and Associates, 1982). These seismic survey lines indicate the presence of a shattered zone several tens of feet wide in the underlying bedrock.

The River Valley Trace is largely covered by rubble and slope wash on the downstream side of the foundation ridge, but it is reasonably well exposed in the ridge to the west. Here the fault appears as a wide zone of dark gouge containing near-vertical calcite veinlets and randomly oriented layers of broken sandstone in a pulverized clayey shale matrix. The margins of this latter exposure are too irregular to allow measurement of an accurate attitude for the River Valley Trace. However, these exposures clearly show that at and directly west of the dam, hard Matilija Sandstone on the northern side of the River Valley Trace has been juxtaposed against softer shale of the Juncal Formation along its southern side.

The southern trace of the Santa Ynez Fault, referred to here as the Alder Creek Trace, trends roughly parallel to the River Valley Trace eastward from Juncal Campground. It is not as linear a feature as its northern neighbor, as it crosses the more rugged terrain along the base of the Santa Ynez Mountains (Fig. 7). It enters the mouth of Alder Creek and continues upstream along the sharp, linear valley of Alder Creek for about one mile. Along this stretch the Alder Creek Trace is separated from Jameson Reservoir and the Santa Ynez River by a relatively low and narrow, but continuous ridge of rocks belonging to the Juncal Formation (Fig. 4 and 6). The northern side of this ridge features shaley rocks while the southern side consists of more resistant sandstone. This ridge diverts northward flowing drainage from the Santa Ynez Mountains westward along the Alder Creek Trace. Near the sharp bend in Alder Creek, the fault trace bifurcates, and as the two breaks trend eastward, the southern one bifurcates again. The more northerly trace continues north of east into and across an area of extremely rugged terrain.

Like the River Valley Trace, the Alder Creek Trace is poorly exposed. Dips of 60 degrees and 70 degrees to the south were measured along the Alder Creek Trace at two sites southwest of the dam (Fig. 7). These exposures are characterized by a series of sharp, parallel breaks in relatively hard rock rather than wide gouge zones. Eastward of these exposures no attitude could be measured along the Alder Creek Trace.

Thus, detailed geologic mapping, supplemented by seismic refraction surveys along the subject stretch of the Santa Ynez Fault, provided discontinuous visual exposures of the two fault traces, spot subsurface locations of the River Valley Trace, and two attitudes on the Alder Creek Trace. Unfortunately, no points were found where the two traces were overlain by a common unit so that a relative degree of offset of that unit along the two traces could be measured. Fortunately, adequate geomorphic evidence does exist at and around Juncal Dam to provide a rather convincing analysis that the most recent fault movement has occurred along the Alder Creek Trace and that movement has not occurred along the River Valley Trace for several tens of thousands of years.

Examination of the topography associated with the two traces of the fault suggests that vertical movement has occurred along the Alder Creek Trace more recently than along the River Valley Trace. Along most of its path the River Valley Trace occupies a wide, topographically subdued valley (Fig. 4). On the other hand, topography along the Alder Creek Trace is sharp and steep, with ground on the southern side of the trace uplifted relative to that on the north (Fig. 5). This uplift is particularly prominent along the southern side of the east-west section of Alder Creek south of the reservoir and dam. Here the tips of several north-trending ridges are cut by a series of triangular facets (Fig. 5). Triangular facets are commonly found along the upthrown sides of relatively young fault scarps and they are the product of rapid erosion of areas where the topography has been truncated and uplifted.

The gradients and forms of the streams associated with the two fault traces also are of significance in demonstrating that vertical movement has occurred more recently along the Alder Creek Trace. The Santa Ynez River is a moderately mature stream occupying a relatively wide valley. Slight meanders, resulting from lateral erosion, have developed in several places, and much of the valley, particularly west of the dam, contains thick deposits of alluvium (Fig. 4). In contrast, drainage courses associated with the Alder Creek Trace are steep, linear, and little alluvium has accumulated along them. The condition of the streams along the Alder Creek Trace are those of streams that have been elevated above local base level and now are attempting to readjust by accelerated downward erosion.

Evidence of a major component of left-lateral movement along the Alder Creek Trace also is evident. The watershed for Alder Creek is located within the northern slopes of the Santa Ynez Mountains. Drainage from this area is generally northward to the Alder Creek fault trace. At this point the course of Alder Creek abruptly turns westward for about 7000 feet, until it merges with the Santa Ynez River west of Juncal Dam. If fault movement on the Alder Creek Trace was entirely vertical, with the Santa Ynez Mountains uplifted, the north-flowing drainage from Alder Creek would, in all likelihood, have incised and breached the low, narrow ridge between the reservoir and the Alder Creek fault trace and joined the Santa Ynez River at that point (Fig. 5 and 6). Instead, Alder Creek turns sharply and flows westward for more than a mile to reach the river. The presence of this intact ridge diverting the upper part of Alder Creek implies that the Alder Creek watershed has been moved eastward for at least 7000 feet along the fault trace to its present position and that the ridge will continue to remain a barrier to northward flow off the Santa Ynez Mountains.

Field evidence along the ridge occupied by Juncal Dam and the ridge to its west indicate geologically recent lateral movement has not been a factor here, and it also provides some feel for the time that has elapsed since the River Valley Trace last moved. Both of these ridges are oriented northwest and both are cut by the River Valley Trace. Both upstream and downstream of these ridges, the Santa Ynez River exhibits a near-linear, east-west course. However, at the upstream side of the foundation ridge for the dam, the river encounters hard Matilija Sandstone. Here it veers sharply northwest and then follows a meandering course through the Matilija outcrop. At first look, it seems odd that the river followed this circuitous path rather than cut directly through the ridges along the less-resistant fault zone. The logical answer is that the existing drainage pattern is antecedent, and that the Santa Ynez River followed the path of least resistance when it was at a higher elevation.

The relative narrowness of the foundation ridge (400 to 450 feet) strongly suggests that it has not been shifted by fault movement during the tens of thousands of years it has taken the Santa Ynez River to incise 150 to 200 feet of hard sandstone. If fault movement in excess of a few feet had occurred during the period of meander downcutting, the foundation ridge would have been breached or offset and the Santa Ynez River would have assumed a more direct course through it.

Thus the weight of geomorphic evidence strongly indicates that at Juncal Dam, movement has not occurred along the River Valley Trace for tens of thousands of years, and that the most recent fault movement, characterized by both vertical and left-lateral displacements, has occurred along the Alder Creek Trace.

ACKNOWLEDGMENTS

Recognition is due the firm of Lindvall, Richter & Associates in general, and Eric Lindvall in particular for the circumstances that enabled me to work in the subject area. Thanks also are due the Montecito Water District, especially Mr. Waymore T. Burrow who was damkeeper at the time my field work took place. Thanks also to Mr. George Ford for the generous use of parts of his unpublished regional geologic map and for use of his excellent aerial photographs. I also owe a vote of thanks to the co-editors of this volume for their patience.

REFERENCES

Dibblee, T. W., Jr., 1966, Geology of the central Santa Ynez Mountains, Santa Barbara County, CA: California Division of Mines and Geology Bulletin 186, 99 p.

Dibblee, T. W., Jr., 1982, Geology of the Santa Ynez-Topatopa Mountains, southern California: In Geology and Mineral Wealth of the California Transverse Ranges, South Coast Geological Society Guidebook No. 10, Santa Ana, CA, p. 41-57.

Lindvall, Richter & Associates, 1982, Investigation of seismic stability of Juncal Dam: Unpublished report on file with California Division of Safety of Dams, Sacramento, two volumes.

Willis, B. E., 1929, Rock conditions at the Juncal damsite: Consultants Report to Montecito County Water District Board of Directors, Montecito Water District, Montecito CA, 20 p.

GROUND FISSURES IN THE TEMECULA AREA, RIVERSIDE COUNTY, CALIFORNIA

Roy J. Shlemon
Consulting Geologist
P.O. Box 3066
Newport Beach, CA 92659

Paul Davis
Consulting Engineering Geologist
711 Main Street
Huntington Beach, CA 92648

INTRODUCTION

In August 1987, several en-echelon ground cracks were observed in an area of new residential grading south of the town of Temecula in southwestern Riverside County, California (Fig. 1). Although initially seemingly quite mundane, it was soon discovered that the cracks passed not only through residential building pads, but also through several recently-constructed houses. Within a few months, similar cracks were observed trending across compacted fill emplaced for a new business park and, in one case, through a large, recently-completed industrial building. Indeed, by December 1987, northwest-trending cracks and fissures were found in a 7-mile long zone in the rapidly-urbanizing Temecula and Wolf Valley areas.

Homeowners, developers, County officials, and the news media soon recognized that the Temecula-area fissure system was a new geological phenomenon, one worthy of great attention, and the subject of litigation. The County of Riverside soon established a "Special Subsidence Zone" that effectively was a moratorium for issuing of building permits until the cause and extent of the fissures had been determined.

Several detailed geological investigations and technical reviews ensued, commissioned by a variety of interested parties, including developers seeking building permits, homeowner plaintiffs looking for redress for alleged physical and emotional distress, and lawyers reaching for technical issues. So began the saga of the "Great Crack of Temecula," one still being played out more than four years later as litigation continues and technical issues are debated. Indeed, at present (early 1992), damage allegations now exceed approximately $50,000,000 and the costs of litigation and technical investigations are probably greater than several million dollars (Corwin et al., 1991).

The extent and cause of the Temecula-area ground fissures are still controversial. Numerous hypotheses abound, but two are generally favored: (1) Aseismic creep along previously unreported faults; and (2) Recent withdrawal of groundwater causing aquifer compaction and local fissuring. Both hypotheses have merit and accordingly, we evaluate them in this paper. Initially, we set forth the geological setting of the Temecula area, pointing out the various geomorphic and subsurface features that apparently control fissure location. We then describe the distribution of fissures and call attention to their appearance and to investigations carried out in two distinct areas, north and south of Temecula respectively, generally deemed the "northern" and "southern" fissure areas (Fig. 1). We also describe the apparent differences in fissures in the two areas based mainly on trench exposures. We also provide examples of pertinent trench logs that bear upon fissure location and causation. We likewise evaluate available seismic and water well data. And finally we present our judgments as to the likely causes of the 1987 ground fissures in the Temecula and Wolf Valley areas of southwestern Riverside County.

GEOLOGICAL FRAMEWORK OF THE TEMECULA AREA

Geomorphic and Tectonic Setting

The Temecula area lies in an approximately one mile-wide trough (graben) bordered on either side by major splays of the Elsinore fault system (Fig. 1). The Elsinore fault zone itself is one of several major northwest-striking faults pertaining to the San Andreas system. In the Temecula area, the trough separates batholithic and other pre-Quaternary rocks on the west from the Plio-Pleistocene Pauba Formation on the east (Figs. 1 and 2). The Temecula area, as described in this paper, includes three contiguous valleys that comprise the Elsinore trough in southwestern Riverside County: The Murrieta Valley on the north, the Temecula Valley in the center, and the Wolf Valley on the south.

The western boundary of the Elsinore fault zone is usually taken to be the Willard fault (Engel, 1959); namely, the several subparallel high-angle faults that bound the Santa Ana Mountains on the west and the Murrieta, Temecula and Wolf valleys on the east. Most Willard-zone faults are traceable only for a few hundred feet where they are often marked by abrupt changes in vegetation reflecting juxtaposition of Bedford Canyon metamorphic rocks on the west and the Plio-Pleistocene Pauba Formation on the east (Fig. 2). Locally, the Willard fault coincides with an abrupt break in slope suggestive of possible Holocene movement. However, recent investigations near the head of the Santa Margarita River show that last displacement of the Willard fault is pre-Holocene, and therefore not active according to present State of California criteria (Schaefer Dixon Associates, 1988a).

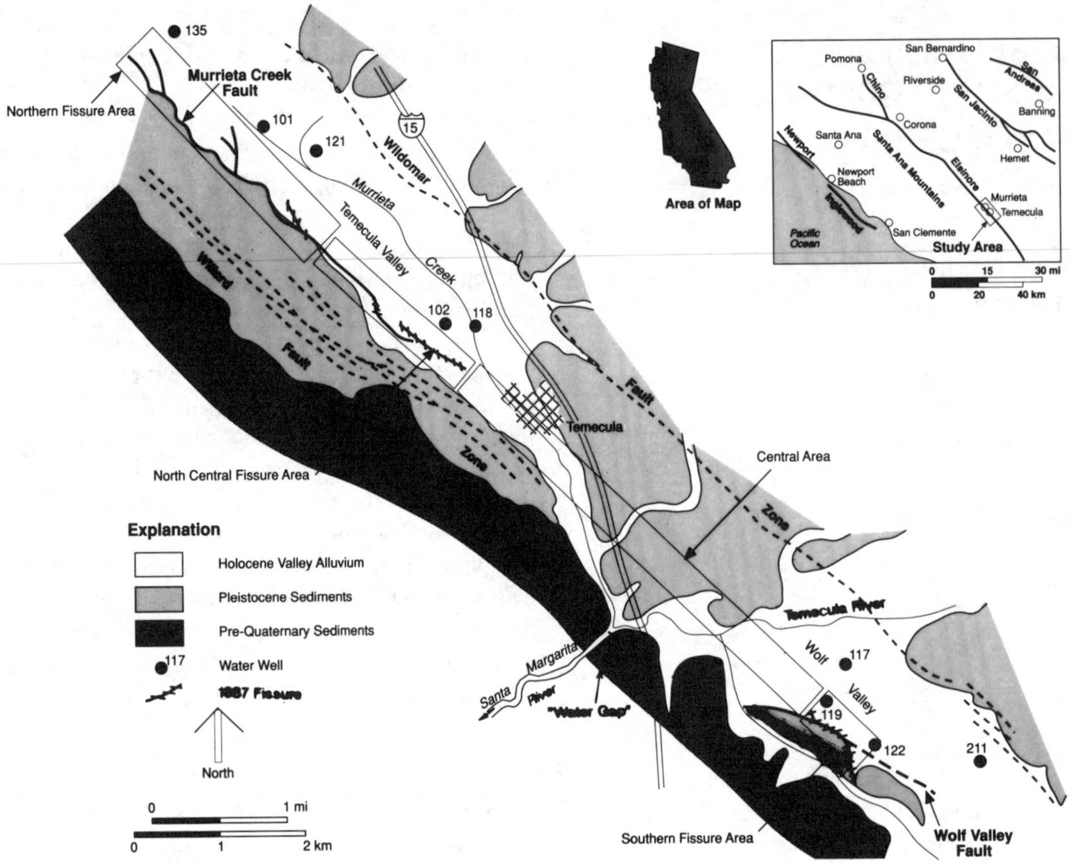

Figure 1. Location and geologic map showing major place names, faults, key water wells, fissure study areas, and related ground cracks in the Temecula and Wolf Valley areas, southwestern Riverside County, California.

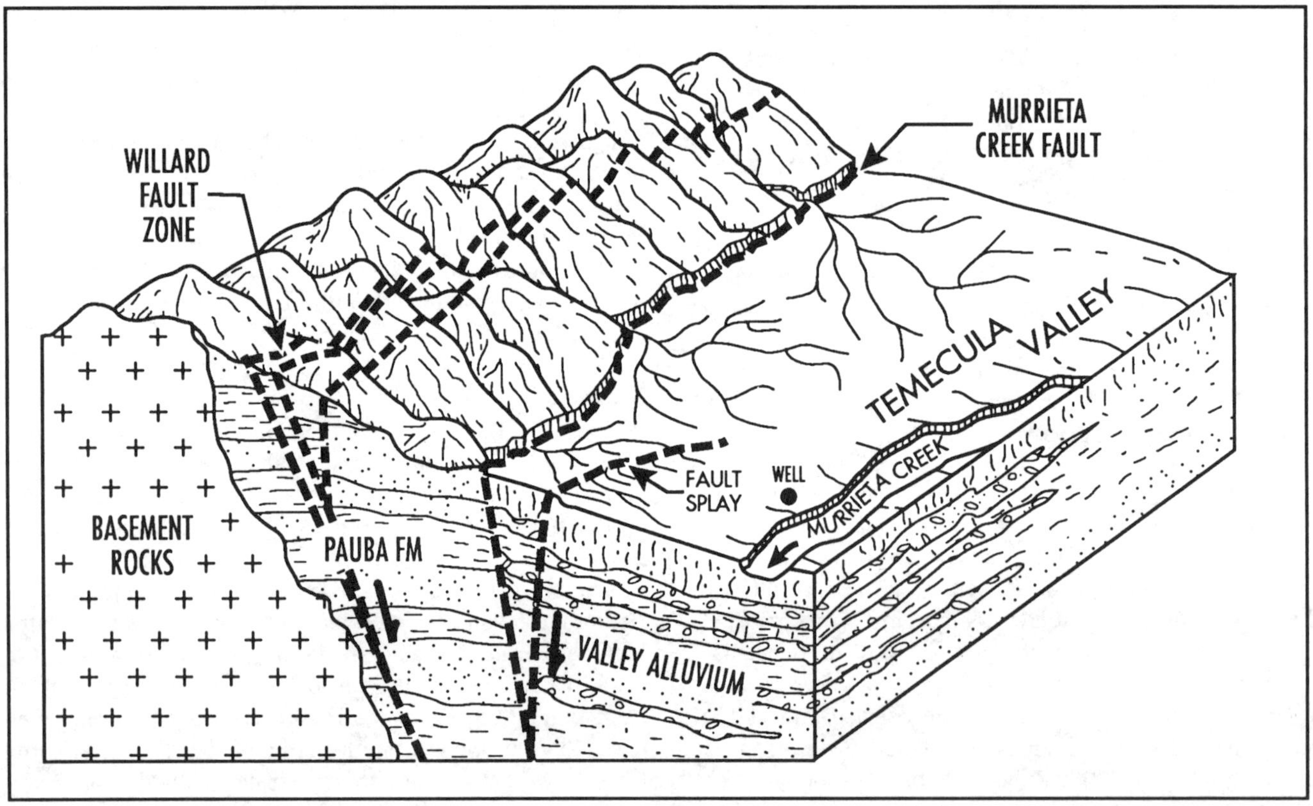

Figure 2. Schematic diagram showing general location of the Willard and Murrieta Creek faults and their relationship to displaced Quaternary sediments.

The eastern boundary of the Murrieta-Temecula-Wolf valleys is the Wildomar fault (Fig. 1). The Wildomar is clearly active (Holocene), for it is well expressed by linear scarps, by offset stream channels, and by abrupt topographic fronts. The fault is photographically visible even where crossing the active floodplain of Temecula Creek, a tributary of the Santa Margarita River on the north side of Wolf Valley (Fig. 1). However, south of Temecula Creek its location is still debateable, despite recent geotechnical investigations (Crisman et al., 1991; Earth Research Associates, 1987).

Two other Temecula-area fault systems bear upon the location and extent of the 1987 fissures. North of "old town" Temecula, in the north-central fissure area, an apparent ground water barrier (Moreland, 1972) was thought to be vertically displaced about 30 feet (Kennedy, 1977). This barrier is of particular interest, for it may coincide in part with some of the northern fissures and the location of a newly-identified active fault now informally designated as the "Murrieta Creek fault" (Bergmann and Rockwell, 1989).

To the south, on the western side of Wolf Valley, Kennedy (1977) delimited the "Wolf Valley fault" as generally occurring at the base of an approximately 38 feet-high and one mile-long ridge. Kennedy (1977) also judged that the Wolf Valley fault was probably active, based mainly on the presence of apparently tilted Holocene sediments, on the occurrence of closed depressions, and on the likely fault control of the linear escarpment in this area. However, unlike the Wildomar fault which defines the east side of the Temecula and Wolf valleys, the Wolf Valley fault was not zoned as "active" by the State of California until post-1987 fissure investigations clearly documented Holocene offset (California Division of Mines and Geology, 1990; Moran, 1989).

Figure 3. Representative 1987 ground cracks in street pavement, north-central fissure area. Relative vertical displacement is less than 1 inch, east side (Valley) down.

Figure 4. Alignment of fissures and "sink holes" in north-central fissure area following 1987 winter rains. The fissures coincide with the underlying Murrieta Creek fault and propagated through about 10 feet of compacted fill.

Stratigraphic Setting

The pre-Quaternary stratigraphy, west of the Willard fault and bordering the Murrieta, Temecula and Wolf valleys, consists mainly of Bedford Canyon metamorphic and Peninsular Range granitic rocks (Fig. 2).

Quaternary units occur mainly within or immediately adjacent to the Elsinore Trough and the Temecula area, namely; the Plio-Pleistocene Pauba Formation, late Quaternary alluvium and colluvium, and remnants of gravel-filled terraces bordering the lower course of Murrieta Creek near its junction with the Santa Margarita River (Kennedy, 1977; Mann, 1955; Schaefer Dixon Associates, 1988a).

The Pauba Formation mainly consists of interbedded fluvial silts, sands and local conglomerates (Mann, 1955). Most deposits were laid down by streams that tapped granitic terrane east of Temecula and debouched into the Elsinore trough, a process that continues today.

The age of the Pauba Formation is unknown; however, it is assumed to be less than 700,000 years old based on vertebrate fossil assemblages (Mann, 1955). Roadcut and trench exposures show that several moderately- to strongly-developed buried paleosols occur within the Pauba Formation. These soils indicate that most Pauba sediments were laid down episodically, separated by periods of regional landscape stability and related soil formation. Presumably, such cyclic deposition and soil formation were caused by regional climatic and related sedimentological change in late Quaternary time (Shlemon, 1985).

Late Pleistocene, stream-terrace sediments occur south of Temecula, where they border the lower Murrieta Creek near its junction with the Pechanga and Temecula creeks and the head of the Santa Margarita River (Schaefer Dixon Associates, 1988a; Fig. 1). Presumably, these terrace gravels were laid down in channels that flowed into an ancestral (antecedent) Santa Margarita River which, in turn, has continued to cut through the rising Santa Ana Mountains that form the western boundary of the Elsinore Trough. This has given rise to the "Santa Margarita Water Gap," a geomorphic enigma still not fully understood (Fig. 1).

Remnant fluvial and distal-fan surfaces about 7 to 10 feet above the modern floodplain of Murrieta Creek are generally deemed "older alluvium," and those clearly laid down by Murrieta, Pechanga and Temecula creeks and their tributaries are informally designated "younger alluvium".

The older alluvium is latest Pleistocene based on approximately 12,000 and 17,000-year old radiocarbon dates for trench-exposed sediments immediately west of Murrieta in the northern fissure area (GeoCon, 1990).

The younger alluvium, affected by the 1987 ground fissures, is mainly Holocene in age and locally exceeds 16 feet in thickness based on recent radiocarbon dates and on interpretation of Cone Penetrometer Test (CPT) soundings in the Temecula area (Bergmann and Rockwell, 1989; Schaefer Dixon Associates, 1988b; 1989a; 1989b). The younger alluvium consists primarily of overbank, channel, and meander-loop deposits laid down by Murrieta Creek. Prior to channelization of Murrieta Creek, many younger alluvial sediments were deposited directly against, and therefore abut, escarpments defining the Murrieta Creek fault. Additionally, however, side-slope, distal-fan deposits often merge almost imperceptibly into the Holocene fluvial sediments and thereby locally "mask" subtle escarpments that mark several Murrieta Creek fault splays.

Two principal aquifers underlying the Murrieta-Temecula-Wolf valley areas yield abundant water for both domestic and agricultural use (Scheliga and McGoldrick, 1984): The upper "Pauba beds" that generally produce water from depths within about 300 feet from the surface; and the lower, more prolific "Temecula sands" generally encountered between 600 and 1200 feet.

DISTRIBUTION OF THE TEMECULA GROUND FISSURES

The Temecula and Wolf Valley ground fissures occur discontinuously within an approximately 7-miles long and 300-foot wide zone. Two generally discrete fissure segments are recognized: a northern and north-central area north of Temecula, and a southern area south of the City (Shlemon and Davis, 1988). No obvious fissures have been reported in the central area, but this may reflect only the present lack of systematic investigations (Fig. 1).

The northern and north-central fissures have been particularly well documented in the past few years, stemming principally from investigations for fault set-backs in various "business parks" that contain new commercial and industrial buildings. The southern fissures, mainly impacting tracts of single-family residences, were investigated soon after the 1987 ground cracks in order to determine appropriate building setbacks, to obtain permits for continuation of construction (GeoWest, 1987; Leighton, 1987), and to ascertain possible cause of the fissures in support of litigation that ensued shortly thereafter (Corwin et al., 1991; Moran, 1989)

The 1987 ground fissures are expressed at the surface by 30-60-foot long cracks in street pavement and adjacent gutters and curbs. Most fissures are en-echelon and left stepping, with horizontal overlap generally less than about 10 feet. No clear evidence of horizontal displacement has yet been seen; and vertical relief is typically less than a few cm with the east (valley) side down (Fig. 3). Damage to residential structures and adjacent streets usually involves slight vertical displacement of curbs and floor slabs. Locally also, especially in the north-central area, incipient ground fissures eventually opened to form discontinuous "sink holes" up to 3 feet wide that increased in size during the winter rains of 1987 (Fig. 4).

Northern Fissure Area

For the purpose of this study, we recognize north and north-central sub-areas within the general northern fissure area. The first fissures north of Temecula were reported on October 2, 1987. This was an auspicious day, for the Magnitude 5.9 Whittier Narrows earthquake had occurred the day before, and the owners of a large industrial structure in a north-central area business park were convinced that the newly-observed ground cracks passing though the eastern part of their building stemmed from either earthquake shaking or from the effects of "loading" caused by a temporary, nearby fill stockpile. Both hypotheses proved incorrect, for it was eventually discovered that the 1987 fissures coincided almost directly with the subsurface expression of a previously undocumented fault (Schaefer Dixon Associates, 1988b). The fault is now informally deemed the "Murrieta Creek fault" after the nearby drainage of that name (Bergmann and Rockwell, 1989).

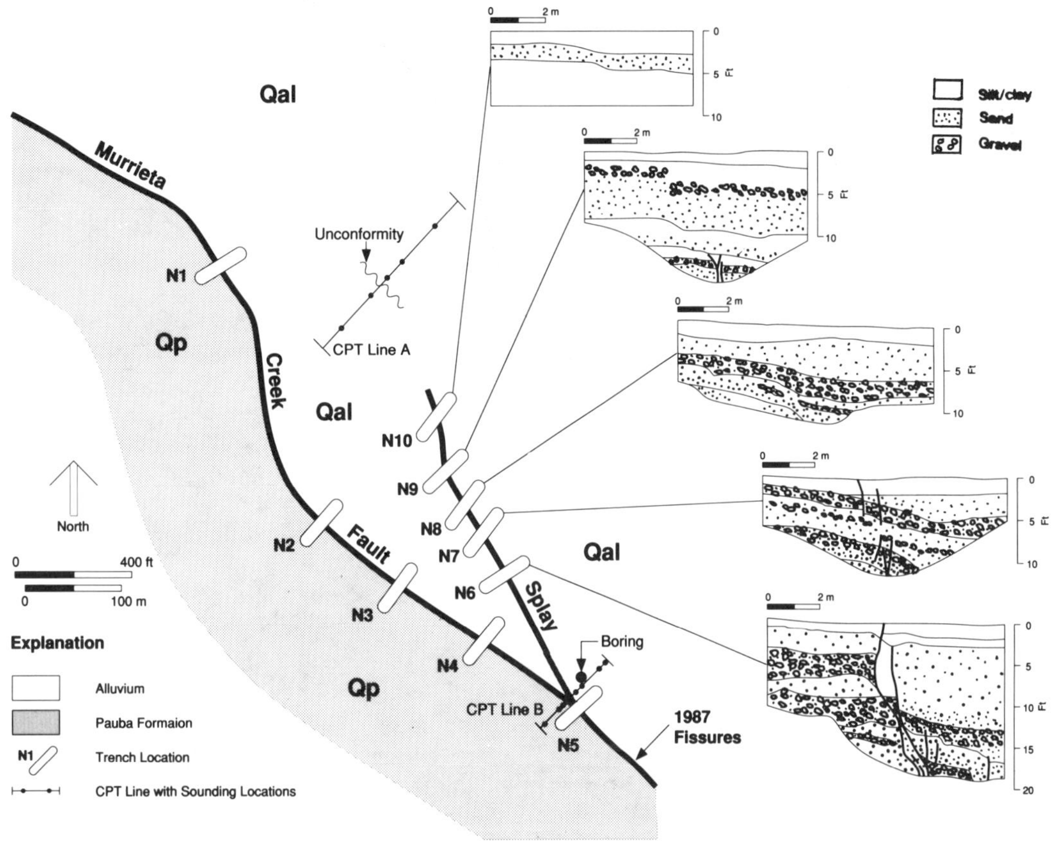

Figure 5. Main trace and splay of the Murrieta Creek fault, northern fissure area; and location of trenches and logs (N-1 through N-10) discussed in text.

The extent and width of fissures in the northern area are well documented; and much subsurface data are available in unpublished consultants' reports (GeoCon, 1990, 1991; Leighton & Associates, 1988a, 1988b; Schaefer Dixon Associates, 1988b, 1989a, 1989b; Soil Tech, 1989). These show that the Murrieta Creek fault generally separates the Plio-Pleistocene Pauba Formation on the west from Holocene valley alluvium on the east (Fig. 2).

Throughout the northern area, the main trace of the Murrieta Creek fault is generally linear to sightly curvilinear and trends northwest. More than 50 trenches, some exceeding 18 feet in depth, have now been placed across the trace of the fault in this area. The excavations show that the Murrieta Creek fault has two components of slip: (1) non-seismotectonic vertical displacement caused by continuing downwarp of graben alluvial sediments on the east; and (2) superimposed seismotectonic offsets with small, recurrent vertical and horizontal movements taking place throughout Quaternary time.

Based on radiocarbon dates from most north and north-central trenches, less than about 7 feet of vertical slip has taken place within the past 12,000 to 17,000 years (GeoCon, 1990, 1991; Schaefer Dixon Associates, 1988b; Soil Tech, 1989). A seeming exception is the 47-feet of vertical seismotectonic displacement reported to have occurred within about the past 12,000 years as deduced from a north-area trench exposure (Bergmann and Rockwell, 1989). However, based on our field observations and on our review of logs for the many other nearby trenches, we propose an alternative hypothesis to explain this "abnormally-high" displacement; namely; that it is only partially of seismotectonic origin, and that most slip stems from relatively slow, continuous downwarping along the western margin of the Murrieta-Temecula Valley graben (Davis et al., 1991). Some specific reasons for ascribing a non-seismotectonic mechanism for much of the reported slip are: (1) the displacement occurs on a short splay fault which, like others nearby, has observable vertical offset dying out within a few hundred feet along strike; (2) there is little, if any, clay gouge associated with the fault zone, in contrast to gouge exposed in nearby trenches emplaced across the main trace of the Murrieta Creek fault; and (3) late Holocene alluvium and cumulic soil profiles thicken to the east (basinward), but are not displaced, where present across Murrieta Creek splay faults (Fig. 5), thus implying that aseismic downwarp is continuing in this part of the graben.

In addition to the Murrieta Creek main trace, the northern fissure area is characterized by at least three distinct arcuate splays that diverge from the main fault and trend north toward the center of the Murrieta-Temecula Valley. Intriguingly, these three splay faults form partial arcs and are nearest to water wells that were pumping in the summer of 1987. These splay faults are thus probably the nearest "geological discontinuities" along which pumping-induced tensile stresses were focused and therefore became the locus of the 1987 ground fissures in this area (Fig. 1).

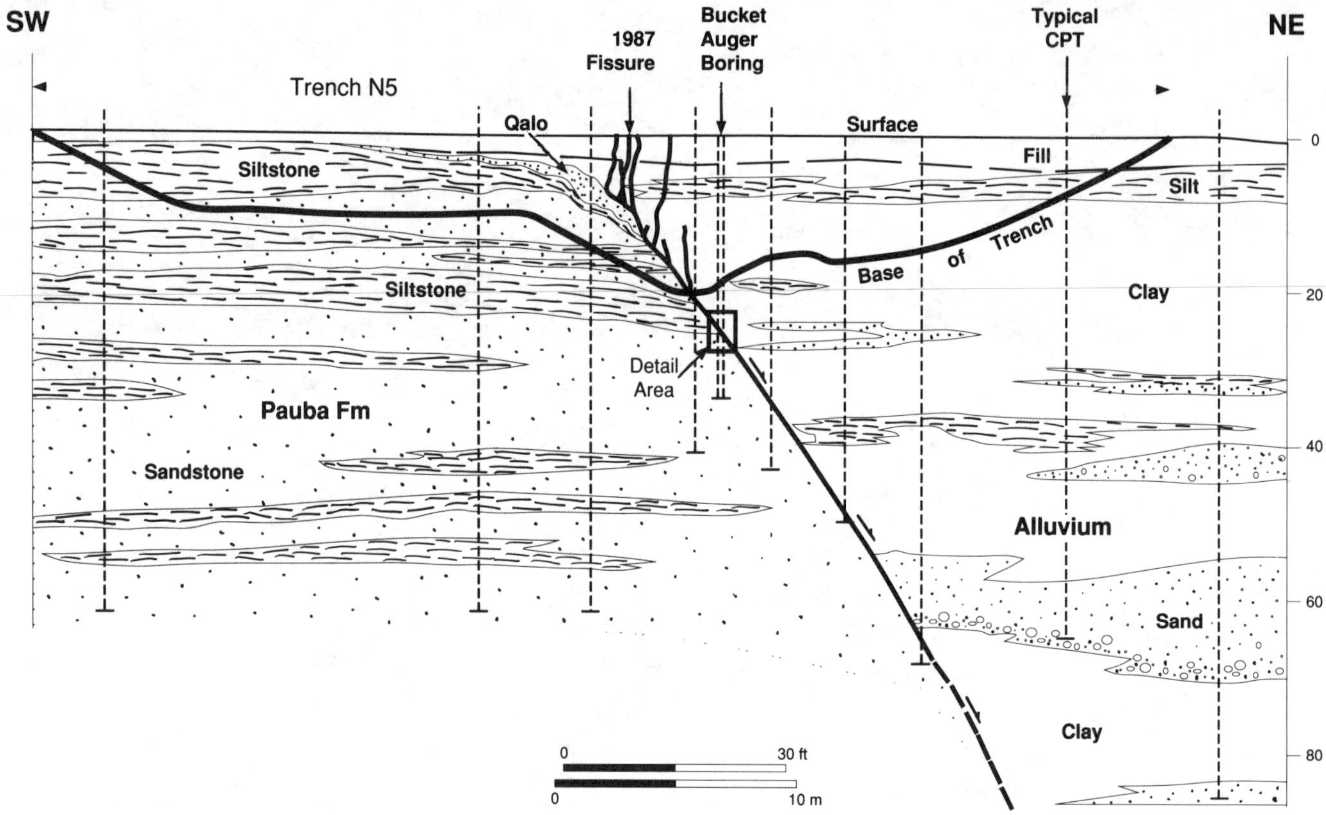

Figure 6 . Log of main-trace Murrieta Creek fault, Trench N-5 and stratigraphic interpretation of underlying sediments based on bucket-auger (Fig. 7) and CPT data.

Representative logs of typical north area trenches emplaced across the Murrieta Creek main trace and a splay fault are shown in Figures 5, 6, 7 and 8. The splay fault trenches show that vertical offset of basin alluvium decreases from about 7 feet at N-6, nearest the main trace, to merely basinward downwarping at N-10 on the north (Fig. 5). The 1987 ground fissures coincide with the splay fault, but is expressed at the surface for only about 650 feet.

The main-trace fault trench, N-5, exposed younger alluvium directly juxtaposed against Pauba-age sediments and downdragged older alluvium (Fig. 6). Supplemented by deep CPT and borehole data, the trench exposures show that the Murrieta Creek main-trace in this area maintains a remarkably constant dip of about 57 degrees to a depth of at least 60 feet (Fig. 6). Further, a bucket-auger exposure at 25 feet depth shows that the slip surface is planar, abrupt, and typified by downdragged beds of the Pauba Formation (Fig. 7). We interpret this large-scale slip and related slip features as indicative of continuous basin downwarping throughout Holocene time. A seismotectonic component probably also governs the form and dip of the slip surface. However, neither Pauba nor older alluvial sediments match on either side of the fault. Therefore, owing to the enormous complexity of determining slip-surface origin in this area, we suggest that hypotheses regarding magnitude of fault slip and related recurrence intervals are presently fraught with uncertainty and hence speculative at best.

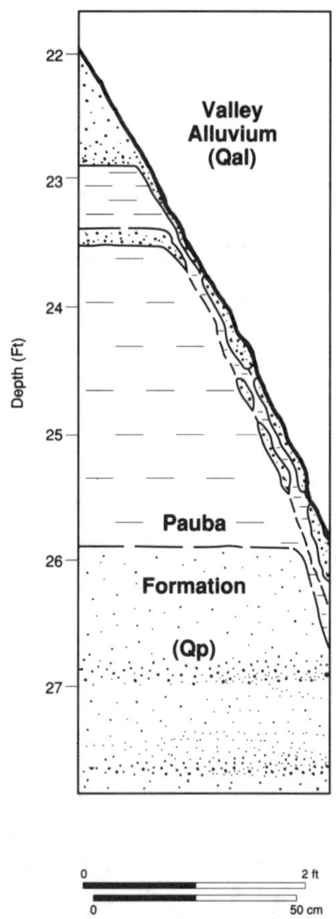

Figure 7 . Detail log of bucket-auger where penetrating main-trace of the Murrieta Creek fault approximately 22 to 25 feet below the surface, north fissure area.

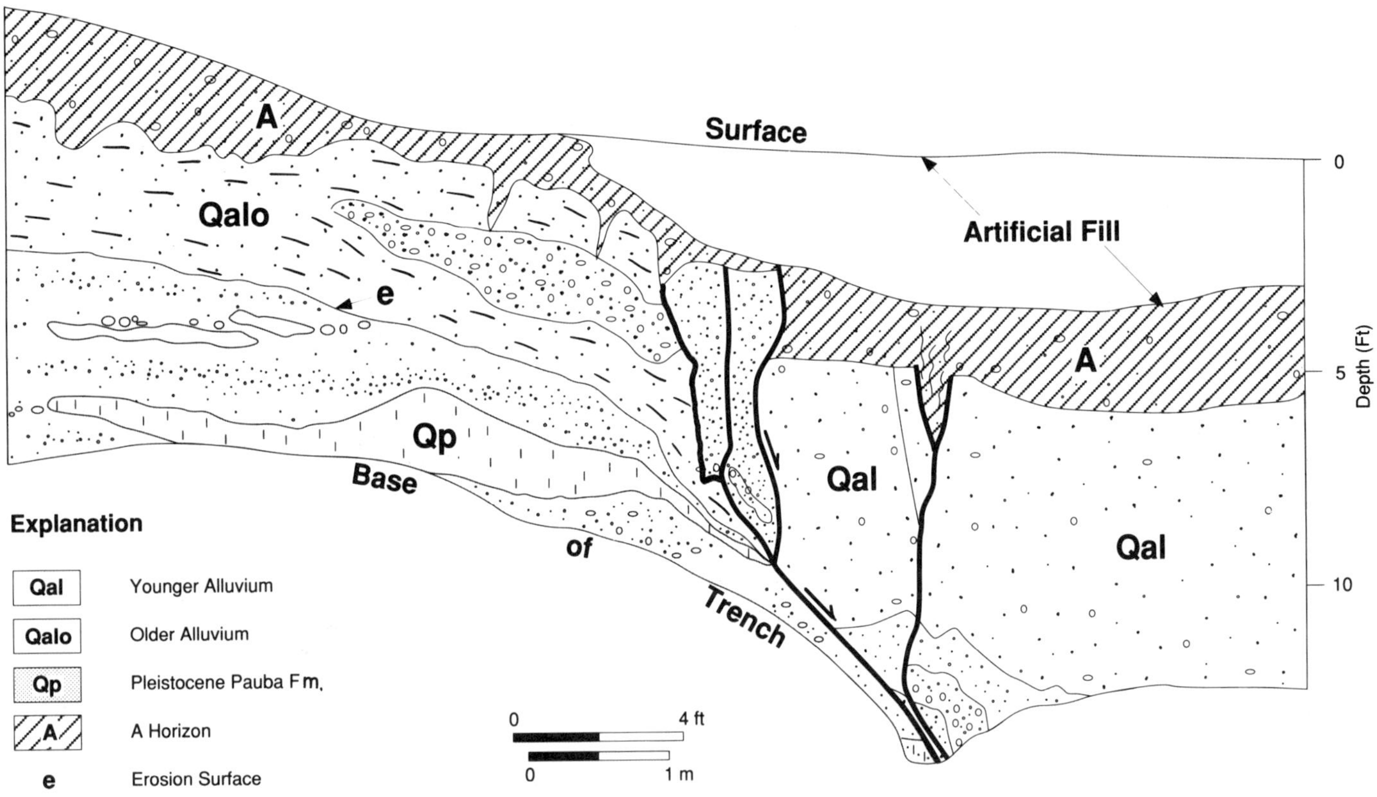

Figure 8. Displacement and "draping" of Pauba formation and older alluvium across the main-trace of the Murrieta Creek fault, Trench N-1, north fissure area (see Fig. 5 for location).

The 1987 ground fissures were also exposed in Trench N-5 (Figs. 1 and 6). On trench walls, the fissures are often reflected by open fractures up to 1/2 inch wide that typically "flower up" toward the surface. But these do not vertically displace sediments; rather, only tension cracks are apparent into which infilling is now taking place (Fig. 6).

Another representative main-trace trench (N-1) crossed an approximately 10-foot high escarpment in the north fissure area (Figs. 1 and 8). Here, too, large-scale vertical slip is apparent with younger alluvial sediments being juxtaposed along Pauba beds and older alluvium. Near-surface alluvium may have well been deposited against the escarpment accounting for some of the apparent vertical displacement. At depth, the slip surface lacks any gouge, but a 7-foot deep "soil infilling" extends to the modern surface. The remnant soil "A-horizon" is only partially preserved owing to agricultural disturbance; but it is probably cumulic in origin, thickening eastward across the fault and toward the subsiding basin.

The 1987 fissures did not propagate to the surface in Trench N-1; however, as in Trench N-5, they "flowered up" into open voids and fractures. We therefore judge that the 1987 ground fissures are but the latest movements taking place on the Murrieta Creek fault system; essentially, an acceleration of long-term natural seismotectonic and downwarping subsidence now induced by recent, rapid ground-water pumpage of nearby wells.

As shown by Trenches N-1 and N-5, Holocene vertical slip on the Murrieta Creek fault in this area exceeds 10 feet. In contrast, measured slip in nearby trenches to the north and south is consistently less than about 6 feet. The reasons for this rapid change in apparent vertical slip are unknown; however, they may relate to segmentation of the graben-bounding Murrieta Creek and Wildomar faults in this area, and hence possibly to differential horizontal slip and the resulting pull-apart of deeply-buried cross faults within the Elsinore trough (A. Hull, personal communication, 1991).

The north area trenches show another phenomenon of interest; namely, that the splay faults tend to "protect" main-trace segments from recent subsidence and fissuring effects. For example, main-trace trenches N-2, N-3 and N-4 exposed neither the 1987 ground fissures nor slip surfaces within the upper 5 to 10 feet of alluvium (Schaefer Dixon Associates, 1989b), features common in the adjacent splay trenches (Fig. 5). We therefore suggest that the splay faults are much younger than the main traces, and perhaps may have formed only in latest Pleistocene or perhaps even in Holocene time.

Following occurrence of the 1987 ground fissures, the north-central "business park" area was investigated to establish fault location and appropriate building setbacks in accordance with provisions of the California Fault-Rupture Special Studies Zones (Hart, 1990). The north-central area is also particularly "sensitive" owing to alleged damage to a large industrial building sited across what was later recognized to be the 1987 fissure and Murrieta Creek fault (Corwin and others, 1991; Schaefer Dixon Associates, 1988b; 1989a; 1989b; Soil Tech, 1989; Fig. 9).

The 1987 fissures propagated upward through approximately 10 feet of compacted fill placed in 1984 (Fig. 4). However, fill laid down in 1986 was cracked only near its base and the fissure did not extend to the surface. These relationships indicate that some ground cracking, although not recognized at the time, may have occurred a few years prior to the summer of 1987. However, wells 102 and 118 (Fig. 1) had been completed in early 1982 and 1986, respectively, and therefore may have been in production prior to 1987 (W. Hardt, personal communication, 1991).

Trenching and interpretation of CPT correlation lines show that the north-central fissures coincide with the now-designated Murrieta Creek fault. Sediments dated by radiocarbon as approximately 9,100 and 12,000 years old are vertically offset less than 2 feet (Schaefer Dixon Associates, 1989a; Soil Tech, 1989). The timing of last north-central fault slip is estimated by the presence of remnant soil-stratigraphic sections preserved in many business park trenches. Generally, non-displaced native sediments below the compacted fill are judged to be about 3,000 to 4,000 years old based on the presence of cumulic soil profiles with slightly-developed argillic (Btj) horizons (Shlemon and Davis, 1988). Accordingly, last seismotectonic movement of the Murrieta Creek fault in this area probably occurred in early to middle Holocene time.

Figure 9 . Representative 1987 ground cracks in street pavement and passing through the large industrial building in background, north-central fissure area.

The Southern Fissure Area

Ground cracks in the southern fissure area were first noticed in August 1987, mainly in recently-constructed residences and in newly-graded building pads in several tracts that "straddled" the previously-mapped Wolf Valley fault (Leighton, 1987; GeoWest, 1987). Subsequent trenching, mapping, and CPT interpretation soon followed, and eventually it was determined that ground fissures generally coincided with a portion of the Wolf Valley fault of Kennedy (1977). There are important exceptions, however. For example, comparable to splays in the northern area, a fissured splay fault branches off into Wolf Valley toward one or more deep wells that were pumping during the summer of 1987 (Fig. 11).

Southern fissure-area trench exposures and radiocarbon dates also show that the related fault system is Holocene in age, thus corroborating Kennedy's (1977) geomorphic interpretation (Leighton, 1987; GeoWest, 1987; Moran, 1989).

Southern-area fissures were typically expressed at the surface by as much as 90-foot long cracks in street pavement and building pads. Street and curbs have since been repaired; and the buildings directly on top or on reasonable projection of the fault have been removed by the developers (Corwin and others, 1991; Fig. 13).

Several trenches to 23 feet deep within the southern area exposed faulted alluvium and Pauba Formation sediments (Fig. 11). Typically, Holocene-age basin alluvium on the northeast has been faulted against Pauba Formation sandstones on the southwest producing a fractured zone several meters wide (Fig. 14). Unlike the Murrieta Creek fault in the north and north-central study areas, however, the fault in the south is characterized by a 5-inch thick, fat, plastic clay gouge as observed in Trench S-3 (Fig. 14). Closely-spaced secondary displacements also occur in the Pauba sands as well as in the alluvial units adjacent to the main fault. Clearly, recurring tectonic faulting has occurred over a considerable time, with large but as yet unmeasured displacements.

The 1987 ground crack closely followed the main fault trace at or near the gouge zone. Portions of the new cracks were infilled with soils and had reopened—indicative of a previous episode of cracking. The 1987 crack ruptured the ground surface through less than 3 feet of engineered fill. The ground crack diverged from the main fault trace to coincide with the short fault splay at its western end (Fig. 11).

Similar to the 1987 northern fissures, the southern area surface cracks followed the splay fault in alluvium opposite the pumping well(s) rather than the main Wolf Valley fault trace. Also, coincidentally, displacements along the splay fault decreased rapidly basinward over a distance of just a few hundred feet. Accordingly, comparable to the northern fis-

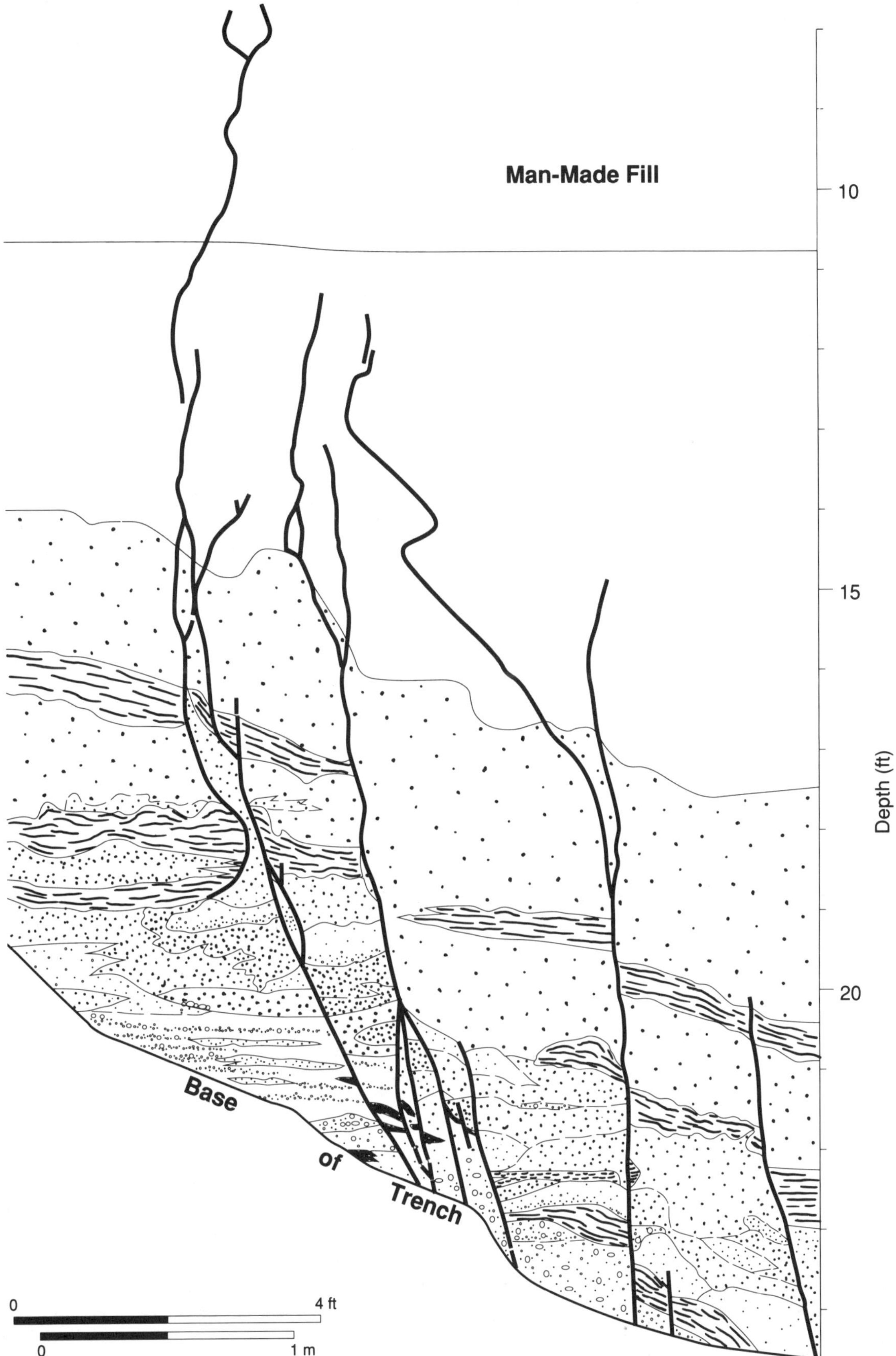

Figure 10. Subsurface expression of open voids and fractures associated with projection of 1987 fissures, north-central area trench. Fractures extend into the base of 1986 compacted fill. Approximately 9,000-year old alluvial units near the base of the trench are vertically displaced approximately 3 feet; horizontal component of slip not determined.

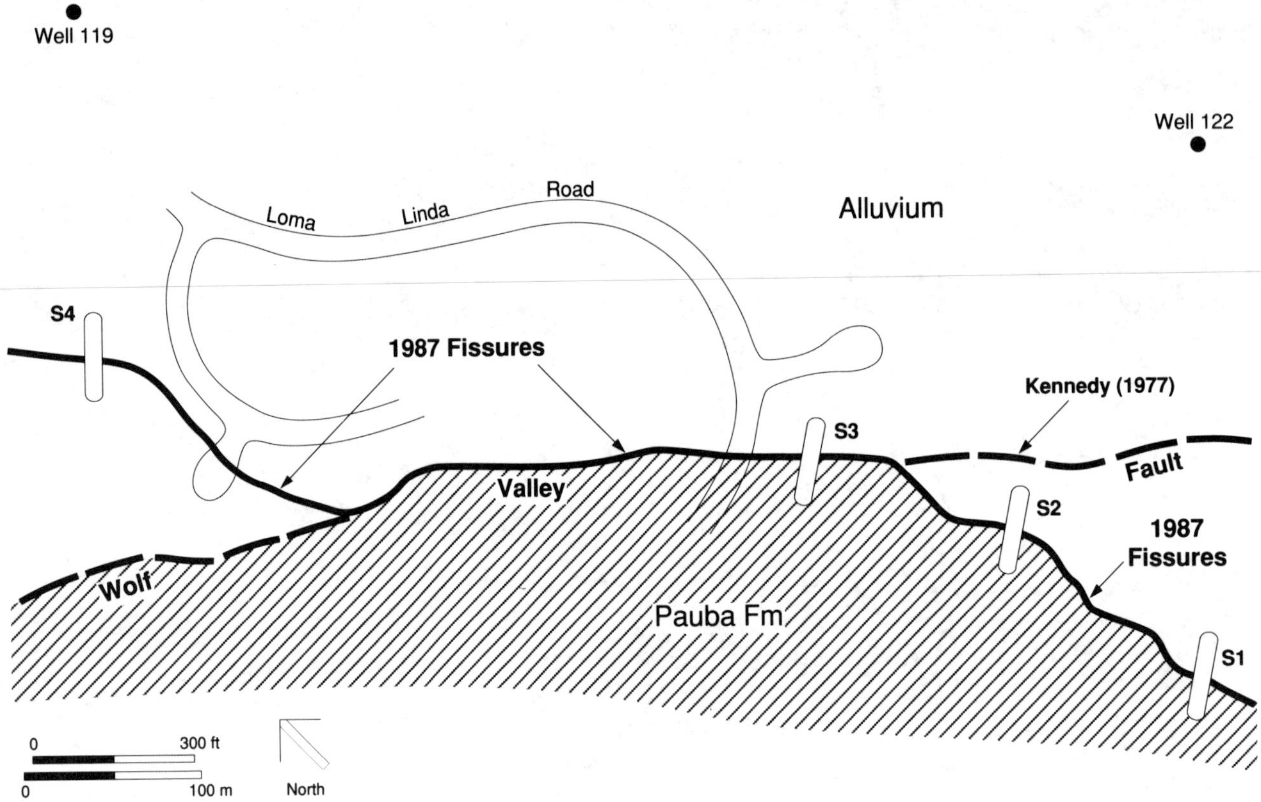

Figure 11. Southern fissure area showing location of representative water wells and trenches placed across the 1987 ground fissures and Wolf Valley fault.

sure area, we believe that southern area seismotectonic events have been probably superimposed on the slow but continuous subsidence of sediments in Wolf Valley.

ORIGIN OF THE TEMECULA GROUND FISSURES

Controversy still surrounds the origin of the 1987 Temecula ground fissures. On the one hand, geotechnical consultants for various homeowner and developer litigants contend that accelerated ground water pumping caused the fissures. On the other hand, consultants for one prominent defendant argue that aseismic creep, on the now-designated active Murrieta Creek and Wolf Valley faults, gave rise to the fissures, and therefore was an "Act of God" (Corwin et al., 1991).

Ground Water Withdrawal

Withdrawal of ground water, hydrocarbons, salt deposits, coal and many other substances have long been known to cause ground subsidence and local fissuring (Holzer, 1984; Johnson, 1991; Johnson and others, 1987; Panel on Land Subsidence, 1991). However, neither hydrocarbon, coal nor salt mining have occurred in the Temecula area; hence these causes for ground fissures are inherently ruled out. Likewise, neither local slope instability nor hydrocompaction of near-surface fluvial or eolian silts, such as well documented in the San Jacinto and San Joaquin valleys (Poland and others, 1975; Holzer, 1980; Lofgren, 1976; Morton, 1977) is likely to be a factor. More plausible explanations, given the local geological setting, are recent and rapid withdrawal of ground water, aseismic creep or some combination of the two.

To date, no systematic leveling survey has been conducted across the Temecula fissures. However, a few data points for a Riverside County road survey in the southern fissure area suggest that very slight but measurable Wolf Valley subsidence had occurred prior to 1987 (Leighton, 1987). Continuing quadrilateral-monument and street-crack surveys show that no measurable movement has occurred since 1987, except for 1989 rejuvenation of cracking in the north-central area. No seismicity was recorded at this time; however, nearby wells (no. 102 and 118) were pumping, but have since been shut down.

The ground water hypothesis for fissure causation is particularly supported by the pumping history of several new wells that coincidentally began heavy production a few days to a few weeks before the ground cracks occurred in the late summer of 1987. For example, three wells in the southern fissure area (nos. 117, 119 and 122), respectively located about 1632, 530, and 960 feet east of the Wolf Valley fault/fissure system, tapped the Temecula Aquifer, a major water-bearing zone typically encountered about 380 to 1300 feet below the surface (Scheliga and McGoldrick, 1984). Two of the nearest wells are shown on Fig. 11. Cumulative production of these wells was essentially nil in June 1987, but dramati-

Figure 12. Southern fissure area ground cracks trending across graded pads and toward newly-constructed homes in background.

cally increased to 230 and 322 acre-feet (af), respectively, during August and September 1987 when the southern fissures were first observed (W. Hardt, personal communication, 1991). When pumping almost ceased in September 1987, after the fissures were discovered, fissure growth apparently stopped (Leighton Associates, 1987). These wells, as others near the newly-designated Wolf Valley and Murrieta Creek faults, have since been shut in based on court order (Corwin and others, 1991).

Similarly, Temecula Valley wells within 1340 feet of the Murrieta Creek fault in the north-central (nos. 102 and 118) and north fissure sub-areas (101, 121), cumulatively produced less than about 100af of water in the several months preceding the observed October 1987 ground cracks in this area. Production "jumped" from about 50 af in July to over 400 af in August and September of that year (W. Hardt, personal communication). Here, too, the temporal relation between ground water pumping and ground cracking seems to be more than mere coincidence. Accordingly, as well documented in Arizona (Schumann, 1974; Slaff, 1990) and in the Las Vegas area (Bell, 1981; Werle and Stilley, 1991), rapid increase in ground water pumpage may well have induced local aquifer compaction and the accompanying tensile stresses that produced the 1987 fissures and ground cracks. Fissure location is likely controlled by nearby geologic discontinuities; in this case, slip surfaces of the Wolf Valley (south) and Murrieta Creek faults that, through processes of

Figure 13. New house astride 1987 ground cracks vacated and being readied for moving, southern fissure area.

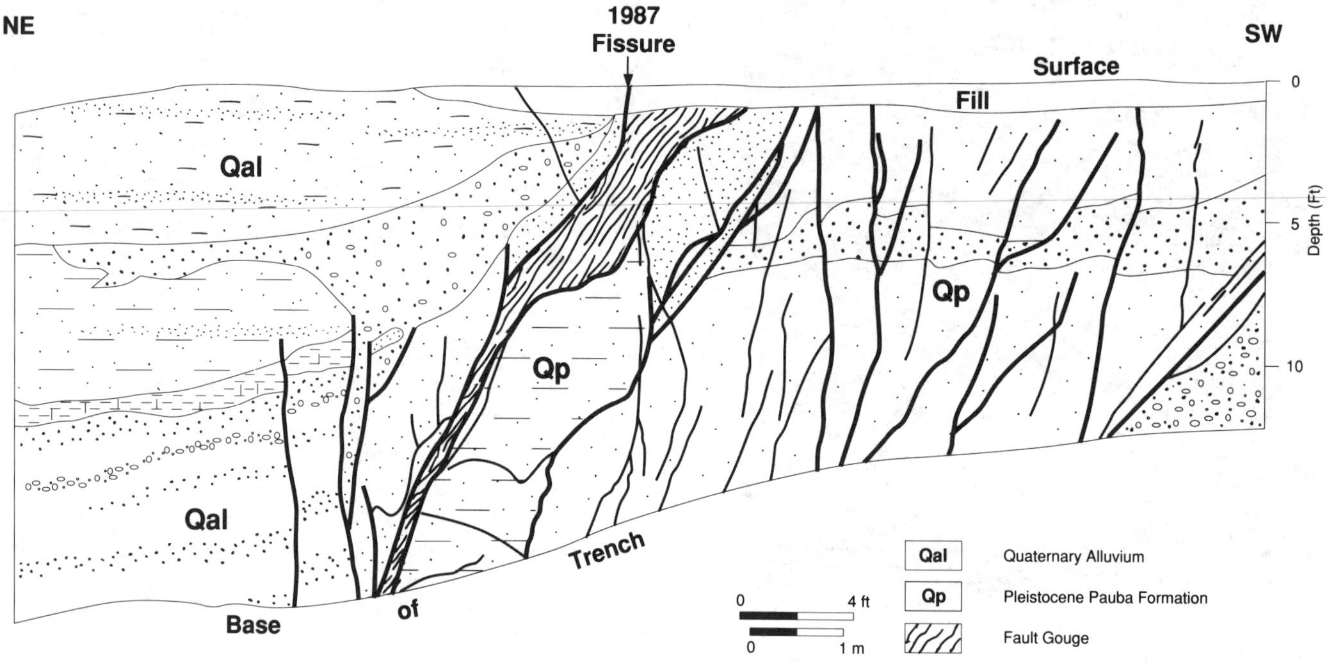

Figure 14. Log of trench S-3, southern fissure area showing fault juxtaposition of Pauba formation and alluvial sediments and coincidence of 1987 ground fissures with gouge of the Wolf Valley fault.

seismotectonic displacement and aseismic downwarping, now juxtapose alluvium against bedrock (Figs. 6 and 14).

Aseismic Creep

The aseismic creep hypothesis has been advanced by consultants for one party to the Temecula ground fissure litigation. Specifically, it has been proposed that the 1987 ground cracks stem solely from tectonic creep along the now-designated, "active" Wolf Valley and Murrieta Creek faults. The hypothesis seems to be difficult to defend, for it is contradicted by the temporal and areal proximity of ground water pumping. Further it is inherently enigmatic, for no measurable seismic events were associated with 1987 fissures (Leighton, 1987), but none might have been expected if movement is truly aseismic.

The aseismic creep hypothesis is not without merit, however, because: (1) no micro-seismic net is present in the Temecula area; (2) because aseismic ground fissures have been documented elsewhere in California; and (3) substantial vertical aseismic downdropping of the Temecula-Wolf Valley graben areas has taken place in Quaternary time.

The presence of aseismic fissures has been documented on the active Garlock fault (Zellmer e. al., 1985). Likewise, en-echelon fissures formed along the Coyote Creek fault immediately prior to, during, and shortly after the April 1968 Borrego Mountain earthquake (Clark, 1972). However, until such time as adequate instrumental records become available, the aseismic creep hypothesis cannot explain the apparent relation between time and amount of ground water pumping and causation of the 1987 Temecula ground fissures.

SUMMARY AND CONCLUSIONS

The 1987 Temecula-area ground fissures are now known to coincide with segments of the newly-designated active Murrieta Creek and Wolf Valley faults. There is no doubt that these faults are active, for Holocene-age sediments are clearly offset. But we judge that this offset reflects two concurrent processes; namely, seismotectonic slip and aseismic downwarping. As exemplified in the north and north-central fissure sub-areas, approximately 9,000- to 17,000-year old radiocarbon-dated sediments are vertically displaced less than 6 feet by the main trace of the Murrieta Creek fault. Exceptions occur in the central part of the northern fissure area where Holocene main-trace displacement may exceed 10 feet. We believe that this separation reflects long-term subsidence and downwarping of adjacent basin sediments on which has been superimposed a lesser component of seismically-induced slip.

The Murrieta Creek and Wolf Valley faults form the western margin of an actively subsiding trough and present the nearest geological discontinuities to several wells that began large-scale production in 1987. We, as several other investigators, believe that the recent and rapid withdrawal of ground water increased subsurface tensile stresses which were focused along the faults, and thus controlled the location of the 1987 ground fissures.

The propinquity of water-well pumping and 1987 ground fissures is seemingly more than a coincidence. Indeed, the start up of several nearby wells in the summer of 1987 closely corresponds with the timing of fissure occurrence. And perhaps more telling is that ground cracking essentially ceased when the wells were shut in. Accordingly, we favor the ground water pumping hypothesis and reject an aseismic creep explanation as the major cause of the Temecula-area fissures.

The 1987 fissures are mostly hairline cracks best expressed in street pavement and in concrete slabs and curbs. To date, no injuries have occurred, no houses have collapsed, and no significant fissure rejuvenation has taken place. Nevertheless, property damage has occurred, an active fault system has been newly recognized, and litigation has ensued. Hypotheses as to fissure origin are still hotly debated by litigants and by generally respected geotechnical consultants. Accordingly, the "Great Crack of Temecula," as deemed in the popular press, will likely prove to be a classic case study. Indeed, it appears that Temecula-type ground fissures, whatever their origin(s), will likely increase as urbanization extends out into the more arid, interior regions of southern California, and will therefore be a subject of economic and geotechnical interest, and perhaps controversy, for many years to come.

ACKNOWLEDGMENTS

The interpretations presented in this paper stem not only from our personal observations and measurements, but also from the technical reports kindly made available to us by the many developers and geotechnical consultants involved in the Temecula area. We therefore thank these developers and consultants, and particularly Bedford Properties (Temecula) for logistical support. We likewise gratefully acknowledge Riverside County Geologist S. Kupferman and consulting hydrogeologist W. Hardt for many helpful discussions; Geologist A. Meeker for unswerving technical assistance; K. Black and J. Siegfried for manuscript typing; and the editors of this volume for their long-enduring patience. We remain, of course, responsible for our interpretations and hypotheses, as well as for any pertinent omissions or other shortcomings.

REFERENCES

Bell, J.W., 1981, Subsidence in Las Vegas Valley: Nevada Bureau of Mines and Geology, Bulletin 95, 84 p.

Bergmann, M. and Rockwell, T., 1989, The Murrieta Creek Fault, a New Branch of the Elsinore Fault, Rancho California, Area, Riverside County, California: Geological Society of America Abstracts with Program, No. 21, p. 5.

California Division Of Mines And Geology, 1990, Special Studies Zones Maps, Pechanga, Murrieta and Temecula Quadrangles: Sacramento, California, scale 1:24,000.

Clark, M.N., 1972, Collapse Features Along the Coyote Creek Fault in the Borrego Mountain Earthquake of April 9, 1968: U.S. Geological Survey Professional Paper 787, p. 190-207.

Corwin, E., Alhadeff, S., Oggel, S., and Shlemon, R., 1991, Earth Fissures, Urbanization and Litigation; a Case Study from the Temecula Area, Southwestern Riverside County, California: in Johnson I. (ed.), International Association of Hydrological Science, Fourth International Symposium on Land Subsidence (Houston, Texas), Publication 200, p. 291-299.

Crisman, R.G., Mcclay, P.L., and Shlemon, R.J., Geologic Bias and Its Affect on Geotechnical Site Evaluations: Old Vail Ranch, Temecula, California; Geological Society of America, Abstracts with Programs, Annual Meeting, San Diego, California, V. 23, No. 5, p. A41.

Davis, P., Meeker, A., and Shlemon, R.J., 1991, Ground Cracks and Surface Displacement, Murrieta Creek and Wolf Valley Faults, Temecula Area, Riverside County, California: Geological Society of America Annual Meeting, Abstracts with Programs, V. 23, No. 5, p. 122.

Earth Research Associates, Inc., 1987, Evaluation of Faulting and Liquefaction Potential, Portion of Wolf Valley Project, Riverside County, County of Riverside; California: Consultants' Technical Report, Job No. 298-87 (Costa Mesa, California) for Great American Development Company, 8 p.

Engle, R., 1959, Geology of the Lake Elsinore Quadrangle, California: California Division of Mines Bulletin 146, p. 1-59.

Geocon, Inc., 1990, Soil and Geologic Investigation for Rancon Business Center, Tentative Map No. 26156, Riverside County, California: Consultants' Technical Report, File No. 04567-07-01 (San Diego, California), for Rancon Financial (Temecula, California), 32 p.

Geocon, Inc., 1991, Supplemental Fault Investigation for Tract No. 25139, Temecula, California: Consultants' Technical Report, File No. 2002-07-01 (San Diego), for 50 Center City Associates/City Associates 2, 11 p.

Geowest Soil Consultants, 1987, Geotechnical Investigation, Tract 19939 Fissure, Pala Road Between Via Gilberto and Via Eduardo, Wolf Valley Area, County of Riverside, California: Consultants' Technical Report, File 742-002, for Jostri Inc. (Temecula, California), 29 p.

Hart, E.W., 1990, Fault-Rupture Hazard Zones in California: California Division of Mines Special Publication 42 (Sacramento, California), 26 p.

Holzer, T.L., 1980, Faulting Caused by Ground-Water Level Declines San Joaquin Valley, California: Water Resources Research, vol. 16, no. 6, p. 1065-1070.

Holzer, T.L., 1984, Ground Failure Induced By Ground-Water Withdrawal From Unconsolidated Sediments: in, Man-Induced Land Subsidence, Geological Society of America Reviews in Engineering Geology, Volume VI, p. 67-105.

Johnson, A.I. (ed.), 1991, Land Subsidence: International Association of Hydrological Science, Fourth International Symposium on Land Subsidence (Houston, Texas), Publication No. 200.

Johnson, A.I., Carbognin, L., and Ubertini, L. (eds.), 1990, Land Subsidence: International Association of Hydrological Science, Proceedings of the Third International Symposium on Land Subsidence (Venice, Italy), Publication No. 151, 939 p.

Kennedy, M.P., 1977, Recency and Character of Faulting Along the Elsinore Fault Zone in Southern Riverside County, California: California Division of Mines and Geology Special Report 131 (Sacramento, California), 12 p.

Leighton & Associates, Inc., 1987, Geotechnical Investigation of "Ground Crack", Tract 19872, Wolf Valley, Rancho California Area, County of Riverside, California: Consultants' Technical Report, Project 6851602-03, for Richmond American Homes (Irvine, California), 19 p.

Leighton & Associates, Inc., 1988a, Preliminary Geotechnical/Fault Investigation 600 +/- Acres, The Pascoe Property, Murieta, Riverside County, California: Consultants' Technical Report, Project 6871384-01, for Rancon Financial Corporation (Temecula, California), 18 p.

Leighton & Associates, 1988b, Revised Geotechnical Report for a Phase II Fissure and Subsidence Investigation and Phase III Geotechnical Investigation for Parcel Map No. 21502, Rancho California, Riverside County, California: Consultants' Technical Report, Project 11860325-02, for Diaz Road Investors (Temecula, California), 22 p.

Lofgren, B.E., 1976, Land Subsidence and Aquifer-System Compaction in the San Jacinto Valley, Riverside County, California — A Progress Report: U.S. Geological Survey Journal of Research, vol. 41 no. 1, p. 9-18.

Mann, J.F., 1955, Geology of a Portion of the Elsinore Fault Zone, California: California Division of Mines Special Report 43 (Sacramento, California), 22 p.

Moran, D.E., Inc., 1989, Significance of Ground Cracks, Riverside County Tract No. 19872, Pala Road at Loma Linda Road, Riverside County, California: Consultants' Technical Report, Job No. 88-23, for Best, Best & Kreiger (Riverside, California), 69 p.

Moreland, J.A., 1972, Maps of the Watersheds of the Santa Margarita and San Ruis Rey Rivers, Riverside and San Diego Counties, California: U.S. Geological Survey Open-File maps.

Morton, D.M., 1977, Surface Deformation in Part of the San Jacinto Valley, Southern California: U.S. Geological Survey Journal of Research, vol. 5, no. 1, p. 117-124.

Panel On Land Subsidence, 1991, Mitigating Losses from Land Subsidence in the United States: National Academy Press, Washington, D.C., 58 p.

Poland, J.F., Lofgren, B.E., Ireland, R.L., and Pugh, R.G., 1975, Land Subsidence in the San Joaquin Valley, California, as of 1972: U.S. Geological Survey Professional Paper 437-H, p. H1-H78.

Schaefer Dixon Associates, Inc., 1988a, Report on Geotechnical Investigation of the Willard Fault, Parcel C-16: Consultants' Technical Report, Job No. 9R1208J, for Rancho California Development Company (Temecula, California), 21 p.

Schaefer Dixon Associates, Inc., 1988b, Geotechnical Investigation, A Portion of Business Park III, Phase 2, Parcels 1, 7, 8, 9, 10 and 11, Rancho California, California: Consultants' Technical Report, Volumes I and II, Job No. 8O-182; for Bedford Properties, Inc. (Rancho California, California), 36 p.

Schaefer Dixon Associates, Inc., 1989a, Fault and Seismicity Investigation, a Portion of Business Park III, Phase 2, Parcels 3 through 6, Rancho California, California: Consultants' Report, Job. No. 9R-128N, for Bedford Properties, Inc. (Rancho California, California).

Schaefer Dixon Associates, Inc., 1989c, Report on Geotechnical Investigation, Assessment District No. 155, Parcel Map 24085, 24086, 21029, 21382, and 21383, Rancho Californif, Riverside County, California: Consultants' Technical Report, Job No. 9R4332C, for Johnson & Johnson, Inc. (Temecula, California), 28 p.

Scheliga, J.T., and McGoldrick, J.P., 1984, Rancho California Water District, Water Resources Master Plan 1984: Consultants' Technical Report, for Rancho California Water District (Temecula, California), 10 chapters.

Shlemon, R.J., 1985, Application of Soil-Stratigraphic Techniques to Engineering Geology: Bulletin, Association of Engineering Geologists, vol. XXII, no. 2, p. 129-142.

Shlemon, R.J., and Davis, P., 1988, Ground Fissures in the Rancho California Area, Riverside County, California: Geological Society of America Abstracts with Programs, no. 20, p. 7.

Slaff, S., 1990 Bibliography on Arizona Earth Fissures and Related Subsidence, With Selected References for Other Areas: Arizona Geological Survey Open-File Report 90-7, 28 p.

Soil Tech, Inc., 1989, Geologic/Seismic Hazards Investigations, Parcel 9 of Parcel Map 19580, Business Park Drive West of Single Oak Drive, Rancho California Area, Riverside County: Consultants' Technical Report, Project No. T3267-FS, for Wescon Properties (Irvine, California), 28 p.

Werle, J.L., and Stilley, A.N., 1991, Effects of Ground Fissures on Development in the Las Vegas Valley, Nevada: Proceedings of the 27th Symposium on Engineering Geology & Geotechnical Engineering, Utah State University, Logan, Utah, p. 32-1 - 32-15.

Zellmer, J.T., Roquemore, G.R., and Blackerby, B.A., 1985, Modern Tectonic Cracking Near the Garlock Fault, California: Geological Society of America Bulletin, vol. 96, p. 1037-1042.

FAULT DEVELOPMENT AND FISSURING NEAR TWENTYNINE PALMS, CA

John H. Foster
Department of Geological Sciences
California State University, Fullerton
Fullerton, California 92634

ABSTRACT

Engineering geology studies near Twentynine Palms, CA, have revealed a change in tectonic style dating at between 100,000 and about 50,000 years ago. The northwest striking West Bullion Mountain fault is a normal fault with over 200 feet of vertical movement, down to the west. Faulting along this structure last occurred prior to about 50,000 years ago, the age of an overlying undisturbed alluvial fan with distinct desert varnish. Recent faulting is occurring further west along the Mesquite Lake fault, a newly formed southern extension of the Calico fault. Several overlapping lines of evidence identify this fault as active, but of particular note is its location within the middle of Mesquite Playa rather than near the edge of the basin. Thus, the faulting most likely succeeded the playa formation. The playa and lake beds formed due to relative motion along the West Bullion Mountain fault. The actual basin low point is west of, but adjacent to, Mesquite Lake fault. It therefore seems likely that the Mesquite Lake fault is a very recent (<100,000 years) extension of the Calico fault; however, displacement in the Twentynine Palms area is only about one-half mile which is less than that documented to the north.

A one-half mile long ground crack occurs north of the Mesquite Lake Fault in the playa basin. This crack trends due north and was thought by earlier investigators to be fault related. However, the discovery of other parallel healed cracks and subsurface detailed studies have shown these features to be desiccation related. Results of drilling show highly plastic clay over most of the playa basin at relatively shallow depth and perched groundwater at about 70 feet below the surface. Although detailed stratigraphy and soil descriptions are not available, it appears that dehydration of the playa clays above about 30 feet has resulted in damp to dry and hard soil conditions. With groundwater present at about 60 feet in clay and the potential for a very large capillary fringe, possibly as much as 25 feet, then the actual control of the ground cracking is likely desiccation of the clays at the 30-foot depth. Where overlain by sandy alluvial playa filling deposits, these may sustain cracks to the surface above the desiccated clay-layer fracture. One such zone east of the runway stood open as a crack in sandy soil and was filled instantaneously by a reddish debris flow which is continuous downwards from a debris flow deposit over the crack.

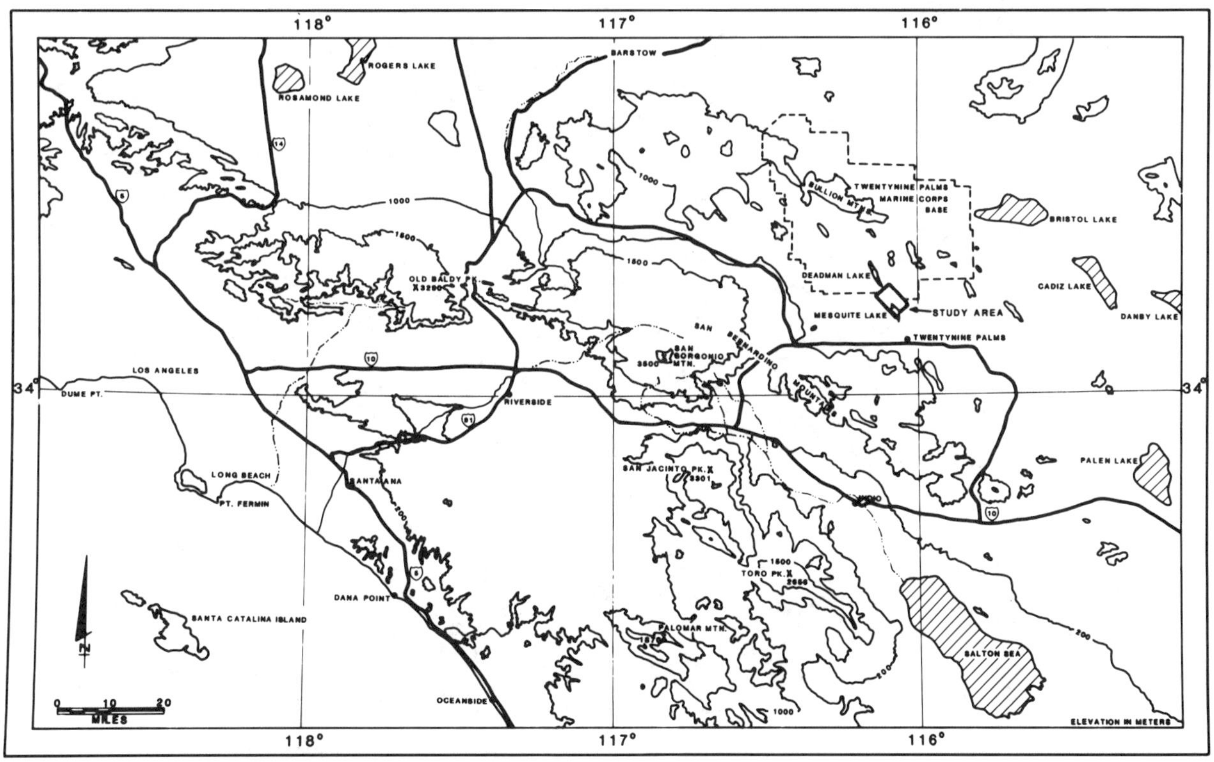

Figure 1. Location map showing outline of 29 Palms Marine Base and study area.

INTRODUCTION

This paper presents the results of on-going work begun in 1982. The studies have been conducted in and around Mesquite Lake Playa, north of Twentynine Palms, California (Fig. 1). This area is a part of the southern Mojave Desert and lies just north of the eastern end of the Transverse Ranges and Joshua Tree National Monument. The geology and structure are typical of the Mojave Desert, that is, rugged granitic or volcanic mountains shaped by northwest-striking faults with mostly right-lateral strike-slip movement. The Mesquite Lake Playa area is unique to the Mojave Desert in that the northwest striking faults terminate or merge with east-west striking left-lateral strike-slip faulting of the Pinto Mountain fault zone east of the town of Twentynine Palms. Mesquite Lake Playa is a low basin between a relatively flat high plains area to the west and the western spur of the Bullion Mountains to the east. Most of the playa and study area is within the boundary of the Marine Corps Air Ground Combat Center at Twentynine Palms. A composite aerial photograph of the study area is presented as Figure 2.

The purpose of initial studies by Wahler Associates (1984) was a concern by the Naval Facilities Engineering Command and Marine Corps about faulting, ground cracking and building foundation distress affecting existing facilities and their potential impact on planned expansion facilities in the study area. Numerous isolated geotechnical studies had been completed but none had attempted to synthesize geotechnical hazards over the whole study area. As a result, planning was uncoordinated and the information base very unorganized. The Wahler Associates studies sought to integrate previous findings, to collect new data and to then establish a data base for base planning. New data were acquired using seismic refraction, gravity and magnetic geophysical studies, trenching, interpretation of aerial photographs, geotechnical borings, and laboratory studies.

Subsequent studies have related these new data to emerging ideas about the structural development of the Mojave Desert. This paper presents some of the results of the above studies as they relate to the general structural setting of the Mojave Desert as now understood.

Acknowledgements

I gratefully recognize Wahler Associates where I worked during the early phases of this study and the NAVFACENGCOM, San Bruno who supported much of the field studies under contract N62474-82-C-0160 to Wahler Associates. I would also like to acknowledge the help and coordination efforts of M.J. Boomer, Chief Engineer at the base Public Works Office. Finally, some of the ideas presented have resulted from conversations and field trips taken with John J. Butelo of Schaefer Dixon Associates, and Dr. Roy J. Shlemon who also provided a detailed log of the West Bullion Mountain fault trench and associated age estimates. I also want to thank my reviewers for their careful consideration of this study: Wes Reeder, County Geologist for the County of

Figure 2. Composite aerial photograph of Main Camp area of 29 Palms Marine Corps Base. North is parallel to road between runway and buildings of the base; scale is approximately 1"=2000'. Mesquite Lake Playa in center; Mesquite Lake fault trace visible as line separating brush covered dunes from the playa, shown by white arrows. Ground cracks visible as dark lines through the runway parallel to road and noted by white arrows. Note arcuate line of vegetation next to road west of playa; this is interpreted to be an old shoreline feature.

San Bernardino; Dr. Michael O. Woodburne, Department of Earth Sciences, University of California at Riverside; and, Dr. Jerry Brem, Chair, Department of geological Sciences, California State University Fullerton. Their comments have greatly improved the quality of this paper.

BASIN DEVELOPMENT

Photogeology

Surficial geology, particularly Quaternary units, was mapped on three different sets of aerial photographs: 1:24,000 scale black and white, 1:12,000 color, and 1:12,000 color transparencies. No field mapping program was done as part of this research effort. Geologic mapping of the regional area was done by Dibblee (1968 and 1967a) at a 15' scale (1:62,500) as well as by Bassett and Kupfer (1962) at a scale of 1:125,000. Dibblee's basic bedrock geology is retained here (Fig. 3 and 4), but some units of later Cenozoic geology have been separated from the broad alluvial categories.

Photogeologic mapping of the study area revealed six Quaternary age surficial geologic units; the basement geology was divided into three units based on cover. and presence or absence of dikes (Fig. 3 and 4). Dissected alluvial and lacustrine deposits (QTa) are present in the elevated areas around the lakebed and exhibit distinct erosional scarring probably due to the incision of the nearby dry stream courses noted as Qsc for the most part. Older intact alluvial fan deposits (Qof) are identified near the base of the western Bullion Mountains. One deposit however, was identified at the playa edge with its apex near the now concealed Western Bullion Mountain fault. Playa and lake deposits (Ql) are identified from high reflectance and are in three discontinuous bodies in the Mesquite Lake area. Areas of warped clay beds and blown sand intersect the once continuous set of playa deposits. Three large accumulations of dune sand (Qd) have developed in the Main camp area. These accumulations are east of Mesquite Lake fault where their presence is due to high groundwater and plant growth on upwarped sections of lake bed; in a crescent of dune sand along the west perimeter of the playa; and as a crescent of dune sand near the southeast perimeter of the playa. Stream channel deposits (Qsc on the map) contain loose sand deposits and some of these narrow areas have been artificially constructed or fortified by the base engineers Undifferentiated alluvium (Qa) is separated out and located north, east and south of the playa. This unit is composed of Recent and near-Recent sheet flood, debris flow and alluvial fan deposits now topographically modified by erosion and more often identified in trenches.

Geophysics

New gravity, magnetic and seismic refraction surveys were run on and off the base to identify basin configuration and fault locations. The investigations weie done to confirm the presence and location of a concealed fault mapped by Dibblee (1968). Also, studies were oriented to identify any

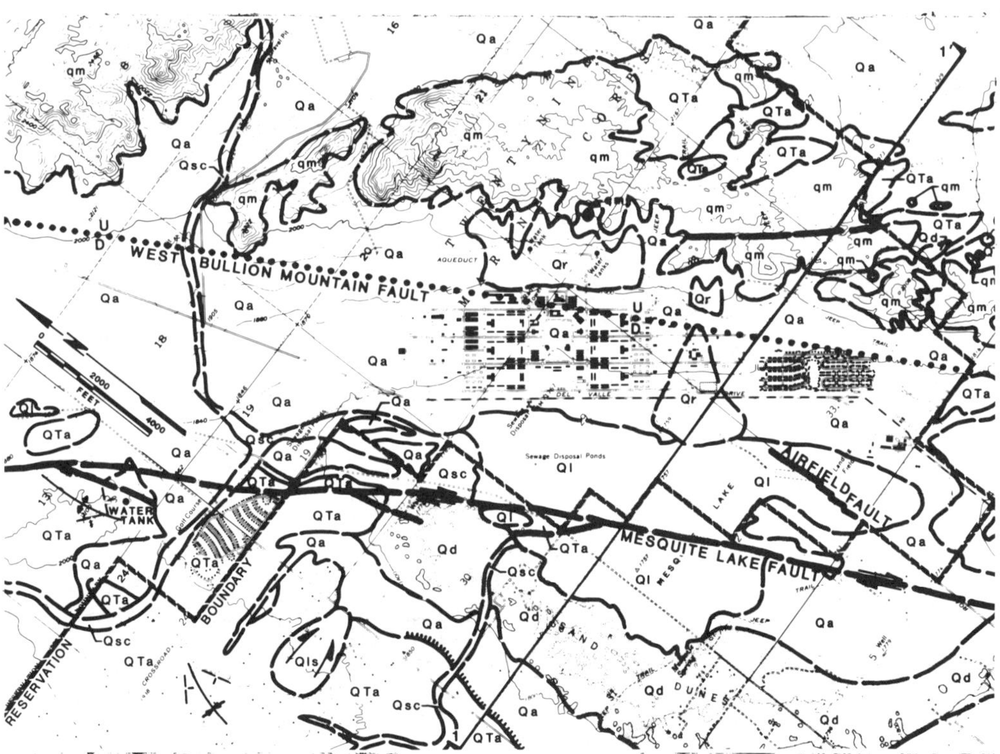

Figure 3. Photogeologic map of Main Camp area. Units are described in figure 4.

subsurface causes, such as a buried basement ridge, for fissuring in the playa. The geophysical data were collected and interpreted by Biehler and Gilpin (1983).

The subsurface geology as interpreted from the geophysical data was resolved into three slow-velocity layers on top of 16,000-feet-per-second quartz monzonite, (Cross section 1-1', Fig. 5). These include an 80-120-foot thick alluvium and playa sediment unit with velocities of 2700 to 3000 feet per second; another 1000-foot-thick section of this same formation that is saturated and has a velocity of 6900-7000 feet per second; and, below that, a 2200-foot thick consolidated formation at 8800 feet per second.

A planar surface was identified on the subsurface quartz monzonite. This surface was interrupted by a 250-foot step, down to the west, interpreted to be a fault. This fault projects upwards to within a few hundred feet west of the concealed fault mapped by Dibblee (1968) and is believed to be the subsurface expression of the West Bullion Mountain Fault as now defined. Basement is again down to the west at the Mesquite Lake fault due to right lateral offset. These relationships are shown in cross section 1-1', Figure 5.

The basin is deepest, and contains over 48,000 feet of sediment, west of Mesquite Lake fault. Gravity and magnetic surveys are interpreted by Biehler and Gilpin (1983) to show a northwest to southeast oriented basin with a central depression about one-half mile west of the Mesquite Lake fault and about one mile south of the center of the playa. Basement

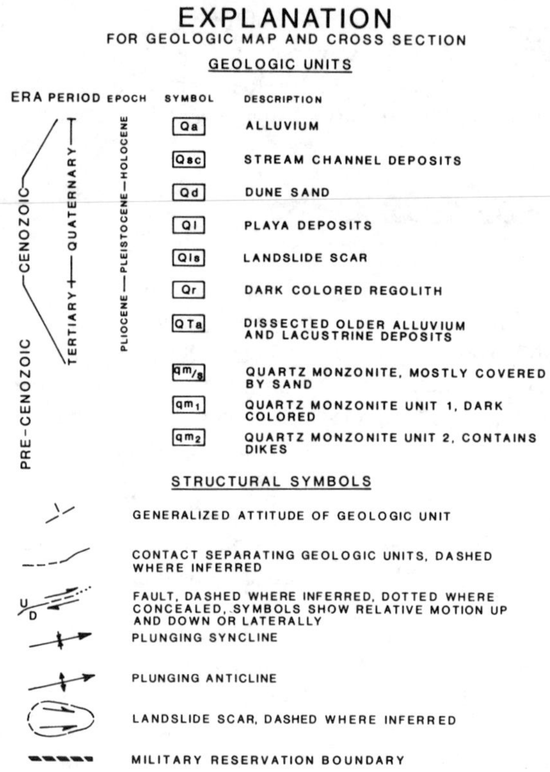

Figure 4. Legend for Figures 3 and 5.

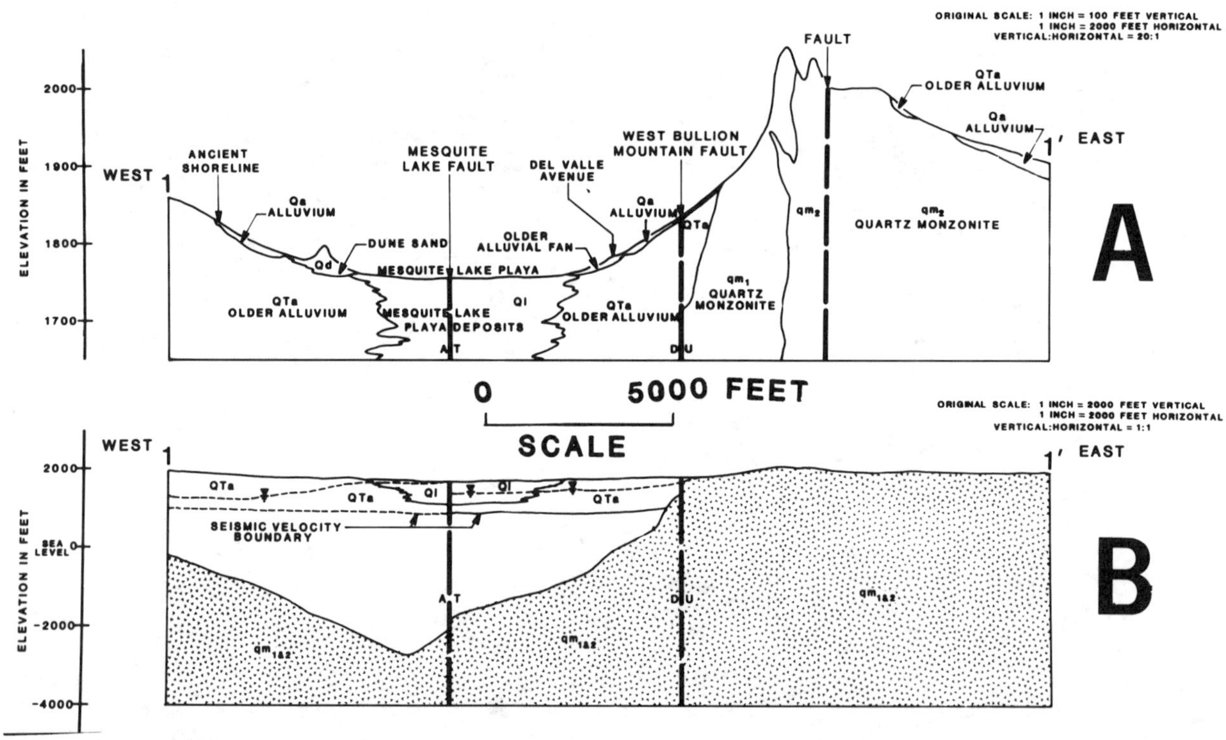

Figure 5. Geologic cross sections 1-1' (see Figure 3 for location). Subsurface interpretation is based on geophysical and borehole data.

rises steeply to the southwest and south and crops out within a kilometer of the Pinto Mountain fault. Seismic refraction data of Biehler and Gilpin (1983) show the contact between the saturated sediment and consolidated sediment sloping downwards and southwards beneath the present playa.

Geotechnical Studies

Drilling, sampling and laboratory analyses of soils were performed in the course of the initial investigation. Of primary interest here are studies conducted in and around the playa area including boreholes BH-1 through 8 as well as BH-57 and BH-58 (Wahler Associates, 1984). Boreholes 1,2,4,6,7,8, and 58 were drilled to depths of 50 feet, and boreholes 3, 5, and 57, were drilled to 100, 75 and 100 feet respectively. Core from boreholes 3, 7, 8, and 57 are composed mostly of clay and lie within the current playa map boundary. The other holes in the area are predominantly sand with some clay and gravel. It appears then that the current basin configuration is the same as that of earlier lake or playa phases. The clays exhibit high swelling potential and are overconsolidated.

The playa basin stratigraphy is similar to that described by Muessig et al. (1957) based on cores drilled in Soda Lake, and by Wells et .al., 1989, based on cores drilled in Silver Lake. These cores reveal finely bedded to laminated sequences of sandy and clayey layers which are hundreds of meters thick. Thick sequences of greenish clay which locally may contain invertebrate fossils are present and are interpreted to be lacustrine sediments from ancient stands of Lake Mojave. Droste (1961) has shown the clay minerals in Mojave Desert playa sediments to be mostly montmorillonite of moderate to high swelling capacity.

Mesquite basin is therefore a typical Mojave Desert lacustrine basin with a record similar to other basins indicating a deep lake phase which dried up after the Mojave pluvial period between 18,400 and 11,400 years ago (Oxygen Isotope Stage 2) into the playa environment of today. This drying is continuing today as groundwater declines regionally creating large desiccation cracks in many of the playa sediments from the shrinking of the expansive playa clays.

GROUNDWATER

Perched groundwater was noted by Wahler Associates (1984) in borehole 57 at a depth of about 70 feet in clay and sand near the south end of the playa, and at about 95 feet in borehole 3 near the north end of the playa. When the difference in elevation of the northern and southern piezometers is accounted for, the water elevation is a flat surface. Therefore the water table of at least a regional perched groundwater body exists at about an elevation of 1687 feet above mean sea level.

Figure 6. Trench wall exposure of West Bullion Mountain fault looking northwest near southeast boundary of Marine Corps Base. Exposure shows 10 cm offset down to the west; mechanical pencil for scale and fault line shown by white arrows.

Figure 7. Trench wall exposure of Mesquite Lake fault due south of southern Marine Corps Base boundary. Offset totals 1.5 m dropping down to the west (away). Note fault steps by displacement of 10 cm thick clay layer in sand indicated by white arrow. Clay layer is frosted white by gypsum coating.

Figure 8. Close-up of paleoliquefaction structure. Sand has blown through carbonaceous layer dated at 1300 100 years before present.

Regional groundwater is in two separate subbasins, Surprise Spring and Mesquite, divided by the Mesquite Lake fault (Schaefer, 1978). Mesquite Spring, west of the fault, does not now flow; however, it has in historic times. Groundwater is about 2 meters deep and generally ponded west of and along the trace of the fault according to Schaefer (1978). East of the fault, water is known to be 260 feet deep in isolated wells (Schaefer, 1978). Flow is generally south and eastwards. Use of the groundwater in the general area of these two subbasins by the military has caused a general regional decline that is affecting the flow patterns and elevations in the Mesquite Lake area (Schaefer, 1978).

EXTENSIONAL FAULTING

West Bullion Mountain Fault

The name West Bullion Mountain fault is herein applied to the concealed fault of Dibblee (1968) and is a modification of West Bullion Fault used by Bacheller (1978). The existence and location of this fault which has been confirmed in this study by geophysical studies and trenching. This fault is a normal fault which strikes north 25° west and dips about 75° west. The fault was observed in trenches near the southeast boundary of the Marine base in line with a surface projection made from the seismic refraction lines (Fig. 6). Offset in the basement averaged about 200 feet down to the west and was identified in all three seismic lines as well as in the gravity and magnetic traverses. These data were collated and interpreted in the ray tracing profiles of Biehler and Gilpin (1983). The fault was only observed in a single trench near the south end of the base. Here, the fault was seen to cut older alluvial sands, fine gravels and clays to within 25 cm of the surface. Two breaks are recorded, one striking northwards and the more westerly striking N25^{o}W.

Age of faulting

The most recent age of faulting was based on field geomorphic and soil stratigraphic relationships (Roy J. Shlemon and Associates, 1985). Shlemon's analysis was based on the sinuosity of the West Bullion Mountains east of the fault indicating pre-Holocene activity and the fact that the fault does not intersect an overlying intermediate-level fan surface which has a moderately developed desert pavement and varnish (patina). This surface is virtually undissected indicating relative youthfulness. Also, the faults cut a depositional (non-pedogenic) clay layer which contains Stage II-III carbonate nodule development (Fig. 6). Finally, the faults do not cut an argillitic soil horizon of B2t and Bk affinity. Therefore, the latest activity on the fault is judged by Shlemon to be greater than 35,000 to 50,000 years old (oxygen isotope stage 3), the age of the undisplaced argillitic horizon. The actual age of fault activity may be as great as, or greater than, 100,000 years because the age of the offset alluvial horizons is only approximately known.

SHEAR FAULTING

Mesquite Lake Fault

Mesquite Lake fault is an active fault which trends generally north 25° west through the center of Mesquite Lake playa and is about one mile west of the West Bullion Mountain Fault. Its trace is easy to follow on the ground and especially on air photographs as a straight line separating dune fields on the west from playa and dissected older alluvium on the east. It was originally mapped by Dibblee (1967a, 1968) and lately it has been related to strike slip faulting in the Mojave Desert by Dokka (1983) and Dokka and Travis (1990).

The nature of movement appears to be right-lateral strike-slip based on offset alluvial units. Trenches excavated by Wahler Associates (1984) showed vertical separation of lake bed and alluvial units that changed relative movement between trenches. Thus, to the north, near Deadman Lake, the offset in the trenches showed uplifted blocks to the west; whereas to the south of Mesquite Lake, blocks are dropped to the west, (Fig. 7). Such a reversal in vertical throw is typical of strike-slip faulting and is extremely unlikely for normal faulting. From borings placed in the lake bed area by Wahler Associates (1984), it appears that green gray lacustrine clay (non-playa) deposits have been offset in a right-lateral sense. The actual offset is indeterminate at this time, but is thought to be on the order of a half-mile or more. Further evidence of activity is the offset of the groundwater table at least 70 feet (and perhaps as much as 260 feet), down to the east. Offset water tables are characteristic features of active faulting in Southern California.

Age of faulting

The age of activity on the Mesquite Lake fault was determined by a carbon-14 date on organic clay from a trench across the fault south of Mesquite Lake (Fig. 7). This clay was sent to two separate laboratories, Beta Analytical in Coral Gables, Florida, and to Dr. Taylor at the University of California Riverside. Samples were prepared by picking out rootlets, dispersing in hot acid to eliminate carbonates, repeatedly rinsing to neutrality and then brought to dryness and given multiple combustions in an enclosed system. Beta Analytical provided a benzene synthesis and normal count on 0.2% carbon for an age of 1220+/-1100 years before present (1950). Dr. R. E. Taylor's lab provided a base insoluble residue age of 1390+/-1150 years before present (1950). Dr. Taylor also provided a base soluble fraction age of 1730+/-100 years before present. Geologically, samples from the carbonaceous horizon are identical and different ages are presumably the result of different lab procedures and methods. Generally the age is considered here to be about 1300 years old or from about 450 A.D.

The clay layer was broken by a fault strand and the same layer was pierced by a liquefaction event (Fig. 8). The fault event is younger than the carbonaceous layer and no other datable units overlie the break, however, this layer terminates against a fault strand a few meters east of the dated location and point of liquefaction. This break extends to the surface. Faulting therefore has occurred at least twice since the dark carbonaceous soil was deposited, presumably the last time swampy conditions prevailed 1300 years ago. As such, a crude maximum recurrence interval of every 600-700 years for earthquakes and faulting capable of producing significant rupture is proposed for the fault.

An interesting aspect of the fault is its relative recency of origination. As detailed above, the West Bullion Mountain fault is believed to be inactive, but may have become so only within the last 100,000 years. Stress relationships for a normal fault are opposed to those conducive to a strike-slip fault forming in parallel. It appears then that a stress field change and, hence, a change in structural style from normal faulting to strike-slip faulting, must have occurred during the latest Pleistocene in this area. The Mesquite lake fault has only recently ruptured lacustrine sedimentary beds formed about 18,000 to 12,000 years ago. Typical basin forming mechanisms of strike-slip faults which create transpressional basins or rhombochasms do not appear to have operated here. Rather, an extant basin has been sheared apart by strike-slip faulting. Other strike-slip faults in the Mojave Desert separate basins and mountain ranges. Older sedimentary rock is exposed north of the Marine base west of the fault and near Twentynine Palms east of the fault. No correlation has been proposed or confirmed for these two stratigraphic sequences and their exposure is most likely related to compressional uplift due to bending of the fault. Certainly more work is required to confirm the slip amount and rate and to relate that to other Mojave Desert faults, but, at this time it seems that the Mesquite Lake fault is quite youthful, having originated only within the last 100,000 years.

Figure 9. Filled ground fissure at airfield; looking north.

Figure 10. Close-up view of ground fissure under deformed temporary runway treads.

SEISMICITY

Seismic History

Twelve earthquakes of Richter magnitude 4.0 or greater have occurred within a 25 mile radius of the main camp area of the Marine Corps Base (Real et.al., 1978; Hill, 1976; and McJunkin, 1980). No earthquakes of this magnitude have occurred historically along the Mesquite Lake fault. Most recently, two earthquakes have occurred near the base: a magnitude 4.6 on November 15, 1975; and a magnitude 4.7 on December 14, 1975. The epicenters of these earthquakes were about a mile apart on a northwest line between the Emerson fault and Goat Mountain, about 17 miles west to northwest of the marine base. The focal depths were 3.5 and 1.1 miles respectively. No surface rupture occurred, however numerous rockfalls were reported. Earlier, in the same year, a magnitude 5.2 earthquake occurred near the Emerson fault at Galway Lake and was accompanied by surface rupture with a right-lateral sense of motion (Hill and Beeby, 1977). The rupture zone was 4.2 miles long and had not previously been mapped. The Galway fault trends northwestward from the Emerson fault and passes along the western edge of Galway Lake. On March 15, 1979, a magnitude 5.2 earthquake struck near Landers on the Homestead Valley fault, also called the Pipes Wash fault, a southward splay of the Emerson fault. The earthquake was accompanied by a swarm of aftershocks and horizontal right-lateral strike-slip movement (Hunter et al., 1980; McJunkin, 1980). The focus of this earthquake and swarm was 3 miles deep.

The proximity and parallel strike of these faults to the Mesquite Lake fault as well as the seismic activity and history, tends to support the idea that the Mesquite Lake fault is active and can be expected to have shallow focus moderate size earthquakes along it with accompanying right-lateral strike-slip motion several miles long. Displacement along the fault could range up to a few inches or more per event if the above examples hold. The Galway event exhibited a maximum displacement of 1/2 inch and the Homestead event exhibited 1 inch of vertical displacement and about 1 cm of right-lateral displacement. Predictions of offset on these northwest striking Mojave Desert faults range suggest offsets measurable in meters based on much larger predicted earthquakes (Greensfelder, 1974; Wesnousky, 1986). Also, the evidence of single event movements of about 2 feet of vertical displacement were observed in trenches placed across the Mesquite Lake fault by Wahler Associates (1984). It therefore may be likely that greater offset and greater magnitude events will occur on the Mesquite lake fault than have been observed recently on nearby similar faults.

Figure 11. Trench exposure of healed ground crack. Note change in units from light sandy deposits below to darker debris flow deposit above. Debris flow is traceable down into filled crack (dark vertical zone in sand; about 1 cm thick).

GROUND CRACKING

History of Ground Cracking

Ground cracks are present in Mesquite Lake Playa trending northwards near the southeast end of the lake bed. These en echelon fissures step to the right in four one-half kilometer segments and were referred to by Fife (1978) as the Airfield Fault (Fig. 3). The cracks are easily visible from the ground and air as narrow filled depressions with abundant vegetation in an otherwise barren environment (Fig. 9 and 10). Trenching of the fissures by Wahler Associates (1984) showed that these features are formed by tension with open linear cavities below the filled depressions as well as large failed blocks in a funnel-shaped downward-narrowing system of cracks. The zone is bounded by single breaks on either side of the main feature which merge at a depth of about 15 feet to a zone no more than a few inches wide. The layered clays and silty sands through which the cracks have propagated are thinly bedded horizontal strata which are not offset across the fissures. A previous study by Fife (1978) described similar subsurface features just after the cracks originated following heavy rainfall in the summer of 1976. The *en echelon* nature of the ground crack suggested to Fife (1978, 1980) that these cracks were essentially the surface expression of a left-lateral fault. However, the subsurface cavities and linear voids suggest tension as the stress mechanism, not compression and

Figure 12. Close up of trench wall showing coarseness of host deposit and vertical walls of thin filled crack. Sandy deposits are slightly cemented; filled crack is about 1 cm thick.

shear, and it appears the fault model is untenable. Neal et al. (1968) have shown that discontinuous *en echelon* "stripes" like those in Mesquite Lake Playa are one of several types normally found in desert playas due to desiccation.

More recent studies in the playa by Schaefer Dixon Associates (1988) have discovered ground cracks outside the playa basin. These cracks are narrow funnel-shaped linear features continuous with and filled by an overlying reddish brown mudflow (Fig. 11 and 12). The unusual feature of these cracks is that they propagate through non-cohesive materials. All other reported ground crack and fissure phenomena have been in cohesive materials undergoing desiccation. The cracks in Mesquite Lake playa appear near the edge of distal portions of alluvial fans and are discontinuous healed features parallel to the playa surface fissures. Trenching did not excavate below the sandy deposits containing the cracks, but it seems safe to assume that crack formation is probably related to the same mechanism as the playa fissure, that is, desiccation of clays below the younger alluvial fan deposits.

Mechanics of Crack Formation

Crack formation in Mesquite Playa is caused by long term desiccation of the underlying clayey playa and lake sediments. This mechanism was first described by Lang (1943). Interpretation of geophysical data described above did not detect subsurface features such as a buried ridge, which might cause the cracking As Neal et al. (1968) and numerous other investigators such as Robinson and Peterson (1963), Holzer and Pampeyan (1984), Jachens and Holzer (1982), and Bull (1964), have reported, desiccation, whether from long-term climatic change or groundwater mining, causes fissuring. The Marine Base has generally not mined groundwater from the playa or nearby areas. Rather, they tend to use water supplied from Surprise Springs many miles northwest of the basin (Schaefer, 1978). Historical records of wells from years back (Moyle, 1960) show groundwater to be at a depth of about 260 feet as it is today. However, long-term desiccation of the desert areas following the pluvial period of the last glaciation has had a major impact on numerous desert basins as reported by Neal et al. (1968). Whereas Fife (1978, 1980) concluded that these features are fault related, their appearance and subsurface geometry are identical to features of ground cracks in other desert basins shown by Neal et al. (1968) and are best explained as being related to long term desiccation.

In other nearby basins such as Broadwell, Lucerne, and Rogers, desiccation has taken the form of large polygons (Fife, 1978,1980; Gary S. Rasmussen and Associates, 1990; Neal et al.,1968). No polygons have formed in Mesquite Playa. It appears that in order for polygons to form, the basin geometry must be wide and unconstricted. Mesquite Playa is apparently not wide enough nor is it sufficiently unconstricted for polygons to form. Where playas begin to narrow, the stripes (fissures) tend to form perpendicular to the long axis and generally parallel to the termination of the playa clay. In Mesquite Lake playa, the stripes have formed in just this manner.

CONCLUSIONS

The Twentynine Palms Marine Corps Base is in an active tectonic region subject to numerous geotechnical hazards including earthquake shaking, faulting and ground cracking. Other soil related hazards such as collapsible soils and expansive soils are present but not discussed here. The geotechnical concerns discussed in this report relate to geological and climatic phenomena that have caused drastic changes in the recent past. The extraordinary finding of this geotechnical study is the relative recency of a change in the tectonic stress regime in this area.

The results of this geotechnical investigation relate to the model of faulting and fault development constructed through the efforts of many researchers relative to widely distributed dextral shear related to the San Andreas fault. Recognition of the Mojave Desert fault system as a series of northwest-striking throughgoing structures was first made by Hewett (1954). The right-lateral strike-slip character of the faults was first described by Dibblee (1961). Since then, the advent of plate tectonics as a viable kinematic structural model has allowed workers deeper insight into the relationship of these faults to the San Andreas system. Garfunkel (1974) proposed a model of simple dextral shear where each fault was bounded by northern and southern faults and the principal compressional axes relative to North American plate motions are north-south. However, compressive stress directions and block rotations predicted by this model are not in accord with the actual field data (Dokka and Travis, 1990). Dokka (1983) and Dokka and Travis (1990) have proposed a model which suggests that the Mojave Desert is an area of six structural domains, each subject to independent rotation above a soft lower crustal zone and separated by one or more of the series of northwest striking strike-slip faults. Their model allows for triangular shaped basins to form near the northern and southern margins of the faults and, most critically, identifies a shift in stress-field dynamics southwards and westwards from the Eastern Mojave shear zone (Death Valley-Granite Mountain fault zones) within the past 1.6-0.7 million years.

The data and interpretations presented here indicate that the Twentynine Palms area is undergoing a very recent adjustment to the older strike-slip rotations recorded in the central Mojave Desert. For example, the northwestern extension of the Mesquite Lake fault is the Calico-Hidalgo strike-slip system. Dokka (1983) maintains that the Calico-Hidalgo fault has slipped about 6 to 10 miles while the evidence, albeit tentative, for slip on the Mesquite Lake fault is for much less slip, about 1 mile, and much more recently activated slip. If, as Dokka and Travis (1990) suggest, the "inter-domain" regions have undergone extension relative to rotation about the central Mojave Desert shear zones, it appears that within the last 100,000 years the shearing has extended through to the borders of the domains, in this instance, the Pinto Mountain fault zone-Eastern Transverse Ranges domain boundary. If these conclusions are valid, it would mean that activity of faulting should be concentrated in smaller earthquakes around the perimeter of the central Mojave Shear Zone. Again this evidence is not overwhelming, but recently, moderate size earthquakes have shaken the Twentynine Palms (M4.6, 4.7, 1975; M5.1, 1977), Johnson Valley (M5.9, 1978), Galway Lake (M5.2, 1975; M5.5, 1984) and Homestead Valley (M5.2, 1979) areas, all of which are near the south and west border of the Mojave Desert province (see Haukson, this chapter).

Desiccation related fissuring is known from most of the playa basins of the southwest. This phenomenon is generally related to climactic change following the last glacial episode 18,400-11,400 years ago resulting in regionally declining water tables and the drying of playa and lake clays. Mesquite Lake Playa contains active and healed fissure-scars. The most recent fissure occurred in 1976 following heavy rains nearby which flooded the playa (Fife, 1978). The cracks are at least 19 feet deep and narrow downwards to the top of the capillary fringe of a perched water table. Cracking takes the form of en echelon right stepping fissures which can remain open for months before they fill up with windblown silt and sand or have water-borne clays. In places, large cavernous voids remain along the fissure trend. These fissures strike about due north and were originally believed to be faults related to the nearby Mesquite Lake fault. However, interpretation of geophysical and geotechnical data indicates that these are extensional features and are not the results of faulting related to the near north-south compressional stress field of the Mesquite Lake fault.

A remarkable feature of the Mesquite Lake Playa desiccation cracks is that some of the older cracks east of the newest crack are in non-cohesive sand and fine gravel to the depth of excavation (16 feet). Such material is highly unlikely to separate with vertical parallel sides the way tension cracks form in clays. Rather, it appears that desiccation of an underlying lake or playa clay bed has caused the overlying slightly cemented sands to separate and stand long enough for a mud flow to fill in the crack. This type of phenomena has not been previously described and warrants further research.

REFERENCES

Bacheller, J., III, 1978, Quaternary geology of the Mojave Desert-Eastern Transverse Ranges boundary in the vicinity of Twentynine Palms, California: Master's thesis, University of California at Los Angeles, 157 p.

Bassett, A.M. and Kupfer, D.H., 1964, A geologic reconnaissance in the southeastern Mojave Desert, California: California Division of Mines and Geology Special Report 83, 43 p.

Biehler S. and Gilpin, B., 1983, Geophysical investigations of the Twentynine Palms area: In Wahler Associates, 1984, Report-volume II-Appendix D, Marine Corps Air-Ground Combat Center geotechnical study, Twentynine Palms, California: Commanding Officer, Western Division Naval Facilities Engineering Command (Code 023), 24 p., plates, figures and appendices.

Bull, W.B., 1972, Prehistoric near-surface subsidence cracks in western Fresno County, California: U.S. Geological Survey Professional Paper 437-C, 85 p.

Cummings, D., 1980, Mechanics of fault deformation and seismotectonic zoning, Mojave desert area, California: In Fife, D.L. and Brown, A.R. (eds.), Geology and mineral wealth of the California desert, South Coast Geological Society, Anaheim, CA, p. 101-120.

Dibblee, T.W., Jr., 1961, Evidence of strike-slip faulting along northwest-trending faults in the Mojave Desert: U. S. Geological Survey Professional Paper 424-B, p. B197-B199.

Dibblee, T.W., Jr., 1967a, Geologic map of the Deadman Lake quadrangle, San Bernardino County, California: U. S. Geological Survey Miscellaneous Field Investigations Map I-488:

Dibblee, T.W., Jr., 1967b, Evidence of major lateral displacement on the Pinto Mountain fault, southern California: Geological Society of America, Cordilleran Section, 63rd Annual Meeting, Abstracts with Programs, p. 32.

Dibblee, T.W., Jr., 1968, Geologic map of the Twentynine Palms quadrangle, San Bernardino and Riverside Counties, California: U. S. Geological Survey Miscellaneous Field Investigations Map I-561:

Dokka, R.K., 1983, Displacements on late Cenozoic strike-slip faults of the central Mojave Desert, California: Geology, vol. 11, p. 305-308.

Dokka, R.K. and Travis, C.J., 1990, Late Cenozoic strike-slip faulting in the Mojave Desert, California: Tectonics, vol. 9, no. 2, p. 311-340.

Droste, J.B., 1961, Clay minerals in the playa sediments of the Mojave Desert, California, California Division of Mines Special Report 69, 19 p.

Fife, D.L., 1978, "Airfield fault", Mesquite Playa, San Bernardino County, California: California Geology, vol. 31, no. 3, p. 51-53.

Fife, D.L., 1980, Giant desiccation polygons and playa features: In Fife, D.L. and Brown, A.R., (eds.), Geology and mineral wealth of the California desert, South Coast Geological Society, Anaheim, CA, p. 414-429.

Garfunkel, Z., 1974, Model for the late Cenozoic tectonic history of the Mojave Desert, California, and its relation to adjacent regions: Geological Soc. of America Bull.,vol. 85, p. 1931-1944.

Gary S. Rasmussen and Associates, 1983, Engineering geology investigation, proposed 20-inch diameter potable water line route between reservoir No. 1 and equalizer tanks west of Deadman Lake, Twentynine Palms Marine Corps Base, Twentynine Palms, California: John Egan and Associates, Inc., San Bernardino, CA, 13 p.

Gary S. Rasmussen and Associates, 1990, Geologic and geotechnical feasibility investigation, proposed residuals repository, Broadwell Dry Lake, north of Ludlow, California: Broadwell Corporation, San Bernardino, CA, 26 p.

Greensfelder, R.W., 1974, Maximum credible rock acceleration from earthquakes in California: California Div. of Mines and Geology, Map Sheet 23.

Hewett, D.F., 1954, General geology of the Mojave Desert region, California, in Jahns, R., Geology of Southern California: California Div. of Mines and Geology Bulletin 170, p. 15-18

Hill, R.L., 1976, Recent earthquakes near Twentynine Palms, San Bernardino County, California: California Geology, vol. 29, no. 6, p. 134-135.

Hill, R.L. and Beeby, D.J., 1977, Surface faulting associated with the 5.2 magnitude Galway Lake earthquake of May 31, 1975, Mojave Desert, San Bernardino County, California: Geological Soc. of America Bull., vol. 88, p. 1378-1384.

Holzer, T.L. and Pampeyan, E.H., 1984, Ground failure induced by groundwater withdrawal from unconsolidated sediment: In Holzer, T.L. (ed.), Man-induced land subsidence: Geological Soc. of America Reviews in Engineering Geology, vol. 6, p. 67-105.

Hunter, L.K., Johnson, C.E., Pechmann, J.C., Ebel, J.E., Given, J.W., Cole, D.M., and German, P.T., 1980, Epicentral locations for the Homestead Valley earthquake sequence, March 15, 1979: California Geology, vol. 33, no. 5, p. 110-114.

Jachens, R.C. and Holzer, T.L., 1982, Differential compaction mechanisms for earth fissures near Casa Grande, Arizona: Geological Soc. of America Bull., Vol. 93, no. 10, p. 998-1012.

Lang, W.B., 1943, Gigantic drying cracks in Animas Valley, New Mexico: Science, vol. 98, no. 2557, p. 583-584.

McJunkin, R.D., 1980, Strong-motion data for the Homestead Valley earthquake swarm, San Bernardino County, California, 15 March 1979: California Geology, vol. 32, no. 1, p. 11-13.

Moyle, W., R.,Jr., 1960, Data on water wells and springs in the Yucca Valley-Twentynine Palms area, San Bernardino and Riverside Counties, California: California Dept. Water Resources Bull. 91-2, 164 p.

Muessig, S., White, G.N., and Byers, F.M., Jr., 1957, Core logs from Soda Lake, San Bernardino County, California: U. S. Geological Survey Bull. 1045-C: p. 81-96.

Neal, J.T.; Langer, A.M.; and Kerr, P.F., 1968, Giant desiccation polygons of Great Basin playas: Geological Soc. of America Bull., vol. 79, p. 69-90.

Powell, R., 1982, Crystalline basement terranes in the southern eastern Transverse Ranges, California: In Cooper, J. (compiler), Geologic excursion in the Transverse Ranges, Southern California: Geological Soc. of America Guidebook, p. 109-136.

Robinson, G.M. and Peterson, D.E., 1963, Notes of earth fissures in southern Arizona: U. S. Geological Survey Circular 466, 7 p.

Roy J. Shlemon and Associates, Inc, 1985, Summary of soil-geomorphic reconnaissance, 29 Palms area.: In Highland Geotechnical Consultants, Inc.,

__________, 1985, Geologic report, West Bullion Mountain fault study, Marine Corps Air Ground Combat Center, Twentynine Palms, California: RGA Consulting Engineers, San Bernardino, CA, 11 p.

Schaefer, D.H., 1978, Ground-water resources of the Marine Corps Base, Twentynine Palms, San Bernardino County, California: U. S. Geological Survey Water-Resources Investigations 77-37: 29 p.

Schaefer Dixon Associates, 1987, Geotechnical investigation report, Branch Hospital Replacement (P-804), Marine Corps Air Ground Combat Center, Twentynine Palms, California: Anderson De Bartolo Pan, Inc., Los Angeles, CA, 9 p.

Wahler Associates, 1984, Report-volume 1, Marine Corps Air-Ground Combat Center geotechnical study, Twentynine Palms, California: Commanding Officer, Western Div. Naval Facilities Engineering Command (Code 023), 54 p.

Wells, S.G., Anderson, R.Y., McFadden, L.D., Brown, W.J., Enzel, Y., ;and Miossec, J., 1989, Late Quaternary paleohydrology of the Eastern Mojave River drainage, souther California: quantitative assessment of the late Quaternary hydrologic cycle in large arid watersheds: Technical Completion Report 14-08-0001-G1312, New Mexico Water Resources Research Institute, 253 p.

Wesnousky, S., 1986, Earthquakes, Quaternary faults, and seismic hazard in California: Journal of Geophysical Research, vol. 91, no. B12, p. 12,587-12,631.

TECTONIC SIGNIFICANCE OF LINEAMENTS ON SATELLITE IMAGES OF THE PENINSULAR RANGES, SOUTHWESTERN CALIFORNIA

Donald L. Lamar
Engineering Geologist
P.O. Box 10784
Eugene, OR 97440

Paul M. Merifield
Engineering Geologist
3411 Wade Street
Los Angeles, CA 90026

ABSTRACT

Prominent lineaments appear on Skylab and Landsat images of the basement terrane of the Peninsular Ranges, southwestern California. Among the most conspicuous are those formed by the active northwest trending Elsinore and San Jacinto fault zones. Field investigations were performed to determine the origin of lineaments on images and the earthquake potential of faults. Several lineaments represent unmapped faults, whereas others resulted from erosion along foliation and joints. The newly identified faults fall into two distinct sets trending north-northeast and east-west. Two faults in the northeast set show right separation. Two of the east-west trending faults are inferred to have left separation. These faults are considered inactive because they are restricted to pre-Tertiary basement rocks, show no geomorphic evidence of movement and are truncated by the active northwest trending faults. This conjugate fault set may have formed in response to west-southwest, east-northeast crustal shortening caused by a Mesozoic subduction zone to the west. The images provide a synoptic overview of regional fault patterns and cross-cutting relations. Analysis of these images is applicable to tectonics and engineering geology investigations of faults for earthquake hazard and the location of water wells

INTRODUCTION

The Peninsular Ranges extend from the western edge of the continental borderland to the Salton Trough, and from the Transverse Ranges to the tip of Baja California. In southwestern California, the Peninsular Ranges Province is underlain by metamorphosed Paleozoic and Mesozoic rocks intruded by late Mesozoic plutonic rocks of the southern California batholith (Jahns, 1954a). Northwest trending faults,

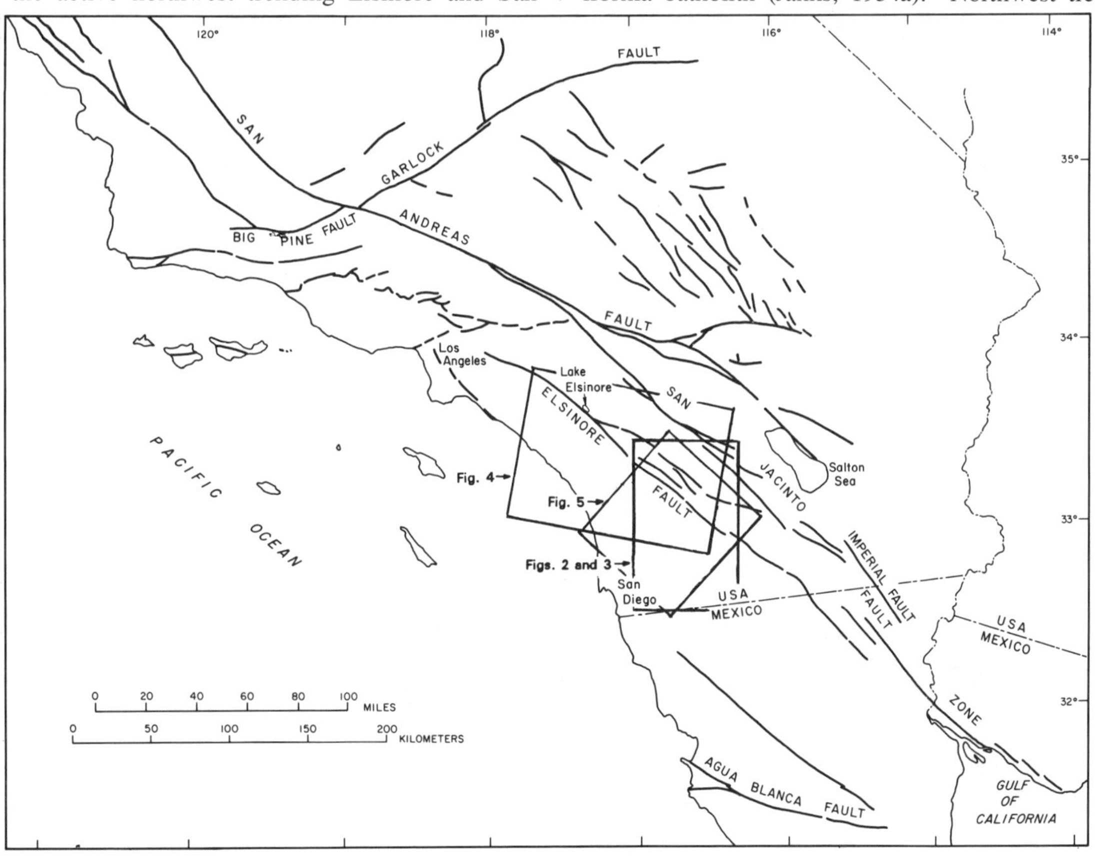

Figure 1. Index map of southwestern California showing major faults, from Proctor (1973), and outlines of Figures 2, 3, 4 and 5.

notably the San Jacinto and Elsinore faults (Figs. 1 and 2), are the dominant structural features. A small-scale Skylab 3 image and Landsat 3 images as well as larger scale color aerial photos from RB-57 and U-2 aircraft, provide a valuable overview of this region. Previous research on geologic features seen on satellite images of the Peninsular Ranges are described by Lowman (1969, 1980) and Merifield and Lamar (1976ab). This investigation considers both the tectonic and the engineering geology application of satellite image analysis. Images and geologic maps in this paper are outlined on Figure 3.

Much of the previous mapping and compilations (Merriam, 1946, 1955, 1958; Larsen, 1948; Everhart, 1951; Gastil and Bushee, 1961; Strand, 1962; Rogers, 1965; Jennings, 1973) were at scales of 1:62,500 or smaller, and some earlier studies did not utilize aerial photography (Richard Merriam, personal communication, 1975). Todd's detailed mapping, with emphasis on petrology, became available after our field work was completed (Todd, 1977; Todd and Shaw, 1979; Todd and others, 1988; these references have information on Todd's 1:24,000 scale U.S.G.S. open-file maps of the area). Lowman (1969) was the first to identify northeast trending lineaments, prominent on Gemini and Apollo photographs of the Peninsular Ranges of southwestern California. These lineaments, the San Jacinto, Elsinore and other faults and linear features are visible in Landsat (Fig. 4) and Skylab (Fig. 5) images.

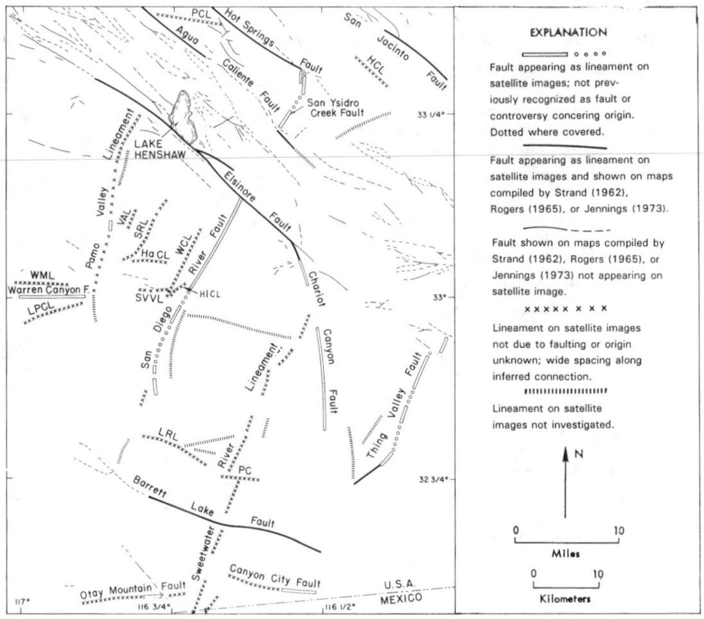

Figure 2. Faults and lineaments recognized on satellite images. Abbreviations: HaCL: Hatfield Creek lineament; HCL: Henderson Canyon lineament; HiCL: Himmel Canyon lineament; LPCL: Los Penasquitos Canyon lineament; LRL: Loveland Reservoir lineament; PC: Pine Creek lineament; PCL: Previtt Canyon lineament; SRL: Sutherland Reservoir lineament; SVVL: San Vicente Valley lineament; VAL: Valle de Los Amigos lineament; WCL: Witch Creek lineament; WML: Woodson Mountain lineament. Figure 1 shows area covered.

NORTHEAST TO NORTH-SOUTH TRENDING FAULTS AND LINEAMENTS

San Ysidro Creek Fault

A segment of San Ysidro Creek and a stream canyon to the south form a prominent 4 mile long north-northeast trending lineament visible on Skylab imagery (Fig. 6). The northwest portion of the San Ysidro Creek area is underlain by Lakeview Mountain Tonalite characterized by prominent, near-vertical east-west trending joints (Fig. 7). Julian Schist to the southeast consists of biotite schist interlayered with gneiss and hornfels; foliation in the metamorphic rocks trends north-northeast, parallel to the intrusive contact and the San Ysidro Creek lineament. A fault is exposed near the northeast end at the dip symbol (Fig. 7).

San Diego River Fault Zone

The northeast trending San Diego River Valley forms a prominent 20 mile long lineament on the Landsat image (Fig. 8). The San Diego River fault zone and other faults and lineaments to the west are shown Figure 9. Although the San Diego River and San Ysidro Creek faults are approximately aligned (Fig. 2), the Landsat image (Fig. 8), Skylab and large scale air photographs show that these features are separated by 6 miles and do not connect as suggested by Lowman (1969). Some previous maps (Sauer, 1929; Miller, 1935) infer a fault on the straight trend of San Diego River

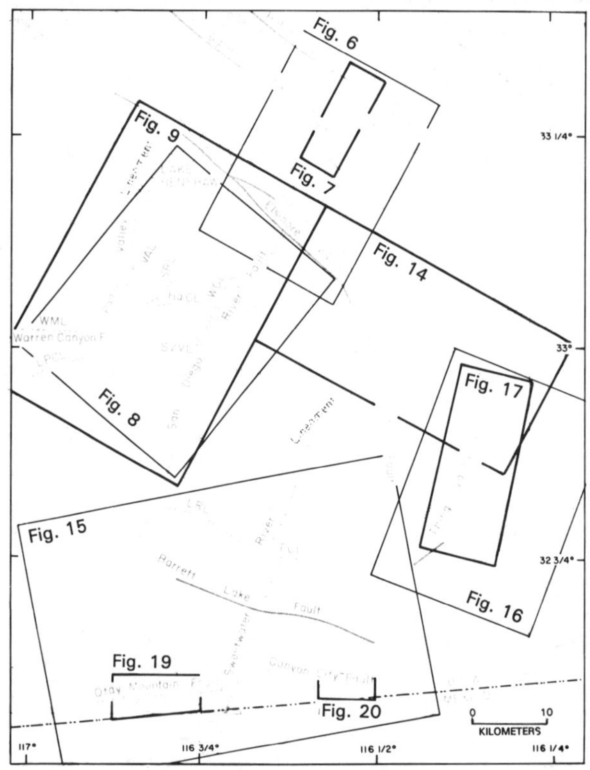

Figure 3. Map showing outlines of geologic maps (heavy lines) and satellite images (thin lines) of areas studied in Peninsular Ranges, southwestern California. See Figure 2 for explanation and abbreviations. Figure 1 shows area covered.

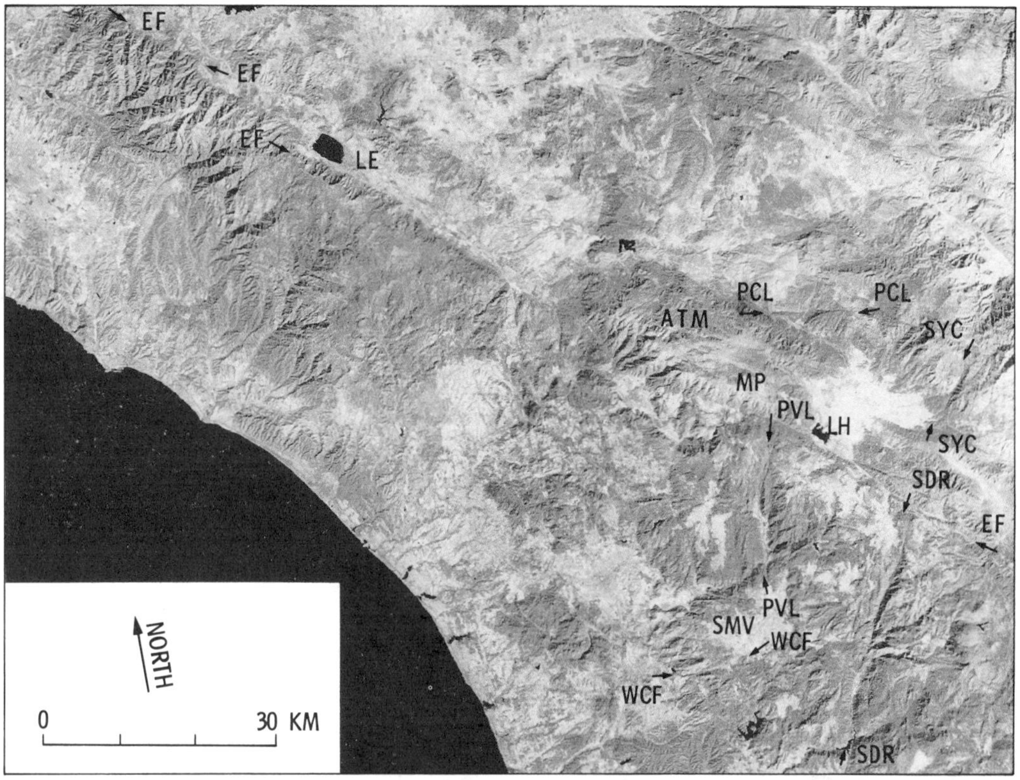

Figure 4. Major faults and lineaments on portion of Landsat image, north central Peninsular Ranges, southwestern California. Abbreviations: ATM: Agua Tibia Mountains; EF: Elsinore fault; LE: Lake Elsinore; LH: Lake Henshaw; MP: Mount Palomar; PCL: Previtt Canyon lineament; PVL: Pamo Valley lineament; SDR: San Diego River fault zone; SMV: Santa Maria Valley; SYC: San Ysidro Creek fault; WCF: Warren Canyon fault. Figure 1 shows area covered.

Valley and the 150 meter (500 feet) difference in elevation between the flat to gently rolling bedrock surfaces on opposite sides of the valley. Everhart (1951) and Merriam (1958) suggest that prominent straight valleys in the region, such as the San Diego River Valley, formed primarily as a result of erosion along joints or foliation. Merriam (1958, Fig. 2) shows the regional northwest trending foliation curving into a northeast trend, parallel to the San Diego River. Our mapping (Figs. 9, 10, and 12) indicates such a change in foliation direction.

The oldest rock unit exposed along San Diego River is the Julian Schist composed of quartzite, quartz-mica schist and gneiss; Stonewall Quartz Diorite and mixed rocks consisting of intimately mixed Stonewall Quartz Diorite and Julian Schist are the next youngest. These rocks are intruded by Cretaceous plutonic rocks: San Marcos Gabbro, Green Valley Tonalite, and Woodson Mountain Granodiorite (Everhart, 1951; Merriam, 1958; Weber, 1963).

The contact between schist and quartz diorite (B, Fig. 10) shows 1400 feet of right separation along the fault. An en echelon zone of fractures is exposed and seen on air photos at E, F, and G. An alignment of saddles and subtle differences in vegetation can be seen on large-scale air photographs (Fig. 11) between H and I. A tabular body of olive green, very fine-grained rock aligned with the canyon and en echelon with the L-M photo lineament is exposed at N (Fig. 12). Individual grains in a photomicrograph (Fig. 13) of rock from N

Figure 5. Skylab image of Peninsular Ranges, southwestern California. Abbreviations: BLF: Barrett Lake fault; CaCF: Canyon City fault; CCF: Chariot Canyon fault; EF: Elsinore fault; EVF: Earthquake Valley fault; GV: Green Valley; HCL: Henderson Canyon lineament; OML: Otay Mountain fault; PVL: Pamo Valley lineament; SDR: San Diego River fault; SMV: Santa Maria Valley; SR: Sweetwater River; SRL: Sweetwater River lineament; SYC: San Ysidro Creek fault; TVF: Thing Valley fault; WCF: Warren Canyon fault. Figure 1 shows area covered.

Figure 6. Skylab image of San Ysidro Creek and north end of San Diego River. Abbreviations: EF: Elsinore fault; EVF: Earthquake Valley fault; HSF: Hot Springs fault; SYC: San Ysidro Creek; SDR: San Diego River. Figure 3 shows area covered.

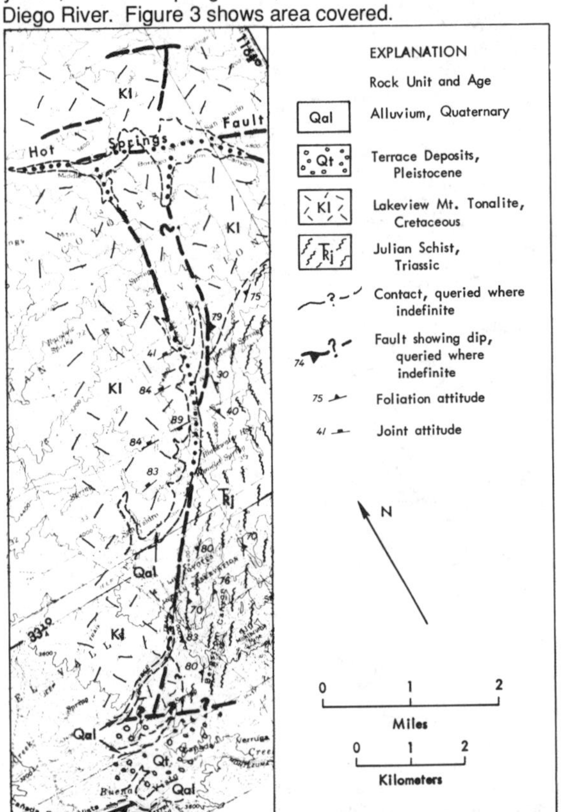

Figure 7. Geologic map of San Ysidro Creek area modified from Merriam (1958), Scheliga (1963) and Weber (1963). Topography from U.S.G.S. Santa Ysabel and Warner Springs quadrangles. Figure 3 shows area covered.

(Fig. 12) are fractured and interpreted to be fault breccia. Bedrock distribution on Figure 12 shows right separation.

An alignment of straight canyon segments and saddles appear on satellite imagery (Fig. 8) along O-P (Fig. 9). The faults on the south end of El Capitan Reservoir trend north-south to north-northwest and dip west as compared to the faults to the north between A and N which strike north-northeast and dip east. This change in strike follows a similar curve in the trend of San Diego River Canyon.

Witch Creek Lineament

The Witch Creek lineament is a 7.6 mile long alignment of remarkably straight canyons (Fig. 8) trending north-northeast parallel to the San Diego River fault zone (Fig. 9). Stonewall Quartz Diorite, Julian Schist and Woodson Mountain Granodiorite are exposed along the Witch Creek lineament (Merriam, 1946, 1958). The San Vicente Valley lineament and the Woodson Mountain Granodiorite - Stonewall Quartz Diorite contact are not displaced by the Witch Creek lineament (Fig. 12). No evidence of faulting was observed and jointing and foliation do not parallel the lineament (Fig. 12).

Himmel Canyon Lineament

A straight segment of Himmel Canyon and canyons to the northeast form a prominent, curving 1.7 mile long lineament (Figs. 8 and 12). The Himmel Canyon lineament does not

Figure 8. Landsat image of San Diego River fault zone and other faults and lineaments in the Woodson Mountain - Santa Maria Valley area. Abbreviations: CC: Chocolate Canyon; DC: Daney Canyon; HaCL: Hatfield Creek lineament; HCL: Himmel Canyon lineament; LPL: Los Penasquitos Canyon lineament; PVL: Pamo Valley lineament; SDR: San Diego River fault zone; SMV: Santa Maria Valley; SRL: Sutherland Reservoir lineament; SVVL: San Vicente Valley lineament; VAL: Valle de los Amigos lineament; WCF: Warren Canyon fault; WCL: Witch Creek lineament; WM: Woodson Mountain; WML: Woodson Mountain lineament. Figure 3 shows area covered.

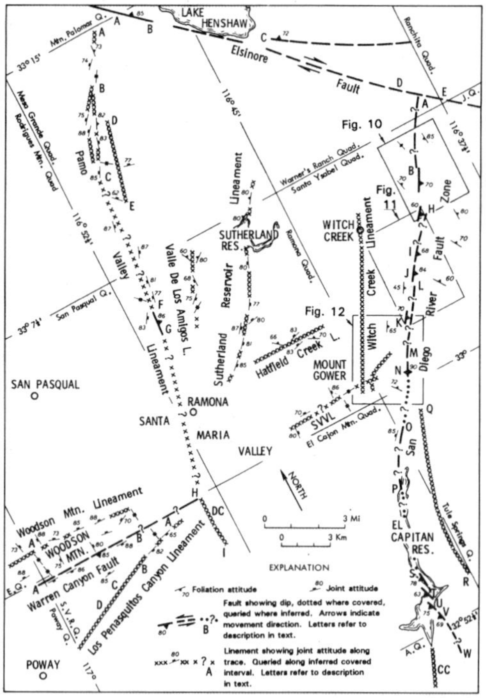

Figure 9. Map showing faults and lineaments between Woodson Mountain and San Diego River and outlines of Figures 10, 11 and 12 showing details of San Diego fault zone. Letters on San Diego River fault zone illustrations are keyed to descriptions on Table 1. Joint and foliation attitudes on figures are usually average of several. Abbreviations: AQ: Alpine quadrangle; CC: Chocolate Canyon; DC: Daney Canyon; EQ: Escondido quadrangle; JQ: Julian quadrangle; SVRQ: San Vicente Reservoir quadrangle; SVVL: San Vicente Valley lineament. Figure 3 shows area covered.

displace the San Vicente Valley lineament or the Woodson Mountain Granodiorite - Stonewall Quartz Diorite contact; neither jointing nor foliation are parallel to the lineament (Fig. 12). Rare bedrock exposures show no evidence of faulting.

Sutherland Reservoir Lineament

The Sutherland Reservoir lineament trends north-northeast for 5.8 miles through Sutherland Reservoir (Fig. 8). Bedrock consists of Julian Schist, Lakeview and Green Valley Tonalite (Merriam, 1946). South of the reservoir joint directions in tonalite parallel the lineament; north of the reservoir foliation in Julian Schist parallels the lineament (Fig. 9). No evidence of faulting was observed.

Valle De Los Amigos Lineament

Straight canyons and saddles form the 2.8 mile long, north-northeast trending Valle De Los Amigos lineament (Figs. 8 and 9). Bedrock consists of Julian Schist, Lakeview and Bonsall Tonalite (Merriam, 1946). Although exposures show no evidence of faulting, foliation in Julian Schist changes strike across the north end of the lineament. Joints parallel the southern portion of the lineament (Fig. 9).

Pamo Valley Lineament

Pamo Valley and other canyon segments and saddles to the north and south form an 11.5 mile long lineament from the north end of Santa Maria Valley, almost to the Elsinore fault

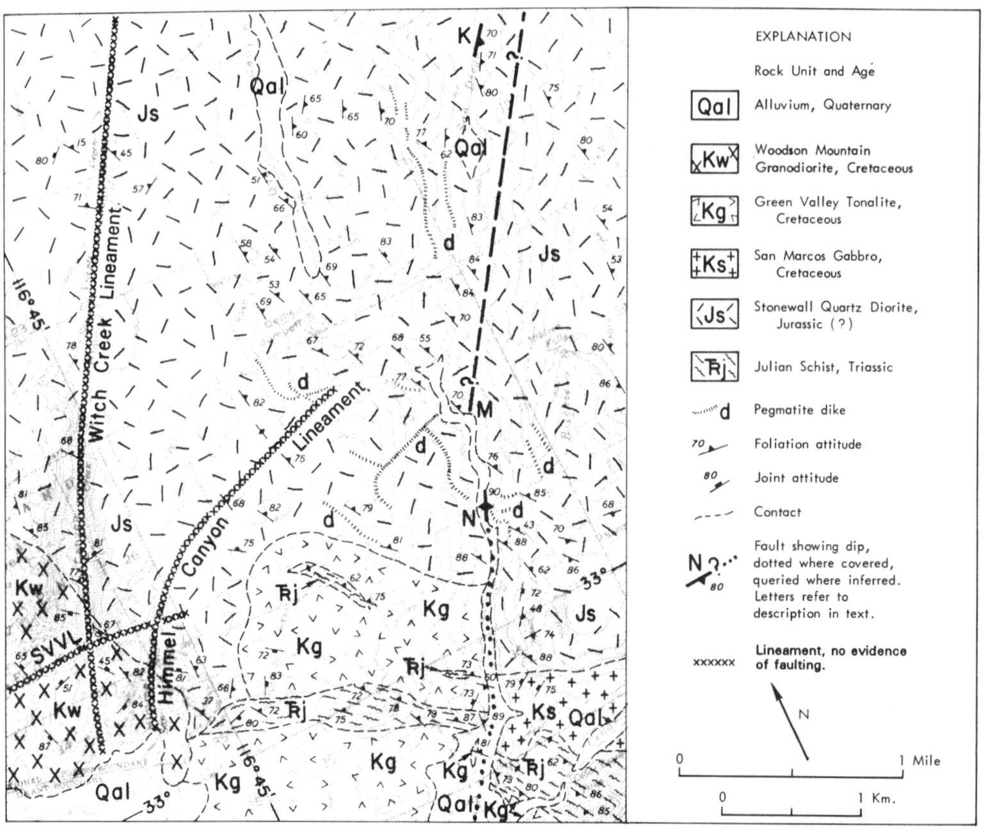

Figure 10. Geologic map of portion of San Diego River fault zone. Topography from U.S.G.S. Santa Ysabel quadrangle. Figure 9 shows area covered.

(Figs. 8 and 9). Bedrock consists of Bonsall, Ramona and Green Valley Tonalite and Woodson Mountain Granodiorite (Merriam, 1946, 1954). Between A and C (Fig. 9) joints and foliation in tonalite parallel the lineament with no evidence of faulting. Between B and C, two parallel lineaments follow straight canyons (Fig. 9). A prominent lineament parallels the Pamo Valley lineament between D and E (Fig. 9). Joints between C and F (Fig. 9) parallel the lineament. A fault is exposed along a straight canyon between F and G; bedrock is similar on both sides of the fault. A similar trending lineament appears south of Santa Maria Valley, between H and I, as a straight segment of Daney Canyon (Figs. 8 and 9).

Los Penasquitos Canyon Lineament

A straight segment of Los Penasquitos Canyon and an alignment of saddles and straight canyons form an 6.3 mile long lineament (Fig. 8). Bedrock consists of Woodson Mountain Granodiorite, Bonsall Tonalite (Gastil and Bushee, 1961) and Black Mountain Volcanics (Strand, 1962). Between A and B (Fig. 9), the lineament is distinct on air photos, but not apparent southwest of B. Joints parallel the northeast end of the lineament (Fig. 9). A north-south trending body of Black Mountain Volcanic rocks (Gastil and Bushee, 1961) between C and D is undisplaced across the lineament.

Figure 11. Portion of U.S.D.A. aerial photo showing lineaments along San Diego River fault zone. Figure 9 shows area covered.

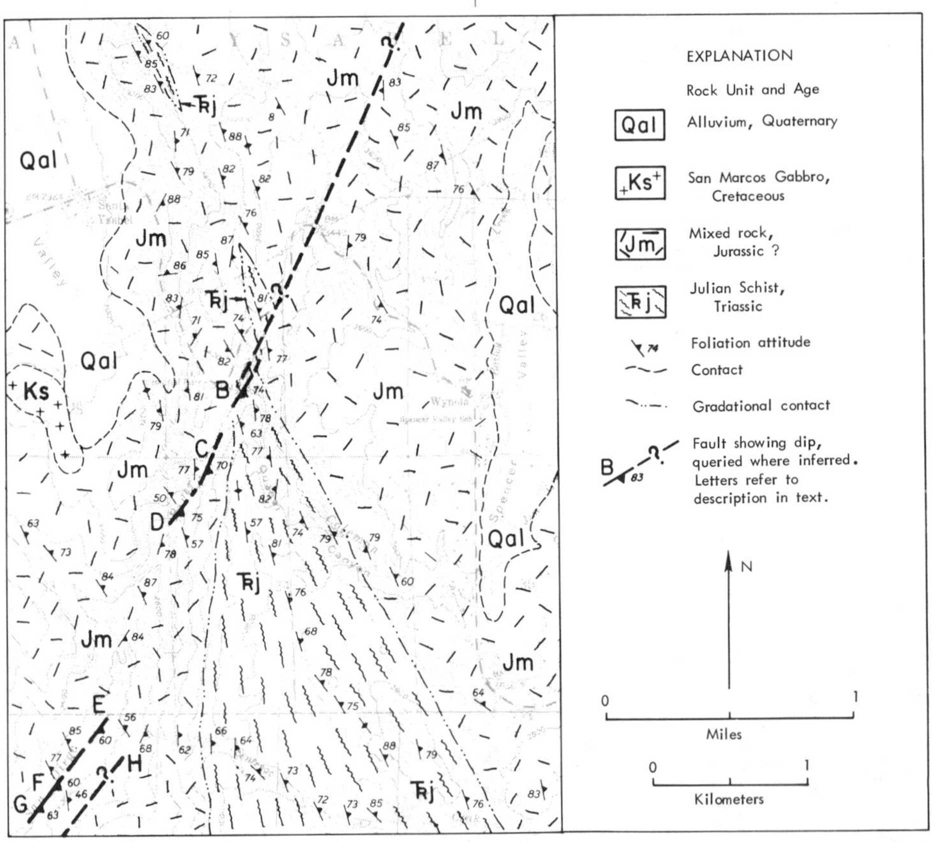

Figure 12. Geologic map of central portion of San Diego River fault zone. Abbreviation: SVVL: San Vicente Valley lineament. Figure 9 shows area covered.

Chariot Canyon Fault

The Chariot Canyon fault branches from the Elsinore fault at the town of Banner and follows the trend of Chariot Canyon; an alinement of straight canyons and saddles extends 10 miles south-southeast of point B (Figs. 5 and 14). A segment at the mouth of Chariot Canyon was mapped by Hart (1964) and Rogers (1965). Bedrock consists of Julian Schist, tonalite and gneiss (Weber, 1963). Based on the distribution of Julian Schist and plutonic rock, Allison (1974ab) reports 5 miles of right separation on the Chariot Canyon fault. Exposures in Chariot Canyon do not reveal a single fault separating Julian Schist on the west from granitic rocks on the east; a broad shear zone occupies the width of the canyon. Foliation in schist and gneiss is undulatory with a general strike conformable with shear surfaces and the canyon trend.

Sweetwater River Lineament

The Sweetwater River lineament consists of a series of discontinuous and en echelon valleys that stretch over 30 miles from 4 miles south of the Elsinore fault zone southwestward to the Mexican border (Figs. 2, 5 and 15). Bedrock consists of Bonsall and Green Valley Tonalite, Woodson Mountain Granodiorite and Cuyamaca Gabbro (Everhart, 1951; Weber, 1963). North of the area we studied, bedrock contacts bend to the right where they cross the lineament and foliation trends east-west, at right angles to the lineament (Everhart, 1951). The lineament in this area may coincide with the axis of a shear fold.

Figure 13. Photomicrograph of fault breccia exposed along San Diego River fault zone at N, Figures 9 and 12. 5 x enlargement.

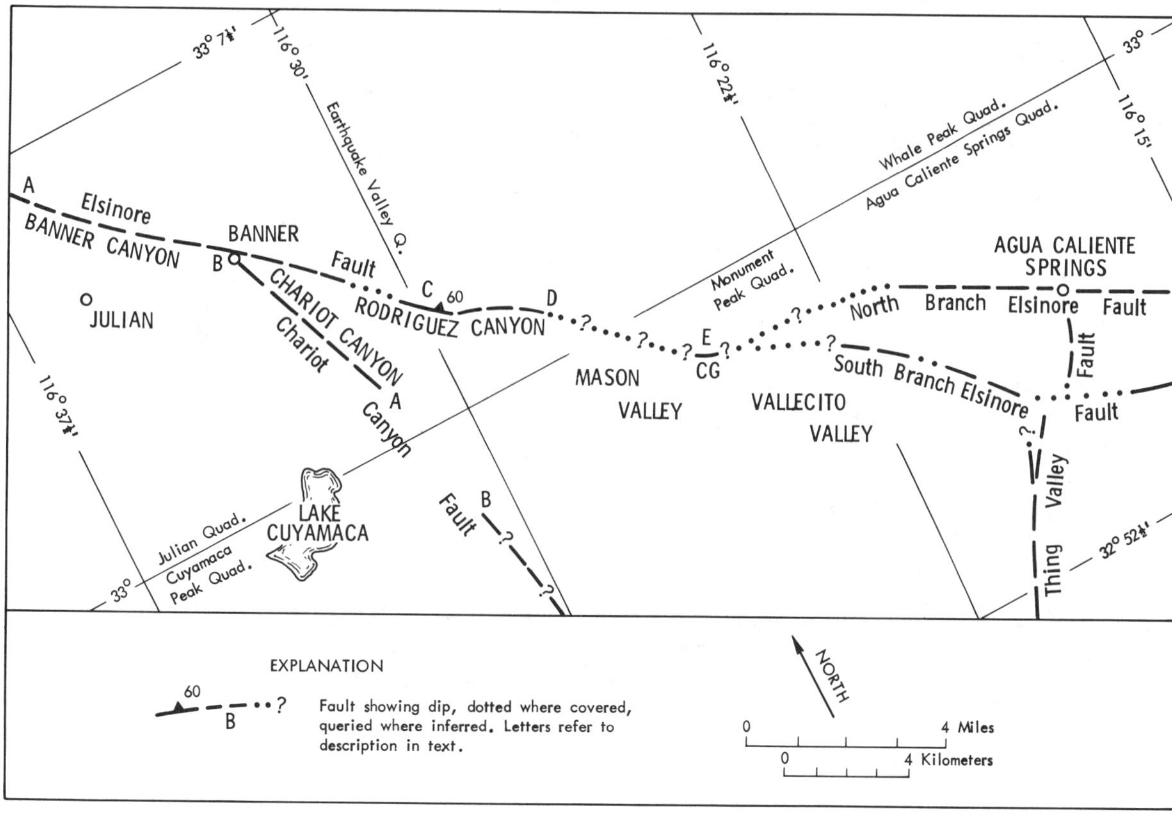

Figure 14. Map showing Elsinore and Chariot Canyon faults in Julian - Vallecito area. Figure 3 shows area covered.

At the north end of the portion we studied, the Horsethief Canyon segment of the lineament (Fig. 15) parallels foliation and slickensided shear surfaces in Woodson Mountain Granodiorite and Bonsall Tonalite. Farther southwest the lineament shows 3/4-mile of left separation on the Pine Creek lineament; the Sweetwater River lineament continues south-southwest along Pine Valley Creek to Barrett Lake (Fig. 15). Foliation along Pine Valley Creek parallels the lineament, which continues south of Barrett Lake as a deep, straight, 2 mile long section of Cottonwood Creek. The origin of the lineament over this segment is unknown; bedrock is massive Woodson Mountain Granodiorite. South of Barrett Junction (Fig. 15) the lineament is recognized for 3 miles as an alignment of canyons and saddles. Joints in Woodson Mountain Granodiorite parallel the lineament.

Thing Valley Fault

Thing Valley and several straight segments of stream canyons to the northeast and southwest form a prominent, fault controlled, north-northeast trending, 12 mile long lineament (Fig. 16). Bedrock consists of Julian Schist composed of quartzite, gneiss, and schist intruded by Lakeview Mountain and Green Valley Tonalite (Merriam, 1955) (Fig. 17). The Elsinore fault between A and B (Fig. 17) was mapped by Merriam (1955) and Buttram (1962) and the Thing Valley fault by Sawicki (1978).

At the north end of the Thing Valley fault (C, Fig. 17), a 23 feet to 115 feet wide, 1000 feet long zone of breccia and fault gouge is exposed. Between points C and D, faults in Lakeview Mountain Tonalite erode as straight canyon segments, aligned saddles, linear depressions, and lines of denser vegetation (Fig. 18). Over most of their length, the faults are covered by boulders accumulated in the low areas eroded in breccia along the fault zone. At point G, segments of the Thing Valley fault show right separation of 2300 to 4300 feet along the south branch of the Elsinore fault (Fig. 17). At point I, a straight, deep canyon is aligned with the fault zone and Thing Valley to the south-southwest is a narrow straight feature along the same alignment. Bedrock contacts across the fault at point I suggest as much as 1/2 mile of right separation. Southeast of Thing Valley (J, Fig. 17) a fault separates Julian Schist from Green Valley Tonalite. To the southwest the fault splays into several branches. Rakes of striations and displaced contacts show reverse right-slip.

ELSINORE FAULT

The northwest trending Elsinore fault zone extends 125 miles from just north of the Mexican border to the northern edge of the Santa Ana Mountains (Strand, 1962; Rogers, 1965). The Elsinore fault zone controls the northeast slope of the north end of the Santa Ana Mountains to the Agua Tibia Mountains (Fig. 4). The fault continues as an alignment of straight canyons and saddles on the southwest slope of the Agua Tibia Mountains (Fig. 4). San Luis Rey River follows a

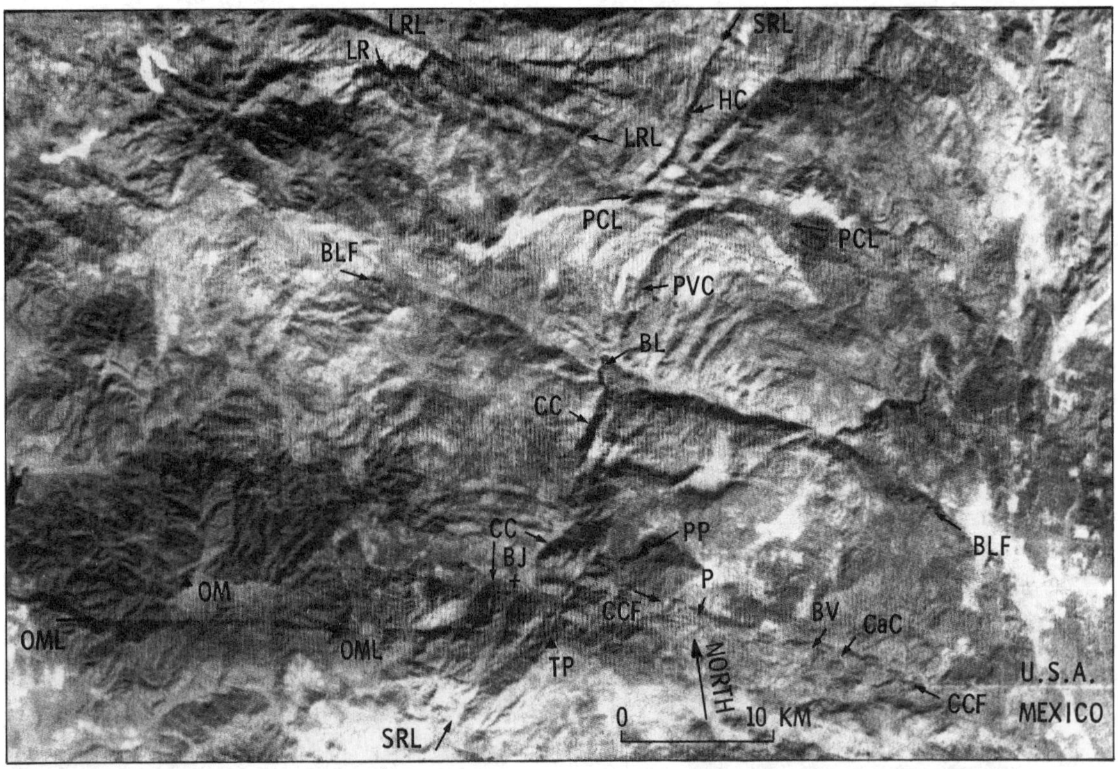

Figure 15. Landsat image of Barrett Lake fault and other faults and lineaments in Peninsular Ranges, southernmost California. Abbreviations: BJ: Barrett Junction; BL: Barrett Lake; BLF: Barrett Lake fault; BV: Bell Valley; CaC: Canyon City; CC: Cottonwood Creek; CCF: Canyon City fault; HC: Horsethief Canyon; LR: Loveland Reservoir; LRL: Loveland Reservoir lineament; OM: Otay Mountain; OML: Otay Mountain fault; P: Potrero; PCL: Pine Creek lineament; PP: Potrero Peak; PVC: Pine Valley Creek; SRL: Sweetwater River lineament; TP: Tecate Peak. Figure 3 shows area.

straight canyon along the fault zone (Jahns, 1954b) and the fault is exposed along Highway 76 northwest of Lake Henshaw (A - B, Fig. 9). The fault is reflected in a northeast facing slope to the southeast end of Lake Henshaw (C, Fig. 9) and a series of straight canyon segments to the southeast (D, Fig. 4). The straight course of Banner Canyon follows the fault (A - B, Fig. 14). The Elsinore fault is exposed at the mouth of Rodriguez Canyon (C, Fig. 14) as a 200 foot wide zone of shattered and slickensided blocks of schist and pegmatite dikes (Hart, 1974).

Rodriguez Canyon (Fig. 14) has a sinuous trace and this segment of the Elsinore fault zone is not obvious on satellite images (Fig. 5). Southeast of Banner (Figs. 5 and 14) the trace is less distinct on satellite images than to the northwest and branches of the Elsinore fault zone are not evident on satellite images of the Agua Caliente area (Fig. 16). The termination of the Elsinore fault as a prominent linear feature on satellite images from Rodriguez Canyon south may be related to a southeastward decrease in right-slip (Weber, 1977; Lamar and Rockwell, 1986) or no right-slip (Lowman, 1980).

Henderson Canyon Lineament

Although Henderson Canyon appears as a prominent lineament (Figs. 2 and 5), there is no evidence of faulting. Foliation in banded gneiss (Henderson Canyon Complex of Sharp, 1967) parallels the lineament.

WEST TO WEST-NORTHWEST TRENDING FAULTS AND LINEAMENTS

Previtt Canyon Lineament

Previtt Canyon and Dodge Valley to the west (Figs. 2 and 4) form the 9 mile long Previtt Canyon lineament. Tonalite and gneiss comprise bedrock (Weber, 1963). Evidence of faulting is lacking and neither joints nor foliation parallel the lineament.

Warren Canyon Fault

A straight, east-west trending segment of Warren Canyon and canyons and saddles to the east follow the 6.6 mile long Warren Canyon fault along the south sides of Woodson Mountain and Santa Maria Valley (Figs. 8 and 9). The east end of the fault terminates against the Pamo Valley lineament (Fig. 8). Bedrock consists of Woodson Mountain and Wolford Granodiorite (Gastil and Bushee, 1961). Joints in granodiorite are predominantly parallel and orthogonal to the fault. A fault along the lineament is exposed in a roadcut on Highway 67 at the head of Warren Canyon. Similar rocks are exposed on both sides of the fault.

Woodson Mountain Lineament

Woodson Mountain is a prominent, elongated topographic high flanked on the south side by the Warren Canyon fault.

Figure 16. Skylab image of Thing Valley and vicinity. Abbreviations: AC: Aqua Caliente Hot Springs; NBEF: North Branch Elsinore fault; SBEF: South Branch Elsinore fault; TV: Thing Valley; TVF: Thing Valley fault. Points G and H are keyed to Figure 17. Figure 3 shows area covered.

Along the north side of Woodson Mountain an 5.5 mile long east-west alignment of straight canyons and saddles forms Woodson Mountain lineament (Figs. 8 and 9). Bedrock consists of Woodson Mountain Granodiorite (Merriam, 1946). No evidence of faulting was observed in the few exposures along the lineament; joints are predominantly parallel and orthogonal (Fig. 9).

San Vicente Valley Lineament

An east-west trending break in slope, straight canyons and saddles form the San Vicente Valley lineament (Figs. 8 and 9). Bedrock consists of Stonewall Quartz Diorite and Wolford and Woodson Mountain Granodiorite (Merriam, 1946). We could not verify the fault along the western portion of this feature shown by Weber (1963) and Rogers (1965). The eastern portion of the San Vicente Valley lineament crosses and does not displace the Witch Creek and Himmel Canyon lineaments or the Woodson Mountain Granodiorite - Stonewall Quartz Diorite contact (Fig. 12). The lineament probably formed by erosion along joints parallel to the lineament (Figs. 9 and 12).

Hatfield Creek Lineament

A fairly straight, east-west trending, 3.8 mile long change in tone occurs along the south side of Hatfield Creek (Fig. 8). Bedrock consists of Lakeview Tonalite (Merriam, 1946).

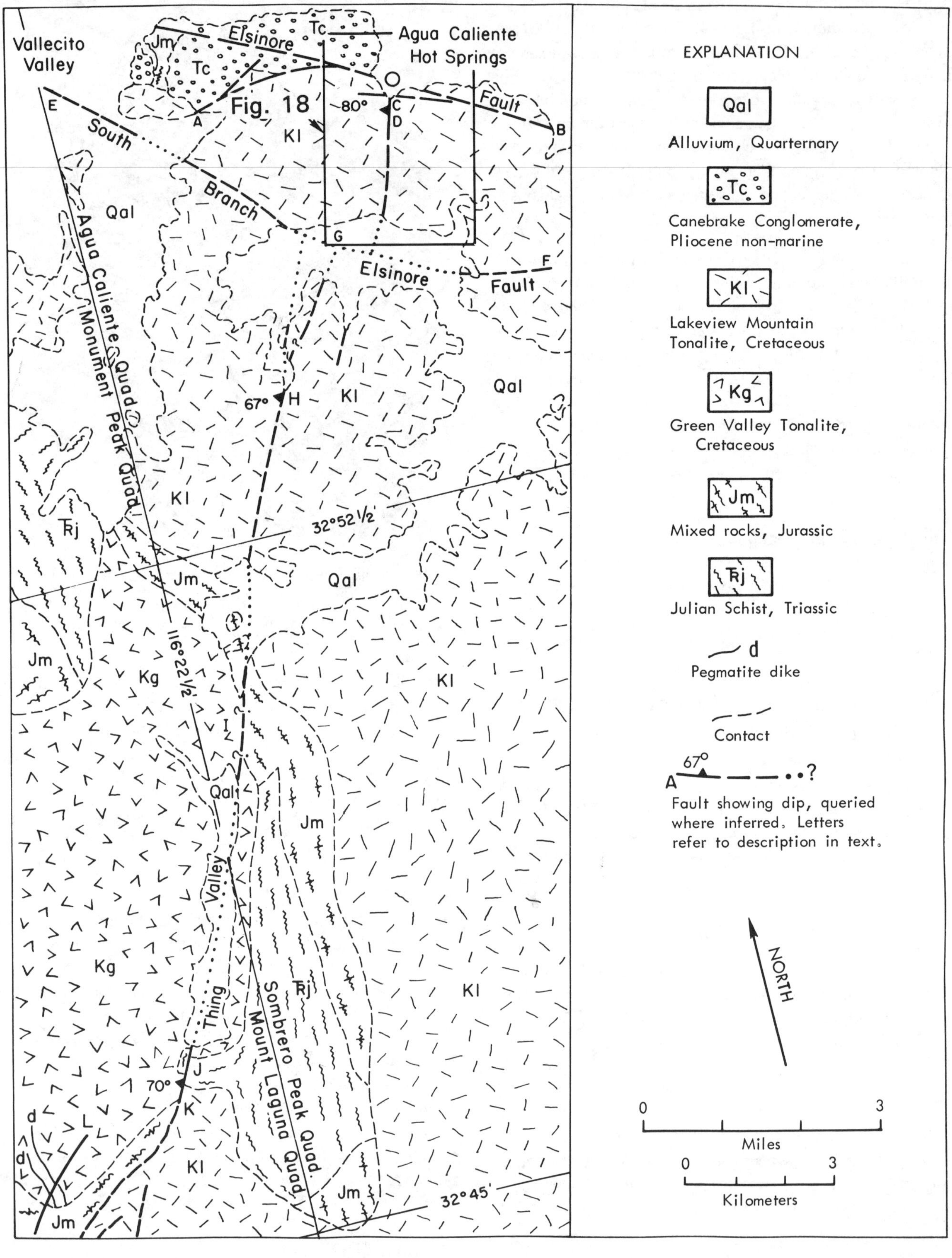

Figure 17. Geologic map of Thing Valley area, modified from Merriam (1955) and Buttram (1962). Figure 3 shows area covered.

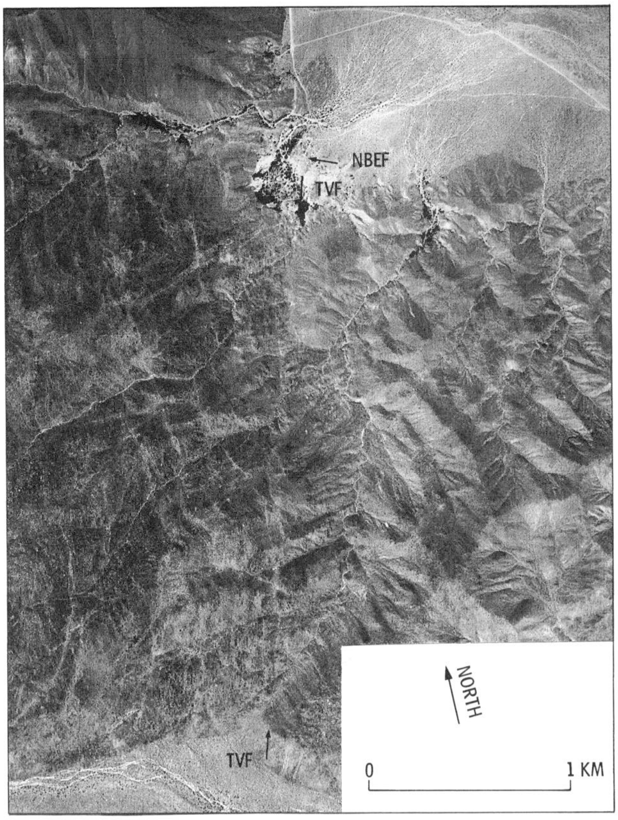

Figure 18. Portion of U.S.D.A. aerial photograph showing Thing Valley fault between north and south branches of Elsinore fault. Abbreviations: NBEF: north branch Elsinore fault; TVF: Thing Valley fault. Figure 17 shows area covered.

Neither foliation nor joints parallel the lineament (Fig. 9) and exposures are not sufficient to establish whether a fault is present. The straightness of the canyon could have been accentuated by boundaries between cultivated fields and natural vegetation; thus neither the existence of a bedrock lineament nor the origin of the tonal change were established.

Loveland Reservoir and Pine Creek Lineaments

The Loveland Reservoir lineament is expressed by 6 miles of west-northwest trending, gently curving valleys. Beginning 2 miles from the east end of the Loveland Reservoir lineament, Pine Creek forms a 3 mile long lineament parallel and en echelon to the Loveland Reservoir lineament (Figs. 2 and 15). These lineaments parallel foliation and the Woodson Mountain Granodiorite - Green Valley Tonalite contact; Bonsal Tonalite and diorite are also exposed (Everhart, 1951). No evidence of faulting was observed, although exposures are poor.

Barrett Lake Fault

Valleys along the Barrett Lake fault (Weber, 1963) form a prominent 25 mile long, west-northwest trending lineament (Figs. 5 and 15). Bedrock is composed of tonalite and gabbro (Weber, 1963). Exposures in the fault zone 0.3 miles west of Barrett Lake consist of gouge and breccia 23 feet wide containing slickensided shear surfaces. Shear surfaces have an average strike of N66°W, a dip of 60°S and rakes of striations of 18°W; the slip is indeterminate.

Otay Mountain Fault

The 6.0 mile long, east-west trending Otay Mountain fault underlies a series of straight, aligned drainage courses and saddles on the south slope of Otay Mountain (Figs. 5 and 15). Black Mountain Volcanics (Strand, 1962) are exposed along the fault which parallels neither foliation nor the dominant joint direction. At the east end of the fault agglomerate shows 600 feet of left separation (A, Fig. 19). In the prominent saddle at B (Fig. 19) thin bedded andesite strikes into massive andesite north of the fault and probable fault breccia is exposed west of the saddle.

Canyon City Fault

A 7 mile long series of valleys are eroded along the Canyon City fault (Fig. 15). Woodson Mountain Granodiorite, Bonsal Tonalite and San Marcos Gabbro (Rogers, 1965) are exposed along the fault. Foliation and joints north of the fault are not parallel to the fault (Fig. 20). South of the fault, foliation is mostly east-west parallel to the fault and joints trend north and northeast at an angle to the fault. Fault gouge and breccia in a zone up to 13 feet wide are exposed at the west end of the fault (Fig. 20). Lithologic units cannot be compared across the fault to determine offset.

APPLICATION OF SATELLITE IMAGERY TO TECTONIC AND ENGINEERING GEOLOGY STUDIES

The perspective afforded by satellite images focuses attention on prominent lineaments not obvious in large scale photography or on the ground. Lineaments on satellite images of the Peninsular Ranges result from erosion along foliation, joints or faults. Cultural features such as major highways and crop boundaries also may produce a linear feature in the images. Field investigations are required to verify their origin. The structural controls of the lineaments described herein are summarized on Table 1. Many prominent lineaments were found to be the result of erosion along joints or foliation rather than faults.

The northwest-trending faults in the north-east portion of the area (Figs. 1 and 2) show geomorphic evidence of recent movement and destructive earthquakes have occurred on the San Jacinto (Allen and others, 1965) and Whittier-Elsinore (Weber, 1987) fault zones in historic time. North of Lake Elsinore, the Elsinore fault zone is comparatively narrow; south of Lake Elsinore the fault may branch into a broad zone of sub-parallel faults in the Lake Henshaw region, notably the Agua Caliente, Aguanga, and Agua Tibia faults (Lamar and Rockwell, 1986). The trace of the Elsinore fault is obvious on satellite imagery at the northwest end where right-slip is significant and indistinct at the southeast where

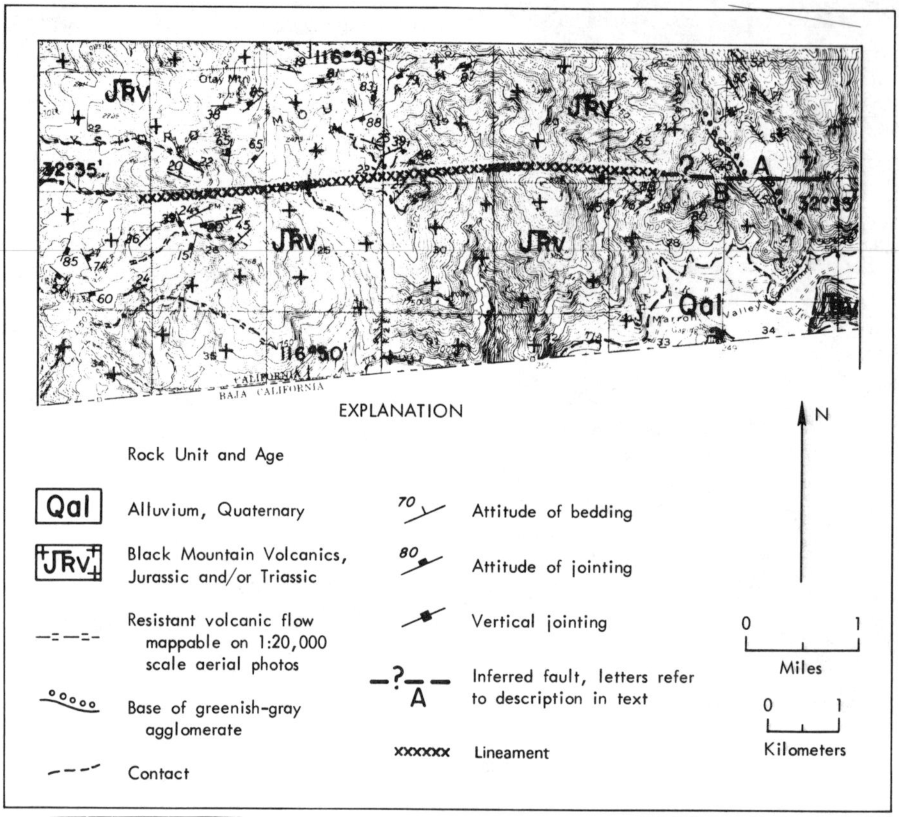

Figure 19. Geologic map of Otay Mountain fault. Topography from U.S.G.S. Jamul quadrangle. Figure 3 shows area covered.

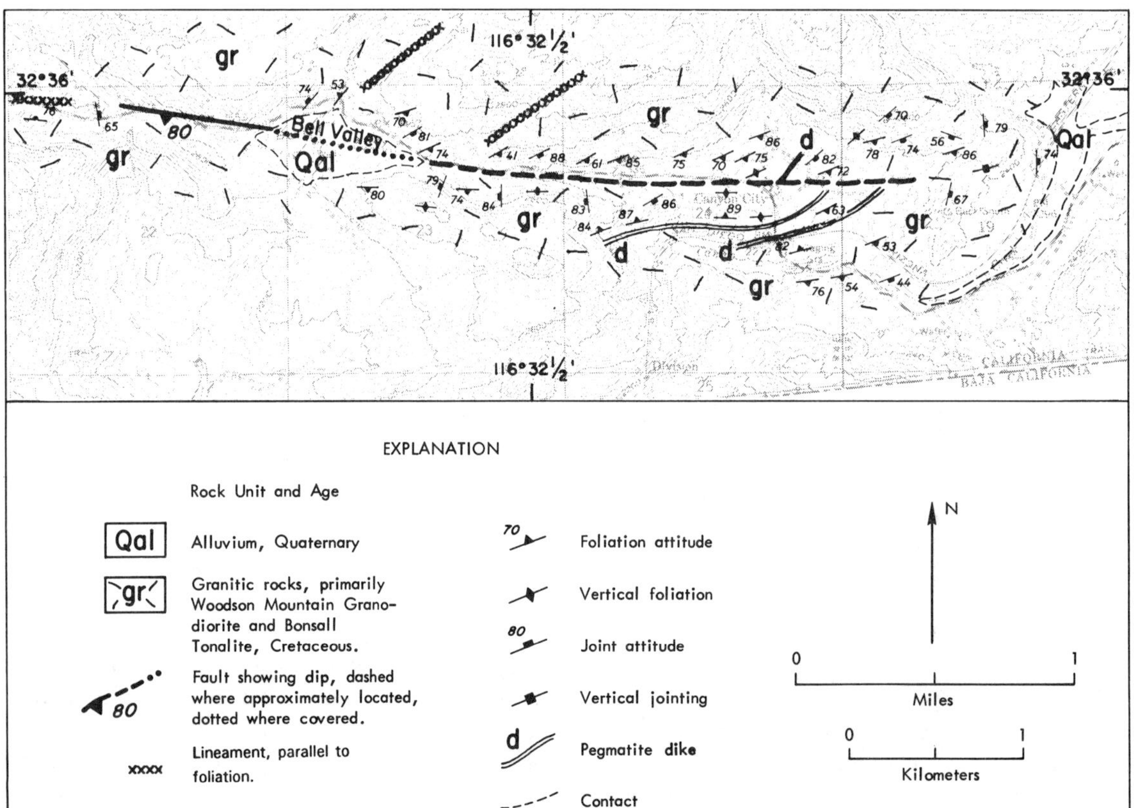

Figure 20. Geologic map of Canyon City fault. Topography from U.S.G.S. Potrero quadrangle. Figure 3 shows area covered.

right-slip is small. Thus, the large right-slip reported along the northwestern segment of the Elsinore fault may be distributed on several branching faults to the southeast (Weber, 1977; Lamar and Rockwell, 1986).

Fault patterns viewed in a single satellite image display regional relationships and may reveal the relative ages of faults. The northeast and east-west trending faults and lineaments in the Southern California batholith and older metamorphic rocks are conspicuous on satellite images of the Peninsular Ranges. The northeast and east-west trending faults are steeply dipping and possess well-developed breccia zones and slickensided shear surfaces. Offsets of steeply dipping contacts by the northeast trending Thing Valley and San Diego River faults indicate right-slip. Left separation of a volcanic unit on the east-west trending Otay Mountain fault could be due to either left-slip or vertical slip. The Sweetwater River lineament appears to be displaced in a left sense by the Pine Creek lineament.

The northeast and east-west faults are inactive: geomorphic evidence of recent displacement is lacking; the northeast-trending faults are truncated by the active northwest-tending faults; right slip on the northeast trending faults is in the wrong sense for these features to be related to right-slip on the active northwest-trending faults.

The northeast and east-west faults may be conjugate shears related to an older stress system. East-northeast, west-southwest crustal shortening is consistent with the fault pattern: right slip on the Thing Valley and San Diego River faults, and conjectural left-slip on the Otay Mountain fault and Pine Creek lineament. On the basis of north-south trending thrust faults and mylonitic zones, Sharp (1968) also suggested an older period (Middle Cretaceous to Eocene?) of east-west crustal shortening in the Peninsular Ranges. The subduction zone parallel to the continental margin during the Mesozoic suggested by Hamilton (1969) and Hill (1971) would have produced east-west shortening.

Satellite images have application to earthquake hazard investigations concerned with the existence and activity of faults. Delineation of faults and associated low strength brecciated rock are also important to foundation studies. The permeability of the bedrock is governed by faults, joints and foliation (Merriam, 1946). Because faults in basement rock may increase permeability by fracturing, preliminary water well locations may be determined by recognition of faults on satellite images. Thus, analysis of satellite images is an important initial phase in engineering geology studies, particularly of large projects, such as the route for the Northern Tier Oil Pipeline (Fritts and others, 1982).

In areas such as southwestern California, where much of the geology has been previously mapped only on a reconnaissance scale, satellite imagery may reveal unrecognized faults, clarify regional tectonic relationships and contribute to engineering geology studies. Several unreported faults were identified in this investigation, and new insights on the tectonic history of the Peninsular Ranges were obtained.

ACKNOWLEDGMENTS

The work was accomplished under NASA Contract NASA 2-7698 and U.S. Geological Survey Contract 14-08-0001-13911. David Amsbury, NASA and Ernest Lathram, U.S. Geological Survey were technical monitors. The project grew out of stimulating discussions with Paul Lowman. Discussions with Richard Merriam, Harold Weber, Victoria Todd, Gordon Gastil, David Amsbury and the late Mason Hill also are gratefully acknowledged. The assistance of students at the University of California, Thomas Troutman, Thomas Davis and Jerry Treiman is also acknowledged.

Table 1 - Summary of field studies of faults and lineaments seen on satellite images (Fig. 2).

Previously unrecognized or unproven fault:
- Canyon City fault
- Otay Mountain fault, probable
- San Diego River fault zone
- San Ysidro Creek fault
- Thing Valley fault
- Warren Canyon fault, joints parallel to lineament

Erosion along joints or foliation:
- Chariot Canyon fault (some evidence of faulting)
- Henderson Canyon lineament
- Los Penasquitos Canyon
- Loveland Reservoir lineament
- Pamo Valley lineament (faulting along one segment)
- Pine Creek lineament
- San Vicente Valley lineament
- Sutherland Reservoir lineament
- Sweetwater River lineament
- Valle De Los Amigos lineament
- Woodson Mountain lineament

Erosion along zone of weakness of unknown origin; not parallel to joints or foliation and faulting improbable:
- Himmel Canyon lineament
- Witch Creek lineament

Unknown origin; not parallel to joints or foliation, exposures not adequate to determine existence of fault:
- Hatfield Creek lineament
- Previtt Canyon lineament

REFERENCES

Allen, C. R., Amand, P. St., Richter, C. F. and Nordquist, J. M., 1965, Relationship between seismicity and geologic structure in the southern California region: Seismol. Soc. Amer. Bull., vol. 55, no. 4, p. 753-797.

Allison, M. L., 1974a, Geophysical studies along the southern portion of the Elsinore fault: M.S. thesis, San Diego State University, 229 p.

Allison, M. L., 1974b, Geologic and geophysical reconnaissance of the Elsinore-Chariot Canyon fault systems: Field Trip Guidebook, December 14 and 15, 1974, San Diego Assoc. of Geologists, p. 21-35.

Buttram, G. N., 1962, The geology of the Agua Caliente quadrangle, California: M.S. thesis, University of Southern California.

Everhart, D. L., 1951, Geology of the Cuyamaca Peak quadrangle, San Diego County, California: Calif. Div. Mines and Geol., Bull. 159, p. 51-115.

Fritts, S. G., P. M. Merifield, D. L. Lamar and J. L. Smith, 1982, Reconnaissance engineering geologic mapping from Landsat imagery along the Northern Tier Oil Pipeline route: Lamar-Merifield Technical Report 82-1, NASA/Jet Propulsion Laboratory, Subcontract 955773.

Gastil, R. G. and Bushee, J., 1961, Geology and geomorphology of eastern San Diego County: Field Trip Guidebook, Geol. Soc. Amer., Cordilleran Section, p. 8-22.

Hamilton, W., 1969, Mesozoic California and the underflow of Pacific mantle: Geol. Soc. Amer. Bull., vol. 80, p. 2409-2430.

Hart, M. W., 1964, The Elsinore fault between Banner Grade and Vallecito Valley, San Diego County, California: San Diego State Univ., Geol. Dept. Undergraduate Research Reports, vol. 8, pt. 3, 14 p.

Hart, M. W., 1974, Field Guide for the December 14 and 15, 1974 field trip: San Diego Assoc. of Geologists, p. 5-9.

Hill, M. L., 1971, Newport-Inglewood zone and Mesozoic subduction, California: Geol. Soc. Amer. Bull., vol. 82, p. 2957-2962.

Jahns, R. H., 1954a, Geology of the Peninsular Ranges province, southern California and Baja California: Calif. Div. Mines and Geol. Bull. 170, ch. 2, p. 29-52.

Jahns, R. H., 1954b, Northern part of the Peninsular Ranges province: Calif. Div. Mines and Geol. Bull. 170, Geologic Guide, 59 p.

Jennings, C. W., 1973, State of California preliminary fault and geologic map, scale 1:750,000: Calif. Div. Mines and Geol., Preliminary Report 13.

Lamar, D. L. and Rockwell, T. K., 1986, An overview of the tectonics of the Elsinore fault: Neotectonics and Faulting in Southern California, Geol. Soc. Amer., Cordilleran Sec., Guidebook and Volume, p. 149-157.

Larsen, E. S., Jr., 1948, Batholith and associated Rocks of Corona, Elsinore, and San Luis Rey quadrangles, southern California: Geol. Soc. Amer., Memoir 29.

Lowman, P. D., 1969, Apollo 9 multispectral photography: geologic analysis: NASA Goddard Space Flight Center, Greenbelt, MD., X-644-69-423.

Lowman, P. D., Jr., 1980, Vertical displacement on the Elsinore fault of southern California: evidence from orbital photographs: Jour. Geol., vol. 88, p. 415-432.

Merifield, P. M. and Lamar, D. L., 1976a, Active and inactive faults in southern California viewed from Skylab: Proc. NASA Earth Resources Survey Symposium, June 8-13, 1975, Houston, Texas, v. 1-B, NASA TM X-58163, JSC-09930, p. 779-797.

Merifield, P. M. and Lamar, D. L., 1976b, Lineaments in basement terrane of the Peninsular Ranges, southern California: Proc. First International Conference on the New Basement Tectonics, Utah Geol. Assoc. Publ. No. 5, p. 94-105.

Merriam, R., 1946, Igneous and metamorphic rocks of the southwestern part of the Ramona quadrangle, San Diego County, California: Geol. Soc. Amer. Bull, v. 57, p. 223-260.

Merriam, R., 1955, Geologic map of Cuyapaipe quadrangle, California, scale 1:62,500: unpublished map (Cuyapaipe quad. presently designated Mr. Laguna quad.).

Merriam, R., 1958, Geology and mineral resources of Santa Ysabel quadrangle, San Diego County, California: Calif. Div. Mines and Geol. Bull. 177, 42 p.

Miller, W. J., 1935, Geomorphology of the southern Peninsular Range of California: Geol. Soc. Amer. Bull., vol. 46, p. 1535-1562.

Proctor, R. J., 1973, Map showing recently active faults in the southern California region: Geology, Seismicity and Environmental Impact, Assoc. of Eng. Geologists Special Publication, Los Angeles (out of print)

Rogers, T. H., 1965, Geologic map of California, Santa Ana sheet, scale 1:250,000: Calif. Div. Mines and Geol.

Sauer, C., 1929, Land forms in the Peninsular Range of California as developed about Warner's Hot Springs and Mesa Grande: Univ. Calif., Pubs. in Geography, vol. 3, p. 199-200.

Sawicki, D. A., 1978, A structural and petrographic evaluation of the Thing Valley lineament, San Diego County, California: M. S. thesis, San Diego State Univ. 98 p.

Scheliga, J. T., Jr., 1963, Geology and water resources of Warner Basin, San Diego County, California: M. S. thesis, Univ. Southern California

Sharp, R. V., 1967, San Jacinto fault zone in the Peninsular Ranges of southern California: Geol. Soc. Am. Bull, v. 78, p. 705-730.

Sharp, R. V., 1968, The San Andreas fault system and contrasting pre-San Andreas structures in the Peninsular Ranges of southern California: Proc. Conf. Geological Problems of San Andreas Fault System, Stanford Univ. Geol. Sci., vol. XI, p. 292-293.

Strand, R. G., 1962, Geologic map of California, San Diego-El Centro sheet, scale 1:250,000: Calif. Div. Mines and Geology

Todd, V. R., 1977, Syntectonic batholithic rocks in south-central San Diego County, A preliminary report: Calif. Div. Mines and Geol., Short Contributions to California Geology, Speial Report. 129, p. 51-59.

Todd, V. R. and S. E. Shaw, 1979, Structural, metamorphic and intrusive framework of the Peninsular Ranges batholith in southern San Diego County, California: in Mesozoic crystalline rocks, Peninsular Ranges batholith and pegmatites, Point Sal Ophiolite, Geological Society of America Meeting, San Diego, California, p. 177-231.

Todd, V. R., G. K. Lee, H. W. Oliver, J. D Causey and S. W. Schmauch, 1988, Mineral resources of the southern Otay Mountain and western Otay Mountain wilderness study areas, San Diego County, California: U. S. Geol. Survey, Bull. 1711-E, p. E1-E18.

Weber, F. H., Jr., 1963, Geology and mineral resources of San Diego County, California: Calif. Div. Mines and Geol., County Report 3, 309 p.

Weber, F. H., Jr., 1977, Seismic hazards related to geologic factors, Elsinore and Chino fault zones, northwestern Riverside County, California: Calif. Div. Mines and Geol., Open-file Report 77-4 LA, 96 p.

Weber, F. H., Jr., 1987, Whittier Narrows earthquakes: Calif. Geol., Sacramento, December, p. 275-281.

FAULT IDENTIFICATION USING MULTIDISCIPLINARY TECHNIQUES, GOLDSTONE DEEP SPACE COMMUNICATIONS COMPLEX, CALIFORNIA

Douglas S. Santo and **Mark B. Schluter**
Senior Geologists
Converse Consultants West
3393 East Foothill Boulevard
Pasadena, California 91107

Roy J. Shlemon
Consulting Geologist
P. O. Box 3066
Newport Beach, California 92659

ABSTRACT

A fault investigation was performed of the Mars/Uranus antenna sites at the Goldstone Deep Space Communications Complex in the Mojave Desert. This investigation was performed for the Jet Propulsion Laboratory, California Institute of Technology, and sponsored by the National Aeronautics and Space Administration. The Goldstone complex is one of three ground stations in the NASA Deep Space Network and includes the Mars/Uranus Facility consists of two large diameter reflector antennas used for communication and control of deep space probes and other missions. The investigation included interpretation of Landsat thematic mapper scene, side-looking airborne radar transparencies, and both color-infrared and black-and-white aerial photography. Four photolineaments suggestive of previously undocumented faults were identified. Three generally discrete morphostratigraphic alluvial fan deposits were also recognized and dated using geomorphic and soil stratigraphic techniques. A total of 14 trenches were excavated across the four lineaments. Trenches show that three of the photolineaments coincide with faults. Last displacement of two of the faults occurred between about 12,000 and 35,000 years ago. The third fault was judged to be older than 12,000 ybp, although uncertainty remains. None of the surface traces of the three faults crosses under existing antennas or structures, however, their potential activity necessitates appropriate seismic retrofit designs and loss prevention measures to mitigate potential earthquake damage to facilities and structures.

INTRODUCTION AND PURPOSE

The Goldstone Deep Space Communications Complex in the Mojave Desert is one of three ground stations in the National Aeronautics and Space Administrations's (NASA) Deep Space Network. With its two sister stations in Spain and Australia, the NASA Deep Space Network is one of the largest and most sensitive telecommunications and radio navigation networks in the world. The primary purposes of the Deep Space Network are to support manned and unmanned space missions and to provide instrumentation for radio and radar astronomy observations in the exploration of the solar system and the universe. Its current uses include tracking and communication with deep space probes including the Galileo, Voyager and Pioneer series, communication support for shuttle missions, monitoring orbital paths of satellites and debris, radio astronomy and the search for extraterrestrial intelligence. The Goldstone complex is managed for NASA by the Jet Propulsion Laboratory (JPL) of the California Institute of Technology in Pasadena, California.

The Goldstone Deep Space Communications Complex operates seven parabolic reflector antennas at five sites called Deep Space Stations (DSSs). Four of the Goldstone DSSs are operational in the Deep Space Network. The fifth station is devoted to research and development activities. The large precision tracking antennas provide the essential link between deep space and ground operations.

Figure 1. The 34-meter Uranus antenna (DSS-15) in foreground, with the newly enlarged 70-meter Mars antenna (DSS-14) in background.

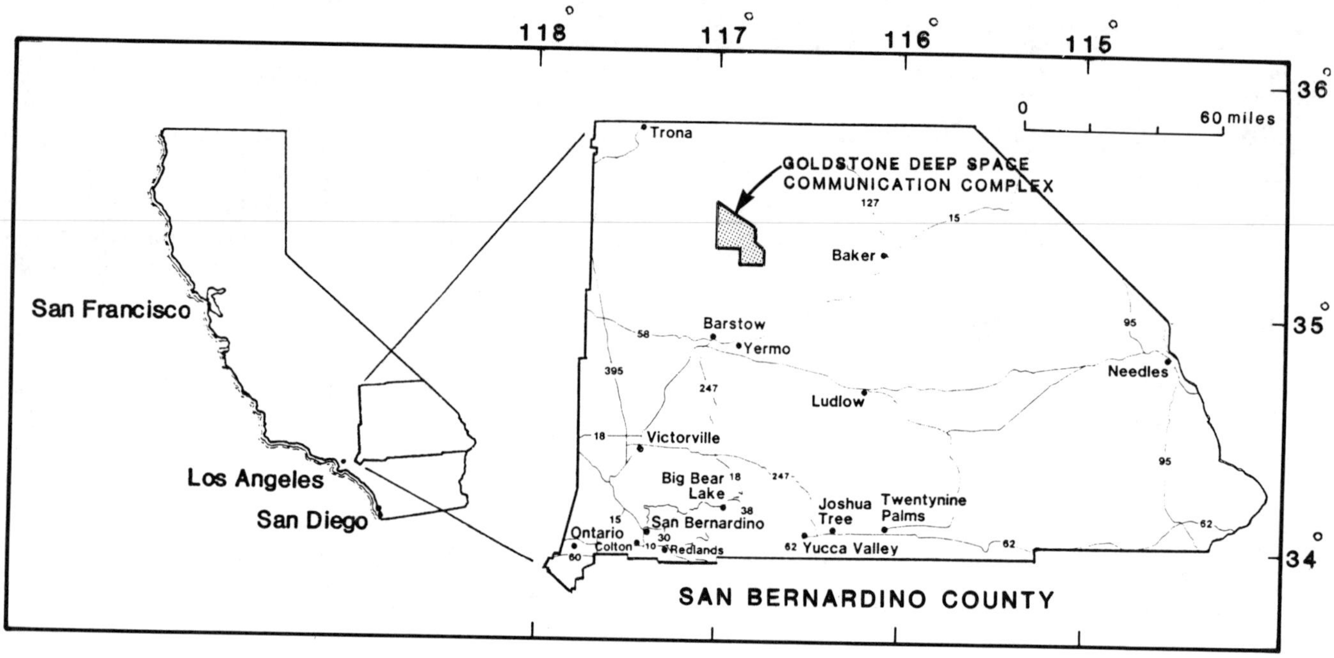

Figure 2. Map showing the Goldstone Deep Space Communications Complex in the Mojave desert, California.

The Mars antenna facility consists of two antennas and surrounding support buildings (Fig. 1). The Mars station antenna (DSS-14), at 70-meters (230-feet) in diameter, is one of the largest antennas of its kind in the world. The antenna was constructed as a 64-meter antenna in 1966 and later enlarged to 70 meters in 1988. Standing more than 235 feet high, this antenna is one of the more striking features of the Goldstone complex.

The Uranus station antenna (DSS-15) is a 34-meter (112-foot), high efficiency, precision-shaped antenna located approximately 1,600 feet southeast of the Mars station antenna (Fig. 1). Built in 1984, this latest antenna addition was first used in January 1986 to support the Voyager 2 spacecraft encounter with the planet Uranus located a distance of 1.8 billion miles from Earth. In the early 1980's plans were developed to construct additional antennas and support buildings at the facility. Geotechnical investigations performed for the proposed expansion revealed several photolineaments within the Mars/Uranus Station valley (Dames and Moore, 1982). The lineaments were thought to be associated with faulting. Refraction seismic lines run through the center of the valley were inconclusive regarding the origin of the photolineaments, and doubt remained about the possible impact of these features and the potential hazard they posed to the facilities. The late 1980's brought a renewed interest in expansion of the facility and a heightened concern among facility designers about the lineaments, and the possible presence of faults within the Mars/Uranus Station valley. A multidisciplinary fault investigation was undertaken to determine the origin of the lineaments and to assess their potential impact on existing and proposed structures.

The purpose of this investigation was twofold: (1) to evaluate the Mars/Uranus Station valley for the presence of faults, including the previously identified lineaments, and (2) to determine the relative age and activity of any identified faults.

LOCATION AND REGIONAL GEOLOGIC SETTING

The Mars/Uranus Station is located in the northwest corner of the Goldstone Deep Space Communications Complex approximately 56 miles north of Barstow, in San Bernardino County, California (Fig. 2). This is a remote area about a 2- to 3-hour drive from major population centers of Southern California. Public access is restricted because the facility is located within the U.S. Army's Fort Irwin National Training Center and east of the China Lake Naval Weapons Center, Mojave Range B.

The Goldstone complex lies in the north-central Mojave Block geomorphic province bounded on the southwest by the San Andreas Fault and Transverse Ranges, on the north and northeast by the Garlock Fault, Tehachapi Mountains, and the Basin and Ranges, and on the east by the arbitrary boundary of the Nevada state line and the Colorado River. The central portion of the block is characterized by potentially active Quaternary and locally active Holocene faults (Hart, 1987). The faults trend northwest/southeast and are predominently strike-slip faults (Fig. 4). The faults cut Tertiary extensional basins separated by low hills of Tertiary volcanic and sedimentary rock or pre-Tertiary intrusive rock (Dokka, 1990). Many basins contain Pleistocene pluvial dry lakes.

TECHNICAL APPROACH

Geologic information within the Goldstone complex is limited and original work was required during the course of the investigation. To cope with the remote location and lack of information in a timely and economical way, analyses were made of remote sensing imagery and aerial photography. Results of the imagery analyses were then checked in the field with ground-truth verification, reconnaissance geologic mapping, geomorphic observations, fault trenching, and soil-stratigraphic analyses. A summary of the technical approach is presented in Figure 3.

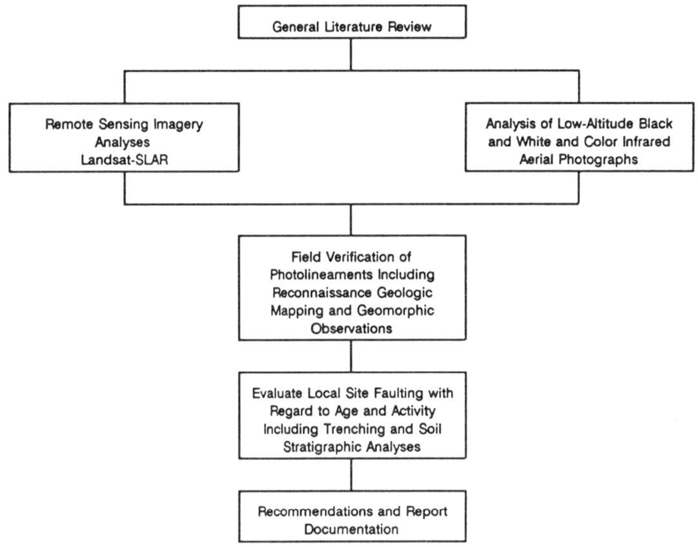

Figure 3. Summary of technical approach.

IMAGERY ANALYSIS

Remote sensing imagery and aerial photography that encompassed 6,000 square miles were reviewed for evidence of possible Quaternary faulting. The imagery included a Landsat thematic mapper scene, side-looking airborne radar (SLAR) transparencies, and both black-and-white and color-infrared aerial photography (Table 1). The imagery was

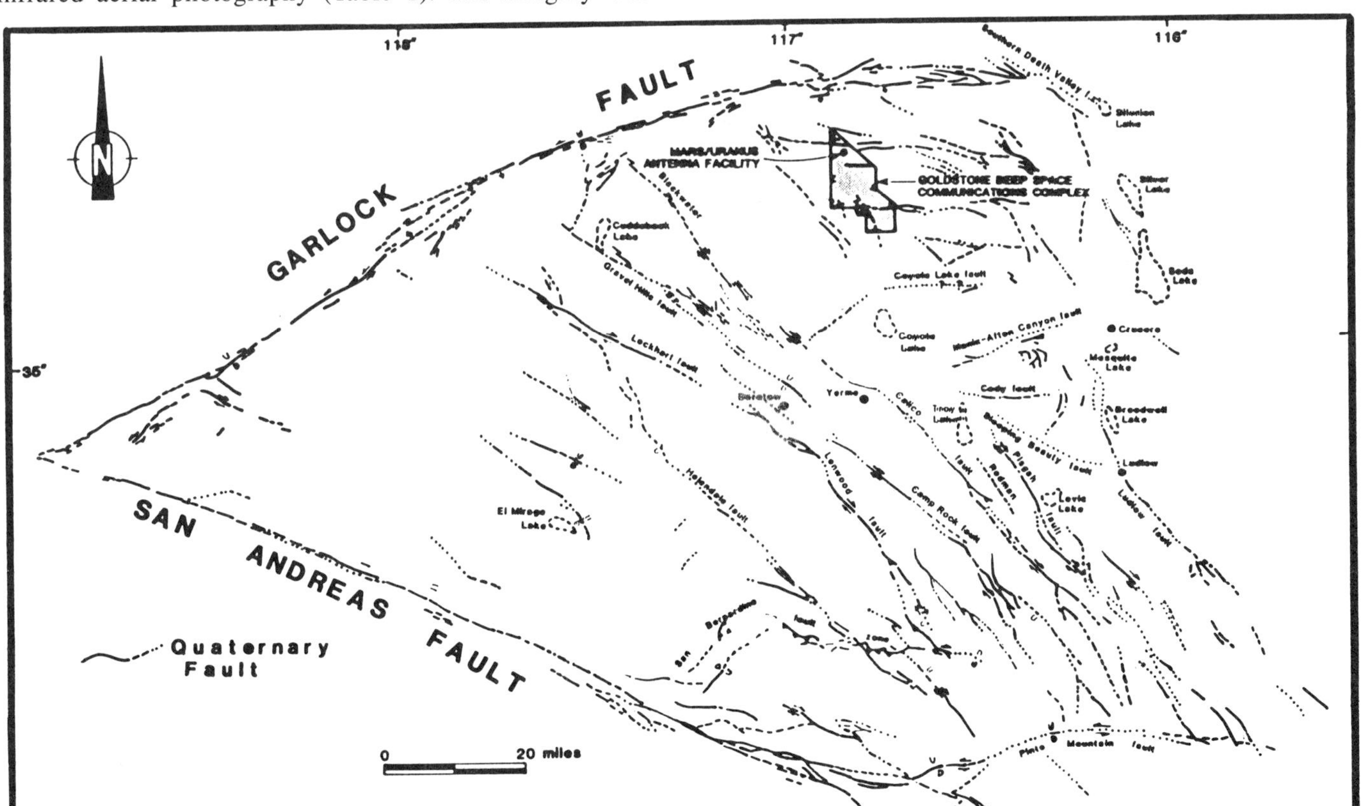

Figure 4. Major Quaternary faults of the Mojave Block Geomorphic Province (Modified after Jennings, 1975).

selected in accordance with the "multi approach" outlined by Glass and Slemmons (1978). This approach optimizes the results of an imagery analysis by reviewing multi-scale, multi-station, multi-band, multi-date, and multi-enhanced imagery. The imagery is then reviewed and interpreted by multiple observers. Geomorphic and photographic features recognized on the imagery and considered to be evidence of faulting included:

- topographic and vegetation alignments
- linear breaks in slope
- rectilinear closed depressions
- linear or uphill facing scarps on alluvial fans
- linear change in drainage or erosional texture
- offset stream channels
- linear change in photographic tone or color
- linear spectral contrasts on Landsat imagery

Field Reconnaissance

After the imagery analysis, photolineaments were checked in the field. The purpose of the field reconnaissance was to differentiate fault related geomorphic features from those formed by man-made, or by natural erosional or depositional processes. In addition, geologic mapping and geomorphic observations were made. Geomorphic observations included relative dissection of fan surfaces, relative development of desert pavement and patina (desert varnish), and fan superposition relationships. In addition early morning, low-angle-sun-illumination 35mm slides were taken of photolineaments.

Results of Imagery Analysis

Based on the photographic evidence and observations made during the field reconnaissance, four principal lineaments were identified and suspected to be associated with Quaternary faulting (Fig. 5). A composite aerial photograph showing these lineaments is presented in Figure 6. The characteristics of each lineament are compiled in Table 2.

EXPLORATION

Field exploration included reconnaissance geologic mapping, siting exploratory trenches, and detailed logging of the trenches including soil-stratigraphic analyses. The trenches served two purposes: (1) to determine if the photolineaments were fault related, and (2) to expose sediments amenable to relative dating by soil stratigraphic techniques. Fourteen trenches were excavated with a backhoe equipped with a 24-inch shovel. (Figure 7).

SITE GEOLOGY

The Mars/Uranus Station is in a linear interior-draining valley bounded on the south by steep volcanic ridges and on the

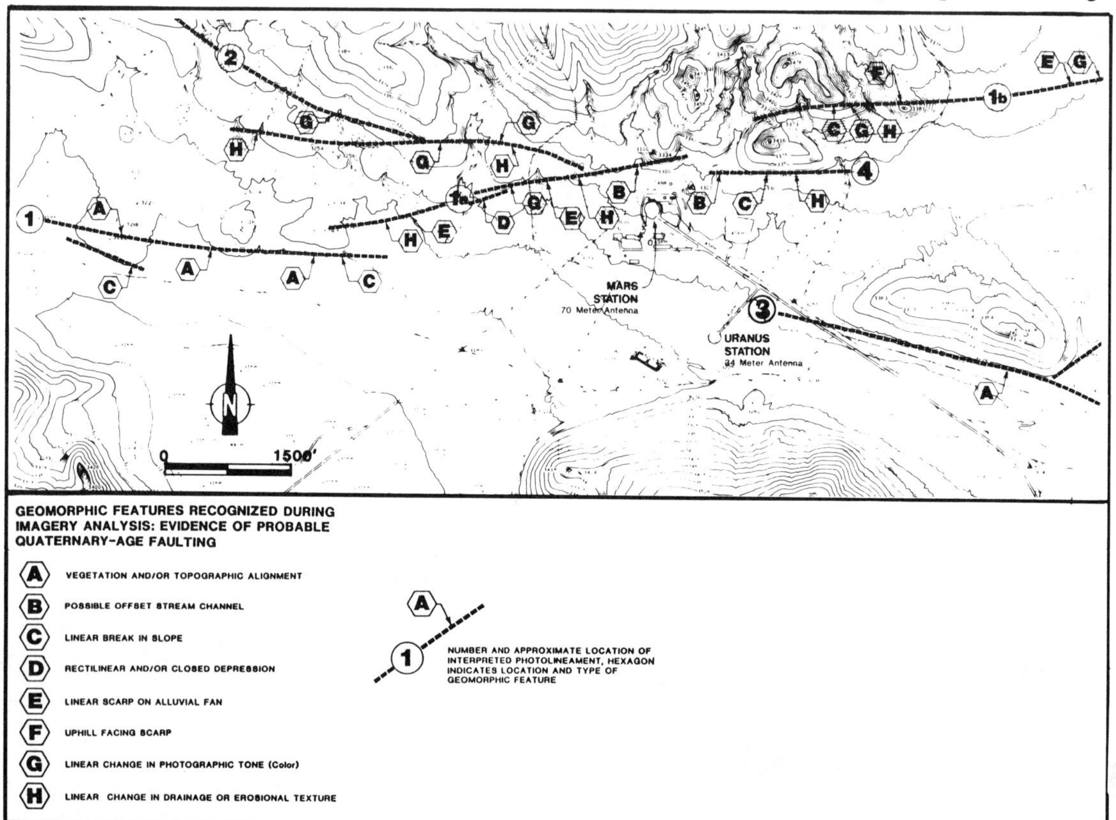

Figure 5. Photolineament map of the Mars/Uranus Station valley

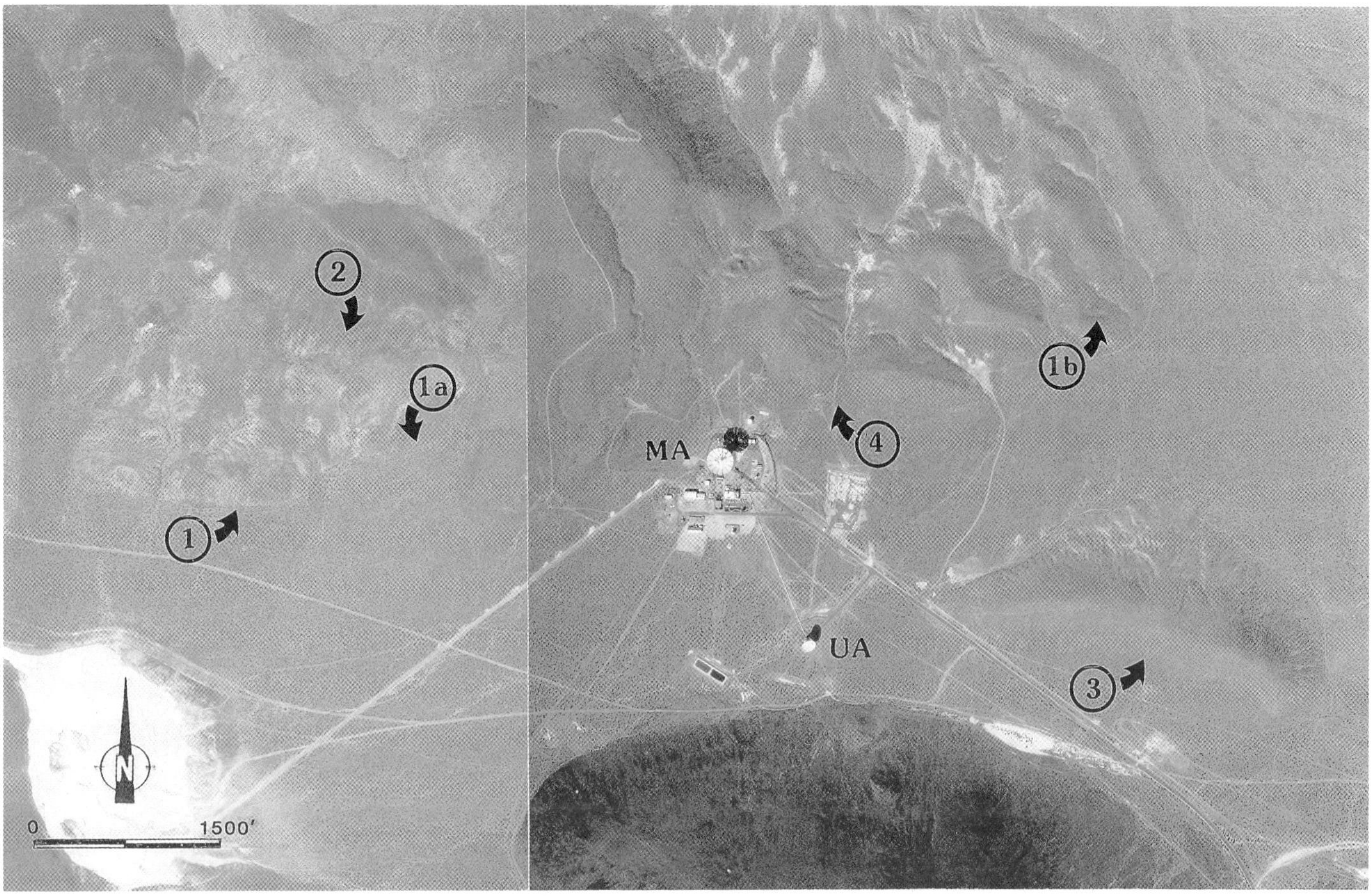

Figure 6. Composite aerial photo of Mars/Uranus Station valley showing photographic expression of lineaments. MA = Mars Antenna, UA = Uranus Antenna. Also note the change in color and morphology on alluvial fan surfaces.

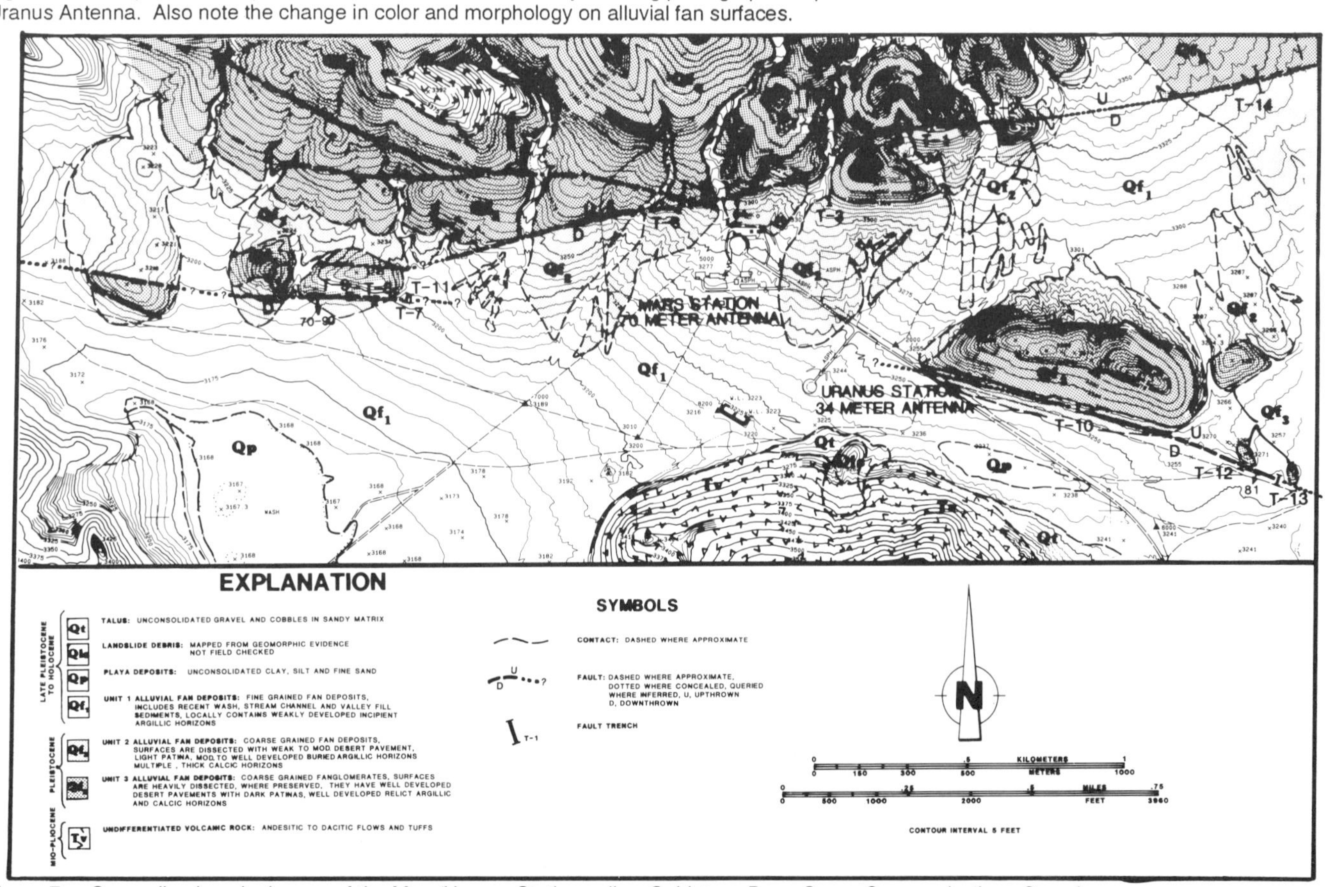

Figure 7. Generalized geologic map of the Mars/Uranus Station valley, Goldstone Deep Space Communications Complex.

north by low-lying alluvial fan capped hills. Southwest of the antennas is a small dry lake. As shown on Figure 7, the major geologic units are: (1) three different-age alluvial-fan deposits; (2) playa deposits; and (3) extrusive volcanic rocks.

Undifferentiated Volcanic Rocks

Andesite forms the steep hills south of the antennas and crops out in two isolated locations in the hills to the north. The rock is light gray to black, massive, and slightly vesicular. It is resistant to weathering and forms steep slopes. Latitic to dacitic tuffs underlie the low hills north of the main valley where they are capped by Unit 3 alluvium. The rock is white to pink, massive to thickly bedded, lithic tuff, and tuff breccia. It is easily eroded and exposures are limited to small stream canyons. The volcanic rocks are probably equivalent to the Pliocene, Lane Mountain Andesite of Dibblee (Dibblee, 1968) or the lower Miocene, Lane Mountain Quartz Latite of Burke (Burke, et al., 1982).

Alluvial Fan Deposits

The alluvial-fan deposits were subdivided into three mapping units based mainly on geomorphic and soil-stratigraphic evidence. Diagnostic criteria included elevation above base level, superposition relationships, fan surface morphology (relative dissection, etc.), development of desert pavement and patina, and soil-profile development. These criteria and the application of soil-profile development for fault activity assessments are summarized in Christenson and Purcell (1985), Machettee (1978) and Shlemon (1978, 1985). Relative ages for site-specific surfaces and alluvial fan sediments are given in Shlemon (1989). The nested geomorphic expression of the three alluvial-fan deposits and their diagnostic surface and soil-stratigraphic characteristics allow them to be mapped as generally discrete morphostratigraphic units designated as Units 1 (youngest), 2 and 3, respectively (Fig. 7).

Unit 3 Alluvial Fans

Unit 3 fan deposits cap the low hills north of the antennas and form the highest level geomorphic surfaces flanking the Mars/Uranus Station valley. Remnant fan surfaces are heavily dissected, but locally well developed desert pavements with dark patinas are still visible on drainage divides. The deposits are coarse gravel/cobble fanglomerates interbedded with finer grained fluvial lenses. On the undissected surfaces these deposits support strongly developed, relict paleosols characterized by truncated argillic horizons and multiple, superimposed, stage III to IV calcic horizons (Btk). Their relative elevation, surface characteristics and pedologic development suggest that these fans are well in excess of 100,000 years old (Shlemon, 1989).

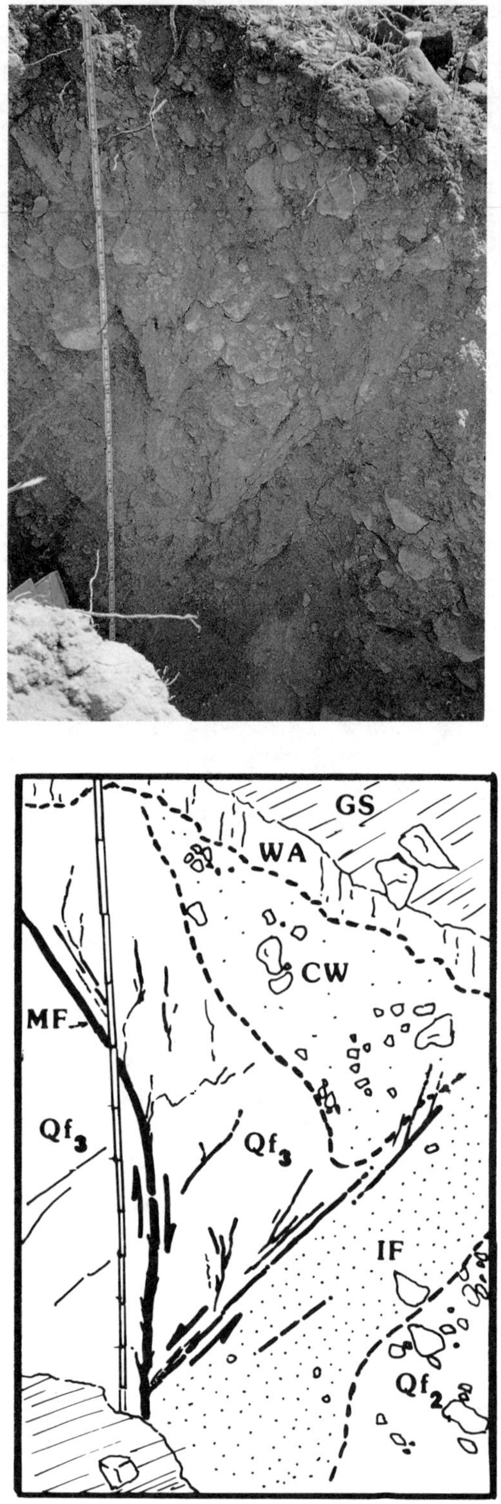

Figure 8. Photo and sketch of fault observed in Trench 8 showing offset alluvial units and possible offset of scarp derived colluvial wedge. Qf_3 = Unit 3 fanglomerate, Qf_2 = Unit 2 alluvium, CW = Colluvial wedge, WA = Weathered Alluvium below active geomorphic surface, IF = Loose, fine-grained infilling, MF = Main Fault Splay, GS = Ground Surface. Folding rule is 6 feet long.

Unit 2 Alluvial Fans

Unit 2 fan deposits form mid-level geomorphic surfaces surrounding the antennas on the west and east. Fan surfaces are moderately dissected with poorly to moderately developed desert pavements. Pavements display a relatively light colored patina. The material is finer grained than Unit 3, but still contains coarse gravel/cobble layers. The soils capping these surfaces are of moderately to strongly developed relict paleosols characterized by argillic horizons and multiple stage II to III calcic horizons. Remnants of moderately developed, near-surface buried soils are also present. Geomorphic and soil-stratigraphic evidence suggest that the Unit 2 fan complexes are at least 35,000 to 40,000 years old, but may be as old as 100,000 ybp.

Unit 1 Alluvial Fans

Unit 1 fans form the lowest level geomorphic surfaces in the Mars/Uranus Station valley. Fan surfaces are smooth, undissected, and represent areas of active deposition. Desert pavement, where present, is thin and patchy. The material consists of relatively fine-grained sands and gravels. Pedogenic profiles are generally absent to weakly developed. Where present, soils are characterized by cumulic profiles with incipient argillic horizons (Btjk) containing disseminated carbonate. Geomorphic and soil-stratigraphic evidence suggests the older portions of these fans are approximately 7,000 to 12,000 years old, but locally may be as old as 35,000 ybp. Recent stream channel and valley-fill sediments were also mapped as Unit 1. These deposits are inherently young, and at the scale of the mapping are difficult to separate from the older portions of the fans. These deposits are estimated to be less than about 5,000 to 7,000 years old.

Other Surficial Deposits

Playa Deposits

Playa deposits consist of unconsolidated, well-sorted clay, silt and fine sand. This material underlies the surface of the dry lake located west of the Uranus antenna. The morphostratigraphic relationship with Unit 1 fans is not entirely clear, but the two are probably intercalated at depth.

Landslides

One small landslide was mapped on the ridge south of the Uranus antenna. The slide was mapped based solely on geomorphic evidence, hence, its age and thickness are yet unknown.

FAULT ACTIVITY ANALYSIS

The origin of photolineaments 1, 2 and 3 were assessed by trenching. Faults were encountered in several trenches placed across the lineaments (Table 3).

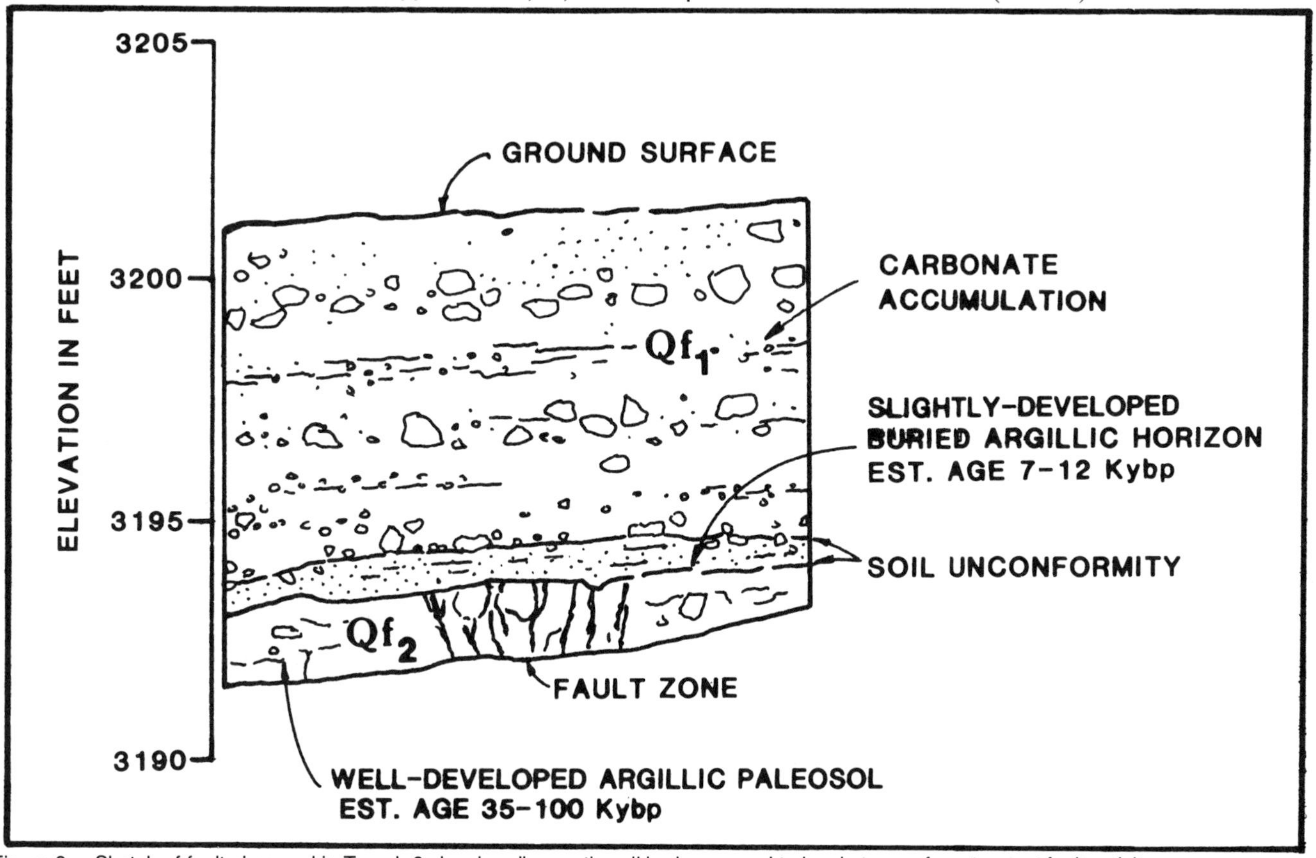

Figure 9. Sketch of fault observed in Trench 9 showing diagnostic soil horizons used to bracket age of most recent fault activity.

Figure 10. Fault exposed in Trench 12 that juxtaposes Unit 2 and Unit 3 alluvium. Note color change at fault: Unit 3 is light colored, Unit 2 is dark colored.

Photolineament 1

The three segments that together comprise Photolineament 1 were explored at trenches emplaced at ten locations within the Mars/Uranus Station valley (Figure 7). Nine of those trenches exposed faults that, near the surface, trend about N75E and comprise a through-going, north-stepping, en-echelon system. The faults form subdued scarps in fan Units 2 and 3, but are not surficially expressed in alluvial fan Unit 1 (Fig. 7).

The Photolineament 1 fault system is characterized by a zone of dislocation several feet wide. Movement has been typically accommodated along multiple slip surfaces and fractures. The slip surfaces generally dip steeply 60° to 90° to the south with the north side-up . Minimum offset was in each case greater than the maximum trench depth which varied from 6 to 10 feet. Further, some fault associated fractures extend into a colluvial wedge presumably deposited on a scarp formed by an earlier faulting event (Fig. 8). This fact, combined with a minimum offset observed of least 6 to 10 feet, suggests that multiple movements have occurred on the Photolineament 1 fault system. However, no site-specific information is presently available from which to judge the recurrence of individual tectonic events.

In Trench 11, the fault offsets a strongly developed buried argillic horizon in fan Unit 2 (Fig. 9). Based on relative profile development this soil required at least 35,000 to 40,000 years to form; however, it may be much older (Shlemon, 1989). Immediately above the fault at this location is a buried cumulic soil profile within fan Unit 1. The cumulic profile is characterized by a weakly developed argillic horizon (Btjk) with disseminated carbonate filaments. This soil is an estimated 7,000 to 12,000 years old (Shlemon, 1989) and is not offset by the Photolineament 1 fault. Accordingly, while not wholly definitive, the non-displaced Unit 1 soil horizon exposed in Trench 11, the lack of any fault-controlled geomorphic expression in Unit 1 fans, and the subdued scarps characteristic of the older fans suggest that last fault displacement most likely occurred in pre-Holocene time.

Photolineament 2

Photolineament 2 was trenched at two locations near the intersection with the fault that forms Photolineament 1 (Fig. 7). Both trenches exposed faults generally coincident with lineament trend. This is exemplified in Trench 6 by a N75E fault trend, and in Trench 5 where the faults trend about N80W. The Photolineament 2 fault has only minor surface expression in the Unit 3 fans and no expression in the Unit 1 fans. In Trench 5, the fault offsets Unit 2 alluvium. These sediments were judged to be in the range of 35,000 to 100,000 years old based on the presence of moderately developed soil profiles. Trench exposures were not sufficient to bracket the most recent offset or to determine the structural relationship between Faults 1 and 2; however, geomorphic evidence suggests Fault 1 truncates and therefore is younger than Fault 2.

Photolineament 3

Three trenches were excavated across the trace of Photolineament 3 (Fig. 7). Trench 12 exposed a fault which displaces Unit 2 and Unit 3 alluvium (Fig. 10). Trench 13, located east of Trench 12, exposed unbroken Unit 2 alluvium and exposed a buried paleosol estimated to be at least 35,000 to 45,000 years old. These fault/paleosol relationships then suggest that last displacement of the Photolineament 3 fault occurred between about 35,000 to 100,000 years ago.

Photolineament 4

Trench 3 was excavated across the trace of Photolineament 4 (Fig. 7). Unit 2 alluvium was exposed along the entire length of the trench. No faults were observed. It is possible that a fault exists below the maximum trenched depth. However, if a fault exists, the soil stratigraphic age assessments of the unbroken Unit 2 alluvium indicates that last displacement occurred at least 35,000 years ago and probably well before that time.

SUMMARY AND CONCLUSIONS

Multidisciplinary techniques were used to determine the presence and age of faults near the Mars/Uranus antenna sites at the Goldstone Deep Space Communications Complex. Photolineament analyses identified four linear trends suggestive of fault control. Site-specific trenching shows that Photolineaments 1, 2 and 3 are fault controlled, but that Photolineament 4 is not. The three, previously unreported fault systems are dated mainly by geomorphic and soil-stratigraphic techniques. Based on the presence of buried and relict paleosols, with relative development of argillic and calcic horizons varying from slightly to strongly developed, three morphostratigraphic alluvial fan units are identified: Unit 1, less than about 7,000 to 12,000 years old; Unit 2, approximately 35,000 to 40,000 years old; and Unit 3, estimated to be in excess of 100,000 years old.

Last displacement of Photolineament faults 1 and 3 occurred between about 12,000 to 35,000 years ago. Subsurface evidence for the existence of Fault 2, which corresponds with Photolineament 2, is less conclusive; however, when combined with geomorphic evidence, clearly suggests faulting. The structural relationship between Faults 1 and 2 is not known. The most recent displacement of Fault 2 is not known, but is believed to be older than Fault 1, based on geomorphic evidence. Review of Landsat and SLAR imagery indicated Faults 2 and 3 may be the northwest and southeast extensions of the same fault.

None of the surface traces of the three faults crosses under or projects into existing Mars/Uranus structures. Seismic risk assessment of local and regional faults utilizing statistical analysis of earthquake data records and computer fault models based on geologic evidence indicated the maximum credible earthquake (MCE) for the site is a magnitude 7.5 event on the Garlock fault located about 12 miles north of the study area. The new fault location and activity assessment will now permit design engineers to proceed with seismic retrofit designs and loss prevention measures for mitigation of potential earthquake damage to support facilities and structures.

ACKNOWLEDGEMENTS

We thank the Jet Propulsion Laboratory in Pasadena for the opportunity to investigate the study area and allowing us to publish this paper. We appreciate the valuable site geologic information, on-site liaison, and technical contributions provided by Tony Riewe of the Jet Propulsion Laboratory. We also appreciate critical reviews of this paper by Howard "Buzz" Spellman, Jr., Chief Geologist of Converse Consultants West, and Perry L. Ehlig, of California State University Los Angeles.

REFERENCES

Burke, D. B., et al., 1982, Cenozoic rocks in the Barstow Basin area of southern California - stratigraphic relations, radiometric ages and peleomagnetism: U.S. Geological Survey Professional Paper 1529-E; 16 p.

Byers, F. M., Jr., 1960, Geology of the Alvord Mountain Quadrangle, San Bernardino County, California: U.S. Geological Survey Professional Paper 1089-A, 71 p.

Christenson, G. E., and Purcell C., 1985, Correlation and age of Quaternary alluvial-fan sequences, Basin and Range Province, Southwestern United States: Geological Soc. of America, Special Paper 203, p. 115 - 122.

Converse Consultants West, 1989, Fault and seismic investigation, DSS-14 Mars Station, JPL/Goldstone Deep Space Communications Complex, San Bernardino, California, Unpublished Report Prepared for Jet Propulsion Laboratory, California Institute of Technology, Pasadena, CA, 28 p.

Dames and Moore, 1982, Geologic and seismic refraction study proposed antenna site Goldstone Tracking Station, Goldstone, California: Unpublished Report Prepared for Jet Propulsion Laboratory, California Institute of Technology, Pasadena, CA, 10 p.

Dibblee, T. W., Jr., 1967, Areal geology of the western Mojave Desert California: U.S. Geological Survey Professional Paper 522; 153 p.

Dibblee, T. W., Jr., 1968, Geology of the Freemont Peak and Opal Mountain Quadrangles, California: California Division of Mines and Geology Bulletin 188, Sacramento, CA, 64 p.

Dokka, R. K., and Travis, C. J., 1990, Late Cenozoic strike-slip faulting in the Mojave Desert, California: Tectonics, vol. 9, No. 2, p. 311 - 340.

Donald R. Warren Co., Engineers, 1961, Final soil study, foundation investigation, advanced antenna system, Goldstone Lake, California: Unpublished report prepared for Jet Propulsion Laboratory, California Institute of Technology, Pasadena, CA, Report No. F61-451.

Glass, C. E., and Slemmons, D. B., 1978, State-of-the-art for assessing earthquake hazards in the United States, in Report 11, Imagery in Earthquake Analysis: Waterways Experiment Station, U.S. Army Corp of Engineers, Vicksburg, MS, 221 p.

Hart, E. W., et al., 1987, Summary Report: Fault Evaluation Program, 1986-1987, Mojave Desert Region and Other Areas: California Division of Mines and Geology Open File Report 88-1; 40 p.

Higgins, C. T, and Streitz, R., 1988, Application of remote sensing to California's geology: California Geology, Sacramento, CA, vol. 41, No. 6, p 123-133.

Jennings, C. W., et al., 1975, Fault Map of California, Map Scale 1:750,000: California Division of Mines and Geology

Kunkel, F., and Riley, F. S., 1959, Geologic Reconnaissance and test-well drilling, Camp Irwin, California: U.S. Geological Survey Professional Paper 1460-F, 38 p.

Machette, M. N., 1978, Dating Quaternary faults in the southwestern United States by using buried calcic paleosols: Journal Research, U.S. Geological Survey, vol. 6, No. 3, p. 369-381.

Nelson, A. R., 1987, A facies model of colluvial sedimentation adjacent to a single-event normal-fault scarp, Basin and Range Province, Western United States: Proceedings of Conference XXXIX, Directions in Paleoseismology, U.S. Geological Survey, Open File Report 87-673, p. 136-145.

Shlemon, R. J., 1980, Quaternary soil-geomorphic relationship, southeastern Mojave Desert, California and Arizona: Geology and Mineral Wealth of the California Desert, Dibblee Volume, South Coast Geological Society, Santa Ana, CA, p. 388-402.

Shlemon, R. J., 1985, Application of soil-stratigraphic techniques to engineering geology: Bulletin Association of Engineering Geologists, vol. 22, p. 129-142.

Shlemon, R. J., 1989, Geomorphic and soil-stratigraphic assessment of fault activity, JPL-Goldstone Deep Space Communications Complex, Mars Station, San Bernardino County, California: Unpublished Appendix Report prepared for Converse Consultants Pasadena on behalf of Jet Propulsion Laboratory, Pasadena, CA, 6 p.

TABLE 1.

Aerial photography, Landsat and SLAR imagery reviewed.

Type	Date	Project	Scale	Roll/Frame	Source
Landsat, TM	07/28/85	—	1:155,000	50514-17515	JPL (EOSAT)
Side-looking Airborne Radar (SLAR)	12/85	TRPA	1:250,000	1/4-7	USGS
Low-altitude aerial photography					
Color Infrared	06/25/83	NHAP 83	1:58,000	89/122, 123 95/136, 137	USGS
Black and White	06/25/83	NHAP 83	1:80,000	62/23124 66/100, 101	USGS
Black and White	10/09/87	87156	1:18,000	1/8-10 2/8-10 3/11-13	JPL
Black and White	01/14/82	—	1:6,000	—/1-8	JPL

TABLE 2.

Characteristics of photolineaments.

Lineament	Length (miles)	Lineament Visible on	Geomorphic Evidence Suggestive of Faulting	Projects Close to Existing Structures	Remarks
1 (1, 1a, 1b)	>3	SLAR, Air Photos	Strong	Yes	Strong lineament defined by a combination of features including obvious topographic alignments linear scarps on alluvial fans, rectilinear depressions and linear changes in erosional texture.
2	>10 (regional)	Landsat, SLAR, Air Photos	Moderate	Yes	Regional lineament visible on high altitude imagery, locally defined by linear changes in drainage/erosional texture and photographic tone.
3	>10 (regional)	Landsat, SLAR, Air Photos	Moderate	Yes	Regional lineament visible on high altitude imagery, probably represents southeast extension of lineament 2.
4	0.4 ±	Air Photos	Weak	Yes	Local lineament (<2,000 feet long) defined by a linear break in slope and an apparently offset stream channel.

TABLE 3.

Summary of fault trenching.

PHOTO LINEAMENT	IDENTIFIED AS A FAULT DURING EXPLORATION	ESTIMATED MINIMUM AGE OF MOST RECENT ACTIVITY (years before present)	REMARKS
1	Yes	~12,000	Through-going fault system dated by geomorphic and soil-stratigraphic methods. Fault trends roughly N75E and is characterized by north-stepping segments up to 1.5 miles long. Minimum offset is 6 to 10 feet, north side up. Some evidence of multiple events.
2	Yes	>12,000	Coincident with regional lineament identified on Landstat imagery. Trenching and geomorphic evidence suggest fault origin. Geomorphically it appears to be truncated by fault 1. However, this was not demonstrated during exploration. The structural relationship with Fault 1 and the approximate age of this fault are yet unclear.
3	Yes	35,000	Fault - coincident with regional lineament identified on Landstat imagery. Dated by geomorphic and soil-stratigraphic methods. Probably forms southeast extension of Lineament 2.
4	No	—	Trenching across the trace of this lineament exposed unbroken 35,000 to 100,000-year-old deposits. Photolineament 4 is not fault related.

Chapter **4**

Legal Aspects

Photo of Bluebird Canyon landslide of 1983, which resulted in multiple lawsuits by the homeowners, and expert testimony by geologists and engineers.

Legal Aspects

Editors Note: Possibly more lawsuits are filed against consulting geologists and engineers in California than in any other region. We suspect it has less to do with the geologists' abilities than with the fact that there are more trial lawyers earning a living here. The first paper is by a seasoned consulting geologist who speaks from personal experience about the pitfalls of consulting resulting in lawsuits. The second paper is an in-depth explanation of the legal system written by an attorney who was also a practicing geologist. The protection and guidelines provided to the practitioner through building codes is reviewed by Mike Scullin. The chapter concludes with a useful checklist for visual inspections for homes in escrow, and a legal bibliography for the professional.

GEOCALAMITIES — THE DO'S AND DON'TS OF CONSULTING IN SOUTHERN CALIFORNIA

By John W. Byer
The J. Byer Group
444 E. Broadway, Suite A
Glendale, CA 91205

CONTENTS

INTRODUCTION

The state of affairs in the geotechnical service industry have deteriorated to the point where my attorney has advised me not to write this article. The fact is, that we are in business makes us a target for liability lawsuits. Many of these lawsuits do not have much substance as to our professional performance. Nonetheless, we all make mistakes and, in legal terms, that makes us negligent. It is the avoidance of mistakes that I would like to discuss here.

I began my career in my college days at the University of California at Santa Barbara taking a geology major. In order to pay expenses, I worked as a soils engineering technician in a laboratory. Here I learned basic soils engineering and the variety of laboratory testing procedures used by soils engineers. The lab also provided materials testing capability, including concrete and steel. Graduation in 1966 was a mild awakening, as employment for geologists was almost non-existent. Traditionally, the two largest employers of graduate geologists are the oil companies and resource companies. At the suggestion of my professor, I turned my attention to the consulting geology business. At that time the field was relatively new and I was advised that an advanced degree would be helpful in obtaining employment. During my graduate studies at USC, I worked as a staff geologist for the Army Corps of Engineers and as the City Geologist for the City of Glendale. After a short stint with a private consulting company, our firm was established and opened for business in 1971. Since 1971 the company has prepared geotechnical reports for over 14,000 projects within the Los Angeles area.

During my college years the term "liability" was not discussed. The possibility that someone could sue me for professional advice was not considered. The purpose of this paper is to identify those mistakes so that others might learn from our experience.

Over the past 20 year,s I have noticed some trends with respect to the geotechnical profession. Registration of geologists, engineering geologists, and geophysicists at the State level eliminate the need for local registration boards. State registration essentially insures that the person has the proper educational background and experience to qualify for his chosen license. Fees charged by geologists remain relatively low with respect to the amount of liability incurred and the value of the finished project. Competition among firms for projects leads to price cutting and minimal exploration budgets. Requesting extra fees because of discovered geologic hazards is common during construction. More and more architects, engineers, lending institutions, and agencies are insisting upon geotechnical studies. They see this as a way of helping to spread liability. Geologic interpretations vary among geologists, especially in areas of prehistoric landslides, rockfalls and debris flows.

In preparing an outline for this article, I decided to break a project down into its various component parts and discuss each part individually. The examples given are from my experience in the Los Angeles area and may not apply to other areas. The components of a project include 1) the land, 2) the client, 3) the Agreement contract, 4) the proposed project, 5) the geologist, 6) the local agency, 7) the intervener, 8) the developer, 9) builder/contractor, 10) the purchaser, and 11) the attorneys.

THE LAND

Engineering geologists and geotechnical engineers would not have assignments if the land were not defective. Southern California has had a history of earth movements and land failures due to geologic defects in the terrain. The land has had a complicated geologic history associated with an emergent coast. Rejuvenation of the mountainous terrain has resulted in rapid downcutting to form steep-walled canyons. Historic and prehistoric landslides are common as these canyons become undercut. Erosion, deposition and ground modification by man has helped to hide these areas.

Renewed movement of the Portuguese Bend landslide in the 1950s alerted the community to the necessity of in-depth geologic studies. Geologists are hired to determine the geologic stability and make-up of the land being considered for a project. Discovery of concealed geologic hazards is not an easy task. The method of exploration used by the geologists is very important with respect to his ability to recognize geologically disturbed terrain. Common types of sub-surface exploration include bulldozer, backhoe, and bucket auger drill rig. The purpose of this equipment is to provide exposures for the geologist to study. These exposures are not available through any other method other than actually digging into the terrain to be studied. Shallow explorations using bulldozers or backhoes are effective where the stable geologic materials are close to the surface. These methods, however, are unsatisfactory where the surface is covered with disturbed geologic materials. Downhole inspection in a bucket auger hole provides the geologist with a much deeper view into the terrain. This is a common practice in Southern California and should be used more in other parts of the United States.

In spite of these exploration techniques, past ground failures and landslide terrain have not been recognized by many geologists working in southern California. For example, there is an area north of Beverly Hills where steep natural slopes consist of large fragments of bedrock mixed with layers of soil and gravel. These materials have been mis-identified by numerous geologists as in-place bedrock. As a consequence, new structures have been built on this material with only conventional foundations, and are a disaster waiting to happen.

Sites can have geologic defects even though these defects are not readily apparent, recognizable, or even discoverable. Development of a site with a hidden geologic defect can lead to exposure of the defect over time. This happened at Portuguese Bend and Big Rock Mesa (see Chapter 8.).

THE CLIENT

The client involved in a project can affect your liability on projects. Several categories of clients are described here with respect to their point of view and the finished product. There is a saying that professionals should not work for speculators. I believe that everyone that develops real estate is in one way or another a speculator. Therefore, we are dealing in a matter of degrees.

The low budget speculator usually deals in marginal property. He is also reluctant to secure the services of surveyors for topographic maps and architects and structural engineers for plans. He will hire a geotechnical consultant only upon the insistence of a regulatory agency. He purchases cheap lots which are difficult to develop and negotiates the lowest possible fee from his design and construction companies. This client will tend to cut corners wherever possible, which can lead to future problems.

The high budget speculator builds in prestigious areas like Malibu, Beverly Hills, Bel Air, West Los Angeles, Hollywood and Encino. This client also tends to cut corners on foundations, walls, and construction and puts the extra money in gingerbread. As a consequence, he sells a very expensive, but cheaply constructed project. These homes are purchased by wealthy, sophisticated people who believe they are buying a quality product. When the defective workmanship begins to show up, all parties involved in the project, whether guilty or not, are threatened with legal action.

Remodelers are very common in the hillside areas of Los Angeles. Many clients remodel property as a hobby or second form of income. Remodelers purchase older homes which were constructed prior to the modern grading and building codes. These older homes usually have non-conforming conditions, including steep slopes, uncompacted fill, and shallow foundations. The remodeler usually requires code variances for his permits and is reluctant to bring the entire property up to the modern building and grading codes.

Mass Builders: Many geotechnical firms derive substantial fees from clients who mass produce homes. These companies are generally large, nationwide companies which build thousands of units a year. Mass builders have an organization which includes the "project manager". Project managers are responsible for the nuts and bolts operation of the project, from the civil engineering design to the earthwork, completion of developed lots and construction of the homes. Project managers are compensated through salary and certain incentive bonuses with respect to the success of the project. Mass builders generally select professional services on the basis of fee biding. Low bidders are forced to cut their exploration budget in order to secure the project, hoping that fees generated during grading will be great enough to cover the less than adequate preliminary exploration.

The personal homebuilder: Clients constructing their own projects, be it their home or other building, are more conservative when it comes to design professionals and construction of the project. These clients willingly hire qualified engineers and design professionals to assist them with their project. They insist on quality work and service and are only disappointed when they don't get what they ask for. These clients tend to accept the recommendations of their team and implement them during construction. These projects have a very low risk of failure or problems. I have customers constructing their own custom homes ask me if our foundation recommendations were strong enough and would it be okay if they put the foundations a little deeper into the bearing material. This professional attitude permeates the entire project.

THE AGREEMENT-CONTRACT

One component of the project that should not be overlooked is that formal agreement between the client and the geotechnical firm. A properly prepared agreement can help limit your liability by describing the terms and conditions of the assignment. A proposal should contain a letter describing the scope of employment, fees, time of delivery, and a paragraph of instructions on how to authorize the project. Combined with this letter should be an agreement to be signed by the owner of the property. Our company's agreement contains 17 terms and conditions, including a limitation of liability paragraph. The agreement also contains an indemnity paragraph which helps insulate the geotechnical consultant from "third party lawsuits". Work should not proceed on any geotechnical assignment with out a signed agreement. The agreement protects the geotechnical firm and is valuable when disputes arise over fees or scope of services. Changes in the scope of the work often occur during exploration of the property. Often additional exploration is required to fully characterize the site. It is important that these changes in scope and fees be documented in writing and agreed to by both parties through signature of amended documents. (Typical Agreement appended hereto.)

THE PROPOSED PROJECT

A proposed project will eventually be described by a set of plans. Plans evolve from the developer's desires and wishes being communicated to the design professionals, the architect, structural and civil engineer. Often, the geologist is asked for input prior to purchase of a property and prior to completion of the plans. The geologist is asked to offer his opinion on the suitability of the site for construction of the desired plan. The geologist making such an opinion exposes himself to tremendous liability. He should not be in the position of deciding the future of a project without adequate subsurface exploration. The initial phase of the design of any project should be a detailed and accurate topographic survey of the site. Poorly prepared topographic maps can lead to serious problems during construction. The most common problem is that designed retaining walls are generally shorter than required to support an excavation. Other problems include a project which does not fit the site boundaries, easements which were not known prior to purchase, and slopes which are steeper than shown on the topographic map. In the City of Los Angeles, hillside projects are subject to a height limitation. Several project have been abandoned because the height of the building exceeded the height limitation. The elevations on the plans, however, showed the proposed projects to conform with the height limitations.

Geotechnical information concerning the project is then submitted to the client, the structural engineer, and the civil engineer. Structural engineers rely on the geotechnical data to design foundations and retaining walls for the project. The civil engineer uses the topographic survey and the plan to design the grading for the project. The civil engineer determines the finished grades, heights of retaining walls, and location of embankments and cut slopes.

THE GEOLOGIST

Armed with his plan, the developer seeks the services of a geologist and soils engineer to determine what geotechnical factors need to be considered in development of the project. The geologist is asked to prepare a written proposal, outlining the scope of services and fees. Many prospective clients will seek bids from several geotechnical firms. Competitive bidding for professional services lead to price cutting. Often the prospective client only looks at the bottom line number and does not determine the scope of the services offered. It is up to the geologist to point out why his proposal is superior with respect to scope of exploration. There is a great difference of opinion among geologists as to what constitutes adequate exploration for any one project. Geologists with experience in a certain area will be aware of potential problems and prepare a proposal according to the anticipated problems. Geologists not familiar with an area can underestimate the exploration necessary to characterize the site. If you are asked to work outside of your normal area, it is advisable to hire a local geologist as a consultant for your project. The local consultants are knowledgeable about the area and can provide important advice on how to proceed with a project.

Most geologists are trained by universities to map geologic structures and identify rock types. Many of the Southern California universities send their students to a desert area where exposures are numerous. They are trained to make detailed geologic maps based upon natural outcrops available. Geologists are also trained in the formation of surficial deposits, including alluvium, talus, soil, and possibly landslide debris. However, geologists are not trained to recognize these surficial deposits from exposures revealed by subsurface exploration. It has been my experience that many large debris-flow deposits have been mis-identified by geologists. In the Malibu area, deposits of debris-flow material contain large boulders of hard sandstone. The surrounding soft soil has been eroded away, leaving the sandstone boulders looking like in-place outcrops.

THE AGENCY

Geologic reports submitted as part of a project are reviewed by the City, County, or State Government that has jurisdiction of the project. Some review agencies, such as the City of Los Angeles, have geologists and engineers on staff for review of reports. Some agencies process the reports through the Building Department where the reports are not actually reviewed. The agency can play an important role in protecting the geologist by providing enforcement of recommendations contained in the report. By requiring the geologist to review building plans before issuing the building permit, the

AGREEMENT

This AGREEMENT is between:

____(Client)

____

(Telephone)________________________and_______________________________

____.

1. Services are limited to providing professional counsel and professional advice regarding subsurface conditions on the job site pursuant to the terms of this AGREEMENT, as follows:

____

____

____

2. Job site: ____

____ Thomas Guide:____________________

Legal Description:___________________________________

3. Fee and Payment:
3.1 Fee for the above services is $________________________________
3.2 Fee shall be paid as follows and shall not be subject to financing or any other contingency:
(a) A retainer fee in the sum of $______________must be paid upon execution of this Agreement before any services are rendered. (b) The balance of the fee shall be paid in full upon rendition of our report at their office.

4. Additional Fees and Payment:
4.1 The above fee does not include fees for services other than set forth in Section 1 of this Agreement. Additional fees for consultation, excavation observations, or site visits during construction will be charged per the attached Schedule of Charges. A new Schedule of Charges is published on the first working day of each year and shall be binding upon the parties.
4.2 Additional fees are due and payable upon presentation of invoice.

5. Submission Not Offer:
Neither the submission of this Agreement nor any quotation shall be considered binding until and unless this Agreement is executed by both parties and any required retainer fee paid.

6. Execution:
This Agreement must be executed by the actual or prospective owner of the Job Site. THE FOLLOWING TERMS AND CONDITIONS APPEARING AND THE ATTACHED SCHEDULE OF CHARGES ARE EXPRESSLY INCORPORATED INTO THIS AGREEMENT AND ARE PART HEREOF.

1. Use of Report:
All reports and work remain our property until all of our fees have been paid in full. Until all fees are paid, such reports and work will not be used for licensing, permits, design and/or construction.

2. Client Information and Responsibility:
2.1 Client will make available all information regarding existing and proposed conditions of the site including but not limited to, plot plans, topographic surveys, hydrographic data, and previous soil data with borings, field or laboratory tests and written reports. Client will immediately transmit any new information which becomes available or any change in plans and shall indicate all property lines, underground utilities and installations and is responsible for the accuracy thereof.
2.2 Client will indemnify against any inaccurate or misleading information furnished by Client.
2.3 Client shall keep the site free and clear of children, animals, persons or property (including all landscaping, shrubs, and flowers) which are likely to or do interfere with access to site or its ability to perform under this Agreement and shall provide free access to site and a safe place to store its equipment and perform its services.
2.4 Client shall provide a representative at the site to supervise and coordinate the job upon 24-hour request by XYZ.

3. Remedies: The following rights and remedies shall be cumulative.
3.1 The right to stop work if any payment shall not be made when due and may keep the project idle until all payments due are received.

3.2 Nothing contained in this Agreement shall preclude from the filing, recordation or foreclosure of any mechanics' lien, stop notice, or similar procedure.

4. Test Holes:
All test holes will be backfilled upon completion of the job. However, if Client requests test holes to remain open after completion of work, Client agrees to pay for all costs in connection with covering and backfilling said test holes at a later date and shall indemnify for all claims, demands and/or liabilities arising therefrom.

5. Limitation of Liability:
5.1 For any injury or damages to persons or property on account of any error, omission, or other professional negligence, liability to Client and to any partner, principal, agent, employee, contractor, subcontractor, or other person who performs work on site and to any person who obtains title to or possession of site or any part thereof (or those claiming under or through any of the foregoing) shall be such that the total aggregate liability to all those named will not exceed fee.
5.2 We shall have the right to post these terms and conditions on site and to notify anyone of said terms and conditions.
5.3 No action or claim arising out of or connected with site or any part thereof or the services to be rendered hereunder may be brought against XYZ by Client, or by any partner, principal, agent, employee, contractor, subcontractor, or other person who performs work on site or obtains title to or possession of site, or any part thereof, or those claiming under or through any of the foregoing, unless such action is commenced within three (3) years from the date of final report to Client.

6. Disclaimer of Warranties:
No warranty, express or implied, of any kind is made or intended in connection with work actually performed or to be performed under this Agreement, or by the fact that XYZ is furnishing oral or written reports or findings.

7. Indemnity:
Client will, during and subsequent to completion by XYZ of the services described therein, hold and save XYZ free and harmless from and indemnify and keep XYZ indemnified against, any and all loss, design defect, damage to property (including, but not limited to, all underground utilities), injury to or death of person, liability, obligation, claim, demand, action, causes of action, cost and expense, including, but not limited to, reasonable attorney fees incurred by XYZ as a result of or arising from any, a combination of or all of the following: (a) Client failing to properly perform any provision of this Agreement required to be performed by Client; (b) XYZ being made a party to any litigation or arbitration proceeding relating to the job described in this Agreement; (c) the rendering by XYZ, or by any agent or contractor of XYZ of services hereunder, regardless if such services are negligently rendered or performed and regardless if such negligence is active or passive; (d) any act or event against which Client is to indemnify XYZ hereunder.

8. Change in Condition or Services:
8.1 Fluctuations in the level of groundwater may occur due to variations in rainfall, temperature, and other factors not evident at the time measurements are made and reported. The level of groundwater may also vary across the site.
8.2 Subsurface conditions, excavation characteristics, and geologic structure will be projected from individual borings or test holes placed on the site. This should not be construed to reflect any variation of subsurface conditions which may occur between the borings or test holes.
8.3 Should any changes be made in the design or location of structures on the site, any report and findings should not be considered valid unless the changes are first reviewed and approved by XYZ.
8.4 XYZ shall not be responsible for injury or damage to the site or any improvements now or hereafter located thereon including any structure, lawn, shrubs, landscaping, walkways, sprinklers, or underground utilities or installations.
8.5 When any or all of XYZ's services are to be on an intermittent or "on call" basis, XYZ will be unable to provide an opinion concerning any such work unless XYZ's field representatives have timely and sufficient opportunities for testing and observation during construction.
8.6 XYZ's services do not include supervision, or coordination or scheduling of work or a review or evaluation of the adequacy of any contractor's safety measures in or near the site.

9. Validity:
Should any provision of this Agreement be held invalid or in violation of any law or regulation, any such provision shall be deemed stricken from this Agreement and shall not affect the validity of the remaining provisions of this Agreement.

10. Merger Clause:
This Agreement, including any Schedule, constitutes a final, complete and total integration of any and all understandings between the parties. Any subsequent change, alteration, addition or modification must be mutually agreed upon, in writing, and signed by both parties.

11. Individual Responsibility:
If Client is a corporation or partnership, the individual(s) signing this Agreement on behalf of Client guarantee that Client will perform its duties hereunder and warrant each is duly authorized to sign this Agreement on behalf of Client.

12. Assignment:
The rights and obligations of Client shall not be assigned or delegated without the prior written consent of XYZ. Subject to the foregoing, each party to this Agreement binds itself, its successors, executors, administrators, and assigns in respect to all covenants and promises of this Agreement._

13. Termination of Agreement:
13.1 In the event either party desires to terminate this Agreement prior to completion of the project, written notification of such intention to terminate must be tendered to the other party and XYZ shall be paid for all services performed to the termination date. In the event that Client notifies XYZ of such intention to terminate XYZ's services, prior to completion, XYZ reserves the right to complete pending analysis, dispose of samples, put equipment in order, and (where considered necessary) to complete a report on the work performed to date. In the event of Client's termination of this Agreement, a termination charge to cover XYZ's cost shall be paid by Client to XYZ, in addition to any other fees or charges.
13.2 If site or any part thereof is destroyed by earthquake or other catastrophe, XYZ will be paid for all work performed theretofore, and both parties will be relieved of any further obligations hereunder.
13.3 If either party should become bankrupt or make an assignment for the benefit of creditors, the obligations of all parties hereunder shall thereupon terminate.

14. Force Majeure:
XYZ shall have no liability for any delay or for any act or event caused by acts of God, Client or Client's agent, weather, labor trouble, acts of public utilities, bodies or officials, acts of contractors or subcontractors, or other contingencies beyond the reasonable control of XYZ.

15. Agreement Solely for Benefit of Parties:
This Agreement is entered into solely for the benefit of Client and XYZ and is in no way intended to benefit or extend any right or interest to any third party. It is the intention of the parties that any benefit derived by any third party is merely incidental to and unintended by this Agreement.

16. Notices:
Any notice required or permitted hereunder shall be given by registered or certified mail, return receipt requested, addressed as indicated, but such address may be changed by written notice given by one party to the other from time to time. Notice shall be deemed received twenty-four (24) hours after deposit in said mail.

17. Miscellaneous Provisions:
17.1 The paragraph or section headings are for purposes of convenience only and are not an integral part of this Agreement.
17.2 This Agreement shall be deemed executed at ____________________, California when signed by an officer of XYZ.

_______________________________ _______________________________

Officer of XYZ Date Client Date

geologist is able to determine whether or not these plans comply with his recommendations. Once the project is under construction, the building department can require the geologist to observe the approve foundation excavations prior to their completion. On some projects a final geologic report is required where the geologist describes the completed project. This type of control from plan check through completion allows the geologist to be involved in the design and construction phases of the project.

Problems arise when the reviewing agency does not allow the geologist to be involved in the review and construction of the project. Many agencies rely solely on the building department to oversee hillside construction. These building officials are not trained in geology, engineering geology, or geotechnical engineering and are not trained to recognize potential geologic hazards. For instance, a geologist may have recognized a hazard, say a landslide, and recommended its stabilization or removal during construction. Unless the agency inspector is aware of this problem, the problem will go unrepaired.

Agency review can be an important step in the process of designing and completing a project. However, there are no standards for reviewers to refer to while performing their work. It is my opinion that many problems could be avoided if reviewers had guidelines for performing their tasks. These guidelines should include a review of the previous reports for an area for which the agency probably has numerous reports on file. Review of the previous geologic reports could reveal geologic conditions differing from those described in the report to be reviewed. A case history comes to mind where a hillside home began slipping following a period of heavy rain. The home was removed from its foundation and a pile and grade beam foundation was constructed on the property. The home was then moved onto its new foundation. Many years later a developer submitted geotechnical reports for construction below this property. The plan called for a cut slope to be excavated to provide room for an access roadway. After completion of the project, the homes above the cut were involved in a slope failure in the same areas as the previous slope failure. Research of the files available to the reviewer would have revealed the fact that the slope above the proposed project had suffered damage in the past.

The reviewer should also determine whether or not the reports before him are based upon adequate exploration. In 1963 the City of Los Angeles Board of Building and Safety Commissioners adopted Resolution No. 636. This resolution outlines hillside subsurface exploratory work and is included here:

CITY OF LOS ANGELES
BOARD OF BUILDING AND SAFETY COMMISSIONERS

RESOLUTION NO. 636
(Adopted April 25, 1963)

SUBJECT:Hillside Subsurface Exploratory Work.

WHEREAS, an Ordinance 123,970 was recently enacted by the City of Los Angeles setting forth new requirements and regulations concerning grading within the City; and

WHEREAS, subsurface exploration will often be necessary as a result of the new requirements set forth in the aforementioned Ordinance; and

WHEREAS, it becomes necessary to outline for professionals the information and studies that should be reported to the Department Staff in conjunction with such subsurface exploratory work; Now, therefore, be it

RESOLVED, that the Board of Building and Safety Commissioners do hereby adopt the following rules concerning Hillside Subsurface Exploratory Work:

Subsurface exploratory work shall be performed by a soils engineer and engineering geologist on hillside grading work when required. Subsurface work shall be performed for the purpose of securing information on which the soils engineer and the engineering geologists shall base recommendations for a grading project. The work shall be conducted at locations most likely to reveal any subsurface weaknesses which may lead to landslide or settlement failures. Particularly, extensive investigation shall be conducted where the stability will be lessened by the grading or where any of the following conditions are discovered or proposed:

1. At fault zones where past land movement is evidenced by the presence of fault gouge.

2. At contact zones between two or more geologic formations.

3. At zones of trapped water or high water table quite often associated with conditions 1 and 2 above.

4. At bodies of intrusive material.

5. At historic landslides or where the topography is indicative of prehistoric landslides.

6.At adversely sloped bedding planes, short range folding, overturned folds, etc.

7. At locations where a fill slope is to be placed above a cut slope.

8. At proposed cuts exceeding 50' in height, unless in extremely competent rock, or of lesser heights in rock of questionable stability.

9. At locations of proposed fills exceeding 50' in height.

10. Where side hill fills on slopes steeper than 5:1 are proposed.

11. Wherever water from rainfall or irrigation or otherp robable sources from both the grading project and adjoining properties is likely to reduce the subsurface stability to a dangerous degree.

Where any of the particular problem zones listed above or other weaknesses are found; subsurface investigation sufficient to thoroughly outline the problem shall be make. The soils engineer and engineering geologist shall submit a written report to the Department of their findings. Wherever feasible, the soils engineer and engineering geologist shall make exploration pits and/or borings of sufficient size to study the subsurface conditions including bedrock. If there are natural exposures on the property which reveal the geologic properties of the underlying bedrock adequately for the purpose of the geologic report, this information may be used in lieu of subsurface investigation. Other aids in exploratory work may be used but subsurface exploratory work sufficient to support the finding shall be performed. The depth of exploration shall be adequate to reveal weaknesses in the soil or underlying bedrock which will affect the proposed grading.

I believe the most important provision of this Resolution is that subsurface openings should be of sufficient size for the geologist to observe the geologic character of the earth materials. This means that the geologist can be lowered into the pits or borings and make direct observations of the earth materials. The Resolution also calls for consultants to perform adequate exploration to support the conclusions and recommendations of their reports.

THE INTERVENER

During the review process, it is possible that an interviewer will question the merits of the project. An interviewer is usually hired by a group opposed to a certain project. The intervener is paid to find defects in the geotechnical reports so that a project can be abandoned, delayed, or made to be more costly. Interveners do not necessarily have to perform exploration or provide new information. In many cases, they need only to send the Building Department a letter indicating that they are questioning the results of the previous studies. The most common complaints of interveners are that the explorations or borings were not deep enough to discover concealed geologic hazards. Interveners also criticize soils engineering laboratory test data by saying that the most critical conditions were not analyzed. By forcing the client to respond to the intervener's questions, the opposition to a project hopes to delay the project and discourage the client from completing it. Interveners can cause friction between the geologist and his client. I have had clients question our professional ability where interveners have created problems. Clients are reluctant to pay for securing the additional information to satisfy the agency and intervener. Some interveners are geologists with honest differences of opinion and an honest objection to a project; some are hired guns who will accept their fee for opposing almost any project.

THE CONTRACTOR

Once the project is finally approved and has cleared all the roadblocks and agency review, the approved plans are sent to contractors for bids. Competitive bidding is designed to get a project constructed for the smallest possible cost. The bidding process results in very tight budgets with very little margin for error. It is in the contractor's best interest to find errors in the plans or geotechnical reports so that he can claim extra work from his customer. "Extras" are a way of life in the construction industry and can lead to disputes between the contractor, geologist, engineer and owner. Probably the most common cause of extras on grading projects is lack of knowledge about depths of removal and recompaction of unsuitable materials. "Removals", as they are commonly referred to in the grading industry, are difficult to define and are usually of much greater volume than anticipated. Discovery of concealed geologic hazards during grading is a large cause of extras. Unrecognized landslide debris or adversely oriented geologic formations require stabilization during grading, leading to huge extras. Developers have come to expect extras during site preparation for projects. Extras occur even with customary geotechnical exploration and reports prepared prior to construction. Developers feel that paying for huge exploration budgets to identify defects in their land is not justified at the beginning stages of a project. They also feel that discovery of these hazards during construction will occur and that remedial work will be done. Therefore, why spend time and money discovering these defects prior to grading? An example: A developer brought preliminary reports and plans to our office for review on a large hillside project designed for residences. The previous reports identified a landslide in the central portion of the project on the basis of one shallow boring. I explained to the developer that I felt that the landslide area was considerably larger than the two reports described, which could lead to huge removals during grading. I requested authorization to drill additional borings to determine the limits of the problem area. In response to my request, the developer told me that we know the landslide was there, we know that the landslide had to be removed, we know that we were going to remove the landslide, and therefore, why should he spend money determining its extent. Removal of the landslide during grading resulted in over 500,000 cubic yards of extra grading. The large extent of the landslide also resulted in revising the grading plans and extending cut and fill slopes well beyond the original plans. The developer was upset about the extras and questioned us on why we had not told him about the magnitude of the problem.

THE PURCHASER

After all of the dust has settled, the project is completed and offered for sale. People who purchase these projects are generally wealthy and expect the final project to be perfect. They are unaware of the construction process which led to the completion of the project. Hidden defects can be present which affect the overall appearance and performance of the project. Most of these construction defects are not related to foundations or geologic conditions, but are the result of construction defects. Nevertheless, the appearance of cracking leads the purchaser to suspect geotechnical problems with his newly purchased project. Apparent distress to buildings can be caused by numerous factors which occur during the building process. For instance, slabs can crack and look distressed through the normal curing process and shrinkage which takes place as the concrete cures. Exterior stucco can crack when applied too wet or when applied too soon after the initial coats are placed. Interior cracks can occur around wall openings as the result of shrinkage of the framing lumber. Larger wood members tend to shrink and twist, causing cracks within wall, above doors, windows, and large openings. The majority of these defects are cosmetic in nature, but have the appearance of structural damage.

THE ATTORNEYS

It is common practice in California for lawsuits to develop over projects which have apparent defects. Attorneys name all parties involved with the project ("100 John Does"), thereby assuming that they will eventually find the guilty party. The geotechnical firm is routinely named in these lawsuits. Attorneys in California are required to obtain an opinion from a professional in the geotechnical field that the proposed litigation has merit. There are members of the geotechnical profession who derive their living by providing these expert testimonials against members of their own profession. The end result of all of this is a long, drawn-out process in which the consultant must defend himself and, eventually, offer a settlement to be dropped from the case. Anyone caught in this poker game should be advised that you should not take your defense to court, no matter how convinced you are that you are right. Remember that attorneys make their reputation by winning cases, not by meting out justice.

CONCLUSIONS

The following is a list of do's and don'ts with respect to performing professional geotechnical services.

Do's

- Do get a signed contract for every professional engagement.
- Do spell out the scope of services to be performed in a formal proposal letter.
- Do insist on a detailed topographic survey of the project and a preliminary plan.
- Do adequate exploration for the scope of the project.
- Do insist on reviewing final plans before permits are issued.
- Do insist on being involved during the construction phase.

Don'ts

- Don't do projects in unfamiliar territory.
- Don't reduce the scope of your exploration to secure a project.
- Don't let the client dictate the scope of your professional work.
- Do not compromise on recommendations and do not change recommendations after the formal report has been issued.
- Don't trade services for a partnership in the project.
- Don't accept conclusions and recommendations of previous geotechnical studies without independent verification of the data presented.
- Don't take over projects where the prospective client is in a dispute with his previous geotechnical firm.

THE NUTS AND BOLTS OF LITIGATION

James H. Patton, Jr.
Attorney and Engineering Geologist
P. O. Box 635
Agoura Hills, CA 91376

CONTENTS

Defendant: *As God is my judge, I didn't do it. I'm not guilty.*

Judge: *He isn't, I am. You did. You are.*

ADMONITION

This offering is intended to be an overview to assist in the understanding of some of the rules which govern your day to day conduct as a professional. It is not intended to be a shortcut to a legal education or the substitute for an attorney. It is based primarily upon California law, but the principles may be generally applicable in other jurisdictions. A particular factual situation may be controlled by statutory or case law which is beyond the scope of this work. If a problem is encountered which is believed to involve legal principles or legal consequences, the advice of an attorney should be sought at the earliest opportunity. It is often said that the person who represents himself has a fool for a client. (The use of the masculine form of a word herein is intended to include feminine form.)

I. INTRODUCTION

In other English speaking countries, a litigious person is said to have *"The American Disease"*. We live in an age and within an area where litigation is looked upon not only as a means to settle disputes, but as a financial enterprise within itself. Many low-bidding contractors plan to use alleged disputes over "extras" and "design errors", in order to turn a marginal bid into a financial windfall. Many individuals seek the opinions of experts not so much for their technical assistance and enlightenment, but as a means of obtaining cheap "insurance." It is comforting to live in a society where, if anything goes wrong, there is someone to whom blame may be passed.

Expediency Rather Than Justice

It is a common misconception that the judicial system is a system of justice. It is actually a system for the resolution of disputes which serves expediency and the ends of the system. Justice is to be served, if at all, in the next life. The judicial system can be thought of as the social lubricant which keeps the machinery of society operating with a minimum of friction, in order that we all can continue to devote our energies to tax generating endeavors. The compromise which the system makes with justice is called *"due process of law"*. It means, at least in theory, that disputes will be resolved according to the established rules and principles for the enforcement and protection of private rights. What those rules and principles are and which interests they serve are determined by the legislature and courts. Legislators are, of course, politicians, and judges are either appointed by politicians or elected. The rules and principles are, therefore, usually the results of the balancing of conflicting interests, altruism diluted by compromise.

Judicial Philosophy

Stare decisis, the policy of courts to adhere to precedent, is intended to give stability to the law and to society in general. Under the doctrine, principles applied by a court to a certain factual situation are applied to all future cases where facts are substantially the same. The old Russian proverb that *"the law is like a sledge, a clever judge can steer it either way"* is, unfortunately, equally applicable to our judicial system. With the passage of time, exceptions to principles are stated, and then exceptions to those exceptions, etc., etc. The conservative judicial philosophy tends to resist change in the law, considering the principles to be like bench marks by which members of society can predict the consequences of their actions in advance and thereby somewhat control their destinies. Those adhering to the liberal judicial philosophy often appear to have less regard for the predictability of consequences and social order. They often seem to consider *stare decisis* to be just another obscure Latin term and to decide a particular case based upon what they perceive to be a just result under the facts, or in furtherance of a particular social goal.

Lawyers and politicians are generally distrusted in our society, but when a lawyer who is a politician slips into judicial robes, he or she is generally assumed to have lost all bias and to be a paragon of fairness and impartiality. Judges are human, and as such are victim to all the weaknesses, prejudices and temptations which corrupt the species.

Money

No discussion of the California legal system would be complete without some reference to money. Money, after all, is the bottom line, the direct or indirect objective in almost every suit. Money is the fuel upon which the engine of justice feeds. Do not be too quick to condemn. Few would choose to labor without pay. Why have accountants risen to such prominence in our society? They are the score keepers in the game of greed, which we must all play in order to survive.

Just how much justice can you afford? Monetary considerations play a major role in litigation. Fees for attorneys, court reporters, and expert witnesses continue to rise at a rapid rate. In deciding whether to go to trial or to settle, one must weight the cost of litigation and the likelihood of a successful conclusion against the potential financial exposure. When considering whether or not to appeal after an unsuccessful trial, these considerations arise again; but then, the possibility of success has diminished and the financial exposure has increased by the cost of the trial transcript, the appeal bond, post judgment interest, and, of course, more attorney's fees. Reevaluation of these factors occurs continually during the course of litigation. Settlement is most likely to occur, when the parties on all sides similarly evaluate the case.

In settlement discussions less attention is paid to the merits of the cases, then to the ability to pay with defendants and cross-defendants being coerced to contributing projected "defense costs" regardless of culpability. The threat of the uncertainty of trial looms over all like the sword of Damocles.

It is generally assumed that professional defendants are (1) insured, and (2) better able to spread the loss then property owners. With the general use of the contingent fee and the inability of victorious defendants to recover more than a nominal amount as costs (attorney's fees are generally not recoverable), there is no down side for the plaintiffs. Some use the system as a State operated and maintained extortion racket. The damaged plaintiff seldom, if ever, recovers his full damages for out of his share comes his attorney's fees. The defendants always lose, since if they win in court, they must still pay their attorneys' fees, which are generally not recoverable. Attorney's fees are paid to the prevailing party in most industrialized countries, except the U.S.A. The greatest beneficiaries of our justice system (if not the only beneficiaries) are, of course, the lawyers. This is certainly not surprising, since they are its creators and operators. There is the story of the young attorney who hung his shingle in a small town which had no other lawyer. He nearly starved before a competitor opened a law office right across the street. Thereafter, they both enjoyed flourishing and financially rewarding careers. Law may be the only field in which business can be created form nothing. For example, a lawyer sitting in his office with little to do can put pen to paper and create a set of interrogatories (written questions which must be answered under oath by the responding party(s)) which he sends to several other attorneys in a pending case. These other attorneys must answer the interrogatories, which in turn provides additional work for the first attorney. If the interrogatories are not answered, then the first attorney can send threatening letters and file motions to which the others must respond. Also, upon receipt of the interrogatories, the other attorneys would be expected to be spurred into interrogatory efforts of their own. Newton's second law finds applicability, at least in part, to law. For every action, there is a reaction, although, in the case of law, it may not be equal.

Selecting An Attorney

Selecting the proper attorney to represent you can present problems. The general practitioner may or may not be suitable. The man who has quite capably handled corporate matters for the last ten years may be totally out of his depth in handling a professional malpractice case. This is the age of specialization. While few attorneys would admit that they are unqualified to perform any legal service, the fact is that some are much more qualified than others. As is the case with most other professions, as legal practitioners develop careers, they tend to gravitate toward some recognized or unrecognized area of specialization. As one specializes in one area of the law, his grasp of other areas tends to slip.

In selecting an attorney some opt for the large, prestigious, high overhead firms with many names on their embossed letterheads and branch offices in exotic places, hoping that this will frighten off the opposition. Actually, few lawyers are frightened off by a large firm because they know that your case is probably being handled by some junior associate, paralegal or clerk in some dank recess of the posh suite. Also, by hiring a prestigious firm, the client signals that there are assets to pay high legal fees and to contribute to settlement. A plea of poverty from a client of a prestigious law firm is not granted much creditability. The high overhead is, of course, paid by the clients.

The most convincing argument for refusing to contribute cost of defense to a settlement package may be made by in-house counsel, whose salary is paid by his employer whether or not the matter proceeds to trial. If your firm cannot afford house counsel, it perhaps could share such an attorney with one or more of your colleagues.

Attorney Fees

The contingent fee, in which the attorney takes a percentage of the recovery, is generally not available to defendants, although it is widely used with plaintiffs. Some attorneys charge a "fair" fee for their services. After the matter is resolved, a determination is made by the lawyer of what would be a "fair" amount to charge based upon the result, the skill, effort, and time involved. By far, the most wide spread method of charging for legal services is by the hour, i.e., time and materials. With this method, periodic statements describing the work performed, the hours utilized, and the amounts expended are sent, with the expectation of payment, to the client. Of course, the client has the opportunity to review the bill and the right to make inquiry into its correctness. Unfortunately, there is little protection against being charged for services which were not performed, or being overcharged for services which were performed. Although the client is forced to rely in large measure upon the integrity of the attorney, the client has the right to have the issues of the fee arbitrated. The lawyer's statement should be more than a tally, indicating how much is owed. It can also serve to keep the client advised of what is happening in the case. When the statement is received, it should be carefully read and analyzed, and if there is anything which is not understood, explanation should be required before payment is made. No payment should be made for any charge which is not understood, or which is disputed.

The organization responsible for the licensing and disciplining of lawyers in California is the State Bar of California. Organizations in other states perform similar functions. If there is a question or a complaint about an attorney, the State Bar should be called for assistance.

Avoiding Suit

The question often is asked, "How do I avoid being sued for my professional work?" Merely not being negligent is not a complete defense. All that is needed to sue is paper, the filing fee, and a typewriter with which to prepare the complaint. How about cause? That may be decided later. Viewed from the prospective of the lawyer, it is much more desirable to initiate suit against everyone who could possibly have even remote exposure, than to take the chance that facts may be subsequently uncovered which point to the liability of someone who was not initially sued. The cost of initially adding additional defendants is just a little ink and the cost of serving the complaint, while the cost of not adding those defendants may be a malpractice suit against the attorney. Unless you have diplomatic immunity, the only real way to avoid being sued is not to perform any professional services, and that does not always work. There are, however, ways to minimize exposure as you traverse the legal mine field, which is a professional career.

Persons Who May Be Sued

A professional owes a duty to exercise the ordinary skill and competence of members of his profession during the course of his activities for the protection of any person who foreseeably and with reasonable certainty may be injured by his failure to do so. This includes the client, the employer, and subsequent purchasers and users of any property which is the subject of professional efforts. A failure to discharge such duty will subject the professional to liability for negligence.

It is a common misconception of professionals who are employed by corporations (or by other professionals) that they are not personally liable to clients of the corporation (or of the professional) and that the corporate entity shields them from liability from claims of third parties. This is not necessarily the case. While the employee may not be liable to his employer's client for breach of contract, he may have liability on theories of negligence, negligent misrepresentation and fraud. One of the purposes of the corporation is to shield its officers, directors and stockholders from liability for the claims of third parties. This protection does not extend to employees of the corporation. Each of us owes a duty to exercise ordinary reasonable care to anyone who could with reasonable foreseeability be injured by our failure to do so. A corporate employee would not only owe such a duty to the client of the corporation, but also to his employer, the corporation. A failure to properly discharge that duty, negligence, could subject the employee to liability both to the client and to his employer, the corporation. For example, if as an employee of a corporation you were negligent in the performance of your duties, you could be sued for such negligence both by the client and by your employer or former employer. Often cases are filed years after the work was done and after circumstances have changed. The employee may have new employment or the corporate employer could be out of business. The former employer could be hostile to the negligent former employee, or the negligent former employee could be the only viable defendant. In a very real sense, the former employee could be left holding the liability bag for a former employer who is no longer viable.

II. THEORIES OF LIABILITY

Although the general rule is that those who sell their services for the guidance of others in their economic, financial and personal affairs are not liable in the absence of negligence or intentional misconduct, this is applicable only in the usual situation where the consultant has no proprietary interest in the developed property, and does not participate as the actual developer. If the consultant also has ownership interest or if he participates as the developer, then additional theories of liability are applicable. Plaintiffs' lawyers tend to attempt to expand the theories of liability upon which their clients may be entitled to recovery under a given set of facts. From the lawyers' viewpoint, it is better to throw all possible theories of recovery at the defendants in the beginning, then to learn that later - discovered facts support an additional theory of recovery which has been barred because it was not asserted earlier. For this reason, the defendant consultant often finds himself or herself confronted with claims for breach of warranty, strict liability, fraud, negligent misrepresentation, negligent and intentional infliction of emotional distress, and breach of contract, as well as the claim for negligence, which is the real basis for the suit.

1. Breach of Contract

A contract is an agreement to do or not to do a certain thing. Each time you promise to perform professional services for a fee you are entering into a contract. Contracts may be oral or written, expressed or implied from the circumstances. The essential elements of a contract are: (1) parties capable of contracting; (2) their consent; (3) a lawful object; and (4) a sufficient consideration. Although, most contracts are enforceable only between the contracting parties, the law recognizes an exception where a contract is made between parties expressly for the benefit of a third person. In this case the third person, who is not a party to the contract, can directly enforce the provisions of the agreement. This device, known as a third party beneficiary agreement, is often asserted by attorneys seeking to enforce provisions of an agreement between the consultant and the developer for the benefit of the injured home owner. This is rarely successful, since such a contract must be expressly for the benefit of the third person.

The contract for professional services, if carefully drafted, can provide substantial protection for the consultant. It should be remembered, however, that with limited exception contractual provisions may be enforced only between the parties to the contract. Typically, the contractual provisions would be enforced between the consultant and his client, while third parties would neither receive the benefits of the contract nor suffer from its burdens nor limitations.

2. Breach of Warranty

A warranty is a written or verbal guarantee that a certain fact is or shall be as promised. In connection with the sale of goods, the law implies warranties of fitness. Warranty imposes liability without regard to fault. It requires a contractual agreement or sale between the party injured and the party from whom recovery is sought. This theory of recovery is generally not viable against professionals. However, some practitioners warrant that their work will be in compliance with generally accepted practices in the profession. Since the law does not impose expressed or implied warranty upon the providing of professional services, they gain nothing by the statement and put themselves in the unfortunate position of guaranteeing their work.

3. Strict Liability

Strict liability, liability without fault for mass produced consumer goods, was a rapidly developing theory of liability in the 1960's and 1970's. It has been applied to almost every type of mass produced product from soap to tract lots and homes. Strict product liability is a large and expanding area of the law. Fortunately, it is not a viable theory of liability against the ordinary professional. Its applicability to developers may provide some benefit for the professional at the trial, where a developer can find himself liable without fault for a construction defect caused by his consultant, against whom negligence must be proved.

4. Negligence

Allegations of negligence form the basis of almost every claim against a professional arising from professional services. Although it is usually accompanied by every other theory the lawyer can dream up, the others are usually just window dressing. Negligence is generally the omission to do something which an ordinarily prudent person would have done under similar circumstances or the doing of something which an ordinarily prudent person would not have done under like circumstances. This is the general definition which applies to all, whether expert or layman. In the case of professionals practicing their profession, negligence is expressed in terms of compliance or noncompliance with a standard of care among like professionals of good reputation practicing in the same or similar community at the same time. The standard of care applicable to a given profession must be determined from the testimony of experts in that field, unless the conduct involved is within the knowledge of laymen.

Reference to a "standard of practice" is not meant to imply that there are established or standard ways of performing professional services with which one must comply regardless of the facts and conditions present. If there were well established standard ways of handling problems, there would be little need of well trained professionals to resolve them. "Standard of practice" has a broader and more generalized meaning. It generally refers to whether the professional in question made use of the tools and information reasonably at his disposal to investigate and analyze the technical problem, and whether he or she reached a reasonable solution to that problem which was based upon generally accepted principals. Perfection is not required. A 1954 California Supreme Court case stated:

> "The services of experts are sought because of their special skill. They have a duty to exercise the ordinary skill and competence of members of their profession, and a failure to discharge that duty will subject them to liability for negligence. Those who hire such persons are not justified in expecting infallibility, but can expect only reasonable care and competence. They purchase service, not insurance."

That case, which established the ground rules for those who sell their services for the benefit of others, involved an error in reporting the depth of fill on a lot to be purchased by a developer. Sound familiar? The developer purchased the lot in reliance on the report and was forced to construct more expensive footings due to the error.

Although failing to comply with a statute or ordinance is normally considered to be negligence as a matter of law (negligence per se), the mere compliance with the letter of the law does not necessarily relieve one of liability, since it is generally recognized that statutes and ordinances set forth only minimum requirements and circumstances may require more than the minimum. If there is a problem, some expert will certainly testify that extra precautions over and above those required by ordinance should have been employed. Further, a professional cannot rely on the approval by an inspector or a governmental entity for relief from liability. Governmental entities and their employees are generally protected from liability by the doctrine of governmental immunity.

Negligence necessarily involves a breach of duty which causes damage. Although we all owe a general duty of care to all members of the community, the duty of professionals in the performance of their services is usually raised by a contract. Therefore, contracts for professional services should be carefully drafted and administered.

Although there is some overlap between the disciplines, registration as a geologist does not empower one to practice civil engineering or any of its various branches. To poach in the areas exclusively reserved for civil engineers is to invite a charge of negligence and fraud and is a crime. Although the definitions of the disciplines are vague and generalized, the statutory definition of civil engineering includes references to specific areas reserved to that discipline. Learn the legal and technical limits of your profession and your skill and practice within them.

Professional education does not end upon receipt of the diploma. The professional disciplines are not static. In order to comply with the standard of care, it is necessary for the professional to be aware of what others in the profession are doing. This may be learned through discussions with colleagues and participation in formal and informal continuing education.

5. Fraud

Fraud, deceit, is: (1) the suggestion, as a fact, of that which is not true, by one who does not believe it to be true; (2) the assertion, as a fact, of that which is not true, by one who has no reasonable ground for believing it to be true (negligent misrepresentation); (3) the suppression of a fact, by one who is bound to disclose it, or who gives information of other facts which are likely to mislead for want of communication of that fact; or, (4) a promise made without any intention of performing it. One who willfully deceives another with intent to induce him to alter his position to his injury or risk is liable for any damage which is thereby suffered.

Actions against professionals based upon intentional fraud are relatively rare, and this theory is often alleged for the sole purposes of attempting to defeat a statute of limitations defense and as a basis for punitive damages. Negligent misrepresentation is a close relative of negligence, and the same facts may often support recovery under both theories. For example, the statement that a particular slope is geologically stable, if untrue, could support an award based upon negligence, or negligent misrepresentation. In the usual case of negligent misrepresentation, it is necessary also to prove the failure to comply with a standard of care, in order to establish the lack of a reasonable basis of belief.

In the case of laymen, the representation of opinion is ordinarily not actionable, however, it becomes actionable when one of the parties possess or assumes to possess, superior knowledge or special information regarding the subject matter of the representation. When one employs a professional, it is for the purpose of securing the assistance of one who possesses superior knowledge and special information in a particular field. It is therefore not surprising that the opinions of such experts, if made without a reasonable basis, provide the basis of liability.

Where material facts are known to one party and not to the other, a failure to disclose them is generally not actionable fraud, unless there is some special relationship which gives rise to a duty to disclose. A client relationship gives rise to such a duty.

An essential element of fraud, regardless of its species, is justifiable reliance. The person injured by the fraud must have reasonably relied upon the material representation or upon the duty to disclose and must have suffered injury as a result of such reliance.

6. Indemnity

In bringing actions against developers, some plaintiffs' attorneys elect not to include the professionals. When this occurs, the developers usually bring cross-complaints seeking indemnity from their professionals. A cross-complaint is a formal complaint made by one who is already a party to a suit seeking some relief from others who may or may not be parties to the action. Indemnity is, in essence, the right of one to have another pay for his loss. Indemnity may be express, such as where it arises from the provisions of a contract, or it may be equitable. The right to equitable indemnity inures to a person who, without active fault on his part, has been compelled by reason of some legal obligation (constructive or imputed liability), to pay damages caused by the initial negligence of another, and for which he himself is only secondarily liable.

> A 1964 case sets forth the doctrine as follows:
>
> "The duty to indemnify may arise, and [equitable] indemnity may be allowed in those fact situations where in equity and good conscience the burden of the judgment should be shifted from the shoulders of the person seeking indemnity to the one from whom indemnity is sought. The right depends upon the principle that everyone is responsible for the consequences of his own wrong, and if others have been compelled to pay damages which ought to have been paid by the wrongdoer, they may recover from him."

A determination of whether or not a party is entitled to equitable indemnity must be determined from the facts of each case, and equitable indemnity is subject to special equitable defenses, such as the doctrines of laches (prejudicial delay in bringing the claim), unclean hands (unconscionable conduct in the transaction) and estoppel (prejudicial conduct restricting proof). Where a claim for express indemnity rests on the provisions of a contract, the claimant's active participation in the wrongful act may preclude recovery.

Express indemnity is a creature of contract and primarily governed by the rules relating to the interpretation of contracts. The intent of the parties as expressed in their agreement should control and where they knowingly bargain for protection, that protection should be afforded.

7. Contribution

Contribution, the sharing of a loss among those responsible for it, is administered in accordance with the principles of equity. Where a money judgment has been rendered jointly against two or more defendants and one of them has paid more than his pro rata share in satisfaction of the judgment, he may seek recovery of the amount paid in excess of his share. The remedy is not available to one who has acted intentionally to cause the injury.

8. Declaratory Relief

Any person who desires a declaration of his rights or duties with respect to another may, in cases involving actual controversy relating to legal rights and duties, bring an action or cross-complaint for that purpose. Although the court may make a binding declaration of such rights and duties regardless of whether or not other relief is sought, declaratory relief rarely travels alone and is usually found in the company of other causes of action. It is a catch-all type of relief which overlaps other more specific relief and which is difficult to challenge procedurally.

III. DEFENSES

The defenses available are varied and their applicability depends upon the nature of the relief sought by the plaintiff or cross-complainant.

A. STATUTES OF LIMITATION

The Legislature has enacted a great number of varied statutes (statutes of limitations) which prescribe the periods beyond which actions may not be brought. Statutes of limitations may be defenses only in judicial proceedings and do not apply to administrative proceedings. The application of a statute of limitations in a civil case acts procedurally only, barring the remedy and not affecting the substantive right or obligation. The obligation still remains, but the remedy is barred by the statute. The significance of this distinction to the professional is that the statute of limitations may be waived, if not properly and timely asserted and that the statute may be tolled (i.e., its operation suspended) by certain acts, circumstances or events. Although the purpose of the statute is to prevent the assertion of stale claims, and it is said to be a favored defense, in reality it is unpopular with judges and particularly with juries who are reluctant to permit a defendant to escape the consequences of wrongdoing merely due to a technicality created by the passage of time. The issue of the statute of limitations is generally a mixed question of law and fact, requiring findings of the judge and the jury (if a jury trial). If there is no factual issue relating to the application of the statute, then the judge acting alone may decide the issue. Court procedure permits the trial of special defenses (e.g., statute of limitations) before the trial of any other issues. In the case where there is a strong statute of limitations argument and where there are substantial or dramatic damages, it is often prudent to litigate this defense first before the trier-of-fact can be prejudiced by impressive damages or a sympathetic plaintiff.

Commencement of the Statutory Period

Civil actions can only be commenced after a cause of action (right of action) accrues. The cause of action generally accrues when the wrongful act is done. The statutory period may begin to run, even though the person injured is ignorant of his cause of action, or of the identity of the wrongdoer. In order to avoid the harsh and unjust result of barring of claims before a person knows he has been injured, the Legislature and the courts have created several of exceptions to the general rule. The most common of these are the rules that the statutory period begins to run when damage results, or when the facts regarding the matter became known or should have become known.

Numerous statutes of limitation relate to almost every aspect of human activity. The statutes of limitations of most concern to actions involving professionals are Code of Civil Procedure sections 337(1)—-four years, written contracts; 337.1—-four years, patent defect in design, etc.; 337.15—- ten years, latent defect, etc.; 338—-three years, trespass or injury to real property, fraud, etc.; and, 339(1)—-two years, oral contracts, etc.

There are situations where more than one statute of limitations may be applicable, or where different periods of limitations run at the same time. If the consultant's activities are limited to giving professional advice and opinions only, the two year period may be applicable, but if the consultant actually participated in activities which caused an injury to real property such as supervising, testing, and inspecting the actual work, the proper statutory period may be three years. The ten year statute applicable to latent defects commences to run from the time of substantial completion of the improvement, while the three year statute for injury to real property runs from the time noticeable damage occurs, or when the plaintiff discovered or with reasonable diligence should have discovered the facts essential to his cause of action. Under certain circumstances, these statutes may be running simultaneously. The ten year statute runs even though the plaintiff may be blamelessly ignorant of his cause of action. If noticeable damage occurred within the first year after substantial completion of the improvement, the plaintiff would have only the balance of the three year period to file suit and would not be entitled to the balance of the ten year period to sue for that damage. If, on the other hand, noticeable damage occurred within the ninth year after substantial completion, the plaintiff would have only the balance of the ten year period to file suit and would lose the balance of the three year period. The ten year statute of limitations does not apply to actions based upon wilful misconduct or fraudulent concealment, and lawyers will often draft complaints to include these elements in attempting to avoid the consequences of the statute.

The period of limitation for causes of action based upon any of the species of fraud is three years, and the period begins to run from the time that the aggrieved party discovers the facts constituting the fraud.

Cross-Complaint For Indemnity

An action for indemnity may be brought by way of cross-complaint in an action involving the same subject matter, or it may be brought later in a separate action. Although the cause of action for indemnity does not accrue until loss has been suffered through payment of an adverse judgment or settlement, a cross-complaint seeking indemnity is permitted in a transactionally related underlying action. For example, a developer who has been sued by a homeowner for damage to property caused by earth movement, may elect to bring his consultants into the main action by way of a cross-complaint, or the developer may elect to wait until he is forced to pay the homeowner in that action and then file a new and separate action against those consultants. The election is usually made in favor of filing a cross-complaint. The advantages of this are (1) the issue of indemnity can be decided at the same time as the issues in the underlying action; (2) the consultants are parties to the action and are thereby forced to defend their own work; and, (3) the cross-complaint may not be barred by the statute of limitations, if the underlying action has been brought before the statutory period runs. If the developer waited to bring a separate action for indemnity, he loses the advantage of the homeowner's filing date and risks having the action barred by the statute.

Substantial Completion

The ten year statute provides, in relevant part, that actions to recover damages for latent deficiency in design, specification, surveying, planning, supervision, or observation of construction or construction of an improvement to or survey of real property must be brought within ten years of the substantial completion of the development or improvement, but not later than the date of (1) final inspection by the applicable public agency; (2) recordation of a valid notice of completion; (3) use or occupation of the improvement; or, (4) one year after termination or cessation of work on the improvement, whichever occurs first. Circumstances may indicate an earlier date of completion. The date of substantial completion relates specifically to the performance by each profession or trade rendering services to the improvement. The statutory period commences to run with respect to a person who has contributed toward an improvement, when such improvement has been substantially completed, and not when the development in which the improvement is a part is completed.

Tolling

Certain acts on the part of potential defendants and cross-defendants can cause the operation of a statute of limitations to be tolled (suspended). Absence from the jurisdiction tolls the statute for the period of the defendant's absence. If after the substantial completion of an improvement the developer or consultant performs repairs, the statutory period may be extended for the period it took to undertake the repairs. This certainly tends to dampen the spirit of cooperation in attempting to remedy problems on a former project. The decision of whether or not to return to a former project to attempt to correct problems requires the exercise of careful judgment, since the statutory period may not only be tolled as to the prior work, but a new statutory period will begin for the repair itself. Assisting in repairs is potentially more risky if the initial statutory period has completely run, since suit may be filed if the repairs are not completely successful and at trial it may be difficult, if not impossible, to determine whether the damages sought in the action were caused by the original (barred) work or by the repair. Under those circumstances, there is little doubt as to where a jury's sympathies will lie.

B. COMPARATIVE NEGLIGENCE

California formerly followed the doctrine of contributory negligence in which any degree of negligence on the part of the plaintiff could completely bar recovery. In 1975, California substituted a judicially created system of comparative negligence which apportions liability between all the parties according to their respective degrees of fault. The trier-of-fact (the jury or the judge, if there is no jury) determines the percentage of fault, if any, for which each party is responsible and the amount of the damages. In theory, each party pays only for that portion of the damages for which he has been found to be responsible, however, reality often produces different results. A defendant who is only minimally responsible can be required to pay damages to a plaintiff who is himself primarily responsible for his injuries. Also, a defendant may be required to pay more than his proportionate share, when his acts combined with those of another to cause the injury and when the other party is for some reason, such as bankruptcy or governmental immunity, shielded from collection. The doctrine has been criticized as producing unpredictable and inconsistent results; requiring judgment beyond the ability of judges and juries; and requiring arbitrary determinations with loss allocation based upon judge and jury sympathies. In spite of the criticism, we are stuck with it.

There is a tendency on the part of the property owners to convert semi-arid sites into a tropical rain forest. Surface drainage is often disrupted and the subsoil saturates. Of course, the property owner will be responsible for that portion of his damage which his negligent acts contributed; however, it is doubtful that a jury of sympathetic homeowners who are doing the same thing will find that it contributed substantially.

C. ASSUMPTION OF RISK

A person who knowingly and voluntarily exposes himself to a known risk and is thereby injured is deemed to have assumed the risk of loss and is barred from recovery. If you stick your finger in a live electrical socket, you have assumed the risk of receiving a shock. The limitations and complications of this defense are beyond the scope of this work.

D. CONTRACTUAL LIMITATION OF LIABILITY

Under certain circumstances, a contractual limitation of liability clause in a construction contract may limit the amount of damages recoverable from a party to the contract. This will be addressed below. It should be kept in mind that this defense may be applicable only between parties to the contract containing the clause. Such a clause would not be a defense to the claims of third parties and would not be a defense against cross-complaints for indemnity brought by a party who was not a party to the contract.

E. MITIGATION OF DAMAGES

Where damage is threatened, the injured party has an active duty to use reasonable care and diligence to protect himself and to minimize his loss. The failure to do so will result in his being unable to recover for the damages enhanced by his failure to mitigate.

IV. DAMAGES

Damage is an element of every theory of recovery that is normally used against professionals. The recovery of damages is the purpose of the litigation. The measure of damages applied depends upon the legal theory used. The measure of damages applied to a cause of action for negligence is different from the measures applied in fraud or breach of contract cases. The measures of damages are set forth in statutes and interpreted by a large body of case law. In general, the purpose of damages in civil actions is to fairly compensate the injured party for the injury actually suffered. Damages must be certain and not speculative, although damages for future harm may be recovered. Under the circumstance where fraud, oppression, or malice of a defendant have been proved, the plaintiff may recovery exemplary or punitive damages which are given for the purpose of punishing the conduct of the guilty defendant, rather than for compensation of the plaintiff.

The measure of damages generally applicable to actions involving property damage is the diminution in market value or the cost of repair, whichever is less. The plaintiff is generally precluded from recovering the cost of restoration in excess of the diminution in value. However, a recent exception (1990) to this rule permits recovery of restoration costs in excess of the decrease in market value, if the owner has a personal reason for restoring the property to the original condition, and where there is reason to believe that the owner will, in fact, make the repairs. Just what personal reasons will justify the additional recovery have not, as yet, been defined.

In those situations where property is purchased in reliance upon incorrect representations made in a geotechnical report, the proper measure of damages for negligence or negligent misrepresentation is the difference in value, if any, between the purchase price and the value of the property at the time of the sale.

In addition to the general damages mentioned above, the plaintiff is entitled to recover special damages, which are those which are not deemed to be a necessary or usual result of the wrongful act but which nevertheless were caused by it. Special damages must be specially requested.

Under some circumstances, a plaintiff is entitled to recover prejudgment interest on the amount awarded. Since the case may take up to five years to get to trial, the amount of prejudgment interest may be substantial. In addition, interest on the amount of the judgment is recoverable from the date the judgment is entered.

V. THE JUDICIAL SYSTEM AND PROCESS

A. COURTS

A detailed description or analysis of the California and U. S. court systems is far beyond the scope of this work. The various courts have monetary and subject matter jurisdictions which control where a particular action may be filed. The Federal system consists of District Courts (trial level), Courts of Appeals, and the United States Supreme Court. The California courts consist of Superior, Municipal, Justice, and Small Claims Courts at the trial levels and District Courts of Appeal and the California Supreme Court at the appellate levels. The Superior Court in each county is the court of general jurisdiction and the court where the majority of cases involving professionals are filed and managed.

B. FILING TO APPEAL

To initiate a suit in the Superior Court, the plaintiff, or his attorney, prepares a formal complaint setting forth the alleged causes of action. This is filed with the Clerk of the Court. A summons is the official notification of the suit and is served along with a copy of the complaint. Various forms of service are authorized, but personal service is most often used. It does little good, in the long run, to avoid service by the process server, since other means of service are available.

Upon being served, it is extremely important to get the summons and complaint into the hands of your attorney at the earliest possible time, since service starts the clock running on your time to formally respond to the complaint, normally thirty (30) days (a shorter period may be applicable). Failure to respond within the allotted time may result in your default being entered. This means that the case is lost without having a day in court. Defaults may be set aside by stipulation (agreement), or by order of the court upon good cause shown. A motion to set aside a default for mistake, inadvertence, surprise or excusable neglect must generally be made within six months. The parties may agree that the time to respond to the complaint (cross-complaint) may be extended, but this should be confirmed with a letter, in order that the extension is not "forgotten".

Upon receipt of the summons and complaint (or cross-complaint), a decision must be made as to the form of the response. If certain irregularities are present, a motion to quash service may be made. The usual options are to prepare an answer, a demurrer, a motion for judgment on the pleadings, or a motion to strike. A decision as to the mode of response involves financial considerations. "A lawyer's time and advice are his stock-in-trade." (A. Lincoln.)

An answer is a formal document (pleading) which admits or denies the allegations of the complaint and sets forth any new matter constituting a defense (affirmative defense). If the complaint is verified (sworn to be truthful), the answer must be verified. A demurrer is a pleading which challenges the legal sufficiency of the complaint. For purposes of pleading, the demurrer assumes that all of the material allegations of the complaint are true, but even so, no case is stated, or it is barred on its face by some doctrine of law such as a statute of limitations. The demurrer essentially says, "So what?" A motion for judgment on the pleadings is quite similar to a demurrer, except that the motion may be made after all the pleadings are in and even at the time of trial. A motion to strike may be made to remove portions of the complaint which are sham or irrelevant or redundant. Such a motion is used to attack an action for professional negligence of an architect, engineer, or land surveyor, where the action was filed without filing the required Certificate of Merit. Such a certificate is not currently required, when filing such an action against an Engineering Geologist.

The court rules on the demurrers and motions as they are brought and normally grants leave to the plaintiff to amend the complaint. Those complaints and portions of complaints which cannot be amended to bring them within acceptable limits are dismissed. Courts are very reluctant to dismiss an entire action without giving the plaintiff every reasonable opportunity to state a case. Fourth and fifth amended complaints are not unheard of.

Eventually, unless the entire action is dismissed, the complaint is answered. The admissions and denials made by the answer to the complaint establish the issues to be determined at trial. Often at the time of filing the answer to the complaint, a defendant will file his cross-complaint setting forth any set-offs he may have against the plaintiff (e.g., unpaid fees), as well as any transactionally related claims (e.g., indemnity) he may have against any other person, whether a party to the action or not. In cases where the plaintiff sues only the developer, the developer's consultants are often brought into the action in this manner. The cross-complaints then go through the answer, demurrer, and motion procedure. When the dust settles, everyone usually is suing everyone else.

C. DISCOVERY

Once the pleading wars have been fought and won or lost, the parties usually embark on voyages of discovery. California has liberal laws permitting discovery of everything which is not privileged and which is relevant to the subject matter of the litigation. In actions involving professionals, discovery usually takes the form of oral depositions; requests for admissions; inspections of documents and things; written interrogatories; and exchanges of expert witness information. With the exception of oral depositions, these discovery tools are available only between parties to the litigation. Documents of non-party witnesses may be obtained through the use of a subpoena duces tecum and their depositions may be taken.

Written Interrogatories

Interrogatories are written questions regarding the litigation which one party sends to another party. The party to whom the interrogatories are sent must object or answer them under oath within thirty (30) days of receipt. Attorneys often agree to extend the time for response. The discovery laws in California were drastically overhauled in 1986. Under the former law, it was not uncommon for large defense firms to inundate their adversaries with burdensome and oppressive interrogatories. With the aid of a computer, a legal secretary could, in a few minutes, generate hundreds of detailed and complex interrogatories which could take hundreds of man hours on the part of opposition attorneys and parties to answer. Under the present law, a party may, as a matter of right, propound any number of official form interrogatories and thirty-five (35) specially prepared interrogatories to another party. No specially prepared interrogatory may contain subparts, or a compound, conjunctive, or disjunctive question. More than thirty-five (35) specially prepared interrogatories may be served, if accompanied with a declaration indicating the justification for the additional interrogatories. The party served may make a motion for a protective order relieving it from the requirement of making responses because the permitted number has been exceeded, or if response would cause unwarranted annoyance, embarrassment, oppression, or undue burden and expense. If the responses to the interrogatories are unacceptable to the party propounding the interrogatories a procedure is available for compelling further answers. The court has the power to impose a range of sanctions on a party who misuses or refuses to make discovery.

Requests For Admissions

Any party may require another party to admit under oath and within thirty (30) days, the genuineness of specified documents, or the truth of specified matters of fact; opinion relating to fact; or application of law to fact. A request for admission may relate to any matter that is in controversy between the parties. Just as in the case of interrogatories, the court has procedures available to enforce compliance and to punish for misuse. The requests for admissions are handy tools which can save the time and expense required for preparing unnecessary proof.

Inspections Of Documents, Things, And Places

Any party may inspect and copy documents and inspect tangible things, and land or other property that is in the possession, custody or control of any other party to the action. A party may inspect, measure, survey, photograph, test, or sample the land, but that land must be in the possession, custody or control of a party to the action. The court lacks the power to order such examination and testing of the land of one who is not a party. Documents of non-party witnesses may be ordered to be produced by subpoena duces tecum. As is the case with other discovery tools, the court has the power to enforce its orders and to issue orders protecting the rights of parties and non-parties.

Oral Depositions

Oral depositions of parties and non-parties are one of the most widely used discovery vehicles. "A deposition is a written declaration, under oath, made upon notice to the adverse party, for the purpose of enabling him to attend and cross-examine." The deposition usually takes place in the informal surroundings of an attorney's office or library. The attorneys, the witness-deponent, and a Certified Shorthand Reporter, who is also a Notary Public, are present. Sometimes parties to the litigation and/or expert consultants are also present. The Reporter-Notary places the deponent under oath and then records on the stenographic machine everything that is said during the proceedings, unless there is an agreement that something is to be "off the record". Only the attorneys can agree to go off the record. Audio tapes and video tapes of the deposition may also be made. A series of questions are asked of the deponent who is required to give responsive answers. From time to time, the attorneys may make objections to the questions. The question must be answered even if objected to, unless the witness for some reason (usually advice of counsel) refuses to answer.

After completion of the deposition, the Reporter's record is transcribed into a booklet which is called a "deposition". The deposition is sent to the deponent who is given the opportunity to review and to change any of the answers to the questions asked. The deponent may not change the questions. After reviewing the deposition and making the necessary corrections, the deponent signs the deposition under oath and returns it. Usually the attorneys agree that the deponent may sign the deposition under penalty of perjury, rather then following the burdensome formal procedure.

The deposition serves three very important rolls. First, it is a discovery tool which provides the means of obtaining information from party, non-party, or expert witness. Second, it is a means of preserving testimony in the event a witness is not available to testify at the time of trial. Third, it provides a means of attacking the credibility of a witness. If a witness changes his testimony in the deposition when he is correcting it, or if he testifies differently at the time of trial then he testified at the time of the deposition, the opposing attorneys can and will bring it to the attention of the trier-of-fact (judge or jury) and argue that the testimony of that witness is not to be believed. This can be very embarrassing to the witness, as well as seriously damaging to the position of one or more parties to the action. Depositions are potent weapons and proper preparation for them is important. Depositions will be discussed further below.

Written Depositions

A procedure exists for taking depositions by submitting questions in writing. (C.C.P. 2028) The procedure is cumbersome and spontaneity is lost. It is rarely used.

Exchange of Expert Trial Witness Information

Normally, the identities and opinions of experts are protected from disclosure on the grounds that they are protected by the attorney's-work-product privilege. However, once the original trial date has been set, a party may demand that all parties make a mutual and simultaneous exchange of information concerning each other's expert trial witnesses. The disclosure not only applies to experts retained for the purpose of the litigation, but also to parties and employees of parties. Failure to properly designate and permit discovery of an expert witness will result in the exclusion of that expert's opinion testimony. The designation of an expert must be accompanied with an attorney's declaration given under penalty of perjury which contains: (1) a brief narrative statement of the qualifications of each expert; (2) a brief narrative statement of the general substance of the testimony that each expert witness is expected to give; (3) a representation that each listed expert has agreed to testify at trial; (4) a representation that each expert will be sufficiently familiar with the pending action to submit to a meaningful oral deposition concerning the specific testimony, including any opinions and their bases that the experts are expected to give; and, (5) a statement of the expert's hourly and daily fees for providing deposition testimony. All discoverable reports and writings may also be demanded.

Within twenty days (20) of the exchange, any party may submit a supplemental list of experts who will testify on subjects listed in the lists of adverse parties, if the supplementing party has not previously designated an expert on that particular subject. There is also a procedure to augment an expert witness list with a subsequently retained expert, if the court finds, upon motion, that prejudice will not result; that the witness is one which the augmenting party would not in the exercise of reasonable diligence have determined to call; or, if the party failed to determine to designate the witness as a result of mistake, inadvertence, surprise, or excusable neglect.

At trial, a party may call as a witness any expert which has been properly designated by another party. An undesignated expert may be called as a witness to impeach the testimony of an expert offered by any other party at trial. The impeaching testimony may include testimony to the falsity or non-existence of any fact used as the foundation for any opinion by an opposing expert, but it may not include testimony that contradicts the opinion.

Of course, experts deposed pursuant to this procedure are entitled to be paid for their services. The party taking the deposition of an expert is required to either send the fee with the service of the deposition notice, or to pay it at the commencement of the deposition. Other arrangements may be made by agreement. The fee required to be tendered is the reasonable and customary hourly or daily fee for the actual time consumed in the examination of that expert by any party attending the deposition. The party designating the expert is responsible for any fee charged by the expert for preparing for the deposition and for travel expenses. If the deposition takes longer than has been anticipated, the attorney taking the deposition must pay the balance of the expert's fee within five (5) days of receiving an itemized statement from the expert witness. Parties or employees of parties are not entitled to such expert witness fees.

It is not unusual for an expert to charge one and one-half to two times the usual rate for time spent giving expert testimony. There seems to be little justification for this gouging. How can a professional justify charging substantially more for stating an opinion at a deposition, than for stating that same opinion orally or in a report? It is clear that lawyers are not the only professionals who use and abuse the system. If there is a dispute, a procedure is in place for having the court determine what the reasonable fee shall be.

Discovery From A Non-Party

One of the biggest complaints of professionals is having to submit to endless depositions and productions of business records of completed projects relating to litigation in which they are not parties. Unfortunately, that is one of the burdens of living in this society. It is possible to include a provision for compensation for future litigation-related services in a contract for professional services. Without such a provision, a professional is limited to the normal witness fees, unless the deposition is taken specifically pursuant to the provisions for discovery of expert witness information.

Discovery from a non-party is limited to oral and written depositions and production of business records and things. The jurisdiction of the court is acquired by service of a subpoena or a subpoena duces tecum requiring the person designated to appear, or requiring the production of specified business records, or both. If the deponent named in the subpoena is not a natural person, (i.e., a corporation) the deposition notice will describe with reasonable particularity the matters on which examination is requested. The deponent is required to designate and produce at the deposition those of its officers, directors, managing agents, employees, or agents who are most qualified to testify on its behalf as to those matters.

The deposition subpoena is effective to require any deponent, who is a resident of California at the time of service, to personally attend, or to produce the business records and things and to attend any court session to consider any issue arising out of the deponent's refusal to be sworn or refusal to answer a question or to produce requested records and things. Unless the court orders otherwise, the deposition must be taken at a place within 75 miles of the residence of the deponent, or within the county wherein the action is pending, and within 150 miles of the deponent's residence.

Procedures are in place to protect the rights of parties and non-party deponents with appropriate protective orders; to require the attendance and meaningful participation of witnesses; and to punish abuses.

Privileges

Society has established a number of privileges against being forced to divulge information. Of these, the Fifth Amendment of the United States Constitution, the privilege against self-incrimination in a criminal case is probably the most well known. Privileges are subject to a number of exceptions and applications which are beyond the scope of this work.

(1) Attorney's Work Product Privilege

It is the policy of the State to preserve the rights of attorneys to prepare cases for trial with that degree of privacy necessary to encourage them to prepare their cases thoroughly and to investigate not only the favorable, but the unfavorable aspects of those cases and to prevent attorneys from taking undue advantage of their adversary's industry and efforts. The attorney's-work-product privilege enforces this policy. In cases involving professionals, the privilege is asserted to protect against unduly intrusive inquiry and is made in response to questions asked at deposition and in interrogatories, as well as to requests for production. The privilege may be waived by making disclosure and is waived as to expert witnesses when mutual discovery of expert information is made.

(2) Attorney-Client-Privilege

In order to encourage and protect full and complete communication between a person and his attorney, society protects such confidential communications from disclosure. The privilege extends to confidential information transmitted between a client and his lawyer in the course of that relationship. The communication may be by a means which, so far as the client is aware, discloses the information to no third persons other than those who are present to further the interest of the client in the consultation, or those to whom disclosure is reasonable necessary for the transmission of information, or the accomplishment of the purpose for which the lawyer is consulted. The privilege includes a legal opinion formed and the advice given by the lawyer in the course of that relationship. "Lawyer" as relates to the privilege is construed broadly and includes anyone reasonably believed by the client to be authorized to practice law in any state or nation.

D. SETTLEMENT

It is not surprising that the vast majority of civil cases settle either before or during trial. The expenses involved, as well as the uncertainties of trial, provide strong motivation for cutting losses. It is said that a poor settlement is always better than a good law suit.

Informal settlement discussions may be held between the parties and/or attorneys at any time. A mandatory settlement conference will be set before a judge approximately one month before the trial date. This takes place close enough to the time of trial that the attorneys and parties have usually completed discovery and are knowledgeable about all aspects of the case. Unless expressly excused for good cause by the judge, all persons whose consent is required to effect a binding settlement and required to be personally present at the mandatory settlement conference. This includes the litigants and authorized representatives of insurance companies which have coverage involved in the case. Normally, the attorneys who will try the case will also be present.

Litigants generally wait in the courtroom, while the case is discussed with the judge in chambers. The insurance representatives observe, while each attorney outlines what he considers to be the important points to consider. The judge next discusses the case in private with the plaintiff's attorney and is advised of the plaintiffs demand and the minimum amount which the plaintiff would accept in settlement. The plaintiff's attorney is generally handling the case on a contingent fee basis with the amount of the fee being a percentage of the amount of recovery. A private discussion is then held between the judge and the defense attorneys in which the judge is advised of the amounts which the various defendants are prepared to offer in settlement. The defense attorneys generally charge by the hour, and part of the contribution of the defense to settlement is usually based in part upon the amount of attorneys' fees and costs which will be saved by not having to proceed further. This "cost of defense" contribution is justified on the grounds that even the most innocent party will have non-reimbursed costs and expenses if the matter proceeds to trial. If the minimum amount for which the plaintiff will settle and the maximum amount which the defendants will contribute toward settlement are reasonably close, the judge will continue to work with the parties until a compromise is reached and settlement achieved. Sometimes additional information is required before settlement may be achieved, and the judge will continue the settlement conference until a later date. If the parties are so far apart in their expectations that the judge sees no hope for settlement, the matter will be returned to its position on the trial calendar.

Normally, the cross-complaints are settled along with the settlement of the complaint, leaving no loose ends. However, in the case where a number of defendants are claimed to be liable to the plaintiff for the same injury, and the plaintiff settles with less than all of the defendants, there remains the possibility that the non-settling defendants will pursue their cross-complaints against the settling defendants. Under these circumstances, the settling defendant, who are also cross-defendants, would gain little by settling, since they save little cost of defense and any judgment against the non-settling defendants could be passed along to the settling defendants through the cross-complaints.

In order to avoid this impediment to settlement, statutes have been enacted which discharge the settling party from all liability for indemnity to any other party, if the settlement is deemed to have been made in "good faith". Most settlements in such cases are made contingent upon the court making a determination that they were made in "good faith". Not every settlement is made in good faith. What constitutes a "good faith" settlement depends upon a number of factors which the judge must take into account at a hearing for that purpose. At that hearing, the court may inquire as to whether the amount of the settlement is within the reasonable range of the settling party's proportional share of comparative liability for the plaintiffs' injuries. Factors to be taken into account include: a rough approximation of the plaintiffs' total recovery and the settler's proportionate liability; the amount paid in settlement; the allocation of settlement proceeds among the plaintiffs; the recognition that the settler should pay less in settlement, then he would if he were found liable after trial; the financial condition and insurance policy limits of the settling defendants; and the existence, if any, of conduct aimed to injure the interests of the non-settling defendants. If the amount offered in settlement is so far "out of the ball park" in relation to these factors as to be inconsistent with the equitable objectives of the statue, the settlement will not have been made in "good faith".

Since the financial condition and insurance policy limits are factors to be considered in determining whether or not a proposed settlement is made in "good faith", consideration should be made to keeping the operating corporation as empty as possible and without insurance coverage. Even where the plaintiff's damages are accepted as great and the liability certain, a disproportionately low settlement figure is often reasonable in the case of a relatively insolvent and uninsured defendant. This requires strong wills, steady nerves and determination on the part of the uninsured defendants.

In recent years, a private judicial system has developed from the ranks of retired judges. Due to court congestion and delay, those of means may choose to employ a private judge to resolve their disputes either by means of settlement, arbitration, or trial.

E. DISMISSAL

Policy requires that the plaintiff shall proceed with reasonable diligence in the prosecution of an action. Time limitations are therefore placed upon the case, and a failure to meet the deadlines subject the suit to the possibility of involuntary dismissal. The summons and complaint must be served upon the defendants within three (3) years of the filing of the action, and the proof of service must be made within the following sixty (60) days. An action must be dismissed if not brought to trial within five (5) years of the date of commencement of the action. The court has the discretionary power to dismiss an action on its own motion or upon the motion of a defendant, if to do so appears to be appropriate under the circumstances of the case. If the case results in a mistrial or a hung jury, a new trial must commence within three (3) years. There are a number of statutes and a considerable amount of case law relating to these restrictions.

F. SUMMARY JUDGMENT

Under the circumstances where a case has no merit or where there is no defense, a party may make a motion for summary judgment. The motion is made upon affidavits, counter affidavits and other documents; and its use is strictly controlled. The motion may be granted only if no triable issue as to any material fact is found to exist.

G. TRIAL

If the matter is not settled, the alternative is a formal trial before a judge and jury or before a judge alone. In those cases where there is a jury, it acts as the trier-of-fact, deciding the factual issues presented. The judge manages the trial and decided issues of law. Where there is no jury, the judge becomes also the trier-of-fact.

Challenge To The Judge

A judge may be disqualified either peremptorily, or for cause. There are time limitations placed upon the exercise of these options, and the challenge must be made before trial commences. Due to court congestion, retired judges are appointed by agreement to conduct a trial as part of a "rent-a-judge" program. These judges, from the "bone pile", are chosen by agreement of the parties, and the challenge procedure is unnecessary.

Pretrial Motions

Before the trial begins, the attorneys generally have an informal conference with the judge in chambers to discuss trial procedure and evidentiary matters. At this time, motions in limine are usually made objecting to evidence in advance, in order to avoid raising objections before the jury when the evidence is offered. In lengthy trials, the court may order that the trial be bifurcated with the issues of liability being determined before evidence of damages is permitted and/or the trial of the cross-complaints are bifurcated from trial of the main action. Trial of special defenses, such as the statute of limitations, may be ordered before trial of the main action. The judge has wide discretion to manage the trial as he deems most appropriate. A motion for judgment on the pleadings may also be made at trial.

Jury Selection-Challenges

Prior to commencement of the trial, a jury is selected from the juror panel. After generally describing the case and identifying the parties and attorneys, the judge usually conducts a preliminary investigation of the prospective jurors, in order to determine if there are any legal reasons why any of them should not become members of the jury. This procedure is called "voir dire". After the judge completes voir dire examination, each attorney in turn is permitted to conduct voir dire. Although the procedure is designed to uncover any prejudice on the part of a prospective juror, many attorneys use it in a manner designed to create prejudice in favor of their clients. The attorneys may challenge (object to) prospective jurors for cause based upon some ground of disqualification, or preemptory where no ground need be stated. The number of preemptory challenges is limited. As a prospective juror is dismissed, another takes his place from the panel. Finally, when twelve prospective jurors are agreed upon, they are given an oath and become the jury. In lengthy cases, alternate jurors are selected in advance to replace any jury member who may be unable to continue to sit. A jury consisting of less than twelve may be agreed upon by the parties.

Opening Statements

After jury selection is completed, the attorneys in turn make their opening statements which for the purpose of acquainting the trier-of-fact (judge or jury) with the theory and nature of the case, in order that the evidence can be better understood and placed in perspective when it is presented. The plaintiff proceeds first, followed by each of the defendants and cross-defendants. Opening statement may be waived or reserved to a later time.

Motion For Non-Suit

If the plaintiff's opening statement does not contain all of the elements necessary to state a prima facie case, any defendants may make a motion for non-suit. The motion raises a defect in the plaintiff's case which cannot be overcome. The same motion is often made at the close of the plaintiff's case and at the close of the case. If the motion is granted, the court may permit the plaintiff to attempt to amend his opening statement to state a prima facie case. If the plaintiff cannot, then the case is terminated.

Plaintiff's Case

The plaintiff next presents the oral and documentary evidence which tends to prove the causes of action alleged in the complaint. A witness is called, placed under oath, and subjected to direct examination by the party calling him. That witness is then cross-examined in turn by the opposing attorneys. The party calling the witness, then has the opportunity to re-direct questions to the witness followed by re-cross examination by the opposition. This procedure continues until the attorneys run out of questions, or the judge exercises his discretion by prohibiting further inquiry. This procedure is generally followed, except in the case where the witness called is adverse, such as where the plaintiff calls the defendant as a witness. Under these circumstances, the party calling the witness is permitted to examination of the witness as if on cross-examination, while the party with whom the witness is identified is limited to re-direct examination. Much more latitude is given on cross-examination, than on direct or re-direct examination. The various objections are made by the attorneys and ruled upon by the judge as the testimony is offered. The purpose of the objections is to permit only the most reliable evidence to be presented to the trier-of-fact. The judge has wide discretion with respect to evidentiary matters, and in a non-jury trial, it is not uncommon for the judge to permit a wide scope of examination without paying too much attention to the usual rules of evidence. It is presumed that in reaching a decision, the judge relied only on the relevant evidence which was properly admitted, and that the prejudicial and irrelevant evidence erroneously admitted was not considered.

Defense Case

Upon completion of the plaintiff's case, the defendants may again make motions for non-suit if there is a jury trial, and a motion for judgment if the case is tried by a judge only. If the appropriate motion is successful, the trial is over. If the plaintiff has presented sufficient evidence to prove a prima facie case, then the defense case proceeds in the same manner as the plaintiff's case.

Rebuttal

Upon completion of the defense case, the parties respectively may then offer only rebutting evidence, unless the judge, for good cause, permits them to offer evidence upon their original cases.

Summation

After the presentation of evidence has been completed, the attorneys for the various parties make their arguments. This is the time when the attorneys are given the opportunity to utilize their persuasive skills to convince the trier-of-fact of the merits of their clients' causes and the shortcomings of the opposition. Because the plaintiff bears the burden of proof in the action, the plaintiff's attorney begins and concludes the oral arguments.

In the case of a jury trial, motions for non-suit may be renewed and motions for a directed verdict may be made as to some or all of the issues. Both motions are in the nature of demurrers to the evidence, and a verdict may be directed in favor of the plaintiff or the defendant. In a non-jury case, a motion for judgment may be renewed at this time.

Jury Instructions

After the completion of oral argument, the judge charges the jury, by reading standard and specially prepared statements which instruct them on the duties and functions of judge and jury; the effect and general nature of the evidence; the use of terminology relating to parties and pleadings; and the specific applicable law. The jury is presumed to have reached a verdict based upon these instructions.

Jury Deliberation And Verdict

After being charged, the jury retires to deliberate until three-fourths of them agree upon a verdict. They may examine the physical evidence and have parts of the transcript of oral testimony read to them. They may have some or all of the jury instructions reread. If, after deliberation, the required number are unable to agree upon a verdict and further deliberation appears futile, a mistrial may be declared and the matter retried. After a verdict is rendered, the jury is discharged. A mistrial may also be declared for any prejudicial conduct at trial which cannot be corrected.

The jury may render a general verdict which decides all of the issues in favor of one or more of the parties, or it may render a special verdict which decides factual issues only and leaves the judgment to the court. The court may instruct the jury to make written findings upon particular questions of fact called special interrogatories.

If the trial has been bifurcated with the jury deciding either the issue of liability or the entire plaintiff's case, the trial process from opening statement to verdict would now be repeated to try the issue of damages or the issues relating to cross-complaints. Often the parties to the cross-complaints will waive jury, so that the jury will be discharged after verdict and the judge will hear what additional evidence the parties to the cross-complaints wish to offer. Judgment is entered on the verdict.

Motion For Judgment N.O.V.

A party who is not satisfied with the verdict may move for judgment notwithstanding the verdict (N.O.V.). The motion is similar to motions for non-suit and for directed verdict, and gives the judge an opportunity to correct the error of not granting one of those motions earlier.

Judge Trial-Statement Of Decision

If the trial is without a jury, the matter is submitted to the judge after oral argument, and in due course a tentative decision is announced. Upon motion made for good cause, a party may be permitted to reopen a case for the presentation of further evidence. The tentative decision is preliminary and is not binding on the court, which may modify or change it. A written statement of decision explaining the factual and legal basis of the decision as to each of the principal controverted issues at trial is prepared upon the request of any party. Parties may make proposals as to the content of the statement of decision and may make objections to it. The court may set a time for hearing of the objections. After resolution of matters relating to a statement of decision, the judgment is entered.

Post Trial Motions

Even after judgment has been entered, procedures are available which permit a party to seek a different result. A party may make a motion for a new trial, or a motion to vacate and enter a different judgment. These motions are designed to permit correction of error or irregularity which occurred during the trial. In general, the grounds for a motion for a new trial are: irregularity in the proceedings; misconduct of the jury; accident or surprise; newly discovered evidence; excessive or inadequate damages; insufficiency of the evidence; and error in law.

Appeal

After the trial and the post trial motions are resolved, an aggrieved party retains the right to have the matter considered on appeal. Appeal is not a retrial of the case, and the scope of the inquiry is limited in general to reviewing errors in law and not to review questions of fact. However, where there is no substantial evidence to support the judgment, it is considered error as a matter of law into which the appellate court may inquire. Error of law is not reversible, unless it appears to have resulted in a miscarriage of justice. An in depth discussion of appellate procedures and review is beyond the scope of this work.

When deciding whether or not to appeal an adverse decision, one must again consider costs. A certified transcript of the trial proceedings must be ordered and the court records to be considered on appeal must be prepared. In the case of money judgments, a stay to enforcement of the judgment may be obtained only upon posting a bond or undertaking in the amount of one and one-half or two times the amount of the judgment depending upon whether or not the bond or undertaking is posted by an admitted surety insurer. Under rare circumstances, a court of appeal may issue a writ staying enforcement of a judgment without the required bond or undertaking.

VI. ALTERNATIVES TO LITIGATION

Arbitration provides a simplified and economical alternative to the standard court procedures. In some superior courts, cases where the amount in controversy is less than $50,000.00, it may be ordered to be arbitrated. The arbitration is final, unless a party files a request for a new trial within a specified period. The arbitrators are lawyers or retired judges who hear cases upon the agreement of the parties.

Due to the congestion in the courts and the delay it causes in bringing matters to trial, a judicial arbitration and mediation industry has developed which provides private judges to parties who can afford to pay for the service. Many retired judges have found a second, parallel career as retired judges. This is viewed by many as justice for the wealthy and indication of the advanced state of decay of judicial institutions.

Many construction contracts contain arbitration clauses requiring disputes to be submitted to arbitrators. These arbitrations are usually decided by arbitrators who are themselves attorneys, contractors, architects or engineers. There is a problem that the arbitrators may be influenced by their own experiences and prejudices. In a jury trial, the attorneys generally like to start with a clean slate and educate the jurors as much of the technical aspects of the action as is believed to be necessary in order for them to reach a knowledgeable decision. Arbitration clauses and awards are enforceable through the courts.

Often the most effective means of avoiding suit is by working with the client to solve problems as they occur. Often a client needs more hand holding than technical advice when a problem arises, and the professional who devotes time and energies to this type of public relations will often be rewarded with additional work, as well as avoiding the cost and aggravation of litigation. Sometimes this approach works and sometimes it does not.

If a dispute with a client degenerates to the point where the consultant is replaced on the project, the consultant not only loses all control over the project and the resolution of the problem, but he can expect the replacing consultant to recommend extremely conservative and much more expensive remedial measures. A consultant does not want to "buy" the potential problems of his troubled predecessor.

VII. SURVIVAL IN THE TRENCHES

Insurance

None of us want to lose what we have worked long and hard to acquire. A lawsuit can wipe out the accumulated wealth of an entire career. Even if the suit is successfully defended, the costs can be staggering. Insurance provides a means of hedging bets. Insurance companies are not in the business to make friends. Risks are evaluated and appropriate premiums are charged. Often premiums are high, exclusions are many, and policy limits are low. The insured pays any excess of a judgment over the policy limits. Often the policy provisions permit the insurance company to control the case and to send it to an attorney who is in its stable of lawyers without regard to the wishes of the insured. The professional may lose control over whether or not to settle. An insurance company, just like any other business, may go out of business and be unable to honor its obligations to the insured. More is usually expected in settlement from those who are insured. Many professionals find that their outlay in legal fees and settlement contributions over the years have been substantially less than the premiums which they would have paid for insurance. Whether or not to obtain insurance overage is an individual policy decision.

Contracts With Clients

The typical contract for professional services should carefully spell out the precise services to be performed, as well as any limitations or conditions. Since most clients have little, if any, actual knowledge or appreciation of the various technical limitations controlling the various disciplines of professionals, these should be carefully spelled out in order that the clients know precisely what they are getting and what they cannot expect. If, for example, the service to be performed is a geologic reconnaissance, care should be taken to explain the limits and limitations of the work and to warn that it will not include geotechnical engineering, structural engineering, civil engineering, or architectural evaluations. Misunderstandings, are best resolved when entering into a contract and not later after problems have manifested. A problem often arises when the initial work done for a client is completed and follow-up services are undertaken pursuant to a verbal authorization or "understanding". To do this is to invite misunderstanding and litigation. It is not unusual that a contract is sent to the client for signature and the work done without its being returned. Under these circumstances, the consultant can expect a difficult time in trying to enforce any of the provisions or limitations of the agreement against the client. No services should be performed without a signed written contract.

The contract should contain a provision that the parties agree that the client may rely only upon the written contents of the reports and may not rely upon any oral representation or statement. With the passage of time, people tend to recall what they wish to recall and what it is in their interests to remember. Perjury is no stranger to courtrooms.

Many consultants include a warranty of compliance with the applicable professional standards to the exclusion of all other warranties in their form contracts. This actually adds little, since the law generally does not impose the law of warranty upon professionals and the inclusion of such a provision merely provides another theory upon which the opposing attorney may rely. It is better to merely include a statement that no expressed or implied warranties are made.

If it can be negotiated, a contractual provision providing that the consultant will be named as an additional name insured on the developer's insurance policy will provide project insurance and coverage for the consultant. It is rare that the developer's insurance carrier will be willing to agree to this, even though they may be in effect insuring the developer against the negligence of the consultant.

A provision providing for attorney's fees is often included in the form contract in the hope of discouraging litigation and paying for the defense of those actions which are filed. In reality, few are discouraged from filing suit because of such a provision, and it tend to help coerce settlement from those who sought to benefit from them. The law interprets such a contractual provision as granting attorney's fees to the prevailing party, even though the provision may have been drafted so as to grant such fees only to the consultant. Serious consideration should be given to whether or not to include an attorney's fees provision in a contract.

A limitation-of-liability provision in a contract is one which limits any recovery to the client to an agreed amount, usually the amount of the professional's fee, or a specified amount. The usual attack on such a provision in a contract is that it is a part of an unconscionable contract of adhesion, in which one's choices are limited to either accepting or rejecting the agreement and the services. Contracts of adhesion are common in our society, and the fine print statements on the back of parking lot claim checks are good examples. A statute permits limitation of liability provisions under certain circumstances in the case of construction contracts. For purposes of the statute, the term "construction contract" is broadly defined and includes most activities of professionals. A recent court decision upheld the validity of a limitation-of-liability provision in the construction contract of civil engineer, where the developer-client had a fair opportunity to accept, reject, or modify the limitation provision prior to entering into the contract. In that case, the court upheld a provision which limited the amount recoverable by the client to the amount of the consultant's fee, $67,640.00, even though the client incurred over five million dollars in expenses to remedy the problem. When parties of relatively equal bargaining position have dealt at arm's length, a carefully drafted limitation-of-liability clause which permits a fair opportunity to negotiate the terms of the contract is enforceable. The parties are bound by the terms of their contract, even if they do not read it. Only the fair opportunity to negotiate and not actual negotiation is required.

Every contract should have a merger clause which states in essence that the parties agree that all of the terms, conditions, and understandings of the agreement between them are merged into the contract which cannot be altered or modified without the written consent of both parties. Such a clause hampers a party from claiming that the agreement was something other than what it says or from attempting to assert additional terms and conditions.

The client should be required to identify, mark and inform the consultant of any subsurface improvements which may be encounter during exploration. The client should be required to secure proper permission to conduct exploration, if the property is actually owned or possessed by a third party.

A clause requiring that the client to indemnify the consultant for any loss or cost incurred as a result of litigation or claims of third parties, where the damage is the result of actions of someone other than the consultant, would be desirable if the client will agree to it.

The main concern when dealing with contracts is that they accurately and specifically described the work to be done for which the fee is to be paid. No services should be performed unless and until the written contract is completed. When the particular services covered by the contract are completed and additional services are required, a new contract relating to the additional services should be prepared.

When dealing with laymen particular care should be taken to explain exactly what can and cannot be expected.

In order to obtain payment for those possible future hours spent in deposition and trial, the contract should contain a provision requiring the client to pay the consultant for any future time spent and expenses and costs incurred in preparing for or giving testimony relating to the project in court or at deposition. The fee should be that applicable at the time the testimony is given.

The contract and the professional reports should contain statements warning of the possibility of changed conditions as part of the warning of the limitations of the service provided.

Field Work

The pre-sale geologic investigation has been a fertile source of litigation, since long before Engineering Geologists became certified by the State. The sole function of these reports is for the purchaser to have someone to blame in the event problems develop with the property. A $500.00 fee for such a project can create legal exposure amounting to hundreds of thousands of dollars. If such employment is taken, particular care should be taken to carefully warn the purchaser in writing of what the report does not contain, and to recommend that the purchaser retain experts in other disciplines to conduct evaluations. When one practices out of his field of expertise, it is negligence.

Always verify the accuracy of the north arrow and scale shown on any plan. Plans prepared by architects and even by surveyors and civil engineers may be in error. Any attitudes of geologic features plotted on a plan containing an incorrect north arrow will be unreliable. Error in north arrows have been found to be as much as 24 degrees off. Occasionally, check the magnetic variation adjustment on your Brunton.

Any field approvals or substantive communications should be made in writing or confirmed in writing immediately. Memories fade with time, and people tend to recall what they wish to recall, or what is in their interest to recall. It is a good idea to develop a field memorandum from which can be filled in at the time of a conference and a copy given to each of those involved.

It is not uncommon for field logs to differ substantially from their counterparts in the published reports. Although a certain amount of editing of details is to be expected when preparing the reports, it is suspicious and casts doubts upon the credibility of the consultant when material changes are found.

In the case where one consultant replaces another on a project, the subsequent consultant should be prepared to accept the full responsibility generally for the entire project and, in particular, for those aspects of the prior consultant's work which was incorporated into the work of the successor. Should litigation arise and the work of the prior consultant be found to be wanting, the subsequent consultant may cross-complain against his predecessor for indemnity, but the statute of limitations may by that time have run as to the work of the first consultant but not as to the replacement. It is doubtful whether the average jury would possess the degree of technical sophistication or interest necessary to make the differentiation and could be expected to find that the damage was caused by both.

Hazardous waste is receiving increasing interest in our society, and the area of the law dealing with it is rapidly growing. In limiting the scope of work in a contract and report, care should be taken to make it clear that the exploration and reporting of the nature and extent of pollution is beyond the scope of the services. When encountering contamination on a site, it should be reported and the recommendation made that appropriate specialists in the field should be retained to investigate and evaluate the site.

You should avoid offering opinions as to legal rights and consequences related to your projects. Legal rights are complex and often require extensive research and consideration even by those who are learned in the law. If you provide legal advice, you will be held to the standards of care applicable to attorneys.

When an injury occurs on the job site, the first concern should be for the well being of the injured. When this is accomplished, the facts and circumstances should be documented and photographs taken if appropriate. Although geologists are not generally responsible for supervision and job safety, the contract should contain provisions which make this clear.

There is little that can be done where the consultant is unaware that the client has failed to comply with recommendations. But where the consultant becomes aware of the failure to comply, letters documenting the non-compliance and warning of the consequences should be sent to the client and possibly to other professionals and governmental agencies whose work may be impacted.

Report Preparation

The technical report should not only accurately advise of all relevant matters, but it should also contain a warning of its technical limitations and its geographic limits. Avoid generalized statements and the use of form or standardized passages which may not be totally applicable. A technical report is no place to soft sell unfavorable aspects of a project. The client that you try to help today may be the plaintiff in your next suit.

Consultants do not have poetic license. Each statement made should be supported by documented facts. One should not become intoxicated with the exuberance of ones own verbosity and thereby drift into untenable positions. Short simple sentences containing words of few syllables are preferred.

If, when making a limited technical inquiry, an adverse condition is observed which affects the client's interests, but does not affect the subject of the limited inquiry, care should be taken in the report to make it clear that the problem condition and the area affected by its influence are excluded from the scope of inquiry, as well as from the conclusions and recommendations. A separate letter may be sent to the client advising of the potential problem and recommending that it be investigated.

Drafts of reports should not be sent to clients for review. There is nothing to justify this practice, and it gives the impression that the client is exercising some technical control over the professional.

Oral reports should be avoided, since they welcome misunderstanding and misinterpretation. In this age of computers there is a tendency to standardize professional reports and portions of reports. Although this is expeditious and convenient, it can lead to problems. Particular care should be used in drafting technical reports, in order that nothing is said which ought not to be said in them, and so that the physical and technical limitations of the project and the services are clearly stated. It should be made clear, for example, that groundwater conditions vary with time and location and that the geologic structure and lithology may change considerably between outcrops or excavations. While writing a technical report, it is good to keep in mind that years later, and after problems have been identified and conditions changed, that report may be dissected by adverse lawyers and your competitors who have a financial interest in finding fault with your work.

The question of the extent of the geologist's responsibility to advise occurs where a geologic hazard is found to extend off the subject property and onto adjacent parcels. Presently, there is generally no duty on the part of the geologist to advise the adjacent owners. They may not want to know, since knowledge would create a duty upon them to advise subsequent purchasers of their properties. The duty to disclose is to the client and would seem to be discharged, if the report property disclosed the extend of the hazard to the client. A gratuitous copy of the report to the adjacent owner is unlikely to be appreciated, and may embroil the consultant in a controversy regarding the adjacent property.

Consultants are not responsible for acts of God, but when such an act is reasonably foreseeable, a consultant may be liable for not advising of its danger.

File Maintenance

It is scientifically correct to keep detailed records of technical projects; it may also be legally suicidal. Embarrassing and legally damaging documents regularly appear with documents produced as part of the discovery process. These are "smoking guns" which wound in the foot, and which can make the difference between winning and losing. In addition, each document gives lawyers something else to talk about and something else to compare. Logs of subsurface exploration, draft reports, lab work sheets, and in-house memoranda present particular problems. Any discrepancies between these documents and information appearing in the reports can be quickly and easily discovered.

File maintenance is a serious problem for any business. Many resolve the problem with a program of file culling which rids the files of unnecessary documentation. If potentially damaging documents are also discarded, then so much the better. Upon completion of a project, the files relating to that project should be reviewed by someone familiar with the project and as much of the file as possible should be discarded. All field logs, laboratory work sheets, draft reports, and draft maps and cross-sections should be discarded. Only final copies of reports and those memos and correspondence which might be of further future use should be retained. Such a program may be justified, if it is part of the regular procedure and done in the ordinary course of business.

Your Deposition

Whether your deposition is taken as an a percipient witness, an expert witness, or a party to the action, the same general procedure and precautions apply. Depositions of persons are occasionally taken before they are added as parties to the suit. The purpose is to secure the testimony before the deponent becomes aware that he is a target in the case, and before he is aware of the need to be represented by an attorney. Because of the important consequences which may result from testimony given at a deposition, it is important that the deponent be represented by an attorney at the proceedings if any of the consultant's work is in question and if there is any possibility that the consultant may be added later as a defendant.

The deposition procedure was outlined above. A few tips on how to conduct yourself at a deposition follow:

1. If the subpoena or notice requires that documents be produced at the deposition, do not produce more than was actually requested. Every scrap of paper gives the lawyers something to talk about, to ponder, and to question. The more documents produced, the larger the scope of the inquiry and the longer the ordeal.

2. Listen carefully to the questions asked. You cannot properly answer any question which you do not fully understand. You may request the reporter to read the question back and you may request that the attorney rephrase the question. Many lawyers lack the technical skill and knowledge to properly structure technical questions. Often a seemingly simple question may encompass a vast and complex technical area.

3. Answer directly, concisely, and honestly. The best answers are often "yes", "no", "I don't know", "I don't remember", or a short narrative answer. The attorneys are evaluating the manner of your responses, in order to predict how effective you will be as a witness at trial.

4. You are required to testify to your best recollection and, if requested, to give your best estimate, but you should not guess or speculate.

5. Do not volunteer anything. The deposition is taken for the benefit of the party whose attorney arranged for it. Nothing that you say at a deposition will accrue to your benefit. State your answer and await the next question in silence. The more you talk, the more you assist the lawyers asking the questions and the longer your deposition will take to complete. Often a stalled interrogation is saved by a verbose deponent.

6. Answer only the question that was actually asked and not the question which you know the attorney intended to ask. When asked if you have an opinion on a particular matter, the answer is "yes" or "no" and not a statement of the opinion. Often totally unintended responses are made to questions simply because the deponent tried to expedite the proceedings by providing the needed information, regardless of the form of the questions asked. It is the job of the attorney to formulate the questions and you have no obligation to assist.

7. The opposition is entitled to examine any document which you have used to refresh your recollection, in order to testify at a deposition or a trial. Be sure that you have not refreshed your recollection from privileged, damaging, incriminating or embarrassing documents.

8. You are entitled to consult records before giving an answer. You should consult records regardless of how well you feel that you know the file. There is a loss of credibility, when a document produced proves an answer to be in error.

9. Often photocopies of documents are shown at depositions. With today's copy equipment, it is easy to modify copies. Often a notation on a photocopy may not appear on the original. Check the original record for any altered or late entry.

10. Do not begin your response to a question until the question is finished, even though you think that you know what the complete question will be. Often attorneys compose questions extemporaneously.

11. The court reporter will not be able to properly record the statements of two or more persons speaking at the same time. To speak when another is speaking is generally considered to be rude, but to do so during a deposition is to invite confusion and error.

12. When dealing with lawyers and particularly on the record, it is wise to use care and deliberation in your choice of words. Often a lawyer will appear unfamiliar with the subject matter and careless in his terminology. To sink into using layman's terms to describe technical matters is to invite embarrassment.

13. Do not jump on the question with an answer. Permit time for the attorneys to make objections. Although objections made at a deposition are made mostly for the purpose of protecting the record, some could provide assistance in formulating your response. Objections made on the grounds that the question is ambiguous, or assumes facts, or lacks sufficient facts to form the basis of a proper hypothetical question should cause you to reflect before you respond. Questions must be generally answered, even though objections are made.

14. Resist trying to educate the attorney in your area of specialization. The attorneys are quite capable of employing their own experts for that purpose.

15. Every question should be considered as calculated to help the opposition. Consider yourself an alien in a hostile environment.

16. If you are asked for your opinion and you have none, do not hesitate to make that fact known.

17. If you are asked to formulate an opinion, make sure that you have available all of the background data necessary to provide a proper basis of the opinion and that the opinion is based on that data. An opinion is worth no more than the facts upon which it is based.

18. Often attorneys attempt to coerce technical witnesses into expressing opinions without compensation. Sometimes an agreement for payment is reached. In those instances where the attorney persists while refusing to pay, it is well to remember the 13th Amendment to the United States Constitution, which prohibits involuntary servitude. To force an unwilling witness to develop an opinion is to force him to perform services. This technique has been successful on occasions in the past. It may suffice to state that you have no opinion; that you do not wish to work for the attorney with or without pay to develop an opinion; and that to force you to develop an opinion would be a violation of your Constitutional Rights.

19. Often lawyers ask questions which present a hypothetical set of facts and request an opinion based upon those facts (hypothetical question). These questions are often long and complex and it is a good idea to write them down. When faced with one of these questions, it is prudent to reconsider it carefully, in order to make sure that all of the information necessary for the formation of an opinion is present within the question. The question stands alone and the deponent should not assume any of the actual facts of the case when developing the opinion. The answer to such a question should be prefaced with a statement making it clear that the answer is based solely upon the facts stated within the question and not upon any of the facts of the case.

20. When asked to state the basis of your opinion, be sure to include the total of your education and experience.
21. Remember that unforeseen results may occur, even if everyone associated with a project has used the highest degree of skill and care.
22. Stay within your area of expertise when testifying and avoid criticizing the work of other disciplines, unless you have expertise in those areas also. Disputes between the defendants can only benefit the plaintiff.
23. If you are a party to an action and are asked questions regarding statements made in the pleadings, it is better to respond that you advised your attorney of the facts in the case and he prepared the pleading which you signed because you believed that your attorney included the proper information in the proper form.

The Expert Witness

Opinion testimony of an expert is limited to subjects that are sufficiently beyond common experience that the opinion of the expert would assist the trier-of-fact. The opinion is based on matter (including special knowledge, skill, experience, training, and education) perceived by or personally known to the witness, or made known to him and which is of a type that may reasonably be relied upon by an expert in forming an opinion upon the subject to which his testimony relates. Expert testimony is limited to those subjects that are beyond the competence of persons of common experience, training, and education. If the fact sought to be proved is one within the general knowledge of laymen, then the expert testimony is not necessary.

Whether or not expert testimony will be required, and whether or not a particular witness will be permitted to testify as an expert in a particular field is within the discretion of the trial judge, and his decision will not be overturned on appeal unless he has clearly abused that discretion. Some judges are quite willing to let almost anyone testify as an expert in any field in which expertise is claimed, stating that the expertise or lack of it goes merely to the weight which should be given to the testimony. Some witnesses are permitted by such judges to testify in areas in which they are not licensed, if they claim to be licensed in an associated or overlapping field. Thus, Engineering Geologists have testified on matters which fall clearly within the purview of the Geotechnical Engineer and vice versa. The courts and the governmental entities responsible for policing the professions seem to show little interest in enforcement. The attorney for the party against whom the testimony is to be offered can ask preliminary questions of the "expert", in an attempt to illustrate lack of appropriate professional credentials.

The need for expert testimony to establish a breach of the standard of care has created stables of forensic experts in every professional field. These become well known in the legal community, where truth is often viewed as a commodity which is subject to negotiation. There is also a tendency on the part of expert witnesses to become advocates for the side who is paying their bill. Objectivity among expert witnesses is often more rare than virtue is in a bordello.

Humans possess a herd instinct which causes them to seek to please, in order to be accepted into the group. Too often those who sell their expert services to attorneys tend to sacrifice their objectivity and become advocates for their client's case. Once on the team, there is an effort, sometimes unconscious, not to disappoint its other members. What results is the rejection, suppression, alteration, and selective use of information which is not consistent with the legal theory. There is often a conscious effort to recommend the most expensive remedial means, in order to maximize the amount of recoverable damages. This makes such consultants popular with the plaintiffs' attorneys and insures future assignments, but they become quickly known and disliked by their colleagues. The prostitution of professional integrity is reprehensible but common.

CONCLUSION

It is hoped that this work is the closest that the reader comes to the judicial system during his or her professional career. The system is the result of innumerable compromises and contains many faults and shortcomings. Perhaps it does not deserve to be referred to as a system of justice. It is, however, the only system for resolving disputes that is readily available and socially acceptable. There is no alternative. It has been said that no one's life, liberty, or property are safe while the courts are in session.

THE STATUS OF GRADING CODE ENFORCEMENT IN CALIFORNIA

C. Michael Scullin
Chief Engineering Geologist
Rogers/Pacific, Inc.
396 Civic Drive
Pleasant Hill, CA 94523

INTRODUCTION

We sent out questionnaires to building officials in California in 1977 and 1991 to try to evaluate the status of grading code enforcement throughout the state. On March 7, 1974, the State of California mandated that all cities and counties adopt Chapter 70, Uniform Building Code (UBC), or an equivalent grading code.

1977 QUESTIONNAIRE

In 1977, to determine the effects of this mandate, 286 questionnaires were sent to building officials throughout the state (Scullin, 1983). While only 31% (88) responded, some interesting data emerged. Whereas 92% of the respondents indicated they had met the state mandate by adopting a proper grading code, only 13% had retained specialized staff grading inspectors. However, 46% of the departments enforcing the grading code had three or more of their personnel assigned to grading. The results of the 1977 survey indicate that there existed a nearly equal division of grading-code enforcement responsibility between building/safety departments and public works departments. The results of this original survey are presented as Table 1.

It is interesting to note that results from the grading-oriented staff section of the questionnaire indicate that in 1977, the majority of those involved with excavation and grading-code enforcement were neither specialized in grading-inspection nor grading-code administration. Administrative grading engineers were generally city engineers or assistant building directors, rather than grading-oriented engineers. Licensed civil engineers, or staff engineering geologists, were primarily involved with the review process prior to issuance of grading permits. At that time, the majority of inspectors were either combined building inspectors or combined public works/engineering inspectors whose primary experience lay in building or street-improvements, rather than the grading aspects of construction.

1991 QUESTIONNAIRE

In 1991, a more detailed questionnaire was sent out to 576 building officials, all Class A members of the International Conference of Building Officials (ICBO) throughout California. Many of the Class A members were fire marshals, State of California Youth Authority, or Office of the State Architect, who are not necessarily involved in grading code enforcement or control. Of this number, 160 responded, most of which were building officials, public works officials, and a small number of planners. A few respondents (10) were fire marshals or from State/Federal agencies. Even though there was only a 21% response, the total number represented nearly twice that had responded 14 years earlier.

In the 14 years interim, there appears to have been a slight increase in the number of agencies that have met the 1974 State mandate (requiring that all cities and counties adopt Chapter 70 UBC or equivalent grading code) from 92% of the respondents in 1977 to 96% in 1991. The number of specialized staff grading inspectors increased slightly, from 13% to 15%. The number of combined building inspector-grading inspectors in creased from 32% to 47% and combined public works inspector-grading inspectors increased from 28% to 38%. This indicates a slight increase in awareness of grading control functions. Agencies that did not enforce any kind of code dropped from 8% to 4%.

Staff geotechnical personnel doubled, from 6% in 1977, to close to 12% in 1991. The use of outside geotechnical consultants increased measurably from 5% to 40%. In 1991, 30% of building officials employed outside geotechnical consultants for peer review. Occasionally, 42% employed outside consultants, while 13% responded that they employed them often. The following peer review purposes were utilized:

Peer review of geotechnical-reports only	13%
Peer review of reports and field pre-inspection	14%
Peer review of grading plans	13%
Peer review on a case-by-case basis	23%
Peer review of field inspection at Dept. request	24%
Peer review of all grading inspection	18%

This represents a tremendous increase in local agency awareness with respect for the need and the application of geotechnical expertise in the hillside grading for development. Hopefully, this awareness will increase in the future as more agencies are faced with development of their hillside areas.

For the most part, miscellaneous queries in the questionnaire were generally divided evenly between yes and no responses:

Statements: Yes or No (1991 Survey)

1. We are generally understaffed for grading inspection - yes 42%, no 48%;
2. We give more attention to the larger grading jobs - yes 43%, no 44%;
3. We basically depend on the applicant's geotechnical/ civil engineers to insure Chapter 70 requirements are met
   - yes 44%, no 38%,
   - during design - 31% yes.
   - during construction - 26% yes.
4. Our inspectors are not particularly well trained in grading operations and testing - yes 45%, no 40%.

Problems experienced in the jurisdictions:

1. Expansive soils problems - yes 54%, no 39%;
2. Inadequate perimeter design - yes 21%, no 58%;
3. Long-term settlement of fill embankments - yes 29%, no 58%;
4. Landslides and slope failures during wet years - yes 31%, no 56%;
5. Drainage problems - yes 68%, no 18%;
6. Drainage/run-off collection problems - yes 62%, no 28%;
7. Stream bank erosion problems - yes 29%, no 56%;
8. Deteriorating infrastructure problems - yes 38%, no 41%;
   corroded culverts - 1%
   broken sewer lines - 2%
   siltation of run-off conveyance - 4%
   pipes/drains - 3%

Litigation Concerns

1. 16% indicated that they have been named in one or more lawsuits in recent years because of alleged ownership of failed improvements or inverse condemnation suits.
2. 16% indicated that developers had threatened lawsuits over plan review or permit issues.
3. Only 3% indicated that the threat of lawsuit made them back down in its concerns in the review/permit/inspection process.
4. 13% indicated that their immediate superiors appear to often be influenced by outside political pressures or job security concerns.
5. 15% indicated they had to collect on performance bonds when grading or paving contractors were unable to complete work.
6. 14% indicated that developers in their area are generally able to exert powerful influence with respect to most technical aspects of their projects.

Over about three-fourths of the respondents indicated that they would be interested in, or would support grading inspection seminars. The author and J. David Rogers routinely conduct two-day grading code administration and enforcement seminars for the ICBO and the Association of Bay Area Governments (ABAG) and have found an increase in their popularity.

ASSIGNMENT OF GRADING SECTION FUNCTIONS

The 1977 survey indicated that grading code enforcement was evenly distributed in California between the building official and the public works official (47% and 44% respectively, with 9% in the Planning Department). However, the 1991 survey shows an increase toward the building official as the primary regulator of grading codes (65% and 28% respectively with 7% in the Planning Department). This shift is significant and merits discussion.

The primary responsibility of all building officials is to protect members of the public from dangers to life, limb, and property through knowledgeable quality-assurance review and inspection of all construction within their jurisdictions. Grading division personnel are specifically charged with review of grading plans for safety and standard of care considerations as well as earthwork operations during construction. To perform this responsibility, grading personnel must be familiar with all aspects of grading construction, be able to accomplish their assigned tasks, and must reflect positive attitudes about the importance of this responsibility. Standard plans and policies exist to guide division personnel toward this goal through Chapter 70 and the standards that have been available since 1963. With experience, grading inspectors perform more efficiently, thereby gaining positive support from their departmental supervisors, the respect of grading contractors over whom they pass judgment, and the community that supports them. Each department member, regardless of status, is responsible for carrying out his or her share of departmental policies and guidelines as well as applying technical skills in performance of tasks. By concentrating on similar phases or groups of activities, the grading inspection staff can fully utilize specialized knowledge and skills. The functional divisions of a grading section are distinctly established here to assure that adequate attention is given to every activity concerned.

The rationale for having a separate grading section has been justified in Southern California along with statistics that support effective grading-code enforcement as a means of reducing losses due to natural geologic hazards (Scullin, 1983). Grading-code adoption has been mandated in California, and its effective implementation will eventually be mandated throughout the nation wherever development pushes into areas of marginal natural stability. Effective grading-code enforcement provides minimum reasonable standards of protection for the health and safety of community members. Such enforcement defines the minimum acceptable standard of safety as a guide for applicants (builders and others) and users or occupants of those sites created or improved through earthwork. It is of little value to adopt a grading code without properly enforcing and implementing it because safe construction sites result from the field application of adequate safety standards, not from administrative adoption of those standards.

The manager who is designing a new organization, or who is modifying an old one, should be familiar with typical ways of grouping grading administrative and inspection activities. Recommendations in this paper are intended as suggestions for developing an organizational structure that can be adapted to the specific needs of individual governmental agencies rather than as a single plan to be universally copied. Activities are grouped in accordance with their characteristics by purpose, process, clientele, and area.

Building departments are normally formed according to purpose, involving either quality assurance inspection and regulation or clientele, such as the construction industry. In contrast, public works organizations are normally organized along lines of process—that process being engineering (e.g. community development, flood control, transportation). However, the subgroups in public works departments are commonly organized along purpose lines. Building departments are also involved with the engineering process; however, the application of this process is oriented toward regulation. A governmental agency must identify its vital operational factors and then combine activities that give optimum results in terms of those factors. Thus, the agency that seeks to give prompt service to the public may find that coordination is a dominant factor; whereas the agency that emphasizes efficient quality inspection at a low cost may stress economies from functional specialization. Grading sections or divisions have a specialized functional orientation.

When grouping or reassigning grading activities into administrative units, these combinations of benefits should be considered:

1. *Take advantage of specialization.* Take advantage of both functional and allied specialists. In addition to grading-code regulation, staff soil engineers and engineering geologists may assist other engineering division sections, such as structural plan-check, while grading inspectors may check single-lot grading plans to assist the grading plan-check section. Grading plan-checkers may also check retaining wall and foundation designs to assist structural plan checkers since the former have a better knowledge of soil-engineering considerations and will have reviewed the site conditions section(s) of consultants reports.
2. *Facilitate control.* Supervision is easier and more efficient in an organization in which the supervisors are knowledgeable about the various activities being performed. They can more efficiently exert control over jobs with regard to the specific construction activities, as well as organizational concepts and policies.
3. *Coordinate activities.* Building department personnel can logically coordinate interrelated activities involving common objectives of various regulatory agencies. Their expertise in earth-science applications would suggest that geotechnical staff could, with great value, coordinate their activities with regard to reviews for and consulting to planning, road, flood control, and water and sewer districts. They may also coordinate with interdepartmental zoning, sand and gravel, and oil-well drilling code sections.
4. *Secure adequate attention.* An important responsibility of the grading section is to earmark earth-science applications with regard to urban development planning, design, and construction. In this manner, the long term safety and welfare of the community is more assured, because developer and design professional liability is only limited to 10 years after the project completion.
5. *Recognize local conditions.* The administrator must recognize local environmental conditions and determine which personnel and what resources would be necessary to accommodate those conditions adequately and efficiently.
6. *Reduce expense.* The administrator may reduce expenses through selective utilization of the available geotechnical expertise, a practice that will increase community support.
7. *Advise.* Grading personnel can advise other departments or interdepartmental sections about earth-science applications and considerations. By providing awareness of earth-science conditions (especially areas of known trouble), they may assist another department to work more economically or effectively. This practice is especially beneficial to small agencies with limited finances, and to large municipalities as well.

The writer's experience with grading-code regulation has been in both a city engineer's office and a county building department. Whether a building department is autonomous or is combined with a public works or planning department, grading-code enforcement belongs within a building department because the functions are similar with regard to enforcing regulations concerning development of private property. This functional regulation activity is consistent with the building department organization structure. The actual enforcement and control procedures are basically the same, for example:

1. They have the same legal authority, and the administrative chapters of the code are consistent.
2. Their ultimate authority and purpose are the same.
3. Their operational processes are the same.
4. Their stage inspections, approvals, and releases are similar;

   a)They can efficiently and effectively hold up building permits until rough grade is approved,

   b)They can hold up releases to utilities of certificates of occupancy until fine grading and drainage devices or protective devices are completed.
5. Their code violation and complaint investigation procedures are the same.
6. Their code compliance or court preparation activities are the same.

It is obvious that grading-code regulations are directly related to building-code regulations in purpose, functional activity, and operational processes.

Another type of problem related to organizational structure deals with the relative merits of including building regulation functions in a public works department or developing building regulations as an autonomous department. Gerald B. Wilson (1964) described the pros and cons of building departments versus public works organizations in the administration of building codes. Wilson solicited opinions of building officials, city managers, public works directors, and planning directors concerning the choice of a building regulatory agency being subordinate to, or included as, a branch of a public works department. He found a number of convincing arguments for inclusion in a public works department:

1. improved coordination in solving problems involving street grading, industrial waste, street use, and off-street parking access;
2. ready access to registered civil engineers;
3. combined inspection of some functions, which reduces processing time for permits;
4. reduced control by city managers;
5. reduced operating expenses;
6. improved efficiency of economics.

However, the arguments against this departmental structure are much stronger from the standpoint of protecting the health and safety of public.

1. The functions of these two agencies are very dissimilar: a public works department is concerned with design, service, and management of public property, while a building and safety department is concerned with enforcing regulations concerning the development of private property, so as to provide for the long term health and welfare of its citizenry.
2. A building department needs direct contact with the city manager by someone familiar with its special needs.
3. Building regulations are generally less-stringently administered under a public works, rather than a building department.
4. Building departments are usually more efficient when combined with planning departments, than with public works departments, because the first two departments share compatible functions.
5. Permit fees intended for building department services may be used in public works departments for unrelated costs of government administration instead of for their intended purpose.

Perhaps the strongest argument against placing a building and safety department within a public works department substantiated by Wilson was his analysis of arguments against this departmental structure. Data from his public works questionnaire indicated that none of the directors who responded to his survey possessed either training or background in building regulation work. And, perhaps understandably, the vast majority indicated that contacts with their own building departments *"were of little or no importance to them in the execution of their duties."* This attitude is in direct conflict with feelings of planning and building officials, who uniformly feel an urgent need for close contact and coordination.

Wilson's research (Wilson, 1964) briefly touched upon grading and excavation regulations. No clear pattern for assignment of this relatively new function had emerged from his study in 1964. Indeed, there is more evidence of confusion related to this function than to any other function investigated. Wilson concludes that, regardless of who may be as signed, ultimate responsibility for grading-code enforcement, the primary enforcement tool in the field will be the control of building permits and inspections. Where the responsibility for enforcement is not in the hands of the building official, close contact should be maintained between his office and the enforcement agency. More often than not, in the cases studied, this function was not assigned to the building official. A review of the basic criteria for the assignment of functions suggests that re-assignment of the grading-code enforcement to the building official should result in more efficient enforcement, provided proper staffing is a part of the transfer.

Standards of Care in Grading Code Enforcement

The California Division of Mines and Geology (CDMG, 1973) indicated that there is a potential $55 billion geologic hazard problem in the state between 1970 and 2000. In this publication, Urban Geology Master Plan for California, they state that the single most effective action that can be taken to reduce losses due to geologic problems is for cities and counties to adopt and diligently enforce modern grading ordinances and building codes. The report goes on to say that, considering anticipated technical advances, it should be possible to reduce property losses by as much as 90 percent. Effective administration and grading-code enforcement alone could reduce or substantially eliminate losses. They estimate that with present technology, strict code enforcement could reduce that projection of loss from $55 billion to $39 billion, at a total enforcement cost of $6 billion for a cost-benefit ratio of 6.2:1. Such an analysis clearly suggests that enforcement of adequate building and safety codes is beneficial.

Slosson and Petak (1987) have noted what appears to be a widening of the gap between standard practice and state-of-the-art or, more succinctly, current capabilities utilizing available knowledge and prudent practices. Studies of case histories of landslides, debris flows, settlement, and other geologic-related types of ground failure have shown in creasing loss of life, as well as property and economic losses that are increasing at a rate greater than inflation. Slosson and Petak (1987) stated that the density of population in hillside areas has increased dramatically since the 1950's. Best estimates on dollar losses from all landslides in the United States range from $1 to $2 billion dollars per year. About $9 billion is attributed to damages to residential structures. They have estimated that losses in California, alone, during the past decade may be easily as high as $2 billion.

Earl Brabb (1987) has estimated that approximately 90% of the landslide-related failures of the past decade could have been avoided through proper engineering solutions and/or land-use management practices. Landslide-related losses were decreased by 97% when a modern grading code was introduced and enforced in the City of Los Angeles in 1963 (Slosson, 1969). The Modern Codes, Standards of Care and the State-of-The-Art available today, are the same as they were in the early 1960's. Slosson and Petak (1987) ask the very basic question - Why does the loss factor increase or why has the gap between "standard practice" and "state-of-the-art" widened if the basic knowledge and methodology for recognizing, predicting, and preventing failure was available at least two decades ago? Their explanation suggests that increased damage levels are due to the following significant factors:

1. The failure of government to adequately review, inspect, and regulate. This, in part, is due to government's reluctance to hire personnel with expertise in geology and geotechnical engineering and, in part, due to government's reluctance and/or outright refusal to provide quality control.
2. Lack of availability of basic geology and engineering classes at most universities. The authors have previously estimated fewer than 5% of graduating students in geological sciences have taken or even had available to them classes in engineering geology, groundwater, marine geology (or costal processes), geomorphology (geomorphic processes), soil mechanics, hydrology, and other subjects pertinent to success in solving problems related to landslides and other types of land failure.
3. Lack of enthusiasm by many practicing geologists to take the time and extend the effort to learn basic fundamentals and/or become involved in continuing education.
4. Lack of commitment to excellence and/or basic scientific truth. Many approach ethics more as a set of guide rules or company/government agency policies rather than as the basic concept of the scientific method of analysis and scientific integrity. Some have forgotten and some may never have been taught. We even hear some suggest that it is acceptable to overlook facts in order to adjust for social and/or emotional needs.
5. Another dimension to the problem is - how many "professionals" are educated and trained with an understanding of what knowledge is available and how to use it? How many professionals are really capable of achieving "state-of-the-art" level of knowledge and expertise? If Slosson and Petak's data base is correct, only 5% to 10% of the practicing engineering geologists and geotechnical engineers have taken necessary university level courses and/or received the proper on-the-job training. *Therefore, the majority of the professional work by which "standard practice" is measured is accomplished by the 90% who have not been satisfactorily educated and trained to accomplish "state-of-the-art" level of professional work. The yardstick used for measurement is short of target. It appears that the level of professional work will drop to the lowest level allowed by government and this level or standard is being established by the median of the 90% that have not been properly trained. Another way of stating this is that those who are setting the standards are incapable of setting these standards.*
6. In our experience while looking at or preparing case histories of failures, we have noticed another common denominator to failure or error. This is related to the often requested assignment by government and/or industry to look for the problem or to provide inspection and supervi-

Figure 1. The compacter on the left is sidewall benching. The water truck is spreading water, the rubber-tired blade is spreading and the scraper on the right is spreading its load. Photograph by C. Michael Scullin.

Figure 2. Grading operation in progress. Scrapers transport earth materials from the cut area and spread it in the fill area where it is spread in thin lifts of four to six inch thickness, watered, and compacted. Photograph by John Shadle, Staff Photographer.

sion - but at the same time remember that, you are being paid to not find rather than to find the problem or defect. Often a governmental agency issues a permit with a condition of approval or variances provided and suggestions that on-site inspection and supervision be provided and that design changes or mitigation will be initiated if any unforeseen conditions are encountered. In this situation, the design and construction professionals are under pressure to "not look" or be replaced.

7. Finally, during the late 1960's we saw the development of "marketing" with the often heard phrase - You know we can do (or could have done) a better job, but if we didn't adjust to the demands of our clients, others would and they would have the marketing advantage.

This has led to a lowering of professional standards with more "rolling of the dice" or "taking a risk" with, of course, the inevitable - more failures. By just combining items #1 and #7, the obvious occurs - the standard of professional work drops to the lowest standard allowed by government.

The author has spent 33 years in grading code regulatory work, and 2nd party or 3rd party review for governmental agencies and for other consultants, as well as expert witness testimony in Standards of Care issues. We have observed additional factors contributing to the widening gap between state-of-the-art and the existing and past Standards of Care in our grading industry. Some of these are as follows:

1. We need to reduce the enormous time lag of 8 to 25 years that exists between the discovery of new technological process and their application by the appropriate public agencies. Slosson and Hauge (1973) describe the lag in three stages.
   (a) They cite a one-to-five year delay between discovering and applying a new technique, which . . . it may take from two to five more years for the technique to be codified.
   (b) An additional delay from five to ten years may occur between codification and effective administration and enforcement.
   (c) Some building officials do not like new technologies because they do not yet understand them; they may hold up the project to the point that the client gives up and fires the consultant in frustration. This clearly requires more communication and training to increase the understanding of the new technology and its merits.
2. Even through the number of staff grading inspectors has in creased in the past 14 years, there is a tremendous difference in effective grading inspection throughout the state. Forty-two percent of the questionnaire respondents stated that they were understaffed for grading inspection while 45% were not particularly well trained for grading inspection activities. Forty-four percent responded that they basically depend upon the site consultants to insure that Chapter 70 mandates have been met. This means that regulatory agencies need more trained inspectors and more 3rd party reviews to assure that the consultants are on the project and doing the job they recommended or promised. The threat of lawsuit if something goes wrong hasn't been successful in skimming the tide of poor quality control on grading jobs. Quality assurance inspection is based upon stage approvals. In grading control, these stage approvals are accomplished through geotechnical observation, inspection, testing and analysis, <u>during each phase of grading construction</u>. If these various stages of grading construction are not approved or acceptable to the licensed geoprofessional or project civil engineer, then they shouldn't be approved by the regulatory agency. A rough grade building site is <u>not</u> considered <u>safe</u> for construction or issuance of the building permit until those responsible professionals (engineers and geologists) indicate that it is safe through their field memos, letters or reports. Since 1964 Chapter 70 called for these stage approvals and the grading/building official inspects according to these stage approvals.

3. Today there is a decreasing standard of care due to:
   (1) Drought periods.
   (2) Retirement/death causing loss of experience.
   (3) Changing political attitudes.
   (4) Relaxation of requirements throughout the state.
   (5) Relaxation of education with more emphasis on computer application and less on field application.
   (6) Difference in enforcement of regulatory codes throughout California.
   (7) Laxness on the part of building officials perpetuated by feeling that responsible professionals will answer to private attorneys if anything ever goes awry on one of their past projects.

The effectiveness of stringent grading codes as deterrents against natural and man-induced landslides has been dramatized and statistically analyzed for the Los Angeles area. However, because heavy rains, moderate to strong earthquakes, paralyzing snows and other natural disasters are unpredictable and sporadic, it is difficult, throughout most of California and perhaps even most of the United States, to sustain public interest in preventive measures long enough to force code changes, implementation, and enforcement before a disaster. In fact, according to Slosson and Hauge (1973) significant changes in public policy implemented by legislation or building codes *occur only in response to a major natural disaster or sequence of disasters.* It is only then that *the press, the public and the legislatures will react.* Consequently, long drought periods create a false security of *don't sweat it while the sun is shining*.

Additionally, this attitude not only allows a relaxation of code requirement and enforcement but aids in changing political attitudes. This is particularly true where professional consultants agree to or are complacent with a lesser standard of care than the basic minimum required by the code. This rescinding of standards is accomplished by those consultants described under the significant factors outlined previously. Those of us who provide third-party peer professional review of geotechnical reports have seen a marked increase in poor quality professional work corresponding with a relaxation of basic code requirements in both northern and southern California. You have to understand that the basic grading control requirements, methods, codes and techniques that we presently operate under have been available to the industry since 1963. The 1963 state-of-the-art grading control has been tested by the heavy rains of 1969, 1978-82, and earthquakes of 1971, 1987, 1988, 1990 in southern California and performed very well.

In hindsight, it would now appear that some of the best grading work was accomplished during the mid-1960's when the consultants, regulatory agencies and contractors were working together to form new standards in order to mitigate storm-related damage and inevitable liability exposure. Thousands of safe building sites were created between the modern 1963 code and late 1960's. These still appear to be performing well. In the interim, we have seen a large body of litigation involving grading prior to 1963 and after the late 1960's. Much of this litigation is due to ground failure or subsidence due to improper grading techniques, relaxation of quality assurance inspection and relaxation of education and training. Many of the 1960's grading inspectors, soil engineers, engineering geologists and civil engineers experienced in grading control and methods are no longer active.

Many of the post-1970 grading officials and professionals have not been afforded adequate training or education with a sound work ethic or commitment to excellence. More often we are finding that there are practitioners in our industry, both public and private, who are *sincerely incompetent*. They want to do a *good job*, but have never been taught what a *good job is* or should be. This is frightening because it creates a distrust among public officials who are dependent upon competent professional expertise. It also creates frustration among competent professionals who are vainly trying to do a good job, but are continuously turned down by the agency during the review process because they are perceived to be overly cautious due to the poor quality reports submitted for review by others. Furthermore, the promoters of poor quality work who are knowingly doing incompetent work are all the while getting more work from the developers while taking work away from the more conservative firms. Thus, the regulatory agency is burdened with the necessity for more scrutiny in the review process and more vigilance in quality assurance inspection. If this is not possible due to insufficient personnel, lack of training or inadequate third party expertise; it becomes obvious that the stage may be set for failure, litigation and large taxpayer endowments in the near future. Some of these *knowingly incompetent consultants* are cavalier in their defense and try to make the building official responsible for their inadequacies. We have known some who have been sued so many times that they have gone bankrupt, formed new companies and are back performing sub standard work under different company names. As Slosson and Petak (1987) noted that, *professional standards drop to the lowest standard allowed by government*.

There are two steps that can and should be taken now to help cure the process. First, the performance of public officials currently engaged in grading-code administration should be improved. They have to either hire more quality control inspectors, undergo additional training in grading code inspection and enforcement and/or retain more third party review expertise to counter the incompetents and control the grading construction. Second, the time lag between codification of change and effective enforcement of that change should be shortened. Intense development pressures in growing urban and suburban areas are forcing development of sites with known geologic hazards. Coupled with demands on developers, builders, engineers, geologists, and governmental agencies to accept more and more responsibilities for their professional activities, liability risks for building-industry personnel are increasing logarithmically. Dewitt Nelson (1965), former Director of the Department of Conservation, State of California, feels that *structures will be built—and in increased numbers, and with vastly increased costs—on sites that will require all the scientific knowledge and engineering skills we can muster to render them safe*.

The State of California Department of Consumer Affairs has tried to instigate sunset legislation of the licensing boards on several occasions due to lack of investigations and enforcement by such boards. It appears that the boards are afraid of being sued for *preventing one from earning a living*. If the state boards relax in policing the professionals, this of course breeds relaxation of requirements by those professionals. However, licensing does and can weed out those who are not current in their education and training, and forces them to gain the required knowledge to pass the tests. This of course, only takes care of the newcomers to the professions, not necessarily the older ones. Many of those in the applied earth science professions need to change their apathetic attitudes and provide better training of their staff personnel. The State also needs to police the professionals more closely as it appears that neither the professional societies nor the professionals are going to do so themselves.

Changes to Chapter 70, U.B.C.
The California Geotechnical Engineers Association, (formerly SAFEA), in southern California made several changes to Chapter 70 in 1984, now published in the 1988 edition of the UBC. The primary change in philosophy is illustrated in changing the definition of approval, (Section 7005) from *shall mean a written engineering or geological opinion concerning the progress and completion of the work, (1976 UBC)*,to *shall mean the proposed work or completed work conforms to this Chapter in the opinion of the building official,* (1988 UBC). In Section 7010 (c) *Preparation of Ground,* the last sentence has been revised so that the soils engineer or engineering geologist accepts instead of approves the site as it is the building official who makes the approvals and not the soil engineer or the engineering geologist, Bush (1988). Additionally, the verbiage in Section 7015 (a) *Final Reports*, has been changed according to the same philosophy. Bush, indicates that Item No.1 has been revised to indicate that the civil engineer is not tacitly responsible for *approval of completed work*, rather to state to the *best of his knowledge* that the work was done in accordance with the final approved grading plans. Bush goes on to say that changes have been made to Items No.2 and 3 so that the consulting soils engineer makes a *finding* rather than providing an *approval*. It is the building official who provided final approval under the requirements of Section 7015 (b). The ICBO Building Standards Proposed Code Changes stated that the reason for this change of philosophy was, *the geotechnical consultants only render an opinion while the building official approves or disapproved the work.* The building officials and the ICBO Code Change Committee has accepted these changes.

We realize that this change of philosophy and transfer of responsibility is a result of much litigation where trial attorneys have attempted to maneuver the professional consultant into a position to *guarantee the stability of the site* when or if he approves the grading work. As part of the professional consultant's defense against liability and loss prevention, he has been advised by his own legal counsel to change the wording in his reports as his test findings and statements of professional opinion do not constitute a guarantee or warranty, expressed or implied. Under California law, the building official cannot be held personally liable if he approved or disapproved a building site, (California Government Code section 816.4) so from a risk or loss prevention standpoint the building official is best suited to approve the grading site construction. However, realistically, the geotechnical consult ant will likely be named in any suit where there is distress litigation and, unfortunately, he will have to pay attorney fees to defended himself even if it is to prove that his actions were completely defensible.

The building official is generally not trained in grading engineering, geotechnical engineering nor engineering geology and, therefore, is not technically qualified to approve the safety of a graded building site unless it is *approved or found to be safe* by the professional consultants. His approval is generally based upon the professional's written approval of the site. It is our opinion that current trends in litigation will eventually bring the professional consultant back to being the person who approves the safety of the site with written opinions concerning the progress and completion of the work rather than expressing a *finding* that is supposed to relieve them of professional responsibility for site safety. In reality, such wording changes have done little to keep geotechnical professionals from being named in lawsuits; if there is a perceived ground failure, the geo-professional will likely be named in a lawsuit if his/her firm is still around.

Other additions or deletions to the 1988 UBC edition were approved by ICBO. There are references in Section 7010 (e) compaction to the standard compaction tests *70-1, Maximum Density Test, 70-2 In-place Density Test, 70-3 Balloon Method, 70-4 Drive-cylinder Method and 70-5 Nuclear Method.* The density tests 70-3 through 70-5 were included into the 1982 Standard Tests and the 70-1, ASTM D1557-70 was revised to the unmodified 5 layer method, in lieu of the 3 layer method, thus providing for greater density. Test method 70-2 was revised to agree with ASTM D1557-78 without changing the testing technique.

Building setbacks from both top and toe of slope were moved to Chapter 29 (See figure 29-1). The purpose of this move was to provide the structural setbacks in case the UBC was adopted without the appendix Chapter 70. During the debate at the ICBO Code Changes Committee in San Antonio, Texas 1983, there were some building officials who preferred having Figure 29-1 left in Chapter 70, as well as Chapter 29, because it applies equally to graded slopes.

Section 7012 (a) General has been changed to delete all drainage facilities and terrace drains on cut or fill slopes of 3:1 or flatter. If this is allowed on slopes composed of sandy silt or clays there will be extensive erosion, regardless of the gradient. Terrace drains were originally devised and required in response to height of slope, regardless of the rock or soil composition to rid the slope of water rapidly to minimize fill erosion. Long, exposed slopes at 3:1 or flatter gradients in sandy silts, silty sands and sandy clays will still erode quicker than natural slopes. Building officials who are planning to adopt a modified version of Chapter 70 UBC should consider this change. You may need a wide access road at the base of the slope to collect the siltation during heavy rainy periods. The following photos illustrate the shortsightedness of the new 3:1 ditch waiver subsection by allowing 150' high cut and fill slopes in landslide-prone terrain without terrace drains, brow-ditch drains, impact walls or diverter walls. When the rains come, and they will, there will be a straight chute for the mudflows to impact the residences built on the graded pads below. This would appear to be absurd in light of the experiences of the 1950's and 1960's. In landslide-prone terrain, more terrace drains are needed to minimize water infiltration into the subsurface instead of less drainage control. Drainage control, surface and subsurface, is basic to the long term stability of a hillside environment! In the cases depicted in the photographs, the regulatory official based his approval solely on the geotechnical consultants recommendations/approval.

Changes made to Chapter 70 in the 1970 edition that do not make sense are the removal of Sections 7007, *Permit Limitations and Conditions (a) through (f) and Section 7008, Denial of Permit, (a) Hazardous Grading and (b) Geological or Flood Hazard.* These subsections were supposed to have been transferred to UBC Section 303 of the Administrative Chapter. Only the validity (liability) jurisdiction of other agencies and time limits were transferred. Storm damage precautions, Conditions of approval, Hazardous Grading, and Geological or Flood Hazard subsections were eliminated. The building official should consider reinstituting them in view of the extensive damage to residences built on ancient landslides, within flood plains, or deep canyons subject to flood damage during periods of intense precipitation.

Another such change was the adding of Section 7012 (d) *Exception A.,B. and C.* These exceptions were slipped into the code after the final draft of October 22, 1968 and were not approved by the Committee. They also do no make much sense! They have nothing to do with the gradient of the pad drainage, which is the subject of this subsection. The Committee suggestion was: *Exception. The gradient from the building pad may be one percent (1%) where building construction and erosion control will be completed before hazardous conditions can occur.* It does not read anything like the Exception: A,B & C in the current Section 7012 (d) Disposal. Building officials and ICBO might consider removing the current exceptions that have nothing to do with drainage facilities or drainage gradients and replace it with the original approved exception.

Deputy Grading Inspector

During 1987, the City of Los Angeles proposed to change the Building Code to enlarge the scope of responsibility of the Deputy Inspection program by adding deputy grading inspector to their *Section 91.0310, Section 1., Subsections (a), (b) and (c) of the Los Angeles Municipal Code.*

The inspector shall be approved by, responsible to, and may be employed by, either directly, or through the architect or engineering firm in charge of the design of the structure, or through the geologic or soils engineering firm providing technical design data for the project, or through an independent approved inspection/test firm.

The engineer's or geologist's *Certificate of Compliance*, Code Sec tion 91.0306 (a), requires the Registered Deputy Grading Inspector be approved by and responsible to the geologic or soils engineering firm providing technical design data for the project. To insure compliance the firm must sign the Certificate of Compliance, (Form B-18), at the completion of the project stating the registered deputy or deputies were responsible to them. Therefore it is important that the geotechnical firm approve the registered deputy grading inspector prior to the time the deputy commences his duties, and also that they know when the deputy first reports to the job.

The above will be in a form of a Memorandum of General Distribution (MGD) originally dated December 29, 1989. Although it is not yet in effect, the City is preparing exams and progressing towards establishing a list of Deputy Grading Inspectors similar to their other Special Inspectors lists. The City currently has a list of 25 tested and approved deputies. The testing covers the grading ordinance, field techniques and terminology, unusual hazardous conditions, crib walls, gravity wall systems, ABC slot cutting, tiebacks, piles, subdrains, backdrains, compaction, dewatering and buttresses. So far there has been about a 30% failure rate.

It would seem that a deputy grading inspector must have a broad knowledge of geotechnical applications, soil testing and procedures, field grading methods and techniques, and code knowledge and enforcement procedures. But, soils engineers and geologists are not normally trained or experienced in code enforcement, much less the soil technicians. Historically, a public agency's grading inspectors have not been fully trained in soil mechanics and soil testing. It seems that the deputy grading inspector must be recruited through the ranks of the private soil technicians, or possibly, the older grading inspectors from various City and County agencies. Some laboratory technicians, engineers, and some geologists have currently been tested. He or she must be a combination of both to provide the soil technician's level of competency in grading control and the grading inspectors' knowledge and judgment in grading code enforcement. The deputy would also need to learn the building department's enforcement procedures. The background requirements are similar to the Class II NICET license, or four years journeyman experience, or 2 years grading inspector.

The deputy will be registered and responsible to the City. However, the geotechnical consultant-of-record will have to approve the owner selection of a particular deputy and sign the deputy's final paperwork. It appears that the City will have a full-time inspector on-site and the consultants will be responsible for their work. The City can withhold grading permits or a grading job until one of their approved deputies are on-site, making the soil engineer or geologist fully responsible for the deputy's work, but the geotechnical firm will have no control or authority over the deputy's qualifications, ability, background, experience, or work integrity. There would also appear to be a conflict of interest for the consultant to be responsible for both the soil technician and the deputy grading inspector.

We realize that this deputy grading inspector method will help relieve city and counties who are understaffed or lack

Figure 3. Looking down a 150' high, 3:1 fill slope toward the building pad. Note the abscence of terrace or brow-ditch drains. Photograph by C. Michael Scullin.

Figure 4. A recently graded 3:1 cut slope, 150' high, constructed without ter race or brow-ditch drains. Photograph by C. Michael Scullin.

Figure 5. Gully erosion after one rain on the 150' high fill slope shown in Figure 3. Photograph by Mitchell D. Wolfe.

Figure 6. Siltation of the rear yard and drainage device after one rain in the same development. Photograph by Mitchell D. Wolfe.

expertise in grading field control or grading code procedures. However, it will likewise tax the geotechnical industry, which is always having to train new technicians and inspectors. It doesn't seem logical or ethical that a geotechnical firm would have to be responsible for someone that they have no control over, either in their training or their work ethic or integrity. Even though the City of Los Angeles is again assuming a pioneering role in this possible grading control methodology, it will be the geotechnical firms who will again have to produce the techniques and define the work parameters for this methodology to work for the safety and welfare of the public.

CONCLUSION

This paper has provided a general review of the development and increasing effectiveness of modern grading codes during the past three decades. However, it clearly emphasizes the need for stronger government control and quality assurance inspection of grading construction, as well as the need for conservative third party review of geotechnical reports submitted to the regulatory agency. The latter is unfortunate indeed, but geotechnical consultants have no one but themselves to blame for the widening gap between the *Standards of Care* and the *State-of-the-Art* as practiced today.

The *modern codes, Standards of Care* and the *State-of-the-Art* available today have been available since 1963. When applied and enforced, they are very effective as deterrents to deleterious earth movement and property loss in our residential environments. In order to accomplish this effectiveness, we must promote the following:

1. We must reduce the enormous time lag that exists between discovery of new technological processes and their application by appropriate regulatory agencies.
2. We must increase the staffing, training, and awareness of grading personnel, both public and private.
3. We must tighten our control of stage approvals during construction and increase responsible action by the geotechnical community, as well as the regulatory agency.
4. We must minimize the differences in enforcement of regulatory codes throughout California.
5. We must eliminate the relaxation of requirements and continuing education that currently prevails.
6. We must change political attitudes toward quality assurance inspection and regulatory code enforcement.
7. We must go back to the basic considerations of long-term public safety and welfare, instead of trying to make money at the expense of the public and through impliance, making the building official responsible for approvals just because he isn't *held legally accountable* under current statues.
8. The State Department of Consumer Affairs should police the *knowingly incompetents* instead of taking the easy way out by sunset legislating the licensing boards.
9. We must not make the building official *responsible* for the *geotechnical consultant's expertise*. Building officials are actually dependent upon the consultant's professional competence in exercising their expertise. If the geotechnical community fails to take charge of their own destiny in this area, socialized engineering will result from forcing government to take full control of the matter.
10. There is no need to continuously amend Chapter 70 to make it less effective or too general for use. The original Chapter 70 was already a minimum performance code that allowed the building official to make use of professional consultants in the enforcement of stage approvals to obtain the quality assurance in hillside development. To lessen the performance aspects of the code increases the opportunity for future ground failure of the graded sites. *The standard of professional work then drops to the lowest standard allowed by governmental regulatory agencies.*

REFERENCES

Brabb, E. E., 1987, Earthquakes and Volcanoes: "Losing By a Landslide", Vol. 18., No. 4.

Bush, V. R., 1988, Handbook to the Uniform Building Code, An Illustrative Commentary: International Conference of Building Officials, Whittier, CA, 533 p.

California Division of Mines and Geology, 1973, Urban Geology Master Plan for California: CDMG Bulletin 198, 112 p.

DeWitt, Nelson, 1965, Role of the Department of Conservation in Landslide and Subsidence Geologic Hazard: Landslide and Subsidence Geologic Hazards Conference, May 26-27, Los Angeles, CA; The Resources Agency Sacramento, p. 9-11.

Kockleman, W. J., 1986, Some Techniques for Reducing Landslide Hazards: Bulletin of the Association of Engineering Geologists, Vol. XXIII, No. 1, p 29-52.

MacFadyen, C.R., 1966, Grading Control Procedures: Seminar on the Importance of Earth Sciences to the Public Works and Building Official, Glendale, CA, AEG Special Publication , p. 65-99.

Olshansky, R. B., 1989, Landslide Hazard Reduction: A Need for Greater Government Involvement: Zoning and Planning Law Report, Vol.12, No. 3, Clark Boardman Co., NY, p 105-112.

Olshansky, R. B., 1990 Landslide Hazard In the United States: Case Studies in Planning and Policy Development": Garland Publishing, NY, 176 p.

Olshansky, R. B., and Rogers, J. D., 1987, "Unstable Ground: Landslide Policy in the United States": Ecology Law Quarterly, Vol. 13, No.4, p. 939-1006.

Scullin, C. M., 1983, Excavation and Grading Code Administration, Inspection and Enforcement: Prentice Hall, Englewood Cliffs, NJ, 405 p.

Slosson, J. E., 1969, The Role of Engineering Geology in Urban Planning and Development: Special Publication No. 1, Colorado Geological Survey, Denver, CO, p. 8-15.

Slosson, J. E.,1985 Encouraging Government Mitigation: The Forensic Geologist/Expert Witness Perspective: presented at the American Bar Association International Symposium, Miami, Florida, October 23-26, 6 p.

Slosson, J. E. and Petak, W. J., 1987, Why Is The Gap Between "Standard of Practice" and "State-of-the-Art" Widening?: AEG News, Vol. 32, No. 2.

Slosson, J. E. and Hauge, C. J., 1973, "The Public and Geology Related Legislation in California, 1968-1972,": Geology, Seismicity and Environmental Impact; Special publication of the Association of Engineering Geologists, p. 23-27.

Slosson, J. E., and Havens, G. W., 1985 "Government Appears to be Failing In Enforcement": presented at Natural Hazards Research Workshop, Univ. of Colorado, July, 5 p.

Slosson, J. E., and Krohn, J. P., 1985, "Effective Building Codes": California Geology, Sacramento, CDMG, June, p.136-139.

Wilson, G. B., 1964, Governmental Organization for Building Regulation in Selected California Cities: M. S. dissertation, University of Southern California 101 p.

TABLE 1 TABULATION OF RESPONSES TO A QUESTIONNAIRE CONCERNING ADOPTION AND ENFORCEMENT OF CHAPTER 70, UBC-TYPE GRADING CODES BY STATE OF CALIFORNIA CITIES AND COUNTIES, 1977[1]

Jurisdiction:Respondents	*Number*	*Percent[2]*
Cities	73	83
Counties	15	17
	88	100
Respondents enforcing Chapter 70, UBC		
Cities	52	59
Counties	12	14
	64	73
Respondents enforcing local grading ordinance		
Cities	15	17
Counties	2	2
	17	19
Respondents not enforcing grading codes		
Cities	6	7
Counties	1	1
	7	8

Department Enforcement	*Number*	*Percentage*
Building departments enforcing Chap. 70, UBC	37	42
Engineering or public works depts. enforcing Chap. 70, UBC	23	26
Building departments enforcing local ordinance	2	2
Engineering or public works depts. enforcing local ordinance	13	15
Split enforcement responsibility: public works department, subdivisions to rough grade, building department finish grade only	3	3
Respondents indicating no department responsibility	3	4
Respondents enforcing no grading code	7	8
	88	100

Staff for Grading-Code Enforcement	*Respondents* City	*Respondents* County	*Total Number*	*Total Percentage*
Administrative Engineer[3]	21	5	26	30
Licensed Civil Engineer	29	6	35	40
Engineering Geologist	3	2	5	6
Use Outside Geotechnical Consultants	4	0	4	5
Grading Plan Checker	37	6	43	49
Specialty Grading Inspector	10	1	11	13
Combined Building (grading) Inspector	24	4	28	32
Combined Public Works, Engineering (grading) Inspector	22	3	25	28
Respondents who staff three or more grading-oriented personnel members	36	4	40	46

[1]The questionnaire was sent to 286 jurisdictions: 40 counties and 237 cities. Responses were returned by 88 jurisdictions (30.77%): 15 counties and 73 cities. The questionnaire was mailed and processed with the assistance of the California Building Officials (CalBO).
[2]Percentages rounded off to nearest whole number.
[3]Administrative Engineer is generally the City Engineer or Assistant Building Director rather than a Specialty Grading-Oriented Engineer.

Source: C. Michael Scullin (personal records)

TABLE 2 COMPARISON OF RESPONSES TO QUESTIONNAIRES SENT OUT IN 1977 AND 1991 REGARDING GRADING CODE ENFORCEMENT IN CALIFORNIA

Jurisdiction:Respondents	*1977*	*1991*
Cities	73	128
Counties	15	22

Grading Code Enforcement	*1977 (Percent)*	*1991 (Percent)*
Building Department	47	65
Public Works Department	44	28
Planning Department	9	7
	100	100
Respondents enforcing Chapter 70, UBC or Modified Version	73	87
Respondents enforcing a local grading ordinance	19	9
Respondents not enforcing any grading code	8	4

Staff for Grading Code Enforcement	*1977 (Percent)*	*1991 (Percent)*
Specialized Grading Inspectors	13	15
Combined Building/Grading Inspector	32	47
Combined Public Works/Grading Inspector	28	38
Staff Geotechnical Personnel	6	12
Use of Outside Geotechnical Consultants	5	40

Number of Personnel in Grading 1991:	*Staff of (1)*	*Staff of (2) or more*
Administration:	Percent	Percent
Building Official	60	5
Public Works Director	28	
Planning Dept.	7	
Grading Plan Check:		
Building Official	45	5
Public Works Director	33	5
Grading Field Inspections:		
Specialty Grading Inspector	13	2
Building Inspector	21	10
Combined Inspector	11	5
Public Works Inspector	28	10
Geotechnical Personnel:		
Staff Geotechnical	3	5
Staff Engineer Geologist	7	5
Outside Geotechnical Engineer	21	5
Outside Engineering Geologist	18	5

[1]Percentages rounded off to nearest whole number.

Source: C. Michael Scullin (personal records)

WALK-THROUGH RESIDENTIAL GEOLOGIC EVALUATION: A HAND-OUT CHECKLIST AND DISCLAIMER

Harry S. Audell
Consulting Geologist
Earthlogics Consultant Group, Inc.
31815 Camino Capistrano, Suite 22
San Juan Capistrano, CA 92675

AUTHOR'S NOTE

For property transfers in today's high liability real estate market the professional geologist is often called upon to evaluate the geologic hazards and risks that may affect or may have a future effect on property or structures. His conclusions and recommendations may be the deciding factor in a homebuyer's decision regarding that major purchase. Because of the liability exposure the geologist receives by being the third party expert, presentation of a standardized handout with checklist and disclosure statement is essential for conveyance of the evaluation results to the homebuyer. Although every site is different, the procedure and methodology can be standardized to better inform the homebuyer and offer maximum liability protection to the geologist.

The standardized checklist itemizes the geologic hazards of southern California and the basic criteria considered in assessing the level of risk of those hazards. It is recognized that the greatest exposure of liability is between the geologist and the homebuyer. The homebuyer is informed and made aware of the existing and potential geological hazards and risks, and is required to acknowledge and agree to, by signature, the results of the evaluation. This procedure offers the buyer a report in writing and the geologist a signed disclaimer stating that the buyer understands the evaluation. The homebuyer keeps the signed handout and the geologist keeps a duplicate copy of the conclusions, which is also signed by the homebuyer.

The following handout begins with the INTRODUCTION which discusses the purpose and intent of the geological evaluation. The SCOPE explains what is and is not included in the investigation and what the homebuyer can expect from the evaluation. The OBSERVATIONS state how the geological walk-through is performed and qualifies the extent of the evaluation. The COMMENTARY addresses the criteria of the geological evaluation and establishes levels of risk to the property and structures. The GEOLOGIC HAZARDS are analyzed and conclusions presented to the client. The LIMITATIONS are presented to make the client aware of the preliminary nature of the evaluation and that other professionals may be needed to evaluate aspects of the property that are not geologically related. The conclusions are presented as a checklist of the observed geologic conditions and risk assessment.

This format has been used in many inspections performed by the author and found effective in offering a service that is easily understood, credible and in keeping with industry standards.

GEOLOGIC EVALUATION OF RESIDENTIAL PROPERTY

INTRODUCTION

Per your request, a Registered Geologist or Certified Engineering Geologist will perform a cursory evaluation of existing geologic hazards and risk assessment at the subject residence and property. This letter attempts to inform the prospective homebuyer as to the technique and methodology used in performing this evaluation. A report summarizing our geologic evaluation, included with a generalized discussion regarding the structural conditions, can be prepared upon your request. If you have any questions during the course of our walk-through evaluation, please do not hesitate to ask.

SCOPE

The purpose of our walk-through inspection is to provide a visual geological evaluation of the existing site conditions and evaluate the hazards and associated risks with the subject property.

The scope of our service is:

1) to perform a walk-through of the residence and property,
2) to review of geological reports produced by the California Division of Mines and Geology (CDMG), the United States Geological Survey (USGS) and consultant's reports, if any,
3) provide a preliminary opinion as to our findings.

However, the scope of our evaluation does not include:

1) a definitive explanation as to the cause and effect of the observed geologic conditions,
2) on-site subsurface exploration or laboratory testing
3) removal of wallpaper, carpet or other floor coverings,
4) a specific prediction of future stability of the residence and property.

The limitations of our evaluation are disclosed so that the prospective buyer is not misled as to the results of this inspection.

OBSERVATIONS

Our geologic observations will be discussed with you. Because of the limited scope of our evaluation, all possible geologic conditions cannot be noted. For this evaluation, the general geologic character of the site will be ascertained by visual examination of the property and review of existing generalized geologic maps of the area. This study does not attempt to make any determinations regarding the structural integrity of the residence or other site improvements. However, upon your request, we can provide professional engineering services regarding structural behavior, performance and integrity.

COMMENTARY

A generalized guideline for assessing the geologic hazard and relative risk to site improvements is provided below. This is a suggested industry standard for determining the geologic hazard-risk assessment for real estate transactions. The following criterion should be considered flexible in application to fit various types of geologic conditions.

1. Identification of geologic hazards affecting the property,
2. The impact of the identified geologic hazard to structures or structural improvements, and
3 Level of risk of the geologic hazard to structures.

For the purpose of this study, geologic risk is assigned to one of two main categories: allowable risk and nonallowable risk. Allowable risk may be low, moderate and high. A house currently in a low level of allowable risk may degrade to an nonallowable level of risk if a geological hazard affecting the structure is allowed to persist, or should future events occur to change the present geological conditions.

GEOLOGIC HAZARDS

A geologic hazard is a naturally occurring or man-made geologic condition or phenomenon that presents a risk or is a potential danger to life, structures or property. The geologic hazards common to the southern California region are numerous and include landslides, expansive soils, soil subsidence, earthquake shaking-liquefaction, flooding, erosion, fault rupture and tsunamis. You will be immediately notified should a geologic hazard impose a high or nonallowable risk to the property or structural improvements.

LIMITATIONS

Without subsurface data this commentary addressing the geological conditions should be regarded as a preliminary evaluation. To fully understand the geological conditions relating to the behavior of structures (or improvements), a detailed investigation with subsurface exploration should be performed. This investigation would involve the combined services of a soil engineer and structural engineer. The causes and effects of the observed conditions, with conclusions and recommendations to correct the subsurface problems, would be presented in a report.

Services performed by this office were conducted in a manner consistent with the level of care and skill ordinarily exercised by members of the profession currently practicing under similar conditions. No other warranty, expressed or implied, is made.

CONCLUSIONS

Our cursory walk-through geological evaluation identifies the geologic hazards that presently affect structures or structural improvements as being:

[ ] landslides   [ ] expansive soils

[ ] soil subsidence   [ ] earthquake shaking

[ ] local site flooding   [ ] erosion

[ ] fault rupture   [ ] tsunamis

Based on the identified hazards, we find that the relative impact to structures may be:

[ ] very low [ ] low [ ] moderate [ ] high [ ] very high

Continued ground movement relative to the applicable hazards may be considered:

[ ] minor   [ ] moderate   [ ] major

It is our opinion that one or more of these hazards may indicate an approximate level of risk to existing structures as being:

[ ] low   [ ] moderate   [ ] high   [ ] nonallowable

Our preliminary commentary addressing the geological hazards and associated risks is not a substitute for a more detailed geologic investigation supported by subsurface exploration.

This opportunity to be of service is sincerely appreciated. If you should have any questions regarding this matter, please contact this office.

Sincerely,

____

Performed By:

____

Date Performed:

____

Location of Property:

AGREEMENT STATEMENT

This residential geological hazard-risk assessment was performed in the presence of the prospective buyer. The undersigned party agrees with the results of this geological evaluation and fully understands its purpose, scope, criteria and limitations. This evaluation is for the sole use the undersigned and is not transferable to other prospective buyers of this same residence.

____ ____________________

Prospective Buyer Date

RESIDENTIAL GEOLOGICAL EVALUATION CHECKLIST

LOT TYPE
[ ] Cut Soil Lot
[ ] Cut Bedrock Lot
[ ] Fill/Soil Lot
[ ] Fill/Landslide Lot
[ ] Fill/Bedrock Lot
[ ] Transition Lot

GEOLOGIC FORMATIONS
[ ] Capistrano Fm
[ ] Puente Fm
[ ] Topanga Fm
[ ] Other _________

EXPANSIVE SOILS
[ ] High Expansivity
[ ] Moderate Expansivity
[ ] Low Expansivity
[ ] Not Applicable

LOCAL FAULT
[ ] San Andreas F
[ ] Newport-Inglewood F
[ ] Whittier F
[ ] Other _________
[ ] Distance ___ miles

SEISMIC SHAKING/DAMAGE
[ ] High Shaking/Damage
[ ] Moderate Shaking/Damage
[ ] Low Shaking/Damage

EROSION
[ ] High Possibility
[ ] Moderate Possibility
[ ] Low Possibility
[ ] Previously Identified
[ ] Not Applicable

SURFICIAL GEOLOGIC UNITS
[ ] Alluvium Deposits
[ ] Colluvium Deposits
[ ] Nonmarine Terrace
[ ] Marine Terrace
[ ] Landslide Deposits
[ ] Artificial Fill

SITE FLOODING/DRAINAGE
[ ] Poor Drainage
[ ] Fair Drainage
[ ] Good Drainage
[ ] Not Applicable

SLOPE STABILITY
[ ] Stable Conditions
[ ] Marginally Stable
[ ] Unstable Conditions
[ ] Not Applicable

SOIL SUBSIDENCE
[ ] High Potential
[ ] Moderate Potential
[ ] Low Potential
[ ] Previous Subsidence
[ ] Active Subsidence

LIQUEFACTION POTENTIAL
[ ] High Potential
[ ] Moderate Potential
[ ] Low Potential
[ ] Previous Liquefaction
[ ] Not Applicable

TSUNAMIS
[ ] High Potential
[ ] Moderate Potential
[ ] Low Potential
[ ] Not Applicable

REFERENCES ON LAW AND THE PROFESSIONAL

American Society of Civil Engineers, 1977, Ethics, professionalism and maintaining competence: ASCE, 345 E. 47th Street, New York, NY 10017.

Association of Soil and Foundation Engineers, 1980, Nine case histories involving geotechnical engineers: ASFE 8811 Colesville Road, Suite 255, Silver Spring, MD 20910. (Evaluation of 9 case histories that detail loss experiences of geotechnical engineers. Themes of case histories are: accepting expert witness assignments, changed conditions, project review, words, duty of the engineer, and Association role.)

_______________, 1985, Guide to forensic engineering, service as an expert witness: ASFE (address above), 47 p.

Bradley, M. D., 1983, The scientist and engineer in court: American Geophysical Union, Water Resources Monograph 8, Washington, D.C., 111 p.

Brown, Glenn A. and Proctor, Richard J. (eds.), 1981, Professional Practice Guidelines: Association of Engineering Geologists Special Publication (Rev. 1985), available AEG Executive Director, 323 Boston Post Road, Suite D, Sudbury, MA 01776, $25.00.

California Division of Mines and Geology, 1975, Recommended guidelines for preparing engineering geologic reports; CDMG Note 44: Free from CDMG, P.O. Box 2980, Sacramento, CA 95812.

Dunn J. R., and Kiersch, G. A., 1974 The professional geologist as an expert witness: American Institute of Professional Geologists, Professional Guide Series, 19 p., available AIPG 7828 Vance Drive, Suite 103, Arvada, CO 80003.

Keirsch, G. A., 1977, Geologist and legal cases; responsibility, preparation, and the expert witness, *in* LeRoy, L., ed., Subsurface geology: Colorado School of Mines Publication, p. 747-752.

Kiersch, G. A., and Cleaves, A. B., eds., 1969, Legal aspects of geology in engineering practice: Geological Society of America, Engineering Geology Case Histories 7, 93 p.

Leighton, F. B., 1992, Mitigation of Geotechnical Litigation in California: Proposal for So. Calif. Contractors Association, Los Angeles (in press).

Loulakis, M. C., 1986, Disclaimers of liability: Civil Engineering, vol. 56, no. 10, p. 32.

Meehan, Richard L., 1981, Getting sued and other tales of the engineering life: MIT Press, Cambridge, MA, 241 p.

Murray, R. C., and Tedrow, C. F., 1975, Forensic geology; earth sciences and criminal investigations: Newark, NJ, Rutgers University Press, 217 p.

Patton, James H., 1973, The engineering geologist and professional liability: p. 5-8 in Moran, D.E. et al. (eds.), Geology, Seismicity and Environmental Impact; Association of Engineering Geologists, Southern California Section, 446 p. (See also Patton paper herein.)

Poynter, Dan, 1987, Expert witness handbook: Para Publishing, Santa Barbara, CA, 243 p.

Shurman, G. and J. E. Slosson, 1992, Forensic Engineering, Environmental Case Histories for Civil Engineers and Geologists: Academic Press, San Diego

Spellman, H. H., 1968, Direct examination of witnesses: Prentice-Hall, Englewood Cliffs, NJ, 306 p.

Sutter, J. and Hecht, M., 1974, Landslide and subsidence liability; Califonia Practice Book No. 65: Calif. Continuing Education for the Bar, Berkeley, CA. (also Nolo Press, Berkeley, CA)

Tank, R. W., 1983, Legal aspects of geology: Plenum Press, NY, 580 p.

Waggoner, E. B., 1981a, The expert witness: Bulletin of the Association of Engineering Geologists, vol. 28, no. 1, p. 29-38.

Waggoner, E. B., 1981b, Construction claims; spurious and justified: Bulletin of the Association of Engineering Geologists, vol. 28, no. 2, p. 147-150.

Chapter 5

Ground Water

Photo of settling basin, Orange County Water District.

Ground Water

Editors Note: The opening papers by Bob Bean and Glenn Brown admirably summarizes the hydrogeology of the entire southern California region, and should be required reading for all practitioners of this specialty. Mike Johnson gives us an update on the ongoing battle against seawater intrusion; and Joe Birman describes a neglected tool used to locate leakage through earthen embankments.

GROUNDWATER AND ITS MANAGEMENT IN SOUTHERN CALIFORNIA

Robert T. Bean
Consulting Geologist
2729 Willowhaven Dr.
La Crescenta, CA 91214

Glenn A. Brown
Consulting Geologist
3212 Whitney Lane
Burbank, CA 91504

CONTENTS

INTRODUCTION

The utilization, development, and management of California's groundwater have been and continue to be key factors in the State's rise to its position as the most populous, and one of the most progressive, states in this country. California's groundwater basins have an essential role in furnishing water supplies during both normal times, and periods of drought such as the five years continuing in 1991. However, statewide groundwater overdraft is estimated at 6 million acre-feet per year during droughts, whereas the overdraft is only about 2 million acre-feet in an average year (Kennedy, California Department of Water Resources, 1990).

Within the last fifteen years, problems of water quality have supplanted problems of quantity as the primary concern in many parts of the State. Volatile organic compounds (VOCs) in concentrations shown as carcinogenic and/or mutagenic by State and Federal standards have been discovered. Many public supply wells have had to be abandoned or their pumpage diluted by blending. Cleanup projects utilizing reverse osmosis or aeration are beginning operation.

Management of groundwater in California has traditionally been on a local level, being handled by water districts and political entities such as counties. However, in recent years, State regulations have increasingly become a framework within which local entities and agencies must operate. Enforcement of both State and Federal laws regarding water quality is the primary responsibility of the State Water Resources Control Board and its nine regional boards. The State Department of Health Services has extensive responsibilities including recommending water quality standards and working with local agencies.

Cooperation in water management among governmental and public agencies has mostly been at a relatively high level in California. It is essential that this cooperation continue at the most advanced technical and economic level possible, if our groundwater and surface water supplies are to bring the greatest good to the greatest number of residents of the State.

The treatment in the present paper is by Hydrologic Study Area (HSA), as defined by the California Department of Water Resources (CDWR, 1975). Principal groundwater basins of the South Coastal HSA from Ventura County to San Diego County are covered first, the counties being handled in alphabetical order. The South Lahontan and Colorado Desert HSA's are then treated in general, with specific coverage of only a few key basins being possible because of limitations of space. In all cases, hydrogeologic conditions are outlined, and past, present, and planned groundwater management activities are covered.

Agencies to which numerous references are made throughout the report are generally referred to by acronym, according to the following listing:

Agency	Acronym
California Department of Water Resources	CDWR
California Regional Water Quality Control Board	CRWQCB
Los Angeles County Department of Public WorksL	ACDPW
Metropolitan Water District of Southern California	MWD
United States Environmental Protection Agency	EPA
United States Geological Survey	USGS

SOUTH COASTAL AREA

Los Angeles County

Coastal Plain

The Coastal Plain of Los Angeles County is the southernmost basin in the county, extending from the sea to surrounding hills and mountains (Figure 1). This 500-square mile area overlies a groundwater reservoir containing almost 32,000,000 acre-feet of storage capacity. The Coastal Plain is now almost entirely urbanized, including a large part of the City of Los Angeles as well as numerous neighboring cities. Nearly 40 percent of the water presently used by this great urban complex is groundwater.

Hydrogeologic Setting: The Coastal Plain is underlain by unconsolidated sands, gravels, silts, and clays laid down by major and minor streams, and as near-coastal marine deposits. The major streams are the Los Angeles and San Gabriel Rivers, and the Rio Hondo, a distributary of the San Gabriel. The coarser-grained deposits occur as a series of flat-lying to gently folded aquifers, separated by aquicludes of clay and silty clay. A detailed description of the aquifer system was prepared by the California Department of Water Resources (1961), and the U. S. Geological Survey (Poland and others; 1956, 1959) described hydrogeologic conditions particularly in the southern and western parts of the Coastal Plain. The Los Angeles County Flood Control District (now Department of Public Works) has prepared pertinent published and unpublished hydrogeologic information, particularly near the coast, where seawater intrusion has been a problem.

Aquifers occur in deposits varying from Holocene to upper Pliocene in age. The most significant aquifer for public water supply is the Silverado Aquifer within the San Pedro Formation, of Pleistocene age. The Silverado is protected from contamination from the surface in most places by overlying strata. However, in some locations toward the southwest the Silverado is merged with the overlying upper Pleistocene Gage Aquifer, described in some of the literature as the "200-foot sand."

Trenches have been cut in the older sediments by the ancestral Los Angeles River in Dominguez Gap and by the ancestral San Gabriel River in Alamitos Gap. These trenches were filled in Holocene (Recent) time, largely by sands and

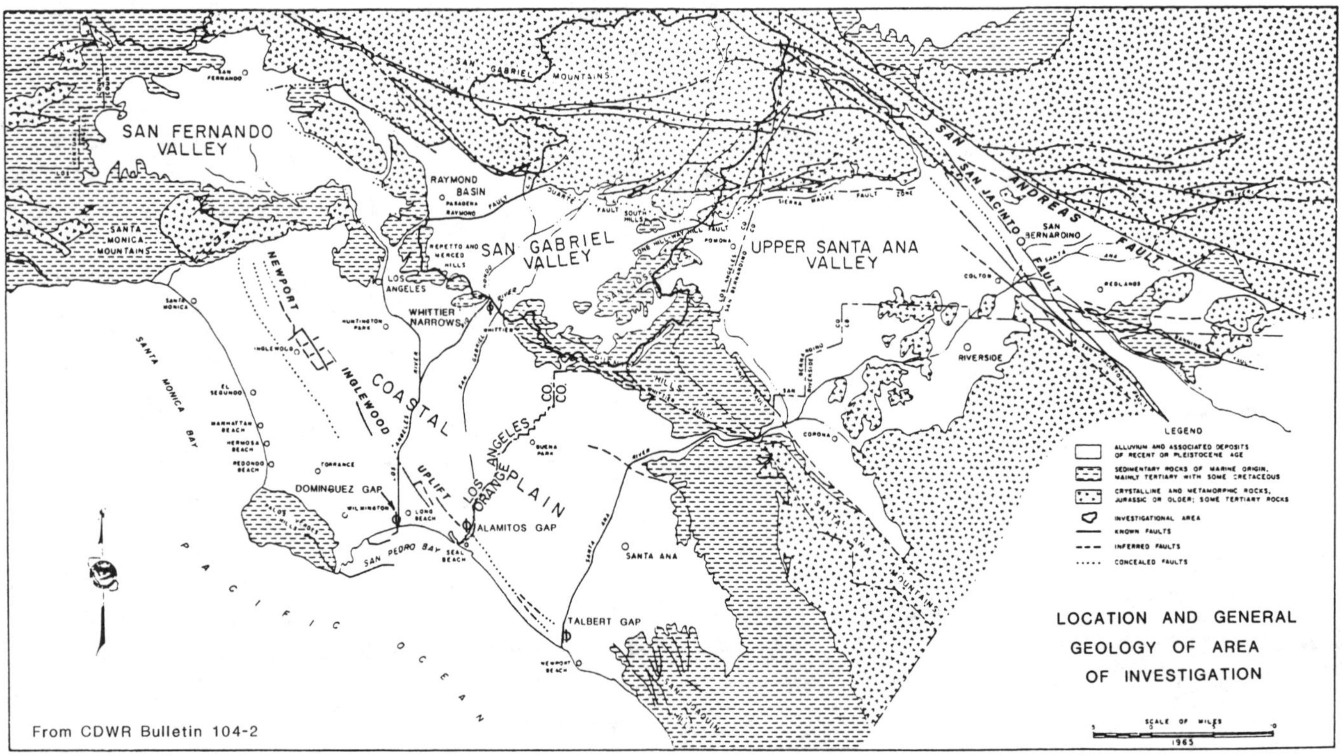

Figure 1. Location and General Geology of Area of Investigation

gravels that presently form the highly permeable Gaspur Aquifer. Where inland groundwater levels are below sea level, seawater readily moves landward through the Gaspur sediments unless checked.

The Coastal Plain is divided into two basins, the Central Basin and the West Coast Basin, by the northwest-trending Newport-Inglewood uplift (Fig. 1). Faults and uplifted older sediments along this zone form a significant barrier to groundwater movement. The uplift extends through Alamitos Gap, where the barrier is present as an anticline in the older sediments. However, there is essentially no barrier effect in the Gaspur sediments filling the trench through the gap.

Principal natural recharge to the aquifers of the Coastal Plain occurs in the Montebello Forebay, just south of Whittier Narrows. Recharge from the surface reaches most major aquifers here, including the Silverado. The most effective artificial recharge facilities, the San Gabriel and Rio Hondo spreading grounds, are located in the Montebello Forebay.

Development and Management: Extensive areas of artesian groundwater flow occurred in the Los Angeles County Coastal Plain under natural conditions, and continued in spite of increasing pumpage from wells until after the turn of the century (Mendenhall, 1904). Pumpage continued to increase, however, and water levels correspondingly have fallen until they have reached nearly 100 feet below sea level in some areas.

When groundwater levels near the coast dropped below sea level, seawater began to move into the aquifers and high-salinity water wells had to be abandoned. This problem, plus the falling water levels, made it imperative that steps be taken to reduce the overdraft and preserve the groundwater reservoir.

Accordingly, court actions were taken to establish long-term availability of groundwater for both the Central Basin and the West Coast Basin. These actions were settled not by traditional courtroom procedures, but by conferences and agreement among the litigating parties, followed by acceptance by the Court of the agreement reached, and a stipulated judgement giving the agreement the force of law. The relative rights of all pumpers were established, and were controlled by the State Department of Water Resources as Watermaster. Beginning in the mid-1960's, annual groundwater withdrawals were limited to 217,367 acre-feet from the Central Basin, and 64,469 acre-feet from the West Coast Basin (Welsh, 1988).

To protect the fresh water aquifers from seawater intrusion, a line of injection wells was installed along the west coast in the mid-1950's. Construction of this fresh water barrier was followed in the 1960's by installation of barriers in Dominguez Gap and Alamitos Gap (Fig. 1). The latter was constructed in cooperation with Orange County, and consisted of extraction wells in addition to wells injecting fresh water. Maintenance and improvement of the fresh water barriers has continued to the present. The water injected is imported by the Metropolitan Water District of Southern California from the Colorado River and also via the State Water Project. The attempt is made to minimize loss to the sea by making the seaward gradient of the injection ridge very gentle, while the landward gradient is much steeper. Currently about 27,000 acre-feet of treated water annually is injected in the barrier system, an estimated 18 percent of which is lost to the sea (Welsh, 1988).

Additional replenishment to groundwater takes place in the Montebello Forebay through spreading of runoff from the San Gabriel River and the Rio Hondo, subsurface inflow from San Gabriel Valley through Whittier Narrows, and artificial recharge of imported water and reclaimed water. The total annual supply to the forebay averages nearly 185,000 acre-feet, consisting of 91,000 acre-feet of local supplies, 56,000 acre-feet of imported water obtained from the Metropolitan Water District, and up to 37,700 acre-feet of reclaimed water (Welsh, 1988).

The use of reclaimed water for artificial recharge is probably the largest such project in the world. The source of the water is three treatment plants located above Whittier Narrows. The desirability of this reclaimed-water recharge project was thoroughly supported by a Health Effects Study carried out by the Los Angeles County Sanitation Districts (1984) in cooperation with other agencies. This study found that there were no undesirable health impacts on the population which had been using well water that included artificially recharged waste water. This conclusion was reached after detailed comparisons were made with a control group of similar socio-economic status, whose water supply contained no reclaimed water.

As a result of controlled pumping, artificial recharge in the barriers to seawater intrusion and in the Montebello Forebay, and current importation of about 60 percent of the water needs, the serious depletion of groundwater supplies in the Coastal Plain has been arrested. During dry periods such as 1986-91, the basin is undoubtedly overdrawn. During wet or normal periods, the Coastal Plain Groundwater Basin is probably being operated at or very near safe yield.

Problems and Plans: The principal groundwater problems presently of concern in the Coastal Plain have to do with water quality. Since the Silverado Aquifer, or the Silverado merged with other aquifers, is the only source presently used for municipal supplies, it is essential that the quality of this source be preserved. The Central Basin Municipal Water District (CBMWD) and West Basin Municipal Water District (WBMWD) conduct a basin-wide groundwater quality monitoring program, and prepare annual reports detailing the results. Any possible deviation from State and Federal drinking water standards triggers evaluation of remedial alternatives.

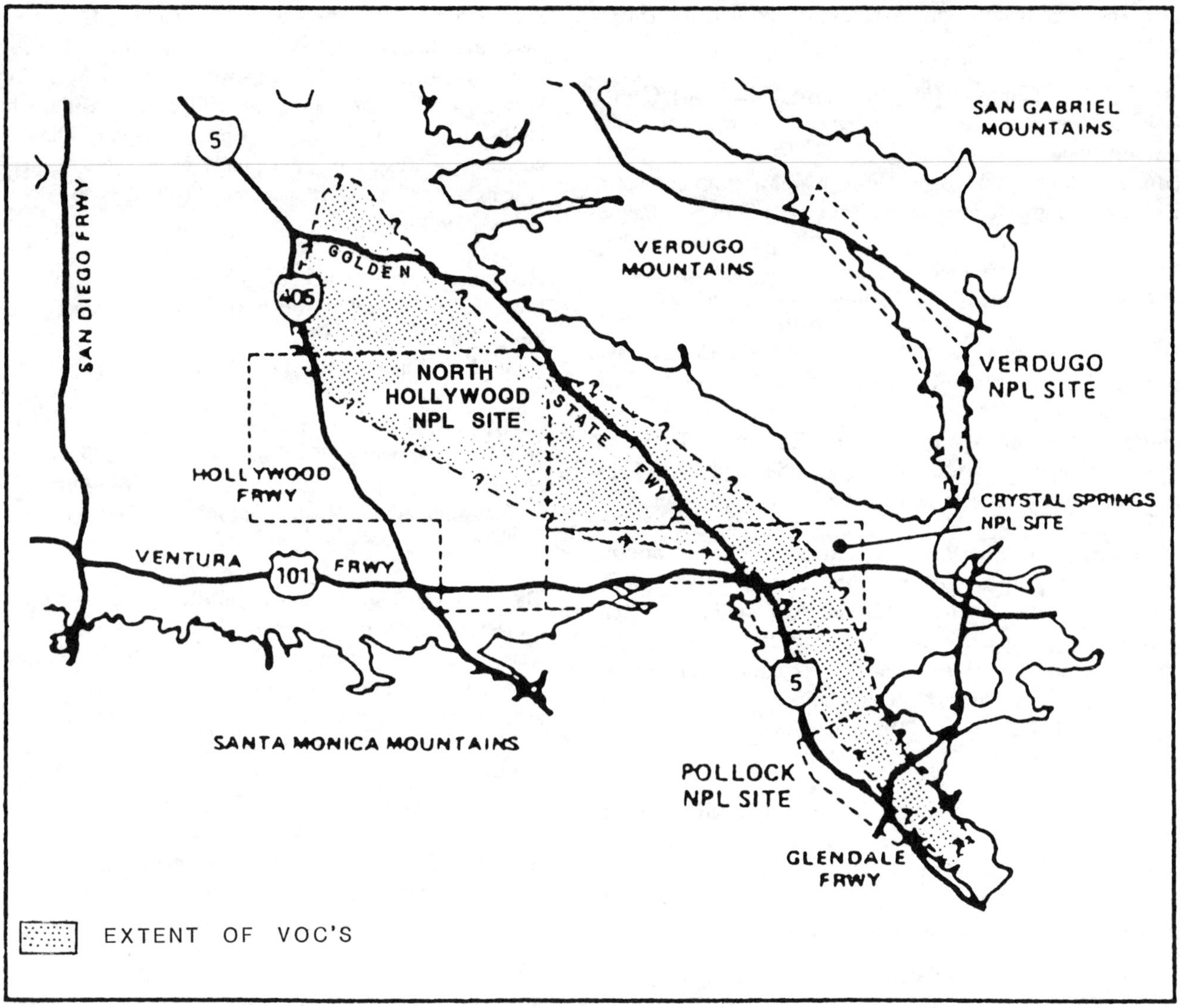

SAN FERNANDO VALLEY

APPROXIMATE EXTENT OF VOLATILE ORGANIC COMPOUND CONTAMINATION ABOVE MCL (EPA July 1990)

Figure 2. San Fernando Valley, Volatile Organic Compound Plumes

Various plans have been proposed for making the fresh-water barriers to seawater intrusion more effective, but necessary funding is difficult to obtain. A related problem involves the fact that when the West Coast Basin Barrier was first constructed, some saline water was already landward of its location. This brackish water has continued to spread, and at present has made groundwater in 20 percent of the West Coast Basin unsuitable for domestic use (CBMWD/WBMWD March, 1991). Various methods of remediation have been considered, but a desalination facility is presently being designed, and is scheduled to begin construction early in 1992. The pilot facility will use the reverse osmosis method and will be funded in part by the U. S. Bureau of Reclamation.

In addition, a mathematical model of the West Coast Basin has been constructed, and is presently being calibrated. This and the model of the Central Basin (see below) will supersede the California Department of Water Resources pioneering model of the Coastal Plain (CDWR, 1961; Tyson and Weber, 1964). A transient program is to be developed for the West Coast Basin model so that flow trends can be simulated and additional remediation measures evaluated.

Contamination of shallow aquifers by hydrocarbons has taken place beneath the many oil refineries of the Coastal Plain. Evaluation of contamination under each refinery was ordered by the State Regional Water Quality Control Board in 1985. Much information on hydrocarbons in groundwater has been obtained as a result of the subsequent investigations and monitoring, as well as more localized investigations of leaking gasoline tanks. This complex subject is treated in an article by Testa elsewhere in this publication.

Subsurface groundwater inflow from San Gabriel Valley enters the Montebello Forebay through Whittier Narrows. The presence of volatile organic compounds (VOCs) in the San Gabriel Basin is a serious concern, as groundwater movement is generally toward Whittier Narrows and the Coastal Plain. As noted above, groundwater in the Montebello Forebay moves into the Silverado as well as other aquifers.

A Groundwater Protection Strategy Study currently under way deals with quality in the Montebello Forebay in three ways: (1) A protection strategy involving evaluation of quality through monitoring; (2) a groundwater model of the Central Basin; and (3) portable wellhead treatment (CBMWD/WBMWD, March 1991).

The groundwater model is being developed under a cooperative agreement between CBMWD and the Metropolitan Water District. They will use the CFEST model utilized by CH2M Hill, Inc. This model is also used to evaluate solute transport as an essential tool in groundwater management strategies.

Wellhead treatment of contaminated groundwater began with the installation and testing of a pilot plant in 1991. The pumpage from a contaminated well is treated by activated carbon and becomes part of the supply for the city of Southgate. Other wellhead treatment programs are under consideration.

Nearly all the programs for various aspects of management of water in the Coastal Plain are cooperative programs involving activity and financing by two or more agencies. In addition, cooperative planning involving the San Gabriel Basin as well has been instituted with a Memorandum of Understanding between the Upper San Gabriel Valley Municipal Water District and CBMWD (CBMWD/ WBMWD, March 1991). Continuation of inter-agency cooperation is obviously essential for optimum management of California's groundwater basins.

San Fernando Valley

San Fernando Valley Groundwater Basin is the northwesternmost of the South Coastal Basins in Los Angeles County (Fig. 1). The Valley is a structural basin bounded on the north by the Santa Susana and San Gabriel Mountains, on the south by the Santa Monica Mountains, and on the west by the Simi Hills. On the east, the Verdugo Hills bound the main San Fernando Basin, but the small Verdugo Subbasin lies north and east of the Verdugos. The Pacoima and Tujunga Rivers enter San Fernando Valley from the San Gabriel Mountains, and are principal tributaries to the Los Angeles River, which leaves the basin on the southeast.

San Fernando Valley has undergone a progressive change from an agricultural to an urban economy during the present century. Groundwater has been a continuing essential component of water supply. However, severe water quality problems have developed in recent years, and the Federal government has designated the Valley as a Superfund site.

Hydrogeologic Setting: Continental sediments of Holocene (Recent) and Pleistocene age constitute the water-bearing fill of San Fernando Basin. Sediments in the eastern part of the main basin are mainly sands and gravels derived from the granitic and metamorphic rocks of the San Gabriels. About two-thirds of the usable groundwater storage capacity of 3.2 million acre feet is in this eastern part of the basin. Transmissivity is high here, and vertical hydraulic conductivity readily permits both natural and artificial recharge, at least to the shallower aquifers.

The San Fernando Basin has been considered as essentially containing a free groundwater body, and has been modeled as such by a finite-difference flow model. However, an extensive layer of fine sediments approximately 250 feet below ground surface probably limits vertical permeability to some extent. In addition, recent well clusters completed to varying depths have shown differing water levels at various depth zones (Mann, personal communication, 1991).

Groundwater in the eastern part of San Fernando Valley was originally a calcium bicarbonate water of very good quality. Unfortunately, contamination from surface sources has degraded this quality severely.

The western part of the basin is characterized by finer grained sediments deposited by the smaller streams draining the principally sedimentary rocks of the surrounding lower mountains. The western sediments of the valley are characterized by high clay content, lower storage capacity, and lower transmissivity. Original water quality in the western valley was a calcium sulfate type generally of poorer quality than to the east, and in some places exceeding sulfate limits of 250 milligrams per liter (mg/l).

Development and Management: Under original conditions, natural recharge to groundwater kept the San Fernando Basin essentially full, and rising water in the southeastern part of the basin flowed out in the Los Angeles River. Agricultural use of the groundwater by deep well pumps became very heavy by the mid-1920's, reducing the outflow. By 1931, pumpage and a series of dry years caused the outflow to cease (Mann and Blevins, 1990).

In the meantime, however, Owens River water arrived at San Fernando Valley in 1913 via aqueduct, and some of this water began to be spread for artificial recharge in 1931-32. Nearly 150,000 acre-feet were spread during the drought years of the early 1930's. A wetter winter came in 1936-37, and in the spring of 1938 a great flood occurred. Natural recharge kept the basin spilling for a number of years, but dry years returned, groundwater levels declined, and by 1949 the basin again stopped spilling.

The City of Los Angeles filed suit in 1955 against the cities of Glendale, Burbank, San Fernando and other pumpers, to establish water rights in the Valley. A Reference was made to the State Water Rights Board, and an investigation carried out which described the hydrogeologic facts of the Valley (State Water Rights Board, 1962). The suit, known as the San Fernando case, was finally decided by the California Supreme Court in 1979 (Mann and Blevins, 1986).

In this landmark case, Los Angeles was given the rights to native water in San Fernando Valley under the Pueblo doctrine, and the right to recapture imported water from groundwater storage. Glendale and Burbank were also granted the right to recapture imported water from storage. The safe yield of the basin was apportioned, and a Watermaster was appointed for the Upper Los Angeles River Area.

Problems and Plans: The improper handling and disposal of organic solvents in eastern San Fernando Valley has resulted in contamination of the groundwater by TCE, PCE, and other organics. Discovery of contaminated wells came within a year of the 1979 judgment settling the San Fernando case. EPA funding was obtained to develop a Groundwater Quality Management Plan, which was completed in 1983 (Mann and Blevins, 1990).

A very large number of small industrial facilities were found without proper sewerage, and as municipal sewers became available, mandatory connections were enforced. Investigation of other actual or potential sources of contamination included storage tanks, sumps, pipelines, and landfills.

By 1987, a cooperative agreement was signed permitting the City of Los Angeles to define the extent of groundwater contamination in four areas designated as National Priority List (NPL or "Superfund") sites. These areas are the North Hollywood, Crystal Springs, Pollock, and Verdugo sites, and are included in Figure 2.

A remedial investigation is currently being conducted under the auspices of EPA and the Los Angeles Department of Water and Power. In Phase 1 of this investigation, 43 shallow monitoring wells plus cluster wells at 14 locations were installed; Phase 2 calls for additional wells. Data from the wells, including borehole geophysics, pumping tests, and groundwater chemistry are being used to construct a multilayered groundwater flow and solute transport model. Utilization of this information is giving a new picture of the hydrostratigraphy of eastern San Fernando Valley, and will be invaluable in helping select the most effective cleanup alternatives.

An aeration facility to clean up TCE-contaminated groundwater has been constructed in the North Hollywood area, and has been in operation, except for some "down time," since 1989. The facility treats the water from a line of 8 wells in an aeration tower, and the treated water in then chlorinated and added to the public water supply. Air- or steam-stripping is also planned for contaminated water in the Burbank area. It is estimated that as much as 80 percent of the safe yield may have to be pumped to clean up the Superfund areas (Mann and Blevins, 1990).

San Gabriel Valley

The San Gabriel Valley Groundwater Basin is an ideal basin for storage and utilization of groundwater. The basin contains over 10,000,000 acre-feet of storage in alluvial sediments that include a large proportion of sand and gravel (CDWR, 1975), and the basin's inland location keeps it free from the danger of seawater intrusion. Furthermore, prolific natural recharge occurs in normal years by percolation into the alluvial fans of the many streams draining the San Gabriel Mountains to the north.

However, in recent years serious problems of water quality have been found, making necessary the closure of some wells and requiring that the pumpage from other wells be blended with higher quality waters. The basin is currently a U. S. EPA Superfund project because of groundwater contamination.

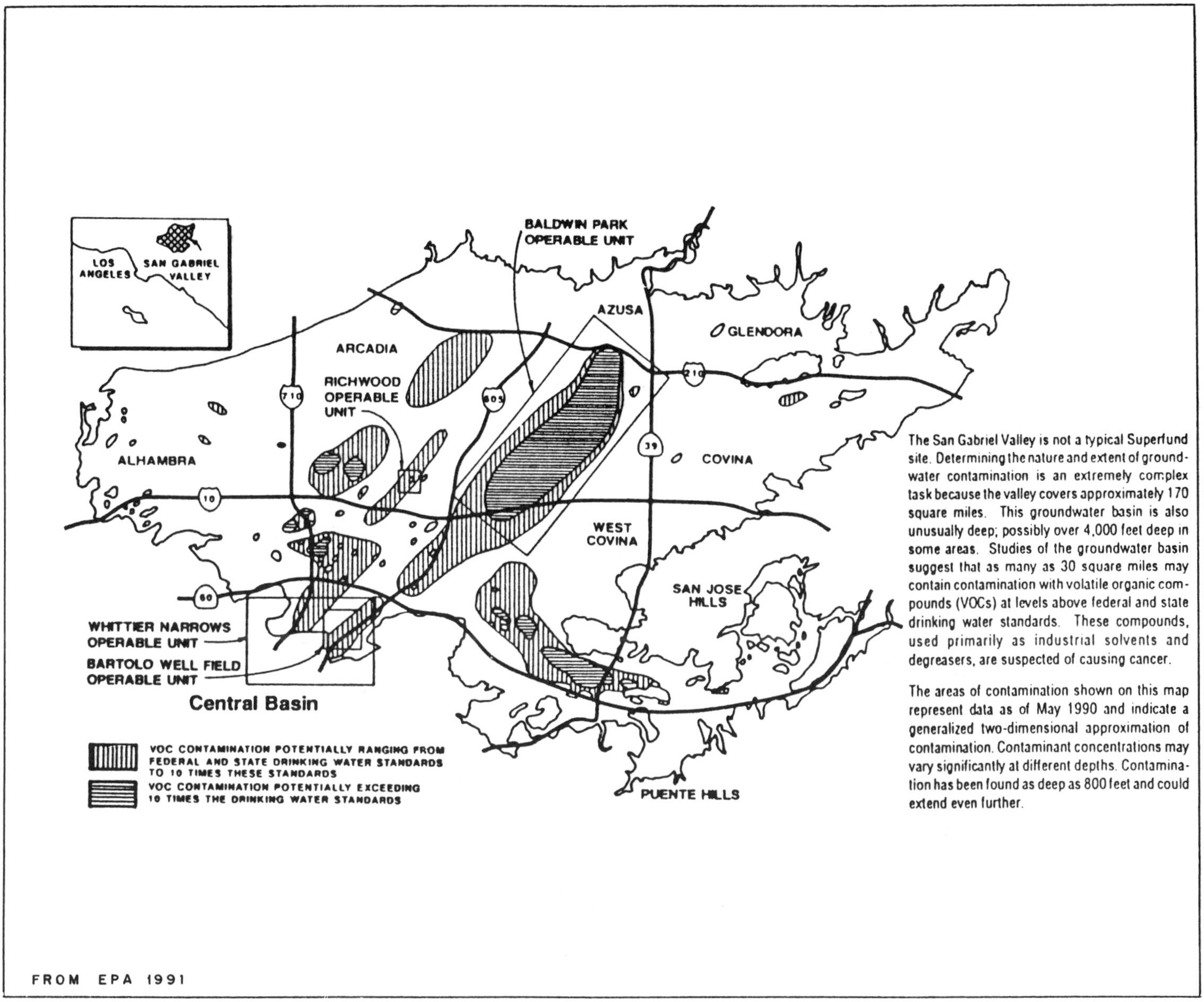

Figure 3. San Gabriel Valley Plumes

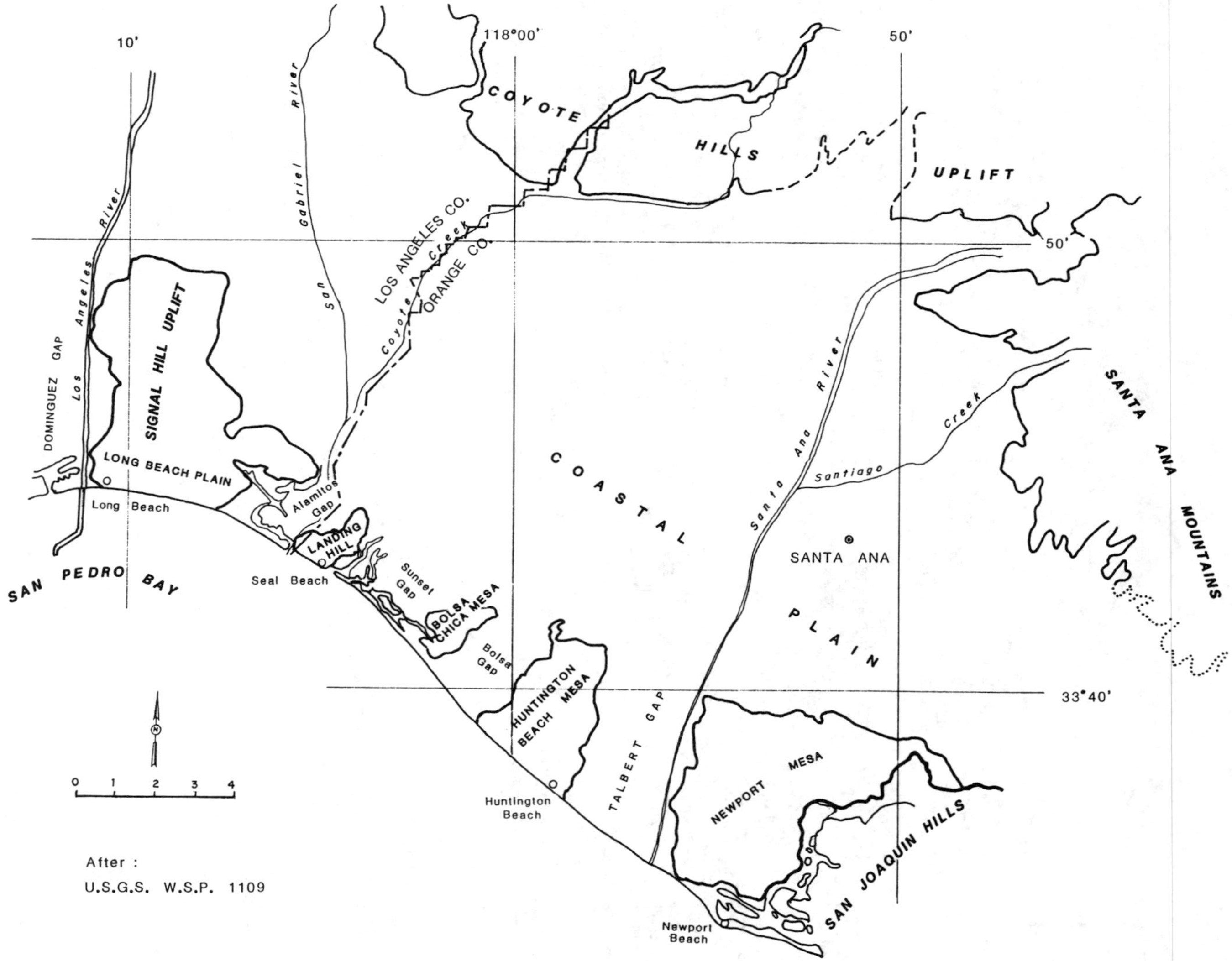

Figuure 4. Orange County Coastal Plain

Hydrogeologic Setting: The San Gabriel Valley Groundwater Basin is a great structural trough filled with alluvial sediments (Fig. 1). The granitic and metamorphic rocks composing the San Gabriel Mountains on the north have contributed most of the sand, gravel, silt and clay forming the groundwater reservoir, particularly in the central part of the valley. In the eastern part of the valley, streams from the bounding San Jose Hills have contributed finer materials to the alluvial buildup, and on the south the Puente Hills east of Whittier Narrows, and the Merced and Repetto Hills farther west, bound the basin and have contributed generally fine-grained sediments. On the northwest, the Raymond Fault separates the San Gabriel and Raymond basins.

The San Gabriel River and its distributary the Rio Hondo are the only through streams crossing the San Gabriel Basin. Both exit the basin in Whittier Narrows. Groundwater flow out of the basin also leaves through Whittier Narrows. Rising water has been an important component of the flow out of the basin.

Sediments in the central part of the San Gabriel Basin become generally finer from north to south, but as noted above the sediments on both the east and west sides are higher in silt and clay. Groundwater in the basin as a whole is considered generally unconfined. However, silt and clay layers throughout the basin cause confinement locally, an effect that shows clearly in most aquifer pumping tests.

The area of the water-bearing sediments in the basin is about 167 square miles, and the depth of the fill is greater than 2,000 feet in places. As noted above, the storage capacity of the San Gabriel Basin groundwater reservoir is over 10,000,000 acre-feet. This is significantly greater than the total gross storage capacity of 6,840,000 acre-feet of the 23 surface reservoirs of the State Water Project, including Oroville Reservoir. Furthermore, at least three-fourths of the San Gabriel storage capacity is clearly useable.

Most of the water-bearing fill of the San Gabriel Basin is "older alluvium" of Pleistocene age, consisting of sand, gravel, silt and clay. Holocene (Recent) alluvium overlies the Pleistocene sediments, transmitting recharge from the surface downward, but mostly lying above the water table. Lower Pleistocene sediments of San Pedro age furnish water to deeper wells north of Whittier Narrows, as do still deeper Pliocene sediments.

Whittier Narrows is a gap about one and a half miles wide, in which the maximum depth of water-bearing materials is about 800 feet. A northeast-trending faulted syncline occurs in the narrows. However, CDWR (1966, Plate 23) shows transmissivities from 300,000 sq. ft./day to 340,000 sq. ft./day through Whittier Narrows.

The Raymond Fault, bounding San Gabriel Basin on the northwest, is quite effective as a groundwater barrier on the west, but becomes less effective toward the east. Faults of the Sierra Madre zone near the base of the San Gabriel

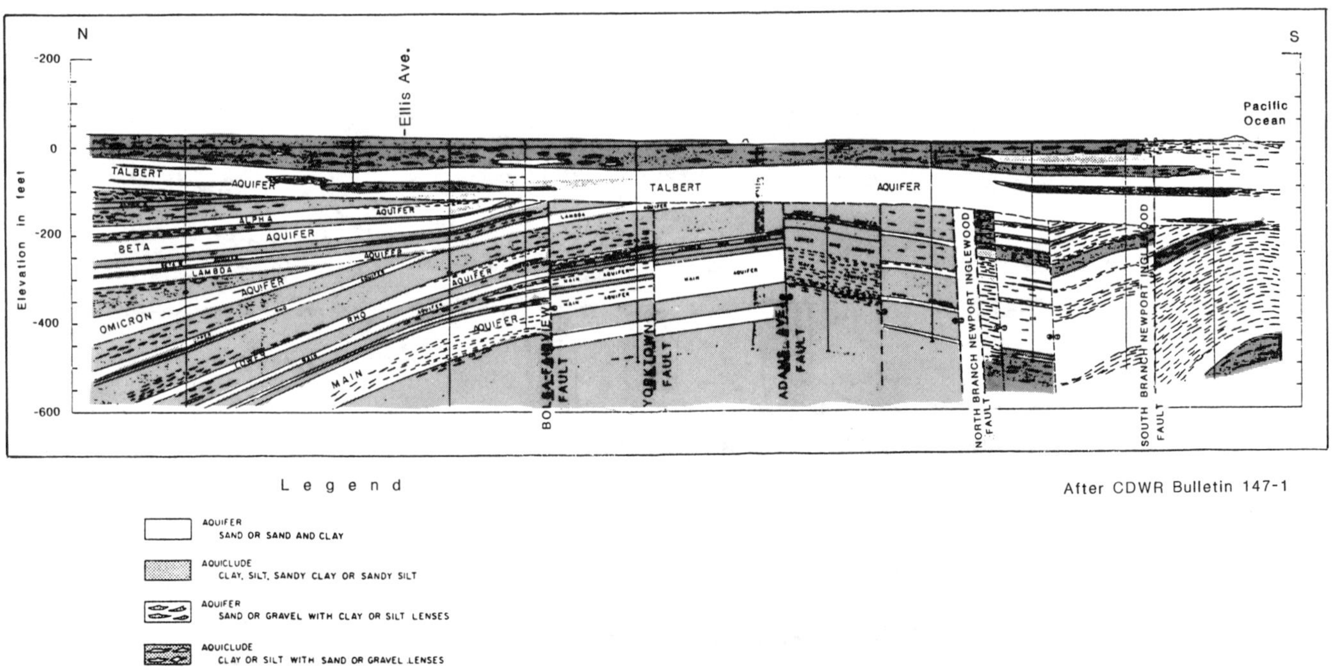

Figure 5. Section through Talbert Gap

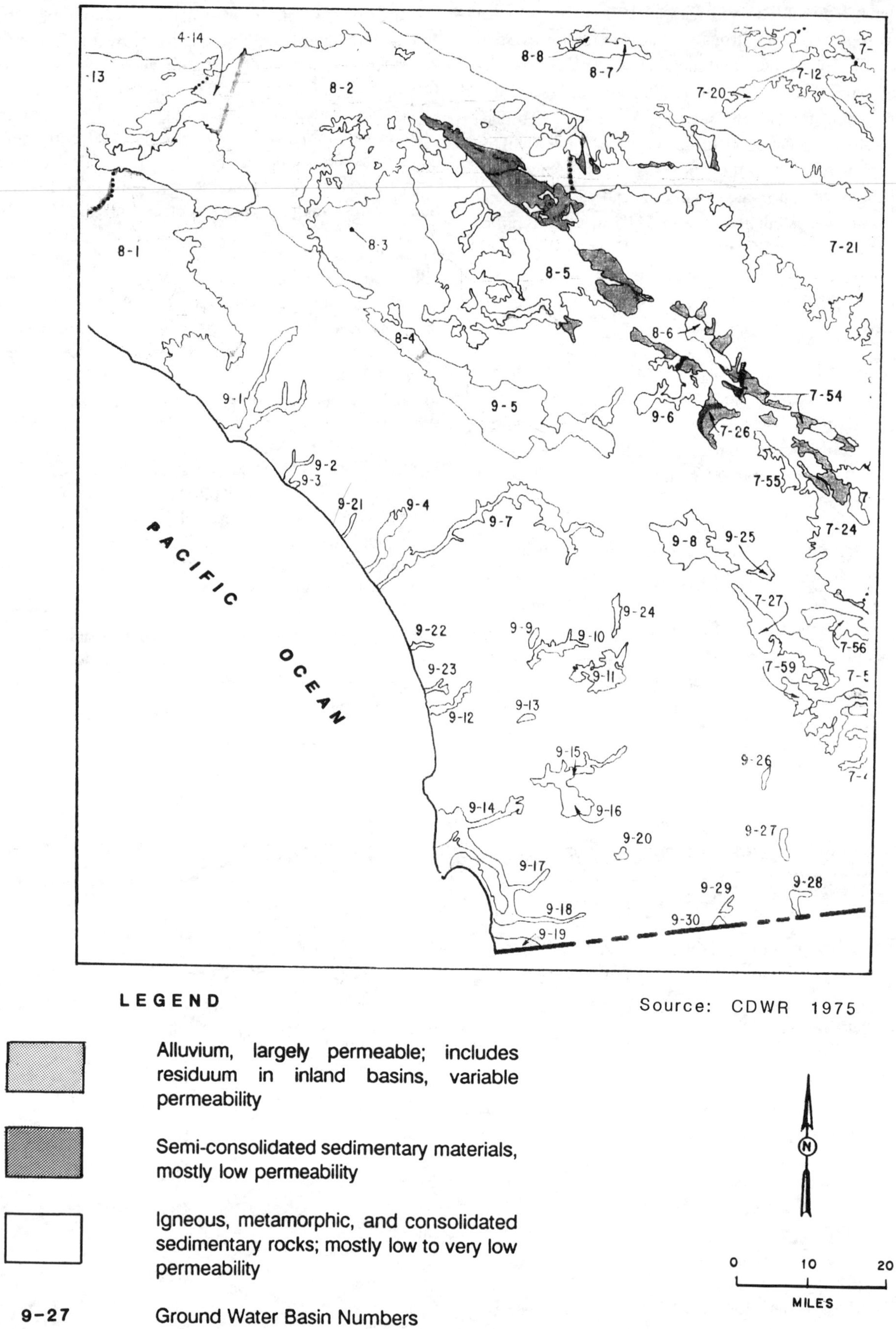

Figure 6. San Diego County, and Western San Bernardino & Riverside Counties, Groundwater Basins

Mountains cause subsurface groundwater cascades in the vicinity of San Gabriel and San Dimas Canyons. A difference in groundwater levels as great as 150 feet has occurred across the Lone Hill-Way Hill Fault, present in a gap at the northeast edge of the basin. However, no faults constituting major barriers to groundwater movement occur except near the edges of the basin, as just described.

Transmissivities throughout San Gabriel Basin were obtained by CDWR by aquifer pumping tests, and confirmed by their mathematical model (1966). In the central part of the basin, transmissivities vary from about 180,000 to 300,000 sq. ft./day, and in the peripheral areas from about 24,000 to 60,000 sq. ft./day. A strong directional variation in transmissivity was found. The higher values are in a north-south direction, generally approximating the direction of original deposition by the ancestral San Gabriel River (CDWR, 1966, Plate 23).

Development and Management: The highly permeable sediments of the San Gabriel Valley Groundwater Basin have been used by water wells of heavy draft throughout the present century. Use was originally primarily agricultural, but urbanization has taken over the valley, with tremendous growth during the middle of the 1900's.

There was early recognition that the significant runoff from the principal canyons of the San Gabriel Mountains to the north should be conserved to recharge the groundwater basin. Fortunately, spreading grounds were then developed in the upper portions of the alluvial fans, and at present 20 spreading grounds are recharging the basin with local runoff and with imported water as well (LACDPW, 1990).

As urbanization progressed in past years, the use of groundwater continued without great change, but instead of irrigation return flow maintaining groundwater levels, sewage began to be exported, and there was concern about falling levels. Users of groundwater in the Central Basin of the Coastal Plain believed that flowage through Whittier Narrows was threatened, and accordingly brought suit against water users in the San Gabriel Basin. A settlement of the suit was negotiated, and became a stipulated judgment of the Court in 1965.

The judgment guaranteed an average annual supply to the Central Basin through Whittier Narrows of 98,415 acre-feet. This includes usable surface flow, subsurface flow, and any water produced in the upper basin and exported to the lower basin. The actual annual entitlement varies in accordance with the previous 10 years' rainfall on San Gabriel Valley. A three-member court-appointed Watermaster administers the judgment (Stetson, 1986).

Pumpage by the various municipalities, water districts, and other entities within the San Gabriel Basin itself has also been established by litigation and a stipulated judgement. The operating safe yield of the basin is established each year by the court-appointed Watermaster, and each producer is allowed to extract a proportional amount of this safe yield. Storage of groundwater in the basin is estimated each year by the level of a sensitive well known as the Baldwin Park Key Well. Groundwater levels of this well show a cyclic pattern that is greatly dependent on rainfall in the mountain watershed to the north. Highs were reached in 1944, 1969, and 1983, with the last high only about 6 feet below that of 1969. During succeeding dry years the level has fallen, and reached a new low in July 1991, when the level was 6 feet below the previous historic low in 1977 (LACDPW; 1990, 1991).

Imported water may be stored in San Gabriel Basin with the right of recovery by the storing party. The Metropolitan Water District has the right to store up to 142,000 acre-feet of imported State water in the basin. However, imported water cannot be spread in the basin if the groundwater elevation in the Baldwin Park Key Well is higher than 250 feet.

Problems and Plans: Problems of water quality have overshadowed problems of quantity in the San Gabriel Basin in recent years. Nitrates in groundwater in excess of the State and Federal limit of 45 mg/l have existed for some time in the easterly portion of the basin, with local occurrences elsewhere. Sources of the nitrates were probably similar to sources in other areas which were first agricultural but then become urbanized, fertilizers and septic tank effluent being primarily to blame. Particularly in the 1980's, it became necessary for numerous water purveyors to shut down some wells with high nitrate content, or to blend well waters with sources whose nitrate contamination was not severe.

In 1979, contamination of groundwater by volatile organic compounds (VOCs) was first discovered in San Gabriel Valley with the detection of trichlorethylene (TCE) in relatively high concentrations in a well in Irwindale (Stetson, 1986). Testing of water from other wells found concentrations of tetrachloroethylene (PCE), carbon tetrachloride, and other VOCs in excess of limits, as widespread contamination was found in the valley.

To date, 30 to 40 square miles of San Gabriel Valley have been found to contain wells having concentrations of VOCs above maximum contaminant levels (MCLs). Plumes of contaminated groundwater exceeding Federal and State standards are shown in detail in Figure 3. The largest plume extends west of south from near the Foothill Freeway, mostly east of the San Gabriel River. Several additional areas of contamination occur as shown, including the City of Industry and vicinity in Puente Valley.

Contamination of the groundwater has undoubtedly been building up for many tens of years. At least 35 landfills have been located in the valley, and have certainly contributed (Lazo, 1983). Only 9 of these remained active in the late 1980's, but nearly all the rest have been buried and the land reclaimed. In addition for many years the San Gabriel Valley has been the site of hundreds to thousands of industries whose waste products evidently have contributed to contamination. Primary among the sources have been degreasers and cleaning solvents, but a great variety of waste products have undoubtedly contributed.

The San Gabriel Valley Groundwater Basin has been designated a Superfund site by the EPA, and the San Gabriel Valley Municipal Water District received a grant in 1985 for a Remedial Investigation/Feasibility Study. However, the many public agencies involved in the Valley, as well as the individual water purveyors, have felt that progress in solving the groundwater problems has been very slow. Concern was also felt that the State Water Resources Control Board might exercise its authority to take over management of groundwater in the basin. Partly because of this concern, the four major water agencies in 1989 adopted a Joint Resolution allocating "responsibility among local water entities for involvement in basinwide water quality protection and remediation" (Kidman, 1990). The Watermaster was designated to coordinate efforts between local, State, and Federal entities to solve the water quality problems. He is given the authority to request pumping or non-pumping of specific wells to control the movement of contaminated groundwater.

Various management procedures for San Gabriel Valley Groundwater Basin have been evaluated during an ongoing engineering investigation by CH2M Hill. The basin is being modeled by a CFEST finite element model that includes solute transport. A water well capture zone in Whittier Narrows has been proposed to protect the Coastal Plain. However, this has not been activated, in part because the principal contamination is still some distance above the Narrows. A water quality monitoring program is under way in cooperation with the Regional Water Quality Control Board. Monitoring wells are being constructed with as many as 15 ports to sample water at selected depths.

The problems of management of cleanup of the San Gabriel Valley Groundwater Basin are very severe. EPA estimates the cost of cleanup at $800 million (Kidman, 1990). Time necessary for cleanup is estimated to be from 30 to 50 years, if sufficient funding is made available.

The plan most favorably considered at this time, according to Harris (1991), is in summary as follows: (1) stop production from uncontaminated wells, in order to stop the movement of contaminated water into the cones of depression caused by pumping; (2) produce water from wells in the highly contaminated areas, to establish a groundwater gradient toward those wells; and (3) treat the contaminated water to make it useable.

The final plan to manage the cleanup of this great groundwater basin, and the necessary funding therefor, remain to be determined.

Raymond Basin: A subbasin of San Gabriel Valley Basin, Raymond Basin lies northwest of the main San Gabriel Valley (Fig. 1), and is separated from the main valley by the Raymond Fault. Raymond Basin was the first large basin subject to significant water rights litigation, and groundwater pumpage became under control of a Watermaster beginning in 1943. The California Department of Water Resources is currently Watermaster.

Santa Clara River Valley, Eastern Basin

The easternmost groundwater basin of the Santa Clara River Valley is in Los Angeles County. Several valleys containing water-bearing materials enter the main valley from the north, and Acton Valley is near the headwaters on the east.

Water-bearing materials include Recent and Pleistocene alluvium, and the Saugus Formation, Plio-Pleistocene in age. All major valleys produce water from alluvium, and the basin containing the Saugus underlies both valleys and highlands north of Newhall. Groundwater movement in the tributary valleys is generally toward the Santa Clara River, and near the river the direction becomes down-valley.

Water quality in the alluvium varies from satisfactory to in excess of EPA's maximum contaminant levels for some constituents. For example, 4 out of 21 wells in Agua Dulce Canyon have exceeded nitrate standards, and certain wells exceeded sulfate and fluoride limits (Barto, Cleath and others, 1982). A well in Mint Canyon exceeded MCLs in sulfate and boron when sampled, and had a total dissolved solids content of 1,022 mg/l. Recent heavy residential development and water use have put a severe strain on groundwater quality, and property owners in certain of the tributary valley basins do not readily release water quality data.

According to Slade (1988), the Saugus Formation contains approximately 1.41 million acre-feet of potentially usable groundwater of acceptable quality in storage. He states that development of additional groundwater supplies from Saugus aquifers is "hydrogeologically feasible."

Orange County

Coastal Plain

The Coastal Plain of Orange County is a continuation to the southeast of the Los Angeles County Coastal Plain, the county line lying approximately along Coyote Creek and the San Gabriel River (Fig. 1). The area of the Plain in Orange County is about 300 square miles. The remainder of the county consists of about 500 square miles of largely hilly to mountainous terrain, except for San Juan Valley, which is treated in a following section.

The Coastal Plain and adjacent parts of the county have undergone an extremely rapid change from an agricultural to an urban economy in the last few tens of years. Groundwater has been essential for both types of development.

Hydrogeologic Setting: The Coastal Plain is underlain by unconsolidated alluvial and near-shore deposits, broken along the coast by a line of low mesas along the Newport-Inglewood structural zone (Fig. 4). The surficial deposits are Holocene (Recent) in the interior lowland areas and in the gaps between the mesas, whereas the mesas are capped by Pleistocene terrace materials.

The most significant gap entirely within the county from the point of view of development and management is Talbert Gap, also known as Santa Ana Gap. The Santa Ana River, the only significant stream flowing through the county, presently enters the sea through this gap, and permeable deposits of the ancestral Santa Ana fill the gap.

Upstream, the Santa Ana River drains the extensive Upper Santa Ana Valley, and enters the Coastal Plain of Orange County through a canyon where flow is now controlled by Prado Dam. For several miles below Prado Dam, the river flows across sands and gravels into which recharge occurs to groundwater, which in turn moves toward the principal pumping areas and Talbert Gap. Existing artificial recharge facilities are located along and near the river in this forebay area.

The detailed hydrogeology of Talbert Gap, to a depth of about 600 feet, was worked out in an intensive investigation by the California Department of Water Resources (CDWR, 1966). CDWR's investigation not only used existing water wells, but included drilling, litho-logging and electric logging of over 12,000 feet of test holes. Also included in the investigation, carried out in cooperation with the Orange County Water District (OCWD), was installation of 23 piezometers (up to 3 in some holes), 10 pumping tests, installation of water level recorders, and obtaining of water quality data.

With this investigation, it was possible to detail the aquifers and aquicludes, faults, and unconformities, in Talbert Gap with the detail shown in Figure 5. As shown, the flat-lying Talbert Aquifer (Holocene) unconformably overlies a sequence of faulted, gently dipping older aquifers and aquicludes.

At the time of the CDWR study, seawater intrusion was a problem of increasing severity. Water with salinity of 500 ppm (mg/l) had moved landward in the Talbert Aquifer almost to Ellis Avenue. Faults in the deeper aquifers, particularly the north branch of the Newport Inglewood Fault, had acted as barriers of varying effectiveness against the landward movement of the salinity. The information obtained was used by the Orange County Water District to install a barrier system against seawater intrusion (see below).

In more recent years, OCWD has carried out a program of drilling and installing monitoring wells throughout the basin and to greater depths. The program is described in District publications, and by Black, Goodrich, and Patton (1988). Wells encounter 10 to 15 aquifers alternating with aquicludes to depths of 1,200 to 1,500 feet. The sediments occur in a synclinal trough plunging to the northwest.

A multi-level groundwater monitoring system has been developed, making possible sampling for water quality and measurement of piezometric levels in each hole. The piezometric head is generally found to increase with increasing depth. As of 1990, seven deep wells and nine shallow wells had been completed, and twenty more deep wells were planned for the next five years. Information obtained is used in management of the basin.

Development and Management: Problems associated with groundwater overdraft were already severe in 1933 when the Orange County Water District was formed by a special act of the State Legislature. Groundwater levels in former artesian areas had dropped well below sea level, and seawater intrusion was advancing inland. Natural recharge had been reduced as urbanization covered forebay areas with impervious buildings and pavement. In addition, diversions of the river flow in the upper basin in San Bernardino and Riverside Counties were increasing.

Imported Colorado River water was first obtained by the District in 1940 and used directly, resulting in some reduction in draft on the groundwater supplies. In the mid-1950's the imported water began to be used in artificial recharge of the aquifers, for by that time groundwater levels in a regional cone of depression about 8 miles from the Coast were more than 80 feet below sea level. As the population continued to explode, additional water sources were clearly needed, and State Project water, a high quality source, was first obtained in 1973 (Richardson, 1988). Furthermore, reclaimed wastewater began to be important as a source of supply.

Present artificial recharge facilities include an extensive replenishment system in the forebay supplying the principal aquifers. This system covers about 1,500 acres, and includes several basins (former gravel pits) along and near the river, and a secondary channel parallel to the main channel of the river. A series of sand and gravel dikes in the secondary channel require a zig-zag flow, and this slow flow recharges groundwater more effectively. Anaheim Lake has been developed from one of the gravel pits, and is currently effective both as an artificial recharge basin and a recreational facility. This lake and certain of the other deep basins are stocked regularly and are thoroughly used by fishermen. Infiltration capacity of the deep recharge basins must be maintained by regular cleaning because of clogging by fine sediments.

At present, the total recharge capacity of the system along and near the Santa Ana River is about 140,000 acre-feet in a dry year (OCWD, February 1990). An additional recharge facility was completed at Santiago Creek on the east, below Irvine Lake, in 1989. Water from the Santa Ana River is piped to the facility, which increases recharge capacity by 20,000 to 30,000 acre-feet in normal and wet years.

In Talbert Gap, the configuration of aquifers developed by the California Department of Water Resources (1966) and shown in Figure 5 was used by the District as a basis for constructing a barrier system to protect the basin from seawater intrusion. The system consists of a line of 23 multipoint injection wells essentially along Ellis Avenue, and a line of 7 extraction wells about two miles away toward the coast. The injection system permits artificial recharge into any specific aquifer needing replenishment. The ridge and trough combination makes it possible to operate the adjacent groundwater basin at a level well below sea level.

The District operates a reclamation facility named Water Factory 21 to furnish water for the injection wells. Wastewater from a nearby treatment plant is given further treatment at Water Factory 21 that includes chemical clarification and settling, ammonia stripping, re-carbonation, and chlorination. Up to 5 million gallons per day (mgd) of this water is then purified by reverse osmosis, and the resulting mixture is further blended with water from wells about 1,000 feet deep, not presently affected by seawater intrusion. The final product is then selectively injected into the pressure ridge. Injection of about 20 mgd or 22,000 acre-feet per year is required to maintain the ridge.

OCWD's operation has been a prime example of progressive water management for several decades. Since 1954 the items of water supply and demand within the basin have been evaluated each year along with groundwater levels, and production amounts accordingly determined for each pumping entity. An assessment is made for the amount to be used by each pumper, and the money is used to purchase imported water from the Metropolitan Water District. Additional funding is provided by ad valorem taxes.

The imported water is used mainly for artificial recharge in the forebay area during the dry season, and for injection in the barrier system in Alamitos Gap. Some imported water is delivered directly to the user as in lieu replenishment. The basin had an accumulated reduction of groundwater in storage of 705,000 acre-feet by 1956 (Richardson, 1988). Artificial recharge and control of pumping reduced this amount to essentially zero by 1964. Since that time, the District has permitted overdraft and dewatering of storage during dry years with the anticipation of recovery during wet years. In 1989, after falling groundwater levels for six years, the accumulated reduction in storage was about 400,000 acre-feet (OCWD, February 1990).

A mathematical model has been developed to simulate both quantity and quality parameters in the multi-level aquifer system of the Orange County Coastal Plain (Williams, 1986).

Problems and Plans: The Orange County Water District describes in considerable detail the District's principal current problems, and the plans to handle these problems (OCWD, February 1990). A principal cause of the problems has been the rapidly increasing urbanization of the upper Santa Ana Basin above Prado Dam. The growing communities there discharge their wastewater into the Santa Ana River, and percolation of this poor quality water in the forebay area below Prado seriously degrades quality of groundwater in the Coastal Plain. Base flow in the river has actually tripled during the last two decades, due to runoff from the many new impervious surfaces in the upper basin, principally buildings and paving, as well as sewage return (OCWD, 1990). Thus more water is available, but the quality is poorer.

The District is enhancing recharge from the deep recharge pits by more frequent cleaning, and has developed the Santiago Creek recharge project, as described above, to take advantage of the additional water available.

Additional water quality problems are due in part to past use of fertilizers and pesticides, as well as septic tanks and cesspools, in the Coastal Plain. As of 1990, about 250,000 acre-feet of groundwater contained nitrate in excess of the 45 mg/l limit for domestic consumption, and 51 wells had been shut down because of this problem. A nitrate removal project, completed in 1990, will add about 33,000 acre-feet of water annually to the supply for the city of Tustin. This project uses both reverse osmosis and ion exchange. Additional research is in progress on biological nitrate removal by denitrifying bacteria, which catalyze the reduction of nitrate to nitrogen gas.

Problems caused by urbanization of the upper basin have resulted in the formation of a joint powers agency by the District and upstream water agencies. A comprehensive water quality control plan for the entire watershed has been formulated, and was adopted in 1975.

Currently, TCE and/or PCE exceed standards in 30,000 to 150,000 acre-feet of water in storage in the Orange County Coastal Plain Basin. A plume with TCE in excess of limits extends westward about 2.5 miles from El Toro Marine Base. A pump-out project begun in 1990 removes about 1,000 acre-feet per year of TCE-contaminated groundwater for use in landscape irrigation. Furthermore, selenium in excess of 10 parts per billion (ppb) renders an additional 175,000 acre-feet of groundwater sub-standard within the basin, and has necessitated the shutdown of 7 wells (OCWD, 1990).

Reclamation and utilization of wastewater are prominent in OCWD's plans for the future. The "Green Acres" project uses water from a new reclamation plant adjacent to Water Factory 21, making up to 7,000 acre-feet of additional water available annually for landscaping and industrial purposes.

The threat of sea water intrusion into the aquifers of Bolsa and Sunset Gaps, the two gaps between Talbert and Alamitos Gaps, increases with increasing pumping draft on the inland aquifers. Injection of treated wastewater is planned for these locations.

San Juan Basin

The alluvium along San Juan Creek and its tributaries is the chief water-bearing formation in San Juan Basin, No. 9-1 on Figure 6. However, the average thickness of the alluvium is only 60 to 90 feet (Woodside and Woodside, 1986). Water quality is variable, but has decreased in recent years, with increasing concentrations of total dissolved solids, particularly in the lower part of the basin.

Rapidly increasing urbanization has resulted in increasing runoff and related problems of degraded quality. Recognition of water problems and the need for management resulted in the formation of the San Juan Basin Authority, a joint powers agency, in 1971.

San Bernardino-Riverside Counties

Upper Santa Ana Valley

Upper Santa Ana Valley, No. 8-2 on Figure 6, is bounded on the north and northeast by the youthful and imposing San Gabriel and San Bernardino Mountains. Thick alluvial deposits have been formed south of these ranges. Bedrock highlands of much lower elevation bound the valley on the south and southwest sides.

Hydrogeologic Setting: Upper Santa Ana Valley as a whole has an estimated groundwater storage capacity of 16 million acre-feet. The size of this basin makes it the largest groundwater resource in Southern California.

The major alluvial basins comprising Upper Santa Ana Valley are Bunker Hill Basin, Chino Basin, and Riverside-Arlington Basin. Bunker Hill Basin lies between the San Andreas Fault, which traverses the south side of the San Bernardino Mountains, and the San Jacinto Fault, which crosses the alluvial valley in a southeasterly direction between San Bernardino and Colton. The San Jacinto Fault, known locally as the Bunker Hill dike, provides closure for the Bunker Hill Basin as it blocks the down-valley movement of groundwater. High groundwater levels in a sedimentary thickness of at least 1,200 feet on the east side of the fault cause problems after years of high precipitation, including a period in the mid-1980's.

Chino Basin is an extensive basin bounded on the north by the San Gabriel Mountains, and terminating on the west at the highlands on either side of the Santa Ana River at Prado Dam. The Riverside-Arlington Basin, on the southeast, has relatively shallow alluvial fill.

Development and Management: Extensive development of groundwater for agricultural purposes took place early in the present century in the Upper Santa Ana region. In more recent years municipal and industrial use has greatly increased, with a corresponding decrease in agricultural use, and a general decrease in water quality (CRWQCB, 1984).

In response to water quality problems, the CRWQCB established an Interim Water Quality Control Plan in 1971, and a complete plan, known as the 1975 Basin Plan, was established four years later. The Santa Ana Watershed Planning Authority (SAWPA) was the prime contractor for basin planning, serving a joint powers agency comprising the Orange County Water District, Chino Basin Municipal Water District, Western Municipal Water District, and San Bernardino Valley Municipal Water District. Objectives were generally based on existing water quality, but were verified by a set of computer models simulating both quantity and quality.

Problems and Plans: In the Bunker Hill Basin, high groundwater levels east of the San Jacinto Fault resulted in a legal complaint after a series of wet years. In 1986 the City of San Bernardino claimed that high groundwater levels within the city were due to artificial recharge activities upstream on the Santa Ana River by the San Bernardino Water Conservation District. This case was settled out of court.

CDWR and MWD have investigated the use of Chino Basin for cyclic storage of imported water from the State Water Project. The investigations led to the preparation of a Draft Environmental Report, which highlighted the water quality problems of the area. High nitrate areas were shown to exist along the Santa Ana River and in the southern portion of Chino Basin. A groundwater model demonstrated that the addition of water in the upper reaches of the basin, where favorable infiltration rates exist, would cause an increase in nitrate problems in the areas already experiencing problems. This created considerable public resistance to the proposed cyclic storage program.

In January 1975, a complaint was filed in the Superior Court of San Bernardino seeking an adjudication of a great part of Chino Basin. This complaint led to an adjudication and physical solution for the management of groundwater which was signed on January 27, 1978. A Watermaster was appointed, and annual reports are prepared documenting extractions and setting forth an operating safe yield for the coming year.

MWD is continuing cooperative studies on Chino Basin management (MWD, 1990). These studies include brackish water reclamation.

San Jacinto-Elsinore Area

The San Jacinto-Elsinore area consists of two principal components: the Perris Block (No. 8-5 on Figure 6), and the Elsinore Basin (No. 8-4) in the Elsinore Trough. The Perris Block lies southwest of the San Jacinto Fault and is composed of scattered crystalline highland masses within a network of alluvial valleys containing groundwater supplies. The Elsinore Trough is a complex graben extending southeasterly from the city of Corona, and containing a significant thickness of saturated alluvial sediments underlying and surrounding Lake Elsinore. Basic hydrogeologic information appears in California DWR Bulletin 15, Appendix B (1959).

The San Jacinto River drains the western side of the mountain of the same name, and after reaching the valley floor recharges the San Jacinto Subbasin of the Perris Block. Groundwater throughout most of the rest of the area receives very limited natural recharge from adjacent highlands. Perris reservoir, located on the valley floor north of Perris, contributes percolation to groundwater in the vicinity. The intermittent flow of the San Jacinto River into the Elsinore Basin and Lake Elsinore is now controlled in Canyon Lake (formerly Railroad Canyon reservoir).

MWD is currently cooperating with local agencies in projects to enhance groundwater management in the Hemet and San Jacinto Subbasins (MWD, 1990). Conjunctive operation of the basins is under consideration. A mathematical model simulates movement of groundwater.

Temecula Valley

Temecula Valley Basin (No 9-5 on Figure 6) lies southeast of Lake Elsinore, generally in the Elsinore Trough. The basin is about 150 square miles in area, and is somewhat complex in stratigraphy and structure. Water-bearing materials include Recent and Pleistocene alluvium, and the Temecula arkose.

A Water Resources Master Plan has been prepared for the Rancho California Water District, roughly coincident with the Temecula Valley Basin (Scheliga and McGoldrick, 1984). This plan reports groundwater storage within the District below 1983 water levels as 1,351,000 acre-feet to a depth of 500 ft, and 2,428,000 acre-feet to a depth of 1,000 feet. Imported water is obtained from MWD, and some artificial recharge has taken place since 1978.

Recent proliferation of housing developments has put a severe strain on both water supplies and wastewater disposal facilities. Land subsidence has occurred, and litigation has taken place on the complaint that the subsidence was caused by groundwater pumpage (see Shlemon, this volume).

Bear Valley

Bear Valley in the San Bernardino Mountains is partially occupied by Big Bear Lake. Recent heavy residential development has required additional water supplies and brought problems of water quality. The need to develop additional water has resulted in the drilling of numerous new test holes and wells. The subsurface information obtained, coupled with recent geologic mapping, is beginning to bring a better understanding of Bear Valley's complex geologic history.

An 1884 map of the valley floor prior to the construction of the first masonry dam suggests the existence of a mountain meadow whose present drainage system was developed by stream capture by Big Bear Creek. Evidence of easterly thickening of older alluvial fans has been found in recent drilling. The total thickness of some of the older alluvial fans is still unknown. Activity on the Helendale Fault has probably had a great effect on the easterly closure of Bear Valley.

Water from some of the area's deeper wells has a high fluoride content, whose source has not yet been determined.

San Diego County

San Diego County is not blessed with the deep structural basins filled with Recent and Pleistocene alluvium such as are present in Los Angeles and adjacent counties to the north. Instead, San Diego County includes three physiographic and hydrogeologic types of areas: 1) coastal semi-consolidated Tertiary sediments; 2) steep-sided coastal valleys containing varying thicknesses of alluvial fill, but less than 200 feet in most locations; and 3) an extensive inland foothill and mountain area in fractured granitic and metamorphic rocks, with intermontane basins where the rocks are mostly weathered to residuum and varying amounts of alluvium are commonly present.

Hydrogeologic Setting

The coastal semi-consolidated sediments have low hydraulic conductivity and storage capacity, and "are generally capable of producing only small quantities of usually poor quality water" (Peterson, 1983).

The four largest alluviated coastal valleys in the county, with their basin numbers for identification on Figure 6, are Santa Margarita River Valley (9-4), San Luis Rey Valley (9-7), San Diegito Valley (9-12), and Mission Valley (lower San Diego River Valley, 9-14). All these coastal basins are subject to seawater intrusion, but pumpage and therefore lowering of the inland groundwater level has been limited by marginal to poor water quality.

The largest intermontane basins, some of which contain significant thicknesses of younger alluvium, are the following: Warner Basin (9-8), Escondido Valley (9-9) and adjacent San Pasqual Valley (9-10), Santa Maria Valley (9-11), El Cajon Valley (9-16), and inland San Diego River Valley (9-15).

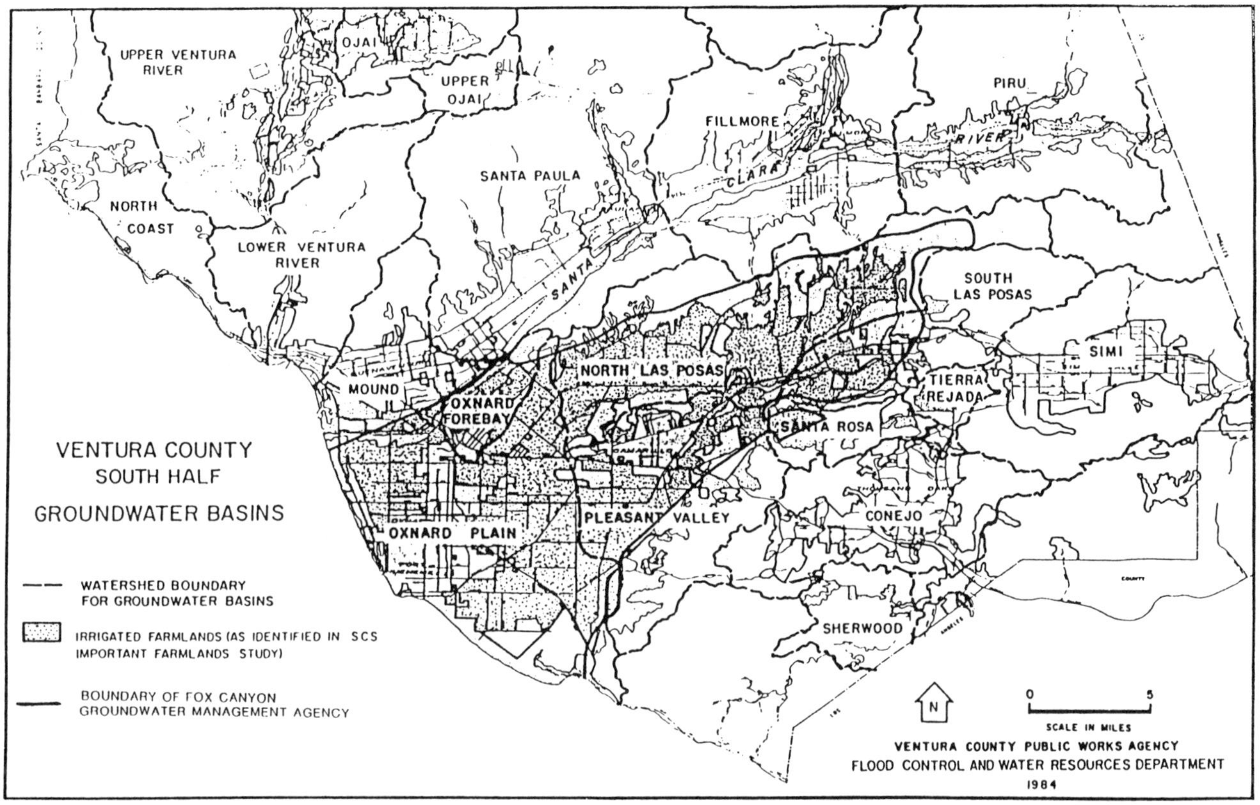

Figure 7. Ventura County Groundwater Basins

Development and Management

San Diego County has experienced explosive population growth, particularly in the last half of the present century. The amount of groundwater available from storage to support this population has been limited, and so has been recharge to replenish groundwater. As a result, local surface supplies have been thoroughly developed, and more importantly, water has been imported through the MWD.

The County of San Diego has attempted to control development and conserve groundwater supplies by a management plan that involves density controls as well as monitoring and continuing investigation (Peterson, 1983). The following table gives the minimum parcel size permitted for single-family dwellings:

Mean Annual Precipitation	Minimum Parcel Size (Gross Acres)
Less than 9"	20
9" to 12"	15
12" to 15"	11
15" to 18"	8
18" to 21"	5
More than 21"	General Plan Density

Problems and Plans

Problems related to groundwater availability have developed in spite of this management plan, and a draft of a revised ordinance to the County Code was prepared in 1991. Problems addressed by this revision include basins with declining ground water levels and a measurable groundwater overdraft, basins having a parcel size smaller than that approved (and in excess of available water resources), and basins where wells have an average yield less than 3 gallons per minute. Any application for development in a problem basin would require a groundwater investigation.

Problems of groundwater quality include nitrates above the 45 mg/l limit, evidently due to fertilizers and/or septic tank effluent. Boyle Engineering (1991, Figure 2) shows large parts of the following basins to have nitrates in groundwater above the stated limit: Escondido Basin, San Pasqual Valley, and El Cajon Valley. Furthermore, the same reference (Figure 3) shows the above three basins with groundwater exceeding 1,000 mg/l in total dissolved solids, plus significant parts of the following basins: Santa Margarita, San Luis Rey, San Elijo (9-23), San Diegito, San Diego River Valley, Sweetwater (9-17), and Tia Juana (9-19). The Boyle report contains specific recommendations for brackish water reclamation.

The San Diego Regional Water Quality Control Board currently has a draft proposal that addresses the water quality problem, and it is proposed that the draft be worked up into a

Basin Plan Amendment (Coe, 1991). Recommendations in this draft include cooperation with county regulatory agencies, cooperative investigation of basin-specific limitations on subsurface disposal, and required hydrogeologic studies under certain conditions in areas of imported water.

Ventura County

Most of Ventura County is occupied by mountains of the Transverse Ranges. The rocks composing these ranges are largely sandstones, shales, and conglomerates of Tertiary age. The strike of the rocks is generally approximately east-west, which is in accord with the direction of the ranges. The largest stream is the Santa Clara River, which enters Ventura County from Los Angeles County on the east. The Ventura River is much shorter, and flows in a southerly direction to enter the Pacific Ocean at the city of Ventura.

The most significant groundwater basins are along the coast, where they are exposed to the Pacific along about 22 miles of coastline. They total about 97 square miles in area. The largest basin for groundwater storage and development is the Oxnard Plain Basin, which covers 73 square miles, and contains a total estimated onshore storage capacity of about 9,000,000 acre-feet (CDWR, 1975). Water-bearing sediments extend inland along the Santa Clara River Valley and into Los Angeles County.

CDWR (1980) includes in the Ventura Central Basin not only the coastal basins noted above, but the basins along the Santa Clara River Valley eastward to the Los Angeles County line, plus Pleasant Valley, Las Posas Valley, and Santa Rosa Valley. Ventura County Public Works Agency subdivides the Central Basin. The Agency shows these various groundwater basins, as well as the other principal basins in the county, as they appear in Figure 7.

Ventura Central Basin

Hydrologic Setting: The sequence of aquifers and aquicludes in the Oxnard Plain Basin was first described by CDWR in 1953. Unconsolidated sediments, about 1,500 feet thick at Port Hueneme, include from the top down the following aquifers separated by intervening aquicludes: a shallow semi-perched aquifer, the Oxnard Aquifer, the Mugu Aquifer, the Hueneme Aquifer, the Fox Canyon Aquifer, and the Grimes Canyon Aquifer. The aquifers are mainly sand and gravel and the aquicludes silt and clay. The Oxnard and Fox Canyon are the most important aquifers; the Oxnard and Mugu are combined in the upper aquifer system, and the other three aquifers in the lower aquifer system. The principal forebay for the aquifers, the Oxnard Forebay, lies in the eastern part of the Oxnard Plain where the Santa Clara River enters the plain (Figure 7).

The Oxnard Aquifer, Holocene in age, is partially confined toward the coast, but open to recharge in the forebay where it also transmits recharge to underlying Pleistocene deposits. The Oxnard Aquifer averages about 110 feet thick in the Oxnard Plain, and generally has hydraulic conductivity of about 225 to 250 feet per day. Off the coast the Oxnard Aquifer is in contact with seawater in the walls of Hueneme submarine canyon, beginning about 2,000 feet offshore near Port Hueneme.

The Mugu Aquifer, upper Pleistocene in age, contains a permeable zone of sand averaging about 175 feet thick and occurring 200 to 300 feet below ground surface. Hydraulic conductivities of this zone are similar to those of the Oxnard Aquifer. The submarine outcrop of the Mugu Aquifer is farther offshore than that of the Oxnard Aquifer.

Thickness of the Hueneme Aquifer (lower Pleistocene) varies from about 300 feet north of the city of Oxnard to about 100 feet at Port Hueneme. The aquifer pinches out to the southeast. Hydraulic conductivity values are estimated to range from about 50 to 75 feet per day.

The Fox Canyon Aquifer, lower Pleistocene, consists of fine-to coarse-grained sand with gravel stringers, interbedded with some silt and clay. Maximum thickness of the Fox Canyon is about 550 feet. The Fox Canyon Aquifer is a major groundwater source, and has given its name to the Fox Canyon Groundwater Management Agency, created in 1982.

The areal extent of the Grimes Canyon Aquifer is approximately the same as that of the overlying Fox Canyon Aquifer, according to CDWR (1970). Few water wells penetrate the Grimes Canyon, but its thickness is listed as 1,500 feet or more.

The sedimentary sequence including the various aquifers is shown by CDWR (1953) to be generally flat-lying to gently dipping beneath the Oxnard Plain proper. However, synclinal structures are shown in Santa Rosa Basin and the two Las Posas Basins. Water-bearing sands and gravels of Holocene and upper Pleistocene age extend eastward along the Santa Clara River valley, and have been significantly developed by wells. In Santa Rosa Valley and Pleasant Valley, groundwater occurs in volcanic rocks in addition to the sedimentary materials.

Development and Management: Seawater began to enter the Oxnard Aquifer many decades ago, when pumping in the Oxnard Plain reduced groundwater levels below sea level. By the early 1930's, seawater intrusion had affected groundwater up to half a mile from the coast (CDWR, 1970), and by 1966 the 500 mg/l isochlor line had advanced almost 3 miles inland. An experimental salt-water extraction type of barrier was put into operation by CDWR in 1966 in an attempt to stem seawater intrusion into the Oxnard Aquifer. The barrier was located in the Port Hueneme area, and pumping continued for about two years. About 9,000 acre-feet of de-

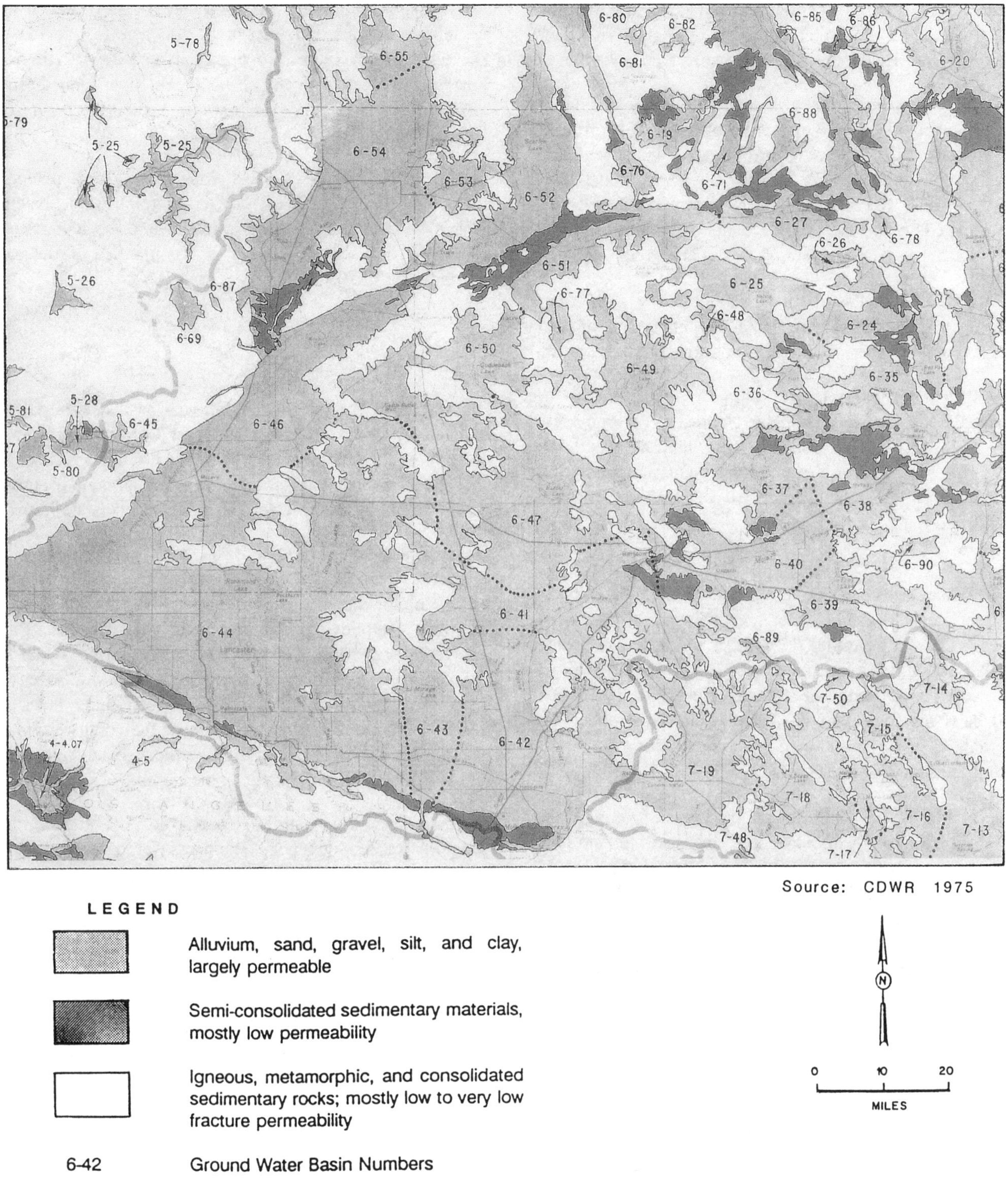

Figure 8. South Lahontan Area Groundwater Basins

graded water was removed during the extraction, and there was "diminution of the sea-water intrusion area" (CDWR, 1970). However, it was decided not to continue this type of barrier.

Additional degradation of water quality in the lower aquifer zone was taking place by downward movement of poor quality water through wells that were gravel packed and/or perforated in the semiperched and Oxnard aquifers and in the lower aquifers as well. To eliminate this problem, the offending wells were plugged in the late 1970's under the direction of the County Resources Management Agency (Farnsworth, 1991).

In 1980, CDWR stated that about 13,000 acres of the upper aquifer were intruded by seawater, and Ventura Central Basin was listed as "Subject to Critical Conditions of Overdraft."

In 1982, after a great many years of disagreement, the various water interests of Ventura County were brought together by the passage of Assembly Bill 2995 Imbrecht, which created the Fox Canyon Groundwater Management Agency (FCGMA). The boundaries of this Agency were to "include all land overlying the Fox Canyon Aquifer" and are shown on Figure 7.

The FCGMA has been active since its formation in taking important steps in groundwater management. Groundwater extractions are now limited in each of the basins delineated by FCGMA, and all pumpers are required to report their extractions. The Groundwater Management Plan (December, 1985) calls for control of the Santa Clara River above Oxnard Plain to bring 43,400 acre-feet annually for groundwater recharge into the Oxnard Forebay.

Problems and Plans: The principal problems in Ventura Central Basin are groundwater overdraft and seawater intrusion. Pumpage in the area of under FCGMA domain was 175,043 acre-feet in 1990. FCGMA plans to reduce overdraft on the groundwater basins to zero and bring the basins to safe yield by the year 2010 (Farnsworth, 1991). In addition to some reduction of pumping, water conservation is promoted, and wastewater reclamation put into practice. Use of reclaimed water is projected to be 11,810 acre-feet annually by 1995 and 12,920 by 2010 in Oxnard Plain and Pleasant Valley combined, and 6,170 acre-feet by 1995 and 6,630 acre-feet by 2010 in North Las Posas Basin (Ventura County Public Works Agency, 1985).

Nitrate concentrations in excess of the 45 mg/l limit have been found in some wells in the Oxnard Forebay, Oxnard Plain, and locally in the Santa Clara River Valley. Probable sources are fertilizers and septic tank effluent. Water in excess of the nitrate limit is useable for irrigation but not human consumption.

Groundwater levels in the lower aquifer system were more than 120 feet below sea level by 1985 in the center of a regional cone of depression in Pleasant Valley Basin, causing serious concern about seawater intrusion. A Seawater Intrusion Abatement Project was formulated, calling for an average annual surface delivery of 11,700 acre-feet to the area affected. This project is operational, and about half of the water is diverted from the Santa Clara River, and the other half from lower aquifer system wells outside the severe cone of depression.

Water quality monitoring wells along the coast are planned to check the effectiveness of the seawater intrusion abatement project by detecting possible intrusion into the lower aquifer system. Additional restrictions are planned if the lower aquifer system is invaded.

Ojai and Upper Ojai Basins

Ojai and Upper Ojai basins are alluvial basins essentially surrounded by Tertiary sedimentary rocks of low permeability, and having areas of about 13 and 3 square miles respectively. Ojai Basin probably discharges a small amount of groundwater into the Ventura River Basin when levels are high.

Groundwater levels in Ojai Basin are very sensitive to seasonal water supply. Wells have supplied water for irrigation and municipal/domestic use for many years, but at present additional supplies are imported from Lake Casitas. Recent dry years and consequent limitation of supplies from both sources have resulted in pumpage exceeding safe yield (Hoffman, 1991). Restrictions on pumpage have been instituted. High nitrates are a problem in the southwestern part of Ojai Basin, and in places high total dissolved solids (TDS) render the water marginal for domestic use.

In Upper Ojai Basin, high TDS also locally render groundwater above recommended standards for domestic use and marginal for irrigation (CDWR, 1975).

Ventura River Basin

The Ventura River Basin, about 10 square miles in area, is separated into upper and lower basins by a constriction near Foster Park. Most development is in the upper basin, but TDS and sulfate are locally high for domestic use and marginal for irrigation.

Simi Valley Basin

Simi Valley Basin is about 25 square miles in area, and contains groundwater in Holocene and Pleistocene alluvium. Groundwater was pumped extensively for agriculture and domestic use until the 1960's, when urbanization began to take over the valley rapidly. State Project water was first imported in the middle of that decade, and problems of water supply soon began to be replaced by high groundwater problems. Water levels in some wells rose about 100 feet, and artesian flow in the western part of the valley took place by 1972 (Hoffman, 1991).

A dewatering project is presently in operation in the western valley. Quality of the water is unsatisfactory for surface disposal, but the water is used by a petroleum company for injection for secondary recovery of oil.

In the Conejo-Tierra Rejada area, groundwater occurs essentially in volcanics of Tertiary age.

SOUTH LAHONTAN AREA

The South Lahontan area extends from the Sierra Nevada and Tehachapi Mountains eastward to the California-Nevada border. The southwestern part of this area is shown in Figure 8.

The South Lahontan is characterized by desert mountains and intermontane basins. All surface drainage is internal. The mountains are composed of a great variety of rocks, most of which have very low permeability and storage capacity.

The basins are largely composed of unconsolidated sedimentary materials eroded from the surrounding mountains. In a typical desert basin, the underlying sediments vary from gravel and sand in the alluvial fans at the base of the mountains to a central playa underlain by clay and evaporites. Groundwater quality is generally best near the mountains, and salinity increases with distance from the mountains, until the water generally becomes unusable for domestic purposes or irrigation in the central part of the basin. Salts may be mined, as in Searles Valley (No. 6-52 on Fig. 8).

The most significant development of groundwater is in the basins along the Mojave River (Nos. 6-40, 6-41, and 6-42), and in Antelope (6-44), Fremont (6-46), and Indian Wells (6-54) valleys.

Antelope Valley

Antelope Valley (6-44 on Figure 8) lies between the Tehachapi Mountains on the north and the Transverse Ranges on the south. Saturated sedimentary deposits, mainly alluvium, in places over 1,000 feet deep, have been developed intensively for irrigation and municipal use during the present century. The greatest use has been in the Lancaster subunit, the area south and west of Rosamond Lake including the city of Lancaster, as delineated by CDWR.

Three principal aquifers occur in the Lancaster subunit: a deep aquifer, which is primarily confined by lacustrine deposits several hundred feet thick; an overlying unconfined zone commonly called the principal aquifer;and an extensive perched water body. Descriptive data on monitoring wells collected by CDWR (Pierotti, 1989) document changes in water level and water quality that have occurred throughout Antelope Valley. Although there is much variation in individual wells, a general decline in water levels has taken place as groundwater has been developed. The decline has been as great as 110 feet according to Pierotti. In a majority of wells, the lowest water elevations occurred in the 1960's or 70's. Since then, water levels have risen in a significant number of wells, due primarily to a change in land use from agricultural to municipal.

Overdraft on the Antelope Valley Groundwater Basin calculated from CDWR Bulletin 118 was about 142,000 acre-feet in 1970. However, Bulletin 118-80 (Fig. 2) does not include Antelope Valley in "basins subject to critical conditions of overdraft."

Mineral quality of groundwater as measured by total dissolved solids is variable in the key wells sampled by CDWR (Pierotti, 1989), but is mostly under 500 mg/l. Variation with time shows no clear pattern of improvement or degradation, although the earliest record shown was 1953. Bulletin 118 of CDWR notes that quality is "locally poor for irrigation and domestic use," and that there are problems, presumably in shallow depth zones, with septic tank failures.

A fairly recent problem evidently due to groundwater withdrawal is land subsidence, notably in the Lancaster area and Edwards Air Force Base. According to the United Nations Educational, Scientific and Cultural Organization (UNESCO), over 460 square miles around Lancaster subsided up to 3 feet between 1955 and 1978 (Los Angeles Times, March 17, 1991). At least 5.5 feet of subsidence has been reported in some places.

Problems associated with the subsidence include the appearance of fissures, about 70 of which have been found in an undeveloped area of Lancaster where home building has been proposed. Additional fissures have caused the closure of a runway at Edwards Air Force Base.

No recent regional investigation of the land subsidence has taken place, but geologic conditions are similar to other areas where the same problem has occurred. As noted above, a deep confined aquifer occurs in the Lancaster subunit. Pumpage accompanied by decline in head of a confined aquifer is the most common cause of regional land subsidence in California, with established examples being Santa Clara Valley and extensive areas in the San Joaquin Valley. In both those areas, land subsidence has been reduced or eliminated by decreasing the pumping draft on the confined aquifers, and substituting the use of surface water. Presumably, the same mitigation method would reduce or eliminate the problem in Antelope Valley. However, complete substitution would depend on the availability of sufficient surface water; which may or may not be at hand.

Mojave River Basins

The Mojave River system begins in the San Bernardino Mountains in the vicinity of Lakes Arrowhead and Gregory. After leaving the mountains, the river flows through three major basins, shown in Figure 8: the Upper Mojave (6-42), Middle Mojave (6-41), and Lower Mojave (6-40). The basins are separated from each other by topographic constrictions. Together they cover a total area of 1,330 square miles. In the lower basin, the river course is easterly beyond Barstow for about 50 miles, where it terminates in Soda Lake, the "Sink of the Mojave."

This ephemeral stream was the only significant source of water, mainly by groundwater recharge, until the advent of the California Aqueduct of the State Water Project (SWP) in 1972. Increasing development of the area resulted in overdrafting of the Mojave basins. The Mojave Water Agency was formed to contract for water from the SWP, and in addition the Agency has provided funding for studies on utilization of groundwater.

In 1967, CDWR published Bulletin 84, "Mojave River Ground Water Basins Investigation." This document served as the basis for an unsuccessful attempt at adjudication of the basins in the early 1970's.

Three reports prepared for the Mojave Water Agency deal with yields of the three Mojave basins (C. M. Engineering and others, 1982 and 1983, and Murk and Crandall, 1985). The U. S. Geological Survey (USGS) has constructed a groundwater model of the Barstow-Daggett area. A body of highly contaminated groundwater, called the "Barstow polluted slug" developed in the 1950's after continuing discharge into the Mojave River channel of wastewater high in phenols and boron. USGS and consulting company studies have focused on the movement of this slug.

Pumpers in the Middle and Lower Mojave Basins have become concerned over their supply of groundwater due to extensive recent development in the Upper Mojave Basin. A lawsuit was filed recently by the City of Barstow against all upstream purveyors of water. The suit claims that Barstow has the right to 30,000 acre-feet per year, and that the heavy growth upstream is depriving the city of its rightful share of water.

The Mojave Water Agency has undertaken studies to determine the extent of the "zone of benefit" from recharge of water in the Mojave River. A proposal has been made to make releases of SWP water from Lake Silverwood for percolation downstream. The zone of benefit study would determine those parties who would receive a supply of SWP water under these conditions, and thereby would share in the cost of the water.

Fremont Valley

Fremont Valley, Basin 6-46 on Figure 8, is essentially an extension to the northeast of Antelope Valley. The Tehachapi and El Paso mountains, with the Garlock Fault at their bases, bound Fremont Valley on the northwest. Drainage of the valley is internal to Koehn Lake.

Use of groundwater includes agricultural and municipal, with the latter principally for California City. Current growing of alfalfa with its heavy use of water is notable.

With some exceptions, decline of groundwater levels in the valley due to pumping has not been excessive in most areas. However, two wells in the western part of the valley, about six miles southwest of Koehn Lake, showed water level drops of 263 and 220 feet from 1974 to 1985 and 1986, respectively (CDWR, 1987).

Indian Wells Valley

Indian Wells Valley, Basin 6-54 on Figure 8, receives major recharge of good quality to groundwater by percolation of streams draining the Sierra Nevada on the west. Groundwater movement is generally eastward, toward a regional cone of depression northwest of the town of Ridgecrest. However, poorer quality water between Ridgecrest and China Lake moves southwesterly toward this same regional low.

A report evaluating all available existing information on groundwater in Indian Wells Valley was completed recently by Bean (1988). He concludes that overdraft under 1985 conditions was slightly over 16,000 acre-feet per year. However, the amount of groundwater remaining in storage is probably close to 2,000,000 acre-feet, based on Dutcher and Moyle's (1973) estimate that 2,200,000 acre-feet were available within 200 feet of the 1921 water table.

Since 1988, drilling in the nearly unused southwesterly part of the valley has developed additional water supplies, and steps have been taken to establish integrated management of the basin.

COLORADO DESERT AREA

The southeastern part of California, Hydrologic Study Area 7 of CDWR, is an area of low precipitation, the Colorado Desert Area. Local groundwater is extremely limited in amount, as it is dependent on percolation on alluvial fans and valley floors from the intermittent rainfall on desert mountains. The mighty Colorado River flows along the eastern margin of this area, and contributes water to various canal and aqueduct systems.

Sixty-one groundwater basins have been delineated by CDWR in this Area, but detailed comments on only two of the largest and most important of these can be made here because of limitations of space.

Coachella Valley

Coachella Valley, shown in part as Basin 7-21 on Figure 6, occupies the northwestern part of a large structural trough that includes the San Andreas Fault in the Colorado Desert region. The trough has received detritus from the adjacent highland areas since Miocene time. Coachella Valley extends from the topographic divide in San Gorgonio Pass to the northern margin of the Salton Sea and covers approximately 600 square miles. The Whitewater River is the principal stream in the Valley. The Valley in the vicinity of, and southerly of Indio, was covered by Lake Cahuilla during upper Pleistocene time. The lake is evidenced by shoreline features such as the deposits at Travertine Rock, and the Cahuilla Lake beds, containing myriads of fresh-water gastropods.

Coachella Valley's early development was related to agriculture. After the turn of the century, date agriculture commenced, experimentally at first. From this meager beginning, the date farming operations of today have succeeded in supplying 90 percent of the date supply of the United States. Citrus orchards and grape vineyards have also developed. The demand for irrigation water led to the construction of the Coachella branch of the All American Canal, which was competed in 1948. The Cahuilla Lake beds now underlie a large semi-perched water zone at shallow depths. To cope with the drainage problems that have developed, a large network of tile drains has been established. Many of these discharge into the lower reaches of the Whitewater River channel and have impacted the surface water quality.

While agriculture was becoming strongly established in lower Coachella Valley, recreational developments were beginning to flourish in the upper portions of the valley. These developments started as small hot spas, but as the magic name of Palm Springs spread nationally, the demand for winter resorts increased. Today, there are many resorts, complete with golf courses.

Investigations by both CDWR (1964) and USGS (1977), the latter stating that groundwater levels declined 100 feet between 1936 and 1974, have furnished the basis for water management. Concern over the adequacy of water supplies was instrumental in the formation of the Desert Water Agency. This Agency contracted with CDWR for supplemental water from the State Water Project. Due to the estimated construction costs of extending the State aqueduct into Coachella Valley, a cheaper alternative was devised. The USGS (1977) had made studies of artificial recharge along the Whitewater River between Windy Point and Palm Springs. Under an exchange agreement with MWD, imported Colorado River water is spread in these previously defined recharge areas. The groundwater thus stored is then available for pumpage.

Imperial Valley

Imperial Valley includes the area of the Salton trough south of the Salton Sea. East and West Mesas occur on either side of the valley. The central trough of the valley is principally underlain by clay deposits, and groundwater supplies are negligible. Recharge to groundwater can occur in the East and West Mesas, but the quality of much of the groundwater is poor. Groundwater was recharged for many years by seepage from canals, particularly from the All American canal near the Mexican border and the Coachella branch along the east side of the valley. However, during the mid-1980's the southernmost 49-mile reach of the Coachella canal was replaced by a lined segment, thus reducing seepage to groundwater by an estimated 120,000 acre feet per year (Angelos, 1991). A pilot project to utilize the old unlined segment for groundwater recharge under controlled conditions recently percolated 18,000 acre-feet to groundwater in one year.

ACKNOWLEDGMENTS

We wish to acknowledge the excellent cooperation received from key individuals within all agencies from which we requested data and information.

Special acknowledgment is given to Law Environmental, Inc., for essential drafting services.

REFERENCES

General

California Department of Water Resources, 1975, California's Ground Water, Bulletin 118, 135 pages; and Map, scale 1:750,000, Valley-Fill Areas and Other Water-Bearing Materials.

——-, 1980, Ground-Water Basins in California, Bulletin 118-80.

Kennedy, David N., 1990, Are California's Water Resources Adequate to Meet the State's Projected Growth? In Coping with Water Scarcity: the Role of Ground Water, 17th Biennial Conference on Ground Water, University of California Water Resources Center, pp. 19-23.

Mann, John F. Jr., 1966, Ground Water in Southern California, In Engineering Geology in Southern California, Association of Engineering Geologists, pp. 237-264.

Tyson, H. N. and Weber, Ernest M., 1964, Groundwater Management for the Nation's Future — Computer Simulation of Groundwater Basins, Journal Hydraulics Division, American Society of Civil Engineers, vol. 90, no. HY4, pp. 59-7.

Los Angeles County

Barto, R. L., Cleath, T. S. and others, 1982, Ground Water Resource Study, Agua Dulce and Mint Canyon, Los Angeles County, California, Phase 1, Hydrogeologic Investigation, James M. Montgomery, Consulting Engineers.

California Department of Water Resources, 1961, Planned Utilization of the Ground Water Basins of the Coastal Plain of Los Angeles County, Bulletin 104, Appendix A, Ground Water Geology.

———, 1966, Planned Utilization of Ground Water Basins San Gabriel Valley, Bulletin 104-2, Appendix A, Geohydrology, 230 pages.

———, Southern District, various years, Watermaster Service in the Raymond Basin.

Central Basin Municipal Water District and West Basin Municipal Water District, March 1991, The Sampler, vol. 1, no. 2, 4 pages.

Harris, Jonathan, June 11 1991, Speech before Association of Engineering Geologists, Southern California Section, Hydrogeologist, CH2M Hill.

Kidman, Arthur G., 1990, San Gabriel Valley Groundwater Contamination as a Cause of Water Scarcity, In Coping with Water Scarcity: The Role of Ground Water, 17th Biennial Conference on Ground Water, University of California Water Resources Center, pp. 129-138.

Lazo, Rodrigo K., 1983, Recent Groundwater Quality Changes in San Gabriel Valley: Unpublished MS Dissertation, Department of Geology, California State University, Los Angeles, 132 pages, 4 appendices.

Los Angeles County Department of Public Works, 1990, Hydrologic Report, 1988-89; and 1991, personal communication.

Los Angeles County Sanitation Districts, 1984, Summary, Health Effects Study, 27 pages.

Main San Gabriel Basin Watermaster, 1991, Eighteenth Annual Report, 132 pages.

Mann, John F. Jr., 1991, personal communication, Consulting Geologist, La Habra, California.

Mann, John F. Jr. and Blevins, Melvin L., 1986, Groundwater Management in the San Fernando Valley Basin, In Hydrogeology of Southern California: 82nd Annual Meeting of the Cordilleran Section, Geological Society of America, California State University Los Angeles, pp. 1-4.

———, 1990, Management of San Fernando Water Rights and Groundwater Contamination, Hazmat West 1990 Conference, Long Beach, California.

Mendenhall, W. C., 1904, Map Showing the Artesian Areas and Hydrographic Contours in the Valley of Southern California, U. S. Geological Survey Water-Supply Paper 142, Pl. VIII.

Poland, J. F., Piper, A. M., and others, 1956, Ground-Water Geology of the Coastal Zone Long Beach-Santa Ana Area, California, U. S. Geological Survey Water-Supply Paper 1109.

Poland, J. F., Garrett, A. A. and Sinnott, Allen, 1959, Geology, Hydrology, and Chemical Character of Ground Waters in the Torrance-Santa Monica Area, California, U. S. Geological Survey Water-Supply Paper 1461.

Slade, Richard C., 1988, Executive Summary Hydrogeologic Assessment Saugus Formation in the Santa Clara Valley of Los Angeles County, California, letter to Robert Sagehorn, Castaic Lake Water Agency, Newhall, California.

State Water Rights Board, 1962, Report of Referee, City of Los Angeles v. City of San Fernando, et. al., LASC Case No. 650079.

Stetson, Thomas M., 1986, Main San Gabriel Basin Ground-Water Management; and Water Quality of the Main San Gabriel Basin, In Hydrogeology of Southern California: 82nd Annual Meeting of the Cordilleran Section, Geological Society of America, California State University, Los Angeles, pp. 5 - 15.

United States Environmental Protection Agency, July 1990, Groundwater Cleanup Studies Continue in the San Fernando Valley Basin, Fact Sheet No. 5.

———, April 1991, San Gabriel Valley Superfund Sites, Fact Sheet No. 10.

Upper Los Angeles River Area Watermaster, 1991, Watermaster Service in the Upper Los Angeles River Area, Los Angeles County, October 1, 1989 - September 30, 1990, 132 pages.

Welsh, James L, 1988, Ground Water Management: Los Angeles Coastal Plain, In Artificial Recharge of Ground Water, A. I. Johnson and D. J. Finlayson, eds., American Society of Civil Engineers.

Orange County

Black, William H., Goodrich, James A., and Patton, Franklin D., 1988, Ground Water Monitoring for Resource Management, In Artificial Recharge of Ground Water, A. I. Johnson and D. J. Finlayson, eds., American Society of Civil Engineers, Pages 446-454.

California Department of Water Resources, 1966, Ground Water Basin Protection Projects, Santa Ana Gap Salinity Barrier, Orange County, Bulletin 147-1.

Orange County Water District, February 1990, Groundwater Management Plan, 32 pages.

——-, 1990, Annual Report 1989-90, 28 pages.

Richardson, Nereus, 1988, Ground Water Recharge at the Orange County Water District, In Artificial Recharge of Ground Water, A. I. Johnson and D. J. Finlayson, (eds.), American Society of Civil Engineers Pages 426-434.

Williams, Dennis E., 1986, Groundwater Modeling in the Orange County Area, In Hydrogeology of Southern California: 82nd Annual Meeting of the Cordilleran Section, Geological Society of America, California State University, Los Angeles, pp. 21-26.

Woodside, Ramon D. and Woodside, Grey D., 1986, San Juan Ground Water Basin in Orange County, California, In Hydrogeology of Southern California: 82nd Annual Meeting of the Cordilleran Section, Geological Society of America, California State University, Los Angeles, pp. 35-41.

San Bernardino-Riverside Counties

California Department of Water Resources, 1959, Santa Ana River Investigation, Bulletin 15, Appendix B, Geology of San Jacinto and Elsinore Units, pp. 101-126.

———, 1983, Chino Basin Ground-Water Storage Program, Final Report.

California Regional Water Quality Control Board, Santa Ana Region, 1984, Water Quality Control Plan for the Santa Ana River Basin (8).

Metropolitan Water District of Southern California, 1990, Metropolitan's Role in the Development of Groundwater in Southern California, Discussion Paper, 45 pages.

Scheliga, J. T. Jr. and McGoldrick, J. P., 1984, Rancho California Water District, Water Resources Master Plan 1984.

San Diego County

Boyle Engineering Corporation, 1991, Brackish Groundwater Reclamation Study Executive Summary, Metropolitan Water District of Southern California.

California Department of Water Resources, 1967, Ground Water Occurrence and Quality: San Diego Region, Bulletin 106-2.

Coe, Art, 1991, Personal Communication, Hydrologist, San Diego Regional Water Quality Control Board.

Peterson, John, 1983, A County Perspective on Groundwater in Fractured Rock, In Proceedings of the Fourteenth Biennial Conference on Ground Water, University of California Water Resources Center, pp. 132-138.

San Diego, County of, 1991, Revised Draft of an Ordinance adding Chapter 7 to Division 7 of Title 6 of the San Diego County Code, Relating to Groundwater.

Ventura County

California, State of, 1982, Assembly Bill 2995, Imbrecht.

California Department of Water Resources, September 1970, Oxnard Basin Experimental Extraction-Type Barrier, Bulletin 147-6.

——-, October 1975, Sea-Water Intrusion in California, Bulletin 63-5.

California Division of Water Resources, 1953, Ventura County Investigation, State Water Resources Board, Bulletin 12.

Farnsworth, Richard B, 1991, personal communication, Hydrogeologist, Fox Canyon Groundwater Management Agency.

Hoffman, La Vern, 1991, personal communication, Hydrologist, Ventura County Public Works Agency.

Thomas, R. G., Marliave, E. C., James, L. B., and Bean, R. T., 1954, Geology and Hydrology of Ventura County, In Geology of Southern California, Bulletin 170, California Division of Mines.

Ventura County Public Works Agency, December 1985, Executive Summary Groundwater Management Plan for the Fox Canyon Groundwater Management Agency.

South Lahontan and Colorado Desert Areas

Angelos, Richard, 1991, Principal Hydraulic Engineer, California Department of Water Resources, personal communication.

Bean, Robert T., 1989, Hydrogeologic Conditions in Indian Wells Valley and Vicinity, California Department of Water Resources and Kern County Water Agency, 51 pages.

California Department of Water Resources, 1964, Coachella Valley Investigation, Bulletin 108.

———, 1967, Mojave River Ground Water Basins Investigation, Bulletin 84.

———, 1969, Water Wells and Springs in the Fremont Valley Area, Kern County, Bulletin 91-16, 157 pages.

———, Southern District, 1987, Antelope Valley Observation Well Data, computer printout.

C. M. Engineering Associates, LeRoy Crandall and Associates, Robert Richey, 1982, Historic and Present Conditions, Newberry Groundwater Basin.

C. M. Engineering Associates and LeRoy Crandall and Associates, 1983, Mojave River Groundwater Basins, Historic and Present Conditions, Helendale Fault to Calico-Newberry Fault.

Duell, Lowell F. W. Jr., 1987, Geohydrology of the Antelope Valley Area, California, and Design for a Ground-Water-Quality Monitoring Network, U. S. Geological Survey, Water Resource Investigations, Report 84-4081.

Dutcher, L. C., and Moyle, W. R. Jr., 1973, Geologic and Hydrologic Features of Indian Wells Valley, California, U. S. Geological Survey Water-Supply Paper 2007.

Loeltz, O. J., Irelan, B., Robison, J. H., and Olmsted, F. H., 1975, Geohydrologic Reconnaissance of the Imperial Valley, California, U. S. Geological Survey Professional Paper 486-K, 54 pages.

Los Angeles Times, March 17, 1991, Pumping Threatens to Sink High Desert's Future, John Chandler, Staff Writer.

Murk, John S. Engineers, Inc., and LeRoy Crandall and Associates, 1985, Historic & Present Conditions, Upper Mojave River Basin.

Pierotti, Bob, 1989, A Network of Key Wells for Monitoring Water Levels and Water Quality in the Antelope Hydrologic Unit, California Department of Water Resources, Southern District, Technical Information Record 1644-IV, 109 pages.

U. S. Geological Survey, 1977, Predicted Water-Level and Water Quality Effects of Artificial Recharge in the Upper Coachella Valley California, Water Resource Investigations 77-29, Swain, L. A., 54 pages.

UPDATE ON SEA WATER INTRUSION

Michael Johnson
Supervising Engineering Geologist
Los Angeles County Department of Public Works
Alhambra, CA 91802-1460

INTRODUCTION

Coastal ground water basins of Los Angeles County have been affected by sea water intrusion for most of this century. As early as 1912, intrusion was detected in the Redondo Beach area of Santa Monica Bay; and by 1932, it was evident that the "plague" had spread to include coastal areas of San Pedro Bay.

The common cause of the intrusion is serious overdraft of ground water to elevations below sea level. Reversal of the natural seaward gradient resultes in sea water intrusion into previously fresh aquifers exposed now both to offshore saline waters and overlying tidal channels. During mostly the 1950's and early 1960's, the problem was studied through detailed geohydrologic and engineering investigations to determine feasibility and develop design concepts for facilities to halt or significantly reduce intrusion.

The purpose of this article is to provide an update on the status of operations at the West Coast Basin Barrier Project. After nearly 40 years of operation and over 800,000 acre-feet of injected water, ground water quality conditions have changed dramatically. The successes and shortcomings of the barrier are demonstrated by the water quality of the Silverado aquifer. Chloride concentrations reveal the barrier's effectiveness as well as the continuing problems due to the landward migration of salt water trapped inland of the barrier.

PHYSIOGRAPHY

The physiographic boundaries of the West Coast Basin are the Newport-Inglewood Uplift on the east; the Pacific Ocean on the south and west; and the Ballona Escarpment on the north. The predominant physiographic features of the Basin interior include the Torrance Plain, the El Segundo Sand Hills, Dominguez Gap, and the Long Beach Plain. Dominguez and Baldwin Hills are the local surface expressions of the Newport-Inglewood Uplift. The Palos Verdes Hills extend about nine miles along the southern boundary, reaching elevations approximately 1500 feet above sea level (Fig. 1).

The Basin covers an area of about 100,000 acres and has a population of about one million people. Urban development, with a vast amount of industrialization, has replaced what was once mostly agricultural activity.

GEOHYDROLOGY

In the West Coast Basin area, Tertiary and Quaternary marine and continental sediments have been deposited on pre-Tertiary non-water-bearing igneous and metamorphic basement rocks. The Tertiary sediments range in age from Eocene to Pliocene and are almost entirely of marine origin. These sediments do not contain fresh water, except for certain late Pliocene aquifers.

The important hydrogeologic units of the Basin are contained within the Quaternary sediments. They are, in order of increasing depth below ground surface, the "200-foot sand" and the Gardena aquifers in the Upper Pleistocene Deposits and the "400-foot gravel" (merged with the Silverado aquifer in the West Coast Basin coastal area), the Silverado and Lower San Pedro aquifers in the San Pedro Formation. Several non-water-bearing units (aquitards) are significant to the local hydrogeology because they provide separation between the aquifers and locally serve as confining layers which interrupt the hydraulic continuity between the aquifer units.

The Silverado zone is the major producing aquifer in the West Coast Basin. It underlies most of the Basin and extends eastward over the Newport-Inglewood Uplift where it is continuous with Pleistocene aquifers in the Central Coastal Plain. The Silverado aquifer reaches a maximum thickness of 500 feet and maximum depth of 1200 feet below sea level in the Dominguez area. Maximum transmissivities reach 150,000 gallons per day-per foot. It is estimated that approximately two-thirds of all the water injected in the West Coast Basin Barrier Project goes into the Silverado zone.

GROUND WATER PRODUCTION

Overdraft conditions have existed in the West Coast Basin for many years. Ground water withdrawals increased substantially in the 1920's and 1930's to meet increasing demands. Recognizing the seriousness of the problem, court action was started in 1945 by several major water agencies in the West Coast Basin. After preparation of a Report of Referee by the California Division of Water Resources (now Department of Water Resources), 46 of the major water users mutually agreed to voluntarily reduce pumping. Soon thereafter, the Department of Water Resources was appointed Watermaster to administer the pumping cutbacks. Under the sponsorship of the West Basin Water Association, a committee of legal advisors proposed stipulated judgement to define the ground water rights of West Coast Basin producers. A decree was issued by the court on August 10, 1961. With some modification in 1966, the total adjudicated allowable pumping in the West Coast Basin is 64,468.25 acre-feet per year. Areas of major ground water production are located in the Carson-Wilmington area, City of Hawthorne, and City of Torrance (Fig. 1).

BARRIER EFFECTIVENESS

The West Coast Basin Barrier consists of 142 injection wells that inject approximately 20,000 acre-feet annually at 104 locations. The wells extend along a nine mile-reach from the Palos Verdes Hills northward to Los Angeles International Airport. The alignment is approximately one-mile inland and parallel to the coastline.

In general, the existing barrier facilities in the "200-foot sand" aquifer are able to maintain water levels above the minimum required for protection. Leakage appears to be fairly well controlled. There are no major problem areas where additional facilities appear to be warranted.

In the Silverado aquifer, the existing facilities are capable of sustaining heads above protective elevations throughout most of the barrier reach. Areas where head conditions are marginal or where intermittent plumes of elevated chlorides have been observed are just north and south of Rosecrans Avenue, at Artesia Boulevard, near 190th Street, and just south of Torrance Boulevard (Fig. 2). These conditions can be controlled by relatively minor operational and physical modification of existing facilities and additional wells are not needed.

A different situation exists in the Lower San Pedro aquifer where facilities have not been completed. Monitoring data indicate that intrusion is occurring through or beneath the partially developed pressure ridge and no barrier wells have as yet been constructed in the 3500-foot northern reach of the aquifer; essentially north of 190th Street. Most of the Lower San Pedro aquifer has elevated chlorides inland of the barrier to about Hawthorne Boulevard. The only obvious fresh water areas are those that have developed around completed and on- line injection wells. Internodal elevations (midpoint between injection wells) are 1 to 14 feet below protective elevations due partly to wells that have been shut down for repair. Camp, Dresser, and McKee, Inc. (CDM), recommended in their 1989 study that 11 new wells be constructed and at least four existing wells be rehabilitated or modified to improve protection of the aquifer. Installation of five of the proposed wells was completed in February of 1991, and it is still too early to determine the full effect of the new facilities.

All of the identified deficiencies in the barrier can be corrected by reducing extractions, adding injection wells, and/or increasing injection rates. However, the most difficult problem to deal with is the cut-off saline wedge trapped inland of the barrier. Substantial intrusion occurred prior to the activation of the barrier leaving significant plumes of high chlorides on the landward side. These plumes continue to migrate slowly under the landward gradient created by the inland extractions and barrier injections. The total estimated quantity of degraded water (greater than 250 mg/l chlorides) inland of the barrier is 460,000 acre-feet.

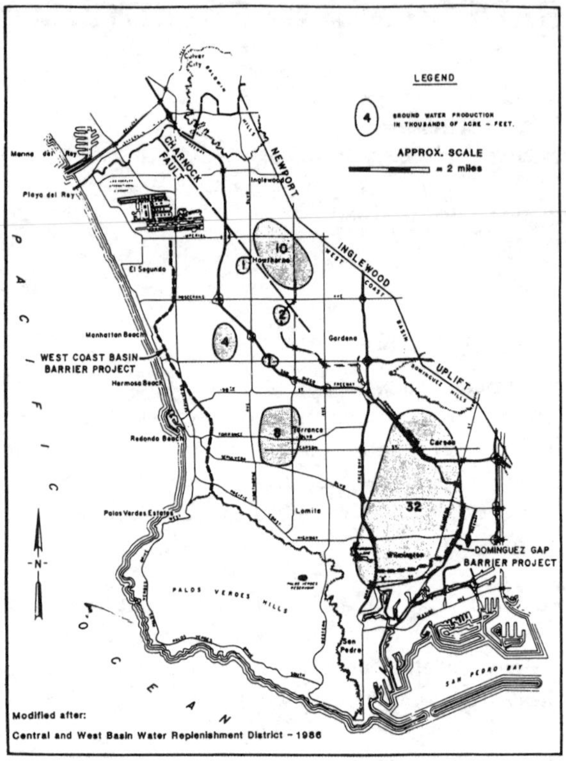

Figure 1. Location map of West Coast Basin sea water barrier showing areas of major ground water extractions circa 1983-84

The major severed plume area in the Silverado aquifer extends southward from Artesia Boulevard to the Palos Verdes Hills and inland to approximately Hawthorne Boulevard (Fig. 2). It is estimated that the leading edge of this saline wedge will reach Crenshaw Boulevard within 10 to 15 years and will impact additional production wells in the city of Torrance. About 200,000 acre-feet of storage is unusable due to this severed wedge.

REMEDIATION

The top priority of work involves completion of the barrier in the Lower San Pedro aquifer. This work includes the addition of injection wells north of 190th Street and rehabilitation and, in some cases, deepening of existing injection wells. Secondly, some reduction of inland pumping is being considered as a way to augment the impact of these new facilities in developing protective elevations.

To mitigate the effects of the trapped saline wedge inland of the barrier, a combination of injection and extraction is being studied where the injection wells are located to drive saline water to the extraction wells.

To eliminate problems associated with the disposal of pumped saline water, desalination and reinjection is the favored option being considered. By employing a system of balanced extraction and injection, the loss of fresh water can

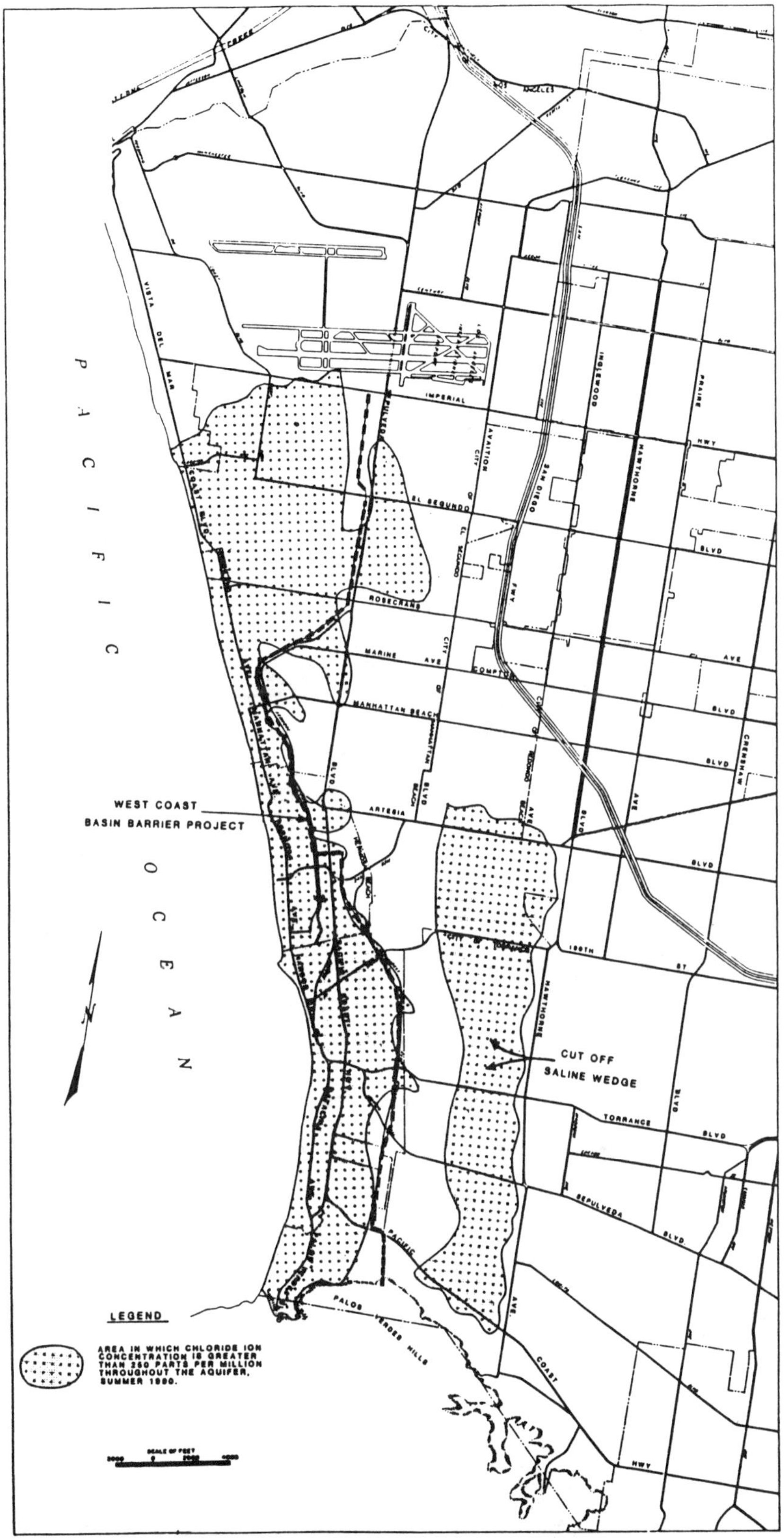

Figure 2. Saline plumes and cut off saline wedge inland of the barrier in the Silverado aquifer

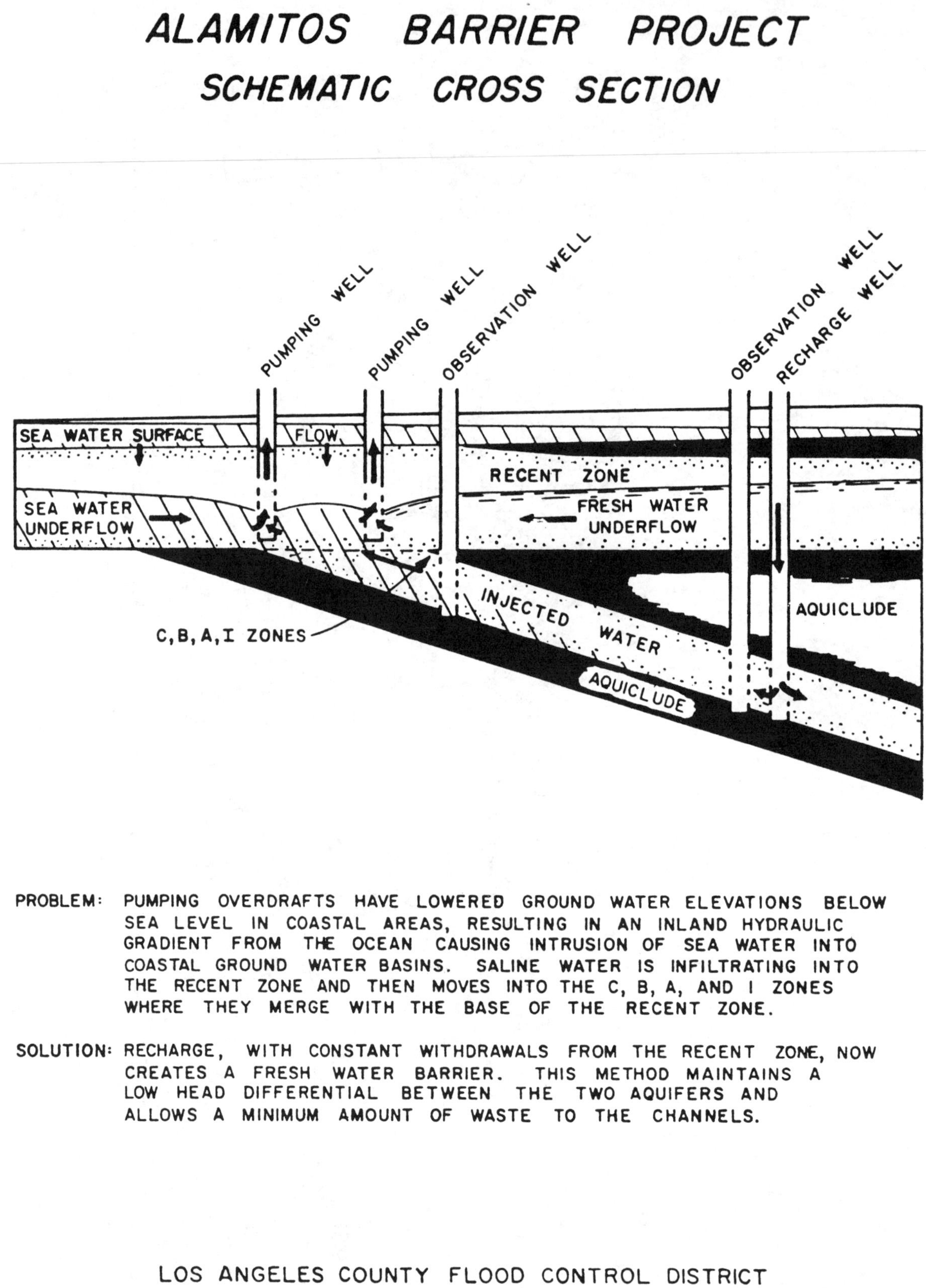

Figure 3. Alamitos barrier project schematic cross section

be minimized and gradients can be more easily controlled (Fig. 3).

In all fairness to the barrier designers, the cut-off saline wedge was a known problem very early on in the design concept. It was apparent that something had to be done to halt the intrusion, however, total containment was not feasible for several reasons but primarily from an economical standpoint. Certainly the number and depth of wells was a consideration, as was the injection water requirement and the length of its supply pipeline.

In retrospect, a much more ambitious plan in the beginning may have been the wiser and cheaper solution to the problem. The goal of total sea water containment with a larger price tag would have been very difficult to "sell" 40 years ago, but in the "20/20 vision rear-view-mirror", it looks like it may have been worth a try.

REFERENCES

California State Department of Public Works, 1952, Draft Report of Referee, California Water Service Company vs. Alexander Abercromby, City of Compton, et al.

California State Department of Water Resources, 1961, Ground Water Geology of the Coastal Plain of Los Angeles County: Bulletin 104.

Camp, Dresser and McKee, 1989, West Coast Basin Barrier Project Study: Prepared for Los Angeles County Department of Public Works.

Central and West Basin Water Replenishment District, 1986, Evaluation of Saline Water Inland of the West Coast Basin Barrier.

Central and West Basin Water Replenishment District, 1991, Annual Survey Report on Ground Water Replenishment.

Rancilio, J. A. and Cristiano, W. E., 1990, West Coast Basin Barrier Project Hydrologic and Operations Report March 1990: Through August 1990, Los Angeles County Department of Public Works.

Zielbauer, E. J., 1966, Sea Water Intrusion and the Barrier Project; in Engineering Geology in Southern California: Special Publication of the Association of Engineering Geologists, Los Angeles (out of print).

THE USE OF SHALLOW GROUND TEMPERATURES FOR TRACING MIGRATING FLUIDS

Joseph H. Birman
GSI/water (Geothermal Surveys, Inc.)
520 Mission Street
South Pasadena, CA 91030

INTRODUCTION

Since 1961, our company has been using shallow ground temperatures to investigate the configuration of migrating underground fluids. The original objective was to help locate best sites for water wells in desert basins. Later we adapted the technique to identify and trace leakage through dams and other impoundments such as mine tailings ponds. The technique is rapid, inexpensive, and highly successful where properly applied and interpreted. As in any other geophysical technique, the thermal technique has advantages and limitations.

The purpose of this paper is to explain the technical basis of the technique as applicable to engineering geology — in this case the investigation of dam leakage — in Southern California.

BACKGROUND

Early use of temperature in exploration was for oxidizing sulfide ore (Lovering, 1948; Lovering and Goode, 1963). Lachenbruch and Brewer (1959) studied thermal effects related to drilling a well in the Arctic. Schneider (1962) presented an early study of thermometry to ground water. Cartwright (1968) used shallow temperatures in a landfill application, and later (1971) in tracing shallow ground water.

Keys and MacCary (1971) compared temperature logging with other geophysical logging in water resources investigations. Woodbury and Smith (1985) quantitatively studied the thermal effects of three dimensional ground water flow. The rise of the geothermal energy industry stimulated increased use of temperature in exploration, but largely in geothermal gradient measurements beneath the reach of surface temperature processes.

Our work began with laboratory research in 1957, and continued with our first field application in 1961, in a Southern California ground water basin (Birman, 1969). In 1963 we began using the technique in dam leakage (Birman, Esmilla, and Indreland, 1971). Over the years, we have continued to refine the technique in a variety of applications (e.g., Roscoe Moss Co, 1990, p. 47-48). We consider the thermal technique as still an empirical tool, to be used with due regard for its limitations, and in the context of a proper understanding of the geohydrologic setting.

BASIS FOR METHOD

In the absence of moving fluid, and beneath the reach of surface temperature processes, the subsurface temperature field is the result of conductive heat tranfer upward from the Earth's interior. The amount of heat conducted to the surface depends on thermal conductivity of the materials through which the heat passes, the cross sectional area under consideration, and the geothermal gradient. These are perfectly analagous to Q, K, A, and i in Darcy's Law describing water flow in a porous medium. Steady-state heat flow sets up a thermal configuration beneath the surface of the ground that can be measured and described.

Water moving beneath the surface of the ground nearly always changes the temperature of the surrounding and overlying materials. The materials are advectively cooled or warmed depending on whether the moving water is acting as a heat sink or a heat source, i.e., whether the arriving water is cooler or warmer than the materials which it enters. Thus, the temperature beneath the surface of the ground results from advective disturbance of the normal conductive thermal configuration.

The temperature disturbances caused by moving water are not restricted to the actual space in which the fluid moves, but extend — by conductive heat transfer — some distance into the surrounding and overlying materials. This provides one of the most important advantages of the thermal technique. Thermal sensors do not have to be in the moving fluid in order to detect it. By thermally mapping areas or profiles, the configurations of anomalies are defined, and their axial portions determined by extrapolation of trends.

Surface and near-surface conditions

The surface of the Earth is affected by time-related temperature processes, of which the two most important to this type of exploration are the diurnal and the annual ambient temperature cycles. The corresponding temperature waves penetrate beneath the surface as determined by frequency and amplitude of the cycle, the values of thermal conductivity and specific heat of the materials. In general terms, in a climate typical of that of Southern California, the depth reach of the diurnal wave is about 1 meter. That of the annual wave is about 20 meters (Jaeger, 1965, p.8).

Where the ground surface is flat, and geologic materials are similar, the isotherms beneath the surface will be horizontal. Where topography is irregular, the isotherms will no longer be horizontal, but will tend to follow the topography for some depth beneath the surface. The rate at which the isotherms flatten downward depends on a number of factors, among which are the thermal parameters of the geological materials.

Another affect of topography is that differences in slope direction result in greater or lesser insolation from the sun, and this complicates the normal subsurface temperature configuration. Natural and artificial differences in vegetation, and urban development add further complications to the transfer of heat into and out of the ground.

These surface-related temperature processes provide a zone of strong thermal noise that effectively masks the small temperature changes related to moving ground water below. If the surface of the ground were not subject to the time-related diurnal and annual temperature cycles; and if there were no topographic, vegetative, or other complexities, small changes in temperature produced by moving ground water a few tens of meters or deeper could be mapped by remote methods such as infra-red surveys. While we ourselves use this technique in very special circumstances, the surface and near-surface thermal noise greatly reduces the applicability of remote infra-red sensing as a viable exploration technique for detecting temperature changes that occur beneath the zone of surface thermal noise.

Subsurface conditions

Two intrinsic thermal parameters of natural materials are thermal conductivity (analagous to hydraulic conductivity) and specific heat (analagous to storativity). These govern the temperature response of materials to thermal processes applied externally. Although all geological materials differ in these parameters, the largest differences relate respectively to: solid bedrock, dry unconsolidated alluvium, and saturated unconsolidated alluvium. These are the three major categories of Earth materials that control geohydrologic processes, and to which we apply the thermal method. If water-bearing bedrock is extensively fractured, this approaches saturated unconsolidated alluvium in its response to thermal processes.

Beneath the reach of the annual temperature wave, except for the rare occurrences of rising steam from geothermal resources, the only short-term dynamic geohydrologic process below the vadose zone is that of moving ground water.

APPLICATION AND EQUIPMENT: FIELD OPERATIONS

We use two modes of thermal exploration: shallow ground temperature measurements to map the lateral distribution of moving ground water; and downhole temperature logging to investigate the vertical arrangement of aquifer/aquitard formations. For both modes, we use precisely calibrated thermistors, and measure to the nearest 0.01 degrees Celsius.

Shallow Ground Temperature Mode

We use the shallow ground temperature technique in reasonably flat, open areas not subject to extensive local variations in surface temperature conditions. The standard depth is 10 feet, with the sensors emplaced in holes backfilled to prevent near surface thermal effects from reaching the probe tips. At this depth, the readings are beneath the reach of the surface diurnal temperature cycle, and within the reach of the annual cycle. Temperatures are recorded after all probes have regained thermal equilibrium disturbed from the drilling and backfilling. The sensors are set directly in the ground, or in small diameter PVC sleeves that can be left in place for long-term monitoring if necessary.

In thermal exploration of large ground water basins, we normally use a grid distribution, with probes as much as a mile apart. For applications on dams and tailings impoundments, the probe spacing may be a few tens or a few hundreds of feet apart. We design the depth, spacing, and geometric pattern of the probe array to fit the geologic and hydrologic setting, structure of the dam, and any specific problems anticipated for the foundation and abutment areas.

The probes may be at constant depth and spacing along the axis of a dam, or at specifically designed spacings in order to investigate an area of suspected leakage. The probes can also be emplaced at several depths within the dam so as to form a vertical monitoring screen.

The probes can be left in place indefinitely and read periodically. Within a reasonable period of time, the normal thermal signature of the dam will be revealed in response to the annual temperature cycle, and to changes in reservoir head. Local anomalies along the profiles are interpreted with respect to leakage or other conditions. When the normal time-temperature response signature for the dam has become known, any departure from that pattern will signal a change in internal conditions, such as development of new leakage or changes in old patterns of leakage.

Downhole Temperature Logging

This is done with a cable-mounted thermistor, also precisely calibrated to respond to changes of 0.01 degree Celsius. Although modern thermistors have very fast response time, this does not apply to the thermistor housing or the cable. Therefore, the best method is to stop at predetermined intervals and allow the equipment to reach equilibrium before taking a reading.

Normally, we measure only within the water column, and in wells or boreholes that are not large in diameter. Large diameter wells permit convective overturn within the borehole, which eliminates the natural thermal gradient. Changes in the downhole thermal gradient may indicate changes in thermal conductivity of the geological formations; active aquifer zones; and vertical crossflow between aquifers. The logging can be done in open holes, or in wells that are blank-cased or screened. The nature of the casing makes no difference: unless there is vertical movement of fluid, the temperatures are indicative of those outside the casing.

ADVANTAGES AND LIMITATIONS

One important advantage of the technique is that it responds to water that is moving. Non-moving water takes on the temperature of its surroundings, and becomes thermally invisible except where it is shallow enough to attenuate the annual cycle.

Another important advantage is that the sensors do not have to be in the moving water in order to detect it. Moreover, because of the nature of conductive heat transfer, the axis of a ground temperature anomaly will occur directly above its source. The technique provides continuous monitoring, costs are low, operation and data collection can be by non-technical personnel, and automation and remote data acquisition are possible.

A serious disadvantage is that a shallow, active aquifer can totally obscure the effects of deeper-lying aquifers; vertical discrimination is not possible in the shallow ground temperature mode. Because the variables are complex, considerable experience and a sound knowledge of the geohydrologic setting are necessary in order to make reliable interpretations.

ENGINEERING GEOLOGICAL EXAMPLES IN SOUTHERN CALIFORNIA

Investigating Leakage in Dams

In Los Angeles, Riverside, and San Diego Counties, thermal investigations have been made on six large dams and one small dam. Four of the large dams have thermal monitoring systems installed and three are being actively monitored by their owners.

The smallest dam, at Seminole Springs near Aqoura, provided the opportunity for long-term research and calibration with hydrological conditions (Birman, Esmilla, and Indreland, 1971) and represents a good example of the use of the thermal technique in an engineering geological application in Southern California.

In 1965, the objective was to locate leakage pathways through the dam that could be grout-injected in order to reduce active leakage at the foot of the dam near the right abutment. Thirty thermal probes were set at 20-ft intervals at 7-ft depth along the crest of the dam. Thermal background was measured by another probe set near the left abutment beyond the end of the dam (Fig. 1).

Figure 2 shows a set of temperature measurements made in 1965 and 1966. Profiles 1 and 2, the first two sets of temperature profiles measured, showed the coolest temperatures at Probe Stations 1, 2, and 3. Part of this zone was grouted by drilling and injection, and the leakage at the downstream toe was significantly reduced. Profiles 3, 4, and 5 show the exothermic effects of the grout at Probe Station 2, the axis of the anomaly.

With the permission of the owner, the probes were left in the dam, and monitored at approximately monthly intervals for the next two years. Much of the leakage pattern is revealed in Figure 2, as described in the following paragraphs. The reservoir became nearly dry during late summer and fall. This is reflected in warmer temperatures with less thermal relief across the entire dam as shown in Profiles 6 and 7.

Surprisingly, a large temperature reversal occurred during the winter months. The portion of the dam that had been the warmest was now the coldest as shown in Profiles 8-11. During this time, the reservoir was at its maximum. To investigate the thermal response, several holes were drilled by auger at Locations A-F. These were logged and bailed in order to gain some insight into relative production rates. The materials at Location C, in the center of the reversed anomaly, showed the wettest conditions of the six drillholes. The temperatures near each drillhole were calibrated with bail tests. The results were interpreted and hydrologically analyzed (1971, p. 2274-2279). This provided a crude analysis of "indicated percolation rates" through the dam at selected times (Fig. 3). The results were then and must still be considered tentative, based on early experience. The study was conceptually sound, however, and is a good example of what can be done with the thermal technique in this application.

Figure 4 shows an example of downhole temperature logging in a deep piezometer downstream of a large dam in Southern California. The logging was made every three months during a year. The effect of the annual wave is clearly shown by the wide variation in temperature near the top of the graph. The normal geothermal gradient is shown in the lower part of the graph, where the temperatures are quite similar, and if extrapolated upward, would pass near the base of the annual wave. A strong reversal in temperature indicates an active aquifer in the depth range of about 60 to 80 feet. An aquitard occurs at about 150 feet, below which aquifer conditions again increase. The dramatic temperature reversal that indicates the strong shallow aquifer is enhanced by the fact that the temperatures continue to decrease during the times of measurement, even though this is beneath the observed range of effect of the annual wave. The interpretation is that discharge in the shallow aquifer was increasing throughout the year from local sources.

Other irregularities and reversals throughout the temperature profiles are probably due to the fact that the piezometer was in fact an unused water well. The fairly large diameter and intervals of screen may have contributed to local irregularities in water movement within the casing.

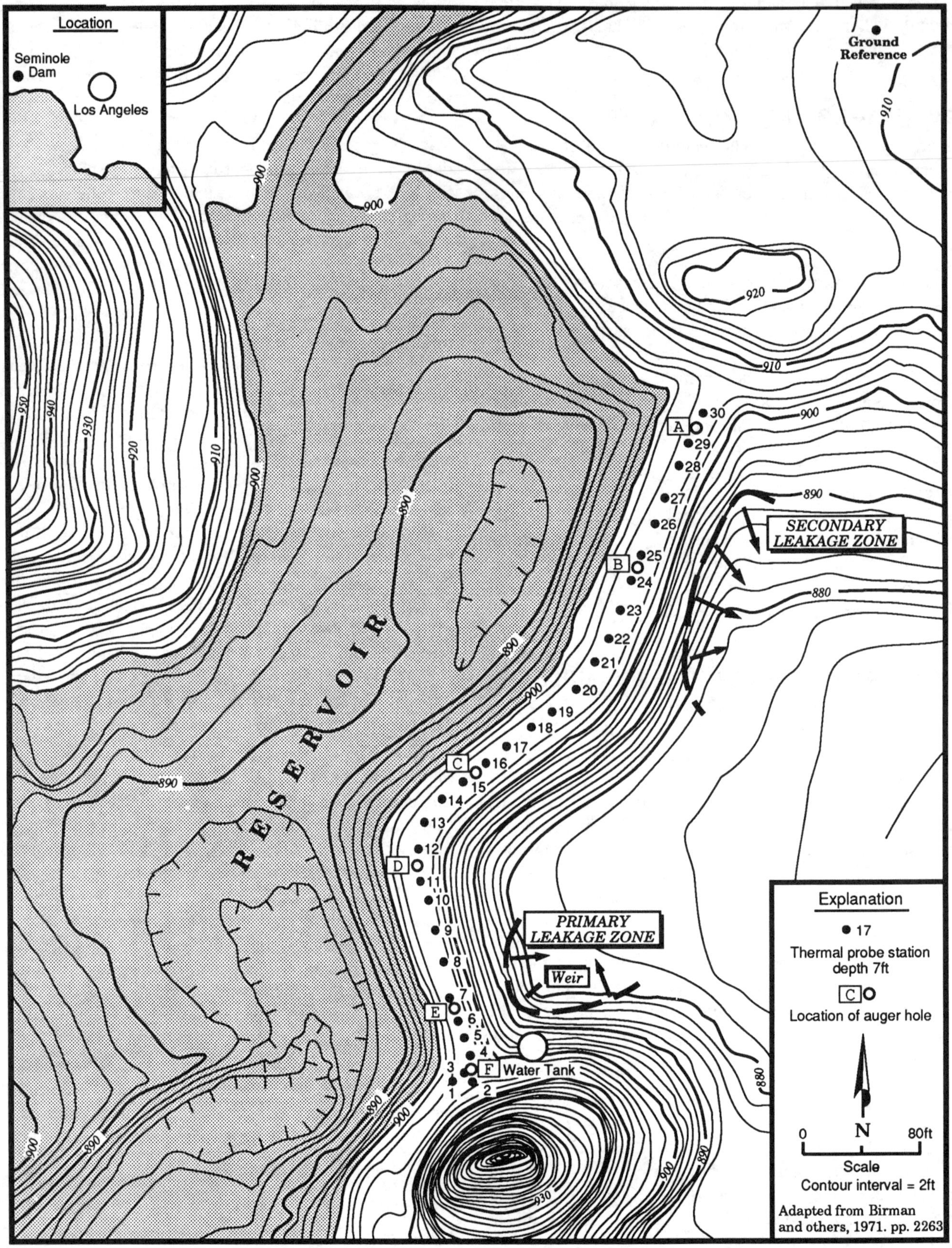

Figure 1. Seminole Dam, showing topographic setting, visible seepage in 1965, and locations of thermal probes and test holes.

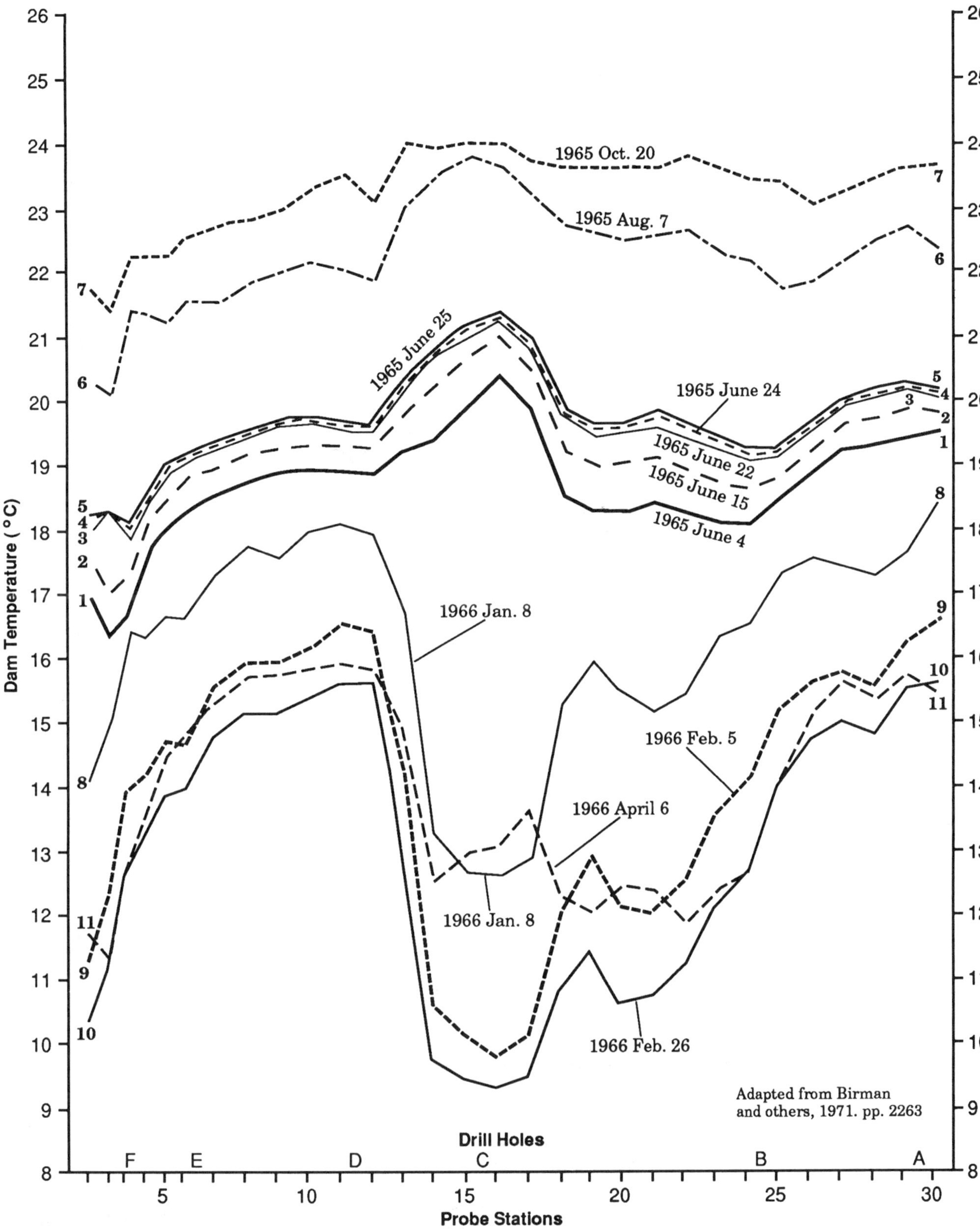

Figure 2. Temperature profiles measured at 7 feet depth along the crest of the dam. June 4, 1965 to April 6, 1966

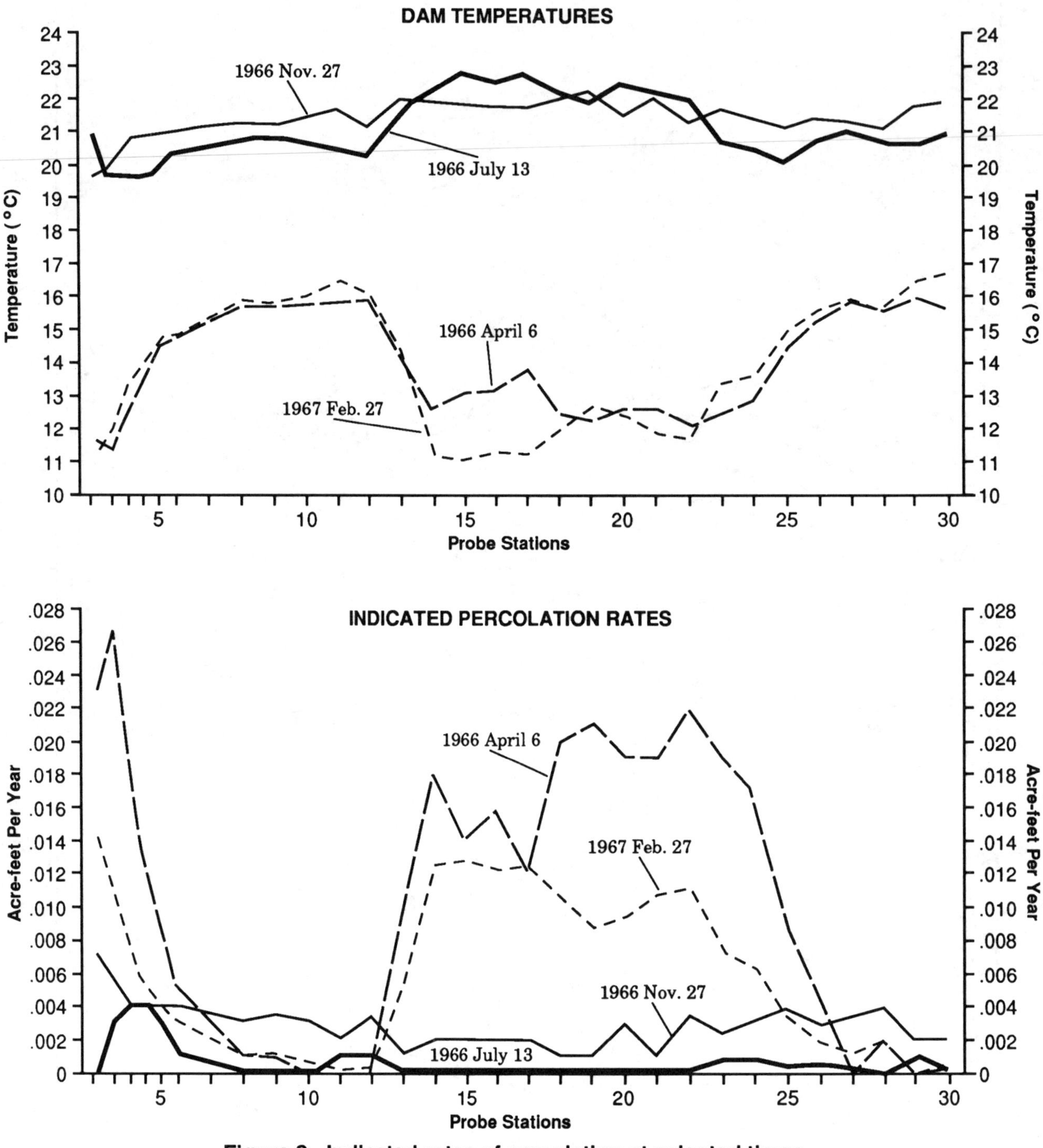

Figure 3. Indicated rates of percolation at selected times in 1966 and 1967, based upon thermal analysis

Adapted from Birman and others, 1971. pp. 2263

Figure 3. Indicated rates of percolation at selected imes in 1966 and 1967, based upon thermal analysis

Tailings ponds and plume migration

We have used the thermal technique to map plume migration from tailings ponds. The technique responds to actively moving fluid, and changes in temperature through time can be related to changes in fill patterns within the impoundments. There is no direct thermal response to water quality. Recognizing a contaminant plume migrating from an impoundment depends on a continuing or changing source of actively moving fluid.

An important contribution of the thermal technique in this application is to map the distribution of normal ground water flow near and beneath the impoundment. This provides the directions in which the contaminants can be expected to migrate.

We do not consider the thermal technique to be applicable to leaky tank investigations (except to determine normal patterns of ground water flow, or the vertical arrangement of aquifers and aquitards). Without continuously moving fluid from the tank, there is no heat sink or heat source effect. While a slug of fluid lost from a tank may remain in the shallow zone and attenuate the annual wave, this is more easily and more directly obtained by shallow drilling.

Methane in landfills.

We have used the thermal technique in four Southern California landfills to locate and track leakage from methane concentrations. Since methane is formed from exothermic processes, it imparts a warm temperature anomaly to its surroundings. Deep-seated methane concentrations are indicated by broad warm temperature anomalies on which gas leaks superimpose much higher-temperature anomaly spikes. Lateral flow of methane produces warm temperature zones.

Bottom temperature surveys.

Crossflow in or out of the bottom of a reservoir produces local anomalies related to the source of the moving fluid and the interface through which it moves. Several surveys of this type have been made with promising results.

An advantage is that the overlying body of water can act as an effective insulating blanket, thus reducing the depth of bottom penetration necessary to derive meaningful results. A disadvantage is that pond liners must not be penetrated. Another is that it is impossible to reoccupy the exact locations in order to obtain repeat readings.

REFERENCES CITED

Birman, J.H., 1969, Geothermal exploration for ground water: Geological Society of America Bulletin, vol.80, p. 617-630.

Birman, J.H., Esmilla, A.B., and Indreland, J.B., 1971, Thermal monitoring of leakage through dams: Geological Society of America Bulletin, vol. 82, p. 2261-2284.

Cartwright, K., 1968, Geophysical surveys in the vicinity of sanitary landfills in northern Illinois: Ground Water, vol. 6, no. 5, p. 23-30.

__________, 1971, Tracing shallow groundwater systems by soil temperatures: Water Resources Research, vol. 10, no. 4, p. 847-855.

Jaeger, J.C., 1965, Application of the theory of heat conduction to geothermal measurements: American Geophysical Union, Geophysical Monograph Series No. 8, Terrestrial Heat Flow, p. 7-23.

Keys, W.S. and MacCary, L.M., 1971, Applications of borehole geophysics to water-resources investigations: Techniques of Water-Resources Investigations of the United States Geological Survey: Book 2, Chapter E1, p. 1-124.

Lachenbruch, A.H. and Brewer, M.C., 1959, Dissipation of temperature effects of drilling a well in arctic Alaska: U.S. Geological Survey Bulletin 1083-C, p. 73-109.

Lovering, T.S., 1948, Geothermal gradients, recent climatic changes, and rate of sulfide oxidation in the San Manuel District, Arizona: Economic Geology, vol. 43, p. 1020.

Lovering, T.S. and Goode, H.D., 1963, Measuring geo-thermal gradients in drillholes less than 60 feet deep, East Tintic District, Utah: U.S. Geological Survey Bulletin 1172, p. 1-48.

Roscoe Moss Co., 1990, Handbook of Ground Water Development: Wiley, Inc., New York.

Schneider, R., 1962, An application of thermometry to the study of ground water: U.S. Geological Survey Water Supply Paper 1544-B, p. 1-16.

Woodbury, A.D., and Smith, L., 1985, On the thermal effects of three- dimensionqal groundwater flow: Jour. of Geophysical Research, vol. 90, p. 759-767.

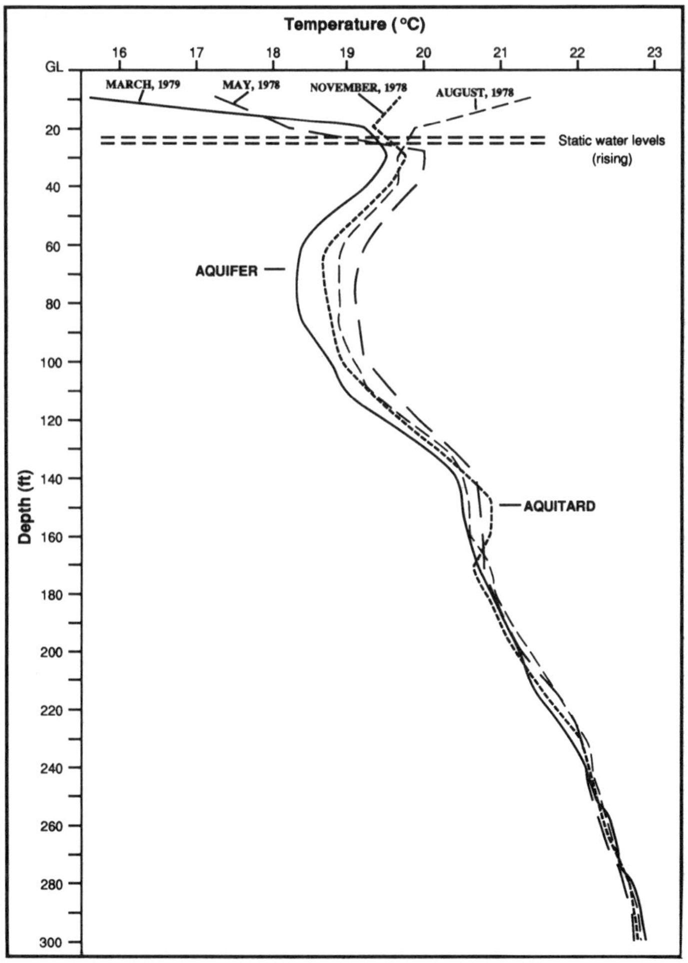

Figure 4. Downhole thermal logs in an observation well downgradient of a large dam in Southern California.

Chapter **6**

Foundation Problems

Settlement of foundation slowly destroying a $3 million home in Brentwood, California. From within the house, daylight can be seen through the cracks.

Foundation Problems

Editors Note: Probably most independent engineering geologists earn their living in southern California from advising clients on the suitability of construction sites—be it a house on a single lot or a vast subdivision involving a thousand future homes. The first paper in this chapter explains the causes of damage to structures and repair methods most utilized. The second paper explains the use of grouting for repair of homes and buildings. The third paper describes a relatively new method of slope repair, and the last paper explains the interesting aspects of constructing the towers for the Palm springs Aerial Tram—from working in 110° F. temperatures at the base of the mountain to flying to the top by helicopter to work in the snow.

CAUSES AND EVALUATION OF RESIDENTIAL DAMAGE IN SOUTHERN CALIFORNIA

Robert A. Hollingsworth, CE, CEG

David J. Grover, CEG

Grover-Hollingsworth & Assocates
31129 Via Colinas, Suite 707
Westlake Village, CA 91263

CONTENTS

INTRODUCTION

Residential structures are a combination of numerous structural elements, products and trades. Each structural element has a specific function, each product a number of performance specifications and tolerances, and each trade certain techniques and practices. The ground upon which the structure rests also has certain properties and performance characteristics. The interaction of the structure and ground can be satisfactory or can lead to distortion and damage.

The level of anticipated residence performance in southern California has risen considerably over the past decade. This rise has resulted largely from three factors. The first is reality disclosure laws, which in California require the disclosure of prior damage and/or defects upon sale of a home. The second is a series of court decisions in the 1960s through 1980s, which led to mass-produced homes being considered manufactured products and the builders being held strictly liable for defects. The third is insurance companies being held responsible for repair of residential damage under "all-risk" homeowners policies. These factors taken together have led to a change in homeowner mentality from "these cracks are just settlement cracks — we patch them every few years" to "these cracks mean my house is defective — who is responsible?"

Engineering geologists and geotechnical engineers have been asked with increasing frequency to evaluate the causes of residential damage. Such evaluation is often a difficult task requiring a knowledge of geology, soils engineering, structural engineering and construction. The type of foundation system utilized has a significant influence on residence performance. "Damage" to a residence can be the result of earth movement, normal construction performance or construction defects. The earth movement processes can be violent as in the case of landslides and seismic events, or subtle and slow as in the case of soil expansion, creep or consolidation. The slow earth movement processes are the subject of this paper.

The geotechnical professional has a number of tools available to analyze earth movement related causes of damage. These tools, typical performance of construction materials and techniques, and many of the potential slow, earth movement causes of residence damage will be discussed. Three case histories will be used to illustrate the less common causes.

RESIDENTIAL CONSTRUCTION

The vast majority of residential structures in southern California are constructed of wood frame, brick or concrete block. The exterior of wood frame structures have stucco, wood or stone siding. Interior wall coverings generally consist of plaster or wallboard. Most residential structures are one or two stories in height.

The floor systems for residential structures can generally be divided into two categories: raised wood and slab-on-grade. Within each category various types of foundation elements are utilized. Raised wood floor systems are the oldest type found in southern California. These systems were used almost exclusively until the advent of slab-on-grade residential construction in the early 1950s. Raised floor systems have, however, continued to be utilized in hillside construction and have become increasingly popular in expensive residential development.

Raised Wood Floor Systems

Raised wood floors can be supported on a number of different foundation elements or combinations thereof. The most common types used in southern California are shallow, continuous concrete footings and shallow, isolated concrete pier footings. Continuous footings can be used exclusively throughout a structure or in conjunction with isolated interior piers. Continuous footings normally support bearing walls and posts which carry second-floor and/or roof loads. The piers generally support the wood floor upon which non-bearing interior, partition walls may rest. The continuous footings normally range in depth from 12 to 24 inches, while the piers range in depth from 6 to 18 inches.

The pier and grade beam system, where small diameter (12 to 18 inch, relatively shallow (5 to 10 feet deep), lightly reinforced piers are connected with structural grade beams, is not very common in southern California. This system, however, is used fairly extensively in northern California. Grade beam systems are used in many hillside homes in southern California, but the grade beams in these applications are supported by larger diameter (18- to 36-inch) belled caissons or straight-shafted friction piles. These hillside systems are typically deeper and more heavily reinforced than the pier and grade beam system.

Slab-on-Grade Systems

Slab-on-grade residential construction has become very common in southern California. Two primary types of slab-on-grade systems are used: non-structural and structural. The non-structural system uses continuous and/or pad footings to support bearing loads and resist deformation or deflection. The slab in these systems transfers live and dead floor loads directly to the underlying earth and is not intended to span any distance without support. Non-structural slabs will support non-bearing partitions and walls and will often be reinforced to resist cracking.

The two structural slab systems used in residential construction are mats and post-tensioned slabs. A mat foundation is a relatively rigid slab which distributes bearing loads more uniformly to the underlying earth. The mat foundation is intended to resist differential movement which can damage the structure (Bowles, 1977). The most common type of mat foundation used for residential construction is a reinforced slab constructed with a grid system of deepened, reinforced beams. These beams provide the stiffness necessary to resist deflection. Mat foundations are only occasionally used in southern California.

The use of post-tensioned slabs for residential construction has become relatively common in southern California. Post-tensioned slabs utilize tendons placed at mid-height in the slab in frictionless sleeves. The tendons are tensioned after the slab has cured to place the slab under compression. Post-tensioned slabs are designed to span over short distances, 5 to 8 feet, where soil support has been lost without excessive deflection. These slabs resist cracking because the concrete is placed under compression by the cables (Weiss, 1985). The slabs are, however, flexible and can be subject to distortion.

The most common type of post-tensioned slab in southern California is a flat slab with thickened edge beams. The slab can distribute post and line loads typically found in one and two story construction to the subgrade without a footing. Pad footings will occasionally be utilized where heavy post loads are present (Weiss, 1985).

GEOTECHNICAL CAUSES OF RESIDENTIAL DAMAGE

Residence damage can result from numerous geologic and geotechnical processes. Many of the processes are relatively slow and have water as a root causal agent. A number of these processes and their potential effects are discussed below. Numerous other geotechnical hazards including fault rupture, liquefaction, landslide/slope instability, subsidence due to fluid withdrawal and ground water rise are beyond the scope of this paper.

Consolidation/Compression

A vertical decrease in soil volume can occur as the result of loading, the introduction of water and shrinkage. The first two causes are often referred to more or less interchangeably as consolidation, compression and collapse. The interchangeability of these terms is not, however, strictly correct. Consolidation is a time-dependent decrease in soil volume associated with the dissipation of positive pore pressure (Lambe and Whitman, 1969). Compression is a vertical decrease in soil volume caused by the application of load (Terzaghi and Peck, 1967). Collapse is a more or less instantaneous volume decrease caused by the introduction of water independent of loading (Dudley, 1970).

Compression and consolidation due to residential loads in southern California generally occur during construction. Consolidation of saturated, fine-grained natural soil deposits due to surcharge by a relatively deep fill mass can, however, require a relatively long time to complete. Consolidation in the lower portions of a deep fill can also occur if compression of the deeper fill due to the increased overburden results in saturation of the deeper portion of the fill mass. The deeper saturated fill is then subject to consolidation when additional loading is imposed.

Residential structures in southern California are normally not damaged by consolidation or compression of a near surface fill or natural soil deposit. Consolidation of the lower portion of deep fills and/or consolidation of the underlying natural soil deposits can, however, lead to residential damage. This is particularly true if the fill depth or thickness of the surcharged natural deposit varies significantly over short distances.

Collapsible Soils

The most common type of vertical soil volume decrease in southern California, aside from shrinkage, results from an increase in soil moisture content independent of a change in loading. This process is commonly referred to as hydroconsolidation or hydro-compression, but may be more properly referred to as collapse.

Soils exhibiting this potential are typically referred to as collapsible. This process can occur in both natural deposits and compacted fill. The amount of collapse is a function of the soil constituents, initial void ratio, stress history and loading (Dudley, 1970). Complete saturation of the soil is not necessary for this process to occur. El-Ehwany and Houston (1990) have shown that 50 percent saturation produced approximately 85 percent of total collapse for a natural soil deposit. Dudley (1970) reports that for collapsible soils in southern California complete collapse occurs by the time the soil reaches 80 percent saturation.

Natural soils subject to collapse upon saturation typically are slightly cemented and exist at low in-situ moisture contents. The soils are often found in aeolian or fan deposits in semi-arid and arid environments. Soils deposited in a rapid fashion by flash floods and mud flows are also prone to collapse. Such soils exist in numerous areas of southern California.

Compacted fill has also been found to be subject to collapse upon saturation. Both improperly and properly compacted fills are subject to this process. Brandon et al. (1990) tested sandy clay fill from a housing tract in San Diego. Fill compacted to a minimum of 90 percent of the maximum density as determined by ASTM D1557 (current compaction standard) was found to be subject to significant collapse (up to 3 percent) when compacted at an initial moisture content slightly dry of optimum. Samples compacted at initial moisture content of at least 1 percent above optimum exhibited very little collapse.

Collapse of natural and fill soils causes significant damage to residential structures throughout southern California. This process is particularly significant in residential damage because it will often result in differential settlement of the structure. Collapse of natural soils and compacted fill in residential areas normally occurs as the result of a moisture increase resulting from irrigation. The irrigation wetting front will progress vertically downward through the soil over time. Collapse will therefore be time-dependent in association with the rate of infiltration. Movement can occur over a number of years in deep fills. Brandon et al. (1990) report up to 12 inches of settlement over a ten year period in a residential tract development in San Diego, with 70 feet of compacted fill.

Compression of Trench Backfill

A relatively uncommon but occasionally significant cause of structural damage is compression of jetted or poorly compacted, deep trench backfill. Utility trenches, particularly sewer and storm drain, are often present in residential neighborhoods and can be as deep as 10 to 30 feet. These trenches, while normally confined to streets, may pass between residences in narrow easements. Portions of the foundations for these structures can fall within the zone of influence of the trench. The zone of influence is defined by planar surfaces projected upward from the base of the trench walls. The planes can range in angle from slightly steeper than a 45-degree angle in granular soils to as flat as 30 degrees in fine-grained soils.

Foundation settlement within the zone of influence occurs as poorly compacted trench backfill compresses laterally. This lateral compression allows lateral, active state/yielding movement of the soils within the influence zone. The lateral soil movement within the influence zone translates into vertical settlement at the ground surface, with the magnitude of settlement increasing toward the trench. Differential settlement of the structure results from the variable magnitude of ground surface movement and the fact that generally only a portion of the residence falls within the influence zone.

Expansive Soils

Expansive clay soils exhibit volumetric changes with moisture content variation. These soils swell when their moisture content increases and shrink when it decreases. When water is made available to a clay soil it is absorbed into the clay structure and strongly bound to the clay minerals. The absorbed water has no mobility and causes an increase in the size of the clay particles and swelling of the soil (Stamatopoulous and Kotzias, 1987).

The primary clay types in order of increasing swell potential are kaolinite, illite, and montmorillonite. The magnitude of the volumetric change is related to the type and percentage of clay, the in-situ density, the initial moisture content, and the loading imposed. The higher the clay percentage in a soil, the more susceptible to volumetric change it will generally be.

The magnitude of potential swell increases with increasing soil density and decreases with increasing initial moisture content. The swell potential decreases with increasing load or overburden pressure (Chen and Ma, 1987). The decrease is generally linear when pressure is plotted on a logarithmic scale. The shrinkage potential increases slightly upon increased loading but can stay nearly constant over fairly wide pressure ranges based on testing by the authors.

The pressures exerted by expansive soils are capable of lifting residence foundations and slabs. Footings are typically subject to only minor lifting. Slabs-on-grade which are lightly loaded, however, are very susceptible to heaving and cracking. Post-tensioned slabs, although much less susceptible to cracking, are moderately flexible and susceptible to distortion which can cause wall damage.

Many one story, slab-on-grade residential structures in southern California use roof trusses to transfer loads to the exterior walls. A single central, interior bearing footing or series of pads is present in other structures to carry a portion of the roof load. All or the majority of the interior walls in these structures are non-bearing and rest on the slab without a supporting footing. These interior walls are highly susceptible to damage in response to slab heaving. Interior walls will be pushed up by slab heaving, producing distortion of the wall/ceiling junction. Walls paralleling the ceiling joists/trusses but falling in between two members can push up through the ceiling board. Walls trending perpendicular to the joists/trusses will develop a regular scallop pattern at the wall/ceiling junction. Extreme pressure can result in the lifting of the roof system off the exterior walls when trusses have been used. The authors have observed such separations which are 1 to 2 inches in width.

The isolated interior piers supporting wood floors are similarly very susceptible to movement. Often slabs-on-grade and wood floors supported by isolated piers will dome upward in relation to the load-bearing footings in response to expansion of the foundation soil. Slabs which are poured adjacent to and not structurally tied to footings will also commonly separate from the footings. Separations can also develop between wood floors and the walls which parallel the floor joists. This type of damage is often incorrectly diagnosed as resulting from settlement of the footings.

It should be noted that swell pressure tests, which measure the pressure required to prevent heave, and swell overburden tests, which measure the percent swell under a particular load, will often overestimate building heave. This fact has been attributed to lateral soil expansion in the field which closes desiccation cracks, and the higher final moisture content and therefore greater swell achieved in laboratory tests (Osman and Sharif, 1987, and Erol et al., 1987). Structures can also tolerate some movement and swelling will generally not cause obvious damage unless the swell potential exceeds 3 percent under a given load (Brown, 1990 and Hollingsworth, 1990).

Creep and Slope Yielding

Creep and slope yielding are processes which result in the slow lateral and downslope movement of soil under a constant external load. These processes are differentiated from slope instability or failure by the absence of a plane or zone of weakness. They are also not subject to a conventional rigid/ plastic failure analysis. Although several processes likely contribute to creep, the most commonly cited is alternate expansion and shrinkage of the near surface soils on a slope which results in their net downslope movement over time. Movement occurs because the unit volume of soil expands perpendicular to the slope face (the direction of least resistance) and shrinks vertically (in response to gravity).

Slope yielding is a process which occurs as a soil, generally fill, undergoes strain as a result of an increase in moisture content. The moisture content increase increases the internal slope load through the addition of the weight of added water, and reduces the soil shear strength. The soil must move or undergo strain in order to mobilize additional shear strength to resist the increase in internal loading.

Creep and slope yielding primarily occur in fine-grained natural and fill soils. The depth of soil subject to creep is highly variable. The depth generally increases with slope height and steepness. Fill slopes constructed at 1-1/2:1 slope gradients, which are common in southern California, are particularly susceptible to creep. Correlation between the depth of soil subject to creep, and soil properties, such as Atterberg Limits or shear strength, have not been successful.

Damage due to creep and slope yielding is very common in hillside residential properties. This is particularly true in mass-graded residential tracts where homes are constructed close to the tops of steep fill slopes. Creep and slope yielding can cause both lateral and vertical foundation movement. The magnitude of movement decreases away from the top of the slope. The affects of these processes have been observed by the authors as far back as 30 feet from the top of fill slopes.

Expansive Bedrock

Expansion of bedrock materials and, in particular, differential expansion over relatively short distances can cause significant residence damage. Bedrock expansion results from four main processes: hydration, chemical alteration, crystal growth and rebound. Bedrock expansion in southern California is typically limited to sedimentary rocks and, in particular, Miocene Age and younger sediments. These sedimentary rocks exist over wide areas of southern California. Sediments of both marine and terrestrial origin are subject to these processes. Differential bedrock expansion is responsible for the greatest incidence of damage. Often the expansion potential will vary markedly over distances of several inches to several feet due to the presence of thin interbeds, shears or faults with variable expansion potential. Damage is typically concentrated at the contact between expansive and non-expansive portions of the bedrock.

The two most common causes of bedrock expansion are hydration of clay minerals and chemical alteration of silt-size particles to clay. Clay minerals within relatively young sediments (particularly Pliocene and Pleistocene) are subject to expansion upon saturation. The clay along fault zones and sheared bedding has been found to be particularly expansive. Silt and sand-size particles within the rock are also subject to chemical alteration as a result of hydration and oxidation. The most commonly altered particles are silt size. The chemical alteration results in a volume increase and increase in the clay constituent. Commonly, grading has removed the weathered profile exposing "fresh" bedrock subject to chemical alteration. The relatively long period of time required for damage manifestation due to bedrock expansion (often eight to twelve years) suggests that weathering is a significant contributing factor (Hollingsworth, 1990). Expansive pressures generated by even narrow beds are capable of heaving foundation elements supporting two story loads. (See Case History #2.)

The second significant cause of bedrock expansion is gypsum crystal growth. Gypsum crystal growth which results in bedrock heaving is generally considered to result from the oxidation of pyrite to sulfate and iron oxide. Gypsum and jarosite form when calcite and/or illite are present. This alteration results in a volumetric increase due to the difference in specific gravities between the minerals (Hawkins and Pinches, 1987). Factors affecting oxidation include the presence of moisture and oxygen (aerobic conditions), biological activity, acidic environment, and temperature.

Gypsum crystallization can take several forms. Prismatic crystals termed selenite are the most common. These generally grow with their flattest side parallel to rock discontinuities. Other forms include laths and acicular crystals. Acicular gypsum crystals have been observed to grow perpendicular to structures and are believed to exert the most force at their growing end (Hawkins and Pinches, 1987). The authors have observed acicular gypsum crystal growth within massive sandstone of the Miocene Topanga Formation in the Santa Monica Mountains. This crystal growth, which is only occasionally visible without magnification, has resulted in significant heave.

Prismatic and tabular gypsum crystal growth have also been observed within shale bedrock. The black shale of the lower unit of the Miocene Modelo Formation exhibits the greatest incidence of gypsum growth in the Santa Monica Mountains, although shale units in the Topanga Formation have also been found to exhibit crystal growth. Black shale in other parts of the country has been found by others to be subject to similar expansion. Heave associated with gypsum crystal growth in shale will be most prominent in flat or relatively flat-lying bedded rocks where the bedding surfaces provide locations of crystal growth.

Another important condition for bedrock expansion due to gypsum growth can be capillary rise which brings the oxidation products close to the surface. Unoxidized, fresh bedrock at the surface is also necessary. Hawkins and Pinches (1987) found in diffractograms of bedrock samples from a cut portion of a hospital site that gypsum was present in samples above 1.5 feet (0.46m) while pyrite was absent. The authors found sulfate concentrations in near surface rock which had weathered since construction to be four to five times higher than in the still unweathered rock below. (See Case History #3.)

Sulfate Attack and Salt Heaving

Sulfate attack on slab-on-grade construction and limited salt heaving have been observed in several areas of southern California. Sulfate attack on concrete results in cracking and spalling of the slab and surface efflorescence along cracks. Slab cracking is caused by the crystallization of salts within the pore space in concrete. Chemical deterioration of the concrete may accompany this process. Crystallization of salts may also occur in the soil directly beneath a slab. The volumetric increase associated with this crystallization causes limited slab heave. Concrete footings suffer spalling, particularly at the slab/footing construction joint. Stucco is also attacked and deteriorated, particularly on the shaded side of the house (Yen and Bright, 1990).

Yen and Bright (1990) studied the problem of sulfate attack in the cities of La Palma, Lakewood and Cypress for the State of California, and found that 80 percent of the damaged homes did not have a moisture barrier beneath the slab. The lack of a moisture barrier allows solute carried to the surface by capillary rise to contact the concrete. Evaporation of water through the concrete results in an increased solute concentration and eventual crystallization.

Water

Water is the root causal agent of all the geotechnical problems previously addressed. Water can itself, however, cause residential damage through non-earth movement processes. Water can carry sulphate to the surface to attack concrete through capillary rise. Water can also migrate through or beneath residential footings and pond in the lowered subfloor areas of raised floor houses or the sunken living areas of slab-on-grade construction. The sources of water leading to foundation movement and residential damage include rainfall, irrigation, shallow ground water, springs and pipe leaks. The periodic heavy winter rainfall which southern California experiences can lead to slope instability, swelling of expansive soils, soil consolidation/collapse, and underfloor dampness problems. Prolonged periods of below-normal rainfall or drought (such as 1986 to 1991) can cause significant damage due to shrinkage of expansive soils.

Irrigation, the process through which southern Californians have turned an arid to semi-arid environment to a lush, temperate environment, is a major factor in residential damage. The amount of irrigation on a residential property averages four to five times the average annual rainfall. (The average annual rainfall in Los Angeles is 14 to 15 inches per year as measured at the Civic Center.) The authors have, however, studied residential properties in the eastern Santa Monica Mountains where annual irrigation exceeds 150 inches per year. Excessive irrigation can lead to lateral movement of moisture beneath slab and raised floor foundations, as well as downward vertical movement of moisture through fill and natural soils. The lateral movement causes heaving of expansive soils, under floor dampness and water seepage through slabs. The vertical movement can result in collapse of compacted fill and natural soils. Vertical movement has proven particularly troublesome in deep, compacted fills.

Damage to residences supported on expansive soils can be caused by a lack of water. Limited or non-existent irrigation and/or drought conditions can lead to drying and consequent shrinkage of expansive soils. When moisture conditions differ around a structure, differential settlement can result from soil shrinkage. Such settlement is often misinterpreted as resulting from consolidation or collapse of the underlying earth materials.

Shallow ground water (the capillary fringe extending to within the zone of foundation influence) and springs can lead to under-floor and slab moisture problems. These conditions, in particular shallow ground water, can lead to chemical attack on concrete when the capillary fringe reaches the foundation.

Both water supply and sewer line leaks can cause damage, but sewer leaks are considerably more difficult to detect. Leaks are particularly troublesome with slab foundations in expansive soil. Even minor leaks can result in serious slab heaving. It should also be noted that moisture from a leak will often migrate through poorly compact, relatively permeable trench backfill, thereby affecting larger portion of the home. Water can also migrate extensively through the sand bedding placed above and below slab moisture barriers.

NON-EARTH MOVEMENT CAUSES OF DAMAGE

Irregularities and cracking occur in residential construction due to typical performance of the building materials, improper construction techniques, poor workmanship, and design deficiencies. The two most common building materials, wood and cement products, both shrink during curing. This shrinkage leads to cracking and distortion. Careless or improper construction techniques can cause distortion of the structure or create conditions susceptible to cracking. Improper or poor conceptual design can lead to excessive structural distortion under design loading and other unfavorable conditions such as underfloor dampness.

Cement Products

Three of the most common building materials in southern California are cement products: concrete, stucco and plaster. Concrete shrinks during curing. The shrinkage coefficient normally ranges from 0.0002 to 0.0006, but occasionally is as high as 0.0010 (Brown, 1990). The shrinkage is primarily a function of the mixing water content and the curing method. Residential slabs, which are normally poured with a medium slump concrete and which often receive little attention during curing, are particularly prone to cracking. Reinforcement placed to control shrinkage and temperature cracking and to resist expansive soil movement can minimize but typically not eliminate shrinkage cracks.

Shrinkage cracks in concrete slabs are generally less than 1/16 inch in width. Crack control joints or expansion joints are usually not provided in interior slabs. Shrinkage cracks generally take five forms: diagonal cracks across corners; somewhat linear cracks through the central one third of large rectangular areas; diagonal cracks extending from inside corners; generally linear cracks parallel to and near exterior walls; and small polygonal cracks (Robertson and Hollingsworth, 1989). The location of these cracks can be controlled by the location of pipes or other utility outlets which partially or wholly penetrate the slab. The majority of shrinkage cracks occur during the first six months after pouring. Shrinkage cracks can, however, continue to occur over two to three years.

Stucco and plaster are also cement products which shrink during curing. Shrinkage cracks in these products are generally hairline to 1/32 inch in width and are again controlled by the water/cement ratio and the application process. The typical stucco cracks extend diagonally from window and door corners and vertically along studs. Diagonal plaster cracks at door and window openings are also common.

Wood

Wood used in framing and decorative treatments shrinks and twists during aging and drying. The amount of shrinkage depends on a number of factors including the wood type, milling process, water content and curing time prior to milling and distribution. Wood shrinkage can result in apparent settlement of over 1 inch in a three story, wood frame construction (Robertson and Hollingsworth, 1989). Gaps resulting from wood shrinkage can develop between exposed beams and plaster or stucco walls. Twisting of wood beams can cause cracking of adjacent walls. Shrinkage of wood mouldings and door frames can leave gaps between the wood and wall.

Wood floors can develop irregularities due to shrinkage and deflection. Floor joists can deflect before and after loading, leading to depressions in the floor. Deflection of cantilevered wood floor sections can result in noticeable floor tilt. Shrinkage of the joists can cause floor irregularities. Floor irregularities can also occur when wood and steel elements are used together.

Slab Construction

Several slab construction techniques and problems can lead to cracking and separations. A common problem with slab on grade construction is the failure to center the steel reinforcement in the slab. This problem is particularly prevalent when wire mesh reinforcement is used. Steel which is at the base of the slab or in the underlying sand cannot adequately resist slab cracking. Cold joints or construction joints in slabs, particularly at steps, are often problematic. Slight movement along these straight line features, even when reinforced, can cause damage to ceramic, stone and linoleum floor coverings.

The slab and footing in non-structural slab-on-grade construction can be poured monolithically or at two separate times. When the second method is used, the slab may cap the footing or be poured adjacent to the interior footing face. Steel reinforcement may or may not be extended from the footing into the slab. Slabs which are not tied into footings with reinforcement or which were not poured over the footing will tend to separate from the interior face of the footing, producing a gap. Gaps with this construction technique of over 1 inch in width have been observed by the authors. The gap results from several processes, including concrete shrinkage, slab heaving and lateral foundation movement.

Wood Framing

Irregularities in wood framing can occur for a number of reasons independent of the performance of the wood product. Framing irregularities can give the appearance that a structure has distorted. Such irregularities have been found to be more common in tract developments. Framing can be distorted by wind during the construction process and, if not corrected, the irregularity will be built into the structure. Improper loading of the roof with roofing material during construction can also result in wall and ceiling distortions.

Drywall Application

Drywall application techniques, as well as inattention to detail, can lead to cracking which can be interpreted to indicate movement. Drywall corners and seams can be poorly taped or left untaped. Drywall seams can be placed at the extensions of window or door frame corners, which are common stress concentration points. Drywall, particularly on ceilings, can be improperly nailed, leading to cracking and distortion.

Lowered Subfloor Construction

It has recently (past ten years) become popular to construct raised floor houses such that the floor surface is only slightly above the exterior grade. The soil beneath the structure must be excavated to provide joist clearance, creating an 18 to 24 inch deep, sump-like area in which water can collect. Water from the ground surface surrounding the house infiltrates into the soil and moves beneath the foundation elements into the lowered area. This movement results primarily from a potential difference due to the ground elevation difference. Occasionally the evaporation of water from the surface soils beneath the structure due to residence heating will induce moisture movement. High moisture conditions beneath a residence can cause foundation movement, dry rot, and odor.

EVALUATION TECHNIQUES

Successful evaluation of residential damage requires careful observation, knowledge of building design and performance criteria, knowledge of geotechnical processes capable of causing damage, and an open mind. A combination of detailed damage observation, site reconnaissance, record and air photograph review, subsurface exploration, laboratory testing and engineering analysis is used. Determination of the cause of residence damage is often difficult. As the old adage goes: things are not always as they seem.

Careful observation, measurement and mapping of the damage is critical. A floor level or manometer survey should be included whenever possible. These steps should preferably be performed before available data on the site and vicinity are researched to avoid any causational bias. This is not always feasible, but a consultant should not try and fit the damage to a hypothesized cause. Rather, the cause(s) should become apparent when the data is collected and reviewed.

Initial observation of a structure should include a cataloging of crack location and width. This information will be useful in determining the direction of movement and will serve as a basis for future monitoring. Site mapping should include the liberal use of a camera. Once detailed observation of the damage is performed, residential damage evaluations follow the same general steps as site evaluations for new construction. Neighborhood reconnais-sance, record research and air photograph review should follow the preliminary observations. The initial observations, reconnais-sance and record review should suggest probable cause(s) of damage and assist in the design of subsurface exploration and laboratory programs to further evaluate the hypothesized cause(s) of damage.

Crack Mapping

Site evaluation should begin with detailed mapping of the interior and exterior crack patterns within a structure.Detailed mapping of cracking within appurtenant improvements should also be performed. The pattern of cracking is often of greater significance than the magnitude of individual cracks. Consideration of the location of load-bearing elements within a structure is important. Interpretation of crack patterns is facilitated by considering the direction of structure movement required to close the cracks.

Floor cracks should be recorded on a plan to aid in the determination of movement patterns. Cracks in slabs with vertical displacement and widths in excess of 1/16 inch (1.6mm) are generally related to earth movement. Abnormal cracking suggestive of slab heaving radiates from a domed area of the slab. Linear, parallel and en echelon cracking with stepped vertical offsets generally indicates differential settlement. Linear cracking along a ridge-like feature indicates heaving of a bed or linear zone within the underlying earth materials.

Various floor coverings respond differently to cracking within the underlying slab. Wood flooring is flexible and can mask even significant slab movement. Ceramic tile, in contrast, is brittle and subject to cracking by even minor movement within the slab, especially if the tile was placed without an adequate mortar bed or slip sheet.

Wall and ceiling cracks should also be mapped and windows, door frames and walls checked using a carpenter's level. Schematic drawings of walls and ceilings are useful in the interpretation of collected data. Wall and ceiling cracks larger than 1/32 inch (0.8mm) may indicate movement in excess of normal adjustment of the structure, especially if associated with distortions in window and door frames. Normal hairline wall cracking radiates away from window and door corners at about 45 degrees. Vertical, regularly spaced stucco cracks are also common.

Cracking due to excessive foundation movement will generally be asymmetrical. Arching of window frames and bowing of window glass indicates compression of the frame. Pronounced cracking at opposite corners of the window will generally accompany the compression. Tension can cause the glass to pull out of its frame, and cracking may only be present at one corner. Shear distortion changes the shape of rectangular window and door openings to that of a parallelogram. Cracking with this movement is generally present at opposite corners.

The exact age of cracks is generally difficult if not impossible to determine. The relative age of cracks and whether or not they have been previously repaired can generally be determined. Plaster and drywall cracks will often exhibit textural differences indicating prior repair. Occasionally the age of wall cracks, particularly exterior cracks relative to prior painting, can be determined.

The relative age of concrete flatwork cracks can generally be determined. Cement products are quite brittle and exposed surfaces subject to slow weathering. The edges of newly formed cracks are generally very sharp but become more rounded and weathered with age. Chemical changes due to contact with the atmosphere over time create a powdery patina on clean, freshly broken concrete surfaces (Michael, 1987).

Manometer Survey

A floor level survey should be performed as part of a residence damage evaluation. Both floors of a two story structure should be surveyed to allow comparison of level patterns. The manometer is a simple and cost-effective tool commonly used to perform floor level surveys. This instrument works on the principle that two interconnected columns of water will seek the same level, regardless of the distance between the columns. One column is connected to a reservoir while the other is attached to a rod which is provided with a scale. The scaled column is moved throughout the residence to record the elevation of the floor relative to the reservoir or datum. Readings should be taken using a systematic grid superimposed on a floor plan. A contour map of relative floor elevation including crack patterns can then be prepared to assist in evaluating movement within the residence. Manometer readings can be duplicated at a later date as a part of a monitoring program.

Careful analysis of crack patterns and floor elevation differences can help differentiate between normal distress and damage due to more serious earth movement. Floor level differences of 1/2 to 3/4 inch showing no obvious pattern are considered normal for slab construction.

Expansive soil heaving results in a generally circular, domed contour pattern with slab-on-grade construction and a series of domes and depressions with raised wood floor construction when isolated piers are present. Heaving of a linear bed or zone will produce a ridge-like contour pattern. Ridge features can also develop in raised wood floors above continuous footings due to settlement away from the footing or joist deflection. Settlement will generally result in a sloping contour pattern toward one wall or corner of the structure.

Site and Neighborhood Reconnaissance

Careful observation of the site and a review of neighborhood conditions should follow residence observation. Observation of the condition of onsite appurtenant improvements should be performed. Cracking in streets and sidewalks, evidence of ongoing or past repairs, and exposed exterior distress on nearby structures should be cataloged as they may indicate a regional problem. Local landforms and geomorphic conditions should be observed as they may reveal the past grading history of the site or suggest the nature of subsurface materials.

Onsite appurtenant improvements, including walkways, patios, pools and decking, can become distressed by a variety of causes including poor workmanship or design, tree roots, poor drainage control, and earth movement. Freestanding block property line walls, which are commonly poorly reinforced and provided with marginal footings, can be good indicators of the direction and magnitude of both vertical and lateral earth movement. Differences in the measured water level across a pool shell indicate relative movement of the shell. Deck separations from rigid fixed improvements, such as pools and residences, can be a good indicator of lateral slope movement. Decks in expansive soils will, however, "walk" away from fixed improvements due to alternate soil expansion and shrinkage. Cracks in concrete decking can indicate earth movement or normal slab peformance. Decks which are inadequately reinforced or not provided with sufficient crack control joints will develop cracks as a result of concrete shrinkage and/or thermal expansion and contraction.

Because water is the primary contributing factor in most geotechnical-related residential damage, evaluation of onsite drainage control and water usage should be performed. Pad, roof and slope drainage control should be reviewed. Drainage devices and irrigation systems should be tested for performance. Pressure testing of water service and pool plumbing lines for leakage and static testing of sewer lines is recommended. Information regarding past maintenance and repairs of drainage devices and plumbing should be obtained, if available. Water use records can also contain valuable information if they are available for a number of years. Care should be taken to include factors such as prior ownership, modification in landscaping, and changes in family size.

Some homes are serviced by private sewerage disposal systems generally consisting of a septic tank and leachfield or one or more seepage pits. The location and type of system should be determined and a review of past maintenance records performed. Testing of the system is also advisable. Many older systems are not constructed in accordance with present-day standards and effluent may be outletted in a manner adversely affecting the movement and stability of near-surface earth materials.

Most linear street cracking parallel to the curb is related to settlement of utility trench backfill, while arcuate patterns with associated curb, gutter and sidewalk damage may represent differential settlement along cut and fill contacts or initial movement of a larger landslide. Local public works departments and utility companies can provide information on the location and depth of utilities and past repair work. Today's minor cracking may be the result of continued long-term movement masked by past street paving repairs.

Understanding the geologic character of a property begins with the recognition of local landforms. Extensive modifications of the natural landscape are common in developed areas. Cut and fill grading operations result in abrupt transitions in natural geomorphic landforms. Landforms created by past grading can often be distinguished by their uniformity when compared to more rounded natural topography. The intersections of graded cut and fill slopes are generally marked by changes in slope gradients. Natural slopes ascend and descend from cut and fill slopes at more gentle gradients. Observation of natural and manmade geomorphic features can lead to an understanding of past processes resulting in the present topography. It is helpful to imagine the area topography prior to modifications by grading. Subsurface conditions and earth materials can then be anticipated prior to subsurface exploration.

Vegetation

Vegetation can also provide clues to the history of the area. Mature oaks and chaparral vegetation, including sumac, toyon, sugar bush, chamise and sage brush, likely indicate natural terrain. However, some of these species may become well established on older, graded slopes. Graded slopes are often landscaped with conifers, eucalyptus, oleanders, and several varieties of ground cover not indigenous to the area, including ivy, dwarf coyote bush, African daisies, grasses and succulents such as sedum and iceplant. Various species of conifer and eucalyptus have been imported since the late 1800s, and can be observed on both natural and manu-factured slopes.

Vegetation can provide clues to slope stability. Well-developed vegetation can be an indicator of long-term slope stability. Conifer and eucalyptus trees in the Santa Monica Mountains with boles of greater than about 12 inches in diameter have been in place on the order of fifteen to thirty years (Michael, 1987). An arcuate bend in tree trunks, however, indicates that soil creep is occurring. The limits of large-scale landsliding can also sometimes be detected by the awkward tilting of trees that normally grow vertically.

Surficial failures or slope repairs are often marked by an absence of vegetation in an otherwise well vegetated slope or by a different vegetation type. Dwarf coyote bush and African daisies were commonly used to landscape slope repairs in the late 1970s and early 1980s. These varieties have fallen out of favor due to disease problems and ivy, periwinkle and oleander are now used extensively.

Record Research

Research of published documents and building department records should be performed following site observation and reconnaissance and prior to subsurface exploration. A review of previous geotechnical reports on file for the site, tract and neighboring properties and tracts should be performed at the city or county building department. Other consultants, familiar with the area, should be contacted when possible. Neighboring property owners may also be contacted, although past experience indicates that people do not like to divulge problems with their properties. Information provided by neighbors should be verified.

Research of permit documents should also be performed. Repair work is often permitted by building officials without requiring geotechnical reports. Data listed on permits, including the names of geotechnical consultants, structural engineers, architects and contractors may also lead to further information not on file in the grading records. The age of construction is also pertinent when considering the construction standards and site development requirements. Data maintained at public works departments are an often-neglected source of available local information. Los Angeles city street and sewer plans dating to the 1920s have provided useful and accurate data to the authors during previous site evaluations.

State subdivision reports also provide information regarding tract grading. The information is generally limited in scope, but is useful when data are missing or unavailable at the local building department. Sometimes data missing from building department files have been transferred to the city attorney offices due to pending litigation in the tract. Efforts to obtain such information can be met with varying degrees of success.

The consultant should become familiar with local building codes, grading ordinances and the standard of care applicable at the time of site development. Standards of enforcement can vary greatly from community to community.

A review of aerial photographs is useful in identifying regional landsliding and recognizing past grading and previous land uses, especially when stereopair photographs are used. Some of the aerial photographic sources available to southern California consultants include the Spence Air Photo Collection maintained at the Geography Department at the University of California, Los Angeles, the Fairchild Collection at Whittier College, Teledyne Geotronics photographs, I. K. Curtis Services, Inc., and U.S. Department of Agriculture stereopair air photographs. Thomas Bros. Maps also produces an aerial atlas compatible with their 1"=2000' Thomas Guide.

Subsurface Exploration and Testing

Subsurface exploration provides critical data in the evaluation and verification of causes of residential distress. Exploration is generally limited by physical constraints such as property lines, onsite improvements, topographic conditions, and access. Subsurface exploration is also an invasive method of study resulting in at least some damage to onsite improvements and landscaping, and disturbance to the homeowner.

The required intensity of exploration will depend upon the type and extent of distress and the geologic complexity of the site. Exploration may consist of shallow and deep excavations utilizing a combination of hand labor and drilling equipment. Graphic logs of test pits and borings should be prepared to assist in the interpretation of data obtained.

Hand labor is often utilized for subsurface exploration and is generally very effective to depths up to 25 feet (7.7m). The size of hand-excavated test pits allows for close observation of exposed foundation elements and earth materials and this method of exploration is often less costly and less damaging than other methods in areas of restricted access. A hand auger can be used to extend hand-dug test pits or to supplement other exploration, but identification of materials is more difficult.

When deeper exploration is required, the use of a drill rig is warranted. Where possible, a full size drill rig is used and can verify information obtained from more limited methods of exploration. Technical advances over recent years have resulted in the development of more sophisticated portable drilling equipment used to explore slopes and other portions of the site with limited access. However, the use of portable drilling equipment is generally quite costly and invasive due to necessary site preparation and clean up.

Adequate sampling and laboratory testing should be performed as a part of the evaluation. Sufficient samples should be retrieved to allow simulation of various conditions which may have led to damage. Testing should attempt to simulate field and historic loading and moisture conditions as closely as possible.

CASE HISTORIES

Case History No. 1

The following case history involves two residential properties on graded lots in the city of Thousand Oaks, Ventura County, California. The properties are generally level and are not adjoined by any significant slopes. Grading and residence construction were performed in the late 1960s to very early 1970s. A 12-inch-diameter sewer line is present in an easement between the two residences. The sewer is 19 to 20 feet below grade.

Damage to the properties consisted of wall and slab cracks, misaligned floors, doors and posts, and out of level floors. Crack mapping and manometer surveys revealed that the damage was concentrated in general proximity to the sewer line. The magnitude of floor movement ranged up to 2 inches. The residence floors sloped down significantly toward the sewer within the limits of the influence zone shown on Figure 1. The driveway directly above the sewer was severely damaged and had settled 6 to 8 inches. The timing of the damage is not known.

Subsurface exploration revealed that the sewer was installed in an approximately 4-foot-wide trench with vertical walls. The upper 4 feet of the trench backfill was found to be compact while the remainder was found to be poorly to moderately compact. The compacted fill consists of silty clay and clayey silt. The alluvium consists of sandy silt and sandy clay over clayey and silty sand. The lower boundary of the zone of influence was found to be approximately 1-1/2:1 (33 to 35 degrees) projected up from the base of the trench.

Case History No. 2

The property is located along the southern flank of the Santa Susana Mountains in the Porter Ranch area of Los Angeles. The tract was graded in 1976 when the subject lot was lowered 25 to 90 feet. A 30-inch-thick compacted fill cap was recommended for all cut lots. The reason for the cap was not specifically identified. The cap was either recommended to control differential bedrock expansion or to control infiltration into buttress fills located along the sides of most cut lots.

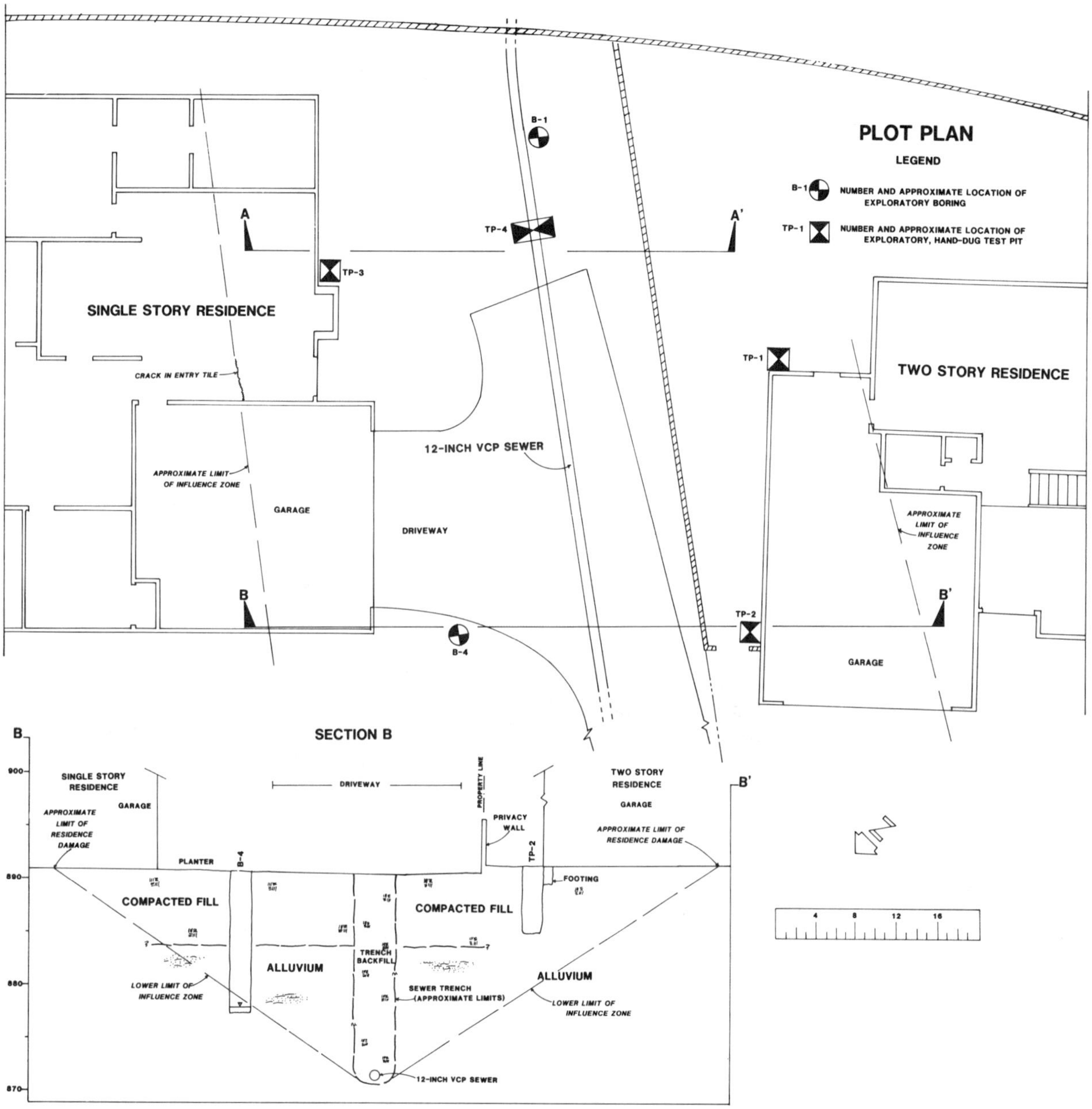

Figure 1 See Case History No. 1

The site is presently developed with a two story single family residence (Fig. 2). The residence is constructed of stucco over wood frame with a tile roof. The residence has a slab-on-grade foundation system. A pool is present in the rear yard. Residence damage consists of interior and exterior wall cracks. Maximum differential elevation across the floor slab is 4 inches. The central bearing footing which carries two-story loading is heaved upward. The pool shell is cracked and out of level. One pool wall has deflected inward and the same wall has heaved upward.

Bedrock underlying the site consists primarily of sandy siltstone included in the Saugus Formation. Occasional claystone beds are present. The claystones are sheared and appear to be associated with faulting. Two trenches were excavated through the compacted fill cap along the line of primary distress. The trenches reveal a 1/2- to 3-inch-thick sheared claystone zone paralleling the axis of primary distress. (Letter J on Trench Log, Fig. 2). The claystone dips 20 to 35 degrees to the north and extends directly below the bearing footing which is heaved upward. The claystone also intersects the northern pool wall near the pool bottom. The rate of change of floor elevation is most severe on the downdip side of the claystone zone. The northern pool wall is noticeably bowed into the pool and the northeast corner has heaved upward.

Bulk samples of the earth materials exposed in the trenches were collected for laboratory testing. Expansion Index tests and Expansion Pressure tests were performed. The Expansion Index of the claystone considered responsible for the damage ranges from 132 to 165. The potential for expansion of this material under one and two story loads ranges from 9.4 to 7.0 percent, respectively. The sheared claystone along the fault zone is therefore capable of lifting the compacted fill cap and causing the residence damage.

Case History No. 3

The property involved in the third case history is located within the central portion of the eastern Santa Monica Mountains in the city of Los Angeles. The tract was graded in 1973. The pad was created by excavating 10 to 150 feet during grading. The least cut was performed at the south corner of the site and the greatest cut at the north corner. A compacted fill cap was not provided on the subject lot, although 30-inch-thick caps were recommended by the tract geotechnical consultant for some lots due to the potential for differential bedrock expansion.

The site is developed with a two story single family residence constructed in 1975 to 1976. The residence had a conventional slab-on-grade foundation system with 12- to 18-inch-deep interior and exterior footings prior to repair. The slab was 3-1/2 to 4-1/2 inches thick and reinforced with 6x6-10x10 welded wire mesh. The foundation system conformed with the consultant's recommendations for lots with a low expansion potential.

Residential damage consisted of a noticeably heaved floor slab, slab cracks, separations between walls and ceiling, and misaligned doors. A manometer survey revealed a slab high trending from the dining room into the family room with a maximum differen-tial elevation of approximately 3 inches (Fig. 3). The damage occurs across a portion of the site excavated 60 to 80 feet. Damage reportedly became noticeable in the residence in 1982 to 1983 and was first observed by the author when it was well developed in 1985.

Bedrock underlying the site consists of sandstone and conglomerate with occasional siltstone interbeds. The bedrock is included in the Miocene Topanga Formation. The sandstone and conglomerate were light grey to blue grey except for the upper 1 to 1-1/2 feet which were weathered brown to orange-brown. The weathering was interpreted to have occurred since site grading. The siltstone was brown to black and gypsum crystalization was observed along fractures. No crystal growth was observed in the sandstone.

The slab high was found to occur above the sandstone and conglomerate rather than above the siltstone in the test trench (Fig. 3). This fact was confirmed during residence repair when the entire heaved slab was removed.

Laboratory testing of samples obtained from test trench 2 revealed expansion indices of 17, 35 and 47 for the grey sandstone/conglomerate, the brown sandstone/conglomerate and the black siltstone, respectively. The expansion potential for these materials is therefore very low to low based on UBC Table 29-C. Chemical tests were performed on samples of the bedrock both during exploration and repair. The tests revealed significant concentra-tions of sulfate in both the sandstone/conglomerate and the siltstone. Sulfate concentrations of 0.008 to 0.220 were measured in the unweathered sandstone/conglomerate while the sulfate concentration in a sample of black siltstone was 1.3 percent. The sulfate concentration in the weathered sandstone/conglomerate bedrock was found to be four to five times higher than in the deeper, unweathered material. The sulfate concentration in the siltstone was, in contrast, nearly identical in the weathered and unweathered zones. Tests also revealed considerable lateral variation in the sandstone/conglomerate sulfate percentage. This variation, however, did not precisely correspond with the slab high.

Damage is considered to be the result of gypsum growth within the bedrock underlying the site. Specifically, the gypsum growth is considered to be occurring within the weathered profile developing above the sandstone/conglomerate bedrock. The weathering process is apparently concentrating the sulfate in the area directly below the slab. The gypsum within the siltstone is not considered a major contributing factor to the damage. This gypsum likely developed in the bedrock at depth prior to grading, since the damage is not centered over this bed and the levels of sulfate in the weathered and unweathered siltstone are similar.

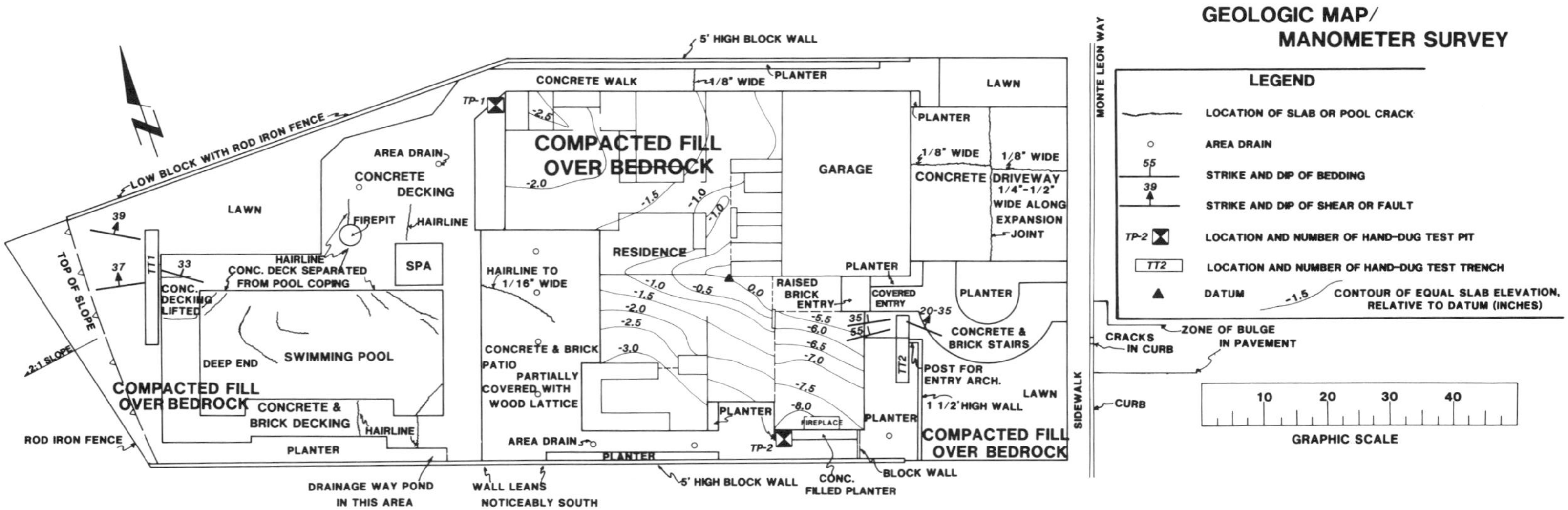

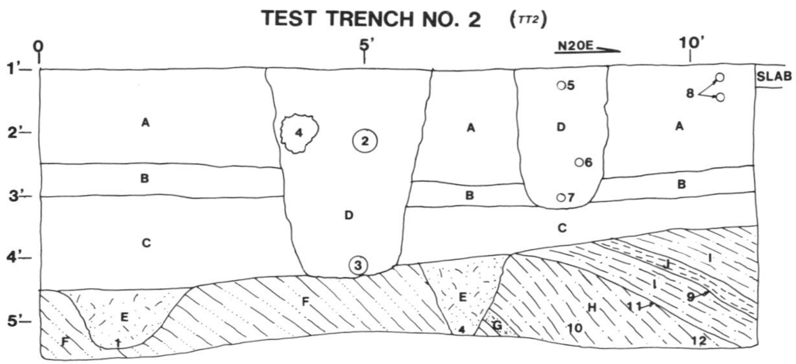

A: Compacted Fill: Sandy Silt, light brown, moist, medium dense to dense, sand fine-grained.

B. Compacted Fill: Sandy Silt, light reddish brown, moist, firm.

C. Compacted Fill: Sandy Silt, light green-brown, moist, dense, sand fine-grained.

D. Trench Backfill: Sandy Silt, light brown, moist, loose to medium dense.

E. Disturbed Bedrock: Sandstone fragments in a Sandy Silt matrix, green-brown, moist to wet, soft.

F. Bedrock: Silty Sandstone, light green-brown, moist, hard, massive, occasional fractures, moderately weathered, sand fine-grained.

G. Bedrock: Sandstone, light gray, slightly moist, hard, friable, thinly bedded, gradational contacts, sand fine- to medium-grained.

H. Bedrock: Sandy Siltstone, light green-brown, moist, hard, thickly bedded, moderately weathered.

I. Bedrock: Possible fault zone, Sandy Siltstone, light green-brown, moist, soft to moderately hard, highly sheared, clay shears ½" to 1" thick present throughout.

J. Shear Zone: Probable fault, Clay, green-brown, moist, plastic, sheared, no slickensides, zone 2" to 3" thick.

1. Water seeping into trench from west side.
2. Drainline: PVC, 4½" diameter, 90° elbow.
3. Drainline: PVC, 3½" diameter.
4. Water seeping into trench from both sides.
5. Pipe: Steel, ¾" diameter.
6. Pipe: Copper, 1" diameter.
7. Pipe: PVC, 1½" diameter.
8. Pipes: PVC, ¾" diameter.
9. Shear: N50W, 20 to 35NE.
10. Bedding: N85W, 55NE.
11. Water seeping into trench along entire contact from east side.
12. Bedding: N80W, 35NE.

Figure 2. See Case History No. 2

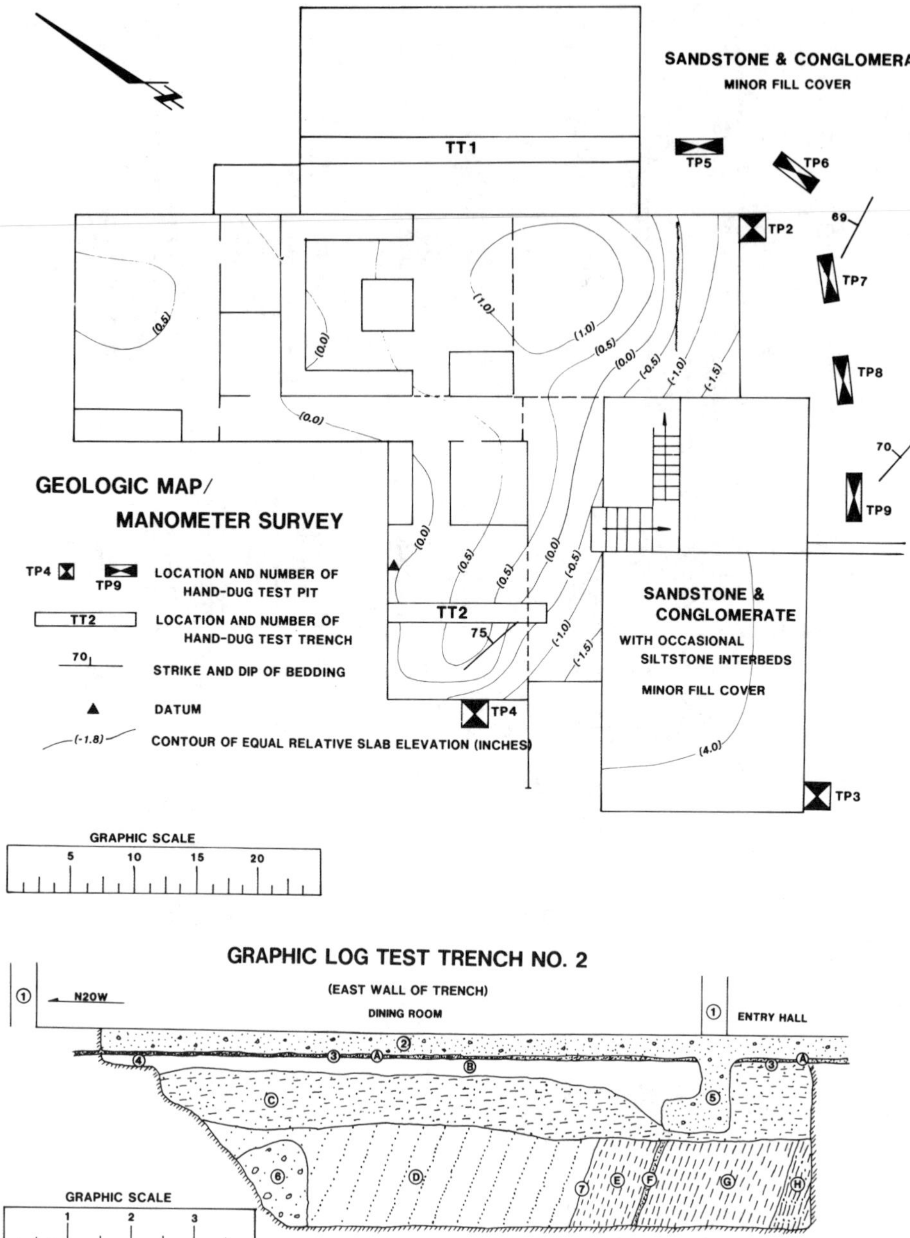

A. Fill: Sand, light gray, moist, medium dense

B. Fill: Sand with Clay binder, mottled gray to brown, moist, medium dense.

C. Weathered Bedrock: Clayey Sandstone, orange-brown to brown to light green-gray, moist, soft to moderately hard, sand fine- to coarse-grained, occasional rounded gravel and cobbles, weathering more extensive toward the surface.

D. Bedrock: Silty to Clayey Sandstone and Conglomerate, gray to blue-gray, moist, moderately hard to very hard with depth, massive, occasional rounded gravel- to boulder-sized clasts, moderately weathered.

E. Bedrock: Sandy Siltstone, blue-gray to black, moist, moderately hard to hard, thickly bedded, moderately fractured, occasional fine gypsum crystallization along fractures, moderately weathered.

F. Bedrock: Sandstone, light blue-gray, moist, moderately hard, laminated.

G. Bedrock: Sandy Siltstone, brown to blue-gray, moist, moderately hard to very hard, thickly bedded, fractured, gypsum crystallization along fractures, moderately weathered.

H. Bedrock: Silty Sandstone, light brown to brown, moist, very hard, thickly bedded, moderately weathered.

1. Framed wall.
2. Concrete slab 3½ to 4½ inches thick, reinforced.
3. Plastic vapor barrier.
4. 1½" copper pipe.
5. Interior concrete continuous footing, reinforced with one #4 steel bar.
6. Boulder.
7. Bedding: N75W, 75NE.

Figure 3. See Case History No. 3

REFERENCES

Bowles, Joseph E., 1977, Foundation and Design, (Second Edition): McGraw-Hill

Brandon, Thomas L., Duncan, J. Michael, and Gardner, William S., 1990, Hydrocompression settlement of deep fills, Journal of Geotechnical Engineering, vol. 116, No. 10, p. 1536-1548.

Brown, Robert Wade, 1990, Design and repair of residential and light commercial foundations, Mc-Graw-Hill

Chen, Fu Hua, and Ma Gui Sheng, 1987, Swelling and shrinking behavior of expansive clays, Proceedings - Sixth International Conference on Expansive Soils, vol. 1, 1-4, New Delhi, India, p. 127-129.

Dudley, John H., 1970, Review of collapsing soils, Journal of the Soil Mechanics and Foundations Division, Proceedings of the American Society of Civil Engineers, vol. 96, No. SM3, p. 925-945

El-Ehwany, Mostafa and Houston, Sandra L., 1990, Settlement and moisture movement in collapsible soils, Journal of Geotechnical Engineering, vol. 116, no. 10, p. 1521-1535.

Erol, A. Orhan, Ohowian, Abdulmohsin, and Youssef, Abdeifahah, 1987,Assessment of Oedometer Methods for heave prediction, Proceedings Sixth International Conference on Expansive Soils, vol. 1, 1-4, New Delhi, India, p. 99-103

Hawkins, A.B. and Pinches, G.M., 1987, Expansion due to gypsum crystal growth, Proceedings Sixth International Conference on Expansive Soils, Vol. 1, 1-4, New Delhi, India, p. 183-188

Hollingsworth, Robert A., 1990, Expansive bedrock of the Mint Canyon and Saugus Formations: Geology and Engineering Geology of the Western Soledad Basin, Los Angeles County, California, southern California Section, Association of Engineering Geologists, Nov. 3, 1990, p. 145-153

Kante, B. A., 1980, Some secrets to building structures on expansive soils: Civil Engineering, ASCE, NY, p. 53-55

Lambe, T. William and Whitman, Robert V., 1969, Soil Mechanics: Wiley, 553 p.

Michael, E.D., 1987, The Role of Geology in Real Estate Appraisal (unpublished), 6225 Bonsall Dr., Malibu, CA 90265

Osman, M.A. and Sharief, A.M.El, 1987, Field and laboratory observations of expansive soil heave: Proceedings - Sixth International Conference on Expansive Soils, vol. 1, 1-4, New Delhi, India, p. 115-120.

Robertson, Hugh S., and Hollingsworth, Robert A., 1989, Differentiation between normal residence performance and landslide-related distress in the evaluation and treatment of slope instability: Publications of the Inland Geological Society, vol. 2, 1989, p . 41-51

Scullin, C. Michael, 1983, Excavation and Grading Code Administration, Inspection and Enforcement: Prentice- Hall, NJ, 405 p.

Stamatopoulos, Aris C. and Kotzias, Panaghiotis C., 1987, Free-Swell test on undisturbed clay, Proceedings Sixth International Conference on Expansive Soils, vol. 1, 1-4, New Delhi, India, p. 111-114.

Terzaghi, Karl, and Peck, Ralph B., 1967, Soil Mechanics in Engineering Practice (Second Edition), Wiley

Weiss, David C., 1985, A simplified design method for the design of post-tensioned slabs-on-grade: Pre-Stressed Concrete Design Lecture Series, the Structural Engineers Association of Southern California, November 12, 1985, 15 p.

Yen, Bing C. and Bright, Robert E., 1990, Residential foundation deterioration, Study for the Cities of Lakewood, La Palma and Cypress, CA, (unpublished)

GROUTING PRACTICES IN SOUTHERN CALIFORNIA

Jack T. Eagen
Senior Vice President

John Tims,
Manager of Grouting Operations

Frank Nonamaker
Grouting Division Project Coordinator

Moore & Taber Geotechnical and Environmental Consultants
1290 N. Hancock,
P.O. Box 19079
Anaheim, CA 92817

Francis Gularte
Hayward Baker, Inc.
P.O. Box 7690
Ventura, California 93006

INTRODUCTION

Pressure grouting is generally divided into two categories: compaction grouting (mechanical stabilization) which improves soil by compaction or densification; and chemical grouting (chemical stabilization) which improves soil by binding soil grains together. For the last forty years, compaction grouting has been used mostly as a remedial measure to repair buildings damaged by differential settlement. A mortar-like consistency soil-cement grout is injected into the foundation soil under pressure at vertical intervals to densify compressible soils and thereby stabilize the soils in place. The grout extends to firm soil beneath compressible materials. Controlled lifting and re-leveling of structures is often a beneficial result of compaction grouting. Chemical grouting is used to provide temporary cohesion to granular soils; to minimize soil liquefaction and densification during seismic vibration; to increase the load-bearing capacity of soil formations, and to control water seepage and flow in certain structures. Chemical grout usually consists of sodium silicate, water, and an organically safe solvent; it resists bacteria growth and resists freeze-thaw cycles. The section on chemical grouting was written by Francis Gularte.

COMPACTION GROUTING

History/Development/Purpose

Compaction grouting has evolved in the United States over the past four decades into a proven method to strengthen and stabilize problematic foundation soils in-situ. Continuing research and development of state of the art equipment, long term monitoring of completed projects and cooperation between the engineering and contracting communities are providing valuable data to substantiate the effectiveness of properly engineered and applied remedial grouting programs.

Compaction grouting is an accepted procedure for treating compressible foundation soils in-situ and correcting structure distress due to settlement. The procedure involves the injection of a soil-cement mixture into the ground under pressure to displace and thereby compact (densify) and stabilize soils. The procedure is most frequently used as a remedial measure to repair buildings damaged by differential settlement, by increasing the strength of underlying supporting soils, and to raise buildings to a reasonable or tolerable level.

The process differs from conventional penetration grouting by its use of a very stiff (low slump) soil cement mix that when emplaced results in a homogenous irregular-shaped mass (grout column) as illustrated in Figure 1. Densification of the soils is accomplished by pressure displacement and compression rather than penetration (permeation) of the grout into the inter-granular voids. The grout tends to move into the weakest soil zone until the force exerted by the pumping pressure exceeds the confining conditions and movement of the ground surface occurs.

Program Development and Layout

Compaction grouting programs are generally designed to treat foundation soils underlying all or portions of structures or pools which have experienced damaging differential movement and exhibit a tilt in excess of some specified amount. In addition, soils influencing the lateral support of the building are also treated. The spacing of grout points are generally pre-determined and dependent on the depth to firm soil beneath the compressible materials. Actual depth of treatment is determined on a point by point basis, utilizing empirical information obtained as the work progresses and/or in accordance with plans, specifications, and recommendations of the engineer.

The location, number and depth of grout injection points may vary depending upon the soil and structure response as the work progresses. Occasionally, grouting operations can be performed entirely from outside by angling beneath structures, but usually it is necessary to work inside the building to provided complete coverage, particularly where the treatment interval is shallow.

The following informational items are valuable and useful in designing a grouting program and preparing a cost estimate:

- A grading plan showing existing pad elevation and original ground contours.
- Boring logs describing or containing classification, depth, consistency, moisture, density, and saturation of the earth materials.

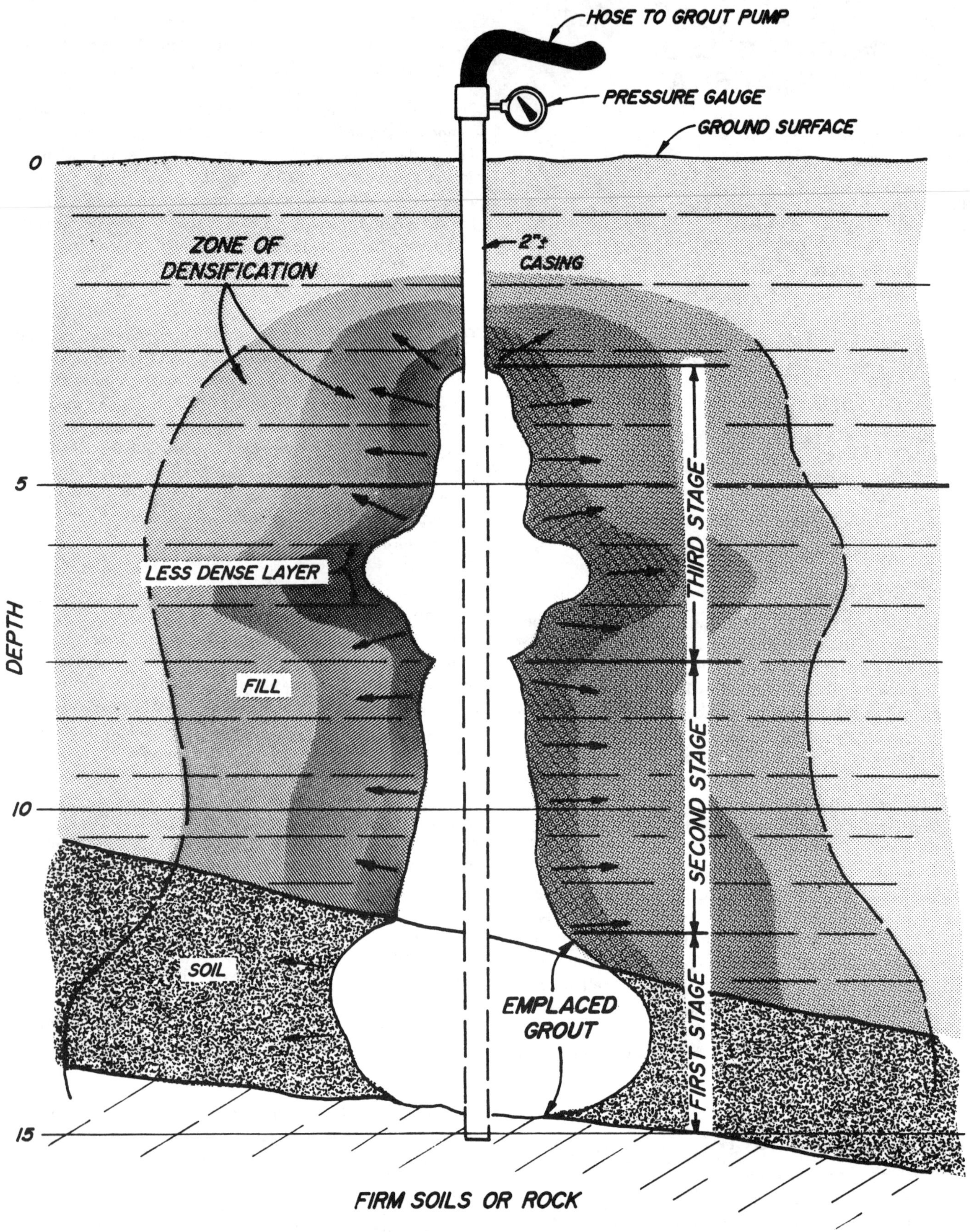

Figure 1. Hypothetical section of idealized irregular grout column showing states and approximate zone of influence.

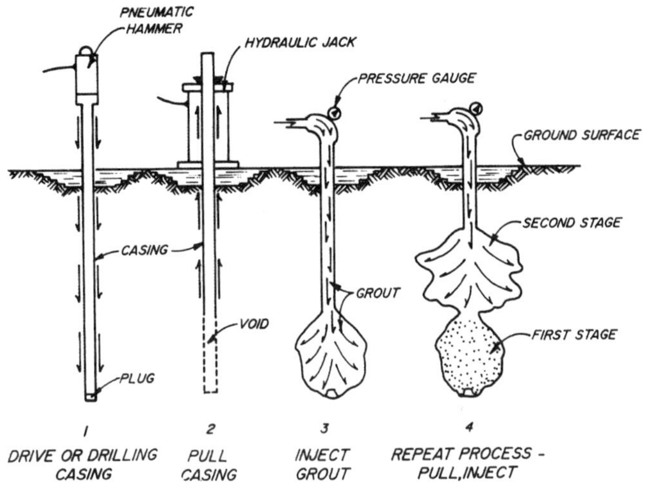

Figure 2. Stage grouting sequence showing drilling or driving to advance casing, pulling casing, injection of grout into void, and repetition of procedure for each stage

- Laboratory test results including maximum density, consolidation characteristics, expansion potential.
- Floor Level Survey; preferably contoured to show pattern, tilt and magnitude of movement on the floor or pool coping.
- House and/or site plan showing locations of structures, appurtences, and subsurface utilities relative to streets and property lines.
- Pre-construction (original) geotechnical, soils and/or geologic reports.
- Geotechnical Investigation or Distress Reports covering the damaged property. Such reports may contain some or all of the above information.

With the above information, the grouting contractor will be able to formulate a detailed grouting program and provide a better estimate or quote on costs. Lacking information on the actual depth or thickness of fill and/or native soil (i.e., alluvium, colluvium, soft sedimentary rock), the moisture, density, saturation, and consolidation characteristics of the soils, and the differential floor elevation (tilt), the contractor/engineer will be limited in their ability to design a comprehensive grouting program. A lack of information forces the contractor to make liberal assumptions on grout take and density increase, to rely on his experience, and to cover himself financially.

Procedures

The primary purpose of most grouting programs is to densify potentially compressible foundation soils which underlie areas experiencing distress and substantially relieve stress on overlying structures by ground lift. Controlled lifting and leveling of overlying improvements is a secondary benefit often achieved during the grouting process. However, lift is not always possible due to ground or structure response, and lifting attempts may have to be terminated to preclude new or additional distress. Additional raising and leveling (if required or desired) can often be attained by mechanical jacking near the completion of the grouting operation.

The grout mix generally consists of a soil-cement mixture having a thick mortar-like consistency. Ongoing controversy exists over the "best" or "ideal" slump for compaction grouting. Some grouting contractors/engineers recommend very low slumps of 0 to 2.0 inches. However, most agree on slumps of less than 4 inches for compaction grouting applications. The grout is pumped through small diameter cased holes (1.5 to 3.0-inch ID) which have been drilled or driven through low density foundation soils into firm natural soil or bedrock. Generally, the grout casing is withdrawn in three to five-foot vertical increments (stages) and the grout is injected at each stage, starting at the bottom and working up, until one of the desired criteria for termination of pumping is achieved (lift, excessive pressures, adverse soil or structure response, grout return). On occasion, a top down procedure is used in the upper few stages to provide a densified "cap" against which deeper stage grouting would act to produce lift. Figure 2 illustrates the general stage grouting sequence, and Figures 3, 4 and 5 show specific steps of the actual operation.

The energy applied to the soil during pumping is transmitted both horizontally and vertically throughout the treated soil mass. The grout tends to displace the weakest soil zone until the force exerted by the pumping pressure exceeds the confining load and surface uplift occurs. In some instances the lack of lateral confinement while pumping near a slope face will limit the lift and effectiveness of densification. Precise monitoring by survey and/or instrumentation of the slope in conjunction with overlying structures is an essential part of the grouting operation.

Compaction grouting has been successfully performed and effective to depths in excess of 100 feet. However, more common applications are intended to treat soils less than 40 feet deep. The effectiveness of compaction grouting is limited at shallow depth (less than 5 feet) due to the reduction in confining overburden loads. Soil improvement and densification are more difficult to obtain at the shallow depths or near slopes and modification of grout point spacing, grout consistency (slump), and pumping rate may be necessary to achieve acceptable results. On occasion, surface "blow-outs" preclude completion of a particular point. Usually relocation or restaging of the grout point is sufficient to accomplish the desired result.

Figure 5. Pumping, batch mixing, and material loading operation.

Figure 3. Driving casing with pneumatic hammer. Casing is driven at slight angle to influence soils directly beneath footing/structure.

Figure 4. Pumping at vertical grout injection point. Hydraulic straddle jack secured to casing for pulling when lift or high pressure is achieved, Technician is observing pressure gauge at inflection point.

Detailed records of casing installation, penetration rates and depth, grout injection intervals (stages), emplaced grout per interval, pumping pressures, and associated ground response are maintained for each grout injection point throughout the project. These records are available to the site or project engineer and when combined with pre-grout investigative information, assist in the empirical development of grouting procedures and techniques to accomplish maximum soil stabilization and lift and permit adjustment of the grouting to accommodate the various soil conditions actually encountered. The grouting data are analyzed to determine the results of the work, such as increased density, structure lift, and ground movement. Density increases of 5% to 10% are common and on occasion, density improvement of as much as 20% might be achieved for some zones or layers.

Application

The major advantage of a designed and properly applied compaction grouting program over conventional remedial measures such as underpinning or caisson and grade beam construction is the injected grouts ability to treat the cause of the problem (compressible soils). This method of repair tends to be site specific rather than structure specific and treats the soils where the problem originates by providing increased strength and bearing characteristics to the soils supporting the structure foundation as well as underground utilities and surrounding flatwork (i.e., sidewalks, patios and decks). Additional advantages and considerations are minimum site disturbance, flexibility and economy.

The grouting operation does not normally require moving or major disruption of normal occupancy, but some noise and landscape disturbance may be experienced during the course of the work.

Grout "take" (volume in cubic feet) can be pre-determined and calculated on some projects if the geotechnical condition of the underlying soils is sufficiently known. A detailed subsurface geotechnical investigation, including in-situ and laboratory testing would be necessary to design a grouting program designating a specified density increase and grout take per each stage. On projects containing a large volume of compressible soil with potential for high grout takes, a test grout program is suggested to define or confirm subsurface conditions and develop parameters for bidding purposes. This procedure provides data on which to design an efficient grout program and to evaluate viability, methods and specifications without submitting to economically unfeasible or unrealistic expectations.

In the absence of engineering recommendations, design plans and specified grout quantities, it will be to the contractor's benefit to pump as much grout as possible into the soil until one of the criteria for termination is achieved. Pumping until lift, high pressure, or grout return occurs (considered refusal) will satisfy the assumptions made for influence, compaction, and coverage of the program. This approach also benefits the contractor financially since most grouting services are performed on a unit cost basis for grout actually emplaced. This thorough procedure is also particularly important if a warranty of the grouting is provided.

With respect to warranties of the soil's performance following grouting, such assurances are becoming more common as compaction grouting gains acceptance, knowledgeable experienced contractors become more proficient and engineers more confident in the effects, capabilities, results, and applications. Warranties might provide for regrouting if some specified amount of subsequent settlement is measured within the area of treatment.

Other innovative applications of compaction grouting have been developed which provide practical and cost effective remediation alternatives, including sub-sealing concrete slab sections of highways and structures, and treatment of sea walls and revetments to mitigate piping of soil. Compaction grouting has also been effectively used to densify loose utility trench backfills in lieu of removal and replacement, to reduce the liquefaction potential of low-density, saturated fine-grained sands, and to increase the uplift capacity of piles. Some examples of the application of compaction grouting are given under Case Histories. An idealized program of compaction grouting to alleviate the effects of settlement is illustrated in Figure 6.

Geotechnical Considerations

Settlement related problems are most often due to compression of the foundation fill or native soil. Thus, compaction grouting by its ability to displace and thereby compact or densify the compressible soils in place effectively treats the cause of the problem and minimizes the potential for future settlement. Although the completed grout columns are composed of material much stronger than the surrounding soils, and may provide some vertical support, they are not relied upon for this purpose.

Artificial fills compacted only to the minimum code requirements are too often relied upon to provide support and perform without future settlement. Even fills compacted to the required code and ordinance values (commonly 90% relative compaction) can be subject to further compression with an increase in soil moisture content (hydro consolidation). Settlement of structures is most often differential with the tilt (deflection) increasing in the direction of increasing fill depth. Approved compacted fills can often be further densified by compaction grouting.

The need for compaction grouting to a large degree will depend on the past performance (how much has it settled) and future performance (anticipated settlement). A structure which has already settled four inches due to compression of the foundation soils may be subject to several more inches of settlement and thus the need for stabilization or treatment by

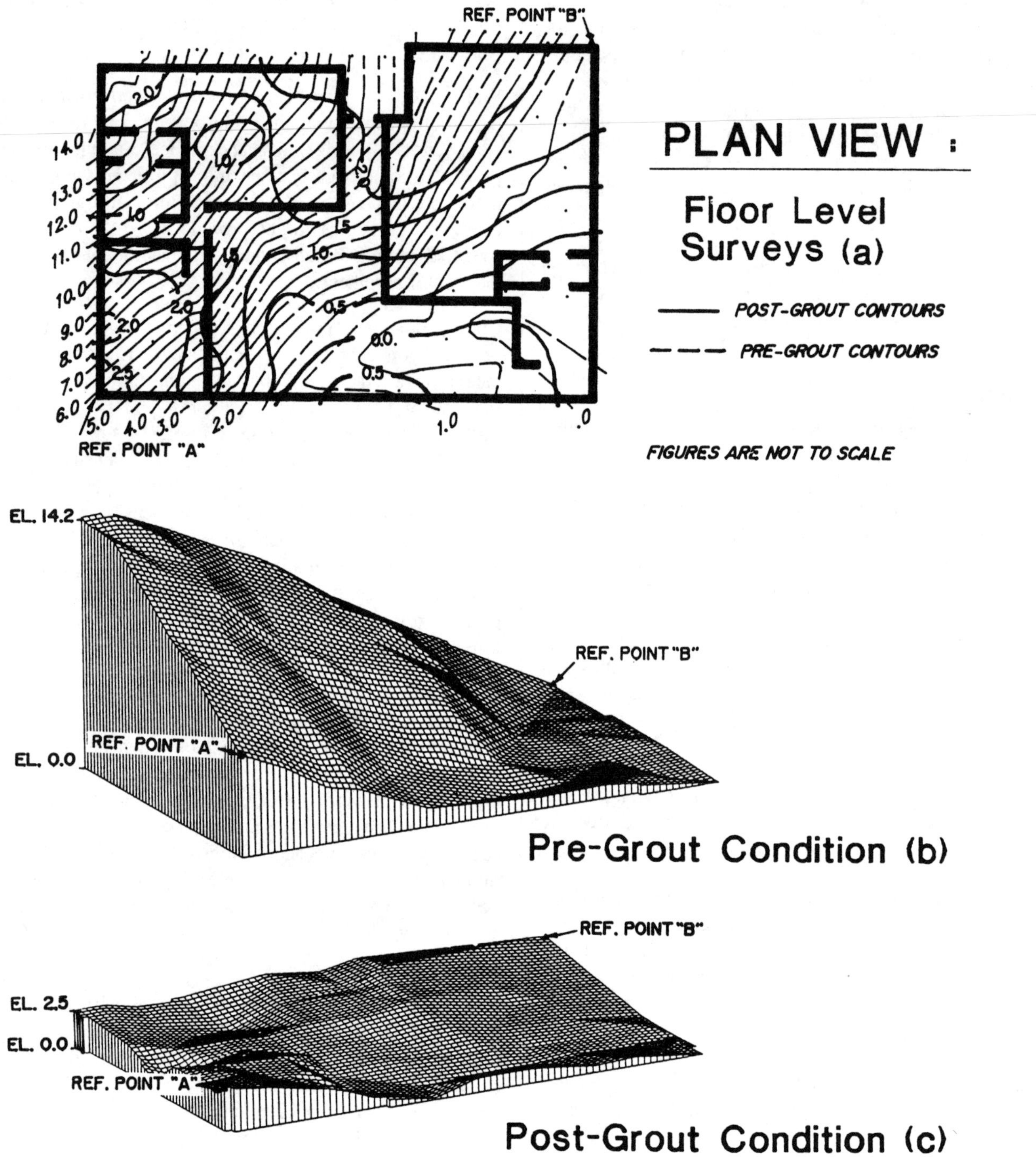

Figure 6. Compaction grouting remedial repair program showing layout of injection points and section view of settlement and subsurface

compaction grouting may be appropriate. On the other hand, most of the compression may have already occurred and only residual or token settlement remains. In such case, a large scale compaction grouting program to achieve a density increase to minimize further compression of already densified soils may in fact be overkill. Strengthening the foundation, mechanical jacking, slab removal and replacement and cosmetic repair may be all that is needed taking into consideration the risk of some slight additional adjustment. The importance and value of competent geotechnical evaluation can not be over emphasized in relation to cause, origin, nature of distress; present condition of foundation materials; need for, type and knowledge of remediation methods.

CASE HISTORIES

Residential Hillside Development, Irvine, California

Densification of fill and alluvial soils beneath four residential lots was accomplished utilizing compaction grouting. Difficult subsurface conditions (large boulders) were encountered throughout the project necessitating rotary percussion drilling for casing installation. The sand cement grout mix (3 inch minus slump) was injected at 339 locations in the areas treated and a total grout volume of approximately 40,500 cubic feet (1500 cubic yards) was injected into the soil. Specifications for the project required that all of the structures conform to a total floor elevation differential not to exceed 1.0-inch at the completion of the grouting work. The structures, which had experienced up to nearly five inches of differential movement, were raised substantially to meet the requirements for the project.

Huntington Beach Pier, Huntington Beach, California

Compaction grouting was utilized to increase the uplift capacity of a 24" diameter pre-cast driven concrete pile which had not achieved the project specified tip elevation. In lieu of pile removal and replacement, approximately 338 cubic feet of sand cement grout was injected into the soil surrounding the embedded pile between -24 and -2 feet (the depth of historic beach scour). Analysis of the grout volume emplaced and assumed area of influence indicated an average 30% increase in densification of the soil treated which met or exceeded the project design requirements.

Repair of Quaywall Sheet Piling, Naval Station, San Diego

Approximately 630 cubic yards of cement/bentonite slurry was injected behind a reconstructed sheet pile quaywall in order to fill voids and seal the sheet piling joints. A portion of the awarded contract for this project included conventional compaction grouting to be used in front of the quaywall (approximately 30 feet below sea level). However, the contractor demonstrated, by means of penetration resistance testing, that the geotechnical conditions were suitable for the planned repairs without modifying the ground seaward of the quaywall by pressure grouting. As a result, a substantial savings in the total contract amount was achieved by this effort.

Treatment of Sewer Line Backfill, Rose Drive, Yorba Linda

Approximately 2400 lineal feet of utility trench backfill soils were densified in place utilizing compaction grouting techniques thereby precluding the need for lane closure, traffic disruption, over excavation and recompaction. A considerable savings in time and money was realized. The process involved treating only the backfill material overlying existing utilities in order to minimize damage to the subsurface improvements. Continual monitoring of the ground surface during the grouting operations revealed that the injected grout mix was confined to the trench excavation area. Subsequent monitoring by the Sanitation District has shown that the areas treated have performed well without settlement.

Remedial Compaction Grouting and Releveling of Structure, Single Family Residence, Monterey Park

Compaction grouting techniques were employed to treat fill and alluvial soils beneath a conventional single story residence in Monterey Park, California. According to the geotechnical consultant, consolidation of the alluvium was the result of loading from the overlying fill. The homeowner reported that distress to the structure worsened considerably as a result of the October 1, 1987 Whittier Narrows Earthquake.

Approximately 10,000 cubic feet (370 cubic yards) of sand/cement grout was pumped into 43 grout injection points ranging in depth from 10 to 43 feet. The structure, which had settled differentially (tilt) in excess of 14 inches, was raised approximately 12 inches by the grout injections. Pre- and post-grout floor level survey data (Fig. 7) show marked improvement from the original out of tolerance condition.

CHEMICAL GROUTING FOR SOIL SOLIDIFICATION

Chemical grouting for soil solidification is used for the following applications:

1. To provide cohesion to granular soils for temporary excavation support for construction of buildings and tunnels in urban areas.
2. To prevent soil liquefaction and or densification during seismic, vibration or shock loading.
3. To increase the load-bearing capacity of soil formations, such as under walls and footings.
4. To make porous soils impermeable to water.

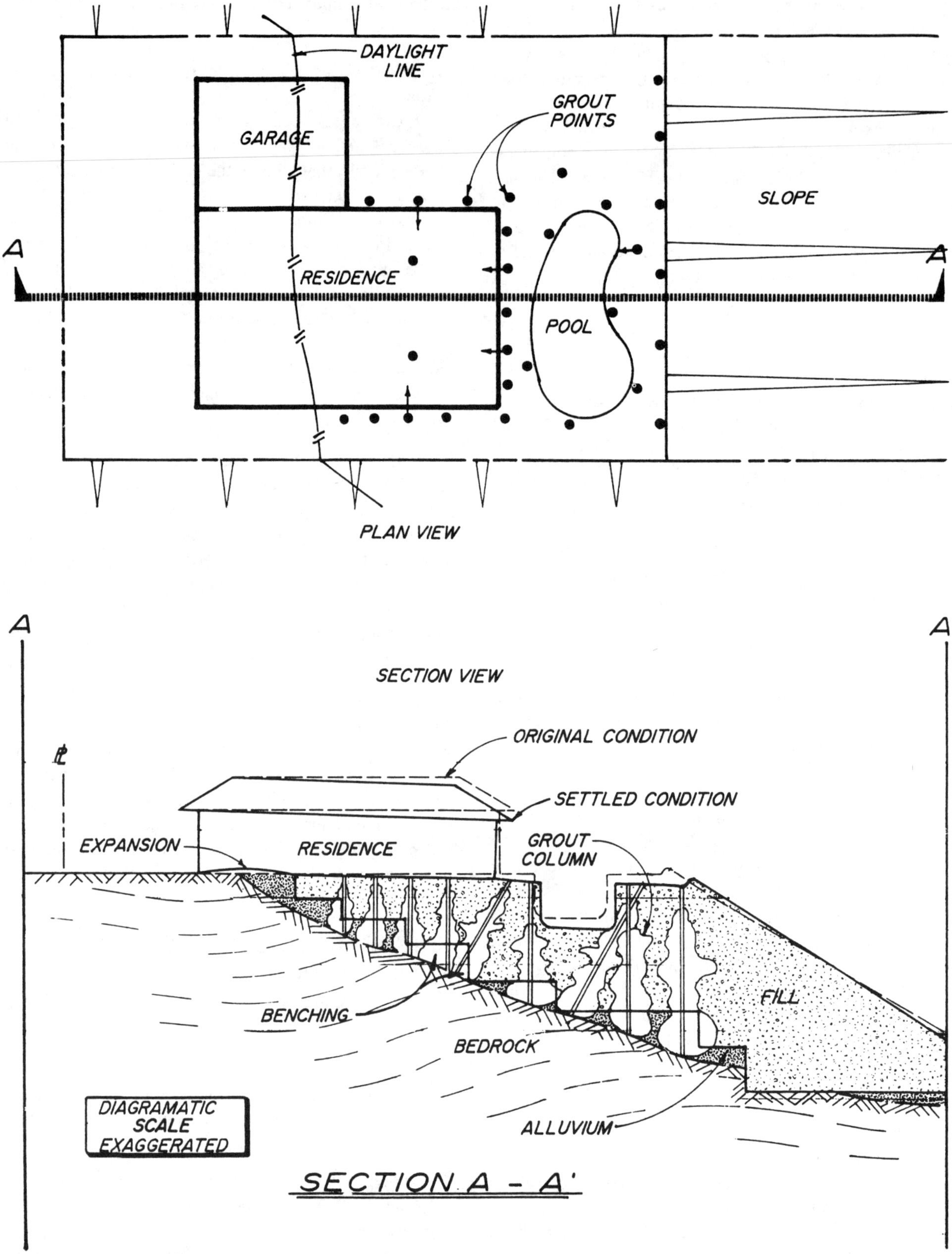

Figure 7. Compaction grouting to stabilize, lift and "relevel" residential structure. Plan view (a) floor level survey (FLS) and three dimensional (isometric) view before (b) and after grouting (g)

5. To prevent flooding of basements, subterranean structures, mines, shafts, and tunnels.

6. To prevent water seepage through dams.

7. To waterproof porous or cracked concrete, block, or brick work - as in airport, highway, sewers, or subway construction - or subterranean concrete masonry structures.

Chemical grout is a pure solution without suspended particles, which allows it to penetrate small pore spaces within the sand mass. Depending upon the chemical used, it will change the sand mass into a solid that has cohesion. The end product approximates a sandstone that will vary from soft and friable to extremely hard. The strength of chemically grouted sands depend upon the percentage of fines present and the type and concentration of chemical grout injected. Most chemical grouts provide unconfined compressive strengths in the range of 10 to 200 psi.

The most commonly used chemical grout consists of sodium silicate, water and an environmentally safe organic solvent. In some cases cement, calcium chloride, sodium aluminate, or sodium bicarbonate are used to change the pH of the solution to fix the gel in the soil pore spaces and provide the desired cohesion. The grout effects the soil in basically two ways: it stabilizes granular soil by providing cohesion which increases the shear strength and load-bearing capacity of the soil, and it occupies the pore spaces thus reducing water flow. Thus trenches or tunnels can be temporarily excavated and shored without movement or loss of the soil. Foundation bearing capacities can be increased for improved loads. Liquefaction or densification settlement that can be caused by earthquakes or heavy operating machinery vibrations can essentially be stopped and where groundwater is present, flow will be reduced.

Several items should be taken into account when designing a chemical grout program:

1. Where grouting improvement is intended to be a permanent solution, an allowance for long term creep with regard to the soils increased strength needs to be taken into account in the design.
2. Where improvement will be permanent, possible changes in the ground water level and chemistry (pH) need to be considered.
3. Where grouting will be used to control water flow, complete and extensive grouting is required.

Sodium silicate - organic solvent chemical grouts provide a more uniform stable grout with controllable strengths in the range of 10 to 200 psi. Sodium silicate - salt or cement chemical grouts are less expensive; however, they are not as stable as the organic solvent mix and provide weaker stabilized soils in the range of 5 to 50 psi. In general the sodium silicate - organic solvent chemical grout:

1. Provides a gel time controllable uniform grout.

2. Is permanent in freeze-thaw cycles and resist frost heave.

3. Resists bacteria growth.

4. Is ecologically safe (depending on the organic solvent or salt used).

Several other types of chemical grouts include arylamide, acrylate, polyurethane, and epoxy. These grouts are not normally used for large scale grouting, however, due to costs and environmental concerns.

The criteria for chemical groutability is based upon soil gradation. When the soil has less than 10% passing the #200 sieve grouting can effectively and efficiently be performed. Grouting gets progressively more difficult as the percentage of fines increases and when over 20% passes the #200 sieve the soil usually is not economically groutable. It should be noted that soil is usually never uniform and the chemical grout will penetrate the mass to varying degrees. The experience of the specialty contractor is the key to determining the type and quantity of grout mix needed because of the tremendous variations in soil structure and composition, and the variable soil environments. The degree of stabilization that can be achieved is generally determined by the grouts chemical properties, density, water content, and overall soil gradation. In determining whether the problem can be solved with chemical grouting each site has to be evaluated on its own merits; grout mix design, gel times, strengths, final permeabilities, ground water, environmental concerns as well as site access for men and equipment.

The best placement of grout is achieved by use of sleeve port grout pipes. These consist of 1 to 2 inch diameter PVC pipes with one way check valves placed on various centers throughout their length. The pipe is installed in the borehole and the annulus cement grouted, thus sealing the multi-ported pipe in the ground. After all of the pipes are installed grouting can be performed in a predetermined sequence (inside to outside or top to bottom, etc.) and grouting can be performed in stages (i.e., 50% the first day, followed by 50% of design volume). In addition, if low pressures and large takes occur, less expensive cement grout can be injected, with the more expensive chemical grout left for last.

Another method of injecting grout is under pressure through open-ended pipes or "needles." These are driven, in a predetermined grid pattern, into the soil and grouted. Then the pipes are driven further and grout injected again. This method only allows grout to be injected once at each location which significantly reduces grouting control if the ground is loose, or grout leaks along the pipe, or a more permeable layer is encountered which consumes all of the grout leaving none for the remaining design zone. This method should only be used for uniform deposits of medium dense to dense relatively clean soils.

Chemical grouting has become an increasingly useful tool in improving granular soil conditions involving many types of construction. No doubt specialty contractors, in coordination with geotechnical consultants, will continue to develop new applications in solving ground modification problems with chemical grouting.

REFERENCES

Bowen, Robert, 1975, Grouting in Engineering Practice: Applied Science Publishers Ltd., London, p. 65-77.

Brown, Douglas R. and Warner, James, 1973, Compaction Grouting: Journal of the Soil Mechanics and Foundations Division, American Society of Civil Engineers, vol. 99, No. SM8, p. 589-601.

GRAF, EDWARD D., 1969, Compaction Grouting Technique and Observations: Journal of Soil Mechanics and Foundation Division. American Society of Civil Engineers, vol. SM5, p. 1151-1158.

Herndon, Joe and Lenahan, Tom, 1976, Grouting in Soils, Vol. 1., A State-of-the-Art Report: United States Department of Transportation Federal Highway Administration Report No. FHWA-RD-76-2,: National Technical Information Service, Springfield, VA.

Herndon, Joe and Lenahan, Tom, 1976, Grouting in Soils, Vol. 2., Design and Operations Manual: United States Department of Transportation Federal Highway Administration Report No. FHWA-RD-76-2,: National Technical Information Service, Springfield, VA.

Littlejohn, G.S., Karol, R. H., Liu Jiacai, et. al., 1982, Grouting in Geotechnical Engineering. In Baker, Wallace Hayward (ed.), Proceedings of the Conference on Grouting in Geotechnical Engineering: The Geotechnical Engineering Division of the American Society of Civil Engineers, New Orleans, LA, p. 35-48; 359-377; 694-707; 923-937.

Littlejohn, G.S., Karol, R. H., Liu Jiacai, et. al., 1982, Grouting in Geotechnical Engineering. In Baker, Wallace Hayward (editor), Proceedings of the Conference on Grouting in Geotechnical Engineering, vol 2 (Panel Discussions): The Geotechnical Engineering Division of the American Society of Civil Engineers, New Orleans, LA

Nonveiller, E., 1989, Grouting Theory and Practice, Elesevier, NY

Prager, M., Bandimer, S., Blacklock, R. and Wright, P., Hampton, D. and Jin, J., Jackura, K. and Graf, E., Vipulanandan, C. and Krizek, R., 1986, Geotechnical Grouting: Transportation Research Board Record 1104, National Research Council, Washington, DC, p. 1-42.

Warner, J., and Brown, D.R., 1974, Planning and Performing Compaction Grouting: Journal of the Geotechnical Engineering Division, American Society of Civil Engineers, vol. 100, No. GT 6, p. 653-666.

APPLICATION OF A GEOGRID-REINFORCED BUTTRESS IN SLOPE STABILIZATION

Gene Custenborder
Southland Geotechnical Consultants
1238 Greenfield Drive, Suite A
El Cajon, California 92021

INTRODUCTION

The application of geogrid-reinforced earth slopes in construction projects is becoming increasingly popular, especially in urban areas where the amount of land available for new development is very limited, and the value of available real estate is relatively high. The geogrid-reinforced buttress was unique and challenging because of the height and steepness of the slope, a limited construction area, and the complexity of the geology within the project area.

The subject slope is within a partially improved residential subdivision located in the Mount Soledad area of San Diego, California. Original mass grading for the subdivision was performed in 1980 and 1981. Grading was a typical southern California-style cut-and-fill operation creating relatively flat building pads on 168 lots for construction of single family homes. Improvements subsequent to grading included paved streets, retaining walls, underground utilities, and landscaping with associated irrigation system.

During the original project mass grading, the project geotechnical consultant identified adverse geologic conditions in the relatively steep natural slope along the southern site boundary. Adverse conditions included an ancient landslide and remolded clay beds within interbedded siltstone and claystone with bedding oriented out of slope. The geotechnical consultant recommended further investigation and analysis to evaluate stability of the slope and its effect on 21 lots located at the top of the slope. The consultant's preliminary stability analyses indicated that portions of the slope had calculated safety factors less than the required minimum safety factor of 1.5. The stability of the slope not only affected the lots at the top of the slope, but also the street on which the lots front. Since this street is the only access into the subdivision, the regulatory agency would not issue building permits for any lots within the subdivision until stability of the slope was increased to meet code requirements.

Subsequently, the developer requested that an investigation be performed by Leighton & Associates to evaluate the slope conditions, and provide recommendations to increase slope stability to an acceptable factor of safety.

FIELD INVESTIGATION AND GEOTECHNICAL CONDITIONS

The field investigation consisted of large-diameter borings and exploratory pits at the top of slope. Exploration was not allowed on the slope areas because they were considered open space, and because of complaints from the very influential and affluent property owners that had views of the area.

Geologically, the Mount Soledad area is an uplifted block of folded and faulted Late Cretaceous, Eocene, and Quaternary marine and non-marine sedimentary rocks. The project area is a south-facing slope situated on the eastern flank of a southeast-plunging anticline. The natural slope ranged from approximately 50 to 150 feet in height with a slope ratio that varied from about 3/4:1 to 2:1 (horizontal to vertical). Bedding attitudes measured during the investigation indicated that the sedimentary units dip generally 20 to 30 degrees to the south, out of slope.

Slope instability at the site was associated with the Eocene Ardath Shale, which consists of thinly bedded siltstone and claystone. The uplift and folding of the area has resulted in localized shearing along bedding planes at frequent intervals within the Ardath Shale. These bedding plane shears (commonly referred to as bedding plane faults) are typically 1/2 to 2 inches thick. The sheared zones have significantly lower strength than the unsheared materials. The low strength creates the potential for slope failures along these surfaces, especially where they daylight in an out-of-slope orientation.

SLOPE STABILIZATION ALTERNATIVES

The following three remedial design alternatives were considered to stabilize the slope:

> Alternative #1 included excavation to flatten the slope to a more stable configuration.
>
> Alternative #2 included construction of a gravity fill to abut the existing slope.
>
> Alternative #3 included excavating the existing materials and somehow reconditioning them to increase their strength properties, then recompacting them as a buttress fill.

Alternative #1, slope flattening, was found to be unacceptable because slope stability calculations indicated that the degree of flattening that could be attained within the limited project area could not provide a sufficient increase in stability to meet code requirements.

Alternative #2, construction of a gravity fill, was an economical and technically feasible solution. However, this alternative was found to be unacceptable to the regulatory agency because it would have involved filling of a canyon which was designated as open space.

Alternative #3, excavating the existing materials and increasing their strength properties, and recompacting them as a buttress fill was the remaining alternative. Evaluations were performed for the use of lime treatment of the existing soils to construct a lime-treated soil buttress, and construction of a geogrid-reinforced soil buttress.

Evaluation indicated that lime treatment of the existing soil could be an economical solution. However, construction of a lime-treated slope of this magnitude was considered somewhat insensitive from an environmental standpoint. Landscaping of the lime-treated materials comprising the slope face would have also been difficult. The construction period for a project of this magnitude would have been lengthy due to the lime/soil mixing process, and because each lift of compacted lime treated soil needs to cure overnight prior to placing additional lifts.

The evaluation of the use of a geogrid-reinforced soil buttress indicated that the geogrid was a cost-effective alternative. The construction period using geogrids was significantly shorter than the lime treatment alternative. It was also favorable from an environmental standpoint, and establishing landscape would be much easier than lime-treated soil.

SUMMARY OF BUTTRESS DESIGN AND CONSTRUCTION

Buttress design consisted of excavating a 3/4:1 (horizontal to vertical) temporary back cut to the designed keyway elevations. Buttress keyways had to be established within high-strength Eocene conglomerate or Cretaceous sandstone units underlying the Ardath Shale. Design key widths ranged from approximately 25 to 75 feet.

Tensar SR-3 geogrid was placed at selected design intervals within the compacted buttress fill. From the factory, SR-3 geogrid rolls are one meter in width and 30 meters long. They were placed with their length perpendicular to the slope face.

Tensar SS-1 geogrid was placed at one foot intervals between SR-3 layers to provide adequate surficial stability for the 1:1 finished slope face. From the factory, SS-1 geogrid rolls are one meter wide and 50 meters long. They were placed with their length parallel to the slope face.

A subdrain system consisting of crushed rock wrapped in filter fabric was placed at the heel of the buttress, and Tensar DC-1100 drainage composite (equivalent to chimney drains) was placed at selected intervals to provide drainage along the back cut.

The buttress design was based on geologic conditions projected from borings located at the top of slope. Because of faulting in two areas, the depth to the high-strength materials targeted for the keyway bottom was found to differ significantly from the proposed design. Consequently, redesign had to be performed in these areas once conditions were exposed during grading.

Construction was planned to proceed in phases to reduce the potential for failures of the excavated back cut. To provide early warning of back slope failures, five slope inclinometers were installed behind the backcut and monitored at regular intervals throughout the construction process. In addition, a monitoring program of the existing improvements at the top of slope was established. No appreciable movement was detected by the monitoring program during construction.

POST-CONSTRUCTION INSTRUMENTATION PROGRAM

An instrumentation program was designed and implemented to monitor performance of the completed buttress. The program consisted of two vertical slope inclinometers, 50 multiple foil-type resistance strain gauges attached to the geogrids, subsurface settlement monitoring using a magnetic probe and sensor rings attached to the slope inclinometers, and surface settlement monuments. Figure 1 is a representation of the instrumention layout.

The results of strain gauge readings indicate a maximum strain of 0.4 percent and an average strain of less than 0.2 percent. These values are well within the design tolerable strain of 2 percent. The results also indicate that most of the strain occurred during buttress construction. Figure 2 presents data from the strain gauge measurements.

Readings of the lower slope inclinometer indicate approximately 1.2 inch of deflection, most of which was caused by surcharging during construction. Readings of the upper slope inclinometer indicate approximately 0.6 inch of deflection which apparently represents post-grading adjustments. Figure 3 presents data from slope inclinometer monitoring.

Settlement on the order of 2 inches over an 11-day period was measured during construction. This settlement is associated with the lateral deflection detected in the lower vertical slope inclinometer during the same period and is believed to be caused by construction activities. A maximum of approximately 1.25 inch of settlement was detected in the upper device. Figure 4 presents data from subsurface settlement monitoring. A maximum of approximately 0.3 inch of post-construction settlement was indicated in surveys of surface monuments installed at the top of the buttress after completion of construction.

CONCLUSION

This project demonstrates the successful application of geogrid-reinforced soil buttress construction. Unstable geologic conditions were stabilized without adverse environmental effects. The total amount of fill placed was approximately 180,000 cubic yards. Approximately 150,000 square yards of

Tensar SR-3, and 40,000 square yards of Tensar SS-1 geogrids were utilized. Total cost for construction of the buttress was approximately $2.1 million. The estimated cost of the lime treated buttress was approximately $3 million.

ACKNOWLEDGEMENTS

I wish to acknowledge the following for their association with the successful completion of the geogrid project: Daniel Chu, Project Engineer, Taipei, Taiwan; Laguna Bernardo Management Company, San Diego, California; Leighton and Associates, Inc., Irvine, California; Sukut Construction, Inc, Santa Ana, California.

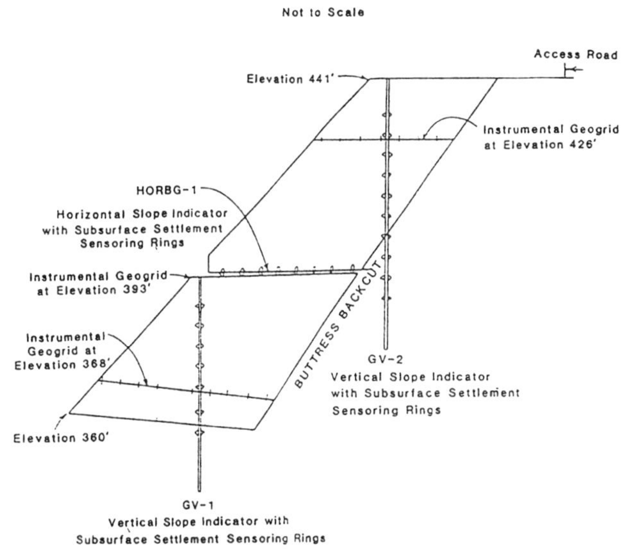

Figure 1. Instrumentation layout

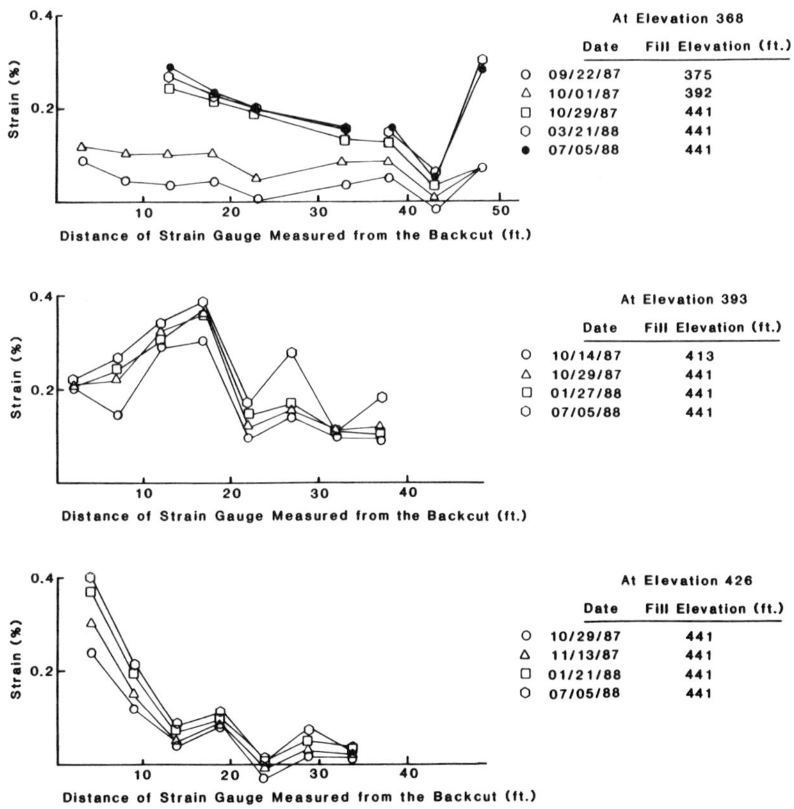

Figure 2. Measured strains from strain gauges

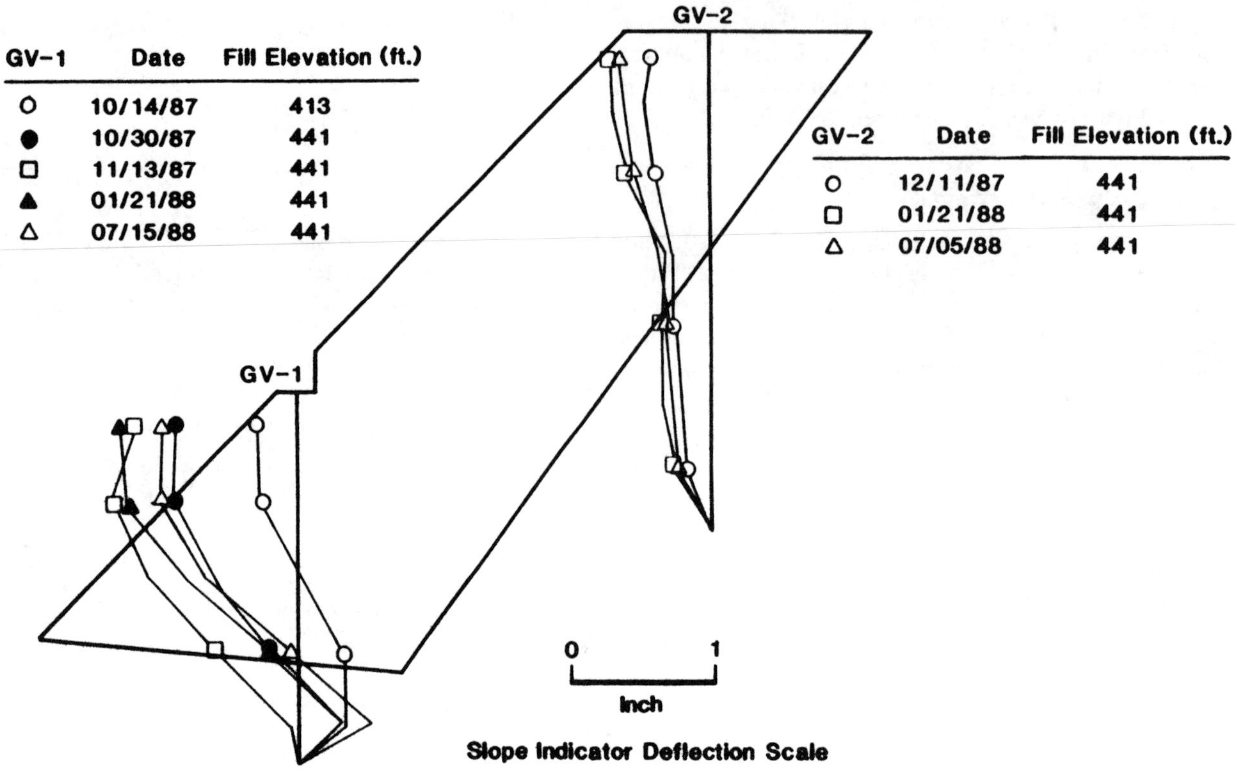

Figure 3. Measured lateral deflections of slope indications GV-1 and GV-2

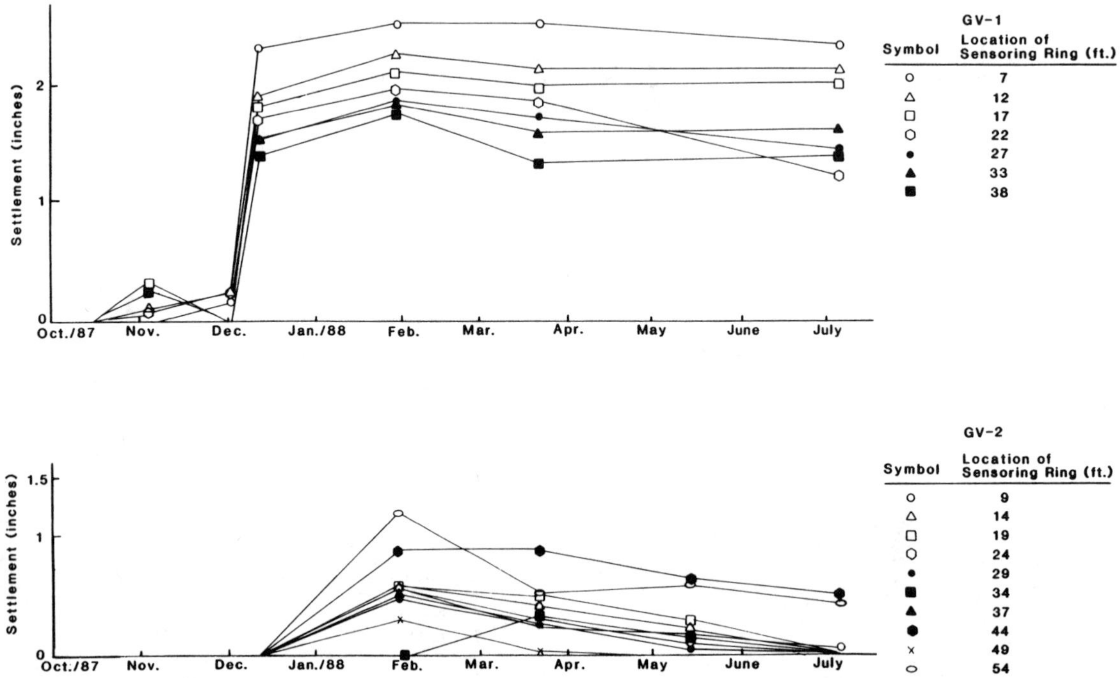

Figure 4. Measured subsurface settlement of sensoring rings at borings

PALM SPRINGS AERIAL TRAM CONSTRUCTION

Russell G. Hood
retired, former Partner
Geotechnical Consultants, Inc.,
Glendale Ca

The Palm Springs Aerial Tramway is located in Chino Canyon just north of Palm Springs, California. It starts at an elevation of about 2600 feet above sea level and rises one mile vertically for two miles horizontally where it terminates at an elevation of about 8500 feet.

The Tramway was built across some of the most difficult terrain over which any major construction project was ever attempted. The lower portion of the Tramway is founded on granite gneiss with typical blocky jointing. Vertical slopes of 500 feet high are very common along the alinement of the Tramway.

Accessibility for construction of the Palm Springs Tramway presented a staggering problem. The entire facility with the exception of the first tower and the drawworks at the bottom, were built from material hauled by helicopter. The contractor used five Bell G-3 supercharged helicopters to transport building materials, workmen, food, and tools over the various areas of the proposed Tramway. To give some indication of the cost of construction, concrete structures at the top of the Tramway were built with concrete that cost approximately $275 per cubic yard. The cost of the Tramway was approximately $8,500,000 and considering the exceptionally rugged terrain, the safety record was near perfect. There were no fatalities on the job, and very few serious injuries.

The facility consists of a drawworks at the lower end; five towers, ranging form about 50 feet high to over 200 feet high; and an anchor works at the top. Two cars, each capable of holding 80 persons, hang suspended from four cables approximately 2 inches in diameter, and one car goes up as the other comes down. Each car is suspended on two such cables with each cable designed with a factor of safety of 3. This factor of safety includes cars fully loaded, a gale blowing over 200 kilometers per hour, an inch of ice on each cable, an earthquake of magnitude of 6 or 7, and an instant deceleration of the cars. All foundations were designed to the same criteria.

The towers were built leaning downslope, some as much as 26 degrees off vertical, in order to give as near a perpendicular support to the cables as possible.

As the design for the towers and the catenaries for the cables were being performed in Switzerland, detailed and precise surveying was necessary to assure that the design fit the topography. Due to the problems presented by exceedingly steep terrain, it was impossible to get any tower base which was more than 40 feet square. Therefore, special design was necessary for the taller towers to assure stability with such a narrow base.

Plane-table surveys with a scale of 1 inch equals 5 feet were made of each tower location. From these, cross sections were made through each side of the proposed tower base, and one through the center. The cross sections showed all the jointing and fracturing of the rock in a three-dimensional presentation. The towers may exert as much as 800,000 pounds in uplift or compressive force on one leg and the maps and cross sections were utilized to make a layout of rock-bolting whereby enough rock mass could be bolted together to assure resistance of these tensional and compressional forces. Rock bolts used in the construction were up to 25 feet long and as much as 200 cubic yards of granitic rock were bolted together to act as a unit to prevent uplift on the leg of the tower. Rock bolts were placed in both a horizontal and vertical direction to assure proper binding of a rectangular block to be tied to a tower leg. Approximately 300 rock bolts and anchor bolts were used in the foundations of the five towers, and each of these bolt locations had to be spotted in the field and the physical location tied to a table showing bolt length, direction, inclination and intended use. This necessitated going over the sides of the cliffs on a scaler's rope and physically marking the exact location of each proposed bolt hole.

Some of the major geological features and conditions that had to be taken in to consideration on this job other than foundation design were:

1. Protection of the towers against <u>rock fall and rock avalanches</u>. Narrow steep-wall gorges occurred throughout the Tramway route. Some gorges have a gradient exceeding 100 percent. Rocks which are broken loose by weathering and frost action slide down these gorges or chutes and tend to wedge in the narrows. Hundreds of tons of boulders then pile up behind these constructions which are held in place by a few key stones. When the stress builds up enough or when the key stone is jarred loose by rolling boulders, the whole mass is released down the gorge in a most spectacular avalanche. During the four to five months of actual geological studies in the field, at least one such avalanche could be heard occurring somewhere within the canyon. Where such rock chutes might jeopardize a tower, special grading and realinement of gorges were necessary.

2. <u>Frost action</u>. The winter weather in Chino Canyon is especially conducive to mechanical weathering of the rocks. Freezing nights and relatively warm days give a daily cycle of freeze and thaw. Water and snow trapped in the joint systems of the granitic rocks work as powerful jacks under the expansion of water changing to ice. This mechanical weathering is very evident, especially in the higher elevations of the Tramway where huge piles of al-

most rectangular boulders cover most of the terrain. To prevent mechanical weathering, all the tower sites were grouted and covered with a blanket of impervious material, usually gunite, to prevent moisture infiltration.

3. Seismic design was based on intensities of earthquakes which had been recorded within the area. The active San Andreas and San Jacinto faults are not too many miles from the Tramway. Magnitudes ranging from 7 to 9 on the Richter Scale were considered in the design of the Tramway facilities.

4. Scour of existing stream channels in Chino Canyon was considered because cloudburst-type precipitation is common in the Palm Springs area. Special precautions were taken by either re-re-routing stream channels or lining them where they were adjacent to any of the structures of the Tramway.

Considering all the problems of alpine terrain, a wide range of climatic conditions, limited areas of use for foundations, and the overseas communication problems with the designer, the planning and construction of the Palm Springs Tramway was truly a challenge for the engineering geologist. His ability to equate geological data to engineering needs played an important role in the teamwork necessary to safely and economically construct such a project.

Reprinted from Engineering Geology in Southern California, 1966, AEG Special publication (out of print).

Figure 1. Palm Springs aerial tram

Chapter **7**

Coastal Erosion

Photo: Large storm waves have removed the protecting beach and are attacking beach homes, Imperial Beach, San Diego County, California

Coastal Erosion

Coastal Erosion in Southern California - An Overview
Bernard W. Pipkin, Hugh S. Robertson and Randall Mills

Applications of Fourier Grain-shape Analysis in the Search for More Accurate Sediment Budgets, Orange and San Diego Counties, California
Robert H. Osborne and Chia-Chen Yeh

Editors Note: Beach front property is among the most desirable and expensive in southern California. When a movie or rock star buys a six million dollar house along the Malibu coast, he or she does not expect to see it partially washed away by a storm surge. The first paper summarizes the status of coastal erosion in southern California; and Osborne and Yeh discuss their work on grain-shape analysis that has yielded new insight into sources of beach sediment.

COASTAL EROSION IN SOUTHERN CALIFORNIA — AN OVERVIEW

Bernard W. Pipkin
Department of Geological Sciences
University of Southern California
Los Angeles, CA 90089-0740

Hugh S. Robertson and
Randall S. Mills
Robertson Geotechnical, Inc.
2500 Townsgate Road (Suite E)
Westlake, CA 91361

CONTENTS

INTRODUCTION

According to the latest U.S. Census population growth has been largest in our coastal states. Stated more succinctly, 50% of the U.S. population and 60% of California's inhabitants live within a one-hour drive of the beach. The Atlantic, Pacific, and Gulf coastal zones are exposed to similar but unique geologic hazards and being developed at an unprecedented rate. Favorable climate, scenic beauty, and proximity to urban centers attracts new residents and has escalated real estate values. Unfortunately, development has been dictated by short-term economic and marketing strategies, rather than long-term planning and environmental considerations. Natural and human intervention in the coastal environment has modified it, mostly with negative impact. Wetlands have been filled or made into marinas, shoreline engineering works have caused siltation and erosion, and lack of early controls led to overcrowded coastal communities. In the past 20 years, planning and control of growth in California has been more aggressive, controlled principally by the Coastal Zone Initiative mandated by the voters.

The study area extends from Point Conception, in Santa Barbara County, to the Mexican border (Fig. 1). It involves about 325 miles of shoreline and 12,000 square miles of watershed, including the Tijuana River drainage basin in Mexico. The shoreline between Santa Barbara and San Diego is about half in private ownership and consists mostly of sandy beaches interrupted by short segments of rocky headlands and cliffs. Because so much of the coastline is suitable for public recreation, there are many State, County, and City beaches. In addition, there are several large military installations, two principal harbor complexes at Los Angeles and San Diego, and many small-craft harbors. Table 1 presents a shoreline inventory and summary of land ownership in each of the 5 coastal counties in this study. Within the study area are six littoral cells (Fig. 2), defined as a segment of coastline that contains a complete cycle of littoral sedimentation including sources, transport pathways and sediment sinks (Inman, 1985). Cells usually terminate at rocky headlands, submarine canyons or both, and the model assumes there is little if any sediment exchange between cells (Emery, 1960).

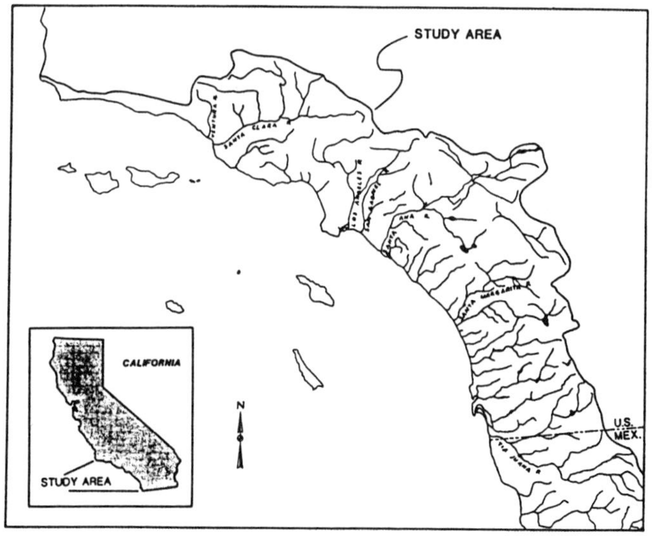

Figure 1. Southern California study area from Point Conception to the Mexican Border.

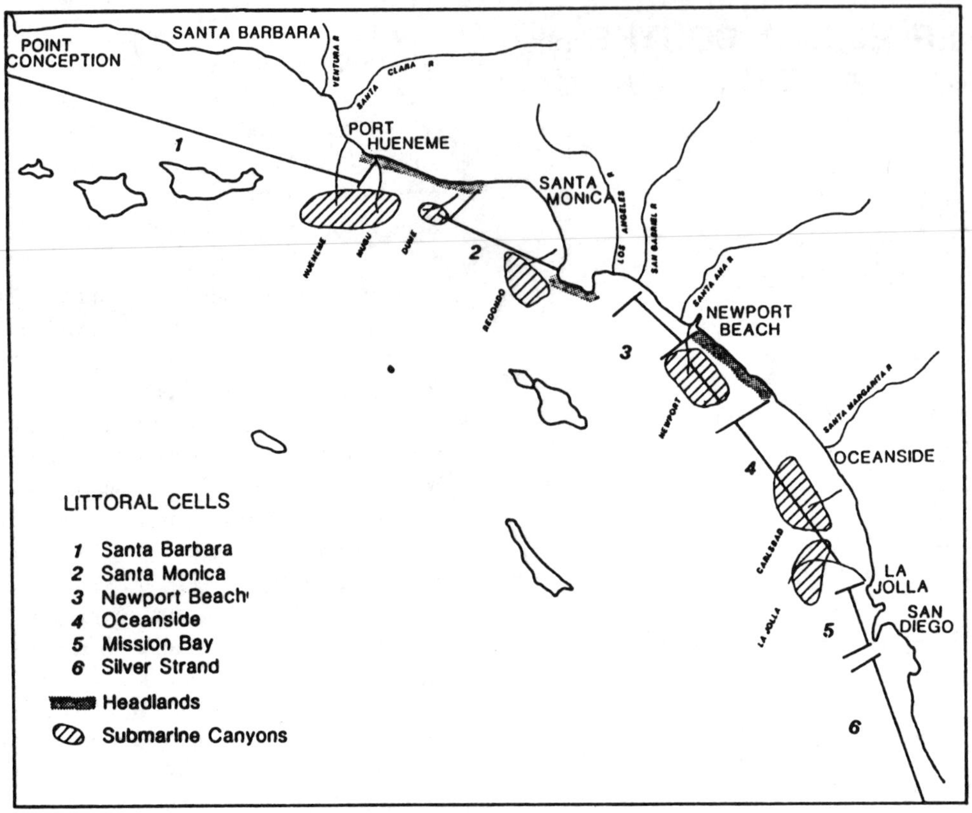

Figure 2. Littoral cells of southern California. (various sources)

Heavy mineral and provenance studies have cast doubt on the self-contained nature of these cells, and recent grain-shape studies suggest that some cells are quite"leaky" (see Osborne, this volume).

Engineers have long recognized the risks associated with development in the dynamic coastal environment. Beach erosion causes loss of recreational areas, property devaluation and increased potential for damage to adjacent land. Thus, good engineering design needs to evaluate long-term and short-term shoreline changes, and the probability and impact of major storm events. As urbanization of the coast expands, so does the development of less stable land masses. The engineering geologist is challenged to recommend means of safely improving what little private coastal land remains, and to repair damage resulting from past mistakes. Many large coastal-engineering projects in southern California were initiated to mitigate or slow coastal erosion. In retrospect, these were rarely solutions but were only temporary protective measures requiring constant monitoring and re-evaluation.

BEACHES AND COASTAL PROCESSES

Long-term natural shoreline changes are the product of (1) sea-level changes, (2) power of the coastal erosion processes acting on a stretch of shore, (3) geomorphology of the shoreline, and (4) sediment supply (Dolan and Kimball, 1988; Henry, Dean, and Olsen, 1987). Man-induced erosion includes interception of littoral drift by jetties and breakwaters, and the arresting of sediment sources by dams, paving, and armoring of headlands. All of the latter are the consequence of urbanization and, to a certain extent, have predictable consequences. Natural events, on the other hand, are much more difficult to predict and mitigate. Long-term changes in coastal erosion may also be difficult to recognize.

Beach erosion processes and design of protective works are covered comprehensively in the U.S. Army Corps of Engineers Shore Protection Manual (1984), the so-called "bible" of coastal engineers. Primers for beach erosion in southern California include Handin (1951), Johnson, (1956), Shepard (1950), and a series of U.S. Congress House Documents and Corps of Engineers publications spanning the period 1940-1991 that treat in detail beach segments from Santa Barbara to San Diego (see Bibliography).

Beaches erode when sediment supplied to a site is less than the wave energy available to move it. Sand starvation may be natural owing to drought or severe storms, or may be due to man-made obstructions to the littoral flow of sediment or sediment sources. Rates of littoral drift based upon measured transects at coastal obstruction are on the order of 10^5 cubic yards per year (cy/y). Using the rule of thumb that a cubic yard of beach material is equivalent to one square foot of beach area, one can see that several acres of beach front are continuously on the move in the littoral stream (Table 2).

Natural Obstructions to Littoral Drift

Twenty nine submarine canyons have been mapped and identified on the continental shelf between Port Hueneme and San Diego. Of these, six extend to within a few hundred yards of the shoreline and act as scavengers of littoral drift. Where canyons are actively intercepting sand there is a significant differences in beach width and drift rates between upcoast and downcoast sides of the canyon head. It has been estimated that Scripps Canyon, in the La Jolla area, may remove 100,000 to 150,000 cy of sand annually from the littoral stream (Shepard and Inman, 1951). Hueneme, Mugu, Dume, Redondo, and Newport Canyons have also been identified as conduits that remove beach sand from the littoral system that eventually finds its way into deep offshore basins by means of turbidity currents or sand flows.

Headlands also compartmentalize the shoreline and act as barriers to littoral drift, so that each segment or cell is self contained and terminates in a headland or submarine canyon. The significance is that canyons or barriers within the cells can potentially affect the shoreline downcoast, but have no affect on adjacent cells. For example, at Redondo Beach, an attached breakwater diverts sand into the canyon head which is very close to shore (Haner, 1971). Redondo Beach is at the end of a littoral cell and should be very wide, however, it underwent severe erosion in the 1960's and was restored only after artificial placement of 2,000,000 cy of sand dredged from offshore sources. The Palos Verdes Hills (headland) and Redondo Submarine Canyon are believed to be the most effective barriers to littoral drift in the study area. Longshore drift rates, canyons, and headlands are shown in Figure 3.

Artificial Obstructions to Littoral Drift

Engineered structures in the study area that have contributed directly or indirectly to beach erosion are listed in Table 3. The City of Santa Barbara was the first to use artificial bypassing of their breakwater by dredging sand from the area of accretion upcurrent to the downcurrent area of erosion. Although the dredge is a permanent feature of the harbor, it does not operate continuously, and some newer structures have a bypass system built into their design. Not all structures have been detrimental, owing in part to their location relative to point sources of beach sand (rivers), or because they are permeable and allow some sand to leak through the structure. Bypassing is usually accomplished by dredging but some dry hauling is done by truck. At the Channel Island Marina in Ventura County a sand trap was excavated upcurrent from the jetty and in the lee of the detached (parallel) breakwater. This facilitates storage of sand that is periodically pumped to feeder beaches downcurrent from the Hueneme submarine canyon.

Changes in Sea Level

Astronomical Tides. Mixed tides, that is two high tides and two low tides of unequal height, characterize the west coast

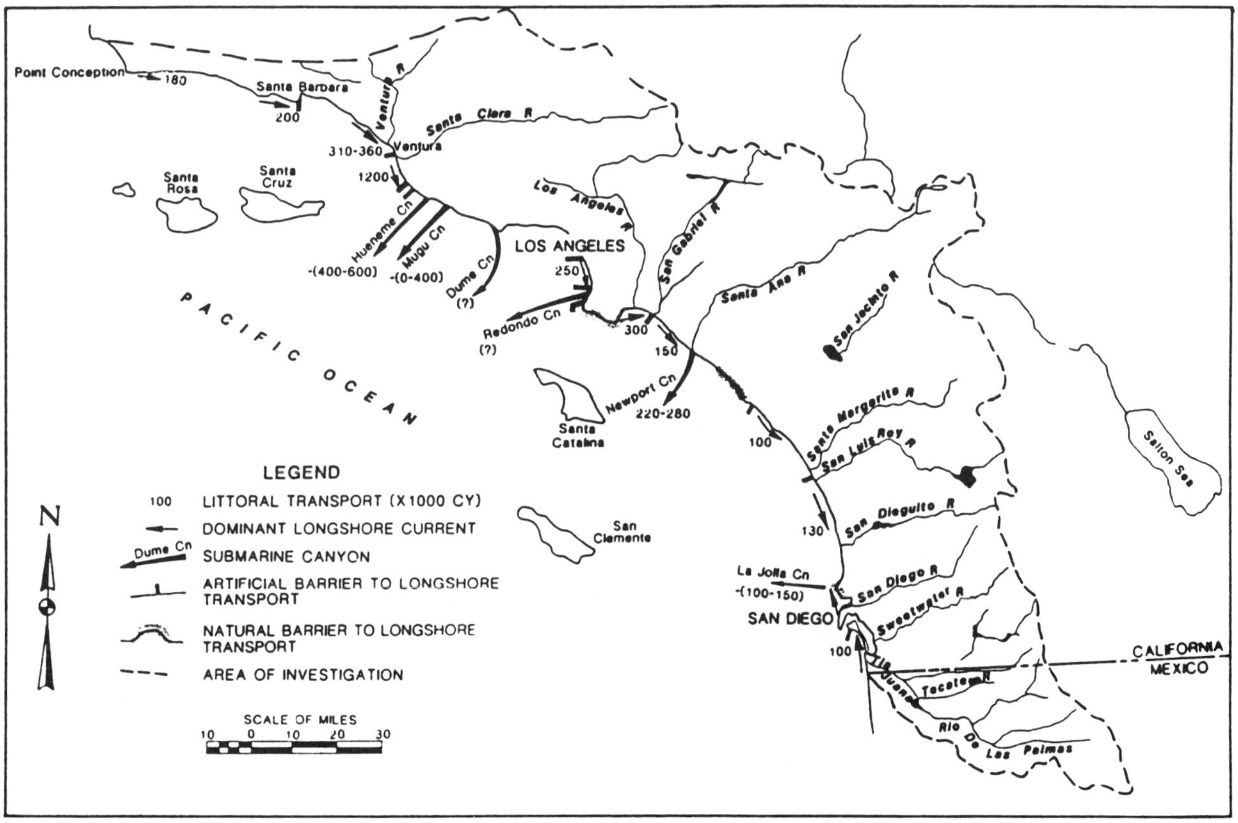

Figure 3. Rates and directions of littoral drift and inferred losses to submarine canyons. (various sources)

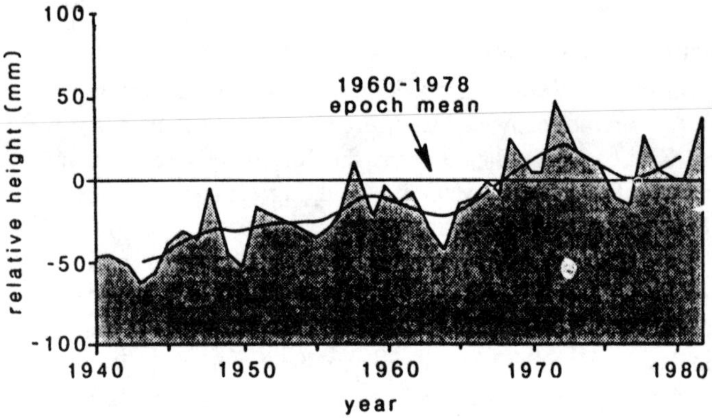

Figure 4 Average annual sea level rise and smoothed curve for United States, excluding Alaska and Hawaii, from tidal gauge records. (Hicks, l983 in Dolan and Kimball, l985)

of the United States. Mean tidal range varies with location and coastal configuration, but is approximately 5 feet for Los Angeles, Orange, and San Diego Counties. The maximum tidal range is close to l0 ft (-2.5 ft to +7.5 feet MLLW datum), and occurs when the sun-moon-earth system is aligned at spring tides. The highest high tides occur when the moon is at perigee, and perigean spring tides compounded by storm surge have been responsible for the l00 most severe coastal storms between 1635 and l976 (Wood, l976). Storm surge in southern California may be roughly approximated as equal to l0 percent of the highest waves anticipated. Assuming a 25-foot wave and perigean spring tides, stillwater level can rise as high as +l0 ft MLLW, or +12.5 ft mean sea level (MSL). This does not take into account the resultant wave runup to even higher elevations and overtopping of protective works. Needless to say, high tides plus storm surge are conditions ripe for extreme beach erosion and structural damage.

Global Warming and Sea Level. Radiocarbon dates on oyster reefs and other organic matter indicate sea level stood about 300 feet (l00 m) lower than present sea level 15,000 years before present (ybp). As Pleistocene glaciers melted, sea level rose rapidly, about 3 feet (1 m) per century (Emery and Uchupi, l972), reaching within a few feet of the a present sea level about 4,000 ybp. Sea level has risen more slowly during the past 2,000 years, and has totaled about .4 ft (15 cm) during the past century (Dolan and Kimball, l988). A recent sea-level rise curve for the United States is shown in Figure 4 (Hicks and others, l983).

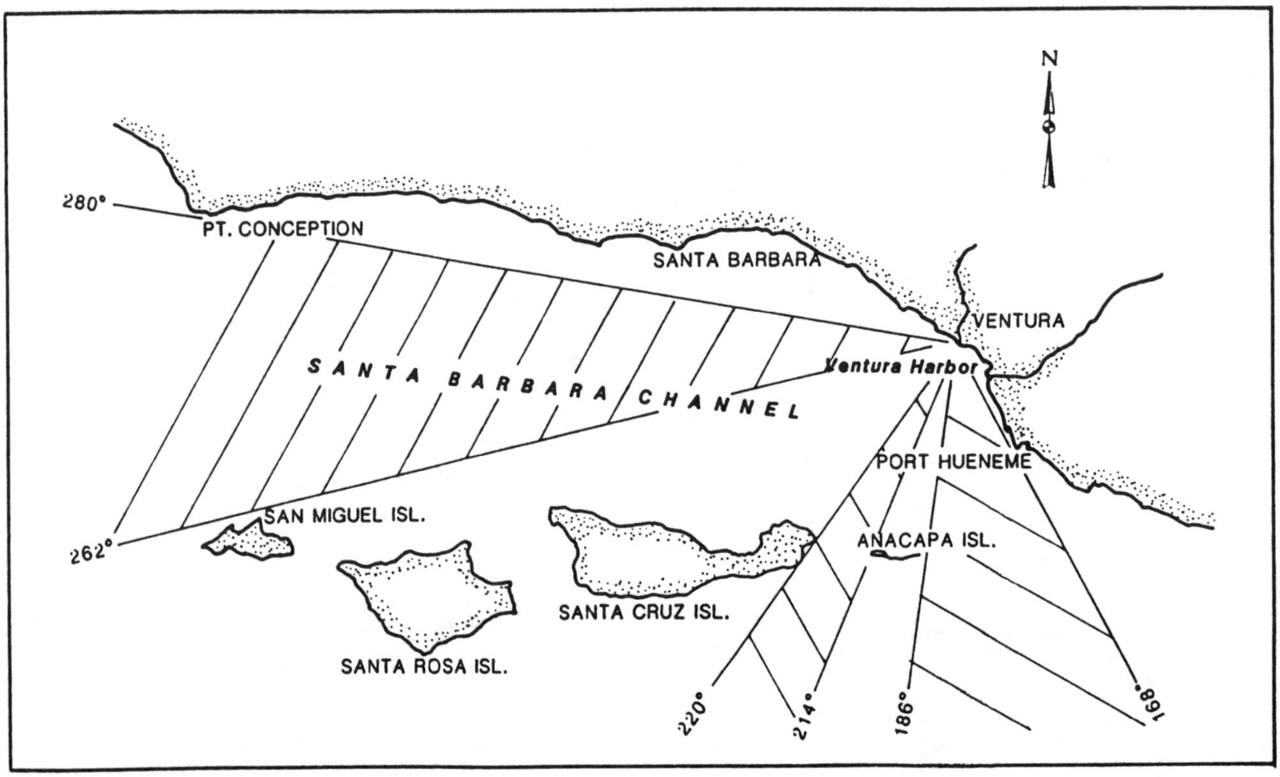

Figure 5 . Storm-wave exposure windows, Ventura Harbor.(Bottin, R., l991, Ventura Harbor, California, Design for Wave and Shoaling Protection; Waterways Exp. Station, U.S. Corps of Engineers)

Global warming due to trapping of long-wave back radiation from the earth by so-called greenhouse gases will increase glacial melting and cause sea level to rise even faster. Best estimates by the Environmental Protection Agency (EPA) using a 1983 model predicts a rise between 4.8 ft (144 cm) and 7 ft (217 cm) by the year 2100 (Hoffman and others, 1983). Many scientists disagree with these figures, however, there is little doubt that sea level is rising and will continue to do so. Tidal data from hundreds of gauges world wide indicate that the rise is by no means universal and is site specific (Emery and Aubrey, 1991). In tectonically active areas land may be rising (or sinking) faster than sea level is changing, thus yielding figures at odds with true eustatic changes. However, if the EPA study is correct, much of the most valuable beach-front property in southern California will be underwater. Approximate methods for estimating the long-term retreat of shorelines resulting from a long-term changes in mean sea level have been discussed by Griese and Aubrey (1987), and Weggel (1979).

Weather Patterns, El Nino, and Coastal Erosion

As noted previously, coastal erosion may be expected when major storms are combined with spring tides. Storms arrive in the southern California coast along four principal tracks, however, the waves actually striking the shoreline from these storms have limited "windows" due to shielding provided by offshore islands. Figure 5 shows these windows for Ventura Harbor as an example (Bottin, 1991). Winter storm waves approach from the west and northwest, whereas high south and southwest swell are associated with disturbances off Mexico or southern hemisphere storms. South swell occurs mostly in the late summer and produces long-period waves capable of producing major damage to south-facing beaches in the study area (Fig. 6).

Kuhn and Shepard (1984) discuss climatic changes and the impact of volcanic eruptions upon these changes. Gases, especially SO_2, produce reflective aerosols and coatings on volcanic dust. The aerosols increase the earth's albedo and may cause cooling on a hemispheric scale. Their study suggests that weather deteriorates and stormy conditions prevail after large eruptions. They cite as evidence the destructive coastal storms of March, 1983 following the eruption of El Chichon Volcano on the Yucatan Peninsula. The storm coincided with spring tides and resulted in extensive property damage along the entire coast from San Francisco to San Diego (Denison and Robertson, 1985; Griggs and Savoy, 1985). Mount Pinatubo in the Philippines erupted in June, 1991, and ranks as the third largest eruption of this century, and the largest in terms of SO_2 emissions. Its impact on climate awaits demonstration.

There is as yet no established link between volcanic eruptions and global climatic perturbations such as El Nino, which occurred in 1972-73, 1976-77, and 1982-83. Knowledge gained from studying the "super" El Nino of 1983 did establish the southern oscillation in atmospheric pressure as the trigger for the event. The southern oscillation is the sudden reduction of the atmospheric pressure differential that drives the southeasterly trade winds between 30° south latitude and the equator. This changing oceanic circulation, formerly called El Nino, is now known as an ENSO (El Nino/ Southern Oscillation) event. The 1983 ENSO resulted in floods and coastal storms in California, Tahiti, Ecuador, and Peru, along with drought and famine in Australia, Africa, and Indonesia.

Source of Beach Sediment in Southern California

Beach sands are terrigenous and derived mostly from coastal watersheds. Heavy-mineral studies of beach sands show that the mineralogy of the northern cells is controlled by the sedimentary Transverse Range provenance, whereas the mineralogy of the southern-cells is dominated by the dioritic Peninsular Ranges provenance (Rice and others, 1976). In other words, nearshore and beach sediments within cells reflect the mineralogy of the rivers draining into that cell. Because of the interrelationship which exists between precipitation, runoff, and sediment production, sediment delivery to beaches varies greatly from year to year. For planning purposes, sediment-rating curves were generated for major watersheds in the study area (State of California, Dept. Water Resources, 1969). These curves reflect the relationship between bed-material (sand) load of these streams and discharge (Fig. 7). These relationships are useful for long-term planning purposes involving artificial beach nourishment. Unfortunately, protracted drought has limited their usefulness.

Urbanization and flood-control structures have severely curtailed delivery of sediment to the littoral system. Table 4 is an inventory by counties of major watersheds and control structures in southern California. For example, the largest watershed in southern California, the Santa Ana River basin, is 80 percent controlled and thus has a limited sand-storage area. It appears that natural source areas of beach sediment are diminishing and in the future we will be required to rely more heavily on artificial nourishment as specific beaches are subject to critical erosion.

SHORELINE CHANGES

Advances and retreats of the shoreline between 1852-1982, from Palos Verdes to Mexico, have been documented using aerial photographs, field surveys, and historical maps (U.S. Army, 1987). A series of 15 Shoreline Movement Maps were prepared showing the interception of the mean high water (MHW) with the land. The shorelines were based upon field surveys in 1852, 1887, 1916-1917, 1933, 1960, and 1972. Aerial photographs were used to establish the 1972 and 1982 MHW line. Errors in data reduction and seasonal fluctuations were the elements of uncertainty in the investigation. The report concludes, surprisingly, that the 19th century

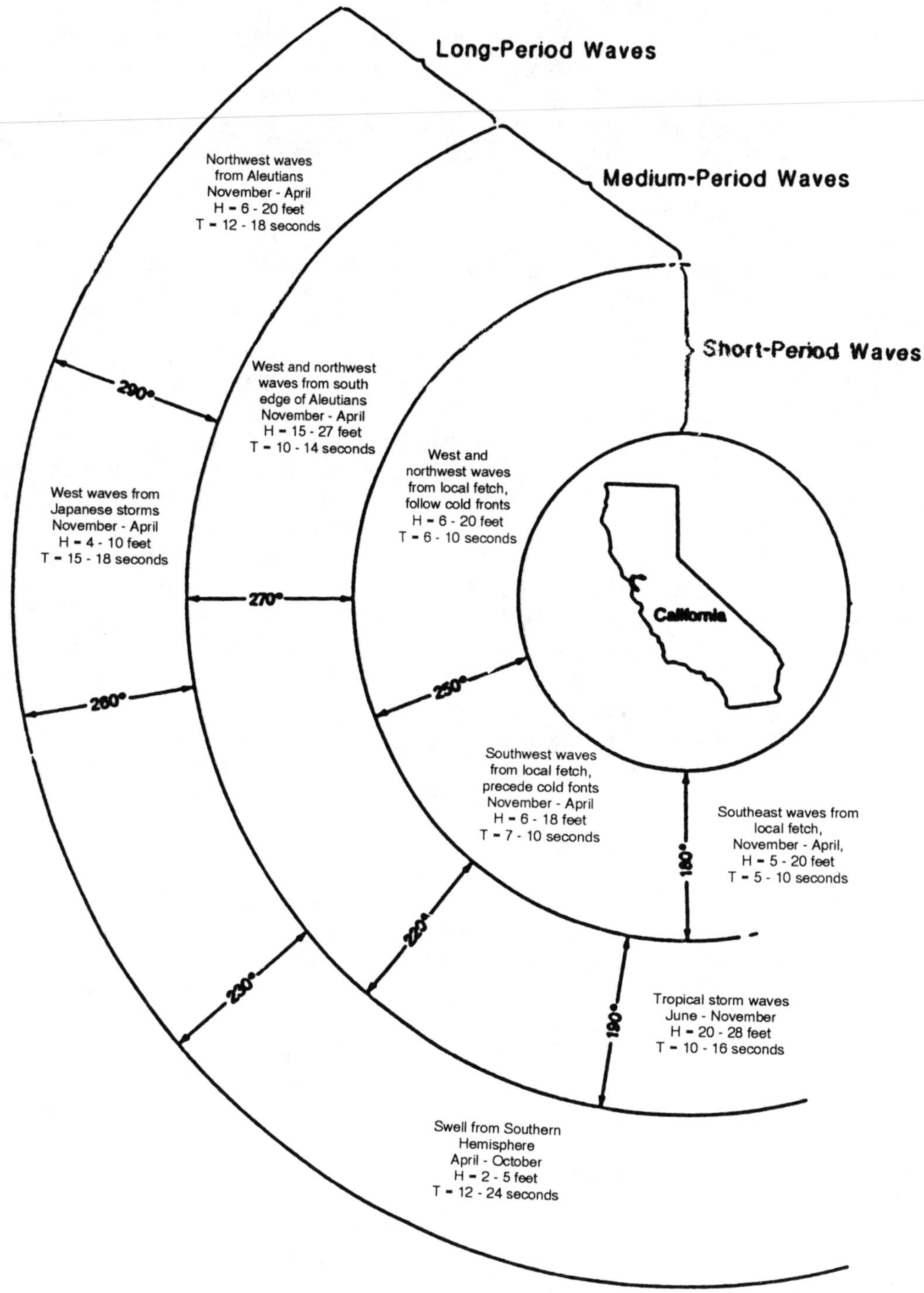

Figure 6. Wave characteristics from different source areas for California (After Inman, Beach Erosion Board Tech. Memo.)

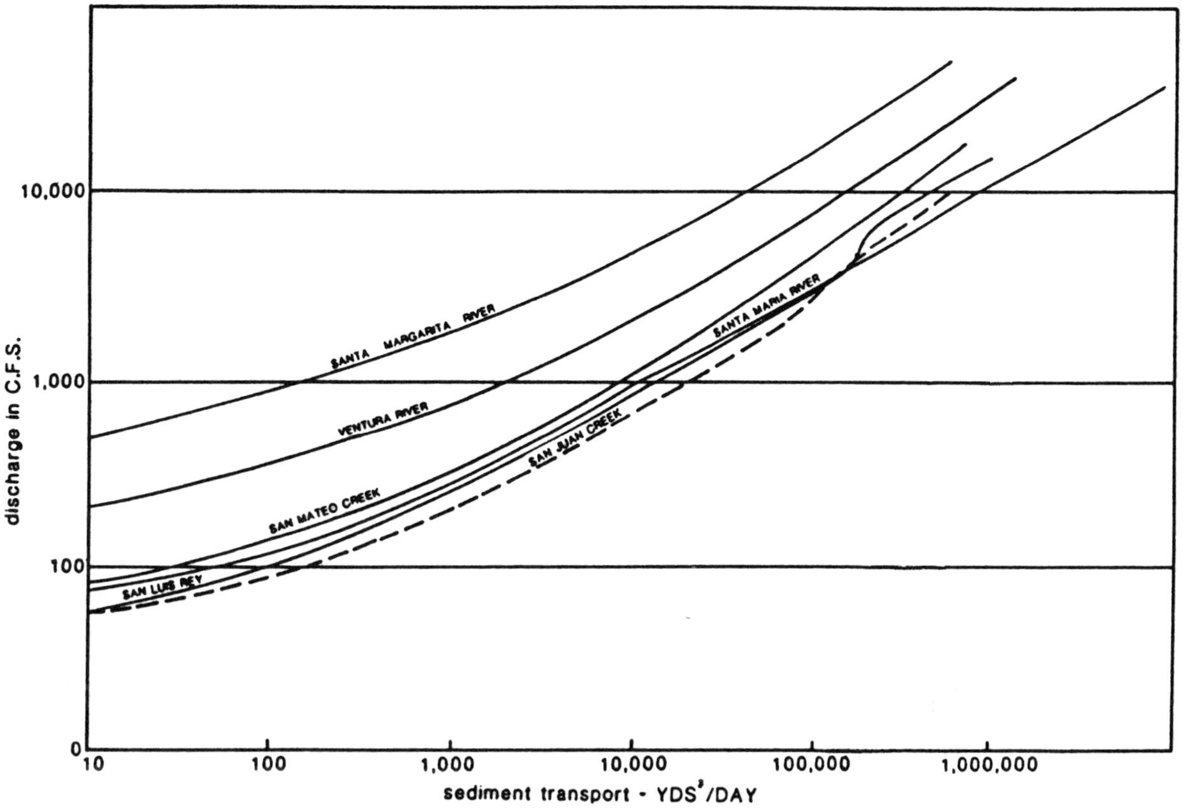

Figure 7. Calculated relationship between discharge and sediment transport using Einstein bed-material load equations. (State of California, l969)

shorelines along most of the coast were landward of those in the later part of this century. This is due to large beach replenishment projects which provided beach fill using material excavated for reasons other than beach replenishment. Such large projects would probably not be possible today because of cost and environmental constraints. As one would expect, breakwaters, jetties, piers, and groins have had a significant effect on advance and retreat of the shoreline, and artificial nourishment and sand bypassing have ameliorated some of the impact of projecting coastal structures. As well as acting like dams, jetties and breakwaters can deflect sand into submarine canyons at the expense of the adjacent shoreline.

Pocket beaches in the study area showed changes unrelated to the distance the adjacent headlands project seaward, the length or orientation of the beach, and the presence or absence of a stream carrying sand to the pocket beach(U.S. Army, l987).

Many large streams in San Diego county flow into lagoons before reaching the ocean. The lagoons act as sediment traps that support marsh vegetation. The study notes that filled areas of lagoons decrease in a northerly direction and may be due to differential land subsidence and/or emergence, with a downward tilt toward the north. Large coastal landslides cause the shoreline to advance which may be seen at La Jolla, Camp Pendleton, and along the Palos Verdes Hills.

STATUS OF THE COASTLINE

The study-area shoreline has been divided into convenient physiographic and littoral units to facilitate discussion of their present condition.

The Santa Barbara Littoral Cell-Point Conception to Point Mugu

Point Conception to Gaviota State Beach. Some fine sand makes its way around Point Conception (Trask, l952), but the stretch is dependent largely upon stream sources on the south side of the Santa Ynez Mountains. Gaviota Creek provides major sand volumes of 300,000 to 350,000 cy/y, which is the same order of magnitude as the volume of sand dredged from Santa Barbara Harbor to the south. The shoreline between Point Conception and Gaviota is relatively straight with narrow beaches and seacliffs ranging from 15-60 ft high. Annual beach cycles of winter erosion and summer building are conspicuous here, with winter-beach profiles exposing cobbles usually covered with sand by early summer. Miocene shales and siltstones support the cliffs and exhibit recessions rates of 2-3 in/y (Griggs and Savoy, l985).

Gaviota State Beach to Rincon Point. This portion of the coast supports cliffs as much as 200-feet high and is backed by seaward-dipping sedimentary rocks of the Santa Ynez Range. Littoral drift is from west to east as south swell is blocked by the northern Channel Islands. The Santa Barbara Breakwater was completed in l929 as a recreational boat facility, and thus began a classic and costly case history of accretion and erosion resulting from man's interdiction into the littoral zone. By 1933 the updrift side of the breakwater had completely filled with sand which began moving along the seaward arm of the structure into the harbor entrance. Severe erosion took place on beaches immediately to the southeast. About 10 miles to the southeast, the beach at Sandy Point was eroded 240 ft inland with $2 million in property damage. Dredging harbor fill to starved beaches downdrift began in 1935, and by 1938 three miles of adjacent beaches were restored. By 1941 sand reappeared on rocky eroded stretches and continued dredging has resulted in a semblance of beach stability for Santa Barbara County. However, sandy beaches in this stretch were badly eroded by the ENSO storms of 1982-83, as were harbor facilities and the waterfront.

Rincon Point to Point Mugu This stretch of shoreline may be divided into two units, the Rincon coast from Rincon Point to the Ventura River, and the Oxnard Coast between the Ventura River and Point Mugu. The stretch terminates at the Hueneme and Mugu Submarine Canyons which intercept much of the drift material. The Santa Clara River is a major supplier of sediment and large volumes of sand-size material is stored in its delta. The heavy rain years of 1969, 1978, 1980, and 1983 produced flows that yielded significant volumes of sand to the littoral stream, as much as 8,000,000 cy in 1969 alone.

The Oxnard coast was originally sandy beaches backed by low dunes, marshes, and lagoons. This area has been so modified by human activity that severe erosion problems have arisen. Levee construction on the Ventura River reduced the volume of sand contributed by the watershed between 1948 and 1959, which was also a period of protracted drought. This combination resulted in a 300-foot recession of the shoreline between the Ventura and Santa Clara Rivers. A field of 7 groins were constructed at San Buenaventura Beach between 1962 and 1967. The beach is presently maintained by sand dredged from the Ventura Marina to the southeast. The present condition on this coastline is in contrast to historic records which suggest that between 1855 and 1933 the shoreline was growing seaward at a rate of 16.5 ft/y. Since the 1960's Oxnard Shores, Mandalay Beach, and Hollywood-by-the-Sea, all downdrift from the Santa Clara River mouth, have periodically undergone critical erosion. Major military construction at Port Hueneme was completed in 1940 and was converted to Navy use in 1942. In 1948, severe erosion progressed past the Ormond Beach power-generating station, 5 miles downcurrent from the south jetty of Port Hueneme. Beach fill and rock revetments had little deterent effect on erosion. Dunes were eroded and breached in an area which had been relatively stable since 1856. Because of this, Channel Islands Harbor was created north of Port Hueneme to trap sand which is bypassed southeastward around the jetties to the eroding beaches (Fig. 8).

Figure 8. Channel Islands Harbor looking southeast toward Port Hueneme and Point Mugu. Sand trapped behind parallel breakwater is periodically pumped to feeder beaches downcurrent.

The Santa Monica Mountains and the Santa Monica Littoral Cell

The Malibu Coast from Point Mugu to Santa Ynez Canyon. The geology of this stretch is dominated by the Malibu Fault which has produced a highly contorted sequence of Cretaceous and Tertiary sedimentary rocks parallel to the coastline (see Drum, this volume). The faulted rocks and complex structures have created unstable conditions resulting in large landslides (see Lownes and Evans, this volume). The fault strikes offshore at the Los Angeles-Ventura County line, and Point Mugu northwest of the fault is supported by both sedimentary and volcanic rocks. Malibu Creek, formerly an important sediment source to the Malibu area, has been controlled by dams and artificial lakes. Cliff erosion, also a source of sediment, has been curtailed by seawalls and rock revetments built to protect cliff-top development. Construction of Pacific Coast Highway in the 1920's, allowed development and curtailed sediment derived from mass wasting of seacliffs.

The Pacific Palisades to Redondo Beach. The Pacific Palisades are at the north end of the Santa Monica coast and are in the transition zone between the Santa Monica Mountains and the Los Angeles Basin. They are in part underlain by deformed and fractured Tertiary rocks, and in part by younger fanglomerates which form unstable and easily eroded high cliffs adjacent to the Pacific Coast Highway. Low bluffs seaward of the highway are reveted and beaches

Figure 9 a. Ballona Creek and Venice prior to construction of Marina del Rey, but after the construction of the north harbor entrance jetty.

Figure 9 b. Marina del Rey in l981. Note difference in beach widths in two photos. (Photos courtesy L.A. County Deptartment of Beaches)

Figure 10. Sand dunes at Playa del Rey and El Segundo in 1947 with Ballona Creek at bottom of photo. A photograph of the same area today would show large facilities for sewage disposal and electrial generation cut into the dunes. (Courtesy Pat Merriam and Pacific Air Industries)

are generally narrow or nonexistent. Where beaches are wider and accessible, groin fields have been installed with varying degrees of success.

An offshore rubble-mound breakwater was built in Santa Monica in 1933. It created a wave shadow which weakened longshore currents and sand accreted behind the breakwater. Now sand must be periodically dredged from behind the breakwater to nourish beaches downcurrent and keep the harbor clear. It is estimated that 259,000 cy/y are deposited behind the breakwater which has deteriorated over the years and was severely damaged during the 1983 storms.

Residential development from Santa Monica to Playa del Rey has occurred in part on a sand spit deposited across the ancient estuary of the Los Angeles River, now called Ballona Creek. Early in this century, parts of the estuary were dredged to form a canal system for the City of Venice, similar to those in Italy. In addition, a detached breakwater was built so close to shore that one can now walk dry-shod to the structure. Much of the canal area and back bay of the estuary have been converted into the largest man-made pleasure craft harbor in the world, Marina del Rey. Dredged in the early 1960's, the marina experienced wave-generated seiche and surge problems. Mitigation required construction of a detached breakwater across the mouth of the entrance jetties to dissipate wave energy. In turn, the decreased wave energy causes shoaling of the marina entrance which in turn requires periodic dredging (Fig. 9 a,b). Accretion has been so extensive against the updrift north jetty that wind-blown beach material moves over the jetty into the harbor entrance creating a shoal in the exit channel.

Wide stable beaches exist between Playa del Rey and Redondo Beach. This is because excavation in coastal dunes for the Hyperion Sewage Disposal Facility and Scattergood Steam Generating Station yielded excess sand which was placed on the beaches at Playa del Rey and El Segundo (Fig.10). The oil refinery and ship-loading facility at El Segundo included an impermeable groin which caused problems downdrift. A permeable groin was built in the 1980's and backfilled with sand updrift to act as a "feeder" to maintain the adjacent shoreline. Although requiring periodic replenishment, a permeable-groin feeder-beach combination appears to be a viable solution to erosion problems connected with groin construction.

The beaches at the cities of Manhattan Beach and Hermosa Beach, have grown wider in the past 50 years, probably in response to the huge amounts of fill placed at El Segundo and Playa del Rey. Redondo Beach, at the end of the Santa Monica littoral cell, cannot make the same claim. Because of the deep water nearshore at the head of the Redondo Submarine Canyon, Redondo Beach was the Port of Los Angeles until the turn of this century when the artificial harbor at San Pedro-Long Beach was built. In 1937 a 1,500 ft attached breakwater was built at Redondo Beach which caused erosion of monumental proportions just downcurrent of the structure (Pipkin, 1985). This was due to wave-energy concentration on the downdrift side of the structure because of the submarine canyon and lack of sand nourishment. Several rows of beach cottages were destroyed by wave action in the 1930's and 1940's (Fig. 11). To solve the problem, an additional 2,800 ft of breakwater was added in a dog-leg or "L" shape to form the marina called Redondo Beach King Harbor. It created quiet water for yacht anchorage and saved the shoreline adjacent to the original 1937 breakwater. Erosion then shifted to the 1-mile stretch of beach between King Harbor and the Palos Verdes Peninsula, eventually requiring 2,000,000 cy of artificially placed sand. Because this part of the coastline is a nodal point for longshore currents, a permeable groin was placed at the north end of the fill area to prevent north-transported littoral drift from being deposited in the canyon head. Today, the harbor breakwater is susceptible to overtopping when large winter storms occur in conjunction with high tides. In addition, wave energy passes through the structure creating adverse current conditions at moorings and flooding of restaurants and recreational facilities (Fig. 12 a-c). In 1988, $14 million in damage was incurred during January storms and resulted in litigation against the City of Redondo Beach. At the request of the Los Angeles District of the Corps of Engineers, a model of the harbor was constructed at the Coastal Engineering Research Center to find a solution (Bottin and Mize, 1991). The model reproduced 8,800 ft of shoreline and included offshore topography to a depth of 60 feet. Optimal remedial measures were found to be raising and sealing 1,300 feet of the north breakwater and extending and raising the south breakwater (Fig. 13 a,b). Model studies like this have proved very useful in solving the problems of overtopping and siltation connected with artificial harbors.

The Palos Verdes Hills consist of a cliffed shoreline capped by 13 marine terraces above to an elevation of 1,200 ft. High cliffs eroded on the seaward side have removed support for the land above resulting in large landslides at Portuguese Bend, Flying Triangle, South Shores, and Point Fermin (see Ehlig, this volume).

The shoreline from Point Fermin to the entrance to Anaheim Bay is controlled by the Los Angeles-Long Beach breakwater. Sediment from the Los Angeles and San Gabriel Rivers are impounded behind the structures and is unaccessible for longshore transport. This segment of the shoreline, formerly part of the San Pedro littoral cell, has been essentially removed from the system.

Newport Beach Littoral Cell

Anaheim Bay to Corona del Mar. The Newport littoral cell extends from the east jetty at Anaheim Bay to the west jetty of Newport Bay. The shoreline along this stretch exhibits disequilibrium, exhibiting both seaward advance and landward retreat. Sediment sources are artificial placement at Surfside-Sunset (340,000 cy/y), West Newport Beach (56,000 cy/y), Balboa Peninsula (190,000 cy/y), and coarse sediments

Figure 11. Storm-wave damage at Redondo Beach in the late 1950's. (Photo courtesy Daily Breeze)

Figure 12 a. Overtopping of breakwater at King Harbor during August, I983, El Nino storms. (Photos a and b courtesy City of Redondo Beach)

Figure 12 b. Flooding of Portofino Hotel due to overtopping and penetration of long-period storm wave energy through the breakwater, I983.

Figure 12 c. Same area of breakwater overtopped during January, l988 storms. Note whitewater inside of the harbor.

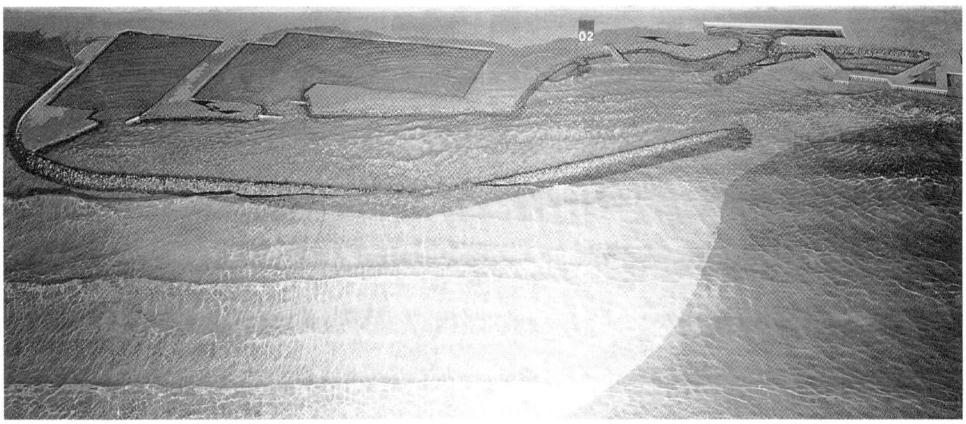

Figure 13 a. Model of King Harbor showing overtopping of breakwater under conditions existing during l983 and l988 storms.

Figure 13 b. Same wave conditions on the model under improvement plan conditions. (photos courtesy Robert Bottin, Jr., see Bottin and Mize,

Figure 14. Newport Beach groin field under storm conditions, 1988. High water extended to first row of houses and did minor flood damage.

discharged at the Santa Ana River (U.S. Army, 1987). At Surfside-Sunset Beach, the shoreline position fluctuated greatly during the replenishment period between 1959-1982, but the beach volume remained relatively constant suggesting beach-fill and littoral-drift rates were the same. This stretch of beach is entirely dependent upon artificial fill which, if terminated, would result in drastic retreat of the shoreline as has occurred in the past.

Bolsa Chica is a coastal marsh of over 1,600 acres, which was partly converted to residential-marina use in the early 1960's. The shoreline at Bolsa Chica Beach has advanced 4 feet over the past 25 years (U.S. Army, 1987). Huntington Beach has experienced only slight erosion since 1874, aided by artificial replenishment, pier construction that broadened the beach, and revetment protection (U.S. Army, 1987).

The Santa Ana River supplies sand to the Newport-Balboa Peninsula, a sand spit. Here, the Newport Submarine Canyon has a strong influence on both beach width and shoreline orientation. It is at the south end of the cell and provides an exit path for sand in the littoral circulation system. Between 1874 and 1926 the beach was stable. Erosion became a problem when the Santa Ana River jetties and Prado Dam were built which severely curtailed sand delivery to this stretch. A field of six groins were constructed in 1968, and 1,000,000 cy of sand from the Santa Ana River was placed between them (Fig. 14). Two more groins were constructed in 1973 and an additional 270,000 cy of sand was placed. Littoral drift rate in this segment is about 130,000 cy/y, but only 72,000 cy/y gets to the Newport Bay jetties, and is lost offshore (U.S. Army, 1987). Canyon scavenging was challenged by Felix and Gorsline (1971) who found the canyon to be silt-filled and inactive during their study. The canyon has been inactive since 1959 (U.S. Army, 1987), and it may be that the canyon only periodically scavenges sand from the littoral system which is rapidly flushed down the canyon into deeper water.

Corona del Mar to Dana Point Headlands. This stretch is characterized by small, picturesque pocket beaches. Differential erosion of Miocene sedimentary rocks and volcanics composing the headland have created a scalloped coastline. The pocket beaches grew between 1875 and 1934, however, beaches south of Laguna Beach have experienced a net retreat since that time. This is due to urban development of south Laguna and to the lack of sediment availability seaward of the foreshore (U.S. Army, 1987).

Oceanside Littoral Cell

Capistrano Beach to Carlsbad. Doheny Beach State Beach Park, and Capistrano Beach at the north end of the cell, have been relatively stable, due mostly to 100,000 cy of sand artificially placed here in the late 1960's. In fact, the beach face has actually advanced slightly between Doheny Beach and the San Clemente pier during the interval 1972-82. San Juan Creek, which discharges at Doheny Beach, is uncontrolled

Figure 15. Camp Pendleton Harbor and Oceanside Marina looking north. San Luis Rey river in lower right and Santa Margarita River in left

and is a potential natural source of sand. One good storm could supply enough sand to nourish Capistrano Beach for several years (State of California, 1967, 1969). San Clemente is built upon a marine terrace and is noted within the engineering geology community for its landslide-prone terrane. Terrace development extends southward to Camp Pendleton and includes the Western White House occupied by ex-President Richard Nixon.

The largest military installation in the United States, the Marine Corps facility at Camp Pendleton, has 15 miles of shoreline and mostly stable or enlarging beaches. This is because the major drainage basins of San Mateo Creek at the north end, and the San Luis Rey and Santa Margarita Rivers toward the south boundary, have not been subject to extensive urban development. At the south end of the Marine Corps base is an artificial harbor with long jetties constructed during World War II (Fig. 15). A sand wedge 5.5 miles long and up to 110-feet wide, has accumulated updrift of the jetties, and the beaches to the south, at Oceanside and Carlsbad, are essentially nonexistent. A small boat harbor was dredged south of Camp Pendleton Harbor specifically to provide fill material for the eroding beaches to the south. This replenishment effort for downdrift beaches has been unsuccessful but continues by dredging fine sediment that accumulate in the harbor entrance (Fig. 16). Studies are ongoing to better characterize the beach sediment so that future fill material will be stable on the beach berm (see Osborne, this volume). The beaches at Oceanside and Carlsbad are now largely cobbles, and some beach homes appear as "stilt" houses because the eroded beach profile is far below the natural one, and their supporting piers are exposed.

Inman and Jenkins (1983) show that a natural balance exists when:

(a) supply = (b) longshore transport = (c) sink,

which for the Oceanside cell under natural conditions was:

(a)273,000 cy/y = (b)254,000 cy/y = (c) 261,000 cy/y.

Currently, 40% of the drainage area is behind dams resulting in a 51% decrease in the supply of sand brought to the coast, or about 136,000 cy/y. Thus an annual sand deficit of about 142,000 cy can be expected to exacerbate any other interdiction in the littoral system, such as jetties and breakwaters that causes erosion.

Leucadia to La Jolla. This segment includes the communities of Encinitas, Cardiff-by-the-Sea, Solano Beach, and Del Mar. It is dominated by narrow beaches backed by bluffs 90-to 110-feet high south of Batiquitos Lagoon. The bluffs have been developed and little erosion occurred until the 1978 and 1983 storms. Shoreline changes at Encinitas between 1888 and 1972 are within the uncertainty limits of the method used, however, the 1982 shoreline was seaward of all other shorelines by an average of 180 feet (U.S. Army, 1987). The City of Del Mar is constructed on the delta of the San

Figure 16. Oceanside-Carlsbad beach looking southward during artificial fill placement, l981. Note the total absence of a beach south of the fill.

Dieguito River. Little shoreline change has been measured here since l888 except for the seaward edge of the San Dieguito River flood plain which has been active. A shoreline advance between l888 and l933 of about 200 feet was measured at the river outlet with little change a mile south.

The shoreline trend between Torrey Pines and La Jolla has shown a seaward advance averaging 0.7 ft/y over the 94 years of record. Most of this is related to cliff instability and slumping. The shoreline is embayed at the heads of the Scripps-La Jolla Submarine Canyons, which also has caused the seacliff to retreat at those locations (Kuhn and Shepard, l984).

Mission Bay Littoral Cell

This cell is 14 miles long between Point La Jolla and the entrance to San Diego Bay. The headlands of La Jolla extend from Point La Jolla to False Point and are characterized by pocket beaches backed by seacliffs. Beach sand is derived locally and there seems to be little transport in or out of the coves. Cliffs, which decrease in height southward, are subject to erosion, and seawalls are conspicuous around the La Jolla Shores development. Three jetties, completed in l950, stabilize the entrance to Mission Bay and the outlet of the San Diego River. Dredging operations in Mission Bay and its entrance have added 1,600,000 cy to local beaches. Because this part of the shoreline is exposed to south swell, the harbor and river inlets come under the influence of both south-and north-moving littoral drift. In l955, a 500-foot long groin was constructed at Ocean Beach to intercept north-moving sand that was shoaling the entrance to Mission Bay, Dredging operations at the entrance during between l951 and l987 have supplied an average of 2.9×10^4 cy/y to both Mission Beach and Ocean Beach. In spite of the additions of beach sand, Ocean Beach showed an average retreat of l00 ft between l960 and l972.

At Point Loma, which separates the Mission Bay Littoral Cell from the Silver Strand Littoral Cell, the Sunset cliffs have eroded an average of 1-3 ft/y for the 35-year period prior to l961 (U.S. Army, l987). Cliff erosion is controlled by joint-sets in flat-lying sedimentary rocks, and deep re-entrants along joints have encroached upon residential streets in the area.

Silver Strand Littoral Cell

The United States Portion. This part of the 16.4 mile-long cell is 13.6 miles and extends from the Zuniga Jetty (east) entrance of San Diego Bay to the Mexican Border. The cell is a beach and sand spit that extends from the Tijuana River to Point Loma. The communities of Imperial Beach and Coronado are on the spit which forms the seaward closure for San Diego Bay. Point Loma and the 1.5 mile-long east jetty entrance to the bay form an effective wave shadow for north swell on this stretch of beach, and net longshore transport is south-to-north. The famous Hotel del Coronado was built near the north end of the spit in l888, and shoreline stabilization devices have operated since 1897, initially to pro-

Figure 17. Rock revetment placed to protect a dwelling at Mandalay Beach, Ventura County. Storm waves commonly bypass the ends of such structures eroding adjacent properties or structures.

Figure 18. A modern coastal development in Malibu utilized a properly designed and constructed bulkhead to resist wave attack and scour. Note the height of the floor of the new home compared to residence to the right.

Figure 19 a. A 1920's vintage home located on Latigo Shores Drive in Malibu was destroyed when the pile foundation collapsed due to battering and sand scour associated with February, 1983 storms.

Figure 19 b. Severe scour in January, 1988 stripped the sand cover from Malibu beaches exposing piles to wave attack scour, and battering by flotsam.

Figure 20. Development along the Malibu coast has produced a variety of designs. Homes with shallow foundations directly on the beach surface were severely damaged by storms in January, 1988. Homes on deep-pile foundations that were elevated above anticipated wave runup faired much better.

tect the hotel. Two miles of revetments, bulkheads, and groins have been built between Imperial Beach and Zuniga jetty to enhance and protect the shoreline. The largest single slug of beach fill (26,000,000 cy) was dredged from San Diego Bay and placed onto the beach south of Coronado in 1946. The beach narrows from north to south, and Imperial Beach at the south end of the spit has undergone critical erosion in the past. This reflects sediment availability as the Tijuana River, the source of sand for the spit, was dammed for flood prevention and water retention. In addition, this part of the U.S. and Mexico has undergone long periods of severe drought, some of the worst in North America.

In recent years, over 30,000,000 cy of sand have been dredged from the bay, of which 16,000,000 cy were deposited south of the Hotel del Coronado, and 1,000,000 cy to the north. This has greatly enhanced the beach in this area but has not helped Imperial Beach, which has a variety of shore protection devices built between 1957 and 1961.

REMEDIAL MEASURES

Site analysis for coastal projects should consider both long-term erosion rates and severe short-term rates for a 100-year storm. If erosion is occurring because of a sand deficit, supplies can be augmented by artificial nourishment with hydraulically equivalent or coarser sediment. This will produce a wider beach but a steeper foreshore and the surf regimen will change. Sand bypassing and harbor-entrance dredging are viable sources of fill material (see Fig. 8). However, if a hydraulic match is not made between fill and native beach material, the restoration program will be unsuccessful.

Concrete block and rock revetments are commonly used protective devices for seacliffs and the backshore areas of beaches (Fig. 17). They require maintenance as they settle, and may be overtopped during infrequent large storms (Fulton-Bennet and Griggs, 1986). They also tend to be bypassed at their ends and erosion may be intensified there. The advantage is they are relatively inexpensive and they absorb wave energy so that the fronting beach may be conserved. Concrete and timber seawalls are more effective, but they reflect wave energy which may transport protective-beach sand offshore. Seawalls are expensive, unsightly, and restrict access to the beach. Also, as with rock revetments, they require maintenance

A very good defensive strategy against damage lies in knowing how a given project will perform under a variety of wave conditions. Model studies can yield this information and have been used to advantage in the past. Unfortunately, models are usually built and operated after a problem has arisen, rather than at preliminary planning and design stages (Bottin, 1991; Bottin and Mize, 1991). The Corps of Engineers Waterways Experiment Station at Vicksburg, Mississippi, has this capability and hopefully this resource will be more fully utilized in the future at early design phases of major projects.

The Design Beach Profile and Foundation Systems

The design wave and beach profiles for coastal projects should be developed by a civil engineer familiar with coastal processes. This utilizes historic records, coastal surveys, bathymetric charts, and knowledge of the performance of similar local projects. He will derive a design beach profile or "scour" profile, which is then used in the design of bulkheads and piling systems. Typically, bulkhead sheeting is placed 2 ft below design beach profile and piles penetrate a minimum of 12 feet below the scour profile. The key element is determining the design wave and the amount of scour that can be anticipated at the site.

The design beach profile is typically shown on a drawing as are various historic beach profiles. The predicted breaker heights and wave runup are then plotted on the profile relative to the proposed improvements. Foundations must be carried below the anticipated scour which typically requires driven or cast-in-place concrete piles. A driven pile is not recommended where beach sands are thin, or where there is a cobble or boulder layer overlying bedrock which can damage piles during driving. Hard bedrock can prevent minimum pile penetration below the scour profile. Drilled piles require casing to keep the hole open and to facilitate drilling into bedrock. Specialized contractors should be used to assure proper placement of reinforcing steel and concrete below the water table. Correctly fabricated, concrete piles are less susceptible to battering damage by wave-thrown debris than wooden piles, and should be incorporated in the design.

BIBLIOGRAPHY

Bottin, Robert, Jr., 1991, Ventura Harbor, California, design for wave and shoaling protection: Coastal Eng. Res. Center, U.S. Corps Engineers, Vicksburg, MS.

Bottin, Robert, Jr., and Mize, Marvin, 1991, Redondo Beach King Harbor, California, design for wave protection: Coastal Eng. Res. Center, U.S. Corps Engineers, Vicksburg, MS.

Denison, F. E., and Robertson, Hugh S., 1985, Assessment of 1982-83 winter storms damage, Malibu coastline, Los Angeles County: Calif. Geology, September.

Dolan, Robert, and Kimball, Suzette, 1988, Coastal erosion and accretion: U. S. Geol. Survey, Nat. Atlas of the United States

Emery, K. O., 1960, The sea off southern California: Wiley, N.Y.

Emery, K.O., and Uchupi, E., 1972, Western North Atlantic Ocean: topography, rocks, structures, water, life and sediments: Amer. Assoc. Petroleum Geol., Memoir 17.

Emery, K. O., and Aubrey, David G., 1991, Sea levels, land levels and tide gages: Springer-Verlag, NY, 237 p.

Felix, D. W., and Gorsline, Donn, 1971, Newport submarine canyon, California: an example of the effects of shifting loci of sand supply upon canyon position: Marine Geology, vol. 10, p. 177-198.

Fulton-Bennett, K. W., and Griggs, G. B., 1986, Coastal protection structures and their effectiveness: State of Calif. Dept. of Boating and Waterways and the Marine Sciences Inst. of the Univ. Calif., Santa Cruz, 43 p.

Griese, G.S., and Aubrey, D. G., 1987, Losing coastal uplant to relative sea-level rise-3 scenarios for Massachusetts: Oceanus, vol. 30, no. 3, p. 16-22.

Griggs, G. B., Littoral cells and harbor dredging along the California coast: Environ. Geol Water Sci., v. 10, no. 1, p. 7-20.

Griggs, G. B., and Savoy, L. E. (Editors), 1985, Living with the California coast: Duke Univ. Press, Durham, N.C., 393 p.

Handin, J. W., 1951, The source, transportation, and deposition of beach sediment in southern California: Beach Erosion Board, U.S. Corps of Engineers, Tech. Memo 22.

Haner, Barbara, 1971, Morphology, and sediments of Redondo Submarine Fan, southern California: Geol. Soc. Amer. Bull., vol. 82, no. 9, p. 2313-2431.

Henry, V. J., Dean, R. G., and Olsen, E. J., 1987, Coastal Engineering: processes, practices and impacts: Assoc. Eng. Geol. Symposium Series, no. 3.

Hicks, S. D., H. A. Debaugh, Jr., and L. E. Hickman, Jr., 1983, Sea level variations for the U.S. 1855-1980: NOAA/NOS Rockville, Md., 170 p.

Hoffman, J. S., Keyes, D., and Titus, J. G., 1983, Projecting future sea level rise: methodology, estimates to the year 2100, and research needs: Office of Policy and Resource Management, U.S. Environmental Protection Agency, Washington D.C., 121 p.

Inman, D. L., 1985, Budget of river sediment in southern California: river discharge versus cliff erosion, in California's Battered Coast: Proceedings Conf. on Coastal Erosion, Calif. Coastal Commission, San Diego, CA, February

Inman, D.L., and Jenkins, S. A., 1983, Oceanographic report for Oceanside Beach facilities: consulting report prepared for the City of Oceanside, 206 p.

Johnson, J.W., 1956, Dynamics of nearshore sediment movement: Amer. Assoc. Petroleum Geol. Bull., vol. 40.

Kuhn, G. G., and Shepard, F. P., 1984, seacliffs, beaches and coastal valleys of San Diego County: Univ. of Calif. Press, Berkeley

Norris, Robert, 1990, Seacliff erosion: a major dilemma: Calif. Geology, vol. 43, no. 8, p. 171-177

Pipkin, B. W., 1985, Santa Monica to Dana Point: p. 307-343 in Griggs and Savoy (Editors), Living with the California coast, Duke Univ. Press, Durham, N.C.

Rice, R. M., Gorsline, Donn S., and Robert H. Osborne, 1976, Relationships between sand input from rivers and the composition of sands from beaches of southern California: Sedimentology, vol. 23, p. 689-703.

Shepard, Francis, 1950, Longshore current observations in southern California: Beach Erosion Board, Corps of Engineers, Tech. Memo. 13.

________, 1963, Submarine geology: Harper, New York.

Shepard, F. P., and Inman, D.L., 1951, Sand movement on the shallow inter-canyon shelf at La Jolla, California: Beach Erosion Board , Corps of Engineers, Tech. Memo 26, 22 p.

State of California, 1967, Progress report on study of beach nourishment along the southern California coastline: Calif. Dept. Water Resources, Southern District.

_______, 1969, Interim report on study of beach nourishment along the southern California coastline: Memorandum Rept., Calif. Dept. Water Resources, Southern District.

Trask, P. D., 1952, Source of beach sand at Santa Barbara, California as indicated by mineral-grain studies: Beach Erosion Board, Corps. of Engineers, Tech Memo. 28, 24 p.

U.S. Army, 1960, Beach erosion control report, coast of southern California, Point Conception to Mexican Boundary; Appendix VII, Interim Report: Corps of Engineers, Los Angeles District.

______, 1961, Special Interim report on Ventura area, beach erosion control report on coast of southern California; Appendix VII: Corps of Engineers, Los Angeles District.

______, 1962, Beach erosion control report, Orange County, California; Appendix V, Phase 2: Corps of Engineers, Los Angeles District.

______, 1984, Shore protection manual, Corps of Engineers, Coastal Engineering Research Center, Waterways Exp. Station, Vicksburg, MS, 2 Volumes

______, 1986, Southern California coastal photography and beach profile index, coast of California storm and tidal waves study: Corps of Engineers, Los Angeles District, Coastal Res. Branch, Ref. No. KCCSTWS 86-2.

______, 1987, Shoreline movement investigation report, Portuguese Point to the Mexican Border (1852-1982): Corps of Engineers, Los Angeles District, Coastal Res. Branch, Ref. No. CCSTWS 87-10.

______, 1988, Ventura Harbor, Ventura County, California: Feasibility Study, Corps of Engineers, Los Angeles District

U.S. Congress, 1940 Beach erosion study, Orange County, California: House Document, 76th Congress, 3rd Session.

_____, 1941, Reexamination of Santa Barbara Harbor, California: House Document 348, 77th Congress, lst Session.

_____, 1952, Appendix I, Coast of California, Carpinteria to Point Mugu, Beach erosion control study: House Document 29, 83rd Congress, 1st Session.

_____, 1952, Coast of California, Point Mugu to San Pedro breakwater, beach erosion control study: House Document 277, 83rd Congress, 2nd Session.

_____, 1956, Oceanside, Ocean Beach Imperial Beach, and Coronado, San Diego County, California, beach erosion control study: House Document 399, 84th Congress, 2nd Session.

_____, 1960, Orange County, California, Appendix V, Phase 1, beach erosion study: House Document 398, Congress, 2nd Session.

_____, 1960, San Diego County, California, Appendix IV, Phase 2, beach erosion control study: House Document 456, 86th Congress, 2nd Session.

Weggel, J. Richard, 1979, A method for estimating long-term erosion rates from a long-term rise in water level: Coastal Eng. Tech. Aid No. 79-2, Corps of Engineers, Coastal Engineering Research Center, Ft. Belvoir, VA.

Wood, F. J., 1976, The Strategic role of perigean spring ties in nautical history and North American coastal flooding, 1635-1976: U. S. Dept. Commerce, Washington, D.C., 538 p.

Table 1

Summary of Shoreline Ownership and Lengths

Santa Barbara County to the Mexican Border

County	Federal	State	Local	Private	Total
Santa Barbara	23.1	6.5	5.2	74.5	109.3
Ventura	7.2	9.5	9.8	15.4	41.9
Los Angeles	1.4	12.6	18.6	29.9	62.5
Orange	0.9	7.8	17.1	15.8	41.6
San Diego	28.3	16.3	10.1	6.7	71.3
Totals (miles)	60.9	52.7	60.7	152.3	326.6
Percent	18.7	16.1	18.6	46.6	100.00

Table 2
Rates of Littoral Drift in Southern California

Location	Rate(xl000 cy/y)	Years of Record	Reference
Santa Barbara	280(b)	1929-46	H.D., 1952(a)
Ventura Harbor	540	Historical	USCE, 1988
Port Hueneme	400(b)	1938-48	H.D., 1957(a)
Port Hueneme	800(c)	1959-62	USCE, 1957(e)
Port Hueneme	1200(d,g)	1948-52	Savage, 1957
Santa Monica	259(b)	1935-	H.D., 1953(a)
Redondo Beach	51.5(d,g)	1935-46	H.D., 1953(a)
Anaheim Bay	300 (d,g)	1961	USCE, 1962(e)
Camp Pendleton	130(b)	1934-52	H.D., 1960(a)
Oceanside	254	----	Inman and Jenkins
Pacific Beach	65(b)	1940-5	H.D., 1960(a)
Imperial Beach	672(f,g)	1946-54	H.D., 1960(a)
Imperial Beach	100(f)	1954-59	H.D.,1960(a)

a. U.S. Congress House Document/year
b. Accretion upcurrent from groin, jetty, or breakwater
c. Combined accretion and loss into submarine canyon
d. Erosion downcoast of artificial structure
e. U.S. Corps of Engineers, Los Angeles District
f. Losses of upcoast (north) littoral drift from Tijuana River.
g. area of periodic critical erosion and/or artificial beach replenishment.

Table 3
Artificial Barriers to Littoral Drift

Locality	Type of Structure	Impact
Santa Barbara(1)	Attached breakwater	Severe downcoast erosion
Ventura Harbor	Jetties	Downcoast erosion
Channel Islands (2)	Jetties and detached breakwater	Downcoast erosion
Port Hueneme (2)	Jetties	Downcoast erosion, increased losses to submarine canyon
Santa Monica(1)	Parallel breakwater	Slight downcoast erosion
Marina del Rey	Jetties and detached breakwater.	Downcoast erosion and nearshore shoaling
Redondo Beach	Breakwater	Erosion, increased losses to submarine canyon
Los Angeles-Long Beach	Breakwater	Detains sediment from two major rivers
Anaheim Bay (1)	Jetties	Severe erosion at Seal Beach
Santa Ana River	Jetties	Downcoast erosion
Newport Harbor	Jetties	Slight erosion downcoast
Camp Pendleton (1)	Jetties	Severe erosion at Oceanside
Agua Hedionda Lagoon	Inlet stabilization jetties	Downcoast erosion
Mission Bay(1)	Inlet stabilization jetties	Down oast erosion
San Diego Bay	Inlet stabilization jetties	Loss of sand to deep water

(1) Maintenance dredging of harbor to feeder beaches downcurrent
(2) Semi-permanent artificial bypassing operations

Table 4
Physical Characteristics
of Coastal Watersheds*

Region	Area (miles^2)	Percent Controlled	Major Dams
Santa Barbara	4,070	39	Whale Rock, Old Creek
Ventura-Los Angeles	4,070	37	Big Tujunga, Rio Hondo, Whittier Narrows, Hansen, Sepulveda basin, Piru Creek
Santa Ana	2,780	80	Prado Flood Control Dam
San Diego	5,024	41	Henshaw, Vail, Hodges, Lower Otay, Tijuana River

*Source--State of California, Department of Water Resources

APPLICATIONS OF FOURIER GRAIN-SHAPE ANALYSIS IN THE SEARCH FOR MORE ACCURATE SEDIMENT BUDGETS, ORANGE AND SAN DIEGO COUNTIES, CALIFORNIA

Robert H. Osborne

Chia-Chen Yeh

Department of Geological Sciences
University of Southern California
Los Angeles, CA 90089-0740

INTRODUCTION

An accurate sediment budget based on comprehensive, long-term study of shoreline changes and quantification of the factors responsible for such changes is required to provide a basis for sound scientific, engineering and management decisions affecting the coastal zone. A sediment budget is a conceptual framework which mathematically describes the transport of sand and coarser-grained sediment into, within and from a defined coastal reach during a specified time interval. The principal objective of sediment budget analysis is to quantitatively predict how previously-identified factors will interact to alter the quantity and distribution of sediment within a given coastal reach (the control volume) from both short- and long-term perspectives. Theoretically, the use of an appropriate budget would make it possible to assess the potential impacts of short-term episodic events such as major storms and floods, of intermediate events such as dam and harbor construction, and long-term events such as sea-level fluctuations and tectonism.

The shoreline segment commonly selected as the basis for sediment budget analysis is the littoral cell. The occurrence of a littoral cell is based on the concept of longshore transport of dominantly fluvially-derived sediment, which is entrapped either by a submarine canyon head or by a point of land extending seaward from the general position of the shoreline. Emery (1960) and Inman and Chamberlain (1960) identified a series of littoral cells along the southern California coast.

Budget analysis also requires accurate short- and long-term estimates for the sediment yields from local sand and gravel sources. Natural and artificial sediment sources and sinks for the littoral zone are shown schematically in Figure 1. Natural sediment sources include river discharge; cliff, bluff and dune erosion; wind and overwash transport; and onshore transport from the inner continental shelf. Natural sediment

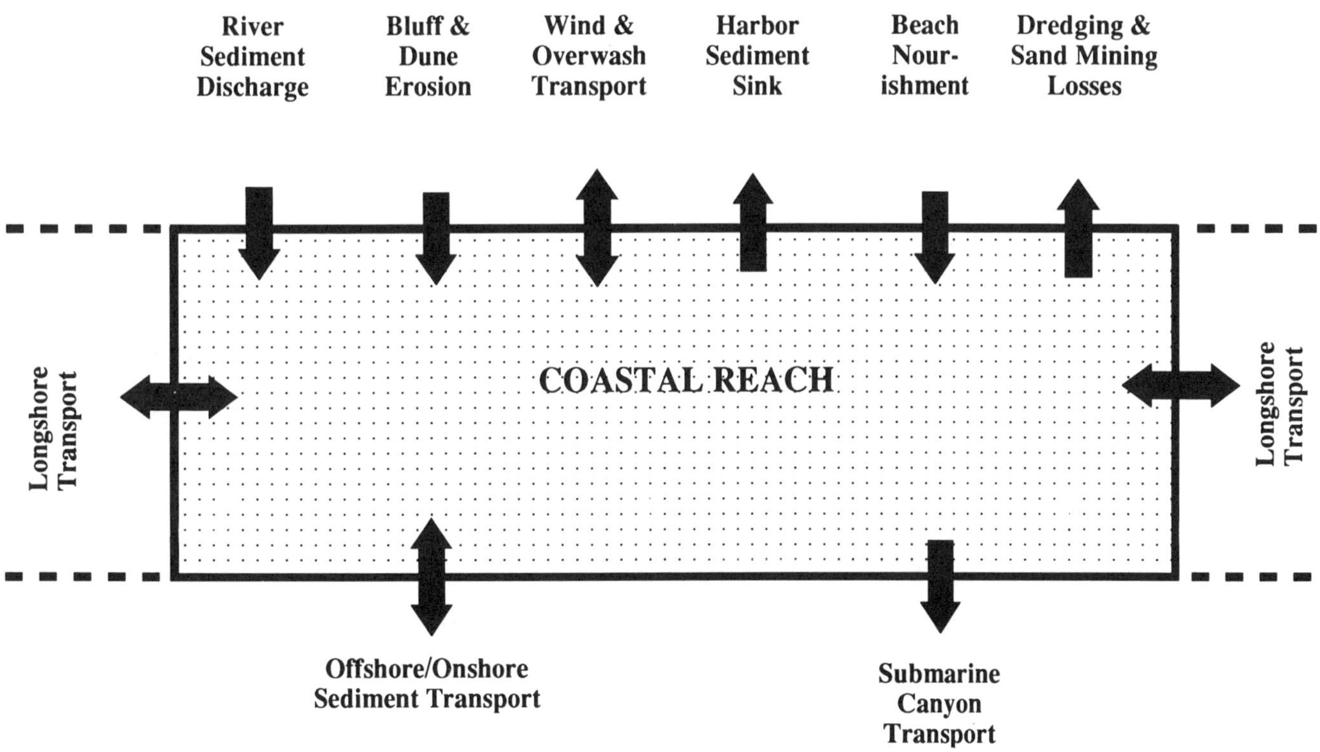

Figure 1. Schematic diagram showing the conceptual sediment sources and sinks as well as idealized transport pathways for beach sand in a defined coastal reach (after Fig. 8, U.S. Army Corps of Engineers, 1987, p. 7).

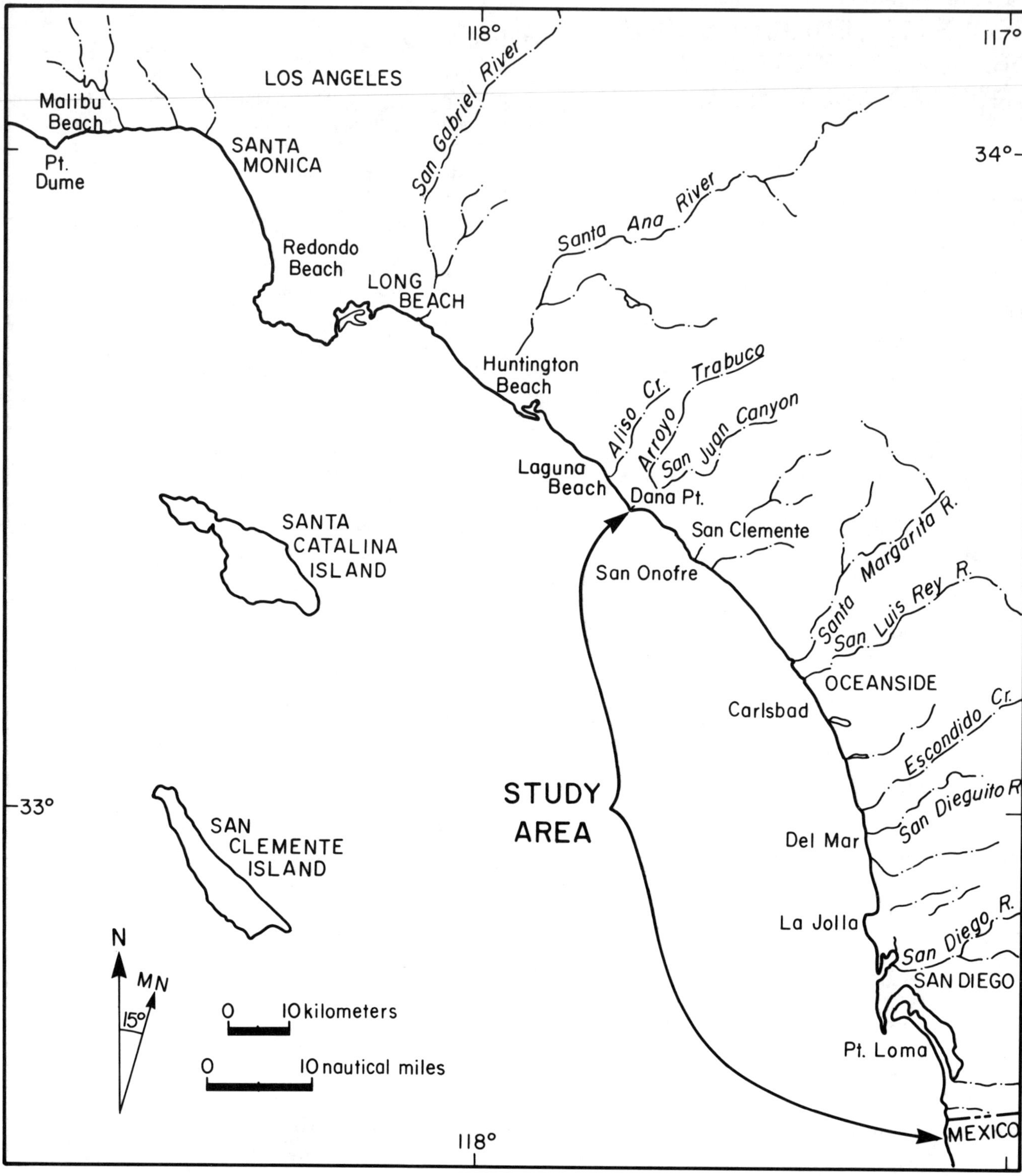

Figure 2. Map showing the location of the study area.

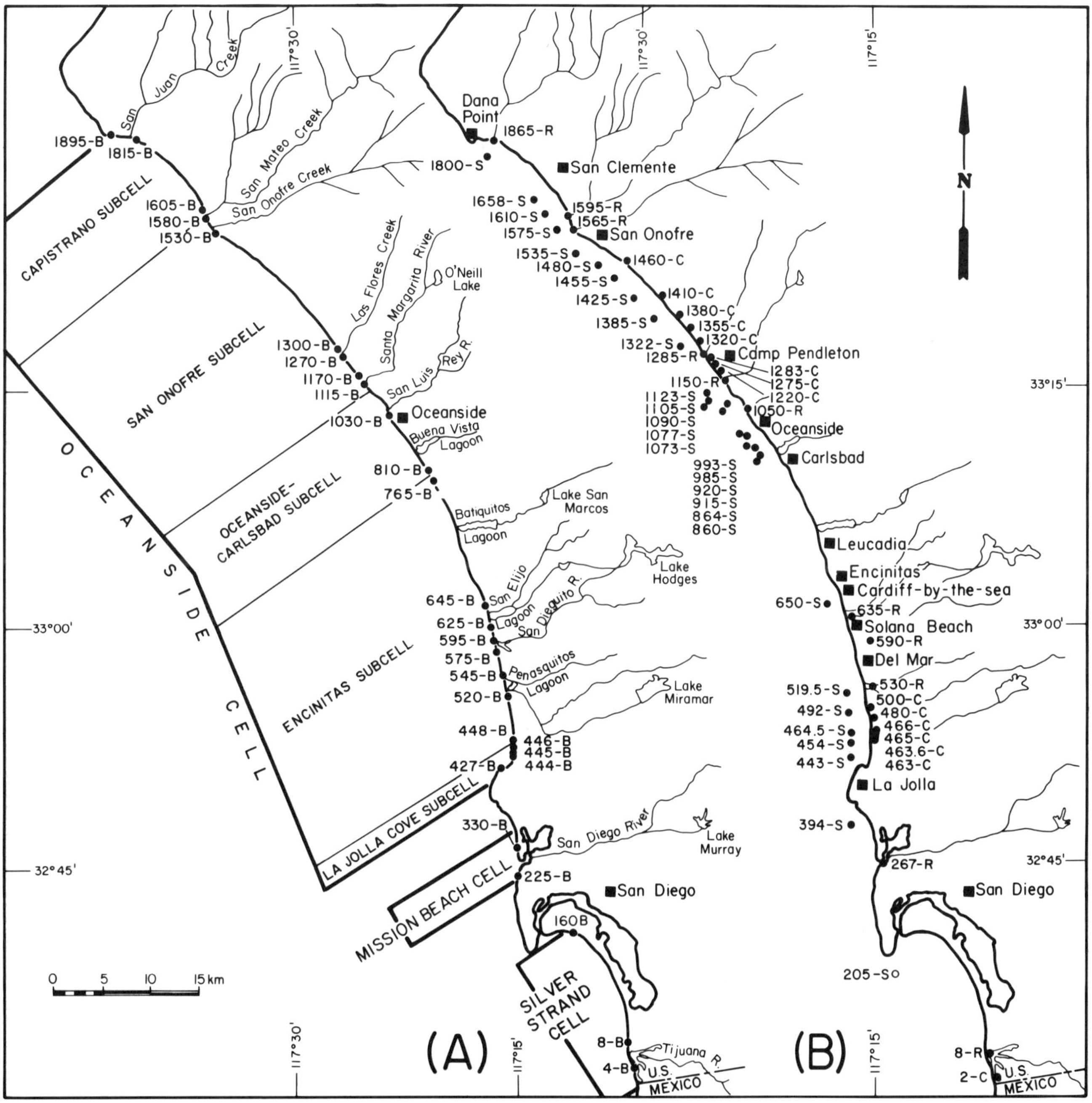

Figure 3. Maps showing the principal geographic features, littoral subdivisions and sample locations from Dana Point to the United States-Mexico border. The three- to four-digit number associated with each sample locality refers to a range line system defined by the U.S. Army Corps of Engineers. The range line values represent the distance in miles along the coastline north of the United States-Mexico border. A decimal point is understood to precede the last digit so that range line 1580 is 158.0 miles north of the border. The letter S, B, R and C following each range line refers to a inner continental shelf, beach, river or cliff sample, respectively. Figure 3A shows the locations of beach samples, and 3B shows the locations of shelf, river and cliff samples.

sinks include losses to submarine canyons and offshore transport to the inner shelf. Beach nourishment (artificial source) and sand mining (artificial sink) may be important locally. Factors normally considered which affect sediment transport within a defined coastal reach include coastal physiography, nearshore wave climate, nearshore currents, tides, sediment characteristics, bathymetry, wind and wave overwash, engineered structures, and short-term sea-level fluctuations.

Although the objectives of sediment budget analysis, including numerical modeling, are extremely important in coastal engineering, state-of-the-art estimates of the sediment volumes and transport rates associated with each source and sink as well as the use of different computational procedures to derive such estimates may produce quite different results. For example, after comparing the estimates for fluvial coarse-grained sediment delivery in the Oceanside area (Figs. 2 and 3), Stow and Chang (1987) show that the results are far from being either consistent or conclusive. They attribute these conditions to the sparsity of estimates, assumptions and inaccuracies associated with the methodologies employed, as well as a general failure to account for river mouth/deltaic processes following major runoff events.

Likewise, after reviewing existing data, literature, sediment budget methodologies and five recent budgets for the Oceanside area (Nordstrom and Inman, 1973; Inman, 1976, 1981; Hales, 1978; Inman and Jenkins, 1983; and Weggel and Clark, 1983) as well as proposing a new preliminary budget, Tekmarine, Inc. (1987) concluded that at present, it is impossible to compute a definitive budget due to uncertainties surrounding the key elements of the budget. The most recent sediment budget for the Oceanside Littoral Cell (Figs. 2 and 3) was computed by Moffatt and Nichol (1989). Their Appendix L summarizes each of the previous budgets for the Oceanside Harbor area as well as the Oceanside Littoral Cell.

There are several key problems in computing accurate sediment budgets that require resolution. One is the fact that wave-hindcasting procedures and conventional monitoring of seafloor topography (profiling) in the littoral zone do not in themselves serve to uniquely define the volumetric contributions from each local source or the exact pathways of the sand in transport. Another key problem is accurately estimating the volume of sediment and the exchange rate associated with the inner continental shelf, i.e. the onshore-offshore or cross-shore vector (Fig. 1). A third is the difficulty of incorporating the effects of intense, natural episodic events such as major storms and floods, which contrast markedly with the relative consistency of seasonally-adjusted fairweather processes. Paradoxically, in southern California the large storm waves which may cause significant coastal erosion and structural damage often are associated with large discharges of sediment and water to the shorezone. Thus regardless of the conceptual complexity and mathematical sophistication of existing sediment budget and numerical modeling procedures, fundamental questions remain concerning the volumetric aspects, transport pathways and rates associated with littoral sediment sources and sinks, Perhaps such questions may be addressed best by appropriate observational data derived from relatively long-term (decades) monitoring experiments.

The purpose of this paper is to address two major and interrelated coastal problems bearing on sediment budget analysis in southern Orange and San Diego Counties using Fourier grain-shape analysis (FGSA). The first problem concerns the identification of local sand sources for seven shoreline segments in the Oceanside, Mission Beach and Silver Strand Littoral Cells (Fig. 3), as well as the estimation of the volumetric contributions from each local source. Of particular importance in this discussion is the role of the inner continental shelf in contributing medium-grained sand to the foreshore. The second problem concerns the short- and perhaps long-term division of a large littoral cell: in this case, the Oceanside Littoral Cell. This problems centers on the question of whether the littoral cell is simply an organizing concept or a physical reality? If real, under what geographic and temporal scales do the sedimentary processes within a given cell operate? The solution to this problem bears heavily on estimates of net longshore transport rates, and is particularly intriguing because the measurement of actual longshore transport rates for sand is often less than that predicted by present wave-hindcasting techniques. The discussion of these topics is prefaced by a brief description of FGSA and the associated statistical procedures employed in this study. The results of FGSA for the Oceanside, Mission Beach and Silver Strand Cells then will be presented followed by an appropriate discussion of substantive issues.

SAMPLE COLLECTION

This study is based on FGSA of 93 sand samples collected from the major rivers, sea cliffs and beaches in the Oceanside, Mission Beach and Silver Strand Littoral Cells as well as the adjacent inner continental shelf. The location of each sample is shown in Figure 3. All of the river and virtually all of the cliff samples used were collected during April 3 to 5 and June 1986. The river samples were obtained from bars in active channels, and were collected at least two kilometers upstream of the river mouths to avoid the possible effects of tidal currents on grain size, grain shape and mineral composition. The tidal effects on the Santa Margarita River are so far reaching that it was necessary to collect this sample about five kilometers upstream of the mouth. The exact geographic and stratigraphic positions of the sea cliff and bluff samples collected in the San Onofre, Camp Pendleton and Torrey Pines areas (Fig. 3) are given in Kuhn and others (1987). The first beach sample set was collected from April 3 to 5, 1986, and the second set on October 8, 1986. The April set approximates the end of the oceanographic winter in southern California, and the October set approximates the end of the oceanographic summer. The beach samples were collected at MLLW at locations about one kilometer northwest and southeast of each major river mouth, as well as other critical locations along the shoreline. Beach samples 765-B and 810-B were collected in February 1989 to provide additional information concerning downcoast transport for the Oceanside-Carlsbad subcell and upcoast transport for the Encinitas subcell, respectively. The La Jolla Cove subcell was defined after the primary sampling program in 1986, therefore the beach samples used for this segment were collected in September, 1989. The beach samples were obtained by inserting a tin can normal to the shoreface to a depth of about 15 cm, which produced a small core with a volume of approximately 500 cm^3.

Inner continental shelf samples were included in this study to better understand the role of cross-shore sand exchange. The seventeen samples with range lines from 443-S through 1123-S (Fig. 3) were obtained from vibracores recovered by Ocean Surveys, Inc., under contract to the U.S. Army Coastal Engineering Research Center, on board the RV *Diamond Lady*. These cores were taken during January and February 1981 at water depths ranging from -11 to -21 m (-37 to -70 ft) MLLW. The 10 shelf samples with range lines from 1322-S through 1800-S (Fig. 3) were collected north of Oceanside Harbor in February 1989 at a water depth of -9 m (-30 ft) MLLW. Depending upon the closure depth used for the Oceanside Littoral Cell, these samples may be considered either lowermost foreshore or innermost shelf.

FOURIER GRAIN-SHAPE ANALYSIS OF SAND SAMPLES

Introduction

Shape is a complex property of sedimentary particles. Many of the attempts to at least semiquantitatively characterize the shapes of such particles have been summarized by Carver (1971). For sand-size grains, sedimentologists traditionally have used charts to visually estimate two shape elements; namely, sphericity and roundness. Sphericity presumably measures the degree to which a grain approaches a spherical shape, whereas roundness measures the sharpness of the edges or corners of a grain. Rittenhouse (1943) published one of the most widely-used charts for estimating grain sphericity. Russell and Taylor (1937) and Powers (1953) published widely-used charts for estimating roundness. In a well-known textbook, Krumbein and Sloss (1963) identified four sphericity classes (very elongate to spherical) and reduced the number of roundness classes from Powers (1953) to form a chart to classify the shape of sedimentary grains in terms of both sphericity and roundness. Conventionally, from one- to several-hundred grains per sample are classified using appropriate visual-estimation charts. The results then are tabulated and statistically reduced to identify grain-shape populations with a given sample set (Osborne and Pipkin, 1982). It is important to note that by adopting and accepting the use of such visual charts, sedimentologists also have accepted the concept that scientifically useful information can be derived from the shape of a two-dimensional maximum projection outline of a grain, i.e., a map view. For example, when measuring sphericity from visual-estimation charts, sedimentologists in fact traditionally have measured the circularity of a two-dimensional grain outline derived from the projection of the maximum grain dimensions onto a planar surface. The use of visual estimation charts also has been adopted by the coastal engineering community to measure both roundness and sphericity. Fourier grain-shape analysis is based on the maximum projection outline of a grain, but improves on visual estimation charts by quantitatively measuring the shape components of the outline by means of a closed-form Fourier series.

Since its introduction by Schwarcz and Shane (1969) and Ehrlich and Weinberg in 1970, Fourier analysis of detrital quartz grain shapes has been applied to a wide variety of natural tracer and petrofacies problems. Mazzullo and Magenheimer (1987) discuss the origin of quartz grain shapes derived from crystalline as well as wind-transported and quartz-cemented sedimentary rocks. The interested reader is referred to the following papers: Brown and others (1980), Ehrlich and others (1974), Ehrlich and Chin (1980), Ehrlich and others (1980), Hudson and Ehrlich (1980), Mazzullo and Ehrlich (1980, 1983), Mrakovitch and others (1976), Riester and others (1982), Mazzullo (1986), Mazzullo and Withers (1984), Van Nieuwenhuise and others (1978), and Young (1980). Fourier grain-shape studies completed in California include Ehrlich and others (1974), Porter and others (1979), Bloom (1979), Clark and Osborne (1982), Gaynor (1984), Osborne and others (1985), Bomer (1985), Ahlschwede (1988), Broadhead (1988), Cho (1989), Osborne and Cho (1989), Osborne and others (1989), Osborne and Yeh (1991), Osborne and others (1991), and Yeh (1991).

Ehrlich and Weinberg (1970) describe a closed-form Fourier method to analyze the observed variation of two-dimensional, maximum-projection, grain-shape area. Grain shape may be estimated by an expansion of the periphery radius as a function of angle about the grain's center of gravity by a Fourier series (Boggs, 1992, p. 53-59). Although conceptually similar to the closed-form methodology described by Ehrlich and Weinberg (1970), the methodology employed in this report makes use of the newer and more widely used Fast Fourier Transform (FFT). The reader is referred to Brigham (1974), Bloomfield (1976), and Bendat and Piersol (1971) for extensive treatments of the mathematically complex FFT. The sample preparation, Fourier methodology and statistical procedures employed are discussed in Osborne and Yeh (1991).

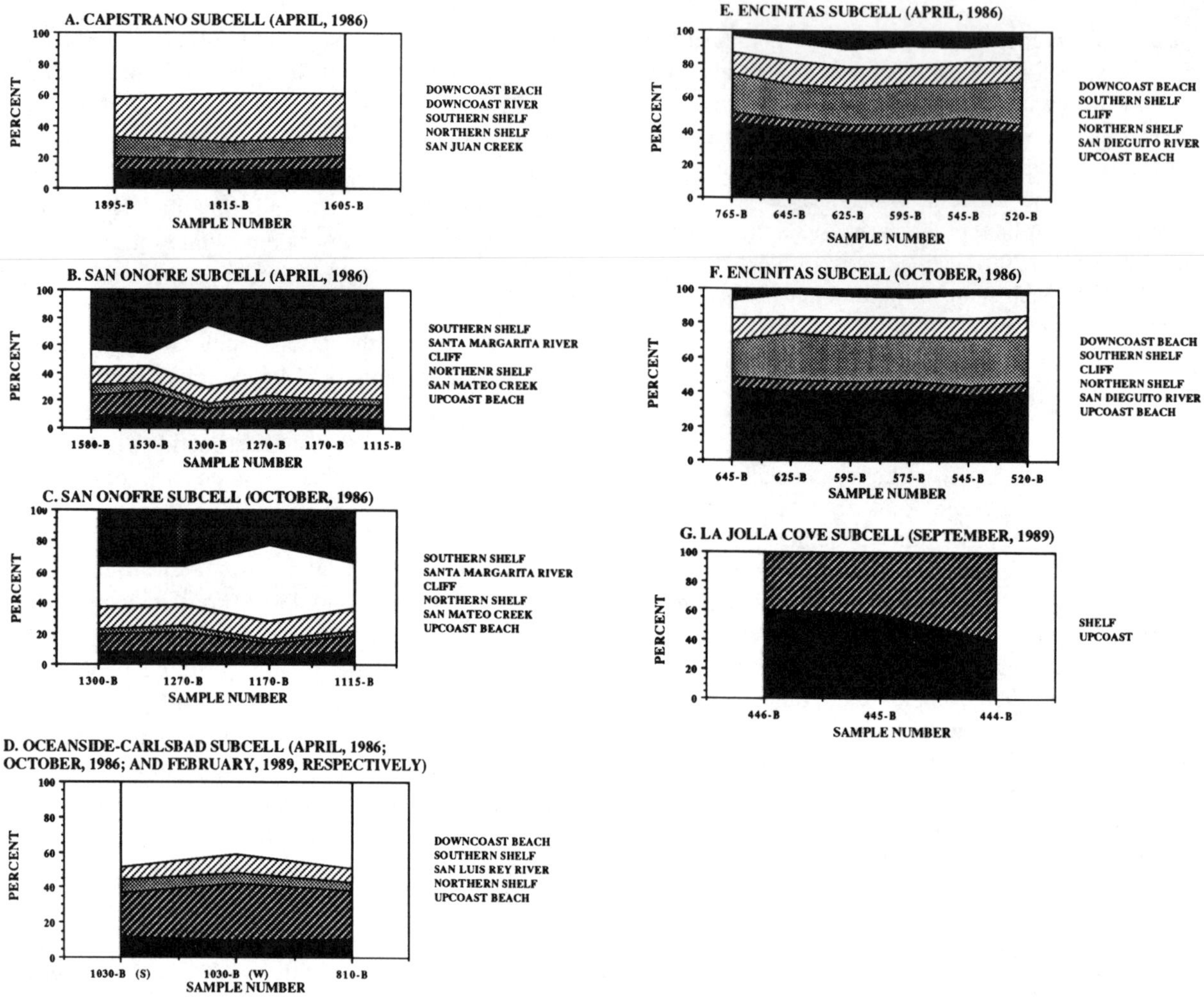

Figure 4. Area plots for the Oceanside Littoral Cell based on the data listed in Table 1. The patterns in each diagram are in the same stratigraphic order as the sources listed along the right margin. For example, in Figure 4A, the percentage of medium-grained quartz from San Juan Creek is shown in solid black, whereas the percentage from the downcoast beach is shown in solid white.

RESULTS AND DISCUSSION

Tables 1 through 3 and Figures 4 through 6 summarize the experimental design and results of FGSA for 93 fluvial, sea-cliff, beach and inner continental shelf samples. Due to the length and sedimentologic complexity of the Oceanside Littoral Cell, a total of 83 samples were analyzed for this coastal reach; whereas only 4 and 6 samples were analyzed for the smaller and less complex Mission Beach and Silver Strand Littoral Cells, respectively.

Oceanside Littoral Cell

The 86.3 km-long Oceanside Littoral Cell (Fig. 3) extends from Dana Point south to Point La Jolla. This is a complex coastal reach backed by cliffs and bluffs in the San Onofre, Camp Pendleton and La Jolla areas. It contains several major rivers (San Juan Creek, San Mateo Creek, San Onofre Creek, Santa Margarita River, San Luis Rey River and San Dieguito River), numerous coastal streams, and several major lagoons (Buena Vista, Agua Hedionda, Batiquitos, San Elijo and Los Penasquitos). The Del Mar Boat Basin and Oceanside Harbor constitute a major small-craft harbor in the Oceanside area. The geomorphologic complexity and length of this coastal reach make it an ideal locality to examine the concept of littoral cell segmentation; i.e. the identification of subcells which may be more meaningful entities for sediment budget analysis.

Although the littoral cell concept is appealing, it has never been demonstrated that sand from a given local source actually follows the suggested pathway through a substantive part of a given cell. This, in turn, gives rise to the possibility of littoral cell segmentation. The distribution of sediment and energy impinging along a given shoreline may be affected by, among other factors, the orientation and physiography of that coastal reach as well as by the presence of engineered structures. From northwest (upcoast) to southeast (downcoast), the occurrence of Dana Point, San Mateo Point, the north breakwater at Oceanside Harbor, Carlsbad Submarine Canyon, Scripps Submarine Canyon and Point La Jolla make it possible to initially divide the Oceanside Littoral Cell into five subcells for testing the concept of littoral cell segmenta-

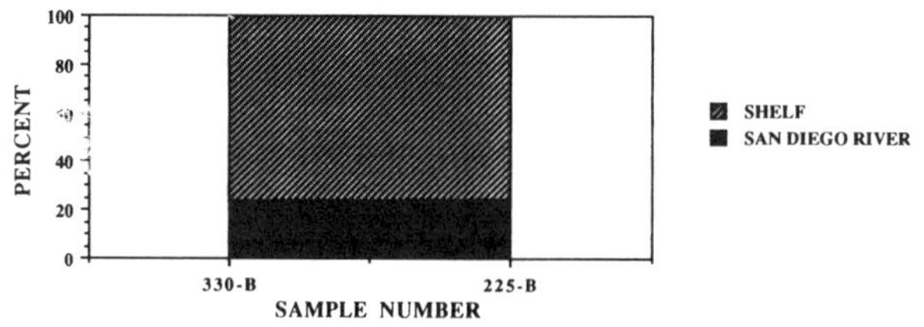

Figure 5. Area plot for the Mission Beach Littoral Cell based on the data listed in Table 2.

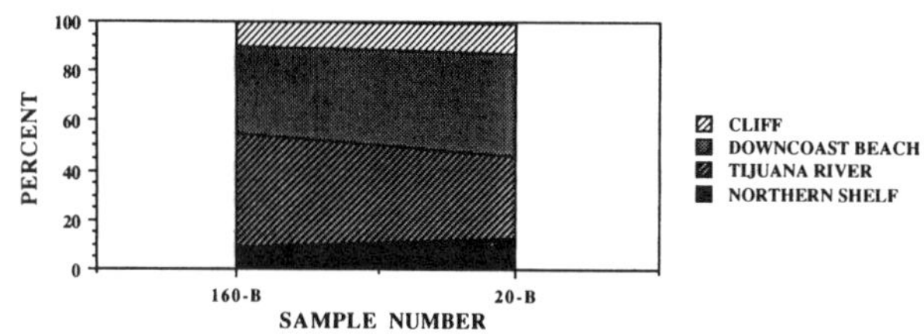

Figure 6. Area plot for the Silver Strand Littoral Cell based on the data listed in Table 3.

tion (Fig. 3). These are named the Capistrano subcell (Dana Point to San Mateo Point), the San Onofre subcell (San Mateo Point to the north breakwater at Oceanside Harbor), the Oceanside-Carlsbad subcell (Oceanside Harbor to the onshore projection of Carlsbad Submarine Canyon), the Encinitas subcell (from the onshore projection of Carlsbad Submarine Canyon to La Jolla Submarine Canyon), and the La Jolla Cove subcell (La Jolla Submarine Canyon to Point La Jolla) (Osborne and others, 1989; Osborne and Yeh, 1991; Yeh, 1991). The average shoreline orientation and average slope from 0 to -18 m (-60 feet) MLLW for each of the five subcells considered are listed in Table 4.

The grain-shape composition of the foreshore samples for each of the five subcells initially identified in the Oceanside Littoral Cell is listed in Table 1 and illustrated in Figure 4. Inasmuch as Osborne and Yeh (1991) discuss these data in detail, only the salient points will be presented here as follows: (1) FGSA indicates that each of the five subcells is distinct with respect to grain-shape composition, which, in part, reflects the occurrence of a characteristic set of local sources supplying each subcell; (2) the inner shelf contribution ranges from 20 percent in the Capistrano subcell to 47 percent in the La Jolla Cove subcell, and averages approximately 35 percent for the entire Oceanside Cell; and (3) with the exception of the San Onofre subcell, the shape composition within each subcell is remarkably consistent. This observed consistency within each subcell is due, in part, to thorough mixing by seasonally-related, bidirectional, longshore and cross-shore currents during an unknown littoral residence time. The peaks of Santa Margarita River-like sand in the San Onofre subcell (Fig. 4B and C) most likely represent pulses of sand from that river delivered prior to 1985-1986 rather than discharge from the similarly-sourced Las Flores Creek (Osborne and Yeh, 1991).

Mission Beach Littoral Cell

The coastal lowland associated with Pacific and Mission beaches occupies the former delta of the San Diego River (Kuhn and Shepard, 1984), which extends almost to Crystal Pier. The only volumetrically important natural sand sources for the Mission Beach Littoral Cell are the San Diego River and the inner continental shelf (Figs. 2 and 3). Simon and Li (1988) estimate an average annual upland yield of only 1,316 cubic meters of coarse-grained sediment for the San Diego River. Sixty percent of the total drainage area of the San Diego River is controlled by El Capitan and San Vincente Dams, therefore only a small part of the annual upland sediment yield is delivered to Mission Bay. Three jetties constructed across the mouth of Mission Bay interrupt the supply of sand from Mission Beach to Ocean Beach, therefore, Ocean Beach may be treated as a pocket beach.

The grain-shape composition of the foreshore samples from the Mission Beach Cell is listed in Table 2 and illustrated in Figure 5. Shapes typical of the San Diego River average only about 24 percent, whereas inner continental shelf shapes average about 76 percent. The one percent variation observed for the shape composition from each source is statistically insignificant. The shelf sample (394-S) was obtained from a water depth of -27 m MLLW, and there is no suggestion that onshore transport necessarily occurred from this depth. Rather, the shelf grain-shape contribution may reflect onshore transport of similarly-shaped sand from somewhat shallower water depths. The high percentage of inner shelf sand observed in the foreshore samples indicates substantive dilution of river-derived sand during an unknown littoral residence time.

Silver Strand Littoral Cell

The Silver Strand Littoral Cell extends from several km south of the United States-Mexico border to Zuniga Jetty at the entrance to San Diego Bay (Fig. 3). According to Tekmarine, Inc. (1987), there is a sediment transport imbalance within this cell. As a result, accretion is expected to continue in the Zuniga Jetty area, and beach erosion will continue to be of particular concern in the following three areas: (1) from Hotel Del Coronado to about 6.4 km south; (2) at Imperial Beach due mostly to slow degradation of the Tijuana River delta; and (3) at Playas de Tijuana (the southern end of the cell), again due to continued destruction of the Tijuana River delta.

The grain-shape composition of the foreshore samples from the Silver Strand Cell is shown in Table 3 and Figure 6. The average grain-shape composition for the two beach samples is as follows: northern shelf, 11 percent; Tijuana River, 40 percent; downcoast beach (south of Tijuana River), 38 percent; and cliff (northernmost extension of Playas de Tijuana), 11 percent. The ranges of values for the northern shelf source (4 percent) and cliff (2 percent) are not statistically significant; however, the range of values for the Tijuana River contribution (13 percent) and downcoast beach (6 percent) requires explanation. One important potential source of variation is the 34-month time interval which elapsed between sampling of the southern and northern beach samples. The southern beach sample (20-B) was collected by the U.S. Army Corps of Engineers in April 1986, whereas the northern one (160-B) was not collected until February 1989. Research in progress will determine whether the observed differences represent systematic trends or relatively short-term temporal variations.

Inasmuch as the two beach samples were collected upcoast of the Tijuana River, the overwhelming dominance of sand derived from the downcoast portions of the Silver Strand Littoral Cell which contain the principal sand sources (Tijuana River and downcoast beach and cliff) clearly demonstrates net upcoast longshore transport. The northern shelf sample (205-S) was obtained offshore of Point Loma at a water depth of -12.8 m MLLW. The fact that the two foreshore samples contain from 9 to 13 percent of their grain-shape composition from this source must reflect onshore and perhaps upcoast longshore transport from this or other shallow inner-shelf sources with similar grain shapes located farther south.

The percentage of medium sand-size quartz derived from each source for the Oceanside, Mission Beach and Silver Strand Cells is summarized in Figure 7.

THE CONTINENTAL SHELF AS A SOURCE FOR LITTORAL SAND

The results of FGSA (Table 5) indicate that an average of 35, 75 and 11 percent of the foreshore sand for the Oceanside, Mission Beach and Silver Strand Cells, respectively was derived from the inner continental shelf. The failure to consider the inner shelf as a sediment source clearly introduces considerable error into sediment budget calculations.

Shepard (1973, p. 133) recognized the likely importance of the inner continental shelf as a source for littoral sand when he stated "The direct source of almost all beach sand is the shallow sea floor, although previously the bulk of the shelf sand was carried into the ocean by runoff from the land. ..... Beaches may also receive their supply from sands that were deposited on the continental shelf during the Pleistocene stages of low sea level when the shelves were largely dry land. Hence, here the terrestrial sources may be remote in time." Evidence that inner shelf deposits constitute a significant source of beach sand may be found in geographically-widespread areas where rivers are not discharging appreciable volumes of sand; where there are no easily-erodable coastal terraces, bluffs, or sea cliffs to supply sand; where sand is actively migrating along shore in a preferred direction, and yet where beaches continue to actively form under natural conditions.

Along the low-gradient Atlantic and Gulf coasts, McMaster (1954, New Jersey), van Andel and Poole (1960, Texas Gulf Coast), Saville (1960, Long Island), Curray (1960, northwest Gulf and Mexico), Giles and Pilkey (1965, southeastern United States) and Pierce (1968, Cape Hatteras) all concluded that much of the beach sand was derived from the erosion of older strata associated with adjacent continental shelves. In one of the more quantitative studies based on long-term changes in shoreline position, Pierce (1968) reported that for a 100-year period, longshore transport and biogenic sediment sources could not account for 337,000 m^3 (42 percent) of an annual input of 796,000 m^3 along a 100-km coastal reach from Hatteras Inlet, North Carolina to Cape Lookout. Thus 42 percent of the annual littoral input was considered to have been derived from unlithified Quaternary sediment or weakly-cemented Tertiary strata exposed on the adjacent shelf.

Along the higher-gradient Pacific coast, Trask (1952, Santa Barbara, California), Bowen and Inman (1966, Sebastian Vizcaino Bay, Mexico), Inman and others (1966, Point Arguello, California), and Clemens and Komar (1988, Cape Blanco, Oregon) all have either suggested or documented net onshore sand transport from the shelf to the adjacent littoral zone. Although Trask (1952) concluded that the adjacent shelf was the only likely source of beach sand in the Santa Barbara area, he did not propose a mechanism to sort and transport relatively coarse grains onshore.

Using Fourier analysis, Riester and others (1982) describe quartz grain-shape populations from beach to outer shelf depths off Long Island. Of interest to this paper is that more abraded beach sand, which is formed at water depths from 0 to -5 m MLW, is mixed with inner shelf sand, which occurs

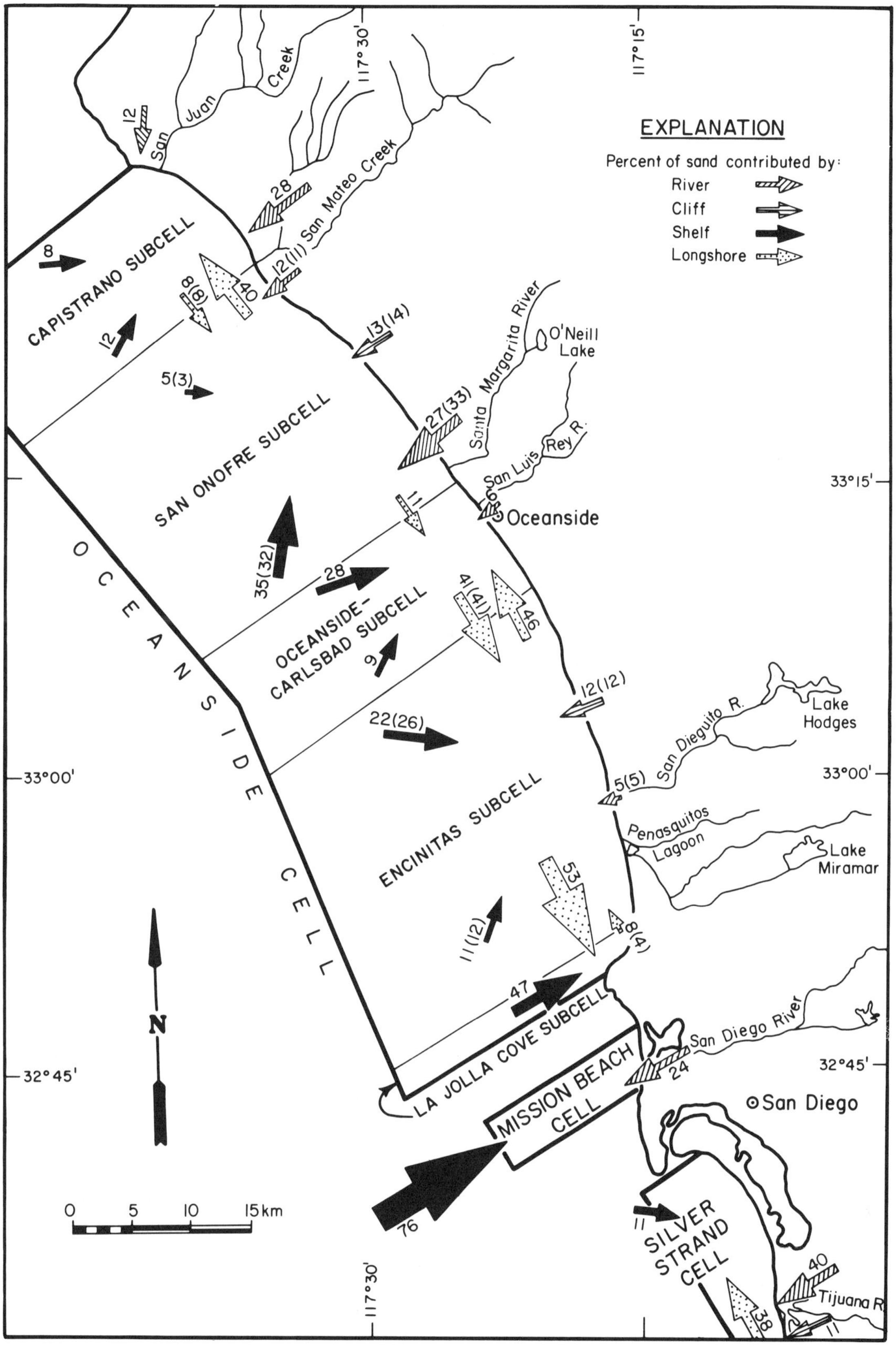

Figure 7. Map showing the percentage of medium sand-size quartz derived from sources contributing to each of the five subcells of the Oceanside Littoral Cell as well as the Mission Beach and Silver Strand Littoral Cells. The average percentage of sand contributed from each source is indicated along the shaft of each arrow. Where one value is given, it represents the composition of the winter sample set. Where two percentages are present, e.g. 22(26), the first value is for the winter and the second for the summer sample set.

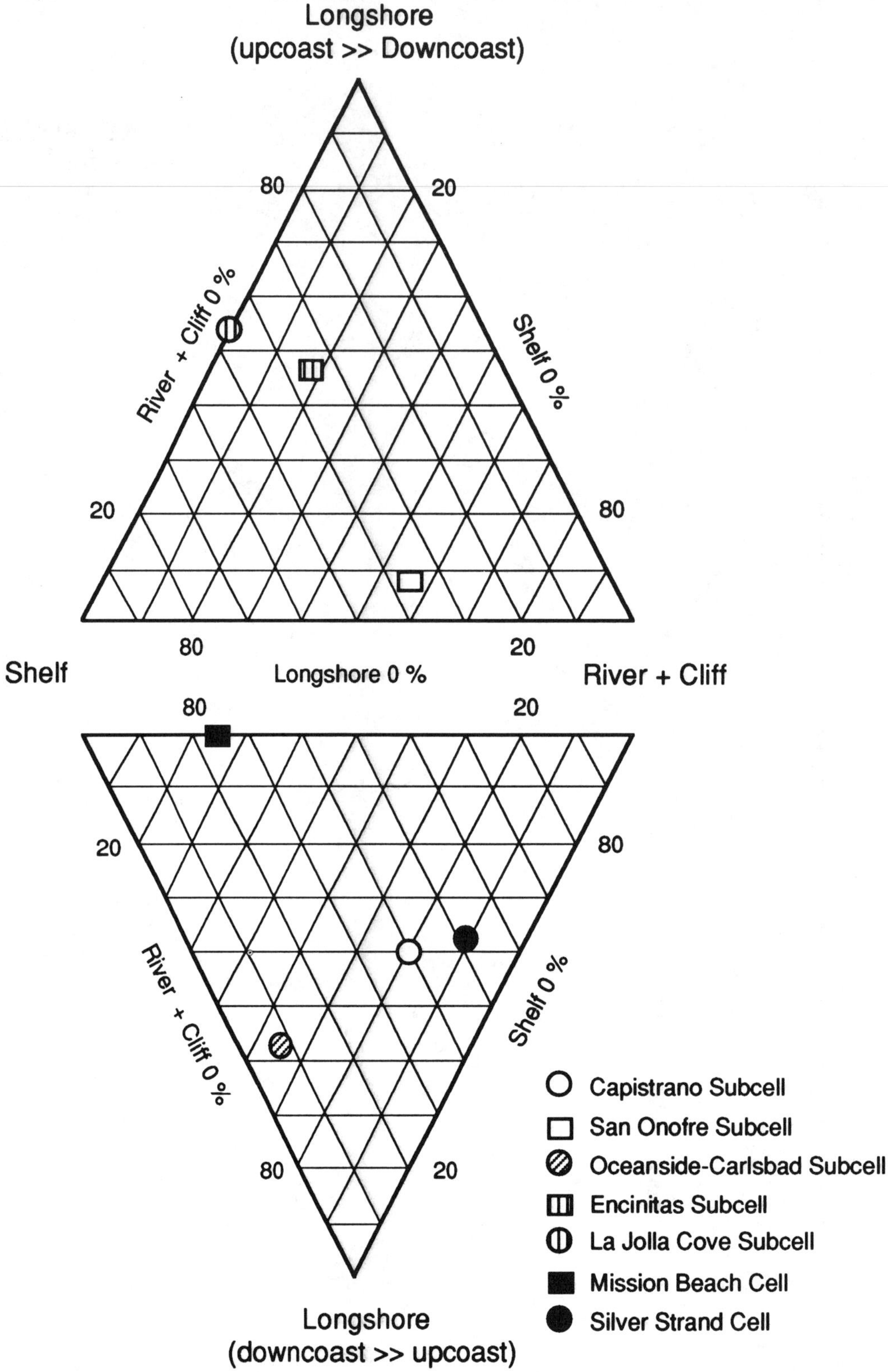

Figure 8. Diamond diagram showing the grain-shape composition of each of the subcells of the Oceanside Littoral Cell as well as the Mission Beach and Silver Strand Littoral Cells. This diagram is based on the data listed in Table 5.

at depths from -20 to -40 m MLW, within a bar complex at water depths from -5 to -10 m MLW. They suggest that during storms, wave base may migrate to water depths deeper than the bar resulting in the mobilization and landward transport of inner shelf sand to the bar complex, where it is mixed with sand eroded from the beach. Riester and others (1982) did not discuss transport or mixing rates, nor was an explicit mechanism proposed to explain the shoreward transport of inner shelf sand.

Dayton and others (1989) describe erosion in San Diego County, which was associated with the exceptionally intense storm of January 17-18, 1988. Seymour (1989a, 1989b) suggested this may have been a 100- to 200-year event. Dayton and others (1989) report that two concrete-filled scuba tanks were transported from a water depth of -12 m to a point inshore of -8 m - a distance of at least 0.5 km from their original location. Likewise, reef-derived boulders were transported from a depth of -15 m as much as 10 m shoreward, and a rock reef projecting about 2.5 m above the substrate at a depth of -18 m was reduced to rubble, and much of this was projected shoreward. Clearly, there is ample evidence of shoreward transport of even boulder-size sediment at inner shelf depths during intense storms. These studies, among others, clearly demonstrate the mobilization and shoreward transport of sand and coarser-grained sediment; however, the fundamental questions of exactly how, when and where remain unanswered.

Although the results of this and other previously-cited studies demonstrate that a volumetrically significant amount of littoral sand was derived from the adjacent inner continental shelf along both low- and high-gradient coasts, the shelf is usually ignored as a source in conventional budget analysis. This unfortunate circumstance is due to uncertainties associated with the volume involved, the rates of supply, the transport pathways, and very limited knowledge concerning the associated transport mechanism(s). In budget analysis, the sediment contributions and losses within a given coastal reach must be balanced so that the mass conservation equation is satisfied:

Sum of volumetric gains from sources = Sum of volumetric losses to sinks

The failure to properly assess the role of the inner continental shelf as a source frequently may lead to considerable overestimation of the role of one or more other sources, and such error may be compounded throughout the computational procedure.

Pierce (1968) considered extremely long surface waves as a potential mechanism to transport inner shelf sand sufficiently far onshore, where it would be entrained by more familiar nearshore wave and current systems to complete the transition from inner shelf to littoral sediment. Dean (1987) discusses several lines of evidence as well as mechanisms concerning onshore sediment transport from shelf depths. Using stream function (nonlinear) wave theory, Dean (1987) shows that the average, nondimensional bottom shear stress would tend to cause shoreward sediment transport across the continental shelf. The shoreward-directed forces require considerable time to transport large volumes of sediment landward; however, the average bottom shear stress increases substantially for shallower water and higher waves, which would accelerate transport rates. Clemens and Komar (1988) demonstrate that part of the present beach sand in the Cape Blanco area was acquired during lower sea-level stands with subsequent onshore transport during the latest transgression. The Clemens and Komar (1988) conceptual model may represent one spectral end member where no inner shelf sand is currently being introduced into the littoral system. The Pierce (1968) and Dean (1987) models may represent another end member where shelf sand is added more continuously to the littoral zone. It also is likely that large storms may either directly or indirectly contribute inner shelf sand in an episodic manner, and therefore may represent a third end member. It seems most probable that all three of these conceptual end-member processes either played or continue to play a role in the introduction of inner shelf sand to our present beaches.

The Sedimentary Petrology Laboratory at the University of Southern California is engaged in coastal research that will lead to a better understanding of the volumetric, temporal and mechanistic factors associated with the contribution of inner shelf sand to the littoral system. Foreshore sample sets collected by K. O. Emery during November 1956 and D. S. Gorsline during June and July 1969 from 55 stations from Point Conception to the U.S.-Mexico border have been augmented by November 1991 (end of the oceanographic summer) and April 1992 (end of the oceanographic winter) sample sets from the same localities. FGSA of these sample sets will permit an evaluation of regional-scale seasonal and relatively long-term (35.4 years) variation in the grain-shape composition of the beach sand as well as an evaluation of the possible change in shape composition as a result of the 1969 floods. In addition, numerous sets of geographically closely-spaced foreshore samples have been collected from beaches in the Oceanside-Carlsbad, Dockweiler-El Segundo and Ventura areas on time scales ranging from bi-monthly to bi-annually from 1986 to present. Beaches in these areas were and continue to be sampled on at least a seasonable basis as well as after each significant storm to evaluate the effects of these short-term events on associated grain-shape populations. The integration of FGSA results for the smaller spatial- and temporal-scale sample sets collected from the Oceanside-Carlsad, Dockweiler-El Segundo and Ventura beaches with those derived from the analysis of the larger spatial- and temporal-scale sample sets collected from Point Conception to the U.S.-Mexico border will permit the reduction of these data by means of a heirarchial analysis-of-variance design. The use of this experimental design will greatly facilitate the required significance testing and interpretation of the massive data set associated with this experiment. It is our belief that relatively long-term, monitoring experiments such as this are essential to better understand the mechanism(s), transport pathways and rates, and volumes associated with the natural introduction of shelf sand into the adjacent littoral system.

DIVISON OF A LARGE LITTORAL CELL

The littoral cell properly remains a fundamental organizing concept in coastal sedimentologic and engineering studies (Peterson and others, 1991; Best and Griggs, 1991). It is generally agreed that littoral cells are defined as zones bounded by prominent headlands and/or submarine canyon heads which greatly impede or terminate longshore sediment transport. However, it also is implicit that longshore transport occurs throughout much if not all of a given cell so that any sediment introduced into or derived from internal sources eventually is delivered to its downcurrent end. To better understand the associated rate and scale of longshore transport within a conceptual cell, it is imperative to trace the movement of sediment from a known point source through a known and substantive time interval. Only then can the net longshore transport rate be determined and used for predictive purposes. The general failure of wave hindcasting procedures and bathymetric profiling to properly identify the local sources and pathways of the sediment in transport renders such methods ineffective in determining longshore transport rates through long distances. The problem of quantifying the flux and net rate of longshore transport in the Oceanside Littoral Cell, among others, is being addressed through the application of FGSA.

The data listed in Tables 1, 2 and 3 were recast to show the percentages of sand derived from subaerial (river + cliff), longshore (downcoast + upcoast), and shelf (northern + southern) sources for each of the five subcells of the Oceanside Cell as well as the Mission Beach and Silver Strand Cells (Table 5; Fig. 8). It is clear from Table 5 and Figure 8 that each of the adjacent subcells as well as the Mission Beach and Silver Strand Cells are compositionally distinct with respect to shape.

In the case of the Oceanside Cell, samples on either side of the defined subcell boundaries have been and are continuing to be collected as part of the heirarchial analysis-of-variance experiment described in the preceding section. If net downcoast longshore transport occurs, then samples on the downcoast side of each subcell boundary should become progressively enriched in upcoast sand. Conversely, if net upcoast longshore transport occurs, then samples on the upcoast side of each boundary should become progressively enriched in downcoast sand. If the net longshore transport rate equals or is close to zero, the grain-shape composition of the samples on each side of the subcell boundaries will remain much the same. Yeh (1991) reports that the compositions of the samples at the subcell boundaries have not changed significantly from 1983 to 1991, therefore it seems that little, if any, sand has been exchanged across these subcell boundaries during this seven-year period. The analysis of sample collected in 1956 and 1969 will extend the considered time interval to 35.4 years. Although this short period of time may not be representative of net transport systems operating at a time scale of centuries to millennia, it is considerably longer than that commonly employed in coastal engineering studies. If these results are corroborated by further research, it would seem that the subcells identified should be the operative littoral unit for sediment budget analysis within the Oceanside Littoral Cell. The same kind of division may be required for other long and geomorphologically complex littoral cells. In any case, weighted local source estimates supplied by FGSA, grain-size and mineralogic analyses must be integrated with mass-balance computations to improve sediment budgets and to constrain sand-transport pathways to and within the littoral zone.

ACKNOWLEDGMENTS

This work is based partially on sand samples and Fourier grain-shape data collected by Kelly S. Ahlschwede, Sean D. Broadhead, Eleanor A. Compton and Kyung H. Cho of the Sedimentary Petrology Laboratory at the University of Southern California. Several additional samples were collected by Gerald G. Kuhn. The vibracores were recovered by Ocean Surveys, Inc., under contract to the U.S. Army Coastal Engineering Research Center, on board the *RV Diamond Lady*. Tim Fogarty wrote the EDGE Fortran Program for digitizing quartz grains. Anthony B. Price, Los Angeles District, U.S. Army Corps of Engineers, contributed several samples and participated in informative discussions regarding the experimental design employed and the interpretation of the results. However, the authors take full responsibility for the content of this paper. The manuscript was typed by Sue Turnbow. Figures 1, 2, 3 and 7 were drafted by Janet Dodds.

Partial funding for this research was provided by the Los Angeles District, U.S. Army Corps of Engineers Contract No. DACW09-87-D-0017.

REFERENCES

Ahlschwede, K. S., 1988, Sources and net littoral transport of sand in San Diego and southern Orange Counties, southern California: Fourier grain-shape analysis [unpubl. M.S. thesis]: Univ. Southern California, Los Angeles, California, 135 p.

Bendat, J. S., and Piersol, A. G., 1971, Random data: Analysis and measurement procedures: Wiley-Interscience, Inc., New York, 407 p.

Best, T. C. and Griggs, G. B., 1991, A sediment budget for the Santa Cruz Littoral Cell, in Osborne, R. H., ed., From Shoreline to Abyss: Contributions in Marine Geology in Honor of Francis Parker Shepard: SEPM (Society for Sedimentary Geology), Special Publication No. 46, p. 35-50.

Bloom, Laurie, 1979, The relationships among river, beach and submarine canyon sands in the southern Santa Barbara littoral cell, Ventura County, California: Fourier grain-shape analysis [unpubl. M.S. thesis]: Univ. of Southern California, Los Angeles, California, 115 p.

Bloomfield, Peter, 1976, Fourier analysis of time series: An introduction: New York, Wiley-Interscience Inc., New York, 258 p.

Boggs, Sam, Jr., 1992, Petrology of sedimentary rocks: Macmillan Publishing Co., New York, 707 p.

Bomer, E. J., 1985, Quartz grain-shape modification in high-gradient fluvial environments: Fourier grain-shape analysis [unpubl. M.S. thesis]: University of Southern California, Los Angeles, California, 139 p.

Bowen, A. J., and Inman, D. L., 1966, Budget of littoral sands in the vicinity of Point Arguello: Coastal Engineering Research Center, U.S. Army Corps of Engineers, Technical Memorandum Number 19, 41 p.

Brigham, E. O., 1974, The fast Fourier transform: Prentice-Hall, Inc., Englewood Cliffs, New Jersey, 252 p.

Broadhead, S. D., 1988, Fourier grain-shape analysis of the sources and littoral transport of sand, San Elijo Lagoon to Point La Jolla [unpubl. M.S. thesis]: University of Southern California, Los Angeles, California, 235 p.

Brown, P. J., Ehrlich, Robert, and Colquhoun, D. J., 1980, Origin of patterns of quartz sand types on the southwestern United States continental shelf and implications on contemporary shelf sedimentation - Fourier grain shape analysis: Journal of Sedimentary Petrology, v. 50, p. 1095-1100.

Carver, R. E., 1971, Procedures in sedimentary petrology: Wiley-Interscience, Inc., New York, 653 p.

Cho, K. H., 1989, Sedimentology of a composite inner-shelf sand body resulting from the resuspension of nearshore sediment by episodic, storm-generated currents, Oceanside California [unpubl. M.S. thesis]: University of Southern California, Los Angeles, California, 104 p.

Clark, R. A., and Osborne, R. H., 1982, Contribution of Salinas River sand to the beaches of Monterey Bay, California, during the 1978 flood period: Fourier grain-shape analysis: Journal of Sedimentary Petrology, v. 52, p. 807-822.

Clemens, K. E., and Komar, P. D., 1988, Oregon beach-sand compositions produced by mixing of sediments under a transgressing sea: Journal of Sedimentary Petrology, v. 58, p. 519-529.

Curray, J. R., 1960, Sediments and history of Holocene Transgression, continental shelf, northwest Gulf of Mexico: in Shepard, F. P., Phleger, F. B., and Van Andel, Tj. H., eds., Recent sediments, northwest Gulf of Mexico, American Association of Petroleum Geologists, Tulsa, Oklahoma, p. 221-266.

Dayton, P. K., Seymour, R. J., Parnell, P. E., and Tegner, M. J., 1989, Unusual marine erosion in San Diego County from a single storm: Estuarine, Coastal and Shelf Science, v. 29, p. 151-160.

Dean, R. G., 1987, Additional sediment input to the nearshore region: Shore & Beach, v. 55, nos. 3-4, p. 76-81.

Ehrlich, Robert, and Chin, Maureen, 1980, Fourier grain-shape analysis: A new tool for sourcing and tracking abyssal silts: Marine Geology, v. 38, p. 219-232.

Ehrlich, Robert, and Weinberg, Bernhard, 1970, An exact method for characterization of grain shape: Journal of Sedimentary Petrology, v. 40, p. 205-212.

Ehrlich, Robert, Orzeck, J. J., and Weinberg, Bernhard, 1974, Detrital quartz as a natural tracer - Fourier grain shape analysis: Journal of Sedimentary Petrology, v. 44, p. 145-150.

Ehrlich, Robert, Brown, P. J., Yarus, J. M., and Przygocki, R. S., 1980, The origin of shape frequency distributions and the relationship between size and shape: Journal of Sedimentary Petrology, v. 50, p. 475-484.

Emery, K. O., 1960,. The sea off southern California: A modern habitat of petroleum: John Wiley and Sons, New York, 366 p.

Gaynor, J. M., 1984, Sources and transport of sand in the littoral zone of Lake Tahoe, California and Nevada: Fourier grain-shape analysis [unpubl. M.S. thesis: University of Southern California, Los Angeles, California, 112 p.

Giles, R. T. and Pilkey, O. H., 1965, Atlantic beach and dune sediments of the southern United States: Journal of Sedimentary Petrology, v. 35, p. 900-910.

Hales, L. Z., 1978, Coastal processes study of the Oceanside, California littoral cell: U.S. Army Corps of Engineers, Miscellaneous Paper H-78-8, 60 p.

Hudson, C. B., and Ehrlich, Robert, 1980, Determination of relative provenance contributions in samples of quartz sand using Q-mode factor analysis of Fourier grain shape data: Journal of Sedimentary Petrology, v. 50, p. 1101-1110.

Inman, D. L., 1976, Summary report of man's impact on the California coastal zone: State of California, Department of Navigation and Ocean Development, Sacramento, California, 150 p. (Reissued by State of California, Department of Boating and Waterways, June 1980).

Inman, D. L., 1981, Evaluation of the budget of waves and sand, Oceanside Littoral Cell and recommendations for monitoring and sand bypassing: Los Angeles District, U.S. Army Corps of Engineers, 24 p.

Inman, D. L., and Chamberlain, T. K., 1960, Littoral sand budget along the southern California coast: Report of the Twenty-first International Geological Congress, Copenhagen, Volume of Abstracts, p. 245-246.

Inman, D. L. Ewing, G. C., and Corliss, J. B., 1966, Coastal sand dunes of Guerrero Negro, Baja California: Geological Society of America Bulletin, v. 77, p. 787-802.

Inman, D. L. and Jenkins, S. A., 1983, Oceanographic report for Oceanside beach facilities: City of Oceanside, California, 149 p.

Krumbein, W. C. and Sloss, L. L., 1963, Stratigraphy and sedimentation: W. H. Freeman and Company, San Francisco, 660 p.

Kuhn, G. G., and Shepard, F. P., 1984, Sea cliffs, beaches and coastal valleys of San Diego County: University of California Press, Berkeley, 193 p.

Kuhn, G. G., Osborne, R. H., Ahlschwede, K. S., and Compton, E. A., 1987, Processes, locations and rates of coastal cliff erosion from 1947 to present, Dana Point to the United States-Mexico border and the stratigraphy of contributing cliffs and bluffs at San Onofre, Camp Pendleton, and Torrey Pines: Los Angeles District, U.S. Army Corps of Engineers, CCSTWS 87-2,298 p..

Mazzullo, Jim, 1986, Sources and provinces of late Quaternary sand on the east Texas-Louisiana continental shelf: Geological Society of America Bulletin, v. 97, p. 638-647.

Mazzullo, J. M., and Ehrlich, Robert, 1980, A vertical variation in the St. Peter Sandstone - Fourier grain shape analysis: Journal of Sedimentary Petrology, v. 50, p. 63-70.

Mazzullo, J. M., and Ehrlich, Robert, 1983, Grain shape variation in the St. Peter Sandstone: A record of eolian and fluvial sedimentation of an early Paleozoic cratonic sheet sand: Journal of Sedimentary Petrology, v. 53, p. 105-119.

Mazzullo, Jim, and Withers, K. D., 1984, Sources, distribution, and mixing of late Pleistocene and Holocene sands on the south Texas continental shelf: Journal of Sedimentary Petrology, v. 54, p. 1319-1334.

Mazzullo, J. M., and Magenheimer, Stewart, 1987, The original shapes of quartz sand grains: Journal of Sedimentary Petrology, v. 57, p. 479-487.

McMaster, R. L., 1954, Petrography and genesis of the New Jersey beach sands: New Jersey State Department of Conservation Series, Bulletin 63, 239 p.

Moffatt & Nichol, Engineers, 1989, Sediment budget report Oceanside Littoral Cell (draft): Los Angeles District, U.S. Army Corps of Engineers, Los Angeles, California, 112 p.

Mrakovitch, John, Ehrlich, Robert, and Weinberg, Bernhard, 1976, New technique for stratigraphic analysis and correlation - Fourier grain-shape analysis, Louisiana offshore Pliocene: Journal of Sedimentary Petrology, v. 46, p. 226-233.

Nordstrom, C. E., and Inman, D. L., 1973, Beach and cliff erosion in San Diego County, California: in Ross, A. and Dowlen, eds., Studies in the geology and geologic hazards of the greater San Diego area, San Diego Association of Geologists, San Diego, California, p. 125-131.

Osborne, R. H., and Cho, K. H., 1989, Sedimentology of a composite inner-shelf sand body resulting from the resuspension of nearshore sediment by episodic, storm-generated currents: Proceedings Volume, Coastal Zone '89, American Society of Civil Engineers, p. 4391-4405.

Osborne, R. H., and Pipkin, B. W., 1982, Geomorphic and sedimentologic analysis for the Oceanside Project: Phase II: Los Angeles District, U.S. Army Corps of Engineers, 121 p.

Osborne, R. H. and Yeh, C.-C., 1991, Fourier grain-shape analysis of coastal and inner continental-shelf sand samples: Oceanside Littoral Cell, southern Orange and San Diego Counties, southern California: in Osborne, R. H., ed., From shoreline to abyss: Contributions in marine geology in honor of Francis Parker Shepard: SEPM (Society for Sedimentary Geology) Special Publication No. 46, Tulsa, Oklahoma, p. 51-66.

Osborne, R. H., Edelman, M. C., Gaynor, J. M., and Waldron, J. M., 1985, Sedimentology of the littoral zone in Lake Tahoe, California-Nevada: California State Lands Commission, 88 p.

Osborne, R. H., Yeh, C.-C., and Lu, Y., 1991, Grain-shape analysis of littoral and shelf sands, southern California: in Krause, N. C., ed., Coastal sediments '91, American Society of Civil Engineers, p. 846-859.

Peterson, C. D., Darienzo, M. E., Petit, D. J., Jackson, P. L., and Rosenfeld, C. l., 1991, Littoral-cell development in the convergent Cascadia margin of the Pacific northwest, USA, in Osborne, R. H., ed., From Shoreline to Abyss: Contributions in Marine Geology in Honor of Francis Parker Shepard: SEPM (Society for Sedimentary Geology), Special Publication No. 46, p. 17-34.

Pierce, J. W., 1968, Sediment budget along a barrier island chain: Sedimentary Geology, v. 3, p. 5-16.

Porter, G. A., Ehrlich, Robert, Osborne, R. H., and Combellick, R. A., 1979, Sources and nonsources of beach sand along southern Monterey Bay, California - Fourier shape analysis: Journal of Sedimentary Petrology, v. 49, p. 727-732.

Powers, M. C., 1953, A new roundness scale for sedimentary particles: Journal of Sedimentary Petrology, v. 23, p. 117-119.

Riester, D. D., Shipp, R. C., and Ehrlich, Robert, 1982, Patterns of quartz sand shape variation, Long Island Littoral and shelf: Journal of Sedimentary Petrology, v. 52, p. 1307-1314.

Rittenhouse, G., 1943, A visual method of estimating two-dimensional sphericity: Journal of Sedimentary Petrology, v. 13, p. 79-81.

Russell, R. D., and Taylor, R. E., 1937, Roundness and shape of Mississippi River sands: Journal of Geology, v. 45, p. 225-267.

Saville, T., 1960, Sand transfer, beach control, and inlet improvements, Fire Island Inlet to Jones Beach, New York: 7th Conference on Coastal Engineering, p. 785-807.

Schwarcz, H. P., and Shane, K. C., 1969, Measurement of particle shape by Fourier analysis: Sedimentology, v. 13, p. 213-231.

Seymour, R. J., 1989a, Unusual damage from a California storm: Shore & Beach, v. 57, no. 3, p. 31.

Seymour, R. J., 1989b, Wave observations in the storm of 17-18 January, 1988: Shore & Beach, v. 57, no. 4, 14-17.

Shepard, F. P., 1973, Submarine Geology: New York, Harper & Row, Publishers, 3rd ed., 517 p.

Simon, Li & Associates, Inc., 1988, River sediment discharge study San Diego region: Technical Report, Los Angeles District, U.S. Army Corps of Engineers, CCSTWS 88-3, 95 p.

Stow, D. A., and Chang, H. H., 1987, Coarse sediment delivery by coastal streams to the Oceanside Littoral Cell, California: Shore & Beach, v. 55, no. 1, p. 30-40.

Tekmarine, Inc., 1987, Oceanside Littoral Cell Preliminary Sediment Budget Report: Los Angeles District, U.S. Army Corps of Engineers, CCSTWS 87-4, 158 p.

Trask, P. D., 1952, Source of beach and at Santa Barbara, California as indicated by mineral grain studies: Beach Erosion Board, U.S. Army Corps of Engineers, Technical Memorandum No. 28, 24 p.

U.S. Army Corps of Engineers, 1987, Consolidated Plan of Study Coast of California Storm and Tidal Wave Study: Los Angeles District, Reference No. CCSTWS 87-1, 369 p.

Van Andel, Tj. H., and Poole, D. M., 1960, Sources of recent sediments in the northern Gulf of Mexico: Journal of Sedimentary Petrology, v. 30, p. 91-122.

Van Nieuwenhuise, D. S., Yarus, J. M., Przxygocki, R. S., and Ehrlich, Robert, 1978, Sources of shoaling in Charleston Harbor: Fourier grain shape analysis: Journal of Sedimentary Petrology, v. 48, p. 373-384.

Weggel, J. R., and Clark, G. R., 1983, Sediment budget calculations, Oceanside, California: U.S. Army Corps of Engineers, Los Angeles District, Coastal Engineering Research Center, Vicksburg, Mississippi, Coastal Engineering Research Center Miscellaneous Paper 83-7, 55 p.

Yeh, Chia-Chen, 1991, Grain-shape composition of coastal and inner continental shelf sand sources from 1983 to 1990: Oceanside Littoral Cell, southern Orange and San Diego Counties, southern California: [unpubl. Ph.D. dissertation]: Univ. Southern California, Los Angeles, CA 249 p.

Young, K., 1980, Fourier analysis of quartz grains, a new method of interpreting sediment movement: Marine Resource Bull., v. 12, no. 2, p. 4-5.

A. Capistrano Subcell (April, 1986): Dana Point to San Mateo Point

	San Juan Creek	Northern Shelf	Southern Shelf	San Mateo Creek	Downcoast Beach
1895-B	12	8	13	26	41
1815-B	12	7	11	31	39
1605-B	13	8	12	28	39

River sample: 1865-R, San Juan Creek
Northern Shelf Sample: 1800-S, Offshore of Dana Point Harbor
Southern Shelf Samples: 1658-S, 1610-S, Offshore of San Clemente
Downcoast River Sample: 1595-R, San Mateo Creek
Downcoast Beach Sample: 1580-B

B. San Onofre Subcell (April, 1986): San Mateo Point to North Breakwater at Oceanside Harbor

	Upcoast Beach	San Mateo Creek	Northern Shelf	Cliff	Santa Margarita River	Southern Shelf
1580-B	9	15	7	13	13	43
1530-B	10	17	6	12	9	46
1300-B	6	8	4	12	45	25
1270-B	8	10	6	13	25	38
1170-B	6	11	2	13	34	32
1115-B	7	10	4	14	37	28

Upcoast Beach Sample: 1605-B
Norhtern River Samples: 1595-R, 1565-R, San Mateo Creek
Northern Shelf Sample: 1575-S, 1535-S, 1480-S, 1455-S, 1425-S, 1385-S, 1322-S, Offshore of San Mateo Creek
Cliff Samples: 1380-C, 1460-C, 1410-C, 1355-C, 1320-C, 1283-C, 1275-C, 1220-C, San Onofre and Camp Pendleton Are
Southern River Samples: 1150-R, 1285-R, Santa Margarita River
Southern Shelf Samples: 1123-S, 1105-S, 1090-S, 1077-S, Offshore of Oceanside

C. San Onofre Subcell (October, 1986): San Mateo Point to North Breakwater at Oceanside Harbor

	Upcoast Beach	San Mateo Creek	Northern Shelf	Cliff	Santa Margarita River	Southern Shelf
1300-B	8	12	3	14	27	36
1270-B	8	13	4	14	25	36
1170-B	6	8	2	12	49	23
1115-B	8	11	3	14	30	34

Upcoast Beach Sample: 1605-B
Norhtern River Samples: 1595-R, 1565-R, San Mateo Creek
Northern Shelf Sample: 1575-S, 1535-S, 1480-S, 1455-S, 1425-S, 1385-S, 1322-S, Offshore of San Mateo Creek
Cliff Samples: 1380-C, 1460-C, 1410-C, 1355-C, 1320-C, 1283-C, 1275-C, 1220-C, San Onofre and Camp Pendleton Area
Southern River Samples: 1150-R, 1285-R, Santa Margarita River
Southern Shelf Samples: 1123-S, 1105-S, 1090-S, 1077-S, Offshore of Oceanside

D. Oceanside-Carlsbad Subcell (April, 1986; October, 1986; and February, 1989, respectively): Oceanside Harbor to Onshore Projection of Carlsbad Submarine Canyon

	Upcoast Beach	Northern Shelf	San Luis Rey River	Southern Shelf	Downcoast Beach
1030-B	12	25	7	8	48
1030-B	10	32	6	11	41
810-B	11	27	6	8	49

Upcoast Beach Sample: 1115-B
Northern Shelf Sample: 1073-S, Offshore of Oceanside
River Sample: 1050-R, San Luis Rey River
Southern Shelf Samples: 985-S, 993-S, 920-S, 915-S, 864-S, 860-S, Offshore of Carlsbad
Downcoast Beach Sample: 765-B

E. Encinitas Subcell (April, 1986): Onshore Projection of Carlsbad Submarine Canyon to La Jolla Submarine Canyon

	Upcoast Beach	San Dieguito River	Northern Shelf	Cliff	Southern Shelf	Downcoast Beach
765-B	45	6	23	13	10	3
645-B	42	5	21	14	11	7
625-B	39	5	22	12	11	11
595-B	39	5	24	11	12	9
545-B	43	5	20	13	9	10
520-B	40	5	25	12	11	7

Upcoast Beach Sample: 810-B
River Samples: 590-R, 635-R, 530-R, San Dieguito River
Northern Shelf Samples: 519.5-C, 650-C, Offshore of Encinitas
Cliff Samples: 500-C, 466-C, 465-C, 463.6-C, 463-C, Torrey Pines Area
Southern Shelf Samples: 454-S, 492-S, 464.5-S, Offshore of La Jolla
Downcoast Beach Sample: 427-B

F. Encinitas Subcell (October, 1986): Onshore Projection of Carlsbad Submarine Canyon to La Jolla Submarine Canyon

	Upcoast Beach	San Dieguito River	Northern Shelf	Cliff	Southern Shelf	Downcoast Beach
645-B	43	6	21	13	10	7
625-B	41	6	27	10	13	3
595-B	41	5	26	12	12	4
575-B	42	5	25	12	11	5
545-B	39	5	28	11	14	3
520-B	41	5	27	12	12	3

Upcoast Beach Sample: 810-B
River Samples: 590-R, 635-R, 530-R, San Dieguito River
Northern Shelf Samples: 519.5-C, 650-C, Offshore of Encinitas
Cliff Samples: 500-C, 466-C, 465-C, 463.6-C, 463-C, Torrey Pines Area
Southern Shelf Samples: 454-S, 492-S, 464.5-S, Offshore of La Jolla
Downcoast Beach Sample: 427-B

G. La Jolla Cove Subcell (September, 1989): La Jolla Submarine Canyon to La Jolla Point

	Upcoast Beach	Shelf
446-B	61	39
445-B	58	42
444-B	40	60

Upcoast Beach Sample: 448-B
Shelf Sample: 443-S, Offshore of Point La Jolla

TABLE 1. RESULTS OF FOURIER GRAIN-SHAPE ANALYSIS FOR BEACH SAMPLES COLLECTED AT THE TIMES INDICATED WITHIN EACH OF THE FIVE INITIAL SUBCELLS. THE VALUES LISTED FOR EACH SAMPLE ARE THE PERCENTAGES OF MEDIUM-SAND SIZE QUARTZ DERIVED FROM EACH OF THE LOCAL SOURCES INDICATED FOR THAT SUBCELL. THE SAND SOURCES CONSIDERED FOR EACH SUBCELL AND ASSOCIATED SAMPLES ARE LISTED BELOW EACH MATRIX.

TABLE 2. RESULTS OF FGSA FOR FORESHORE SAMPLES COLLECTED FROM THE MISSION BEACH LITTORAL CELL. VALUES LISTED FOR EACH SAMPLE ARE THE PERCENTAGES OF MEDIUM-SAND SIZE QUARTZ DERIVED FROM EACH OF THE LOCAL SOURCES LISTED BELOW THE MATRIX.

	San Diego River	Shelf
330-B	24	76
225-B	25	75

River Sample: 267-R, San Diego River

Shelf Sample: 394-S, Offshore of San Diego River

TABLE 3. RESULTS OF FGSA FOR FORESHORE SAMPLES COLLECTED FROM THE SILVER STRAND LITTORAL CELL. VALUES LISTED FOR EACH SAMPLE ARE THE PERCENTAGES OF MEDIUM-SAND SIZE QUARTZ DERIVED FROM EACH OF THE LOCAL SOURCES LISTED BELOW THE MATRIX.

	Northern Shelf	Tijuana River	Downcoast Beach	Clif
160-B	9	46	35	10
20-B	13	33	41	12

Northern Shelf Sample: 205-S, Offshorn of Point Loma

River Sample: 8-R, Tijuana River

Downcoast Beach Sample: 4-B

Cliff Sample: 2-C, Border Field State Park, South of Downcoast Beach Sample

TABLE 4. AVERAGE SHORELINE ORIENTATION AND AVERAGE SLOPE FROM 0 TO -18 M (-60 FT) MLLW FOR EACH OF THE FIVE SUBCELLS CONSIDERED.

Subcell	Trend	Average Slope (0 to -18 m MLLW)	
		m/km	Degrees
Capistrano	N42°W	5.9	0.34
San Onofre	N45°W	6.0	0.34
Oceanside-Carlsbad	N39°W	11.6	0.66
Encinitas	N12°W	18.0	1.03
La Jolla Cove	N28°E	40.2	2.30

TABLE 5. AVERAGE COMPOSITION BY SUBCELL FOR THE OCEANSIDE LITTORAL CELL AND FOR THE MISSION BEACH AND SILVER STRAND LITTORAL CELLS

	River and Cliff	Longshore	Shelf
Capistrano Subcell	40	40	20
San Onofre Subcell	55	8	37
Oceanside-Carlsbad Subcell	6	57	37
Encinitas Subcell	18	47	35
La Jolla Subcell	0	53	47
Mission Beach Cell	25	0	75
Silver Strand Cell	51	38	11

Chapter 8

Landslides

Photo of earthquake-induced landslide of new cut slope at Interstate 5 north of Sylmar, 1971.

Landsides

Major Landslide Repairs in Southern California
Lawrence R. Cann and Edward A. Steiner

Portuguese Bend Landslide
Perry Ehlig

Big Rock Mesa Landslide and Litigation, Malibu
Dennis A. Evans and Richard E. Lownes

Mega-Landslides in Southern California
Martin L. Stout

Paleolandslides in the San Gabriel Mountains
J. David Rogers, Lawrence H. Herber and Thomas H. Hibner

L. A. County's "Drainage Benefit Assessment Areas" Landslide Mitigation
Robert A. Larson

State-of-the-Art Shear Key for Stabilization of a Large Landslide
Richard M. Ramirez

Vegetation to Prevent Surficial Slope Failures
Paul M. Merifield

Editors Note: Coping with slope instability is a way of life for many in southern California. With the flatlands mostly built-up, new construction has been moving into the hills. Here we sometimes find subtle geologic problems that need to be recognized and remediated. This is exemplified by the extensive methods used to stabilize the Portuguese Bend landslide. The Big Rock Mesa Landslide of the 1980s resulted in the largest legal award of this type in history—$92 million. The next two papers demonstrate the importance of basic geomorphology and surficial geology in the recognition of gigantic ancient landslides. The chapter concludes with three papers on methods, both technical and statutory, to prevent catastrophic slope failures.

MAJOR LANDSLIDE REPAIRS IN SOUTHERN CALIFORNIA

Lawrence R. Cann
Principal Engineering Geologist

Edward A. Steiner
Senior Project Geologist

Leighton and Associates
2121 Alton Parkway
Irvine, California 92714

CONTENTS

INTRODUCTION

As property values and new home construction costs have risen, mitigation of landslides has become increasingly attractive in contrast to abandonment. In many cases, it is more cost effective to design and implement remediation for a landslide than to abandon and raze a damaged home, and rebuild elsewhere. This is especially true in southern California where costs have made it possible to use remediation techniques which would otherwise be too expensive for a single family home or small structure.

This paper uses four case histories for landslides in Southern California which have been repaired since the 1973 AEG Annual Meeting, the last one held in southern California. These case histories demonstrate a unified approach to landslide stabilization that, in our opinion, provides a high level of benefits when compared to the costs of repair.

The case histories include the Bluebird Canyon landslide in Laguna Beach, the Cove Road landslide in Dana Point, the Seagate landslide at Niguel Shores, and the Big Rock landslide in Malibu.

STAGES OF LANDSLIDE REPAIR

The repair of a complex landslide requires the integrated efforts of people with a wide variety of training and expertise, working as a well-organized team. Like any successful team, there must be considerable advanced planning prior to beginning the repair, and then implementation and coordination of the plan during the repair. A complex landslide repair generally progresses through five stages: the emergency response stage, the geotechnical investigation and interpretation stage, the planning and design stage,the construction stage, and the completion stage.

Emergency Response Stage

Most major landslide repairs begin with a series of emergency earthwork measures taken within a few days after the event to minimize potential for injury or additional property damage. In the interests of expediency, these measures are usually performed before enough information is available to adopt a permanent repair plan. The most important issue at this stage of the slide repair is to halt the landslide by providing temporary stability while a permanent solution is designed. During the emergency the Engineering Geologist and Geotechnical Engineer help to determine the extent of areas which may be influenced by the sliding, help to assess the damage to public and private property, and evaluate the need for heavy equipment. Heavy equipment is used to build temporary access roads, move earth to increase temporary stability, break down dangerously over steepened headscarps, fill in ground fissures and grabens, or construct temporary drainage or run-off control devices. It is important to deal honestly and sympathetically with the affected residents. The loss of a home is also the loss of the largest investment many people will make. The loss of a home is a major crisis for people and the respect and support of all parties is appreciated.

Investigation and Interpretation Stage

Before any landslide repair can begin, the landslide must be thoroughly understood or characterized. Coordination meetings between the client, design civil engineer, engineering geologist, geotechnical engineer, grading contractor and agency are of paramount importance. The main purpose of these meetings is to communicate the geotechnical conditions in sufficient detail and clarity to be understood by the non-technical members the team. Pre-construction planning is absolutely necessary and, if the repair is to involve mass grading, it is very important to develop an earth work logistics plan. Illustrations, including maps and cross-sections depicting the geotechnical conditions, utilities, and easements must be prepared, and thoroughly understood by all the team members prior to commencing the landslide repair.

Time is of the essence because the slide area is in a "metastable" condition, and additional failures, such as headscarp collapses, are common. The investigation stage of a major landslide repair requires quick response and, in some cases, may involve working around the clock. The geotechnical investigation phase of the landslide repair should include, at least, the following tasks:

- Background research of all available pertinent geologic reports and maps of the specific area. This should be supplemented by collection and review of stereoscopic pairs of aerial photographs taken before and after the slide. Both vertical and oblique photographs in color, black and white, and false color infrared should be reviewed. The infrared photographs are particularly useful to define areas of disrupted drainage, and the margins of the landslide.
- Detailed geologic mapping of existing exposures. The purpose is to understand the root cause of the failure, and geologic conditions are often a factor. This should include existing landslide morphologic features. The surface cracks can be useful indicators of the kinematics of the landslide.
- Subsurface exploration consisting of backhoe or dozer trenches, and large diameter and small diameter borings, if needed. The best results are usually obtained from the large diameter borings which allow the geologist first hand observations of the nature of the landslide rupture surface. The investigator should also collect samples for laboratory testing to determine the strength of the earth materials and the chemical composition of ground water, if present. Care must be taken in this exploration program because the active slide mass is still unstable and may create dangerous subsurface conditions.
- Preparation of maps and cross sections to depict the three dimensional geometry of the landslide. The geometry is critical to accurate analysis of remediation options.
- Detailed geotechnical stability analyses to evaluate the remediation options for stabilization.
- Preparation of a report that summarizes the conditions describes the elements needed to stabilize the slide area.

The Planning And Design Stage

Repair of a landslide is dependent upon the type, size and geometry of the slide, topographic conditions in the vicinity of the landslide, the extent of damage, the surface and ground water conditions, the final stability requirements, the availability of construction material and equipment, and economic constraints.

The planning and design effort is critical to the landslide repair. It requires a synthesis of all of the available data, and it is the point at which the engineering geologist, geotechnical engineer, design civil engineer, and owner must have a clear exchange of ideas. Technical, social and logistical problems must be approached with an open mind and some compromises may be necessary.

The geotechnical issues and design options must frequently be discussed with those trained in other disciplines. These include attorneys, contractors, insurance companies, homeowners, and community leaders.The best results are achieved with a combination of visual and verbal techniques. The geotechnical consultant should make a detailed, written outline of the summary of the findings so that there is a written as well as an oral summation. Sometimes the data and interpretation can be depicted on 3-dimensional models as well as maps and 2-dimensional cross sections. Pictures of the site are beneficial and shorten the otherwise long-winded and detailed descriptions. The best aids to understanding are:

- Maps, drawings, pictures or video tape recordings of existing conditions encountered at the sites.
- Cross-sectional or 3-dimensional views to show the geometry of the landslide and alternative mitigation measures.
- Overlays on maps and cross-sections or sequential models to demonstrate the various remedial alternatives. This makes selection of design options easier and less-time consuming.

The grading contractor can play an important role in this preliminary stage. This is particularly true when repairing landslides in populated areas, such as at the Bluebird Canyon or the Seagate slides discussed below. Both of these slides were successfully repaired with minimal loss to additional property, and at a cost much lower than what would have occurred without good advanced planning.

The land owner is the ultimate decision maker because he/she reaps the rewards or bears the cost of the repair. To have a successful project, the landowner must know and understand the geotechnical problems and their potential impacts. For example, if the landslide material has a high moisture content, it may require specialized and costly treatment before it can be reused as compacted fill. Logistical planning is critical to the success of the project. It should include:

- Frequent meetings between the entire project team
- A proactive, positive attitude toward problem solving.
- Good topographic and other survey data from the design Civil Engineer.
- Clearly illustrate and well understood sequence of the repair project at each anticipated stage including the location of off-site disposal and borrow areas, temporary stability conditions, dewatering requirements, etc.
- A thorough study of the types of equipment and manpower that will be needed on the project.
- A stockpile of supplies needed to complete the repair

- A thorough study of the quantities and topographic conditions involved so that construction traffic patterns can be established to maximize efficiency and minimize handling of earth materials.

Contingency planning for conditions at the site which differ from the interpreted conditions.

Construction Stage

The most critical time during a hillside repair is during the removal of the landslide and construction of the slope stabilization elements. The most common stabilization method is the removal of portions or all of the landslide and their replacement with an earth fill buttress. This requires the excavation of buttress keys and manufacture of temporary cut slopes. In most cases the back slope excavation is only marginally stable with a factor of safety at or less than one. Special precautions must, therefore, be taken to minimize the failure of these back slopes. Great care is needed during this period. A back cut failure may radically increase the total yardage of earth materials moved or threaten adjacent properties.

Methods of Achieving Temporary Stability

A failure of a temporary buttress back slope can easily double or triple the total buttress yardage quantity and can significantly prolong construction and increase grading, geotechnical and civil engineering costs. The potential for failure can be reduced by:

- Excavation of the back slope of the key at the shallowest possible angle, usually no steeper than the dip of bedding or the dip of the landslide rupture surface. In most cases the upslope location of buildings, utilities, and topographic conditions dictate the shallowest back cut slope angle possible.
- Excavation of the key in sections so that the entire key and back slope are not open and unsupported at the same time.
- Time is critical. Subdrain installation and buttress construction should begin and be completed as quickly as possible. If the buttress keyway is wider than about 25 feet, fill may be placed against the back slope first to provide additional temporary support for the slope before placing the fill in the outer portion of the keyway.
- Construction of steel soldier pile walls or reinforced concrete caisson walls may be needed if temporary back slope failures can not be tolerated because existing structures are close to the top of the landslide. Temporary soldier piles can be extracted after the landslide is removed or the back slope is supported by the buttress fill. Or the piles can be designed to be left in place as a part of the permanent fix.
- Ground freezing is a method for strengthening the soil on a temporary basis. It was utilized with limited success in the Seagate slide area in 1981. It has some potential for certain types of excavations in the appropriate soil types.
- Soil cement columns can also be considered for temporary stabilization of back slopes.

Cost of Landslide Repairs

It is difficult to determine the repair costs for a major landslide because of the many variables involved. However, the cost of simply removing slide debris and replacing it as compacted fill is much higher that normal hillside grading costs. For example, to grade a baseball field of compacted fill, in August 1991 cost approximately \$1.25 /cubic yard. In contrast, earthwork for landslide repairs can cost from \$5 up to 25 or \$30/cubic yard.

CASE HISTORIES

Selection Criteria For Accompanying Case Histories

The following case histories demonstrate the methods previously discussed. They all contain a common purpose—the restoration of an unstable area to an acceptable degree of safety, to minimize additional damage to nearby properties, and to repair the area as efficiently as possible. The repair efforts described in these case histories all share the following common attributes:

1) Thorough planning of the repair effort;
2) Complete and unabiguous communication:
3) Positive attitude among all partners toward the issues and contributions of others;
4) A sense of teamwork by all members;
5) A flexible and creative approach to problem solving; and
6) The resources (equipment, materials, and manpower) to complete the project despite changed conditions.

Each case history was selected because it demonstrates a combination of unique conditions and an innovative method of repair.

Big Rock Landslide, Malibu

Introduction

Repair of the Big Rock landslide required an innovative repair to maintain access along the heavily traveled Pacific Coast Highway in Malibu. The repair incorporated a variety of monitoring methods to warn of incipient failure during construction (Clark, 1986; Evans and Lownes, this volume).

The Big Rock landslide, not to be confused with its much larger active neighbor to the west, the Big Rock Mesa landslide, began as a slow moving rock and debris slide during the winter of 1979. The landslide included the upper portion of a natural-over-cut slope excavated for the original Pacific Coast Highway (PCH) some 60 years ago. This designed cut slope was more than 200 feet high with a gradient of about 1:1 (horizontal to vertical). The name "Big Rock" was derived from the highly visible 100-ton boulder perched at the upper portion of the slope. The landslide presented a major challenge to CALTRANS as they sought to protect the row of homes on the Pacific Ocean across PCH from the slope. Adverse conditions included the weather, steep terrain, the limited area in which to work, and the need to keep an important highway open (Fig. 1).

First movement was noted by displacement of the "Big Rock" a few inches to a few feet downhill. After a number of frustrating attempts, CALTRANS finally removed the "Big Rock" by pulling it down onto PCH with a skip loader and cable system. In April, following removal of the rock, three major head scarps with lateral displacements of up to 1.5 feet appeared which, along with sloughing of small rocks and debris onto PCH indicated involvement of a much larger area. The debris was raining down from the toe region of the moving mass, approximately at the top of the cut slope. On April 13, PCH was closed to all traffic. Two lanes were reopened on May 8 after concrete berms were placed on the roadway to protect the open lanes from falling and rolling debris. The containment zone formed behind the berm filled rapidly with debris. A 20-to 30-foot high wall of steel H-beams with heavy wood lagging was driven into the road along its center lane to protect the remaining two open lanes. Within a week the wall was breached by a 30-ton boulder. The barrier remained in place as slide movement slowed, and proved to be an effective but not fail-safe method of keeping PCH open.

Geologic Setting

Preliminary geologic mapping of the slope face showed the slope was breaking across the predominant bedding and was following old fracture surfaces or the contacts between the diabase dikes and country rock. The bedrock consisted of Tertiary sediments of the Vaqueros and Sespe formations which was overlain by the Topanga Formation. These rocks were intruded by irregular shaped bodies of the diabase . All rocks near the ground surface were highly fractured and weathered, especially the diabase dikes.

The geometry of the slide was poorly known. It was estimated to be only 30 feet thick, based on the pattern of scarps and absence of rotational movement. Visual examination of the failure surface by large diameter drilling was precluded by high cost, the very steep slopes, and hardness of the rock. Small diameter core drilling adjacent to the slide mass located several zones of poor quality rock at depths of more than 100 feet. Coring on the slide mass itself could not be completed due to the unsafe slope conditions. Therefore the

Figure 1. Aerial photograph of the Big Rock landslide, Malibu, California. Pacific Coast Highway and existing homes in foreground.

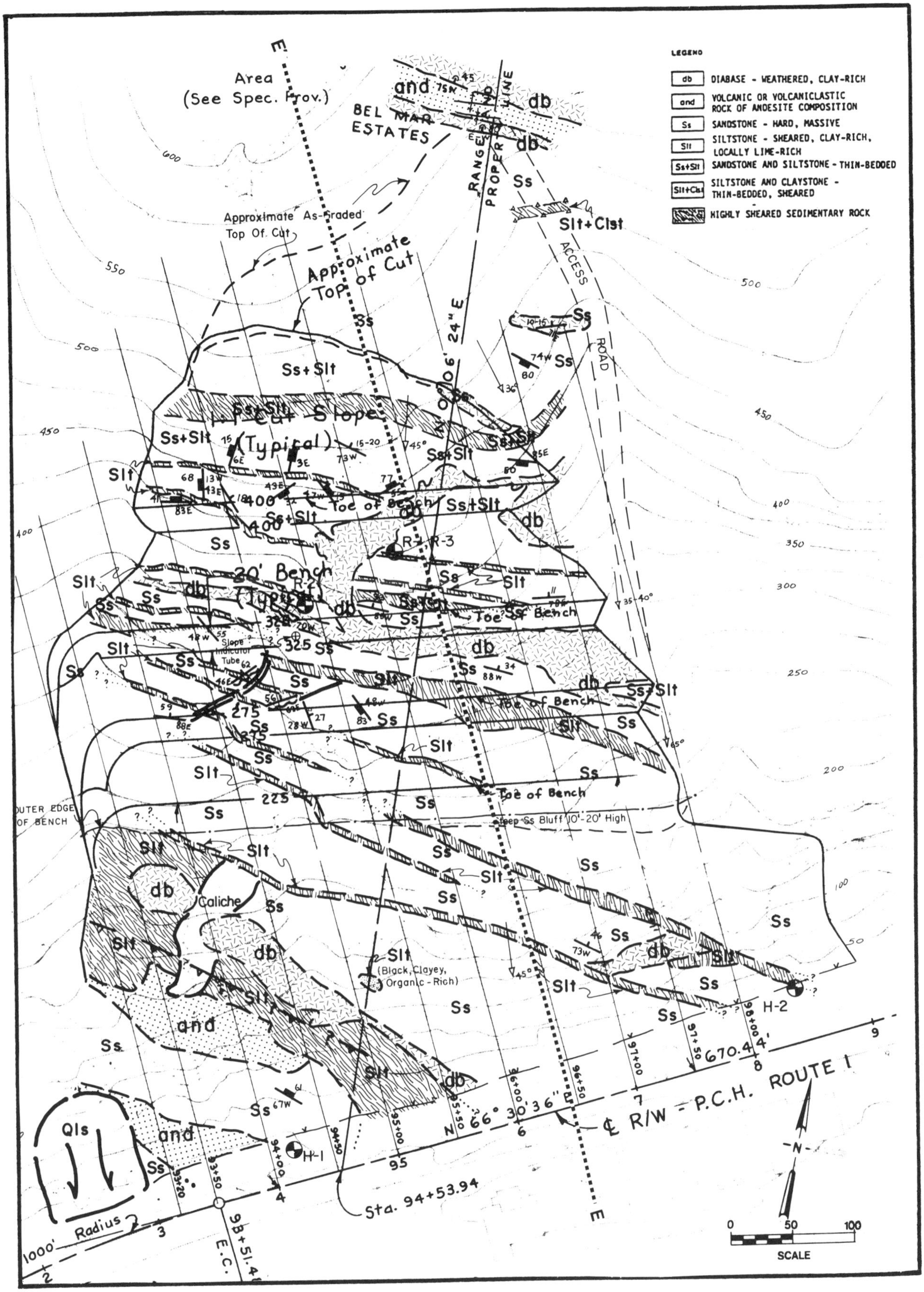

Figure 2. Schematic geologic map of the Big Rock landslide.

limits of the landslide were based primarily on surficial mapping. The generally high degree of weathering and abundant clay content of the diabase bodies made them the likely hosts of the rupture surfaces (Figs. 2 and 3).

Monitoring

The slide presented a continuing hazard to the adjacent highway and homes. As a result, a three-part monitoring program was begun almost immediately. The program was designed as an early warning system for renewed or catastrophic movement. It consisted of: (1) geophones installed at shallow depths in the surface of the slide to detect subaudible rock noises (SARN). The SARN system could effectively detect high levels of micro-seismic activity in the slide, but was less effective in predicting imminent large scale movement. (2) CALTRANS installed a series of long baseline (10 to 25 meter) wire-cable extensometers with an early warning arrangement at highway level. Large movements in the head scarp would set off the signal, allowing closure of PCH to traffic. The system worked well, but the correlation between slide movement and sloughing of debris onto the roadway was poor. In fact, the road was closed several times when birds landed on the cable causing it to stretch. (3) Finally a network of points on the slope face was established and resurveyed to determine which portions of the slope face were actually involved in the slide and to provide a warning of any change in the rate of movement (on a scale of weeks) which could threaten residences along PCH as well as the highway itself. The bulk of the slope face was not involved in the slide, but the uppermost points on the face were slowly moving downslope.

This was an important slide to monitor due to the potential impact of continued or accelerated movement on public safety and property. While it was not possible to predict micro-movements which released debris directly to the base of the slide, the three systems did demonstrate the larger movement pattern and gross changes in rate of movement.

Remediation

Remediation consisted of excavating the loose slide debris and broken rock to a fresh slope face with 1:1 slopes, with wide drainage and slough benches to the top of the existing rock face. Even then designed fresh slope would have had questionable surficial stability because of the deep fracturing. Therefore, the plan included a rock anchor system of #11 steel rebars with an average length of 30 feet drilled perpendicular into the slope face and grouted in place. These were placed at 7 1/2 feet on center over the entire freshly excavated face ands the areas in between dowels were covered with a chain link mesh, welded to the tops of the rebar, to reduce the rockfall hazard.

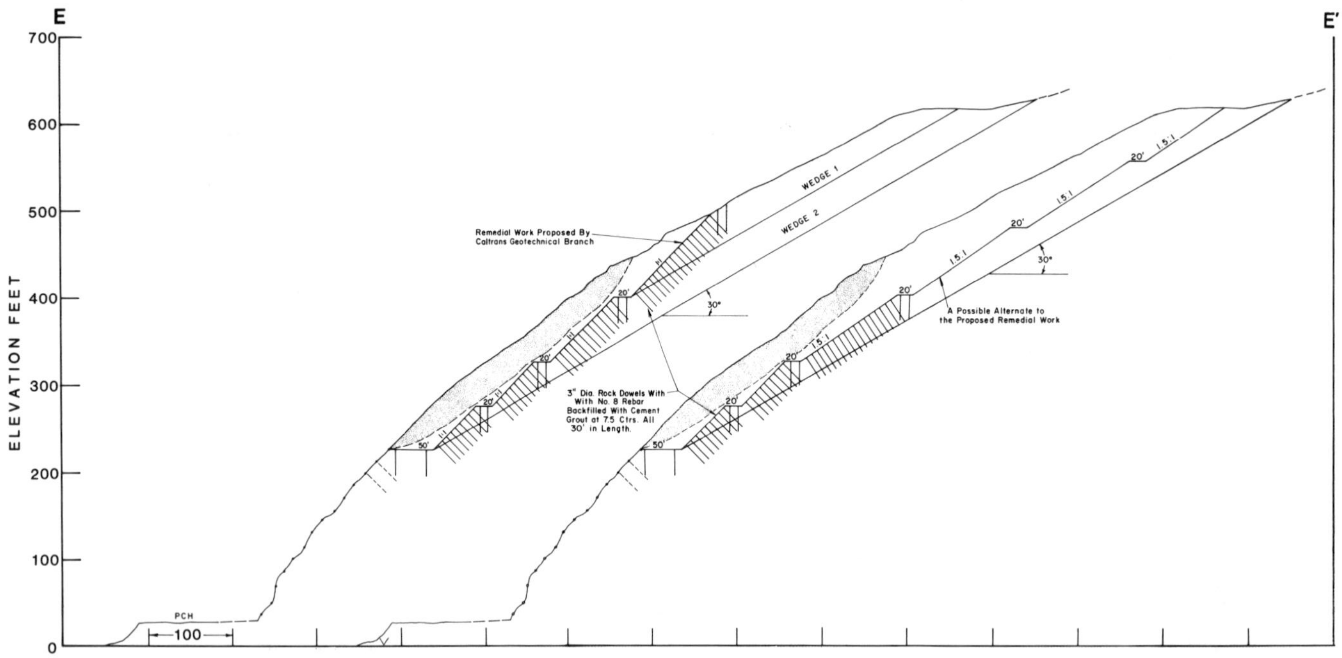

Figure 3. Geologic cross-section of the Big Rock landslide.

Figure 4. Aerial photograph of the partially completed Big Rock landslide repair.

The grading began in September 1979 with removal starting at the top of the proposed cut slope. The slope cut was begun far enough upslope to remove approximately 60 vertical feet of material in the landslide zone. Since the slide was estimated to be about 30 feet thick, the designed cut would remove the active slide and the weakest, highly broken surficial bedrock.

Steep, locally vertical cuts were made where resistant sandstone layers were encountered. The 1:1 design was continued downslope to near the toe at the road shoulder where a vertical cut 8 feet high was made.

Grading was slowed by loose rock spilling down slope, becoming airborne during descent, and falling onto the highway and adjacent properties; difficult ripping and rock breakage in the cemented sandstone and some diabase. Construction rockfall was mitigated by further lengthening and raising the temporary catchment barrier along the highway. Grading was completed in June 1980 and the lower slopes were sprayed with a hydroseed compound (Fig. 4).

The Bluebird Canyon Landslide, Laguna Beach

Introduction

The successful repair of the Bluebird Canyon landslide (Leighton et al., 1986) required the active participation of city, county, state, and federal officials including the President of the United States. The slide was notable by the support of the entire community for the slide victims, and the efforts of the repair team. This was exemplified by the absence of litigation and hold-harmless agreements signed prior to the repair.

The Bluebird Canyon Landslide (Fig. 5) occurred on October 2, 1978 in a hillside section of the City of Laguna Beach, California (Fig. 5 and 6). This landslide resulted in major devastation and property damage, but no loss of life or injury. The slide dammed the Bluebird Canyon drainage course; ultimately, 24 houses had to be demolished; all major utilities and 775 lineal feet of public roads were destroyed; and nearly 100 residences around the periphery of the slide were threatened or adversely impacted (Fig. 6).

Emergency measures were initiated with reestablishment of access to the slide area which began on October 3, 1978, and the commencement of the installation of a 72-inch diameter storm drain on November 18, 1978 in the Bluebird Canyon drainage course. This storm drain was completed one day before a torrential rainstorm that could have caused major down-stream flooding. All stabilization measures were essentially completed by June 29, 1979, approximately eight months following the occurrence of the landslide. Many phases of the stabilization measures were constructed around the clock on a 6 to 7-day workweek and commencing on October 3, 1978. The remedial stabilization measures included the construction of five principal-engineered devices as fol-

Figure 5. The Bluebird Canyon landslide occurred during the early morning hours of October 2, 1978. Ultimately 24 homes were lost. Miraculously, there were no deaths or serious injuries. (photo courtesy Woody Higdon)

lows:

1) A storm drain in Bluebird Canyon to alleviate damming of the channel.
2) A horseshoe-shaped shear key buttress consisting of a 10-celled prism of compacted fill.
3) Sixty-six soldier piles placed vertically around the perimeter of the buttress excavation, and then interconnected with welded grade beams.
4) A gravity butress, an earthen dam of compacted fill placed in Bluebird Canyon in front of the landslide.
5) A subdrainage system consisting of perforated pipes and permeable materials.

General Geologic Relationships

The Bluebird Canyon Landslide of October 2, 1978 was a block glide that reactivated a prehistoric slide with a history of at least three different episodes of movement. It occurred on a dip-slope that contained a faulted synclinal warp (Figs. 7, 8a, 8b, and 9). The basic conditions for landslide failure were a combination of (1) rock weakness, (2) canyonward-dipping bedding and an ancient slide structure, (3) saturation of the graben area and basal rock units of the landslide resulting in buildup of hydrostatic pressure in at least the graben area and pore pressures along the basal rupture sur-

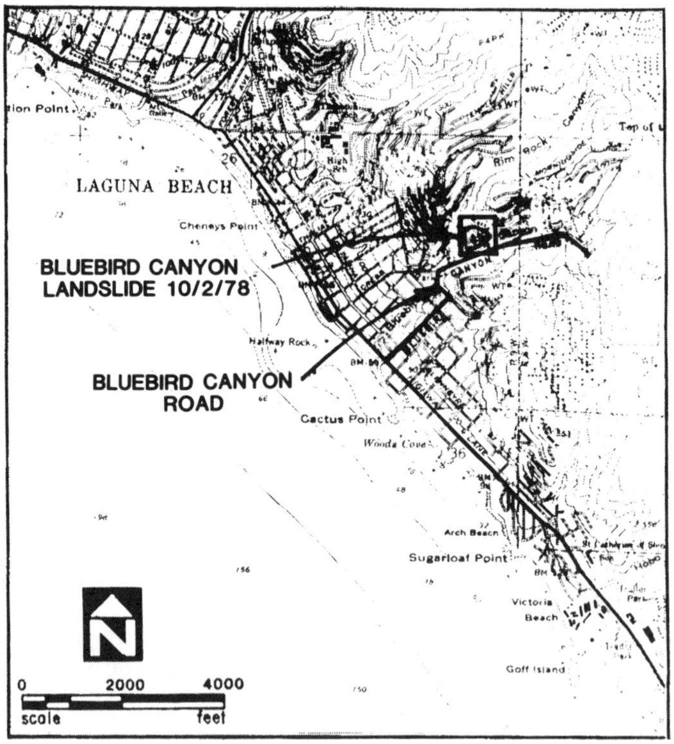

Figure 6. Index Map

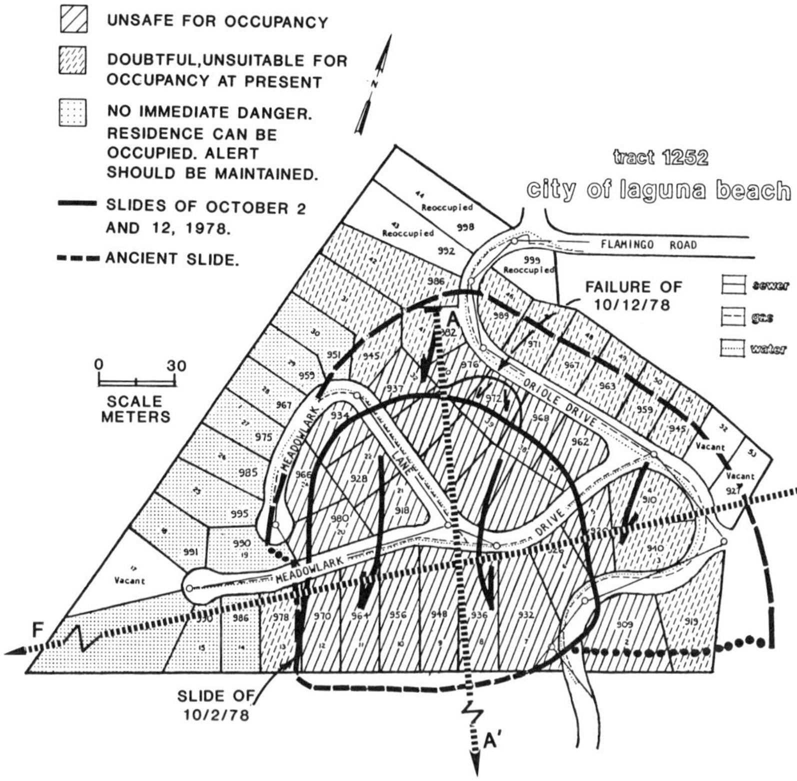

Figure 7. Site Index Map

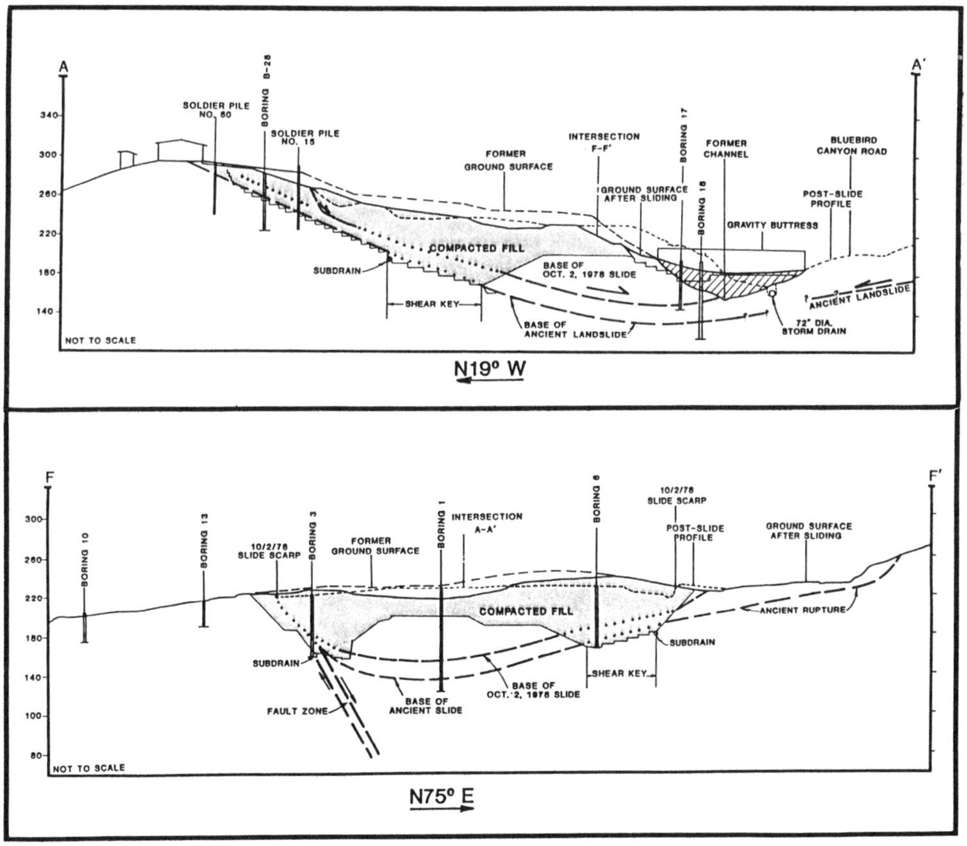

Figure 8. Geologic Cross-sections

face and (4) canyon entrenchment and erosion at the foot of the dip-slope and ancient slide (Figs. 7, 8a, and 8b).

Phase I Emergency Stabilization Measures-Storm Drain Installation

The location and size of the required storm drain were determined by the design civil engineer, Robert Bein, William Frost and Associates (RBF). This determination was based on the area topography, rainfall data, and availability of CSP (Corrugated Steel Pipe). The design called for a 72-inch diameter, 600 foot long storm drain (Fig. 10) to be located at the modified bottom of the Bluebird Canyon channel (see Figs. 11 and 12 for general location).

The main geotechnical concern during the required trenching was reactivation and extension of the slide. The Oriole Drive embankment between Bluebird Canyon Road and Tract 1252 produced a buttressing effect that probably prevented the eastern portion of the October 2nd slide from extending to the eastern margin of the ancient rupture surface. Trenching along the toe of the slide and across Oriole Drive would have undermined the October 2nd slide and breached the buttress created by the Oriole Drive embankment. Consequences of reactivation and extension of the slide would have been:

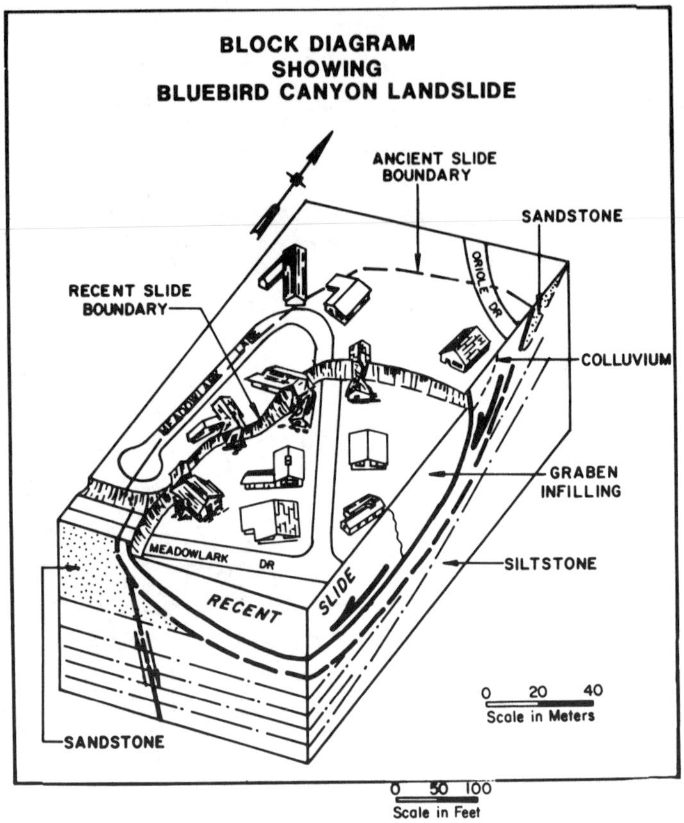

Figure 9. Block diagram showing the Bluebird Canyon landslide.

Figure 10. A 72-inch diameter, 600 foot long storm drain pipe was installed during Phase I stabilization construction. Workers were tethered and the city Fire/Rescue squad stood by to provide prompt assistance if needed.

A. Jeopardize the safety of the construction and inspection team.

B. Destruction of approximately 500 additional feet of Oriole Drive, the previously damaged residence at 909 Oriole Drive and three other previously undamaged residences (910, 919 and 940 Oriole Drive).

C. Formation of a lake upstream of the slide and possible breaching of the landslide dam.

To safeguard against the above threats the following was recommended:

A. Trenching for pipe installation be done in small increments (30 feet in length).

B. Prior to trenching across Oriole Drive, a buttress fill be constructed in the canyon bottom, east of Oriole, in order to compensate for the breaching of the buttressing effect of the Oriole Drive embankment.

C. Trenching, pipe installation and backfill of trench across Oriole Drive be accomplished on a round-the-clock schedule to minimize the length of time available for failures.

Installation of the storm drain was completed with no reactivation of the slide.

Phase II Of Stabilization Measures

A very conservative approach was used for evaluation of strength parameters of the slide materials and rupture surface. As a result, a much higher than normal level of confidence could be established in the adopted factor of safety of 1.2 rather than the normally accepted safety factor of 1.5. The main stabilization elements recommended were a drained shear key buttress at the head of the slide and a gravity buttress (Figs. 11 and 12) at its toe. These buttresses, the associated sub-drainage system and ingrading backslope failures are discussed below.

Shear Key Buttress

The shear key buttress consisted of a massive zone of compacted structural fill keyed into firm bedrock underlying the disturbed and slide-involved rock units. The approximate plan location of the buttress key is indicated in Figures 11 and 12. Typical cross-sections of the buttress are shown on Figures 8a and 8b. Earlier geotechnical reports recognized the high potential for backslope failures during the construction of the shear key buttress, and in order to reduce this potential, recommended:

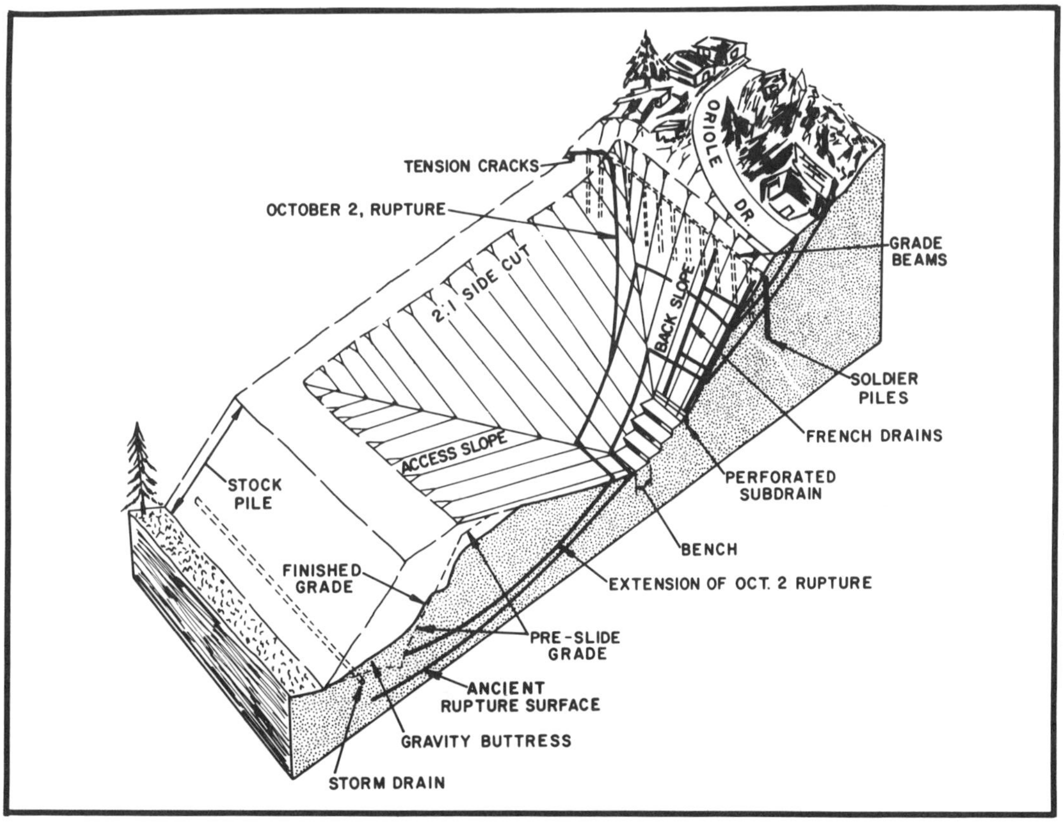

Figure 11. Block diagram showing design and construction elements of the stabilization plan. View is to the northwest.

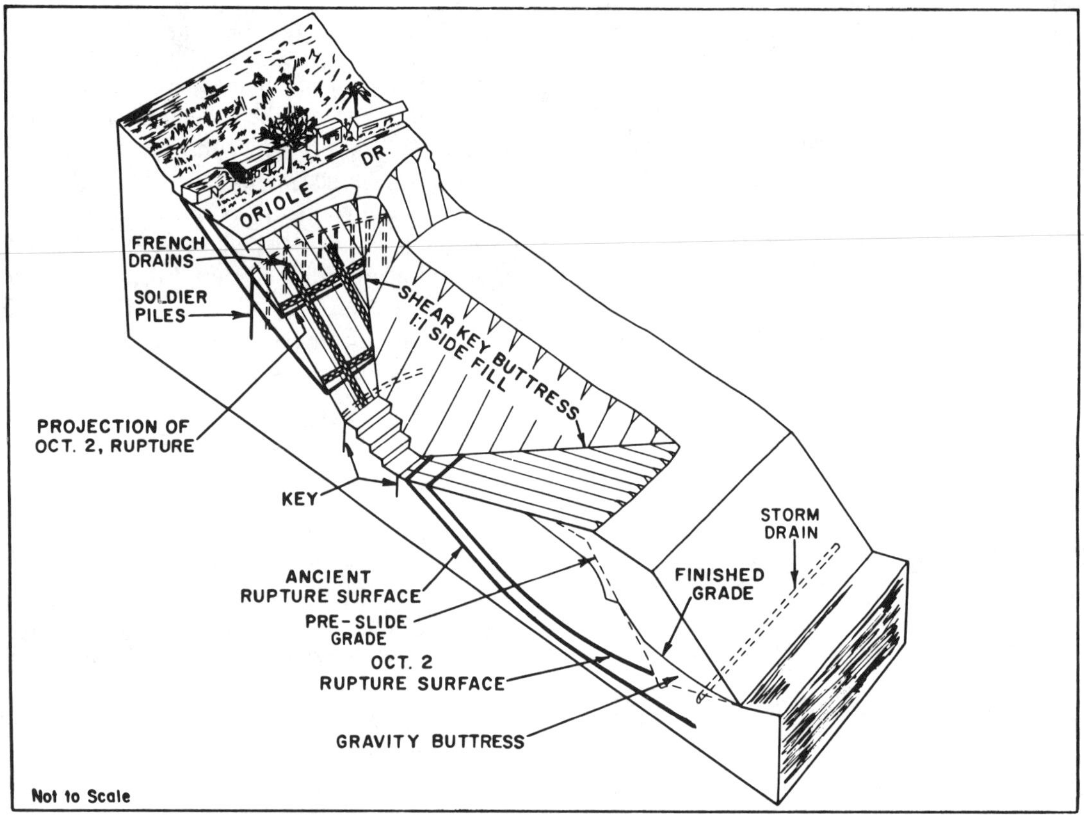

Figure 12. Block diagram showing design and construction elements of the stabilization plan. View to the northeast.

A. Installation of a system of 2-foot diameter reinforced concrete soldier piles.

B. Excavation of the backslope in relatively small sections with round-the-clock grading as necessary.

Forty-eight soldier piles were installed between January 22 and February 6, 1979, at locations indicated on Figures 11 and 12. Availability and economics governed the choice of steel beams used in soldier piles. In order to maximize effectiveness of the soldier piles, they were interconnected at the surface by welded grade beams. Sukut Construction Company, the project contractor, devised a ten-phase grading scheme that combined the geotechnical recommendation of phased backslope excavation with the earthwork constraints. Figures 11 and 12 depict how the combination of soldier piles and the phased backslope excavation reduced the probability of major backslope failures.

Backslope Failures

Geologic mapping and strain monitoring were useful in predicting the location and time of occurrence of backslope failures. As anticipated, numerous backslope failures occurred as a result of the removal of lateral support during excavation of the shear key buttress, particularly during the wet season. Only the failures of April 23, May 12 and May 15 slid on the "ancient rupture surface." All other slides occurred on shallower surfaces. For the most part, the combined effect of soldier piles and phased backslope excavation was successful in containing these failures to the area in front of the soldier piles, preventing further upslope failure in previously undamaged areas. Only the April 23rd failure extended significantly upslope beyond the soldier piles. Despite the extensive counter measures, one residence was destroyed and an 80-foot section of Oriole Drive was lost. As a result of the April 23 backslope failure, eighteen additional soldier piles were recommended and installed, so that buttress stabilization could continue.

Gravity Buttress

The recommended gravity buttress consisted of a compacted earth embankment placed at the slide-modified canyon bottom (Fig. 8a, 8b, and 11 through 14). This feature improved the gross stability of the slide mass by elevating the finish grade of the canyon bottom and effectively reducing the overall height of the south-facing slope. The initial design of this buttress was modified near the end of the construction phase of the shear key buttress. About 15,000 cubic yards of excess soil was placed as compacted fill in the canyon bottom east and west of Oriole Drive.

Subdrainage System

High ground water levels within the ancient graben zone and its role in elevating pore water pressures within ancient rupture zones and triggering the October 2, 1978 slide were documented by subsurface probes. Perched water in the ancient graben zone was also confirmed by numerous seeps and high moisture content observed during the excavation of backslope of shear key buttress. The likelihood of future ground water buildup was reduced by installation of a subdrainage system.

The subdrainage system consisted of a main subdrain and a lacework of french drains (Fig. 11 and 12). The main subdrain is located at the foot of the backslope and consists of a perforated ABS pipe with both 6-inch and 4-inch diameter sections. It was embedded in a 24-inch- wide trench filled to a minimum height of 24 inches with Class 2 filter material. The subdrain is horseshoe-shaped in plan view, with half the drain flowing to the east side of the area and half to the west side. Flow gradient is a minimum of 2 percent toward the 72-inch storm drain to which the subdrain is connected on both the east and west sides of the landslide area. Terminal sections of the subdrain are unperforated in order that water cannot enter the fill and slide materials at the foot of the slide. Class 2 filter material was placed beneath the storm drain to provide subdrainage of the slide material not removed during emergency grading.

At 30-foot intervals french drains radiate from the main subdrain up the backslope. They consist of 2-foot wide trenches filled with a 1-foot minimum thickness of filter material. The radial pattern of french drains is interconnected at 20- to 25-foot vertical intervals by near-horizontal lateral drains. A total of 2,840 feet of 4- and 6-inch diameter ABS pipe and 1,614 tons of Class 2 filter material were used in the subdrainage system.

Earthwork and Quality Control of Fill Compaction

The fill materials used for stabilization consisted of silty clays, clayey silts and fine clayey silty sands. They were spread in thin lifts in approved areas (shear key buttress, gravity buttress and storm drain backfill), moisture-conditioned to near-optimum moisture content and compacted by sheepsfoot rollers and heavy construction equipment. At the base of the shear key buttress, the fill was keyed into firm undisturbed bedrock beneath the basal rupture surface of the ancient slide (see cross-sections). The October 2nd and ancient slide materials were entirely removed from the shear key. All loose slide debris resulting from backslope failures were removed by stripping and/or benching. Brush and loose surficial slide debris were cleared from the base of the gravity buttress; however, most of the October 2nd slide debris and alluvium at the canyon bottom were left in place.

Deep-Seated Stability

The recommended emergency stabilization measures were implemented. The three principal measures for deep-seated stability are: (1) a horseshoe-shaped shear key buttress, (2) a

SITE STABILIZATION IN PROGRESS ON BLUEBIRD CANYON LANDSLIDE, MARCH, 1979

Los Angeles Times Photo by Deris Jeannette

Figure 13. Buttress construction in progress, Bluebird Canyon.

Figure 14. Stabilization of the landslide was essentially completed by June 29, 1979; after only 8 months. Nearly all destroyed homes were

gravity buttress in Bluebird Canyon, and (3) an elaborate integrated subdrainage system.

Analyses of data showed that the adopted criterion of achieving a factor of safety of 1.2 could be met using conservatively estimated strength parameters based on the post-failure geometry of the slide mass. In addition, analyses were also presented which indicate that the factor of safety is 1.5 when strength parameters were back-calculated from pre-failure geometry. In light of the close correlation between strength parameters determined in the laboratory and strength parameters back-calculated from pre-failure geometry, it was the professional geotechnical opinion that the stabilized area meets the normally accepted static stability state of 1.5. Pseudostatic analysis using a lateral earthquake loading of 0.15g (gravitational force) yields a dynamic factor of safety of 1.13 which is normally acceptable.

Conclusions

The Bluebird Canyon Landslide of October 2, 1978 has been successfully stabilized and the blocked drainage course restored. The safety hazards posed to public right-of-way and private properties upslope of the area and downstream of the drainage course have been alleviated. Geotechnical conditions encountered during grading were essentially as anticipated. The major refinements to previous findings were: (1)

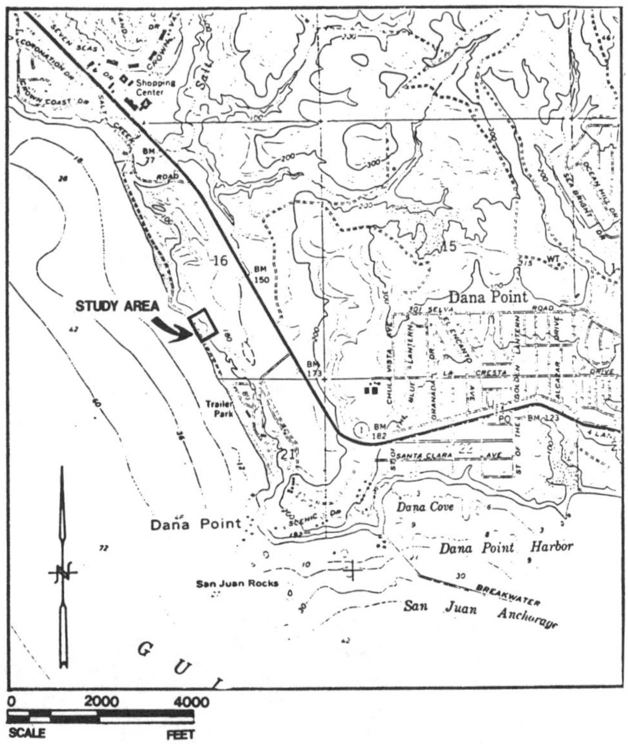

Figure 15. Index Map, Seagate landslide

the "ancient rupture surface" was found to be locally slightly lower than previously inferred, and (2) a north-trending, east-dipping normal fault was found to bound and control the western side of the ancient and October 2nd landslides (Fig. 8a and 8b).

The emergency repair measures completed within the scope of this report have been performed in accordance with FDAA guidelines. Since these guidelines do not allow permanent improvements, further measures are needed to render the emergency stabilization measures permanent.

Seagate Landslide Repair, Niguel Shores

Introduction

The Seagate landslide repair is unique in that it attempted to use a frozen-ground method to stabilize the buttress front cut and back cut. The repair constantly battled the pounding storm surf, unusually high ground water conditions, and geologic mapping while excavating a buttress key more than 20 feet below sea level. Environmental constraints along the Pacific Ocean required the use of a conveyor-belt system to transport the excavated landslide material to a stockpile and return it to the buttress cut. Temporary soldier piles were later installed at the toe and at the head of the landslide to provide emergency backslope stability support. Dewatering was necessary to control subsurface ground water flow upslope of the landslide area and from the Pacific Ocean. Extensive sandbagging and temporary restraining devices were necessary to minimize flooding of the site by high tides and storm surf.

The Seagate landslide occurred in 1977 shortly after the site was rough graded (Fig. 15). The slide occurred during the heavy rains of the winter of 1977 to 1978. Although, the slide took place before expensive coastal view homes had been built, the landslide severely damaged six of the residential lots and a portion of the road. The toe of the landslide extended onto the beach at the toe of the previously graded slope (Fig. 16). The movement of the landslide was temporarily halted by excavation and removal of material at the head of the slide; thus, reducing the driving force.

The slide area was extensively investigated and analyzed prior to commencement of the landslide repair. Numerous large-diameter and small-diameter borings were drilled and geologically logged in addition to the excavation and logging of backhoe trenches. Extensive soil engineering testing and stability analyses were performed in order to arrive at an effective repair plan.

The grading contractor as well as the owner provided valuable input during the design process to assure that the repair alternative selected could be constructed and would be as cost effective as possible. The Seagate landslide repair commenced in mid-September 1980 and was completed by the end of February 1981.

Figure 16. Near-vertical aerial view of the Seagate Landslide repair. The beach and Pacific Ocean are at the bottom of the picture. The conveyor-belt system utilized to remove slide debris from the excavation is shown on the left side of the picture. The buttress key was approximately 75 percent completed at this stage.

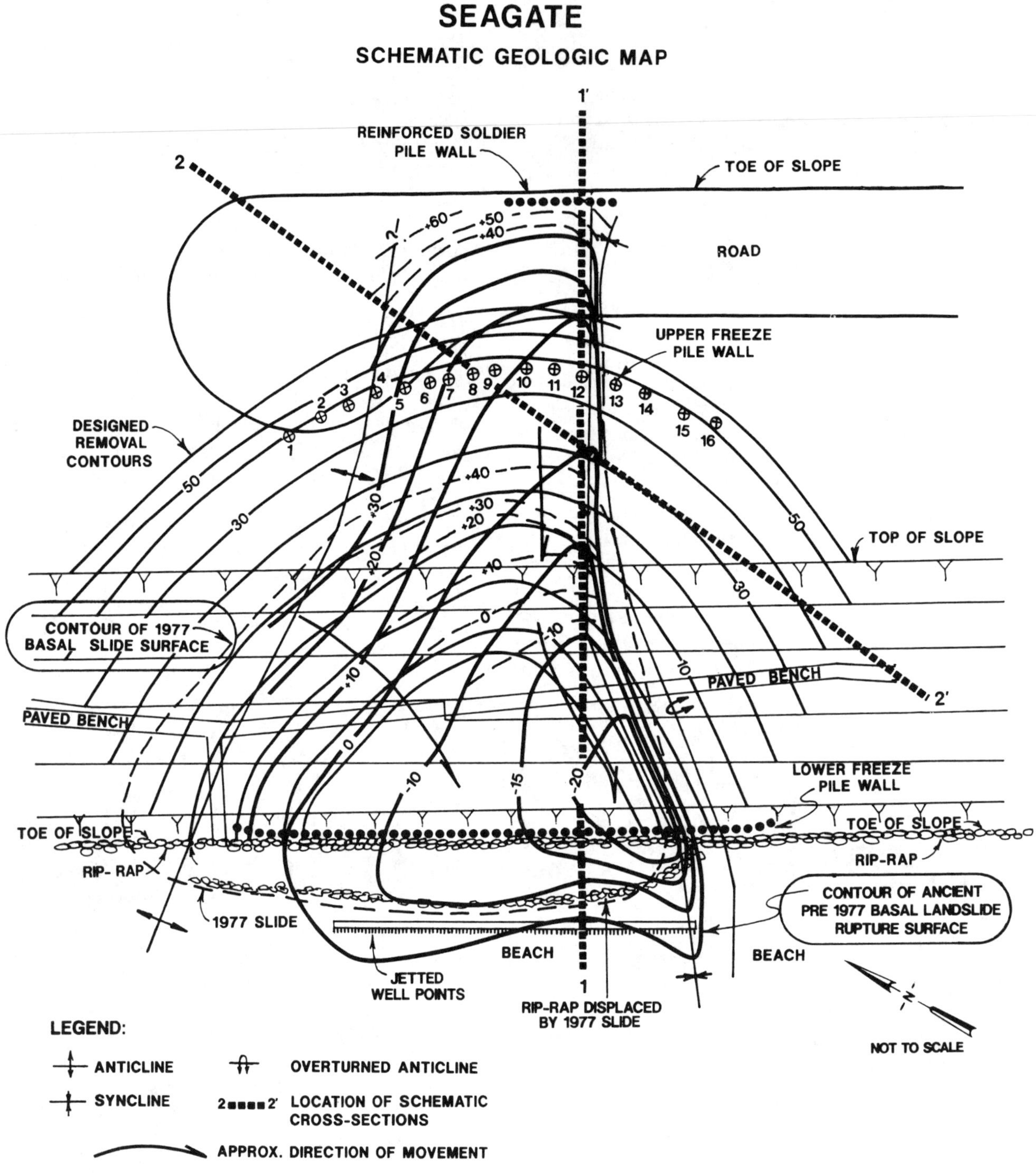

Figure 17. Schematic Geologic Map, Seagate

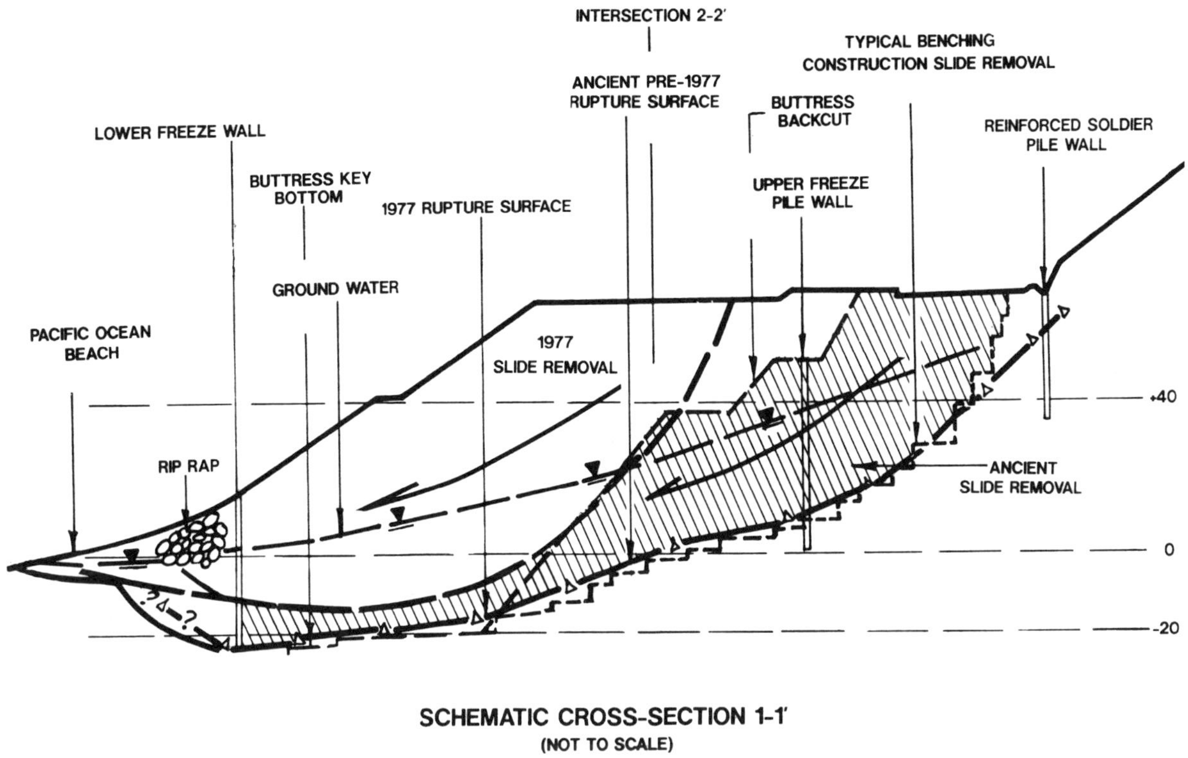

Figure 18 a. Geologic Cross-Section 1-1'

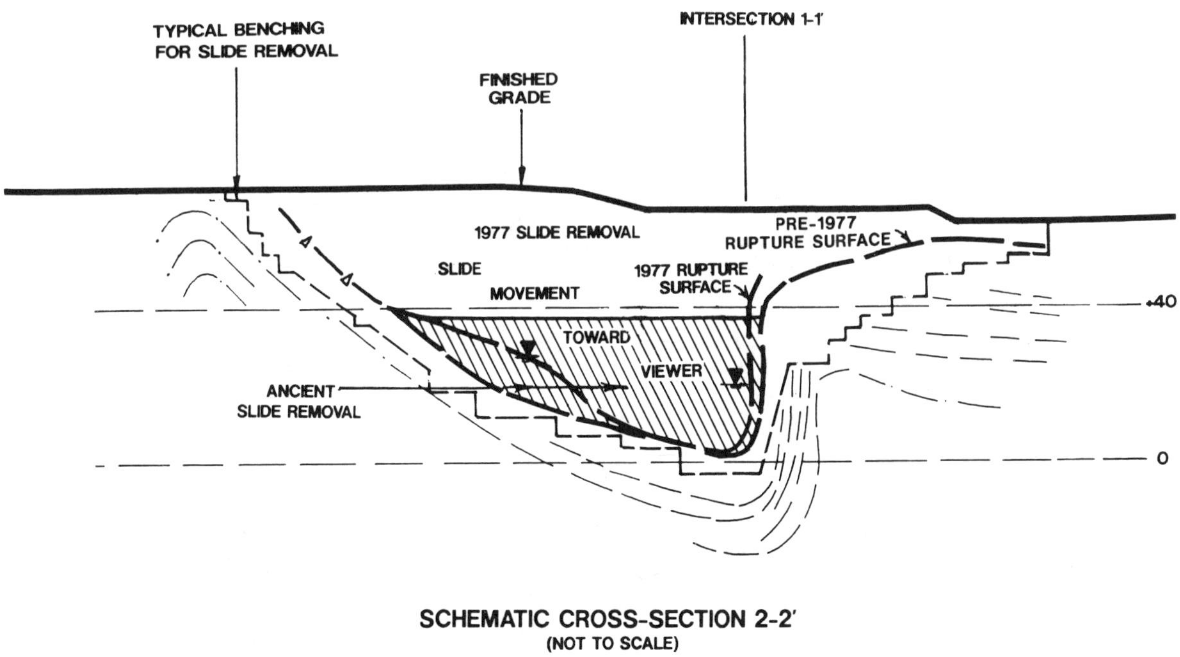

Figure 18 b. Geologic Cross-section 2-2'

Summary Of Landslide Conditions-Rupture Zones

The projected 1977 rupture surface of the landslide and the older rupture surface are shown on the Schematic Geotechnical Map (Fig. 17) and Cross-Sections 1-1 and 2-2 (Figs. 18a and 18b). The location of the 1977 rupture surface was mapped essentially as anticipated during shear key excavation. The rupture surface was slightly deeper in the southern area. The rupture surface involved both older terrace deposits and artificial fill as originally interpreted. Additional slide materials encountered during grading consisted of lenses of broken, crumbled siltstone and shale in a silty clayey matrix in various thicknesses up to 6 feet, overlain with terrace deposits. The older rupture surface was parallel to bedding with a 6-foot thick, near-vertical sheared zone forming the southern boundary (Figs. 17, 18a, and 18b).

Several prominent potential rupture surfaces were observed within the bedrock during shear key excavation. These planes are parallel to bedding, 2 to 3 feet beneath the older basal rupture surface. They are polished and slickensided clay surfaces probably formed during past movements (Fig. 19). Because these zones are susceptible to future movement, they were essentially removed during the excavation of the key. These weak layers as well as the 1977 rupture surface were considered during the shear key and buttress design.

Groundwater

The moisture content of earth materials excavated during shear key construction ranged from below optimum in sandier fill and upper sandy landslide debris to saturated in landslide debris below the ground water level. Numerous seepage zones were mapped in the back-cut slopes and landslide debris above the 1977 rupture surface. These occurred near the contacts of: (1) sandier fill layers over silty clayey fill layers, (2) artificial fill over shale and siltstone bedrock, (3) landslide debris above the 1977 rupture surface, and (4) highly jointed and fractured siliceous shale overlying siltstone. Seepage zones below the ground water level generally flowed at 10 to 20 gallons per minute. The seepage zones were confined to porous sands and permeable, fractured siliceous shales. While the porous sands acted as aquifers in the fill and landslide debris, the prominent seepage in bedrock is controlled by secondary structural features (i.e., jointing and folding). Consequently, the wedge-shaped synclinal trough would promote the buildup of pore pressure.

During shear key excavation, ground water infiltration from temporary construction cut slopes presented substantial delays to grading. The phreatic surface was encountered at approximately the 5-foot elevation near the toe and the 8- to 10-foot elevation near the back-cut slope. Numerous sumps, six deep-well points, and a horizontal rock drain were eventually installed within the excavation to remove ground water (Fig. 20). In addition, 60 or more manually jetted well points were installed seaward of the riprap to intercept sea-

Figure 19. Fat blue-grey clay rupture surface of the ancient landslide. (See Figure 7 for approximate location; left side of backhoe).

Figure 20. Temporary steel pile wall was constructed at the toe of slope after failure of the lower freeze wall. This was later removed once the

water infiltration into the shear key buttress excavation since the lower freeze wall never froze into a solid impermeable wall (Fig. 21). These were installed to minimize ground water circulation and aid the ground freezing.

The deep well points were drilled below the rupture surface into bedrock, cased with perforated 12-inch ABS pipe, and the bottom backfilled with Class 2 material. Installation of the six well points occurred at various periods during grading (i.e., Deep Wells 1-3, November 11-13, 1980; 4-6, November 25-26, 1980; and 7-8, attempted and abandoned December 2, 1980) as ground water problems continued. Throughout shear key excavation, numerous sumps were established to collect ground water (Fig. 21). All sumps were installed with backhoes (either rubber-tired or track-mounted) in saturated flowing sands using perforated corrugated metal pipe and Class 2 permeable material. The sumps were usually installed less than 10 feet in depth.

A 35-foot-long horizontal french drain consisting of 2-inch rock backfill in a 4-foot-wide trench was excavated below the rupture surface along the southern boundary. Flowing sand required the utilization of a steel shield (4 x 12 x 8 feet) for excavation to a depth of elevation -23 feet. Consecutive segments trending N28°E were excavated, end-to-end, beginning opposite Well Point 5 toward the lower freeze wall (December 17-20, 1980). Near the lower freeze wall, a CMP pipe was installed to an elevation of -33 feet in the final segment to pump the collected ground water. All segments of the trench were immediately covered with existing sands.

The manually jetted well points were installed December 10-13, 1980. Each well point was installed using existing water hydrant pressure to force the pipe through the sands into the shale and siltstone. All well points were connected to a manifold system and a single-suction vacuum pump. Effluent output from all dewatering systems equaled approximately 100 gallons per minute but varied from 100 to 600 gallons per minute depending upon tidal fluctuations. The phreatic surface was eventually lowered 12 to 22 feet below sea level, allowing phased shear key excavation.

Design Considerations And Implementation-Slide Mass Stabilization

Recommendations for slope stabilization called for an all-earth shear key buttress with an associated subdrainage system. This stabilization system was installed in multiple phases to minimize backslope failures and headward progression of the landslide. The buttress, temporary stabilization measures, subdrainage system, and backslope failures during grading are discussed below.

Excavated earth materials were removed from the subject site by means of a system of conveyor belts and trucks (Fig. 22). A hopper was attached at the excavation end to allow nearly

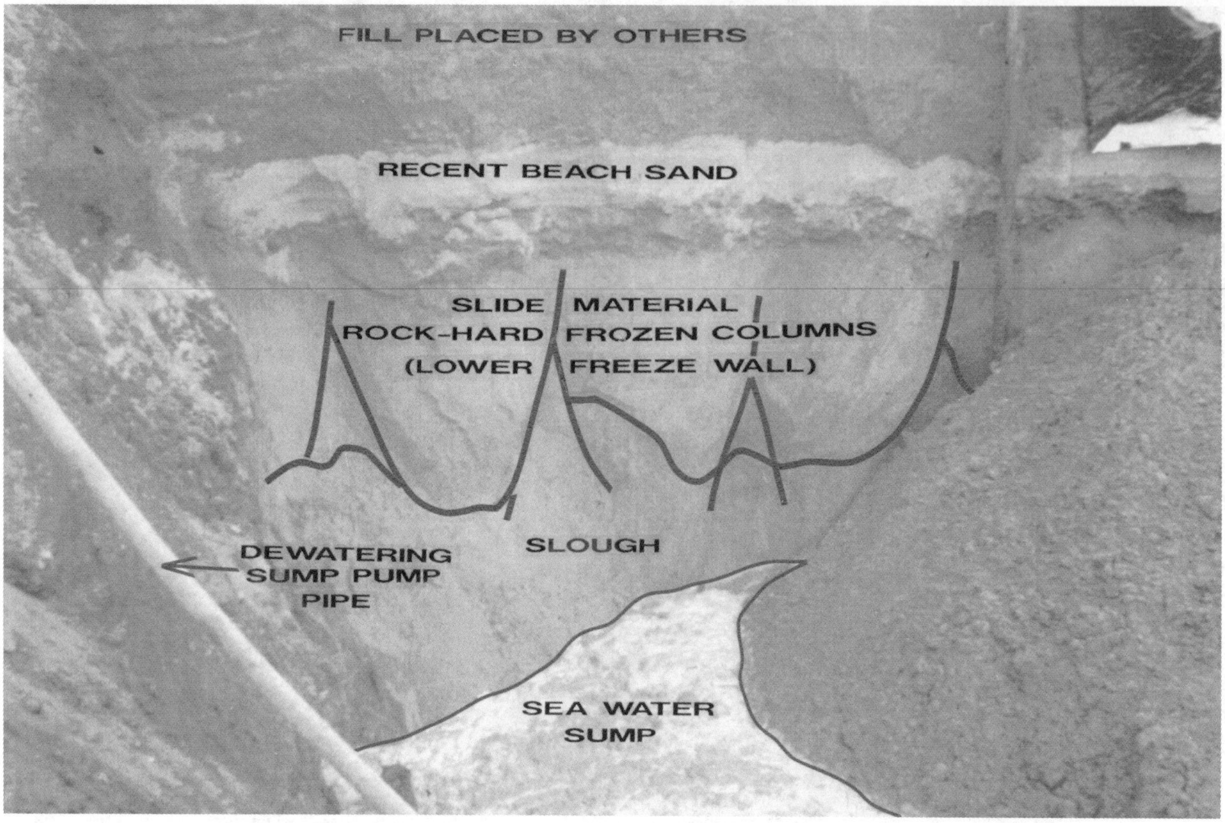

Figure 21. Lower freeze wall that was constructed along the beach at the toe of slope. Upper portions of the frozen columns were rock-hard but these narrowed with depth and sea water seeped into excavation.

continuous feeding of the conveyor-belt system. Earth materials traveled up the interconnected belts along the path of an existing terrace drain bench, finally being stockpiled on the northerly seacliff bluff. Approximately 640 lineal feet of conveyor belt was utilized to lift the material over 150 vertical feet from the excavation to the stockpile area. Each belt section was 40 feet in length (16 total) and each section was powered by an electric motor serviced by an onsite generator. Upon completion of the shear key excavation, the conveyor belts were reversed and the materials were returned to the excavation by the belts and trucks.

Shear Key Buttress

The shear key buttress consisted of a massive zone of compacted structural fill keyed into bedrock once the landslide materials were removed. In both plan and cross-section, the approximate extent of the shear key buttress is shown on Figures 17, 18a, and 18b. The high potential for backslope failures during construction of the shear key had been recognized. Shear key excavation in small successive portions was recommended to reduce this failure potential. Round the clock construction was recommended as an additional option to minimize the period of time the shear key excavation was left unsupported. To accomplish this, Sukut Construction, Inc., the project contractor, devised a multiple-phase grading scheme. This scheme combined geotechnical recommendations with necessary earthwork constraints for phased backslope and toe slope excavations. A temporary system of freeze piles was constructed in a portion of the backslope and the toe area opposite the riprap to provide temporary support for the construction excavations (Fig. 23). In addition to ground freezing, concrete and steel piles were to be installed as an emergency contingency measure should the frozen piles fail. Excavation began in the deeper southwest portion of the subject slide and progressed phase-by-phase along the toe of the subject slide and toward the backslope phase. The base of the shear key buttress was stair-stepped from southwest (-22 feet) to northwest (-4 feet), and southwest (-22 feet) to east (+23 feet) in order to remove the ascending rupture surface.

Strain Monitoring

A system of strain monitoring devices was installed above the back-cut slope to provide early detection of the location and extent of potential failures in the back wall of the excavation. Utilization of these devices improved the general safety of workers in the shear key area and provided information to adopt alternative recommendations for the safety of the surrounding residences above the excavation. The system included three inclinometers and three extensometers installed along the top of the back-cut slope to monitor minute ground movements during excavation. The extensometer con-

Figure 22. Slide material removed was lifted nearly 150 vertical feet to an offsite stockpile area by conveyor-belt system shown in the picture. Once the key had been completed, the earth material was returned to the key by the conveyor and trucks and compacted to form the earth

Figure 23. A large compressor unit was situated at side top of the excavation during slide removal. A chilled brine solution was utilized to create the upper and lower free walls. The upper freeze wall is located downslope of the compressor.

sists of a rigid rod, extended between two pins, driven vertically into the ground. Any movement between the pins is detected on a dial indicator connected to the rigid rod. The system is capable of measuring changes as small as a few thousandths of an inch over a baseline length of 10 to 30 feet. Movements of this magnitude are known to precede many backslope failures and serve as good early warning systems.

The inclinometer consists of a longitudinally grooved vertical tube installed in the ground through the depth where the potential for sliding exists (approximately 60 feet at the Seagate landslide). The tube is accurately and regularly monitored at appropriate time intervals to detect minute changes in the position of the tube, reflecting subsurface lateral movements. During shear key grading, the inclinometer was an inexpensive method of detecting the activation of those slide planes prior to back-cut slope failures.

Both the extensometer and inclinometer measurements indicated movement along the potential slide planes beneath the road during shear key excavation prior to backslope failure. They also defined approximate lateral limits of potential backslope failure.

Failures of Temporary Construction Cut Slopes

Several failures occurred in the temporary construction cut slopes during shear key excavation. All failures occurred in the 1977 rupture surface or its projection into the slope from previously activated slide debris. The primary cause of temporary cut slope failures was the necessary removal of lateral support during excavation for the shear key buttress. Locations most susceptible to instability which could endanger the neighboring properties were recognized, and recommendations for temporary shoring were made by Leighton and Associates. A system of frozen ground columns backed up by preparedness for installation of soldier piles was selected by the project contractor for shoring.

One backslope failure was of major proportions, extending almost to the easterly curb line of the road. On January 26, 1981, a portion of the upper frozen pile wall collapsed and the ground upslope of the system moved downward and outward from the road and into the buttress excavation (Figs. 24 and 25). This failure involved the reactivation of old slide material and structural fill that had been placed over the old slide materials by others. The excavated phase was temporarily backfilled to halt the movement of this slide mass, while a concrete and steel soldier pile wall was installed above the developing scarp (Fig. 20). Headward progression of the slide mass continued during shear key excavation but did not breach the soldier pile wall. Consequently, shear key excavation and buttress filling was continued on a dawn-to-dusk work schedule to complete the compacted buttress to a stabilizing level as quickly as possible.

Figure 24. The upper freeze wall failed eventually but provided good temporary support for several months.

Figure 25. Failed frozen columns and freeze pipes of the upper freeze wall. These were found to be of large diameter and rock-hard in some areas and of small diameter and only partially frozen in other areas.

Temporary Frozen Ground Stabilization Measures

Temporary stabilization of the toe cut and a portion of the back-cut slope was attempted, utilizing a system of frozen ground piles. This unique method of temporary stabilization consists of steel casings installed into the ground on appropriate spacings and a specific coolant circulated inside the casings. Progressively, an increasingly larger column of frozen ground (earth materials with frozen pore moisture) forms around the steel pipes. Potentially, the individual frozen columns of earth materials can grow in diameter until they are bonded together as a continuous, solid frozen wall.

Each freeze pipe consists of 3-5/8-inch steel pipe (53 feet long in the upper pile wall and 40 feet long in the lower pile wall), with 1-1/2-inch PVC plastic tubing inside, and capped at the surface. Individual freeze pipes are interconnected and coolant (calcium chloride) is circulated under pressure from the refrigeration unit. The as-built locations of the frozen ground pipes are shown (Fig. 17).

During shear key excavation, both the upper frozen piles and lower freeze wall partially collapsed (Fig. 25). The lower freeze wall bent and failed on December 31, 1980. Excavation halted until a temporary steel soldier pile wall was installed within the excavation (Fig. 17). The upper freeze piles bent and later failed on January 26, 1981 (Fig. 24 and 25). Excavation again halted until a permanent concrete and steel soldier pile wall was installed above the slide scarp. Approximate locations of the upper (16) and the lower (48) freeze piles are shown in Figure 17.

The following three key factors are believed to have contributed to the failure of the frozen columns: (1) overestimation of thermal conductivity of earth materials, (2) insufficiency of soil moisture in some strata, and (3) presence of a high rate of seepage in other strata.

Temporary Soldier Pile Stabilization Measures

Temporary stabilization of a portion of the toe cut and a portion of the back-cut slope was necessary after collapse of the initial freeze wall stabilization. This was achieved by installation of soldier pile walls within the excavation opposite the lower freeze wall and east of the road adjacent to the curb. The soldier pile wall consisted of 34-foot-long steel H-beams (13W120) installed into the ground. Upon the decision to install steel soldier piles, actual installation began immediately. Drilling commenced in the late afternoon and continued throughout the night despite intermittent heavy rain. In less than 48 hours, the holes had been drilled, steel beams placed, concrete poured, and bond beams were being welded.

The lower steel soldier piles were installed January 5-8, 1981, east of the failed freeze pile wall on 5-foot centers. Each steel soldier pile was rooted a minimum of 5 vertical feet below the basal rupture surface in shale and siltstone bedrock and backfilled with local sands. A grade beam and terminal, lateral braces were welded near the top of the soldier pile wall to maximize strength. An additional H-beam was buried on the beach to act as a dead man. Two 5/8-inch cable tiebacks were connected to the middle of the welded grade beam and the dead man for added support. Wooden planks (6 x 12 inches) were inserted in the webbing between successive H-beams to retard sand flow into the excavation during shear key excavation (see Fig. 20). The soldier pile wall allowed vertical toe cut slopes into bedrock parallel to the pile wall for the deepest phases of shear key excavation. When the level of compacted structural fill had risen to a safe elevation, the dead man, cables, grade beams, and lateral supports were removed and all but one soldier pile beam were extracted from the partially backfilled excavation. The majority of wooden planks were removed as the compacted fill rose; however, some of the planks were not retrievable.

The upper soldier pile wall was installed January 27-28, 1981 along the eastern curb of the road. The 3-foot-diameter concrete reinforced soldier piles were on 6-foot centers with welded grade beams between the tops of individual soldier piles to maximize the integrity and effectiveness of the wall. The soldier pile wall prohibited additional headward progression of old slide debris, allowing excavation of the expanded shear key buttress and placement of the compacted structural fill.

Subdrain System

The subdrain system consisted of a main subdrain and a series of collector french drains. The main subdrain was installed near the base of the backslope in undisturbed bedrock below the projection of the pre-1977 basal rupture surface. It consisted of a perforated 8-inch ABS pipe laid in a 36-inch-wide trench with a minimum of 18 inches of Class 2 permeable material both above and below the pipe. The main subdrain is curved in plan view and drains to both ends. At these two points, terminal sections of solid 8-inch ABS drain the system to the toe of the buttress. A minimum flow gradient of 4 percent is maintained throughout the system to drain water to the riprap. At 30-foot intervals, french drains radiate from the main subdrain up the backslope. They consist of 24-inch trenches excavated into either bedrock or compacted structural fill with a minimum 2-foot thickness of Class 2 permeable material. Two radial extensions were added at the center of the main subdrain system to drain the extended back-cut slope generated during back-cut slope failure. The radial extensions consist of perforated 8-inch ABS pipe laid in a 24-inch trench with a minimum of 12 inches of Class 2 filter material above and below. The radial extensions are installed in compacted structural fill with french drains at their distal termini, installed as described above. In addition, the french drains are interconnected at approximately their mid-height on the back-cut slope by auxiliary drains. The auxiliary drains consist of 24-inch trenches excavated into bedrock with a minimum 2-foot

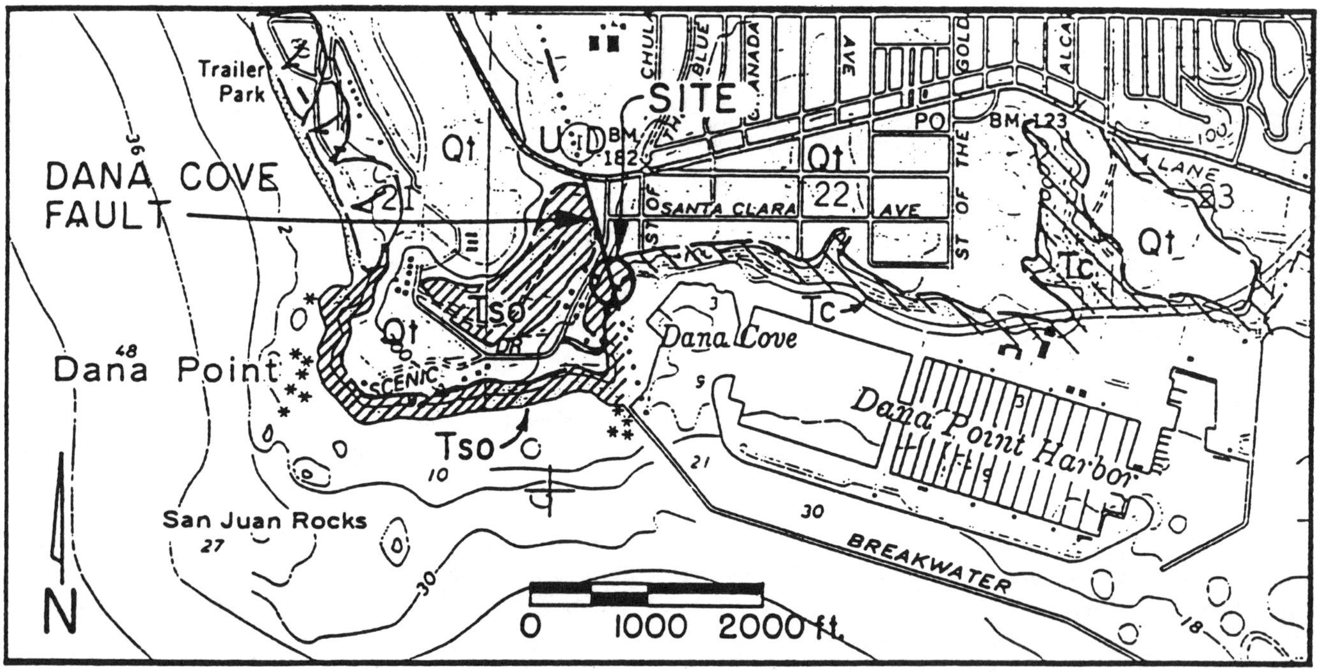

Figure 26. Site Location Map

thickness of Class 2 filter material. From the midpoint, the auxiliary drains maintain a minimum flow gradient of 4 percent toward the interceptor french drains.

A second subdrain system consisted of a nonperforated 4-inch ABS pipe installed in compacted structural fill along the property line of Lots 22 and 23. This subdrain system collected effluent from two existing 4-inch subdrain systems east of the road and drained at the midslope bench drain.

Cove Road Landslide, Dana Point

Introduction

In 1980, a wedge failure in a nearly 150 foot high, 60 degree sea cliff at Dana Point Harbor, California (Fig. 26) undermined a portion of a restaurant on the top of the bluff. Slide material blocked Cove Road, the county road, at the toe of the slope, blocking access to the western end of the harbor and marina. The complicated geologic conditions and a twenty-two million dollar lawsuit filed by the owner of the restaurant required use of sophisticated geotechnical stabilization measures while the restaurant continued to serve its customers. The landslide has been summarized in Kerwin (1989).

The Investigation Phase:

- The limited working area, access problems, and steep terrain necessitated the use of cranes to position drilling and excavation equipment
- The potential failure block extended approximately 40 feet beneath the restaurant structure (Fig. 27).
- The slide block consisted of a weakly cemented, massive silty sandstone with an average unconfined compression strength of approximately 100 psi.
- Shear values of the weaker siltstone bed had an angle of internal friction of 28 degrees and a cohesion of 300 psf. The clay gouge fault and rupture surface clays had an angle of internal friction of 14 degrees and a cohesion of 350 psf.
- Stability analyses of the critical wedge using strength parameters computed from the original failure geometry (i.e., Ø = 28° and c = 230 psf) verified the near equilibrium condition of the critical failure wedge. Construction of a restraining structure of adequate size and orientation to apply the rock anchor stabilizing loads necessitated trimming portions of the sandstone scarp to a 1:1 slope. The driving and resisting force for the trimmed configuration of the potential failure block were assumed equal (F.S. = 1.0). In accordance with local practice, a factor of safety of 1.5 against sliding was required for stabilization. Therefore, additional resistance of 1650 kips was required.

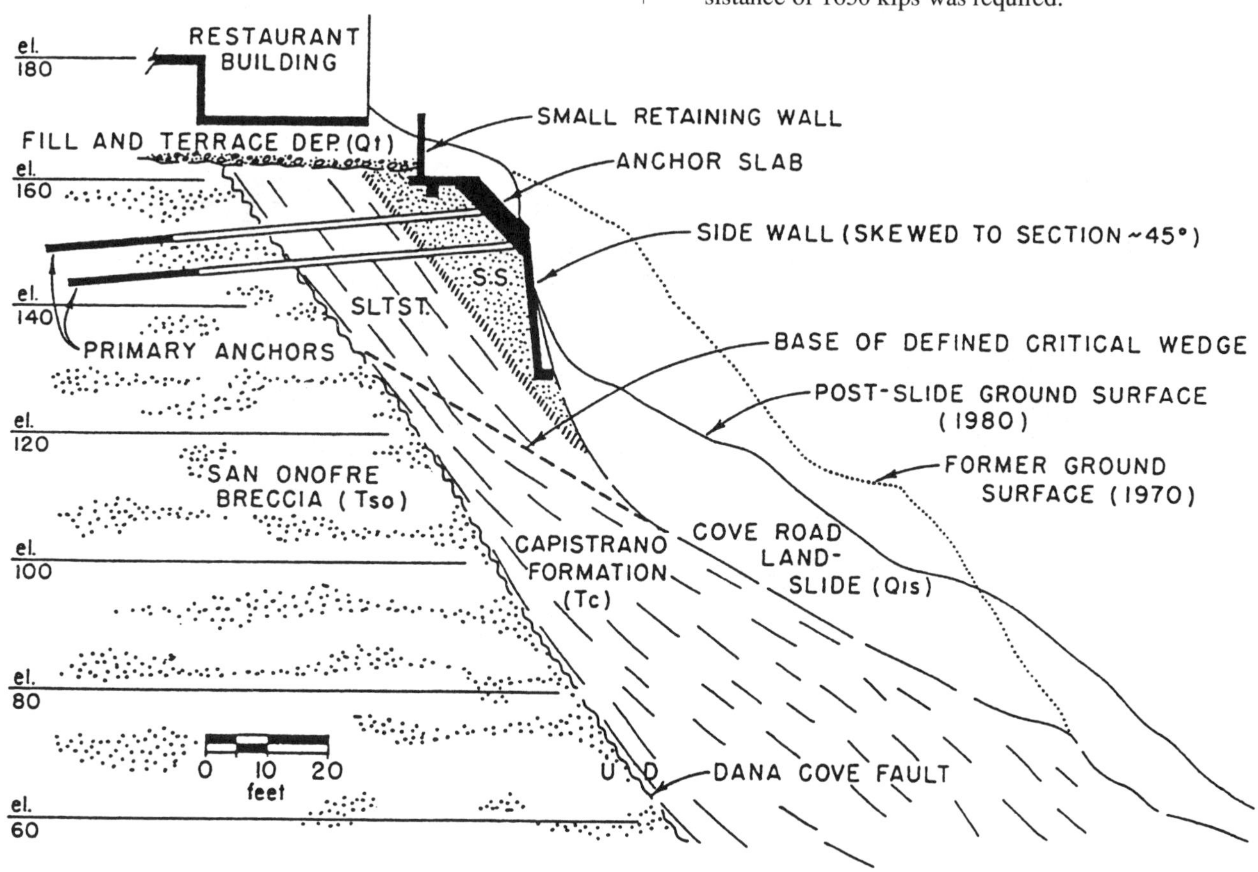

Figure 27. Geologic section of the critical headscarp wedge viewed looking northeast, perpendicular tot he alignment of the primary anchors. (From Kerwin, 1989).

Figure 28. View from the Dana Cove Pier. Construction of a pile supported, temporary crane pad was necessary to provide access and support for drilling and other construction equipment on the site. the 16 primary anchors were drilled from a wooden deck supported by an adjustable steel frame (center of photo). Drilling of the 18 secondary anchors was accomplished from a crane-held platform locked in place with chain binders on shallow, temporary deadmen (not shown). Phase I Construction -January, 1984. (Photograph from Kerwin, 1989).

The Monitoring Phase:

The restaurant owners desired to continue at full capacity during the repair of the failure even though patrons at a few tables appeared to be situated in mid-air. To conclude, a catastrophic failure and automatic alarm system consisting of borehole monitoring system was installed beneath the restaurant. Constant monitoring was needed to evaluate minor movement during construction and to mitigate life threatening smaller failures. Up to 0.06 inch of slopeward displacement was measured on two separate occasions following heavy rains and during the construction of the key structural elements of the repair. No significant damage, deaths or serious injuries occurred during this three year landslide repair.

The Construction Phase (September 1983 to August 1986):

The construction phase was divided into two parts: 1) the repair of the head scarp area; and 2) the repair of the toe of the slide. Construction site access was difficult, because there was virtually no right of way except for foot traffic, a temporary pad for a 65 ton capacity friction crane had was constructed (Fig. 28). Six steel "H" beam columns were built to support the outriggers of the crane.

- The headscarp of the slide had to be trimmed and laid back prior to the construction of any supporting elements. A small dozer and rubber tired backhoe were lowered onto the head of the slide up by the crane. This work had to be done carefully so as to not undermine the restaurant. Much of the work was done by hand and pneumatic clay spades.
- Sixteen anchors with design modes of 175 kips each drilled at 5 degrees from the horizontal were installed from a portable platform lowered and held in place by the crane.
- The length, relatively high loads, and difficult access conditions favored steel cable anchor tendons rather than solid rods. Six 0.6-inch diameter tendons were required to accommodate the 175 kips design load for each primary anchor. Loads of 40 to 105 kips for the secondary anchors necessitated installation of two to four tendons per anchor.
- The first and, possibly most important, is a "side wall" constructed to adjoin the south edge of the primary anchored slab. Supported by anchors bonded in Capistrano Formation sandstone, this wall provides confinement and erosion protection to foundation rock underlying the anchor slab. A second wall covers the headscarp along the exposed fault plane. This "headscarp wall" is supported by anchors penetrating the breccia in a fashion similar to that of the primary system. The "headscarp wall" and a "small retaining wall" to the east provide access around the south corner of the restaurant and prevent possibly damaging degradation of the foundation materials. Open, unreinforced joints were constructed between each of the four elements of the restraining structure to accommodate possible elastic deformation of the rock from anchor loads. The construction joints were not filled and sealed until all anchors had been tested, approved and locked off at the specific loads.
- The aggregate design capacity of the 34 anchors installed for the restraining structure was 2,010 tons. However, the applied stresses are greater because anchors were tensioned and locked off at loads equivalent to 120 percent of the design working capacity to compensate for possible long term losses. The potential for unacceptable future load loss was evaluated prior to final tensioning by routine proof testing of all anchors to 150 percent of design load. Selected anchors were also "lifted off" to determine whether residual loads after selected time intervals were within specified limits (Littlejohn, 1981).
- Surficial stability of the clay siltstone had to be achieved by driving shallow rebar anchors into the material and covering the surface with an interim coating of shotcrete.
- The San Onofre Breccia bedrock was difficult to drill and very abrasive. To help minimize this hard drilling, a larger drill rig had to be utilized and the drilled holes had to be provided with a casing in order to maintain the alignment of the hole and to be able to purge the hole.
- Ultimately, a standard horizontal boring machine with a variable jacking capacity up to 75 kips was used to drill the primary anchor holes. Heavy duty continuous flight auger and a specially fabricated rock drilling bit with retractable "wing" cutters excavated material immediately ahead of the casing, including annular clearance. Casing was advanced the full depth of the hole and was subsequently pulled back approximately 23 feet for placement of the anchor assembly (2 to 3 feet for a packer and the remaining 20 feet for bonding). All anchor borings were initially drilled "dry" in accordance with limitations established during settlement conferences before construction. Subsequent to anchor failures during load testing, holes were redrilled through the casing using essentially the same equipment adapted to rotary wash drilling methods and hardware.

The relatively high jacking forces utilized to drill and advance the casing necessitated design and erection of a larger, stronger drilling structure which incorporated an adjustable platform. A temporary framework of steel "H" beams was constructed above the primary anchor slab. Bearing support was provided by a single pile cast in bedrock at the toe of the critical wedge. Lateral resistance to the jacking force was attained by anchoring the uppermost steel members of the framework to the concrete retaining wall footing cast in sandstone at the top of the anchor slab. Analyses indicated the jacking force applied by the drill rig should be limited to 20 kips to maintain design stability.

- Because of the corrosiveness of the environment, double corrosion protection was specified for the anchor tendons. In addition, the tendons had to be encapsulated in corrugated polyethylene type.
- Ground water and grouting problems necessitated replacing some anchors. It was found that separate feed lines to both the inside and outside of the anchor capsules were needed. It was found that packer assemblies failed to seal out minor accumulations of ground water seepage.
- The final phase of repair was the construction of a 45 foot high, precast concrete crib type retaining wall at the toe of the slope.
- Crucial aspects of this phase of the project included removal and recompaction of the landslide debris and installation of an extensive subsurface drainage system. This portion of the slope restoration work could not have been safely accomplished without prior stabilization of the bluff top.

The entire project required approximately three years to complete and Cove Road was officially re-opened in August 1986.

ACKNOWLEDGEMENTS

The authors wish to thank Scott Kerwin and Jack Eagen of Moore and Taber, Anaheim, California, who kindly provided the photographs of the Cove Road landslide and allowed us to use their report.

REFERENCES

Clark, B.R., 1986, The Big Rock slide, Malibu, California: in Ehlig, P.E., (Compiler), Landslides and landslide mitigation in southern California, Geol. Soc. America, Cordilleran Section Guidebook, p. 71-75.

Kerwin, S.T., 1989, Sea cliff stabilization using long rock anchors: in Cann, L.R., and Steiner, E.A., (compilers), Engineering geology along coastal Orange County: Assoc. of Engineering Geologists, Southern California Section Guidebook, p. 103-116.

Leighton, F.B., 1969, Final geologic report on the general plan study for the City of Laguna Beach.

_____, 1969, Engineering geologic report of storm damage, City of Laguna Beach.

Leighton, F.B., Cann, L.R., and Poormand, I., 1986, Emergency repair operations for the Bluebird Canyon landslide of October 2, 1978: in Ehlig, P.E., (Compiler), Landslides and landslide mitigation in southern California, Geol. Soc. America, Cordilleran Section Guidebook, p. 35-41.

________, 1984, Verde Canyon landslide, San Clemente, Orange County: Calif. Geology, Calif. Div. Mines and Geology, August.

Leighton and Associates, 1975, Background data and geotechnical information for the seismic safety and safety element: Presented to the City of Laguna Beach.

Littlejohn, G.S., 1981, Accepted criteria for the service behavoir of ground anchorages: Ground Engineering, v. 14, 26-29, Brentware England, Foundation Publishers Ltd.

Morton, P.K., and Others, 1973, Geo-Environmental maps of Orange County, California: Calif. Div. Mines and Geology Preliminary Report 15.

Rex, Robert W., 1972, Origin of the salt of the Imperial Valley and surrounding watershed areas: Report for Bureau of Reclamation Contract #14-06-300-2258, University of California at Riverside (IGPP-UCR-72-33), 38 p.

Tan, S.S., 1976, Geology and engineering geologic aspects of the Laguna Beach Quadrangle, Orange County, California: Calif. Div. Mines and Geology Special Report 127.

Varnes, D.J., 1978, Types of slope movements: Transportation Research Board Committee A2T58, from Spec. Report 176, Chapt. 2.

Vedder, J.G., and Others, 1957, Geologic maps of the San Joaquin Hills-San Juan Capistrano Area, Orange County, California: U.S. Geological Survey Oil and Gas Investigations Map OM 193.

Ziony, J.I., and Others, 1974, Preliminary map showing recency of faulting in coastal southern California: U.S. Geological Survey Miscellaneous Field Study Map MF 585.

EVOLUTION, MECHANICS AND MITIGATION OF THE PORTUGUESE BEND LANDSLIDE PALOS VERDES PENINSULA, CALIFORNIA

Perry L. Ehlig

Department of Geological Sciences
California State University
Los Angeles, CA 90032
and City Geologist
City of Rancho Palos Verdes

ABSTRACT

The 260-acre Portuguese Bend landslide is part of a two-square-mile Pleistocene megaslide on the south flank of Palos Verdes Peninsula. Sliding occurs in weak bentonite beds near the Portuguese Tuff in the middle Miocene Altamira Shale member of the marine Monterey Formation. The megaslide occupies a seaward plunging structural trough in which bedding undulates but dips an average of 7 degrees in the direction of slide movement. The megaslide originated about 120,000 years ago when the sea occupied a marine terrace whose shoreline angle is 140 feet above sea level. Sliding probably started on an uphill ramp where beds dip 15 to 20 degrees seaward and propagated downhill to the beach where the slide base was as much as 200 feet below sea level. The toe propagated to the surface by folding and subsequent thrusting. The head of the megaslide migrated uphill by arcuate slumping on the west and sliding down a west-dipping ramp on the east. The megaslide is broken into a complex of subslides controlled by undulations in the underlying bedrock. Sliding was probably promoted by ground water perched above the Portuguese Tuff in the main part of the slide and by wave erosion and artesian pressure beneath the tuff near the coast.

The megaslide's entire coastal part has reactivated since the start of residential development in this area. The Portuguese Bend landslide reactivated in 1956 when Los Angeles County placed road fill along its head. Its unabated movement destroyed 130 of 160 homes within it. Part of it has moved 800 feet. The 80-acre Abalone Cove landslide started in response to high rainfall in 1978 and part of it moved 30 feet. Several homes were damaged before it was stabilized. The 50-acre Klondike Canyon landslide began in 1980 but moved less than three feet and caused no major damage. The Abalone Cove and Klondike Canyon landslides were stabilized by removal of ground water and control of surface drainage. Mitigation was paid for by establishment of Geologic Hazard Abatement Districts, the first two in California.The Rancho Palos Verdes Redevelopment Agency was established in 1984 to stabilize the Portuguese Bend landslide. A plan was developed to control the slide in phases starting near the head

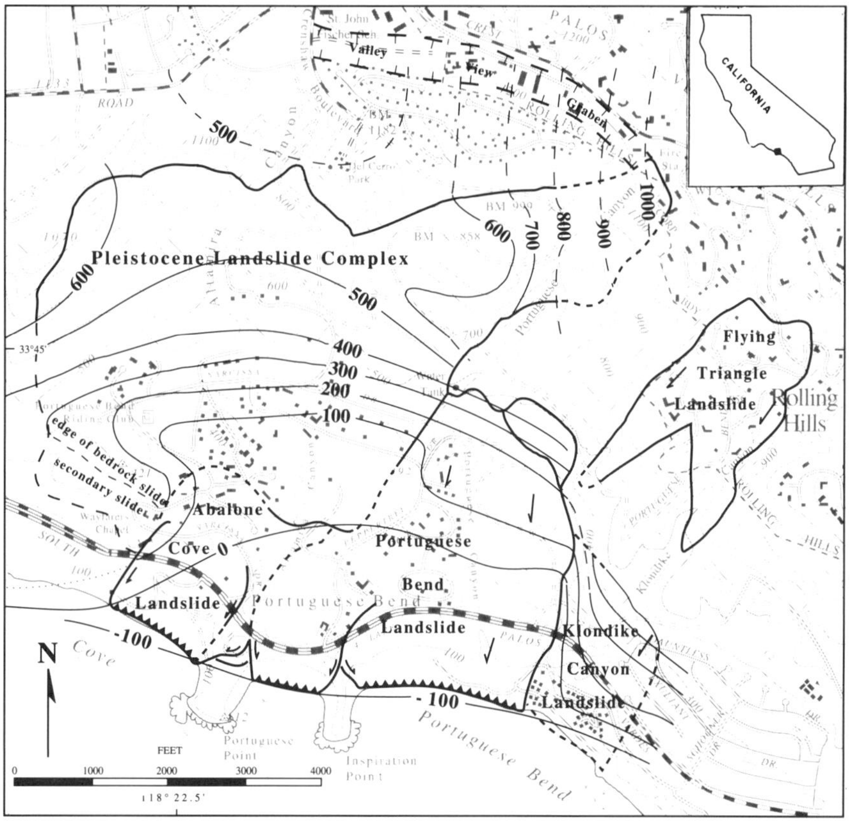

Figure 1. Map showing the Pleistocene landslide complex and the Portuguese Bend, Abalone Cove, Klondike Canyon and Flying Triangle landslides. Structure contours in feet are on the base of the Portuguese Tuff.

and working seaward. Control measures include pumping ground water from wells, restoring surface drainage, and moving earth from where it exerts excess driving force to where it exerts a net resisting force. Because low permeability made ground water removal slower than anticipated, grading was increased to immediately improve stability. The landward part of the slide appears stable and the central part is nearly stable. The seaward part remains active because of wave erosion. Gabions were installed along the toe for temporary shoreline protection but have broken because of percussion and corrosion. A rock revetment and toe berm are needed for permanent stabilization. The project has cost about $3,000,000 to date. After stabilization is complete, undeveloped areas will probably be used for a golf course and public recreation.

HISTORICAL BACKGROUND

The Los Angeles metropolitan area contains few settings more attractive for residential development than the large topographic amphitheater overlooking Portuguese Bend and Abalone Cove on the south flank of the Palos Verdes Peninsula. This area has a generally mild, sunny, smog-free, marine climate and an excellent view of the Pacific Ocean and Catalina Island, located 23 miles offshore. However, its gently rolling terrain, so ideal for residential development, is within the two-square-mile Portuguese Bend landslide complex (Fig. 1).

Development of Portuguese Bend began in 1946 even though the slide complex was recognized by Kew (1926) and Woodring and others (1946). The landslide was ignored because building codes did not require geologic investigations and, even if the landslide had been investlgated, its existence is unlikely to have stopped development. Stability analyses at that time would have shown the area was stable because laboratory tests would have yielded much higher shear strengths for bentonite than the 6° angle of internal friction and 75 psf cohesion obtained by back calculations on the Portuguese Bend landslide. Even today, most test results are too high.

While homes were being built at Portuguese Bend, Los Angeles County was extending Crenshaw Boulevard southward from the crest of the peninsula to Palos Verdes Drive South. In August 1956, cracks appeared in road fill as much as 70 feet thick, which was being placed across the northeast corner of the ancient landslide. Within a month, the cracks propagated westward and seaward to define the part of the Portuguese Bend landslide which moves into Portuguese Bend. During the next several months, movement progressed westward to include the segment of the Portuguese Bend landslide which moves into the cove between Inspiration Point and Portuguese Point. The fully developed slide covered 260 acres and contained about 160 homes within it. All but 30 of the homes were ultimately destroyed.

Studies were performed to determine the feasibility of stabilizing the landslide. Two plans were suggested, one of which involved the building of a buttress along the toe of the slide. This plan required working within the ocean and was rejected because of cost. The other plan involved the installation of shear pins near the toe of the slide where the slide mass was relatively thin. Although this plan received poor reviews, 23 reinforced concrete shear pins, each 20 feet long, were installed. All failed by plucking, tilting or shearing. The unearthed remains of several shear pins are visible in the surf zone at low tide (Fig. 2).

Based on our current knowledge, sliding could not have been stopped solely by installation of shear pins along the toe of the slide. Even if enough pins had been installed to stop movement along the basal slip surface, the thrusting action of the slide would have developed a new slide base above the tops of the shear pins.

In 1961, a legal precedent was set when Los Angeles County was held responsible for starting the slide and, by reason of inverse condemnation, was found liable for the destruction of about 130 homes. In an effort to protect itself from additional liability, Los Angeles County declined further involvement with the slide. Homes which might otherwise have been condemned were allowed to remain in the slide and be repaired at the discretion of the homeowner.

In 1973, the City of Rancho Palos Verdes was incorporated. The Portuguese Bend landslide was included within its boundaries. At the time of incorporation, the city agreed that arrangements within the slide area would remain status quo and that no action would be taken to stabilize the slide. However, the situation was altered by an unusually wet period beginning in 1978.

In April 1978, following the rainiest winter since 1890, the Abalone Cove landslide adjacent to Portuguese Bend began moving within an 80-acre area containing 25 homes. The City of Rancho Palos Verdes took two actions in response to the slide. First, it placed a building moratorium on the entire area containing the Portuguese Bend landslide complex pending investigations to determine which parts, if any, were sufficiently stable to permit development. Second, it commissioned a study to determine what could be done to stop the Abalone Cove landslide.

The study of the Abalone Cove landslide was supervised by the writer, and found a high water table throughout the slide mass. It was anticipated that the slide could be stabilized by removing ground water from as few as six wells. The Abalone Cove Homeowners Protective Association (ACHPA) was formed to install the wells and a pipeline to transport the water to the ocean. The dewatering system was installed using $250,000 in donated funds and production started on March 2, 1980, near the end of the third rainiest winter since 1890. The rate of slide movement abruptly slowed in response to dewatering. High rainfall during the 1982-83 rainy

Figure 2. Reinforced concrete caissons exposed by wave erosion at the toe of the Portuguese Bend landslide in 1981. Pins were installed through the base of the landslide with their midpoints 42 feet below sea level in 1957.

season caused a slight resurgence but only creep has occurred since 1985.

In the meantime, the City of Rancho Palos Verdes sought and received legislation that permitted the establishment of Geologic Hazard Abatement Districts (State of California Public Resources Code, Division 17, Sections 26500 through 26600). Such districts can abate geologic hazards by taxation of the benefiting properties. In 1981, the Abalone Cove Landslide Hazard Abatement District (ACLAD) was established as the first Geologic Hazard Abatement District in California. It includes all properties within the Abalone Cove landslide and all properties within the inactive Portuguese Bend landslide complex located up slope from the active landslide. ACLAD, which took over ACHPA's abatement facilities and activities, has worked well to this time.

Another important event was the filing of a lawsuit by properties owners (Horan et al. vs. County of Los Angeles et al.). Funds were requested to permanently stabilize the Abalone Cove landslide. A panel of experts (David Cummings, Perry L. Ehlig, Richard T. Frankian, John F. Mann, Richard L. Meehan and James E. Slosson) was established to evaluate the feasibility of permanently stabilizing the Abalone Cove landslide and to make recommendations regarding methods and costs of stabilization. In December 1984, the Panel concluded that permanent stabilization was possible and could be accomplished for about ten million dollars. In the subsequent settlement, Los Angeles County agreed to provide ten million dollars in low interest bonds to pay for stabilization of the Abalone Cove landslide and an additional five million dollars in bonds to aid in the stabilization of the Portuguese Bend landslide. The previously established technical panel was retained to recommend and oversee remedial work within the Abalone Cove landslide. An Improvement District has been formed to handle the bond indebtedness. The bonds will be repaid from the tax increment obtained from appreciation of property values as a result of slide abatement.

While dewatering was proving successful in 1980 in abating the Abalone Cove landslide, two other landslides began to move. One was the 50-acre Klondike Canyon landslide adjoining the east side of the Portuguese Bend landslide. Although the area was thought to be underlain by stable bedrock, investigations demonstrated it was an ancient slide whose maximum displacement was only a few tens of feet. In May, 1981, artesian water burst forth from a hole being drilled to install a slope indicator near the toe of the slide. The hole was converted to an artesian dewatering well and became the main method of abating the slide. The Klondike Canyon Geologic Hazard Abatement District was formed to handle abatement activities and is still operating. A survey monument adjacent to the Portuguese Bend slide has moved only 2.5 feet since it was installed in 1957; most of the movement occurred between 1977 and 1981. Another monument near the center of the slide has only moved 1.6 feet since it was installed in 1956. The slide has moved no more than 0.2 foot since 1981 and no homes have been lost because of the movement.

PLEISTOCENE PORTUGUESE BEND LANDSLIDE

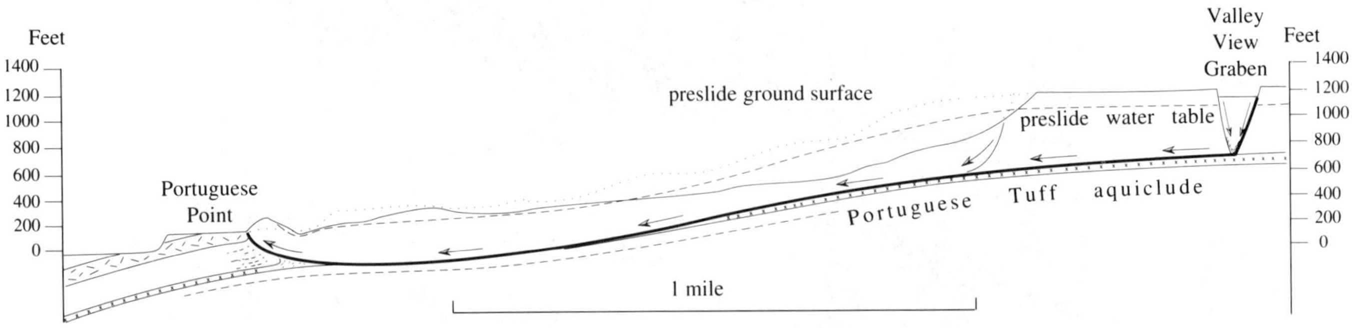

Figure 3. Vertical cross section through the Pleistocene landslide complex showing Portuguese Tuff at slide base and conjectured position of preslide ground surface and water table.

The Flying Triangle landslide also began moving in 1980, initially within an area of 5 acres in the City of Rolling Hills. In late 1983, movement expanded to an area of nearly 90 acres. The down slope part of the slide splits in an inverted V-shaped pattern at the up slope edge of a stable hill called "Flat Top" (Fig. 1). One branch moves southwest into Paintbrush Canyon. The other branch moves south towards Klondike Canyon. Its base passes beneath the bottom of Klondike Canyon to the east of Flat Top but rises and uplifts into the canyon further south. The down slope parts of both branches are in the City of Rancho Palos Verdes. The Flying Triangle landslide is a reactivated ancient slide but is not physically a part of the Portuguese Bend landslide complex. The only abatement activity consists of two horizontal drains installed by the City of Rancho Palos Verdes in order to relieve artesian pressure at the toe of the Klondike Canyon branch.

Starting in 1978, above average rainfall caused a large increase in the rate of movement within the Portuguese Bend landslide. High rates of movement lasted until the start of remedial grading in October 1986. For example, survey monument R-l, located in the east central part of the slide, moved 237 feet horizontally during the 20 year period from October, 1956, to October, 1976. But it moved 382 feet horizontally in the 10 year period from October, 1976, to October, 1986. The accelerated movement greatly increased the cost of repairing the 0.8 mile segment of Palos Verdes Drive within the slide. Prior to 1978, the city spent an average of less than $40,000 per year repairing the road. During the period from 1983 to 1986, the average cost was $250,000 per year and consumed most of the city's total road maintenance budget. This, and the potentially adverse effect of the Portuguese Bend slide on the stability of the slides on either side of it, provided the incentive to evaluate the feasibility of abating the Portuguese Bend landslide. Clearly, the city would benefit if the slide could be slowed or stabilized at a reasonable cost. Success in abating the Abalone Cove landslide gave hope that stabilization was feasible.

During the summer of 1983, I analyzed available data on the Portuguese Bend landslide to learn as much as possible about its mechanics and potential methods of stabilizing it. The results of my study suggested that stabilization of all but the coastal part of the slide could be achieved at a modest cost by removing ground water, reestablishing surface drainage, and moving earth from areas where it produces a high driving force to where it would produce a neutral or negative driving force. At the request of the city, the data was reviewed by a volunteer panel of eighteen geotechnical experts, most of whom had extensive experience in the Palos Verdes Peninsula. The group included several geotechnical engineers, a specialist in ground water geology and a coastal engineer. The panel divided into subcommittees to make specific recommendations on things that might be done to improve stability. The panel concluded that stabilization appeared feasible but more data were needed, specifically a current topographic map and ground water information.

In 1984, the city funded the work requested by the technical panel. It also established the City of Rancho Palos Verdes Redevelopment Agency to carry out the stabilization effort. The City Council serves as the agency's Board of Directors and the agency's boundaries coincide with those of the Landslide Moratorium Area. Thus began the effort to stabilize the Portuguese Bend landslide. The city provided about $200,000 to start the project. A $2,000,000 grant from the state has covered the cost of most grading and additional funds have been provided by the city and small grants to the city. Total expenditures to date approach $3,000,000.

GEOLOGIC SETTING

The Palos Verdes Peninsula is a northwest-trending domical ridge, 9 miles long and up to 5 miles wide. Its crest has gently rolling topography at elevations ranging from 1480 to 1100 feet above sea level. Below this upland, remnants of a flight of Pleistocene marine terraces ring the peninsula and demonstrate it was an island during most of its geomorphic evolution. The peninsula is currently bounded by the Los Angeles coastal plain to the north, by Los Angeles Harbor on the east, and by the Pacific Ocean on the south and west.

The peninsula's shape reflects its Quaternary uplift as a doubly plunging anticline within the hanging wall of the southwest dipping Palos Verdes fault. Wells drilled for oil have established the subsurface position of the Palos Verdes fault along the northeast margin of the peninsula but the fault has not been located at the surface and may not penetrate late Quaternary deposits.

The Mesozoic Catalina Schist forms the core of the Palos Verdes anticline. Schist crops out near the center of the peninsula and is at shallow depth beneath much of its northeast flank. In early middle Miocene, the schist formed an irregular, southwest-sloping sea floor on which the Monterey Formation was deposited. The Monterey Formation is divided into the middle Miocene Altamira Shale, the late middle to early upper Miocene Valmonte Diatomite and the late upper Miocene to early Pliocene Malaga Mudstone. The Altamira Shale is further divided into the tuffaceous, cherty and phosphatic members. The tuffaceous member is of special interest because it contains weak bentonite beds that host the slip surfaces of most large landslides within the peninsula, including the Portuguese Bend landslide complex.

The tuffaceous member comprises all but the uppermost part of the Altamira Shale and represents the infilling of a tectonically formed submarine basin centered south of the peninsula. Its thickness increases in a distance of four miles from as little as 300 feet in the area of schist outcrops in the northeast part of the peninsula to about 4000 feet where penetrated by an exploratory oil well at Long Point to the west of Abalone Cove. It includes cherty and phosphatic shale of largely biogenic origin; siltstone and fine-grained sandstone derived from Catalina Schist and lessor volcanic sources; basalt in the form of pillows, sills and irregularly shaped intrusions; tuff turbidite, and ash fall and debris flow tuff which are largely altered to bentonite; and dolostone formed by diagenesis and geothermal activity. Much of the member contains thin parallel beds giving the false impression that beds should be continuous over long distances. Most beds lens out over short distances and parts of the lithologic sequence, in units as much as 100 feet thick, are affected by syndepositional slumping and debris flow.

In the vicinity of the Portuguese Bend landslide complex, the Portuguese Tuff is the only easily recognized stratigraphic unit within the tuffaceous member. It occurs about 450 feet below the top of the tuffaceous member and consists of about 60 feet of irregularly bedded tuff that appears to have been deposited by turbidity currents during a single eruptive event. Most of it was converted to a nearly pure waxy bentonite. The Portuguese Tuff functions as a zone of low shear strength and as an aquiclude. The base of the Portuguese Bend landslide complex ranges from 50 feet above the top of the tuff to the top of the tuff in most places. Except, along the coast it is near the base of the tuff in the Abalone Cove and Portuguese Bend slides and 70 feet below the tuff in the Klondike Canyon slide.

The cherty member of the Altamira Shale overlies the tuffaceous member along the head of the Portuguese Landslide complex and occurs in rotational slump blocks within the central and western parts of the complex. It is about 50 feet thick and consists of thin bedded chert and porcelanite derived from diatomite. A zone of distinctive chalcedony concretions, unique to this member, makes it easy to identify the cherty member wherever it crops out within the peninsula. The cherty member was deposited across a flat sea floor after cessation of local volcanism and tectonism.

The phosphatic member of the Altamira Shale and the Valmonte Diatomite are present along the head of the landslide complex and highest slump blocks but are not present within the main body of the complex.

During Pleistocene uplift, wave erosion removed the upper part of the Monterey Formation from most of the Palos Verdes Peninsula. The Malaga Mudstone was removed from all but the northern margin. Erosion was greatest along the peninsula's seaward-facing south flank. As a result, the tuffaceous member crops out across the entire south flank. Bedding within the tuffaceous member dips seaward subparallel to the slope of the ground surface. This creates conditions suitable for translational sliding along weak bentonite beds. In addition, wave erosion cut notches into the south flank creating steep wave-cut cliffs in which beds were inclined unsupported out of slope, further favoring formation of landslides.

For a comprehensive coverage of the geology of the Palos Verdes Peninsula refer to Woodring and others (1946). Rowell (1982) presents a detailed chronostratigraphy of the

Monterey Formation; Conrad and Ehlig (1983) provide details on the lithostratigraphy of the Monterey Formation, and Bryant (1987) provides recent information on the marine terraces and Quaternary tectonics of the peninsula.

PORTUGUESE BEND LANDSLIDE COMPLEX

Setting

The Portuguese Bend landslide complex occupies a broad structural, coastal and topographic embayment on the south flank of the peninsula. The structural embayment is shown by contours on the base of the Portuguese Tuff (Fig. 1). It probably formed by differential compaction of sediments deposited in a basin bounded on the east, south and west by thick intrusions and extrusions of basalt, and on the northeast by a basement high of Catalina Schist. The schist and basalt experienced negligible compaction while biogenic and pyroclastic-rich sediments experience extensive compaction. The compaction created a structural bowl beneath the area of the landslide complex. The bowl was subsequently tilted southward during folding of the Palos Verdes anticline. This created steeper south dips beneath the landward part than the seaward part of the complex. The structure is further modified by gentle folds which partly control the boundaries of the slide complex and the development of subslides within it.

The coastal embayment and part of the topographic embayment reflect the greater ease with which the tuffaceous member of the Altamira Shale was eroded in this area compared to the areas on either side. Relatively soft siltstone and shale predominate in the embayment whereas hard basalt and dolostone predominate on either side. Along the present coast, resistant basalt and associated dolostone form the sea cliff to the west of Abalone Cove and resistant dolostone of geothermal origin forms the promontory east of Portuguese Bend. A basalt sill buttresses Portuguese and Inspiration points within the embayment but Holocene wave erosion breached the sill on either side of the points.

In the uphill part of the complex, the basal slip surface is at or slightly above the top of the Portuguese Tuff. At the toe, the basal slip surface is near the base of the tuff. The Klondike Canyon landslide is an exception. Its basal slip surface is 70 feet below the tuff because a monoclinal flexure has raised the bedrock beneath it (Fig. 1). The western edge of the complex occurs where wave erosion removed the Portuguese Tuff from the coastal area prior to development of the complex. The eastern edge occurs where the Portuguese Tuff becomes deeply buried beneath overlying strata near the coast and the seaward part is buttressed by an anticline.

Origin of the complex

In order to properly evaluate the mechanics and stability of the Portuguese Bend landslide complex, it is necessary to know whether the complex originated as a single large megaslide or as a coalescing group of semi-independent slides of variable age. As seen today, the landslide complex is broken into numerous subslides. Based on geomorphic expression, individual subslides might have separate origins from their neighbors, or they might represent parts of a megaslide that started moving as a unit but fragmented as movement progressed.

Jahns and Vonder Linden (1973) believe the complex consists of semi-independent slides formed during three separate time intervals. As interpreted by them, slides in the up-slope part of the complex formed before development of the seventh marine terrace whose shoreline angle is reported to be 650 feet (Jahns and Vonder Linden, 1973, Table 1); slides within the intermediate part of the complex formed before final development of the first marine terrace whose reported shoreline angle is 140 feet, and slides whose toes are along the present coast formed after the first marine terrace. Their interpretation that the headward part of the complex predates the seventh marine terrace is based on several observations. Remnants of a landslide overlie the Portuguese Tuff and occur up slope from the seventh terrace in a bowl-shaped area west of the Portuguese Bend landslide complex. Within the complex, a topographic embayment extends across the slide at the approximate elevation of the seventh terrace. They also report the presence of marine deposits of the seventh terrace resting on inplace bedrock down slope from the exposed base of the landslide on the west side of Portuguese Canyon. I have observed stream gravels which contain reworked marine cobbles but have not found a marine terrace in this area.

Although the south flank of the peninsula contains slides of various ages, I interpret the Portuguese Bend landslide complex to have originated as a single megaslide at the time when the ocean occupied the abrasion platform which caps Inspiration and Portuguese points (terrace 1 of Jahns and Vonder Linden, 1973, but terrace 2 of Woodring and others, 1946, and terrace 2a of Bryant, 1987). My interpretation is based on the following evidence:

1. Subsurface data indicates the deepest slide plane is a relatively smooth surface that extends continuously beneath the entire complex. The slip surface is near the top of the Portuguese Tuff in the upper part of the complex but varies from above the tuff to slightly below the tuff in the central and seaward parts of the complex.
2. The slide base is more than 100 feet below the surface in much of the area where the ground surface is near the elevation of the seventh marine terrace. The topographic embayment is a reflection of the bedrock structure beneath the slide. To the best of my knowledge, no marine terrace deposits occur within the slide in the vicinity of the projected position of the seventh marine terrace.
3. In the western part of the slide, slump blocks containing the cherty member of the Altamira Shale are distributed

from the head of the slide to the coast. The inplace cherty member crops out at an elevation of about 1000 feet in the head scarp of the complex. On the ridge at the west edge of the slide, the cherty member was removed by erosion prior to emplacement of the eighth marine terrace whose shoreline angle is about 950 feet. These observations are in agreement with those of Jahns and Vonder Linden (1973). However, I conclude that the slump blocks containing the cherty member originated near the head of the slide and were transported to the toe as part of a megaslide; whereas, they appear to believe that transport was accomplished by having younger slides develop beneath the toes of older slides.

4. There is no apparent difference between post-sliding landform modification or soil development throughout the parts of the slide not involved in the Holocene Abalone Cove and Portuguese Bend landslides. These two landslides contain the same features as the upslope part of the complex, including a thick soil mantle over old slump blocks and graben deposits, except that these features are disrupted by Holocene sliding.

5. Regarding the timing, the seaward edge of this Pleistocene slide is exposed on the east side of Inspiration Point, as previously recognized by Jahns and Vonder Linden (1973). Here, the Portuguese Tuff was bent upward to a nearly vertical position by up thrusting along the toe of the Pleistocene slide and then unconformably overlain by marine terrace deposits. A similar relationship is present on the inland edge of Portuguese Point but subsequent sliding has thrust the toe of the slide over the seaward part of the terrace and uplifted and back tilted the part of this terrace which overlapped the toe. The age of this terrace is about 120,000 years old (Bryant, 1987).

An important aspect to the difference between my perspective of the evolution of the landslide complex and that of Jahns and Vonder Linden (1973) is that I am concerned with the mechanics of sliding whereas they were concerned with the space-time relationships between sliding and sea level and climatic changes. There is general agreement regarding the interpretation of observable features; of course, more information is available now than when their paper was written. In the case of the Portuguese Bend and Abalone Cove landslides, there is agreement that both are reactivated parts of the Pleistocene landslide. The toes of the two slides are a short distance seaward of the Pleistocene toe and have developed in response to loss of support along the Holocene shoreline angle.

From a mechanical point of view, it is difficult to account for the development of the seaward part of the complex if the complex originated as a series of slides rather than a single slide. The seaward part has a gently sloping base and, at the time it formed, its toe ramped up from as much as 200 feet below sea level. The initial shearing resistance should have exceeded driving force generated within the seaward part of the slide. However, the landward part of the complex could

Figure 4. View of Portuguese Bend area looking north on September 3, 1947. Notice the landslide scarp on the slope inland from the pier. (Spence Aerial Photos)

have provided the extra driving force needed to cause the initial failure (Fig. 3).

The initial slide was probably destabilized by wave erosion and ground water. Its toe is near the axis of a gentle arch in the bedrock. Wave erosion removed weight from the top of the arch. This allowed beds above the incipient slip surface to yield by bending upward in response to the thrusting force produced by sliding further uphill. Instability was probably promoted by development of a perched water table above the Portuguese Tuff in the uphill area, and development of artesian pressure beneath the tuff near the coast. During the early stages of sliding, the toe developed as a slide-propagation fold. Uplift would have been aided by artesian pressure from ground water trapped beneath the Portuguese Tuff. As movement progressed, the slip surface propagated through the fold to form the ramp which is preserved inland from Portuguese Point (Fig. 3).

During initial movement, the head of the megaslide was probably a short distance up slope from the present day 600-foot ground elevation contour (Fig. 1). As movement reduced lateral support, rotational slide blocks developed up slope from the head scarp and slid downhill to form the stepped terraces visible in the northwest part of the complex (Fig. 4 and 5). The northeast arm of the complex and Valley View graben (Fig. 1 and 3) probably also developed at this time due to the combined effects of loss of lateral support and a high water table.

The Pleistocene landslide probably stopped moving because of the combined effects of loss of driving force by displacement of mass away from its head and by drainage of ground water from its up slope area as a result of slide-induced permeability along joints and fissures. Self buttressing by accumulation of debris along the toe after the latest Pleistocene glacio-eustatic lowering of sea level was probably not a factor. Had it been, the entire complex should have reactivated in response to Holocene wave erosion.

The present Portuguese Bend and Abalone Cove landslides coincide with a prehistoric Holocene landslide as interpreted from scarps, closed depressions and displaced landforms visible on aerial photographs that predate historic movement. The Holocene landslide developed from the coastal part of the Pleistocene landslide complex. Movement occurred after wave erosion largely removed strata that buttressed the toe of the Pleistocene slide. Thus, the Portuguese Bend and Abalone Cove landslides are a reactivated part of the Pleistocene landslide complex.

The Holocene landslide is split into three segments by the buttressing effect of Inspiration and Portuguese points. The eastern segment moves into Portuguese Bend; the central segment moves into the unnamed cove between the two points, and the western segment moves into Abalone Cove. The two points are supported by a seaward dipping basalt sill which ranges from about 50 to 100 feet thick within the confines of the points. The sill extends beyond the points but was removed by wave erosion before the start of Holocene sliding. Because Inspiration Point is narrow, the thrust force of the slide mass caused its inland edge to rupture along steeply inclined, conjugate tear faults that diverge at an angle of about 65 degrees. The slide mass cleaves as it pushes against the leading edge of these faults. Because Portuguese Point is wide, the Holocene slide mass initially ramped up onto the point. Subsequent movement caused tear faults to develop on either side of the point but the main tear faults do not join at their inland edge. There is a belt of slide material that is still being thrust up the ramp at the inland edge of Portuguese Point. This belt forms the division between the Portuguese Bend and Abalone Cove landslides.

The most recent prehistoric movement of the Portuguese Bend landslide was less than 4,800 æ 180 years ago based on the C-14 age of carbonized wood taken from pond deposits that were displaced and tilted by prehistoric movement (Emery, 1967). The Abalone Cove landslide moved at the same time as shown by the continuity of a head graben that extends between the two slides, and by the similar youthfulness of scarps, closed depressions and other geomorphic features within the two slides. The Klondike Canyon landslide probably moved at the same time, as suggested by black adobe soil which fills prehistoric fissures, but its displacement was too small to leave well defined geomorphic features.

The part of the Portuguese Bend landslide located between Inspiration Point and Portuguese Canyon was affected by sliding during the period between 1931 and 1947. A well-defined head scarp is visible in a 1947 photograph (Fig. 4) but absent from a 1931 photograph. The scarp is at the same location as the head of the seaward subslide, as described later. This indicates the seaward subslide had a safety factor near 1.00. It may have been creeping when the main slide reactivated in 1956. MacKintosh and MacKintosh (1957) report that the pier, which was built 400 feet east of Inspiration Point in 1946, required repairs before 1956 at the same place where it was offset by sliding in 1956.

PORTUGUESE BEND LANDSLIDE

Boundaries

As described above, Inspiration Point splits the Portuguese Bend landslide into two parts. The eastern part which moves into Portuguese Bend is hereafter referred to as the main Portuguese bend landslide. The western part which moves into the cove between Inspiration Point and Portuguese Point is hereafter referred to a the western segment of the Portuguese Bend landslide. The main landslide has an area of about 190 acres and the western segment has an area of about 70 acres.

Figure 5. View of Portuguese Bend area looking north on September 26, 1956, one month after reactivation of the Portuguese Bend landslide. Road fill where sliding began is on the right central side of photo. (Whitehurse Aerial Surveys)

The main Portuguese Bend landslide has a lobate shape with a length to width ratio of about 1.5:1. Its greatest width is 2700 feet and occurs in the vicinity of Palos Verdes Drive. Its greatest length is 3900 feet.

The head of the main slide occurs where the base of the slide rises to the surface along a structural ramp in the underlying inplace bedrock. After reactivation in 1956, the head was initial marked by a well-defined head scarp and graben but subsequent movement uncovered seaward dipping bedrock along the central part of the head. In 1986, grading uncovered bedrock for a distance of 1,000 feet along the head. Bedrock was also exposed at the northwest corner of the slide but was subsequently covered by fill to provide surface drainage and eliminate a vertical scarp. Bedrock is not exposed along the eastern part of the head because Paintbrush Canyon was incised through the base of the slide prior to placement of fill during construction for Crenshaw Boulevard. The head of the slide has been greatly modified by grading since 1986 and is no longer active.

Along most of the slide's east edge, bedding is upturned by a west-facing monoclinal flexure. As a result, the basal slide plane rises to the surface and forms a sharply defined trace that extends nearly 3000 feet from the head to the toe of the slide. Along much of this length, movement is nearly horizontal and parallel to the trace. Thus, where the trace crosses Palos Verdes Drive, it produces a left lateral offset in the

Figure 6. Offset in Palos Verdes Drive at east edge of slide in Oct., 1991.

roadway without noticeable vertical offset or convergence or divergence across the break (Fig. 6).

The western edge of the main slide has a poorly defined boundary. In-between the head of the main slide and the head of the western segment, the gently inclined basal slip surface continues westward beneath the inactive part of the slide complex. Displacement dies out across a broad zone which contains a series of northeast to east trending slices bounded by steeply inclined en echelon faults, fissures and discontinuous half-grabens. The westward decrease in displacement is accommodated by clockwise hinging and rotation of these slices.

Between the head of the western segment and the point where the slide splits into two segments at the inland edge of Inspiration Point, the boundary between the main slide and the western segment is poorly defined. The buttressing effect of Inspiration Point caused a narrow belt of land located directly inland from the point to move less than the land on either side. Since 1956, total horizontal displacement along the axis of this belt ranges from 85 feet at its inland edge to nearly 100 feet at Palos Verdes Drive.

The toe of the main slide is along the beach where the base of the slide rises to the surface. Movement causes the slide mass to thrust over the beach and wave cut bench but wave erosion continually removes material from the toe. Consequently, the position of the toe varies through time due to variations in the rate of slide movement and the rate of wave erosion. In its most advanced position, the toe was as much as 400 feet seaward of its initial 1956 position. Wave storms and the slowing of movement due to slide mitigation have caused the toe to retreat landward as much as 150 feet in recent years.

The western segment of the Portuguese Bend landslide is 1000 feet wide along its toe in the cove between Inspiration and Portuguese points. It extends inland 2000 feet giving it a length to toe-width ratio of 2:1. The east side is transitional into the main Portuguese Bend landslide as described above and is interpreted to be nearly straight. The west side is complex. It is about 300 feet wider at the inland edge of Portuguese Point than in the cove between the two points. The widening is accommodated by thrusting and westward displacement along the slide ramp at the inland edge of Portuguese Point. A survey monument on the hill above this ramp was displaced 2.2 feet west toward Abalone Cove between 1957 and 1963 (elevation changes were not determined during this period). No survey data are available for the period from 1963 to 1976, but it is reasonable to assume that movement continued during this period. Since 1976, the monument on this hill has moved 7.3 feet west and has risen 1.3 feet in elevation. North of Palos Verdes Drive, the west

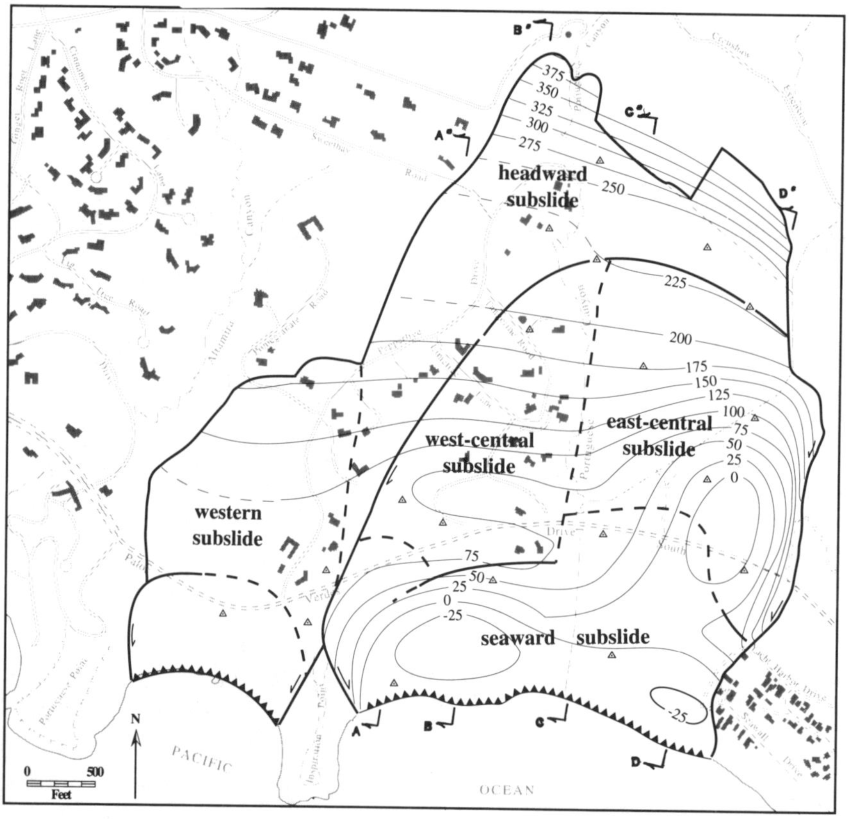

Figure 7. Map of Portuguese Bend landslide showing boundaries of subslides and structure contours in feet on base of slide.

side is marked by northeast-trending fissures and small grabens similar to those along the northwest edge of the main slide. After the Abalone Cove landslide began moving in 1978, movement occurred throughout the zone between the two landslides. Thus, the boundary between the western segment and the Abalone Cove landslide is transitional.

Subslides

Because of its large area and large length to thickness ratio (30:1), and variations in conditions which affect the rate of movement from one part of the slide to another, the Portuguese Bend landslide behaves as a group of semi-independent subslides. As define here, each subslide is controlled by undulations in the slide's base and/or lateral boundary conditions. Therefore, each subslide maintains a fixed location through time as the slide mass translates from one subslide to another. Recognition of these subslide is an essential element in the plan to permanently stabilize the Portuguese Bend landslide.

The main slide is divisible into four subslides. The four are referred to as the headward, east-central, west-central and seaward subslides. The western segment of the slide, which is itself a subslide, is further divided into the landward part of the western segment and the seaward part of the western segment. The subslides and structure contours on the base of the slide are shown on Figure 7.

The headward subslide is crescent shaped and has an area of about 45 acres. It forms the up slope and western parts of the main slide. Its up slope half is controlled by undulations in the underlying bedrock to which its basal slip surface conforms. Along its head, the slide base dips 15 to 25 degrees seaward. The slide mass on this ramp generates a large excess driving force. Further down slope the slide base flattens at the crest of an anticlinal undulation (see Fig. 3) to a seaward dip of only 4 to 5 degrees. In this area the slide mass generates a net resisting force providing the water table is low. The base of the western part of headward subslide has an average dip of only 5 degrees. As a result, its resisting force exceeds its driving force providing the water table is low.

The east-central subslide is nearly rectangular in shape and has an area of about 60 acres. Most of its driving force is generated on the south flank of the anticlinal undulation that controls the adjoining down slope edge of the headward subslide. Here, the slide's base dips between 10 and 18 degrees seaward. Because movement accelerates along this structural ramp, a broad graben has formed by extension across the crest of the anticlinal undulation (C-C', Fig. 8). The graben is included as part of the east-central subslide. The seaward part of the subslide is controlled by a spoon-shaped structural depression where the base of the slide descends to a few feet below sea level and then rises gently on the seaward side. On either side of the depression, the slide's base is subhorizontal in the direction of movement. Consequently, the downhill part of the subslide generates a net resistance to sliding.

The west-central subslide includes about 40 acres within the residential area west of Portuguese Canyon. Its shape is roughly a right triangle with the hypotenuse on the northwest. It is a westward continuation of the east-central subslide except it is partly supported on its downhill side by a subsurface ridge. The slide mass must rise 30 to 40 feet as it moves to the crest of the ridge; this produces a large force resisting movement. The origin of the ridge is undetermined but it may consist of steeply inclined beds along the upturned toe of a deep-seated abandoned part of the Pleistocene landslide. The east end of the ridge is probably terminated by a tear fault. This is suggested by the presence of a steeply inclined right-slip fault which extends for a few hundred feet along the southeast edge of the west-central subslide. The fault is parallel to the direction of slide movement and regenerates at the same location as the slide sheet translates through this area.

The seaward subslide has an area of about 45 acres. Its toe extends eastward from Inspiration Point for a distance of 2,300 feet along the beach. West of Portuguese Canyon, it consists of a single major rotational slump block (cross sections AA' and BB', Fig. 8). Here, its head scarp is about 700 feet inland and is above the ramp on the seaward side of the anticlinal ridge that buttresses the west-central subslide. The head scarp continually regenerates at this location and is in the same place as the scarp visible on the 1947 photograph (Fig. 4). The steep slip surface exposed in the head scarp flattens downward to merge with the base of the main slide near the top of the ramp. From there the slip surface follows the arcuate path of the main slide base. At the base of the ramp it flattens into a bowl-shaped structural depression whose center is about 45 feet below sea level and then rises to the surface at the beach (B-B', Fig. 8).

In the vicinity of Portuguese Canyon, the seaward sublide extends over 1,000 feet inland. Its base dips about 5 degrees seaward in the inland area but flattens to subhorizontal near the coast. The eastern part of the seaward subslide extends about 900 feet inland. Its base has a similar but more subdued shape to that of the western part.

The separation of the western segment of the Portuguese Bend landslide into two subslides is controlled by the buttressing and constricting effects of Inspiration and Portuguese points. Movement of the landward subslide is retarded by these effects while the part of the slide within the slot between the two points is not. Consequently, the seaward subslide moves faster than the landward one and has its head marked by an arcuate graben in the vicinity of Palos Verdes Drive.

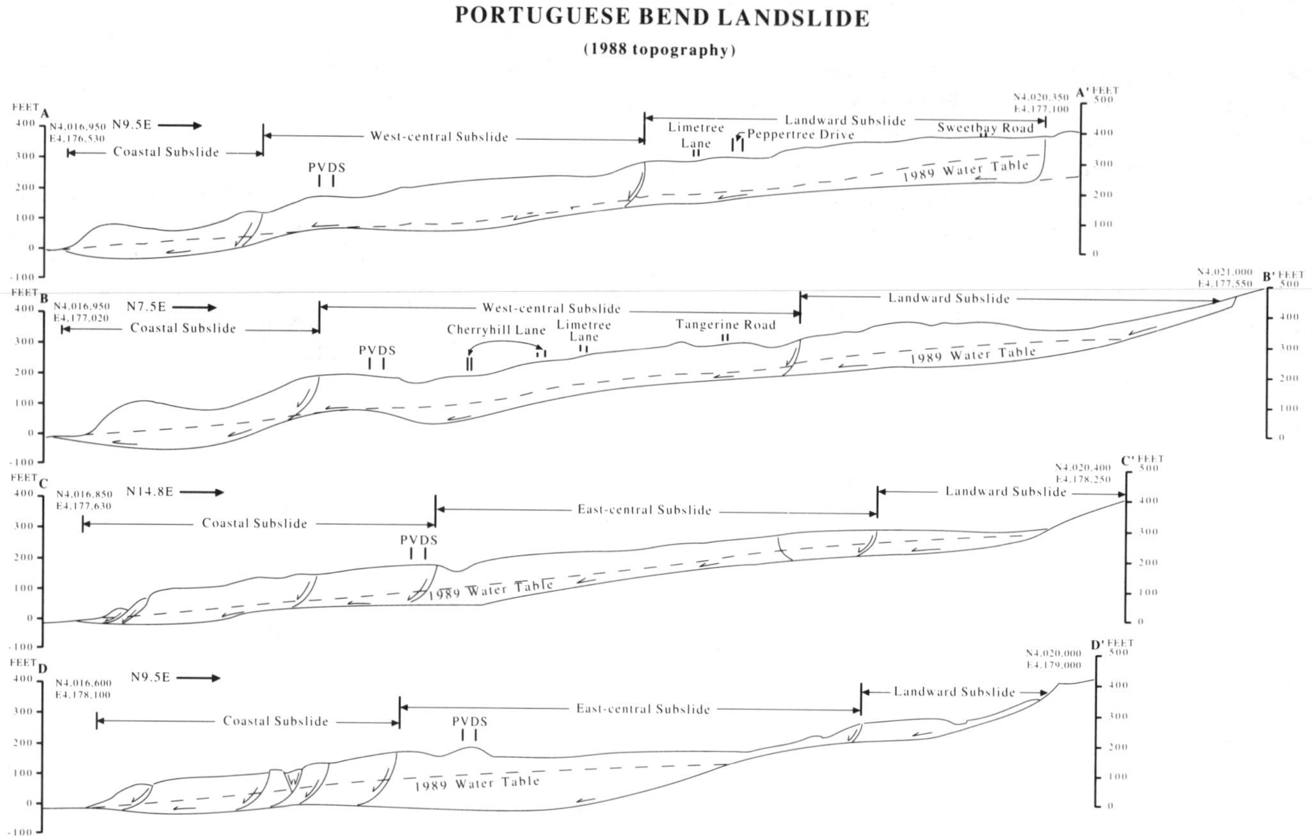

Figure 8. Vertical cross sections through the Portuguese Bend landslide. See Figure 7 for locations of sections.

Mechanics

The main Portuguese Bend landslide probably originated as a simple translational glide sheet driven by excess driving force generated along structural ramps. The earliest movement may have developed along the ramp which extends uphill from the headward subslide. Here, bedding dips as much as 30 degrees seaward. Assuming slippage occurred along a tectonically sheared bentonite bed whose angle of internal friction was 6 to 8 degrees and cohesion near zero, the slide sheet overlying this ramp would have generated three to four times as much driving force as resisting force. As excess driving force was transmitted downhill, it was sufficient to overcome a net resisting force in areas of gentle or reverse dip, at least as far downhill as the shoreline. The toe developed near the then existing shoreline angle because the slide sheet was thinnest at this location. Uplift at the toe was probably promoted by artesian pressure beneath the Portuguese Tuff which is a perfect aquiclude. This is suggested by a change in the stratigraphic position of the slip surface from above the Portuguese Tuff beneath the main body of the slide to within or at the base of the Portuguese Tuff in the vicinity of the toe. The artesian condition could result from surface water entering bedrock beneath the Portuguese Tuff in the area northeast of the slide and along the bottoms of Paintbrush and Klondike canyons. If artesian pressure exceeded overburden pressure, anticlinal uplift could have begun along the shoreline prior to sliding. This would have promoted sliding by reducing support along the toe.

The Portuguese Bend landslide is no longer a simple translational glide sheet. Much of the original slide sheet was displaced to the coast and removed by wave erosion, particularly in the area east of Portuguese Canyon. The present slide mass is a mixture of broken and disoriented bedrock blocks, slide debris, alluvial and pond sediments which were deposited in flat areas and closed depressions, masses of Portuguese Tuff and other slump deposits derived from above the head of the main slide, and fill and debris from human activity. This melange, which has an average thickness of about 125 feet, is weak and parts of it deform like a mass of incoherent earth. The slide still gains a large part of its driving force by translation down inclines in the basal slip surface. However, driving force is also generated by the lowering of mass along normal faults that merge with or terminate at the base of the main slide. There are two styles of normal faulting. One creates nearly symmetrical grabens. The other creates a stair-step arrangement of rotational blocks. Both are capable of causing distention above a subhorizontal slip surface.

Symmetrical grabens form in areas where the slide base and ground surface are subhorizontal and the downhill part of the slide is moving faster than the uphill part. The graben at the head of the east-central subslide is an excellent example (C-C', Fig. 8). This graben is bounded by steeply inclined normal faults that dip toward each other. Numerous normal faults occur between the bounding faults. These create secondary grabens and horsts. The driving force created by lowering of wedge-shaped blocks helps drive the downhill part of the slide. This type of faulting is capable of causing lateral spreading above a weak subhorizontal slip surface.

The stair-step arrangement of rotational blocks develops where surface elevations rise conspicuously in the uphill direction. The orientation of the main slide base is of minor important. The necessary conditions are well developed within the seaward subslide. Here, as the slide mass moves forward, wave erosion continually regenerates a steep cliff along the toe of the slide. Because of an unbalanced rotational force, a seaward dipping listric normal fault develops landward from the cliff (C-C', D-D', Fig. 8). The fault has a steep inclination near the surface but flattens downward to merge with the base of the main slide. As the block is lowered by backward rotation, a cliff is created on its landward side. This promotes the formation of another rotational block further uphill. Directly east of Portuguese Canyon the series of rotational blocks extends as much as 1,000 feet inland from the coast. The retrogressive growth of rotational blocks has an important impact on the stability of the main slide. Seaward from where the most landward listric normal fault joins the basal slip surface, the rotation of blocks may provide all of the driving force needed to overcome frictional resistance to sliding. Under this condition, the seaward part of the slide does not buttress the landward part, even where the base of the slide dips landward. Only the viscous component of friction resists the thrusting action of the landward part. Because viscous friction is a function of velocity, the landward part will continue to move so long as it has an excess driving force at its downhill edge.

Both absorbed water and ground water play a major role in the mechanics of the Portuguese Bend landslide and were the main causes of temporal variations in the rate of slide movement between 1956 and the start of remedial grading in late 1986. This is shown in Figure 9 by the relationship between rainfall, measured 1 mile northeast of the Portuguese Bend landslide, and the rate of slide movement at Los Angeles County survey monument R-l, located in the east-central subslide. In the vicinity of R-l all water comes from natural sources. From 1956 to 1963, frequent survey measurements show movement accelerated during rainy months and slowed during dry months. Since 1962, the time between measurements does not permit direct comparison between rate of movement and monthly rainfall but it does permit a direct comparison between the rate of movement and the deviation of rainfall from a long-term average of 12.9 inches per year. Notice that between 1959 and 1978, rainfall deviated only slightly from the long-term average and that the rate of movement has the same pattern as the cumulative deviation of rainfall. Starting in 1978, high rainfall caused a large cumulative deviation from the long-term average and was matched by a large increase in the rate of slide movement. The fact that movement did not slow greatly during periods of below average rainfall in 1981 and 1984, indicates the natural depletion of water is slow.

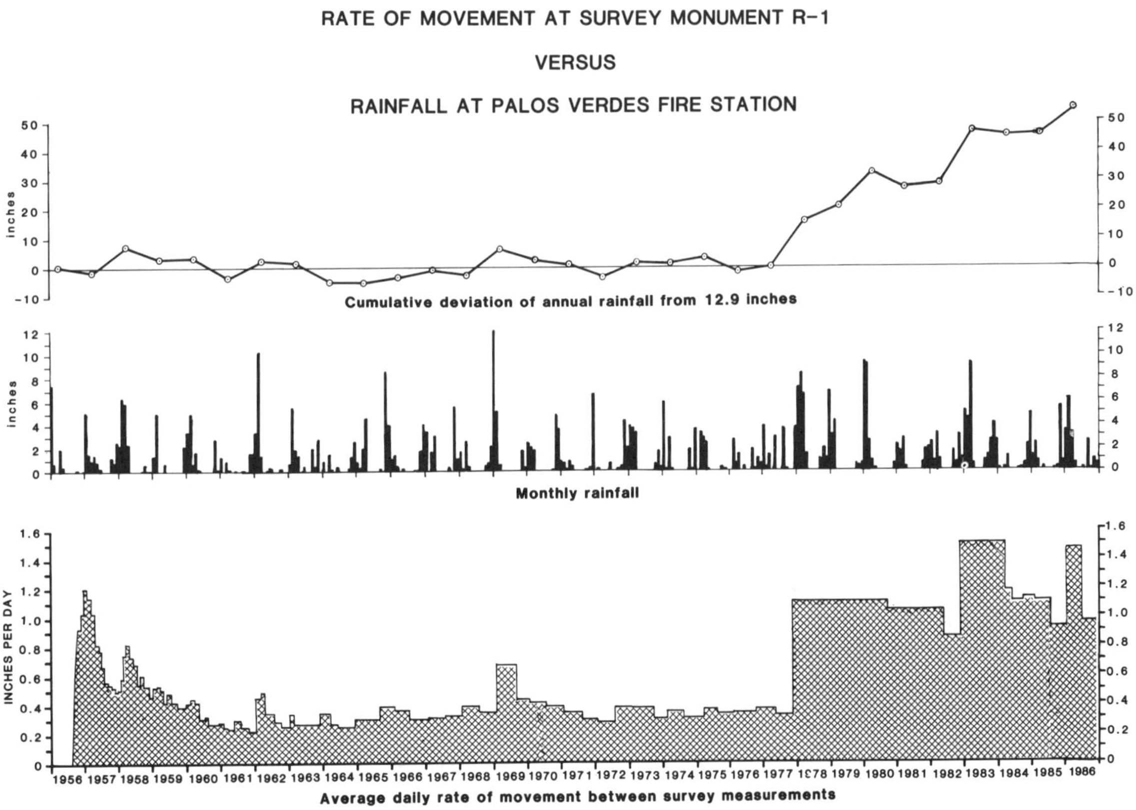

Figure 9. Graph showing relationship of rainfall to rate of slide movement.

Most water within the slide mass is absorbed water. In the zone above the water table, the amount of absorbed water varies seasonally and also annually. Where the slide base is above the water table, the shear strength of bentonite along the basal slip surface varies as a function of its absorbed water content. The strength decreases as the bentonite absorbs water and increases as it loses water. Within the body of the slide, an increases in the amount of absorbed water causes a decrease in the bonding strength between particles and allows deformation to occur more freely. This is especially important in areas where normal faulting and other types of deformation are occurring. Loss of absorbed water makes the slide mass more rigid and resistant to internal deformation. Variations in the absorbed water content account for most of the seasonal variation in the rate of slide movement.

Absorbed water may have played an important role in triggering the Portuguese Bend landslide during placement of road fill in 1956. The bentonite at the base of the slide was probably nearly saturated prior to placement of the fill. Compaction of the bentonite under the added weight of fill probably caused the development of excess pore water pressure, thereby reducing the shear strength of the bentonite. Such an effect was observed in September 1990, when 60,000 cubic yards of earth were moved from the upper part of the slide and placed as a fill in the most easterly part of the slide north of Palos Verdes Drive. Slide displacement is nearly horizontal in the area of the fill and, therefore, must consume more energy than it generates. However, the local rate of movement increased from 0.042 inches per day prior to placement of the fill to 0.077 inches per day upon completion of the fill. Movement subsequently slowed at a decelerating rate for several months until the rate fell below the original rate. This pattern can be explained by a reduction in shear strength due to development of excess pore water pressure during placement of the fill followed by a gradual increase in strength as the pore pressure dissipated.

Ground water reduces the stability of a landslide through its buoyancy effect. In the Portuguese Bend landslide, material below the water table exerts only about half as much normal force causing frictional resistance along the base of the slide as material above the water table. Consequently, in an area where the slide mass is 100 feet thick, a 2 foot rise in the water table will causes a 1% decrease in frictional resistance to sliding. For this reason, ground water played a major role in sustaining the movement of the Portuguese Bend landslide from 1956 to 1978 and was responsible for accelerated movement commencing with high rainfall in 1978. Ground water may also have been an important factor in the enlargement of the original slide beyond the immediate vicinity of the road fill in 1956. Movement in all but the seaward part of the landslide can be stopped by lowering the water table.

Data obtained from monitoring wells installed in 1984 and subsequent years indicate ground water is present above the base of the Portuguese Bend landslide in all but a narrow zone along the slide's northern and eastern margin. The slide mass acts as an unconfined aquifer perched on an impervious slide base. In 1984, the water table was a relatively smooth seaward sloping surface with an average height of about 45 feet above the slide base. The thickness of the saturated zone varied because of undulations in the slide base; it was greatest (as much as 110 feet) above troughs and depressions in the slide's base and least along the crests and seaward flanks of ridges. Since 1984, the water table has been lowered an average of about 10 feet as a result of pumping and low rainfall. Progress in lowering the water table has been slow because of low permeability, particularly in the area east of Portuguese Canyon.

Little reliable data are available regarding ground-water conditions prior to 1984. Based on current knowledge, ground water was present above the slide base in most areas, even in 1956. Unfortunately, no monitoring wells were installed during this period. Data was obtained during the drilling of borings. Most borings which penetrated the water table were filled or abandoned without allowing the water to reach its static level. Monitoring wells drilled since 1984 have shown that it may take many days or even weeks for the water level to stabilize in borings because most slide material has a low storage coefficient (about 5%) and low permeability. Seepage was encountered above the slide base in several bucket auger borings logged by me in 1957, including one located at the northwest edge of the slide. Based on monitoring data and well production, we now know that between 10,000 and 15,000 gallons of water per day enter the northwest corner of the slide as part of the regional ground-water flow. Similar flow probably existed in 1956. If it did, the 1956 configuration of the water table was similar to that in 1984 except much lower.

One question which is often asked is whether artesian pressure exists beneath the slide. If it did, it could explain the unusually low friction angle (6 degrees) determined by back calculations of stability. I have observed no evidence of artesian pressure beneath the main part of the slide. Wherever borings penetrate the slide base, they typically encounter unoxidized bedrock which appears to be impermeable. In 1985, two piezometers were installed beneath the base of the Portuguese Bend landslide, one in the east-central subslide and the other in the seaward subslide. Neither recorded artesian pressure. A multistaged piezometer is installed in the transition zone between the Portuguese Bend landslide and seven are distributed within the Abalone Cove landslide. All extend far below the slide base. None has recorded artesian pressure.

I am aware of only one boring that has encountered artesian water within the Portuguese Bend landslide. It was drilled on the east side of the slide about 200 feet inland from the beach in 1957. Artesian water flowed at a rate of several gallons per minute from beneath the base of a thick zone of sheared bentonite. The water flowed from the hole for many days. At the time, I assumed the water was coming from beneath the main slide base but it may have come from beneath a seaward dipping listric slip surface associated with a secondary rotational slide. At that time, houses near the beach were disposing of their sewage in pits located up slope from the boring.

In 1981, artesian water was encounter 700 feet further east at the toe of the Klondike Canyon landslide. The water was trapped beneath the Portuguese Tuff. Subsequent pumping has lowered the water table below sea level with no observable effect on the Portuguese Bend landslide. Thus, artesian pressure may occur locally near the beach but there is no evidence of artesian pressure affecting movement of the slide.

PORTUGUESE BEND LANDSLIDE STABILIZATION PROJECT

Plan of Control

During planning of the stabilization project, it was clear that remedial activities would be limited by several constraints. Total funding would be meager and stretched over a period of many years. Therefore, stabilization would have to be carried out in phases. Each phase would have to survive any subsequent slide movement, or be inexpensively repaired, until final stabilization was achieved. It would be impractical to use the common method of stabilizing a large slide by keying a buttress into undisturbed bedrock at its toe and supporting the uphill part by one or more shear keys locked into bedrock beneath the slide. The keyway for a buttress would require excavating below sea level. Any excavation for a shear key would have to extend through 50 to 150 feet of highly unstable slide material with the deep parts well below the water table. Funds would be inadequate to remove a substantial part of the slide and replace it with compacted fill founded on stable bedrock. Stabilization of the seaward part of the slide would require permanent shoreline protection but prerequisite studies and the obtaining permits and funds would take many years. Therefore, uphill stabilization would have to progress without permanent shoreline protection. A final constraint was that abatement activities must be as nonintrusive as possible in developed part of the slide.

A plan of control was developed in accord with these constraints. It had two objectives: (1) permanent lowering of the water table to increase resisting forces, and (2) redistribution of the slide's mass to decrease driving forces and increase resisting forces. The water table would be lowered by removing ground water from wells and restoring surface drainage. A temporary system of culverts would be installed to conduct water to the ocean. Installation of a sewer system was recommended for the developed area but was deemed impractical while the land was moving. Mass-grading techniques were to be used to recontour the undeveloped part of the slide in order to reestablish drainage and redistribute earth to decrease driving force and increase resisting force. Ultimately, the shoreline would be protected by a rock revet-

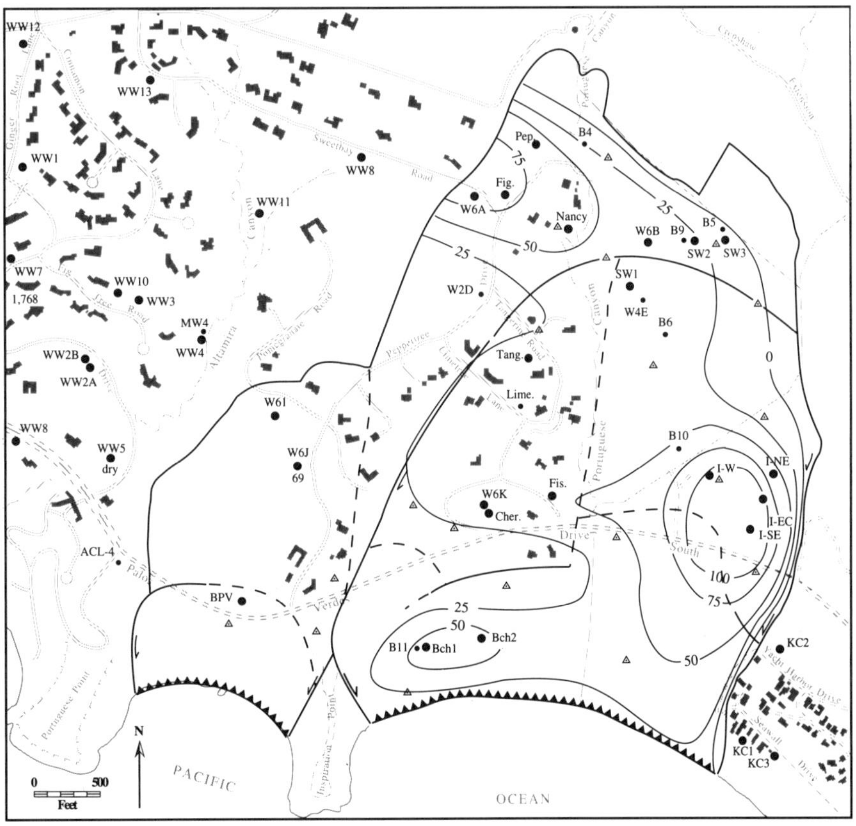

Figure 10. Map of producing and monitoring wells. Isopachs show thickness in feet of saturated zone above slide base in August, 1991.

ment, or other permanent structure, and the stability of the seaward subslide would be improved by construction of a toe berm.

Dewatering System

In 1984, the Redevelopment Agency funded the installation of monitoring wells to determine ground water conditions within the Portuguese Bend landslide. A total of thirteen wells were installed in the area north of Palos Verdes Drive. No wells were installed in the seaward part of the slide because of rapid movement and secondary sliding. Well spacing was nearly uniform but specific sites were chosen in areas most likely to support good producing wells. These included grabens and fissured ground where extension was likely to have increased porosity and permeability, and areas above troughs and depressions in the slide base where the water table was likely to have its maximum height above the slide base. Of the thirteen wells, five showed good potential for production and were fitted with six-inch PVC casing and gravel pack to permit their future use as dewatering wells. The others were fitted with two-inch PVC casing for monitoring.

Of the five wells suitable for production, the three most westerly wells were connected to the Abalone Cove dewatering system and placed into production near the end of 1984. (The water is piped to the ocean because its high sulfate content makes it unusable for domestic or agricultural purposes.) The other two wells and three new wells were placed in production during 1985 and 1986. Additional wells were installed between 1986 and 1991 to increase the total to 19 currently operational wells (Fig. 10). Thus far, only three wells have been rendered unusable for production - one from slide deformation, one from vandalism and one from clogging by bentonite.

Initial production from most wells exceeded 1,000 gallons per day (g/d) and several produced over 10,000 g/d. Production generally declined rapidly during the first few days or weeks and then stabilized to a gradual decline. Total well production peaked at about 30,000 g/d in 1987 due to above average rainfall in 1986 and an increase in the number of producing wells. Production declined below 20,000 g/d in 1989 and has subsequently stabilized at about 17,500 g/d (August, 1991). Of this, over 11,000 g/d comes from two wells at the northwest edge of the slide and an additional 2,000 g/d come from two nearby wells. This production is from perennial inflow of ground water from the northwest and represents interception rather than depletion. Production from the two wells located above the trough in the base of the west-central subslide is currently 2,000 g/d but oscillates through time in a pattern unrelated to rainfall. Most of this water is of domestic origin. Eight wells are present in the undeveloped area east of Portuguese Canyon and north of Palos Verdes Drive. Production from these wells has declined through time to a current total of only 1,700 g/d. Most of this water represent depletion of ground water stored in this area.

The wells are not as effective as originally anticipated. When I first evaluated the possibility of stabilizing the Portuguese Bend landslide, I assumed wells within it would have about the same productivity as wells within and up slope from the Abalone Cove landslide. In the latter, each of the first six wells averaged more than 5,000 g/d during their first two years and two exceeded 30,000 g/d. The present system of eleven wells in the Abalone Cove system currently produces 65,000 g/d of which 50,000 g/d comes from three up slope wells which intercept perennial inflow of ground water from the crest of the peninsula. Based on past performance, the Abalone Cove dewatering system has a proven capacity of over 200,000 g/d.

There are three reasons for the differences in well performance between the two areas. First, the Abalone Cove wells had higher initial production because its water is stored in permeable breccia and course alluvial deposits which form compartments between intact blocks of displaced bedrock. The blocks act like dams that impede the flow of ground water. In the Portuguese Bend landslide, breccia and alluvial deposits are abundant but generally have low permeability because they contain clay derived from Portuguese Tuff exposed within and uphill from the Portuguese Bend landslide. Surface runoff picks up clay from the tuff (bentonite) and transports it down fissures to the water table where it is deposited in pore spaces.. Permeability is exceptionally low in the area east of Portuguese Canyon where most of the slide mass has the properties of clay-rich earth. This material also contains numerous steeply inclined secondary slip surfaces which impede the flow of ground water. Bedrock blocks are present west of Portuguese Canyon but they are highly fractured and form ineffective barriers to ground water migration. A second reason for lower initial production in the Portuguese Bend landslide is that much of the water is stored in basins above troughs and closed depressions in the slide's base. The slide mass is under compression in these areas and removal of ground water promotes compaction which reduces porosity and permeability. The third reason is that the Portuguese Bend landslide receives less than one-third as much inflow from up slope ground water sources and from local sewage disposal. This is beneficial; it means less recharge from these sources.

The stabilization strategy has changed to accommodate the poor performance of wells. The last 4 wells were installed in the headward part of the slide to complete a belt of 8 wells arranged to intercept ground water flowing into the slide and to deplete ground water within the headward part of the slide. This eliminates a major source of recharge into down slope areas. Greater emphasis has also been placed on moving earth to achieve a rapid improvement in stability. This has the added benefit of permanently improving stability over and above that achieved by depletion of ground water.

Figure 11a. and 11b. Before grading and after grading view of Lake Ishibashi area showing filling of closed depression. Telephone pole at lower right in Figure 11a is near same location as pipe casting shadow in Figure 11b.

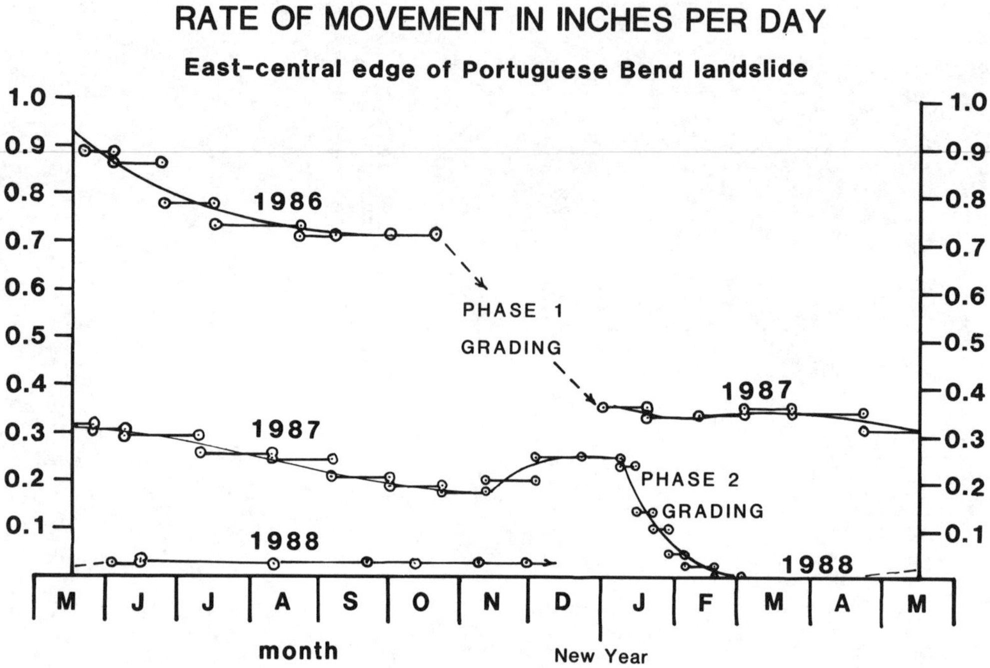

Figure 12. Rate of slide movement at northeast corner of east-central subslide, May 1986 to May 1988. Horizontal bars are average rate of movement between measurements. Continuous line is synthetic curve simulating instantaneous rate of movement.

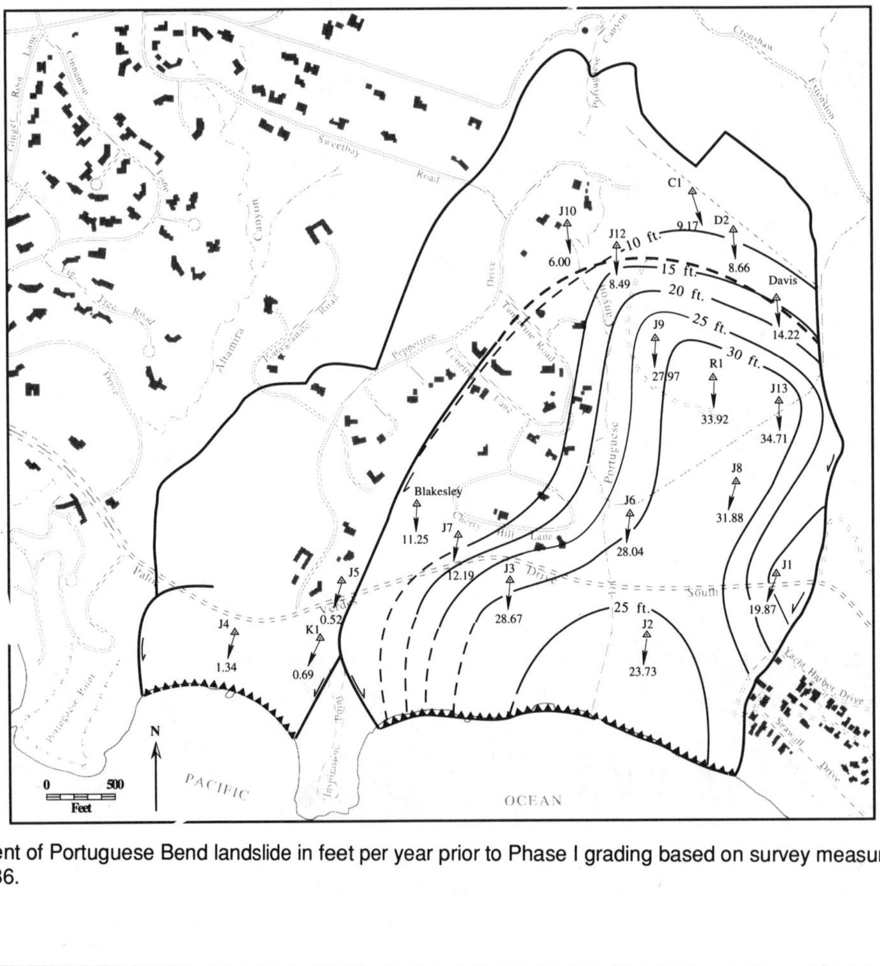

Figure 13. Displacement of Portuguese Bend landslide in feet per year prior to Phase I grading based on survey measurements of 16 January 1985 and 1 October 1986.

Figure 14. Displacement of Portuguese Bend landslide in feet per year during period between Phase I and Phase II grading based on survey measurements of 25 January 1987 and 4 November 1987.

On a long-term basis, great emphasis will be placed on actions designed to minimize ground water recharge. Phase I Grading and Drainage Improvements

The plan for Phase I grading was devised by me, reviewed by the Panel of Experts and finalized in mid 1986. The plan had two goals: (1) to slow the movement of the headward and east-central subslides by moving earth from above slide ramps where it produces a large net driving force to areas where the slide base is flat, and (2) to reduce ground water recharge by restoring drainage within the undeveloped area north of Palos Verdes Drive and conducting the runoff to the ocean. For aesthetic reasons, the grading plan was designed to create smooth natural-appearing topography and to eliminate steep hazardous slopes.

Grading occurred between mid October and mid December, 1986. A total of 560,000 cubic yards of slide material were moved. Of this, 200,000 cubic yards were moved along the up slope edge of the slide and 350,000 cubic yards were moved from the northeast part of the east-central subslide. The remaining 10,000 cubic yards were moved to extend a drainage trough from the main grading area to the ocean. Most of the earth was placed in the southeast part of the east-central subslide to fill a large depression and restore drainage (Fig. 11).

Prior to Phase I grading, Portuguese Canyon drained into a graben at the northwest corner of the slide where the runoff recharged ground water within the slide. This was corrected by filling the graben and excavating a broad southeast sloping trough along the head of the slide so as to conduct drainage to the east side of the slide. The trough is located where the slide base dips 15 to 25 degrees seaward; thus, the excavation greatly reduced the force driving the headward subslide. On the uphill side of the trough, slide material was removed to inplace bedrock in a belt 1000 feet long with an area of 4 acres. The bedrock surface was ripped and tamped so vegetation could be planted and the bedrock would not act like a paved surface directing runoff to the slide. Where the slide base came to the surface on the downhill side of this area, the slide mass was over-excavated to a depth of 10 feet below final grade and replaced by compacted fill. This provided a stable base for a corrugated steel drainage culvert that was installed to conduct runoff from Portuguese Canyon to a culvert at Paintbrush Canyon on the east side of the slide.

Prior to 1986, divergent slide movement and erosion by runoff down Paintbrush Canyon partially exposed bedrock in a belt about 20 feet wide and 1000 feet long on the northeast side of the slide. During grading, this belt was widened to about 100 feet by removing slide material from inplace bedrock. The bedrock was ripped and tamped to eliminate potential slip surfaces and then overlain by a minimum of 10 feet of compacted fill. This provided a stable surface for the drainage culvert and reduced the likelihood that runoff from the ridge east of the slide would permeate to the slide base.

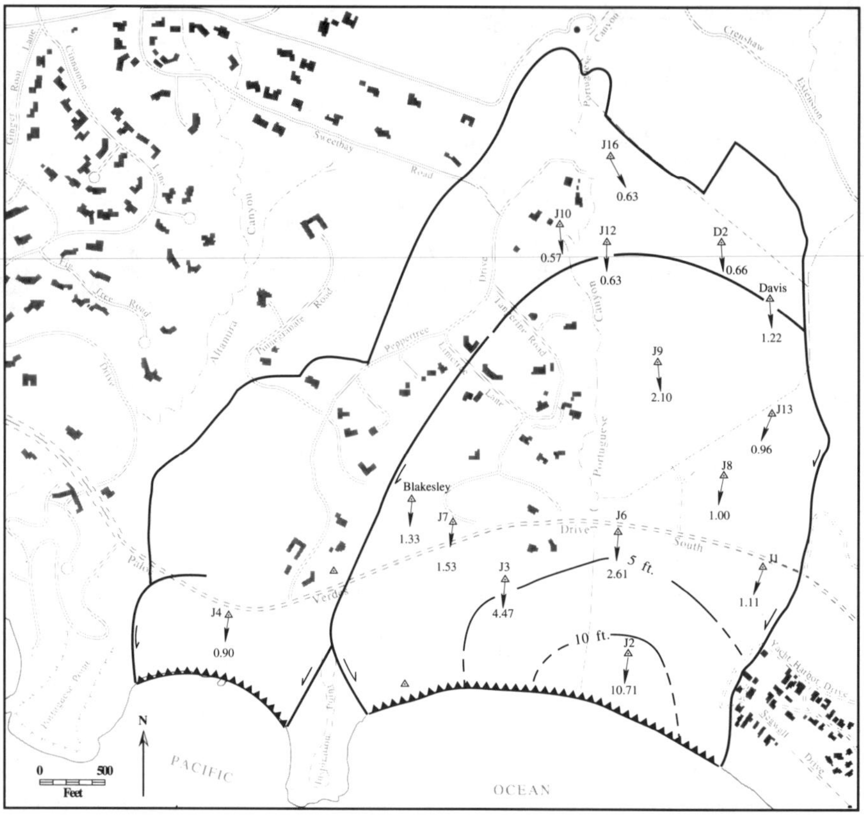

Figure 15. Displacement of Portuguese Bend landslide in feet per year during Phase II grading based on survey measurements of 7 March 1988 and 25 January 1987.

The addition of fill also addressed a public concern that grading would reduce support from the ridge east of the slide.

In order to reduce driving force from the east-central subslide, 350,000 cubic yards of earth were removed from the ramp on its northeast side. Here, the slide base has an average dip of 16 degrees in the direction of slide movement. The earth was moved a short distance seaward to fill the depression and restore drainage in the southeast part of the subslide. The need to reduce driving force was not entirely compatible with the need to restore drainage. Although most fill placed in the depression added resisting force, the northern part did not.

Phase I grading, aided by removal of ground water from wells and low rainfall, caused the headward subslide to nearly stabilize. Monitoring along the northeast edge of the east-central subslide shows movement decreased from 0.72 inches per day (in/d) before grading to 0.36 in/d after grading (Fig. 12). The two central subslides slowed more than 50% as shown by a comparison of Figures 13 and 14.

Phase II Grading and Drainage Improvements

The success of Phase I grading made it desirable to perform additional grading as soon as feasible in order to minimize damage to Phase I drainage improvements and to make drainage improvements along Palos Verdes Drive. At the time, Palos Verdes Drive was in chaotic condition. All storm runoff from the roadway went into nearby fissures. Although the roadway had been moved landward on several occasions, it had migrated into the seaward subslide where secondary rotational slides continually disrupted it. No amount of maintenance work could have restored drain for more than a few days at a time. Therefore, the road would be rebuilt near its 1956 position in order to place it where movement could be controlled and where road fill would be beneficial because of a subhorizontal to landward slope on the slide base.

A cut of 500,000 cubic yards was designed at the head of the east-central subslide in order to reduce its driving force. About 200,000 cubic yards of this earth would be used to reconstruct Palos Verdes Drive and the remainder would be placed as a toe berm in the area east of Inspiration Point where the bases of the slide rises in a seaward direction (cross sections AA' and BB', Fig. 8). The initial plan indicated 40,000 cubic yards of earth would be removed from the headward subslide in an effort to complete its stabilization; however, this was eliminated from the final plan to reduce costs. In fact, the initial bids exceeded available funds. It was necessary to reduce the width of Palos Verdes Drive in order to lower costs. Grading was planned for late fall and early winter to obtain low bids during the slack part of the grading season without excess risk of high rainfall. Rebidding delayed grading but fortunately it was an unusually dry rainy season.

Phase II grading was performed between early January and mid March, 1988. During the first three months after grading, the east side of the slide did not appear to move in the area northward from Palos Verdes Drive. In June, cracks reappeared near a monitoring station at the northeast corner of the east-central subslide. As far as I could tell, the hiatus was real and represented relaxation of compression within the down slope part of the east-central subslide. During the remainder of 1988, movement averaged 0.03 in/d at this station (Fig. 12).

Figure 15 shows the annual rate of movement during the year after Phase II grading. Movement in the central subslides slowed to about 20% of what it was prior to Phase II grading.

September 1990 Grading

During 1989, the central part of the headward subslide continued to move at an average rate of 0.02 inches per day. The rate was gradually decreasing due to removal of ground water but movement would likely increase should rainfall be above average. Therefore, the decision was made to remove 60,000 cubic yards of earth from the up slope edge of the subslide, essentially as originally planned for Phase II grading. The slide base has an average dip of 14 degrees in the removal area. Grading was performed in September, 1990. My monitoring indicates no movement has occurred along the head of the slide since completion of this grading. The headward subslide is not expected to reactivate in the future, even during rainy periods. But only experience will prove if this is correct.

The earth which was removed from the headward subslide was placed as fill on the east side of the slide north of Palos Verdes Drive. Slide movement is nearly horizontal in this area; therefore, the fill should produce a net resisting force. However, the rate of movement increased during emplacement of the fill, as previously described. The rate subsequently slowed asymptotically to 0.040 in/d in mid 1991. The probable cause of the accelerated movement was consolidation of bentonite along the base of the slide and the consequent build up of pore-water pressure along the slip surface. My analysis of movement recorded at my monitoring station indicates the east side of the slide moved about two inches more than it would have if the fill had not been placed. But the increased movement should be compensated for by a decrease in movement after completion of consolidation.

Temporary Shoreline Protection

The need for permanent shoreline protection was recognized during planning for the stabilization of the Portuguese Bend landslide, but temporary protection appeared unnecessary. On January 17 and 18, 1988, while Phase II grading was getting underway, the most severe wave storm of this century eroded an estimated 15 to 25 feet of land from the toe of the slide and removed much of the rock that tended to armor the beach and retard erosion. During the next few months, wave erosion continued at a rapid rate. East of Inspiration Point, the seaward face of the toe berm was oversteepened by erosion and its lower part began to collapse into the surf. Further east, secondary rotational slides accelerated in response to the maintenance of a nearly vertical cliff by wave erosion. This caused a rapid decrease in surface elevation near the coast. At the same time, the slowing of slide movement by Phase II grading resulted in a net landward migration of the shoreline. These two factors had the combined effect of decreasing support to the slide mass further uphill and caused secondary sliding to propagate toward Palos Verdes Drive. Therefore, temporary shoreline protection was recommended.

Gabions were chosen as the least expensive and most environmentally acceptable method of temporary shoreline protection. They consist of plastic coated, galvanized wire mesh in the form of rectangular cages which are filled with beach rock and stacked in a stairstep pattern to form a protective wall along the beach. Near the end of 1988, gabions were installed along the western 800 feet of beach to protect the toe berm. They performed well during the first year and funds were obtained to extend the gabions 1,400 feet to the east side of the slide. However, just as construction of the new gabions began, a storm destroyed many of the original gabions. As soon as the gabion wall was breached, waves eroded behind the gabions and undermined them. Also, the impact of stones cut the plastic coating the wire and, wherever the plastic ruptured, saltwater corrosion weakened the wires and ultimately caused them to break. The 1,400 feet of gabions were installed but are being damaged by impact of stones and by tilting and toppling from upthrusting along the toes of secondary slides. They are likely to last no more than two years.

I have recommended against further use of gabions unless their shortcomings can be corrected. At present, it is unlikely that permission or funding can be obtained to install other types of temporary shoreline protection.

Future Mitigation

Minor funding is available for two projects at the end of this year (1991). First, the channel along Portuguese Canyon will be opened from Palos Verdes Drive northward to the head of the central subslides so runoff from the residential area west of the Canyon can be directed to the canyon and hence to the drainage system installed during Phase I grading. Residents are asked to conduct their runoff to the canyon. At present, most of the residential area drains into fissures. Second, about 30,000 cubic yards of earth will be removed from the head of the east-central subslide in order to reduce its rate of movement. The subslide is not expected to stop but should slow to about 0.01 in/d.

Five major mitigation projects remain: (1) the installation of permanent shoreline protection; (2) construction of a toe berm; (3) regrading of the seaward subslide; (4) redistribution of mass within adjacent parts of the central subslides, and (5) installation of a sewer system in the residential area. Upon completion of the project, each subslide should have a safety factor above 1.00. Thus, uphill subslides will not be pushing on downhill subslides.

The feasibility of constructing permanent shoreline protection is under study by the Army Corps of Engineers. Protection will most likely consist of a rock revetment constructed on inplace bedrock seaward of bedrock disturbed by sliding. A separate but coordinated project will be the construction of a toe berm behind the revetment. The berm will resist the forward movement of the slide and function as a counterweight in areas where the base of the slide ramps up from below sea level. Earth for its construction will come from the inland part of the seaward subslide. The ground surface will be lowered in areas where the base of the subslide dips seaward so as to reduce force driving the main slide. The ground profile between the toe berm and the head of the seaward subslide will be flattened as much as possible to minimize forces tending to cause rotational slumping. Thus, the three projects will be coordinated. These projects will also establish proper drainage seaward of Palos Verdes Drive.

An estimated 200,000 cubic yards of earth will be removed from the ridge north of Palos Verdes Drive and east of Portuguese Canyon in order to reduce the force driving the west side of the east-central subslide and to minimize forces tending to cause rotational slumping. Most of this earth will probably be used in construction of the toe berm. Part of it may be used to fill the hollow on the north side of Palos Verdes Drive to the west of Portuguese Canyon. This area is on the uphill side of the subsurface ridge in the slide base and the added weight of fill will resist slide movement. However, surface elevations will be kept below those existing prior to 1956 as a precaution against subsurface shearing through the ridge.

After slide movement stops, sewers will be installed in the residential part of the slide. Homeowners will pay for the sewers by means of a sewer district. Sewers will eliminate about 8,500 gallons of water per day currently being injected into the landslide. This will eliminate an important source of ground water recharge and will permit wells to lower the water table below existing levels.

Future Land Use

The future use of land within the Portuguese Bend landslide will depend upon the success in completing the proposed mitigation measures. If all proposed mitigation is completed, it should be possible to maintain the slide's safety factor at or above 1.2. This should permit restricted development. If mitigation is not completed and the slide remains unstable or in a quasi-stable condition, the slide area will most likely remain much as it is today. Assuming the slide is stabilized, land users will have to pay the cost of maintaining wells and drainage improvements and permanent monitoring of the slide to assure its continued stability. For this reason, it would be desirable to permit types of land use which generate the tax revenues or user fees needed to maintain slide abatement activities.

The city-owned land seaward of Palos Verdes Drive will most likely be used for recreational purposes. The use is not known at this time but, when determine, can be taken into account during the design of remedial grading. Slide abatement funds could be obtained from users fees.

The undeveloped area north of Palos Verdes Drive is currently under consideration for a golf course. I support such a use because surface drainage and ground water conditions are easily monitored and controlled in a golf course and there is good access for drilling and maintaining wells. The northeast part of the slide provides a suitable area for construction of a club house and related facilities. Here, the slide mass is thin and underlain by gently dipping bedrock. It should be feasible to remove and recompact the slide material to provide a stable base for construction.

Some individuals urge that the undeveloped land north of Palos Verdes Drive be left unimproved so that it can revert to native vegetation for wildlife habitat. Patches of native habitat can be designed into a golf course but I am opposed to leaving the land idle because it will generate no revenues for slide abatement and the uncontrolled growth of brush will create a fire hazard.

Future land use within the existing residential area of the landslide is clouded by several factors. This area, located west of Portuguese Canyon and north of Palos Verdes Drive, was divided into residential lots prior to the start of sliding in 1956. Many of the original homes were destroyed. Most of the 32 residences in this area are on grade beams or other special foundations that permit adjustments to accommodate ground deformation. All present homeowners purchased their homes long after movement began. They moved into the slide because prices and property taxes were low. Most are opposed to anything which will increase their taxes, including sewers. None really know where their land is. Most contend that their houses and landscaping mark their land and, therefore, their land has moved with the slide. But this creates impossible situations. Not only has the land surface been distorted by rotation, tilting, strike-slip faulting, dilation and compression, but, in addition, most of the original boundary monuments have been destroyed. When, in 1988, Palos Verdes Drive was relocated northward toward its preslide position, lot owners sued the city for placing the road on their property. The city won the lawsuit but the dilemma remains. If the homeowner's property has not moved from its original location, most homes are on someone else's land.

The only satisfactory way to resolve this issue is to void the original land descriptions and establish new ones. First priority must go to reestablishing the boundaries of developed lots, but much negotiation will probably be necessary. Lots cannot be restored to their original shaped because the east side of the residential area has moved as much as 400 feet further seaward than the west side. Also, vistas have changed. Some lots that originally had ocean views are now in hollows.

The problem of land ownership is more than academic. A sewer cannot be installed without easements and easements can't be granted without knowing who owns the land. Landowners might be asked to grant a blanket easement but they are unlikely to do so without a financial incentive. The same applies to drainage improvements.

An important unresolved issue is whether the owners of vacant lots will ever be able to build on their lots. At present, the city building code will not permit construction of habitable structures on sites that have a safety factor of less than 1.50 (forces resisting sliding are 1.5 times greater than shearing force tending to cause sliding). It is unlikely the safety factor will be above about 1.25 upon completion of slide mitigation. However, unlike other areas where the true safety factor is imprecisely known and no actions are taken to maintain stability after development occurs, the safety factor of this area will be known with good precision and factors affecting the slide will be monitored and abated when necessary. Therefore, if the slide is stabilized and the factor of safety raised to about 1.20 or higher, the carefully controlled development of vacant lots appears feasible. Most lots would require regrading, and special foundations would be needed. In addition, architectural designs would have to be approved by the Portuguese Bend Homeowners Association. Two things would be accomplished by permitting development. First, it would give the owners of vacant lots an incentive to make drainage improvements and cooperate in all other aspects of slide abatement. Second, it would increase lot values by at least ten times and in some cases as much as 100 times. This would provide the Rancho Palos Verdes Redevelopment Agency with tax funds to carry on abatement activities. Their funding comes from taxes generated by the increase in property values resulting from slide abatement.

Of course, only time will tell what really happens. Excellent progress has been made toward stabilizing the Portuguese Bend landslide but many fiscal, technical and political hurdles remain before the project can be completed.

ACKNOWLEDGMENTS

Special thanks are due to Scott Thompson for preparing computer-graphic illustrations and to Bernard Pipkin for reviewing the manuscript and making helpful suggestions.

REFERENCES

Bryant, M. E., 1982, Geomorphology, neotectonics, and ages of marine terraces, Palos Verdes Peninsula, in Cooper, J. D., comp., Landslides and landslide mitigation in southern California: Assoc. of Engineering Geologists, Southern California Section, p. 15-25.

Conrad, C. L., and Ehlig, P. L., 1983, The Monterey Formation of the Palos Verdes Peninsula, California - an example of sedimentation in a tectonically active basin within the California Continental Borderland, in Larue, D. K., and Steel, R. J., (eds.), Cenozoic marine sedimentation, Pacific margin, U.S.A.: Pacific Section, Soc. Econ. Paleontologists and Mineralogists, p. 103-116.

Easton, W. H., 1973, Earthquakes, rain, and tides at Portuguese Bend landslide, California: Assoc. Engineering Geologists Bull., v. 10, p. 173-194.

Emery, K. O., 1967, The activity of coastal landslides related to sea level: Rev. Geog. Phys. et Geol. Dynam., V. 9, p. 177-180.

Jahns, R. H., and VonderLinden, K., 1973, Space-time relationships of landsliding on the southerly side of Palos Verdes Hills, in Moran, D. E., Slosson, J. E., Stone, R. O., and Yelverton, C. A., eds., Geology, seismicity, and environmental impact: Assoc. Engineering Geologists, Special Pub., p. 123-138.

Kerr, P. F., and Drew, I. M., 1969, Clay mobility, Portuguese Bend, California: Calif. Div. Mines and Geology Special Report 100, p. 3-16.

Kew, W. S., 1926, Geologic and physiographic features in the San Pedro Hills, Los Angeles County, California: Oil Bulletin, v. 12, p. 513-518.

MacKintosh and MacKintosk, 1957, Report of earth movement, Portuguese Bend, California: unpublished technical report, 16 p.

Merriam, R., 1960, Portuguese Bend landslide, Palos Verdes Hills, California: Jour. Geology, v. 68, p. 140-153.

Rowell, H. C., 1982, Chronostratigraphy of the Monterey Formation of the Palos Verdes Hills, in Cooper, J. D., (comp.), Landslides and landslide mitigation in southern California: Assoc. of Engineering Geologists, Southern California Section, p. 7-13.

VonderLinden, K., 1972, An analysis of the Portuguese Bend landslide, Palos Verdes Hills, California: Stanford Univ., unpubl. Ph.D. thesis, 260 p.

Woodring, W. P., Bramlette, M. N., and Kew, W. S. W., 1946, Geology and paleontology of the Palos Verdes Hills, California: U. S. Geol. Survey, Prof. Paper 207, 145 p.

BIG ROCK MESA LANDSLIDE AND LITIGATION, MALIBU

Dennis A. Evans
Geotechnical Engineer & Engineering Geologist
Shepardson Engineering Associates, Inc.
5810 El Camino Real, Suite A
Carlsbad, CA 92008

Richard E. Lownes
Engineering Geologist
Lownes Geologic Services
651 S. Hudson Avenue
Pasadena, CA 91106

INTRODUCTION

The destructive movement of the Big Rock Mesa landslide began in September 1983 and prompted a local neighborhood association, the Concerned Citizens for Water Control, to retain E.D. Michael and D.A. Evans, Inc. to reduce or stop the land movement. At that time there were 216 homes on and in the vicinity of the slide utilizing private sewage disposal systems. Most owners tended gardens and landscaping which required extensive irrigation. Our work, with Richard Lownes, retained by Evans, eventually included drilling 18 vertical dewatering wells, 30 horizontal drains, several exploration borings, surface geologic mapping, weekly and monthly progressive crack mapping, placement of 42 slope indicators, and many piezometers. The work took approximately two and onehalf years to complete culminating in a seven-volume report in 1986 prepared by D.A. Evans, Inc. Following that time several lawsuits occurred which involved the State of California, the County of Los Angeles, insurance companies and the homeowners. This litigation established landmark laws and legal precedent. Since the global settlement of the lawsuits the County of Los Angeles has retained other consultants to monitor the landslide cracks, the slope indicators and the water wells.

This report was prepared mainly from Evans (1986) and maps and illustrations are from that report. The Hydrogeology section was written by E. D. Michael for the Evans report and is condensed in this paper by Lownes. Evans wrote the sections on litigation and engineering. All photographs are courtesy of E.D. Michael and were taken during the investigation. (Stout in this chapter also discribes the Big Rock Mesa landslide.)

LITIGATION

The Big Rock Mesa landslide is certainly not the largest in recorded history. The volume of the slide mass was less than 100 million cubic yards and involved less than 250 acres of hillside. But, it occurred in southern California in an urban area inhabited by people with above average economic means and political power, including several celebrities. Equally important, it occurred when defects in the design and construction of real estate improvements is a popular subject of debate in our courtrooms. It is very likely that the moment the landslide started, it was destined for litigation. Indeed, it ultimately became the largest and most complex lawsuit in California history.

The discussion of the litigation offered here will probably not withstand the scrutiny of a lawyer. It is solely based on the authors observations and impressions after spending literally hundreds of hours giving deposition and courtroom testimony. The mass of litigation that resulted from the landslide forever changed the role of the engineering geologist and geotechnical engineer in our society and the way liability for damaged real estate improvements is determined and measured by the courts.

The Big Rock landslide generated two very distinct areas of litigation. The area that got the most publicity was the litigation which pitted the homeowners as plaintiffs against the County of Los Angeles, the County Waterworks District and the State of California as defendants. That litigation started in late 1983, involved three separate trials, and ended up in a negotiated settlement in late 1989. When the settlement was finally reached, both sides were preparing for what had, at that time, been affectionately nicknamed "the mega-trial". The County was in the process of converting the Hollywood Palladium into a special multimillion dollar courtroom. The design of the courtroom involved a unique and complex electronic communication and television system capable of allowing a judge, jury and some 250 lawyers to function collectively for an estimated year long trial involving 400 parties. Concurrently, the so-called "mini- mega-trial", the third trial of the plaintiffs over 70 years of age had started in Los Angeles County Superior Court. In 1985, a three-month trial in Los Angeles County Superior Court involving one elderly homeowner couple as plaintiffs resulted in a two million dollar judgment against the County of Los Angeles for inverse condemnation. The Waterworks District and the State of California were found not liable. The judgment against the County was, however, reversed in 1987 by an appellate court. Another three-month trial by other elderly homeowners resulted in a jury verdict against the State of California. That verdict resulted in a $25 million settlement with the State.

One of the most significant elements of this facet of the litigation was economic. A substantial majority of the applied earth science consulting community was involved in one way or another. It has been conservatively estimated that the consulting fees paid to applied earth science firms directly related to this area of litigation greatly exceeded the money spent by the Improvement District established by the homeowners to initially stop landslide movement. Moreover, it was estimated that if the "mega-trial" proceeded to fruition, the combined legal and expert witness fees would have exceeded the litigated damages caused by the landslide.

The other area of litigation involved the so-called "third-party" suits. Those were lawsuits filed by homeowners against their homeowners insurance carriers, several of which were tried in Federal courts. One lawsuit filed in Superior Court resulted in an important legal precedent involving the extent of repair for which an insurance carrier can be liable. The "Strickland vs. Federal" appellate decision held the insurer liable for the cost of stabilizing the property which supports the insured structure, not just the structure.

One final note about the litigation provides an interesting insight into the power of the real estate market in southern California and to the length of people's memories. In the aftermath of such a disaster, knowledgeable people offered the opinion that the 92 million dollar "global settlement" of all Big Rock Mesa landslide cases (an amount less than the estimated loss of value due to the "taint" of the landslide) was agreed to by the homeowners because resales of the homes on the landslide were better than predicted. As a result, it was becoming increasingly more difficult for the homeowners to prove diminution of value.

SOCIAL IMPACTS

There were a number of people, including families that were over 70 years of age, living in homes on the landslide. Some had purchased their homes in the 1960s and could not afford to purchase another home once their Big Rock homes were destroyed. Another family, a retired couple, purchased their home with cash only months before landslide movement became apparent. They were forced to abandon their home when the County declared it unsafe. One home was demolished and removed by the homeowner's insurance carriers to avoid future liability. Many of the homeowners were neither affluent nor well-known and dozens of homes were posted by the County of Los Angeles as being unsafe for occupancy. Some affected homeowners, in a show of defiance or because of lack of funds, refused to leave their homes. Others simply packed their belongings and left. Reportedly, one family was so emotionally affected, they refused to acknowledge the existence of the landslide.

The common thread running through the Big Rock Mesa litigation was that everyone who lived on or near the landslide, or owned property in the area, was substantially affected and their lives were changed catastrophically.

PREVIOUS INVESTIGATIVE WORK

The Big Rock Mesa area has been the subject of a number of geotechnical investigations, the earliest of which was included in a report by Moran, Proctor, Mueser & Rutledge (1959) based on literature review, theoretical stability analysis, analyses of data from subsurface instrumentation and surface geologic mapping. That study included the Malibu coast from Santa Monica to west of Big Rock Mesa. The applicable portions of that study briefly mention minor landslides in roadcuts at Big Rock.

Numerous reports by E.D. Michael, Moore & Taber, Inc. and others were submitted to the County of Los Angeles as part of the data required by the Regional Planning Commission for its review of the Tentative Map for Tract 26263 at the time of its development.

Final interpretation by the USGS of geologic conditions in the Big Rock Mesa area was presented by Yerkes and Campbell (1980).

In 1973, Dr. Paul Merifield prepared a letter-report for the Malibu Mutual Drainage Company which summarized information regarding observed ground cracks within Tracts Nos. 26263 and 27463.

Also in 1973, the County of Los Angeles prepared a report prior to the formation of an improvement district to outline the need to dewater the Big Rock Mesa area. E. D. Michael prepared reports in 1973 summarizing consulting work performed for the Malibu Mutual Drainage Company, and in 1978 Paul Merifield and Jack Eagan reported on the level of ground water and evidence of instability in a paper published by the Association of Engineering Geologists.

Following the major sliding in 1983 and 1984, D.A. Evans, Inc. prepared several interim reports of geotechnical evaluation for the Big Rock Mesa landslide which summarized preliminary findings, conclusions and recommendations derived from the investigation of the then-active landslide to those dates.

A number of miscellaneous consulting reports and technical papers were prepared by various geotechnical firms concerning the Big Rock Mesa area from 1983 through 1986.

PROJECT DESCRIPTION

The purpose of the D.A. Evans work was to satisfy the terms of contracts dated February 17, 1984 and December 6, 1984 between D.A. Evans, Inc. and Los Angeles County Improvement District No. 2629 and No. 2629R, describing possible stabilization of the slide. The work culminated in Evans (1968) Final Report which set forth a comprehensive discussion of the geotechnical evaluation and includes a summary of findings regarding the landslide. It also includes conclusions as to the effect of the dewatering system on ground movement and future stability of the area as well as recommendations regarding maintenance of facilities in the District.

It was determined at the outset that if the slide was to be stabilized at all, it would be by dewatering. Initial attempts

Figure 1. Oblique photo of Big Rock Mesa, circa 1952.

at dewatering involved the reactivation of existing dewatering wells and the construction and operation of new wells. This was judged at the time to be the most feasible means to begin dewatering because hydraugers, also referred to as horizontal drains, were then believed to produce at relatively low rates. Furthermore, water well contractors were available on short notice whereas contractors qualified to construct horizontal drains of sufficient length were not.

Extensive modification of the surface drainage system, particularly in the western part of the Big Rock Mesa area, was considered necessary to make the dewatering program as efficient as possible and was undertaken in late 1983 to construct a temporary surface drainage system which would, in the short term, alleviate some of the more serious problems of infiltration presented by the existing system. Its effectiveness over its intended one-year life was never seriously tested because of the fortuitously low rainfall of the 1983-84 and 1984-85 storm seasons.

Geologic mapping was carried out by D. A. Evans, Inc. personnel as well as by R. E. Lownes and E. D. Michael under contracts to D. A. Evans, Inc. as consultants. Installation, maintenance and construction of the dewatering wells were under the direction of Michael and Lownes who prepared the geologic data for the report. The entire project geotechnical evaluation was under the direction of Dennis A. Evans.

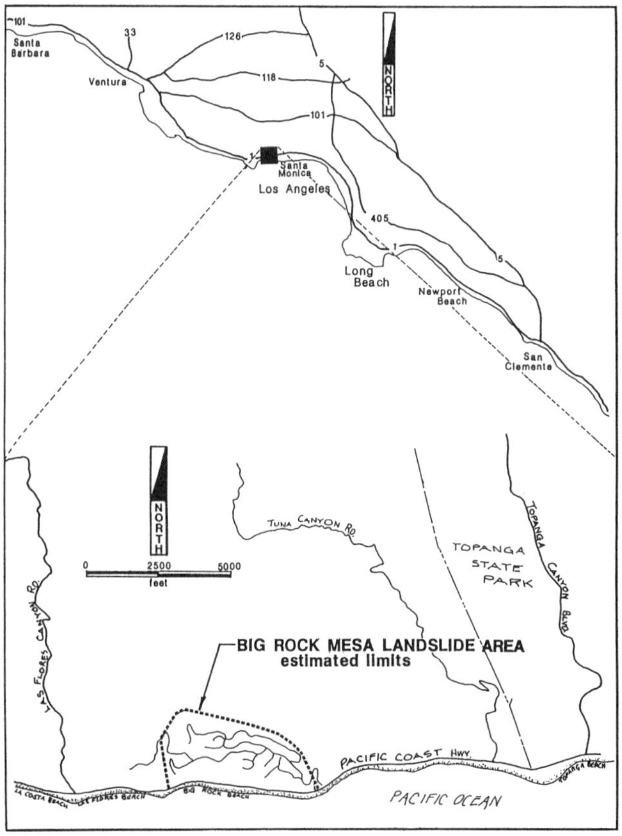

Figure 2. Area Location Map

LOCATION

The active, modern Big Rock Mesa landslide occupies about 220 acres of mountainous terrain along the Malibu coast which has for a number of years been known locally as Big Rock Mesa (see Figs 1, 2 and 3). It is within the city of Malibu, which incorporated in 1990.

SCOPE

To the extent that the principal and over-riding goal of the geotechnical evaluation was to develop a technique to arrest slide movement on an emergency basis, the scope of work was continually re-examined and revised, as appropriate. The means to attain this goal included dewatering wells and horizontal drains, drainage modification and reduced water use by the residents.

GEOLOGY OF BIG ROCK MESA AREA

The geology of the Big Rock Mesa study area was mapped in detail by R. E. Lownes and E. D. Michael. Prior to that, the most reliable and complete geologic work immediately available was that of Yerkes and Campbell (1980) which was not sufficiently detailed for the purposes of this study but was helpful in establishing regional geology.

In the Big Rock Mesa landslide area, important contacts are masked by landslide debris and man-made fill. The exposures of geologic conditions in Piedra Gorda Canyon are particularly complex and difficult to interpret.

Geologic Units

Based primarily on Yerkes and Campbell (1980) six bedrock formations were defined Fig. 4). The distribution and thicknesses of these various units are indicated on Plate II-2 Geology Map. Detailed descriptions are in the D. A. Evans report. All fossil identification were by Dr. Takeo Susuki of the Department of Earth and Space Sciences, UCLA.

Tuna Canyon Formation. The section of the Tuna Canyon Formation, as defined by Yerkes and Campbell (1980) is exposed in the northeastern corner of the mapped area in the eastern slope of Piedra Gorda Canyon where it is thrust over Sespe beds. It consists of highly fractured, olive-brown siltstone and shale interbedded with massive, resistant siltstone and fine-grained sandstone.

Coal Canyon Formation. Rock believed to be of Paleocene to Eocene age is exposed in the area northwest of the Big Rock Mesa landslide area consisting of alternating sections of conglomerate, sandstone and siltsone.

Sespe Formation. Earlier investigators who have worked in or near the Big Rock study area (e.g., Simonson, 1936; Soper, 1938; McGill, 1948) have recognized that the upper Eocene to lower Miocene Sespe Formation of the Santa Monica Mountains consists of a lower red bed sequence in which sandstone

Figure 3. Site Location Map

and conglomerate predominates, a middle unit of gray, light tan and light green, coarse-grained sandstone with minor pebble conglomerate, and a finer-grained upper unit of variegated shale, siltstone and fine-grained sandstone which Yerkes and Campbell (1979) designate the Piuma Member of the Sespe Formation. A similar Sespe sequence is present in the Big Rock Mesa study area. Only the upper few hundred feet of the lower coarse-grained red beds are present. The best exposures, consisting of massive, coarse-grained, arkosic sandstone and cobble conglomerate are in the seacliff between Rockpoint and Eagle Pass canyons.

Vaqueros Formation. Outside the Big Rock Mesa area, the lower Miocene Vaqueros Formation is exposed as a tongue within the Piuma Member along Piuma Road, (Yerkes and Campbell, 1979). The Vaqueros can be observed just east of the mouth of Eagle Pass Canyon eastward to the eastern side of Rockpoint Canyon and in a mass of prehistoric landslide debris north of the Moonshadows restaurant.

Topanga Canyon Formation. Yerkes and Campbell (1979) give group status to middle Miocene rock formerly referred to as the Topanga Formation. The lowermost unit of their Topanga Group is the Topanga Canyon Formation which is divided into three members, from youngest to oldest, the Cold Creek, Fernwood and Saddle Peak Members.

Intrusive Rock. Samples of relatively unweathered dark igneous rock near Moonshadow channels (probably incorporated in landslide debris) are medium-grained, altered, porphyritic, olivine diabase and altered ophitic diabase. Rock of this type is widespread outside the study area in the nearby Santa Monica Mountains, particularly in association with the Sespe and Topanga Canyon formations. In the Big Rock Mesa area they are a dull olive-brown color due to weathering.

Zuma Volcanics. Yerkes and Campbell (1979, p. 27) have assigned the name Zuma Volcanics to lower and middle Miocene "...basaltic and andesitic flows, breccias, pillow lavas, mudflow breccias, aquagene tuffs, minor amounts of volcanic sediment, and locally, interbeds of mudstone and siltstone..." that occur only south of the Malibu Coast fault. Zuma volcanics are well exposed at lowest tide within about 250 feet of shore. Unfavorable surf conditions and poor underwater visibility made a detailed examination impractical. The outcrops, isolated knobs protruding from the bottom sand, suggest an overlying conformable section. Samples recovered are mostly orthoquartzite.

Landslide Debris. Landslide debris masses which appear in the 1928 Fairchild aerial survey photographs are defined as prehistoric while other debris is historic. In addition to this fundamental classification, all landslide units except the recent episode, the subject of this report, are further classified as indicated on the Geology Map (Plate II-2). For clarity, the boundaries of the recent episode of movement are delineated on Plate II-4 Ground Cracks. Most of the landslide debris masses can be recognized and correlated with their source rock by inspection of the topography and geomorphology. Major masses of the prehistoric debris are described on Plate II-2 according to both the phase and the bedrock foundation from which the masses are derived, i.e., either units of the Sespe, Vaqueros or Topanga Canyon Formations, although technically phase-4 prehistoric debris appears to be entirely derived from phase-3 debris.

Minor prehistoric debris masses occur in steep slopes of Piedra Gorda Canyon, the seacliff, and Rockpoint Canyon. Historic debris masses are generally confined to south-facing slopes. Several of the larger masses are derived from prehistoric debris. Two of the historic slides caused structural damage. One, below the Inland Lane cul-de-sac, resulted in loss of support for a patio area at the rear of 20282 Inland Lane about 1972. The other, located just south of 20676 Rockpoint Way destroyed a residence there in 1978.

Terrace Deposits. Marine and continental terrace deposits occur on abrasion platform remnants located just east of Big

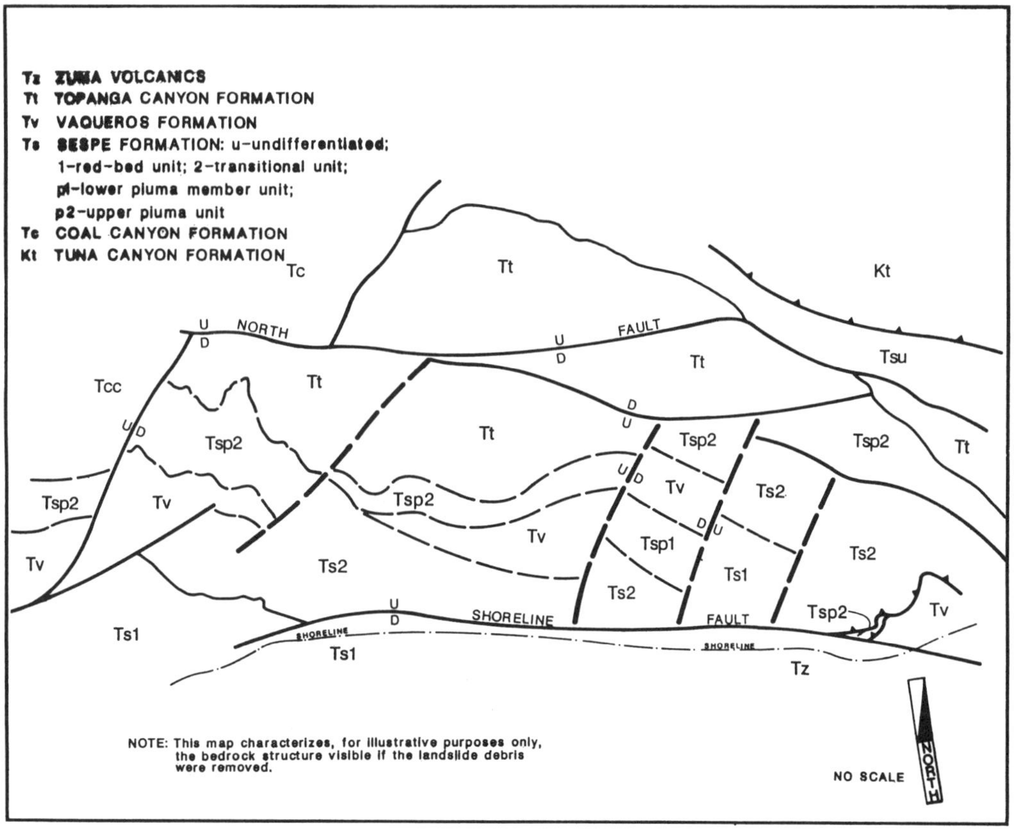

Figure 4. Geologic Map

Rock Drive and just west of Rockpoint Canyon. The deposits are probably no more than a few feet thick. A thin zone of gravel and a few cobbles occur at the base where it unconformably overlies sandstone of the Saddle Peak Member of the Topanga Canyon Formation. Scattered, molluscan fragments are present in the loose surface sand of this deposit.

Stream Alluvium. Deposits of unconsolidated stream alluvium were mapped only where they are sufficiently well developed to be reflected in the topographic expression of the area. However, most of the stream channels in the Big Rock Mesa area are underlain by alluvial deposits including boulders several feet in diameter.

Man-Made Fill. Man-made fill is widely distributed in the Big Rock Mesa study area. Most roads have a certain amount of fill on the downslope side as the result of the typical grading operations for road development which involves cutting on the upslope side and depositing fill on the downslope side. With the possible exception of certain stretches along Big Rock Drive in the lower part of the study area, it is reasonable to conclude that none of this fill was placed as engineered fill. Engineered fill masses have been constructed in the eastern part of the Big Rock Mesa area as a result of residential development. The largest of these fill masses are delineated on the Geology Map (Plate II-2).

Geologic Structure

Structural deformation in the vicinity of Big Rock Mesa appears to be almost entirely the result of faulting. Folding is monoclinal exhibiting only slight warping, with the main geologic structure a system of faulted blocks. Geologic Sections A- A' through D-D' (Plate II-3) illustrate, in a general way, the block-faulted character of the area.

Shoreline Fault. According to Yerkes (1985), this fault along the shoreline forms the northern limit of a zone bounded on the south by the Anacapa fault. The two meet at the coast to become the Santa Monica fault. Yerkes and Campbell (1980) have mapped the Malibu coast fault as being offshore east of Las Flores Canyon, but do not indicate how far seaward it occurs. Work to date suggests that a branch of the Malibu coast fault, if not the main trace itself, is just onshore along the base of the seacliff in the vicinity of the Big Rock Mesa study area. Evidence for this conclusion includes (1) presence of the reef rock, here correlated with the Zuma Volcanics, (2) the presence of a section of extrusive rock penetrated by a CalTrans exploratory boring located just south of Pacific Coast Highway about 2,500 feet west of Big Rock Drive, and (3) the presence of a coarse-grained sandstone and conglomerate unit with unusual diagenetic chlorite (presumably the product of low-grade metamorphism associated with the faulting) which is exposed in the beach area 1,200 feet east of Rockpoint Canyon. This lithology, entirely different from the on-shore lithology, strongly suggests the existence of a major discontinuity.

LANDSLIDE HISTORY

The Big Rock Mesa area has a long history of landsliding. Two landslide episodes, one involving a series of major prehistoric slides and the other involving the movement which was on-going between mid-1983 and mid-1985, have special significance in considering the relative stability of the Big Rock Mesa landslide area.

Major Prehistoric Slides

Six major prehistoric landslides are believed to have occurred in phases in the Big Rock Mesa area. Plate II-6 Prehistoric Landslides, depicts the estimated limits of these features. Their location and relative ages have been interpreted both from the remnants of the crowns as they appear in aerial photographs and from the manner in which some debris masses apparently truncate others. The earliest phase began just west of the mouth of Piedra Gorda Canyon with later phases progressing westward at successively higher elevations.

Carbonaceous samples recovered in an exploratory boring which penetrated third and fourth phase debris at depths of about 90 and 100 feet have been dated by Professor C. R. Berger at UCLA using the radiocarbon method (Roy J. Shlemon, pers. comm.). The samples indicate ages of 30,000± and 26,670, ± 2,580 years, b.p., respectively. On the northern side of Pacific Coast Highway, about 400 feet east of Moonshadows, there is an exposure of subsidiary, prehistoric slide debris deposited on a steep bluff slope developed in second phase debris. This subsidiary debris is overlain by a section of silty sand in which is preserved charcoal and clam shells of a Chumash Indian midden (Meigan, 1985). The charcoal and shells have radiocarbon dates of 1875 ±50 years and 1925 ± 50 years, b.p., respectively.

Current Movement

The current Big Rock Mesa landslide includes about 220 acres as shown on Plate II-4. This includes the 155± acre area of initial movement, the 65± acre possible westerly enlargement of movement and two areas of known subsidiary movement, the seacliff and the headscarp. A scarp just south of the North Peak has persistently marked the northern extent of movement of the current Big Rock Mesa landslide since it was first noted in 1983. Within a period of about six months, it increased in height from a few inches to about five feet. Thereafter, except for minor adjustments, this feature and associated tension cracks maintained roughly the same configuration until grading for the site of an exploratory boring in 1985 destroyed much of it.

Movement below the Inland Lane cul-de-sac in 1972 was at that time thought to be the result of a localized seacliff failure (Merifield, et al., 1978, p. 142-148). However, the extensive system of tension cracks mapped in 1972 by Merifield may have been the surface manifestation of dilation over a much larger area and the precursor of the current failure. Fractures definitely correlated with the current movement were noted about 1979 in the eastern portion of the landslide area where highway pavement and a section of adjacent roadcut were offset.

A theory was offered at that time by other investigators that the toe of the current slide extends several hundred feet offshore. This was based upon an interpreted irregularity in the offshore abrasion surface and an apparent offset in overlying Holocene sediments in certain of McClelland "boomer" profiles. However, these features are indistinguishable from many others in the profiles, strongly suggesting alternative explanations. Moreover, it seems unlikely, because of wave action, that a break in the loose bottom sediments could be maintained at a water depth of only 15-20 feet for a year or more. In any event, compelling evidence that this theory is incorrect is indicated by detailed photogrammetric surveys of reef rock exposed in the surf zone in both 1971 and 1985. These surveys (horizontal accuracy of one foot or less; vertical accuracy of 0.1 foot) indicate no horizontal or vertical movement during that period. Judging from the manner in which the Pacific Coast Highway surface and adjacent structures have deformed, it is logical to conclude that the toe of the current slide is confined to the shore and is in fact controlled by resistant extrusives and quarzites of the Zuma Volcanics which are seaward of the shoreline fault. Slope indicator data indicate the landslide is translational with base failure at the toe. The main mass was stable (i.e. inactive) from about the middle of 1985 at least until the Evans report was produced in 1986.

Fractures in various structures in the vicinity of McAnany Way, north of Rockpoint Canyon, were noted early in 1984. Some months later, large tension cracks in the slab of a house in Rockpoint Way and a series of tension cracks in the Pacific Coast Highway pavement east of Rockpoint Canyon had developed. These fractures appeared to be caused by dilation of the slope. Initially, this dilation was thought to be a precursor movement. The penetration of a slide plane at a depth of 220 feet in an exploratory drill hole on the Stone property at 20826 West Big Rock Drive (GeoSoils, pers. comm., 1986) indicates that the dilation may have been part of a western enlargement of the landslide. Boundary evidence for the westerly enlargement however, is lacking leaving the question of that extension suspect.

Seacliff Landslide. The seacliff landslide extends from just east of the east contact in a westerly direction to a point about 400 feet west of Roca Chica Drive cul-de-sac and then with less certainty another 600 feet to the west contact. The total horizontal distance along the upper edge is 3,400 feet. It was an active subsidiary slide entirely within the main landslide mass. It appeared to be a steep rotational, basal failure which did not extend far south of the base of the seacliff at Pacific Coast Highway. The seacliff slide includes about 40 acres. Within it, 27 homes have been posted by the County of Los Angeles as being unsafe. It was possible that another 10 or 12 homes could have been jeopardized if the movement continued. Its present status is unknown.

Headscarp Failure. The steep slope immediately below the Northeast Ridge and southwest of the North Peak by about 1,200 feet, is referred to as the headscarp. Continued movement in, and adjacent to, the headscarp, was attributed to both localized adjustment and movement of the main mass. It destroyed a residence directly below the North Peak and threatened several others. The boundaries of this movement were not clearly manifested in the form of ground cracks, but appeared to be confined to the scarp and immediately adjacent areas.

Causes of Landsliding at Big Rock Mesa

Two scenarios are possible with respect to the age of the prehistoric debris. If the carbonaceous samples recovered by Shlemon were incorporated as living material at the time of failure, the third or fourth phase of debris is between 24,090 and 30,000+ years old. If it was dead at the time of incorporation, the debris is younger. In either case, the second phase debris is at least 1,825 years old inasmuch as the undisturbed midden deposits rest upon a steep seacliff slope developed in that debris.

The coastal terrace remnants predate the prehistoric slides. Since the age of the youngest of the coastal terraces in Malibu, the Dume, is about 100,000 years b.p., there may have been a steep coastal slope in the area that is now the Big Rock Mesa area for a period of at least 1,825 years and as much as 70,000-75,000 years without massive failure. It seems likely that exposures of relatively weak Piuma Member beds in the eastern part of the study area gradually became oversteepened by wave erosion and failure began 25,000-30,000 years ago. The sequential occurrence of the prehistoric slides from east to west through the block-faulted section of Piuma Member rock support this speculation even through the possibility exists that a high ground water level or seismic excitation, or both, contributed to the failures.

It is reasonable to conclude that a critical level of ground water was not reached in the previous 1,825-year period inasmuch as the midden deposit is undisturbed. It is also reasonable to conclude that during this period of relative stability, there must have been numerous occasions when naturally induced ground-water recharge brought the level of ground water up but did not cause a massive failure. The coincidence of the current failure with (1) a rise in the level of ground water that started about 1961, (2) the lengthy prior period during which no failure occurred, (3) the extent of the current slide, (i.e., it includes all major prior phases of failures), and (4) the concentration of private waste water disposal effluent, is convincing evidence that the current landsliding is a result of an artificially induced high groundwater level.

The headscarp failure is most logically due to loss of support caused by movement of the main mass of debris. The sea-cliff failure is a result of the dilation and oversteepening of an already steep slope in prehistoric debris resulting from movement of the main mass.

Although rainfall from the period 1978 through 1980 may have given impetus to the current movement, fractures observed by Merifield (op. cit.) in 1972 indicate that the slide may have started before 1978. It seams clear that, but for the artificially high level of ground water that accompanied development beginning around 1960, normal increments of natural recharge would not have been sufficient to cause the failure.

SEISMICITY

It is generally recognized that the Malibu Coast fault (MCF) is of comparatively recent age and active by any reasonable definition of the term. Based the Alquist-Priolo Special Studies Zone Act definition of an active fault the MCF meets the criteria.

The occurrences of earthquakes that have caused widespread damage along faults which previously were not considered to be even potentially active (e.g. the 1971 San Fernando event) emphasize the limited significance of the "active and "potentially active" distinctions. The active character of the Malibu Coast fault zone is suggested by offsets in Quaternary deposits (e.g., Hoots, 1931, Greene, et al., 1975, and Drum, herein). Perhaps the most significant, however, is the occurrence of seismic events within the zone itself. For example, Yerkes and Lee (1979) report more than 100 events within the southern boundary fault of the MCF in the period 1970-75 ranging from a Richter Magnitude of less than 1.0 up to the 5.9 event of February 21, 1973 at Point Mugu. For purposes of this study, a seismic event on the Malibu Coast fault with a Richter Magnitude of 6.0 to 6.5, a period of strong motion of about three seconds (i.e., a frequency of three hertz), and a duration of strong motion of about 15-20 seconds has been used to define the Maximum Probable Earthquake for the site.

HYDROGEOLOGY

Domestic Production

Water for domestic use and a limited amount of agriculture and stock watering was initially provided from privately owned wells in and immediately adjacent to Big Rock Mesa. Records held by Los Angeles County Waterworks District No. 29 indicate that by 1960, at least 20 wells had been drilled in the vicinity. All of the supply wells had been constructed by about 1955. Water was supplied to the Big Rock Mesa area and consumers along the beach by the Big Rock Beach Water Company, a private entity.

Non-domestic Production

The wells constructed for this study performed basically as expected through 1987. Initial well capacity in some cases exceeded 70 gallons per minute (gpm), gradually dropping as the volume of local dewatered rock enlarged and the hydraulic gradient declined. In those cases where surging developed as the pumping level neared the sump intake, the pumping was modified to maintain maximum production without air entrainment. Well capacity varied widely. Of the 18 wells utilized for dewatering, daily production ranged from about 5,000 to 100,000 gallons per day (gpd). In 1986 only seven wells were operating. Total production to January 1, 1986 was estimated to be 246 acre-feet (af), or about 85 million gallons. Production from the better wells initially was limited by the capacity of the pump.

Production from our horizontal drains also varied widely, up to about 125,000 gpd. Of the 30 drilled, 25 were still producing in 1986. Total production to January 1, 1986 from the horizontal drains was estimated to be 140 acre-feet (af), or about 48 million gallons.

Vacuum Pumping. A testing program was undertaken to determine the performance of certain of the horizontal drains when subjected to a partial vacuum. Test equipment consisted of a special pump driven by a diesel-powered generator, a vacuum-pressure meter, and a water meter. The pump produced a partial vacuum gage pressure of up to 15 inches of mercury, i.e., about 0.8 atmosphere. The test consisted of the recording of production for a period of about eight hours while maintaining more or less constant negative pressure. The results in all cases indicated an essentially constant flow rate for the period of the test.

The data were analyzed by producing graphs relating production and time to obtain a representative test flow rate. The test rate was compared with gravity flow rate of several weeks before and after the test.

The results indicate that the total net gain in the existing gravity flow rate would not be more than 10 gpm. This, together with the cost of installing and operating a vacuum system, and the uncertainty regarding the period over which the increased vacuum flow would be effective, is a reasonable basis for the conclusion that vacuum pumping of the horizontal drains is not advisable except under emergency conditions.

Water Quality. Samples of the water were periodically collected from most of the dewatering wells and horizontal drains constructed. The biological quality of the water was of concern because of the necessity for surface disposal. The primary purpose in developing a data base for the chemical character of the water was to determine if various sources could be identified and related to the geological history of the study area.

The biological quality of the water was monitored by the Los Angeles County Department of Health Services. The fecal

Figure 5. Oblique air photo of Big Rock Mesa, 1984

Fiure 6. Example of cracks in Big Rock Drive

Figure 7. Portion of Seacliff slide on Pacific CoastHighway near Moonshadows Restaurant

Figure 8. Drilling near the headscarp

Figure 9. Flowing horizontal drains along Pacific Coast Highway

coliform count during the initial stages of production of several of the wells and drains probably was high. However, after a few days, it dropped significantly. In 1986, none of the water from the wells or drains, registered a coliform count greater than 20 per 100 milliliters. Extensive growth of algae at some outlets during the first year of dewatering indicated the presence of a significant amount of nutrient in the water. In 1986, however, there was very little algae growth. Rust stains at the outlets of some of the wells and drains, and a fine patina of iron oxide on pumps removed from some of the wells strongly suggest the presence of iron bacteria, specifically Crenothrix is suspected.

The first known measurements of the chemical character of the ground water in the Big Rock Mesa area are recorded in nine analyses by the Smith-Emery Company dated May 28, 1959. Copies of these analyses probably were made as a requirement for transferring the interests of the Big Rock Beach Water Company to the Ocean Mutual Water Company. A summary of these results is contained in Table III-B-Chemical Constituents of Ground Water in the Evans Report.

Aquifers

Water supply aquifers in the Big Rock Mesa area are poorly developed, and many of the early wells were apparently abandoned because of low yield. Nevertheless, for purposes of dewatering, there are at least three aquifers of geotechnical significance. These are the bedrock, landslide debris, and channel-fill aquifers.

The bedrock aquifer includes portions of the Sespe, Vaqueros, and Topanga Canyon Formations. Its base is at an unknown level but is probably slightly below sea level. It is probably saturated within the study area except for an upper, unsaturated zone near the surface at the base of the seacliff. Hydraulic conductivity in the bedrock aquifer is low, probably no greater than 0.1 foot per day (fpd). Specific yield may be no more than two percent. There are insufficient data to estimate transmissibility.

Landslide Debris Aquifer. The large, complex mass of prehistoric landslide debris has hydraulic characteristics distinct from the bedrock due to the more intense fracturing caused by landsliding. An upper zone of debris derived from the Topanga Canyon Formation provides the highest specific yield. It is a massive breccia with fragments no more than a few inches in longest dimension in some locations. At other localities boulders 20 feet or more in diameter occur. A lower zone, presumably Sespe and Vaqueros rock, seems to have yielded water only from isolated fractures.

Ground water in the Topanga Canyon formation- derived debris occurs in a relatively thin zone at its base and is perched above less permeable Sespe- or Vaqueros-derived debris.

Falling water was common in wells and exploratory drill holes which penetrate this aquifer. In most wells, a perched condition also was indicated by a sharp break in the recovery curve. Hydraulic conductivity probably was in the range of 0.1 - 10 feet per day. The specific yield may have been on the order of 10 percent. There was insufficient data to estimate transmissibility.

Channel Fill Aquifer. The ancient debris-filled channel recognized by Moore (1961) extends from the seacliff near Inland Lane cul-de-sac northward at least to the vicinity of the intersection of Seaboard Road and Roca Chica Drive. The debris has in the past stored a substantial amount of ground water and has a relatively high transmissibilty. Dewatering wells which have the highest capacity are usually located close to the channel axis. Judging from well production, the specific yield was probably on the order of 10-20 percent and transmissibility is in the range of 10-500 square feet per day.

Ground Water Movement. Ground water in the Big Rock area moves generally southward to the ocean. Prior to residential development, the regime must have reached a condition of long-term equilibrium. The sparse vegetation shown in 1928 aerial photographs provides some basis for the conclusion that under an equilibrium condition, the level of ground water was relatively low. Initial construction for the Pacific Coast Highway resulted in the local steepening of the seacliff, and the placement of relatively impermeable fill for the roadway. This probably inhibited spring flow and evaporation at the base of the seacliff, but had little effect on the overall subsurface drainage.

Ground-water contour maps were produced in Reference 5 at intervals of about three months. By the end of 1985, the level of the piezometric surface had lowered about 100 feet along the upper edge of the seacliff, about 70 feet at the northern side of the lower mesa (in the area of Big Rock Drive), and about 20 feet below Little Rock Way. At higher elevations, there appears to have been little change except in the immediate vicinity of the dewatering well at Cool Oak Way and Big Rock Drive. The decline in the piezometric surface between mid-1984 and mid-1985 is characterized on Plate IV-6, Volume IV of Evans' (1986) final report.

The contour maps clearly show that significant subsurface outflow to the ocean occurs below the base of the seacliff. An attempt was made to define the zone of outflow and the range of hydraulic conductivity in that section by packer testing. Both four-and six-inch-diameter packers were tried without success. In practically all attempts, the hole could not be sealed because of the weakness of the surrounding rock material. The resulting loss of fluid pressure (by-pass) vitiated the results.

Dewatering

Four hydrauger drains were apparently installed by others in the seacliff about 1965 as a condition of approval of Tracts 26263 and 27463 by the Los Angeles County Regional Planning Commission. These drains (see Plate III-4 Dewatering Facilities for location) were to be 700 feet long, 360 feet on center, with a positive gradient of six percent. There is no record of when these drains were actually installed or whether the design criteria where actually achieved. Moreover, it appears that they were not monitored after they were installed. Reportedly, Drain H-3 was producing 8,650 gallons per day and Drain H-4 was producing 5,240 gallons per day in November 1974. By the time the eastern portion of the study area had been fully developed, they were producing a total of no more than about 5,000 gallons per day (gpd).

Dewatering was accomplished with both wells and horizontal drains to its conclusion in March 1986. Detailed records of production and other relevant data were maintained.

Well Design and Construction

Initially, a low yield was expected from the horizontal drains; therefore, a program of existing well rehabilitation and new well construction was immediately instituted. Priority was given to maximizing production even at the expense of efficiency.

Large rotary wash drilling equipment was utilized with satisfactory results obtained down to 250 feet. Below that depth canvernous landslide debris caused heavy drilling fluid loss. Details of well construction and equipment are found in Evans (1986).

Horizontal Drain Design and Construction

The construction of horizontal drains in the Big Rock Mesa Landslide area, was accomplished with a fourth- generation modification of an Aardvark drilling machine, designed by Soil Sampling Service, Inc. of Puyallup, Washington. The drill utilizes a specially fabricated four- inch-diameter rod, driven hydraulically. The drilling operation is similar to that of a normal rotary-wash rig with the drilling fluid pumped through the rod to carry the cuttings out the annulus. The drilled hole is completed by inserting 1-1/2-inch diameter slotted PVC pipe in the drill rod and subsequently withdrawing the rod. This technique theoretically permits the installation of drain pipe more than 1500 feet long. Lost circulation and caving generated by high water flow limits boring lengths.

HYDROLOGIC EQUILIBRIUM

The basis for applying the concept of hydrologic equilibrium to the Big Rock Mesa study area is the assumption that the direction and containment of both surface and subsurface flow for a specified block of mountain slope is essentially constant. Therefore, the use of the concept in analyzing a complex mass of bedrock and landslide debris is as valid as the more conventional application to a ground-water basin, even though there may be less probability of achieving accurate results. The observation period used for this study was July 1, 1984 to June 30, 1985.

Direct Estimation of Subsurface Outflow

The direct measurement of hydraulic conductivity of rock sections along the northern side of Pacific Coast Highway was attempted utilizing a special packer tool designed to isolate five-foot-thick increments at any depth in the boring. No satisfactory results were obtained because of leakage (by- pass) around the packers. In every case attempted, down to a depth of about 50 feet, a chamber pressure sufficient to force fluid into the formation could not be reached before leaking or "by-pass" developed. Since only a statistical analyses was possible, any successful result below 50 feet would not have been significant, and the packer tests were abandoned.

Limited geothermal testing was undertaken on Pacific Coast Highway near the toe of the slide to determine if that technique could be used to define the sea water/fresh water interface, since that interface is the lowest elevation that overflow can occur. The results were promising, however, funds for more detailed testing were not available.

Implications

The ability of the Big Rock Mesa study area to transmit ground water to the ocean is the fundamental limiting factor in determining relative stability. Above a certain subsurface outflow rate, buoyancy and seepage forces become high enough to cause movement. Stability depends upon keeping the subsurface outflow rate below a critical value. Preliminary hydrologic equilibrium analysis suggests that, under the conditions which prevailed prior to the current failure, this critical value was about 15 acre-feet per year.

Reduction of the various increments of recharge is therefore extremely important. Adequate surface drainage can reduce the quantity of rainfall infiltration but little can be done to reduce the rainfall infiltration rate, which can be greater than the five percent value assumed for the period of observation. High-intensity rainfall is now better able to infiltrate the fractured debris. More promising is the possibility of reducing infiltration due to irrigation. It appears that far too much water is used in the study area for this purpose. Moreover, the present surface drainage system does little to reduce infiltration. The subsurface inflow rate probably fluctuates only slightly because of a relatively constant gradient which is not easily controlled. The eventual need for a public sewer system will probably depend upon the future infiltration rate, yet to be observed, which will depend upon the management of high-intensity storms and whether irrigation is reduced.

Future Performance

There was no opportunity during this study to test the dewatering system under conditions of increased natural recharge because there was relatively little rainfall since it was installed. Nevertheless, the system had significantly lowered the level of ground water in a period of about two years through 1986.

It is unknown whether or not the level of ground water can ever be reduced to such an extent as to increase the relative factor of safety to 1.5. Reduction in the capacity of the dewatering system was expected as the level of ground water lowered. The level decreased at a much lower rate in 1986 than it did at the start of dewatering. In fact, ground-water contour maps for the last quarter of 1985 show very little change, suggesting that the dewatering system had reached its capacity. This, and the continued increment of artificial recharge coupled with an eventual increase in natural recharge, implies that an equilibrium condition will be reached. However, performance of the dewatering system to 1986 indicated that additional dewatering devices, either drains or wells, could be effective to further lower whatever equilibrium level may develop.

Additions to the dewatering system, to that time seemed most likely to reduce the level of ground water significantly, included two or three deep wells located north of the North Fault in the ridge between Piedra Gorda and Rockpoint Canyons. This fault may be a partial ground-water barrier and, hence, act as a reservoir in effect, helping to maintain a relatively high hydrodynamic head to the south. It is unknown if these wells were built.

GEOTECHNICAL ENGINEERING

Introduction

The Big Rock Mesa landslide was a reactivation of an older, prehistoric landslide complex. The focus of movement of the episode covered by this study was essentially that of at least one of the earlier episode(s) of movement. It is not clear if the original movement occurred along discrete geologic planes of weakness. What is clear is that movement did occur along a geologic plane of weakness which consisted of presheared soil or rock. For purposes of engineering investigation and analysis, the landslide was considered as one mass; no attempt was made to evaluate earlier episodes of movement. Separate analyses were performed, however, for both the seacliff and the headscarp areas.

Investigative Techniques

At the inception of the study, the geotechnical consultants involved agreed that, from a practical standpoint, dewatering of the slide mass represented the only possible method to slow or arrest movement. As a result, the investigative process focused on installing dewatering devices and ground movement instrumentation as quickly as possible delineating the vertical and lateral limits of the landslide, developing a mathematical model of the landslide and determining the relative factor of safety under static and seismic loading conditions that could be achieved by the dewatering.

Instrumentation

One of the first elements of the investigative process in the Big Rock Mesa study area was the installation of slope indicators. Data obtained from these devices were used in conjunction with the then existing network of ground survey points and the surface geologic mapping of group cracks to initially verify the lateral limits of movement established by the County of Los Angeles and to initially establish the vertical limits of movement.

Slope Indicators. The County of Los Angeles reportedly installed slope indicators on Pacific Coast Highway in the Big Rock area prior to about 1980. Private consultants installed other comparatively short (less than 100 feet) inclinometer casings prior to and during 1983 as part of investigations of individual properties. During the investigation of the Big Rock Mesa landslide which began in late 1983, on the order of 40 inclinometer casings were installed. The length of the casings varied from less than 100 to 450 feet. Most of the tubes were standard, aluminum, 1.79 inch O.D. Digitilt casing, although some plastic tubes were used. The casings were installed in both small-and-large diameter holes and, in most cases, the holes were backfilled with sand.

Piezometers. Throughout the investigation, the level of ground water in the study area was monitored by means of piezometers. In lieu of installing a separate set of tubes specifically designed to measure piezometric surface in an active landslide, the slope indicator tubing installed was modified to function as an open stand-pipe for measurement of standing water. Specifically, the indicator tube was perforated by drilling 1/4-inch-diameter holes at regular intervals (± five feet) in the wall of the tube between the sensor tracks. In addition, the annulus was backfilled with clean sand and the bottom of the tube plugged prior to installation. The devices were read using an electronic probe each time the slope indicator sensor was inserted in the tube.

In an effort to minimize in-filling of the tube with backfill sand, some of the earliest tubes installed were wrapped with synthetic fabric. However, the unevenness of the drill hole caused installation difficulties and tearing of the fabric, particularly in holes deeper than about 200 feet. As a piezometric device, the tubes have the advantage of being economic to install and to read. They have the disadvantage of a somewhat limited useful life and producing data of limited accuracy in areas where the ground water is perched. The estimated locations of all known piezometers are shown on Plate IV-1 Ground Features Map.

Surface Geologic Mapping. At the time the investigation was initiated, the County of Los Angeles had performed some investigative work including the preliminary delineation of the lateral boundaries of movement. That boundary was used for some time as a general limit for study purposes. Throughout the study, surface geologic mapping was performed for two primary purposes: (1) to understand and depict the geologic framework of the study area, and (2) to delineate ground cracks resulting from the ground movement. All surface mapping of ground cracks was performed at a scale of one inch equals 100 feet using the topographic map which serves as a base for Plate IV-1 Ground Features Map. The results of the geologic mapping are shown on Plate II-1 Geology Map: the results of the mapping of ground cracks are summarized on Plate IV-3 Geotechnical Map.

Plane Ground Surveys. In 1980, the County of Los Angeles installed a network of ground survey points within and adjacent to the study area. The network was apparently installed because both data from County slope indicator installations and repeated damage to the Waterworks District No. 29 water transmission main in Pacific Coast Highway suggested the possibility of landslide movement. The network consists of more than 80 individual points on Pacific Coast Highway (both in and outside the study area) and in the residential area on the mesa. The location (both vertical and horizontal) of these points with respect to a "fixed" datum was re-determined by the County of Los Angeles at a more or less regular interval.

Subsurface Exploration

A number of exploratory holes were excavated in the Big Rock Mesa study area during this investigation. The estimated locations of the drill holes for which either driller's or geologic logs were prepared are shown on Plate VII-3 Drill Hole Location Map.

The relative accuracy of the logs varies widely. The least accurate are the driller's logs since they are prepared by nonprofessional personnel. Next, in increasing order of accuracy are the geologic logs prepared by trained, professional geologists. In this case, variation is limited to the relative ability of the individual geologist and the method of obtaining cuttings from the drill hole during rotary-wash operations. The next most accurate logs are those made from observation of cores and the most accurate logs are those prepared by direct, down-hole observation of the large-diameter holes by professional geologists.

Laboratory Testing

Comparatively little laboratory testing was performed. At the onset of the investigation, it was assumed that the relative factor of safety against landslide movement was near unity at the time movement started. This permitted the mathematical determination of the effective shear strength along the slide plane at the inception of movement. Since these calculated values were reasonable and compatible with the generally accepted range of values for residual shear strength under similar stress conditions, the need for laboratory derived values of shear strength was diminished. As the investigation proceeded and the slide plane was penetrated by drill holes, it became apparent that it would be very difficult to obtain an undisturbed sample of the rock or gouge within the slide plane. Three of the exploratory drill holes excavated by D. A. Evans, Inc., produced samples of a slide plane which were considered suitable for laboratory testing.

Strength tests were performed on samples taken from each of the three sites. In each case, the sample was subjected to a cyclic direct shear test using a motorized, automatically-reversing, single shear apparatus. The rate of strain was set at 0.0001 inches per minute with a total displacement of 0.4 inches per cycle. The shearing was repeated until the shear stress became a constant value. The samples were all submerged for several days prior to shearing. The results of the tests are tabulated below:

Sample	Angle of Internal Friction, Degrees	Cohesion, Pounds Per Square Foot
DH-1	40	0
SP-4A	26	100
SP-26	12	0

The sample taken from DH-1 was quite sandy and not particularly representative for the slide plane. However, the samples from SP-4A and SP-26 appeared to be representative of clay gouge and the test results for these samples are regarded as being representative (i.e., the right order of magnitude) for the effective shear strength along a pre-sheared surface such as a slide plane.

Landslide Model

Much of the ground data regarding the Big Rock Mesa study area were digitized using a computer program. This included the topography, the piezometric contour maps, the lateral limits of the recent episode of the landslide and the contours on the slide plane. This permitted not only the rapid computation of the volume of the landslide (more than 90 million cubic yards) but the volume of rock within the landslide below the piezometric surface at any one time. It also made the computation of a comparatively accurate hydrologic balance possible. As a result of the computer- assisted modeling, it was also possible to prepare three- dimensional perspective views of the landslide for illustrative purposes.

All of the analyses to determine the relative stability of the landslide and its components were computer assisted. The earlier analyses were performed using the SLOPE program. Later, analyses of relative stability were performed using the TSLOPE program.

Relative Stability

Main Slide. For purposes of our investigation, two simplifying assumptions were made. It was assumed that, at the time the failure started, the relative factor of safety was unity, i.e., the forces causing movement were in balance with the forces resisting movement under static loading conditions. Further, it was assumed that the piezometer head extant when the landslide started could, for practical purposes, be represented by the head in September 1983. These assumptions were considered reasonable inasmuch as (1) the landslide did not have the characteristics associated with episodic, "creep" failure at a calculated factor of safety of greater than unity, (2) the buildup of ground water storage in the slide mass had occurred over a long period of time (15 to 20 years), and (3) the dilation associated with initial slide movement would tend to slow or stop the continued buildup of ground water.

Using the assumptions described, several conventional analyses of the relative stability of the landslide were performed to determine the shearing resistance along the slide plane necessary to provide a factor of safety of unity. The Morgan-stern-Price, limit-equilibrium technique (available in both the SLOPE AND TSLOPE computer programs) was used. These analyses were repeated as more information about the configuration of the slide plane was obtained. Using the "equilibrium" values of shearing resistance, analyses were then performed for several sections of the landslide to determine the change in factor of safety due to changes in the piezometric surface. The following is a summary of the values for the "equilibrium" shearing resistance and the calculated relative factors of safety under static loading conditions for three piezometric surfaces. The locations of the section analyzed are shown on Plate IV-3 Geotechnical Map. The geotechnical sections are included as Plates IV-4 Geotechnical Sections. The positions of the slide plane on the geotechnical sections is based on Plate IV-5 Slide Plane Contours.

Geo-technical Section	Angle of Internal Friction[1] Φ, degrees	Factor of Safety (3/84)	Factor of Safety (1/85)	Factor of Safety[2] (dewatered)
A-A'	181.11.21.3			
B-B'	191.01.11.3			
C-C'	201.11.21.3			
D-D'	191.21.21.3			
E-E'	161.11.21.3			
F-F'	101.11.11.2			
G-G'	121.21.21.2			
H-H'	141.11.31.4			
I-I'	151.31.31.5			
J-J'	201.01.11.2			

1 - The calculated effective angle of internal friction, F, assuming the effective cohesion, c', is zero needed to produce a relative factor of safety of one in September, 1983.

2 - The calculated relative factor of safety under static loading conditions if the piezometric surface in the landslide area is lowered to the elevation of Pacific Coast Highway.

These data indicate that, for a relative factor of safety of unity, the strength parameters on the slide plane varied between F = 10 degrees, c'=0 and F = 20 degrees, c' =0, an average of F = 16 degrees, c' = 0. These values were not unreasonable for the residual strength along a pre-sheared failure surface through the rock which comprised the study area. Further, the data indicated that, for static loading conditions, a dewatering system which had the capability of lowering the piezometric surface to the elevation of Pacific Coast Highway would elevate the relative factor of safety about 20 to 50 percent (an average of 30 percent).The landslide mass as of 1986 remained sensitive to seismic loading. Conventional pseudostatic analyses indicated that the relative factor of safety against seismically-induced movement would fall to less than unity for even a comparatively small magnitude earthquake. Because of the size of the landslide (the axial length is the same order of magnitude as the wavelength of the shear wave generated by an earthquake), the pseudostatic method of analysis tends to produce an unrealistically low value for the factor of safety. Nonetheless, the yield acceleration (i.e., the lowest ground acceleration needed to produce dislocation) at that time was comparatively low. Specifically, the following values were appropriate.

Geotechnical Section	Yield Acceleration (ky)
A-A'	0.04
B-B'	0.03
C-C'	0.04
D-D'	0.04
E-E'	0.04
F-F'	0.03
G-G'	0.04
H-H'	0.06
I-I'	0.06
J-J'	0.03

Assuming a Maximum Probable Earthquake with a Richter Magnitude of 6.0 to 6.5 (ground acceleration of 0.17g to 0.23g), the calculated displacement would be on the order of one and nine inches.

Structure Damage

Although not all of the homes within the Big Rock Mesa landslide area were inspected internally and externally during the investigation, it is reasonable to conclude that all of the improved properties were damaged in some fashion by the landslide movement. Distress was most visibly pronounced along the brow of the seacliff, in the area below North Peak and along the original defined western boundary of the landslide. At or about the time that the investigation started, several homes had already been abandoned by their owners. During the winter of 1983-84, the relative risk of damage or destruction of a number of homes at the brow of the seacliff (and consequent danger of injury to their inhabitants) increased to the point where the County of Los Angeles ordered the homes posted as unsafe.

REFERENCES

Anonymous, Final Report, Pacific Palisades Landslide Study: unpublished report, Moran, Proctor, Mueser & Rutledge, Consulting Engineers, New York, pursuant to Agreement Number 13616, dated May 26, 1958 with the State of California as authorized by California Statistics, 1957, Chapter 2009, 2 volumes, 13 tables, 30 plates.

County of Los Angeles, 1973, Engineering Geology Report Big Rock Mesa Proposed Ground Water Improvement Maintenance District, Malibu: Engineering Geology Section, Department of County Engineers, May 13.

Evans, D. A., 1983, Interim Report of Geotechnical Evaluation - Big Rock Mesa Landslide, Malibu, California, for the Concerned Citizens for Water Control: D. A. Evans, Inc. consulting report, Job No. L83-102-02, December 30.

Evans, D. A., 1984, Second Interim Report of Geotechnical Evaluation - Big Rock Mesa Landslide, Malibu, California, for the County of Los Angeles, Job. No. L83-102-02.

Evans, D. A., 1986, Final Report - Geotechnical Evaluation, Big Rock Mesa Landslide, Malibu, California: for Los Angeles County Improvement District 2629R, 7 volumes and 2 appendices.

Greene, L. G. Clarke Jr., S. H., Field, M. E., Linker, F. I., and Wagner, H. C., 1975, Preliminary Report on the Environmental Geology of Selected Areas of the Southern California Continental Borderland: U. S. Geological Survey, Open-file report. p. 75-596.

Hoots, Harold W., 1931, Geology of the Eastern Part of the Santa Monica Mountains, Los Angeles County, California: U. S. Geological Survey, Professional Paper 165-C, p. 83-134, 19 plates.

McGill, John T., 1948, Geology of a Portion of the Los Flores Quadrangle, Los Angeles County, California: unpublished M.A. Thesis, University of California, Los Angeles

Merifield, P. M., Michael, E. D., and Eagan, J. T., 1978, Dewatering Big Rock Mesa, Malibu; Geologic Guide and Engineering Geology Case Histories, Los Angeles Metropolitan Area: Association of Engineering Geologists; 1st Annual California Section Conference, p. 142-148.

Meighan, Clement, 1985, A Midden Remnant in the Big Rock Slide Area: unpublished consultant report, July 31.

Michael, E. D., 1960, Engineering Geology Report, Tract 26263: unpublished consulting report for Jennings Engineering Company, December 28.

Michael, E.D., 1961, Supplement, Engineering Geology Report, Tract 26263: published consulting report for Jennings Engineering Company, February 23.

Michael, E. D., 1973, Progress Report, Dewatering Project, Big Rock Mesa: unpublished consulting report for Malibu Mutual Drainage Company: November 15, 11 p.

Moore, R. F., 1961, Stability Investigation - Tentative Tract 26263, for Roberts Enterprises; Moore & Taber consulting report Job No. 1142F, July 24.

Simonson, Russel R., 1936, Conglomerates of the Sespe and Topanga Formations of the Dry Canyon Quadrangle, Santa Monica Mountains, California: unpublished M. A. thesi, sUniversity of California at Los Angeles .

Soper, E. K., 1938, Geology of the Central Santa Monica Mountains, Los Angeles County, California: California Journal of Mines and Geology, vol. 34, no. 2, p. 131-180.

Yerkes, R. F., 1985, Earthquakes and Surface-Faulting Sources, Geologic and Seismologic Setting, in Evaluating Earthquake Hazards in the Los Angeles Region - An Earth- science Perspective: U. S. Geological Survey Professional Paper 1360, p. 25-41.

Yerkes, R. F., and Campbell, R. H., 1979, Stratigraphic Nomenclature of the Central Santa Monica Mountain, Los Angeles County, California, U. S. Geological Survey Bulletin 1457-E.

Yerkes, R.F., and Campbell, R.H., 1980, Geologic Map of East-Central Santa Monica Mountains, Los Angeles County, California: U. S. Geological Survey Miscellaneous Investigations, Map I-1146.

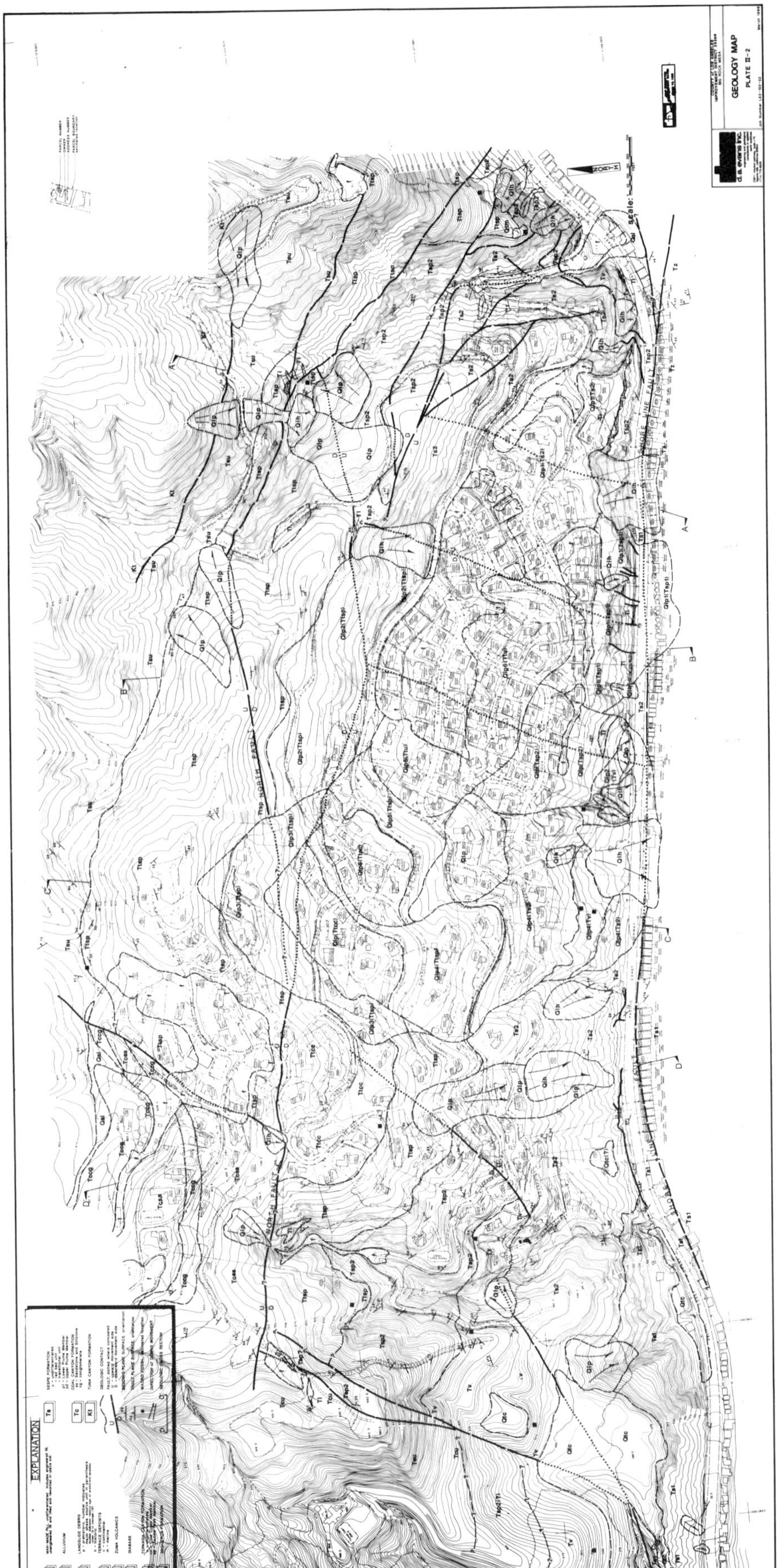
GEOLOGY MAP
PLATE II-2
EXPLANATION

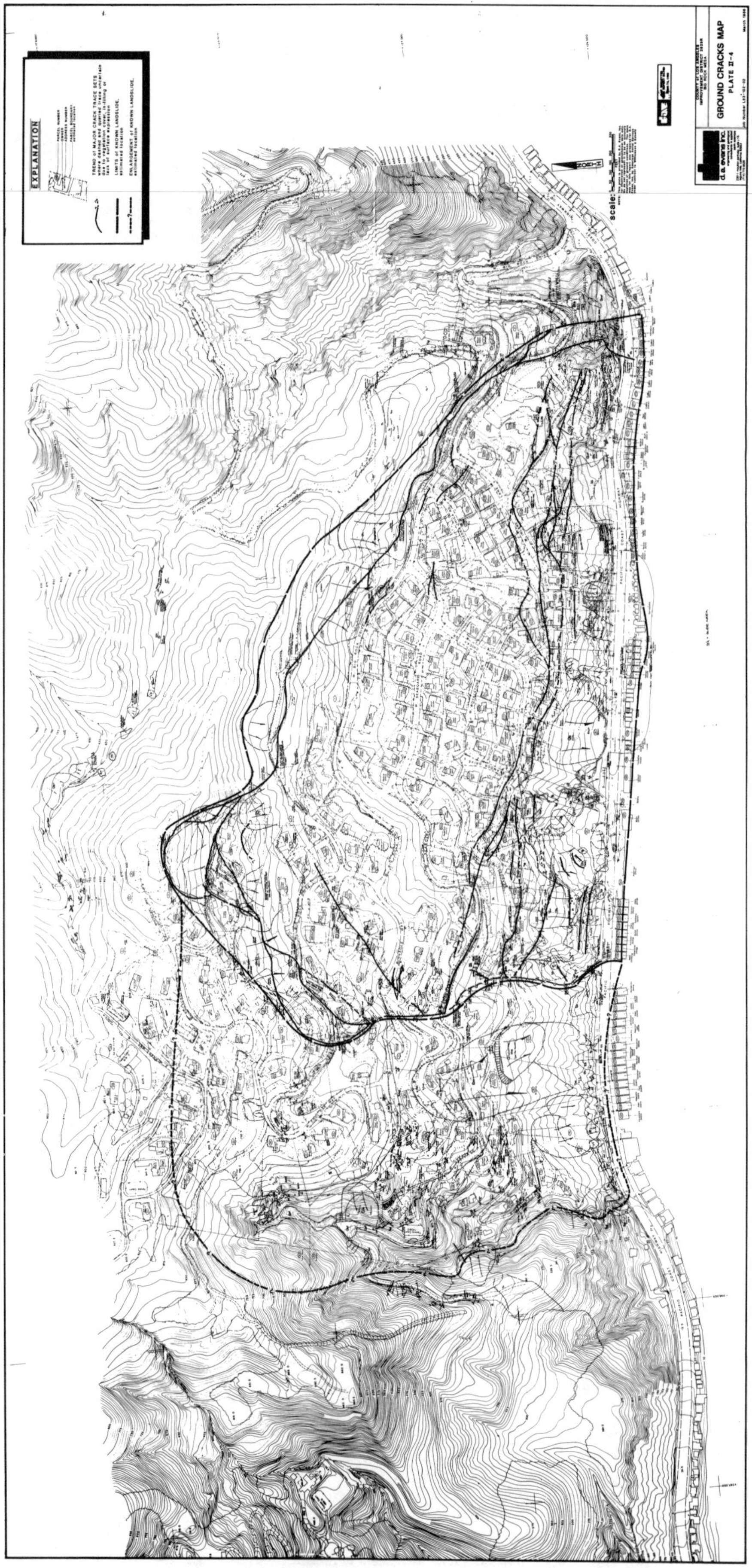
EXPLANATION
GROUND CRACKS MAP
PLATE II-4
scale

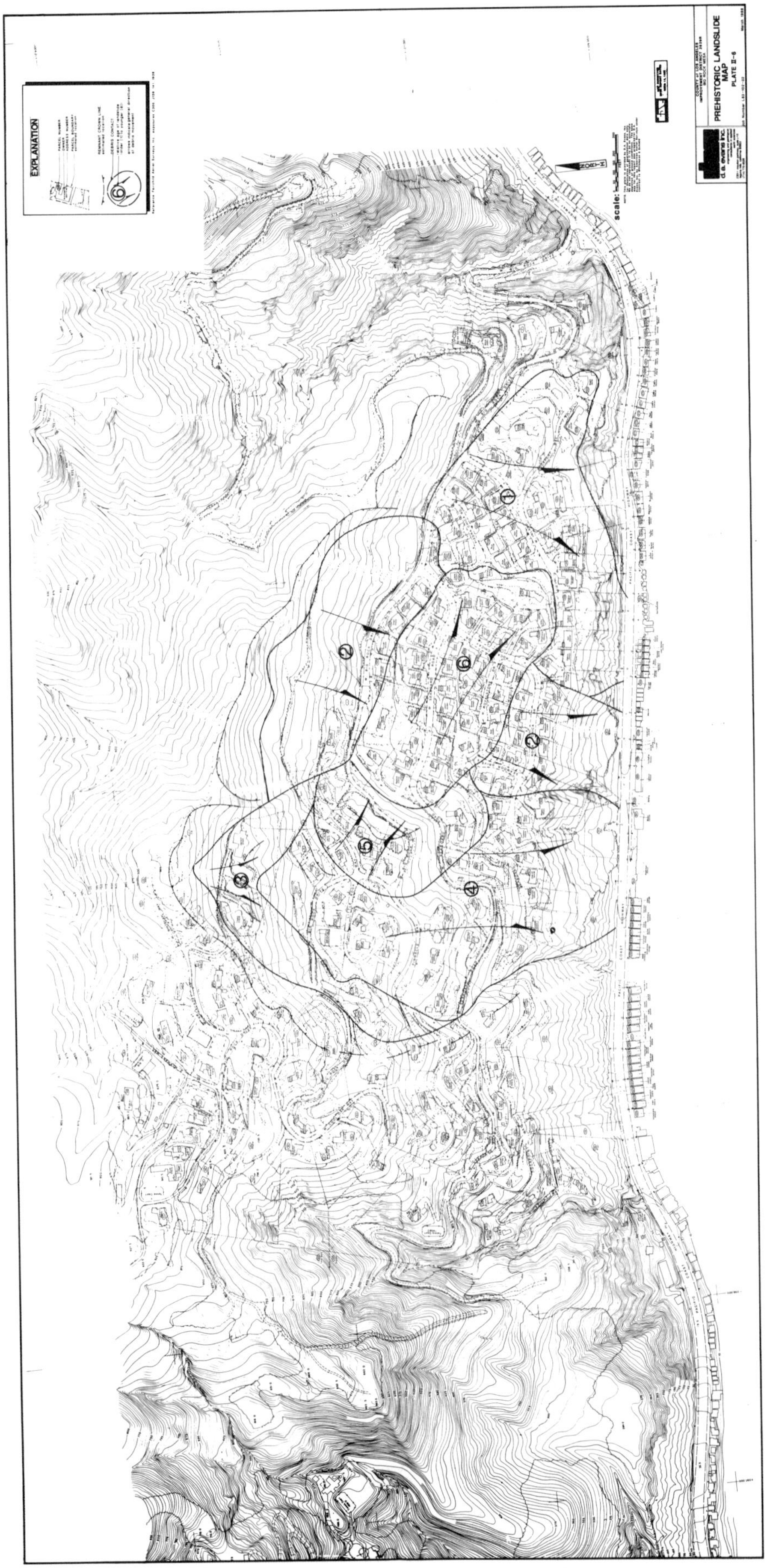
EXPLANATION
PREHISTORIC LANDSLIDE MAP
PLATE II-6

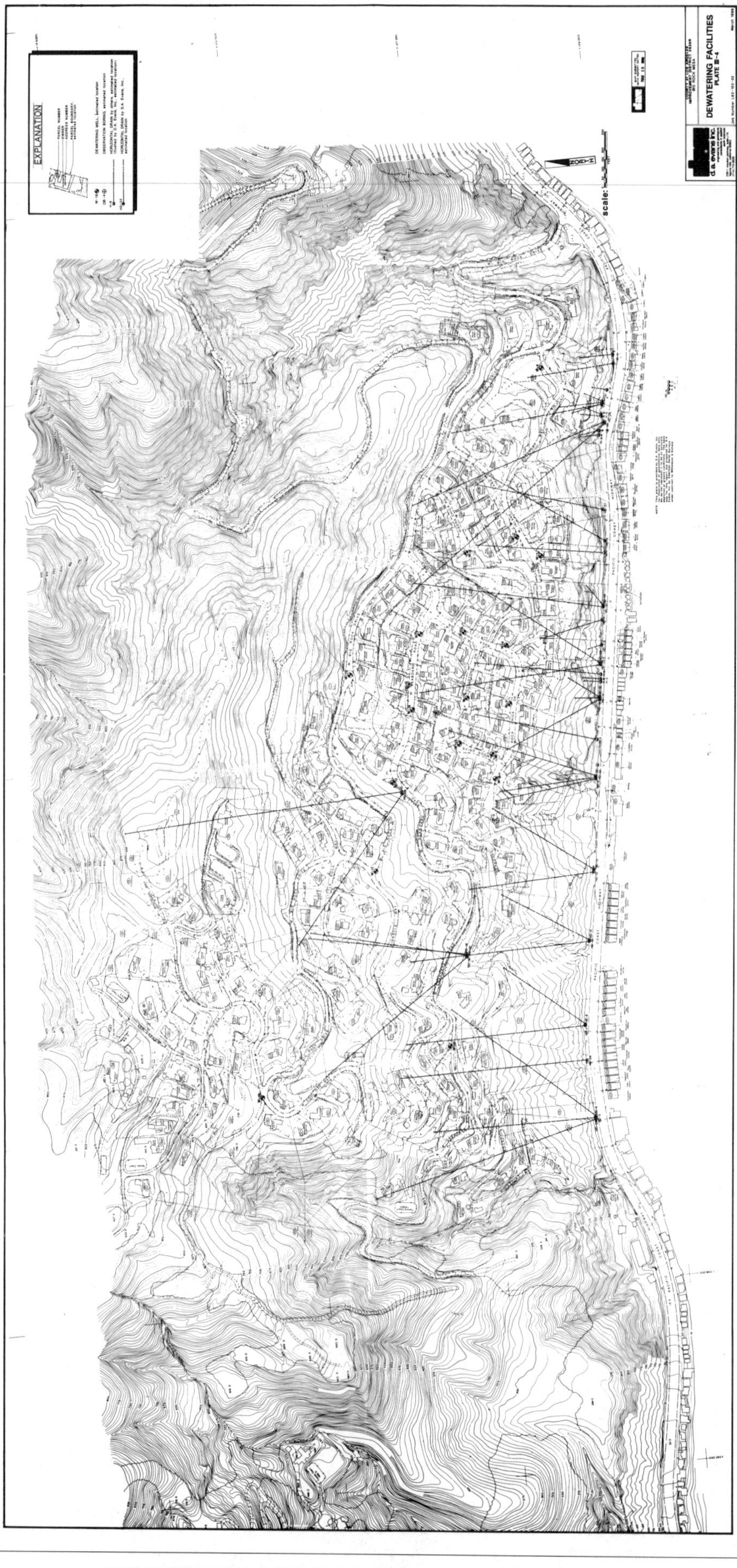
EXPLANATION
DEWATERING FACILITIES
d.a. evans inc.
scale:

MEGA-LANDSLIDES IN SOUTHERN CALIFORNIA

Martin L. Stout
Department of Geological Sciences
California State University
Los Angeles CA 90032-8203

INTRODUCTION

Mega-landslides, or megaslides, are words that do not now appear in the 3rd edition of the AGI glossary. In the field mega-landslides are more common than recognized, and in many ways, are similar to detachment or listric normal faults within areas of extensional tectonics.

For this paper, mega-landslides, or megaslides for brevity, are defined here as covering areas at least 2-3 km^2 (490 to 740 acres), or a minimum of 3/4 square mile, and are bottomed with known or inferred sheared clays. This definition excludes the deposits involved with turbulent flow characteristics such as debris flows. There is no upward limit to the size of a megaslide. The largest slides are still to be discovered on the earth—albeit they may not necessarily be Quaternary in age. The 2-3 km^2 figure was chosen only to emphasize that megaslides are not always the "easy to spot", "throw a stone across", garden variety of landslide, many of which are commonly dealt with on a routine basis by engineering geological consultants. Obviously the larger the area involved in landsliding, the more difficult it becomes to recognize as an identifiable and mappable feature—and equally applicable, the older a megaslide, the more difficult it is to interpret the deposits accurately, and ultimately to identify the mass positively as a landslide.

Once identified, megaslides make engineering decisions tenuous at best because it is difficult, if not impossible, to design reasonable mitigation measures for stability. In a sense, this is like an engineering geologist finding an active fault on a client's property in California—some land is almost automatically going to be lost if the fault is determined to be active, and similarly, if a megaslide is identified, there may be some ground lost because of questions regarding stability.

PRIMARY FAILURE MECHANISMS

Primary failure mechanisms, or those mechanisms wherein friction is reduced considerably to allow failure to occur, include water, clay, air, and seismicity, or combinations of these. Some unique variations have been shown to include volcanic gas/water/magma intrusions. The author feels that water in all its various forms, is the most important primary ingredient in earth-based megaslides.

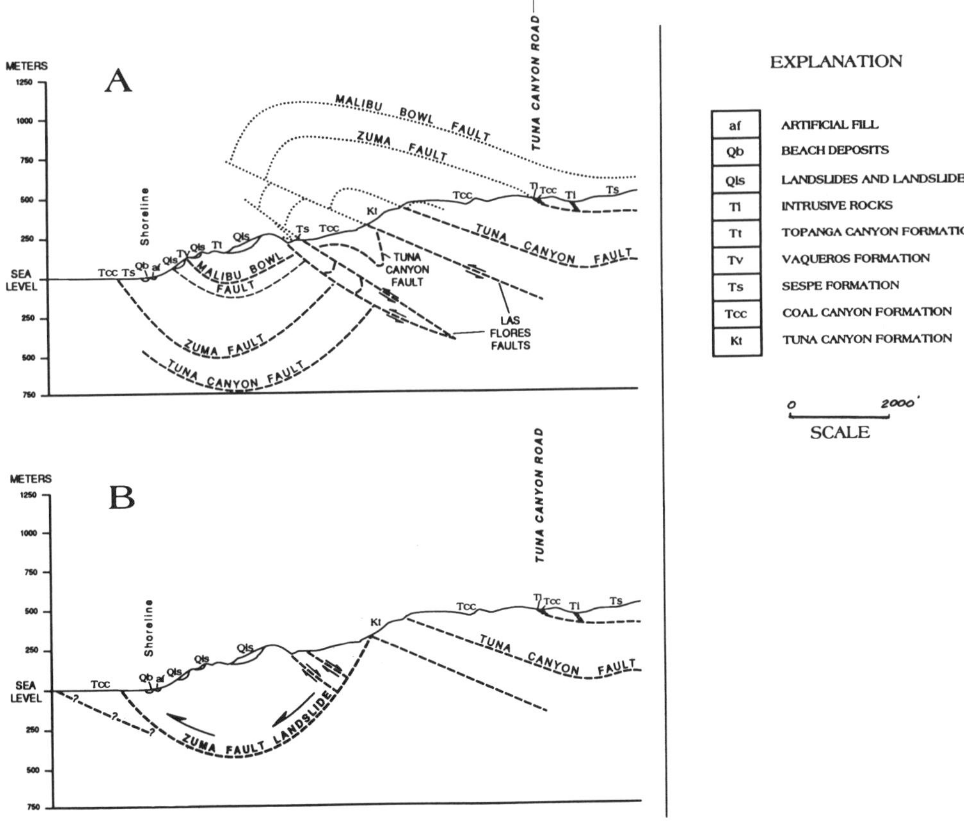

Figure 1. Geologic sections through the Big Rock Mesa landslide complex; 1-A is modified from Yerkes and Campbell (1980); 1-B shows a possible megaslide interpretation utilizing the same geologic section.

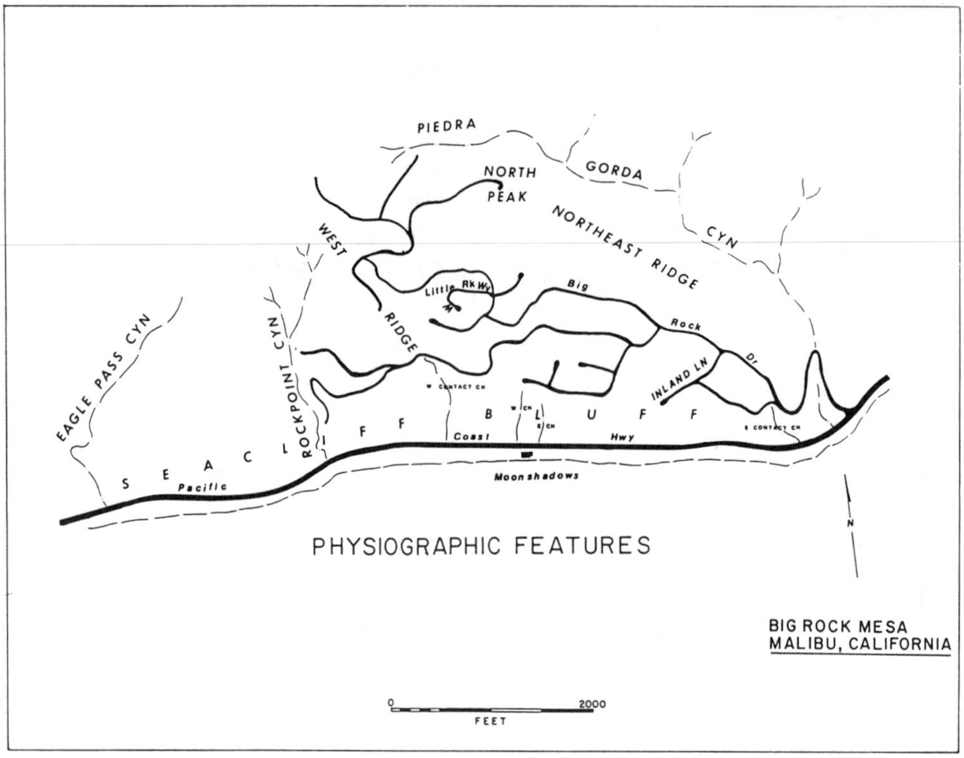

Figure 2. Map showing physiographic features of the Big Rock Mesa area, including Piedra Gorda Canyon. (from Michaels, 1986).

SECONDARY FAILURE MECHANISMS

Secondary failure mechanisms usually involve structural or stratigraphic controls which maximize the friction loss due to a primary mechanism, and generally are a mappable feature in the bedrock. The geometry of many failure surfaces in megaslides suggests that the dominant secondary mechanisms of failure are controlled by one or a combination of the following:

1) low-angle faulting, either detachment-style or thrusting, with lateral-slip possible, particularly in "flower structures"; any combination of displacement which places relatively more competent rocks over weaker rocks.
2) unconformities, particularly where an extended period of time has weakened, by weathering, the rocks just below the unconformity.
3) bedding/foliation/joint planes, again particularly important are lower-angle dips (°), in heterogeneous shear strength rocks; dips are not as critical in homogeneous shear strength rocks.
4) high angle fault surfaces.

These weaker surfaces generally contain one of the above primary materials of low-shear strength allowing failure to occur, or they allow accumulation of water to develop higher than expected pore pressures. Low-transmissivity zones in any strength rock allows collection of subsurface waters in fractures and joints, which explains failures in some relatively homogeneous crystalline rocks. The author recommends that the usual mapping approach should be to look for the "weak link" first; then, if not found, possibly resort to another interpretation. The critical part of this field assessment is when does a geologist give up in the search for "weak links", because they are known for their poor exposure or total absence in outcrop, and can be extremely thin in the subsurface, so can be missed in cores or large diameter borings.

TIMING

Megaslides are not actively occurring on the earth's surface today, and the author feels that most megaslides were controlled by the same massive water transfers which also caused glacial periods (Stout, 1991). This is not to say that there are many megaslides on the surface of the earth, portions of which are being reactivated on almost a daily basis because of climate, people, or seismicity. Within California, the more famous recent landslides such as those affecting State Highway 1 along the Monterey coast, Love Creek landslide in Santa Cruz County, Portuguese Bend and Big Rock Mesa landslides in Los Angeles County, Bluebird Canyon and near-coastal landslides of San Clemente in Orange County, and Soledad Mountain in San Diego County have all occurred within larger and older landslide debris, although the original movement of material may not have been large enough to qualify the mass as a megaslide. More recently, the large-scale ground cracking accompanying the 1989 Loma Prieta earthquake, has been found to be largely in older landslide deposits, although not all areas had been identified as such prior to the earthquake.

SOUTHERN CALIFORNIA STRUCTURAL SETTING

In southern California, the Transverse Ranges provide an ideal structural setting for low-angle faulting, and hence the generation of megaslides (e.g., Fig. 3). Regions experiencing this style of deformation are the southern side of the Santa Monica Mountains; southern side of the San Gabriel Mountains; northern and southern sides as well as within the San Bernardino Mountains; both sides of the Santa Clara River Valley in the vicinity of Santa Paula and Fillmore, to cite a few. In several of these sites, basement or crystalline rocks have been placed over sedimentary or alluvial sequences, whereas in others, sedimentary rocks have been placed over other sedimentary rocks.Prehistoric large landslides that have characteristics of flowage, such as the Blackhawk landslide and the Martinez Mountain rock avalanche, have left older slide debris in the source area (Stout, 1988; Baldwin 1981, 1986). Regions experiencing extensional tectonics with young detachment faults are in and around the Salton Trough, continuing into Baja California, and this region is another area of megasliding.

PROBLEMS OF IDENTIFICATION

Large, currently active coastal landslides in southern California are almost always formed in reactivated landslide debris—the legacy of an earlier climatic period of more intense landslide activity. One of the more visited and famous of these is the Portuguese Bend landslide, which actually involves only about 20% of the 5.3 km^2 (1280 acres) surface area of the older landslide mass. The older slide mass is marked by an arcuate headscarp. Problems of interpretation of landslide boundaries in this area take place where features are not included within that arcuate scarp, such as a linear valley north of the arcuate head scarp (Ehlert, 1986). If such prominent geomorphic features as headscarps, bulbous toes, hummocky topography, etc. are subdued by other processes such as secondary landsliding and/or erosion, which generally implies age, identification by surface mapping becomes much more difficult, and data from the subsurface is easier to miss or misinterpret.

Figure 3. Vertical aerial photo of south front of San Gabriel mountains with cities of Altadena at left and Sierra Madre at right. Center of photo shows prehistoric Henninger Flats landslide with its toe along Eaton Wash (NW trending white line) and its head scarp east of the "P" at left-center of photo. The maximum width of the slide is 5200 feet. Note the traces of the Sierra Madre fault zone appear across the landslide, suggesting fault movement since the slide occured. (Photo courtesy of Richard Proctor.)

BIG ROCK MESA LANDSLIDE

Problems of identification are exemplified by the well known Big Rock Mesa landslide in Malibu, California, which has now been studied by numerous consultants from southern and northern California, and is the subject of another paper in this volume. Ironically, many consultants argued in the 1960's about whether the area was a landslide or an elevated and faulted marine terrace, but once movement started, this question was moot. However, once the slide was active, attention is directed toward factors of mitigation, and not much effort is expended trying to "produce" other problems by enlarging the landslide boundaries and mass, and ultimately the cost of engineering structures. The debate now relates to whether or not there is additional landsliding in the area, and if so, the size of the ancient landslide. The author is in a distinct minority for including the Big Rock Mesa area as part of a megaslide, for even though the present active slide covers only about 0.9 km^2 (217 acres), the proposed megaslide to the west covers more than 4 km^2 and extends beyond Las Flores Canyon.

The Big Rock Mesa landslide is about 2.1 miles due west of the Los Angeles Civic Center, along the east-west portion of the Malibu Coast. Mapping by Yerkes and Campbell (1980) of the U.S. Geological Survey, showed numerous thrust faults in this area, and the faults themselves are currently the subject of intense debate as to whether to include them in the California's Alquist Priolo Special Studies Zones. Figure 1A shows a geologic cross section reproduced from Yerkes and Campbell, (1980) which coincidently passes through the Big Rock Mesa landslide (Qls on seaward facing slopes). Figure 1B shows my suggested interpretation indicating that even though the sheared surfaces appear to be real, the last movement in this area along the surface of thrusting may have been that caused by landsliding. In this case, normal faulting, as opposed to reverse, could be expected in the pull-away of the megaslide.

The difficulty of recognizing landslide masses and mapping in complex areas such as Big Rock Mesa suggests evidence for and against this theory. Many consultants have noticed the geomorphically enigmatic arcuate course of Piedra Gorda Canyon (Fig. 2), and most would agree that there does not appear to be any slip surface exposed in or along the present stream course. However, the arcuate configuration of Piedra Gorda Canyon may suggest much earlier subsequent control of stream evolution, possibly along a landslide break. This pattern can also be produced by thrusting. This activity must predate the Holocene. In the lower reaches of Piedra Gorda Canyon, a fault has been mapped along the valley bottom because the geology does not match across the valley, yet north of the active Big Rock Mesa landslide, sheared zones appear to cross the valley bottom without interruption (B. Yen and Associates, 1992).

CONCLUSIONS

At present, most mega-landslides, with the exception of slides in sensitive clays, appear to be in quasi-static equilibrium suggesting that the dominant triggering mechanisms are not currently active. Where smaller portions of megaslides have been reactivated, the initiation of movement can almost always be traced to groundwater conditions and/or activities of people, and to a lesser degree, natural undercutting of slopes. Megaslides have not shown a tendency to be reactivated by large seismic events. Consolidation of ancient landslide debris is a common event during ground shaking, particularly where high groundwater levels prevail, and this may be misinterpreted as reactivation. In areas previously undisturbed by landsliding, megaslides have been triggered seismically where sensitive clays are involved, or where conditions are prone to liquefaction. Southern California has no historic accounts of megaslides triggered under these conditions.

REFERENCES

Baldwin, J.E. II, 1981, The Martinez Mountain rock avalanche: MS thesis, California State University, Los Angeles, 58 p.

__________, 1986, Martinez Mountain rock avalanche: *in* Guptil, P.D., Gath, E.M., Ruff, R.W., eds., Geology of the Imperial Valley, Annual Field Trip Guidebook #14, South Coast Geological Society, Santa Ana, p.37-48.

Ehlert, K.W., 1986, Origin of a mile-long valley located northerly of the ancient Portuguese Bend landslide, Palos Verdes Peninsula, southern California: *in* Ehlig, P.L., ed., Landslides and landslide mitigation in southern California, Cordilleran Section, Geological Society of America, Field trip guidebook, p. 167-172.

Michael, E.D., 1986 Progress in stabilizing Big Rock Mesa, Malibu, California: *in* Ehlig, P.L., ed., Landslides and landslide mitigation in southern California, Cordilleran Section, Geological Society of America, Field trip guidebook, p. 81-96.

Stout, M.L.,, 1988, Geology of Blackhawk Canyon, San Bernardino Mountains, southern California: Geological Society of America, Abstracts with Programs, v. 20, p. A361.

__________, 1991, Landscape evolution during the last 35,000 years—Perspective of a southern California engineering geologist: *in* Abbott, P.L. and Elliott, W.J., eds., Environmental Perils, San Diego Region, San Diego Association of Geologists, p 161-164.

Yen, Bing, and Associates, 1992, A geotechnical engineering evaluation, Big Rock Mesa, Malibu, California: Los Angeles County Improvement District 2629R2.

Yerkes, R.F., and Campbell, R.H., 1980, Geologic map of of the east-central Santa Monica Mountains, Los Angeles County, California: U.S. Geological Survey Miscellaneous Investigations Series Map I-1146.

PALEOLANDSLIDES IN SAN ANTONIO CANYON, EASTERN SAN GABRIEL MOUNTAINS, CALIFORNIA

J. David Rogers
CE and Engineering Geologist
Rogers/Pacific, Inc.
396 Civic Dr.,
Pleasant Hill, CA 94523

Lawrence J. Herber
Department of Geological Sciences
California State Polytechnic Univ.
3801 W. Temple Blvd
Pomona, CA 91768

Thomas H. Hibner
GSI/Water (Geothermal Surveys, Inc.)
520 Mission St.
South Pasadena, CA 91030

INTRODUCTION

San Antonio Canyon lies in the eastern San Gabriel Mountains, immediately north of Upland, California (Figure 1). The Canyon's trend appears to be fault-controlled, cutting across the structural grain of the San Gabriels. Its source lies at the foot of "Old Baldy", or Mt. San Antonio, at 10,064 feet, the highest point in the range. From its source, it drops steeply over a 10 mile run to San Antonio Dam, a flood control basin at the mouth of the Canyon (Elevation 2150 feet).

The steep slopes of San Antonio Canyon have spawned numerous paleolandslides. Near Mt. Baldy Village (Figure 2, upper) the volumes of these slides range from 7 to more than 170 million cubic yards. The Canyon's side slopes record successive rejuvenation: the slopes of the inner gorge are approximately 1:1, with upper canyon slopes between 1.5:1 and 2:1, and relief of 1200 to 4200 feet (Figure 2, lower). The paleoslides are sourced in Precambrian gneiss (San Gabriel Terrane) on the west side of the gorge, and in mixed granodiorite (Cretaceous)- quartzite-marble-gneiss rocks of likely Paleozoic age on the opposing (east) side of the canyon (May and Walker, 1989). The slides appear to be of late Quaternary age.

This article grew from on-going investigation of paleoslides in crystal line rock undertaken by students and faculty in the Geological Sciences Department at California State Polytechnic University, Pomona over the past 25 years. Results to date suggest that toppling and progressive failure initiated by keyblock removal at slope toes could be common failure mechanisms for these ancient slides in otherwise, seemingly competent crystalline rock.

Toppling failure is of special interest because it provides a mechanism for slope failure in "high strength" crystalline rock. Failure is induced by slow slope deformational processes such as creep, which promote progressive dilation of the rock mass fabric; which in-turn, allows for discontinuities to open up and for discrete inter-block translations (slip) to occur. Creep-induced failure mechanisms are most pronounced in long, high slopes, where incremental strains are carried up through the mass if the toe area is unrestrained (or undercut). In simple terms, the higher the slope, the greater the amount of creep-induced strain we could expect to see accumulated near its crest. As creep-induced strains continue to accumulate, a shear zone develops at the base of the creeping mass, initially near its toe. This zone of macro-shear strain then progresses upslope until a failure threshold is reached, unleashing the entire affected mass of rock.

Temporary pore water pressures associated with snowmelt or heavy precipitation, moderate seismic events of at least 0.15 g acceleration and 25 seconds duration, or long-term chemical weathering of the dilated mass may work together to trigger the apparently catastrophic failures evidenced in the Canyon area.

Published work on paleolandslides in the immediate vicinity includes geologic reconnaissance (Baird, 1956; Morton and Streitz, 1969; and Morton, Sadler and Minnich, 1987) and geomorphic studies (Herber, 1976, 1987; Shelton, 1966; and Vrabel, 1988). More detailed work has focused on the origin of failure mechanisms of these slides (Archer, 1979; Hibner, 1990; Hibner and Herber, 1991; Herber et al., 1991; Rasplicka, 1991; and Ressel, 1989). Much of this latter work was successful in back-analyzing the block-glide style failure at Brown's Flat (Archer, 1979) as well as the toppling and keyblock controlled failures discussed in this article.

GEOLOGIC SETTING

The Sunset Peak and Hog Back slides originated from east-southeast facing Sunset Ridge, approximately 1.3 and 2 miles southwest of Mt. Baldy Village (Figure 1). In this area Sunset Ridge is underlain by a southwest dipping homoclinal structure, as defined by the foliation of the gneiss (Figures 3 and 4).

The Sunset Peak and upper Hog Back evacuation scars show well-layered (banded) biotite-quartzo-feldspathic gneiss with well-defined, continuous interlayers of hornblende-plagioclase gneiss of Precambrian age. These rocks overlie a moderately well to poorly layered hornblende-plagioclase gneiss of Cretaceous age. The contact is vividly seen as a dull green, locally epidotized conformable mylonite zone in some areas, but is locally faulted and intruded with basaltic dikes, adjacent to, and within the Hog Back scar area (Figure 3).

The rock mass within and on the periphery of both slide scars is usually fresh, but locally slightly weathered; and moderately to highly jointed and fractured. Joints, fractures and layering rarely show calcite, clay or iron oxides, but most exhibit a thin (0.05 in.) epidote filling that is often slicked.

Joint and foliation data used in the slope stability analyses were mapped in the slide evacuation scars, or projected on-strike from the side scar areas. These data were then projected with the appropriate apparent inclinations onto two-dimensional maximum cross sections developed for computerized back-analysis.

Sunset Peak Slide

As presented in Figure 4, the scar and margins of the Sunset Peak Slide exhibit well-defined foliation/layering and joint suite orientations. The foliation of the banded gneiss is generally continuous, thinly spaced (1 to 6 inches), near planar, and consistently dips southwest, an average of 31 degrees, into the slope. The rock mass is thoroughly perturbed by joints, which are continuous for 10 to 40 feet, of moderate spacing (3 to 6 feet), near planar, and frequently open (0.05 to 0.30 inches). One set of joint dips an average of 50 degrees southwest, into the slope, while a second dominant set dips an average of 62 degrees, downslope, to the northeast (depicted schematically in Figure 5).

Backanalysis of the slide was effected along the maximum cross section developed by Hibner (1990), shown in Figures 4 (location) and 5 (the section). By estimating the slide surface area (82 acres), volume (9 million cubic yards) and apparent thickness (averages 69 feet, reaching a maximum of 200 feet), initial pre-slide conditions could be estimated, as shown in Figure 5.

Three principal sets of discontinuities were modeled utilizing progressive block deformation analysis (or BDA). Progressive block analysis is accomplished by making various assumptions as to the frictional strength between blocks, depending on the infilling observed in the field. For the Sunset Peak analyses, the following block-to-block friction criteria were utilized:

1. Rock to rock contact, discontinuous, slightly undulatory joints, with no mineralogic infillings: $\Phi = 40$ degrees
2. Rock to rock contact, smooth, planar joints, with no mineralogic infillings: $\Phi = 30$ degrees; and
3. Discontinuity filled with slickensided epidote infilling: $\Phi = 25$ degrees;

After these interblock friction values are assigned to the respective discontinuity suites, gravity is applied through the program and strains predicted. In this manner, permanent deformations are then input into the next analysis, and the process attempted again. With each succeeding run, plastic deformations are progressively fed into the analysis, and the long- term, plastic distortions of the mass of discontinuity-bordered blocks can be analyzed. Such a system of analysis allows for the progressive modeling of an inelastic system with progressive, permanent deformations, such as slope creep.

The first few analyses on the Sunset Peak Slide indicated that toppling would initiate at the toe of the model slope, controlled by steeply-inclined foliation and jointing (Figures 5 and 6). As the toppling progressed, its effects become increasingly magnified, leading to the progressive undercutting of larger and larger blocks (Figure 6). The block deformational analyses suggests that, occasionally, wedge-shaped pieces could have been expected to drop into the toppling voids, temporarily retarding further movement (shaded block in Figure 6). The stalling effect of such "wedges" was probably abated by the continued creep of the rock mass, downslope of the wedges. This is because inter-block strain is accumulated upslope, as an integral function (Goodman and Bray, 1977). In this manner, the inter-block strain (or degree of block tilt) increases upslope, causing the greatest deformations at the highest points of the slope. In this manner the "wedges" sketched in Figure 6 would be turned on their sides, negating their buttressing or stalling effect. The later stages of block deformation are sketched in Figure 7.

Eventually, the creep-deformed mass would have creeped as far as it could (Figure 7), and thereby exert enormous shear stress along the "hinge line" of the toppling. Catastrophic failure of the entire mass could then ensue due to accumulation of strain along the hinge line (Figure 8); or be suddenly triggered by environmental extrema, such as precipitation or seismic excitation.

The slide's geometry (Figure 4) suggests that it initially moved in a easterly direction. The current configuration of the scar area is presented in Figure 9. A photograph of the slide scar and the topographic bench created by its debris run-out is shown in Figure 10. Physical characteristics of the slide mass suggest that it represents the latest late -Pleistocene failure within an older, larger slide complex.

Hog Back Slide

The Hog Back Slide is located just east of the Sunset Peak Slide, as shown in Figures 4 and 11. The Hog Back slide is the most obvious of those exposed in the Canyon, having blocked the channel sometime in late Pleistocene time (Shelton, 1966). Today, Mt. Baldy Road, built in 1954-55, cuts through the head of the slide, replacing the old highway that ran parallel to the old overflow gorge, cut over 90 feet into the mixed hornblende-plagioclase gneiss-granodioritic rocks, comprising the east wall of the Canyon (Figure 12).

The discontinuity suites perturbing the Hog Back scar are generally similar to those of the Sunset Peak slide. However, layering in the Cretaceous hornblende-plagioclase gneiss is somewhat less well developed. In addition, the joint orientations appear to be somewhat more scattered. Stereographic analyses indicate an apparent adversely oriented foliation dip of 5 to 9 degrees on the lower slopes, changing to northwesterly, in-slope dips of 18 to 40 degrees on the upper slopes, along the line of maximum cross section (shown in Figure 13). Joints dip adversely 18 to 38 degrees to the southeast along the cross section of the lower slope, and steepen to 62 to 83 degrees on the upper slope, as shown schematically on Figures 13-16 and 18-19.

Backanalysis of the Hog Back Slide was effected in much the same manner as that of Sunset Peak, discussed previously. Figure 13 presents the initial pre-failure cross section, as deduced from field mapping and mass balance computations by Herber (1987) and Herber et al., (1991). Discontinuity suites mapped in the side scarps and evacuation scar of the slide were projected as the fabric of the failed mass. In this instance, the three-dimensional characteristics of the discontinuity field were input into a computerized keyblock analysis (Goodman and Shi, 1985) because of the sheer number and geometric complexity of the problem.

Figure 14 presents a graphic portrayal of the keyblock analysis at the toe of the model slope, where we presume it was undercut by the proto San Antonio Creek. In this case the interblock friction between blocks was assumed to be 35 degrees, a fairly conservative number. The keyblock analysis predicted pull-out of a tapered wedge, subjected to principal stresses aligned at approximately 45 $\Phi/2$ degrees from the sloping surface of inclined valley-side (stress relief) joints, seen in the canyon wall. The keyblock program cannot predict the actual size of the block, only that it should theoretically exist and whether or not it is inclined to fail. If an assumed block size is chosen, block deformational analysis can then proceed, as shown in the analysis for the Sunset Peak slide. If progressive block deformational analysis is applied (Step 1c in Figure 15), a chain- reaction of block displacements is predicted, as shown in Step 1d, shown in Figure 15.

Progressive block deformational analysis indicates that the slippage of the toe keyblock would initiate further creep-induced strain of the now-unsupported mass, as shown in Figure 16. As the mass continued to creep and dilate, progressively larger strain is built up. At this juncture we decided to run a finite element analysis of the dilated, joint-bordered mass of blocks to evaluate what sort of loads would be generated at the boundaries of the dilating mass. Figure 17 presents the results of this study, utilizing the program JTROCK (Hittinger and Goodman, 1981; Rogers and Figuers, 1982). The JTROCK analysis predicts that the toe keyblock would be in compression (shown shaded in Figure 17). It also indicates that the dilated mass lying above the displaced keyblock goes into tension, due to loss of constraint. Just below this tension zone, however, is a sort of stress reversal, with very high shear stresses, presumably induced by traction of the dilated mass over the intact country rock. Lines of equal tensile and shear stress predicted by the finite element analysis are presented in Figure 17.

We can infer from these analyses that slippage of keyblocks in the toe area of the slope may have initiated creep-induced deformations of the slope above it, extending a considerable distance above the initial perturbation, depending on how far the initial keyblock(s) were displaced. If the keyblock(s) actually began to slide into the undercut channel, a chain reaction of upslope events was likely set into motion. No frame of time is suggested herein, but we can assume that the stress redistributions assumed herein took many thousands of years to occur.

Eventually, the traction-induced shear stresses would grow to become sufficient to induce the catastrophic failure of the mass, likely along adversely dipping block interfaces, above the keyblock, as depicted in Figure 18. The Hog Back Slide failed as a largely coherent block glide with minor internal shearing, as seen in the major lithologic contacts that are preserved within the slide mass, as indicated in Figures 4 and 19. The slide does appear to have come downslope in discrete segments, each one thrusting upon the one preceding it. San Antonio canyon was thereby dammed, creating a lake which we can assume soon overtopped the landslide dam near the extreme limits of its run-out, on the opposing east walls of the Canyon. The comparison of slide profile and that now exposed at the toe is shown in Figure 19.

The Hog Back Slide also holds a special place in the developmental history of California and that of the world at large. Because of the landslide dam, the gradient of San Antonio Creek increases dramatically through the overflow gorge carved at the slides eastern foot. In 1892, a 1,320 foot long tunnel was excavated through the Hog Back Slide to feed a powerhouse penstock to fall 300 feet to the creek bottom below the landslide dam. This spot had been chosen for its favorable topography by Dr. Cyrus Baldwin, President of the fledgling Pomona College. The problem then be came not one of generating the electricity, but how to transport it the 14 miles to Pomona, the longest distance yet attempted for commercial electric service up to that time. A young engineer named A.W. Decker had recently moved to Southern California because he suffered from tuberculosis. Baldwin engaged Decker's services as engineer for his hydropower scheme, which seemed impractical because of the long distance between the source and the proposed destination of the power. Decker formulated the idea of oil-filled "step- up" transformers at the powerhouse to transmit the electricity the long distance to Pomona, then "stepping down" at the destination for safe distribution to customers. So began the "step up/step down" system of electrical transmission in use all over the world today. A more complete synopsis of the story is contained in Myers (1986). The original powerhouse has since been destroyed by flooding, but the old tunnel still carries water across the Hog Back slide, to a small generator further down the Canyon-maintaining a secure place in the history of technology few appreciate.

CONCLUSIONS

Paleolandslides blanket much of the rugged San Antonio Canyon area, some of these with volumes in excess of 170 million cubic yards (such as the Cow Canyon Slide). Despite the seemingly favorable structural orientation of fabric discontinuities such as foliation, seemingly resistant slopes comprised of granitic gneiss appear prone to failure by mechanisms which are just beginning to become understood. In part, the lack of appreciation is partly due to the relative infrequency of such slide events, when compared to statistically frequent slides which perturb the softer Tertiary-age rocks comprising the southern California basins. Simple application of frictional stereographic analyses may be insufficient to explain the observed failures. Three dimensional analyses incorporating progressive displacements of discrete, discontinuity-bordered blocks suggests that such failures may initiate over considerable periods of time due to normal gravitational stresses.

The presence of trapped interstitial water, such as the condition presented with rapid snowmelt, could serve to buoy effective interblock friction and thereby, trigger failure. Alternatively, seismic loading could also serve as a trigger mechanism for catastrophic failure. San Antonio Canyon certainly lies within a zone of heightened seismicity, being defined by the active left-lateral San Antonio fault, its mouth crossed by the Cucamonga fault zone, and the San Jacinto and San Andreas faults lying but a few miles to the northeast.

Inexpensive iterative computer simulations presently exist to aid in the assessment of complex rock slope assessments. Even so, a tremendous amount of judgment and area experience is necessary in evaluating what frictional characteristics to assign to the various discontinuity suites. An area can only be evaluated with efficacy after first studying and back-analyzing its paleoslide features in sufficient detail so as to gain an insight to the modes of past failures. This, in of itself, may prove more elusive than it would appear, as the two examples we have presented are but a fraction of those that remain to be analyzed in the immediate area.

ACKNOWLEDGMENTS

The authors wish to thank the many students at Cal Poly Pomona who have spent countless hours mapping in the San Antonio Canyon area as part of class projects and senior thesis studies. Richard E. Goodman strongly influenced the methodologies employed in back-analyzing crystalline rock slope failures. Conversations with and articles by David M. Cruden and Barry Voight on slide mechanisms in crystalline rock terrain were also particularly influential in the analytical approaches taken by the authors in the course of their studies.

Fred H.P. Chin of Rogers/Pacific performed the keyblock analyses while Sands H. Figuers of Rogers/Pacific performed the finite element assessments utilizing joint elements with transversely anisotropic (layered) rock masses. Shi Gen-hua kindly reviewed the keyblock analyses results and provided helpful input. John Shelton provided the excellent aerial photographs of the Hog Back Slide, while W.H. Freeman kindly granted permission to reproduce one of the figures from his classic text, Geology Illustrated, which was the first publication to recognize and assess the Hog Back Slide, in 1966. It is still in publication after 25 years.

REFERENCES

Archer, R.L., 1979, Browns Flat Landslide - A study of a landslide in gneissic terrain: Senior Thesis: Dept of Earth Sciences, Calif State Polytechnic University, Pomona, CA.

Baird, A.K., 1956, Geology of a Portion of San Antonio Canyon, San Gabriel Mountains, California: Master's thesis, Pomona College, Pomona, CA, 91 p.

Goodman, R.E. and Bray, J., 1977, Toppling of Rock Slopes: Proceedings of the Specialty Conference on Rock Engineering for Foundations and Slopes: American Society of Civil Engineers, vol. 2, p. 201- 234.

Goodman, R.E., and Shi, Gen-hua, 1985, Block Theory and Its Application to Rock Engineering: Prentice-Hall, 338 p.

Hibner, T.H., 1990, Failure Analysis of the Sunset Peak Rock Avalanche in San Antonio Canyon, CA: Senior Thesis, Dept of Geological Sciences, Calif State Polytechnic University, Pomona, CA.

Hibner, T.H., and Herber, L.J., 1991, Failure Analysis of the Sunset Peak Rock Avalanche, Southern California: Abstracts with Program, Cordilleran Section Meeting: Geological Society of America: San Francisco, vol. 23, no. 2, p.35.

Herber, L.J., 1976, Potential Geologic Hazards on the South Side of the San Gabriel Mountains: faults, landslides and flooding: Spring Field Trip Guidebook, Far Western Section, National Association of Geology Teachers: 21 p, 10 plates.

Herber, L.J., 1987, Hog Back: a grossly stable prehistoric translatory rock block slide, San Antonio Canyon, Southern California: Centennial Field Guide to Cordilleran Section, Geological Society of America: p. 201-202.

Herber, L.J., Hibner, T.J., Ressell, M.W.Jr., Vrabel, C., and Rogers, J.D., 1991, Failure Mechanisms of Catastrophic Subalpine Paleolandslides in Crystalline Rock of the Central San Gabriel Mountains, San Antonio canyon, Southern California: Abstracts with Program, Annual Meeting, Geological Society of America, San Diego, vol. 23, no. 5, p. A126.

Hittinger, M., and Goodman, R.E., 1978, JTROCK - A Computer program for Stress Analysis of Two-Dimensional, Discontinuous Rock Masses: Report UCB/GT/78-04, Dept Civil Engineering, University of California, Berkeley, 147 p.

May, D.J., and Walker, N.W., 1989, Late Cretaceous Juxtaposition of Metamorphic Terrains in the Southeastern San Gabriel Mountains, California: Bulletin of the Geological Society of America, vol. 101, p. 1246-1267.

Morton, D.M., and Streitz, R., 1969, Preliminary map of Major Landslides, San Gabriel Mountains, California: Calif Division of Mines and Geology, Map Sheet 15, 1:62,500.

Morton, D.M., Sadler, P.M., and Minnich, R.A., 1987, A Field Guide to Landslides in the Inland Valleys and Adjacent Mountains of Southern California - Examples from the Central and Eastern San Gabriel Mountains of Southern California: Inland Geological Society, University of California, Riverside: p. 25- 27.

Myers, W.A., 1986, Iron Men and Copper Wires: Trans-Anglo Books, Glendale, 272 p.

Rasplicka, Alan, 1991, Cow Canyon landslides in San Antonio Canyon: Senior Thesis, Dept of Geological Sciences: Calif State Polytechnic University, Pomona, CA.

Ressel, M.W., Jr., 1989, Failure Analysis of the Cow Canyon Slide, Mt. Baldy, California: Senior Thesis, Dept of Geological Sciences, Calif State Polytechnic University, Pomona, CA.

Rogers, J.D., and Figuers, S.H., 1982, JTROCK II - Corrected version of JTROCK to run on IBM computer system: Dept Civil Engineering, University of California, Berkeley.

Shelton, John S., 1966, Geology Illustrated: W.H. Freeman, San Francisco, 434 p.

Vrabel, Craig, 1988, Cow Canyon Landslide: Senior Thesis, Dept of Geological Sciences, Calif State Polytechnic University, Pomona, CA.

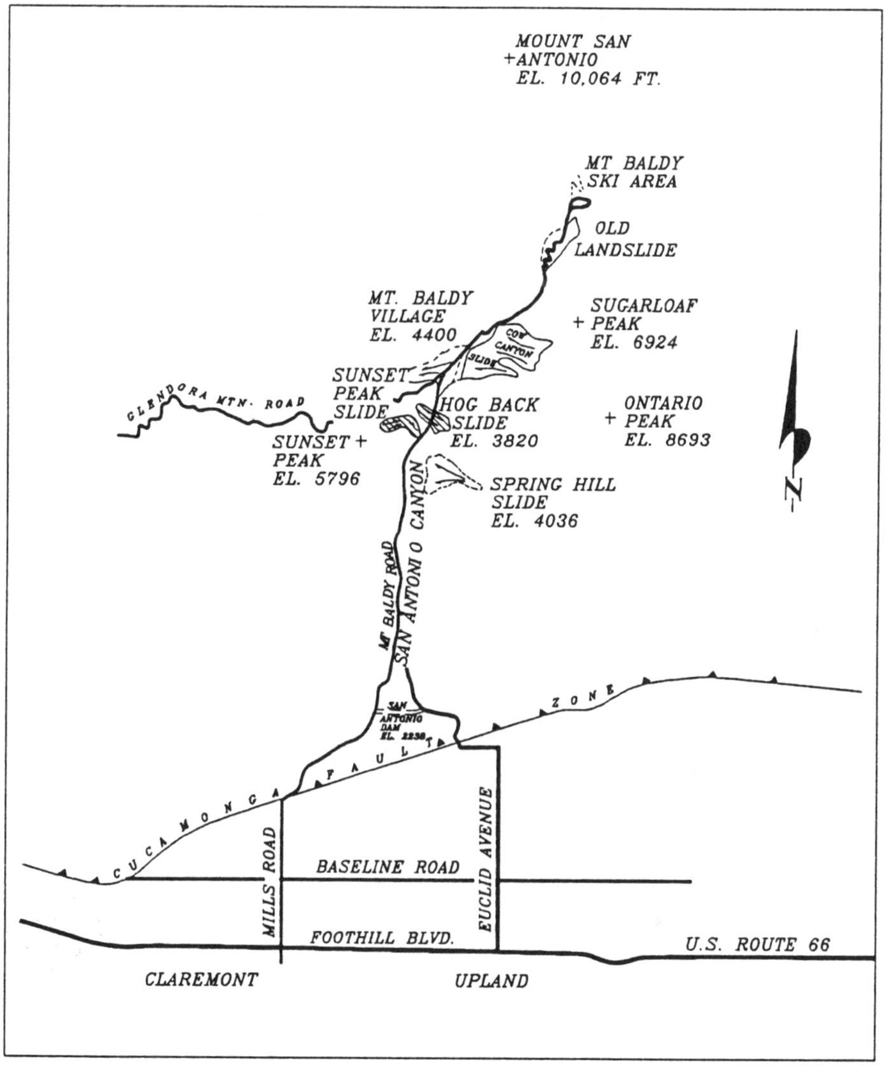

Figure 1. Vicinity map showing the approximate location of the Hog Back and Sunset Peak slides in the rugged San Antonio Canyon of the eastern San Gabriel Mountains. Rising to 10.064 feet, Mount San Antonio is the highest peak in the San Gabriels, a range bounded by the San Andreas and San Jacinto faults on the north and Cucamonga fault on the south.

Figure 2. (upper) View looking up San Antonio Canyon in the vicinity of Mt. Baldy Village (elevation 4250 feet). The valley floor has been widened by channel aggradation behind the Hog Back slide, which blocks the Canyon a mile downstream. The slopes in the background average 50 degrees and are comprised of a mixture of quartzite, marble, and granitic gneiss.

Figure 2. (lower) View looking down San Antonio Canyon, at the Hog Back slide (center foreground). The eroded surface of Spring Hill looms in the background, a half-mile downstream of the Hog Back (arrows indicate the direction of paleomotion). Considerably higher and older than the Hog Back slide, Spring Hill is typical of the massive paleoslides which must have periodically dammed the gorge.

Fig.ure 3. Basaltic dike cutting thru the Precambrian-age hornblende-plagioclase gneiss, as seen along Mt. Baldy Road a half mile south of the Hog Back slide. The basalt dikes parallel an older suite of joints perturbing the ground mass.

Figure 4. Blocky character of the hornblende-plagioclase gneiss, as seen in outcrop along Mt. Baldy Road, just south of the Hog Back slide. Although the foliation dips into the slope, these outcrops show numerous block pull-outs which appear to be joint-controlled.

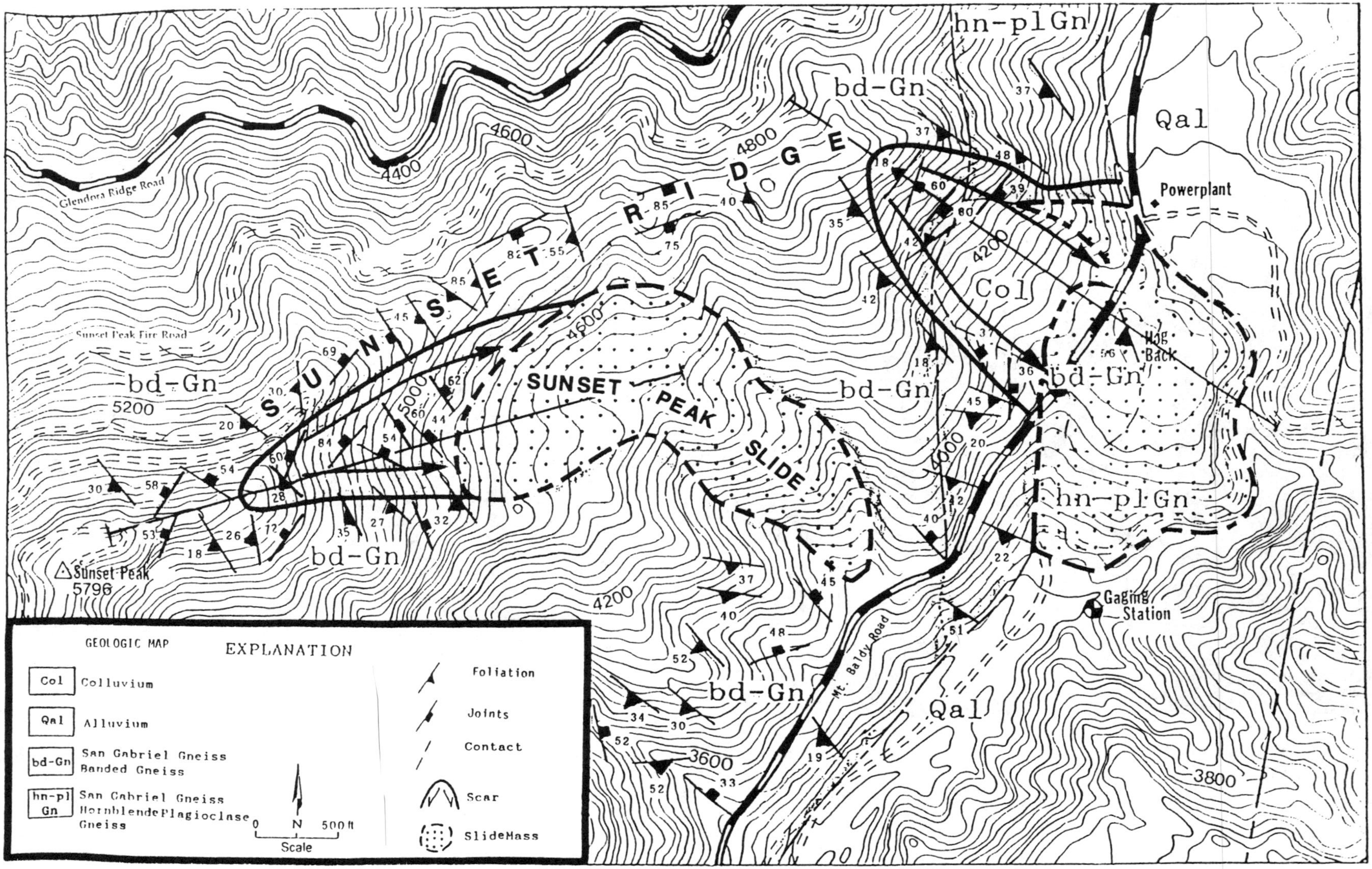

Figure 4. Geologic and topographic site map of the Sunset Peak and Hog Back slide areas along the western flank of San Antonio Canyon. Cross section lines delineate the trends of the sections used in the back-analyses presented in Figures 5-10 and 14-20.

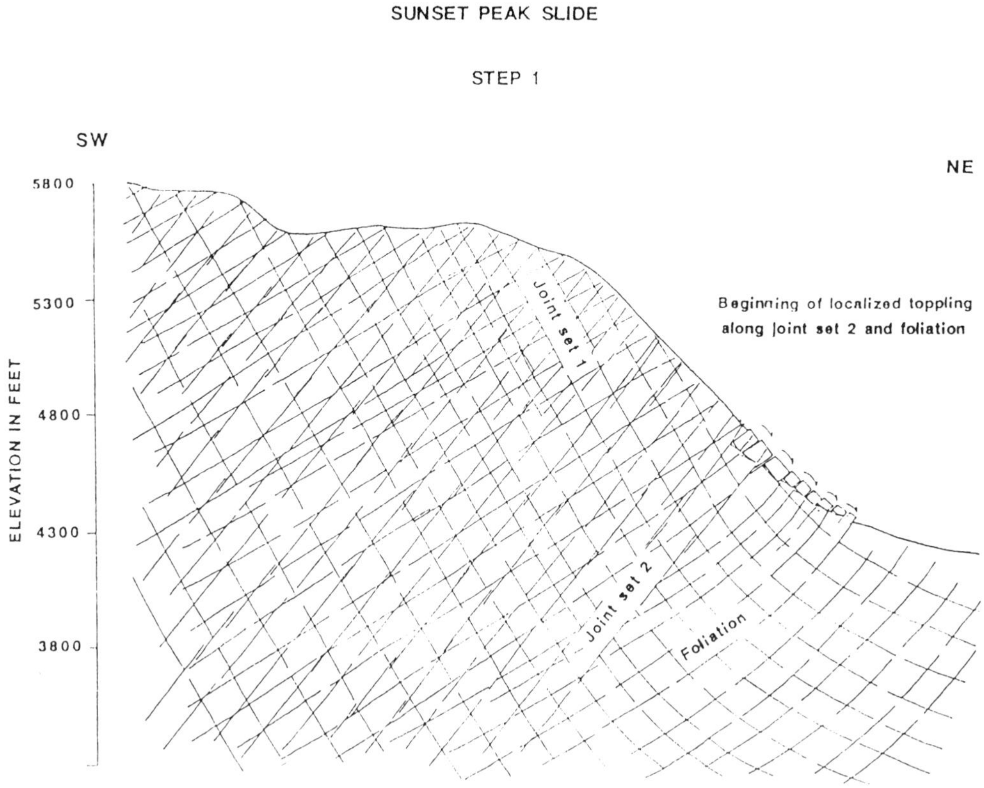

Figure 5. Initial pre-slide condition of the Sunset Peak slide, as con structed by Hibner (1990). The apparent inclinations of two primary systematic joint suites and regional foliation are presented, based upon projection of mapping along and within the slide scar. After applying gravity forces, initial interblock deflections are predicted at the toe of the slope, as shown.

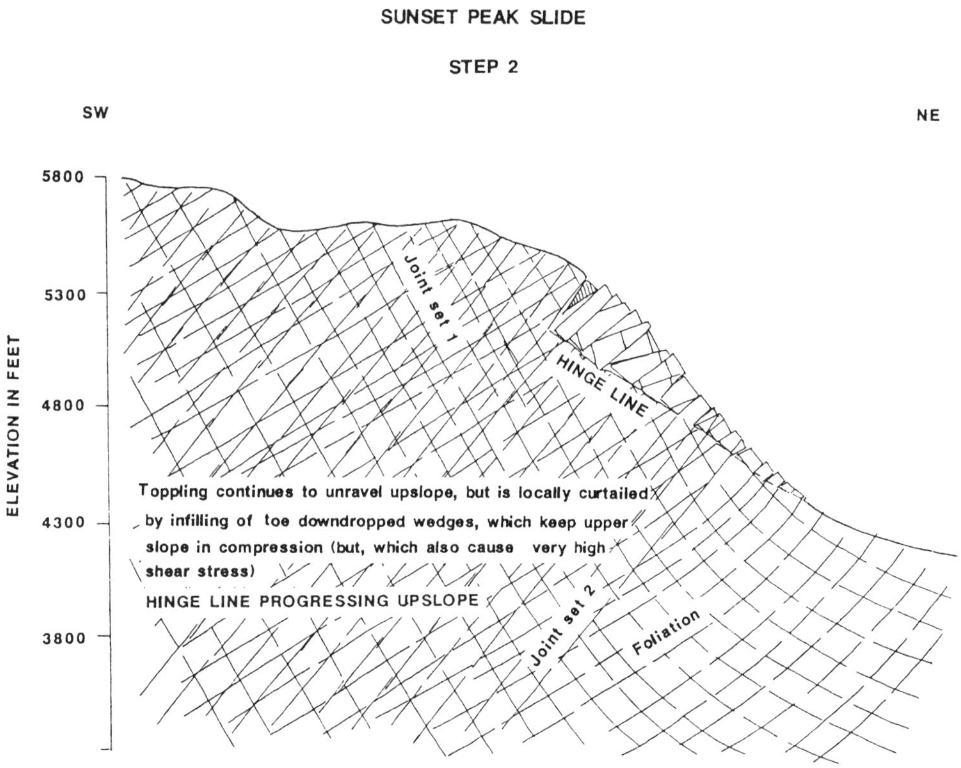

Figure 6. After applying successive load increments, a toppling mode of failure is predicted along joint set 2. Blocks bordered by joints sets 1 and 2 tend to wedge into the interblock depressions, even though the analysis is somewhat biased by the joint spacings selected. A plastic “hinge line” eventually forms at the base of the affected zone which even tually develops very high shear stresses due to mechanical translation.

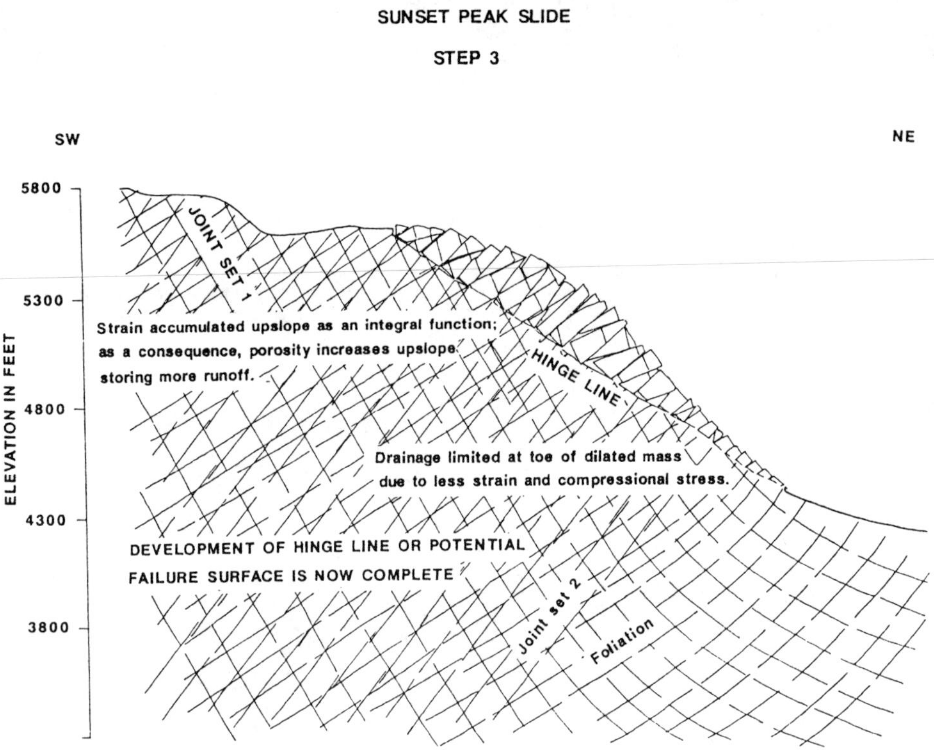

Figure 7. As toppling continues, interblock strain is translated upslope as an integral function. In this manner, the sum of all interblock strain is continually accumulated upslope, causing the largest deflections to occur at the crown of the affected area. Such an affected mass could be expected to dilate appreciably, with the greatest porosity being towards the head and the least towards the toe. In this manner, mass infiltration at the upper end could be expected to outpace that at the lower, less-deformed end, thereby leading to net accumulation of moisture during periods of intense precipitation.

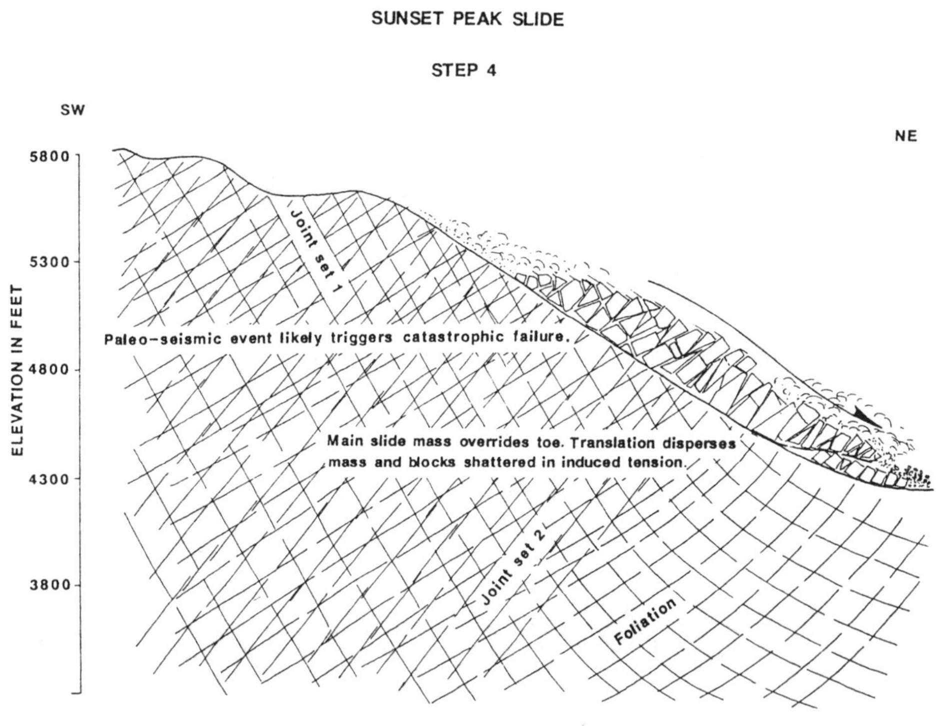

Figure 8. When sufficient moisture were accumulated or excitation by seismic paleo seismic events, a net rise in pore pressures would eventually exceed the normal effective strength mobilized along the hinge line. A catastrophic failure appears to have ensued, disaggregating and displacing the entire mass a considerable distance downslope.

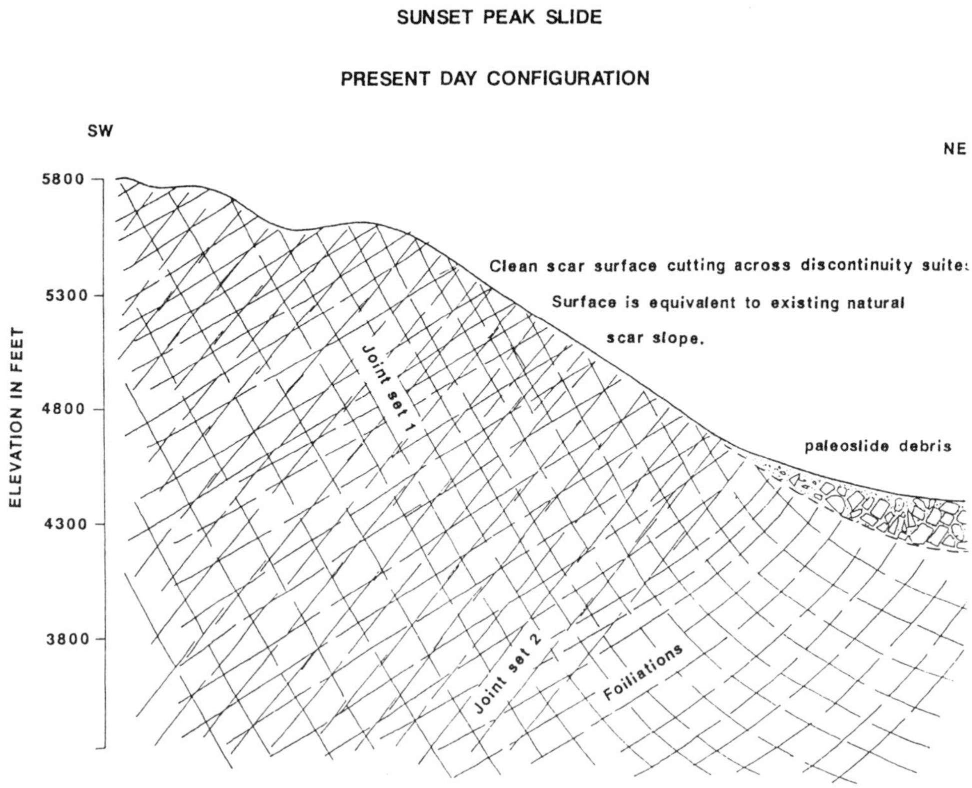

Figure 9. Present-day configuration of the Sunset Peak slide in cross section by Hibner (1990). A clean evacuation scar is observed and paleoslide debris is neatly distributed along a curved, lenticular corridor. Minor colluvial infilling of the upper debris area masks details of those materials and testifies to the relative old age of the slide.

Figure 10. View looking westerly, at the Sunset Peak slide and evacuation scar, from across San Antonio Canyon, on Spring Hill. The slide scar extends almost to the crest of the peak, extending to an elevation of approximately 5600 feet. The debris run-out turned downslope, and is now found as low as elevation 3650. White dots trace the approximate limits of sliding.

Figure 11. Aerial oblique view of the Hog Back and Sunset Peak slides by John Shelton. White dashes indicate the approximate boundaries of the slide mass while the dots delineate the limits of the evacuation scar. The white arrow indicates the approximate thrust of the movement, as discerned from the structure preserved in the slide mass at the lower center of the photo. The Hog Back slide dammed San Antonio Canyon sometime during late Pleistocene or early Holocene time, and the new channel is incised into the granitic gneiss exposed on the eastern side of the Canyon.

Figure 12. The Hog Back slide with evacuation scar at right (west), show ing existing Mt. Baldy Road and the old road along the overflow gorge at left (east) of the slide. White arrows indicate the approximate motion track of the debris. The flat area in the lower right portion of the photograph is built of gravels retained in the lake that formed behind the landslide dam. Photo by John Shelton, used by permission of W.H. Freeman & Co.

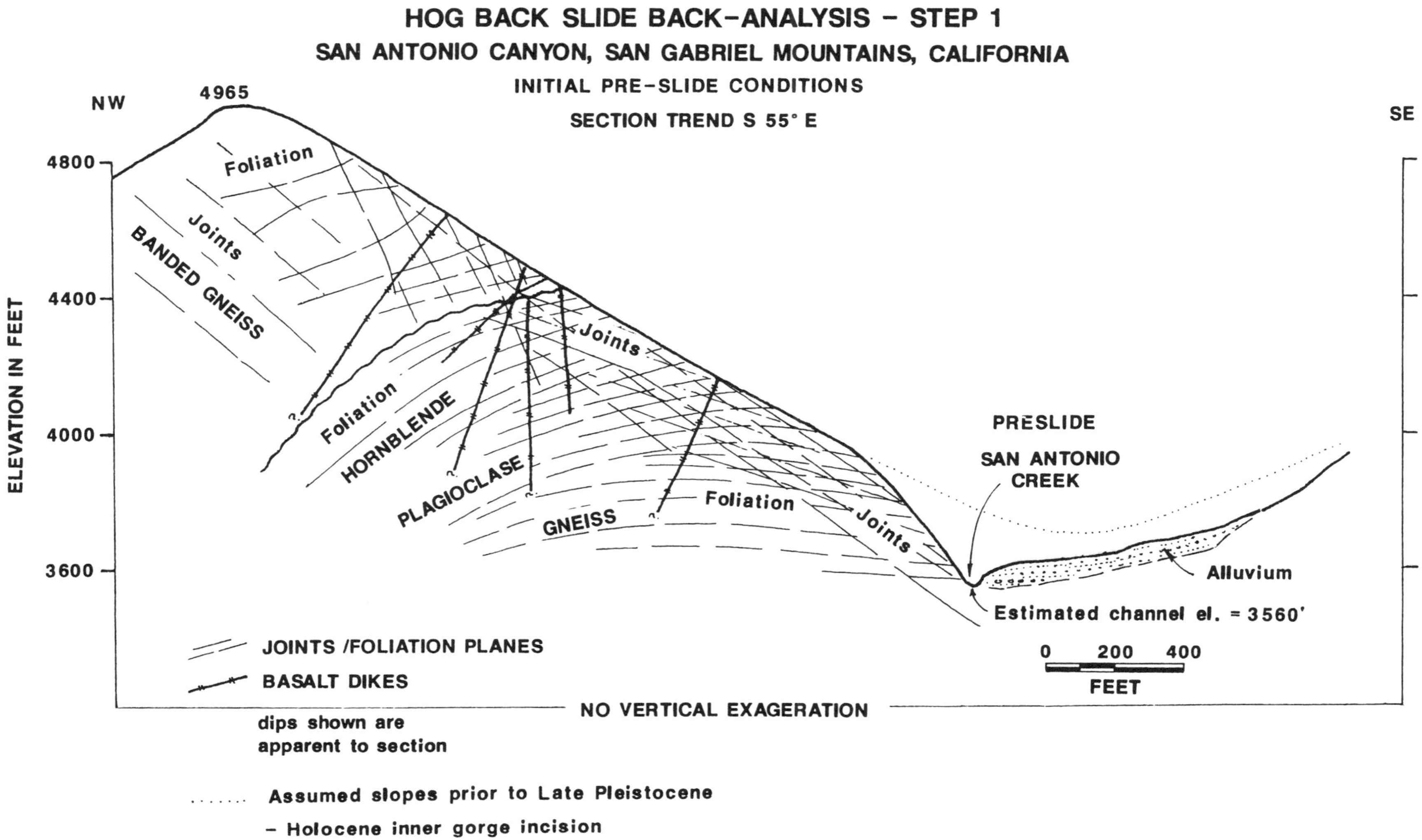

Figure 13. Initial pre-slide conditions at the Hog Back slide, as developed by Herber (1987) and Herber et al (1991). A complex structure is suggested, with foliation of the hornblende-plagioclase gneiss progres sively shifting out-of-slope as the foot of the original slope is approached. Though less numerous, secondary valley-side joints also perturb the ground mass, particularly in the inner gorge area.

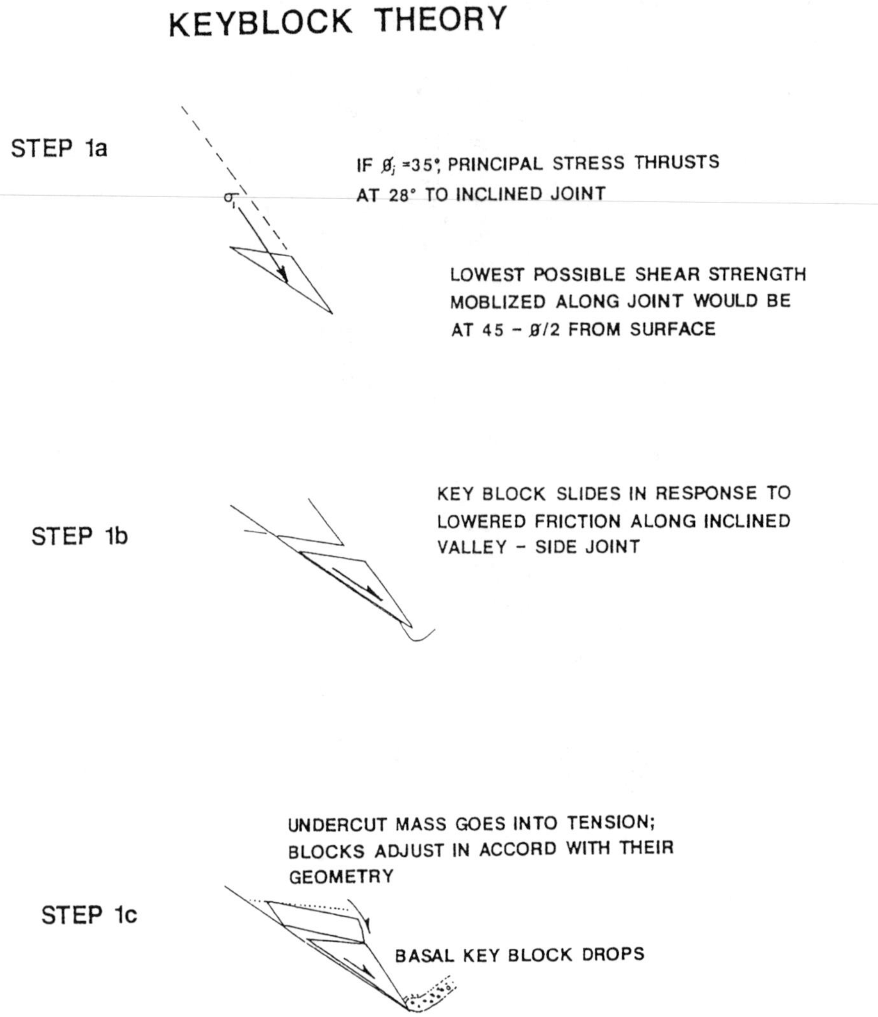

Figure 14. Initial block movement assessments revealed by the computerized keyblock analysis. Keyblocks bounded by out-of-slope dipping foliation and valley-side joints were identified in the analysis. Discrete interblock deflections are allowed to rationally predict the each succeeding step.

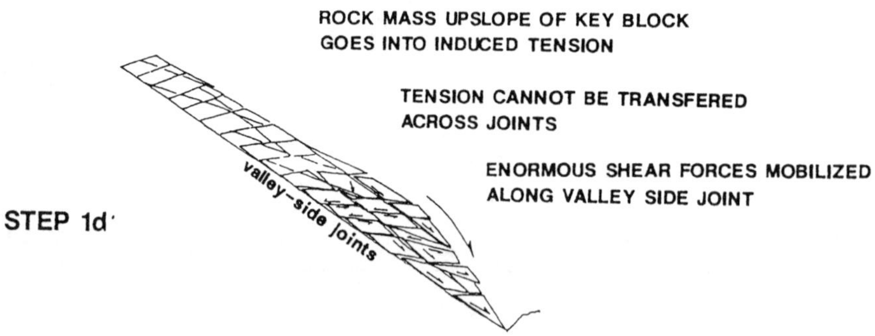

Figure 15. As the boundary constraint of each block is removed, block deformational analysis tracks the progressive distortion of the mass as a whole. Very large point-to-point stresses are predicted between the tilted blocks, presumably causing spalling, shattering and smoothing of asperities as the mass dilates and begins to creep downslope.

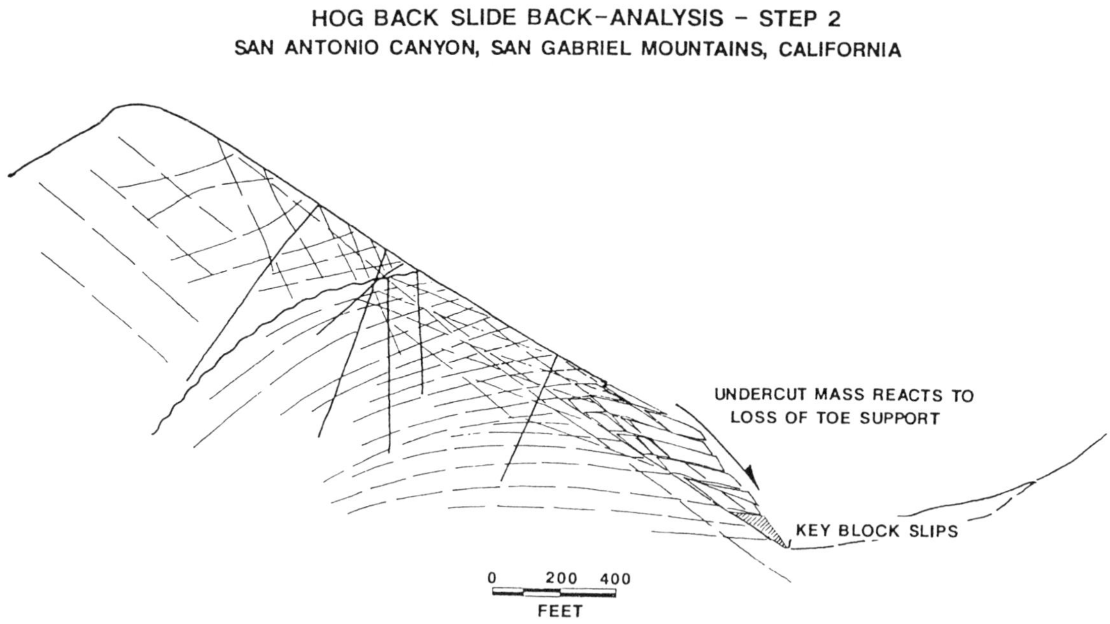

Figure 16. Block deformational analysis of the Hog Back slide mass as the basal keyblock is allowed to slip into the incised channel of a paleo-San Antonio Creek. If the keyblock is sufficiently large or strong, it may not shatter, and could serve to buttress the remaining slope, depending on how far it travels before becoming wedged. The early keyblock analyses used in this study cannot model block scale on more than just a gross, qualitative scale. New forms of the analysis are now being developed which can accommodate discontinuity spacing variances.

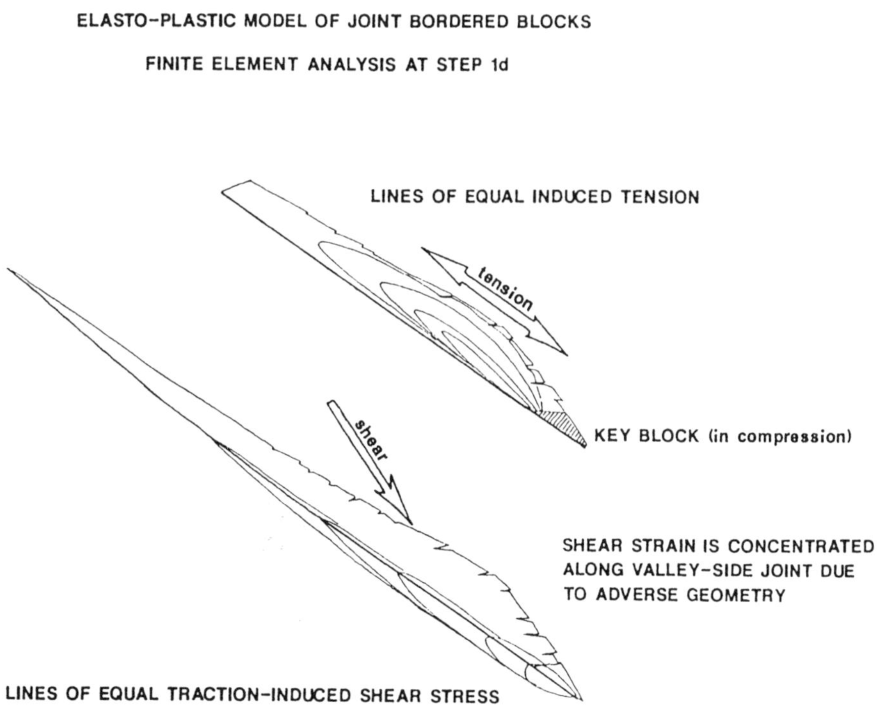

Figure 17. Finite element model predictions of stress distributions after the basal keyblock was allowed to translate an appreciable distance (one half the block length). Although the keyblock is in compression, the dilated mass above it appears to be in tension, an logical result within such an intensely jointed media. Shear stresses are predicted along the valley-side joints forming the "hinge line" between the dilated creeping mass and the underlying, undisturbed rock. These shear stresses are generated by traction along the creep boundary.

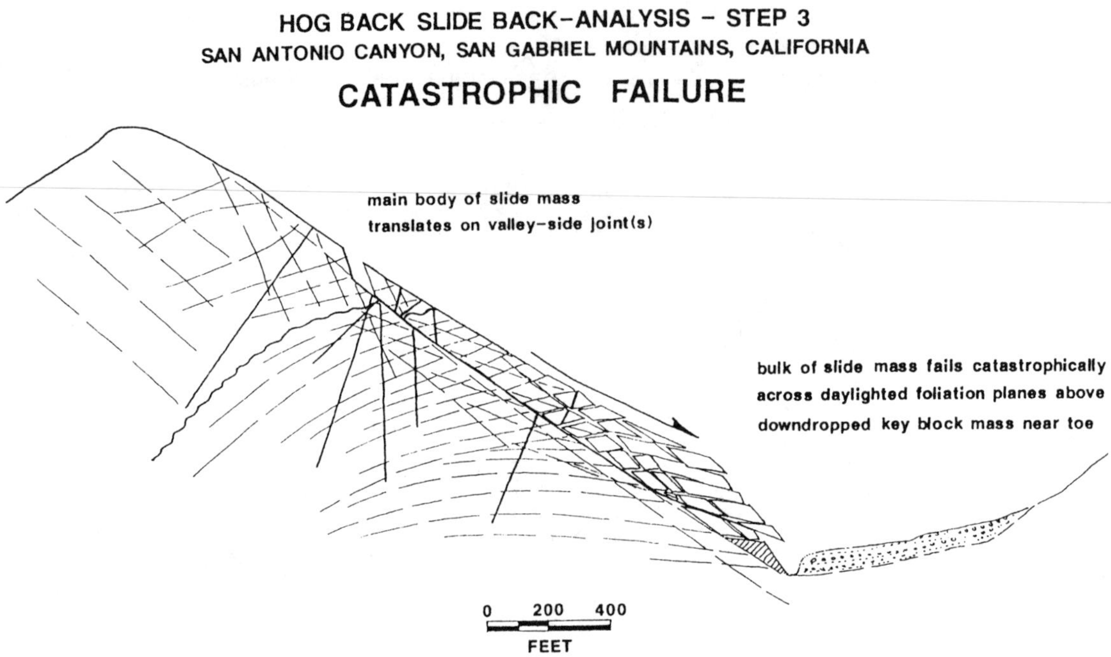

Figure 18. Hypothesized catastrophic failure of the Hog Back slide, based on progressive block deformational analysis. The bulk of the slide mass appears to have failed across daylighted foliation planes above the downdropped keyblock mass near the slope toe.

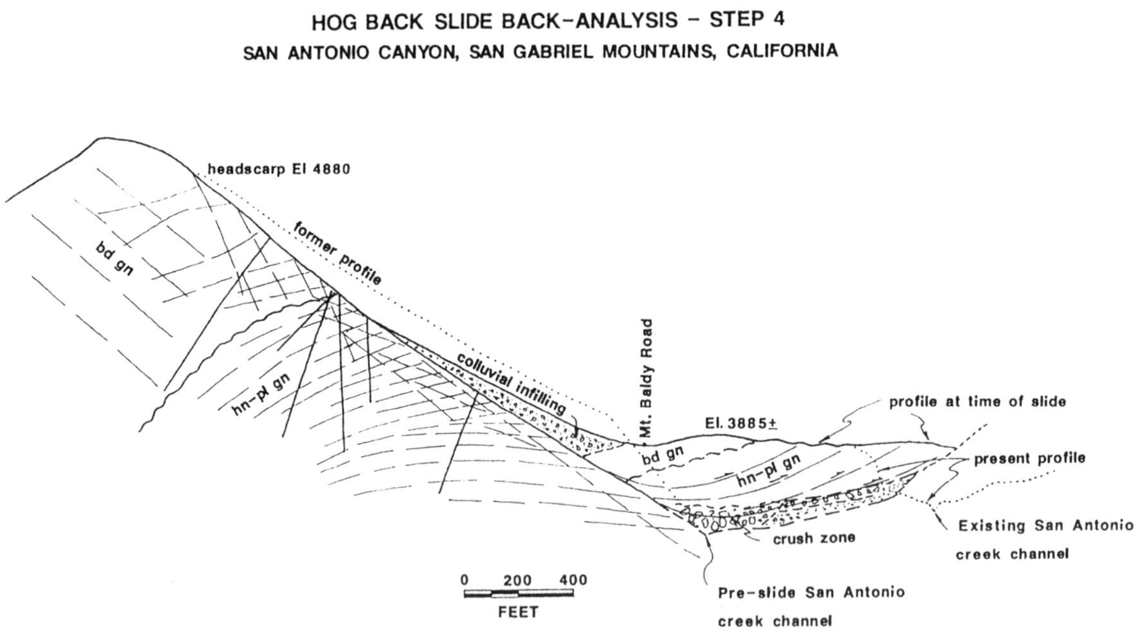

Figure 19. Existing conditions at the Hog Back slide, as determined by Herber (1987). The hypothesized profile at the time of failure can be compared with the existing channel at the extreme eastern toe of the slide. Note how the contact between the banded San Gabriel and hornblende- plagioclase gneiss was conformably translated downslope, suggesting a block glide mechanism of failure. The thick breccia zone at the base of the slide debris attests to the rapid and catastrophic nature of the failure.

DRAINAGE BENEFIT ASSESSMENT AREAS IN LOS ANGELES COUNTY: A CASE HISTORY OF MITIGATIVE DESIGN FOR LANDSLIDES

Robert A. Larson
District Engineering Geologist
Los Angeles County Department of Public Works
900 S. Fremont Ave., Alhambra, CA 91803

INTRODUCTION

Drainage Benefit Assessment Areas (DBAA) are used in Los Angeles County to tax property owners for operation and maintenance of surface or subsurface drainage facilities. Four of the existing twenty-eight DBAA in the County consist of subsurface facilities utilized to mitigate an increase in groundwater elevation. This paper discusses these four DBAA, their histories, and the problems and resulting solutions associated with them. A DBAA is utilized when essentially an entire subdivision is affected by a potential landslide hazard and passive mitigative design is either not feasible or sufficient. The DBAA is an active mitigative design that includes periodic monitoring, maintenance, and repair of drainage facilities, and if necessary, additional geotechnical analysis. A manual of operation and maintenance procedures tailored to the requirements of a DBAA is prepared and utilized to run each program. Preliminary guidelines for a monitoring and maintenance manual are included in Appendix A. The system is proposed by the geotechnical consultant and reviewed by the Department of Public Works prior to approval of the project.

L.A. County Improvement Districts are also property owner tax assessment areas, but they are established to maintain, repair, or dewater an existing, active geologic hazard. Improvement Districts have been established for Big Rock Mesa and Calle de Barco landslide areas. A DBAA is used to maintain a drainage system in order to preclude the occurrence of a hazard.

The four DBAA, numbered 3, 6, 18, and 24, are located in Canyon Country, northwest of Solemint Junction (Fig. 1 and 2). DBAA #3 is situated near the intersection of Whites Canyon Road and Nadal Street. DBAA #6 and #18 are situated east of Whites Canyon Road and north of Shangri-la Drive. DBAA #24 is presently proposed in Plum Canyon (Fig. 3).

Formation of a DBAA

The DBAA is a property owner tax assessment program created under the Benefit Assessment Act of 1982, Chapter 6.4 of Part 1, Division 2 of Title 5 of the Government Code of the State of California. The DBAA is formed by resolution of the local legislative body. Each year, a report must be submitted to that body that discusses the operation and costs of the preceding year, and estimates costs for the next year of operation. The cost of operating the DBAA varies from year to year, depending on monitoring levels, repair costs, pumping requirements, and inflation. A public hearing is held after appropriate announcements are given, usually in local newspapers. After the meeting is held, a resolution must be passed in order to assess the property owners and continue the functioning of the DBAA. Property owners are assessed a fee directly on their property tax each year. Initially, the developer consents to form the DBAA and pays the tax until the dwellings are constructed and occupied, and the new owners receive their first tax bill. Owners of common areas are also assessed, based on the size of the parcel. The money received from this tax pays for the operation (including maintenance, repair, and administration) of the assessment district, and cannot be used for any other purpose.

Regulation of a DBAA

The Department of Public Works, the Department of Regional Planning, and the Board of Supervisors (or City Council) are involved with the establishment of a DBAA in Los Angeles County. The units within the Department of Public Works that have responsibility for regulating the DBAA include: 1) Geology and Soils Development Review Section of the Materials Engineering Division; 2) Construction Division; 3) Flood Maintenance Division; 4) Planning Division; and 5) Subdivision Section of the Land Development Division.

The L. A. County Geology and Soils Development Review Section is the lead unit in establishing the need for and administering the creation of a DBAA. The geotechnical aspects of the dewatering systems and monitoring requirements are subject to review and approval by this section. The Flood Maintenance Division reviews plans for the system as the agency responsible for maintaining the system over the life span of the project, presumably 50 years. Their recommendations are included in the plans to ensure feasibility of maintenance. The Construction Division reviews the plans for type of construction materials, manhole and vault design, and electrical requirements. The Planning Division processes an application and provides legal documentation for the assessment area. The Subdivision Section reviews the Tentative Map and Final Map for compliance with the subdivision code. The Department of Regional Planning reviews the Tentative Map for compliance with the planning policies of the County and establishes the conditions of approval, including the requirement to form a DBAA. They present the Tentative Map and its conditions of approval to the Board of Supervisors for final approval. The formation of the DBAA is one of the many conditions set by Regional Planning.

The DBAA were originally formulated assuming that the County of Los Angeles, through it's Department of Public

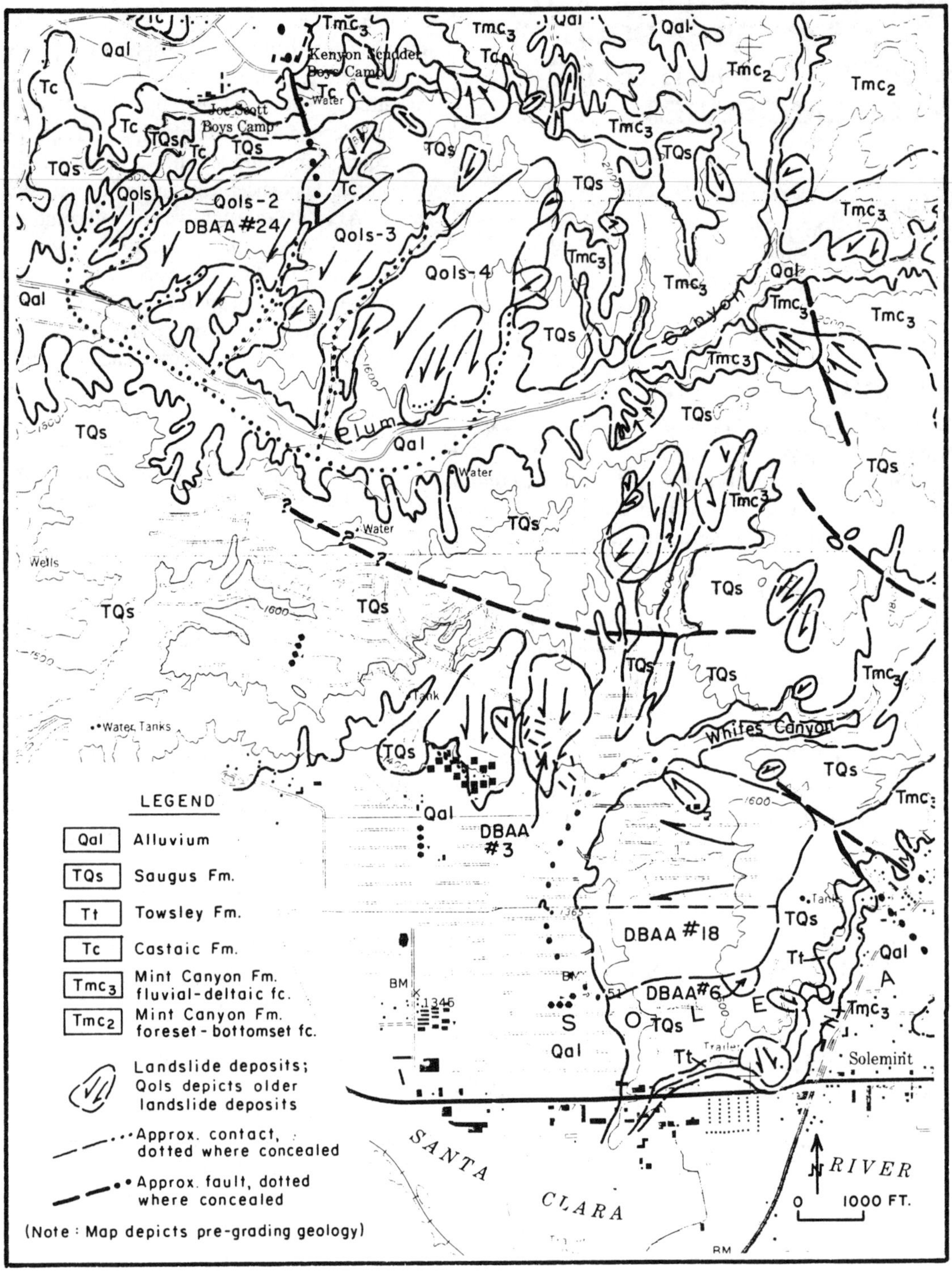

GENERALIZED GEOLOGIC MAP OF THE PLUM CANYON–WHITES CANYON AREA, CANYON COUNTRY, LOS ANGELES COUNTY, CALIFORNIA

(Modifed from Saul, 1990, Larson, 1990, landslides interpreted by author; Base map is U.S.G.S. Mint Canyon Quadrangle, 1988)

Figure 1. Geologic map of Canyon Country, depicting the landslides discussed in the paper. The Plum Canyon landslide is Qols 1, 2, 3, and 4 in the northwest corner of the map. Note that much of the boundary is buried by alluvium. The other landslides are depicted by their DBAA number. The dashed line on the DBAA #18 landslide (Shangri-la) approximates the northern boundary of the shear key.

Figure 2. Aerial photo looking east at the Shangri-la landslide. The crown of the slide is the prominent slope below the water tanks in the center of the photo. The toe (present contact with alluvium) corresponds to the toe of slope of the graded area of DBAA #18 and to the prominent break-in-slope to the north. The approximate side boundaries correspond to the northern limit of houses and to the southern limits of the recently-graded area. The locations of DBAA #3 and #6 are also depicted.

Figure 3. Aerial photo looking northwest at the Plum Canyon landslide complex. The landslide corresponds with the flat, erosional surface on the north side of Plum Canyon. The houses at the edge of the photo, west of Qols-2 are situated on a buttress remnant of Qols-1. The proposed Whites Canyon-Plum Canyon road will extend from the dead end road, just below and east of the water tanks, across the ridge and into Plum Canyon, where it will cross the landslides and join with the existing road.

Works, would retain responsibility for the programs. The County, through its Flood Maintenance yards, is experienced at maintaining a wide variety of civil engineering works and the maintenance required for a DBAA is well within their capabilities. However, the City of Santa Clarita was formed in 1987 and assumed responsibility for DBAA #3 and #6 on December 15, 1987 and then proposed DBAA #18 on March 2, 1989. The Santa Clarita City Council formed DBAA #18 by resolution on June 27, 1989. Presently, the boundary between the City of Santa Clarita and L.A. County coincides with the western boundary of DBAA #24, which has been proposed for annexation to the City. The City subcontracts their public works administration to a private consultant and has subcontracted the County Flood Maintenance Division to perform the monitoring functions for the DBAA. Thus, a more complex chain of responsibility for the DBAA now exists than originally intended.

Advantages and Disadvantages

There are several advantages and disadvantages inherent in the existing system. Advantages include the following: 1) the property owners directly affected by the potential geologic hazard pay for the mitigation; 2) the property owners are directly assessed on their tax bill each year; 3) a secure source of funds are available to the responsible legislative body to administer the programs; 4) the programs cannot be terminated by the property owners, but only by the ruling legislative body; an action that is not likely; 5) an active program exists to ensure stability of the property.

Some of the disadvantages are: 1) in the event of unanticipated or changed conditions that require additional funding above the standard yearly level, the Flood Maintenance Division or other entity responsible for the system would have to request the funding increase from the ruling legislative body. The funding decision thus becomes political, since the affected property owners may express their objections to a funding increase at the required public hearings; 2) the increase in funds would not be available until the next tax collection period, which could be a year away; 3) the governmental body cannot relinquish responsibility for the systems and would have to fund the increases until tax revenue is available; and 4) the governmental body in whose jurisdiction the DBAA is situated becomes responsible for the maintenance and repair of the drainage system, and thus, the stability of the site.

GEOLOGY

Geologic Setting

All the sites are situated in major drainages north of the Soledad Canyon between Mint Canyon and Bouquet Canyon (Fig. 1), some 26 miles northwest of the City of Los Angeles. Three of the DBAA are located in the Soledad Canyon drainage; the fourth is in the Plum Canyon drainage. The sites are underlain by Quaternary to Oligocene sedimentary rocks of the Soledad Basin. The crystalline basement rocks are Precambrian gneiss and Mesozoic granitoids (Ehlig, 1981).

The sedimentary units in the area of concern, from youngest to oldest, are the Saugus, Towsley, Castaic, and Mint Canyon Formations (Saul and Wootton, 1983). These formations have never been deeply buried and retain much of the engineering characteristics of a moderately-consolidated sediment, except where cementation has occurred. Lacustrine, shallow marine, brackish-water, fluvial, and alluvial fan depositional environments are represented. The fine-grained sediments associated with these environments appear to have provided the weak plane of resistance along which the landslides have failed.

Structurally, the sites are located near the axis of the post-Miocene Soledad Basin synclinorium. Generally the formations dip to the west at Solemint Junction (southeast corner of Fig. 1), changing dip toward the south at the junction of Plum-Bouquet Canyons. East-west folds disrupt this general trend and dips range from about 5°-30° whereas slopes in the area vary from 3°-35°. Daylighted-bedding conditions occur in many of the slopes.

The sites are situated between the San Andreas Fault and the Sierra Madre-San Fernando-Santa Susana Fault zones and periodically experience strong ground shaking from earthquakes. Both of these fault systems have generated significant earthquakes that could have induced initial failure of the landslides. Epicenters for the 1971 San Fernando earthquake and aftershocks as well as the 1991 Castaic Lake earthquake occurred near the sites. The San Gabriel and Elkhorn (Green Ranch) Faults are situated closer to the sites, but appear to be less active than other major faults in the area. The potential for seismically-induced reactivation of the landslides was evaluated during the design of the projects.

Quaternary History

A broad piedmont sloping gently southward from the Sierra Pelona to the Saugus basin, formed, since onset of significant climatic fluctuations some 2.5 ma (Morrison, 1991), which received sediment from the north and east. Since cessation of Saugus Formation deposition some 400 ka (Levi and others, 1986) the area has been uplifted, degraded, and aggraded to it's present form. The major drainages within the Soledad Basin took on the approximate form we see today, controlled by antecedent drainage from the San Gabriel mountains and Sierra Pelona. Tectonism in the Ventura Basin as well as along the San Andreas, San Gabriel, and the Sierra Madre faults in conjunction with the largest Quaternary climatic changes associated with the last four glacial-interglacial cycles resulted in repeated incision and deposition that left a series of strath and alluvial terraces in the major drainages. Evidence of frequent stream capture, the presence of wind gaps, thick alluvial and slope lag deposits, and a multitude of old landslides characterize the western Soledad Basin.

McMillan and others (1991, their Fig. 1), based on eustatic sea-level curves, suggest that latest incision of canyons near the Ventura coast occurred 70-18 ka, followed by aggradation from 18-13 ka. The Soledad Basin is 50 miles inland, thus the geomorphic response may have temporally lagged behind the response at the coast. Bull (1991) provides evidence of a major aggradational event in the Mojave Desert from 11-9 ka. The toes of the two largest landslides, Plum Canyon and Shangri-la, both overlie alluvium and are overlain by alluvium, thus failure appears to have occurred prior to the last major aggradational event that filled the canyons.

This suggests an age of major landslide movement similar to that of many other landslides in southern California (Stout, 1991), probably 18-11 ka. Based on similar arguments, Oborne and others (1990) suggest failure of the Plum Canyon landslide between 15-12 ka. However, the landslides may have failed during an earlier cycle of channel downcutting, lateral planation, and aggradation. Bull (1991) documents a very complicated Quaternary history of incision and deposition in the drainages of the San Gabriel Mountains just south of the sites. Since failure, the climate has fluctuated greatly, but overall changed from temperate (?) to semi-arid, with a corresponding decrease in groundwater storage.

Description of the Landslides

The DBAA #3 landslide (Figures 1, 2, and 4) is some 2000 feet long, up to 1000 feet wide, and is much narrower at the toe then the crown. The landslide apparently failed southward along bedding planes, resulting in backward rotation of the beds in the toe area. The slide plane in the vicinity of the structures is essentially flat. The toe of slope was probably eroded by drainage from Whites Canyon, leading to failure.

The DBAA #6 landslide (Fig. 1, 2 and 5) is some 400 feet long and 700 feet wide. This landslide failed northward along bedding into the graben area of the Shangri-la landslide (DBAA #18). The disrupted landslide materials appear to have a higher permeability than the buttress fill, resulting in the perching of groundwater within the retained landslide despite the heel drains in the shear key.

The DBAA #18 and #24 landslides have been informally named the Shangri-la Landslide and the Plum Canyon Landslide, respectively (Larson, 1990). Both landslides occur in similar geologic environs, yet demonstrate vast differences in morphology and internal structure. Both slides involve displacement of the Saugus Formation with the main failure plane at the contact or just within the underlying Towsley or Castaic Formations. The morphological differences arise from the amount of displacement that each slide experienced. The Shangri-la landslide is shown in Figure 2 and the Plum Canyon landslide is shown in Figure 3.

The Shangri-la landslide is situated on the western slope of a northerly trending, terrace-capped ridge, adjacent to and northwest of Solemint Junction. Ramirez (1990, and this chapter) provides a detailed description of the development of this landslide which is some 3500 feet long and 3200 feet wide. The slide moved westward over alluvium into the partially-filled erosional meander of the Santa Clara River. The main slide plane, up to 135 feet beneath the surface, occurs in blue-grey siltstone and sandstone, some ten feet below the oxidized-unoxidized contact within the Towsley Formation. An alternate hypothesis places the failure zone within the Sunshine Ranch Member of the Saugus Formation. The undulatory shear zone, which commonly consists of 4 inches of fat clay, dips westerly at 6°-7°.

Deformation is characterized by a prominent headscarp, closed depressions on the surface, internal thrusting, back rotation of beds, radial horst- and graben-style shearing, and extension over alluvium. Secondary, compound landsliding is significant. The main graben was filled with up to 20 feet of colluvium (California GeoSystems, 1989). The closed depressions contained lacustrine sediments.

The Plum Canyon Landslide complex is situated on the northern slope of Plum Canyon, a tributary to Bouquet Canyon. The landslide complex, approximately 1 mile long and 2 miles wide, moved southward toward and locally into the westerly-draining channel of Plum Canyon. The slide

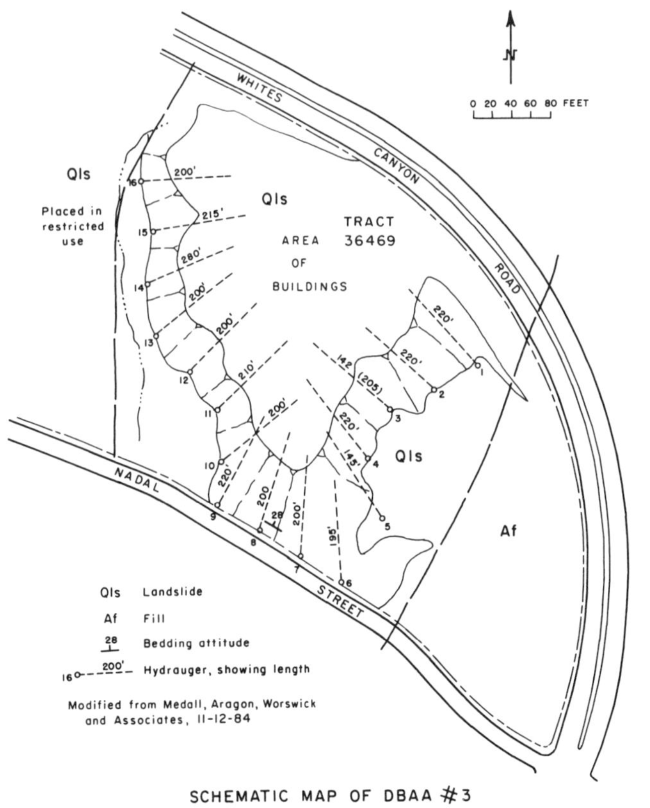

Figure 4. Map of DBAA #3 depicting location of the horizontal drains and general physiographic features.

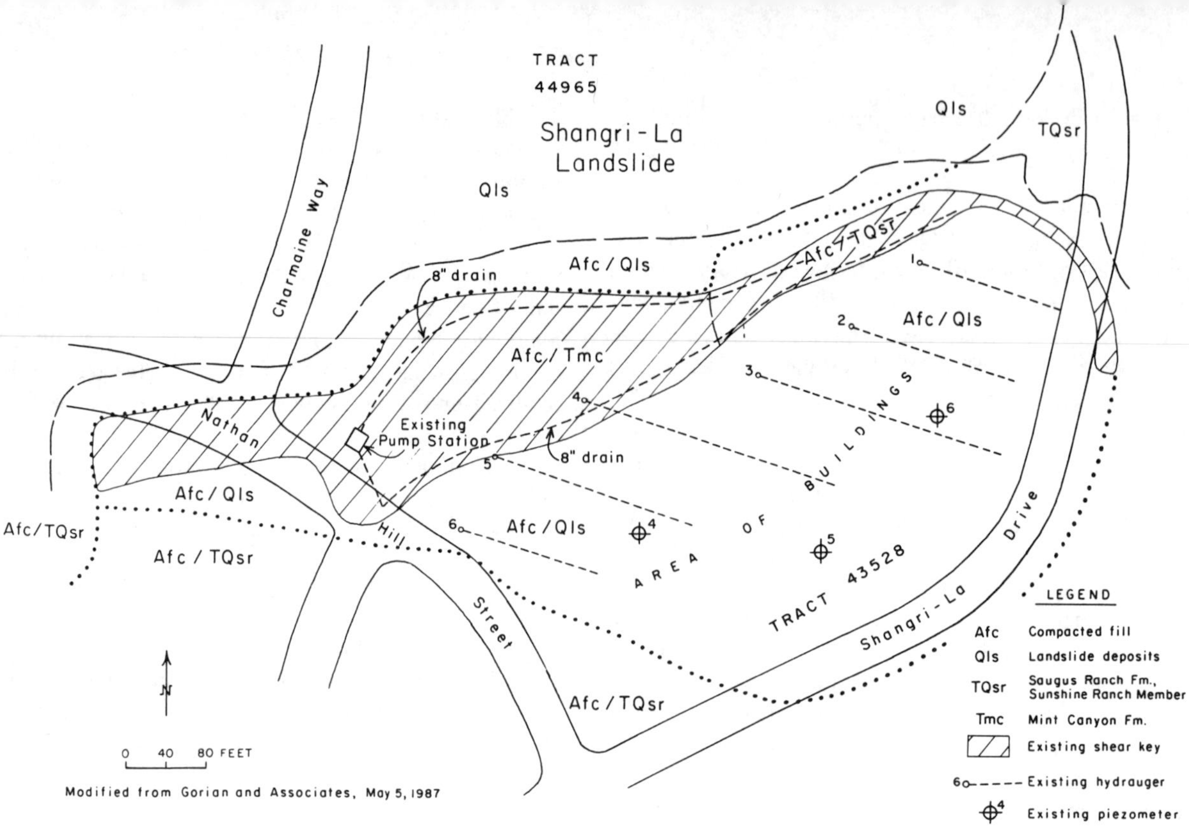

SCHEMATIC MAP OF DBAA # 6

Figure 5. Map of DBAA #6 prior to recent grading on Tract 44965 north of the shear key. The outlets of horizontal drains #1, #2, and #3 have been connected to a 4-inch drain and are now under fill. A new monitoring well has been installed in the shear key some 40 feet north of horizontal drain #5 outlet.

plane, exposed near the crown is a 1-inch thick, brown, clay that dips southerly at approximately 10 degrees. The slide has a maximum thickness of 300 feet and has locally overridden alluvium (Leighton and Associates, 1985).

Oborne and others (1990) present a detailed report on the landslide. Bedding within the slide mass is remarkably homoclinal, undeformed and essentially parallel to the failure plane. A syncline beneath the toe of the slide brings the failure plane to the surface in Plum Canyon. Northward dipping beds at the toe support the presence of the syncline, but could also indicate local rotation. The crown is morphologically subdued, and graben-filling sediment has not been found. The complex is divided (by hypothesis) into four large landslides (Qols - 1,2,3,4) along secondary, alluvial-filled drainages. Erosion has resulted in removal of most of the crown area of the slide, which is presently represented by the terminus of the flat-topped ridges. Plum Canyon is bowed southward around the toe of the slide suggesting somewhat greater movement than indicated by internal deformation and geomorphic features. Abundant secondary landslides occur on and around the margins of the slide complex which appears to have failed as a coherent block.

The Groundwater Problem

The sites are currently situated in a semi-arid environment; the corresponding groundwater levels are low. Some perched water was observed in the Plum Canyon and Shangri-la landslides. Water levels in the Shangri-la landslide appeared to decrease from the time of initial exploration to the grading phase of development, resulting in reduced in-grading dewatering from what was anticipated. Much of the subsurface exploration for both landslides was conducted during the drought conditions of the late 1980's.

Water introduced into the subsurface subsequent to development, principly from irrigation, often perches on low-permeability layers including slide planes within landslides. Irrigation can be anticipated to add 60-100 inches per year of equivalent precipitation in unpaved areas in addition to normal annual rainfall. The resulting rise in the elevation of the water table results in hydrostatic pressure creating an upward buoyant force that lowers the resisting force and the factor of safety. The policy of the County of Los Angeles requires an applicant, through their geotechnical consultants, to demonstrate a static factor of safety of 1.5 and a pseudostatic factor of safety of 1.1 for all slopes for the life of the project. County geologists have recognized that groundwater storage often increases in landslides due to irrigation after development and that standard subdrains can and do fail. Recognition of these two principals have resulted in the extensive groundwater monitoring and dewatering systems of the DBAA.

The prediction of groundwater conditions in "dry" landslides is not trivial and requires estimation based on experience in other landslides, an understanding of the physical phenomenon, and laboratory analysis of representative samples. Consideration of anticipated groundwater conditions in the design of the drainage facilities increased with each new project. The following discussion is based on proposed DBAA #24.

A hydrologic model is constructed. Since the landslides are presently "dry", and were once presumably saturated with water, subsurface discharge must have occurred in conjunction with decreased annual rainfall as the climate changed. The discharge may have occurred over a long period of time (1-10 ky) when compared with the lifespan of the project, therefore, the assumption of no subsurface discharge appears reasonable with regard to anticipating future groundwater conditions. Hydraulic conductivity can be estimated from grain size analysis and from field observations. Variation of permeability and porosity, based on sediment-type and fracturing, can be observed and mapped in the field. An estimate of recharge can be made assuming irrigation and rainfall infiltrate in the areas of the landslide that will not be covered with concrete and structures. The maximum discharge rates of subdrains and wells can be calculated. Potential recharge is compared to maximum discharge to determine the sensitivity of the system. The monitoring and dewatering system must be designed to detect the rise in groundwater in sufficient time to allow for effective dewatering of the slide mass prior to groundwater rising above the factor of safety = 1.5 surface. The location of monitoring/pumping wells is sensitive to the predictions of the model since a cone of influence, the primary spacing parameter for dewatering, is not derived from well tests.

ASSESSMENT AREAS

Drainage Benefit Assessment Area #3

Assessment Area #3 is Tract 36469 and is situated between Whites Canyon Road and Nadal Street (Fig. 1 and 4). Drainage facilities consist of 16 horizontal drains with lengths from 142 feet to 280 feet. Lots 1 and 2 of Tract 36496 were deeded in their entirety as easements for installation and maintenance of the drains. The horizontal drains were installed to preclude the water level from rising above a diezometric surface assumed in the static stability calculations in order to demonstrate a factor of safety of 1.5. There are 232 unit parcels in Tract 36496. The estimated annual cost in 1984 for maintenance of the system was $10,000 with an estimated family dwelling unit price of $43.00. The estimated annual cost in 1990 was $10,900 with a unit cost of $46.98.

Tract 36469 is a unit tract of Tentative Tract 36383, which was submitted to the Department of Regional Planning on January 4, 1978. Rough grading of the tract was approved on January 9, 1985. The DBAA was formed by resolution of the Board of Supervisors on September, 13, 1984. An operation and maintenance manual was submitted which outlined anticipated parameters for running the DBAA (Medall, Aragon, Worswick and Associates, 12/26/84). Prior to Final Map approval a performance bond covering the cost of installation of the horizontal drains was required for the amount of $57,600.

The drains were first serviced by Flood Maintenance in July, 1987. Maintenance consisted of rodding and flushing the drains on a yearly basis. Drain #2 had a measured flow of 18.2 gpm and drain #12 had a flow of 11 gpm. All other drains were dry at this time. In 1991, drains #1, #3, and #13 were found to be blocked at depths of 143 ft., 160 ft., and 180 ft., respectively. An attempt to redrill drain #13 resulted in a lost drill stem. Alternatives are presently being considered. Drains #1 and #11 drip water and drains #2 and #12 had measured flows of 0.25 gpm and 0.06 gpm, respectively, on August 20, 1991 (Wayne Snively, personal communication, 1991). These drains were constructed without benefit of a paved, surface drain to direct water to a street drain. A few of the drains regularly intercept irrigation water and as a result the grass field at the toe of slope is often wet and muddy. Groundwater levels within the landslide can only be estimated, since monitoring wells were not included in the design.

Drainage Benefit Assessment Area #6

Assessment Area #6 is Tract 43528 and is north of the intersection of Shangri-la Drive and Nathan Hill Street (Fig. 1 and 5). Drainage facilities consist of 4 monitoring/pumping wells, 6 horizontal drains, and a below-grade pump station for the shear key back drains. The drainage system was installed to preclude the water level from rising above a phreatic surface assumed in the pseudostatic stability calculations in order to demonstrate a factor of safety of 1.1. There are 122 units within Tract 43528. The estimated annual cost in 1987 for maintenance of the system was $7,000. The estimated annual maintenance cost in 1991 was $34,200 with a unit cost of $206.

The shear key (Fig. 5) supports landslide materials left in place on Tract 43528 and separates this remnant landslide from the larger Shangri-la landslide to the north. Eight-inch diameter heel and toe drains were constructed in the keyway and connected to a subterranean pump station. Two low-flow pumps direct water to an adjacent storm drain. One of the pumps may be removed for servicing without affecting the system. Six monitoring/pumping wells were originally planned, three in the shear key fill and three in landslide materials. Only the three wells in the landslide materials were constructed at the time of grading. The developer was given permission to install the other wells in the following year; however, additional grading to the north improved the stability of the area such that the wells were not deemed necessary by the consultant.

The three wells were constructed to a total depth of 10 feet below the slide plane and consist of 4-inch PVC pipe screened 5 feet below and above the slide plane. Six horizontal drains were constructed into the landslide materials. These drains were not deemed necessary by the consultant, but were required by the County. The drains were constructed with metal pipe outlets and paved surface drains that directed water to a storm drain. An access road and associated easement was required to provide maintenance access to the drains. The drains consisted of solid 2-inch PVC pipe within the fill and slotted pipe within the landslide materials.

Tract 43528 is a unit tract of Tentative Tract 32929, which was submitted to the Department of Regional Planning on August 1, 1983. Rough grading of the tract was approved on December 19, 1986. The DBAA was formed on November 21, 1985 by the Board of Supervisors. An operation and maintenance manual dated May 5, 1987 was submitted that outlined anticipated parameters for running the DBAA (Gorian and Associates, 1987). This manual was approved by the Geology and Soils Section on June 4, 1987.

The system was first serviced by Flood Maintenance in October, 1988. Water was measured in all three monitoring/pumping wells. Apparently, within a month of initial irrigation of the vegetation on the tract, water had infiltrated and perched on the slide plane. By May, 1989 the water level in well #5 had risen more than 10 feet above the base of the well, and the first "red flag" was reached. The "red flag" was a water level that required increased frequency of monitoring.

Because the monitoring wells had not been installed within the shear key, adherence to the monitoring and maintenance manual could not occur.

Flood Maintenance decided to take a proactive approach and to pump well #5. However, it was found that it was very difficult to lower the submersible pump into the well, due to slight bends in the alignment. It was also discovered that there was no way to determine the amount of water that the pump station was directing into the storm drain from the shear key drains. The water levels have not risen to the red line (factor of safety 1.1) and grading to the north on Tract 44965 improved the stability of the area. This grading placed additional fill against the existing shear key and covered horizontal drains #1-3. These drains were connected to solid 4-inch pipe constructed within the fill with an outlet at the slope face.

The water surface has continued to rise. In 1991, the water surface was 20 feet above the bottom (at the second "red flag" 10 feet above the slide plane) in wells #5 and #6, with less water in well #4. The pump station occasionally turns on to pump water out of it's gallery. In December of 1991, horizontal drains #4, #5, and #6 were flowing water. Because of the rising water table, the Flood Maintenance Division has installed one more monitoring/pumping well into the shear key and propose two more horizontal drains (Wayne Snively, pers. comm., 1991).

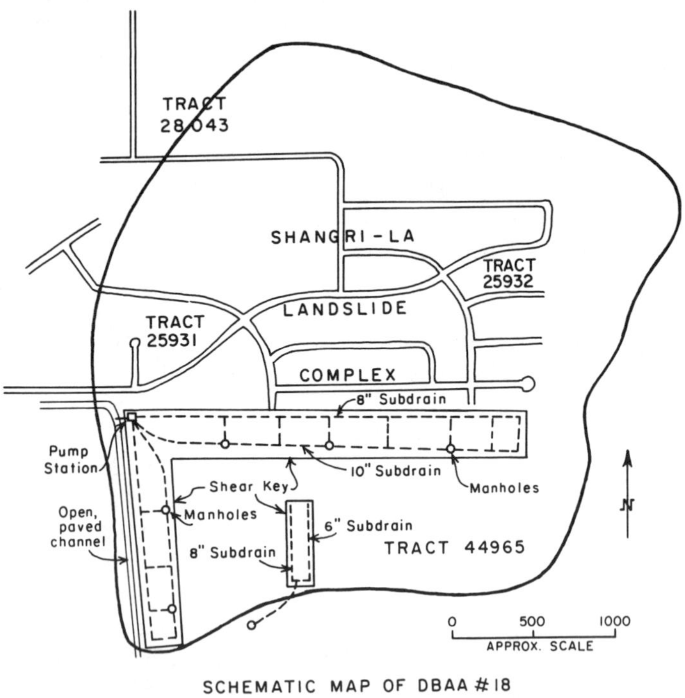

Fiure 6. Map of the Shangri-la landslide, depicting the general location of graded tracts. The landslide failed westward. The area of the slide encompassed by the shear key generally coincides with the boundaries of DBAA #18 on Tract 44965. The western boundary of the landslide is buried under alluvium. The northern shear key isolates the tract from the remaining landslide mass that has not been stabilized.

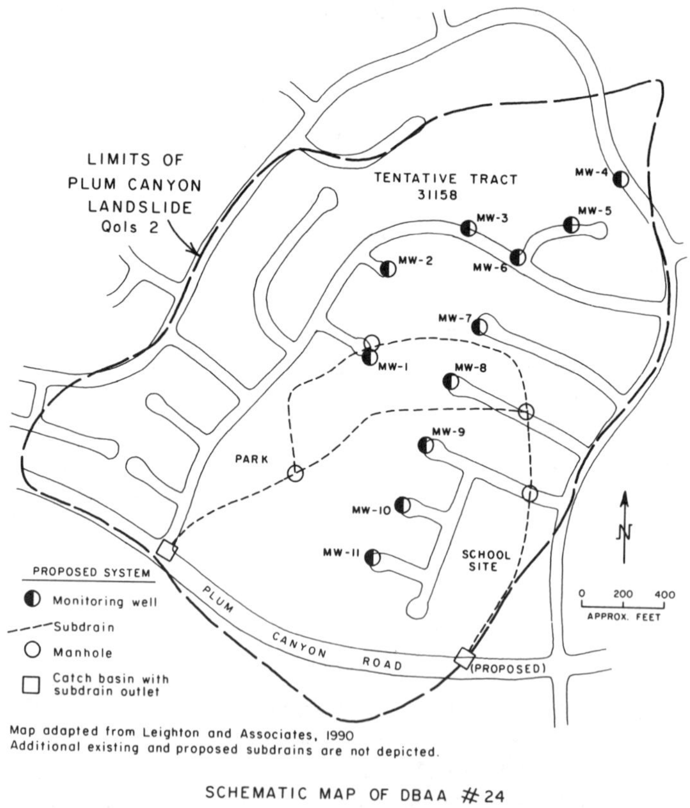

Figure 7. Map of the Plum Canyon Qols-2 landslide. The landslide failed toward the southwest. The Plum Canyon drainage coincides with the southwest boundary of the slide. Much of the landslide boundary is buried by alluvium.

The practical knowledge gained from this experience was put to use in the design of drainage facilities for DBAA #24 where 6-inch wells and accessible flow meters at pump stations were required. In addition, xerophytic landscaping has been recommended in order to decrease the amount of irrigation. It is my understanding that a revised monitoring and maintenance manual has not been completed for DBAA #6.

Drainage Benefit Assessment Area #18

Assessment Area #18 is Tract 44965 and is situated north of the Tract 43528 (Fig. 1, 2, and 6). The assessment area covers the south one-third of the Shangri-la landslide. Drainage facilities consist of an extensive subdrainage system within the shear keys and a monitoring well. Ramirez, (this chapter) discusses the construction of the 2.5 million yard shear key and related 1.3 million yard grading for condominium development. When condominium construction is completed there will be 293 dwelling units. The estimated annual cost for maintenance in 1991 was $21,400 for a predicted unit cost of $73.04, assuming all the units were occupied.

The drainage system was installed to preclude the water level from rising above a water surface assumed in the pseudo-static stability calculations in order to demonstrate a factor of safety of 1.1. The stability was more sensitive to water rising in the shear key than in the landslide mass. A 20-foot rise in the water level from the base of the shear key would require initiation of active dewatering measures.

Tract 44965 is the unit tract of Tentative Tract 44965 (formerly a part of Tentative Tract 32929), which was submitted to the Department of Regional Planning on October, 2, 1986. A revised map was approved by the Geology and Soils Section on April 27, 1987. Grading of this tract was conducted in two phases. This first grading plan was for the mass grading of the shear keys. The second plan was for surface grading to create suitable pads. The grading plan for the shear keys for this tract was approved by the Geology and Soils Section on February 22, 1988. Prior to completion of the project, regulatory agency control of geotechnical review was transferred on March 2, 1989 to the newly-created City of Santa Clarita. The second grading plan and rough grading of the tract was approved by the City of Santa Clarita. The DBAA was formed on June 6, 1989 by resolution of the City Council. It is my understanding that a monitoring and maintenance manual was submitted to the City and approved.

Development on the landslide occurred in two phases. Single family dwellings and a school were constructed during the 1960's on the northern half of the slide prior to its recognition. This area included Tracts 25931, 25932, and 28043 (a unit of Tract 27493). Grading and construction on these tracts occurred from 1963 through 1966. A low-angle fault, dipping east into the slope at 5-10 degrees, was observed during grading of Tract 27493 (Geotechnical Consultants, Inc., 1966). A buttress was constructed to mitigate the adversely fractured rock above the fault. It now appears that this fault may have been an exposed branch of the thrusted toe of the landslide.

The southern half was graded in 1987-89, after extensive geotechnical investigation that started in the late 1960's. Stabilization of the southern half of the slide required construction of a shear key with elaborate subdrainage. Some condominiums have been constructed and others are proposed.

The drainage system consists of subdrains in two separate shear keys and monitoring/pumping wells. In the small, central shear key (Fig. 5), an 8-inch toe drain and a 6-inch heel drain were constructed with a solid lateral drain directing water by gravity to the slope face. The main shear key consists of some 10 miles of pipes that directs water to a subterranean pump station in the northwest corner of the project. The pump station, some 50-60 feet below adjacent grade at the bottom of a 5-foot diameter manhole, directs water to the adjacent storm drain channel, immediately west of the tract.

The main drain in this key is a solid 10-inch pipe with access available from five 5-foot diameter manholes. Flood Maintenance Division required the 10-inch pipe and maximum spacing of 660 feet between manholes in order to maintain the system. The sizing and spacing standards are required so that a television camera can be utilized for inspection of the drain to allow for flushing and rodding. The manholes can also be converted to pump stations.

Connected to the 10-inch pipe is the 8-inch perforated heel drain and associated backdrains. The heel drain is situated some 5 feet below the slide plane. A christmas-tree-like arrangement of backdrains extends up the backcut for the shear key. The backdrains consist of 6-inch solid pipe, spaced every 200 feet along the shear key, and connect to horizontal filter-wrapped backdrains spaced every 25 vertical feet up the backcut. A toe drain, consisting of 8-inch perforated pipe directs water to the pump station. Cross drains between the heel and toe drains were constructed across the bottom of the key. One monitoring/pumping well has been installed.

The drainage system appears to be functioning as anticipated. The submersible pump, which cycles on when water elevation in the shaft is 1.5 feet, has a flow of about 5 gpm. Manholes were constructed without ladders, and therefore entry is difficult. Recently, safety concerns of a regulatory nature has made it difficult for County personnel to enter shafts, and the effect this could have on inspections has not yet been determined. Apparently, no one has yet entered the manholes in DBAA #18 for inspection (Wayne Snively, pers. comm., 1991).

Drainage Benefit Assessment Area #24

Proposed Assessment Area #24 is Tentative Tract 31158 and is situated north of the Plum Canyon (Fig. 1, 3, and 7). The assessment area covers essentially all of landslide Qols-2 and adjacent areas to the north and northeast. Landslide Qols-2 is part of the Plum Canyon landslide complex. Oborne and others (1990) describe this landslide complex in detail. Proposed drainage facilities consist of 8-inch perforated pipe placed at appropriate grade with manhole access, and 11 monitoring/pumping wells (Leighton and Associates, 1990). The drainage system is designed to preclude the water level from rising above the phreatic surface calculated to demonstrate a factor of safety of 1.5. The tentative tract proposes 537 lots, including a school site and a park. The estimated annual cost for maintenance of the system is currently $18,000. Thus the dwelling unit cost would be about $32.

Part of the proposed drainage facilities have been constructed. The 8-inch perforated pipe has been placed in deep excavations in the graben area of a secondary slide within Qols-2. Prior to excavation it was hypothesized that the secondary graben slide planes would act as aquicludes, subdividing the potential water-bearing rocks of the larger landslide. The excavations allowed detailed observations that indicated the absence of continuous clay shears. The subdrains were placed in the excavations and extended into the adjacent alluviated canyons where they were temporarily terminated until grading resumes on the next phase.

Construction remaining to be completed consists of extending these subdrains to catch basins along Plum Canyon Road, construction of manholes and the monitoring/pumping wells, and completion of grading. The proposed design includes manholes that will allow flushing and rodding of sections of the subdrain and can be converted to pump stations, if needed. The monitoring wells will consist of 6-inch perforated pipe completed to a depth immediately above or below the slide plane. The manholes and the wells will have subsurface drains connected to the storm drain system, such that if pumping is initiated in the future, water will not be directed onto the street. Each well and manhole will have a subsurface electrical system, such that if pumping is required power will be available without external hookups.

A storm drain discharge permit is required from the State Water Quality Control Board to discharge subsurface water into the storm drains. A permit cannot be applied for now since pumping is not anticipated in the near future and the quality of the water that may be pumped cannot be known. During project planning it was assumed that a permit could be obtained in the future, if pumping is required. If the permit cannot be obtained, expensive remedial measures would have to be implemented.

Tentative Tract 31158 consists of 6 unit tracts and was first submitted to the Department of Regional Planning on April 23, 1986. Grading for this tract is divided into three main phases. Grading of phases 1 and 2 commenced and was completed while approval for the remaining phase was being sought. The third phase will consist of grading unit tracts 46270, 46269, and 31158, as well as completion of the drainage facilities for the Assessment Area. The developer is not allowed to build on the completed tracts until the drainage facilities are completed and accepted by the County. The grading plans for phase 3, including plans for the drainage facilities and a proposed monitoring and maintenance manual are currently under review and approval is imminent at the time of writing this paper (January, 1992).

Development on this slide complex will occur over a period of more than a decade, and possibly two. The slide was recognized prior to development, thus extensive geotechnical investigations have been completed. Houses were built on the western portion of the slide complex in the early 1980's, and have been incorporated into the City of Santa Clarita. Construction of the northward extension of Whites Canyon Road over the ridge to connect with Plum Canyon Road (overlying the slide) is proposed in 1992-93 as part of a larger, area-wide, traffic management plan.

CONCLUSIONS

A DBAA provides a proactive approach to development on large landslides. The groundwater elevation within the landslides shall be monitored for the lifespan of the project. If groundwater levels rise, such that a stability hazard may occur, measures to lower groundwater are in place and others are ready to be activated. Considering that the homes in these tracts will probably be occupied for many decades, a DBAA can be considered a long- term experiment in graded tract hydrogeology. Assuming that monitoring is not discontinued at sometime, the record will be invaluable to future geotechnical practitioners.

The County has learned valuable lessons from the establishment of each DBAA, and has, in turn, applied this experience to the administration and approval of the next DBAA. The procedures for regulating the approval of a DBAA, as well as the requirements for maintaining the systems have improved with each project. This is the normal course of development for new technologies.

ACKNOWLEDGEMENTS

This paper is based on the author's participation as a regulator for DBAA #18 and #24, discussions with L.A. County geologists and many of the consultants over a period of 4 years, and the public files of the Department of Public Works. Robert Chezem and Wayne Snively kindly shared their knowledge of these DBAA. Review of a draft of this paper was provided by Gay Haven, Chris Sexton, James E. Slosson, Ray Waldbaum, and the editors; their critique is appreciated. Photos were taken in 1990 by the author. Geo-Soils, Inc. supported the aerial photography. Drafting services were kindly provided by Gay Haven. The responsibility for the data and opinions presented lies solely with the author. This paper does not represent the opinions or policies of, nor endorsement by the Los Angeles County Department of Public Works.

APPENDIX A

General Guidelines for a Monitoring and Maintenance Manual

A. Provide general introduction of dewatering system; description, purpose, methods to evaluate effectiveness, and critical water elevations.

B. Provide plans, drawings, as-built reports, and well data:
 1) As-graded plan depicting surveyed location and depth of wells, subdrains including horizontal drains, backdrains, manholes, etc.
 2) As-built geologic map utilizing the as-graded plan as a base map.
 3) Regional geologic map.
 4) As-built geologic cross-sections depicting the safe line (factor of safety =1.5 or1.1, whichever is most critical) and other action levels (i.e., increased monitoring frequency).
 5) Geologic log and well construction inspection report for each well including detailed lithology and structure, duration of drilling, time of completion; drillers name and type of rig, well construction materials actually used, location of perforations and solid pipe, thickness and type of impermeable seal, any construction difficulties encountered, any change in anticipated geologic conditions, written certification that wells were completed according to the specifications and recommendations of the consultant, etc.

C. Pump and flow meter specifications, storage, availability if pumping is necessary, a flow meter must be operational at wellhead for dewatering monitoring.

D. Operation and maintenance procedures:
 1) Frequency and method of inspection of subdrain, wells, and associated components.
 2) Method of rodding, flushing, repair of system including frequency.
 3) Inspection frequency and procedure for water pumps and flow meter.
 4) Service manuals, drawings, source of parts, etc., for water pump and flow meter.
 5) Allowable downtime for each pump or group of pumps.

E. Monitoring procedures for wells and subdrains:
 1) Method
 2) Reaction levels of groundwater
  a) Levels for increased frequency of monitoring for each well - frequency of observation should be linked to initial and continuing irrigation as well as high rainfall years.
  b) Levels for pumping.
  c) Levels for emergency construction of new wells.

F. Provide recommendations for a geologic review and yearly report.

REFERENCES

Bull, W.B., 1991, Geomorphic responses to climatic change: Oxford Univ. Press, New York, 326 p.

California GeoSystems, Inc., 8/2/1989, Final geologic report and compaction report on shear key construction, Tract 44965, Santa Clarita, CA: Public files, City of Santa Clarita.

Ehlig, P.L., 1981, Origin and tectonic history of the basement terrane of the San Gabriel Mountains, central Transverse Ranges: in The Geotectonic development of California, Ernst, W.G., (Ed.), Prentice-Hall, p. 253-283.

Geotechnical Consultants, 5/27/1966, Report of Tract 27493: Public files, Los Angeles County, Department of Public Works, Land Development Division.

Gorian and Associates, 5/5/1987, Horizontal drains and observation wells, Tract 43528, Saugus, California: Public files, Los Angeles County, Department of Public Works, Land Development Division.

Larson, R.A., 1990, Landsliding in the western Soledad Basin, Los Angeles County, California: Geology and Engineering Geology of the Soledad Basin, Los Angeles County, California; Buckley, C.I., and Larson, R.A. (Ed.), Guidebook for the Southern California Section of the Association of Engineering Geologists Annual Field Trip, October 2, 1990, p. 77-97.

Leighton and Associates, Inc., 10/21/1985, Geotechnical investigation of landslide Qols - 2, Tentative Tract 33987 and 37081, Saugus area, County of Los Angeles, California: Public files, Los Angeles County, Department of Public Works, Land Development Division.

Leighton and Associates, Inc., 10/8/1990, Monitoring and maintenance of dewatering system for landslide Qols-2, Tracts 44960, 37081, 46268, 46269, 46270, and 31158, Plum Canyon, Saugus area, County of Los Angeles, California: Public files, Los Angeles County, Department of Public Works, Land Development Division.

Levi, S., Schultz, D.L., Yeats, R.S., Stitt, L.T., Sarna-Wojcicki, A.M., 1986, Magnetostratigraphy and paleomagnetism of the Saugus Formation near Castaic, Los Angeles County, California: Neotectonics and faulting in southern California, Ehlig, P.L., compiler: Guidebook for the 82nd Annual Meeting, Cordilleran Section, Geological Society of America, March 25-28, 1986, Los Angeles, California, p. 103-108.

McMillan, K., McNamara, J.E., Kile, M.B., Butler, G., Lopez, W.B., and Shlemon, R.J., 1991, Tectonic geomorphology of the westernmost Camarillo Hills, Ventura County, California: Engineering geology along the Simi-Santa Rosa fault system and adjacent areas, Simi Valley to Camarillo, Ventura County, California; Blake, T.F., and Larson, R.A. (Ed.), Guidebook for the Southern California Section of the Association of Engineering Geologists Annual Field Trip, August 24, 1991, p. 223-229.

Medall, Aragon, Worswick and Associates, 11/12/1984, Revised hydroauger operation and maintenance recommendations, Tract 36496, Newhall-Saugus area, Los Angeles County, California: Public files, Los Angeles County, Department of Public Works, Land Development Division.

Morrison, R.B., 1991, Introduction: in Morrison, R.B. (Ed.), Quaternary Nonglacial Geology - Conterminous U.S.: Geological Society of America, DNAG volume K-2, p. 1-12.

Oborne, M., Khiabani, R., and Hancock, R., 1990, Geology of the Plum Canyon landslides, north half of the Mint Canyon quadrangle, Soledad Basin, Canyon Country, California: Geology and Engineering Geology of the Soledad Basin, Los Angeles County, California; Buckley, C.I., and Larson, R.A. (Ed.), Guidebook for the Southern California Section of the Association of Engineering Geologists Annual Field Trip, October 2, 1990, p. 116-131.

Ramirez, R.M., 1990, Shear key stabilization of a large, ancient landslide, Santa Clarita, California: Geology and Engineering Geology of the Soledad Basin, Los Angeles County, California; Buckley, C.I., and Larson, R.A., (Eds.), Guidebook for the Southern California Section of the Association of Engineering Geologists Annual Field Trip, October 2, 1990, p. 109-115.

Saul, R.B., 1990, The Mint Canyon Formation and associated rocks in the Soledad Basin, Los Angeles County, California: Geology and Engineering Geology of the Soledad Basin, Los Angeles County, California; Buckley, C.I., and Larson, R.A. (Eds.), Guidebook for the Southern California Section of the Association of Engineering Geologists Annual Field Trip, October 2, 1990, p. 26-34.

Saul, R.B., and Wootton, T.M., 1983, Geology of the south half of the Mint Canyon Quadrangle, Los Angles County, California: California Division of Mines and Geology, Open-File Report 83-24 LA, 139 p.

Stout, M.L., 1991, Landscape evolution during the last 35,000 years- Perspective of a southern California engineering geologist: Environmental Perils-San Diego Region, Abbott, P.L., and Elliott, W.J. (Eds.), San Diego Association of Geologists, p. 161-164.

STATE-OF-THE-ART SHEAR KEY FOR STABILIZATION OF AN ANCIENT MEGA-LANDSLIDE

Richard M. Ramirez
Engineering Geologist
California Geo/Systems, Inc.
312 Western Avenue
Glendale CA 91201

INTRODUCTION

A stabilization system of three state-of-the-art compacted fill shear keys extending 3/4 mile in total length with a maximum depth of 150 feet, and involving 2.5 million cubic yards in volume was constructed in 1989 for permanent stabilization of a part of an ancient mega-landslide. A 310-unit residential subdivision was constructed on the stabilized landslide. The site is located in northern Los Angeles County in the City of Santa Clarita, California, which is about 30 miles north of downtown Los Angeles.

The unique features of the stabilization shear key design and construction include the following:

1. Construction of the 3/4-mile-long shear key was performed by grading 22 separate 200-feet-long segments or bays, each bay being excavated and partially backfilled before the adjacent bays were graded.

2. The shear key stabilization system is comprised of three separate keys; an upper and lower key and one separation key along the east longitudinal property line. (Fig. 1).

3. An extensive system of subdrains and collector drains were placed along the bottom and backside of the shear key to drain into a deep sump where a pumping station was constructed.

4. Construction of five access shafts, each 5 feet in diameter and spaced approximately 660 feet apart and located along the alignment of the main subdrain collector line or "heel" drains.

5. Three groundwater observation wells were constructed to a depth of 20 feet below the basal slide plane along the upslope side of the shear keys to monitor groundwater levels.

6. Five television observation chimneys or camera access tubes were constructed to monitor the water flow along the extensive system of 10-inch diameter subdrain collector pipes.

7. Survey monuments for vertical and lateral monitoring of ground movement were utilized along the perimeter of the deep lower and separation shear key excavations.

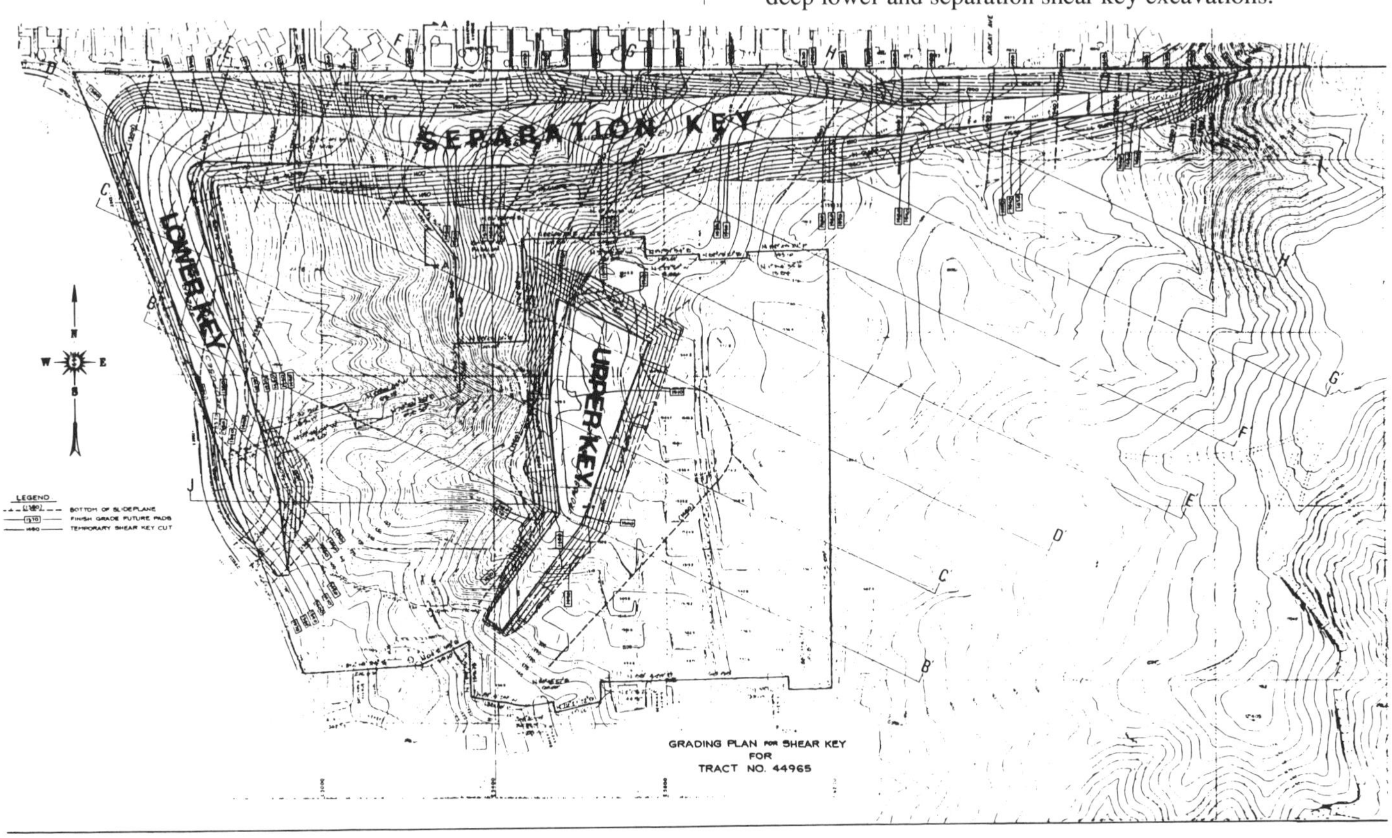

Figure 1. Shear-key locations

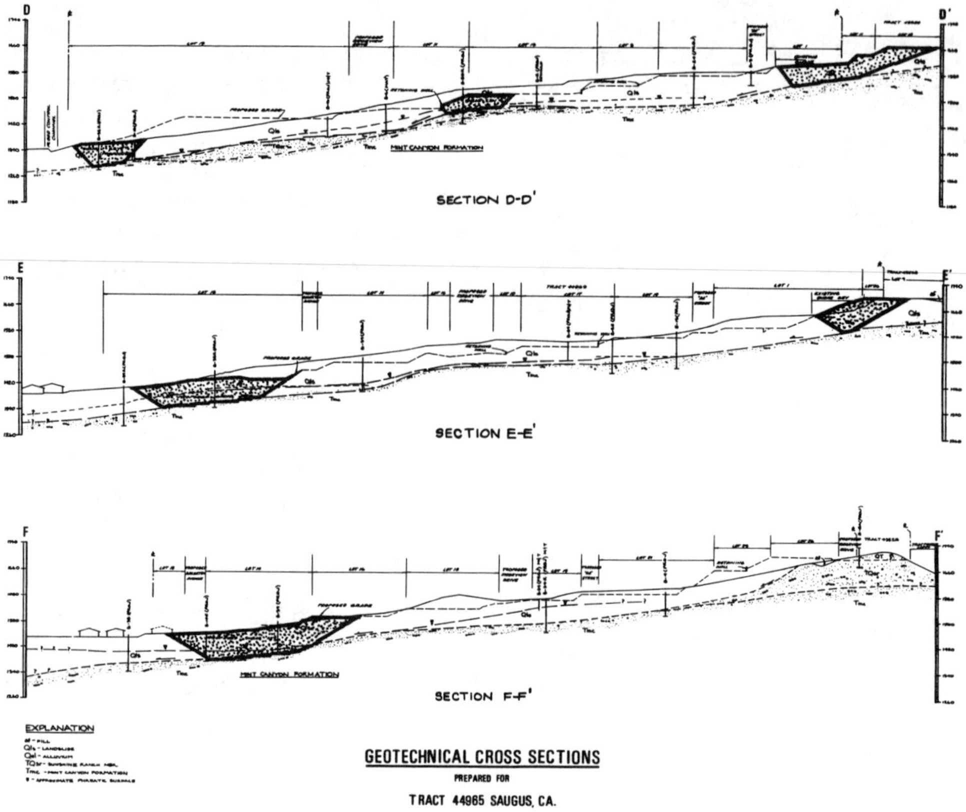

Figure 2. Geotechnical cross sections

8. Ground vibration monitoring was conducted during grading to record equipment vibrations.

9. A Groundwater Maintenance Manual was prepared by the geotechnical consultant to assist in the long-term monitoring of the groundwater.

10. A Drainage Benefit Assessment District was formulated to continuously fund the monitoring and maintenance of the subdrain system (see Larson this chapter).

GEOLOGY

The site is in the Mint Canyon area in the eastern portion of the Soledad Basin of Los Angeles County. The underlying pre-slide geologic structure is characterized by westerly dipping sedimentary strata comprised primarily of coarse sandstone, conglomeratic sandstone with thin beds of mudstone and bentonitic clays. In the site area, the strata are typically daylighted as the beds are low angle and vary in dip from 10° to 25° west.

Ancient massive translational sliding (Quaternary Age) affected a 300-acre area of this daylighted natural slope. Down-cutting by a tributary of the Santa Clara River resulted in the prehistoric landsliding of a major portion of the west-facing ridge slope during Quaternary time. Prior to identification of the mega-landslide residential pads were graded and homes were constructed in the middle 1960's on the northern portion of the landslide without stabilization. This existing developed area is to the north of this shear key project.

EXPLORATION

A total of 45 borings were drilled in the study area by various private geotechnical consultants. During the grading operations an additional thirteen borings were drilled at various bottom of key locations to assure that the basal slide was identified.

The equipment used for the exploratory drilling included 24-inch diameter bucket augers and 8- to 10-inch diameter rotary wash borings ranging in depth from 26 to 133 feet. Where possible, the 24-inch borings were downhole inspected and logged by an engineering geologist. In-place and bulk samples of the landslide debris, clay rich layers, slide plane, and underlying in-place bedrock materials were obtained with either a "Pitcher" type core barrel or a "California" thin wall sampler during the various stages of exploration.

In-place oriented samples of the various slide planes were obtained during the grading operations to verify the design residual shear strengths of the basal slide plane. Adjustments of the shear key widths were made during grading where the shear strengths were found to be lower than the design values.

Figure 3. Aerial view of completed grading of shear key and residential pads.

Figure 4. Construction slope failure in landslide debris on a 1 1/2 horizontal to 1 vertical temporary cut for shear key. Slope was re-graded to a 1 1/4:1 ratio.

Figure 5. Incipient slope failure on a temporary 1:1 construction cut slope in clayey siltstone slide debris. Note man on slope.

Figure 6. Exploratory drilling at bottom of 200-foot long bay for verifying that basal slide plane had been encountered in excavation.

Figure 7. Base of 5-foot diameter access shaft (left) and 10-inch chimney sweep for television access being constructed. Six of these observation-maintenance access shafts were in the project. Basal slide plane about 10-feet above access shaft base.

Figure 8. Typical grading-compaction equipment used in one of the 22 construction bays.

ANCIENT LANDSLIDE CHARACTERISTICS

Because of the age of sliding (Quaternary), the landslide materials were found to be very dense, tight, free of open voids and generally suitable for accepting additional compacted fill loads and suitable for supporting residential foundations.

The exploration indicated that the site was underlain by one primary deep slide and at least three overlapping shallow, younger slides. During grading it was determined that the basal slide was unusually undulatory and that the maximum basal slide depth encountered in one of the shear keys was approximately 135-feet deep. The basal slide plane was typically a clay-rich claystone or mudstone, locally bentonitic and highly plastic, striated in a northwest direction, and ranged in thickness from a few inches to about 24 inches.

Perched groundwater was found localized in only a few locations of the key excavations just above the basal and secondary slide planes. The maximum groundwater rate pumped was about 15 gpm.

GEOTECHNICAL ANALYSIS FOR SHEAR KEYS

Based on the original geological data obtained and supplemented by drilling and testing, several geotechnical cross-sections were prepared to provide the grading engineer with a final design of the shear keys. The key designs were modified as grading progressed as a result of additional in-place sampling and residual shear testing of the basal slide plane. Also the temporary front and backcuts for the shear keys were modified from 1:1 or $1^{1}/4$:1 to $1^{3}/4$ due to construction slope failures (primarily on the backcuts).

The minimum safety factors utilized for the shear key design were 1.5 static and 1.1 for pseudostatic conditions (0.15g). For temporary shear key excavations a 1.25 safety factor was utilized. A total of 2.5 million cubic yards of compacted fill was placed to construct the shear keys. An additional 1.3 million cubic yards of compacted fill was placed to construct the residential building pads and streets.

The constructed shear key widths at the level of the basal slide plane varied from 50 to 135 feet and the shear key depths varied from 55 to 150 feet. The total length of the three shear keys is 4400 feet with the separation key along the north side being the longest at 2400 feet.

The lower 15 feet (minimum) of each shear key was placed at a relative compaction of 95 percent and the remaining fill was placed at 90 percent relative compaction.

Shear Strength Parameters Utilized In Shear Key Analyses

Type of Soil or Rock	Unit Weight (pcf)	Static Values (psf)	Static Values (deg)	Seismic Values C (psf)	Seismic Values Phi (deg)
Cross Bedding	130	400	36.0	1000	34
Along Bedding	130	200	6.5	400	9
Compacted Fill	130	400	36.0	1000	34

Slope Stability

A slope stability computer program based on "Stability of Earthen Slopes" by Huang (1983) was utilized in the slope stability analysis. This program considers pore pressure at various water levels in the slices selected for analysis.

The water level in the slices through the landslide debris was taken into account, but was not considered within the compacted shear key as any groundwater will be intercepted by the extensive subdrain system in front and under the shear keys. The shear keys were designed assuming that the groundwater level at the bottom of the shear key rises to a maximum of 10 feet above the key bottom. The backup observation/pumping well system will be utilized if the groundwater rises to a level greater than 10 feet above the shear key bottom.

DEWATERING

Prior to the initial grading a primary sump was constructed at the northwest corner of the property. Groundwater encountered during construction of each shear key bay was less than 15 gpm and was transferred to the primary sump where the water was periodically pumped to a nearby flood control channel. The total volume of groundwater encountered during grading was less than the anticipated volume based on exploratory boring data.

SUBDRAINS

State-of-the-art features characterizing the shear key construction include the elaborate system of subdrains, collector wells, pump station, subdrain access shafts and television monitoring chimneys.

Continuous and interconnected subdrains were placed into bedrock at both the heel and toe of the shear key for their entire lengths. Toe drains consisted of 8-inch diameter perforated pipe and heel drains consisted of 10-inch solid pipe. Lateral drains consisting of 6-inch diameter perforated pipe were placed every 25 vertical feet against the backcuts on the upslope side of shear keys. Outlet pipes for the lateral drains were placed every 100 feet, and are connected to the 10-inch diameter solid pipe at the heel of the shear keys. Bottom transverse subdrains consisting of 6-inch diameter perforated pipe were placed every 150 to 300 feet of shear key length, across the width of the shear key bottom, and connected to the solid 10-inch heel drain. Additional subdrain pipes were installed in areas that experienced heavier flows during construction.

All subdrains consist of Schedule 40 PVC pipe and are surrounded by 3 cubic feet per linear foot of Caltrans Class 2 gravel. Mirafi filter fabric was wrapped around the Class II drain material for all of the subdrains.

Pump Stations

Groundwater collected in the subdrains flows through the collector pipes to a 5-foot diameter sump-pump station in the northwest corner of the property. The drained groundwater is then pumped out and released into a nearby concrete flood control channel.

Access Shafts and Television Observation Chimneys

Five separate concrete-lined access shafts 5 feet in diameter were constructed every 660 feet of shear key length to provide physical access to the 10-inch diameter collector pipes (heel drains) for inspection and maintenance. Chimneys consisting of 8-inch diameter PVC Schedule 40 solid pipe are located adjacent to each access shaft, providing passage for rodding equipment and television cameras from the ground surface to the 10-inch diameter collector pipes (heel drains) in the shear key bottom.

The 5-foot diameter access shafts allow for additional pump stations should blockage or damage occur in any portion of the collector system.

Observation/pumping Wells

In addition to the six access shafts which may also be used as "pumping wells", three additional groundwater observation wells were constructed in the public right-of-ways within the subdivision. These 8-inch diameter wells were drilled with a rotary wash drill rig to a minimum depth of 20 feet below the slide plane and are also designed as pumping wells.

SHEAR KEY CONSTRUCTION PROCEDURE

Three compacted fill shear keys for stabilizing the prehistoric landslide were completed during a one year period. Two main interconnected compacted fill shear keys were constructed along the north and west boundaries of the subdivision, creating an L-shaped configuration. An isolated third (upper) shear key was constructed near the center of the property. Construction of the shear keys was performed in 22 separate 200-feet-long "segments" or "bays" for a total length of 4400 feet. Initially, the excavation and shear key placement was conducted on alternating bays to preclude slide activation. The temporary shear key excavation slopes varied from 1:1 to 2:1 depending on the temporary stability of earth materials and on the shear key depth.

The deepest shear key excavation was 150 feet near the central portion of the site. The total shear key volume was 2.5 million cubic yards and the total grading construction cost was approximately 5.4 million dollars; an average cost of $1.75 per cubic yard for the shear key construction.

Settlement monuments were placed at final grade at various locations to monitor the settlement characteristics of the shear key fill. Construction survey monuments were placed along the perimeter of the long west and north key for the purpose of monitoring ground movement. Periodic vibration monitoring was also conducted during the grading operations to record equipment vibrations.

After the shear keys were completed, the "normal" grading for the construction of streets and building pads was initiated. Typically, all cuts within the ancient slide mass required stabilization by buttressing or by equipment width stabilization fills. Standard buttress subdrains and terrace surface drains were employed.

Maintenance Manual

A manual for monitoring and maintenance of the main pump station, the subdrain system and the observation wells was prepared to assist the City of Santa Clarita and the Homeowners Association. The City of Santa Clarita is designated as the agency responsible for long term monitoring and maintenance of the elaborate subdrain and observation well system. The City currently subcontracts the monitoring and maintenance operations to the Flood Maintenance Division of the Los Angeles County Department of Public Works.

Drainage Benefit Assessment Area

In order to provide long term assurance that the subdrain system is functioning as intended, a Los Angeles County Drainage Benefit Assessment Area (DBAA) was established to provide initial and on-going funding for its monitoring and maintenance. (See Larson, this chapter).

CLOSURE

The successful construction of the elaborate shear key system demonstrates that well-designed massive shear keys allows for the stabilization of ancient mega-landslides which previously were considered economically unfeasible.

Because minor changes in groundwater levels could have a major adverse effect on the gross stability of the stabilized landslide, several backup safety features were incorporated into the design and construction of the stabilization system.

REFERENCE

Huang, Y.H., 1983, Stability Analysis of Earth Slopes: Van Nostrand Reinhold Company, N.Y., 305 p.

SURFICIAL SLOPE FAILURES: THE ROLE OF VEGETATION AND OTHER LESSONS FROM RAINSTORMS

Paul M. Merifield
Consulting Geologist and Adjunct Professor
Dept. of Earth and Space Sciences
University of California, Los Angeles 90024

ABSTRACT

Widespread damage in hillside areas of southern California resulting from major rainstorms during 1978 and 1980 focused attention on risks to life and property from surficial slope failures, including debris slides, debris flows, and debris floods. Both man-made and natural slopes suffered damaging surficial failures, however mountain and foothill areas below burned watersheds were affected by considerably more debris than unburned areas. Slopes vegetated with grasses or other ground cover lacking deep root systems were particularly susceptible to failure.

Preventive measures include the following: (1) Stricter enforcement of existing grading codes (e.g., the requirement for a continuous cover of deep-rooted vegetation). (2) Stability analyses of surficial failures. (3) Regional studies by state and local governments to identify hazardous areas. (4) Site-specific investigations by consultants when properties change hands, or when significant grading or structural changes are accomplished. (5) Increased efforts to educate the public about slope maintenance to mitigate slope failures.

INTRODUCTION

Debris slides, debris flows and debris floods, brought on by devastating rainstorms in 1978 and 1980, caused widespread damage to hillside residential properties in southern California. These events, as well as the 1982 rainstorms in northern California and the spring runoff in Utah in 1983, have increased awareness of the damage and loss of life that can result from surficial failures.

Natural slopes in southern California are commonly 1-1/2 horizontal to 1 vertical or steeper. The mantle of soil and colluvium, which is generally thinner than 3 feet but accumulates to depths up to about 10 feet in ravines, is therefore near its angle of repose.

Vegetation consists mostly of chaparral: shrubs that grow primarily in winter and form a nearly continuous canopy over the slopes. Sediment moves downslope principally during infrequent heavy rainstorms; the production of sediment increases enormously in burned areas when heavy rains follow brush fires within one to three years.

OVERVIEW OF 1978 AND 1980 STORMS

Precipitation, which occurs almost entirely in the winter months, averages about 15 inches in the southern California coastal plains (but more than twice that in some mountain and foothill areas). Annual rainfalls of twice the average are not uncommon: Since 1877 the Los Angeles Civic Center has recorded nine years of rainfall with seasonal totals between 26 and 38 inches.

In terms of intensity and severity of damage, the rainstorms culminating February 10 and March 4, 1978, and the mid-February storms of 1980 stand out prominently. Figure 1 shows the 1978 rainfall figures for Mt. Wilson in the San Gabriel Mountains, which commonly records the highest precipitation in southern California. The storm culminating on February 10 began five days earlier; on the night of February 9-10, more than 8 inches of rain was recorded at some localities within a 24-hour period. One station at Crystal Lake in the San Gabriel Mountains recorded 13 inches on February 10. Several southern California stations exceeded 100-year return periods of high-intensity rainfall (Pyke, 1982). However, these rainfall intensities for a given 24-hour period are not unique: On New Year's Eve and New Year's Day, 1933-34, Pasadena recorded 12 inches; and on January 25, 1969, over 20 inches of rain was recorded at some stations in the San Gabriel Mountains. The highest rainfall recorded in southern California for a single 24-hour period was 26.12 inches at Hoegees station in the San Gabriel Mountains on January 22, 1943.

The other heavy rainstorm period of 1978 occurred from February 27 to March 4, 1978: Mt. Wilson recorded more than 24 inches during this period (Fig. 1). The rains of March 4 were particularly intense: La Crescenta, at the foot of the San Gabriel Mountains, received 6.79 inches of rain on that day.

In mid-February 1980 a series of six storms occurred over a nine-day period. Mt. Wilson received over 30 inches of rain during these storms (Fig. 2) and Topanga Canyon, a residential community in the Santa Monica Mountains, received 17 inches (Fig. 3). The Topanga Canyon station recorded 8.3 inches in 24 hours on February 17, most of which fell in a 10-hour period. At Bel Air, 5 miles away, 3.5 inches of rain fell in the same 24-hour period, exemplifying the wide variation in precipitation as these storms moved across southern California in relatively narrow belts.

STORM DAMAGE

The City of Los Angeles reported damage at 3,102 addresses during the 1978 winter storms. Los Angeles County, in a less comprehensive survey of unincorporated areas, reported damage at 1,796 individual street addresses (Weber et al., 1979). Thirty-eight storm-related deaths and $220 million in damages were attributed to the 1978 storms (Brooks, 1982).

Much of the flood damage in mountain and foothill areas occurred in burned watersheds, which produced far more debris than unburned areas (Shuirman et al., 1985). In burned areas, several debris basins—some recently constructed—overflowed, while in unburned areas basins proved quite adequate (Brooks, 1982). Extensive areas of the San Gabriel Mountains had been burned over in preceding dry periods, and the residential communities below south-facing watersheds were especially vulnerable when the intense storms arrived. In 1978 and 1980, forty percent of the estimated 2,300,000 cubic yards of debris deposited in the debris basins operated by the Los Angeles County Flood Control District was attributed to burned watersheds (Davis, 1982). Zachau, Rubio, and Shields Canyons, all within burned watersheds in the southern San Gabriel Mountains, overtopped their debris basins in 1978.

The effect of debris flooding in Shields Canyon is noteworthy. The watershed in this canyon above La Crescenta is only about 0.3 sq mi, with slopes averaging 1:1. In the 1960's new homes were constructed in the canyon above an older debris basin; a smaller debris basin was built above the new development along with a concrete channel to convey flow through the development to the old basin (Davis, 1982). During the intense storm of February 9-10, the smaller basin filled to capacity and debris overtopped the structure. The channel, which was too narrow to accommodate debris up to boulder size, became clogged (Fig. 4) and flow was diverted out of the channel and onto the street. The high volume of flow and the steep gradient of the street enabled the debris flood to transport automobiles parked along the street. Because the street made a right-angle turn just above the old basin, the debris flood left the street and dammed up in the front yard of a single-family residence (Fig. 5).

An event at Hidden Springs on the Middle Fork of Mill Creek in the San Gabriel Mountains was the costliest result of the February 1978 storm in terms of lives lost (Shuirman et al., 1985). A fire in July 1977 burned the entire 3.5 sq. mi. watershed of the Middle Fork of Mill Creek. The small resort community and 13 persons were swept away by an estimated 14 feet deep flow of water and debris produced by the rains of February 8-10, 1978.

Grass-covered slopes were particularly susceptible to surficial failure during the 1978 storms. Extensive areas of the urbanized foothills are underlain by Miocene marine shales which weather to clay-rich soils that support primarily shallow-rooted grasses (Fig. 6). Debris flows originating in these slopes did extensive damage to residences lacking adequate protective devices or sufficient setbacks. Even where structures were undamaged, yards were commonly inundated with debris, requiring costly removal.

Debris flows on wooded or chaparral-covered slopes originated chiefly in colluvium-filled ravines. Relatively small failures, starting as slides high on slopes near the heads of natural drainage courses, often became enormous debris flows by the time they reached the base of the slopes. One such failure in a tributary of Benedict Canyon above Beverly Hills incorporated large volumes of ravine-filling material as it moved more than 600 feet downslope. The watershed has a relief of about 375 feet and an area of about 10 acres, and slopes at the heads of the ravines are 35-45°, decreasing to 20° at the confluence of the two main branches.

Most of the debris came from the northernmost branch of the tributary (Fig. 7), where failure was observed to have originated on or about February 23, 1978 (Michael, 1980); the debris had moved about 50 feet on a failure surface apparently controlled within bedrock just below the overburden mantle. The 40-foot-wide, 10-foot high scarp was about 80 feet below the top of the slope, within an area modified by scarifying the slope of vegetation (Fig. 8). Water seeping from the scarp and the quick condition of a portion of the debris indicated a near-saturated condition. Saturation and initiation of movement of the slide can probably be attributed to the earlier storm period from February 4-10.

The next series of storms began on February 28. For the period February 28 to March 4, the nearby upper Stone Canyon station recorded 19.24 inches, of which 8.26 inches fell on March 4 (Kauper, 1982). Debris flows, which had moved part way downslope during prior rains, became reactivated by the intense downpour of March 4, and additional ravine fill up to 13-feet deep, was scoured from the channels (Fig. 9). An inadequately protected residence, constructed in the axis of the tributary canyon, was inundated with debris (Fig. 10).

Brush removal required to reduce the fire hazard on natural slopes below residences apparently was a significant contributing factor to the initial failure. A bulldozer moving parallel to the slope scarified the brush on the upper 100 feet or so of the slope, but also created a ribbon of uncompacted fill surcharging the natural soil and ravine fill along the margin of the scarified zone. The fill was unstable on a slope of 35° to 45°; moreover, the colluvial mantle on the slope above was deprived of stabilizing chaparral. Replanting consisted of shallow-rooted grasses and flowering plants, and surface drainage was uncontrolled.

The 1980 rains caused even more damage than those of 1978, primarily because concentrated, heavy rainfall occurred over a larger portion of southern California. Thirty deaths were attributed to the storms of February 1980, and damages in southern California were estimated at $400 million. One hundred eleven homes were destroyed and 1,350 were reported damaged (Weber, 1980). The foothill town of Monterey Park on the eastern edge of the Los Angeles basin suffered extensively from debris flows. In older residential areas, many slopes, both natural and man-made, planted

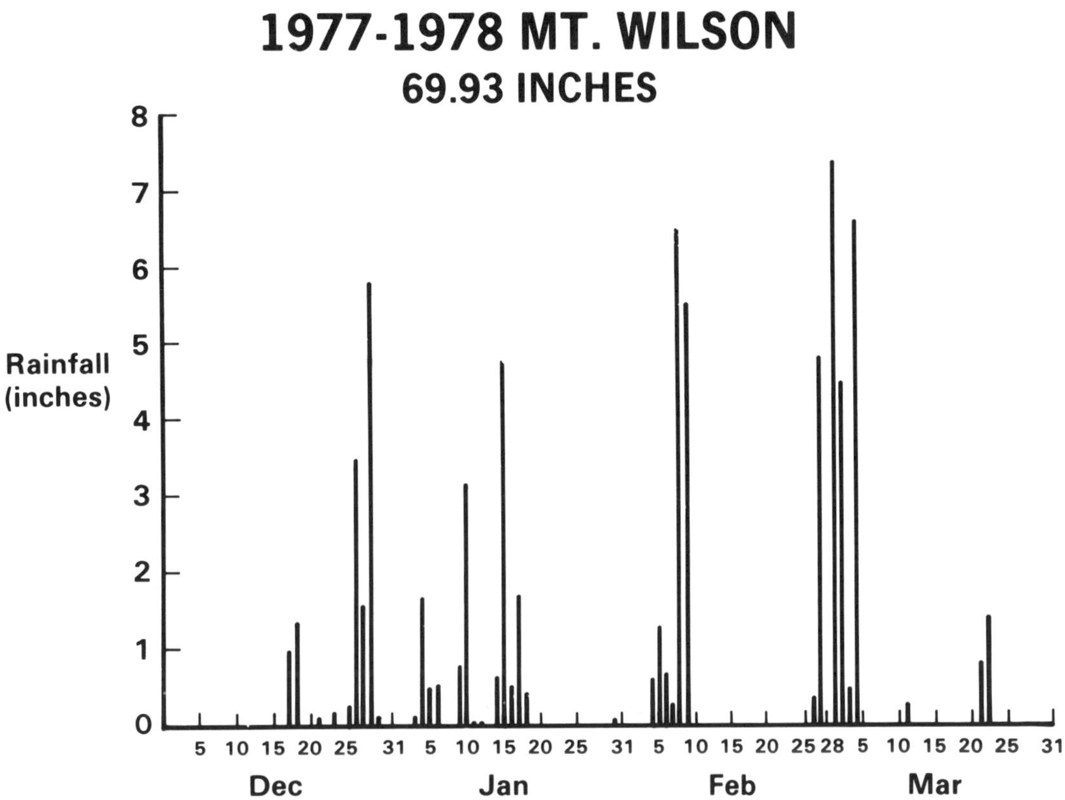

Figure 1. Daily rainfall, December-March, 1977-1978, Mt. Wilson station, NOAA Climatological Data.

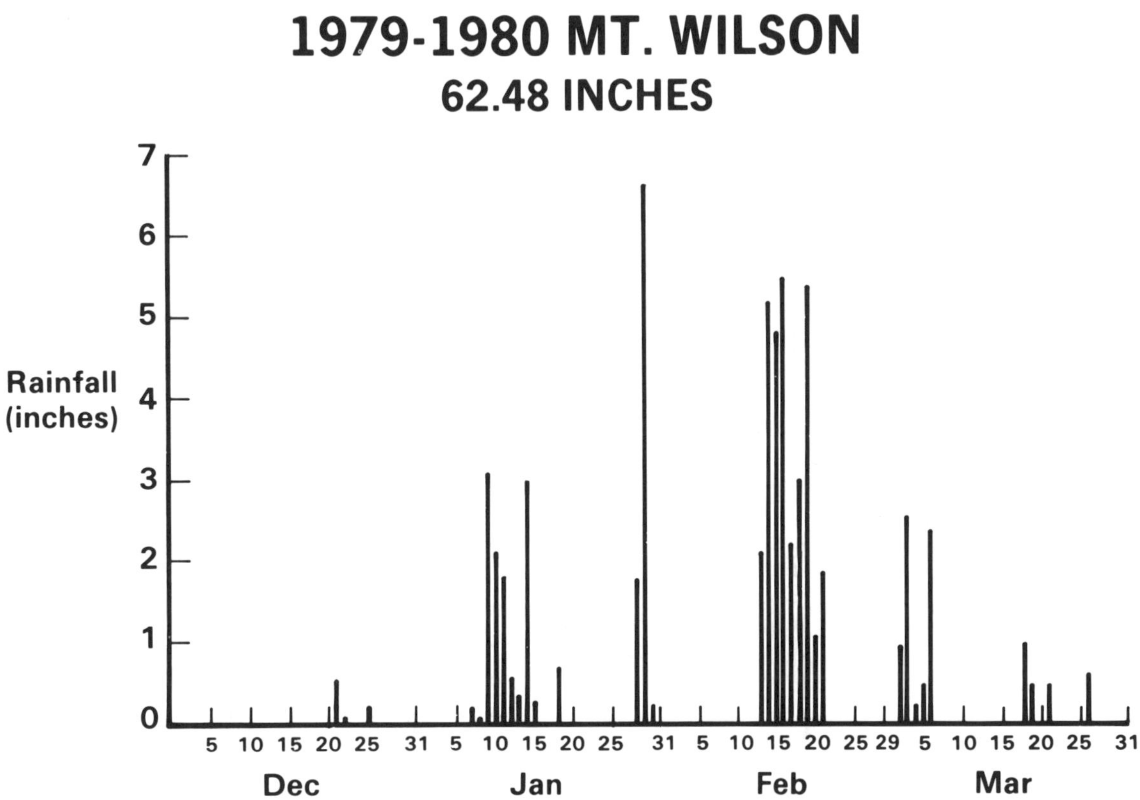

Figure 2. Daily rainfall, December-March, 1979-1980, Mt. Wilson station, NOAA Climatological Data.

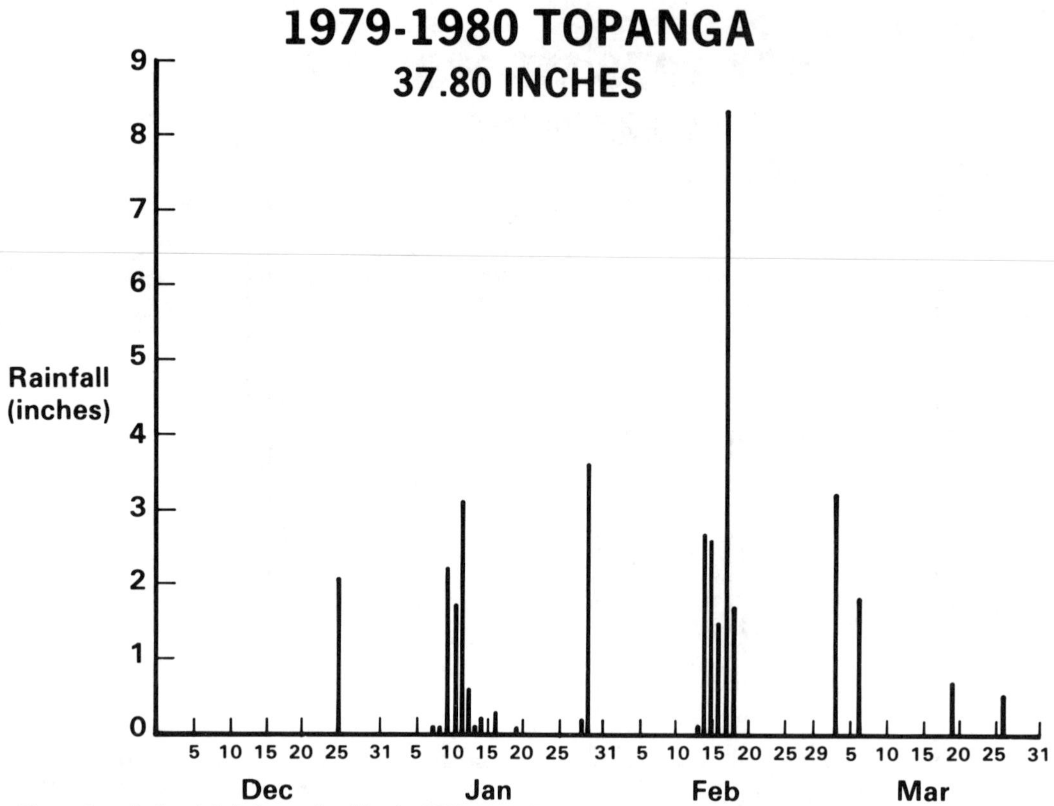

Figure 3. Daily rainfall, December-March, 1979-1980, Topanga station, NOAA Climatological Data.

with common ice plant failed; new man-made slopes with shallow-rooted ground covers were also susceptible to failure.

The rains and runoff were particularly severe in Topanga Canyon, washing out the principal roads and isolating the canyon community for several days. Low bridges and imprudent construction practices by local homeowners impeded high discharge and caused flood damage when natural channels overflowed. In Riverside Conty, the San Jacinto River overflowed, and portions of the town of San Jacinto near the river were flooded. Lake Elsinore, receiving waters from the San Jacinto River, rose to unprecedented heights, inundating shoreline homesites.

Greater than normal precipitation also occurred in southern California during the winter of 1982-1983 (over 30 inches at Los Angeles Civic Center). But rainfall was distributed over the season, and short, intense storms comparable to 1978 and 1980 did not occur. A histogram of daily rainfall for Topanga Canyon is shown in Figure 11. Debris flows and debris flooding were uncommon, but 1983 was characterized by deep-seated landslides presumably caused by abnormally high ground-water levels.

GRADING CODES

The first grading code for the City of Los Angeles was established in 1952, however geological and soil engineering investigations for residential construction were not required and rarely sought. Los Angeles County established its first grading code in 1957, Orange County in 1962. Deficiencies in these early codes have periodically been corrected; major revisions in the City of Los Angeles code were made in 1963, establishing stricter standards and better defining the roles of the geologist and soil engineer. In 1974, California state law required that all permit-issuing agencies within the state adopt Chapter 70 of the Uniform Building Code (UBC) as a minimum; some agencies, notably the City of Los Angeles, maintain even stricter standards. The California division of Mines and Geology (CDMG) has also issued a series of guidelines outlining the elements that should be covered in geologic reports related to development (Slosson, 1984).

Of significance to the following discussion is the treatment of surficial failures in the codes. In 1979, following the severe damage wrought by the 1978 storms, the City of Los Angeles began requiring analyses of surficial failures. For slopes exceeding 2:1, a minimum factor of safety of 1.5 against surficial failure was specified, to be derived by calculations for an infinite slope with seepage parallel to the slope. The use of other methods is subject to approval by the Department of Building and Safety. The minimum assumed depth of soil saturation is 3 ft (0.9 m) or the depth to firm bedrock. Soil strength parameters are to be derived from the testing of representative samples under conditions approximating saturation. The required stability is to be achieved by appropriate mitigating measures.

Figure 4. Clogged storm drain below debris basin, Shields Canyon, La Crescenta.

Figure 5. Effects of debris flooding, Shields Canyon, February 1978.

Figure 6. Surficial failures on grass-covered slopes, foothills of the San Gabriel Mountains at Big Tujunga Wash, 1978. (Photo courtesy of H.F. Weber, Jr.).

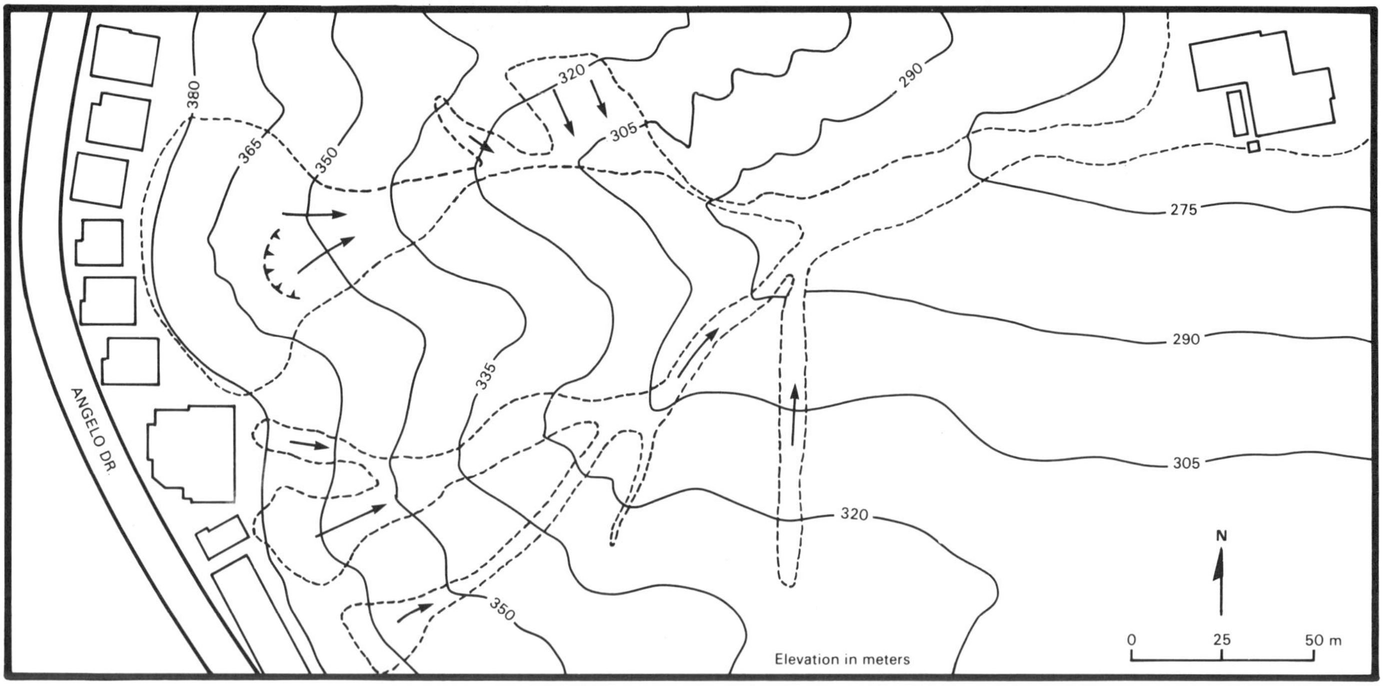

Figure 7. Debris flows originating above Liebe Drive, Santa Monica Mountains (after Michael, 1980). Scarp of initial failure shown in northern tributary.

Slope vegetation requirements for man-made slopes are set forth in Chapter 70 of the UBC; the City of Los Angeles has the same requirements. Slope vegetation should consist of trees with a minimum spacing of 20 ft (6 m), or shrubs spaced 10 ft (3 m), or a combination of the two. Vegetation on natural slopes is not covered in the code. In practice, natural vegetation is commonly cut off at ground level, removing the leaf conopy, for fire protection or replaced by shallow-rooted ornamental species.

Mitigation of water and debris flooding is called for in general terms in Chapter 70 of the UBC. Specific provisions are not mentioned, and mitigating measures are expected to fall within "good engineering practice." In recent years developers of residential tracts have been made responsible for flood-damage control, but the Los Angeles County Flood Control District and the Army Corps of Engineers have assumed responsibility for protecting many urbanized areas by constructing dams and debris basins.

DISCUSSION AND CONCLUSIONS

The conclusions that can be reached from observing surficial failures during the rainstorms of 1978 and 1980 reinforce those of Campbell's (1975) study of slope failures caused by the 1969 rainstorms in the Santa Monica Mountains. Prerequisites to initiation of debris flows are slopes of 2:1 or greater and a sufficient thickness of soil, colluvium, or poorly consolidated artificial fill. Many fill slopes constructed before updated grading codes were in effect were poorly compacted or placed on improperly prepared surfaces. Even well-compacted slopes become less well consolidated in time owing to physical and chemical weathering processes and the activity of organisms. In the case of slopes constructed of compacted clay soil—and most compacted slopes in California have a significant clay fraction—weakening of the slope face is most apparent owing to the process of soil expansion. Natural slopes also undergo continuous change as new soil forms, moves downslope, and accumulates in ravines.

Precipitation sufficient to bring the soil to field moisture capacity (about 10 inches), followed by a rainstorm with an intensity of greater than 0.25 inches per hour, constitutes the critical failure criteria (Campbell, 1975). Slosson and Krohn (1982) observed that surficial failures are most severe when heavy rains (over 10 inches) occur for five or more days and especially when the most intense rainfall occurs near the end of the storm period.

When the intensity of rainfall is greater than the ability of the soil or fill to drain, the slope-forming material becomes saturated and seepage proceeds parallel to the slope. An analytical model describing the initiation of shallow sliding, which is derived from the Mohr-Coulomb equation, has been available for many years and was discussed by Campbell (1975) in relation to soil failures in the Santa Monica Mountains during the 1969 rainstorms. The factor of safety for a

Figure 8. Scars from debris flows above Liebe Drive, Santa Monica Mountains. Note upper slopes scarified of most large plants. (Photo courtesy of City of Los Angeles, Dept. of Building and Safety)

Figure 9. Main channel scoured by debris flow above Liebe Drive, Santa Monica Mountains. (Photo courtesy of City of Los Angeles, Dept. of Building and Safety)

Figure 10. Residence at cul-de-sac, Liebe Drive, Santa Monica Mountains, March 1978. (Photo courtesy of City of Los Angeles, Dept. of Building and Safety).

saturated infinite slope (Fig. 12) is a function of the effective cohesion, the angle of internal friction, the slope, the depth of sliding, and the unit weight of the material.

In utilizing the model for compacted clay soils, Day and Axten (1989) caution against using the same shear strength parameters (cohesion and friction angle) as those for gross stability (deeper arcuate failures), as currently allowed by permitting agencies. Their research and that of others indicates that the cohesion of clay soils approaches zero, and the friction angle increases to 40-45°, at low normal stresses simulating low confining pressures near the slope face. These parameters yield lower factors of safety than those extrapolated from higher normal stresses.

To illustrate the application of the model to field conditions, the factor of safety calculated using two sets of shear strength data from tests on saturated samples of shallow failures is plotted in Figure 13; the data suggest that a slope is unstable when the factor of safety is less than 1.5. The sensitivity of the factor of safety to cohesion has been emphasized by Gray and Leiser (1982). Figure 14 plots factor of safety as a function of cohesion for the two sets of data wherein the set of data for artificial slopes includes both failed and unfailed slopes. For this set, an excellent correlation between cohesion and factor of safety exists, largely because the other parameters (slope, unit weight, depth of failure and friction angle) are the same or nearly the same for different samples. The set for the natural slopes contains a much wider variation between these parameters—especially slope, which varies from 2:1 to 1:1, and depth of failure, which varies from 2 feet (0.6 m) to 15 feet (4.6 m). Nevertheless, for a factor of safety of 1.5, a minimum cohesion of about 200 lbs/sq ft (0.10 kg/cm^2) is indicated.

Observations of surficial failures in 1978 and 1980 suggest that shallow-rooted vegetation—such as grasses and ice plant—is relatively ineffective in preventing surficial failures; in fact, such ground covers may promote soil failures by increasing the infiltration of surface water, thereby leading to "a more rapid and thorough saturation of the soil mantle" (Campbell, 1975).

Deep-rooted woody vegetation, on the other hand, increases soil strength by root reinforcement and also by the buttressing and soil arching between plants; its removal by fire or clear cutting increases the potential for failure. This cause-and- effect relationship is well documented in the literature from studies around the world (Gray and Leiser, 1982; Greenway, 1987). The negative effect of brush fires on surficial slope stability is well known in California (Cleveland, 1973; Davis, 1982). Root reinforcement increases the cohesion and can be added to the numerator of the factor of safety equation (Gray and Leiser, 1982).

Woody plants providing a high concentration of deeply penetrating flexible roots with high tensile strength are most effective for slope stability. Soil shear strength increases

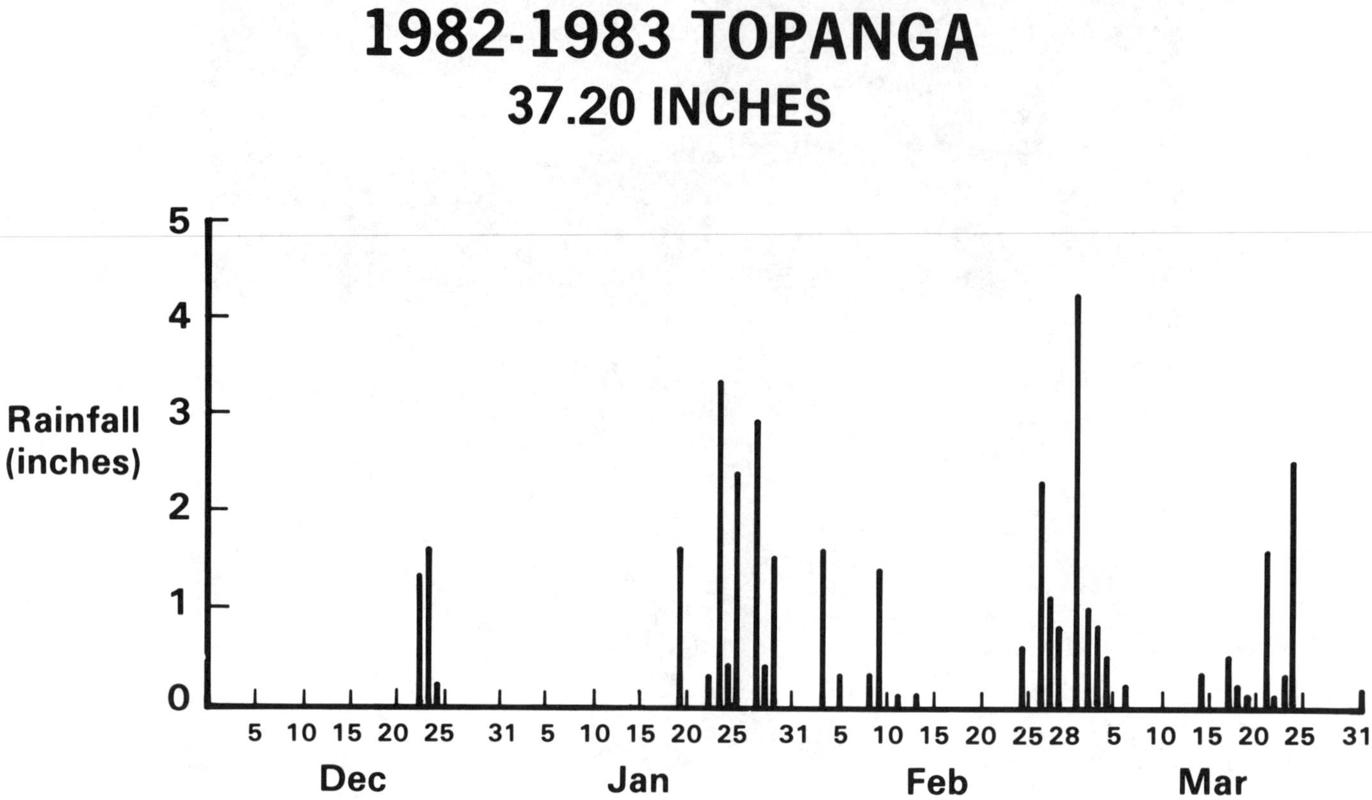

Figure 11. Daily rainfall, December-March, 1982-1983, Topanga station, NOAA Climatological Data.

SURFICIAL STABILITY ANALYSIS

$$\text{F.S.} = \frac{c' + (\gamma_t - \gamma_w)\, h \cos^2 \alpha \tan \phi'}{\gamma_t\, h \cos \alpha \sin \alpha}$$

γ_t = total, saturated density of soil
γ_w = density of water
h = vertical depth of slide
α = slope angle
c' = effective cohesion
ϕ' = effective friction angle

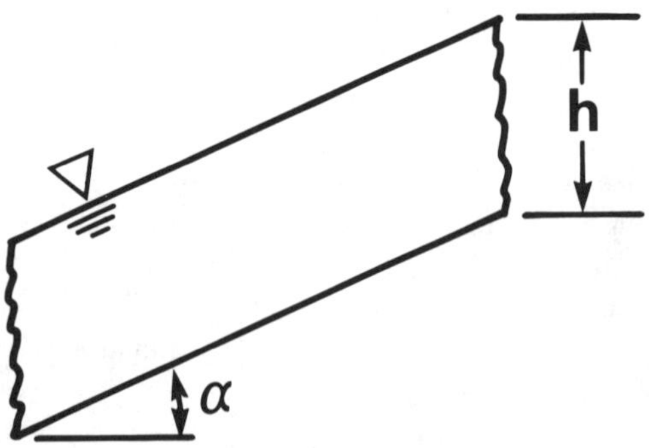

Figure 12. Stability analysis for saturated surficial slab of infinite length.

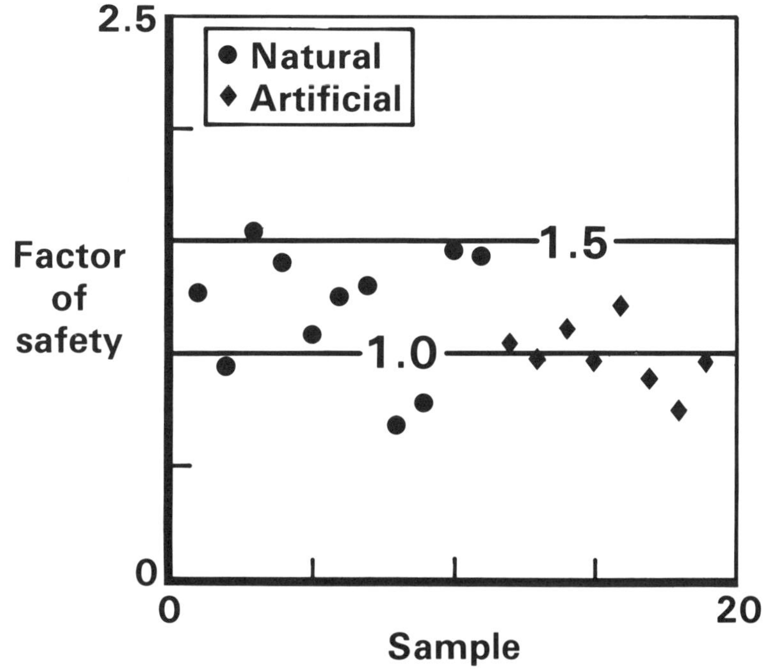

Figure 13. Factor of safety calculated from equation in Fig. 12 for failed slopes. Dots: natural slopes (Pascucci, 1985); diamonds: artificial slopes, 2 cuts and 6 fill slopes (unpublished data).

with the bulk weight of roots per unit volume, which can also be expressed as the ratio of the area of the roots to the area of the sliding surface. Based on in situ shear test data for sandy residual soil on granitic slopes forested with Douglas Fir, Gray and Leiser (1982) concluded that root-area ratios in the range of 0.05 to 0.10 percent increased soil shear strength 1-2 pounds per square inch (0.07-0.14 kg/cm^2 or 144-288 lbs/ft^2). An increase in cohesion of 1-2 psi will increase the factor of safety for the slopes in Fig. 14 to over 1.5.

In another study near Cincinnati, Ohio, tree roots increased the factor of safety against shallow sliding ninefold; the sliding surface in this case was the contact between bedrock and colluvium. The average shear strength contributed by tree roots penetrating the contact was determined to be about 0.06 kg/cm^2. In forested areas, colluvium-mantled slopes were stable up to 35°, whereas deforested slopes were subject to sliding at 12° to 14° (Riestenberg and Sovonick-Dunford, 1983).

Studies in the San Dimas Experimental Forest near Glendora (Rice et al, 1969) have demonstrated that soil failures are 3 to 5 times more frequent for grass-covered slopes than for brush- covered slopes, and the minimum angles for failure were less for grassy cover than for most chaparral.

Review of the literature indicates that most trees are deep rooted; among the large shrubs a number are cited as being deep- rooted and drought resistant. Table 1 is a selected list of shrubs with slope-stabilizing characteristics compiled from Emery (1967), Radtke (1983), Pecoff (1983), Perry (1987), and lists provided by the City and County of Los Angeles. The California scrub oak (Quercus dumosa) (Fig. 15) has one of the deepest and most extensive root systems of the southern California chaparral species, penetrating as much as 25 feet (8 m) into fractured bedrock but also spreading laterally, thus reinforcing the colluvial layer.

Buttressing is provided by firmly anchored tree trunks or large diameter shrubs. Arching occurs when soil attempts to move through and around a row of trees (Gray and Leiser, 1982). If the rows are staggered, the buttressing and arching affects a large portion of the slope (Fig. 16).

Damage from surficial failures during the 1978 and 1980 rainstorms occurred to homes built close to the base of steep, soil-mantled slopes. Swales or natural ravines concentrating flows toward home sites produced the most hazardous situation. Failed slopes were commonly vegetated with grasses, ice plant, or other shallow-rooted ground covers. In some cases, deep—rooted native growth was removed for fire protection and replaced with shallow-rooted ornamental species.

Radtke (1983) has emphasized the inadvisability of completely removing chaparral because of its slope-stabilizing characteristics. Alternatively, chaparral should be thinned

FACTOR OF SAFETY AS A FUNCTION OF COHESION

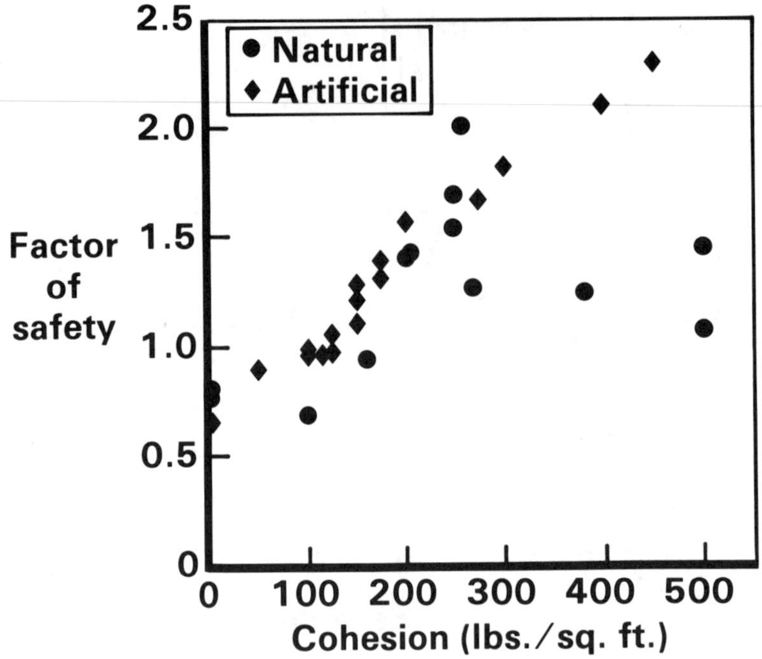

Figure 14. Factor of safety as a function of cohesion. Samples from artificial slopes include failed and unfailed slopes. Highest and lowest values of each data set not used.

and pruned to reduce fuel volume and interspersed with carefully selected deep-rooted, drought-tolerant, low-fuel species, at least until the new plants are mature. Plants with a low growth profile gener ally have lower fuel volume than taller species.

The effect of strengthened grading codes has been encouraging. A survey of 37,000 sites graded before 1963 in the City of Los Angeles showed 2,790 failures or 7.5 percent, whereas 36,000 sites graded after 1963 had 210 failures or 0.07 percent (Slosson and Krohn, 1982). Until recently, stability analyses were not performed to investigate the potential for surficial failures. The City and County of Los Angeles now require that the potential for surficial slope failures be assessed as part of the geotechnical investigation. Some other permit-issuing agencies, however, do not even require geological studies. The vegetation requirements in Chapter 70 of the UBC are not generally adhered to; hydromulching of fill slopes with grass seed, which is not in compliance with the code for slopes exceeding 15 feet (4.6 m) in height, is common practice. More over, it is questionable whether the code, which calls for trees at 20-foot (6-m) or shrubs at 10-foot (3-m) spacing, is adequate. Spacing should be such that the slope is covered with a continuous canopy of trees or shrubs with root systems sufficient to reinforce surficial unconsolidated materials and bind them to firmer substrate.

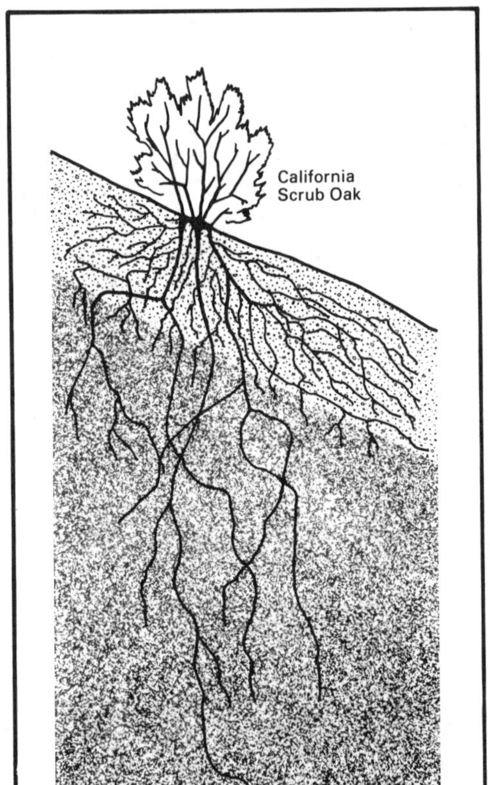

Figure 15. Root system of California scrub oak (Quercus dumosa), after Hellmers et al, 1955.

Burned watersheds yield far greater amounts of water and debris runoff than unburned areas. The vegetal canopy and some of the shallower root systems are destroyed, resulting in greater erosion by raindrop impact and surface flow. Moreover, increased soil creep between rains produces deeper ravine fill for movement during ensuing storms (Simpson, 1969). The intense heat of fires produces a hydrophobic layer just below the surface, an impervious waxy substance formed from organic matter in the soil which is believed to contribute significantly to increased runoff (Cleveland, 1972)

Observations of the 1978 and 1980 rainstorms showed that debris basins were not designed to accommodate the increased debris that results when brush fires precede such storms within two to three years. In addition, many privately engineered flood control provisions in residential tracts proved to be under designed, and drainage provisions were commonly neglected. Developers of housing tracts are required to install bench drains, collection basins, berms and other devices to convey water quickly away from slopes; however, when the homes are completed, responsibility for maintaining the devices passes to the homeowners, who seldom make a coordinated effort to maintain devices common to several properties. The devices become clogged and fail to function when the need arises.

The 1978 and 1980 rainstorms have made geologists and soil engineers more aware of the causes of surficial failures. Slopes with the potential for surficial failure can be more clearly identified; grading codes need to be updated to reflect this knowledge and enforced. For new development, stability analyses for the surficial slab should be performed on both natural and engineered fill slopes.

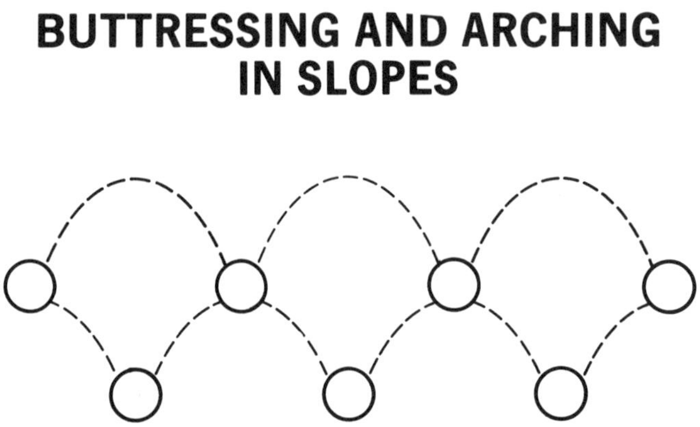

Figure 16. Buttressing and arching effect of trees or large shrubs (adapted from Gray and Leiser, 1982).

Many older residences were built in the axes of drainage courses with little or no protection from debris flows and debris flooding. Structural mitigative measures, such as debris fences and deflection walls (Gray and Leiser, 1982; Chin, 1985; Hollingsworth and Kovacs, 1981), have proven to be effective and can be constructed at relatively low cost. Planting trees or deep-rooted shrubs, such as those listed in Table 1, is even less costly.

Special slope stability studies of existing developments in areas known to be sensitive to slope failure are appropriate tasks for state and local governments. Site-specific investigations by private consultants when property ownership is transferred alerts sellers and buyers to potential geologic hazards that can usually be minimized or abated at modest cost in comparison to the value of the property. Recommendations to bring older residences up to modern standards can also be incor porated into geological/soil engineering reports for structural additions or new grading. Further education of the public on slope stability, including the maintenance of mutually owned drainage devices and the planting of appropriate slope vegetation, is recommended.

ACKNOWLEDGMENTS

The writer is grateful for the comments of D.L. Lamar, E.D. Michael, J.W. Cobarrubias, A.G. Keene, J. Rowlands, and G.W. Axten.

Table 1. Deep-rooted shrubs

Acacia cyclopsis (Acacia cyclops)
Acacia longifolia (Sidney golden wattle)
Acacia pycnantha (Golden wattle)
Acacia rodelens cultivar ongerup (Prostrate acacia)
Acacia verticillata (Star acacia)
Achillea millefolium (White yarrow)
Arctostaphylos gladulosa (Eastwood manzanita)
Atriplex canescens (Four-wing saltbush)
Atriplex halimus (Sea orach)
Atriplex lentiformis (Quail bush)
Atriplex nuttlei
Atriplex rhagodioides (Silver saltbush)
Atriplex undulata (Wavy leaf saltbush)
Baccharis pilularis prostrata (Prostrate coyote bush)
Baccharis pilularis, subsp. consanguinea (Coyote brush or chaparral broom)
Baccharis pilularis, subsp. pilularis (Dwarf chaparral broom)
Baccharis viminea (Mulefat)
Callistemon species (Bottlebrush)
Ceanothus crassifolius (Hoaryleaf ceanothus)
Ceanothus crisseus (Carmel ceanothus)
Ceanothus griseus horizontalis (Carmel creeper)
Ceanothus gloriosus (Point Reyes ceanothus)
Ceanothus leucodermis (Chaparral white thorn)
Ceanothus megacarpus (Bigpod ceanothus)
Ceanothus oliganthus (Hairy ceanothus)
Cercidium microphyllum (Parkinsonia microphyllum) (Littleleaf Palo Verde)
Displacus longiflorus (Mimilus longiflorus) (Bush monkeyflower)
Eriodictyon californicum (Yerba santa)
Eriodictyon trichocalyx (Hairy yerba santa)
Eriogonum giganteum (St. Catherine's lace)
Fremontodendron californicum (Fremontia californica) (Flannelbush)
Garrya elliptica (Coast silktassel)
Grevillea banksi (Crimson coneflower)
Heteromeles arbutifolia (Photinia arbutifolia) (Toyon or California holly)
Isomeris arborea (Bladder pod)
Juglans california (Southern California walnut)
Juniperus species (Junipers)
Keckiella cordifolia (Penstemon cordifolius) (Honeysuckle penstemon or climbing penstemon)
Lotus scoparius (Deerweed)
Lupinus arboreus (Tree lupine)
Mahonia nevinii (Berberis nervinii), Nevin barberry
Nerium oleander (Oleander)
O'Conners leguma (Strawberry clover)
Prunus ilicifolia (Hollyleaf cherry)
Prunus lyonii (Catalina cherry)
Quercus dumosa (Scrub oak)
Quercus wislizenii var. frutescens (Dwarf interior live oak)
Rhamnus crocea (Redberry)
Rhus integrifolia (Lemonade berry)
Rhus laurina (Malosma laurina) (Laurel sumac)
Rhus ovata (Sugar bush)
Rhus trilobata var. pilossissima (Squaw bush)
Rosmarinus officinalis (Rosemary)
Salvia apiana (White sage)
Salvia leucophylla (Purple sage)
Salvia mellifera (Black sage)
Salvia sonomensis (Creeping sage)
Spartium junceum (Spanish broom)
Sambucus mexicana (Elderberry)
Symphoricarpos mollis (Creeping snowberry)
Viguieria lacineata (San Diego daisy)
Yucca whipplei (Chaparral yucca)
Zauschneria cana (California fuchsia)

REFERENCES

Brooks, N.H., 1982, Overview and summary, in Storms, floods, and debris flows in southern California and Arizona 1978 and 1980: Proceedings of symposium, Sept. 17-18, 1980, National Academy Press, Washington, D.C., p. 1-24.

Campbell, R.H., 1975, Soil slips, debris flows, and rainstorms in the Santa Monica Mountains and vicinity, southern California: U.S. Geological Survey Prof. Paper 851.

Chin, A.E., 1985, Landslide and debris flow hazard mitigation measures for existing development, in Bowles, D.S. (ed.), Delineation of landslide, flash flood, and debris flow hazards in Utah, General Series UWRL/G-85/03: Utah Water Resources Laboratory, Utah State University, Logan, Utah, p. 511-531.

Cleveland, G.B., 1973, Fire + rain = mudflows, Big Sur, 1972: California Geology, vol. 26, p. 127-135.

Costa, J. E., and Wieczorek, G. F., (eds.). Debris flows, avalanches: process, recognition and mitigation: Geol. Soc. Amer. Reviews in Engineering Geology, vol. 7, 248 p.

Davis, J.D., 1982, Rare and unusual postfire flood events experienced in Los Angeles County during 1978 and 1980, in Storms, floods, and debris flows in southern California and Arizona 1978 and 1980: Proceedings of a symposium, Sept. 17-8, 1980, National Academy Press, Washington, D.C., p. 243-256.

Day, R.W., and G.W. Axten, 1989, Surficial stability of compacted clay slopes: Journal of Geotechnical Engineering, vol. 115, no. 4, p. 577-580.

Emery, D.E., 1967, Native plants for erosion control in southern California: Leaflets of the Santa Barbara Botanic Garden, vol. 1, no. 1, p. 97-111.

Gray, D.H., and A.T. Leiser, 1982, Biotechnical slope protection and erosion control: Van Nostrand Reinhold, NY, 271 p.

Greenway, D.T., 1987, Vegetation and slope stability, in Anderson, M.G., and K.S. Richards (eds.), Slope stability: geotechnical engineering and geomorphology: Wiley, p. 187-230.

Hellmers, H., J.S. Horton, G. Juhren and J. O'Keefe, 1955, Root systems of some chaparral plants in southern California: Ecology, vol. 36, no. 4, p. 667-678.

Hollingsworth, R.A., and G.S. Kovacs, 1981, Soil slumps and debris flows: prediction and protection: Bull. Assoc. Engineering Geologists, vol. 28, p. 17-28.

Kauper, E.K., 1982, Meteorological report on conditions existing over the eastern Santa Monica Mountains, February 28 through March 4, 1978: Metro Monitoring Services, Covina, California, October 25.

Los Angeles, City of, 1978, Guide for erosion and debris control in hillside areas: L.A. Dept. Building and Safety, 27 p.

Michael, E.D., 1980, Source of mudflow debris originating between Angelo Drive and Liebe Drive cul-de-sac: Consulting Geologist (private report), Malibu, California.

National Geographic, 1969, California's trial by mud and water: October 1969

Pascucci, V.N., 1985, 475 landslides in the City of Pacifica, California, in Bowles, D.S. (ed.), Delineation of landslide, flash flood, and debris flow hazards in Utah: General Series UWRL/G-85/03, Utah Water Resources Laboratory, Utah State University, Logan, Utah, p. 99-112.

Pecoff, R., 1983, Innovative erosion control plant materials and planting procedures, in Scullin, C.M., Excavation and grading code administration, inspection, and enforcement: Prentice-Hall, p. 255-266.

Perry, B., 1987, Trees and shrubs for dry California landscapes: Land Design Publishing, San Dimas, California.

Pyke, C., 1982, Return periods of 1977-80 precipitation in southern California and Arizona, in Storms, floods, and debris flows in southern California and Arizona 1978 and 1980: Proceedings of symposium, Sept. 17-18, 1980, National Academy Press, Washington, D.C., p. 77-86.

Radtke, K.W.H., 1983, Living more safely in the chaparral-urban environment: U.S. Forest Service, Pacific Southwest Forest and Range Experiment Station, General Technical Report PSW-67, 51 p.

Rice, R.M., E.S. Corbett and R.G. Bailey, 1969, Soil slips related to vegetation, topography and soil in southern California: Water Resources Research, vol. 5, no. 3, p. 647-658.

Riestenberg, M.M., and S. Sovonick-Dunford, 1983, The role of woody vegetation in stabilizing slopes in the Cincinnati area, Ohio: Geol. Soc. Am. Bull., vol. 94, p. 506-518.

Shuirman, G., J.E. Slosson and D. Yoakum, 1985, Relationship of fire/flood to debris flows, in Bowles, D.S. (ed.), Delineation of landslide, flash flood, and debris flow hazards in Utah: General Series UWRL/G-85/03, Utah Water Resources Laboratory, Utah State University, Logan, Utah, p. 178-194.

Sidle, R. C., Pierce, A. J., and O'Loughlin, C.L., 1985, Hillslope stability and land use: Am. Geophys. Union Water Res. Monogrpah no. 11, 140 p,

Simpson, L.D., 1969, Hydrologic report on storms of 1969: Los Angeles County Flood Control District.

Slosson, J.E., 1984, Genesis and evolution of guidelines for geologic reports: Bull. Assoc. Engineering Geologists, vol. 31, p. 295-316.

Slosson, J.E. and J.P. Krohn, 1982, Southern California land-slides of 1978 and 1980, in Storms, floods, and debris flows in southern California and Arizona 1978 and 1980: Proceedings of symposium, Sept. 17-18, 1980, National Academy Press, Washington, D.C., p. 291-320.

Weber, F.H., 1980, Landsliding and flooding in southern California during the winter of 1979-1980: California Div. Mines and Geology, Open-file report 80-3LA.

Weber, F.H., J.A. Treiman, S.S. Tan and R.V. Miller, 1979, Landslides in the Los Angeles Region, California - effects of February-March 1978 rains: California Div. Mines and Geology, Open-file report 79-4LA.

Chapter 9

Dams

Photo of Copper Basin thin arch dam in San Bernardino County.

Dams

Editors Note: Imported water is the life-blood for semi-arid southern California. Therefore, most dams are built here to store water in their reservoirs, compared to other areas in the U.S. where the main purpose of dams may be flood control, developing hydroelectric power, and recreation. The papers in this chapter on dams describe their purposes, unique design, construction, and some failures.

OVERVIEW OF MAJOR DAMS IN SOUTHERN CALIFORNIA

Richard J. Proctor
Consultant
327 Fairview Avenue
Arcadia, California 91007

Water is the single most important commodity to semi-arid southern California. Seven local public water agencies store and supply potable water to more than half of California's population of 31 million. Three aqueducts each bring water more than 300 miles from three different sources to the greater Los Angeles and San Diego metropolitan areas.

The oldest aqueduct is the City of Los Angeles' Owens Valley Aqueduct, completed in 1913 and enlarged in the 1960s, with more than 30 reservoirs on line. Next is the Colorado River Aqueduct of the Metropolitan Water District of Southern California (MWD), whose construction was completed in 1939 and enlarged in the 1960s, with eleven reservoirs on line. Most recently the California Aqueduct was completed in 1974 with four major reservoirs in southern California (see Table of Major Dams).

The City of San Diego operates many storage reservoirs, eight with capacities greater than 20,000 acre-feet. More than 90% of the city's water is derived from the Colorado River and California Aqueducts.

The U. S. Bureau of Reclamation has constructed and operates six large water supply dams and reservoirs in southern California, mainly in Santa Barbara and Ventura Counties and in the San Joaquin Valley. Several small water agencies own and operate individual dams and reservoirs, some also used extensively for recreation, such as Arrowhead and Big Bear lakes.

Water supply engineers and planners realized long ago that emergency reservoir storage was needed south of the San Andreas fault, because all three aqueduct sources are north and east of the fault. As a result, more than 1,000,000 acre-feet of storage is available to the southern California metropolitan areas should future movement on the San Andreas fault disrupt the aqueducts. This contingency allows several months water supply while the aqueducts are being repaired.

The MWD's Domenigoni Dam (UC) in Riverside County will create the largest storage reservoir in southern California—800,000 acre-feet. The reservoir will consist entirely of imported water because the dam will close a U-shaped range of hills, similar to the Perris Reservoir setting, with essentially no natural drainage into the reservoir.

Four dam failures in southern California are worth noting, because two failures were caused by earthquake shaking and two were caused by poor foundation geology: Sheffield Dam was built by the City of Santa Barbara in 1918, and failed during the Santa Barbara earthquake of 1925. Sketchy records show it was constructed of improperly (by today's standards) compacted earth in relatively thick layers with a minimum of compactive effort (Sherard and others, 1963). Investigations indicated that during earthquake shaking, the central part of the 35-foot high embankment liquified and slid downstream about 100 feet. No one was injured.

St. Francis Dam, 45 miles north of Los Angeles, failed suddenly in 1928, drowning more than 400 people (see Rogers in following paper). This failure was largely responsible for the realization of the need for thorough geologic investigations prior to siting a dam. This dam failure also greatly fostered the creation of the specialty of engineering geology. The dam was sited on a fault that separates sandstone from adversely dipping schist. The many prominent geologists who investigated the failure are listed in the following Table of Early Project Geologists.

Baldwin Hills Reservoir was constructed on a hill in Los Angeles by scooping out the hilltop and creating a dam by filling a ravine. It was constructed on splays of the Newport-Inglewood fault zone. Failure is believed to be related to both poor site location and dam design. Differential subsidence along one of the faults was caused by local petroleum pumping and the weight of the water on the lining overlying the fault (see Baldwin Hills Reservoir Failure References at end of this chapter). Water traveled along the fault, largely undetected by the subdrain monitoring system, and caused piping failure of the embankment. Five people drowned.

Lower San Fernando Dam was built in 1918 by the hydraulic fill method of consolidating earthfill by ponding. During the strong shaking of the 1971 M6.5 San Fernando earthquake, the embankment failed. The nearest surface rupture traces were only a few hundred feet east on the Mission Wells fault segment, though the epicenter was about 17 miles northeast. The dam failed by liquefaction causing the central part to slide into the reservoir (see References at end of chapter). The headscarp extended downslope beyond the dam crest. The reservoir water level reached to just four feet below the lip of the headscarp, thus preventing overtopping. Fortuitously, the reservoir level was 14 feet lower than normal due to a recent inspection of the spillway, and to an apparent misreading by the caretaker who lowered the reservoir a few feet more. The inundation area of the dam is the central San Fernando Valley, where 80,000 people were evacuated for three days, until the reservoir was lowered to a safe level. Thus, what would have been the worst disaster in the U.S. was narrowly averted. The dam was emptied, the embankment reshaped, and a new smaller dam (Los Angeles Dam) was constructed in 1977 within the old reservoir.

The following **Table of Major Dams in Southern California** provides some statistics on 65 of the largest dams, including two under construction, in the categories of water supply, flood control, and hydro power.

Some early pioneering geologists became employees and consultants to various public agencies that were planning and building large dams and tunnels in Southern California. Some of these early engineering geologists are listed in the **Table of Early Project Geologists**.

TABLE OF MAJOR DAMS IN SOUTHERN CALIFORNIA

Owner and Name	County	Year	Capacity acre-feet	Type
WATER SUPPLY DAMS				
California Department of Water Resources				
Castaic	L.A.	1973	350,000	Earth
Cedar Springs (Silverwood)	S.Bern.	1973	131,000	Earth
Perris	Riv.	1973	179,000	Earth
Pyramid	L.A.	1971	78,000	Earth
Los Angeles Department of Water and Power				
Bouquet Canyon	L.A.	1936	36,000	Earth
Encino	L.A.	1924	10,000	Earth
Haiwee	Inyo	1913	58,000	Hydraulic
Long Valley (Crowley)	Mono	1941	183,000	Earth
Los Angeles	L.A.	1977	10,000	Earth
Lower San Fernando	L.A.	1918/77	10,000	Hydraulic
Mulholland (Hollywood)	L.A.	1924	4,000	Gravity
Tinemaha	Inyo	1928	16,000	Earth
Stone Canyon	L.A.	1924	10,000	Hydraulic
Metropolitan Water District				
Copper Basin	S.Bern.	1938	22,000	Arch
Gene Wash	S.Bern.	1937	6,000	Arch
Domengoni (UC)	Riv.	1998	800,000	Earth
Mathews	Riv.	1938/61	182,000	Earth
Morris	L.A.	1935	34,000	Gravity
Skinner	Riv.	1973	43,800	Earth
City of San Diego				
Barrett	S.Diego	1922	44,000	Gravity
El Capitan	S.Diego	1934	116,000	Hydraulic
Lake Hodges	S.Diego	1918/36	34,000	Mult-Arch
San Vincente	S.Diego	1943	90,000	Gravity
Savage	S.Diego	1919	56,000	Gravity
Sutherland	S.Diego	1954	29,000	Mult-Arch
Sweetwater	S.Diego	1888	28,000	Gravity
Morena	S.Diego	1895	50,000	Rockfill

South Bay Irrigation District

Lake Loveland	S.Diego	1945	25,000	Arch

Vista Water District

Henshaw	S.Diego	1923	203,000	Hydraulic

U. S. Bureau of Reclamation

Bradbury (Cachuma)	S.Barb.	1953	205,000	Earth
Casitas	Ventura	1959	254,000	Earth
Friant (Millerton)	Fresno	1947	520,000	Gravity
Imperial	Imperial	1938	85,000	Earth
Parker (Havasu)	S.Bern.	1938	648,000	Arch
Twitchel	S.Barb.	1958	240,000	Earth

United Water Conservation District

Santa Felicia	Ventura	1955	100,000	Earth

Rancho California Water District

Vail	Riv.	1949	51,000	Arch

Lake Arrowhead Company

Lake Arrowhead	S.Bern.	1922	48,000	Hydraulic

Bear Valley Mutual Water Company

Bear Valley (Big Bear)	S.Bern.	1911	54,000	Mult-Arch

FLOOD CONTROL DAMS

L. A. County Flood Control District

Big Tujunga	L.A.	1931	4,000	Arch
Cogswell	L.A.	1935	8,000	Rockfill
Pacoima	L.A.	1929	4,000	Arch
Puddingstone	L.A.	1928	17,000	Earth
San Gabriel No. 1	L.A.	1938	44,000	Earth

Irvine Company

Santiago Creek (Irvine)	Orange	1933	25,000	Earth

U. S. Corps of Engineers

Hansen	L.A.	1940	27,000	Earth
Isabella	Kern	1953	570,000	Earth
Mojave	S.Bern.	1971	157,000	Earth
Pine Flat	Fresno	1954	1,000,000	Gravity
Prado	Riv.	1941	314,000	Earth
Seven Oaks (UC)	S.Bern.	1999	113,000	Earth
Santa Fe	L.A.	1949	32,000	Earth
Sepulveda	L.A.	1941	17,400	Earth
Success	Tulare	1961	850,000	Earth
Terminus (Kahweah)	Tulare	1962	150,000	Earth
Whittier Narrows	L.A.	1957	67,000	Earth

HYDRO POWER DAMS

Southern California Edison Company

Big Creek No. 7 (Redinger)	Fresno	1951	35,000	Gravity
Florence Lake	Fresno	1926	64,000	Mult-Arch
Gem Lake	Mono	1917	17,000	Mult-Arch
Huntington Lake	Fresno	1917	89,000	Gravity
Mammoth Pool	Fresno	1960	123,000	Earth
Shaver Lake	Fresno	1927	135,000	Gravity
Vermillion (Lake Edison)	Fresno	1954	125,000	Earth

Pacific Gas & Electric

Courtright	Fresno	1958	123,000	Rockfill
Wishon	Fresno	1958	125,000	Rockfill

TABLE OF EARLY PROJECT GEOLOGISTS

California Department of Water Resources

compiled by Laurence B. James

Geologist	Date	Job Status
Chester Marliave	1930s-early 1940s	Chief Geologist
Rollin Eckis	1932-1934 (Los Angeles Office)	Chief Geologist (later President of ARCO)
Elmer Marliave (Charter M AEG)	1940s and early 1950s	Chief Geologist
Laurence James (Charter M AEG)	1956-1976	Chief Geologist
Wilferd Peak (M AEG)	1976-1984	Chief Geologist
Charles P. Berkey	1920s	Consultant on Dam Safety Board
F.L. Ransome	1928	Consultant on St. Francis Dam failure
Bailey Willis	1928	Consultant on St. Francis Dam failure
George Louderback	1920s-1930s, 1953 - 1954	Consultant on Dam Safety Board Tehachapi Tunnels

Geologist Consultants employed by DWR 1946-19746

Roger Rhoades	Geologist member of the State Water Project Consulting Board and of the boards for Land Subsidence in San Joaquin valley and for investigation of the failure of Baldwin Hills Reservoir.
Hugo Benioff	Seismologist, Consulting board for Earthquake Analysis.
Clarence Allen (M AEG) Analysis.	Geologist, Consulting Board for Earthquake
Bruce Bolt	Seismologist, Consulting Board for Earthquake Analysis.
George Kiersch (HM AEG)	North Coast tunnels and Pyramid Dam.
Frank Nickel	North Coast tunnels and State Water Project.
Elmer Marliave (Charter M AEG)	State Water Project (1950s-1960s).
Claire Holdredge (Charter M AEG)	North coastal studies (1950s).
Pierre St. Amand	Seismologist, Baldwin Hills Reservoir failure investigation, 1963.
U. S. Grant IV	Geologist, Safety of Dams, Malibu damsite.
David Todd	Geohydrologist, underflow at Castaic damsite
Robert Loufbouro	Mining geologist, North coastal tunnel studies

TABLE OF EARLY PROJECT GEOLOGISTS (cont.)

Metropolitan Water District of Southern California

compiled by Richard J. Proctor

Geologist	Date	Job Status	Project
F. L. Ransome	1930-1935	Consultant	Colorado River Aqueduct
V. P. Pentegoff, (M AEG)	1930s &1964-1976	Safety Inspector; later a consultant to MWD	Colorado River Aqueduct; various tunnels, dams
T. F. Thompson, (M AEG)	1930s &1964-1976	Safety Inspector; later a consultant to MWD	Colorado River Aqueduct; various tunnels, dam
E. J. Zielbauer, (Past Pres. AEG)	1930s	Safety Inspector; (became Ch. Geol. LAFCD)	Colorado River Aqueduct
John P. Buwalda	1933-1954	Consultant	Colorado River Aqueduct; San Juan Tunnel
Richard J. Proctor, (Past Pres. AEG)	1958-1980	First full-time geologist hired by MWD with a geologist title; currently a consultant to MWD	Colorado River Aqueduct; various tunnels, dams
Edwin B. Eckel, (HM AEG)	1960s	Consultant	Bolsa Island Desalting Plant
William I. Gardner, (Charter M AEG)	1960s	Consultant	Bolsa Island Desalting Plant
Clarence Allen, (M AEG)	1960s	Consultant	Bolsa Island Desalting Plant & currently on proposed projects
Richard H. Jahns, (HM AEG)	1964-1976	Consultant	Various tunnels, dams
John F. Mann, (M AEG)	1968	Consultant	San Fernando Tunnel, groundwater
Mason L. Hill, (M AEG)	1972	Consultant	Newhall Tunnel oil seeps
Charles F. Richter, (HM AEG)	1978	Consultant	Seismic hazard to pipelines

Los Angeles Department of Water and Power

compiled by Henry S. Mayeda

Geologist	Date	Job Status	Project
John C. Branner	1925	Consultant	St. Francis Dam.
Robert T. Hill	1928	Consultant	Bouquet Dam.
F.L. Ransome	1929-32	Consultant	Bouquet Dam; Bell Cyn. and Las Virgenes aband. dam sites.
Karl Terzaghi	1930s	Consultant	Mono Craters Tunnel
A.E. Sedgwick	1930s	Consultant	Bouquet, Tinemaha, Mono Basin dams.
Charles P. Berkey	1931-33	Consultant	Bouquet, Lower San Fernando Dams.
Rush T. Sill	1932-33	Consultant	Bouquet, Lower San Fernando Dams.
Chester Marliave	1934	Consultant	Haiwee Dam.
Robert R. Wilson, (M AEG)	1947-67 & 1967-71	Chief Geologist; Consultant	Many projects. Baldwin Hills Reservoir, failure litigation.
U.S. Grant IV	1967-71	Consultant	Baldwin Hills Reservoir failure litigation.
John P. Buwalda	1952	Consultant	Eagle Rock Dam.
John F. Mann , (M AEG)	1961- present	Consultant	Ocean Wood Terrace landslide; groundwater litigation San Fernando Valley and Mono Basin.
James Richmond	1965-66	Consultant	Eagle Rock, Santa Ynez Dams.
Charles F. Richter (HM AEG)	1971-80	Consultant	Member, Dam Safety Board.
V.P. Pentegoff, (M AEG)	1971-75	Consultant	Member, Dam Safety Board.
Robert T. Bean, (Past Pres. AEG)	1975-80	Consultant	Member, Dam Safety Board.
Richard H. Jahns, (HM AEG)	1972-73	Consultant	Chatsworth, Fairmont, Haiwee, Los Angeles, Lower Franklin Dams.

TABLE OF EARLY PROJECT GEOLOGISTS (cont.)

Los Angeles County Flood Control District

compiled by Michael Johnson

Geologist	Date	Job Status	Project
Louis T. Hill	1925	Consultant	Several dams
John P. Buwalda	1926	Consultant	Several dams
Andrew C. Lawson	1925	Consultant	Santa Anita, Pacoima a nd Puddingstone Dams
F. L. Ransome	1928,1929	Consultant	Big Tujunga Dam
Chester Marliave	1929,1935	Consultant	Mainly Big Tujunga Dam
C. F. Tolman	1933	Consultant	Cogswell Dam
R. F. Cuthbert	1935,1936	Consultant	Big Dalton, Pacoima, and Cogswell Dams
Bailey Willis	1930s	Consultant	San Gabriel Dam
A. E. Sedgwick	1935	Consultant	San Gabriel Dam
Warren N. Thayer	1937-1950s	First full-time geologist hired by LAFCD	
E. J. Zielbauer, (Past Pres. AEG)	1950s-1970s	Chief Geologist	

Photo 1. Atomic Energy Commission's "Blue Ribbon" committee for MWD's proposed Bolsa Island Desalting Plant, 1967. Left rear Mike Agbabian, CE; William Diment, USGS Geophysicist; William Gardner, deceased, former USBR chief geologist; Clarence Allen, Caltech emeritus professor of geology; lower left: R.McFerren (?); Ed Eckel, deceased, former USGS Geologist; Harry B. Seed, deceased, former professor of Soils Engineering, UC Berkeley; and Stanley Wilson, CE, deceased, founder of Shannon-Wilson. MWD photo.

Photo 2. Part of the prominent Caltech geology faculty circa 1947. Five geologists/seismologists who were also consultants, front row from left: John P. Buwalda, Ian Campbell, Richard H. Jahns, Beno Gutenberg, Charles F. Richter.

REASSESSMENT OF THE ST. FRANCIS DAM FAILURE

J. David Rogers
CE and Engineering Geologist
Rogers/Pacific,Inc.
396 Civic Drive
Pleasant Hill, CA 94523

INTRODUCTION

St. Francis Dam was built by the City of Los Angeles Bureau of Water Works and Supply in 1925-26 as a curved concrete gravity dam, approximately 200 feet high, in San Francisquito Canyon, about 5 miles northeast of what is now Magic Mountain, California. The stated purpose of the dam was to provide 38,000 acre-feet of emergency storage for Los Angeles-Owens River Aqueduct water should slippage along the San Andreas fault sever the aqueduct. The dam failed catatrophically upon its first full filling, near midnight on March 12/13, 1928, killing 450 in the San Francisquito and Santa Clara River Valleys. It was the greatest man-made disaster in California history.

No less than a dozen separate investigations of the failure followed, most notably, that of the State of California, which convened on March 19th, made one site visit, and issued their report 5 days later (Committee Report, 1928). The State's consulting board concluded that the dam's failure emanated from a fault contact between the Pelona Schist and Sespe formation on the right, or west abutment. This somewhat simplistic explanation was offered after observing that blocks from the dam's right abutment were found further downstream than those of the opposing, east side.

The failure of St. Francis Dam still serves to remind us that disaster awaits those who do not employ sound engineering geologic input, regardless of their respective engineering abilities. The public outcry following the disaster led to the humiliation and demise of William Mulholland, the dam's principal sponsor and the architect of Los Angeles' water supply system. It also led to the formation of the State's Division of Safety of Dams (DSOD), which retains regulatory authority over all but federally-owned dams in California.

Detailed geomorphic assessments suggest that St. Francis Dam was constructed with its eastern abutment against paleomegaslides, comprised of the Pelona Schist. The balance of this article explores what is currently being learned about the St. Francis disaster by reassessing its failure with modern forensic analytical techniques, many of which were unavailable to our professional contemporaries in 1928. Research-to-date suggests that St. Francis was not designed with a modern appreciation of uplift theory; the dam's base width was not as thick as previously assumed; and the designers were not aware that the left abutment was a paleomegalandslide or that the Sespe beds comprising the opposing abutment would slake upon submersion. A review of the available evidence suggests the dam failure sequence was brought about by the partial reactivation of the paleomegaslide, within the schist comprising the east abutment. As the slide mass was saturated to a depth of some 200 feet, "keyblocks", or discontinuity-bordered wedges beneath the sloping east abutment, appear to have been hydraulically "lifted", thereby breaking the dam and initiating a rapid chain-reaction failure, similar to the style of failure subsequently experienced at Malpasset Dam in France in 1959.

THE SITING OF ST. FRANCIS DAM

The St. Francis Dam was originally conceived by Bill Mulholland, who served as Chief Engineer of Los Angeles' Bureau of Water Works and Supply (later the Department of Water & Power) from 1886 until his untimely retirement in 1928. Mulholland was among the first in America to conceive and construct a long-distance municipal water system, the 300-mile long Los Angeles-Owens River Aqueduct, completed between 1906-1913.

Mulholland visualized a dam site along San Francisquito Creek, adjacent to one of the camps built during the aqueduct's construction. He had excavated exploratory adits into the Sespe beds and filled these with water to evaluate the adequacy of the Sespe beds to support a reservoir (Mulholland, 1928). Mulholland had also recognized the tendency of the Pelona Schist slip "along its slaty cleavage" in the outcrops along the southeast side of the canyon (City of Los Angeles, 1911). Mulholland's attraction to the dam site was linked to what he perceived as favorable topography: a natural narrowing of the canyon downstream of a broad, upstream platform, thereby creating a large water storage area (Mulholland, 1928). Unbeknownst to Mulholland, the reasons for this favorable topography lay in the fact that the site had already served as a natural reservoir, due to damming of San Francisquito Creek by ancient megalandslides in the Pelona Schist. The seemingly intact Pelona Schist had rotated downward, onto the opposing bank of Sespe conglomerate, thereby blocking the Canyon and creating a landslide dam (Figure 1). The waters of San Francisquito Creek had eventually overtopped the landslide dam and reexcavated a channel. The broad flat area, seen by Mulholland as an excellent reservoir site, had actually been created through sedimentation behind the paleolandslide dam. Evidence of the paleoslides and the impounded lake is clearly seen today, in the form of stepped terraces along the Pelona escarpment (Figures 1, 13 and 14).

DESIGN AND CONSTRUCTION

The Los Angeles Aqueduct was completed in 1913, a major engineering marvel in its time, bringing Mulholland a great deal of notoriety and respect. But, to the ranchers of the

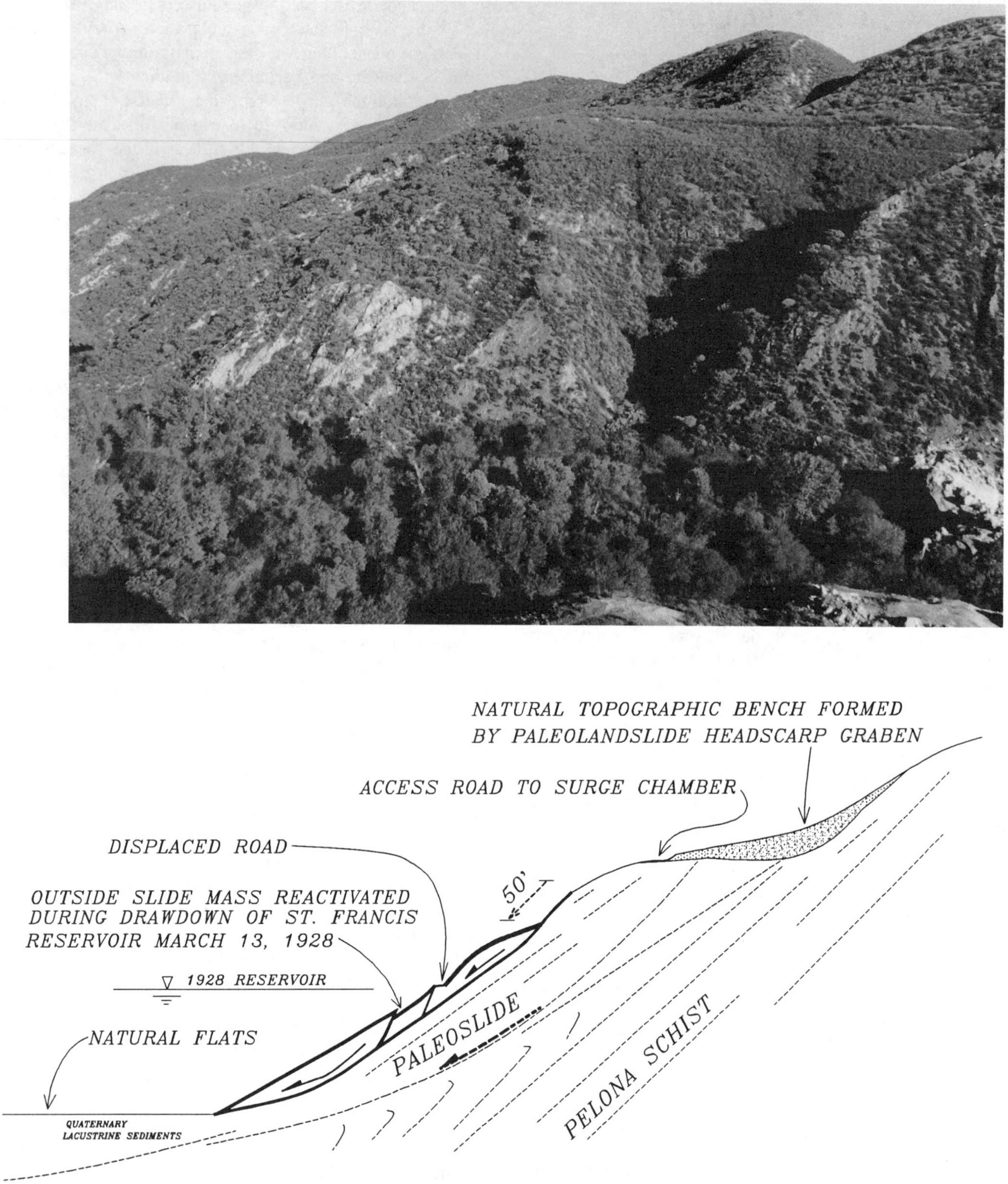

Figure 1. Present day view looking upstream at Pelona Schist escarpment along the southeastern side of San Francisquito Canyon. The remains of St. Francis Dam lie in the right foreground, while a drawdown-induced landslide, which slid approximately 50 feet into the reservoir, is at left center. The bench at the top of the slope appears to be the headscarp graben of an extensive paleomegaslide complex developed within the Pelona Schist, likely during the Pleistocene Epoch. The Los Angeles Aqueduct was built upon this surface in 1908-13. A schematic cross section through the 1928 slide and the surrounding slope is presented below.

Owens Valley, Mulholland was a fiend with no equal, as he was the point man for Los Angeles' tangled web of water rights acquisition that took away the water from their homesteads. The ranchers had been outmaneuvered by slick city lawyers who lobbied for favorable legislation in far-away Washington, D.C., long before the ranchers realized what the Angelinos were up to. The ranchers' sense of frustration often came to a head, and in mid-1920's, they struck back by dynamiting the aqueduct at vulnerable points along its remote traverse through the Owens Valley. The plight of these rural folks drew considerable sympathy at the time, even within newspapers, such as the Los Angeles Record, which never failed to take its shots at the City's Departments of Water and Power. These circumstances combined to create an air of secrecy and suspicion amongst Mulholland's Water Works people, which likely prevented much of St. Francis Dam's engineering records from ever being evaluated by outside entities (for example, no correct cross section of the dam had ever been released or published prior to this paper).

In 1916 the Los Angeles Department of Public Services bifurcated a new companion agency to Mulholland's Bureau of Water Works and Supply, called the Bureau of Power and Light. This new bureau constructed San Francisquito Powerhouse Number 1 along the aqueduct a year later (about 5.5 miles upstream of what would later become the St. Francis Dam site). In 1920, Power and Light constructed Powerhouse No. 2, six miles down the canyon, approximately 7000 feet downstream of the future damsite. Both of these facilities remain in service today.

Because of the political sensitivity of water rights acquisition, Mulholland appears to have proceeded with the planning of St. Francis Dam in a virtual veil of secrecy. Los Angeles Department of Water and Power (LADWP) records show that the City surveyed the entire dam and reservoir area in detail in 1922. The City filed a condemnation petition for the reservoir inundation area with the Federal government in April 1924. Their original plan called for a concrete gravity dam, virtually identical to the design then being utilized for the Hollywood Reservoir (which after the failure of the St. Francis Dam, was buttressed with an earthen berm against its downstream face, ENR, 1934). As originally conceived, the dam was to extend some 175 feet above the bed of San Francisquito Creek and provide storage for 30,000 acre feet of water. In July 1924, the lake's capacity was increased to 32,000 acre feet by adding a small wing dike, approximately 10 feet high, extending from the west abutment. Construction began in the spring of 1924, with the first concrete being placed the following August.

In July 1925, after eleven months of placing concrete, Mulholland's engineers apparently decided to increase the reservoir capacity once again (to 38,168 acre feet), by adding an additional 10 feet to the dam's height (but, without changing the base width). This increased storage was made possible by raising the dam to a height of 185 feet above the

Figure 2. St. Francis Dam as it appeared just after its completion in May 1926. It was a curved concrete gravity section, standing approximately 200 feet high and was comprised of 130,000 cubic yards of concrete without benefit of any contraction joints, drainage galleries, cut-off walls, or grout curtain. The Pelona Schist made up the east abutment, on the right side of the photo, while the Sespe formation red beds comprised the upper two-thirds of the west abutment (shown at left).

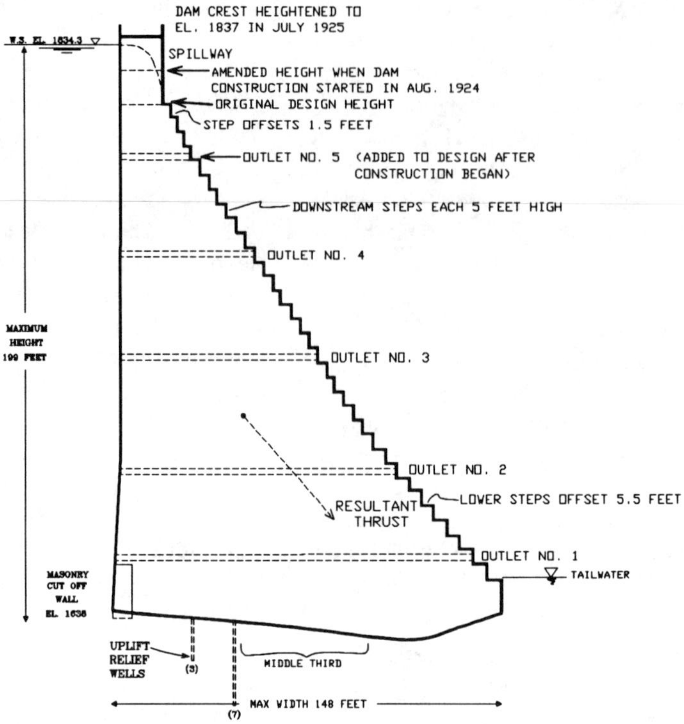

ELEVATION VIEW OF ST. FRANCIS DAM

TAKEN THROUGH SURVIVING BLOCK 1

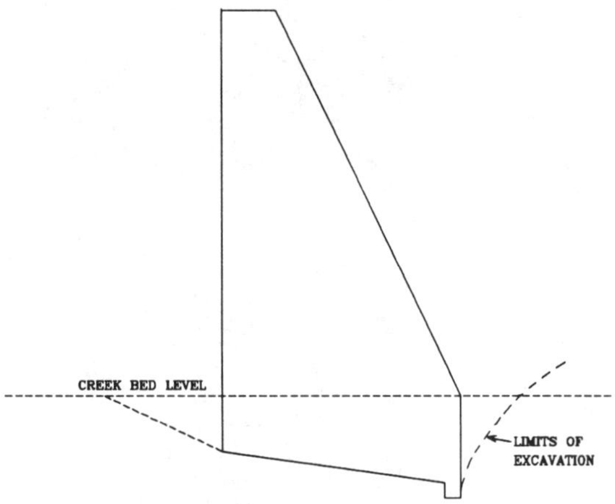

DESIGN SECTION THRU STA 1 + 25.00 FROM LA DWP FILES. LIKELY THE ORGINAL DESIGN, THE SECTIONS ARE CONSISTENT IN SHOWING THE DOWNSTREAM TOE CHOPPED OFF VERTICALLY, BELOW THE LEVEL OF THE STREAM BED.

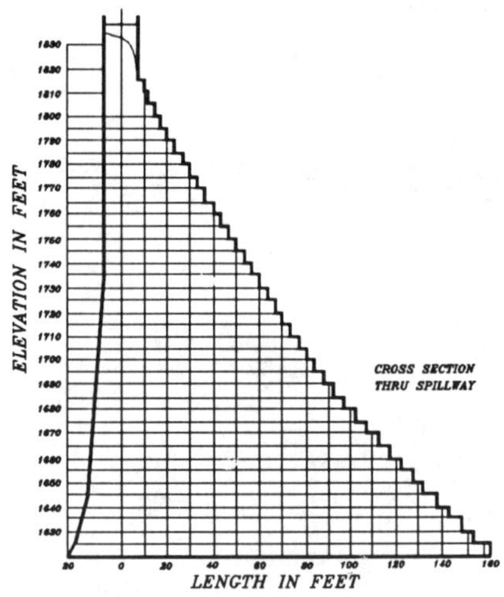

CROSS SECTION PROVIDED BY THE LOS ANGELES BUREAU OF WATER WORKS AND SUPPLY TO GOVERNOR'S INQUIRY BOARD, SHOWING A FLARED DOWNSTREAM TOE. MULHOLLAND TOLD THE CORONER'S INQUEST THAT THE DAM WAS 170 FEET THICK AT THE BASE (WHICH, IF TRUE,APPEARS TO HAVE BEEN ADDED LATER)

Figure 3. Three cross sections of St. Francis Dam, through the outlet works contained in Block 1, the only section of the dam to survive the failure. The upper section represents a reconstruction by the author. That at the lower right is taken from LADWP files, and likely represents the original dam design. Note how the downstream toe was clipped, a feature Outland (1963) had discovered when examining construction photos. The section at lower right was the one provided to the various boards of inquiry into the failure. A few uplift relief wells had been provided for the central core of the dam, within the stream channel. No such protection was afforded the sloping abutments.

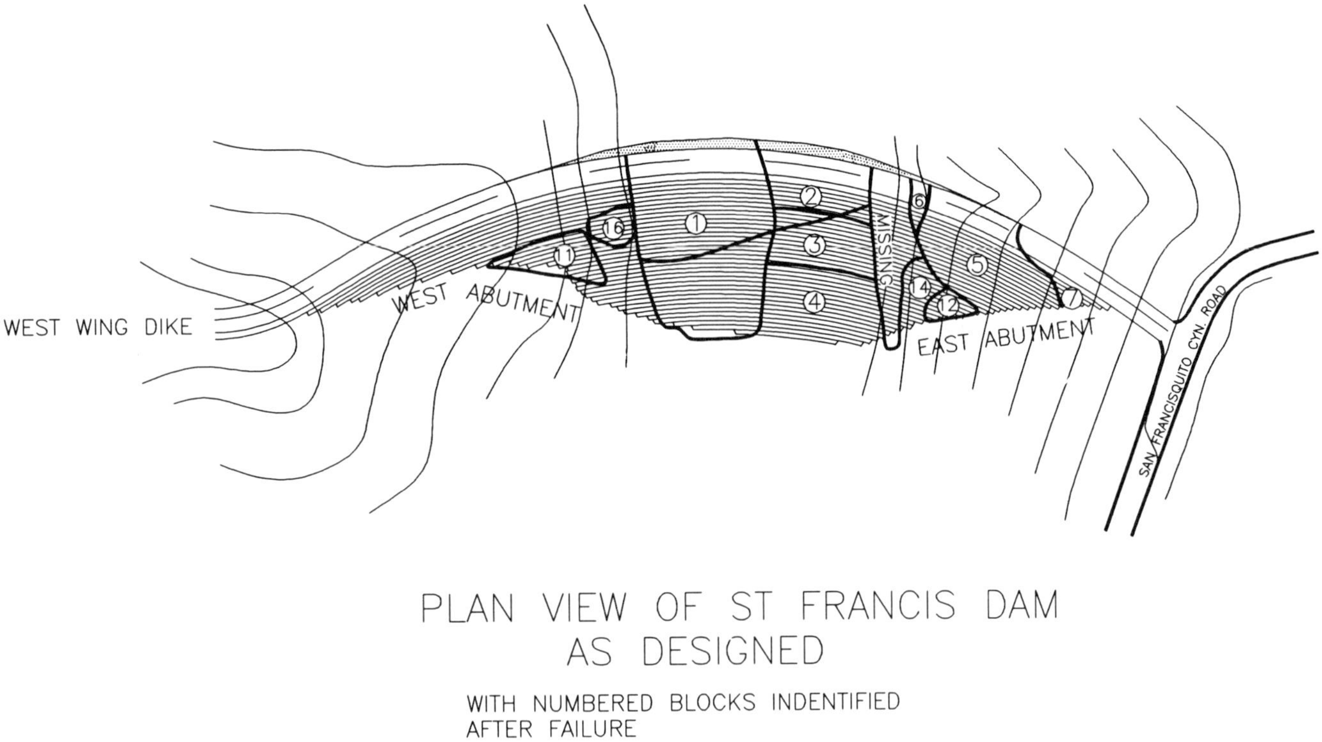

Figure 4. Plan view of the main section of St. Francis Dam as orginally designed, from the files of the Los Angeles Department of Water and Power. Diagonal lines and numbers delineate the blocks identified in the post-failure analysis by the State (shown in Figure 8). The portion of the dam between Blocks 2/3/4 and the East Abutment (Blocks 5/6/7) was never found, this portion of the dam becoming the so-called "missing section". The dam's horizontal steps were each five feet high; and varied in width, from 5.5 feet in the lower part of the structure, to as little as 1.5 feet in the uppermost part of of the dam. The unique spacing of the steps enables forensic assessment of where the blocks displaced downstream came from within the structure.

creekbed, and extending the west wing dike some 1300 feet along a natural ridgeline in the Sespe.

Owens Aqueduct water was first diverted into the reservoir on March 1, 1926, and the structure was completed the following May. The main dam section across San Francisquito Creek was designed as a free-standing gravity section, as shown in Figure 3. Very little seepage relief was afforded in the design, only a few uplift relief wells beneath the central core were employed (as shown in the upper half of Figure 3). No cutoff walls, grout curtains, or drainage inspection/collection galleries were employed in the structure's design, a fact that later investigative boards would repeatedly mention. In addition, no provision was made for emplacement of expansion joints within the main structure.

The main portion of the dam was arched upstream, on a 500 foot radius, as depicted in the plan view, shown as Figure 4. Additional load-carrying capacity due to arch action was neglected in the dam's design, a common and seemingly conservative decision. During the Los Angeles County Coroner's Inquest following the dam's failure, Mulholland would testify that the dam was designed with a "factor of safety of three or four" (Mulholland, 1928, p. 10).

INITIAL OPERATION AND FILLING

Filling of the St. Francis reservoir was nothing short of dramatic, the level rising at an average rate of 1.8 feet per day over the first three months (March 1 to June 1, 1926). The St. Francis reservoir pool reached elevation 1832, three feet below spillway crest, by May 10, 1927 (177 feet above the creekbed). This elevation was held until May 26, at which point the spring run-off ceased, and the reservoir lowered. Modest spillway capacity was afforded the structure through the employment of narrow slits, only 18 inches high, that were built into the main dam section (from elevation 1835 to 1836.5, as shown in Figure 3). In all, there were eleven 20-foot wide spillway bays (seen to good effect in Figure 2). An additional emergency overflow weir was built into the end of the wing dike section, approximately 1100 feet northwest of the main dam section.

During the initial filling of 1926-27, several cracks appeared, transverse to the dam's axis, in the downstream face of the main structure. These cracks appeared to Mulholland to be "transverse contraction cracks", presumably caused by the thermal stresses associated with the curing of the mass concrete (the main structure was comprised of 130,500 cubic yards of concrete, placed in just 16 months). Two sets of cracks appear to have formed, those on the steeply rising flanks of the structure, and two within the maximum section. The approximate locations of these fractures, deducted from verbal descriptions and photographs, are depicted schematically in Figure 5. The flank cracks were observed to be wider at their juncture with the abutments, narrowing upwards; while the two longer, main section cracks, appeared widest at the top, near the parapet wall. The transverse

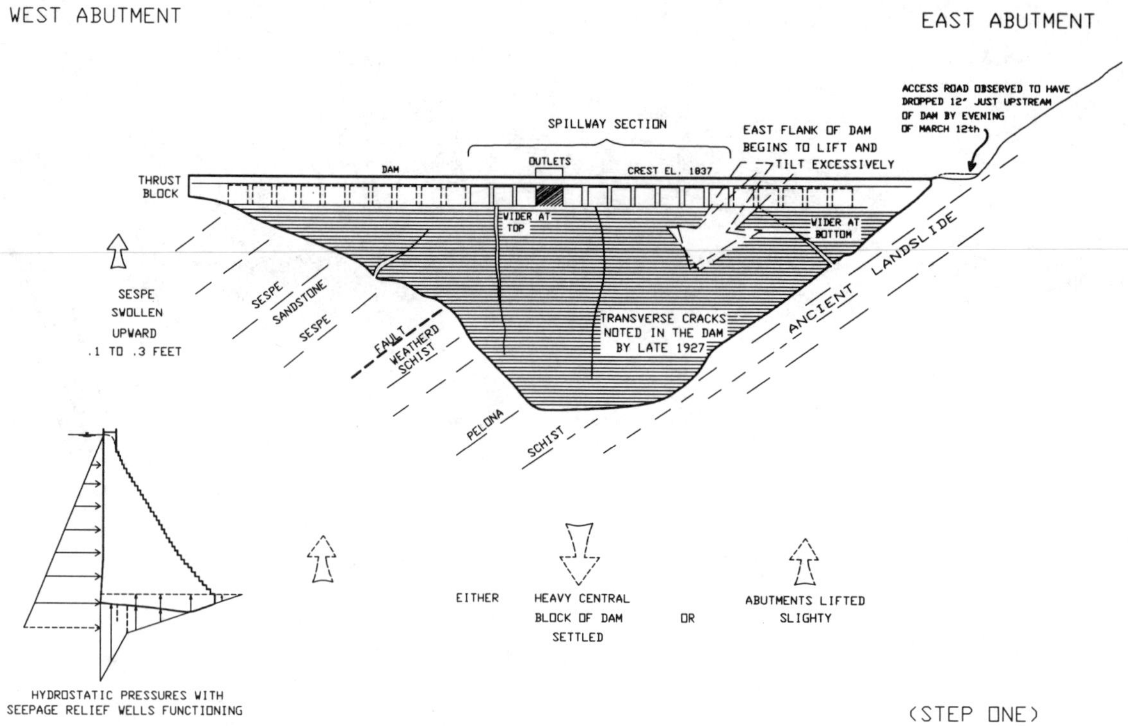

Figure 5. Schematic elevation view of transverse cracks which appeared on the downstream face of the dam prior to its failure. Th four cracks most often described are portrayed, their having been formed by late 1927. Both abutment cracks were angled obliquely, and widest at their junctures with the foundation rock, suggesting either abutment uplift or main section settlement. The two longest cracks, in the dam's central section, were later found to be those bordering Block 1.

cracks within the main dam block were wedged with oakum on the downstream face and backfilled with cement grout.

In February 1928 the lake level again rose to the dam crest and a number of leaks developed in the foundation, mostly on the west side, within the Sespe conglomerate. One of these leaks was in the vicinity of the fault separating the Sespe from the Pelona Schist. It would appear that the Sespe was little deformed in this location, having been conformably deposited upon an eroded surface of the Pelona, nearly coincident with the latter units' regional foliation. The paucity of structural distortion more than a few feet into either unit, suggests that inter-formational slippage along this boundary was responsible for the fault developing (due to the dramatic stiffness variance between the two units; see Clements, 1929).

When the spring run-off began arriving in January 1928, the reservoir was allowed to fill to maximum capacity, reaching elevation 1934.75 (three inches below spillway) by March 7, 1928, whereupon no additional aqueduct water was diverted into the lake. The previous years' leaks gushed forth with additional discharge and new leaks developed on both abutments. During the first week of March, a substantive leak developed along the wing dike, issuing artesian flow of about 0.60 cubic feet per second (cfs). Mulholland ordered a work crew to install a 8" concrete pipe underdrain from this point, eastward along the base of the dike, to discharge along the west abutment contact of the main dam section, giving an appearance that there was seepage emanating along this juncture (Figure 6).

INCIDENTS LEADING UP TO THE FAILURE

By Monday March 12th, the reservoir pool had been held at Elevation 1934.75, just three inches below spillway, for five days. Wind-driven waves were lapping up, over the spillways, complicating the task of discerning between last-minute leaks and spillage (Figure 6). With all downstream storage facilities similarly filled to capacity, excess water from the Los Angeles Aqueduct were released into San Francisquito Canyon for the first time in almost two years that morning; the 30 cfs discharge emanating from the aqueduct's overflow gates where it crossed Drinkwater Canyon (about 3000 feet downstream of Powerhouse 2). Some downcanyon residents, noticing water in the usually dry creekbed, wondered if something wasn't wrong at the dam.

That same morning, the damkeeper, Tony Harnischfeger, had telephoned Mulholland to say that a new, larger leak, had developed on the Sespe abutment, and that the discharge was "dirty", cause for concern to any dam engineer, as this would be an alarm to the possibility of piping. It was a Monday morning, and Mulholland, very much the personification of "the man in charge" (Mulholland made a practice of visiting all of his 19 dams once every 10 to 14 days according to his testimony before the Coroner), chose to personally inspect the dam immediately with his chief assistant engineer, Har-

Figure 6. View of downstream face of St. Francis Dam, taken around noon on Monday March 12, 1928, the day of the failure. Note the windswept spillage coming over the spillways, thereby complicating the task of discerning last minute leakage from spillage. At this time approximately 2 cfs of water was running down the west abutment contact, shown at left center

vey Van Norman. Mulholland and Van Norman arrived by chauffeured car at the dam at 10:30 AM, whereupon they inspected the dam for the next two hours (Figure 7).

Wave-whipped water was issuing over the spillways, even though the lake had dropped to 1834.3 over the past two days (Figure 6). Their inspection revealed that about 2 cfs outflow was now emanating from the the western abutment, but the water was decidedly clear, which would preclude the possibility of piping (but not that of uplift). The volume of water issuing from the abutment was observed to be somewhat inconsistent, with a surging style of flow. Further investigation showed that the water had only become muddy where it washed over uncompacted sidecast fill, on the west abutment access road (which had a 18% uphill grade). Though not appreciating its significance at the time, the pair also noticed "a small stream of clear water" cascading down the exposed (downstream) juncture of the dam's east abutment, against the schist. Mulholland and Van Norman left the dam around 12:30 PM, assuring Harnischfeger of its apparent soundness. Within 12 hours, Harnischfeger and his six year old son would be the first victim's of the dam's collapse, their bodies never being found.

The apparent conditions at the dam site on Monday evening, March 12, 1928, are depicted schematically in Figure 5. No less than three separate parties drove by the dam within an hour of its apparent failure, just before midnight. All of these people were Bureau of Power employees stationed at Powerhouse No.1, five miles above the dam. The first of these was a family quoted by Charles Outland (1963, p.208) in his book, *Man-Made Disaster*, who had refused to be publicly identified. They told Outland that they had driven the San Francisquito access road, between Powerhouses 2 and 1, the evening of the failure. The unidentified source(s) related they had observed "the road had dropped at least 12 inches, just upstream of the dam's east abutment" (in the Pelona Schist). The location identified would seem to correspond to the upstream lateral scarp of the 1928 east abutment slide, shown in Figures 13, 14 and 15. Such an observation would be expected if the east abutment landslide was beginning to drop, and thereby, thrust against the back of the dam.

Two other employees stationed at Powerhouse 1, Elmer Steen and Katherine Span, had left their friends' home at Powerhouse 2 around 11:35 PM, climbing the canyon's southeast wall, and passing the dam's east abutment crest around 11:40 to 11:45 PM. In their testimony before the Coroner's inquest, they stated they had not seen "anything unusual", only that the dam was blacked out (which was normal) and appeared "spooky" in the moonlight as they passed along the rough San Francisquito Canyon Road (which was unpaved). Either of these individuals may also have been Outland's "mystery witness", as he had tracked them down for interviews in the 1950's. At the time of the interviews, Steen was on an LADWP pension.

Figure 7. (upper) Bill Mulholland, Harvey Van Norman and damkeeper Tony Harnischfeger walking across the crest of the swollen dam, around noon on March 12, 1928, approximately 12 hours before the failure. The west dike, running across the top of a natural ridge in the Sespe formation red beds, lies at right center

The last person to see the dam and live to tell about it was Ace Hopewell, another Powerhouse 1 employee, who traversed the canyon road by the dam sometime between 11:50 and 11:55 PM. Driving alone on a motorcycle with sidecar, he should have been able to notice if anything significant (such as concentrated, orifice outflow) was occurring, but he did not. He did recall seeing car headlights upcanyon, presumably those of Steen and Span's vehicle, but noticed nothing else that was unusual. About a mile above the dam, Hopewell stopped suddenly, sensing an unusual sound or shaking. He pulled over, but kept the engine running on his motorcycle, smoking a cigarette while listening to the strange crashing sounds about a mile behind him. He then continued up the canyon, reaching Powerhouse 1 and learning of the disaster he had so narrowly missed.

What happened next has been subject to considerable speculation since the time of the failure. A Stevens Water Stage Gage, situated upon the deep, central core of the dam (Figure 17), was retrieved after the failure. An examination of the pencil trace on the recording graph paper indicated that, in the 40 or so minutes prior to the catastrophic failure, the lake level appeared to have lowered about 3.6 inches, in an accelerating manner. The dam was presumed to have failed at 11:57:30 PM, as that was the time the Southern California Edison (SCE) Lancaster power line, running along the east abutment, was suddenly cut. This power line extended the length of San Francisquito Canyon, on a series of poles placed in tandem, and was located about 90 feet above the

Figure 7. (lower) Approximately the same view, as seen the following day, after the dam failed, leaving the west dike in place. Water had been leaking from beneath this dike the previous day. Note the widespread occurance of shallow earthflows on the Sespe slope, likely due to rapid drawdown-induced pore pressure differential as the reservoir emptied in about an hour (photo from Ventura County Museum of History and Art).

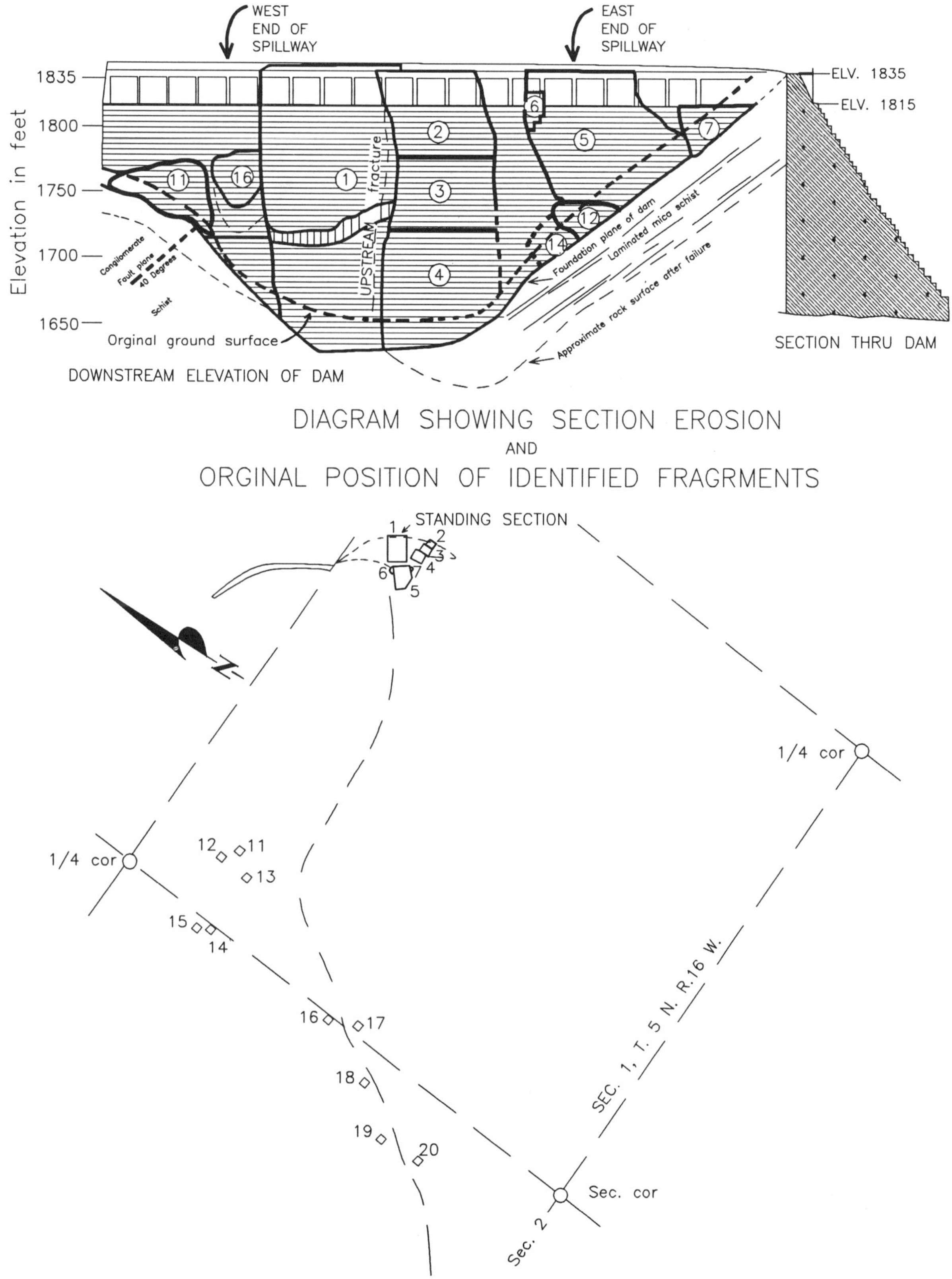

Figure 8. Distribution of fragments from the St. Francis Dam, as delineated in the State report of the disaster, issued 11 days after the failure. Some fragments from the west abutment were found as much as half a mile downstream, thereby influencing the Board's conclusion that the failure initiated on that side of the dam.

Figure 9. One of the better exposures of the Pelona Schist, seen today in the scar of the 1928 east abutment slide (shown in plan in Figure 15). Planes of foliation form the planar face, with numerous joints cutting the ground mass. A set of parallel shears cuts into the slope, while another, older set of joints is infilled with milky quartz. Discontinuity data from this and another, adjacent series of outcrops, was incorporated into the keyblock anaysis of the east abutment (photo by author).

east abutment crest of the dam (depicted in Figures 10, 13L, and 17).

The Stevens Gage consisted of a 12-inch diameter pipe affixed to the upstream face of the dam (Figure 17). Water fed into this so-called "stilling well" through a 1-inch diameter hole at its lower extremity, with the purported intent of filtering out the oscillatory effects of waves on the back of the dam. In the failure's aftermath, just about everyone attempted to correlate the apparent drop in lake level recorded by the gage with the theoretical outflow that could have lowered the lake by 3.6 inches. Predictably, this turned out to be a considerable quantity of water, beginning with around 1,000 cfs, and building up to over 15,000 cfs, just before the failure; an enormous amount of water in tiny San Francisquito Canyon not to be noticed by the 11:30 PM shift change at Powerhouse 2 or by any of the three aforementioned parties that drove along the creek, leading up to the dam just before the failure. The road connecting the dam to Powerhouse 2 paralleled the creek to a point about 2600 feet downstream of the dam, then climbed the east wall of the canyon, joining the east abutment crest at Elevation 1837. We can catch a glimpse of this road in Figure 26.

KINEMATIC ANALYSIS OF THE EAST ABUTMENT

In attempting to understand the potential failure mechanisms of either abutment, we first need to study the geometry and final positions of the dam's identifiable pieces (Figure 8). These positions can then be combined with LADWP survey data and geologic mapping to enable preliminary rock mechanics analysis to evaluate the abutments (Figure 9). As-built surveys indicate that at the base of the east abutment, a steep "step" was cut into the schist, leaving a slope inclined at approximately 55 degrees (from horizontal). By combining the excavated abutment profile survey (from LADWP files) with outcrop measurements, a computerized keyblock analysis (Goodman and Shi, 1985) was performed on the schist exposed in the east abutment (which was once part of the paleomegaslide). Figures 10 thru 12 present the results of such an analysis. Wedge-shaped keyblocks A, B and C were identified beneath the projection of the dam's original east abutment contact, in the vicinity of the steep cut shown on the construction surveys. The keyblock program serves to identify the geometry of these wedges, not their actual size. The representations made herein only approximate those that likely existed beneath the actual structure (since 40 vertical feet of schist, comprising the east abutment, was washed downstream during the 1928 failure, we are forced to presume that the fabric of the removed mass was similarly configured).

The original position of the dam's east side was then superposed over the identified keyblock wedges, as presented in Figure 11. Figure 12 graphically depicts an apparent imbalance of uplift forces and the dam's effective weight, absent any grouted seepage cutoff (in the style of analysis made

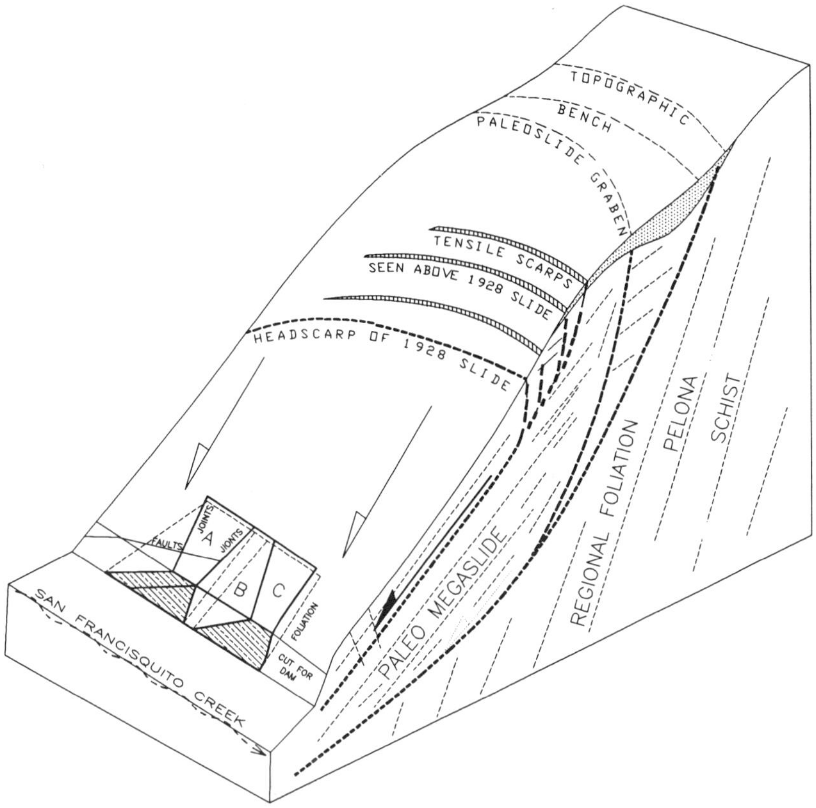

Figure 10. Schematic block diagram view of the St. Francis Dam east abutment, showing rock wedges A, B and C, identified in the keyblock analysis. The blocks are identifed by their geometry, the sizes portrayed here are only for purposes of illustration. Block B would present particularly unfavorable geometry with respect to uplift. The approximate positions of the paleomegaslide and the smaller, 1928 east abutment slide are shown.

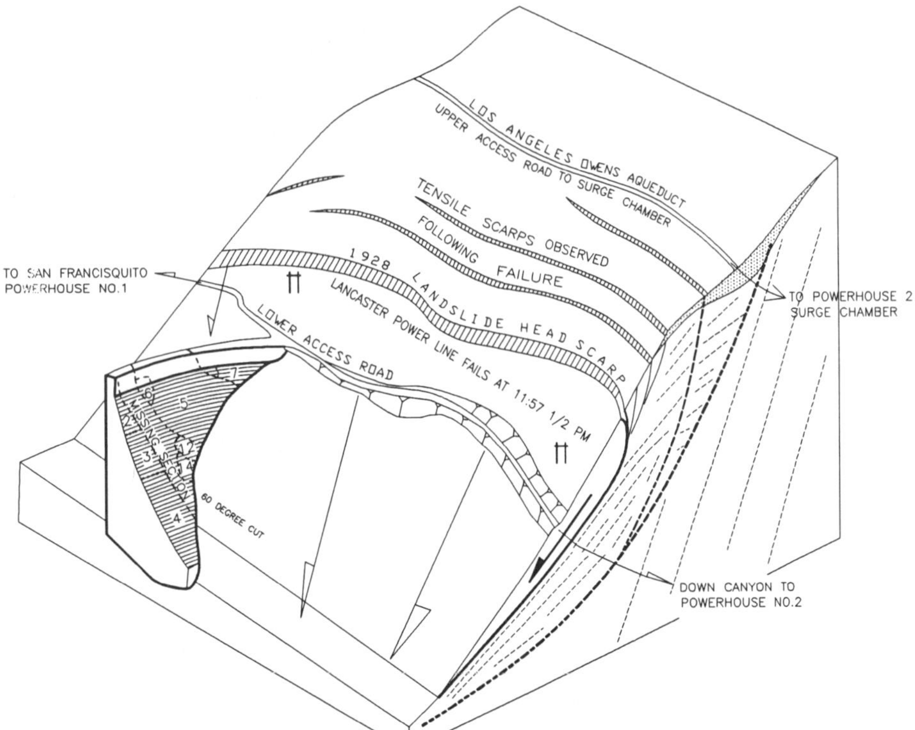

Figure 11. Schematic block diagram of the conditions on the east abutment of St. Francis Dam, just prior to its failure in March 1928. The "missing section" of the east abutment lies upon keyblocks A, B and C, shown in the previous figure, which coincide with the steepest excavation made for the dam. The approximate aereal limits of the east abutment slide are shown, as well as that of the Edison Company power lines that went down at 11:57:30 PM on the evening of the disaster. These lines were situated 90 feet above the dam crest.

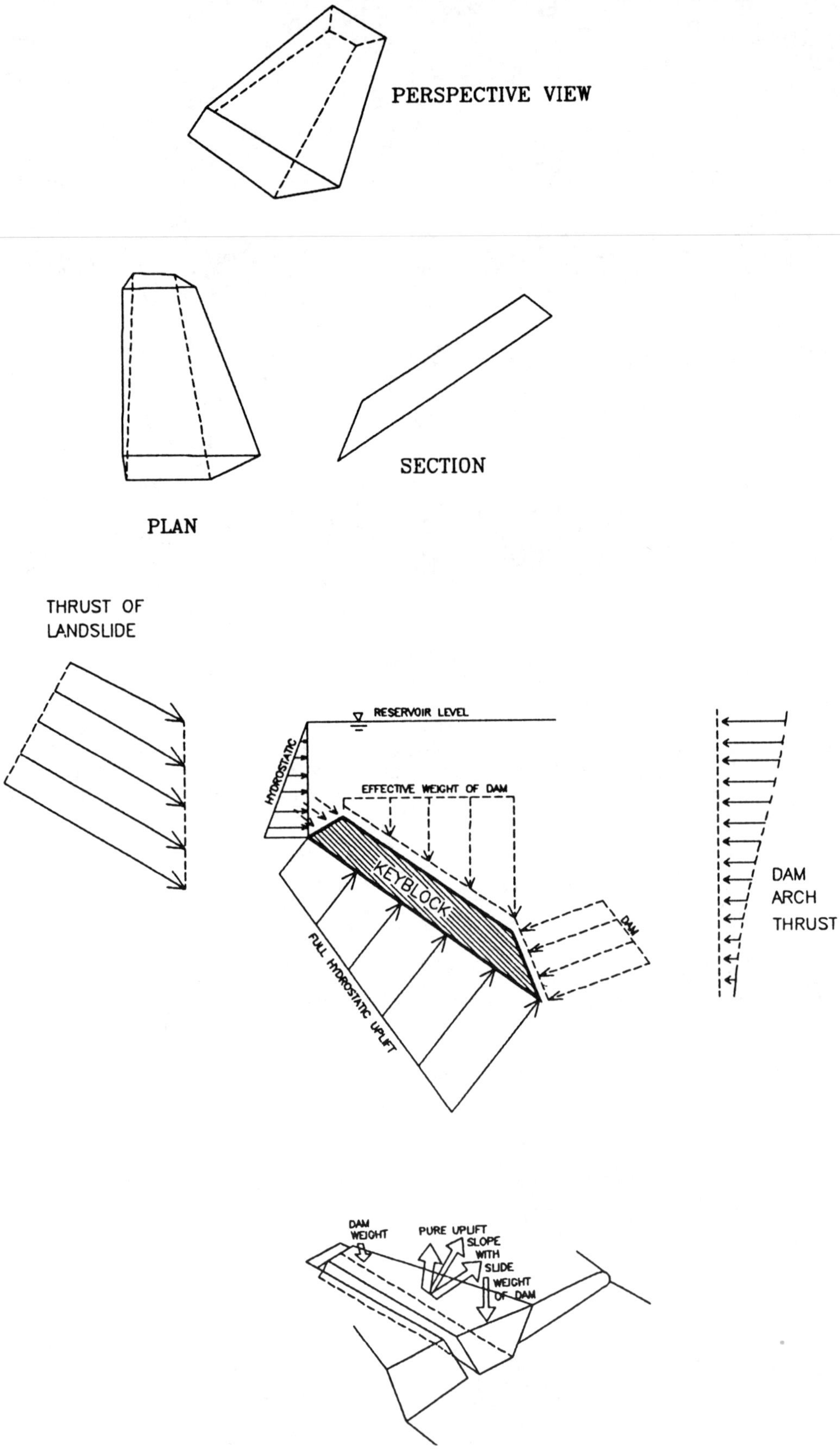

Figure 12. Plan, section and perspective views of keyblock "B", shown schematically in Figure 10. The section view parallels the schist foliation, the lower face that of the abutment excavation. The lower view shows an approximation of the loads that were likely acting on this and adjacent wedges, beneath the east abutment after the maximum reservoir pool was attained, five days preceding the failure. A simple stability analysis suggests that two-thirds of the lake's full reservoir head would have been sufficient to have lifted this block, with the dam upon it, the same manner of failure by which Malpasset Dam later failed in 1959.

by Pacher, 1964; and Londe, 1968). If two-thirds of the full reservoir hydraulic pressure were developed beneath any of the three wedge configurations ("B" being the worst of the three), it can be theoretically demonstrated that the dam abutment would have been lifted (as shown in the lower part of Figure 12), throwing parts of the sloping abutment into tension. Even with only 50% uplift, the normal effective weight of the dam, between Blocks 2/3/4 and 5/6, would appear prone to imminent failure (utilizing the Governor's Board of Inquiry block nomenclature, shown in Figure 8). Though slightly more complex (due to the sheer number of discontinuity suites present), the east abutment failure would appear similar to that experienced (due to a tetrahedrally-shaped wedge) at Malpasset Dam, a concrete arch structure in France, which collapsed catastrophically during initial filling in 1959 (Londe et al.,1968, 1969,1970; Pacher, 1964; Terzaghi, 1962).

There is no reason to openly discount Outland's (1963; p.208) "mystery witness" with regard to incipient motion of the east abutment, some 30 to 40 minutes preceding the failure. Such an observation would be expected, in light of the slide's enormity, seen the following morning (Figures 13 and 14). Realizing that landslide motion on the east abutment, regardless of how slight, preceded the dam's imminent demise, is a critical piece of information in unraveling the failure puzzle. We could reasonably expect that the incipient motion of such a large slide (877,500 tons/500,000 cubic yards of schist) would impose enormous loads on the dam

Figure 13 (upper) View looking southeast, at east abutment slide scar, from across canyon, atop the west abutment thrust block. The enormity of the slide can be vividly appreciated, the affected mass involving well over 500,000 cubic yards. The oakum patching of the diagonal crack in Block 5 can be clearly seen (arrow). Blocks 1 and 5 are indicated by their respective numbers

Figure 13. (lower) The same view as seen at the damsite today. "A" indicates the position of the old upper road along the LA Aqueduct (and within the paleoslide headscarp graben); "B" is the outcrop shown in Figure 9; "C" are the remains of dam Blocks 2 and 3; "D" is the remains of Block 1, demolished in 1929.

Figure 14. (upper) View looking uphill, into the evacuation scar of the 1928 east abutment slide, as seen from the floor of the canyon, just downstream of the dam's former position. This photo shows the guardrail of San Francisquito Canyon Road, parallel to the slide's west headscarp (thereby fixing its relative vertical position, shown in Figure 15). Photo reproduced by permission of Ventura County Museum of History and Art.

Figure 14. (lower) Post-failure view looking up-canyon from the higher, surge chamber road, above the east abutment slide headscarp (position "A" in lower portion Figure 13), a few days after the failure. Retrogressive tension scarps reached 200 feet uphill of the slide, encraoching on the high road as shown, dropping it by about 10 feet. The Edison Power line that went down at 11:57 PM is shown as "A" and the slide depicted in Figure 1 is delineated as "B".

(251,000 tons dry weight/130,000 cubic yards of concrete), thereby progressively distorting the distribution of stresses within the structure (and negating most of the design's conventional loading assumptions).

Figure 15 presents a present-day topographic plan map of the dam site, prepared by geology students at Cal Poly Pomona in 1980. The former position of St. Francis Dam was overlain on the present-day topography utilizing as-built surveys of the dam (some common points of fixity still exist, such as the base of the thrust block, atop the west abutment). From Figure 15, it can be appreciated that the upstream portion of the slide must have been dropping and thrusting against the lowest portion of the east abutment wall, in the vicinity of keyblocks A, B and C. As mentioned previously, this steeply sloping east abutment had not been provided with any manner of foundation cutoff, drainage, or uplift pressure relief. Given the fact that the eastern abutment was comprised of schist that had been displaced by paleolandsliding, it likely possessed sufficient porosity to develop extremely high pore pressures (a facet discussed in some detail by Willis, 1928).

From the pattern of transverse cracking, it would appear that uplift of the east abutment may have occurred with the first filling of the reservoir in 1926-27. If this were the case, excessively high toe pressures would have been generated at the base of the inclined section, between Blocks 2/3/4 and 5/6.

As the outermost piece of the paleomegaslide continued to creep downslope, it would have loaded the dam obliquely (Figure 15). Such lateral loading, much higher than that imposed by reservoir water, would have brought the upstream face of the dam into tension, induced by the downstream bending of the narrow abutment section (Blocks 5/6/7). The sudden creation of tensile stresses in the heel or upstream face of a brittle concrete dam is a requiem for disaster, as full reservoir hydraulic pressures would enter into such cracks and negate the dam's stabilizing dead load through bouyance (upper half, Figure 27). A simple calculation can demonstrate that, were full reservoir pressures allowed to leak into a transverse crack (or series of cracks within the "missing section", between Blocks 2/3/4 and 5/6/7), the dam's resultant thrust would be deflected far downstream, theoretically causing overturning failure.

Excessive overturning forces, brought about by tilting, could then be expected to initiate high bending stress at the downstream toe of the dam. As deflection continued, the downward concentration of force on the downstream toe usually initiates the failure process through one of two modes: bearing capacity failure of the dam foundation (Fishman, 1979), or induced tensile spalling, bringing about the "cleaving" of the sloping toe Preliminary calculations reveal that either of these initial failure modes would have been expectable results at St. Francis; the concrete tensile strength was fairly low (100 to 130 psi) and the displaced Pelona Schist would have been subjected to something around

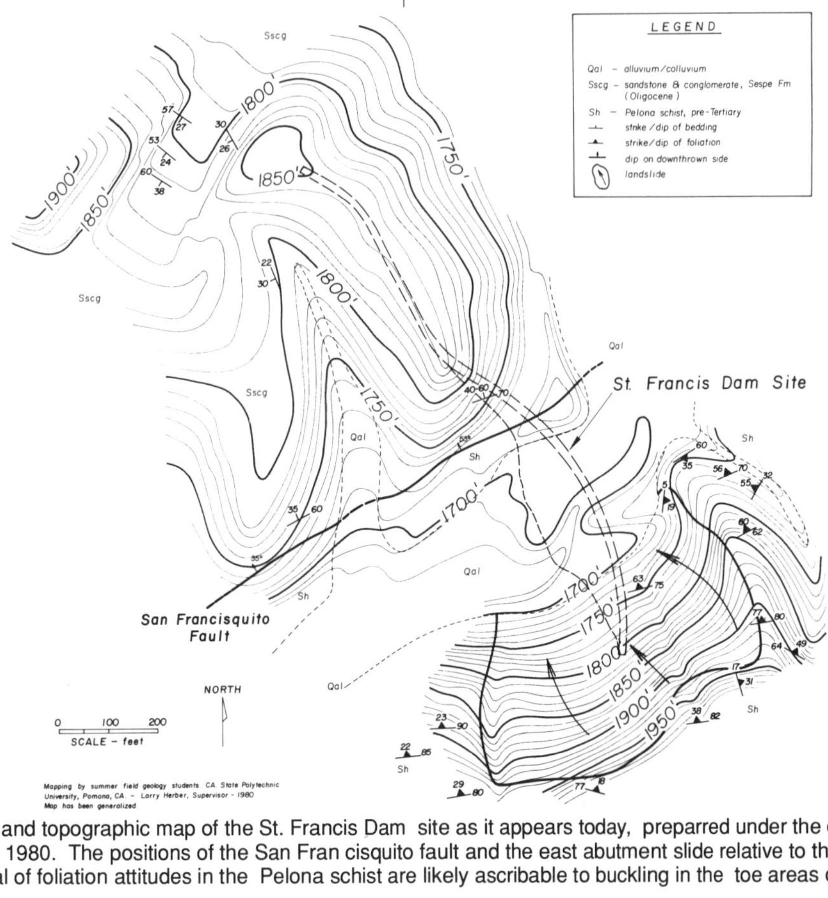

Figure 15. Surface geologic and topographic map of the St. Francis Dam site as it appears today, preparred under the direction of Professor Larry Herber at Cal Poly Pomona in 1980. The positions of the San Fran cisquito fault and the east abutment slide relative to the main section of the dam are easily discerned. Reversal of foliation attitudes in the Pelona schist are likely ascribable to buckling in the toe areas of the paleomegaslide.

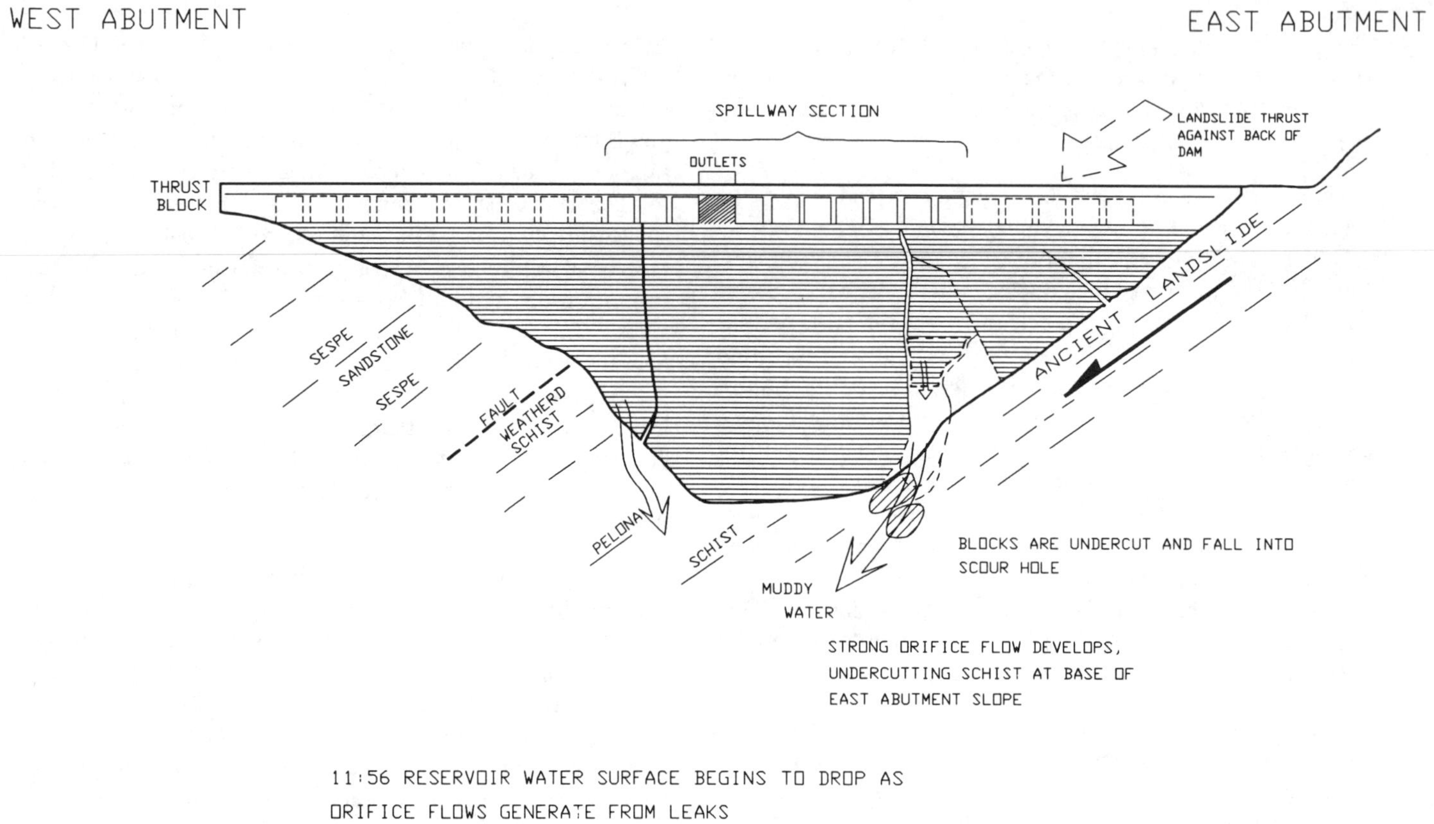

Figure 16. (upper) Schematic elevation view depicting the onset of under cutting beneath the "missing section" of the east abutment slide (see Figures 8 and 11), between Blocks 2/3/4 and 5/6/7. This sequence likely began around 11:50 to 11:55 PM, and would have been hiden from the view of people passing along the road leading to the east abutment crest.

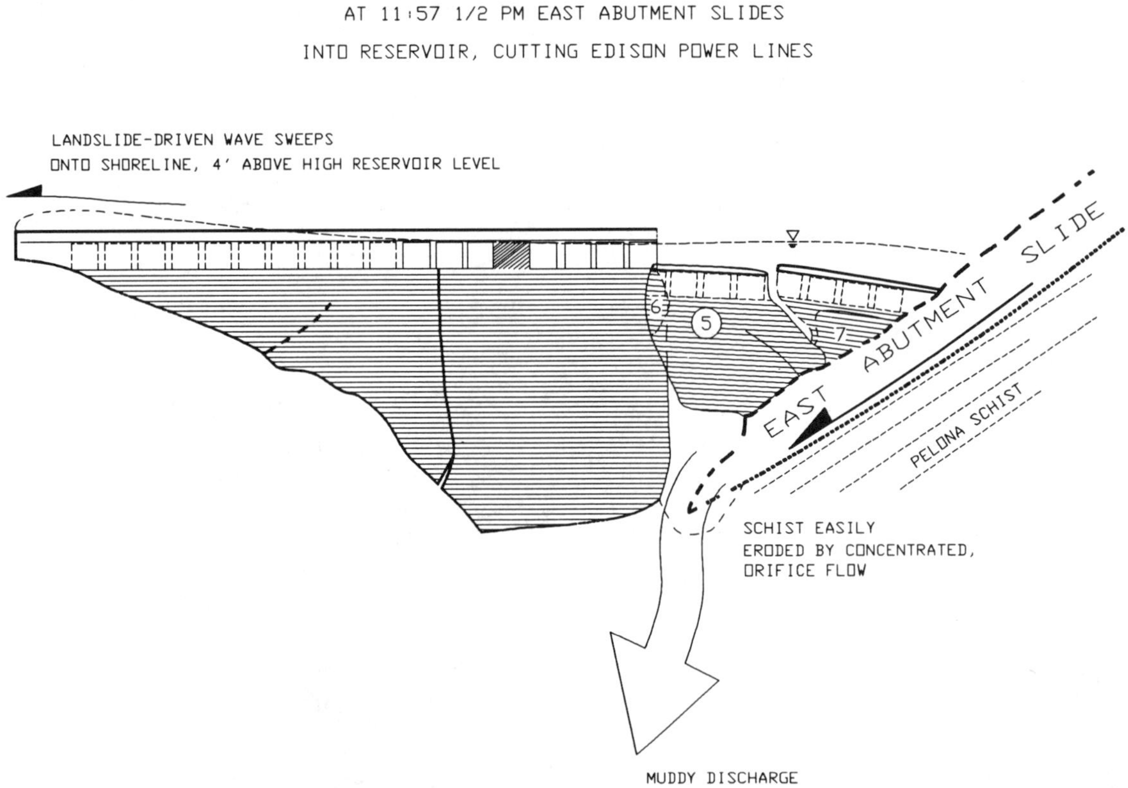

Figure 16. (lower) Schematic portrayal of the initial mobilization of the east abutment slide, around 11:57:30 PM. A sudden mass movement is suggested by several facts, including, but not limited to, the dropping of the Edison power lines some 90 feet above the dam crest at this instant (arguing against a toe-headward, progressive style of failure); the creation of wave floatsom 4 feet above the maximum reservoir level; and the apparent movement of Blocks 5/6/7 completely across the downstream face of the dam.

Figure 17. Post-failure view looking downsteam, at the upstream face of Block 1, the only remaining section of the dam. The thin black lines run ning vertically down the back of the dam are the gate controls for the dam's five outlet conduits (shown in Figure 3). The snapped portion of the 12-inch diameter housing for the Stevens Water Stage Gage is clearly seen to the left of the uppermost outlet gate, pointing towards the east abutment, between elevation 1800 and 1815 (arrow). The lake had been at elevation 1834.4 just prior to the failure. Photo used by permission of the Ventura County Museum of History and Art.

30,000 psf toe pressure, far above what we could expect it to accept without ample distortion, due to its accurate anisotropy and the fact that it must have dilated during translation by sliding.

If the dam's failure initiated in the lowermost east abutment, we could expect full hydrostatic pressures would then be exerted on the opposing faces of any transverse cracks, reaching the upstream face of the dam. Once this occurred, arching action towards the east abutment would have ceased. This loss of arching would then increase the cantilever loads on the remaining section, likely causing it to deflect excessively, lift and tilt. Such a scenario offers an alternative hypothesis to explain the 3.6 inch shift in maximum pool level faithfully recorded by the Stevens Gage: the dam could have been lifting, as opposed to the reservoir dropping just prior to the failure. Grunsky (1928) had been the only engineer to assert such a possibility in the days following the failure.

If reservoir water began to escape from either of the dam's lower abutments, its erosive effects would have been most acutely felt against the moving mountain of schist (as none of this material remained on the canyon walls following the failure). Orifice flow, emanating from an ever-enlarging breech would be expected, as depicted schematically in Figure 16 upper. The "missing section" of east abutment, between Blocks 2/3/4 and 5/6, can theoretically account for up to 400,000 cfs of escaping water with a near- full reservoir head (Figure 16 lower). But, if a hole that large were developing, it is almost certain that the unrestrained abutment (Blocks 5/6/7) would have been moving into such a void. It is reasonable to assume that some significant amount of concentrated, orifice flow must have preceded the catastrophic failure in order to create the "misty haze" that suddenly descended upon the Powerhouse 2 community near midnight, awakening Lillian Curtis and her husband (as discussed later).

Regardless of just how large a hole developed, a massive chain reaction failure must have ensued shortly after a substantive portion of the "missing section" (between Blocks 2/3/4 and 5/6/7) was removed (Figure 16 lower). The apparent outpouring of water thru the "missing section" (Figures 8 and 11) must have undercut the incipient slide mass; thereby setting about its apparent catastrophic failure (Figure 16). Blocks 12 and 14 were later found a half mile downstream (Figures 8 and 16 lower), about 20 feet above the channel level, and offset 50 yards to the outside of the first bend in the San Francisquito channel (Figure 8).

We are left to deduce that catastrophic sliding of the dam's east abutment occurred before the reservoir had experienced many appreciable drainage for six key reasons:

1. The morning after the failure, flotsam and debris was observed along the reservoir shoreline at a level 4 feet higher than the maximum level of the reservoir pool;

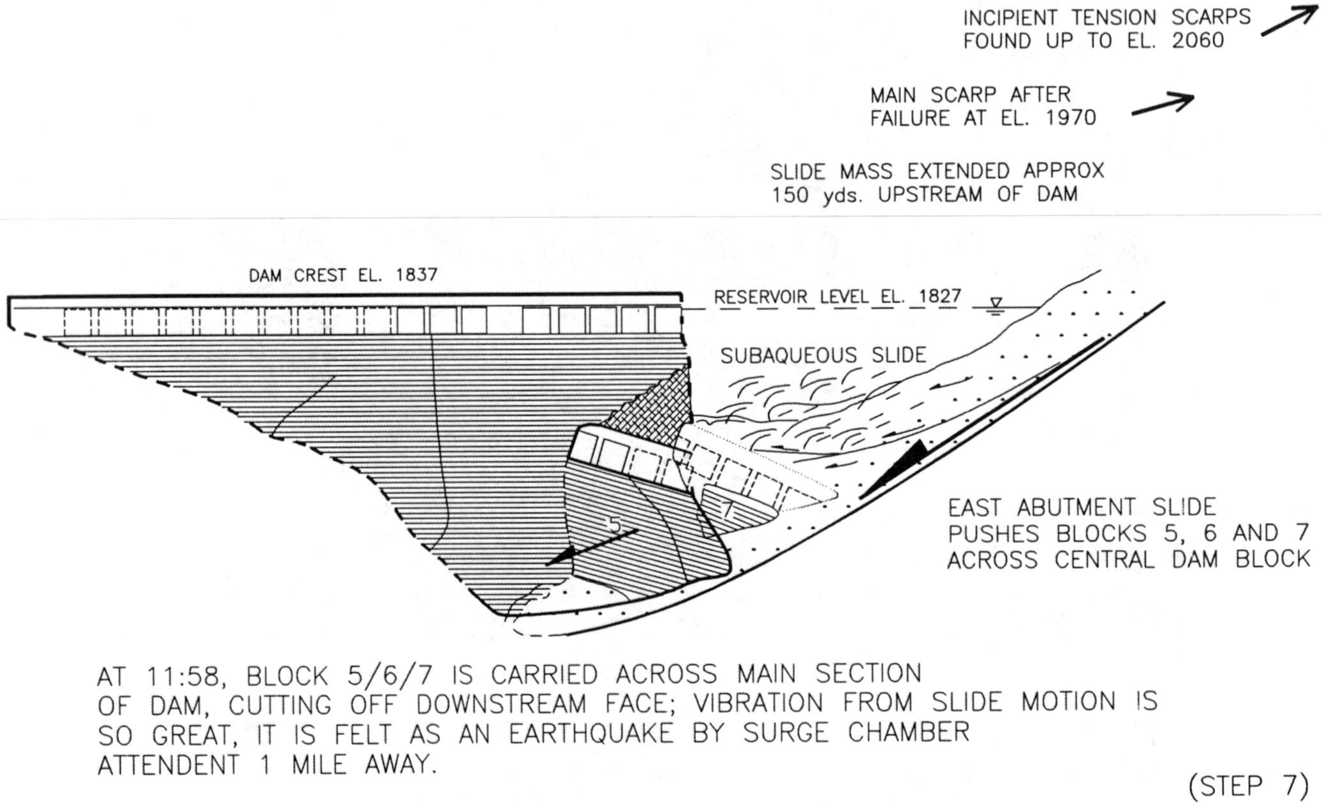

Figure 18. (upper) Schematic portrayal of the east abutment slide carrying Block 5/6/7 across the downstream face of the dam's central section (Blocks 1/2/3/4) early in the failure sequence. At this time, the lake level had likely only dropped 3 to 7 feet, to around elevation 1827. The outer 20 feet of the dam's face was abraided by the movement of Blocks 5/6/7.

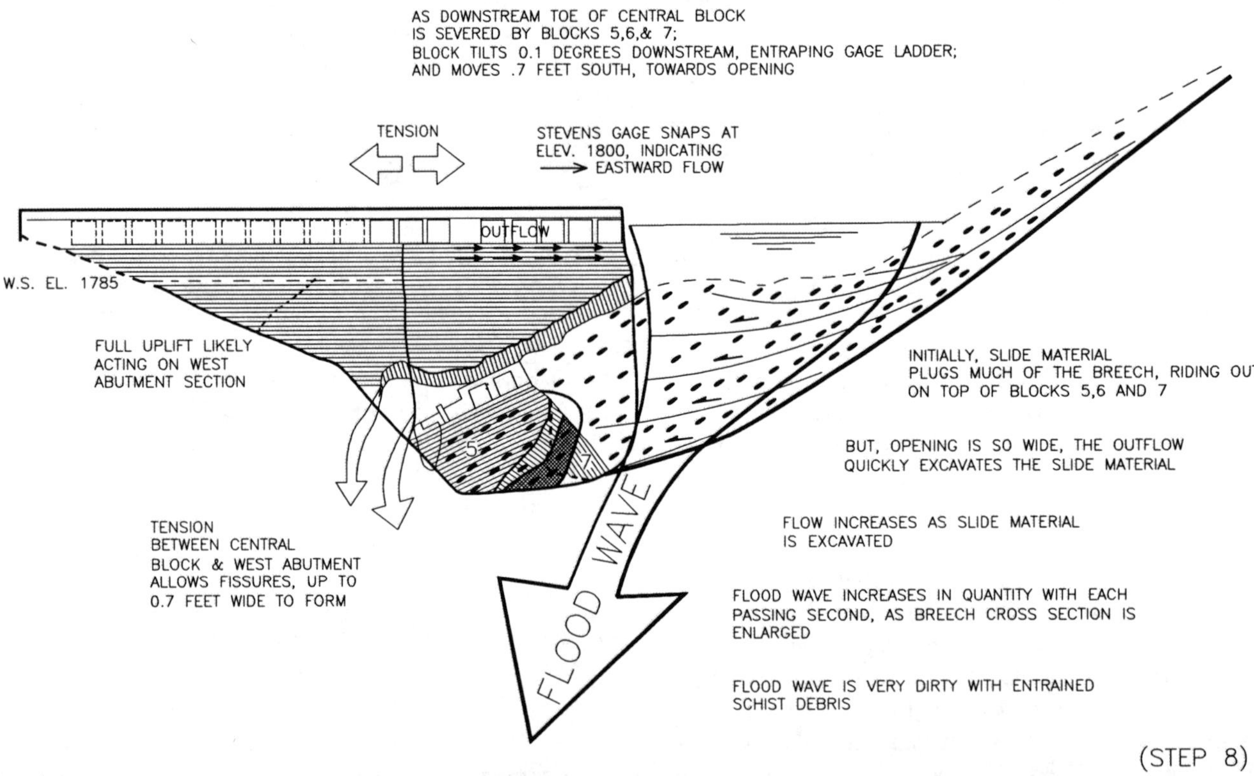

Figure 18. (lower) Schematic portrayal of the landslide dam, temporarily formed across the east abutment breech. Debris from this translational slide probably overthrust upon itself, building up a short-lived impedence to outflow. The formation of a temporary dam with increasing discharge accounts for the 30 to 90 seconds of precursory outflow related by survivors at Powerhouse 2 before the maximum discharge arrived, killing 120.

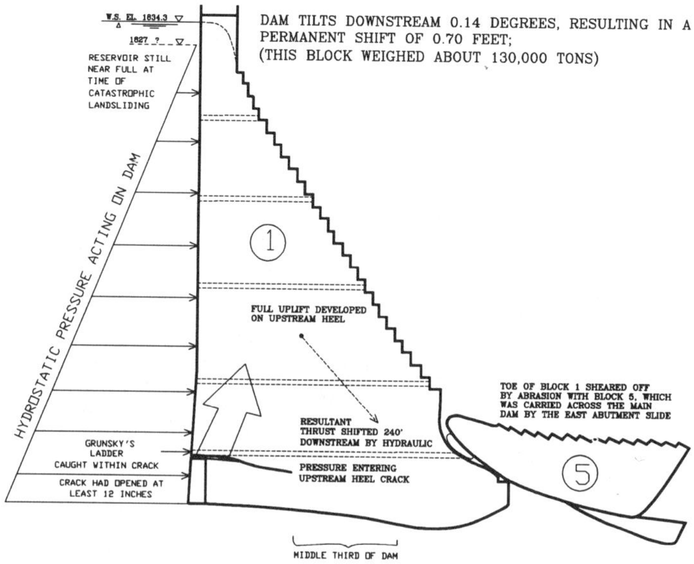

Figure 19. Loading conditions likely exerted upon Block 1, based on physical evidence observed immediately following the failure. The snapping of the Stevens Gage around elevation 1815 suggest that the reservoir was still full when maximum outflow conditions were developing through the east breech. When Block 5 sheared off the toe of Block 1, the dam's resultant thrust would have been shifted downstream of the structure, leaving the upstream heel in tension. The heel crack opened up at least 12 inches, allowing full hydrostatic pressures to develop under this section. The uplift wells placed beneath the dam's main section likely saved it from disaster. The sloping abutments were not afforded any such protection.

meaning that a sizable wave had been generated while the lake was still at a relatively high level;

2. The Stevens Gage stilling well pipe, affixed to the upstream face of Block 1 (on the west side of the dam's central core), was found to have been buckled between elevation 1800 and 1815, towards the east abutment (Figure 17); indicating that a massive outpouring was occurring to the east side of the central dam block when the lake was still high.

3. Blocks 5/6 and 7, comprising the majority of the east abutment, were carried across the entire main dam section, shearing off approximately 20 feet of the downstream toe of this main section (Figures 18 - 21). Enormous lateral forces had to have been acting on the dam when this occurred, in order to overcome the hydrostatic pressures of the reservoir (or its remnants), which would have been pushing normal to the downslope motion of these blocks. The fact that piles of schist were found atop Blocks 5 and 7, some 40 feet above the creekbed (Figure 20), also indicates that they were subsequently covered with schist detritus;

4. Lillian Curtis and Ray Rising, the only adult survivors of a population of 123 DWP employees/family members at Powerhouse 2, were both awakened by extra-normal sounds of high water discharge and an unusual "foggy haze" that had settled over their community. Curtis had sufficient warning to get out of her home with her 2-year old son, before the water reached its maximum level of 110 feet, whipping away the Powerhouse 2 community at 12:02:30 AM. She described the flood water as a sort of "liquid mud", indicating an extremely high suspended bed-load in the initial flood wave. That this initial event was sweeping and occurred early on in the failure sequence is collaborated by E.H. Thomas, the surge chamber attendant atop the mesa above Powerhouse 2, who upon hearing and feeling what seemed to him like seismic tremors, ran down the canyon slope to see what had happened. He reached the high water line at 12:15 AM, only to find that the water had already receded 20 vertical feet.

5. There simply was not sufficient time for an appreciable lowering of the reservoir between the earliest time Ace Hopewell drove by (11:50 PM) and when the Edison power lines went down, on the east abutment slide at precisely 11:57:30 PM. The fact that these lines were located almost 100 feet above the crest level of the east abutment seems to have been overlooked in the post-failure assessments (Figures 11 and 14 lower).

6. The fixed times of known events suggests that the failure sequence was rapid and catastrophic, arguing against a conventional, piping style of failure. The dam looked fine shortly before the failure, between 11:50 and 11:55 PM (Hopewell); failed at 11:57:30 (downed Edison lines on east abutment); destroyed the Powerhouse 2 area at 12:02:30 (Curtis); and had already receded 20 feet by 12:15 AM (Thomas). The reservoir was emptied within an hour (when first reached by DWP employees Spainhower and Lindstrum at 1:09 AM). All the accounts are consistent with respect to time, but spaced very closely.

The upper and lower halves of Figure 18 present what likely occurred, given the verifiable parts of the puzzle (both physical evidence and eyewitness accounts). After having its toe rapidly undercut and eroded by strong orifice flow (which would produce a misty "fog"), the toe of the east abutment slide was severely undercut (Figure 16). Such undercutting should have been sufficient to trigger the east abutment landslide, which was already in a stage of incipient failure if the road had indeed dropped 12 inches by the evening of March 12th. The east abutment slide was a simple translational failure, along slippery planes of foliation in the mica schist. We can infer that the slide must have began its descent around 11:57:30, dropping the Edison Company Lancaster power lines. Depending on the speed at which the slide dropped, the moving mountain of schist should have infilled the tremendous void, at least temporarily (lower half, Figure 18). An alternative hypothesis could have the toe of the east abutment slide gradually being eaten away by the outpouring water, but such a scenario would then be unable to account for the very long trip taken by Block 5; which was shifted all the way over to the opposing abutment. We are left to conclude that the translation of Blocks 5/6/7 to the opposing side of the channel, below Block 1, argues for a rapid, dramatic slide motion, so large that it presented an inexorable force; temporarily greater than that posed by the force of the outpouring reservoir water. Mass balance assessments comparing the effective force of the slide, a saturated dam (Figure 27) and the reservoir water support such a premise.

In addition, if the east abutment blocks came down slowly, the outpouring water should have been capable of translating them some distance downstream (Figure 18). Years after the failure, journalist Outland (1963) had been the only person to correctly deduced this key part of the puzzle. The tremendous "grinding motion" of the massive concrete blocks (which sheared off a 20 foot thick section of the downstream face) was the likely source of the "seismic tremor" described by Ace Hopewell and surge chamber attendant Thomas and his mother, all atop the schist ridgeline. The block kinematics involved in such a massive translation can be better appreciated when viewing the dam in plan, as shown in Figures 4 and 15.

As Block 5 was carried upon the translating landslide, it sheared off the downstream face of Blocks 2/3/4 and Block 1, as depicted in Figures 19 and 20. By removing the downstream toe of the dam's central block, the resultant thrust would have been projected even further downstream, thereby leaving the upstream heel in tension (as the dam tried to overturn). Post-failure assessment by Grunsky (1928) showed that Block 1 had separated at its upstream heel, between the masonry cut-off wall and the dam proper (Figure 19). Such tilting would have progressively worsened, due to the development of full hydrostatic uplift pressures, within the crack. This must have occurred, as the gage ladder was found swallowed into the crack at the base of the remaining section, suggesting that the dam MUST have tilted at least 12 inches to allow complete access into this crack (Figures 17 and 22).

The approximately 500,000 yards of schist landslide debris would have been taken into the massive void created by the removal of the entire east abutment. Temporarily, the breech would have been partially blocked by the sheer volume of slide debris (as depicted schematically in the upper-half of Figure 23). The outpouring waters of the reservoir, some 200 feet deep, could be expected to rapidly excavate this temporary "landslide dam", issuing forth a progressively larger discharge, as the schist was quickly removed by the outpouring water. Short-lived, but dramatic increases in overflow discharge typify the breaching of erodable landslide dams (Lee and Duncan, 1975). Peak flows observed from landslide dams have been the largest recorded channel flows in modern time (Schuster, 1985).

The massive landslide failure sequence likely took only 30 to 90 seconds, just enough time for the few survivors at Powerhouse 2 to realize something was amiss and gain a brief chance to escape (Mrs. Curtis' husband had awakened her, pushed her out the uphill window, and told her to "run for

Figure 20. The remains of Blocks 5 and 6, as seen immediately after the failure. The two lowermost steps of Block 1 appear to have been left in tact at the west end. Note the accumlation of schist detritus upon Block 5, about 20 to 30 feet above the creek bed.

the hills" with their two year old son. He then went into an adjacent bedroom to retrieve their two older daughters, but did not escape; so there was precious little time to react). Mrs. Curtis and other survivors of the flood are consistent in describing the maximum wave front as consisting of "liquid mud", likely resulting from the 500,000 cubic yards of weathered schist that was rapidly being scoured and picked up in the east abutment breech. In the flood's aftermath, the Los Angeles County Coroner (1928) would testify as to the large amounts of silt found in the victim's bodies, attesting to their drowning in extremely turbid water. Based on a hydraulic assessment of the flood wave scour depth and channel cross section, the maximum flow quantity appears to have been approximately 1.7 million cfs, a much larger number than that supposed by the various boards of inquiry, as the mechanics of critical flow were little understood at the time. The height of the flood wave at Powerhouse No. 2, 7000 feet downstream of the dam, was later measured at 110 feet above the floor of the Canyon (the maximum water level of the dam being only 179 feet above Canyon floor). It would appear, therefore, that the reservoir emptied very quickly by excavating a "mega-breech" through the slide debris (Figure 23 upper). As the eastern side of the dam was torn away by the massive landslide, there would have been a marked relaxation of arch stresses to that side of the dam. The main dam section could be expected to relax slightly, towards the gap. This relaxation would have allowed the pre-existing transverse crack, bordering the right (west) side of Block 1 (depicted in Figure 5), to open itself up, thereby subjecting the fissure to full reservoir hydraulic pressure. The west abutment was now beginning to break up, for the same reasons experienced on the east side.

As the great quantity of water issued forth from the massive east abutment breech, the central core of the dam was being severely undercut by back-eddy flow (Figure 23 upper). We know from post-failure surveys that a hole 60 feet deep was cut beneath the dam's maximum section, undermining Block 2/3/4. As this occurred, Blocks 1 and 2/3/4 must have tilted towards the undercut/void, or towards the south. Post-failure surveys showed that Block 1 had moved 5.5 inches towards the east abutment and 6 inches downstream (a total of 0.70 feet on a bearing of south 3 degrees west). Such motion would release even all constraint on the west abutment (above the fault), thereby leaving it to absorb the remaining reservoir load as a free-standing gravity section.

As the central core of the dam moved towards the east abutment breech, the stability of the west abutment must have quickly deteriorated, likely initiating failure beneath Block 16 (Figure 25), in vicinity of the fault contact. We can deduce that the west abutment failed after the east side for two key reasons:

1. The Stevens Gage was pulled towards the east abutment at when the reservoir pool was at an elevation of between 1800 and 1815 feet (as seen in Figure 17); and
2. Post-failure photos clearly show that the west abutment construction road was not damaged above the 1785 elevation; despite being comprised of erodable sidecast fill from the Sespe formation (see photos in Figure 24). Undisturbed brush is clearly seen above this level in all the post-failure photos.

The west abutment was probably removed in rapid fashion in a chain-reaction sequence of progressive undercutting. Blocks 11 and 16 (shown in Figure 8), the two largest found more than a half mile downstream (Figure 25), were probably carried that distance by the maximum flow emanating from the dam, which could only have occurred after a substantial removal of the slide debris (in order to create a large opening, akin to the failure of a natural landslide dam). This brief passage of time allowed the reservoir to drain to a level between Elevation 1800 (the bottom of the weakened Stevens Gage casing) and Elevation 1785 (the top of the scour line on the west abutment).

The glaring separation of the west abutment pieces far downstream (Figures 8 and 25) could be ascribable to their smaller relative size (when compared to the blocks on the east side), and the fact that they were probably loosened at about the same time as the schist landslide debris was being excavated from the eastern breech. The tremendous amount of sediment entrained in the reservoir outflow would cause the discharge to have been of much greater relative density, than a comparable discharge of clean water. The relative submerged, or effective weight, of the concrete blocks would be directly proportional to the density of the fluid they displace. In this manner, such enormous blocks might weigh only a fifth of their dry weight, as shown diagrammatically in Figure 27. Back-analyses utilizing conventional, "clear water channel hydraulics", would not replicate the eyewitness descriptions of "moving mud" and the simple fact that 500,000 cubic yards of disintegrated Pelona Schist quickly vanished from the scene.

Towards the end of the failure sequence, Block 2/3/4 fell into the scour hole excavated beneath it by the massive east breech (Figure 21). This hole was some 60 feet deep, according to the LADWP post-failure surveys! As the east side of the dam's central core toppled over, it separated along near-horizontal cold/pour joints, markedly observed in the pieces remaining on site today. This final sequence is sketched in the lower half of Figure 23, and likely occurred when the reservoir had emptied to a level of 40 to 60 feet above the creek bed (based on the schist detritus left on Blocks 4 and 5, Figure 21).

POST-FAILURE NOTES

In the failure's aftermath, surveys by the City's Bureau of Water showed the enormity of the east abutment slide, as well as numerous shallow slides induced by rapid drawdown (Figures 1 and 7 lower). Precise post-failure surveys showed that the standing section (Block 1) had been permanently displaced 0.70 feet (Figure 26). These same surveys also found that the west wing dike (Figure 7 lower) had lifted 0.10 to 0.30 feet, presumably due to swell of the Sespe beds. This swelling may partly account for the many leaks observed in that area prior to the failure. However, nobody ever tried to tie the wing dike leaks to the collapse of the main dam.

A new road had to be constructed through San Francisquito Canyon as the east abutment slide had forever erased the old route. The new road was built shortly after the failure, cutting through what had been the west abutment (Figure 15). Over the past few years the City has constructed a new highway alignment which bypasses the dam site, its' cuts slashing through the former position of the overflow dike, approximately 1200 feet northwest of the dam (also shown on Figure 15). Were it not for engineering geology classes, the memory of California's greatest man-made disaster will simply lie as a footnote in the annals of the State's civil engineering history.

CONCLUSIONS

Many lessons can still be gleaned from the St. Francis failure. Foremost of these is the critical importance of soliciting sound engineering geologic input from more than one

Figure 21. Photo looking upstream, showing the final resting positions of the east abutment and main section blocks. Note the schist detritus left upon Blocks 4 and 5. The oakum patch in Block 5 is also very apparent. Blocks 2/3/4 separated very late in the failure sequence, when much less water was available to disloge them (they toppled upstream).

Figure 22. Grunsky's (1928) photo showing the upstream gage attendant's ladder wedged into a large crack at the west upstream heel of Block 1. Block 1 would have to have tilted at least 12 inches to swallow up the lad der, which was 18 inches wide. There is little doubt, therefore, that the upstream heel of the dam's central section was in tension and that, at some point, either before or during the failure sequence, full hydrostatic pressures must have acted upon the block's interior, significantly negat ing its effective dead load

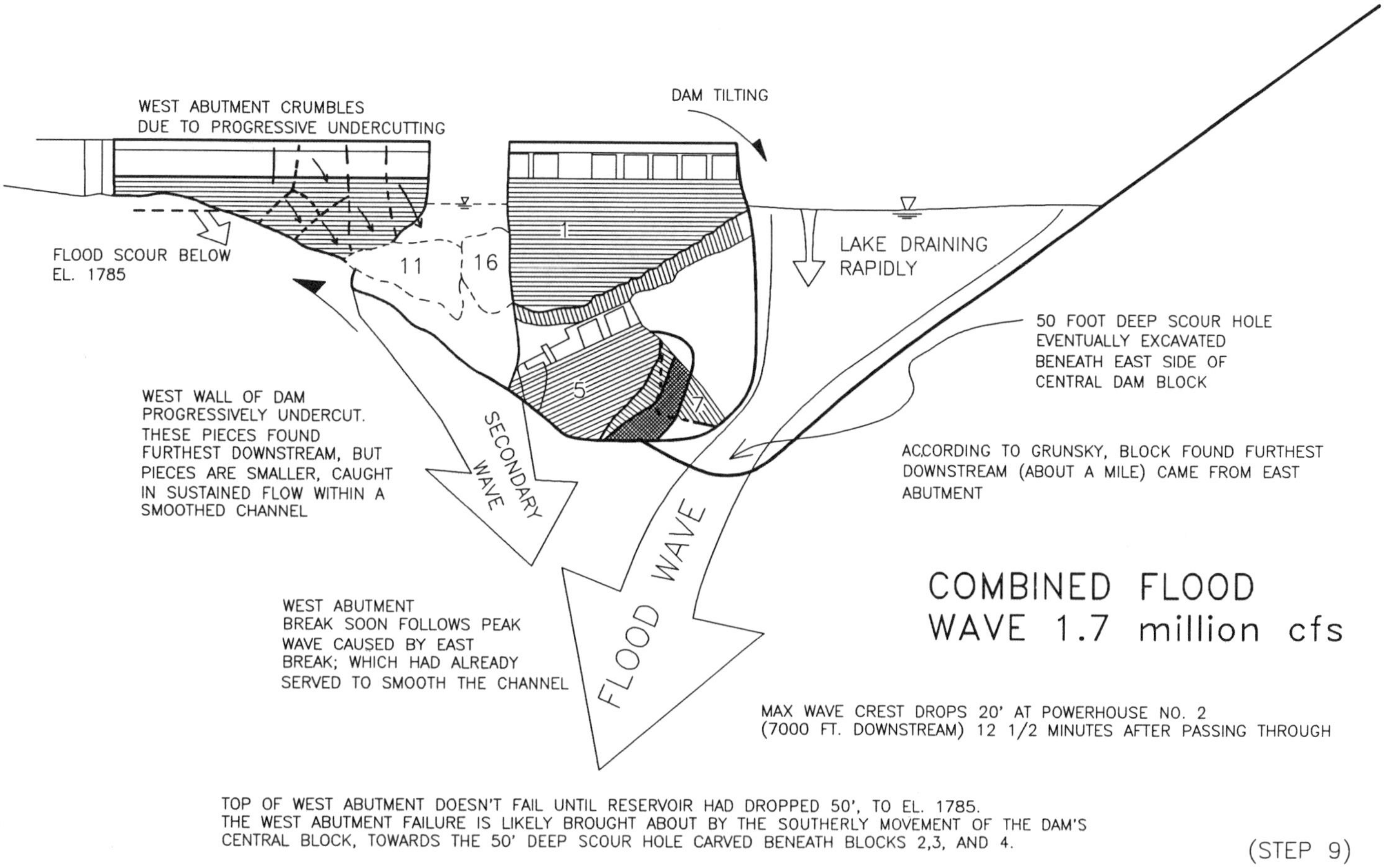

Figure 23. (upper) Schematic portrayal of east and west breeches combining to form the maximum discharge of 1.7 million cfs, sometime shortly after midnight on March 12/13, 1928. This wave crested at Powerhouse 2, 7000 feet downstream, sometime between 12:02:30 and 12:08; receeding 20 feet by 12:15 AM. The maximum outflow was likely caused by the rapid ero sion of the debris plug created by the east abutment slide at 11:57:30 PM.

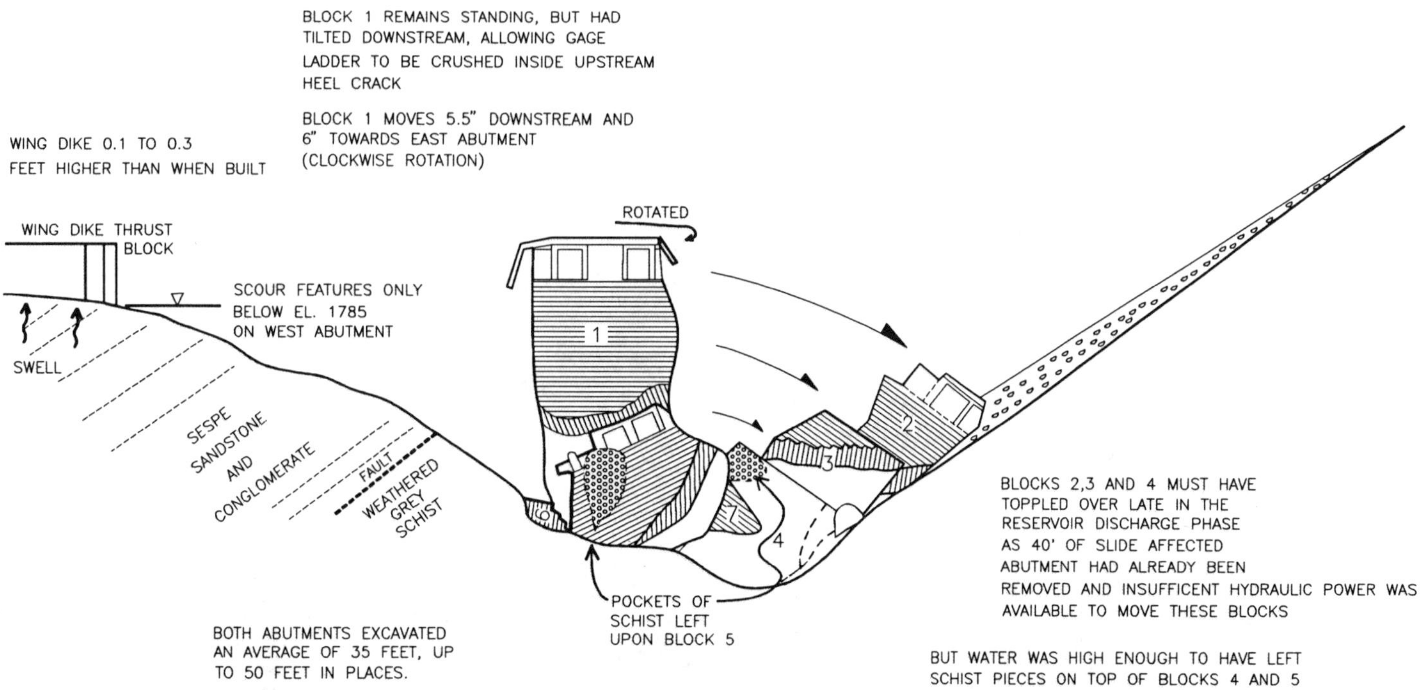

Figure 23. (lower) Schematic portrayal of the final stages of the St. Francis failure. The west abutment was scoured away very quicky once the initial piercement occurred, just west of Block 1 (see Figure 24). Block 2/3/4 was undercut by the outpouring through the east breech, eventually toppling over into a hole 50 feet deep, falling backward. The cutting of such a deep hole on the east side of the main structure also argues for more discharge emanating from this side of the structure. This is how the scene looked the following day (Figure 25).

Figure 24. (upper) View looking down on the dam site from the upper surge chamber road (Figure 14 lower) the morning after the failure. Note how the original construction road, leading up to the wing dike, is not damaged above elevation 1785 (arrow). This suggsts that, at a minimum, the upper portion of the west abutment did not fail until the reservoir had drained at least 30 to 50 feet (photo courtesy of Ventura County Museum of History and Art).

Figure 24. (lower) Detail of failed west abutment, looking downstream, with people circled for scale. Only a portion of the west dike access road side-cast fill was excavated by the outpouring water (arrow), with scrub brush left upon the cut slope, down to elevation 1785, 50 feet below the reservoir level. Grunsky's ladder, caught in the upstream heel crack as Block 1 tilted downstream, is annotated (photo courtesy of Ventura County Museum of History and Art).

expert source. No less than four prominent geologists viewed the site (Branner, Louderback, Ransome and Willis), but only Willis (1928) recognized the geomorphic indicators of ancient landsliding. Somehow, in the failure's aftermath, too much emphasis was placed on the fault, which was so vividly exposed on the west abutment (Figures 17 and 24).

It appears that William Mulholland simply didn't appreciate the concepts of effective stress and uplift, precepts that were then just beginning to gain universal acceptance (LaRue, 1928). Harza's (1949) award-winning paper on uplift theory had been written 15 years before it was accepted for publication, due to the reticence of some of the senior members of the dam engineering profession in the 1930s and '40s. The other weak link in Mulholland's design process had been the apparent omission of any outside consultants to review the dam's design, a curious decision considering that he had previously convened a consulting board to review the Los Angeles-Owens Aqueduct plans. During the Coroner's Inquest, Mulholland (1928) stated that he had brought Stanford Professor John Branner to the dam site to view it before construction had commenced. The Governor's Board of Inquiry concluded that the owners of all dams should submit their plans for review by an outside board of consultants; a recommendation almost identical to that made by Idaho's Teton Dam Inquiry Board a half century later (1976). The Coroner's Inquest reached a similar verdict, quoted by

Figure 25. - Close up view of dam Block 16, which at 10,000 tons, was the largest piece of the dam moved any appreciable distance downstream (as shown in Figure 8). This block came from the lower part of the west abutment, adjacent to Block 1, which remained standing. The small holes visible on the upper cold joint (arrow) were for the rakers supporting the stepped forms for the dam's downstream face. Blocks of this size can be most easily transported in a dirty fluid or slurry, which greatly diminishes their effective weight, as suggested in Figure 27. Photo courtesy of Ventura County Museum of History and Art.

Figure 26. Overview of the flood's aftermath, immediately downstream of the dam. The flood wave scoured the canyon's slopes to a maximum depth of 120 feet above the stream level. The Edison power line crossed the channel, but was situated above the maximum scour line (arrow). San Francisquito Canyon road began climbing east wall of the canyon, at upper left. Photo courtesy of Ventura County Museaum of History and Art.

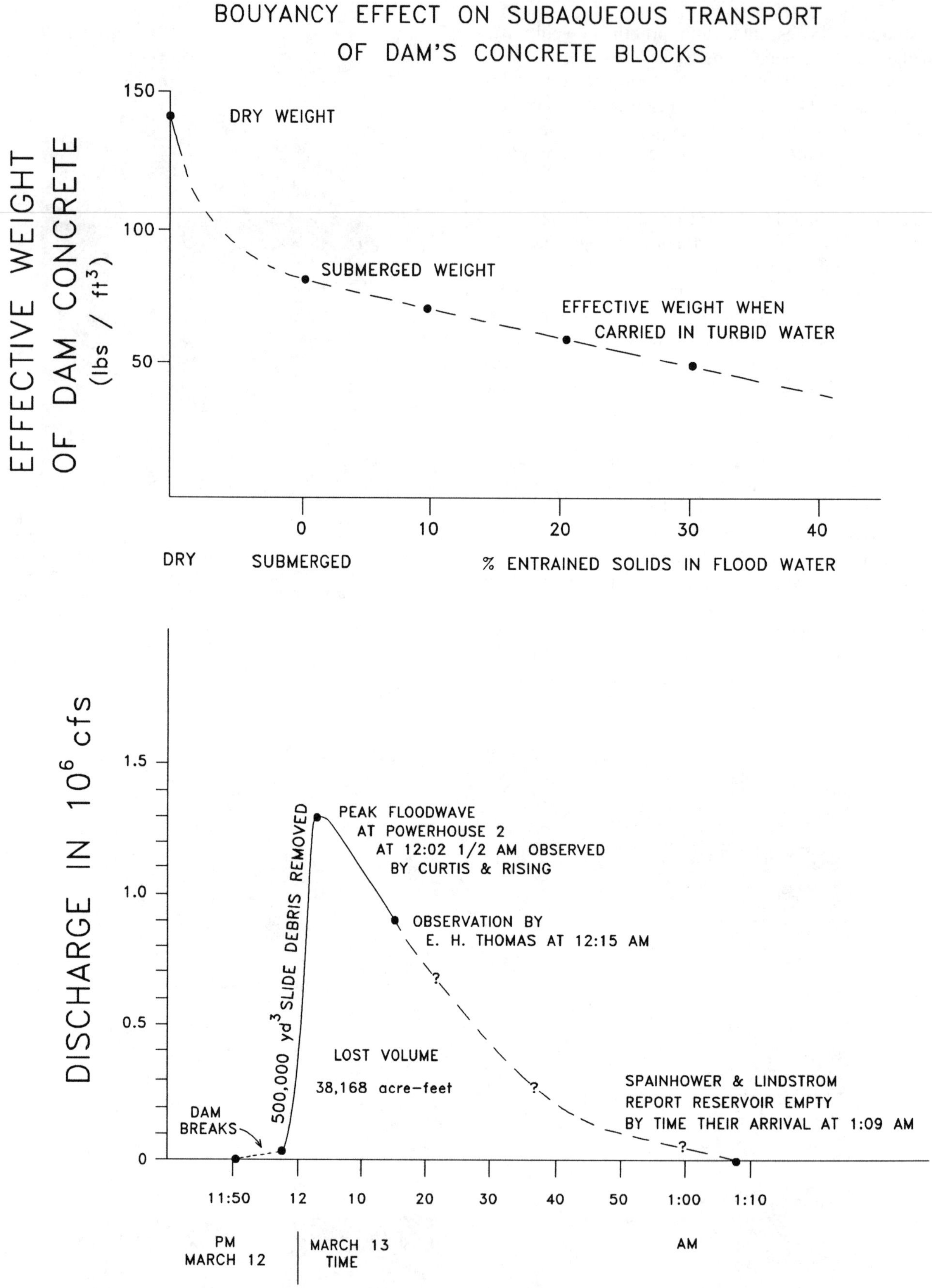

Figure 27. Preliminary evalation of flood wave turbidity effects on bouyancy of the dam's concrete blocks. Although heavy when dry, tests of the dam's concrete show it to have a void ratio of around 13%. Post- failure photos also suggest that the blocks were saturated, as they continued to weep moisture out of their cold joints for days following the failure. If saturated, and immersed in turbid flood waters, the effective weight of the blocks was reduced by as much as 67 percent

Raphael (1988), stating:

> *A sound policy of public safety and business and engin eering judgment demands that the construction and operation of a great dam should never be left to the sole judgment of one man, no matter how eminent, without...checking by independent experts."*

With those conclusions expressed, the State legislature enacted legislation to bring about a dam control law the following year. This new law mandated that the owner of proposed dams pay for a review of their projects by a board of eminent civil engineers and geologists retained by the State Engineer. This body subsequently became the Division of Safety of Dams (DSOD) within the State Department of Water Resources, and was one of the first agencies created strictly for dam safety review in the world.

Looking back, its particularly sad that the engineering profession's opinions were almost wholly predicated on little, if any, actual field investigation. Everyone seems to have accepted the City's cross sections and Mulholland's testimony with respect to the dam's maximum width, without having verified them through proper field measurements. With the exception of Grunsky and Willis, the published discussions focused on the fault, the propensity of the Sespe beds to slake under submersion, and the Sespe's low shear strength, which was about one fourth that of the dam's concrete. But, the more significant culprits appear to have been a proper appreciation of uplift theory and the incorporation of sound geomorphic and rock mechanics assessments. A similar failure mechanism befell Malpasset Dam 32 years later, and it was only after years of studying that failure that the importance of complex rock mechanics analyses were demonstrated. Had people kept working on St. Francis, the actual complexity of the failure might have been discovered.

It is of more than passing interest to note that most of what we can discern from the St. Francis failure didn't come from civil engineers or geologists, but through the efforts of a single journalist. The late Charles Outland (1910-1985), a long-time Ventura County historian and native of Santa Paula, spent the better part of his lifetime researching the failure. As a high school senior, Outland watched the flood decimate his hometown in the early morning hours of March 13, 1928. It was a sight he was never to forget. For 30 years he researched the St. Francis failure. This effort culminating with the publication of his first book, Man-Made Disaster, which appeared in 1963. It continues to be the only definitive work ever published on the construction and failure of the St. Francis Dam.

For engineering geologists, St. Francis will forever exist as a warning beacon to those who do not heed the importance of geologic input. A structure is only as strong as its foundation. No student of geology in southern California has completed their college curriculum without a visit to the imposing waste and heartfelt tragedy of the St. Francis site. Might we never forget it.

ACKNOWLEDGMENTS

The author wishes to thank his colleagues Robert B. Rogers and David L. McMahon, for their many constructive technical discussions, photographic assessments and hydraulic calculations which have greatly aided in unraveling the failure. Professor Lawrence J. Herber of Cal Poly Pomona provided me with my first visit and critical assessment of the dam site. Jerome M. Raphael, Professor Emeritus of Structural Engineering at U.C. Berkeley, provided me with his files on the dam's failure which included many original materials from State and Federal agencies. Richard E. Goodman, the late Richard Jahns and James E. Slosson periodically rekindled my interests with their long-held assertions that no one had really figured out how and why St. Francis actually failed on both abutments simultaneously. Dr. Fred Chin preparred the computerized keyblock analyses and tested samples of the foundation materials in the Rogers/Pacific lab. Charles Johnson, Curator of the Ventura County History Museum, provided invaluable assistance in gaining access to unpublished photographs and the personal research files of Charles Outland.

REFERENCES

Bailey, F.S. and Griffith, J.R., 1929, Discussion of Essential Facts Concerning the Failure of the St. Francis Dam: Proceedings, American Soc. Civil Engineers: vol. 94, December, p. 2729-2731.

Bryan, Kirk, 1929, Problems involved in the geologic examination of sites for dams: American Institute of Mining, Metallurgical and Petroleum Engineers, Technical Publication no. 215.

California, Division of Water Resources, 1931, Report of the Consulting Board on the Safety of the Proposed Pine Canyon Dam, Los Angeles County, California: State Printing Office, 22 p.

Clements, Thomas, 1929, Geology of a Portion of the Southeast Quarter of the Tejon Quadrangle, Los Angeles County, California: Master's thesis in Geology: California Institute of Technology, Pasadena.

Clements, Thomas, 1966, St. Francis Dam Failure of 1928: in Engineering Geology in Southern California, R.Lung and R. Proctor, ed, Southern California Section, Association of Engineering Geologists, p. 88-91. (Out of print)

Committee Report for the State, 1928, Causes Leading to the Failure of the St. Francis Dam: Calif. State Printing Office, Sacramento, 77 p.

Engineering News Record, 1934, Mulholland Dam Backed by Earthfill Against Downstream Face: vol. 112, May 3, p. 558-560.

Fishman, Yuri A., 1979, Investigations into the Mechanisms of Failure of Concrete Dam Rock Foundations and Their Stability Analyses: Proceedings of the 4th International Congress on Rock Mechanics, International Soc. Rock Mechanics, Montreal, vol.2, p. 147-152.

Floris, A., Grunsky, E.L., and Morris, S.B., 1930, Discussion of article on Essential Facts Concerning the Failure of St. Francis Dam: Transactions, American Soc. Civil Engineers: May, p. 1023-1032.

Forbes, Hyde, 1928, Geological Foundations at the St. Francis Dam Site: Engineering News Record, vol. 102, p. 596-597.

Godfrey, E., Turner, C.A.P., and Jorgenson, L.R., 1930, Discussion of article Essential Facts Regarding the Failure of the St. Francis Dam: Proceedings of the American Soc. Civil Engineers, January 1930, p. 163-174.

Goodman, R.E. and Gen-hua Shi, 1985, Block Theory and Its Application to Rock Engineering: Prentice Hall, 338 p.

Goodman, Richard E., 1990, Rock Foundations for Dams: A summary of exploration targets and experience in different rock types: Keynote Address, 10th Annual USCOLD Lecture, U.S. Committee on Large Dams, New Orleans, March 6, 1990, U.S. Committee on Large Dams, Denver, p. 1-31.

Grunsky, C.E. and Grunsky, E.L., 1928, St. Francis Dam Failure: Western Construction News, May 25, 8 p., 10 plates.

Harza, L.F., 1949, The Significance of Pore Pressure in Hydraulic Structures: Transactions, American Society of Civil Engineers, vol. 114, p. 193-289.

Hill, L.C., Evans, H.W. and Fowler, F.H., 1929, Essential facts concerning the failure of St. Francis Dam: Report of the Investigating Committee: Proceedings of the American Society of Civil Engineers: vol. 94, October, 1929, p. 2147-2163.

Jakobsen, B.F., Gerry, M.H., Jr., Marsh, F.B., and Finkle, F.C., 1930, Discussion regarding the article Essential Facts Concerning the Failure of the St. Francis Dam: Proceedings of the American Soc. Civil Engineers: April 1930, p. 827-846.

Jansen, Robert B., 1980, Dams and Public Safety: U.S. Dept Interior, Water and Power Resources Service (Bureau of Reclamation), Denver, 332 p.

LaRue, E.C., 1928, San Francis Dam failure: unpublished manuscript, E.C. LaRue, consulting civil engineer, Los Angeles, 3 p.

Lee, K.L. and Duncan, J.M., 1975, Landslide of April 25, 1974 on the Mantaro River, Peru: National Research Council, Committee on Natural Disasters, Washington, D.C., 72 p.

Londe, P., Vigier, G., and Vormeringer, R., 1969, Stability of Rock Slopes, A Three Dimensional Study: Journal of the Soil Mechanics and Foundation Engineering Division, American Society of Civil Engineers, vol. 95, no. SM-7.

Londe, P., Vigier, G., and Vormeringer, R., 1970, Stability of Rock Slopes, Graphical Methods: Journal of the Soil Mechanics and Foundation Engineering Division, American Society of Civil Engineers, vol. 96, no. SM-4.

Los Angeles, City of, 1911, Sixth Annual Report of the Bureau of the Los Angeles Aqueduct: Dept of Public Services, Bureau of Water Works and Supply.

Longwell, C.R., 1928. Lessons from the St. Francis Dam: Science, vol. 68.

Los Angeles County Coroner, 1928, Transcript of Testimony and Verdict of the Coroner's Jury in the Inquest over Victims of the St. Francis Dam Disaster: Book 26902, 874 p.

Mead, Elwood C., ed, 1928, The St. Francis Dam Failure: New Reclamation Era: May, p. 66-71 (report of the committe appointed by the City of Los Angeles to investigate the failure).

Mulholland, William, 1928, Testimony Offered Before the Inquest of the Los Angeles County Coroner: in Los Angeles County Coroner, cited above: Mulholland's testimony is contained in p. 6-37; 87-126; 373-386, 582-589, and 620-624.

Outland, Charles F., 1963, Man-Made Disaster: the story of St. Francis Dam: Arthur H. Clark Co., Glendale, CA, 249 p.

Pacher, F., 1964, The Influence of Fissure Water on the Stability of the Rock Abutments of Arch Dams: International Journal of Rock Mechanics and Mining Science, vol. 1, p. 327-339.

Ransome, F.L., 1928, Geology of the St. Francis Dam Site: Economic Geology, vol. 23.

Raphael, Jerome M., 1988, Concrete Gravity Dams: in Development of Dam Engineering in the United States, Kollgaard and Chadwick, ed., Pergamon Press, NY, p. 22-23.

Schuster, Robert L.,ed, 1986, Landslide Dams: Processes, Risk, and Mitigation: American Society of Civil engineers, Geotechnical Special Publication No. 3, 164 p.

Terzaghi, Karl, 1962, Stability of Steep Slopes on Hard Unweathered Rock: Geotechnique, vol. 12, no. 3, p. 251-270.

Willis, Bailey, 1928, Report on the Geology of St. Francis Damsite, Los Angeles County, California: Western Construction News, June 25, 4 p., 3 plates.

HYDROELECTRIC DAMS IN THE SIERRAS OF THE SOUTHERN CALIFORNIA EDISON COMPANY

Patrick Hamilton
Senior Engineering Geologist
Southern California Edison Company
Rosemead, California 91770

INTRODUCTION

The Big Creek-San Joaquin River hydroelectric development of the Southern California Edison Company is located on the west side of the Sierra Nevada, which flank the eastern side of the San Joaquin Valley of California (Figure 1). It has attained a premier place among the power projects of the world. The system is a composite of six major reservoirs, sixteen tunnels driven through solid granite and steel penstock to nine powerhouses.

The development had its inception in the late 1800s by the young engineer, John S. Eastwood. His early surveys led to the hydroelectric scheme for the Big Creek Project. The initial development was completed in 1912 with the construction of three dams forming Huntington Lake and the associated powerhouses No. 1 and 2. Over the years, the Project has been expanded to meet the electrical needs of Southern California. After World War I, increased generation was required, and the Huntington Lake Dams were raised to increase the storage and, hence, generation capacity. New storage dams were built in 1926 and 1927, Florence and Shaver dams, respectively. Vermilion Dam, originally studied in 1917, was constructed in 1954. The most recent of the major dams, Mammoth Pool, was completed in 1960.

The story of the construction of the Big Creek-San Joaquin River development is a significant chapter in the history of reclamation of waters for the development of hydroelectric energy and irrigation. The initial development of the system required the voluntary isolation of construction men behind winter snow barriers of the High Sierras, reached only by dog teams; the conquering of seemingly insurmountable engineering difficulties in a country of deep canyons and towering peaks; the construction of a 56-mile railroad through difficult mountain terrain in less than a year to provide a means of transporting personnel, equipment and material into the area; tunnels of record lengths driven in record times; the installation of impulse turbines with the highest hydraulic heads in the world at the time of their installation.

The initial development of Huntington and Florence dams and the connecting tunnels is still considered a near-miracle in heavy construction. Construction was accomplished by the use of steam power, mule teams, a large work force, and ingenuity. The drill jumbo, control blasting, mucking ma-

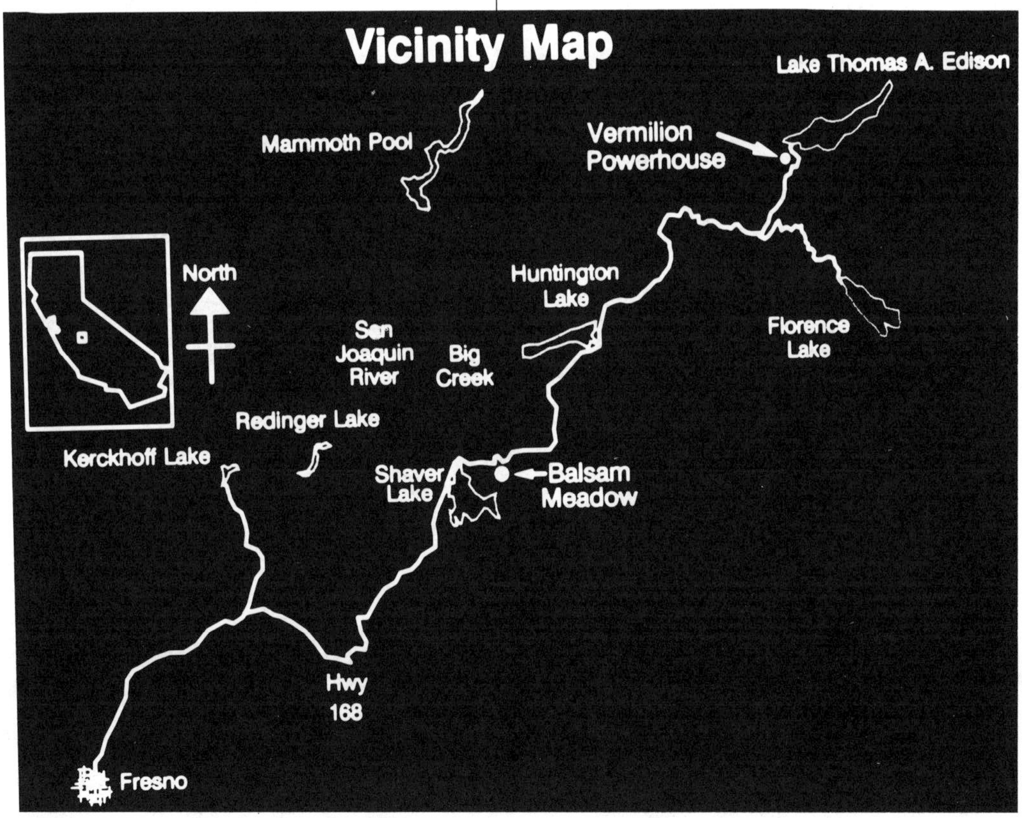

Figure 1. Vicinity Map

chines and lazer surveying commonly used today did not exist, but the jobs were completed in record time with excellent results.

In the present Big Creek system, the total operating capacity of the 23 generating units in the nine powerhouses is 999,600 kilowatts. With the exception of amounts distributed locally, the kilowatts are transmitted to the load centers in the San Joaquin Valley and the southern California area. With large drainage areas at high altitudes and with reservoirs filled during April through June principally from melting snow, the addition of powerhouses at Mammoth Pool, Big Creek 4, the Portal plant, No. 5 unit at Big Creek 3, Eastwood, and the recent turbine repower brought the annual production of the Big Creek project to approximately 3.5 billion kilowatt-hours under average hydrological conditions.

This paper describes the five major dam projects in the Big Creek system. It begins with a look at the path along which power is generated by the hydroelectric facilities and the regional geology of the Sierra Nevada Mountains. A brief description of the characteristics of each dam is then given with a short local geologic description. As with any major construction, each has unique features such as the multiple arch design at Florence Dam, the foundation treatment at Vermilion, or the size of the embankment fill at Mammoth Pool. These unique features are addressed.

BIG CREEK OPERATIONS

Six man-made lakes are used to collect and store water. They are Lake Thomas A. Edison, Florence Lake, Huntington Lake, Shaver Lake, Redinger Lake and Mammoth Pool. A series of tunnels and penstocks connect the system so water can be diverted from upper to lower lakes through powerhouses as it is needed.

The first turbine-spinning assignment for the water is at Portal Powerhouse at the end of Ward Tunnel. This tunnel diverts water from Lake Edison on Mono Creek and Florence Lake on the South Fork of the San Joaquin River into Huntington Lake. Water from Huntington Lake then drops 2,100 feet through Powerhouse 1 or 1,700 feet through Eastwood Powerstation into Shaver Lake. From Powerhouse 1, the water drops another 1,900 feet to Powerhouse 2.

Directly beside Powerhouse 2 is Powerhouse 2A, which receives water from Shaver Lake. Water from these two powerhouses is discharged into a common pool. From here the water drops another 800 feet to Powerhouse 8. Mammoth Pool Reservoir feeds water through an eight-mile tunnel to Mammoth Pool Powerhouse. Water discharged from Mammoth Pool Powerhouse is combined in a common pond with water from Powerhouse 8 and is then sent through Powerhouse 3. It is collected at Redinger Lake and discharged through Powerhouse 4.

Figure 2. Huntington Lake Dams

REGIONAL GEOLOGY

The Big Creek hydroelectric facilities are located on the western slope of the central Sierra Nevada in California. The dominant topographic feature of the region is the San Joaquin River drainage basin. The entire 1,300-square-mile extent of the river basin is cut by deep canyons and is crossed by major ridges. Broad upland benches, dotted with meadows, lie at the base of the jagged peaks that rim the basin. The highest elevation of the drainage divide is 13,972 feet at Mt. Humphreys. Water from the melting snows of the high ranges flows through upland valleys and then unites to form the main branches of the San Joaquin River. The branches then join into one main river that flows out to the foothills in a deep narrow canyon.

The Sierra Nevada is a massive, north-south trending, tilted fault block mountain range with asymmetric flanks. The tilting and uplift of the block occurred in several stages during the Miocene to Pleistocene epochs. Prior to these relatively recent stages of diastrophism, there occurred many periods of massive intrusion, crustal deformation, and volcanic activity. More recently, recurrent volcanic and glacial activity, together with constant erosion, has shaped the Sierra into its present form.

The dominant rock types in the Sierra are granite, quartz-monzonite and granodiorite. In addition, there are some roof pendants of metamorphic rock and a few remnant flows of volcanic rock remaining.

DESCRIPTION OF 5 MAJOR DAMS

Huntington Lake Dams

General Description — Huntington Lake Reservoir, with a drainage area of 97.7 square miles, has a storage capacity of 89,166 acre-feet. There are four dams that form this reservoir with an overall length of 4,126 feet and a maximum height of 170.5 feet above the foundation contact line. The spill elevation of the reservoir is 6,950 feet. Figure 2 is a photo of Dam 1 in the foreground, with Dams 2 and 3 visible in the background. Water from this reservoir may be transported by conduits to Big Creek No. 1 Powerhouse or diverted to Shaver Lake via the Eastwood Powerstation.

Dams 1, 2 and 3 at Huntington Lake, part of the initial development in 1911-1912, are curved in plan and originally formed a reservoir with a capacity of 45,000 acre-feet. They were founded on good granite and constructed of concrete made from local aggregate. The dams were subsequently raised 38 feet to provide increased reservoir capacity, and Dam 3A, straight in plan, was constructed at this time. Spillway provisions originally consisted of overflow sections supplemented by a battery of siphon spillways which have since been converted to overflow sections. The overflow section at Dam 1 was eliminated and replaced by an auxiliary spillway constructed adjacent to the dam. This spillway has manually operated sliding gates that can be operated between the elevations of 6,945 and 6,950 feet.

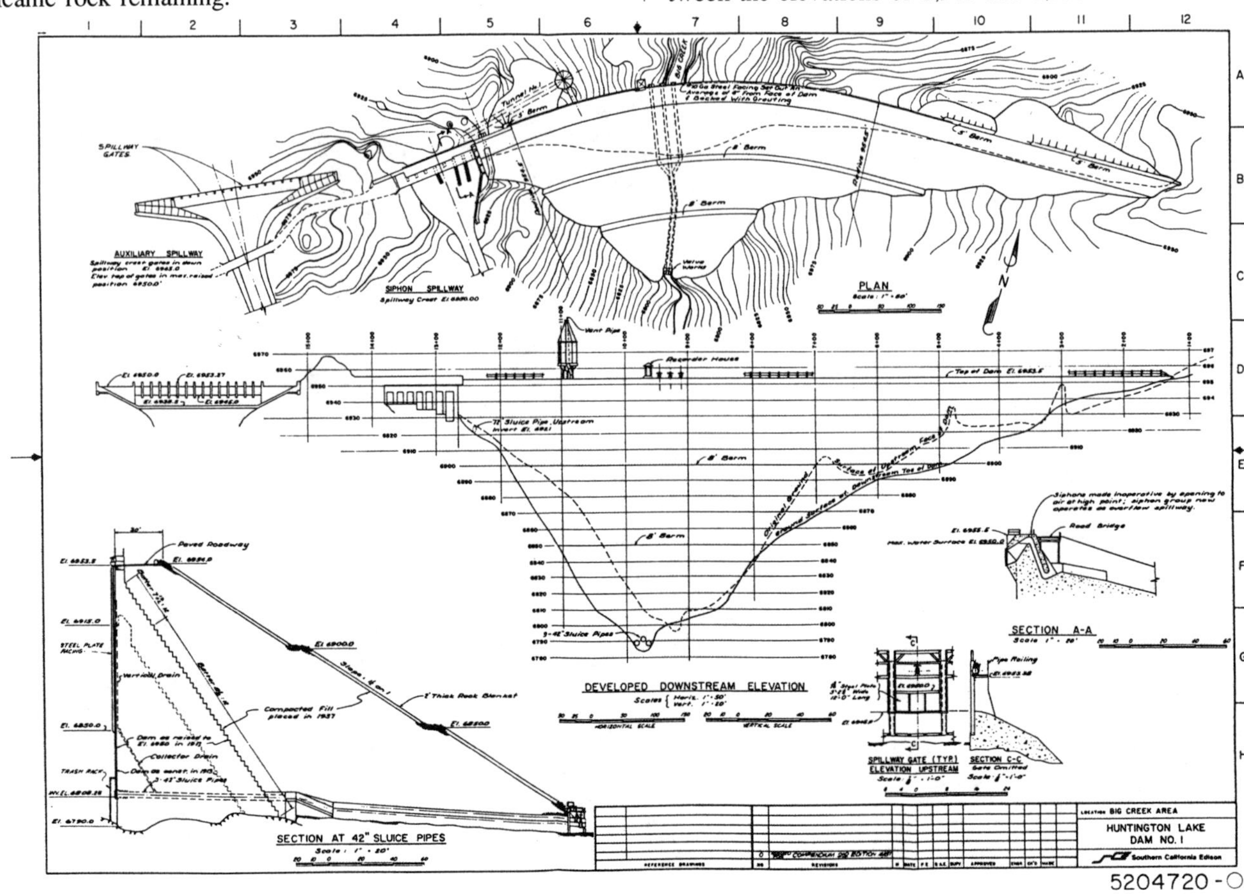

Figure 3. Plan and section of Huntington Lake Dam Number 1

Geology — The Huntington Lake basin is located within a single intrusive body of the Sierra Nevada batholith, the Dinkey Creek pluton. The segment exposed in the Huntington Lake area consists of relatively coarse-grained granodiorite containing numerous small xenoliths (inclusions) of diorite. The granodiorite has a moderately developed foliation produced by viscous flow during crystallization.

The granodiorite contains joints spaced at broad, irregular intervals. Joint orientations tend to form a rectilinear pattern consisting of three sets of joints with each set oriented nearly 90 degrees to the other two. The most prominent joint set trends northeast and is nearly vertical.

The condition of the exposed granodiorite ranges from decomposed, where it is easily disintegrated into coarse sand and rock fragments, to hard, unweathered rock. Weathering is most intense where the quartz diorite is mantled by soil and forest litter. However, the presence of large natural exposures of hard, unweathered rock in many places and the common occurrence of unweathered rock beneath weathered rock observed in road cuts suggest that the base of the weathering zone is shallow in most places.

The shape of the Huntington Lake basin was caused by the repeated movement of glaciers in the Pleistocene Epoch aralle to the strike of the dominant joint, N60°E. The movement of glacier ice during the Tahoe and again in the Sherwin stages is evident by morainal deposits along the basin.

Construction — Dam 1 is situated on Big Creek, a tributary of the San Joaquin River. It is a 170-foot high, arched, concrete gravity dam with a crest length of 1,335 feet at an elevation of 6,953.5 feet. The original dam was built in 1913 and raised about 38 feet to the present height in 1917 (Fig. 3).

The dam has been extensively modified by placing embankments against the downstream face and on parts of the upstream face, and by sheathing a large part of the upstream face with concrete-backed 10-gauge sheet steel. The embankments or buttresses were constructed to improve the stability of the arch dam and to protect the concrete against freezing. The steel sheets were used to protect the concrete in areas where the embankment could not be placed. The underside of the steel facing is drained by a pattern of half-round conduits which discharge to a low level outlet pipe.

Dam 2 is on a small unnamed tributary of Big Creek. It is a 120-foot high, arched concrete gravity dam, and like Dam 1, is buttressed fully on the downstream face and extensively on the upstream face by embankments. The crest length is 1,862 feet. The downstream buttressing fill extends for the full length of the crest and was topped out by a row of granite blocks. The blocks were recently integrated to form a parapet by facing their upstream side with gunite.

Dam 3, situated on a small tributary of Big Creek, is about 165 feet high, with a length of 640 feet. As with Dams 1

Figure 4 Florence Lake Dam

and 2, it was originally constructed in 1913 as a curved concrete gravity dam, was raised in 1917, and was later extensively buttressed both upstream and downstream by embankments.

The spillways for Huntington Lake are located at Dam 1. For the original 1913 reservoir, spill was permitted over the crest of the dams. When the dams were raised in 1917, a long overflow section was provided in the crest, and a concrete siphon spillway about 100 feet long was constructed near the right abutment of Dam 1. The siphons were reported as having a history of unsatisfactory operation. Therefore, an additional and independent spillway was cut through granite rock to the right of the right abutment. This spillwayconsists of two overflow concrete ogee weir sections, totalling 110 feet long, with crests at elevation 6,950 feet, and a gated section 203 feet long with the sill at elevation 6,945 feet. The gated section is provided with 15 manually operated, vertical-lift slidegates.

Florence Lake Dam

General Description — Prior to the construction of Florence Lake Dam, there was no storage available to supply water to Huntington Lake although a small natural glacial lake existed in the valley drained by the river. The present reservoir, with a drainage area of 171 square miles and a storage capacity of 64,406 acre-feet, was created by the construction of a multiple-arch dam 3,156 feet in length with a maximum height of 154 feet above the foundation contact line (Fig. 4). Fifty-six thousand cubic yards of concrete were used in the construction of the dam, which was completed in 1926. The spillway elevation is 6,327.5 feet.

Geology — Florence Lake Dam is constructed transverse to a broad U-shaped valley created by glaciation along the South Fork of the San Joaquin River. The head branches of the repeated glacial influence to the site were Goddard Canyon, Evolution Valley and Piute Canyon. Glacial ice moved from these canyons, carving the U-shape at Florence Lake. The glacial aspect of the valley is pronounced and unmistakable.

Upland shoulders have developed on both sides, thereby delimiting the U-shape more sharply. Associated with these shoulders are many small hanging valleys, nearly all of which contain typical glacial tarns. At Blaney Meadows along the reservoir shoreline, the U-shaped trough is still further accentuated by the presence of lateral moraines that extend like continuous embankments along the upland shoulders. These shoulders stand about 2,300 feet above the valley floor, and the circling peaks rise 3,500 to 4,600 feet.

A distinguishing feature of the valley is a ridge of glacially smoothed, almost flawless, massive granitic rock rising 200 feet above the floor and projecting squarely across the valley below the natural Florence Lake. Only a small notch has been eroded by the river. It is this ridge line on which the

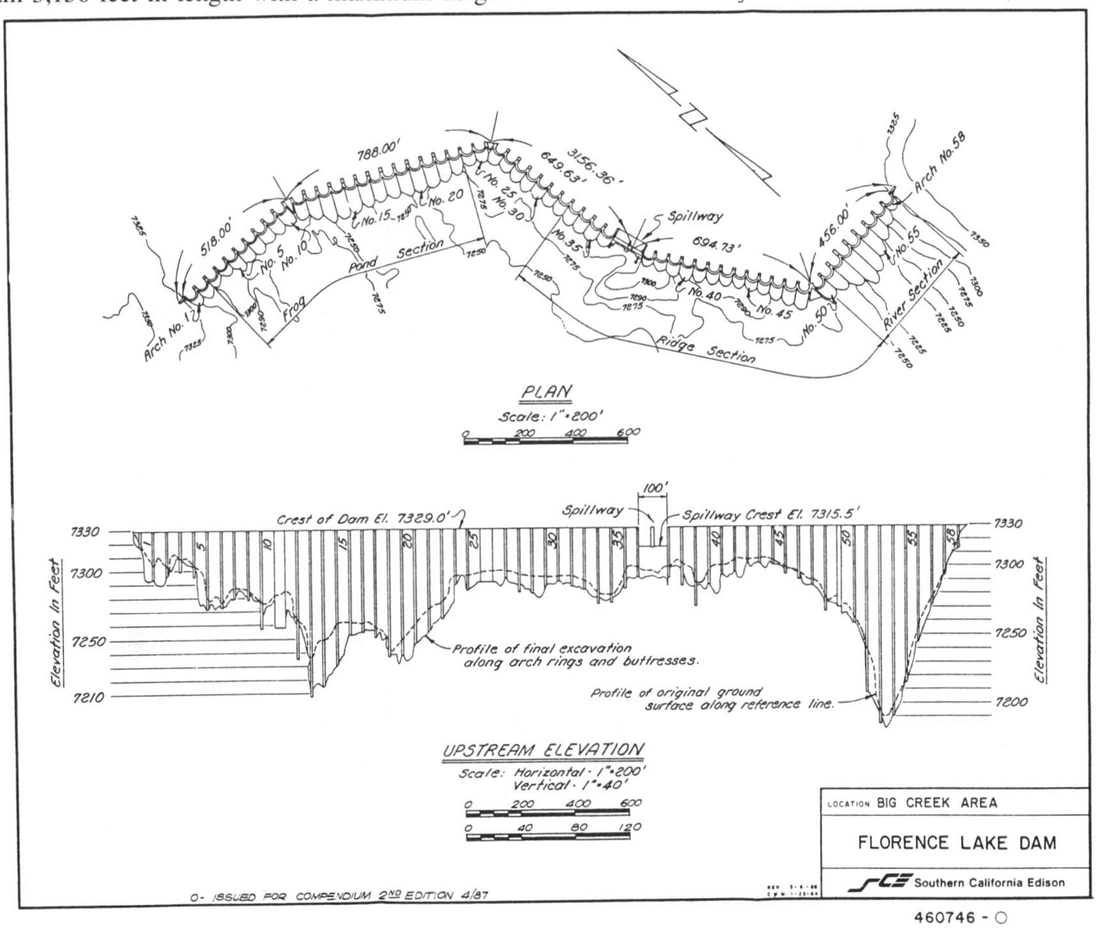

Figure 5.

dam was founded. Typical glacial landforms are evident from the smooth, striated rock surfaces, with erratics to the domal aspect of several outcrops, indicating exfoliation. The rock outcrops at the site are massive and composed of foliated granodiorite of the Sierra Nevada pluton. Locally, it has been mapped as the Mount Givens Granodiorite consisting of highly competent material with a widely spaced joint pattern.

The site for the dam was selected after extensive field studies beginning in 1923. Various types of dams were investigated to find one suitable for the site, with strong emphasis on constructability and economic factors such as expense of hauling and relative inaccessibility. A rockfill structure, using material from the Florence Lake Tunnel and faced with either earth or asphalt-covered planking to make it impervious, was given extensive study. Various disadvantages were foreseen with such a dam, and test pits in nearby meadows indicated a possible shortage of material for the earth covering. Because estimates showed it to be approximately 10 percent lower in cost than any other, the multiple arch form was adopted. John S. Eastwood is credited with having originated and developed the first designs for a multiple arch dam.

Florence Lake Dam is one of the longest multiple-arch dams in the world. The 58 arches comprising the dam were constructed on a five-tangent alignment in order to benefit from the irregular rock ridge forming part of the basin (Fig. 5). It was found that temperature, shrinkage and rib-shortening stresses could be accommodated by a modification of the shape of the arch from the conventional normal circular section. Buttresses, which are on 50-foot centers, are stiffened by counterforts at the upstream head, the downstream edge and intermediately. The extrados surface is a compound cylinder, and the intrados an inverted compound cone.

The foundation for the dam is in excellent quality rock and was exposed for the entire length of the structure except for a short distance across the San Joaquin River and a secondary channel. The arches were trenched into the granitic rock by light blasting to prevent shattering of the rock. All overburden was removed from the river and stream sections with excavation of more than 100 feet deep. Exfoliation jointing was evident in the foundation excavation by the exposure of open seams which were completely filled by a grouting program. Grout holes were drilled on 10-foot centers to a minimum depth of 10 feet around all of the arches for the entire length of the dam and grouted at a pressure of 90 psi. Only 15 holes accepted more than 50 sacks of neat cement grout while most of the holes took very little grout.

The spillway is controlled by two 50 x 14-foot drum gates with controls arranged so that the crest can be maintained at any elevation desired within their range. This type of gate was selected for ease in handling trash and logs and because no electric power is available at this location for power-operated gates.

Figure 6. Shaver Lake Dam

On the downstream side of the arches, sufficient backfill was added and drainage provided to keep water from standing against the concrete in freezing weather. Some backfill at the base of the arches on the upstream side was placed for the same reason when the lake would be empty in the winter. In spite of efforts to produce concrete of highest quality possible, it was not long after completion of the dam before the effects of freezing were recognized.

Several types of coating for the purpose of waterproofing were placed on various arches for experimentation. The first coating, which was applied to the upstream face in 1926, was totally ineffective. It was a coal tar preparation which caused very large daily temperature variations due to its black color.

In 1941, the application of a compound called Asbestile resolved the problem. Before it was applied, the surface of the arches was prepared by removing disintegrated concrete and restoring the section with gunite. Also, reflecting paint was applied to reduce large temperature fluctuations. These measures were so successful that there appears to have been little deterioration of the concrete since it was applied. Modest maintenance has been required every year and probably will continue so for the life of the structure. The prompt correction of any leaks which develop and the practice of keeping the reservoir empty during the winter are necessary to hold the line on deterioration.

Shaver Lake Dam

Water from the Shaver Lake Reservoir is transported through Big Creek Powerhouse 2A by means of a conduit which is comprised of a tunnel and penstock. The elevation of the reservoir is 5,370 feet with a storage capacity of 135,283 acre-feet. Drainage area for this reservoir is comparatively small, being only 30 square miles; and since there is not enough water from natural drainage to fill the lake, all excess water which cannot be stored in the upper reservoirs is diverted into Shaver Lake via the Huntington-Pitman-Shaver conduit/tunnel through Eastwood Powerstation.

Shaver Dam is located on Stevenson Creek, a tributary of the San Joaquin River (Fig. 6). It is a concrete gravity dam 1,760 feet long with a compound downstream slope of 0.63, 0.67 and 0.80 to 1. In plan, it consists of two nearly equal tangents, intersecting at 26 degrees, connected by a600-foot radius curve. The maximum height above the streambed is 183 feet, and maximum height above the foundation contact line is 198 feet. A saddle dike, designed as a concrete core earth fill, extends the left end of the dam by about 409 feet. Reservoir spill, which has historically never exceeded about 100 cfs, is permitted to occur directly over the crest of the dam at a one-foot deep by 250-foot long notch in the parapet wall. Completed in 1927, the dam is constructed of over 281,000 cubic yards of concrete made with native quarried granitic rock.

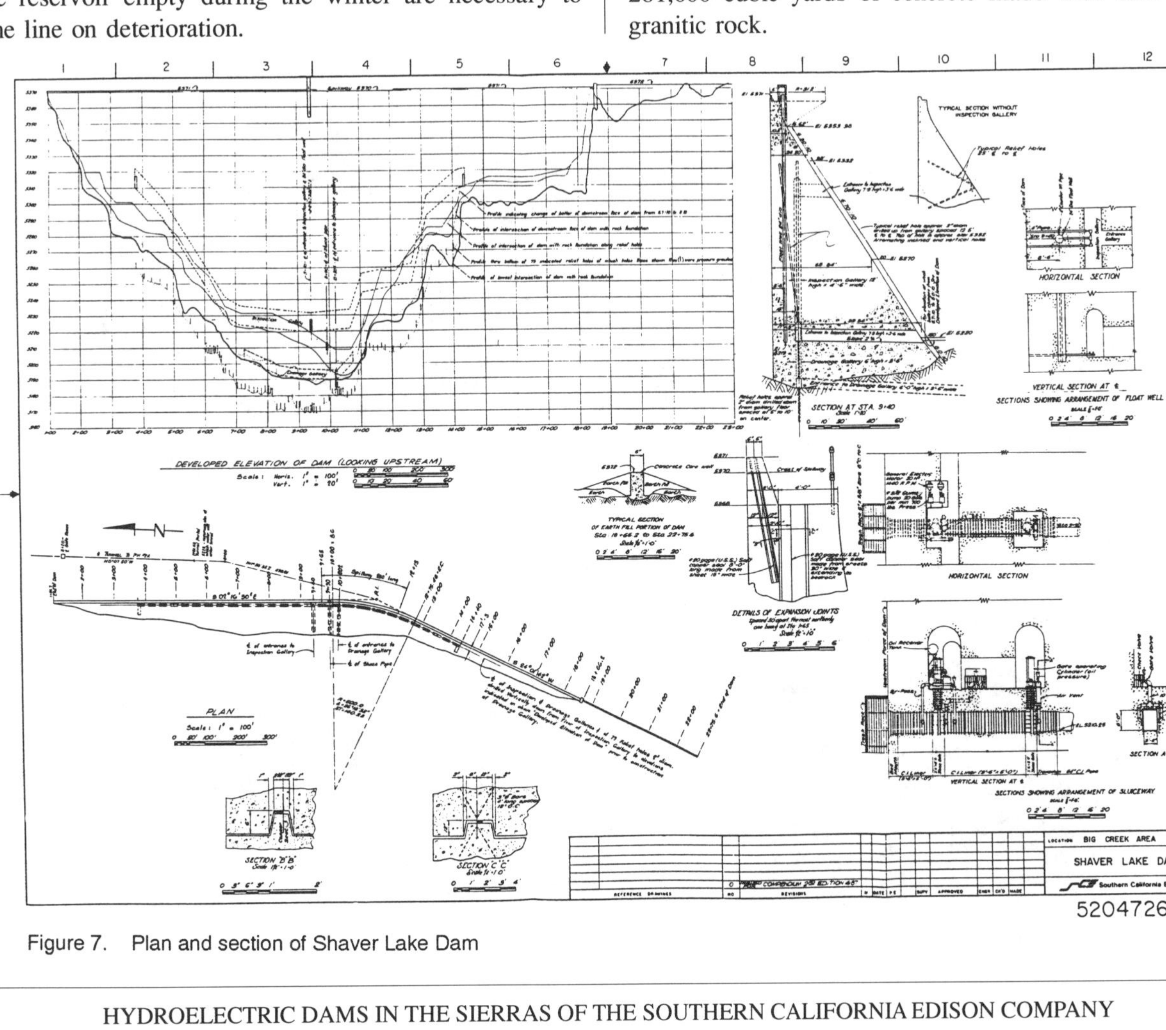

Figure 7. Plan and section of Shaver Lake Dam

The Shaver Lake basin is underlain primarily by Mesozoic intrusive rocks of the composite Sierra Nevada batholith. The dam is situated entirely within a single intrusive body, the Dinkey Creek pluton. The segment exposed at the dam consists of relatively coarse-grained quartz diorite containing numerous small xenoliths of diorite. The quartz diorite has a moderately developed foliation produced by viscous flow during crystallization.

The quartz diorite contains joints spaced at broad, irregular intervals. Joint orientations tend to form a rectilinear pattern consisting of three sets of joints, with each set oriented nearly 90 degrees to the other two. The most prominent joint set trends northeast and is nearly vertical. Foliation has a northwest trend and a nearly vertical inclination. The condition of the exposed quartz diorite ranges from decomposed, where the quartz diorite is easily disintegrated into coarse sand and rock fragments, to hard, unweathered rock. Weathering is most intense where the quartz diorite is mantled by soil and forest litter. However, the presence of large natural exposures of hard, unweathered rock in many places within the area and the common occurrence of unweathered rock beneath weathered rock observed in roadcuts suggest that the base of the weathering zone is shallow in most places.

Shaver Dam is situated on land acquired from the Fresno Flume and Lumber Company in 1919. In 1920-21, while the exploration, design and final location for the dam were being considered, a tunnel was driven to utilize the water of Stevenson Creek through Big Creek Powerhouse 2. A small, temporary 40-foot long, rock-fill dam was constructed upstream of the new dam site, forming a reservoir of between 5,000 and 6,000 acre-feet. A flume was then constructed, connecting this temporary impoundment to the newly constructed tunnel.

Excavation for Shaver Dam began in the spring of 1926. Some sluicing of loose material was performed on the abutments, with most of the heavy excavation accomplished by use of steam shovels. Badly broken, weathered and partially decomposed granitic rock was removed and the site excavated to hard, firm rock. In addition, a trench approximately five feet wide was excavated along the upstream toe to such depth as necessary. No heavy shooting was allowed, and shattering the rock or opening up seams was avoided as much as possible. About 125,000 cubic yards of material were excavated for the dam foundation (Fig. 7).

To control leakage through the jointed rock, holes were drilled and grouted on 10-foot centers near the upstream face of the dam. In this program, approximately 220 holes from 10 to 20 feet deep were grouted. The total grout take was about 700 sacks of cement, of which about 350 were taken in only five of the holes. The grout program, in terms of current practice, is judged to be modest; however, it is clear from seepage records that the grouting was adequate to reduce seepage to an unimportant quantity.

Figure 8. Shaver Lake Dam

To construct a gravity dam, it was essential that uplift pressures be kept to a minimum. A drainage gallery was formed in the dam structure, and drain holes were drilled from the gallery into the foundation. Water flowing into the gallery from these drain holes is discharged at the downstream face. To lower uplift pressures further, a lower drainage gallery was subsequently excavated along the foundation line, half in concrete and half in rock foundation. This gallery, about half as long as the upper one, presented an unusual opportunity to examine the effectiveness of the original grouting and the bond between the dam and bedrock. The contact was very good wherever encountered as evidenced by only a small amount of leakage.

A siphon system was also installed in the upper gallery to further lower the water table beyond the range of the lower gallery by drawing water from the drain holes. This siphon system, which creates a vacuum on the line by the water ejection principle, has worked effectively since its installation. The ejectors are supplied water directly from the reservoir. The combination of the upper and lower galleries and the siphon system has successfully reduced uplift in and under the dam to very small values.

Vermilion Dam

Lake Thomas A. Edison, with a capacity of 125,000 acre-feet, was completed in 1954 by the construction of Vermilion Dam. The dam is a rolled, earth-fill structure, 161 feet high with a crest length of 4,243 feet located across Mono Creek, a tributary to the San Joaquin River (Fig. 8). When filled, the reservoir has a capacity of 125,000 acre-feet.

The economic construction of Vermilion Dam became feasible with the improvement in construction and engineering skills between 1917, when the dam was first conceived, and the 1950s, when it was designed and constructed. Early studies envisioned a concrete structure, but the first borings showed deep glacial deposits over the foundation area, making excavation to bedrock excessively expensive and hence, such a dam impractical. The new techniques in soil mechanics and foundation design permitted an economical and practical design using local materials.

Before the Pleistocene ice age, the present location of Mono Creek Valley was occupied by an erosion valley carved out of granitic rock. At the dam, the bottom of the old riverbed is situated at a depth of approximately 250 feet below the deepest point of the present valley.

During the Pleistocene ice age, this region of the Sierra Nevadas was invaded repeatedly by valley glaciers descending to an altitude of a few hundred feet below that of the dam site. The glaciers deposited subglacial, marginal and fluvio-glacial materials in the valleys. At the dam site, the youngest member of the glacial deposits consists of a set of low, terminal moraines resting on impervious glacial till. At

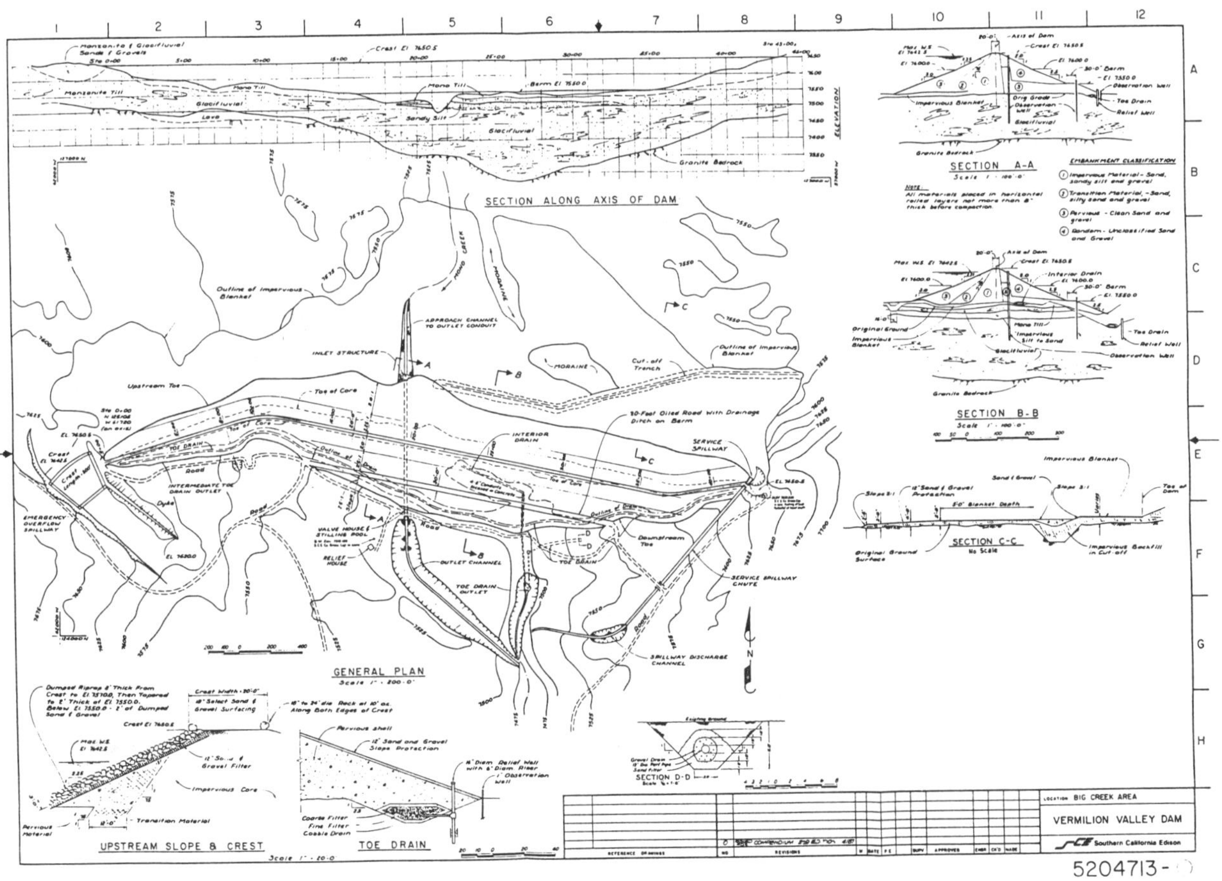

Figure 9. Plan and section of Vermilion Valley Dam

the west end of the dam site, the remnants of a kame terrace were encountered.

The upper part of the fluvio-glacial sediments beneath the youngest till sheet contains one large lense and several smaller lenses of silt with a variable thickness. The continuity of at least the youngest fluvio-glacial and glacial lake deposits was disrupted by small erosion valleys carved out by streamlets and sand-gravel mixtures. Therefore, the pattern of stratification of the subsoil at the dam site is so erratic that it was impractical to determine its average coefficient of permeability on the basis of the results of laboratory tests.

Upstream from the dam, the subsoil consists of coarse-grained, somewhat silty, fluvio-glacial and alluvial deposits with an almost horizontal surface, underlying an area of the valley floor with a length of several miles and a width of several thousand feet. Formerly, Mono Creek meandered on this floor. In recent geological times, the meandering creek began to erode, forming incised meanders with steep side slopes. At the dam site, the creek has cut to a depth of 40 feet below the old valley floor.

Vermilion Dam is a compacted earth dam consisting of four zones for a total of over 5,000,000 cubic yards including the impervious blanket (Fig. 9). Zone 1 is an impervious section composed of fine silty sand and silt with minor amounts of gravel. The transition material of Zone 2 is the siltier variety of the Zone 3 pervious section. Zone 3 is clean sand and gravel. Zone 4, the downstream stability section, is constructed from unclassified sand and gravel with a 12-inch protective slope cover of selected sand and gravel.

The foundation of the dam represents an irregularly stratified, layered system of over 200 feet of glacial deposits of varying permeabilities. Complicating the system are the cut, then backfilled, trenches of Mono Creek, its tributaries and glacial streamlets. Because of the impracticality of excavating a deep cutoff trench, it was necessary to cover the ground surface some 1,300 feet upstream of the dam with a 6- to 13-foot thick impervious blanket. By connecting the blanket to the impervious core and using the same material, the path of percolation through the foundation was greatly increased.

During geologic mapping and subsurface exploration, an apparently continuous silt stratum was located from a point about 1,000 feet downstream to a distance of more than 2,000 feet upstream of the dam site. This stratum, at a depth of 10 to 25 feet, was considered continuous in the design, so a cutoff trench, to a maximum depth of 25 feet, was constructed at the upstream limit of the impervious blanket. Even if the assumption was in error, the added cutoff would again lengthen the seepage path.

Besides providing the increased path for seepage, adequate drainage facilities along the downstream toe of the dam were also constructed to eliminate the possibilities of a failure due to piping. Water entering the pervious downstream section

Figure 10. Mammoth Pool Dam

of the dam through its base flows into a reversefilter toe drain running the full length of the dam. An extensive system of relief wells was installed below the dam and tied to the toe drain. This would reduce any hydrostatic pressure which could conceivably occur within the predominantly horizontal lenses of materials with varying permeabilities. It is apparent that the cutoff and impervious blanket have been affective in controlling seepage since the maximum seepage flow for the toe drain system in the past 10 years has not exceeded six cubic feet per second.

The service spillway with a discharge capacity of 1,200 cubic feet per second is situated at the east abutment of the dam. It is a concrete weir with a crest elevation of 7,634.5 and is equipped with one 15-foot by 8-foot radial gate. Floods in excess of 1,200 cubic feet per second can escape over a broad emergency spillway at the west abutment, through a gully and into Mono Creek.

Mammoth Pool Dam

The Mammoth Pool Dam, completed in 1960, is one of the most recent additions to the hydro storage facilities. It is located across the main San Joaquin River above Big Creek Powerhouse No. 8 and is one of the highest earthfill dams in the United States. The dam is 820 feet long with a height above streambed of about 330 feet and a gross height above bedrock of about 420 feet. The embankment required approximately 5,250,000 cubic yards of fill, which was obtained from borrow areas within two to three miles of the dam site. An overflow spillway was constructed in a natural saddle west of the dam site and has a discharge capacity of 170,000 cubic feet per second (Fig. 10).

A large part of the runoff of the main San Joaquin River and its western and northern tributary streams flows into the Mammoth Pool Reservoir. The reservoir is approximately eight miles long with a capacity of 119,940 acre-feet. A continuous flow of water is maintained below the dam in order to preserve the fish life in the river. To recover the energy from the release, a 1,250 kVa unattended generator was built at the end of a power tunnel diversion located at the base of the dam.

The dam site is situated in a narrow, steep-walled gorge carved by the San Joaquin River. The drainage area behind the dam is over 91 square miles of dendritic patterned tributary streams emptying into the river. The valley occupied by the reservoir is characteristic of the Sierra Nevada with youthful steep slopes and canyon walls. Along the reservoir, there are moderately wooded slopes alternating with barren granitic domes.

Prior to construction, massive blocks of granitic rock, some over 25 feet in diameter, have fallen from the adjacent cliffs, forming a natural "dam." A stretch of deep, quiet water about 1.5 miles in length occurred upstream of the "dam," which was locally known as the Mammoth Pool. The original

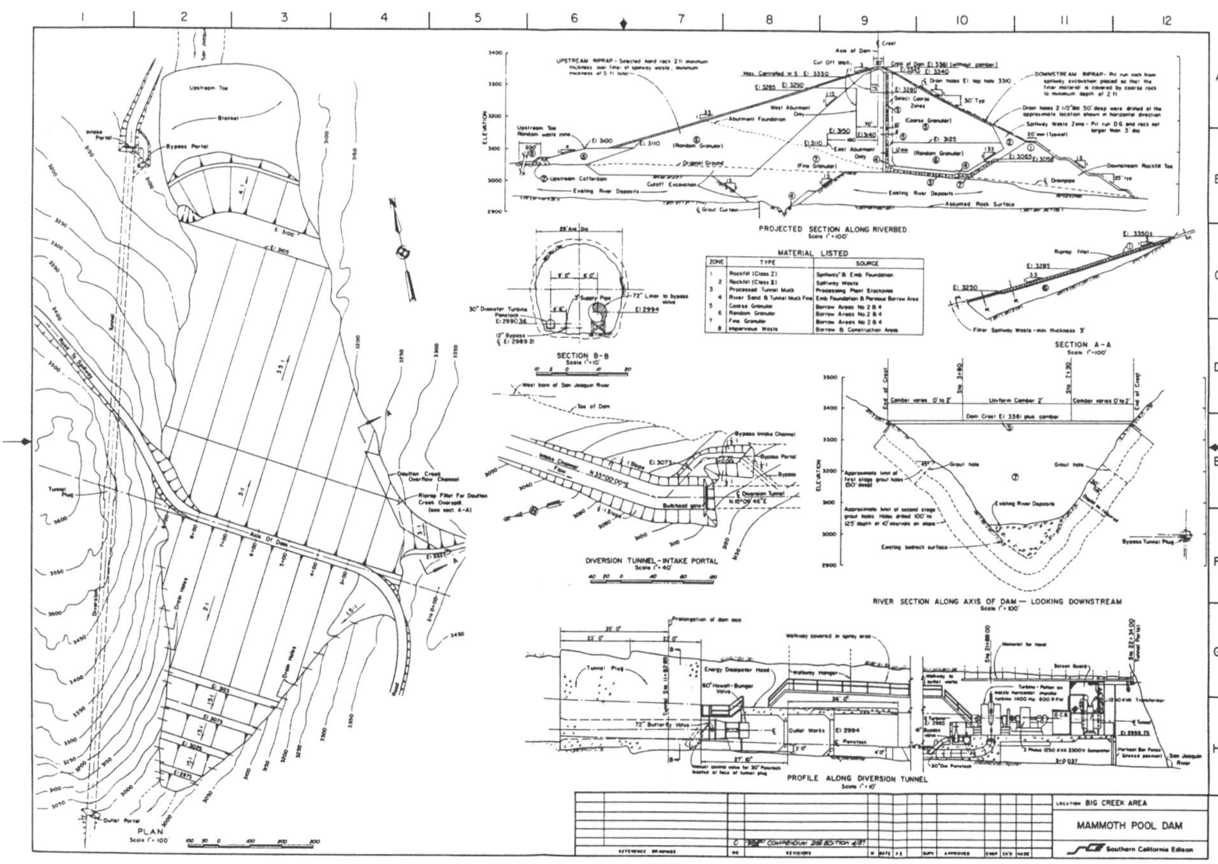

5204729-O

Figure 11. Plan and section of Mammoth Pool Dam

1955 geologic investigation describes the site: "Topographically and as regards to the strength, homogeneity, imperviousness and insolubility of the major portion of the rock, the site is virtually ideal." The dam and reservoir is situated entirely within a single Mesozoic intrusive body, the Mount Givens Granodiorite. The granodiorite is generally medium grained, but contains scattered crystals of feldspar up to 1.5 inches in length.

The most obvious structural feature of the region is the jointing pattern of the granodiorite. There are three readily discernible sets. One is the set of relatively flat to steeply curved joints which result from sheeting and exfoliation and which may be expected to decrease with depth beneath the cover. The second is a set of nearly vertical joints with varying strikes, the most prominent of which have a nearly north-south trend paralleling the segment of the river at the dam site. A third, minor vertical joint set trends nearly east-west at the dam site, more or less perpendicular to the river.

The most conspicuous joints in the region are the curved concentric fractures which divide the granite into great shells or domes. Weathering, by attacking these shells successively, produces the domes so characteristic of the Sierra in general and other regions of massive granite. This structure and its topographic consequences are well illustrated by the dome that forms the west abutment at the dam site. In the vicinity of the Mammoth Pool site, these shells range from 6 to 50 feet in thickness.

The dam embankment is essentially a homogeneous impervious type section having a central cutoff extending approximately 100 feet below the huge boulders, gravel and sand of the stream level to bedrock (Fig. 11). The dam has an internal drain system, an upstream drawdown zone, and an upstream and a downstream rockfill toe. Both abutments were cleaned and stripped in the central section of the dam to the bottom of the cutoff trench to provide a continuous seal between the impervious embankment and the bedrock. The massive granitic formation was an excellent foundation. However, because of the joint pattern in the rock, extensive consolidation and cutoff curtain grouting were required principally to minimize leakage.

A vertical drain downstream from the axis of the dam was constructed to control seepage within the embankment. The drain is a precautionary measure against the effects of transverse cracking in the fill. In the event such cracks should form as a result of post construction consolidation or arching, the downstream portion of the vertical drain was constructed of very pervious materials capable of yielding high localized water flow.

In the portion of the foundation between the cutoff trench and the downstream rock toe, the boulders were removed from the channel. A pervious drain and filter zone was provided to prevent the migration of fine material from the embankment fill into voids in the natural foundation materials. Perforated drain pipes connecting to outfall lines adjacent to each abutment have assured a positive outlet for seepage intercepted by the vertical drain and the horizontal basal filter zone.

All of the rock fill in the upstream "drawdown zone" and the downstream rock toe was obtained from required excavation at the dam site and spillway. Transition or filter zones were placed between both of these rockfill sections and the adjacent earthfill of the embankment. The drawdown zone or rockfill shell substantially increased the stability of the upstream slope of the dam and assured an adequate safety factor with respect to sloughing or shear slides under severe drawdown conditions.

Tunnel muck obtained from the excavation of the diversion tunnel was used for the construction of the upstream cofferdam and rockfill toe, which served to divert the San Joaquin River through the diversion tunnel. The downstream slope of the dam between the crest and the rockfill toe is protected against erosion by a layer of tunnel muck at least 12 inches thick.

CASE HISTORIES OF DAMS IN SOUTHERN CALIFORNIA

MORRIS DAM - The Dam with a Built-in Crack

From V. L. Peugh Report, City of Pasadena, 1935

Morris Dam, near Azusa, was built by the City of Pasadena in the period 1932 to 1935 and sold to the Metropolitan Water District by agreement with Pasadena as one of the original 13 member cities of the District. It is a concrete gravity dam rising 328 feet above a diorite foundation in San Gabriel Canyon. The current reservoir capacity is 27,500 acre-feet, which is several thousand acre-feet less than originally designed owing to heavy silting, mainly from the record floods of 1938 in southern California.

A vertical, zigzag, manmade "crack" (construction joint) was built into Morris Dam as a result of uncovering a minor fault which intersects the axis of the dam at a near-right angle. The open joint is 3 feet wide for the full height of the dam, and is offset in four places by the contact of two concrete blocks of the dam; the contacts are packed with several layers of plywood, bituminous filler and copper water stops. The object of this sliding joint is to allow up to 6 1/2 feet of movement in the dam should this fault become active, which is highly unlikely, as undisturbed old alluvial gravels were later found overlying the fault; the gravels were estimated to be 10,000 to 20,000 years old.

The Morris reservoir spillway is a unique design created from a hydraulic model made at the defunct California Institute of Technology Hydraulics Laboratory. During spilling the water shoots far downstream to prevent undercutting of the structure

PARKER DAM - A dam like an Iceberg

From MWD History and First Annual Report, 1939

Parker Dam is located just below the junction of the Bill Williams and Colorado Rivers, about 16 miles northeast of Parker, Arizona. It raises the water surface about 72 feet to elevation 450, creating Lake Havasu with an original storage capacity of 717,000 acre-feet, the largest potable water reservoir in southern California. The dam was built and is maintained by the U. S. Bureau of Reclamation; the Metropolitan Water District, however, paid the cost of Parker Dam which amounted to more than $12,000,000 in 1938. It is a variable radius arch type, utilizing 380,000 cubic yards of concrete. The dam height is 320 feet above gneiss bedrock, which was exposed after removing a remarkable 233

Figure 1. Morris Dam, spillway and reservoir, near Azusa. Built by the City of Pasadena in 1935, it was sold to the Metropolitan Water District in 1941. Note in the background the U.S. Navy's variable angle test launcher, used extensively during World War II for testing torpedo entry angles. (MWD photo)

Figure 2. Parker Dam construction circa 1936 on the Colorado River. The foundation is at the base of a 233 feet deep alluvial channel, possibly the world's deepest dam with only 87 feet above the river bed. One-half of the power generated operates the five MWD pumping plants along the Colorado River Aqueduct. (USBR photo courtesy of MWD)

Figure 3. Construction of Lake Mathews Dike (auxillary dam) in 1960, adding to the reservoir capacity of the original dam built in 1938. Note the drain and filter zones in the center of the embankment. Equipment: Earth scraper/hauler at left, water truck spraying fill to obtain optimum moisture, and dozer with trailing disks to mix layer of newly placed fill. (MWD photo)

vertical feet of alluvium; therefore, like an iceberg, only 87 feet of the dam extends above the present riverbed.

An interesting sidelight event occurred on November 10, 1934 before construction actually began. The Governor of Arizona proclaimed martial law in this part of the state and dispatched an armed detachment of the Arizona National Guard (on board the "Arizona Navy") to take possession of the dam abutment area in Arizona, contending that the Congress of the United States had not authorized construction of the dam. The United States brought suit in the Supreme Court seeking to enjoin Arizona from interfering with the construction of the dam, and on February 13, 1935, under a temporary restraining order, the Secretary of the Interior authorized a resumption of construction. The court, however, rendered a decision in the case on April 29, 1935 sustaining the Arizona Contention, and all work on the dam was immediately stopped by the contractors, Six Companies, Inc. (the builders of Hoover Dam). President Roosevelt approved the Rivers and Harbors Act on August 30, 1935, which specifically authorized construction of Parker Dam, and construction resumed again.

Early in February 1937, when the fill of the upper cofferdam was nearly completed but not yet faced with riprap, a flash flood occurred on the Bill Williams River. During a 72-hour period the run-off averaged 20,350 cubic feet per second with a peak discharge of 92,500, causing the Colorado River to flow upstream and create a lake above the upper cofferdam which extended several miles up both streams. The ponded water rose to elevation 405, forty-five feet over the diversion tunnels invert, at which stage the tunnels discharged 42,200 cubic feet per second. The lower cofferdam was threatened by the scouring action of the flood waters issuing from the tunnels, but was saved by strenuous and timely work on the part of the subcontractor, J. F. Shea Company.

LAKE MATHEWS DAM

Richard J. Proctor
Consultant
327 Fairview Avenue, Arcadia, CA 91007

Lake Mathews is the terminal storage reservoir of the 244-mile long main Colorado River Aqueduct of the Metropolitan Water District. It is in Riverside County, 9 miles southwest of the city of Riverside. The reservoir was originally constructed in the period 1935 to 1938 as Cajalco Reservoir, formed by Cajalco Dam and Dike. Each structure was constructed of rolled earthfill embankment, with slopes of 2-1/2:1 to 3:1.

By raising the original dam and dike crests 33 feet and constructing an additional dike in the period February 1959 to March 1961, Lake Mathews Reservoir was increased from 104,000 to 182,000 acre-feet. The dam is now 264 feet high and just over a mile in length, and Dike No. 1 is 127 feet high and a mile and a third long. Total embankment fill in these two structures, plus the smaller Dike No. 2, exceeds 13,000,000 cubic yards. Before reservoir enlargement could begin, the downstream rubble blanket of the dam and the downstream ice plant growth on the dike were removed. A new toe drain averaging 12 feet deep was required at Dike No. 1 to intercept and save reservoir seepage through decomposed diorite.

Geology: The reservoir rests on a portion of a vast stable granitic block, represented on the surface by the Perris peneplain. The rocks in the eastern part of the reservoir are mixed medium-grained diorite and San Marcos Gabbro which have been injected by northwest striking pegmatite and aplite dikes. Decomposed diorite was used for fill material in Dikes 1 and 2. Weathering in the diorite averages greater than 20 feet under 3 feet of topsoil. In the western part of the reservoir the main rock is fine-and coarse-grained biotite-rich diorite. Almost exclusively within the quartz monzonite are swarms of black tourmaline dikes and several massive tourmaline-quartz plugs. One of these, a mile west of the reservoir, contained the famous Temescal Tin Mine, which is now depleted.

Several quartz latite porphyry dikes were intruded into the quartz monzonite in the late Cretaceous. Genetically related to these volcanic intrusions are dikes and plugs of dark blue quartz diorite porphyry which appears to represent the last igneous activity in the region. The southeast portion of the dam rests on a roof pendant of Triassic Bedford Canyon Formation, which in this locality is quartzite and siliceous slate.

More than 1,000 test holes by the Metropolitan Water District in the 1930s revealed lake and stream sediments at least 80 feet thick which have filled the old depressions in the ancient topography. Meager outcrops of this conglomeratic green sand and siltstone south of the reservoir have not been described by geologists who have worked on the western Perris block. The borrow pits for embankment fill material for the dam were stripped of their mantle of soil and terrace gravels in 1959 to expose more of these sediments, including the bones of two large early Pliocene camels and an oreodont. The sediments have been named the Lake Mathews Formation and have an areal extent of six square miles. The fossils were donated by the author to the Los Angeles County Museum. In addition to these sediments being used for fill material in Mathews Dam, deposits of decomposed diorite and quartz monzonite compose about 20 percent of the volume of the embankment.

Grouting: The grouting at the extensions of the original dam was mainly on 20-foot centers with an initial eight to one water-cement grout ratio, and stage grouted at 5 feet and 30 feet below cutoff wall concrete.

Five rock types were encountered in the cutoff wall of the dam—quartz monzonite, diorite, quartz diorite porphyry, quartzite, and conglomerate. Their average grout "take" results are as follows: Least "take" of all was the Lake Mathews Formation conglomerate in the form of a 200-foot wide, 20-foot deep channel. It comprises 6 percent, by length, of all rock types encountered in the cutoff trench. Probably owing to its silty matrix, it took only 5.7 sacks of grout per hole. In addition to grouting, the upstream surface of this old alluvial channel was covered with a 5-foot thick compacted earth blanket to prevent future foundation seepage.

The quartz monzonite comprises 54 percent of the rocks in the cutoff trench and took 8.4 sacks of grout per hole. Well-fractured quartzite of the Bedford Canyon Formation comprises 8 percent of the rock types and took 9.2 sacks of grout per hole. Together, the quartz diorite porphyry and decomposed fine-grained diorite comprise 32 percent of the rock types and took 10.4 sacks per hole. Only the east abutment of Dike No. 1 required grouting. The rock is medium-grained decomposed gabbroic diorite. Twenty-five 30-foot stage grouted holes took an average of 15.7 sacks of grout per hole. Diorite, then, is the most porous rock type at Lake Mathews.

CEDAR SPRINGS DAM (Silverwood Lake) A dam designed against potentially active faults

Steve Stankov
Consulting Engineering Geologist
29371 Buckhaven Street, Laguna Niguel, CA 92677
Former DWR geologist.

Introduction

Silverwood Lake, impounded by Cedar Springs Dam, is an integral feature of the California Aqueduct and serves primarily to regulate aqueduct flows. It provides a method of meeting peak water demands and a source of emergency water during periods of aqueduct shutdown. Silverwood Lake also helps to satisfy the recreational needs in southern California, and serves as a forebay for the operation of Devil Canyon Powerplant in San Bernardino.

Geology Of The Area

This dam, completed in 1971, is located in the northwestern portion of the San Bernardino Mountains, near the northerly boundary of the Transverse Ranges Geomorphic Province and the Mojave Desert.

The regional structure is dominated by the San Andreas Fault which forms the southern and western boundary of the San

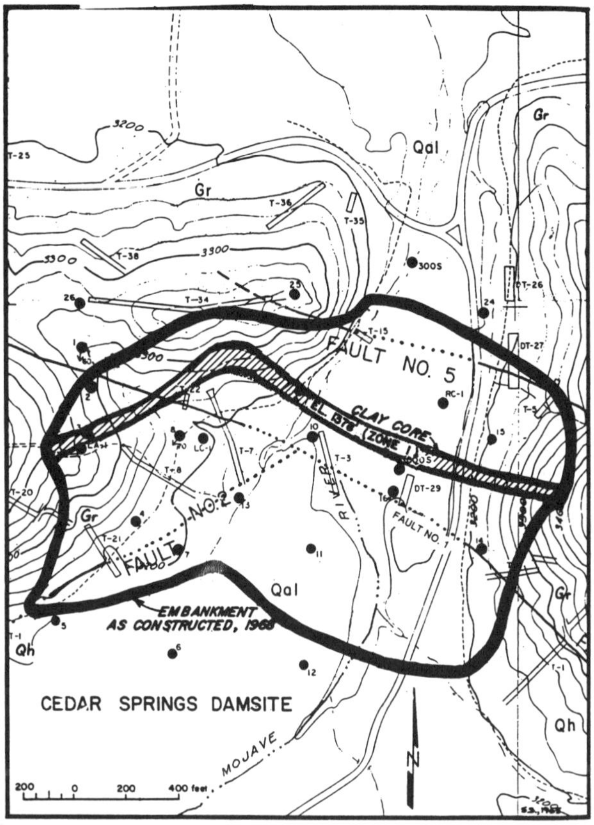

Figure 4 Geologic map of Cedar Springs damsite in 1965 showing exploratory work. Qh = Harold Formation sandstone, Gr = Granitic rock, Qal = Recent alluvium.

Bernardino Range. Many of the valleys are fault controlled, and they are subparallel to the San Andreas fault system.

The primary rock types of the area are gneissic granodiorite and quartz monzonite of Mesozoic age. Tertiary to Quaternary continental sediments, namely, the Harold Formation, and older and Holocene alluvium are preserved along structural troughs and valley floors (Meisling and Weldon, 1982).

The Harold Formation occupies the valley floors and dips generally to the north. On their northern flanks, the valleys are terminated by faults. Silverwood Lake lies in one of these valleys. As much as one thousand feet of vertical displacement is believed to exist in the lower reservoir area. The contacts of the Harold Formation with the plutonic rocks along the formation's south side are at least in part depositional.

At the dam, the granitic rock is a light gray quartz monzonite. Much of the rock is medium-grained and exhibits poorly to well defined foliation. In the lower reservoir area, the foliation has a regional dip of 5° to 15° to the north-northeast and strikes about N70^{o}W. An elongate body of gneissic granodiorite trends northwest across the dam site. Also, minor amounts of chlorite schist and marble have been found at the dam site and the lower reservoir area. Joint directions are variable, with steeply dipping planes, but set approximate (1) N70^{o}E, (2) N20^{o}E, (3) N15^{o}W, and (4) parallel with foliation.

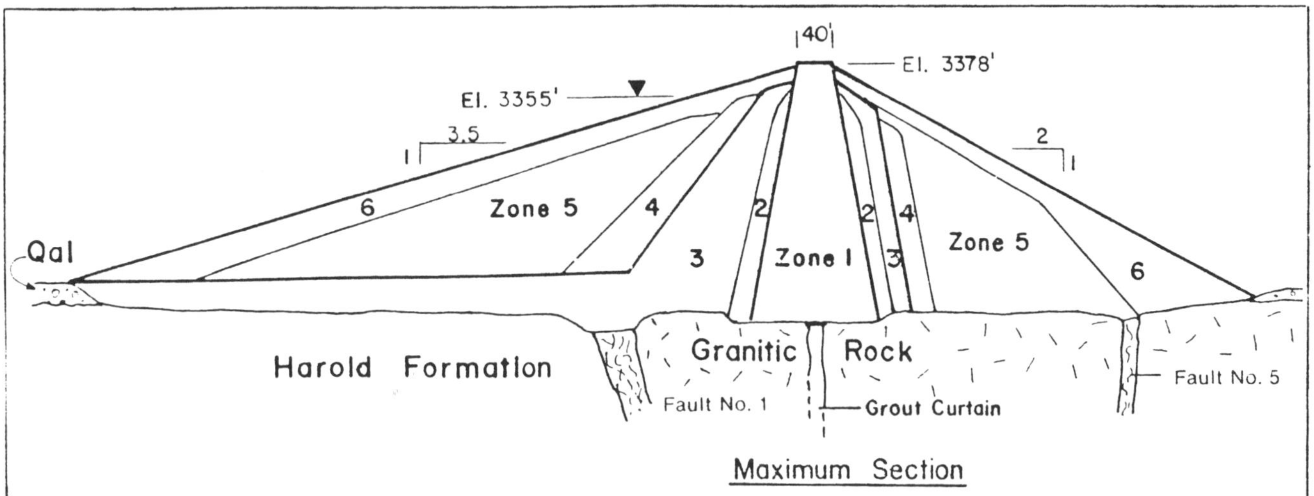

Embankment Materials

ZONE		MATERIAL	SOURCE
Impervious	1	Lean Clay	Apple Valley
Filter	2	Silty Sand	Harold Fm.
	3	Gravelly Sand	Stream Bed
Stability	4	Rolled Rockfill	Quarry Spalls
	5	Rolled Rockfill	Quarry
	6	Dumped Rockfill	Quarry

Figure 5. Maximum geologic section showing faults in foundation.

The Harold Formation and the alluvium are preserved in the lower reservoir area (Fig. 4 and 5). The Harold Formation consists predominantly of indistinctly bedded to massive, indurated, white arkosic sandstone. Claystone and siltstone lenses to one foot thick were found during exploration. Beds of clean sand were also encountered, but for the most part, this unit was found to be high in silt content. Rounded, hard pebbles and cobbles of quartzite and granitic rocks were found disseminated throughout the beds in the formation. Older and Holocene alluvium occupies the reservoir area floor and consists of fairly clean sands and gravels.

Exploration

Preliminary exploration at this site was performed intermittently from 1960 to 1964. During this time, the exploration activities included aerial photographic studies, field mapping, drilling 30 diamond core holes at the dam site, and drilling auger holes at the borrow areas, as well as under-flow measurements using pump tests and installation of observation wells.

Final exploration using bulldozer and dragline trenches commenced during the final design of the dam. This was accomplished between July 1964 and March 1968. In October 1964, evidence of Holocene displaced alluvium was uncovered in Trench T-21 during trenching in the dam foundation (Fig. 4 and 6). Although the preliminary investigations delineated the faults and the rock distribution in the foundation area, nothing was found to suggest Holocene displacement.

Potentially Active Faults

With the found offset of Holocene alluvium, as shown in Trench Log T-21 (Fig. 6), the Department of Water Resources called in its consulting boards; the Earth Dam Consulting Board and the Consulting Board of Earthquake Analysis. Both boards were asked to review the geologic conditions at Cedar Springs Dam. As a result to the Department's own investigative efforts and the reviews of the consulting boards, the conclusion was reached that the faults, Fault No. 1 and Fault No. 2, at the dam site (Fig. 4) must be considered potentially active. Maximum fault movement of 3 to 5 feet was determined possible (with the direction not specified) by the consulting boards. This was based upon the fault offset of alluvium in Trench T-21 and configurations of faults at the dam site (California Department of Water Resources, 1968).

Design For The Safest Possible Structure

Consequently, various supply alternatives were studied. The Department considered at least eight viable alternatives to the construction of the 200,000 acre-feet storage reservoir as originally planned during the preliminary design. These studies, together with the DWR consulting board's reviews, led to the conclusion that a lowering of the dam and reservoir

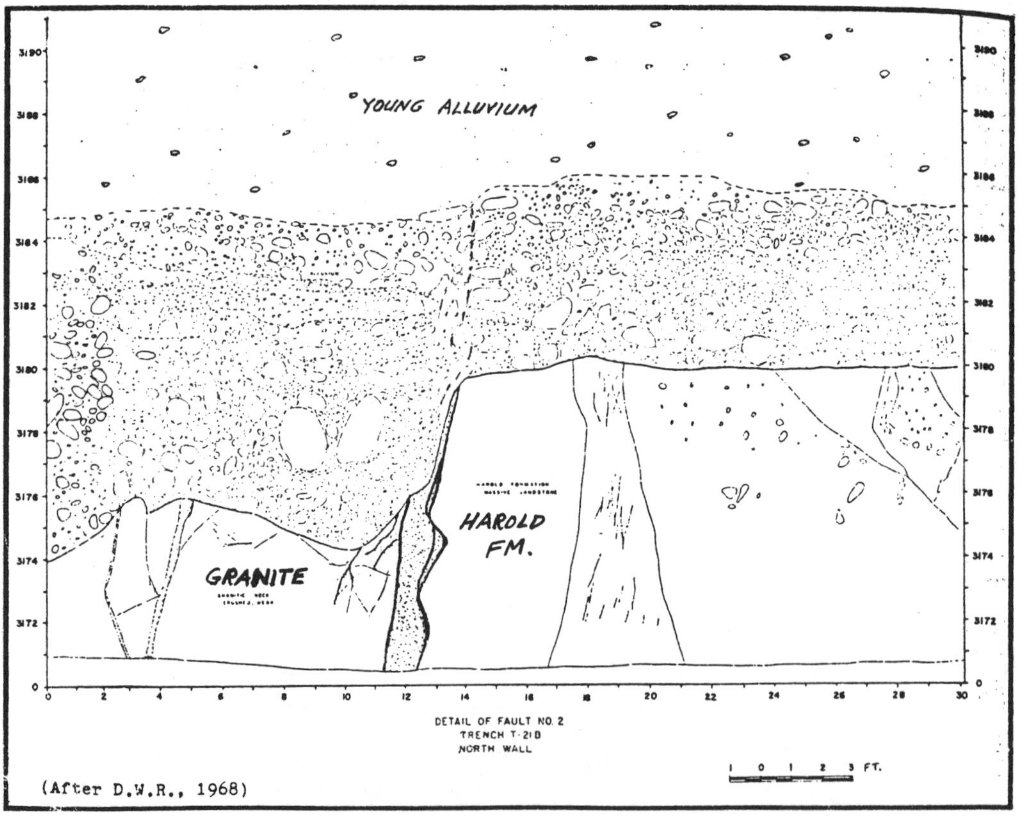

Figure 6. Fault No. 2 in Trench T-21B extending through alluvial gravels.

capacity was the most prudent of the options. However, many design changes had to be implemented.

The most significant modifications were: (1) the lowered dam crest elevation from 3,460 feet to 3,378 feet, and decreasing the maximum water surface by 3,355 square feet, thereby reducing the reservoir capacity to 78,670 acre-feet; (2) changing the design of the embankment zoning, including a large rolled-rock shell; and (3) shifting the alignment and location of the dam's axis so that all of the impervious core is on granitic rock.

The Embankment

The cross section shown on Figure 5 shows the conservative embankment design. The design includes a thick impervious clay core, zones of cohesionless filters designed to prevent piping of materials from one zone to another, and a thick rock filled zone for maximum stability.

The clay core was designed to accommodate the possible offset of the foundation of more than 5 feet. To guard against material migrating through open crack sand eroding the embankment, a lean clay was imported from Apple Valley along the Victorville-Lucerne Valley Road (Ely, 1982). This material was placed at more than optimum moisture content so the clay would deform plastically and heal readily. Thick filter zones and rolled rock fill zones (Zones 2, 3, and 4) were also to mitigate against piping of the materials and their migration from one zone to another.

The shape of the core in plan is concave-downstream so that any deflections of the dam embankment would tend to compress because of this extreme thickness. The core is even more enlarged through the left abutment. The alignment of the core was moved as far as possible from the faults in the foundation. In addition, the weighted stability sections, Zones 5 and 6, were thickened so that the embankment could withstand full reservoir pressure on the core.

Grout Curtains And Control Of Foundation Leakage

Two grout curtains were used in this dam. The downstream curtain is up to 150 feet deep, and the upstream curtain is 25 feet deep. Additional blanket grouting was provided in the unsuitable rock areas as well as the fault zones, using a maximum depth of 25 feet and pressures up to 50 psi. The Harold Formation was not grouted.

In addition, to insure against excessive foundation leakage and possible erosion of the foundation materials, the abutment areas upstream from the core was blanketed with Zone 1 lean clay material (Fig. 5). Zone 3 blanket material was placed over Fault No. 1 to protect against piping of any eroded foundation materials at the left abutment downstream into the rock fill sections.

Design Against Possible Settlement Due To Seismic Shaking

The large sections of rolled rock fill placed in Zone 5 of the embankment were compacted in thin lifts to reduce any future settlement. All of the alluvium was removed from under the dam section. This material was about 30 feet thick in the upstream portion of the dam foundation, while it thinned to about 5 feet thick at the toe of the downstream portion. Thus, any possible liquefaction potential of the native materials was eliminated.

Substantial Freeboard Designed Against Seiche

The dam crest elevation of 3,378 is 23 feet above the maximum water surface. A paved road over the dam crest guards against erosion if overtopping should occur. The spillway is located away from the embankment and is cut through granitic rock in the left abutment. The construction of the Cedar Springs Dam and its appurtenances was completed in 1971, and the reservoir was filled in 1972.

Acknowledgments

The author is grateful to the California Department of Water Resources for the opportunity to participate in this project. A special thanks to Jack Marlette and Guy Hanagan for support during the preliminary investigation. The outstanding efforts of Art Arnold and the late Vadim Voloshin during the final exploration of this project is fondly recalled. Frank Kresse and others on the various design teams are also thanked for the opportunity to assist them. Thanks are due to Karen Faris for drafting and Donald Fife and Mark Bryant for their review of the manuscript.

This is a condensed version of a paper that appeared in *Geology and Mineral Wealth of the CA Transverse Ranges*, in 1982; printed with permission of the South Coast Geological Society.

References

California Department of Water Resources, 1963, Status of Geologic exploration at Cedar Springs Dam.

California Department of Water Resources, 1968, Office Report, Engineering Geology of Cedar Springs Dam and Reservoir.

Ely, M. II, 1982, Robertson Clay Mine, Fifteen Mile Valley Quadrangle, California: in Fife, D. L. and Minch, J. A. (eds.) Geology and Mineral Wealth of the California Transverse Ranges, South Coast Geological Soc., Santa Ana, CA.

Meisling, K. E. and Weldon, R. J., 1982, The Late-Cenozoic structure and stratigraphy of the western San Bernardino Mountains: Geol. Soc. America Guidebook, April 19-21, 1982, p. 75-81.

Sherard, J. L., Cluff, L. S., and Allen, C. R., 1974. Potentially active faults in dam foundations: *Geotechnique*, vol. 24, no. 3, p. 367-428.

GARVEY RESERVOIR, SPRINGS AND LEAKAGE

Richard J. Proctor
Consultant
327 Fairview Avenue, Arcadia, California 91007

Garvey Reservoir is a cut-and-fill type reservoir impounding 1,600 acre-feet of Colorado River water in the City of Monterey Park. Constructed in 1954, it is located in the Repetto Hills in the Repetto Formation marine sandstones. Although carved out of the top of a hill, the main embankment is 150 feet high with 3:1 slopes.

Ironically, within a year and one half after the reservoir was filled, springs began to occur in a new subdivision beyond the base of the embankment fills. However, chemical analysis showed that the water was not Colorado River water, and fluorescein dye placed in the reservoir failed to show up in the springs. In addition, old photographs show a wet area existed before the subdivider filled the gullies. It appeared to be a coincidence in timing that the old springs broke through the subdivision fill soon after the reservoir was filled. A new subdrain system in the subdivision eliminated the seepage problem.

However, in the late 1980s, some thirty years later, water seeps were again noticed. This time the water was traced to foundation leaks at the bottom of the reservoir. Speculation was that perhaps the Whittier Narrows earthquake of October 1, 1987 (M6), only 5 miles east, may have contributed to the cracks on the reservoir bottom. The reservoir was drained, the leaks repaired, and a thick clay lining was placed in the reservoir bottom and sides

Figure 7. Overview of Garvey Reservoir in Monterey Park nearing completion in 1954. Reservoir leakage 30 years later required major lining repairs. (Steve Barrett photo, courtesy of MWD)

REFERENCES ON OTHER DAMS IN SOUTHERN CALIFORNIA

Arnold, A. B., 1966, The California Aqueduct in Southern California, p.57-62 in R. Lung and R. Proctor, (eds.), Engineering Geology in Southern California: Assoc. Engrng. Geol. Special Publication, Los Angeles, 389 p. (out of print) (Pyramid, Castaic, Silverwood and Perris Dams.)

Arnold, A. B. and Hanegan, G. L., 1974, Rebound and foundation settlement at Castaic Dam: Bulletin Assoc. Engrng. Geol. vol. XI, no. 1, p. 15-27.

Bauman, P. A., 1960, Cogswell and San Gabriel Dams: Am. Soc. Civil Engrs. Transactions, vol. 125, part II.

California Department of Water Resources, 1988, Dams Within Jurisdiction of the State of California: DWR Bull. 17-88, 121 p. (Statistics on 1212 dams in California.)

California Division of Water resources, 1931, Report of the Consulting Board on the safety of proposed Pine Canyon Dam, Los Angeles County: 22 p. (Morris Dam).

Dowlen, R. J. and D. L. Schug, 1983, Field Guide to selected dams in San Diego County: Assoc. Engrng. Geol. Ann. Mtg. Guidebook, 157 p.

Jackson, D. C., 1988, Great American Bridges and Dams: The Preservation Press, Wash. D.C., 357 p. (Friant, Littlerock and Mulholland Dams.)

Lund, LeVal, 1974, Silver Lake Reservoir reconstruction, p. 814-828 in Inspection, Maintenance and Rehabilitation of Old Dams: Am. Soc. Civil Engrs., NY, 946 p.

Mayeda, H. S. and J. B. Weldon, 1978, Los Angeles Reservoir and Dam, p. 122-128 in Guidebook of the Los Angeles Metropolitan Area: Assoc. Engr. Geol. So. Calif. Section, Los Angeles, Don Lamar, Field Trip Chairman

Morris, S. B., and C. E. Pearce, 1934, Concrete gravity dam for faulted mountain area: Eng. News-Record, vol. 113, p. 823. (Morris Dam).

Morris, S. B., 1954, Morris Dam construction: Engineering Geology Packet, Geol. Soc. America Ann. Mtg., Los Angeles, 5 p.

Morrison-Knudsen Company, 1947, River tamers, dams and their builders: P. D. Nations (ed.), Boise, ID, 109 p. (Imperial, Parker and Santa Fe Dams).

Proctor, R. J., 1966, The Colorado River Aqueduct, p. 72-80 in R. Lung and R. Proctor, (eds.), Engineering Geology in Southern California: Assoc. Engrng. Geol. Special Publication, Los Angeles, 389 p. (Parker Dam, Mathews Dam and 9 others.)

Peugh, Verne L., 1935, Technical report on the construction of Morris Dam: Pasadena Water Dept., 218 p., 72 plates.

Roche, R. B., 1974, Program for the investigation and analysis of dams of the L.A. County Flood Control District, p. 804-813 in Inspection, Maintenance and Rehabilitation of Old Dams: Am. Soc. Civil Engrs., NY, 946 p.

Sarkaria, G. S., 1974, Stress studies in three old arch dams (including Pacoima Dam): p. 418-439 in Old Dams: Am. Soc. Civil Engrs., NY.

Schuyler, J. D., 1888, The construction of Sweetwater Dam: Am. Soc. Civil Engrs. Trans., vol. XIX, p. 201-220.

Scott, R. F., 1971, San Fernando Earthquake, 9 February 1971 preliminary soil engineering report, p. 299-331 in P. C. Jennings, (ed.), Engineering Features of the San Fernando Earthquake of February 9, 1971: Caltech Report No. EERL-71-02, 512 p. (San Fernando Dam and Pacoima Dam.)

Seed, H.B., 1975, Dynamic analysis of Lower San Fernando Dam during the earthquake of February 9, 1971: Jour. of Geotechnical Engineering, Am. Soc. Civil Engrs., NY, vol. 101, p. 889-912.

Sherard, J.L., Woodward, R.J., Gizienski, S.F., and Clevenger, W.A., 1963, Earth and earth-rock dams: Wiley, NY, 725 p. (Sheffield Dam.)

Terzaghi, K., and T. M. Leps, 1960, Design and performance of Vermillion Dam: Am. Soc. Civil Engrs. Transactions, vol. 125.

Walker, F. C., 1958, Development of Earth Dams Design in the Bureau of Reclamation: U.S. Bur. Reclamation, Denver, CO (Cachuma and other dams.)

Wolfe, C. G., 1974, Seismic safety of Henshaw Dam, p. 871-899 in Inspection, Maintenance and Rehabilitation of Old Dams: Am. Soc. Civil Engrs., NY, 946 p.

Yerkes, R. F., M. G. Bonilla, T. L. Youd and J. D. Sims, 1974, Geologic environment of the Van Norman Reservoirs: U. S. Geological Survey Circular 691-A, 35 p.

Baldwin Hills Reservoir Failure References

California Department of Water Resources, 1964, Investigation of Failure, Baldwin Hills Reservoir: Report of Findings, State Engineering Board of Inquiry, April 1964, 171 p., 48 fig., 12 tab., 92 photos.

Hamilton, D. H., and Meehan, R. L., 1971, Ground rupture in the Baldwin Hills: Science, April 23, vol. 172, p. 333-344.

Hudson, D. E., and R. F. Scott, 1965, Fault motions at the Baldwin Hills Reservoir site: Bull. Seism. Soc. of Am., vol. 55, p. 165-180.

James, L. B., 1968, Failure of Baldwin Hills Reservoir, Los Angeles, California, in Kiersch, G. A., ed., General case histories on dams, tunnels, highways, and underground construction: Geological Society of America Engineering Geology case Histories 6, p. 1-11.

Jessup, Walter E., 1964, Baldwin Hills Dam failure: Civil Engineering, vol. 34, no. 2, p. 62-64.

Kresse, F. C., 1966, Baldwin Hills Reservoir failure of 1963: p. 93-103 in Lung, R., and R. Proctor, eds., Engineering Geology in Southern California: AEG Special Publication (out of print) 389p.

Leps, T. M., 1972, Analysis of failure of Baldwin Hills Reservoir, in Proceedings, Specialty Conference on Performance of Earth and Earth Supported Structures, Purdue University: Am. Soc. Civil Engineers, NY, p. 507-550.

Paris, B. R., Hume, N. B., May 1964, Aftermath of an Emergency — The failure of the Baldwin Hills Reservoir: Public Works, p. 103-107.

Warne, W. E., Feb. 1964, The Baldwin Hills Dam Failure: Western Construction, vol. 39, no. 2, p. 78-80.

Wilson, R. R., 1949, Geology of the Baldwin Hills Reservoir Site and Immediate Vicinity: Unpublished report, Los Angeles Department of Water and Power.

Chapter **10**

Tunnels

Photo of 26-foot diameter TBM that excavated the Newhall water tunnel.

Tunnels

Editors Note: Even with New York City's 242 miles of subways, southern California probably has more miles of tunnels than any similar area in the world—more than 300 miles. Two of the most recent are described—the first part of the long-awaited L.A. Metro subway, and an 8-mile sewer tunnel that extends beneath LAX. Other tunnel case histories describe geologic problems including large water inflows, methane, a large fault crossing, and open joints deep in the Sierras. Dr. Bakhtar's paper explains several rock mechanics classification systems and provides an example of how one system was used in a test tunnel/chamber.

OVERVIEW OF TUNNELING IN SOUTHERN CALIFORNIA

Richard J. Proctor
Consultant
327 Fairview Avenue
Arcadia, CA 91007

Southern California contains just over 300 miles of civil tunnels, possibly more than any similar area in the world. At least 280 individual tunnels make up the total. Most were constructed to convey water to this semi-arid area, which is the second most populous in the U. S., containing more than 18 million people.

Three main aqueducts supply potable water to Southern California from different sources. The Los Angeles City Owens Valley Aqueduct was the first built and initially completed in 1913. Major additions were the St. Francis Dam in 1928 (see paper herein), the Mono Craters Tunnel (see paper herein), Lake Crowley in the 1930s, and the Second Aqueduct Project in the 1960s. This aqueduct system contains more than 100 tunnels with a combined length of 76 miles.

The Colorado River Aqueduct was completed in 1940, with 92 miles of main aqueduct tunnels and 43 miles of distribution tunnels. The Metropolitan Water District of Southern California (MWD) is a six-county governmental body created in 1928 by the state legislature; it owns and operates the world's most extensive facilities which, along with their distribution of California Aqueduct water, supplies two billion gallons of water per day to 220 cities.

The California Aqueduct of the California Department of Water Resources (DWR) began supplying water to the MWD and other public agencies in 1974. It has 9 tunnels totaling 19 miles in southern California, mainly in the Tehachapi Mountains, the Ridge Basin of the western San Gabriel Mountains, the San Bernardino Mountains, and at Perris Reservoir.

Other significant local tunnel projects include 27 miles of sewer tunnels and 19 miles of flood control tunnels.

The following table lists the dimensions, date completed, and geology of all tunnels in Los Angeles County, plus future planned tunnels.

Major tunnels planned for southern California in the future include:

1. The expansion of the MWD's water distribution system, especially in the San Bernardino Mountains and into Riverside, Orange and San Diego Counties, with 23 miles of tunnels scheduled for construction 1993-98. If the extension of the MWD Foothill Feeder in the eastern San Gabriel Mountains is approved, then an additional 32 miles of tunnels will be constructed after the turn of the century.
2. The Los Angeles Metro subway (see paper herein) will continue expanding, with 28 miles mainly from downtown through Hollywood and into the San Fernando Valley, planned to be completed by 2004.
3. The Southern California Edison Company's Bicep Project in the Sierras will include 8 miles of tunnels.
4. The on-again-off-again California-Nevada High-Speed Train (magnetic levitation) would mostly be underground, between Los Angeles and Las Vegas, and thus would require more than 200 miles of tunneling.

TABLE OF TUNNELS IN LOS ANGELES COUNTY

Compiled by Richard J. Proctor, 1973 and 1992

The following compilation lists more than 200 civil tunnels in Los Angeles County in the categories of transportation, and aqueducts and utilities. Seventy-five tunnels are within the city limits of Los Angeles. Existing tunnels are separated from a list of 12 Future Planned Tunnels. The oldest tunnels were built in 1876 and one used redwood staves for lining. Adits are not listed except for the cities of Pasadena and Sierra Madre where "water tunnels" are unusually long and of historic interest. Nineteen geologic formations have been excavated; the geology of each tunnel is keyed to the footnote on the last page. TBM means Tunnel Boring Machine excavation.

TRANSPORTATION TUNNELS

Tunnel Name	Length	Bore	Completed	City	Rock Type*
California Division of Highways					
Pasadena Fwy. (Figueroa St.)	1700' (4 tunnels)	56'	1931	L.A.	ss,sh (11)
Angeles Crest Hwy. (Kratka Ridge)	1100' (2 tunnels)	33'	1950	- -	gneiss
Los Angeles City Streets					
2nd Street	1502'	50'	1924	L.A.	sh (11)
3rd Street	1045'	35'	1901	L.A.	sh (11)
Griffith Park	300'	23'	1933	L.A.	gr, volcanics (18)
Mulholland OC (Sepulveda Blvd.) (built by Calif. Div. Hwys.)	655'	49'	1930	L.A.	ss, sh (6)
Grand Ave. (Music Center)	107'	9'x16'	1965	L.A.	sh (11)
Pershing Square Pedestrian	140'	9'x15'	1968	L.A.	Qal
Los Angeles County Roads					
Angeles Forest Hwy.	686'	26'	1940	- -	gr(anorthosite)
Malibu Cyn. Road	300'	30'	1955	- -	ss (18)
Seminole Bypass, N. bound	1225' (2 tunnels)	30'	1968	- -	volcanics (18)

Tunnel Name	Length	Bore	Completed	City	Rock Type*
Seminole Bypass, S. bound	1250' (2 tunnels)	32'	1973	- -	volcanics (18)
Mt. Baldy Road	700' ±	26'	1954	- -	Qal, gneiss
Soledad Canyon (Lang)	550'	32'	1954 (enlarged)	- -	gr(anorthosite)
East Fork San Gabriel River Road	400'	26'	1961	- -	gneiss
East Fork San Gabriel River Road (aband.)	350'	26'±	1930s	- -	gneiss
Dume-Kanan (Newton Ridge)	970'	46'	1975	- -	volcanics (18)
Southern Pacific Railroad					
Soledad Cyn. (Lang) Nos. 18, 19	266' & 328' (2 tunnels)	17'x22'	1876	- -	gr(anorthosite)
Newhall, No. 25	1.3 mile	16'x21'	1876	- -	ss, sh (10, 19)
Hasson-Chatsworth, Nos. 26, 27, 28	1.7 mile (3 tunnels)	14'x18'	1904	- -	ss (2)
Unknown mining company					
Railroad tunnel 2 miles N. of Azusa for gold mine in East Fork San Gabriel Canyon	500'	18'±	1890s	- -	gr(diorite)
Pacific Electric Railway (defunct)					
Hollywood Line Subway (aband.)	4840'	18'x28'	1925	L.A.	sh (11, 12)
Hill Street No. 1 (aband.)	530'	18'x28'	1911	L.A.	sh (11)
Hill Street No. 2 (aband.)	930'	18'x28'	1911	L.A.	sh (11, 12)
Los Angeles County Transportation Commission (was SCRTD)					
Red Line subway (Union Station to Alavarado Street)	8.6 miles (2 tunnels)	21' (TBM)	1992	L.A.	Qal, sh (11,12)
AQUEDUCT & UTILITY TUNNELS					
Pacific Telephone Co.					
Olive Street	750'	7' (TBM)	1969	L.A.	sh (11)

Tunnel Name	Length	Bore	Completed	City	Rock Type*
Callequas Water District					
Santa Susana	1.5 mile	10'	1968	L.A.	sh (11)
California Department of Water Resources					
Angeles	7.1 mile	36'	1970	- -	sh, ss (1,8)
Castaic Diversion	0.6 mile	24' & 32'	1968	- -	sh (1)
Los Angeles City Dept. Water & Power					
Owens Valley Aqueduct	22.9 miles (50 tunnels in L.A. County)	12'	1913	1 mile in L.A. City	gr, gn, schist, ss, sh, (1, 4, 5, 9, 10, 15, 16, 19)
2nd Barrel Owens Valley Aqueduct	1.3 mile (4 tunnels)	9'	1968	- -	schist, ss, (9, 10)
Ascot Res. Inlet	465'	5'	1926	L.A.	sh (11)
Stone Canyon Res. Bypass	3296' (4 tunnels)	7'	1952	L.A.	ss, sh, slate (2, 6, 14, 18)
Stone Canyon Res. Outlet	1000'	7'	1954	L.A.	slate (14)
Upper Franklin Can. Res. Inlet	3803'	7'	1914	L.A.	ss (2, 18)
Upper Franklin Can. Res. Outlet	175'	5'	1914	L.A.	gr, slate (14)
Lower Franklin Canyon Res. Outlet	175'	6'	1954	L.A.	slate (14)
Bouquet Res. Inlet-Outlet	1027'	11'	1933	- -	ss, sh (4)
Bouquet Res. Spillway	1350'	9'	1932	- -	schist (9)
Encino Res. Inlet-Outlet	1117'	7'	1963	L.A.	ss (18)
Drinkwater Res. Inlet	1100'	11'	1923	- -	schist (9)
Chatsworth Res. Outlet	945'	6'	1931	L.A.	ss, sh (2, 6)
Chatsworth High Line	4432' (4 tunnels)	9'	1931	L.A.	ss, sh (17)
Santa Inez Outlet	572'	12'	1966	- -	ss (16)
Upper Hollywood Res. Spillway	291'	5'	1933	L.A.	ss (18)
Upper Hollywood Res. Inlet	2500'	6'	1923	L.A.	ss (18)
Lower Hollywood Res. Old Spillway	125'	6'	1923	L.A.	ss (18)

Table 1 (cont.)

Tunnel Name	Length	Bore	Completed	City	Rock Type*
Lower Hollywood Res. New Spillway	100′	6′	1933	L.A.	ss (18)
Silverlake Res. Outlet	276′	6′	1933 & 1952	L.A.	ss (11)
Silverlake Res. Spillway	2500′	6′	1908	L.A.	ss (11)
Rowena	1002′	6′	1942	L.A.	ss (11)
Eagle Rock Res. Inlet	1036′	10′	1953	L.A.	gr, gneiss
Eagle Rock Res. Outlet	564′	6′	1953	L.A.	gr, gneiss
Elysian	3875′	7′	1943	L.A.	ss, sh (11)
Elysian Res. Inlet	680′	4′	1937	L.A.	ss, sh (11)
Elysian Res. Outlet	500′	4′	1937	L.A.	ss, sh (11)
Buena Vista	1300′	7′	1943	L.A.	ss, sh (11)
Dry Canyon Res. Bypass	3410′	10′	1957	- -	sh (1)
Chatsworth Storm Drain Bypass	1307′	14′	1971	L.A.	sh (2)
Maclay Highline	7000′	7′	1918	L.A.	ss (10, 15)
Metropolitan Water District of Southern California Upper Feeder (Colorado River Aqueduct water)					
Monrovia Tunnel No. 1	1.5 mile	12′	1937	- -	gr
Monrovia Tunnel No. 2	0.2 mile	12′	1937	- -	gr
Monrovia Tunnel No. 3	6.1 miles	12′	1938	Duarte, Monrovia	gr, gneiss, marble
Monrovia Tunnel No. 4	1.5 mile	12′	1938	Monrovia	gr, gneiss
Sierra Madre	1.3 mile	12′	1936	Sierra Madre	Qal
Pasadena	3.6 miles	12′	1937	Pasadena	Qal
San Rafael No. 1	0.8 mile	12′	1937	Pasadena	gr, gneiss, quartzite
San Rafael No. 2	1.1 mile	12′	1937	L.A., Glendale	gr(diorite)
Hollywood	3730′	8′	1940	L.A.	ss, sh, volcanics (18)

Tunnel Name	Length	Bore	Completed	City	Rock Type*
M.W.D. Palos Verdes Feeder (Colorado River Aqueduct water					
Oak Hill	600′	8′	1939	So. Pasadena	ss (18)
Ascot	1600′	8′	1936	L.A.	sh (11)
Palos Verdes Res. Outlet	730′	8′	1939	Rolling Hills Estates	diatomite (7)
M.W.D. Second Lower Feeder (Colorado River Aqueduct water)					
Coyote Creek Crossing	506′	8′ (TBM)	1967	Long Beach	Qal
605 Freeway Crossing	1277′	8′ (TBM)	1967	Long Beach	Qal
M.W.D. Foothill Feeder (Feather River Project water)					
Castaic No. 1	1.0 mile	26′ (TBM)	1970	- -	sh, ss (1, 15)
Castaic No. 2	2.5 miles	26′ (TBM)	1969	- -	ss (15)
Saugus	1.2 mile	26′ (TBM)	1969	Saugus	Qal, ss (15)
Placerita	0.2 mile	26′ (TBM)	1969	Newhall	ss (15)
Newhall	3.3 miles	26′ (TBM)	1970	Newhall, L.A.	ss, sh (10, 15, 19)
San Fernando	5.5 miles	22′ (TBM)	1973	L.A.	Qal, ss, sh, (10, 15, 17)
Glendora	6.2 miles	19′	1968	Glendora, San Dimas	Qal, sh, gr, gn, volcanics (11, 18)
M.W.D. Sepulveda Feeder (Feather River Project water)					
Balboa Inlet	0.9 mile	14′	1968	L.A. (Sylmar)	ss, sh (10, 15)
Balboa Outlet	0.7 mile	16′ (TBM)	1969	L.A. (Sylmar)	ss, sh (15, 17)
M.W.D. Sepulveda Feeder (Feather River Project water)					
Sepulveda	1.4 mile	12′	1970	L.A. (Encino)	slate, ss, sh (6, 14)
M.W.D. Yorba Linda Feeder (Feather River Project water)					

Table 1 (cont.)

Tunnel Name	Length	Bore	Completed	City	Rock Type*
Tonner Tunnel No. 1	0.9 mile	11' (TBM)	1975	- -	ss, sh (11)
Tonner Tunnel No. 2 (1.2 mile in Orange County)	2.5 miles	11' (TBM)	1975	- -	ss, sh (11)
Rialto Pipeline-Live Oak (5 tunnels)	0.3 mile	12'	1974	LaVerne	Qal, gr
Los Angeles City Sewers					
Central Outfall	2.1 miles	9'	1903	L.A., Inglewood	Qal, ss (13)
La Cienega-San Fernando Relief	2.9 miles	9'	1956	L.A. (Hollywood)	Qal, volcanics gr, ss, sh (2, 4, 18)
North Central Outfall	5.3 miles (3 tunnels)	12'	1959	L.A., Culver City	Qal, ss (13)
Westwood Relief No. 1	1700'	8'	1961	L.A.	Qal
NORS	8 miles	12' (TBM)	1992	L.A. (under LAX)	Qal, ss (13)
Los Angeles County Sanitation District					
White's Point (Palos Verdes Hills)	5.0 miles (2 tunnels)	8' & 12'	1936 & 1954	L.A. (San Pedro)	ss, sh, schist (3, 7,13)
Long Beach Interceptor	1.4 miles	7'	1956	Long Beach	Qal
Los Angeles County Flood Control District					
Storm Drain #1102, No. 3 & 4	1728'	16'	1975	L.A.	Qal
Storm Drain #1109, No. 2	2254'	17'	1974	Montebello	ss, sh (11)
Storm drain #1105 (Redondo Beach)	64,183'	13' & 7'	1972	Redondo Beach, Torrance	dune sand
Storm drain #5241 (La Tijera)	1529'	13'	1971	L.A.	Qal
Storm drain #59 (Central)	3586'	9'	1961	L.A.	Qal, ss, sh (11)

Tunnel Name	Length	Bore	Completed	City	Rock Type*
Storm drain #146 (Glendale Ave.)	264'	12'	1954	Glendale	Qal
Storm drain #60 (Tropico)	2804'	14'	1954	L.A.	Qal
Storm drain #67 (Hollenbeck)	2257'	16'	1961	L.A.	Qal
Storm drain #286 (Manhattan Beach)	5516'	5'	1955	Manhattan Beach	dune sand
Storm drain #40 (Sycamore-Scholl)	2029'	16'	1959	Glendale	gr(diorite)
Storm drain #57 (Sacatela)	475'	6'	1963	L.A.	Qal
Storm drain #84 (Slauson Ave.)	114'	17'	1959	L.A.	Qal, ss (13)
Storm drain #152 (Gardena Ave.)	1047'	7'	1955	L.A., Gardena	Qal
Storm drain #482 (8th Street)	2665'	7'	1962	L.A.	Qal
Santa Anita Dam Debris Sluiceway	1513'	11'	1969	Monrovia	gr(diorite)
Devil's Gate Dam Outlet	495'	13' x 6'	1920	Pasadena	gr(diorite)
Thompson Creek Dam Outlet	340'	6'	1927	- -	gr(diorite)
San Antonio Wash-Chino Creek	577'	18'	1973	Pomona	Qal
San Gabriel Dam Outlet	1930'	30'	1932	- -	gr(diorite
San Gabriel Dam Sluiceway	1225'	7'	1932	- -	gr(diorite)
Brown's Gulch Drainage	580'	9'	1935	- -	gr
Cogswell Dam Outlet	1220'	14'	1932	- -	gr
Puddingstone Dam Outlet	725'	7'	1927	- -	ss, volcanics (18)
Big Dalton Dam Outlet	500'	6' x 3'	1930	- -	gr(diorite)
Pacoima Dam Spillway	70'	8'	1929	- -	gneiss
Ybarra Cyn. Res. Inlet-Outlet	1800'	12'	1977	- -	ss (15)
Ebey Cyn. Res. Inlet-Outlet	1400'	12'	1977	- -	ss (15)
Pasadena Water Dept.					
Eaton Canyon No. 1	890'	5' x 3'	1895	Pasadena	gr(gabbro)
Eaton Canyon No. 2	440'	5' x 3'	1887	Pasadena	gr
Eaton Canyon No. 3	150'	5' x 3'	1888	Pasadena	gneiss

Table 1 (cont.)

Tunnel Name	Length	Bore	Completed	City	Rock Type*
Arroyo Intake	180'	5' x 3'	1890s	Pasadena	gr
Devil's Gate Dam area	4200'± (includes adits)	5' x 3'	1890s	Pasadena	gr, Qal
Azusa Power Conduit (built by Pacific Gas and Electric Co.)	4.8 miles	7' x 3'	1876	- -	gr, gneiss, marble
Morris Dam-Pasadena Pipeline	4800'	7' x 3'	1933	- -	gr, gneiss
H. Huntington Estate					
Utility Tunnel	200'±	7' x 5'	1911	San Marino	Qal
City of Sierra Madre					
Sierra Madre (water adits built by "Lucky" Baldwin)	1668' (3 "tunnels")	5½' x 3'	1880s	Sierra Madre	gr(diorite)
U.S. Army Corps of Engineers					
Ft. MacArthur Utility	?	?	1914±	L.A. (San Pedro)	ss, sh (7)
Unknown water companies					
Aband. water tunnels located 2 miles N. of Azusa	600'± (2 tunnels)	5'±	1880s	- -	gr(diorite)
Aband. water tunnels, Elysian Park-downtown L.A.	1000'±	5'±	1880s	L.A.	ss, sh (11)

FUTURE PLANNED TUNNELS

Tunnel Name	Length	Bore	Estimated Completion	City	Rock Type*
Metropolitan Water District of Southern California					
Foothill Feeder (Feather River Project water)					
Sunland No. 1	2.5 miles	18'	2005	L.A.	ss, sh (6, 10, 19)
Sunland No. 2	1.9 miles	18'	2005	L.A.	sh (6)
Verdugo	7.0 miles	18'	2005	L.A., Glendale	gr, gneiss, quartzite, marble, ss, sh (6,10,18)
Flintridge	3.5 miles	18'	2005	Glendale, Pasadena	Qal, gr, gneiss
Arroyo Seco Cross-Feeder	0.7 mile	8'	2005	Pasadena	Qal, gr
Altadena	4.0 miles	16'	2005	Pasadena, Altadena	Qal
Santa Anita No. 1	4.5 miles	16'	2005	Pasadena, Arcadia	gr, gneiss, gabbro
Santa Anita No. 2	7.4 miles	16'	2005	Arcadia, Monrovia, Duarte	gr, gneiss, marble
Los Angeles County Roads					
East Fork San Gabriel River Road	2300' (5 tunnels)	26'	1990s	- -	gr, gneiss
California Division of Highways					
Crystal Lake (State 39)	3± miles	?	1990s	- -	gr, gneiss
California-Nevada High-Speed Train					
Rapid Transit	15 miles in L.A. Co.	?	?	Sylmar to Acton	gr, gneiss, anorthosite
Los Angeles County Transportation Commission					
Rapid Transit (Alvarado St. to North Hollywood)	28 miles (2 tunnels)	21'	2004	L.A.	Qal, sh, ss, gr (6,11,12,18)

*Rock type abbreviations:
ss = standstone, sh = shale or siltstone, Qal = alluvium or terrace deposits, gr = granitic rocks
Formation names: (1) Castaic, (2) "Chico", (3) Franciscan (Catalina) Schist, (4) Martinez, (5) Mint Canyon, (6) Modelo, (7) Monterey, (8) Peace Valley, (9) Pelona Schist, (10) Pico, (11) Puente, (12) Repetto and Fernando, (13) San Pedro, (14) Santa Monica Slate, (15) Saugus, (16) Sespe, (17) Sunshine Ranch, (18) Topanga, (19) Towsley.

Table 1 (cont.)

ROCK MECHANICS FOR DESIGN OF TUNNELS

Khosrow Bakhtar
ARSM, Ph.D., President
Bakhtar Associates
2429 West Coast Highway,
Suite 201, Newport Beach, CA 92663

CONTENTS

INTRODUCTION

There is a growing demand for tunneling at shallow depth in urban areas for transportation, underground storage, and utilities. This demand grows with the population growth in metropolitan areas. The deciding factors for construction of tunnels in modern cities is because tunnels provides a less disruptive alternative to construction on the surface, and they can be very cost-effective. Excavations are made under buildings, streets and utility lines without requiring their relocation.

Depending on the local geological conditions, groundwater, physical and mechanical properties of soil or rock, depth of cover, and dimensions of the tunnel, the best excavation method and support system can be recommended. Ground stabilization techniques may be required prior to excavation of a tunnel to minimize the risk of excessive surface settlement which may result in structural damage and expensive lawsuits.

Following any excavation activity in rock, the question of stability and support requirements are of paramount importance. Most of the methods used for improving properties of soil are applicable to weathered rock mass. Additionally, rock reinforcement schemes such as steel or wooden props, brick or concrete arches, concrete or steel linings, are used as support elements. These schemes are designed to provide additional strength to the rock mass in order to prevent failure. The most economical and useful forms of reinforcement are rock bolts and stressed cables. These are used to load the rocks in such as way that resistance to shear along planes of weakness is improved, and additional tensile strength is given to the rock mass. Mechanically anchored, friction anchored, and resin grouted rock bolts are considered support elements which fall under rock reinforcement categories. Applications of gunite or exotic shotcrete mixtures to the excavated surfaces before any significant deformation of the rock can occur may also be considered as reinforcement.

By and large, the support requirements are dictated by the engineering characteristics of the geologic formations through which the tunnel excavation is made. For hard ground tunneling, applications of rock mechanics principals to the rock mass can aid in the safe and economic design and construction of the structures. In southern California, using various kinds of shield techniques, soft ground tunneling has been practiced for many decades. However, hard ground tunneling within city limits has not been practiced as much. The recent planned extension of the Metro Rail through the Santa Monica Mountains in Los Angeles, and extensive plans for water tunnels in San Diego has directed attention to the practice of rock mechanics principals in hard ground tunneling.

This paper describes the state-of-the-art approaches for rock mass classification, joint characterization and modeling for design and construction of underground facilities. The methodology outlined and described in a case history can be used for site characterization and economic design of support systems for tunnels and large underground openings. The field index tests can be performed by engineers or geologists on recovered cores or in situ if access to the excavation is available.

ROCK MECHANICS

Excavation in a rock mass causes redistribution of the stresses around the opening and subsequent disturbance in the local equilibrium conditions. The successful design of a structure is contingent on the ability to predict the load-deformation characteristics of a rock mass for a given structural configuration, resulting from these excavation induced disturbances. Rock mechanics principals are used as empirical tools for assessment of the rock mass behavior, determination of the appropriate safety factor, and design of a support system. The material characteristics of the rock mass and associated discontinuities (joints) provide the pertinent design parameters needed for estimation of the required factor of safety.

Generally speaking, the strength, deformation, and conductivity coupling of discontinuities effect the behavior of rock mass hosting an underground structure. Because a rock mass is anisotropic and heterogeneous, the properties vary as a function of location. Ideally, the local stability of the rock mass, during and post-excavation, is restored by means of applying a suitable support system. The nature and extent of the support requirements depend on the local geology, rock mass conditions and characteristics, and the purpose for which the excavations are made. In mining, the long-term stability of the openings usually are not of prime concern. Following the depletion of the ore, the mined areas are closed. In civil projects, such as transportation tunnels and underground facilities, the long-term stability of the structures are of prime interest.

Rock mass classification and field index testing can provide the necessary tools for identification of the level of support requirements and assist in economic and safe design of the structural components. The following discussion summarizes the most widely used schemes for rock mass classification, provides a basis for comparison, and shows the practical applications of the most common scheme with the aid of an example.

MAJOR ROCK MASS CLASSIFICATIONS

Rock classifications are typically based on the results of strength tests and the conditions of the core samples. Geologic features such as joints, faults and folds play important roles. Table 1 is an abbreviated listing of most of the rock mass classification systems currently in use internationally in the field of underground excavation and tunneling.

Table 1. Major rock mass classification systems

Name of Classification	Originator and Date	Country of Origin	Applications
Rock Load	Terzaghi (1946)	U.S.A.	Tunnels with Steel Support
Stand Up Time	Lauffer (1958)	Austria	Tunnelling
Rock Quality Designation (RQD)	Deere (1963) Deere et al., (1969)	U.S.A.	Core Logging, Tunneling
RSR Concept	Wickham et al., (1972)	U.S.A.	Tunnels with Steel Support
Geomechanics (RMR System)	Bieniawski (1973)	South Africa	Tunnels & Mines
Q-System	Barton et al., (1974)	Norway	Tunnels, Large Chambers

Terzaghi Rock Load Classification (1946)

The Terzaghi Rock Load Classification has been used extensively in the U.S. for the last 50 years. It is used primarily to select steel supports for rock tunnels. However, it is unsuitable for modern tunneling methods in which rock bolts and shotcrete are used.

The main features of Terzaghi's classification are shown in Figure 1. During the construction of a tunnel, some relaxation of the rock mass will occur above and on the sides of the tunnel. The loosened rock within the area abcd will tend to move in towards the tunnel (Fig. 1). This movement will be resisted by friction forces along the lateral boundaries ac and bd. The friction forces transfer the major portion of the overburden weight, W, onto the materials on either sides of the tunnel. The roof and sides of the tunnel are required only to support the balance which is equivalent to a height H_p. The width B_1, of the zone of rock in which movement occurs will depend upon the characteristics of the rock mass and the tunnel dimensions H_t and B.

Lauffer Stand-Up Time Classification (1958)

Lauffer introduced the concept of an unsupported span and its equivalent stand-up time, which was a function of rock mass quality. It appears to be excessively conservative when compared with present day tunnelling methods. Stand-up time is a period that a tunnel will stand unsupported after

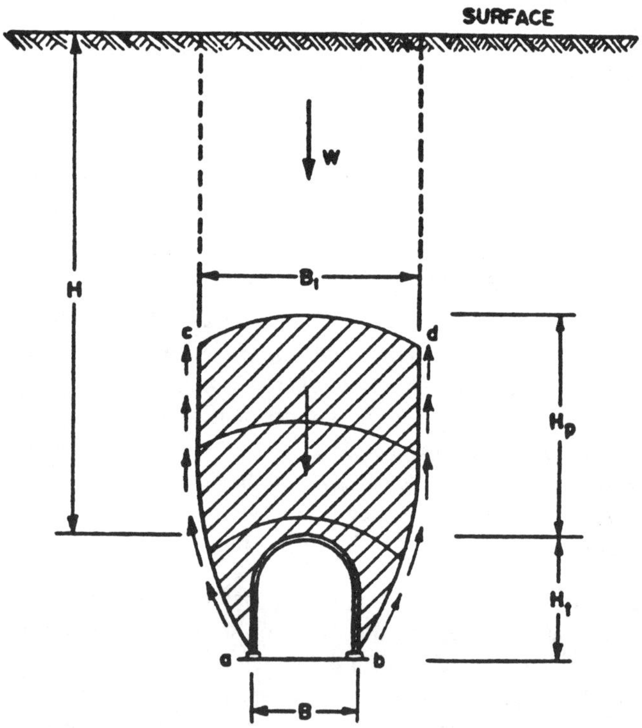

Figure 1. Simplified diagram of tunnel rock load (after Terzagi, 1946)

excavation. It should be noted that a number of factors may affect the stand-up time, such as orientation of tunnel axis, shape of cross-section, excavation, and support methods. Lauffer's classification is no longer used in its original form because it has been modified several times by other Austrian engineers.

Deere Rock Quality Designation - "RQD" (1963)

Deere's RQD is a simple description of conditions of the recovered drill cores. It has been successfully adopted as a part of subsequent classification systems. On its own, it does not account for the conditions of joint surfaces and filling materials, and may be overly sensitive to orientation effects. RQD is based on the modified core recovery procedure. This, in turn, is directly based on the number of fractures and amount of softening or alteration in the rock mass as observed in the rock cores from a drill-hole.

Instead of counting the fractures, an indirect measure is obtained by summing up the total lengths of the recovered core, but counting only those pieces of cores which are 10 cm (4 inches) in length or longer, and which are hard and sound. The following relationships between RQD index and the engineering properties of rock was proposed by Deere:

RQD (percent)	Rock Quality
Less than 25	Very Poor
25 - 50	Poor
50 - 75	Fair
75 - 90	Good
90 - 100	Excellent

If the cores are mechanicaluy broken by handling or by the drilling process mechanical breakage (i.e., the fracture surfaces are fresh irregular breaks, rather than natural joint surfaces), the fresh broken pieces are fitted together and counted as one piece, provided that they form the requisite length of 4 inches. Some judgement is necessary in the case of sedimentary rocks and foliated metamorphic rocks.

Wickham, Tiedemann and Skinner, RSR rock Structure Rating, (1972)

The RSR concept introduced numerical ratings and weighing to relate rock mass quality, excavation dimensions, and steel support requirements. The method was an immediate forerunner to the two methods now used most frequently on an international basis - the RMR and Q-System.

Bieniawski "RMR" Geomechanics Classification (1973)

The RMR evolved from several earlier system, and has undergone several changes (1974, 1975, 1976, and 1979) since its first introduction in 1973. The method was eventually based on 49 case records, though details of these cases with their relations to support recommendations have not been published. Recent applications of RMR system have been made in mining, which has extended the data base considerably.

The engineering classification of rock masses, based on RMR, utilizes the following six parameters which are measured in the field and can also be obtained from borehole data:

(a) Uniaxial compressive strength of intact rock material
(b) Rock Quality Designation (RQD)
(c) Spacing of Discontinuities
(d) Conditions of Discontinuities
(e) Groundwater Conditions
(f) Orientation of Discontinuities

To apply the Geomechanics Classification, the rock mass along the tunnel is divided into a number of structural regions, Bieniawski (1984), i.e., zones in which certain geologic features are more or less uniform within each region. The above six classification parameters are determined for each structural region from field measurements and entered into the standard input data sheets and, final assessment, engineering classification of the rock mass are made. An excellent and detailed description of this scheme is outlined in a recent book by Bieniawski (1984) which is a valuable reference for engineers and geologists involved in hard ground tunneling.

Barton, Lien, and Lunde - Q-System (1974)

The Q-System was developed independently of Wickham et al., (1972) and Bieniawski (1973) methods, but relied extensively on the RQD method of Deere (1963). In introduces five additional parameters to modify the RQD value to account for the number of joint sets, the joint roughness and alteration (filling), the amount of water, and the various adverse features associated with loosening, high stress, squeezing, and swelling. The classification method and associated support recommendations were based on an analysis of 212 case records.

In the method for estimating rock mass quality the size parameters chosen to describe the Q are combined in the following way:

$$Q = (RQD/J_n)\ (J_r/J_a)\ (J_w/SRF) \qquad \text{Eq. 1}$$

where: RQD = Rock Quality Designation
J_n = Joint Set Number
J_r = Joint Roughness Number (least favorable)

J_a = Joint Alteration Number

J_w = Joint Water Reduction Factor

SRF = Stress Reduction Factor

It is important to note that values of J_r and J_a relate to the joint set or discontinuity most likely to allow failure to initiate. The important influence of orientation relative to the tunnel axis is implicit. Detailed descriptions of the six parameters and their numerical ratings are given in Table 2. The range of possible Q-values (approximately 0.001 to 1000) encompass the whole spectrum of rock mass qualities from heavy squeezing ground to sound unjointed rock. The case records examined included 13 igneous rock types, 26 metamorphic rock types, and 11 sedimentary rock types. More than 80 involved clay mineral joint fillings of various kinds, including 12 swelling clay occurrences. However, most commonly the joints were unfilled and the joint walls were unaltered or only slightly altered.

The Q-System is more detailed than any of the other methods regarding the inter-related factors; roughness (or degree of planarity), alteration (filling), and relative orientation. The classification of "least favorable features" (for J_r and J_a) represent one of the strongest features of the method. It also seems to be a factor which is virtually ignored in the other classification schemes. For example, in Bieniawski's RMR method, although data for all joints and discontinuities are collected, only the average data are incorporated in the numerical ratings. Furthermore, in the RMR it is impossible to separately vary the degree of joint roughness and the degree of infilling, as obviously may occur in practice. Also, Bieniawski's ratings for Parameter #4, roughness and filling thickness, cannot be verified independently.

Figure 2 shows the relationship between Q-value, the tunnel span, and the type of excavation (ESR) for approximately 200 successfully supported excavations.

The Q-value can also be used to estimate the equivalent stiffness or the overall modulus of deformability of the rock mass. Knowledge of the overall modulus of deformability can assist in economic design of the concrete liner. This knowledge is particularly useful when designing support liner for pressure tunnels where internal loading conditions should also be accounted for.

In cases where the geology is non-uniform, the rock mass along the tunnel horizon is divided into a number of structural regions where the geological features are more or less uniform within each zone. The Q-value is then evaluated for each section based on the six parameters shown in Equation 1. The resulting Q-values are used to determine the support requirements and the modulus of deformability for each section. Estimates for ranges of deformation modulus are made based on the compilation of in-situ test data by Barton (1980) and Bieniawski (1974). Large-scale deformation modulus

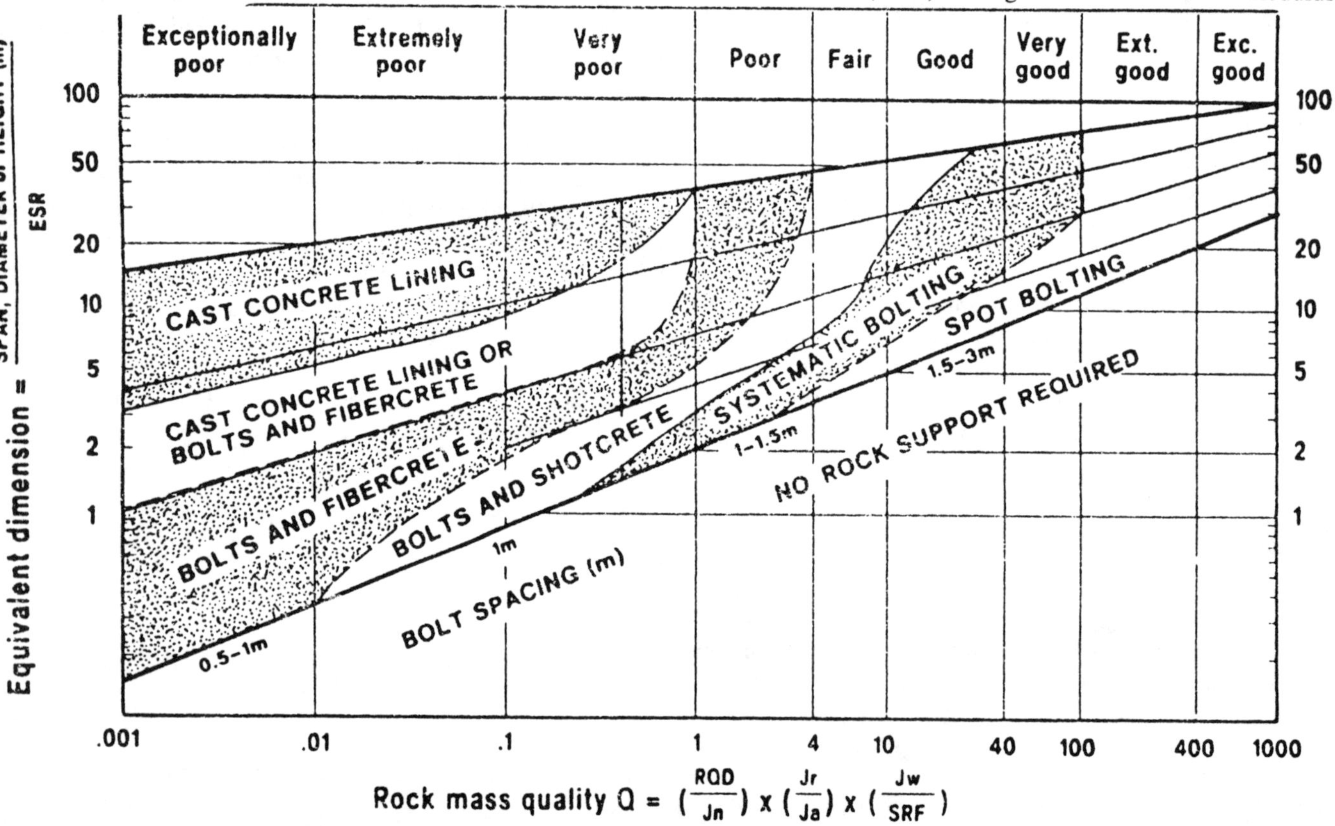

Figure 2. 200 case records plotted in terms of rock quality and capital excavation dimensions. These cases form the bases for support dellection. (Barton, et al., 1974)

measurements that have been correlated with Bieniawski's (1974) RMR rock mass quality ratings are shown by open squares in Figure 3. The filled circles are values that have been correlated with Q-values by Barton, et al. (1982). The relationship between RMR and Q was obtained by Bieniawski. As indicated in the figure, approximate upper and lower bound values of modulus are given by:

$E_{upper} = 40 \log Q \text{ (psi)}$ GPa Eq. 2

$E_{lower} = 10 \log Q \text{ (psi)}$ GPa Eq. 3

ROCK JOINT BEHAVIOR

In designing rock mass structures, one of the most important factors to consider is the shear strength of the joints or the rock mass discontinuities through which a failure is likely to occur. The behavior of joints has been of traditional interest to the geomechanicians for many years. Chief among the variables studied has been the basic influence of effective normal stress (or confining pressure) on the observed shear strength.

In rock mechanics, some pioneering work was performed by Patton (1966 a, b) who suggested the likelihood of scale effects on shear strength. This was followed by concrete experimental evidence of actual scale effects, reported by Pratt et al. (1972). The influence of scale effects was quantified by Barton and Choubey (1977) and conclusively confirmed in a major experimental test program conducted by Bandis (1980). Recently, the influence of shear rate on joint strength has been added to the variables affecting behavior, with important contributions by Crawford and Curran (1981) and Gillette et al. (1983). Studies reported by Bakhtar (1985) further confirms the influence of shear rate on frictional characteristics of joints.

Generally, the basic joint shear behavior is affected by the following factors:

- Normal stress
- Basic friction angle (Φ_b) and residual friction angle (Φ_r)
- Peak dilation angle ($d_{n}o$) - associated with joint roughness
- Size
- Temperature
- Chemical effects of water
- Gouge composition

Rock joint characterization, based on index testing, usually yields the important parameter required to assess the joint behavior. Methods of quantifying roughness and wall strength and utilizing them in shear strength relations were developed

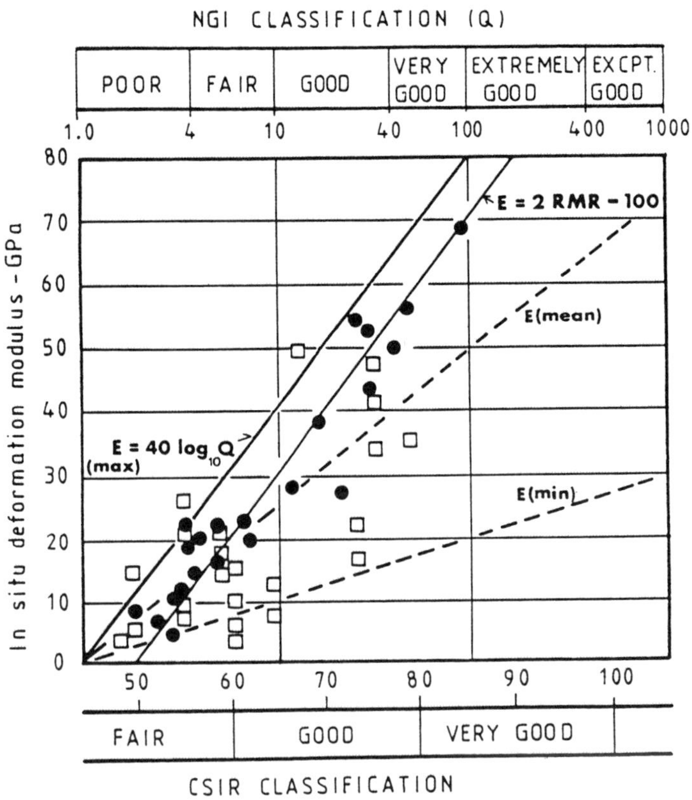

Figure 3. Estimation of in situ deformation modulus from two classification methods (Barton et al., 1980)

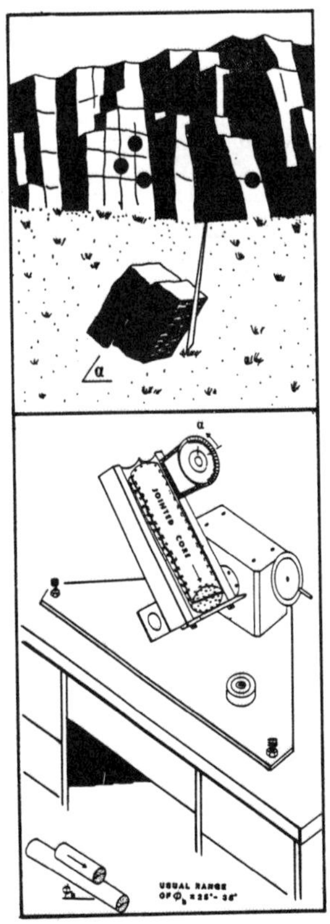

Figure 4. Characterization tests for quantifying joint rougness (JRC) and basic friction angle using axially jointed core.

by Barton and Choubey (1977). These methods were applied by Bandis (1980) and recently by Bakhtar (1985) in studies of joint deformation and strength. Barton and Bakhtar (1983) and Barton et al. (1985) showed how the strength, deformation, and conductivity coupling of rock joint for construction of dams, tunnels, and slopes in jointed water bearing rock can be determined based on the index test data.

Joint Characterization

The parameters required for complete joint characterization can be defined as follows:

JRC	Joint Roughness Coefficient
JCS	Joint Wall Compression Strength
σ_c	Unconfined Compression Strength (Rock Adjacent to Joint Wall)
Φ_r	Residual Friction Angle
e	Conducting Aperture
E	Mechanical Aperture

Two of the parameters listed above can be obtained indirectly from extremely simple tilt tests using pieces of intact and jointed core, as shown in Figure 4. The ideal sample would be jointed axially, but routine testing can. also be obliquely jointed samples, as typically recovered from a drilling program. The joint roughness coefficient (JRC) is obtained from such tests as follows:

$$JRC = (\alpha - \Phi_r) \;/\; \log(JCS / \sigma'_{no}) \qquad \text{Eq. 4}$$

where:

α = tilt angle when sliding occurs;

σ'_{no} = corresponding value of effective normal stress when sliding occurs (weigh upper sample, correct for cos , measure joint area).

The value of JRC typically varies from zero to about 15, the former value corresponding to residual nondilatant joint surfaces for which $\alpha = \Phi_r$.

The residual friction angle, Φ_r, may be lower than the basic friction angle, Φ_b, obtained from the tilt tests on core cylinders (lowest sketch Fig. 4) due to weathering or alteration effects. A simple empirical relation was developed by Barton and Choubey (1977) to estimate Φ_r from Φ_b, using Schmidt hammer. It is based on rebound tests on both the unweathered, dry rock (rebound R) and on the weathered and saturated joint wall (rebound r):

$$\Phi_r = (\Phi_b - 20) + 20r/R \qquad \text{Eq. 5}$$

The Schmidt hammer test is also used to estimate the joint wall compression strength (JCS) of the fresh or altered joints in the saturated state, if this is appropriate to in-situ conditions. The relationship derived by Miller is used to convert the rock density and the rebounds (r) and (R) to estimates of the compression strengths JCS and %c respectively. Full details of the above characterization tests were given by Barton and Choubey (1977).

The parameters JRC, JCS, and Φ_r are all that are needed to develop shear strength, displacement, dilation, and normal stress-closure curves for any given joint. However, coupling conductivity with these processes requires additional information concerning the initial joint aperture, since closure or dilation resulting from stress changes are superimposed on these initial apertures. The joint deformation and conductivity coupling studies reported by Barton et al. (1985) provides the detailed description and methodology for evaluation of the hydromechanical response of fractures.

Shear Strength Based on Rock Joint Characterization

Shallow excavations are most strongly influenced by the geologic discontinuities such as joint sets, faults, and the degree of weathering of the rock mass. The parameters discussed previously can be obtained from extremely simple tilt tests using pieces of intact or fractures cores. The fracture or joint roughness coefficient (JRC) is obtained from the tilt test (Fig. 4) using the relationship given by Equation 4.

In the absence of cores, the joint roughness coefficient is determined in the field using a profile gage. Field activities also include:

- Determination of joint dips and dip direction, supplemented with photographs.
- Spacing of joints for major joint sets.
- Indication of the typical continuity of major joint sets.
- Estimation of block sizes within each major joint set.

Joint Modeling

The parameters discussed above can be used to generate shear-strength-displacement and dilation-displacement plots for relevant block size, using the empirical joint modeling technique described by Bakhtar and Barton (1984). The results of in-situ rock mass characterization are used as the input, and the model predicts, for each through-going joint, how much shear mobilization can occur before peak shear strength of the discontinuity is reached.

The conceptual method which is used to investigate the shear strength-displacement and dilation-displacement along the joint surfaces, was the empirical model developed by Barton (1973). In this model the peak drained friction angle (Φ') of a continuous (cohesionless) joint or fracture is given by the following relation:

$$\Phi' = JRC \log(JCS/\sigma'_n) + \Phi_r \qquad \text{Eq. 6}$$

where

JRC = peak value of the joint roughness coefficient

JCS = joint wall compression strength

Φ_r = residual friction angle

σ'_n = effective normal stress

At shear displacements smaller than or larger than those required to mobilize peak strength, the JRC value is expressed as JRC (mobilized). Thus, at a given shear displacement, the strength is given by:

$$\Phi'_{mob.} = JRC_{mob.} \ \log (JCS/\sigma'_n) + \Phi_r \qquad \text{Eq. 7}$$

The mobilized value of the roughness coefficient, which ranges in value from 0 to 20 for smooth to rough joints, determines the angle of dilation, which results from overriding or partial shearing through the aspirities. This in turn depends on the ratio JCS/σ_n , with high values of the ratio causing strong dilation and limited fracture surface damage and low values causing the opposite effects. The approximate angle of dilation (d_n) at any given displacement is given by:

$$d_n \text{ (mob.)} = .5 \ [\ JRC_{mob.} \ \log (JCS/\sigma'_n)] \qquad \text{Eq. 8}$$

Estimates of Joints Peak and Residual Shear Strengths Based on Index Tests

1. Poisson's relationship, relating overburden pressure to horizontal stress (for isotropic horizontal stress field), can be used to determine estimates of the in-situ stress field, where the horizontal components of an isotropic stress field is estimated from

$$\sigma_h = (\nu / 1 - \nu) \ \sigma_v = \qquad \text{Eq. 9}$$

where:

σ_v = vertical stress, psi (MPa)

ν = Poisson's ratio (assumed value of 0.25)

σ_h = horizontal stress, psi (MPa)

Knowing the magnitude of horizontal and vertical stresses, the transformation equation (Equation 10) can be used to compute effective normal stress (σ_n) on joint plane.

$$\sigma_n = 1/2(\sigma_v + \sigma_h) \ - \ 1/2(\sigma_v - \sigma_h) \ \ \cos 2\beta \qquad \text{Eq. 10}$$

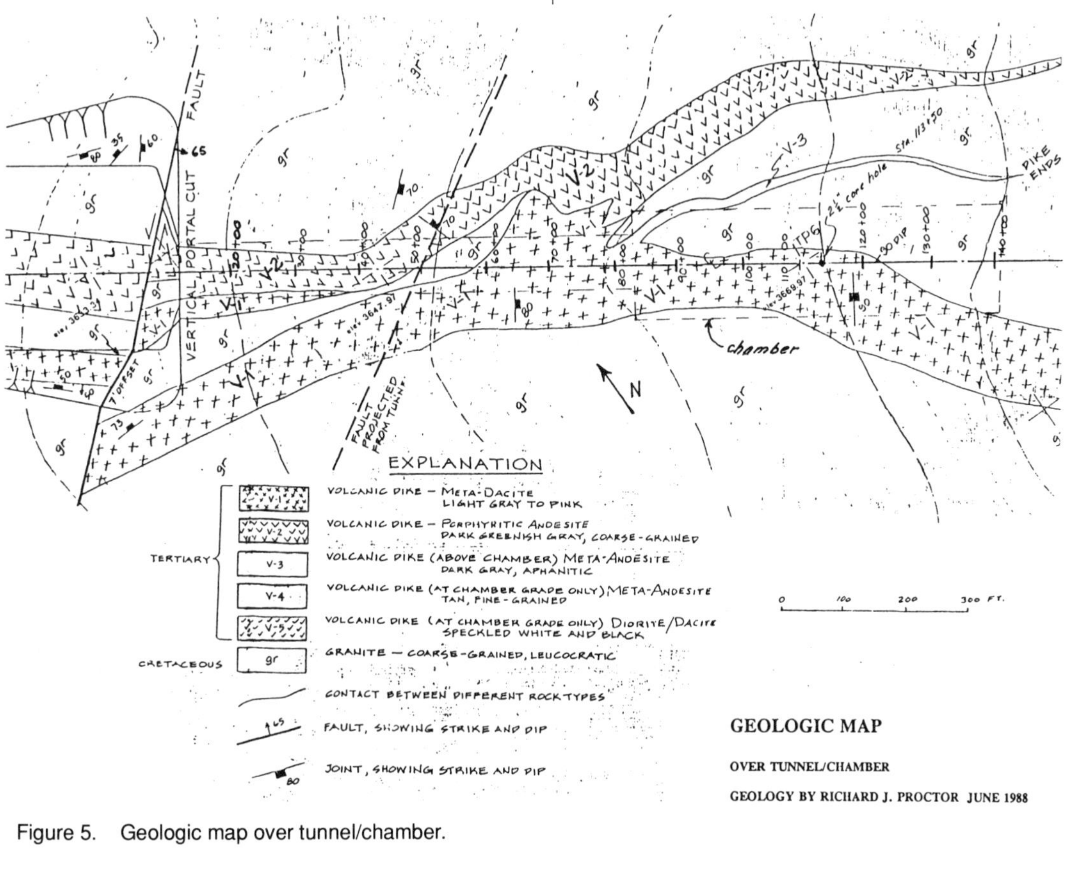

Figure 5. Geologic map over tunnel/chamber.

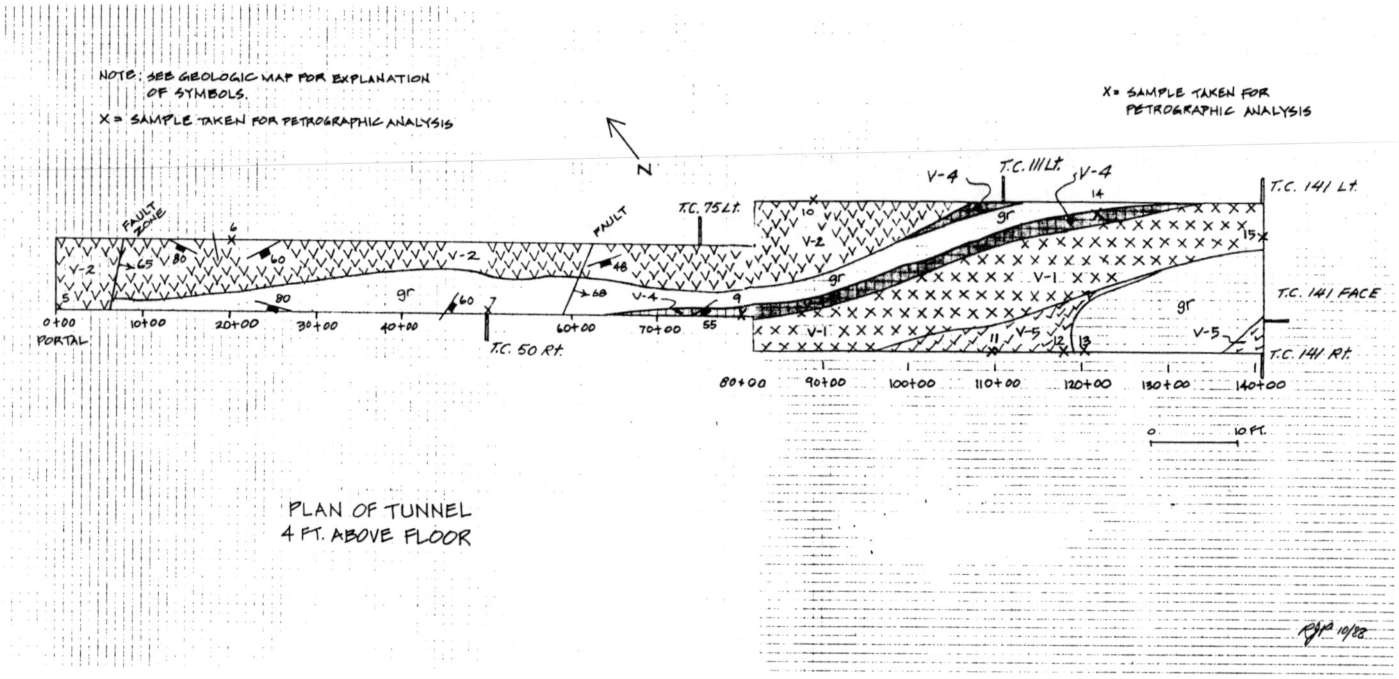

Figure 6. Geologic plan of tunnel/chamber 4 ft. above floor.

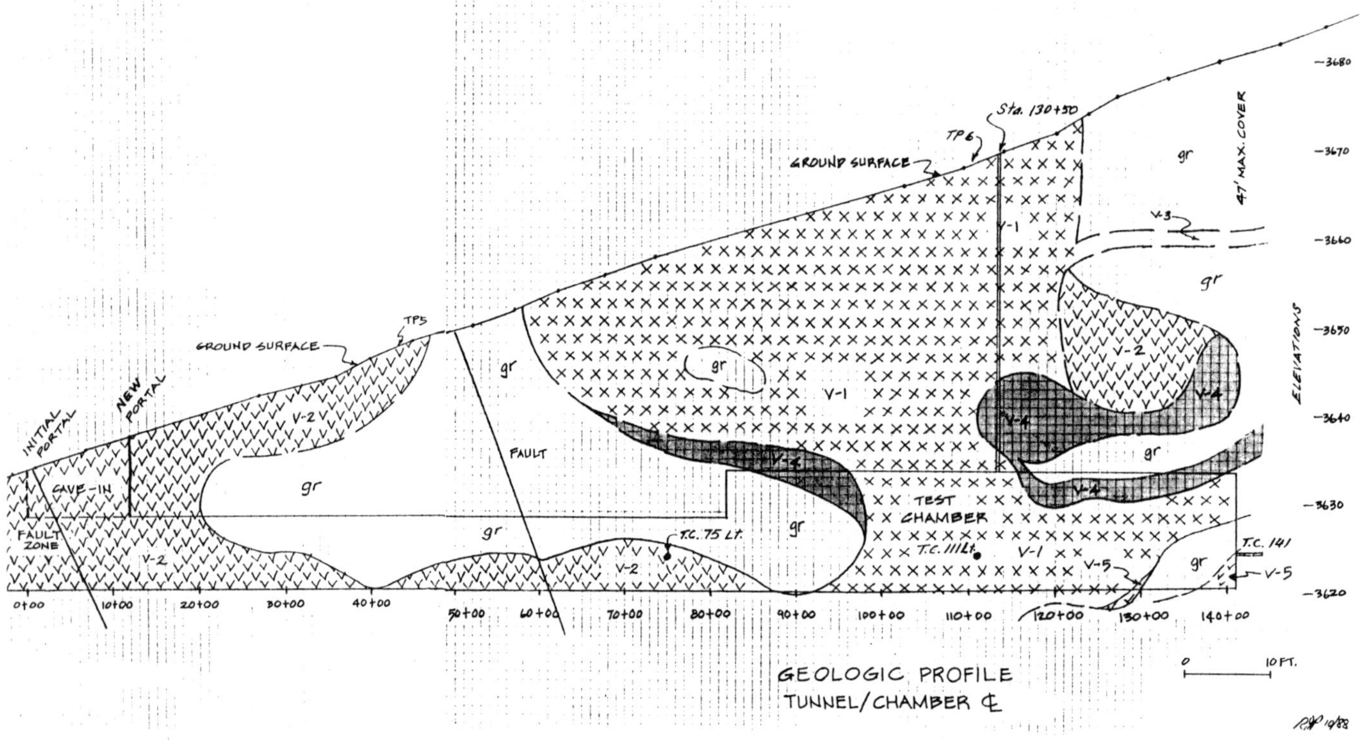

Figure 7. Geologic profile along tunnel/chamber centerline.

where β is the angle between vertical stress and fracture plane for each joint set and defined as

$$\beta = 90^{o} - \alpha \qquad (\alpha = \text{measured dip angle})$$

2) The "Joint Model", based on the Barton JRC-JCS concept (Barton and Bakhtar, 1984) is used to generate plots showing shear stresses (shear displacement and dilation), shear displacement curves for the major joint sets at initial stages of shear mobilization of blocks.

The simple procedure described in the preceding pages allows the important properties of major discontinuities such as peak and residual shear strength values to be determined based on major tests performed in the field.

CASE HISTORY APPLICATION OF Q-SYSTEM

Introduction

A short tunnel with an attached chamber was recently excavated in the southern California desert. It provided an opportunity to apply the Q-System of rock mass classification. The site geology and tunnel/chamber were mapped by Richard Proctor and the tunnel/chamber joint characteristics were logged by Khosrow Bakhtar, under separate contracts

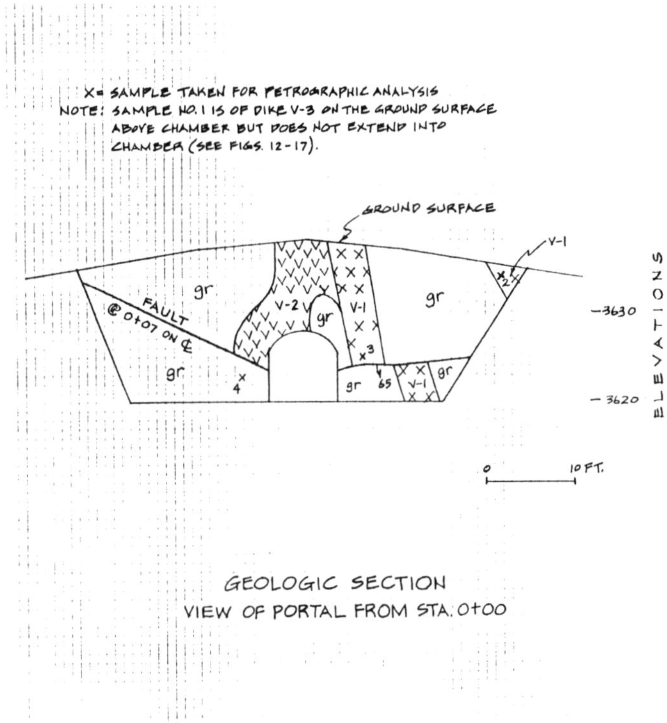

Figure 8. Geologic section, view of portal from Sta. 0+00.

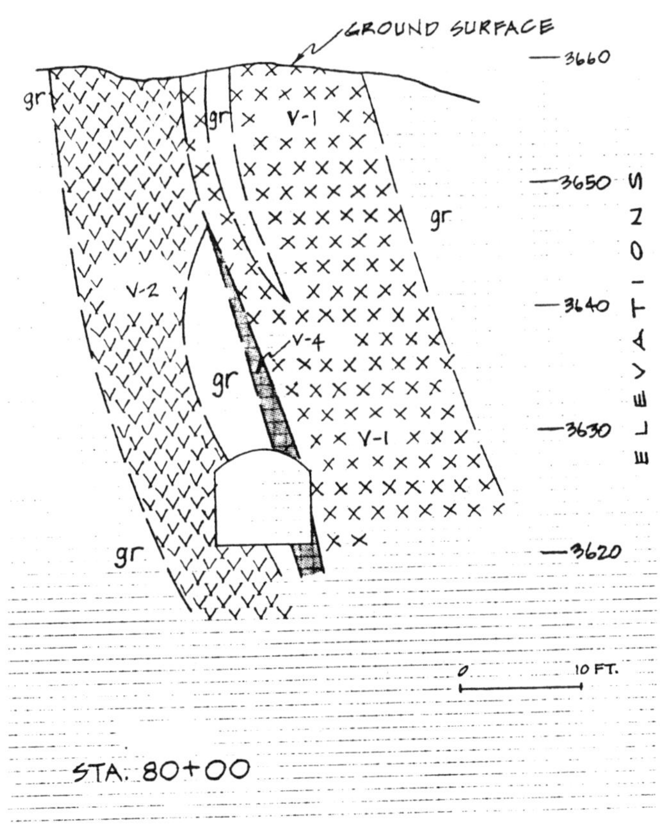

Figure 9. Geologic section of tunnel at Sta 80+00.

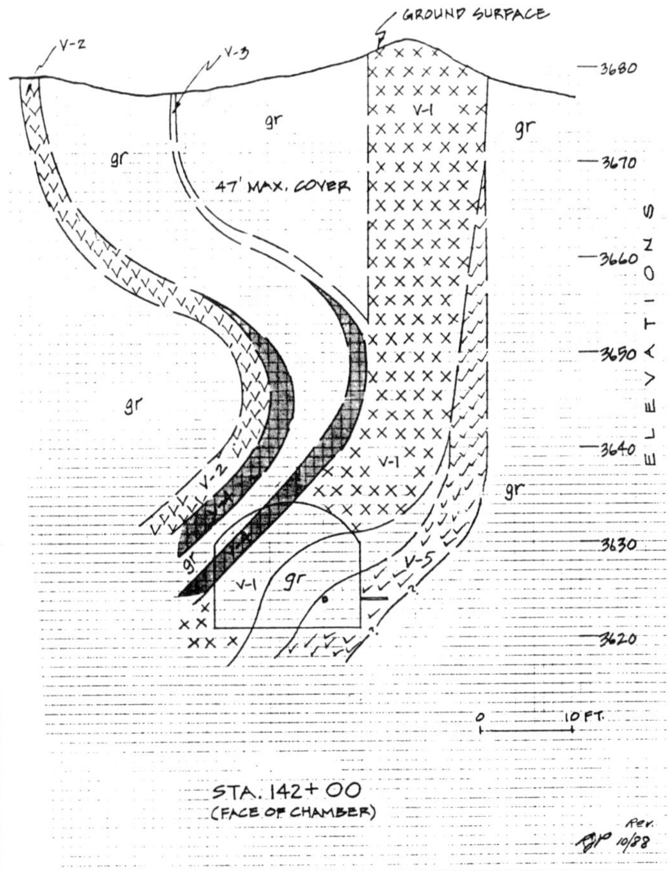

Figure 10. Geologic section at end of chamber, Sta. 142+00.

with the owner. Proctor's geology was not available to Bakhtar when he made the following evaluation. Thus, Figures 5 through 10 are by Proctor and the remaining figures and text are by Bakhtar.

The dimensions of the tunnel were 8 x 8 x 82 feet, and for the attached chamber were 13.2 feet high by 16.5 feet wide by 59 feet long. The entire tunnel/chamber was excavated by drill and blast. The maximum cover over the chamber was about 49 feet which is a shallow depth for rock characterization using the Q-System. Because of the unevenly weathered nature of the rock mass, the tunneling contractor was unable to excavate smooth uniform walls in the tunnel and chamber. Blasting operations resulted in mechanical breakage along the walls. In the absence of pre-construction drill cores, it was impossible to estimate the extent of weaker rock areas. The blocky nature of the rock caused wedge and block failures in the tunnel and chamber. Extensive rock bolting was required around the portal area using split bolts and plates. No specific bolting pattern was followed and the bolts were installed at the contractor's discretion.

Geology

The tunnel/chamber is in deeply weathered, coarse-grained granite that is cut by five volcanic dikes. Figure 5 shows the surface expression of these dikes and their composition. Figures 6 and 7 (geologic plan and profile) show that the dikes comprise the majority of the excavation. This is fortunate because the owner wanted a hard rock site and the dikes are harder than the surrounding decomposed granite. Had the tunnel centerline been located only 30 feet eastward, none of these dikes would have been encountered. (Figs. 8 - 10).

All rocks are strongly fractured with several prominent joint sets which enabled the following Q-System rock mass analysis to be performed.

A large fault became exposed at the portal which revealed 7 feet of horizontal offset of a volcanic dike. the portal area caved requiring a new portal to be constructed 12 feet southward (Fig. 5 - 8).

Rock Mass Analysis

Figure 11 shows the locations, dip and strike of three major joint sets with observed in the tunnel. In the chamber, two major joint sets and one minor shear were logged. Figure 12 shows the locations, dip and strike of these features. The thickness of the clay in the shear was 2 inches at the invert, but became very tight and disappeared at the crown of the chamber. All joint planes were smooth-planar with clay coatings.

The Q-System developed by Barton et al. (1974) was used for engineering characterization of rock masses in the tunnel and chamber. The system is essentially a weighing or ranking process in which the negative and positive aspects of the rock mass are assessed. The method is based on a numerical assessment of rock mass quality using the six different parameters given by Equation 1.

Geological engineering judgement and information in Table 2 were used to estimate the in-situ values for several parameters needed to determine the Q-values. Details of characterization and justification for assigning numerical values for each parameter in Equation 1 are as follows:

Rock Quality Designation (RQD) was estimated (in the absence of drill cores at the site) from the number of joints per unit volume. Equation 11, a simple relationship developed by Palmstrom (1975), was used to estimate the number of joints in a 1 cubic meter volume wedge observed in the tunnel-chamber wall. Therefore,

$$RQD = 115 - 3.3\ J_v \text{ (approximately)} \qquad \text{Eq. 11}$$

where

J_v = total number of joints per m^3

In Equation 11 J_v is about 17, which gives an estimated RQD of about 59, representing fair quality rock mass. Figure 13 shows the location of the joint within the volume of the wedge. Attempts were made to differentiate between the natural joints and blast induced fractures.

Number of Joint Sets (J_n) for the tunnel three sets of joints (plus several random) were noted. The locations, dips, and strikes of these joints are shown in Figure 11. A joint set number (J_n) of 12 based on the Q-System was assigned to the tunnel.

For the chamber, two joint sets and one minor shear were observed. The shear intersected the axis of the chamber almost perpendicularly. Because the joint sets represented the most unfavorable conditions, they were accounted for in the analysis and the shear was ignored. However, if the shear contained expansive clay with shrink/swell potential and a uniform thickness, it would have been accounted for in the analysis based on the Q-System. A value of 6 was assigned for the joint set number in the chamber. Therefore,

Tunnel: $J_n = 12$

Chamber: $J_n = 6$

Joint Alteration Number (J_a) applies to a joint set or a single discontinuity most likely to initiate failure. Silty or sandy clay coatings with zones of altered clay, continuous in appearance but less than 0.2 inch (5 mm) thick, were noted in the tunnel and chamber.

The joints were not healed and contained altered clay infillings which are obviously unfavorable because of their in-

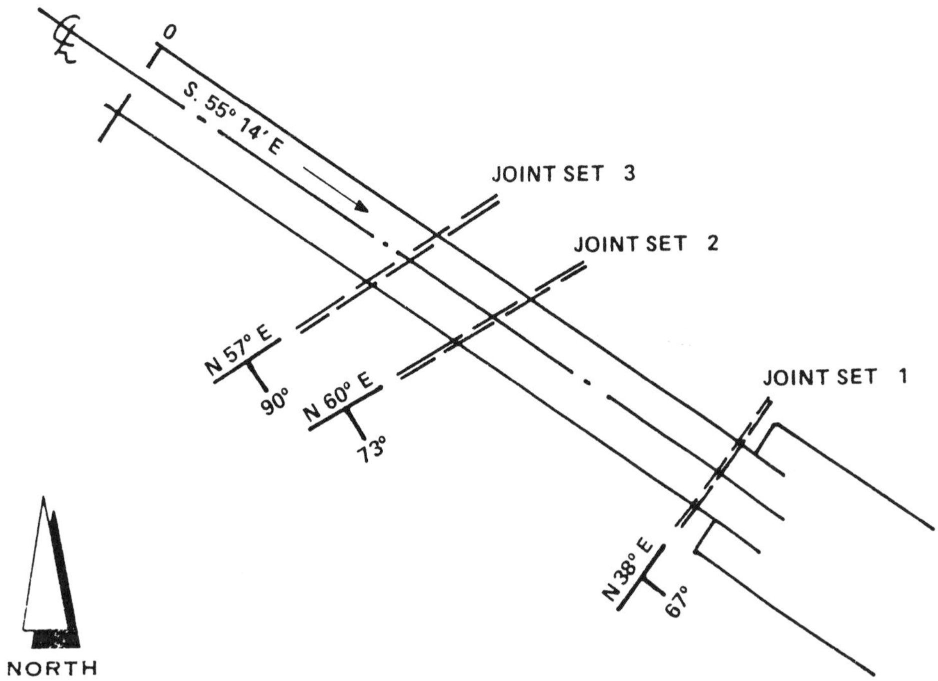

Figure 11. Plan of tunnel showing prominent joint sets used for Q-System analysis.

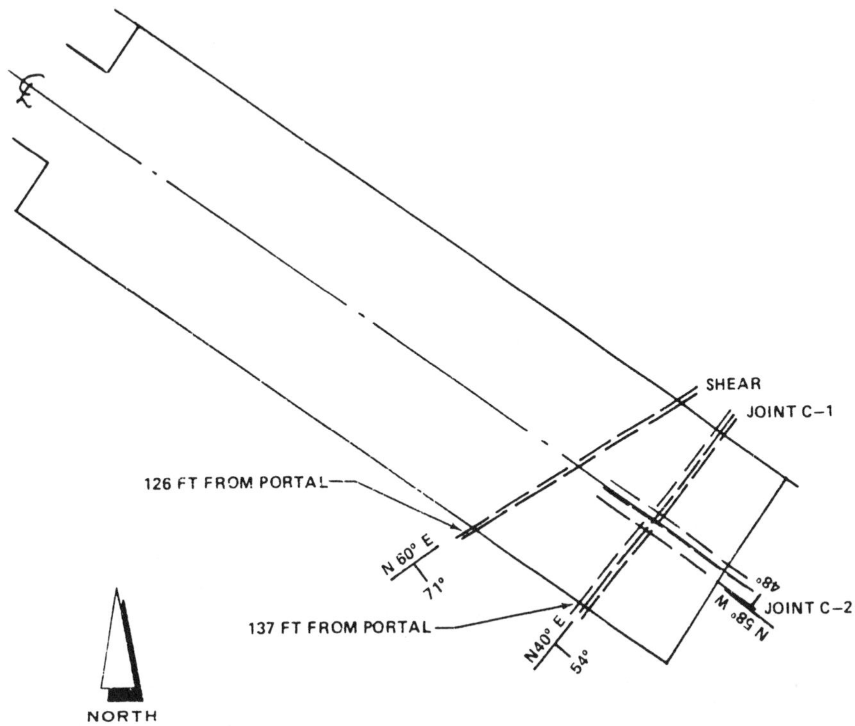

Figure 12. Plan of chamber showing prominent joint sets used for Q-System analysis.

stability. A value of 3 was assigned to J_a for joints in tunnel and chamber. Therefore,

$J_a = 3$ for tunnel and chamber (see Table 2)

Joint Water Reduction Factor (J_w) The excavation of tunnel and chamber was conducted under completely dry conditions. Therefore,

$J_w = 1$ for tunnel and chamber

Stress Reduction Factor (SRF) has 16 classes which are divided into four broad groups (Barton et al., 1974)

a) Weakness zones causing loosening or fall-out
b) Rock stress problem in competent rock
c) Squeezing; flow of incompetent rock
d) Swelling; chemical effects due to water uptake

A value of 2.5 was assigned to SRF for both the tunnel and the chamber. This value refers to the case where the structure is close to the surface ($\sigma_1/\sigma_2 > 200$). Therefore,

SRF = 2.5 for tunnel and chamber (see Table 2)

Results of Q-Analysis

Values assigned to parameters in Equation 1 are used to estimate Q for the tunnel and chamber. The results are summarized in Table 3.

Table 3. Summary of results based on Q-System

	Parameters					
Location	RQD	J_n	J_r	J_a	J_w	SRF
Tunnel	59	12	1	3	1	2.5
Chamber	59	6	1	3	1	2.5

$Q = (RQD/J_n)\ (J_r/J_a)\ (J_w/SRF)$

Q (tunnel) = 0.65 Q (chamber) = 1.30

In general, the host rock through which the excavations were made consist of highly weathered granite with near-vertical dike intrusions. Three major sets of joints in the tunnel and two sets of joints in the chamber were observed. The major joint sets contained altered clay coatings. The calculated Q-values for rock at the tunnel and chamber were 0.65 and 1.30, respectively. These values represent very poor to poor rock mass quality. Therefore, the long-term stability of the tunnel and chamber require a permanent support system. The Q-values with appropriate ESR values were used to determine the best choice for support requirements and estimates for range of the modulus of deformability for tunnel and chamber based on Figures 2 and 3. These indicate that rock bolts and shotcrete were the proper choice for support for the tunnel and chamber.

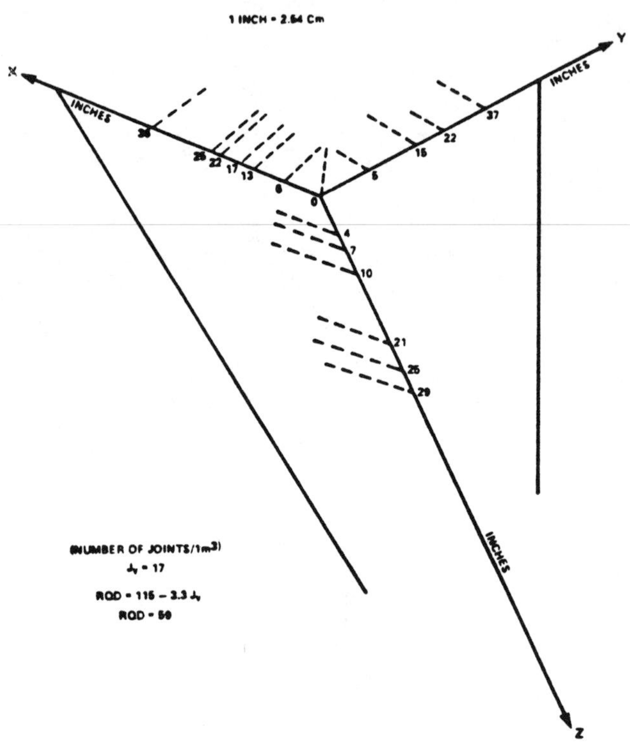

Figure 13. Sketch of the wedge used to estimate RQD

Estimates for the range of the modulus of deformability for the chamber based on the compilation of in-situ test data (Barton, 1980 and Bieniawski, 1974) presented in Figure 3 were found to be

Estimate of $E_{max} = 40 \log Q = 6.61 \times 10^5$ psi (4.65 GPa)

Estimate of $E_{min} = 10 \log Q = 1.65 \times 10^5$ psi (1.14 GPa)

Joint Characterization and Modeling

The results of in-situ rock mass characterization and index testing were used to characterize the joints and determine their initial behavior based on the Barton JRC-JCS Concept. The initial normal stress on the fracture plane for each joint set was computed using a transformation equation. Typical block sizes were 17 inches (0.43 m) to 22 inches (0.65 m) in length. Emphasis is placed on modeling mobilization of shear resistance for the blocks to reach the peak and residual stages. The basic input parameters to the Barton joint model were obtained from simple index tests performed on several

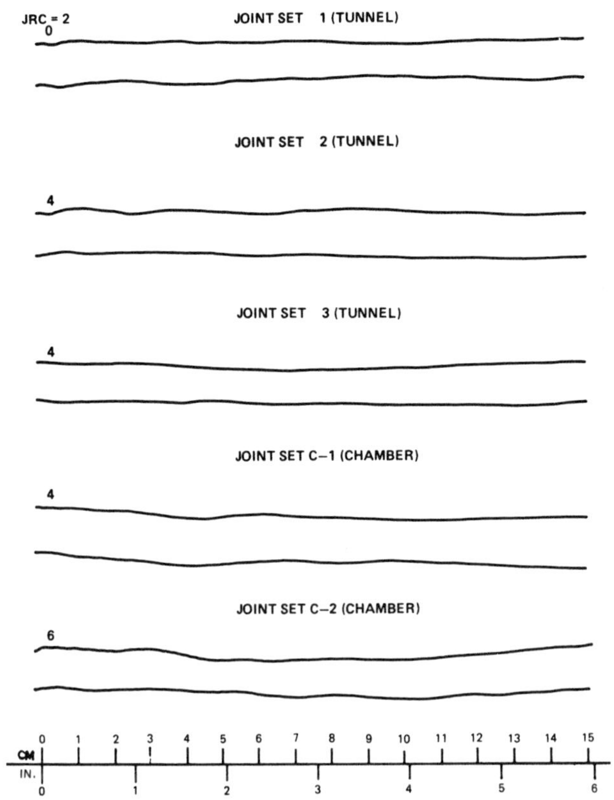

Fig ure 14. Small scale roughness profile measured on joints (profiles are measured in direction of dips)

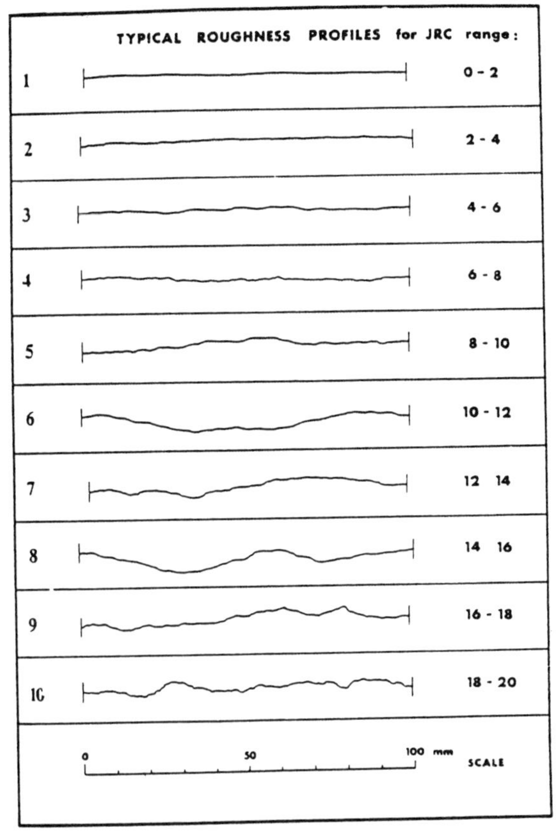

Figure 15. Typical JRC values for joint samples of different roughness (Barton and Choubey, 1977)

joint surfaces and intact rock in the tunnel and chamber. The information required for complete joint characterization and input to the Barton model consists of the following (Barton and Bakhtar, 1983):

- Joint Roughness Coefficient, JRC_o (laboratory scale)
- Block Length, L_n
- Laboratory Sample Size, L_o (10 cm)
- Joint Wall Compression Strength, JCS_o (laboratory scale)
- Effective Normal Stress on Joints, σ_n
- Smooth Hydraulic Aperture, e_o
- Residual Friction Angle, Φ_r
- Basic Friction Angle, Φ_b
- Unconfined Compressive Strength of Intent Rock, σ_c

The manner in which the above parameters were determined in-situ is described as follows:

Joint Roughness Coefficient (JRC_o)

The small scale profiles of the joint roughness taken from the major joint sets are shown in Figure 14. These profiles were taken using a 6.1 inch (15 cm) profile gage. The small scale roughness profiles obtained using the profile gage were then compared with a standard set of roughness profiles, Figure 15, published by Barton and Choubey (1977). For the joint sets under consideration, this comparison revealed values of 2, 4, 4, 4, and 6, representing an average value of 4.

Dimensions of the blocks encountered within the five joint sets were measured. Their overall measured dimensions are

Joint Set #1: 17 in (43 cm) x 16 in (41 cm) x 17 in (43 cm)

Joint Set #2: 22 in (56 cm) x 16 in (41 cm) x 10 in (25 cm)

Joint Set #3: 22 in (56 cm) x 15 in (38 cm) x 27 in (69 cm)

Joint Set C-1: 22 in (56 cm) x 12 in (30 cm) x 15 in (38 cm)

Joint Set C-2: 17 in (43 cm) x 10 in (25 cm) x 16 in (41 cm)

Figures 14 and 15, presented earlier, show the locations and strike/dip directions of the various joint sets. The block length used for modeling refer to the side parallel to the dip of the joint sets. The following block length from joint sets (parallel to dip) were considered:

Joint Set #1: L_n = 17 in (43 cm)

Joint Set #2: L_n = 22 in (56 cm)

Joint Set #3: L_n = 22 in (56 cm)

Joint Set C-1: L_n = 22 in (56 cm)

Joint Set C-2: L_n = 17 in (43 cm)

Average Block Length: 20 in (51 cm)

Laboratory Sample Size (L_o)
For modeling purposes, small scale sample size (L_o) is considered to be 4.1 in (10 cm), Barton and Bakhtar (1983).

Joint Wall Compression Strength (JCS_o) - Small Scale

The joint planes were weathered and their unconfined compressive strengths were different from those of intact rocks. A Schmidt hammer was used to estimate the unconfined compressive strength of the joint walls. For each joint set, the average for more than 15 rebounds of the hammer was taken, and after making the necessary correction for hammer orientation, the unconfined compressive strength values were computed using the relationship developed by Miller (1965):

$$\log 10\,(\sigma_c) = 0.00088\,\gamma R + 1.01 \qquad \text{Eq. 12}$$

where (σ_c) = unconfined compression strength of surface (MN/m_2)

γ = unit weight of dry rock

R = rebound number

The density measurements gave an average value of 166 lb/ft^3 (2.65 gm/cc) for the rock. The average heights of rebound for the Schmidt hammer test performed at the joint walls were 34.2, 17.9, 23.3, 15.1 and 8.4. Equation 3 was used to compute the corresponding joint compressive strength (JCS_o) as follows: 9,000 psi (62.0 MPa), 3,810 psi (26.3 MPa), 5,063 psi (35.0 MPa), 3,287 psi (22.7 MPa), 2,310 psi (15.9 MPa). The average JSC_o for joint walls was calculated to be 4,695 psi (32.38 MPa).

Effective Normal Stress on Joints (σ_n)
The average depth of about 30 feet (18.29 m) was used to estimate the stress fields around the joint planes. For an average density of 166 lb/ft^3 (2.65 gm/cc), the unit weight of the rocks becomes about 26 KN/m^3 which results in an average vertical stress of about 69 psi (0.476 MPa). Poisson's relationship, Equations 9 and 10, were used to estimate the isotropic horizontal stress fields and effective normal stress on joint plane.

Based on the measured dip angle during in-situ rock mass characterization, the following values were computed for the effective normal stress for the five joint sets: 0.29 psi (0.002 MPa), 8.12 psi (0.056 MPa), 14.28 psi (0.098 MPa), 10 psi (0.069 MPa) and 32 psi (0.220 MPa). The average effective normal stress was then calculated to be 12.9 psi (0.089 MPa).

Smooth Hydraulic Aperture (e_o)
The smooth hydraulic aperture is an important input parameter to the "Shear Displacement" joint model, (Barton and Bakhtar, 1983). The joint apertures on the walls of the tunnel and chamber were extremely disturbed as a result of blasting operations. The "initial" exposed apertures were measured at several locations along the tunnel and chamber walls using a feeler gage. Because of the shallow depth and extent of weathering, the values obtained for e_o using the feeler gage were used for modeling, and no attempts were made to determine the extent of damage by using the "Joint-Normal Closure" model, Barton and Bakhtar (1983). Average values of 0.2 in (0.4 mm) and 0.01 in (0.279 mm) were measured in-situ for the exposed apertures in the chamber and tunnel, respectively.

Residual Friction Angle (Φ_{ro})
A simple empirical relationship developed by Barton and Choubey (1977), Equation 5, was used to estimate (Φ_{ro}). The following values were obtained for the residual friction angle of the five joint sets: 25°, 19°, 22°, 21°, 21°.

Basic Friction Angle (Φ_{bo})
It should be noted that (Φ_{bo}) is usually determined from tilt tests on saw-cut rock surfaces or smooth cores. For the case under study, in the absence of cores, engineering judgement was used to estimate Φ_{bo} at 30° based on tilt tests.

Unconfined Compressive Strength of Intact Rocks (σ_c)
A Schmidt hammer was used to estimate the unconfined compressive strength of intact rocks at the location of major joint sets. The average rebounds were 45.2, 39.8, 38.4, 27.8, and 25.6. Using Equation 12 and an average value of 166 lb/ft^3 (2.65 gm/cc) for density, the estimates of unconfined compressive strength of rocks were obtained as: 16,132 psi (111.25 MPa), 12,074 psi (83.27 MPa), 11,216 psi (77.35 MPa), 6,417 psi (44.26 MPa), 5,715 psi (39.41 MPa).

Summary of Joint Characterization Results
Table 3 represents the summary of joint characterization results. In the absence of pre-construction drill cores and good quality rock sample, the values for Φ_b (basic friction angle) were estimated based on field observations, testing experience, and engineering judgement. Several pieces of intact rocks were collected during excavation and attempts were made to recover intact cylindrical samples for laboratory testing by the tunnel contractor. Difficulties were encountered during coring the blocks and eventually 1 inch diameter (2.54 cm) plugs were used for tests to determine unconfined compressive strength of the rocks. No deformation measurements were made during uniaxial stress testing; therefore, no information was available on elastic properties of the intact rock in the tunnel.

The laboratory test results from plugs taken at five locations yielded the following values for unconfined compressive

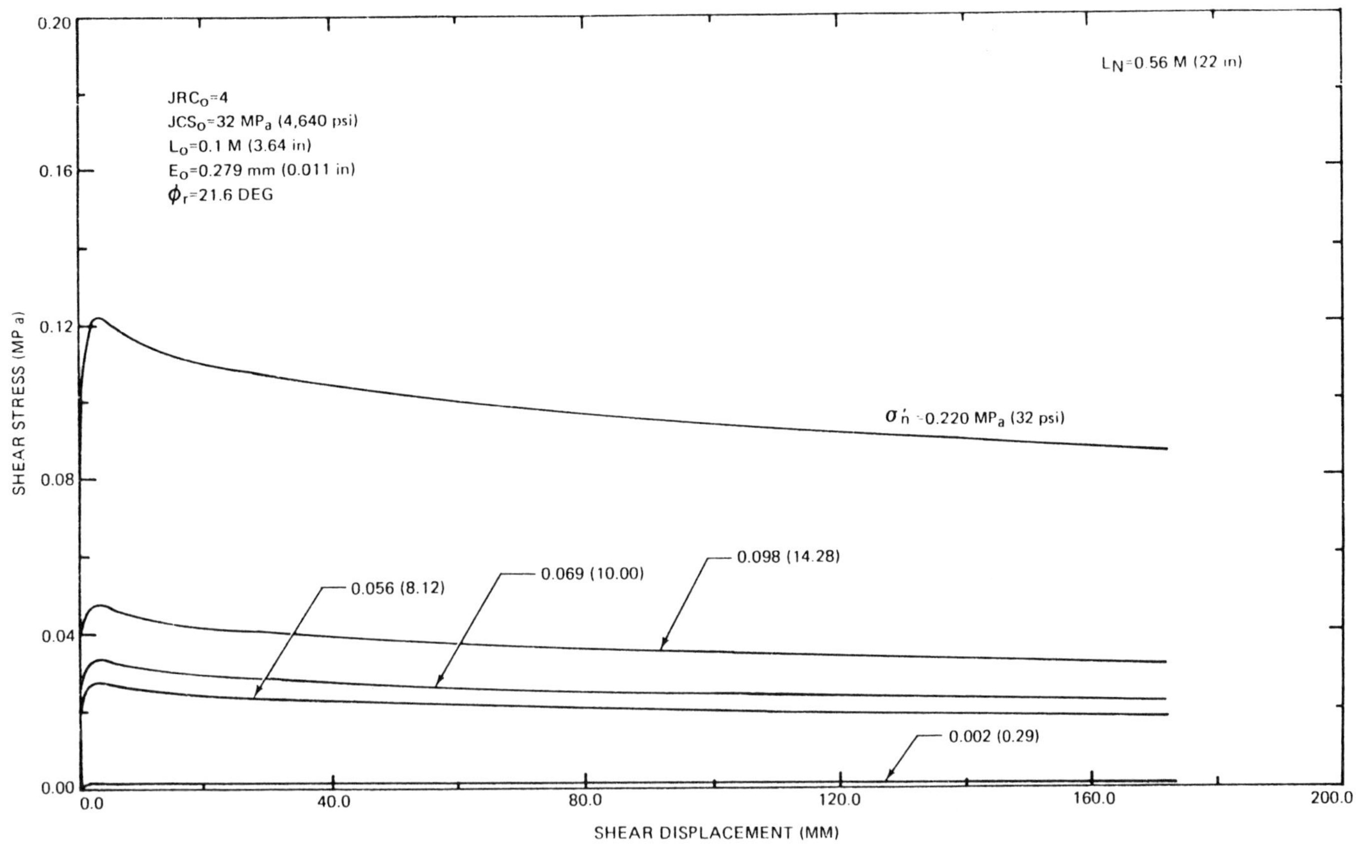

Figure 16. Shear stress-displacement curves generated under various levels of effective normal stress for block size 22in (.56 m).

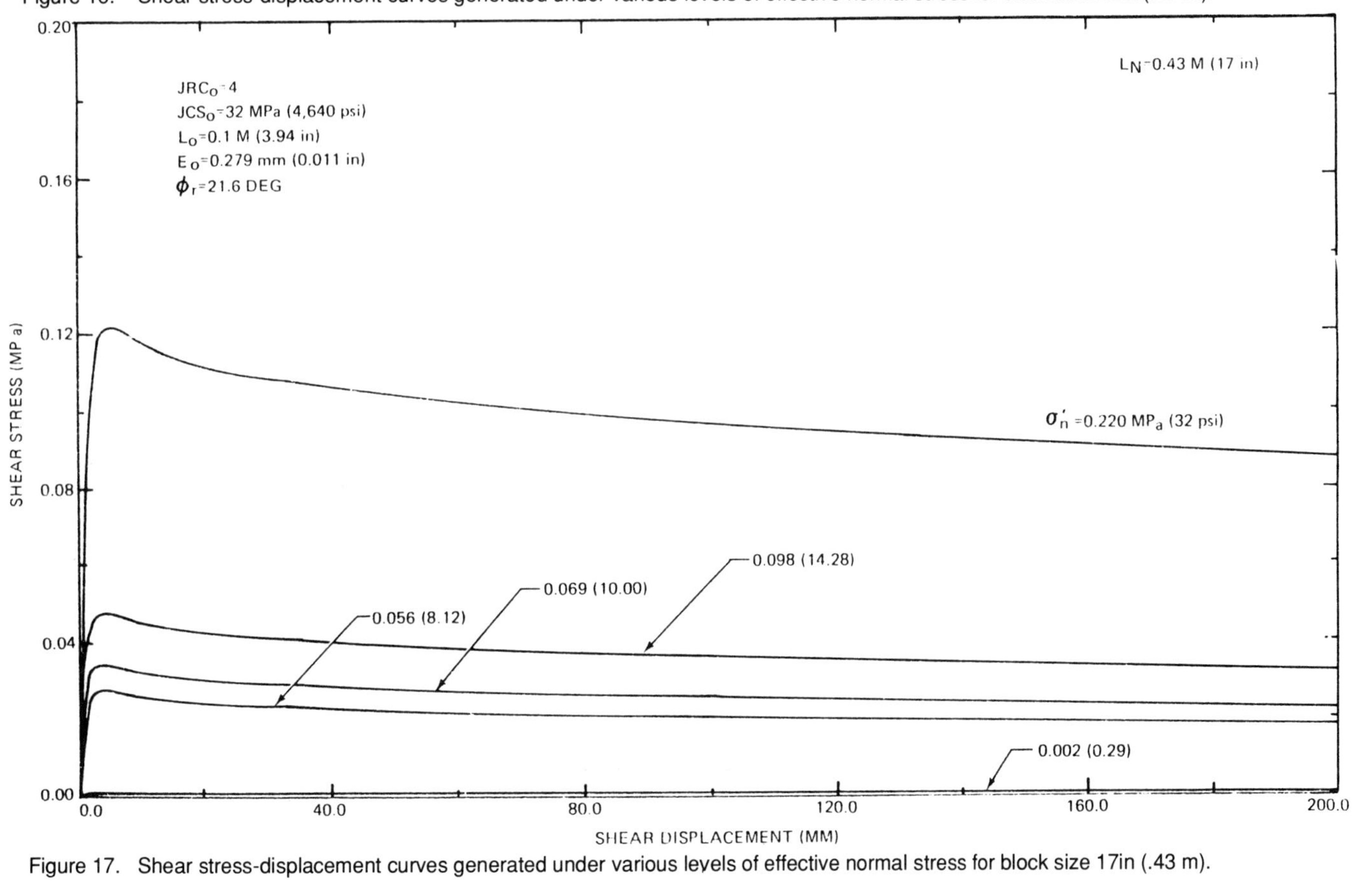

Figure 17. Shear stress-displacement curves generated under various levels of effective normal stress for block size 17in (.43 m).

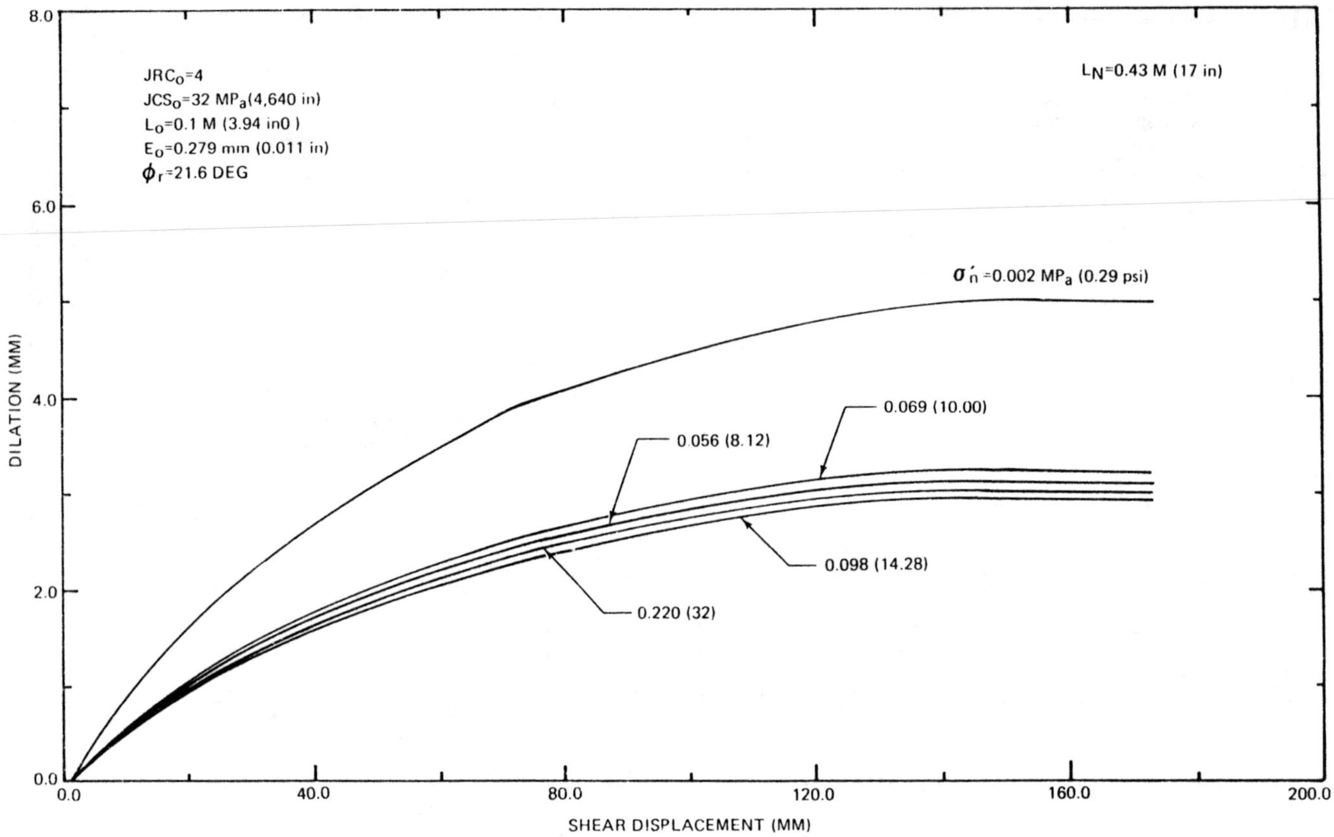

Figure 18. Dilation-displacement behavior fro block size 17in (0.43 m).

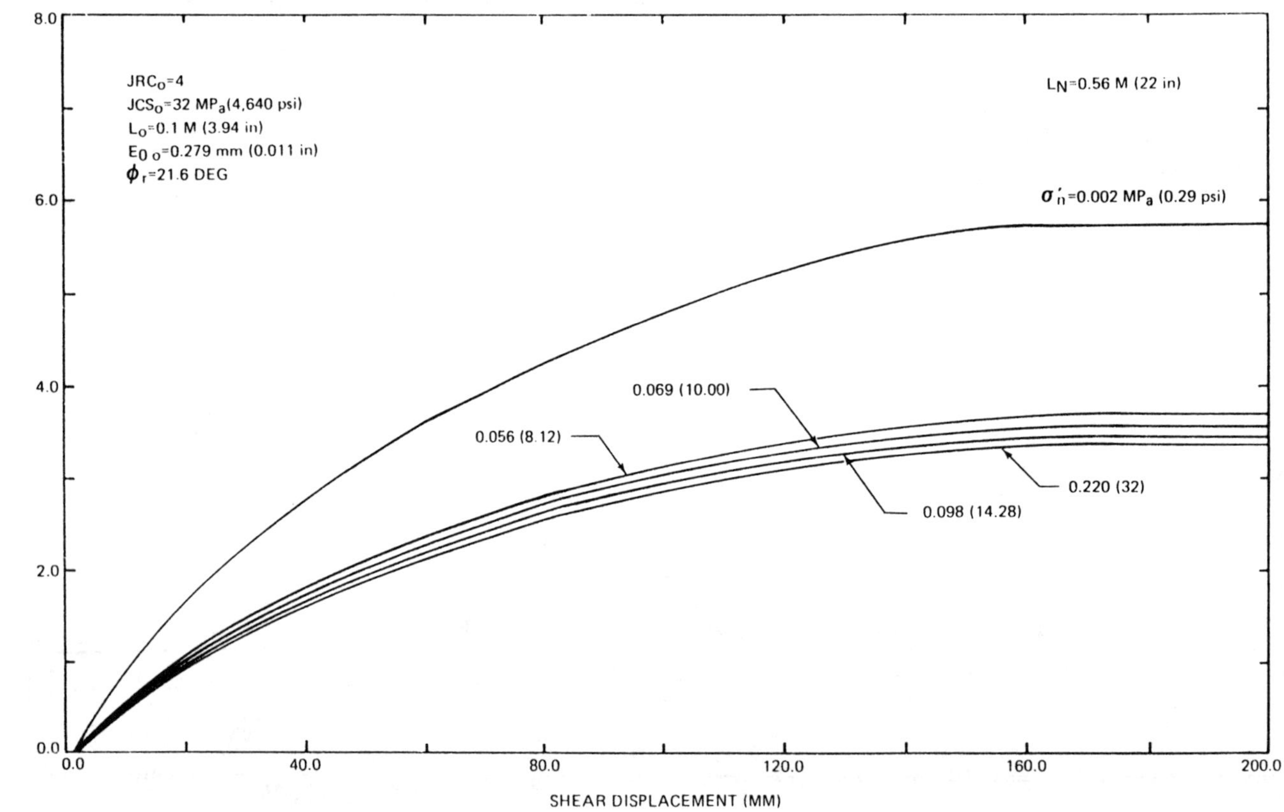

Figure 19. Dilation-displacement behavior fro block size 22in (0.56 m).

strength: 40,666 psi (280 MPa), 18,062 psi (124.57 MPa), 22,012 psi (151.80 MPa), 15,991 psi (110.28 MPa), 3,664 psi (25.27 MPa), 3,475 psi (23.96 MPa). Because information on dimensions of specimens tested and details of testing procedures were unavailable, it was difficult to compare the above values with those reported in Table 4, obtained from Schmidt hammer. However, the range of values obtained from laboratory tests were consistent with those measured in-situ using a Schmidt hammer.

Comparison of Shear Stress-Displacement Behavior

Figures 16 and 17 illustrate the difference in shear resistance between block lengths of 17 in (0.43 m) and 22 in (0.56 m). The curves shown were generated at normal stress levels computed for each dip angle. Because of the nature of joints at the site, one major joint orientation can be assumed and the stress levels used represent the different depths. For the two block sizes considered, the initial peak shear resistances were almost identical, 17.69 psi (0.122 MPa) and 17.52 psi (0.1208 MPa), and post-peak behavior curves were similar.

As a result of coupling between dilation and effective normal stress, the actual block resistance will follow a complex path of effective normal stress with a fairly uniform reduction in stress as shearing occurs. The peak shear resistance of the smaller block is mobilized faster than that of the larger block.

Comparison of Shear Displacement-Dilation Behavior

Figures 18 and 19 show the variation of shear displacement-dilation curves for the two size blocks considered. Note the finite displacement required to mobilize dilation. Peak shear resistance corresponds to the parts of the dilation curves showing maximum dilation angles, because of the peak value of JRC (mobilized). Dilation continues post-peak, but at a gradually diminishing rate. Higher dilation values were observed for the larger block.

CLOSURE

The preceding pages describe techniques for rock mass classification, joint characterization and modeling. The methods outlined can be applied in the field to completely characterize a rock mass hosting an underground facility for the purpose of estimating support requirement.

REFERENCES

Bakhtar, K., 1985, Large scale behavior of rock joints: Ph.D. Dissertation in Rock Mechanics, University of Utah, Department of Mining Engineering.

Bakhtar, K., Barton, N., 1984, Large scale static and dynamic experiments: 25th U.S. Symposium on Rock Mechanics, Northwestern University, Evanston, Illinois.

Bakhtar, K., Martin, W., Dietl, B., and Byrnes, P., 1991, Ground stabilization in shallow depth soft ground tunneling: Special Session, Am. Soc. of Civil Engineers, Orlando, Florida.

Bandis, S., 1980, Experimental studies of scale effects on shear strength and deformation of rock joints: Ph.D. dissertation, University of Leeds, Department of Earth Sciences.

Barton, N., 1973, Review of the new shear strength criterion for rock joints: Engineering Geology, vol. 7, p. 287-332.

Barton, N., 1980, Estimation of in situ joint properties - Nasliden Mine: International Conference on Application of Rock Mechanics to Cut and Fill Mining, Luelea, Sweden, Institute of Mining and Metallurgy, London.

Barton, N., and Bakhtar, K., 1983, Rock joint description and modeling for hydrothermomechanical design of Nuclear Waste Repositories: Parts 1-5, CANMET, Mining Research Laboratories, Ottawa, Ontario.

Barton, N., and Bandis, S., 1982, Effects of block size on shear behavior of jointed rock: Keynote Lecture, 23rd U.S. Symposium on Rock Mechanics, Berkeley, California.

Barton, N., Bandis, S., and Bakhtar, K., 1985, Strength, deformation and conductivity coupling of rock joints: Inter. Jour. of Rock Mechanics and Mining Sciences and Geomechanics Abstract, vol. 22, no. 3, p. 121-140.

Barton, N., and Choubey, V., 1977, The shear strength of rock joints in theory and practice: Rock Mechanics, vol. 10, p. 1-54.

Barton, N., Lien, R., and Lunde, J., 1974, Engineering classification of rock masses for design of tunnel support: Rock Mechanics, vol. 6, no. 4.

Bieniawski, Z. T., 1973, Engineering classification of jointed rock masses: Transactions, South African Institution of Civil Engineers, vol. 15, No. 12.

Bieniawski, Z. T., 1974, Geomechanics classification of rock masses and its application in tunneling: Proceedings, Third International Congress on Rock Mechanics, Inter. Soc. of Rock Mechanics, Denver, Colorado, vol. IIA, p. 27-32.

Bieniawski, Z. T., 1975, The Point Load Test in geotechnical practice: Engineering Geology, vol. 9, p. 1-11.

Bieniawski, Z. T., 1976, Rock mass classification in rock engineering: Proceedings, Symposium on Exploration for Rock Engineering, Editor Z. T. Bieniawski, A. A. Balkema, Rotterdam, p. 97-106.

Bieniawski, Z. T., 1979, Tunnel design by rock mass classification: U. S. Army Corp of Engineer, Waterways Experiment Station, Tech. Report GL-79-19, 133.

Bieniawski, Z. T., 1984, Rock mechanics in mining and tunneling: A. A. Balkema, Rotterdam.

Crawford, A. M., and Curran J. H., 1981, The influence of shear velocity on the frictional resistance of rock discontinuities: Inter. Jour. of Rock Mechanics and Mining Sciences, vol. 18, p. 505-515.

Deere, D. U., 1963, Technical description of rock cores for engineering purposes: Rock Mechanics and Engineering Geology, vol. 1, no. 1.

Deere, D. U., and Miller, R. P., 1969, Engineering classification and index properties of intact rock: Technical Report, No. AFNL-TR 65-116, Air Force Weapons Laboratory, Albuquerque, New Mexico.

Gillette, D. R., Stur, S., and Ko, H-Y, 1983, Dynamic Behavior of rock joints: Proceedings, 24th Symposium on Rock Mechanics, Texas A&M University.

Lauffer, H., 1958, Gerbirgsklassifizierung fur den Stollenbau: Geologie und Bauwesen, vol. 24, no. 1.

Miller, R. P., 1965, Engineering classification and index properties for intact rock: Ph.D. Dissertation, University of Illinois, Department of Civil Engineering.

Palmstrom, A., 1975, Characterization of degree of jointing and rock mass quality: in Norwegian, Internal Report, Ing. A. B. Berdal A/S, Oslo.

Patton, F. D. 1966a, Multiple modes of shear failure in rock and related materials: Ph.D. Dissertation, University of Illinois, Department of Civil Engineering.

Patton, F. D., 1966b, Multiple modes of shear failure in rock: Proceedings, 1st Congress, Inter. Soc. of Rock Mechanics, Lisbon, vol. 1, p. 509-513.

Pratt, H. R., Black, A. D., Brown, S. W., and Brace, W. F., 1972, The effects of specimen size on the mechanical properties of unjointed diorite: Intern. Jour. of Rock Mechanics and Mining Sciences, vol. 9, p. 513-529.

Terzaghi, K., 1946, rock defects and loads on tunnel support: Rock Tunneling with Steel Supports, R. V. Proctor and T. White, eds., Commercial Shearing Company, Youngstown, Ohio.

Wickham, G. E., Tiedemann, H. R., and Skinner, E. H., 1972, Support determination based on geologic predictions: Proceedings, Rapid Excavation and Tunneling Conference, AIME, New York.

Table 2. Details of Rock Mass Parameters (Barton et al., 1974)

Table 1. Descriptions and ratings for the parameter RQD.

1. ROCK QUALITY DESIGNATION (RQD)	
A. Very poor	0 - 25
B. Poor	25 - 50
C. Fair	50 - 75
D. Good	75 - 90
E. Excellent	90 - 100

Note: (i) Where RQD is reported or measured as ≤ 10, (including 0) a nominal value of 10 is used to evaluate Q in equation (1).

(ii) RQD intervals of 5, i.e. 100, 95, 90, etc. are sufficiently accurate.

Table 2. Descriptions and ratings for the parameter J_n

2. JOINT SET NUMBER	(J_n)
A. Massive, no or few joints	0.5-1.0
B. One joint set	2
C. One joint set plus random	3
D. Two joint sets	4
E. Two joint sets plus random	6
F. Three joint sets	9
G. Three joint sets plus random	12
H. Four or more joint sets, random, heavily jointed, "sugar-cube" etc.	15
J. Crushed rock, earthlike	20

Note: (i) For intersections use $(3.0 \times J_n)$

(ii) For portals use $(2.0 \times J_n)$

Table 3. Descriptions and ratings for the parameter J_r

3. JOINT ROUGHNESS NUMBER	(J_r)
(a) *Rock wall contact* and (b) *Rock wall contact before 10 cms shear*	
A. Discontinuous joints	4
B. Rough or irregular, undulating	3
C. Smooth, undulating	2
D. Slickensided, undulating	1.5
E. Rough or irregular, planar	1.5
F. Smooth, planar	1.0
G. Slickensided, planar	0.5

Note: (i) Descriptions refer to small scale features and intermediate scale features, in that order.

(c) *No rock wall contact when sheared*	
H. Zone containing clay minerals thick enough to prevent rock wall contact	1.0
J. Sandy, gravelly or crushed zone thick enough to prevent rock wall contact	1.0

Note: (ii) Add 1.0 if the mean spacing of the relevant joint set is greater than 3 m.

(iii) $J_r = 0.5$ can be used for planar slickensided joints having lineations, provided the lineations are orientated for minimum strength.

Table 4. Descriptions and ratings for the parameter J_a

4. JOINT ALTERATION NUMBER	(J_a)	(ϕ_r) (approx.)
(a) *Rock wall contact*		
A. Tightly healed, hard, non-softening, impermeable filling i.e. quartz or epidote	0.75	(-)
B. Unaltered joint walls, surface staining only	1.0	(25-35°)
C. Slightly altered joint walls. Non-softening mineral coatings, sandy particles, clay-free disintegrated rock etc.	2.0	(25-30°)
D. Silty-, or sandy-clay coatings, small clay fraction (non-soft.)	3.0	(20-25°)
E. Softening or low friction clay mineral coatings, i.e. kaolinite or mica. Also chlorite, talc, gypsum, graphite etc., and small quantities of swelling clays	4.0	(8-16°)
(b) *Rock wall contact before 10 cms shear*		
F. Sandy particles, clay-free disintegrated rock etc.	4.0	(25-30°)
G. Strongly over-consolidated non-softening clay mineral fillings (continuous, but <5 mm thickness)	6.0	(16-24°)
H. Medium or low over-consolidation, softening, clay mineral fillings (continuous but <5 mm thickness)	8.0	(12-16°)
J. Swelling -clay fillings, i.e. montmorillonite (continuous, but <5 mm thickness). Value of J_a depends on percent of swelling clay-size particles, and access to water etc.	8-12	(6-12°)
(c) *No rock wall contact when sheared*		
K, L, M. Zones or bands of disintegrated or crushed rock and clay (see G, H, J for description of clay condition)	6, 8, or 8-12	(6-24°)
N. Zones or bands of silty- or sandy-clay, small clay fraction (non-softening)	5.0	(-)
O, P, R. Thick, continuous zones or bands of clay (see G, H, J for description of clay condition)	10, 13, or 13-20	(6-24°)

Table 5. Descriptions and ratings for the parameter J_w

5. JOINT WATER REDUCTION FACTOR	(J_w)	Approx. water pres. (kg/cm²)
A. Dry excavations or minor inflow, i.e. <5 l/min. locally	1.0	<1
B. Medium inflow or pressure, occasional outwash of joint fillings	0.66	1-2.5
C. Large inflow or high pressure in competent rock with unfilled joints	0.5	2.5-10
D. Large inflow or high pressure, considerable outwash of joint fillings	0.33	2.5-10
E. Exceptionally high inflow or water pressure at blasting, decaying with time	0.2-0.1	>10

Table 2. (Cont.)

	(J_w)	Approx. water pres.
F. Exceptionally high inflow or water pressure continuing without noticeable decay	0.1-0.05	>10

Note: (i) Factors C to F are crude estimates. Increase J_w if drainage measures are installed.

(ii) Special problems caused by ice formation are not considered.

Table 6. Description and ratings for parameter SRF

6. STRESS REDUCTION FACTOR

(a) *Weakness zones intersecting excavation, which may cause loosening of rock mass when tunnel is excavated.*	(SRF)
A. Multiple occurrences of weakness zones containing clay or chemically disintegrated rock, very loose surrounding rock (any depth)	10
B. Single weakness zones containing clay or chemically disintegrated rock (depth of excavation ≤ 50 m)	5
C. Single weakness zones containing clay or chemically disintegrated rock (depth of excavation >50 m)	2.5
D. Multiple shear zones in competent rock (clay-free), loose surrounding rock (any depth)	7.5
E. Single shear zones in competent rock (clay-free) (depth of excavation ≤50 m) ..	5.0
F. Single shear zones in competent rock (clay-free) (depth of excavation >50 m) ..	2.5
G. Loose open joints, heavily jointed or "sugar cube" etc. (any depth)	5.0

Note: (i) Reduce these values of SRF by 25-50% if the relevant shear zones only influence but do not intersect the excavation.

(b) *Competent rock, rock stress problems*	σ_c/σ_1	σ_t/σ_1	(SRF)
H. Low stress, near surface	>200	>13	2.5
J. Medium stress	200-10	13-0.66	1.0
K. High stress, very tight structure (usually favourable to stability, may be unfavourable for wall stability)	10-5	0.66-.33	0.5-2
L. Mild rock burst (massive rock)	5-2.5	0.33-.16	5-10
M. Heavy rock burst (massive rock)	<2.5	<0.16	10-20

Note: (ii) For strongly anisotropic virgin stress field (if measured): when $5 \leq \sigma_1/\sigma_3 \leq 10$, reduce σ_c and σ_t to $0.8\,\sigma_c$ and $0.8\,\sigma_t$. When $\sigma_1/\sigma_3 > 10$, reduce σ_c and σ_t to $0.6\,\sigma_c$ and $0.6\,\sigma_t$, where: σ_c = unconfined compression strength, and σ_t = tensile strength (point load) and σ_1 and σ_3 are the major and minor principal stresses.

(iii) Few case records available where depth of crown below surface is less than span width. Suggest SRF increase from 2.5 to 5 for such cases (see H).

(c) *Squeezing rock plastic flow of incompetent rock under the influence of high rock pressure*	(SRF)
N. Mild squeezing rock pressure	5-10
O. Heavy squeezing rock pressure	10-20
(d) *Swelling rock chemical swelling activity depending on presence of water*	
P. Mild swelling rock pressure	5-10
R. Heavy swelling rock pressure	10-15

ADDITIONAL NOTES ON THE USE OF TABLES 1 TO 6.

When making estimates of the rock mass quality (Q) the following guidelines should be followed, in addition to the notes listed in Tables 1 to 6:

1. When borecore is unavailable, RQD can be estimated from the number of joints per unit volume, in which the number of joints per metre for each joint set are added. A simple relation can be used to convert this number to RQD for the case of clay-free rock masses (Palmstrom, 1975):

$$RQD = 115 - 3.3\,J_v \text{ (approx.)}$$

where

J_v = total number of joints per m^3

$(RQD = 100 \text{ for } J_v < 4.5)$

2. The parameter J_n representing the number of joint sets will often be affected by foliation, schistocity, slatey cleavage or bedding etc. If strongly developed these parallel "joints" should obviously be counted as an complete joint set. However, if there are few "joints" visible, or only occasional breaks in bore core due to these features, then it will be more appropriate to count them as "random joints" when evaluating J_n in Table 2.

3. The parameters J_r and J_a (representing shear strength) should be relevant to the *weakest significant joint set or clay filled discontinuity* in the given zone. However, if the joint set or discontinuity with the minimum value of (J_r/J_a) is favourably oriented for stability, then a second, less favourably orientated joint set or discontinuity may sometimes be of more significance, and its higher value of J_r/J_a should be used when evaluating Q from equation 1. *The value of J_r/J_a should in fact relate to the surface most likely to allow failure to initiate.*

4. When a rock mass contains clay, the factor SRF appropriate to *loosening loads* should be evaluated (Table 6a). In such cases the strength of the intact rock is of little interest. However, when jointing is minimal and clay is completely absent the strength of the intact rock may become the weakest link, and the stability will then depend on the ratio rock-stress/rock-strength (Table 6b). A strongly anisotropic stress field is unfavourable for stability and is roughly accounted for as in Note (ii), Table 6b.

5. The compressive and tensile strengths (σ_c and σ_t) of the intact rock should be evaluated in the saturated condition if this is appropriate to present or future in situ conditions. A very conservative estimate of strength should be made for those rocks that deteriorate when exposed to moist or saturated conditions.

Table 3. Summary of joint characterization results

Joint Set	JRC_o	L_n m	L_n (in)	L_o m	L_o (in)	JCS_o MPa	JCS_o (psi)	σ_n' MPa	σ_n' (psi)	e_o mm	e_o (in)	ϕ_r°	ϕ_b°	σ_c MPa	σ_c (psi)
#1	2	0.43	(17)	0.10	(3.94)	62.0	(9,000)	0.002	(0.29)	0.279	(0.011)	25	30	111.25	(16,132)
#2	4	0.56	(22)	0.10	(3.94)	26.3	(3,810)	0.056	(8.12)	0.279	(0.011)	19	30	83.27	(12,074)
#3	4	0.56	(22)	0.10	(3.94)	35.0	(5,063)	0.098	(14.28)	0.279	(0.011)	22	30	77.35	(11,216)
C-1	4	0.56	(22)	0.10	(3.94)	22.7	(3,287)	0.069	(10.00)	0.432	(0.017)	21	30	44.26	(6,417)
C-2	6	0.43	(17)	0.10	(3.94)	15.9	(2,310)	0.220	(32.00)	0.432	(0.017)	21	30	39.41	(5,715)
AVERAGE	4	0.51	(20)	0.10	(3.94)	32.4	(4,695)	0.089	(12.90)	0.355	(0.014)	21.6	30	71.11	(10,310)

JRC_o - Joint Roughness Coefficient

L_n - Block Length

L_o - Laboratory Sample Size

JCS_o - Joint Wall Compression Strength

σ_n' - Effective Normal Stress on Joint

e_o - Smooth Hydraulic Aperture

ϕ_r° - Residual Friction Angle

ϕ_b° - Basic Friction Angle

σ_c - Unconfined Compressive Strength of Intact Rock (Schmidt hammer)

LOS ANGELES METRO RAIL PROJECT: HISTORY AND CONSTRUCTION OF PHASE I

Richard F. Escandon
Law/Crandall, Inc.
200 Citadel Drive
Los Angeles, CA 90040

Anthony F. Stirbys
Dames and Moore
911 Wilshire Blvd., Suite 700
Los Angeles, CA 90017

Charles W. Daugherty
Parsons Brinckerhoff
5610 Redbird Center Drive, Suite 400
Dallas, TX 75237

INTRODUCTION

The Los Angeles Metro Rail rapid transit system is one of the largest single construction projects ever undertaken in the southern portion of the Golden State. As a primarily underground project, it has faced some issues and encountered some problems that make it a good case history for the enlightenment of engineers, engineering geologists and tunnelers contemplating similar underground works. This paper is intended to provide an appreciation for where Metro Rail now stands in its development from an interesting concept to a viable construction project to a functioning system.

This paper is also intended to provide a summary of some of the important technical developments experienced on the project. As a case history study, it skips over excess theory to emphasize descriptions of how things actually get accomplished. It is hoped that this publication will help disseminate the information to the widest possible audience and serve those seeking solutions to the types of problems faced by Metro Rail.

A BRIEF HISTORY

With the final demise of the famed "Big Red (trolley) Cars" in the late 1950s and early 60s, the populace had become almost totally dependent on the internal combustion engine to traverse a sprawling network of communities that make up the greater Los Angeles area. As years passed, traffic on L.A.'s city streets and highways became more and more congested gaining national attention not only from the resultant snarled traffic conditions, but also from the air pollution generated. It was clear to some transportation experts that the area's love affair with the private automobile needed to be reined in to some degree. Clean running electric trains were an obvious partial solution to such a problem. If some of them were designed to be high speed and grade separated from other traffic by encasement in subsurface alignments, they would be capable of carrying vast numbers of people and would encourage the removal of a large number of automobiles from surface roads.

As early as 1953 a study was undertaken to determine the feasibility of beginning the development of a new L.A. County rapid transit system with the construction of a monorail line between Long Beach and the San Fernando Valley. Twenty-seven years later in 1980, Disneyland was the only "municipality" with a monorail, but the citizens had finally — after many rejections — backed the visionaries through approval of a referendum for a 150-mile rapid transit system. This was to consist of a mix of heavy rail, light rail and dedicated busways. The heavy rail system, which is the focus of this paper, came to be known as Metro Rail. It is a conventional steel wheel on steel rail arrangement, with trains traveling mostly in twin tunnels slightly more than 20 feet in excavated diameter. The preliminary engineering for the system began in December 1981 and actual excavation of the first phase began on September 29, 1986. The finish and systems contracts (architecture, rails, electrification, signaling, etc.) are presently nearing completion for the first phase of construction and trains are expected to be running by September 1993. The second phase of Metro Rail construction is just getting underway and is expected to be completed by 1998.

The original concept for the Metro Rail subway was that it was to start at Union Station, the existing railroad station, then swing through the Central Business District, proceed west under Wilshire Boulevard, north under Fairfax Avenue, thence through Hollywood to North Hollywood in the San Fernando Valley (Fig. 1). Although the approach from the yard site on the flood plain of the Los Angeles River would be at grade, the remainder of the 18.5-mile route was planned to be below ground with most of the running lines contracted as mined tunnels and the 17 stations as cut- and-cover boxes. The system was divided into phases so as to create convenient funding packages. The first phase was dubbed the Minimum Operable Segment No. 1 (MOS-1) and is the portion for which the structure is now complete. West of MOS-1 the alignment led straight through the heart of the La Brea Tar Pit area, world renowned for the large number of Pleistocene age fossils that have been recovered from its oil and gas laden sediments.

Gas proved to be the most critical factor for Metro Rail. Design proceeded apace in 1983 and 1984, but came to an almost abrupt halt on March 24, 1985. On that date, a methane gas explosion in a department store basement within the abandoned Salt Lake Oil Field near the planned Fairfax/Beverly Station destroyed the store, made national TV news, and got the attention of nearly everyone who cared about underground construction in the area. Metro Rail de-

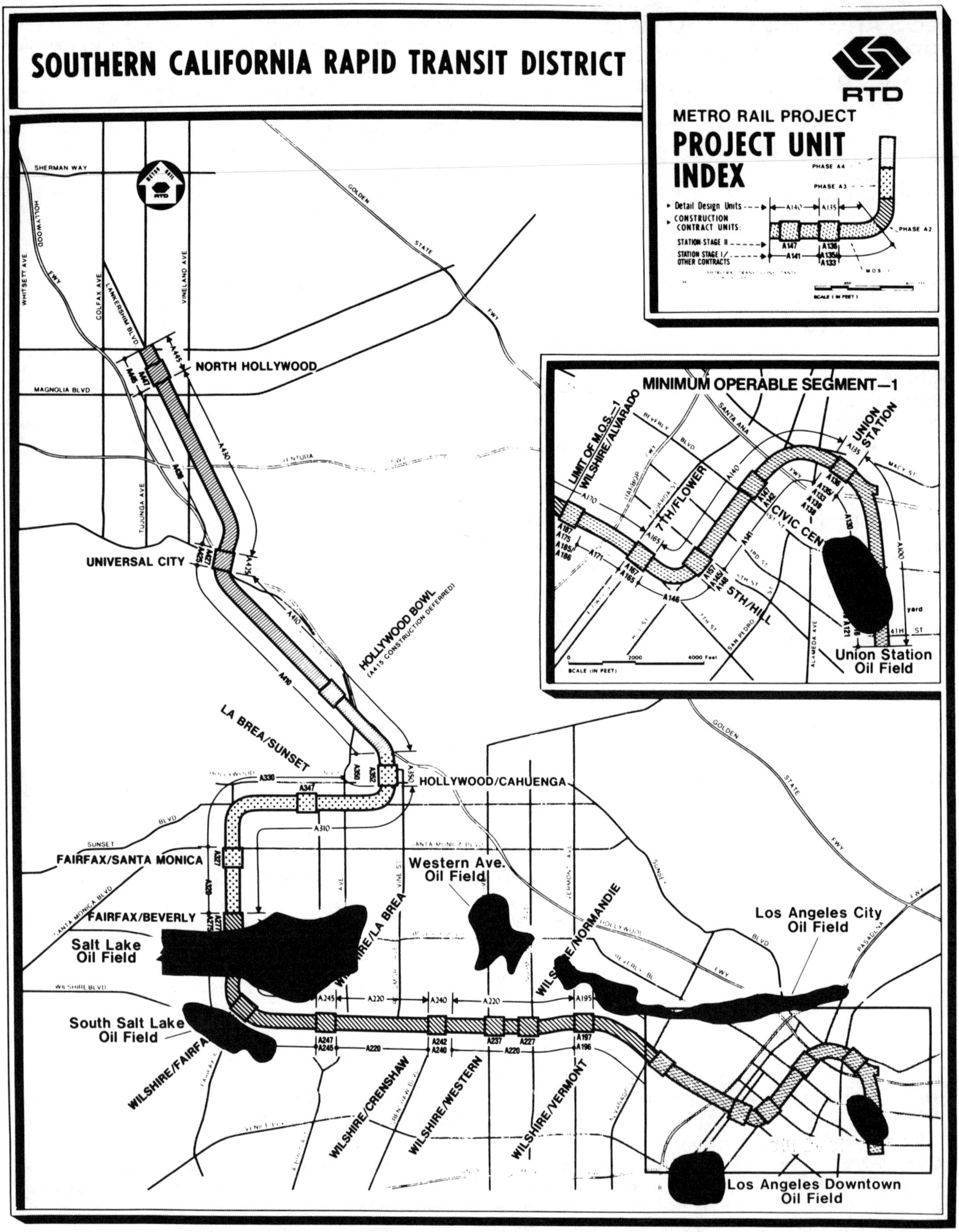

Figure 1. Original Metro Rail route showing oil fields and MOS-1

signers had known of the gas related hazards all along and had laid plans to circumvent them during both construction and operational phases. However, regulatory agencies and many members of the public were unconvinced. In December 1985, a bill was enacted by Congress to prohibit tunneling through the "potential risk area" in the vicinity of Wilshire and Fairfax.

After much study, Metro Rail planners re-evaluated the route west of Wilshire/Vermont Station and instated a modified system now comprised of 16 stations spaced along a 22.5-mile alignment that reaches Hollywood by following Vermont Avenue northward and Hollywood Boulevard to the west. There is still a short southern leg of alignment extending westward along Wilshire Boulevard. If funding ever becomes available, it can be extended to Santa Monica by safely skirting the the south boundary of the potential risk area. Pressure to bring the trackage up into an aerial structure along Vermont and Hollywood was successfully resisted. Except for the yard approach, Metro Rail is still completely underground. MOS-1 was untouched by the change in plans. The revised second construction/operational package now extends from the Wilshire/Alvarado (Westlake/MacArthur Park) Station to the Hollywood/Vine Station. As in the original system, the revised Metro Rail alignment still involves mostly soft ground construction, with only the reach through the Topanga Formation of the Santa Monica Mountains between Hollywood and the San Fernando Valley suggesting hard rock tunneling methods.

MOS-1 DESCRIPTION

MOS-1 involves construction of 4.4 miles of subway routed through the heart of the downtown area of Los Angeles (See inset on Fig. 1) and includes at- grade storage yard and maintenance facilities, five subterranean subway stations and approximately 13,000 feet of twin tunnels. The alignment crosses two major freeways at three locations and passes along side or beneath several major buildings along its route. The stations are spaced about 1600 to 5000 feet apart at locations strategic to ridership and are 60 to 65 feet wide and 55 to 81 deep.

MOS-1 was divided into a number of construction and procurement contract units involving both Stage I structural contracts and Stage II finish and systems contracts. Eight of the contracts involved underground station and/or tunnel construction and are the primary focus of this paper. Of these eight contracts, two are strictly tunnel line construction (Contracts A146 and A171) and four involve mainly station construction (Contracts A135, A145, A165 and A175). Contract A141 is the only contract which included both station (Civic Center Station) and tunnel line construction. Contract A130 is also multifaceted involving tunnel, open-cut and cut-and-cover construction.

DESIGN CONSIDERATIONS

The majority of underground works in the United States face the classic tunneler's problem: how to excavate and support the ground and how to handle an excess of groundwater. Thanks to its geographic location, the L.A. Metro Rail system is blessed with the kind of geology that goes a long way toward minimizing these particular problems. However, that same location — in combination with the chosen alignment — forced Metro Rail to contend with a combination of circumstances and hazards that may never be faced in their entirety by any other U.S. tunneling project. Before addressing all of the deeper technical details of design and construction, it may prove fruitful to set the stage by summarizing most of the unusual problems faced and overcome by Metro Rail designers and builders.

Seismicity and Faulting

For the public at large, seismicity is probably uppermost in the minds of those contemplating underground construction and travel in southern California. However, a properly designed tunnel is one of the safest places to be in an earthquake because the whole structure can move with the ground rather than being whipped back and forth like a building with its bottom tied to the ground and its top unrestrained. San Francisco's BART system took the lead in designing in this environment in the 1960s and Metro Rail followed that lead in the 1980s (see Assoc. of Engineering Geologists, 1981). Seismicity and faulting forced onto L.A. Metro Rail a series of designs and actions which may be unique in their scope:

- Designs for both permanent structures and initial excavation supports that absorb rather than resist the shaking (strong ground motion) and associated distortions from tremors generated from active faults in the area. The design's success is attested to by the lack of damage in spite of the system having been subjected to two moderate earthquakes during the first construction phase.

- Metro Rail is equipped with an extensive network of computerized strong motion detectors to shut down train operations in case shaking exceeds a certain threshold.

- The seismic regime of Los Angeles is also characterized by a network of major and minor faults, some active, some inactive. For the first year of tunnel construction, until the threat of fault encounters could be properly put into perspective, Metro Rail took the action of having mined tunnel face mapping continuously performed by engineering geologists. This effort was backed up by contingency plans for tunnel lining enhancement in case an encountered fault posed a threat to the tunnel's integrity.

Gassy Ground

As already noted, gas — mostly methane and hydrogen sulfide — was the second geology-related problem of great concern to designers and to the public. Although the oil fields had at one time been a generator of great revenue for Los Angeles, the associated natural gas was more a source of concern for Metro Rail. Buildings and underground structures in certain localities have to be protected against gas intrusion and designers have to be aware that changing migration patterns can cause the insidious mixtures to pop up in unexpected places. Because Metro Rail was to be constructed within this more or less gas-laden ground, practically its entire soft ground length was designated "gassy" or "potentially gassy". And because the completed tunnels would carry electrified cars and people, not simple utilities, the system had to be made safe for long term operations as well as during the construction phase. This resulted in a unique combination of features and activities including:

- The complete encasement of all underground structures in a 100-mil thick high density polyethylene membrane (HDPE) to restrict gas flow into the tunnel to less than 1 percent of what it would be through bare concrete.
- Popularization of a two-pass, cast-in-place concrete inside pre-cast concrete segment type of tunnel lining by those contractors who choose not to use steel ribs and wood lagging as the initial support elements (Fig. 2). The two-pass system was necessitated by the requirement for the HDPE and concrete sandwich. The arrangement was later adopted by the Washington, D. C. Metrorail System, which had a need to create a PVC and concrete sandwich for waterproofing purposes. Although the two-pass concrete lining had first been tried in Seattle, when it migrated to D.C., it was informally referred to as the "Los Angeles Tunneling Method."
- L.A. Metro Rail will have one of the world's most sophisticated sniffing and ventilation systems to protect patrons in case the first line of gas defense, the HDPE membrane, is breached. Computer controlled sensors mounted throughout the tunnels willsignal Rail Control Center if there is an unexpected build-up of gas, and the operator can remotely alter ventilation arrangements so as to cut the gas flow/concentration to an acceptable level. He may also prevent trains from entering the affected area.
- The improperly abandoned oil wells that dot the L.A. landscape are also a problem in that they could contain gas under pressure, able to flood a heading with an asphyxiating or flammable mixture if encountered by a tunneling shield. This led to the requirement that the tunnel contractor keep a probe hole and magnetometer survey (to detect abandoned metallic oil well casings) a minimum of 50 feet ahead of the advancing tunnel face. The magnetometer requirement is accompanied by specified methods of properly plugging and abandoning any well casings encountered by the mining operation.

Environmental Design Issues

Metro Rail designers and builders have also had to cope with several environmentally related problems that have ranged from the unusual to the unique. At one point, Contracts A130, 135 and 141 were practically designed and ready to be bid when the belated discovery of hazardous wastes in parts of the Los Angeles River flood plain put everything on hold. Newly dug piers for the elevated busway paralleling the Hollywood Freeway encountered wastes from a very old coal gasification plant. This resulted in the stoppage of busway construction and Metro Rail bidding as ways were sought to tiptoe through the pockets of carcinogenic materials. Months later, the rail system was again ready for bidding, but only after the realignment of trackage that moved the structure as much as 140 feet to the northeast in the area of Union Station.

Although water is a relatively minor concern for most of the Metro underground construction, the groundwater affecting portions of the three contracts in the Union Station area proved to be a definite problem. The ground in the area is pervious, the water table high, and there is a high rate of recharge due to the presence of the Los Angeles River. Worse yet, the groundwater in this area is so heavily charged with pollutants, primarily hydrogen sulfide, that treatment was required before the product of dewatering could be released into the storm drains or the river. In order to expedite construction, the Metro Rail owner decided to handle the prob-

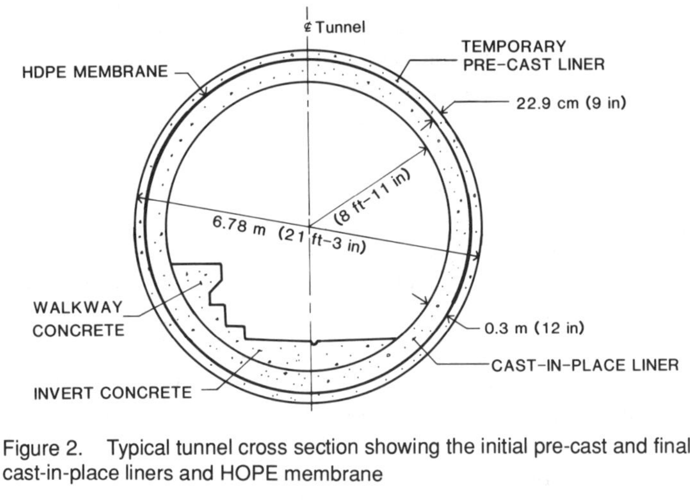

Figure 2. Typical tunnel cross section showing the initial pre-cast and final cast-in-place liners and HOPE membrane

lem for all three contracts and so set up and operated a water treatment plant capable of handling up to 7 million gallons per day on a continuous basis to the end of construction. The treated water was pumped into the river and the solid waste was dried out and routinely disposed of as a non-toxic material.

INSTRUMENTATION

Prior to the subway's conception a large comprehensive program of geotechnical instrumentation such as that used on Metro Rail had never been carried out as part of any Los Angeles project. Metro Rail designers completed plans for the first 4.4-mile segment without definite proof that the ground would react to construction activities in precisely the manner predicted. With the desire for more data, they provided MOS-1 with an array of instrumentation in excess of that needed strictly for construction control monitoring. The program was intended to create a body of data that would be used to check original design assumptions and refine future designs. The 17,600 feet of underground construction was monitored through a program that entailed hundreds of strain gages and load cells on cofferdam struts and tiebacks, 40 inclinometers, 180 borehole extensometers, 45 shallow settlement indicators, 20 deep settlement indicators, 35 deep bench marks, 300 liquid tube monitoring devices, 5200 building settlement reference points, and 3600 ground surface settlement reference points.

PROJECT GEOLOGY

The MOS-1 alignment is located at the northern end of the northwest trending Peninsular Ranges geomorphic province, near the southern boundary of the east-west trending Transverse Ranges province. Major structural elements in the vicinity of the project include the Santa Monica-Hollywood fault zone to the north, the Whittier fault zone to the east, and the Newport-Inglewood fault zone to the southwest. On a more local scale, the MOS-1 alignment is located near the northern margin of the Los Angeles basin just west of the Los Angeles river and skirts the southeast tip of the Elysian Park hills as it traverses through downtown Los Angeles. Structural trends along the alignment are associated with the southwestern flank of the southeast trending Elysian Park anticline and several northeast to northwest trending faults of limited extent to the north and east of the project. The trace of one of these faults, the MacArthur Park fault, is projected to cross near the vicinity of the Wilshire/Alvarado station (Converse et al. 1981) but was not encountered during construction.

Geologic conditions along the alignment are shown on the geologic profile (Fig. 3) and include both younger and older alluvial deposits of Quaternary age as well as soft-sedimentary bedrock units of Tertiary age. Holocene younger alluvium associated with the Los Angeles river floodplain underlies most of the downtown area and was encountered in over 33 percent of all tunnel and station excavations. The younger alluvial deposits consist of dense granular sands and

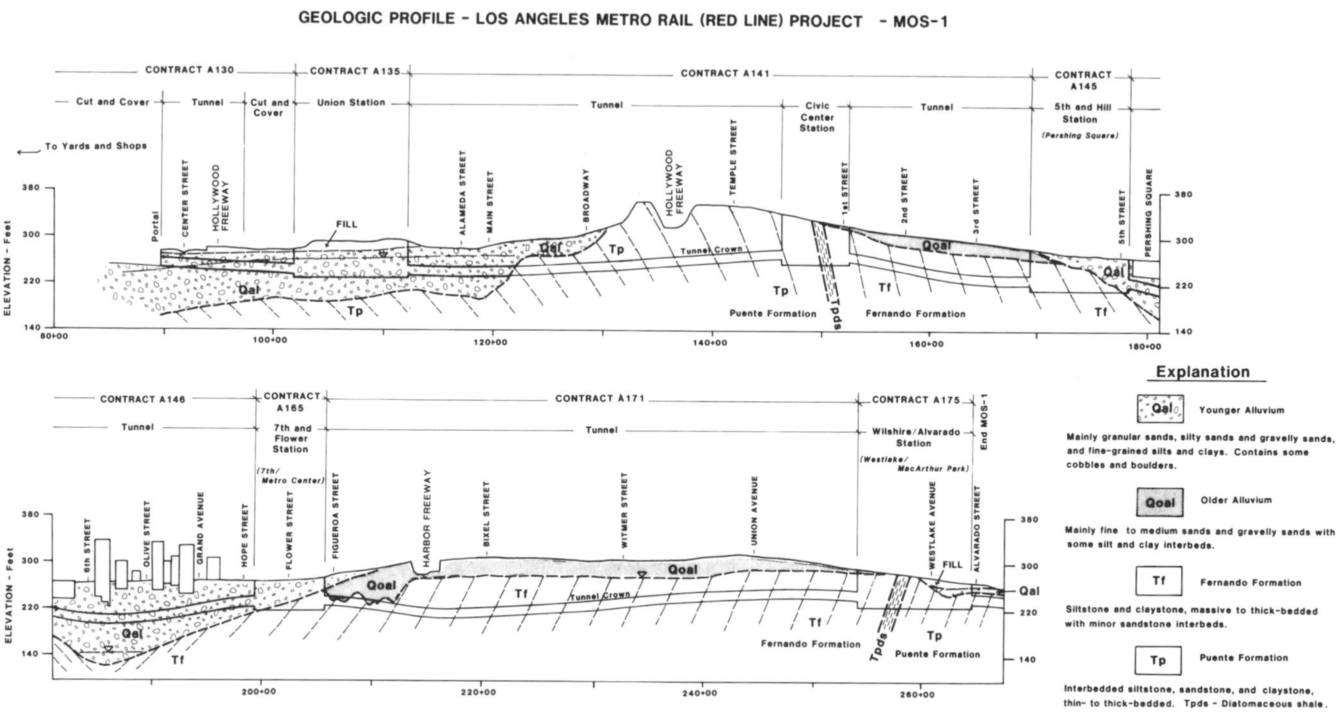

Figure 3. Geologic profile along the MOS-1 alignment

gravels with cobbles and boulders near the Los Angeles river and numerous stiff fine-grained silt and clay interbeds. Pleistocene older alluvium, consisting of dense fine to medium sands and gravely sands with some silt and clay interbeds, also blankets portions of the alignment but was encountered in less than 2 percent of all excavations.

Siltstone, claystone and sandstone bedrock of the Pliocene Fernando and Miocene Puente Formations underlie the alluvial deposits in the downtown area and are exposed in outcrops within the low lying Elysian Park hills just north of the alignment. The bedrock was encountered in approximately 65 percent of the all tunnel and station excavations and is generally thick- to thin-bedded and contains local hard cemented beds and concretionary nodules. Both bedrock units have the engineering properties of hard cohesive soils and are considered soft-ground tunneling materials due to their general excavation character and relatively low unconfined compressive strength. The transition from Fernando to Puente Formation is recognized by a nearly 100-foot-thick zone of diatomaceous shale with interbedded siltstone at the stratigraphic top of the Puente Formation.

Several small faults were observed in bedrock exposures at the Civic Center and Wilshire/Alvarado Station excavations. Known active or potentially active faults were not encountered along the alignment.

Ground Water

Ground water encountered on the project is associated with the Los Angeles Forebay Area of the Central Ground Water Basin and occurs primarily within the younger and older alluvial deposits. The alluvial deposits serve as a pervious free ground water reservoir resting upon the underlying relatively impermeable Fernando and Puente Formations. Ground water levels at the eastern portion of the project near the Los Angeles river were measured within 20 to 25 feet below the ground surface prior to construction and, as stated earlier, site dewatering was performed in this area prior to tunnel and station excavation. Ground water levels throughout the remainder of MOS-1 were generally well below the tunnel invert elevation except for perched ground water at the base of the older alluvium on Contracts A171 and A175 and local seeps within the bedrock along joints and sandy interbeds.

Documentation Of As-built Geologic Conditions

A challenging exercise on the project was the documentation of as-built geologic conditions during tunnel excavation. Conventional tunnel mapping techniques usually involve detailed mapping of tunnel walls as they are exposed during construction and have been discussed in detail in previous papers (Jack, 1968; Proctor, 1971). These techniques are most applicable in hard rock tunnels where walls are open for observation for some period of time. However mapping of walls in soft-ground tunnels can be very difficult if the walls are never exposed for view, as was the case in the shield-driven tunnels used on the Metro Rail project because they were supported by a continuous lining immediately behind the tunneling shield, the geology is never exposed except at the face.

In order to document conditions as they were encountered, a program was developed to map the face of the excavation during tunneling. This would be analogous to mapping the exposed bottom of a 24-inch diameter boring rather than logging the geology exposed on the walls. Mapping of the face was performed at a scale of 1:60 with emphasis placed on soil classification or bedrock descriptions, engineering characteristics, Terzaghi (1950) ground classification, and ground water. Mapping proved to be a difficult task in view of the cramped conditions at the working face and/or the limited time available at the face between or during each four-foot advance. Approximately 2000 face maps were made for the over 26,000 feet of tunnels resulting in an average of about one face map for every 13 feet of tunnel. Photographs were also taken on a regular basis. The as-built conditions were compared to the anticipated conditions on a daily basis and plotted on an as-built geologic profile.

TUNNEL CONSTRUCTION

The tunnel portion of MOS-1 involved construction of nearly 13,000 feet of side by side tunnels with bore diameters on the order of 22 feet. Taken together, the twin tunnels account for over 400,000 cubic yards of excavated soft rock and soils. Ground cover along the entire alignment ranges from as shallow as 12 feet near the eastern portal to over 90 feet in the vicinity of Hill Street and the Hollywood Freeway. The twin tunnels are separated by 7 to 20 feet at the springline and are joined together at regular intervals by emergency cross passageways spaced about 750 feet apart throughout the entire alignment.

Tunnel operations were divided among four major contracts ranging in tunnel length from as little as 730 feet (Contract A130) to about 5300 feet (Contract A141). Access to the level of tunnel operations was provided by a number of access shafts and adits located at strategic locations along the MOS-1 alignment.

Like other similar tunnel projects, tunnel operations on the Metro Rail project can be divided into three major activities — excavation, support of excavations, and protection against ground loss and associated damage to existing structures. The scope of each of the three major operations is dictated largely by ground conditions, proximity to critical structures, and length of tunnel contract. The following sections provide a description of excavation and tunnel support operations as well as a brief summary of some of the more interesting aspects of tunneling and ground control experienced on MOS-1.

Shield Tunnels and Tunnel Support

Soft-ground tunneling was performed using a variety of sophisticated state-of-the-art tunneling machines such as the one shown on Figure 4. Each machine was custom designed or selected for the project based on the anticipated ground conditions and tunnel diameter. The "shield" tunneling machines used on the Metro Rail project consist of cylindrical casings made of thick structural steel to protect the miners during excavation and a backhoe or spade-type excavator for digging at the front or leading edge of the machine. The machines operate by digging at the face with the excavator and dragging the excavated material (muck) onto a belt conveyer located near the face. The excavated soils are then transported on the conveyer belt to a series of muck cars attached to a diesel-powered locomotive which in turn transports the excavated soils to the access shafts where the muck is hoisted to the surface and dumped. The process is repeated throughout the course of mining.

The rear or trailing end of the shield tunneling machines are also provided with a canopied tail section under which the initial tunnel lining can be safely assembled and erected for tunnel support. Once the segments or tunnel linings are in place within the tail section, the shield propels itself forward by expanding a series of hydraulic "shove" jacks against the just installed tunnel lining set. The tunnel lining in turn reacts against the previously installed set and so on. Once the machine has advanced another four feet into the ground, the shove jacks are retracted and another four- foot set is installed in the tail section and the process is repeated.

Tunnel support design calls for construction of both initial (temporary) and permanent support systems with the initial lining later incorporated into the final permanent lining. Initial tunnel support, which is used for temporary support of the excavation during tunneling operations, consists of either pre-cast concrete segments (Fig. 2) or conventional full-circle ribs and lagging. A cast-in- place concrete liner is later added for the final or permanent finish. A 100 mil thick high density polyethylene gas resistant membrane (HDPE) is also incorporated into the lining between the initial and permanent supports. Figure 3 shows a typical tunnel cross section showing the pre-cast concrete and cast-in-place liners with the HDPE sandwiched between the two.

Tunnel Construction History

A171 Tunnels

Contract A171 was the first tunnel section to get underway and involved nearly 5000 feet of twin tunnel alignment extending from the 7th/Flower Station on the east to the Wilshire/Alvarado Station on the west. Tunneling was performed using a 22-foot diameter open-face Mitsubishi digger shield (Fig. 5) custom designed for the project and equipped with a massive backhoe-type excavator for digging at the open end of the machine. Initial tunnel support was provided

Figure 4. Mitsubishi digger shield being prepared for start of A171 tunnels.

Figure 5. A171 tunnel showing initial pre-cast concrete liners. (Photo courtesy of Rail Construction Corporation.)

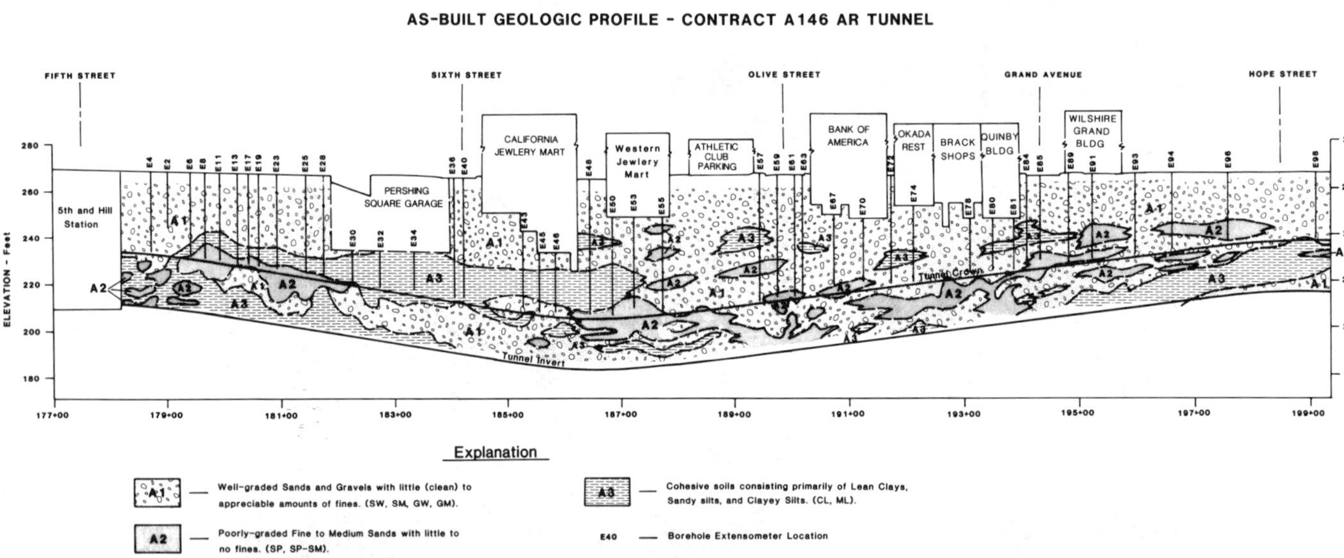

Figure 6. As-built geologic profile for the right or outbound AR tunnel - contract A146

by pre-cast concrete segments which were installed in the tail of the shield with the aid of a ring-gear type segment erector capable of lifting each of the four 7000-pound segments into place. Ground conditions along the tunnel section were nearly ideal for soft-ground tunneling with over 95 percent of the section in soft bedrock and less than 5 percent in mixed-face alluvium and bedrock. The soft bedrock provided a near ideal medium for excavation with the massive excavator while providing stable "firm ground" working conditions at the face.

Tunneling was performed on two eight-hour shifts per day with the third shift used for maintenance, survey work and magnetometer probing. Daily tunnel advance rates for A171 ranged from about 65 to 100 feet per day and were about 33 percent higher than anticipated based on some previous estimates for the Los Angeles area (Yerkes et al. 1977). The AL (left or inbound tunnel) was completed on May 18, 1988 after 182 calendar days of mining while the AR tunnel (right or outbound tunnel) was completed in only 83 calendar days on September 17, 1988.

A146 Tunnels

The A146 tunnel section was the second tunnel contract to get underway and involved 2100 feet of twin tunnels routed through the heart of the downtown area. The A146 tunnels were also excavated with a Mitsubishi shield and excavator similar in design to that used on the A171 tunnels. Unlike the A171 tunnels however, the entire section was excavated in recent alluvial river deposits above the water table. The alluvial materials consisted largely of relatively clean granular sands and gravels with interbedded silt and clay layers as shown on Figure 6. Ground cover along this section ranges from 30 to 60 feet as the tunnels passed under or along side eighteen major structures including several four- to thirteen-story buildings and the subterranean Pershing Square parking structure.

Analyses performed prior to construction based on geologic conditions, ground cover and anticipated tunneling techniques suggested that 2 to 3 inches of surface settlement could be expected where tunnels pass directly beneath buildings if no special protective measures are taken (Hansmire, 1985). Anticipated damage to buildings due to these settlements could range from minor cracks in walls and floors to seriously distressed structural support elements. In order to limit damage due to tunneling, a number of critical buildings were underpinned with deepened foundations prior to construction and a program to perform compaction grouting during mining was planned. The compaction grouting technique makes use of a highly viscous sand-cement grout injected at high pressures to expand and form a grout bulb which recompacts loosened soil thus minimizing building settlements. The grout is injected through pre-drilled pipes which terminate between the tunnel crown and building foundations (see Fig. 7). Over

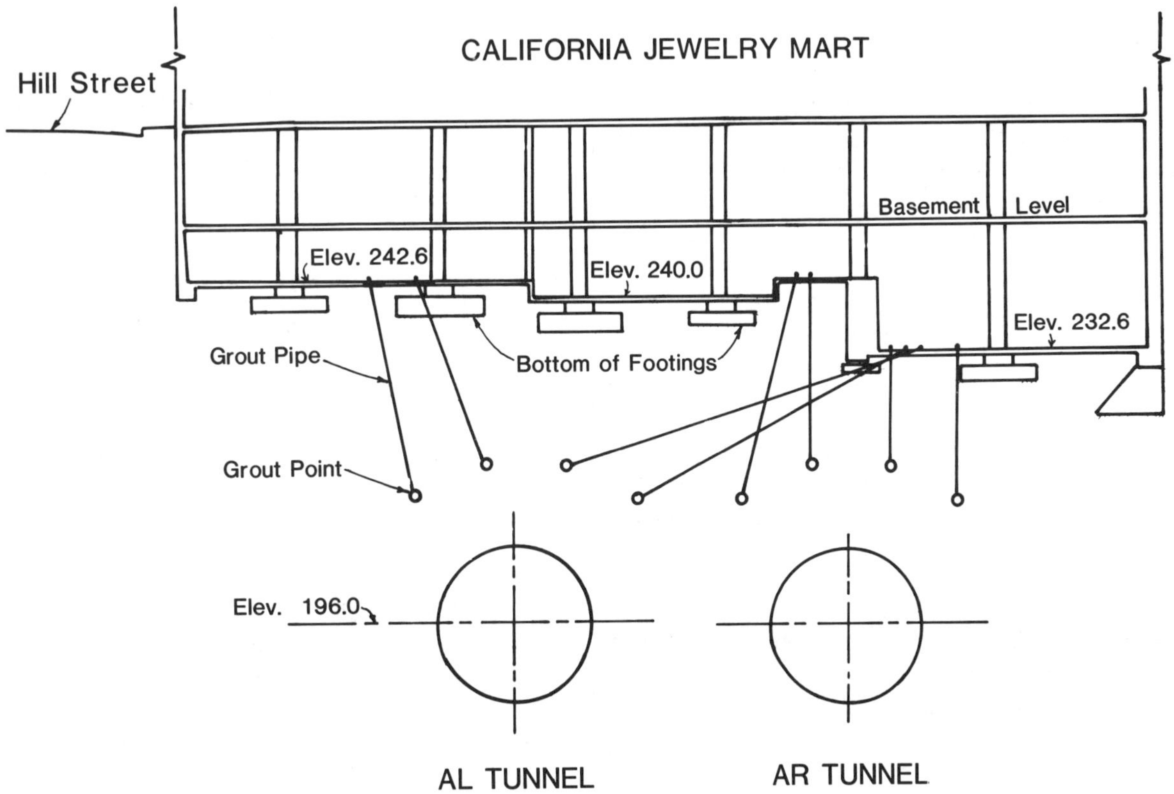

Figure 7. Typical compaction grout scheme - Contract A146

700 grout holes were drilled from the surface to within 5 to 10 feet of tunnel crown.

Chemical grouting was also used in conjunction with the compaction grouting program on Contract A146 where excessive ground losses through the tunnel face were observed to be occurring in cohesionless sand deposits at or above the tunnel crown. The chemical grouting program involved injecting a sodium silicate mixture into the sandy soils ahead of the tunnel through grout pipes drilled through the tunnel face to form a canopy of weakly cemented sandy soils above the tunnel crown. The canopied zone was maintained a distance of 20 to 60 feet in front of the tunneling machine at all times while in the sandy stratum. Over 290,000 gallons of chemical grout was use for both the AL and AR tunnels (Robison and Wardwell, 1991). Drilling and grouting were performed on a daily basis during the swing and graveyard shifts while mining followed during the day shift.

The combined chemical and compaction grouting programs kept structural settlements to acceptable levels and also recompacted loosened ground so there will be little chance of future settlements if the area should be subjected to seismic shaking. Daily tunnel progress for A146 averaged about 40 feet per day.

A141 Tunnels

Contract A141, at 5300 feet, was the longest tunnel contract and also the only contract awarded including both tunnel line and station construction. The tunnel section begins at Union Station heading first northwest and then southwest beneath Hill Street. The tunnel alignment is interrupted by Civic Center Station before continuing southwest beneath Hill Street to the 5th and Hill station. Because excavation for the Civic Center station was completed prior to tunneling, it was necessary for the tunneling machine (and all of its associated trailing gear) to "break" into the station excavation through the northern bulkhead wall and then be ferried along on tracks through the station prior to continuing towards its destination at the 5th and Hill station.

Excavation was performed using an open-faced digger shield, manufactured by the Robbins Company, similar to that used on the A146 and A171 tunnel contracts. In contrast to the two earlier tunnel contracts, the contractor choose to use the more conventional full- circle steel ribs and wood lagging for the initial tunnel support.

Of the four tunnel contracts, geologic conditions along the A141 alignment were the most varied, beginning in saturated younger alluvium near Union Station before transitioning into soft bedrock of both the Puente and Fernando Formations. With the exception of some unexpected sagging of the crown in one of the cross passage excavations, tunneling on A141 was performed without excessive surface settlements or damage to existing structures.

A130 Tunnels

The A130 tunnels, the last tunnel contract to get underway, was one of the shortest tunnel sections on the project but involved some of the most challenging ground conditions. The tunnel section included twin 21-foot diameter tunnels extending some 730 feet northwest from the portal entrance beneath the Hollywood Freeway and the intersection of Commercial and Center Streets. Ground cover ranged from 12 feet near the portal entrance to 20 feet beneath the freeway. The freeway is underlain by about 8 feet of fill which in turn overlies recent alluvium and high groundwater at the tunnel elevation. Steel ribs and wood lagging was the chosen initial support.

Due to the shallow ground cover, potentially unstable conditions, and concern for the Hollywood Freeway a program to stabilize a 530 section of the alignment with chemical grout was implemented prior to tunneling. The grouting program was performed by Hayward-Baker, Inc. of Ventura and included over 2,000,000 gallons of grout (Gularte and others, 1991) and over 400 horizontal and vertical grout holes beneath the freeway as well as the Commercial and Center Street intersection. A 200-foot long stretch of vacant ground between the freeway and intersection was left ungrouted. Site dewatering was also performed in conjunction with the grouting program.

The effectiveness of the grouting program was demonstrated by the limited settlements observed in the grouted zones as well as by the tunnel performance after a major fire erupted in the left or inbound tunnel on July 13, 1990 destroying all the wooden tunnel lining between the steel ribs. The grouted portions of the tunnel remained intact while the 200-foot section of ungrouted ground sections suffered complete collapse. In the aftermath of the fire, regulatory agencies decreed that wood lagging would no longer be permitted in the construction of mined tunnels on the Metro Rail Project in California.

STATION CONSTRUCTION

Construction of the 4.4-mile segment of MOS-1 included five stations as indicated in Figure 1 (see inset). The Civic Center, (Contract A141), Pershing Square (A145) and 7th/Flower (A165) stations were constructed using cut-and-cover procedures while Union (A135) and Wilshire/Alvarado (A175) stations were constructed using open-cut techniques.

Union Station

Union Station, as indicated in Figure 1, is located at the eastern end of the Metro Rail alignment near the Los Angeles River. The site geology is characterized by fill underlain by permeable alluvium to a depth of some 90 feet. The alluvium in turn is underlain by the relatively impermeable Puente Formation. The groundwater level in the surficial alluvium is contaminated and is located approximately 20 to 25 feet below the surface. Contract specifications required

Figure 8. Excavation of Union Station showing exposed slurry walls

that dewatering be performed to five feet below the lowest construction elevation and that the contaminated water be treated prior to discharge. As a result of the site geological conditions and the expected large volume of contaminated water, the Contractor elected to design and construct a deep concrete slurry cut-off wall embedded five feet into the Puente Formation for groundwater control and ground support.

Individual slurry wall panels, each measuring 18 feet long by 30 inches wide, were excavated through the alluvium and into the impermeable bedrock 86 feet to 98 feet below the surface. The panels were excavated using 150-ton cranes outfitted with 29 inch wide, 12 ton clam shell buckets. A bentonite slurry was pumped into the excavated trenches to maintain wall support during the excavation process. Cages, constructed of rebar, were then installed into the completed trenches and the bentonite muds were displaced by 4000 psi concrete using tremie techniques. Alternate slurry wall panels were excavated to allow for the adjacent concrete panels to set. Following completion of the slurry wall support system, the alluvial materials within the perimeter of the excavation were removed using front- end loaders and dump trucks (Fig. 8). As the excavation proceeded downwards, 45-foot to 60-foot long cable type tiebacks spaced vertically 15 feet were installed and tensioned through openings in the slurry wall.

Three "instrumentation load zones" monitored the performance of the support system. Instrumentation included load cells on tiebacks and two inclinometers per "instrumentation load zone". In addition, observation wells both inside and outside the excavation perimeter, and ground and building survey settlement reference points were regularly monitored. The load cell and inclinometer data collected indicated no significant movement of the excavation walls during construction or load build-up within the tiebacks. The slurry wall performed well within the limits specified in the contract.

The use of the tieback/slurry wall system allowed construction clearance within the excavation and, with the assistance of dewatering wells, minimized the inflow of water. No construction delays were incurred due to water problems.

7th/Flower Station

The 7th/Flower Station (A165) located in downtown Los Angeles is a dual purpose station in that the upper level is the terminus for the North-South running downtown to Long Beach light rail Blue Line, and the lower level includes the Metro Rail Red Line. The upper level was completed in February of 1990 and the Red Line is scheduled for completion in 1993. The A165 station was excavated in alluvium underlain by the Puente Formation (Fig. 3). The excavated material was removed to the surface using a mobile vertical belt conveyer system located on the floor of the excavation.

As the material was brought to the surface, it was dumped into a chute and funneled into waiting trucks. Ground support was provided by soldier piles and wood lagging, struts and local tiebacks. Three load zones were instrumented with inclinometers, load cells, strain gauges and observation wells. Ground and building settlement survey reference points were also monitored on a regular schedule. The instrumentation data collected indicated that all movement and load levels were well within design limits.

Civic Center and 5th/Hill Stations

Civic Center Station, (A141) located at Hill and First Street, is 700 feet long by 65 feet wide situated at a depth of 65 to 81 feet below the surface (Fig. 9). This station was excavated through a thin alluvium/fill cover underlain by bedrock materials of the Puente and Fernando Formations using cut-and-cover methods. The temporary support consisted of soldier piles on six foot centers inserted into pre-drilled holes. The piles were installed in structural concrete from the bottom of the excavation to 12 feet below subgrade. From the surface to the subgrade, lean concrete was used to secure the piles in the holes. Deck beams were installed across the excavation for traffic maintenance and were supported on every other pile. The station was excavated using bulldozers and dump trucks using a ramp at the southern end of the excavation. As the excavation proceeded, the lean concrete surrounding the soldier piles was exposed and removed from the face and treated timber lagging was installed between the piles.

Lateral support was provided by two levels of pipe strut cross- bracing in the upper portion of the excavation and three levels of post-tensioned cable tieback anchors in the lower portion (Fig. 10). The two pipe strut levels were installed horizontally at 18 foot spacings and at depths of approximately 10 and 25 feet below the surface. The struts were used for the upper level support in lieu of tiebacks in order to avoid drilling laterally into existing building footings. The 30-inch diameter struts were pre-loaded to 50 percent of the design load and welded to stiff horizontal steel wales using horizontal 1/2 inch thick gusset plates. The lower level tiebacks anchors were installed on six foot vertical centers below the strut levels to complete the support.

Three sections of the excavation were designated "instrumentation load zones" (Fig. 11) and selected to monitor the performance of the temporary support system. Two sections were located near the ends of the excavation and one section was located near the middle. Instrumentation in each load zone included two inclinometers (one on each side of the excavation) installed in the soil immediately behind the soldier piles and three vibrating wire strain gages spaced 120 degrees apart on three consecutive struts on both B and C levels (six in total), two diameters from the ends. Load cells were installed on all three (D, E, and F) tieback levels on both sides of each load zone (18 tiebacks per load zone).

Figure 9. Civic Center Station looking southwest. (Photo courtesy of Rail Construction Corporation.)

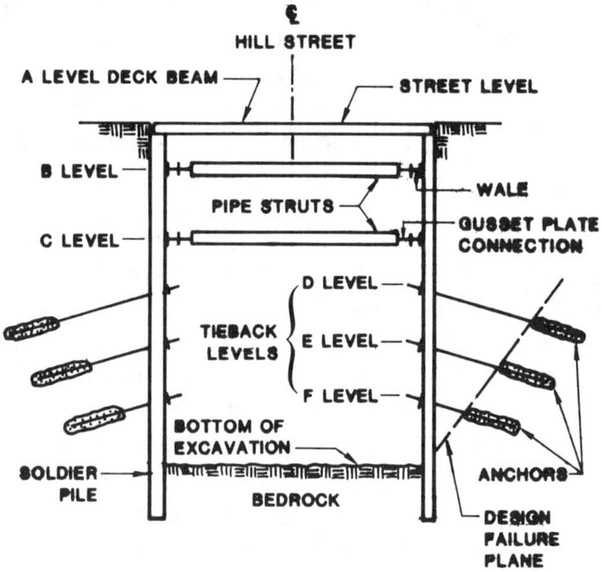

Figure 10. Contract A141 - Typical Cross Section

The deck beams (level A) were not instrumented. In addition, building and ground survey settlement reference points within 100 feet of the perimeter of the excavation were installed and monitored on a regular basis.

The temporary support system was designed by the contractor using specified loading criteria provided in the contract documents. Soil pressure diagrams were specified with lateral earth pressures of 20H (H is the depth of the excavation). In selected areas, 24H was used in order to limit movements of adjacent structures.

During the course of the excavation, a review of the strain gage data revealed that higher than normal loads were accumulating. In addition, inclinometer data indicated movement of walls was occurring. Several struts were accumulating loads close to the design levels well before completion of the excavation. Near the end of the excavation, four C level pipe strut plate connections located between Load Zones No. 1 and No. 2, buckled upwards. The wales they were attached to also were deformed. This observation prompted an immediate closure of Hill Street and an investigation program was initiated. Following much analysis, a solution which included adding an additional pre-loaded auxiliary row (B') of struts, between levels B and C, was implemented. The buckled connection plates and deformed wales were also repaired. It should be noted that the building survey data indicated that little or no settlement had occurred during the load build-up. Following installation of the B' auxiliary level of struts, only minor changes in the load levels were observed in any of the struts. By this time most of the excavating was completed.

The high load built-up in the struts at A141 station was unexpected and a verification study of the phenomenon was implemented. A review of the local area construction data revealed that a combined support system, i.e., struts and tiebacks, in Los Angeles bedrock materials, with instrumentation, was unusual if not unprecedented. Several literature studies were undertaken in order to research any such occurrences from across the country and to provide background data for such phenomenon.

As a part of the verification study, the remaining station excavations along the Metro Rail alignment were examined to see if the observed phenomenon at A141 could be duplicated. Contract A145 (5th/Hill Station) was selected. The northern geologic profile at A145 (Fig. 4) appeared similar to A141 and exhibited up to 70 percent soft bedrock. In addition, struts being installed could be mounted with additional strain gages welded to the struts and monitored closely during the excavation procedure.

The A145 excavation is about 820 feet long, 64 feet with and the depth varies from 68 feet at the north to 58 feet at the southern end. The temporary support system consisted of soldier piles on six foot centers installed into 36 inch predrilled holes. These piles were embedded in structural con-

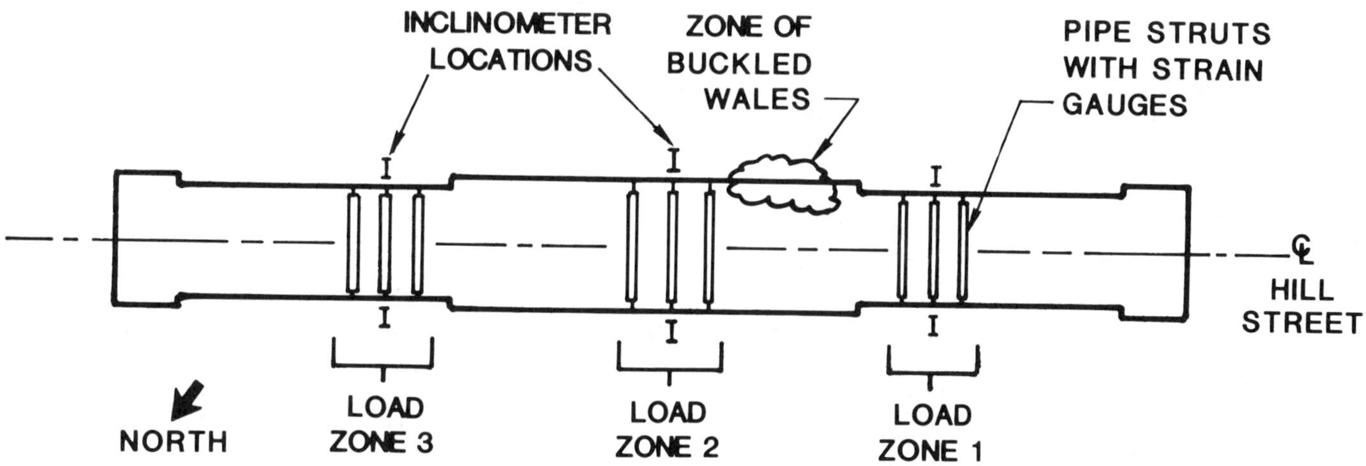

Figure 11. Plan view of Civic Center Station showing load instrumentation zones. Only instrumented pipe struts are shown.

crete 12 feet below the bottom of the excavation. Excavated materials from this station were also removed using a mobile conveyer system located on the floor of the excavation. As the excavation proceeded, four inch by 12 inch timber lagging was installed between the piles. Three levels of pipe struts were installed during excavation and were spaced vertically about 18 feet and horizontally about 15 feet.

Unlike A141, the pipe struts at A145 were connected to continuous horizontal wales with direct strut end bearing (shimmed) connections at all levels. The 28 struts of the northern end (study section) of the station were instrumented with vibrating wire strain gauges. Three strain gauges were welded two pipe diameters from the strut ends at 120 degrees apart. In addition, inclinometers, observation wells and settlement survey points installed previously in the "load zones" were closely monitored on a regular schedule during the excavation.

Although there were similarities between the two sites, it should be noted that A145 had several distinct differences:

- thicker alluvial cover
- 1800 feet south of A141 and up section geologically
- temporary support was provided entirely by struts
- the strut pre-load levels were specified at 25 percent of the design load-except for several struts selected in the northern section which were pre-loaded to 50 percent of the design
- total earth loading per design criteria was less than for A141 due to the lesser depth of excavation

During the course of the excavation, survey data did not indicate any significant movement. However, the strain gauge and inclinometer data observed in the northern end (study section and north load zone) was significant.

The two bottom levels of struts (six struts) in the north load zone were located in the bedrock. Upon completion of the excavation, the inclinometer data indicated movement up to 0.5 inches at the 40 to 45-foot depth and strain gauge data indicated that two struts on the second level had exceeded the load design criteria. Of the 28 struts in the study section, in bedrock, more than 80 percent exceeded the design load criteria, although no noticeable strain in the overall support system due to overloading was observed.

Several hypotheses have been postulated for the higher than normal loads observed at A141 and A145:

- Wedge failure hypothesis
- Creep hypothesis
- Swelling hypothesis
- Load transfer hypothesis

The "Wedge Hypothesis" suggests that adversely oriented bedding or joint planes could cause higher than "normal" earth pressures. Bedding and other structural features were mapped in detail in both excavations and dips were determined to be generally oblique to the excavation walls. This hypothesis was determined to be an unlikely cause of the observed phenomenon.

The "Creep Hypothesis" suggests that a high in-situ horizontal stress regime present in the bedrock due either to compressional tectonic stresses or over-consolidation could be responsible for the load build-up.

The "Swelling Hypothesis" suggests that swelling clays as a result of local water intrusion could be responsible for the high loads. The phenomenon observed at A141 and A145 was not a localized event.

The "Load Transfer Hypothesis" suggests that loads from lower tier tieback anchors at less than 100 percent of design, could transfer loads to the above struts and thereby create an overload situation. However, the strain gauge data collected at A141 during the excavating phase indicated higher than normal loads on some struts occurred even before the tieback levels were reached. The load transfer hypothesis should be ruled out as a possible mechanism at A141 and at the same time does not address the high loads experienced in an all strut situation at A145.

From the above discussion, no real definite conclusion or hypothesis can be accepted without further studies or specific investigations. But at this time, the "Creep Hypothesis" appears to best explain the phenomenon observed. The Los Angeles area is a well documented active tectonic region and high horizontal stress conditions due to tectonic activity should not be considered unusual. In addition, residual stresses from over consolidation would only aggravate the situation further.

Wilshire/Alvarado Station

The instrumentation data observed at the A175 Wilshire/Alvarado Station (Fig. 11) further corroborates the phenomenon noted at A141 and A145. This station is an elongated open-cut excavation of varying width and depth and was constructed in mainly the Puente and Fernando Formations. Temporary support was provided by soldier piles braced entirely by tiebacks at all levels. Three "load instrumentation zones," which included inclinometers and load cells on tiebacks, were utilized to monitor the performance of the excavation. Some of the inclinometers in the load zones indicated up to one inch of inward movement at the 20 to 30 foot depth level. Inward movements of similar magnitudes were observed at the A141 and A145 station excavations.

SUMMARY

Ground conditions encountered along MOS-1 ranged from nearly ideal for soft- ground construction to those requiring numerous innovative techniques to facilitate construction and protect against ground loss and damage to existing structures. The tunneling and station excavation techniques used on MOS-1 have proved to be effective overall in providing for both safe construction operations as well adequate implementation of the design concepts. With relatively few exceptions, the ground response to tunneling and station excavation has been within tolerable levels and damage to existing structures were kept to a minimum.

Many questions still remain unanswered regarding the ultimate cause or causes of the buckled strut connections at Civic Center Station and future studies may be needed to reevaluate the engineering characteristics and expected performance of soft-bedrock type materials, such as the Puente and Fernando Formations, in relation to deep station wall excavations. A review of the performance of differing combinations of lateral support systems may also be warranted.

ACKNOWLEDGMENTS

The authors wish to thank the Rail Construction Corporation and the Los Angeles County Transportation Commission for permission to write this paper.

REFERENCES

Association of Engineering Geologists, 1981, Rapid transit construction costs related to local geology: Proctor, R. J., (ed.), AEG Bulletin Issue of D.O.T. Symposium held at Caltech,, vol. XVIII, no. 2, p. 133-220 (8 papers including LA Metro and BART)

Chase, A.P., Heuer, R.E., Proctor, R.J., and Lindwall, C.E., 1978, Feasibility of tunneling for L.A. Metro subway: report to Chief Engineer, So. Calif. Rapid Transit District, 21 p.

Converse Consultants, Earth Sciences Associates and Geo/Resource Consultants, 1981, Geotechnical Investigation Report, Metro Rail Project: for the Southern California Rapid Transit District, Two Volumes.

Converse Consultants, Lindvall, Richter and Associates, 1983, Seismological Investigation and Design Criteria: for the Southern California Rapid Transit District.

Escandon, R.F., R.C. Lindsay, S. S. Navid, and M. A. Piepenburg, 1989, Tunneling performance and geotechnical aspects of L.A. Metro Rail Contract A-171: Proceedings, Rapid Excavation and Tunneling conference, ASCE/AIME, N.Y., p 641-660.

Gularte, F.B., Taylor, G.E., Monsees, J.E., and Whyte, J.P., 1991, Tunneling performance of chemically grouted alluvium and fill, Los Angeles Metro Rail, Contract A130: Proceedings, Rapid Excavation and Tunneling Conference, AIME/ASCE, NY, p.291-307.

Hansmire, W.H., 1985, Supplemental Report, Influence of tunneling and requirements for building protection, Section A140: for the Southern California Rapid Transit District.

Jack, H.A., 1969, Discussion: Tunnel mapping methods: Bulletin of the Association of Engineering Geologists, Vol. 6, no. 2, p.151-156.

Lamar, D.L., 1970, Geology of the Elysian Park-Repetto Hills Area, Los Angeles County, California: Calif. Div. Mines and Geol., Special Report 101, p. 46

Mayo, R.S., 1982, Shield Tunnels: in Bickel, J.O. and Kuesel,T.R., (eds.) Tunnel Engineering Handbook, Van Nostrand Reinhold Company, NY, p.93- 122.

McCusker, T.G., 1982, Soft Ground Tunneling: in Bickel, J.O. and Kuesel, T.R., Tunnel Engineering Handbook, Van Nostrand Reinhold, NY, p.70-92.

Metro Rail Transit Consultants, 1984, Supplemental Criteria for Seismic Design of Underground Structures: for the Southern California Rapid Transit District.

Murthy, K.N. and Monsees, J.E., 1991, Los Angeles subway - a triumph despite tribulations: Proceedings, Rapid Excavation and Tunneling Conference, AIME/ASCE, NY, p.584.

Monsees, J.E., 1988, Los Angeles Metro Rail project-engineering challenges: World Tunneling, London, March, p.68-73.

Navin, S. J., 1991, HDPE as a gas and water barrier, L.A. Metro Red Line expansion: RETC conference proceedings, AIME-ASCE, N.Y. p, 253-265

Polacek, M.L., Navin, S.J., and Daugherty, C.W., 1990, Use of magnetometer to detect abandoned oil well casings in L.A. Metro Red Line project, APTA 1990 Rapid Transit Conference.

Proctor, R.J., 1971, Mapping geological conditions in tunnels: Bulletin of the Association of Engineering Geologists, vol. VIII, no. 1, p.1-43.

Proctor, R.J., 1973, Geology and urban tunnels - including a case history of Los Angeles: in Geology, Seismicity, and Environmental Impact, Association of Engineering Geologists, Los Angeles, CA, p.187-199.

Proctor, R.J. and Monsees, J.E., 1985, Metro rail project design issues related to gassy ground, Proceedings, Rapid Excavationand Tunneling Conference, AIME/ASCE, NY, vol.1, p.488-505.

Proctor, R.J., Monsees, J.E., and D. G. Hammond, 1985, L.A. Metro rail project, overview and challenges: Underground Space, Pergamon Press, vol. 9, p. 120-125

Robinson, M. J., T. J. Lyman, and W. A. Ruemmele, 1989, Ground response to tunneling, L.A. Metro Rail Contract A-171: Proceedings Rapid Excavation and Tunneling Conference, AIME/ASCE, NY, p.624-640

Robison, M.J., and Wardwell, S.R., 1991, Chemical grouting to control ground losses and settlements on Los Angeles Metro Rail Contract A146: Proceedings, Rapid Excavation and Tunneling Conference, AIME/ASCE, NY, p.179-195.

Stirbys, A.F., Navin, S.J., Louis, S.K., and Roth, W.H., 1989, Design and construction of access shaft and cross adit, Los Angeles Metro Rail Tunnel Contract A146: Proceedings, Rapid Excavation and Tunneling Conference, AIME/ASCE, NY, p.604-623.

Terzaghi, K., 1950, Geologic aspects of soft ground tunneling, in Applied Sedimentation, Wiley, NY, Chapter 11.

Yerkes, R.F., Tinsley, J.C., and Williams, K.M., 1977, Geologic aspects of tunneling in the Los Angeles area: U.S. Geol. Survey, Map MF-866, 67p.

NORTH OUTFALL REPLACEMENT SEWER (NORS) TUNNEL

Timothy P. Latiolait
Principal Engineering Geologist

Richard F. Escandon
Project Engineering Geologist

Arnie K. Hammock
Project Engineer

Law/Crandall, Inc.
200 Citadel Drive
Los Angeles, CA 90040,

Keith G. Farrell
Director of Geological Services
EMCON Southwest
3300 N. San Fernando Blvd.
Burbank, CA 91504.

ABSTRACT
The North Outfall Replacement Sewer (NORS) is an 8-mile-long tunnel extending from the Baldwin Hills to the Hyperion Treatment Plant in El Segundo. NORS generally parallels the North Central Outfall Sewer constructed in the mid-1950s. It extends beneath Los Angeles International Airport (LAX) where tunneling was carefully controlled. NORS encountered continental and marine deposits ranging in age from early Pleistocene San Pedro Formation to Holocene alluvium and man-made fill. Geologic structural features in the area of NORS consist of northwest-trending faults and folds, including the Newport-Inglewood fault zone, the Overland Avenue and Charnock faults, and the Gardena syncline. Geotechnical studies for NORS were conducted from 1987 to 1989 and included drilling 144 borings to define geologic, soil, and ground water conditions along the tunnel alignment. Additionally, a preliminary environmental site assessment was conducted to identify areas of potential soil and/or ground water contamination. Construction of the NORS tunnel project began in November 1989, and was conducted from two primary staging areas, or portals. Soft ground tunneling was performed using various machines and ground support systems depending upon tunnel diameter and anticipated ground conditions. NORS was completed in 1992.

INTRODUCTION

The North Outfall Replacement Sewer (NORS) is an 8-mile-long tunnel project that consists of a main tunnel conveying sewage to the Hyperion Treatment Plant, plus four diversion lines connecting the main tunnel to major existing sewers. All tunnels are in soft ground and are lined with precast or reinforced concrete.

The NORS tunnel begins at the north end of the Baldwin Hills, in Culver City, and extends south to southwesterly through the Baldwin Hills, under the San Diego Freeway (Interstate 405) and Centinela Creek, beneath the residential and business districts of the Westchester District of the City of Los Angeles, beneath Los Angeles International Airport (LAX), and terminates at the Hyperion Treatment Plant at Imperial Highway and Vista Del Mar. The sewer alignment is shown on Figure 1.

The sewer diameter varies from 90-inch to 150-inch internal diameter. Construction of NORS included 4 diversion units (designated Diversions 1 through 4); 5 junction structures; 4 access structures; 18 manholes to tunnel (42 to 105 feet deep); and numerous permanent and temporary ancillary structures and related facilities.

All tunnel construction operations were performed through two portal areas consisting of 72- to 75-foot-deep shafts. One portal, designated South Portal, was located just north of LAX and Lincoln Boulevard. The second portal, North Portal, was located in the Inglewood Oil Field north of West Los Angeles College.

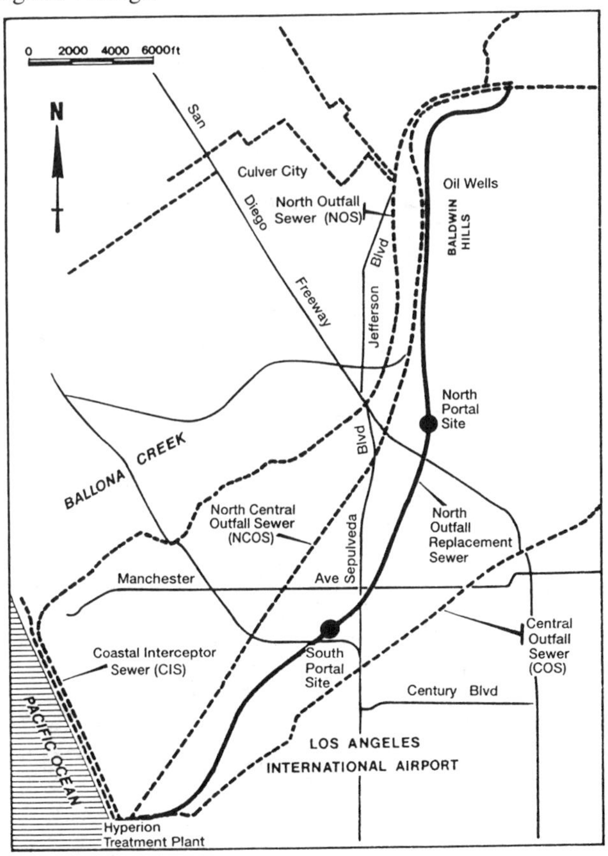

Figure 1. Location Map

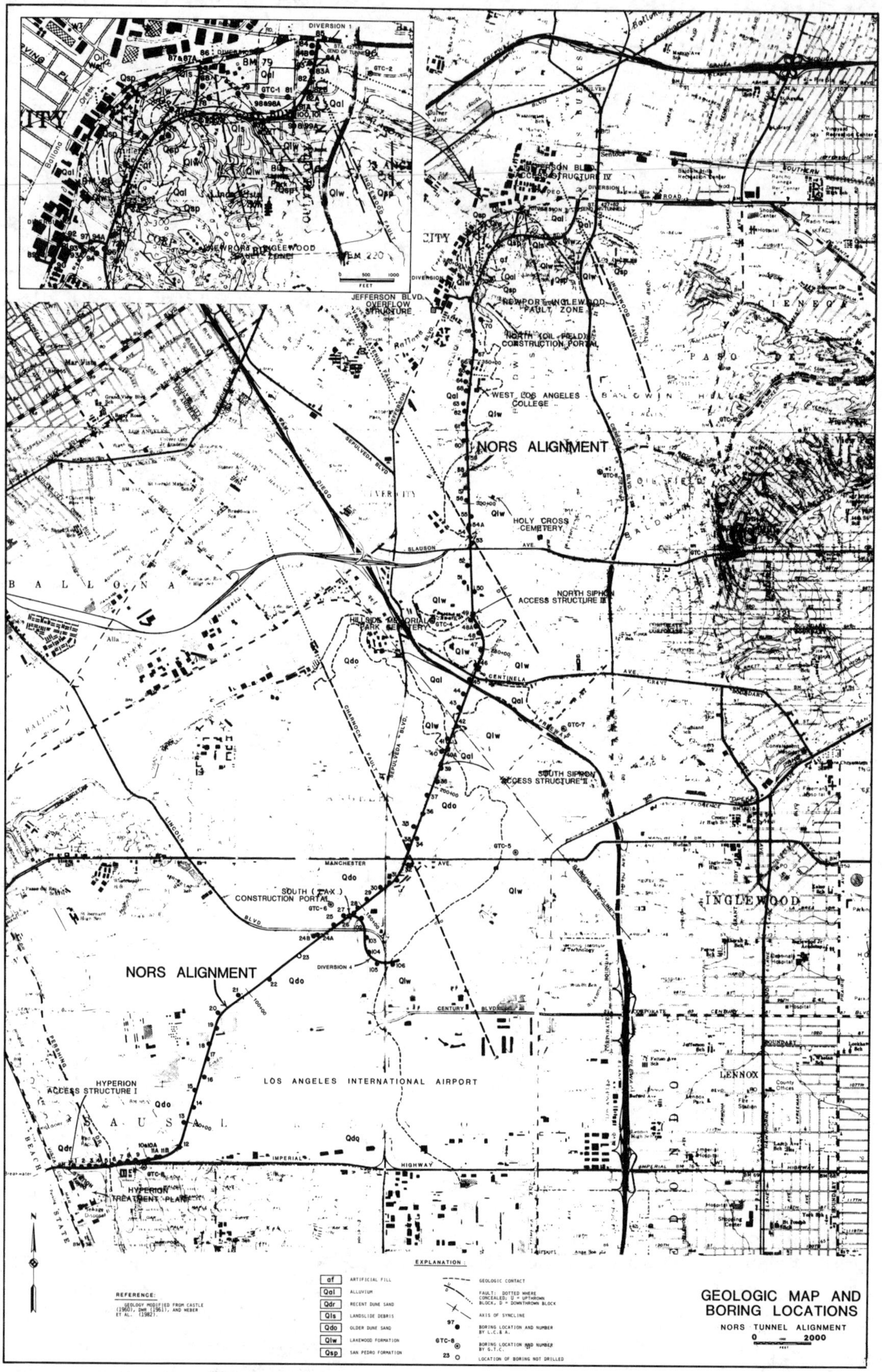

Figure 2. Geologic Map and Boring Locations

GEOLOGIC SETTING

NORS is in the Los Angeles basin, a northwest-trending alluviated lowland, sometimes called the coastal plain (Mendenhall, 1905). The Los Angeles basin, located at the northern edge of the Peninsular Ranges and abutting the southern margin of the central Transverse Ranges, is a deep structural depression into which an estimated 32,000 feet of marine and fluvial sediments have been deposited since Late Cretaceous time. The present form and structural relief of the basin was largely established during late Miocene through early Pleistocene time, when sediments were uplifted and folded along the north, east, and southwest flanks of the basin (Yerkes et al., 1965).

The topography in the vicinity of NORS is dominated by the Baldwin Hills, an uplift along the Newport-Inglewood fault zone. These hills rise 513 feet above sea level and 400 feet above the adjacent alluviated plain. The north and west sides of the Baldwin Hills are marked by a moderately high escarpment, whereas the south and east sides slope more gently. Other landform features in the area of NORS include Ballona Gap and the El Segundo Sandhills. The gap, 1.2 to 3 miles wide, was cut into a late Pleistocene marine surface by the westward-flowing ancestral Los Angeles River.

The El Segundo Sandhills consist of a coastal belt of recent and ancient, or older, dunes and sandhills that extend 3 to 5 miles inland (Poland et al., 1959) from Ballona Gap to the Palos Verdes Hills. The recent (active) dunes, about 0.5 mile wide, lie adjacent to the coast and attain a maximum height of 185 feet above sea level. The stabilized older dunes constitute the main part of the El Segundo Sandhills and rise 245 feet above sea level.

Geologic Units

The NORS tunnel extends through geologic units that range in age from early Pleistocene to Holocene. The areal extent of these units is presented on Figure 2. The subsurface distribution of the units along the alignment is presented on Figure 3.

Pleistocene Units

The early Pleistocene age San Pedro Formation constitutes the oldest materials along the NORS alignment. This formation underlies most of the Los Angeles Basin as well as the Santa Monica and San Pedro shelves offshore. Within the project area, the San Pedro Formation outcrops in the Baldwin Hills and is composed primarily of marine grey to blue-grey silt, clay, and fine-grained sand with occasional lenses of coarse sand and gravel. These units are generally massive and well-consolidated.

Unconformably overlying the San Pedro Formation are late Pleistocene deposits of the Lakewood Formation. The Lakewood Formation is exposed in the Baldwin Hills and the Fox

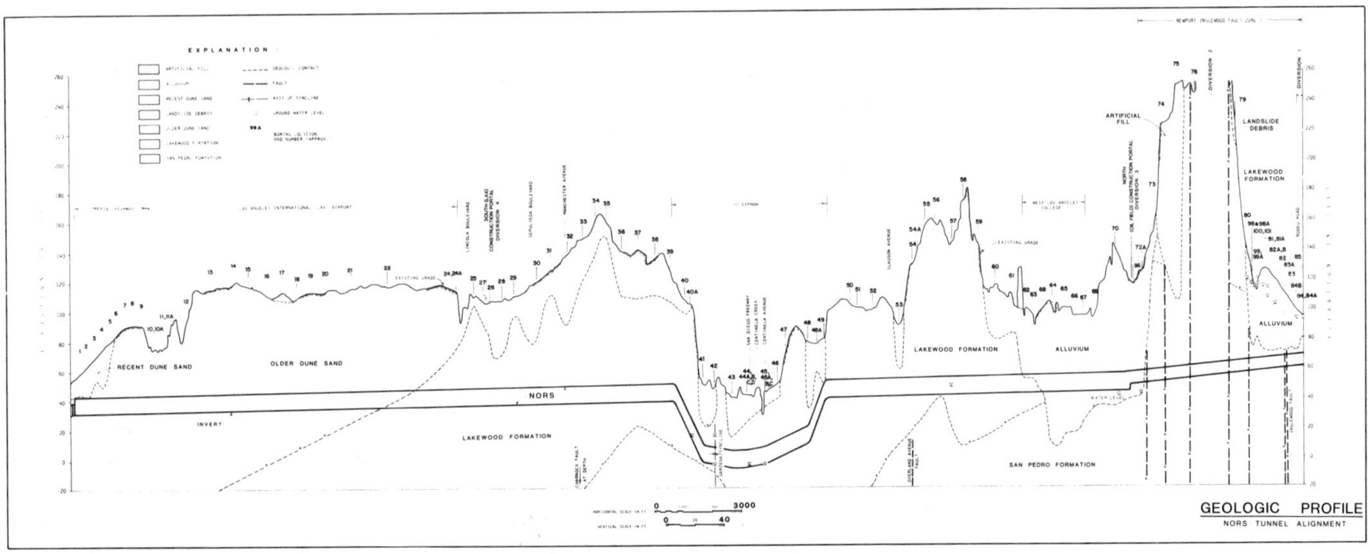

Figure 3. Geologic Profile

Hills area of Culver City and underlies the El Segundo Sandhills beneath Westchester. This formation constitutes the primary material encountered in the tunnel and is composed of marine to non-marine, poorly consolidated, crudely stratified to occasionally cross-bedded sand and gravel with lenses of silt and clay.

Late Pleistocene age dune sand, referred to as older or ancient dune sand, occurs in the Los Angeles Basin along the Santa Monica Bay coastline, south of the Ballona Escarpment. The older dune sand, composing the majority of the El Segundo Sandhills, formed as ancient offshore sand bars that were later modified by wind and stream action. These deposits consist of fine to medium sand, silty sand, and where weathered, clayey sand. Additionally, minor silt, clay, and gravel lenses, as well as zones containing abundant shell fragments, have been identified in the older dune sand.

Two ancient landslides, presumably Pleistocene age, have been identified at the northerly tip of the Baldwin Hills (Weber et al., 1982). The smaller of the two landslides, located above the tunnel, was not explored during our investigation. The other landslide occurs upslope of Jefferson Boulevard in the vicinity of Diversion 2 and is composed of highly oxidized grey-brown clayey silt derived from the underlying San Pedro Formation.

Holocene Units

Holocene units include recent, or active, wind-blown dune sand, alluvium, and artificial fill. The recent dune sand occurs as a narrow strip along the coast from Playa del Rey to Redondo Beach. It unconformably overlies the older dune sand and occurs along the tunnel alignment beneath the Hyperion Treatment Plant and in the southwest corner of LAX. The recent dune sand consists of fine to medium sand. Holocene alluvium occurs in the flood plains of Ballona and Centinela Creeks, and in canyons draining westerly from the Baldwin Hills. In general, the alluvium is composed of combinations of sand, silt, and clay with some lenses of gravel, cobbles, and boulders, particularly near the base of the alluvium.

Artificial fill materials occur sporadically along the tunnel alignment. The materials are typically derived from near source materials and, therefore, are similar in composition to the respective underlying units. One fill area of significant depth and extent occurs in the Baldwin Hills beneath the athletic field of Culver City Park. This section of fill is composed of mixtures of sand, silt, and clay with large amounts of construction trash debris, such as wood and concrete.

Geologic Structure

The major geologic structural features in the area of NORS reflect the northwest structural trend that dominates in the Peninsular Ranges. They include the active Inglewood fault, the potentially active Overland Avenue and Charnock faults, and the Gardena syncline.

Inglewood Fault

The Inglewood fault constitutes the northernmost 9 miles of the Newport-Inglewood fault zone. This fault was crossed by the tunnel and trends northwest-southeast through the Baldwin Hills. North of the Baldwin Hills, the fault is buried by Holocene deposits in the Ballona Gap. Exposures of the fault mapped in Ballona Creek, prior to channelization, and in a storm drain excavated along Rodeo Road in the 1930s, indicate the Inglewood fault is marked by a 40-foot-wide shear zone, with a principal "broken zone" approximately 8 feet wide (Holman, 1943). As described by Holman, the fault dips 65 to 85 degrees toward the east, is confined to "argillaceous siltstone" units (presumably of the San Pedro Formation) and overlain by about 8 feet of unfaulted Holocene alluvial sediments.

Overland Avenue Fault

The Overland Avenue fault was first named by Poland et al. (1959) and was located based on water well log correlation and water level information in the Ballona Gap area of the Los Angeles Coastal Plain. The Overland Avenue fault extends northwest-southeast across the West Los Angeles/Culver City area, roughly parallel to Overland Avenue, for an approximate length of 6 miles. The Overland Avenue fault is buried by Holocene age deposits for most of its mapped length and is only approximately located in the Culver City/Los Angeles area. As mapped, the fault parallels the active Inglewood fault. Ziony and Yerkes (1985) classify the Overland Avenue fault as a "capable" fault, or a fault that is considered a candidate for future fault rupture. This criterion was based on the evidence of offset along the Overland Avenue fault within the last 750,000 years.

Charnock Fault

The buried Charnock fault was recognized as a ground water barrier as early as 1939 by the Los Angeles County Flood Control District (Baumann and Laverty, 1940). Poland et al. (1959) subsequently assigned the name "Charnock fault" to the barrier because it presumably passes west of the Charnock well fields in the City of Santa Monica. The fault trends northwest-southeast from approximately the intersection of La Grange Avenue and Sawtelle Boulevard in West Los Angeles, to south of Redondo Beach Boulevard in the vicinity of Gardena. The trace of the Charnock fault is roughly parallel to the trace of the Overland Avenue fault and the Newport-Inglewood fault zone, located 1 mile and 3 miles northeast of the Charnock fault, respectively. For the most part, the Charnock fault is concealed by Holocene deposits.

Gardena Syncline

The Gardena syncline is an elongated, northwesterly-trending, southeast-plunging fold extending from Culver City southeasterly to Long Beach. This fold parallels the Newport-Inglewood fault zone, and the fold axis crosses beneath the NORS tunnel in the area of Centinela Creek. The syncline is best developed in Tertiary and lower Pleistocene sediments and is moderately developed in upper Pleistocene sediments.

Ground Water

Ground water in the vicinity of the NORS project occurs within Pleistocene aquifers associated with the San Pedro and Lakewood Formations, in Holocene aquifers, and as isolated areas of perched water. The aquifers include the Silverado aquifer of the Lakewood Formation (previously known as the "200-foot sand" of Poland et al., 1959), and the Ballona aquifer ("50-foot gravel" of Poland et al.) in the Holocene alluvium. Sediments of the Gage aquifer occur along the tunnel alignment in the siphon area and below Westchester and LAX. The Ballona aquifer materials are present in the area of Diversion 3 and below West Los Angeles College. Silverado aquifer deposits are not present within the depth of the tunnel invert.

Along a major portion of the NORS alignment, the tunnel sections are above any anticipated ground water surface. Exceptions include a portion of the siphon area where perched water and ground water occur within the proposed tunnel zone; the north end of the tunnel alignment where ground water is about 20 to nearly 50 feet above the tunnel crown; and along Diversion 1 and the north end of Diversion 2 where the ground water level is above the top of the tunnel crown. Localized perched zones occur within the Baldwin Hills, but are not easily predicted because of lithologic changes with discontinuous permeable zones in the Lakewood and San Pedro Formations. However, perched water typically occurs near the contacts of the permeable alluvium and Lakewood Formation with the less permeable San Pedro Formation. Perched water above clay lenses occurs in the older dune sand, but is generally seasonal and in small quantity.

GEOTECHNICAL INVESTIGATION FOR NORS

The NORS tunnel for the most part parallels the nearby North Central Outfall Sewer (NCOS), which was constructed in the mid-1950s. As in the case of the NORS study, LeRoy Crandall and Associates was the geotechnical consultant for the 1955 NCOS investigation. We are, therefore, afforded the opportunity to compare the two studies to see how the "state-of-the-art" for tunnel investigation has changed over the approximately 30-year period.

NCOS Investigation

The geotechnical studies for the NCOS sewer, conducted between June 1955 and January 1956, consisted of a preliminary study to evaluate three proposed sewer routes and additional studies once the final alignment was selected. As part of the preliminary study, 28 borings were drilled to obtain soil samples for laboratory testing, with the boring spacing varying from approximately 800 feet to 10,000 feet. After selection of the final sewer alignment, measuring 8.05 miles, samples were obtained from an additional 26 borings that were drilled. Boring spacing for the final alignment, including borings from the preliminary study, was typically 1,000 feet or less; the maximum drilled depth was 348 feet.

Six water level observation wells were installed in the westerly half of the alignment (between the Hyperion Plant and Sepulveda Boulevard) to determine ground water conditions within the planned tunnel. Although perched water conditions were identified in borings drilled in the northerly portion of the alignment (Sepulveda Boulevard to the north end of Baldwin Hills), no observation wells were constructed.

Chemical analyses of three water samples were performed to determine the deleterious effects of the ground water on concrete. The analyses were limited to a determination of the levels of chloride and sulfates in the water.

The geologic conditions along the sewer alignment were addressed in a 10-page preliminary geologic study prepared by Bear and Gardett (1955) for LeRoy Crandall and Associates. The study, conducted before the field exploration, described known fault locations, areas of subsidence within the Baldwin Hills, ground water levels, locations of oil fields, and the nature of the geologic units anticipated along the alignment. Although the geologic report acknowledged the potential for damage to the sewer at its crossing with the active Inglewood fault, no consideration was given to designing the tunnel for possible fault rupture.

NORS Investigation

The geotechnical investigation for the NORS tunnel was conducted from July 1987 through March 1989. It included the drilling and logging of numerous borings, geophysical logging, installation of several water level monitoring wells, aquifer testing, measurement and analysis of in-situ gases, laboratory analyses of soil samples, and chemical analyses of ground water and gases. A Board of Geotechnical Consultants advised on the geotechnical construction aspects of the project. The Board members were Ralph Peck, David Hammond, Wayne Clough, Richard Proctor, and John Witte.

Field Exploration

The NORS sewer alignment was explored by drilling 144 borings to depths of 19 to 300 feet. Twelve of the borings were drilled by Geotechnical Consultants (1988), as part of the preliminary investigation prior to the selection of the final NORS alignment. The typical spacing for the borings was about 500 feet. The borings were drilled with rotary-wash drilling equipment, with some selected borings drilled with bucket auger and hollow-stem auger equipment. Specifically, bucket auger rigs were used to define caving potential, retrieve samples unaffected by drilling fluid, define seepage and water levels, and/or facilitate observations of subsurface conditions. Undisturbed and standard penetration tests were taken at various depths in the borings, with frequent sampling performed within the "critical zone" (defined as being from two tunnel diameters above the tunnel crown to one tunnel diameter below invert). The materials encountered in the borings were classified according to the Unified Soil Classification System and were assigned a geological formation designation (such as alluvium, San Pedro Formation, etc.).

Geophysical logging was conducted (gamma ray and E-logs) in 16 selected borings to more precisely define layering and geologic contacts, to characterize the subsurface conditions, and to supplement the information obtained during drilling of the borings. The data from the geophysical logs included natural gamma ray, spontaneous potential, and resistance curves.

Ground Water Data

The occurrence of ground water along the alignment was evaluated using previous ground water data developed for the area and by direct measurement of water levels in monitoring wells established along the tunnel route. Previous data included information gathered by the Los Angeles Flood Control District (1982, 1985, and 1987), the California Department of Water Resources (1976 and 1988), Poland et al. (1959), and by LeRoy Crandall and Associates (1955) during exploration for the NCOS. The Los Angeles County Flood Control District's data included water level measurements for both shallow and deep aquifers that were regularly measured prior to 1978, at which time water level measurements in the shallow aquifers were discontinued. Water level information presented in Poland et al. (1959) was limited to data gathered prior to 1945 for the Silverado aquifer, and data presented on Water Level Contour Maps of the Torrance-Santa Monica Area for the years 1933, 1941, and 1945. The California Department of Water Resources data consisted of water level measurements obtained from existing water wells located in the vicinity of the NORS project.

To evaluate the ground water conditions along the planned tunnel alignment, 53 wells (consisting of monitoring, pumping, and observation wells) were installed in previously drilled exploratory borings at selected locations along the tunnel route. Included in the 53 wells were three pumping wells and seven observation wells installed for aquifer testing.

Because of anticipated ground water at or above tunnel invert for the siphon area and for the northern end of the alignment, we conducted aquifer testing at three well sites to obtain hydrologic data for dewatering evaluation purposes. The siphon area tests were performed at the locations of Borings 44 and 45; the third test was performed at Boring 100. Each test site consisted of one pumping well (installed in Borings 44, 45, and 100) and at least two additional observation wells constructed no farther than 50 feet from the pumping well. Drawdown data from the wells were used to approximate the transmissivity and storativity of the aquifers.

Seismic Considerations

Research of earthquake engineering literature indicates that tunnels perform very well during earthquake shaking, with only minor damage occurring in tunnels from ground accelerations less than 0.4g and no damage observed when accelerations are less than 0.2g. The major exception is where tunnels have been displaced by surface fault rupture or by earthquake-induced landslide movement. NORS extends over three known fault traces: the Inglewood, Overland Avenue, and Charnock faults, as well as other lesser fault splays of the Newport-Inglewood fault zone. The Overland Avenue and Charnock faults have no evidence of fault movement during the Holocene epoch; accordingly the possibility of these two faults affecting the NORS tunnel during its design life would be very remote. The Newport-Inglewood fault zone, however, could possibly affect the NORS tunnel if a significant earthquake on the Inglewood segment in the Baldwin Hills occurred.

To address concerns related to possible fault rupture problems at the crossing of the Newport-Inglewood fault zone with the tunnel, Dr. Clarence Allen of Caltech was retained by the NORS design engineers. Allen concluded that any near-surface displacement on the Inglewood fault during the life of NORS is very unlikely. However, there is some finite possibility that the NORS tunnel could be offset in association with movement on this fault. For an earthquake with a moment-magnitude of 6.5, Allen postulated that, on the average, such an event would occur about once every 800 years, and that there could be about 8 inches of horizontal displacement associated with fault rupture, with about twice as much horizontal as vertical displacement expected. He also stated that other branches of the Newport-Inglewood fault zone crossing the alignment would be much less likely to rupture than the main Inglewood fault (Allen, 1987).

To accommodate the anticipated offset, the northernmost 300 feet of the tunnel were designed as a 90-inch-diameter sewer line inside the 144-inch-diameter (12-foot-diameter) tunnel bore. The annular spaces between the 90-inch segments and the 144-inch bore were filled with light-weight compressible concrete.

Environmental Concerns

Because of increasing societal awareness of environmental problems and the resulting regulations in the last few years, real and perceived problems at a construction site can result in tremendous financial costs and even bankrupt the project. The majority of larger construction projects now have environmental assessments performed at project sites in addition to geotechnical feasibility and design investigations.

Historically, the principal environmental concern with tunneling projects in Los Angeles has been the presence of methane or other noxious gases. While this was still a major concern for the NORS project, other potential problems also existed due to the urban nature of the 8-mile alignment. The tunnel alignment traversed commercial and industrial areas, a closed landfill, two active oil fields, residential neighborhoods, and an airport, as well as two areas that would encounter ground water and require some amount of dewatering during construction.

The environmental assessment for the project was performed in two phases. The first phase included a historical review of the property along the alignment. This included a review of historical aerial photographs and topographic maps, as well as a site reconnaissance and conversations with long-time residents of the area. After identifying various historical uses, as well as present day use, local building records, fire department records, and other available sources of information regarding past site use were checked for industrial site use or underground storage tanks that could have affected the subsurface area. Because of the depth of the majority of the tunnel, minimal affects were anticipated from contamination by most surficial sources.

Fourteen potential problem areas were identified along the alignment. These included an area of commercial/industrial buildings and a Chevron distribution facility at the north end of the alignment, a landfill, two oil fields, an oil separator facility, historic aboveground tanks (contents unknown), four retail gasoline service stations, and 3 acres of aviation fuel storage at LAX.

The second phase of investigation included selective drilling, sampling and chemical testing in these identified areas, or further review of available investigation data gathered by others at the respective sites (such as the aviation fuel problems at LAX). In addition, ground water sampling and chemical testing was performed in the two general areas of anticipated dewatering. A suite of general mineral and other chemical tests were done in accordance with the California Regional Water Quality Control Board requirements for temporary discharge and National Pollutant Discharge Elimination System (NPDES) permits. During the second phase of investigation, the potential for methane gas in the identified areas was also explored. The significant potential problem areas are summarized in the following sections.

Oil production facilities: The planned tunnel alignment passes through two active oil production fields, one at each end of the project. These are the Inglewood and Hyperion Oil Fields, of which the Inglewood is the more significant of the two fields. Active production in the Inglewood Oil Field began around the turn of the century and continues today. Historic aerial photographs from Whittier College and UCLA depict the oil field with derricks and drilling sumps along the alignment as far back as 1929.

Abandoned oil wells were considered a potential hazard along the alignment. Older, potentially uncharted well locations were considered a hazard due to past abandonment practices that did not require cement seals in the producing zones or near the surface. The old wells represent a potential pathway for methane migration, as well as a potential obstacle during construction. Many older wells did not contain deep cement plugs or grouted intervals in the producing zones. Older wells were commonly filled with drilling mud only and sealed at the surface with a steel well cap. Any encountered wells would have to be reabandoned, and this would stop progress on tunnel construction.

Oil production methods that could affect the subsurface soils as a result of past use include old sumps, drilling ponds, areas of past spills, storage tanks, and separation facilities. In addition, the presence of crude oil in the subsurface (resulting from past production activities, leakage, or spillage) has the potential to produce methane gas as a by-product. Separation facilities associated with oil production separate the pumped product into three phases: water, petroleum, and natural gas. Separation facilities commonly use refined products such as diesel fuel as a cleaning solvent during maintenance operations. Consequently, it is not unusual to find abundant soil contamination from both crude and refined products around such operations.

The tunnel alignment passes close to a separation facility in the Inglewood Field near West Los Angeles College. Historic photographs indicate this facility has been in the same approximate location since at least 1938. No indications of hydrocarbons were found in the boring drilled along the alignment in the vicinity of the separator facility.

Landfills: One landfill is present along the alignment in the athletic field area of Culver City Park. Based on a review of historic photographs, the landfill area was formerly used as an oil well drilling location.

Abundant debris and trash fill were encountered in one boring to a depth of 83 feet. Although no significant hydrocarbons, sludge, or sump pond materials were noted in the borings, the oil field surrounding the site indicates that oil field related materials may have been placed within the landfill. Within the landfill, one boring was converted to a gas monitoring well by installing two vapor wells with perforated intervals at separate levels within the landfill materials. Gas vapor monitoring indicated the presence of low concentrations (25 to 800 parts per million methane) in the vapor monitoring wells.

Commercial/industrial sources: Various commercial businesses and industries along the proposed tunnel alignment posed potential environmental problems to tunnel construction, including four gasoline stations located near the alignment. No evidence of contamination was discovered in borings drilled along the alignment near these potential sources.

The LAX airport fuel dispensing facilities have experienced past leakage. As much as five feet of fuel product has been reported floating on the ground water surface, approximately 1,000 feet from the alignment. The NORS alignment passes under LAX at a point between two known spillage areas within the airport.

There was also a known contamination problem resulting from leaks along a north-south fuel supply line that intersects the alignment. No evidence of hydrocarbon contamination was encountered in the exploration borings at LAX, however. There are a few other industrial buildings and commercial facilities along the alignment that were identified as potential sources of environmental impairment, including eight commercial buildings and a Chevron Dispensing Facility located at the north end of the alignment.

NORS CONSTRUCTION

Construction of the NORS tunnel project began in November 1989 and was conducted from two primary staging areas, the South and North Portals. The purpose of the two portal sites was to allow simultaneous tunneling operations to proceed at both ends of the project.

The South Portal, located just north of LAX, includes two 72-foot-deep, 34-foot-diameter circular shafts linked by 250 feet of hand-mined connector tunnel. The two shafts and starter tunnel permitted tunnel driving in two opposite directions along the main tunnel alignment.

Tunnel operations at the North Portal were conducted primarily through a single 75-foot-deep, 31-foot-diameter vertical shaft. This shaft, along with 285 feet of hand-mined starter tunnel, also provides access to two tunnel headings in opposite directions along the main tunnel alignment. Two secondary 34-foot-diameter shafts, designated Access Structures II and III, were also located at opposite ends of the inverted siphon section (see Figure 3) and were 88 and 61 feet deep, respectively. These secondary shafts provided vertical access for shield modification and retrieval as the tunneling machine entered and exited the siphon section.

Shaft and Starter Tunnel Construction

The early stages of the project involved construction of vertical shafts to provide access to the level of tunnel operations.

Figure 4. Robbins Tunneling Machine with Trailing Gear

The upper portions of the shafts were provided with a 2-foot-wide, 4-foot-deep concrete collar installed around the perimeter to aid in stability and to prevent surface water from entering the shaft. Earth materials were excavated using a backhoe and loader, and hoisted to the surface with a crane and muck box. Circular steel ribs were then lowered in sections into the excavation by crane and assembled around the periphery of the shaft with each successive set hung with tie rods from the previously installed set. Expanded metal linings were then placed between the flanges of the steel ribs to act as a form for backfill concrete and a high-slump pea gravel concrete was pumped between the metal lining and the shaft wall to fill any voids. The process was repeated until the bottom of the shaft was reached.

Starter tunnels were manually excavated using top heading and bench methods, and hand-mining techniques. The horseshoe-shaped tunnels were supported using a lattice-type framework of steel arch ribs, and vertical posts with wood lagging and concrete backfill between sets.

Tunnel Operations

The main tunnel from Hyperion Treatment Plant to the North Portal consisted of a 150-inch-diameter section that makes up over 36,000 feet or 87 percent of the main tunnel alignment. This portion of tunnel was mined in three drives: the drive south from the South Portal to the Hyperion Treatment Plant; the drive north from the South Portal to Access Structure III; and the drive south from the North Portal to Access Structure III. The drive north from the north portal completed the remaining 5,871 feet, or 13 percent, of the main tunnel alignment, and extended north through the Baldwin Hills and then northeast below commercial and residential developments to a junction structure at Diversion 1. The drive north was divided into two sections: the drive north to the "Earth Pressure Balance" (EPB) section, and the EPB section itself. The two sections differed in tunneling methods and tunnel support systems employed.

Tunnel Support and Equipment

Tunnel support for the 150-inch section was provided by both initial and permanent linings. The initial liner provided temporary support during tunneling and consisted of precast concrete segments. Each 4-foot-wide set included four 7-inch-thick segments weighing approximately 4,200 pounds each. The final or permanent liner consisted of a 150-inch-inside-diameter precast reinforced concrete pipe. The interior of the pipe was covered with a polyvinyl chloride (PVC) sheet liner designed to prevent disintegration of the concrete from sulfuric acid. The pipe was installed in 24-foot sections inside the initial liner, with the annulus between the initial and the final liner filled with backfill grout.

Soft ground tunneling was performed using two open-faced excavator shields (see Figure 4), designated Shields No. 1 and No. 2, manufactured by the Robbins Company and

Figure 5. Lovat "Earth Pressure Balance" Tunneling Machine

selected on the basis of anticipated ground conditions. The tunneling machines weigh about 125 tons each and measured 20 feet long with a 17-foot bore diameter. Each machine was designed with articulated joints to aid in steering and was equipped with a paddle-type excavator for digging at the face, twin-arm erectors for lifting and installing the precast segments into the tail of the shield, and 200 feet of trailing gear including conveyer belt and electrical and hydraulic power units. The tunneling machines were propelled by 18 shove cylinders connected to a thrust ring that reacted evenly against the previously installed concrete segments. The thrust systems was capable of up to 7,000 pounds per inch of thrust at the cutting edge.

Ground control at the face was provided by a fixed breasting table at springline and six hydraulically operated breasting doors above springline. Ground below springline was controlled primarily by free raveling of the excavated materials to their angle of repose. Where difficult ground conditions were encountered, the breasting system was modified to include a second breasting table above springline and reconfiguration of the breasting doors.

Initial tunnel support for the drive north to the EPB section was provided by 4-foot-wide sets constructed of steel ribs and full-circle wood lagging. Initial tunnel support for the EPB section was provided by prefabricated, gasketed metal liner plates. The final or permanent liner consisted of 90-inch- and 96-inch-inside-diameter precast reinforced concrete pipe similar in design and construction to that described in the 150-inch section. A crushable cellular concrete, consisting of a light-weight portland cement product, was used as backfill where the EPB tunnel crossed the projected Inglewood fault.

Tunneling was performed using a 12-foot-diameter Lovat closed-face tunnel boring machine (see Figure 5) especially designed to convert to EPB mode when entering into water-bearing ground conditions (EPB section). The machine was 20 feet long and included an articulated cutterhead to aid in steering, and a rotating cutting head equipped with spade-type teeth for excavation. The machine was propelled by twelve 85-ton shove cylinders with a maximum thrust capacity of 850 tons. The Lovat tunneling machine was also equipped with built-in gas detectors, a rib erector and expander, and associated conveyer belts and trailing gear.

Ground control was performed by hydraulically operated flood control doors within the rotating cutting head. The doors were fully closed when the tunnel was not being advanced or individually opened during tunneling. The machine was convertible to EPB mode by activating a system of "pressure relieving gates," which act as a second closable bulkhead within the cutting head. The pressure relieving gates are hydraulically adjusted to open only when pressures exceed a preset earth pressure within the cutting head and in front of the machine. As the machine was thrust forward, the pressure increases, allowing the pressure relieving gates to open and muck to pass from the cutting head to the conveyer belt.

Diversion 2, 3, and 4 Tunnels

The remaining tunnels on the project included the Diversion 2, 3, and 4 tunnels. These tunnels were 970, 1,413, and 2,810 feet long, respectively, and were excavated using a 12-foot-bore-diameter open-face shield. Excavation was performed with both a small backhoe-like digger similar to the Robbins machine and a rubber-tired mucker. Tunnel support was provided by steel ribs and wood lagging similar to that used in the north drive from the North Portal and a reinforced concrete pipe for the final liner.

Comparative Geologic Conditions

Geologic conditions encountered during tunneling were documented by periodic mapping of the tunnel face and logging of instrumentation holes drilled along the alignment. Over 1,100 face maps were prepared for the main and diversionary tunnels. A typical face map is presented as Figure 6. Geologic data contained on the face maps were plotted on a continuing, as-built geologic profile and compared with anticipated conditions as the tunneling progressed.

Instrumentation

The instrumentation program used on the NORS project was designed to inform the contractor and the engineer of ground movements within the vicinity of the tunnels. The instruments measured settlement in the soil mass within the influence of the tunnel as the ground movements propagated toward the surface or structures.

Settlement measurements were used to verify whether the tunneling procedures were maintaining ground movements within the specified criterion and to provide early warning of adverse trends. Measurements were also used in determining the source of excessive lost ground and the degree to which protective or remedial measures were limiting damage to structures. The performance and structural integrity of the tunnel lining system were also evaluated with instrumentation measurements.

A preconstruction survey was conducted to document the existing condition of all buildings and structures within the zone of influence of construction. Diagrams, photographs, and videos were made with details being given to the number, length, and width of cracks, condition of door and window jams, locations of water marks, and condition of paint.

Instruments were installed to monitor ground movements as tunnel construction passed beneath or near selected locations. These locations included critical structures and utilities, where major changes in the soil stratigraphy existed, and critical alignment conditions. Test sections were also in-

CONSTRUCTION INSPECTOR'S DAILY JOB REPORT
TUNNEL FACE/GEOLOGY REPORT

JOB TITLE		JOB No.
NORTH OUTFALL REPLACEMENT SEWER	SOUTH DRIVE FROM NORTH PORTAL	EXX-31392

DAY: THURS	SHIFT: GRAVEYARD	TIME: Start: 0200 End:	ACTIVITY ID: 09600

Description of Work Done, Contractor, Limits, Quantities, Sampling, Instructions to Contractor (See Bureau Manual, B1-3 Records & Reports)

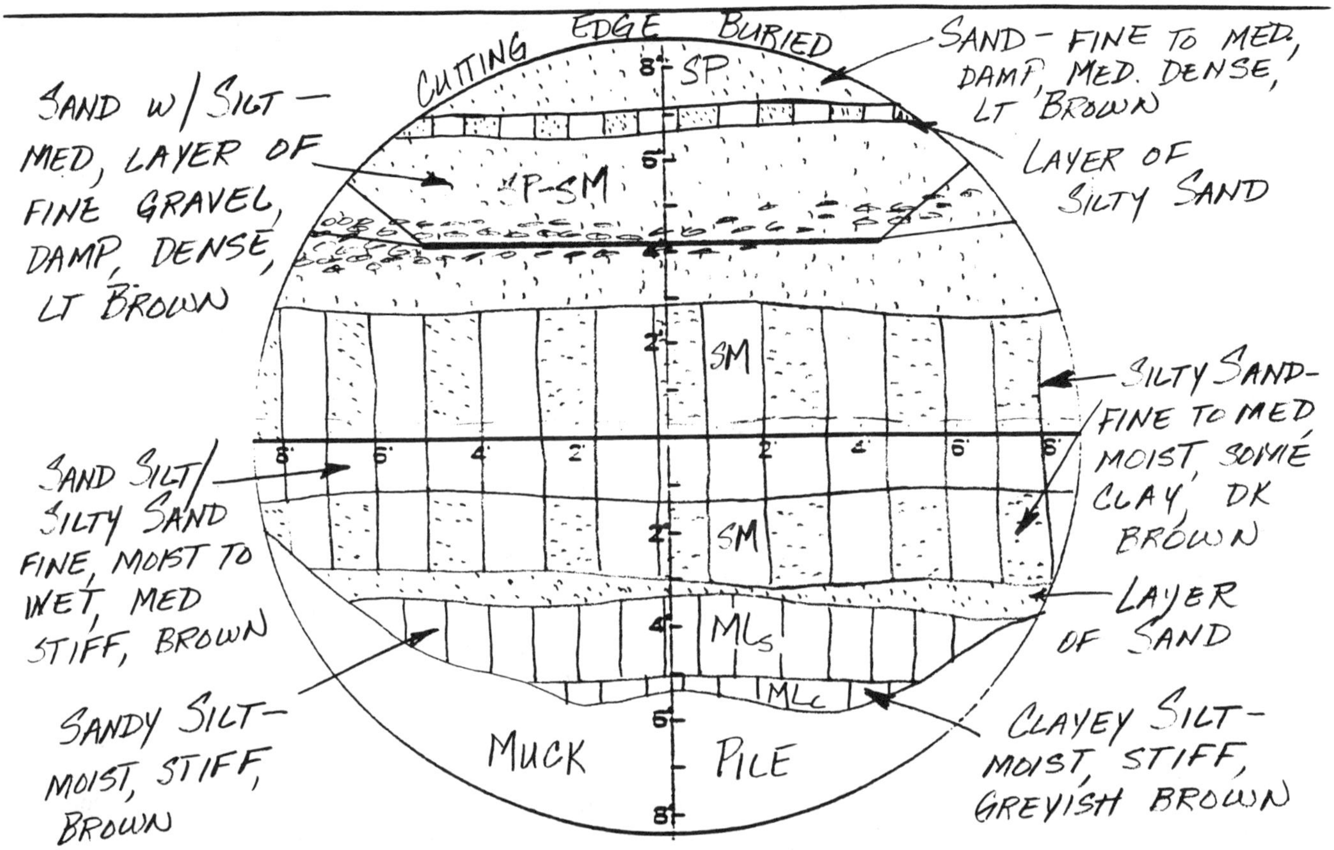

FACE STATION: ______+______ LAST RING SET: # 131

• FACE STANDS NEAR VERTICAL: UPPER TABLE HAND MUCKED

• NO GROUNDWATER OBSERVED

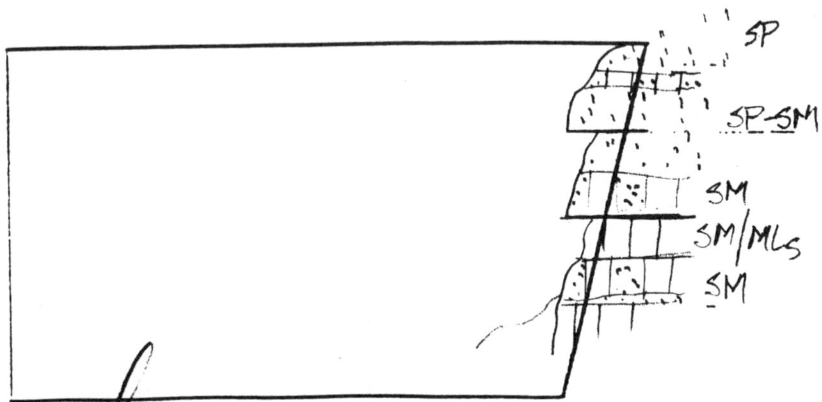

PREPARED BY: A. King Hammack

Figure 6. Typical Face Map

stalled at the South Portal for development of soil response data during start-up operations. Surface-type monitoring devices used on the NORS project included pavement settlement markers, building markers, and vibration monitoring devices. The pavement settlement markers were usually installed on centerline and offsets with extensometers. Subsurface-type monitoring devices used included extensometers, extensometer/inclinometers, and piezometers.

In all, 58 extensometers were installed at selected locations along the centerline of the NORS project. Of these, 57 were the Sondex model extensometers, and 1 was a Boros-Head multiple position borehole extensometer. The Sondex extensometer was chosen for its ability to measure vertical movement at many points within the soil mass above the tunnel from a single location. Induction rings of the Sondex extensometers were located at elevations of 3, 5, and 7 feet above the tunnel crown (the highest point of soil excavation), and thereafter at 5-foot intervals for the next 30 feet, then at 10-foot intervals to the ground surface. The Boros-Head multiple position borehole extensometer was used as an additional instrument when its mechanical type of anchoring systems was needed. The anchors were located at 3, 13, 23, and 33 feet above the crown.

Combined extensometer/inclinometer instruments were also installed at ten selected locations at 10-foot offsets from the tunnel centerline. These instruments were used to obtain measurements of both vertical and horizontal movements. Induction rings were located at springline elevations and at 5-foot intervals to the surface. Inclinometer data were measured at 10 feet below invert elevation and at 2-foot intervals to the surface.

Instruments were monitored at intervals relative to the position of the shield as it approached and passed instrument locations. Baseline readings were obtained at least 30 days prior to the shield passing within 100 feet of the instrument locations. Readings were then obtained when the shield was 50 feet and 20 feet away, when the leading edge of the shield passed the instrument location, when the middle of the shield passed, when the tail of the shield passed, and when the tail of the shield was 5, 10, 20, 30, 50, and 100 feet past the instrument location. Instruments located within the test section at the South Portal to obtain start-up information were read after every shove when the leading edge or the tail of the shield was within 40 feet of the instrument's location.

DISCUSSION

The NORS investigation, like the preceding NCOS investigation, provided the information necessary for design and construction of the "soft ground" tunnel driven from the north end of the Baldwin Hills to the Hyperion Treatment Plant in El Segundo. As well as defining the engineering geologic and geotechnical conditions along the tunnel alignment, the investigations have aided in a further understanding of the geologic environment and Quaternary stratigraphy in the northern Los Angeles basin.

The NCOS investigation preceded the NORS investigation by approximately 30 years. In that time, the methodology for deriving the geotechnical and design data by means of exploratory borings has not changed significantly. There has been an increased level of concern, however, related to seismic, ground water, and environmental conditions. Although both the NCOS and NORS tunnels cross the active Newport-Inglewood fault zone, only the NORS tunnel has been specially designed to withstand fault rupture. The heightened awareness of seismic risk can probably be somewhat attributed to the 1971 San Fernando earthquake, which was also the impetus behind the passage of state legislation (such as the 1972 Alquist-Priolo Special Studies Zones Act) and more stringent building design criteria.

Evaluation of the ground water conditions along NCOS was limited to the installation of only six monitoring wells, all located between Sepulveda Boulevard and Hyperion Treatment Plant. No attempt was made to delineate perched water conditions within the Baldwin Hills. In contrast, evaluation of the ground water along the NORS alignment included the installation of 53 wells (spread out along the entire alignment) and aquifer testing to provide data for dewatering.

Within the last few years, soil and ground water contamination has become a major issue. Governmental regulations related to environmental problems, and the high costs of treating contaminated soil/ground water, necessitated an environmental assessment of the NORS alignment. The assessment included chemical analyses of water and gas samples; the majority of the chemical testing was of ground water samples to demonstrate that the ground water quality met the requirements for discharge of water to sewers or storm drains. Soil and ground water contamination was not a concern during the investigation and subsequent construction of NCOS.

ACKNOWLEDGEMENTS

The authors are grateful to Mr. Kao Liao, Project Manager of NORS, and the City of Los Angeles for allowing us to publish this paper. We would also like to thank Ms. Carol Van Natta, who provided invaluable editorial and constructive comments, as well as typing the manuscript; constructive review by Mr. Mike Robison; and preparation of illustrations by Mr. Todd Tostado.

REFERENCES

Allen, C. R., November 26, 1987, letter to Daniel, Mann, Johnson, and Mendenhall, (LosAngeles), pertaining to possible fault offset on the Inglewood fault.

Barrows, A. G., 1984, A Review of the Geology and Earthquake History of the Newport-Inglewood Structural Zone, Southern California: California Division of Mines and Geology Special Report 114: 115 p.

Baumann, P., and Laverty, F.B., 1940, Annual Report of Hydrologic Data, Season 1938-39: Los Angeles County Flood Control District, 215 p.

Bear, T. L., and Gardett, P. H., 1955, Baldwin Hills-Hyperion Sewer Project, Phase I - Geological Study: report for LeRoy Crandall and Associates, 10 p.

California Department of Water Resources, 1961, Planned Utilization of the Coastal Plain of Los Angeles County: Bulletin 104, Appendix A - Ground Water Geology, 181 p.

California Department of Water Resources, 1964, Investigation of Failure, Baldwin Hills Reservoir: 64 p.

California Department of Water Resources, 1976, Hydrologic Data 1974, Southern California: Bulletin No. 130-74, vol. V, 481 p.

California Department of Water Resources, 1988, Hydrologic Data 1985, Southern California: Bulletin 130-85, vol. V, 208 p.

Castle, R. O., 1960a, Geological Map of the Baldwin Hills Area, California: U.S. Geological Survey Open-File Report.

Castle, R. O., 1960b, Surficial Geology of the Beverly Hills and Venice Quadrangles, California: U. S. Geological Survey Open-File Report.

Castle, R. O., and Yerkes, R. F., 1976, Recent Surface Movements in the Baldwin Hills, Los Angeles County, California: U.S. Geological Survey Professional Paper 882, 125 p.

Desai, D. B., Merritt, J. L., and Chang, B., 1989, Shake and Slip to Survive - Tunnel Design: Proceedings, Rapid Excavation and Tunneling Conference, AIME/ASCE, N.Y., p. 13-30.

Geotechnical Consultants, Inc., 1988, Preliminary Geotechnical Report, North Outfall Replacement Sewer, for the City of Los Angeles, 2 vol.

Hauksson, E., 1990, Earthquakes, Faulting, and Stress in the Los Angeles Basin: Journal of Geophysical Research, vol. 95, no. B10, p. 15,365-15,394.

Holman, W. H., 1943, Fossil Localities in Southwest Part of Hollywood Quadrangle, California: unpublished data.

LeRoy Crandall and Associates, 1955, Report of Foundation Investigation, Phase II, North Central Outfall Sewer, for the City of Los Angeles: 9 p.

LeRoy Crandall and Associates, 1955, Report of Foundation Investigation, Phase III, Unit I, North Central Outfall Sewer, for the City of Los Angeles, 14 p.

LeRoy Crandall and Associates, 1956, Report of Foundation Investigation, Phase III, Unit II, North Central Outfall Sewer, for the City of Los Angeles, 13 p.

LeRoy Crandall and Associates, 1989, Report of Geotechnical Investigation, North Outfall Replacement Sewer, for the City of Los Angeles, Books A and B.

Los Angeles County Flood Control District, 1982 and 1987, Hydrologic Report, Hydraulic and Water Conservation Divisions.

Los Angeles County Flood Control District, 1985, Ground Water Contour Map for Deep Water Aquifers, Fall 1985: map scale 1:63,350.

Mendenhall, W. G., 1905, Development of Underground Waters in the Western Coastal Plain Region of Southern California: U.S. Geological Survey Water Supply Paper 139.

Poland, J. F. and Piper, A. M., 1956, Ground Water Geology of the Coastal Zone of Long Beach-Santa Area, California: U.S. Geological Survey Water Supply Paper 1109, 162 p.

Poland, J. F., Garrett, A. A. and Sinnott, A., 1959, Geology, Hydrology, and Chemical Character of Ground Waters in the Torrance-Santa Monica Area, California: U.S. Geological Survey Water Supply Paper 1461: 425 p.

Proctor, R. J., 1973, Geology and Urban Tunnels - Including a Case History of Los Angeles: Geology, Seismicity, and Environmental Impact: Association of Engineering Geologists Special Publication, Los Angeles, p. 197-199.

Smith, M., 1991, North Outfall Replacement Sewer: North American Tunneling, World Tunnelling and Subsurface Excavation, vol. 4, no. 4, London, p. N4-N11.

Weber, F. H., Jr., Hsu, E. Y., Saul, R. B., Tan, S. S. and Treiman, J. A., Slope Stability and Geology of the Baldwin Hills, Los Angeles County, California: California Division of Mines and Geology Special Report 152, 93 p.

Yerkes, R. F., McCulloh, T. H., Schoelhamer, J. E., and Vedder, J. G., 1965, Geology of the Los Angeles Basin, California - An Introduction: U.S. Geological Survey Professional Paper 420-A, 57 p.

Ziony, J. I., and Yerkes, R. F., 1985, Evaluating earthquake and surface faulting potential. In Ziony, J. I. (editor), Evaluating Earthquake Hazard in the Los Angeles Region - An Earth Science Perspective: U.S. Geologic Survey Professional Paper 1360, 505 p.

CASE HISTORIES OF TUNNELS IN SOUTHERN CALIFORNIA

SAN JACINTO TUNNEL

Thomas F. Thompson
Deceased.
Formerly Consulting Engineering Geologist,
Burlingame, CA

The driving of the 13-mile San Jacinto Tunnel, located near Banning, California, proved to be the most difficult project involved in the construction of the Colorado River Aqueduct of the Metropolitan Water District of Southern California.

Pre-construction geologic studies of the San Jacinto Tunnel alinement in the early 1930s consisted of surface reconnaissance of the area through which the line was projected using the only topography then available — U. S. Geological Survey quadrangles at a scale of one inch equals one mile - supplemented by core borings spaced at comparatively wide intervals along the line. Resistivity geophysical studies were made on the vicinity of both ends of the tunnel to determine depths of landslide materials or other overburden as would influence the establishment of portals.

These early investigations led to the conclusion that hard, mostly massive granitic rocks of Cretaceous age, including granodiorite, quartz monzonite and quartz diorite, would be the predominant type through which the tunnel would be driven, but that at least one long section of Paleozoic or older metamorphic rocks (mica schists, quartzites, and crystalline limestones) and possibly several smaller bodies of these materials would be encountered. Although some suggestion of faulting was noted, it then was thought that the faults would not represent serious obstacles to advancement of the work since no one visualized the tremendous volumes of water that were later found to be associated with these structural features. (Fig. 1,2).

Early Water Difficulties

In early construction stages, excavation had advanced only 160 feet east from the bottom of Potrero shaft when a shear zone was encountered. After penetrating a short distance into this fault a heavy flow of water, estimated at 7,500 gpm, suddenly surged into the heading accompanied by over 1,000 cubic yards of rock debris. The limited sections of tunnel east and west from the shaft were soon flooded and the water filled the 815-foot shaft to within 150 feet of the surface.

Detailed Geological Studies

As a result of the flooding of the Potrero headings, a detailed geologic study of the tunnel line was initiated in 1935 to de-

Figure 1. Typical working conditions at a top-bench heading in the San Jacinto Tunnel. Note 5-piece crown set.

Figure 2. Sudden inflows of water at a heading similar to this one attained a maximum volume of 16,000 gpm in the San Jacinto Tunnel.

termine whether additional faults of similar nature lay ahead of the various headings.

Aerial photos were obtained, initially giving a coverage of about one mile in either side of the line, and later expanded to include an area of about 600 square miles. Office stereoscopic studies of the contact prints were made, followed by field location and verification of geologic features on the photos. The results of field mapping were transferred to mosaics at the same scale, and geologic sections were prepared illustrative of conditions to be encountered ahead of the tunnel faces.

As a result of these studies, 21 faults, striking northwesterly across undriven portions of the tunnel line, were located and several segments of east-west faults were mapped, all of which represented potentially difficult ground at their tunnel-level projection. The attitude of the faults, as determined on the surface, proved to be unusually consistent with depth. As the tunneling was advanced, it was found that water occurred in open interconnected fractures and joints in the hanging (northeast) wall of the northwest-striking faults. Out in the hanging wall the rock became increasingly tight away form the actual zone of shearing. When tunnel headings approached northwest-striking and easterly-dipping faults from the east, water under high pressure was generally developed at some distance away from the fault plane in hard but somewhat fractured rock. As the heading neared the fault the open fractures were more numerous and volume of inflow increased. Normally by the time the footwall was reached, where gouge and crushed material was present, the hydrostatic head had generally been reduced to the extent that little difficulty was encountered in driving through the soft, faulted material characteristic of the actual planes of slippage. Conversely, when headings were being advanced from the southwest side of the northeasterly-dipping faults, (through their footwalls) large flows of water under very high pressures were freed instantaneously, resulting in strong surges that carried large quantities of broken rock, sand and clay gouge into the tunnel. This feature is best exemplified by the experience in penetrating the Goetz fault zone from the west (footwall) side in May 1937. A heavy surge of water carrying some 3,000 cubic yards of sand poured through the ruptured diaphragm represented by fault gouge, and appreciably delayed tunnel progress. As indicated on the total inflow hydrograph the peak flow from all headings approximated 40,000 gpm, with a maximum flow from one point of 16,000 gpm. Water pressures of 600 psi were measured in a few instances and pressures ranging from 150 to 300 psi were common.

Heroic Measures Taken

Advance knowledge obtained from combined geologic and hydrologic studies of the surface trace of the faults known to lie ahead of tunnel faces materially assisted in planning methods of attack as would reduce tunneling costs, and even more important, prevent excessive delay. "Heavy ground" tunneling procedures and methods of contending with underground water in unprecedented volume were initiated or largely owe their development to the ingenuity of those who were in charge of the driving of the San Jacinto Tunnel. Some of the more outstanding of these developments were:

a. The use of "feeler" holes drilled horizontally out ahead of the face to determine the presence of high pressure water or soft ground in advance of the heading. These were particularly useful for "pin-pointing" the location of faults that had been projected form the ground surface far above.

b. The use of pioneer tunnels, smaller in section than the main bore, which were driven parallel to and ahead of the main tunnel when difficult driving ground was known to lie ahead of the face. These reduced water pressures which otherwise would have been difficult to hold on the face of the larger main heading and also permitted attacking the faulted section from both sides (by means of crosscuts to the main tunnel line after the fault was pierced). A further advantage was gained from the two added faces that would result from the crosscut.

c. The use of "detours" whereby the alinement would be temporarily departed from to enable crossing a particularly obnoxious faulted zone at right angles to the strike, after which the original alinement would be resumed and the tunnel line straightened.

d. Techniques for grouting off water ahead of the driving faces through the injection of neat cement grout in long diamond or percussion-drilled holes under exceptionally high pressures (up to 1,500 psi). By so doing the open fissures in which the water was present were sealed off and the rock was consolidated so that driving difficulties were reduced.

e. The use of steel reinforced gunite "rings" in zones of heavy ground where inflows continued after the face had been advanced. These were constructed in circular section (in contrast to the conventional horseshoe shape of the tunnel) and permitted the shutting off of water and stabilization of heavy ground by use of high pressure consolidation grouting before placing the conventional tunnel lining.

f. The development of procedures for grouting through bulkheads (either concrete or sand bag) when water inflows and the character of the rock were such as to obviate the possibility of the customary means of grouting (through drill holes drilled from a rock face).

g. The extensive use of gunite where the rock had tendencies to air-slack and spall, thereby eliminating need for costly steel support and wood lagging. Over 1,600 cubic yards of gunite was placed in approximately 7,600 feet of tunnel.

In addition to the foregoing described measures for detecting and correcting rock defects and adverse water conditions, the constructors of the San Jacinto Tunnel are credited with bringing about innumerable other advances in the art of tunneling, such as methods of drilling, blasting, mucking and timbering as well as making important contributions to underground safety practices.

Reprinted from Engineering Geology in Southern California, 1966, Association of Engineering Geologists publication (out of print).

References

Burkholder, J.L, 1935, Methods and progress of tunnel construction: Civil Engineering, vol. 5, no. 2.

Corner, D. C., 1938, Modern speed in tunnel driving as demonstrated on Colorado River Aqueduct: Explosives Engr., vol. 16, p. 241-245.

Henderson, L. H., 1939, Detailed geological mapping and fault studies of the San Jacinto Tunnel line and vicinity: Jour. Geol., vol. 47, p. 314-324.

Leadbetter, B. C., 1938, Driving an extremely difficult tunnel: Engr. News-Record, vol 121, p. 669-672.

Smith, L. D., 1939, Boring through a rainstorm: Scientific America, vol. 160, p. 140-141.

Thompson, T. F., 1954, The San Jacinto Tunnel on the Colorado River Aqueduct of The Metropolitan Water District of Southern California: Engineering Geology Packet, Geol. Soc. America Ann. Mtg., Los Angeles, Nov. 1954, and in Engineering Geology in Southern California, 1966, AEG Publication (out of print).

MONO CRATERS TUNNEL

Robert Wilson and Henry Mayeda
Formerly Chief Geologists,
Los Angeles Dept. Water and Power

The Mono Craters Tunnel, completed in 1940, was the most challenging and difficult of all the LADWP Los Angeles Aqueduct construction projects. The tunnel pierces the volcanic necks underlying Mono Craters and the high mesa between Mono Basin and Long Valley. Its length is 11.3 miles with an inside diameter of 9 feet 7 inches. Working from six headings, the excavation required 5 1/2 years. Sixty-seven percent of the tunnel required support, some of which was very close, with ribs on 9- to 12-inch centers. Progress varied from extremely slow to as much as 90 feet in 16 hours in cemented volcanic ash. A large variety of rocks were penetrated, including rhyolite, tuff, volcanic ash, granitics, metamorphics, calcareous sandstone, glacial moraines, alluvium and lake beds (see Figure 4). One section of the tunnel penetrated small spurs of granite with intervening valley filled with glacial gravels, volcanic ash and detrital materials.

Dr. Karl Terzaghi (1946), after reviewing the various types of formations penetrated, wrote: "It is not surprising that nearly every difficulty inherent in rock tunneling was encountered in driving this tunnel." Large volumes of water under high pressures constituted one of the greatest difficulties. When an abundantly fractured volcanic plug was penetrated, two

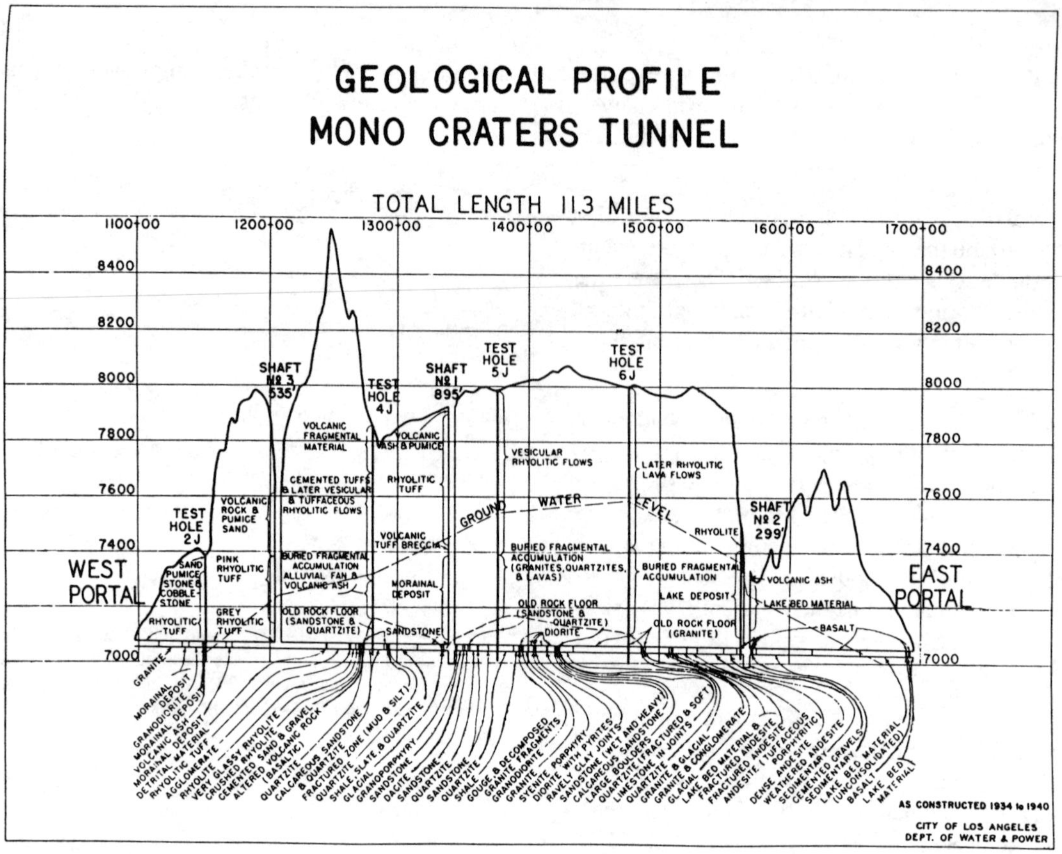

Figure 3. Geological profile - Mono Craters Tunnel

pump stations were installed with a capacity of 18,000 gpm. This obstacle was compounded by the fact that the water was charged with CO_2 gas which was liberated into the tunnel in volumes as high as 800 cu. ft./min. "It was necessary to force air at the rate of 30,000 cu. ft./min. into the tunnel to keep the CO_2 content of the tunnel air below 2 1/2 percent (Gresswell, 1940). The sedimentary rock through which this plug intruded contained calcareous sandstone, quartzite and shale, with carbonate coatings in the joints. "The source of carbon dioxide is most likely in these calcareous formations which have been broken down by volcanism and the action of sulphurous and other acids commonly found in active volcanoes: (Jacques, 1940). CO_2 gas is still a problem in this tunnel during inspection and maintenance work.

Other difficulties included flowing and squeezing ground where the floor would heave and the sides and heading were extremely difficult to hold. Spiling and often breastboarding was required. Some of the worst squeezing was encountered at contacts between metamorphic and granitic rocks. "Each contact was accompanied by much faulting and weathering with the feldspatic mineral in the rocks in many instances highly kaolinized, causing the ground to squeeze from the sides and arch, and raising of the floor (Gresswell, 1940).

Reprinted from Engineering Geology in Southern California, 1966, AEG Publication (out of print).

References

Gresswell, W. K., 1940, Short report on the geological formations encountered during the Mono Craters Tunnel: Calif. Jour. of Mines and Geology, vol. 36, no. 2, p. 199-204.

Jacques, H. L., 1940, Mono Craters Tunnel construction problems: Journal of the American Water Works Association, vol. 32, no. 1.

TECALOTE TUNNEL

Robert E. Trefzger
Retired Engineering Geologist
U. S. Bureau of Reclamation
10200 Mt. Vernon Road, Auburn, CA 95603

Summary
This tunnel was characterized by (1) hot water (to 117°F), (2) water pressures as high as 375 psi, (3) inflows as great as 9,200 gpm, (4) squeezing ground, and (5) methane and hydrogen sulfide gases causing two explosions. The hot water inflows were so irritating to the miners (especially their eyes) that they filled muck cars with cool water and submerged themselves when passing through certain reaches.

The lesson to be learned (again) is that adequate geologic investigation prior to tunnel construction is a must, if surprises as encountered in the Tecolote Tunnel are to be avoided. In

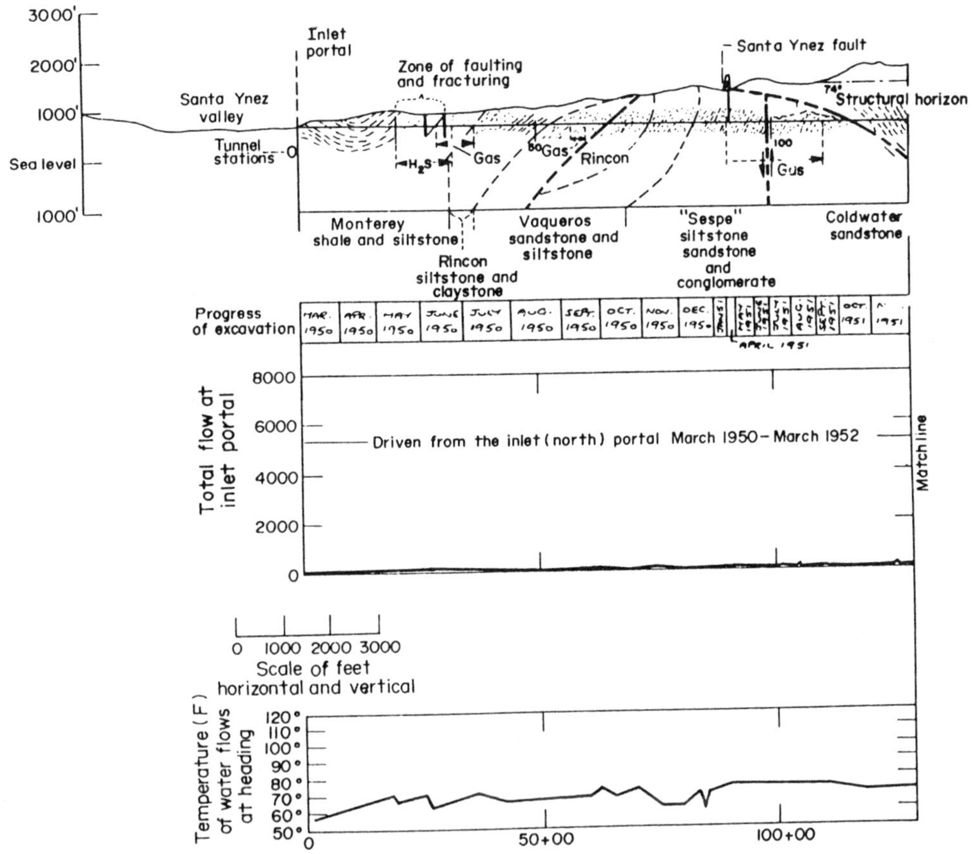

Figure 4a. Tunnel Profile - North portion

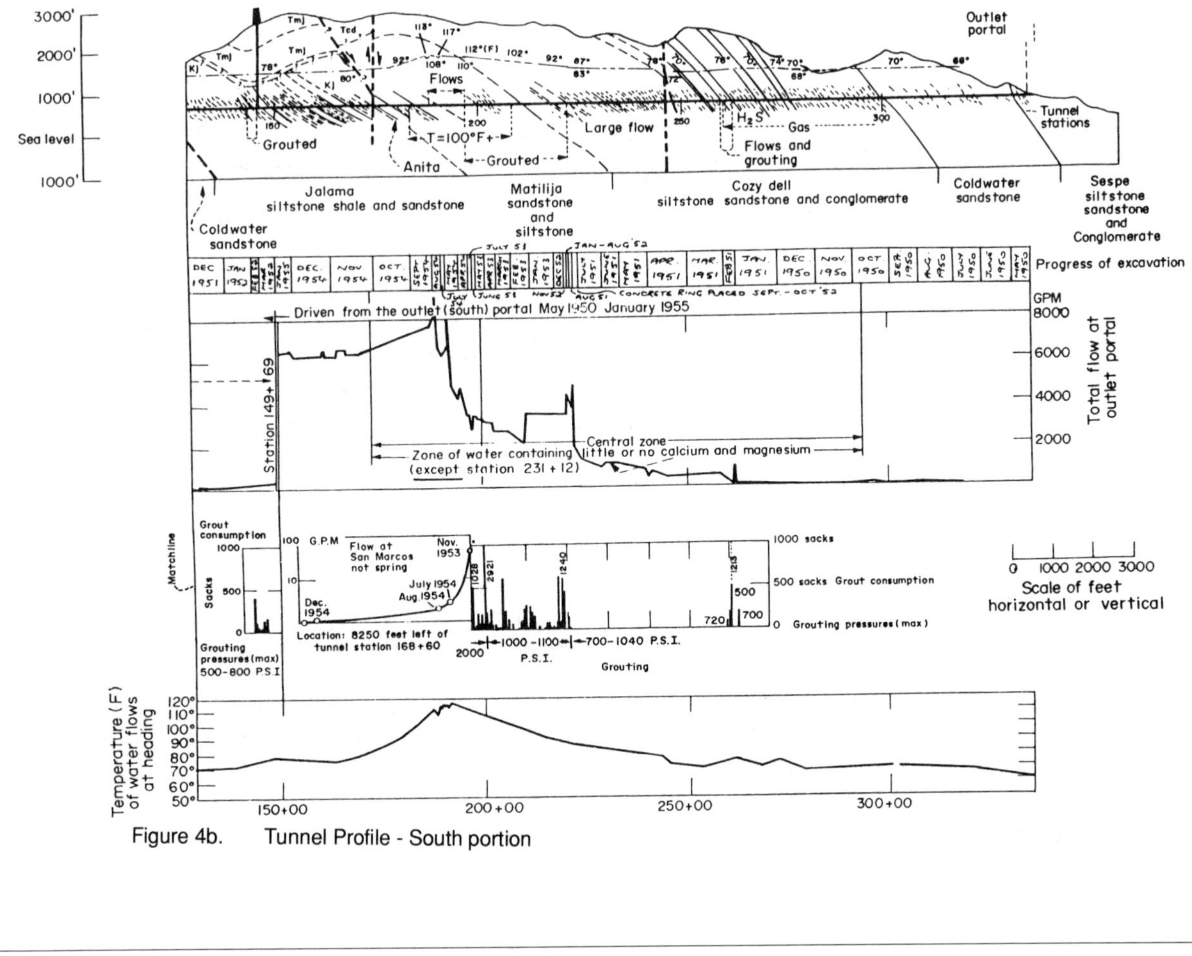

Figure 4b. Tunnel Profile - South portion

none of the references listed are the results of any exploratory borings given.

Although hot springs were known in the area, the paucity of geologic investigations led to significant increases in hazard and cost. In fact, the net payments were $12,301,236 and the bid price was $4,764,853, for a cost overrun of 260% (USBR, 1959, p. 108).

The Tecolote Tunnel penetrates the Santa Ynez Mountains about 12 miles northwest of Santa Barbara, California, and carries water from Cachuma Reservoir, on the Santa Ynez River, to the narrow coastal area east and west of Santa Barbara. The tunnel is 6.4 miles long and 7 ft. in diameter. Excavation began from the inlet and outlet portals in 1950. At the inlet heading, excavation was stopped at Station 149 + 69 in order to line this section with concrete before Cachuma Reservoir covered the portal. Progress continued irregularly at the outlet heading until the tunnel "holed through" on January 15, 1955.

During construction, Bureau of Reclamation geologists mapped the tunnel geology on a scale of 1 inch = 10 feet. Large rock samples were collected every 15 to 20 ft. through most of the tunnel for a cooperative committee of 18 to 20 oil companies. Detailed stratigraphic and paleontologic data from these samples were given to Bureau geologists by several of the companies.

Geologic setting.

The Santa Ynez Mountains are one of the east-west trending Transverse Ranges and consist of a thick sequence of Tertiary and Cretaceous sedimentary rocks. Most of the range consists essentially of the southerly limb of a major anticline which has been cut near its structural crest by the Santa Ynez fault (which has a combination left-lateral and thrust displacement), traceable for at least 90 miles.

The Santa Ynez fault divides the tunnel line into two major structural units: (1) a sequence of younger, north-dipping beds complicated by faulting and a minor syncline near the inlet portal; and (2) toward the outlet portal, a sequence of older, south-dipping beds cut by minor faulting. Small faults and shear zones are common throughout the tunnel but do not greatly affect the overall structural picture. The joints are the principal sources of the waterflows.

Problems during construction.

In the inlet leg many steel H-beam ribs and invert struts were deformed and pushed out of position by "squeezing ground". This trouble usually occurred in and near fault zones and in areas of deformed and somewhat sheared rock. In the Santa Ynez fault zone and in the zone of major faulting at Station 99 + 00, the distorted ribs often had to be replaced two or three times, in some cases with 6-inch H-beam ribs and invert struts spaced 18 inches apart.

Minor seeps of natural gas (essentially methane) were encountered at various places in the tunnel. In the lower Monterey, Matilija, and Cozy Dell Formations, the gas came from fractures and joints in the otherwise impermeable rock; and in the Vaqueros, Sespe, and Coldwater Formations, the gas occurred in the in the interstices of poorly-cemented, soft sandstones. A minor explosion of accumulated gas burned several workmen in the inlet heading area near Station 89 + 87. A much stronger explosion took place in the outlet leg during a work stoppage, about 26 hours after the ventilating fan was shut off. Between Station 272 + 00 and Station 293 + 00 the blast was very severe, and over 2,000 ft. of 12-inch ventilating pipe were collapsed and deformed.

Minor quantities of hydrogen sulphide were dissolved in most of the waterflows throughout the tunnel. Large amounts were encountered only in one small section, between Station 260 + 00 and Station 262 + 00. Concentrations of gas emanating from the highly charged water into the heading area occasionally became severe enough to discomfort the miners and force a temporary halt of work.

Waterflows were the principal construction problem in Tecolote Tunnel. At the inlet heading only very minor flows and seeps were encountered prior to a flow of sand and gas-bearing water at Station 89 + 87. Further to the south the inlet heading encountered only minor flows until it reached Station 143 + 42. Between this station and Station 149 + 69, where excavation was stopped, several flows of from 50 gpm to 200 gpm occurred, in each case from 4 to 6 grout holes. These flows all came from fractures and joints in well-cemented sandstones and siltstones of the Jalama Formation and were successfully grouted off.

In the outlet section of the tunnel, all flows came from fractures and joints in the well-cemented sandstones and siltstones and were very minor for the first several thousand feet. The hydrogen sulphide-rich flows, between Station 260 and Station 262, were the first sizable flows, about 400 gpm and 1,000 gpm, and although under the highest static pressures (375 psi) found in the tunnel, were successfully sealed off by grouting. Continuing progress gradually picked up more water until at Station 222 + 55 the total flow at the outlet portal was about 2,000 gpm. At this point a large flow of water and finely fractured and crushed siltstone poured into the heading area. This flow varied from 1,200 gpm to 2,800 gpm during the following 16 months which were spent in grouting attempts, advancing drainage drifts 30 ft. on either side of the main face, and driving the main face through the loose material to solid rock. This area was by far the most time-consuming and troublesome spot in the entire tunnel. A 245-ft section was concrete-lined to confine and control the water and muck flows before tunnel driving resumed.

Further progress through the lower Cozy Dell and upper Matilija Formations to Statio 196 + 85 was slow, due to the

Figure 5. Looking downstream from tunnel face at station 188 + 69

extensive, time-consuming grouting efforts and to the gradually increasing temperatures of the rock and water. The object of the grouting was to reduce the water flows to a practical minimum, and much success was realized in achieving this end. In the 2,445 ft of tunnel between Station 221 + 30 and Station 196 + 85, over 19,600 sacks of cement were pumped as grout into a total of more than 725 grout holes. Pre-grouting water flows from these holes totaled over 12,000 gpm. An estimated 2,000 gpm were flowing into this 2,445-ft grouted interval when work was suspended on July 21, 1953 for renegotiation due to large inflow and high temperature.

Progress at the outlet heading was at a standstill at Station 196 + 85 from July 21, 1953, to April 12, 1954, when a new sub-contractor took over and excavation was resumed. During most of this time, total flow at the outlet portal stayed at about 3,600 gpm. No grouting of water flows was attempted when progress resumed at Station 196 + 85; and in the 825 ft. of tunnel excavated to Station 188 + 60, total flow increased from 3,600 gpm to 9,100 gpm. This peak flow tapered off to about 7,400 gpm in a month's time, and when the tunnel holed through, flows at the outlet portal had decreased to about 6,500 gpm.

Water and rock temperatures rose steadily as the heading advanced from the vicinity of Station 221 + 00 (88°F) to Station 196 + 85 (112°F). Chemical analyses of flows in this area show sodium and bicarbonate to be the principal ions in solution with minor amounts of carbonate, sulphate, chloride, and silica, and with very low to zero values for calcium and magnesium. Total dissolved solids in the warm and hot waters ranged from 260 to 320 parts per million. It is believed that the hot water flows were meteoric waters which were heated at greater depth and subsequently ascended up the east-southeastward plunging anticline, through joints and fractures, and reached a thermal equilibrium with the surrounding rocks. Highest water temperatures recorded were 117°F at Station 191 + 70. [The combination of excessive temperature and high humidity greatly reduced labor efficiency, although partly alleviated by increasing the ventilation and by dehumidifying the air supply.]

Condensed from Engineering Geology in Southern California, 1966, Association of Engineering Geologists publication (out of print).

References

Crocker, E. R., 1955, Hottest, wettest tunnel holed through: Civil Engr., vol. 25, p. 142-146.

Engineering News-Record, September 1951, "Tecolote Tunnel is advanced four miles in tough going through Coast Range," vol. 26, No. 9, p. 78-79.

Engineering News-Record, February 1954, "Work changes hands, cost upped", p. 25.

Engineering News-Record, June 1954, "Heat being conquered at Tecolote Tunnel:, p. 45-46.

Engineering News-Record, June 1954, "History of hot tunnel". p. 46.

Gardner, Wm., 1957, Tecolote Tunnel: A discussion on p. 32-34, in "Conference on Geological Engineering Problems of Water in California", issued by Univ. Calif. Berkeley, Committee on Research in Water Resources, 162 p.

Rantz, S. E., 1961, Effect of tunnel construction on flow of springs and small streams in the Tecolote Tunnel area of Santa Barbara County, California: U. S. Geol. Survey Professional Paper 424-C.

Thomas, C. F., 1956, The Tecolote Tunnel: Explosives Engr., vol. 34, p. 40-47.

Trefzger, R. E., 1966, Tecolote Tunnel, p. 201-205 in Lung, R. and Proctor, R., eds., Engineering Geology in Southern California: AEG Special Publication, 389 p. (out of print).

U. S. Bureau of Reclamation, 1959, Technical Record of Design and Construction of the Tecolote Tunnel: U.S.B.R., Denver, Colo., 109 p.

Western Construction, August 1954, Fighting 6,800 gpm hot water flow in Tecolote Tunnel: vol. 29, No. 8, p. 59-62.

GLENDORA TUNNEL CROSSING OF THE SIERRA MADRE FAULT

Richard J. Proctor,
C. Marshall Payne
D. C. Kalin

formerly,
The Metropolitan Water District of Southern California, Los Angeles; now consultants.

Summary
Well-exposed sections across major fault zones are uncommon. An excellent opportunity to observe and study such a zone was provided in 1966 during the construction of the 6.2 mile Glendora tunnel, part of a construction program to distribute imported northern California water within the southern part of the state.

The Sierra Madre fault zone is the major break that extends for a distance of 55 miles along the steep south front of the San Gabriel Mountains. As mapped in the tunnel, this fault zone has an apparent thickness of 520 ft. Within it are oil-stained, brecciated shale of the Puente Formation (Upper Miocene) and fault slivers of Precambrian gneiss and Cretaceous quartz-rich granite, which have been thrust at least 700 ft. over Quaternary terrace deposits.

Geologic mapping and core hole information proved a longer tunnel route would be several million dollars cheaper to build because of poor rock conditions along the shortest route.

Tunnel construction was not adversely affected by the fault materials encountered, but was affected by bentonite shales adjacent to the fault zone. Interpretation of core borings made prior to tunnel construction predicted the nature of the fault zone because of an unusual drill log sequence showing terrace gravels 37 ft. beneath the gneiss and granite "basement rocks."

Introduction
The internal structure of major fault zones rarely is exposed for study. Thus, whenever a major fault is freshly exposed, geologists should promptly undertake a detailed recording of its features for posterity. In California, good exposures of breaks within the active San Andreas fault system are rare (e.g. Proctor, 1962), but new descriptions can be expected from new exposures in large construction projects.

Construction of the 6.2-mile Glendora Tunnel of the Metropolitan Water District of southern California began in November 1965, and excavation was completed in December 1968. This bore is one of 11 pressure tunnels totaling 52 miles that are being constructed to distribute surplus northern California water from the State Water Project (Lung and Proctor, 1966).

With a finished inside diameter of 15 1/2 ft. and a bore diameter of 18 1/2 ft., the tunnel was being constructed by the J. F. Shea Co., Inc., at a bid cost of $22 million.

The tunnel route extends easterly under the steep front of the San Gabriel Mountains from a point below Morris Dam in San Gabriel Canyon to San Dimas Canyon Park. The tunnel was driven westward from the East Portal, and both eastward and westward from the Dalton Adit located approximately midway in the tunnel route.

All prospective tunnel bidders were issued a copy of a geologic report written by the three consulting geologists retained by the Metropolitan Water District: V. P. Pentegoff, R. H. Jahns and T. F. Thompson. The report included the logs of the 31 borings made during exploration, plus the geologic portion of the Historical Record of the nearby Monrovia Tunnels No. 1, 2 and 3, previously completed by the District in 1939.

Rock types and conditions
The San Gabriel Mountains comprise a set of irregular fault blocks that have been uplifted during Pleistocene and recent times to their present general position. The topography is characterized by sharp ridges and deep, V-shaped canyons. Slides are numerous, and weathering of the pervasively fractured rocks is very active. Locating faults in the field is difficult here because of dense brush cover, steep terrain, similarity of many rock types, and a mantle of soil and debris that blankets most of the hills.

The mountains are composed chiefly of Precambrian gneisses, schists, and other metamorphic rocks, along with Cretaceous granite, diorite, and other plutonic igneous rocks. Middle Miocene volcanic rocks, consisting of andesite flows and dikes, welded tuff, and tuff breccia, were encountered in the eastern one-third of the tunnel. From the East Portal 2,500 ft. of Upper Miocene Puente Formation shale, fault zone material, and terrace deposits of probable Late Pleistocene age were penetrated. Several eroded highs of tuff and conglomerate of the Topanga Formation (Middle Miocene) cropped out under the terrace deposits at the floor of the tunnel.

In general, most rock types encountered in tunneling were moderately to highly jointed and fractured. Alteration zones and small faults were common throughout. Minor water inflows appeared intermittently along the entire length. Tunneling conditions were generally fair to good but included some locally squeezing and swelling ground (Terzaghi, 1946; Proctor, 1971).

Sierra Madre fault zone

The 55-mile long Sierra Madre fault is the major break along the steep south front of the San Gabriel Mountains. The zone is unusual in that it is obviously a right lateral fault at its western end in Big Tujunga Canyon where it merges with the known right-lateral San Gabriel fault; it then becomes a dominantly reverse and thrust fault in its center portion, and finally appears to abut the known right-lateral San Jacinto fault at Cajon Pass. This suggest orogenic activity and dip-slip movement are now more active than strike-slip movement at the south side of the San Gabriel Mountains.

There are many locations in the 30 miles between the communities of Sunland and Glendora where the fault plane has been measured indicating basement rocks thrust over terrace deposits. One of the most accessible exposures is on the west wall of Santa Anita Canyon just north of the road onto Wilderness Park of the City of Arcadia. Here brecciated gneiss overlies old alluvium with the fault dipping 30° north. (See Plates in Crook et al., 1987.)

Twelve hundred feet from the East Portal the Sierra Madre fault zone, or Cucamonga fault, was encountered. The tunnel intersected the fault nearly perpendicular and exposed an excellent section of the entire fault zone, 520 ft. thick, as described later. As observed in the tunnel, the Sierra Madre fault consists of brecciated shale of the Puente Formation, containing slivers of gneiss and quartz-rich granite, all thrust at least 700 ft. over terrace deposits. The fault is a true thrust with the dip of the undulating plane averaging approximately 12°.

Earthquake risk

It should be realized that it is practically impossible to avoid founding major linear engineering structures on faults in southern California. According to R. H. Jahns (personal communication, 1965) "the Glendora Tunnel would extend across the Sierra Madre fault zone, but relatively low known seismicity in this area, fortified by geologic evidence indicating little or no fault displacement for thousands of years before present (based on locally undisturbed older, elevated terrace gravels), leads us to the conclusion that possibilities for serious earthquake damage during the useful life of a given structure — that is, 50-100 years. Tunnels are among the civil engineering structures least affected by ground shaking resulting from great earthquakes (e.g., Duke and Leeds, 1959). On this theme, a disquieting fact on the occurrence of great earthquakes is that of the four largest earthquakes recorded in the United States in historic time, excluding Alaska, two occurred on the west coast (Fort Tejon, California, in 1857, and San Francisco, in 1906) and two occurred in the supposedly "quiescent" eastern United States (New Madrid, Missouri, in 1811, and Charleston, South Carolina, in 1886).

Geologic Exploration Demonstrates Longer Route Is Cheaper

All preliminary routes involved a tunnel about 14,900 ft. long between San Gabriel Canyon and the mouth of Dalton Canyon. From here, one alternative route involved a pipeline extending about five miles to an existing water treatment plant at La Verne. The pipeline alternative had two drawbacks: Right-of-way would extend through expensive residential areas, and the pipeline would have to be constructed to withstand up to 200 ft. hydraulic head in these residential areas.

The other alternative routes included combinations of pipe siphons across Dalton Wash below the junction of Little Dalton and Big Dalton Canyons, with various shorter or longer tunnels through the lower frontal part of the Glendora Hills, and thence a relatively low-pressure pipeline to La Verne.

In all, 24 core holes totaling 5,078 ft. and seven power auger holes totaling 476 ft. were drilled, plus several trenches excavated, during the investigation period of 1964-1965. For all these exploratory holes, detailed geologic logs were prepared, initial elevations of groundwater were established, and perforated plastic pipe installed or periodic measurements of depth to the groundwater surface. In addition, hydrologic data were collected of the 49 identified springs, wells, and water tunnels in the general vicinity.

The originally proposed Glendora Tunnel alinement was 6.0 miles long (31,500 ft.) consisting of two tunnels with a 2,000-ft. long inverted siphon connecting them across Dalton Wash. Several core borings were made along the eastern part of this alinement, where 1 1/2 miles of soft, wet tuffs and one mile of hard andesite flows would have been penetrated by the tunnel. The remainder of the tunnel would have been in hard granitic rock and gneiss, plus several hundred feet in a shale section common to all alternative routes. The cores of the tuffs and tuff breccias oozed from the core barrel like toothpaste, denoting very incompetent rock conditions. A preliminary cost estimate indicated \$1,400/ft. for tunneling in this material and \$700/ft. for the hard volcanic, gneissic, and granitic rocks. The estimated cost of this tunnel was \$27 1/2 million.

By moving the alinement one-half mile northward into the mountains, the siphon route could be eliminated and access

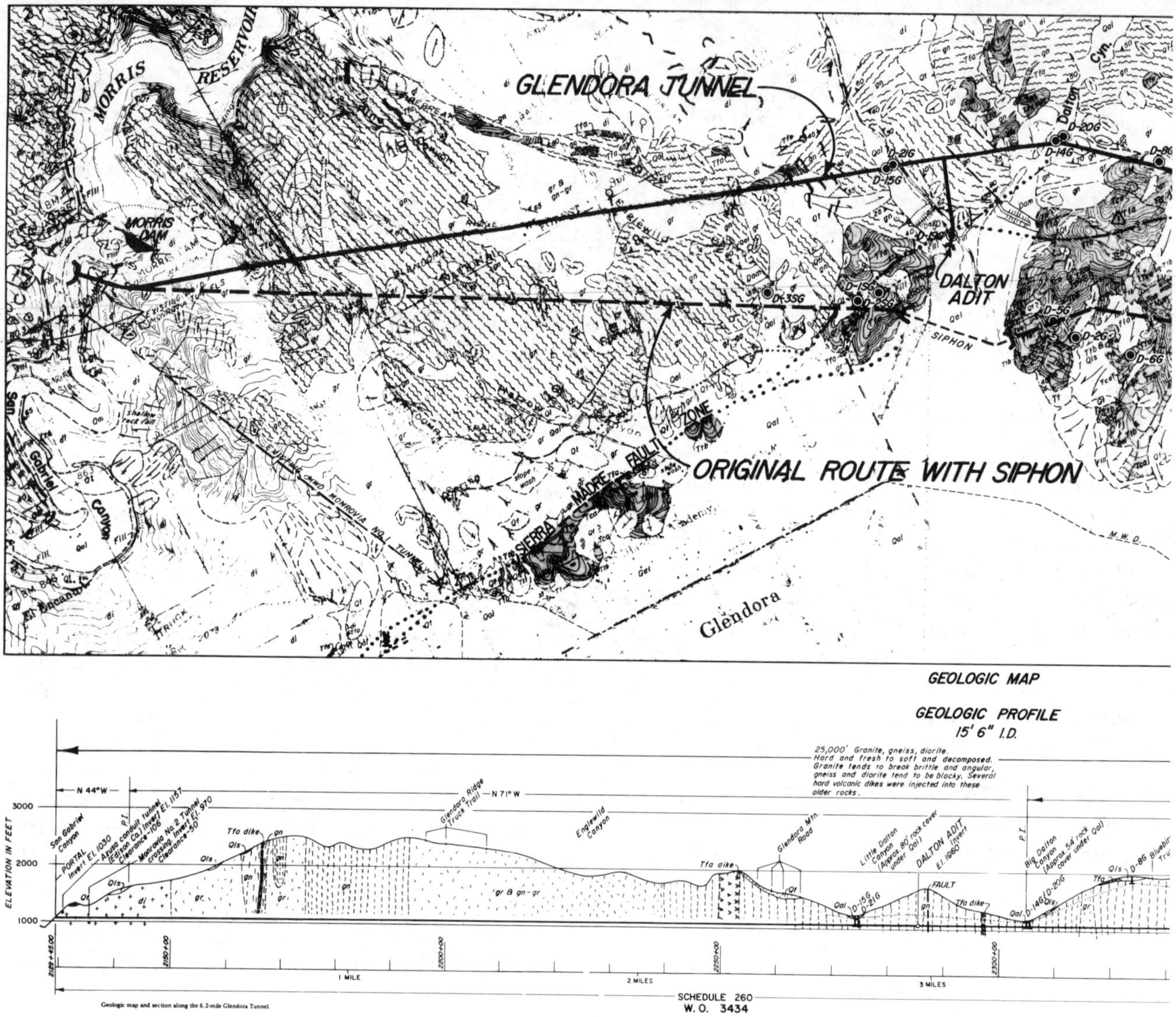

Plate 1 Geologic map and section of Glendora Tunnel

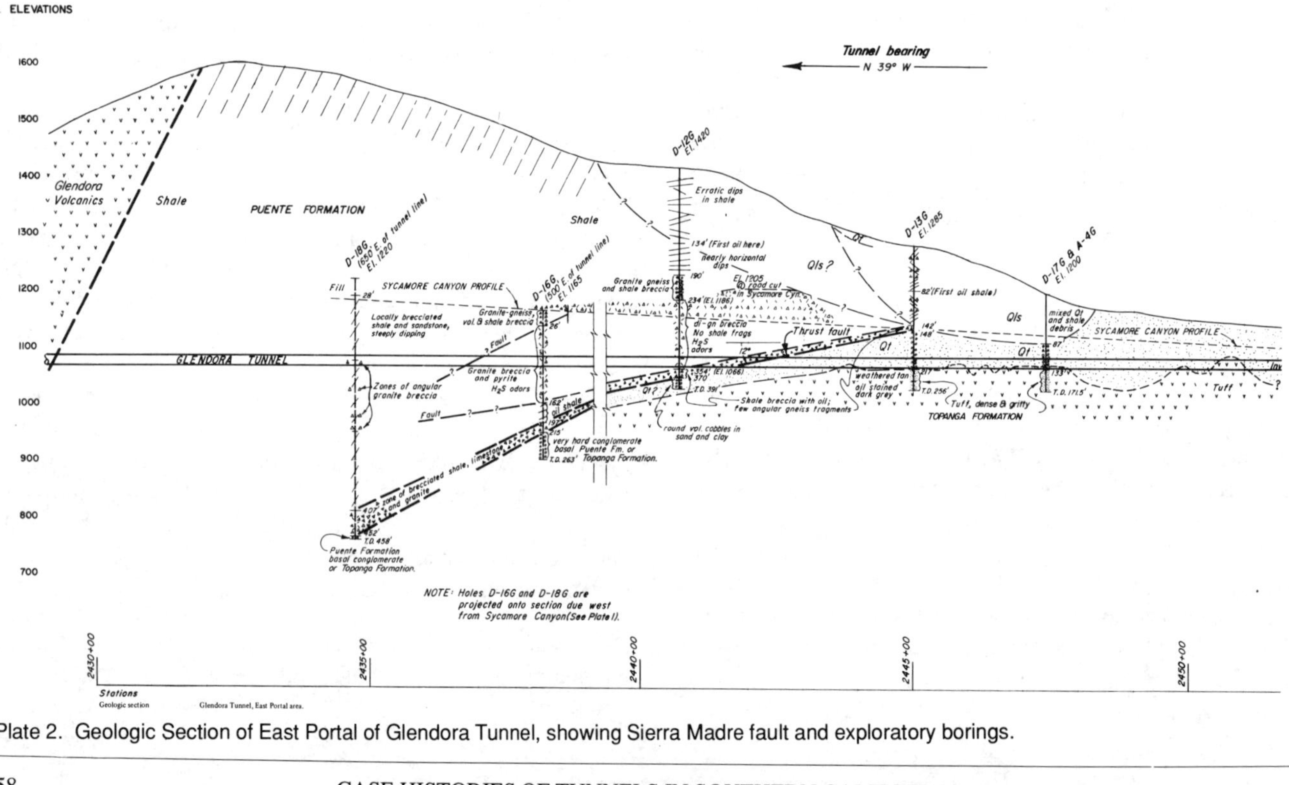

Plate 2. Geologic Section of East Portal of Glendora Tunnel, showing Sierra Madre fault and exploratory borings.

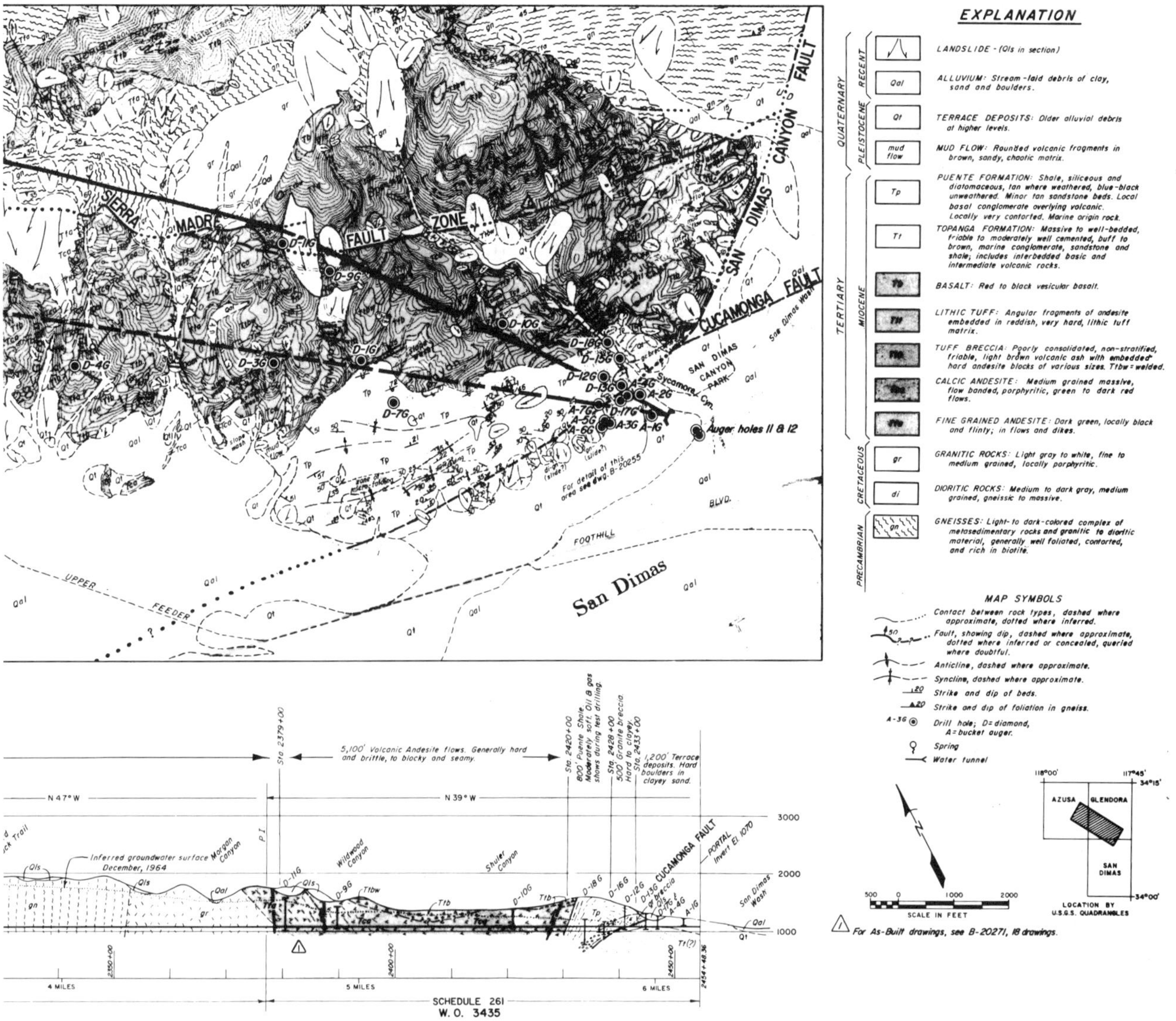
EXPLANATION
QUATERNARY
RECENT
PLEISTOCENE
LANDSLIDE - (Qls in section)
Qal ALLUVIUM: Stream-laid debris of clay, sand and boulders.
Qt TERRACE DEPOSITS: Older alluvial debris at higher levels.
mud flow MUD FLOW: Rounded volcanic fragments in brown, sandy, chaotic matrix.
TERTIARY
MIOCENE
Tp PUENTE FORMATION: Shale, siliceous and diatomaceous, tan where weathered, blue-black unweathered. Minor tan sandstone beds. Local basal conglomerate overlying volcanic. Locally very contorted. Marine origin rock.
Tt TOPANGA FORMATION: Massive to well-bedded, friable to moderately well cemented, buff to brown, marine conglomerate, sandstone and shale; includes interbedded basic and intermediate volcanic rocks.
Tb BASALT: Red to black vesicular basalt.
Ttf LITHIC TUFF: Angular fragments of andesite embedded in reddish, very hard, lithic tuff matrix.
Ttb TUFF BRECCIA: Poorly consolidated, non-stratified, friable, light brown volcanic ash with embedded hard andesite blocks of various sizes. Ttbw = welded.
Tca CALCIC ANDESITE: Medium grained massive, flow banded, porphyritic, green to dark red flows.
Tfa FINE GRAINED ANDESITE: Dark green, locally black and flinty; in flows and dikes.
CRETACEOUS
gr GRANITIC ROCKS: Light gray to white, fine to medium grained, locally porphyritic.
di DIORITIC ROCKS: Medium to dark gray, medium grained, gneissic to massive.
PRECAMBRIAN
gn GNEISSES: Light- to dark-colored complex of metasedimentary rocks and granitic to dioritic material, generally well foliated, contorted, and rich in biotite.
MAP SYMBOLS
Contact between rock types, dashed where approximate, dotted where inferred.
Fault, showing dip, dashed where approximate, dotted where inferred or concealed, queried where doubtful.
Anticline, dashed where approximate.
Syncline, dashed where approximate.
Strike and dip of beds.
Strike and dip of foliation in gneiss.
Drill hole; D= diamond, A= bucket auger.
Spring
Water tunnel
SIERRA MADRE FAULT ZONE
SAN DIMAS CANYON FAULT
CUCAMONGA FAULT
SAN DIMAS CANYON PARK
Sycamore Cyn.
San Dimas Wash
Auger holes 11 & 12
For detail of this area see dwg. B-20255
FOOTHILL BLVD.
UPPER FEEDER
San Dimas
SCALE IN FEET
AZUSA
GLENDORA
SAN DIMAS
LOCATION BY U.S.G.S. QUADRANGLES
For As-Built drawings, see B-20271, 18 drawings.
5,100' Volcanic Andesite flows. Generally hard and brittle, to blocky and seamy.
Sta. 2379+00
Sta. 2420+00 800' Puente Shale. Moderately soft. Oil & gas shows during test drilling.
Sta. 2428+00 500' Granite breccia. Hard to clayey.
Sta. 2433+00 1,200' Terrace deposits. Hard boulders in clayey sand.
N 47° W
N 39° W
Inferred groundwater surface December, 1964
Morgan Canyon
Wildwood Canyon
Shuler Canyon
PORTAL Invert El. 1070
4 MILES
5 MILES
6 MILES
SCHEDULE 261
W. O. 3435

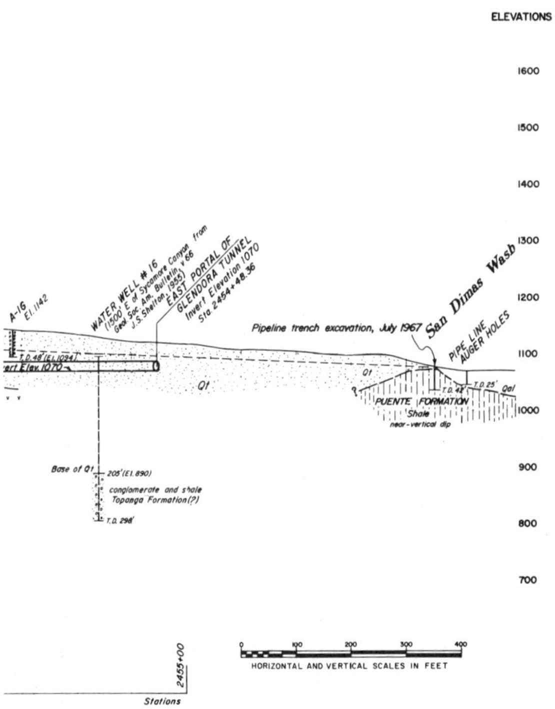
ELEVATIONS
WATER WELL # 16 (1500' E. of Sycamore Canyon from Geol. Soc. Am. Bulletin, v. 66 J. S. Shelton, 1955)
EAST PORTAL OF GLENDORA TUNNEL Invert Elevation 1070 Sta. 2454+48.36
Pipeline trench excavation, July 1967
San Dimas Wash
PIPE LINE AUGER HOLES
PUENTE FORMATION Shale near-vertical dip
Base of Qt 205' (El. 890)
conglomerate and shale Topanga Formation(?)
T.D. 298'
HORIZONTAL AND VERTICAL SCALES IN FEET
Stations

to the tunnel provided by an adit. Thus the eastern part of the alinement would be in gneiss and granitic rock with only one mile in soft to hard volcanic rocks. This alinement was adopted; it is 6.2 miles long, or 1,200 ft. longer than the original route, but cost of construction was estimated to be almost $5 million less. Although some squeezing ground was encountered in the tunnel causing a comparatively small "changed conditions" payment to the contractor, we are certain that great hazard and expense were prevented by avoiding the shortest route.

Test Borings Reveal Thrust Fault Or Sole Of A Large Landslide

The dilemma

Field mapping in Sycamore Canyon, in the east portal area of the proposed tunnel, revealed a small outcrop of brecciated quartz-rich granite surrounded by shale of the Puente Formation. The occurrence of this granite can be explained alternatively as: (1) bedrock underlying the Puente Shale; (2) a fault sliver of basement rock forced into the Puente Shale; or (3) a remnant in a large landslide derived from the granitic highlands one mile north.

A D-4 Caterpillar tractor was used to try to expose a contact between the Puente Shale and the granite breccia. However, debris from slope wash and a small landslide obscured the contact on the west side of the canyon; roadfill obscured it on the east side. It therefore became necessary to core-drill the area to gain more information.

Five NX holes were drilled, three on the ridge on the west side of the canyon and two from points in the canyon bottom. The first hole drilled was No. D-12G located on top of the ridge. The following remarkable log resulted:

0-134 ft. =	Landslide debris. (Erratically dipping tan Puente Shale).
134-190 ft. =	Puente Shale. (Well bedded, oil-stained).
190-234 ft. =	Breccia. (Angular pebbles of diorite gneiss, quartz-rich granite, and chips of Puente Shale).
234-354 ft. =	Gneiss. (Hard, fresh, angular fragments with interstitial soft clay, and a strong odor of hydrogen sulfide, H_2S).
354-370 ft. =	Oil shale breccia. (Recemented angular pieces of soft, oil-stained shale with a few scattered angular gneiss fragments, minor gouge).
370-391 ft. =	Terrace gravels. (Rounded volcanic cobbles and bleached shale chips in a brown clayey sand).

The bore hole data negated the bedrock outcrop possibility (1) in Sycamore Canyon.

The conclusion

The numerous small landslides in the Glendora Hills led to the original assumption that this could be a large landslide. However, the presence of oil and gas in the bore holes suggested that these volatiles might have escaped had landsliding occurred. The particular type of granite in the outcrop at the bottom of Sycamore Canyon was one of the major factors in calling this feature a fault; no outcrop of similar granite could be found in the nearby mountains. Further, a major fault (the Cucamonga fault) has been mapped in places along the front of the range by others (Eckis, 1934, Shelton, 1955; Baily and Jahns, 1954; Dibblee, 1968). Their mapping placed the buried fault trace at the toe of the Glendora Hills. Considering the above evidence, it was generally agreed that faulting must have taken place with subsequent minor landsliding covering the fault scarp.

Deeply incised canyons across the buried fault trace indicate that the major disturbance occurred at least several thousand years ago. This leads to the probability that further fault movement here will not occur within the relatively short life of the project.

Details Of The Sierra Madre Fault Zone

The fault has been mapped at a scale of ten feet to the inch. The zone can be divided into three distinct fault materials or units: (1) Shale breccia, composed of both "bedded" shale masses (megabreccia clasts), and "structureless" material. (2) Granitic breccia, composed of highly fractured quartz-rich granite and diorite with irregular-shaped blocks of gneiss and scattered fragments of pegmatite. (3) Shale gouge, composed of yellow and tan, soft, moist clay. The fault zone materials were encountered for a distance of 520 f in the tunnel, where its true thickness is about 140 ft.

The sequence and thickness of each unit are as follows: the shale gouge unit rests directly on terrace conglomerate in one place and on fossil soil in another (Fig. 6). This suggests that locally the thrust plate partly scraped off the fossil soil, and also slid out over it. The fault plane undulates but averages about 12° northward dip. The gouge is 9 ft. in maximum thickness, but in several places it is entirely absent. The gouge is so soft that the hammer could be completely imbedded into it with little effort. It is noteworthy that five steel support ribs rest entirely upon soft gouge but did not deform. This fact can be explained by blocking and wedging behind the horseshoe-shaped sets, little or no overburden pressure, and absence of groundwater.

Overlying the gouge is 70 ft. of oil-stained "bedded" shale breccia, followed by 110 ft. of intermixed granitic and shale breccia, and finally 60 ft. of "structureless" shale breccia containing a patchy zone of quartz diorite fragments (Fig. 7).

Figure 6. Photo and sketch of Sierra Madre fault gouge and Puente Formation shale breccia overriding fossil soil on terrace deposits. Rib No. 248

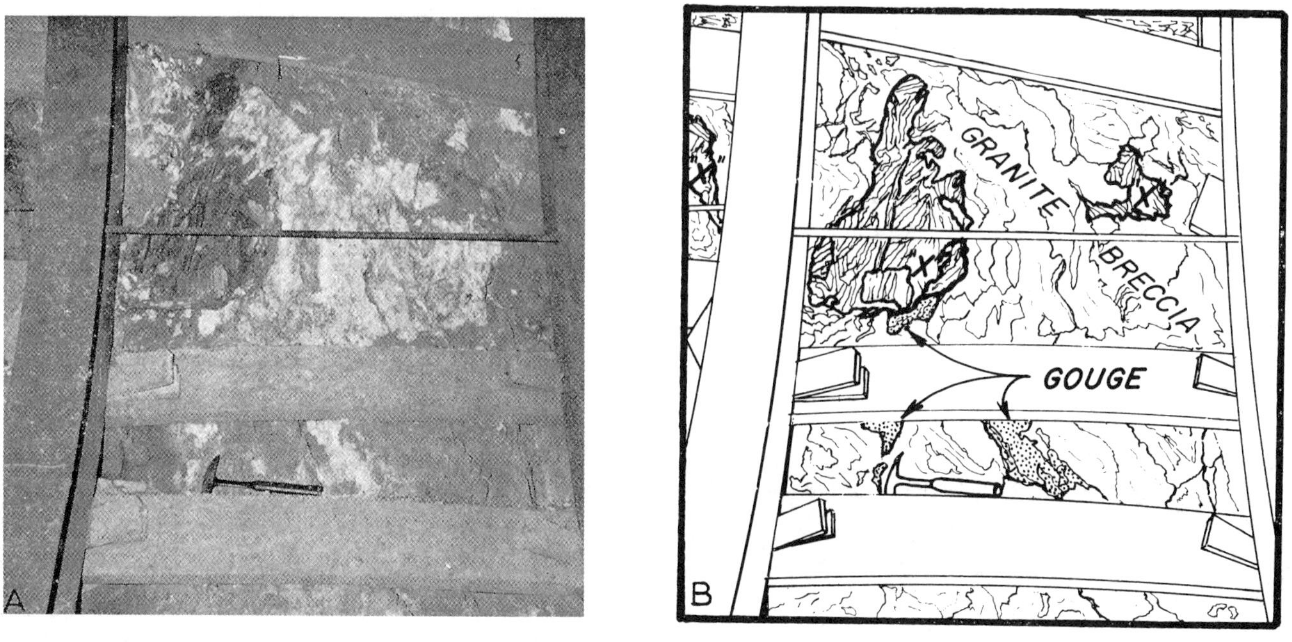

Figure 7. Photo and sketch of granite-diorite breccia with large inclusions of shale breccia "X". Rib No. 294 shown at right.

Overlying these fault units is well bedded shale of the Puente Formation which appears undisturbed except for isolated moderately dipping faults marked by one or two inches of gouge.

The strong hydrogen sulfide odors encountered in the borings in the gneiss and shale breccia were not noticeable in the tunnel owing to good ventilation.

Tunneling Progress

No unusual tunnel excavation problems occurred in crossing this fault zone, mainly because of low cover and absence of groundwater. The contractor experienced his most difficult tunneling several hundred feet farther into the tunnel in the Puente Shale. For a distance of about 300 ft the shale is interbedded with thin bentonite-rich layers that caused a swelling ground condition. The contractor elected to excavate a portion of this shale by the top heading and bench method. Top heading advance was by steel channel spiling and blasting. Progress in the bentonite shale zone averaged 5 ft./day, compared to 14 ft./day through the 520 ft. The best daily progress was 40 ft., in hard diorite near the West Portal.

Groundwater occurred as minor seeps in most parts of the tunnel. In one 20-ft section of tunnel, a flow of 30 gpm was encountered in loose lapilli tuff, which caused the only running condition noted.

Acknowledgements

The following geologists were consulted in 1966 by the District concerning the existence and recency of a massive landslide as might affect the Glendora tunnel: Clarence R. Allen, Professor of Geology and Geophysics, California Institute of Technology; George B. Cleveland, Senior Geologist, California Division of Mines and Geology; John C. Crowell, Department of Geology, U.C.L.A. (now at University of California, Santa Barbara); Thomas W. Dibblee, Jr., Geologist, U. S. Geological Survey; Rollin Eckis, President of Atlantic Richfield Oil Corporation; Richard H. Jahns, Dean, School of Earth Sciences, Stanford University; Vladimir P. Pentegoff, Consulting Engineering Geologist, Burlingame, California.

Condensed from 1970 article in Engineering Geology: vol. 4, p. 5-63, Elsevier Publishing Co. Reprinted with permission.

References

Bailey, T. L. and Jahns, R. H., 1954, Geology of the Transverse Range Province, Southern California: Geology of Southern California, Calif. Div. Mines Geol., Bull. 170.

Crook, R., C. R. Allen, B. Kamb, C. M. Payne, R. J. Proctor, 1987, Quaternary geology and seismic hazard of the Sierra Madre and associated faults, Western San Gabriel Mountains: p. 27-63 in USGS Professional Paper 1339.

Dibblee Jr., T. W., 1968, Displacements on the San Andreas fault system in the San Gabriel, San Bernardino, and San Jacinto Mountains, Southern California: Proceedings of Conference on Geologic Problems of San Andreas Fault System. School of Earth Sci., Stanford Univ., Stanford, CA p. 260-278.

Duke, C. M. and Leeds, D. J., 1959, Effects of earthquakes on tunnels: Protective Construction Symposium. Rand Corp., Santa Monica, CA , p. 337-365.

Eckis, R., 1934, Geology and ground water storage capacity of valley fill: Calif. Div. Water Resources Bull. 45, 279 p.

Proctor, R. J., 1962, Geologic features of a section across the Casa Loma fault, exposed in an aqueduct trench near San Jacinto, California: Geol. Soc. Am. Bull., vol. 73, p. 1293-1296.

Proctor, R. J., 1970, Mapping geological conditions in tunnels: Assoc. Engineering Geologists Bull., vol. 8, no. 1, p. 1-42.

Shelton, J. S., 1955, Glendora volcanic rocks, Los Angeles, Basin, California: Geol. Soc. Am. Bull., vol. 66, pl. 45-90.

Terzaghi, K., 1946, Introduction to tunnel geology: R. V. Proctor and T. L. White (Editors), Rock Tunneling with Steel Supports. Commercial Shearing and Stamping Co., Youngstown, Ohio.

BALSAM MEADOW HYDROELECTRIC PROJECT, OPEN JOINTS AT GREAT DEPTH

Patrick Hamilton
Project Geologist
Southern California Edison Company
Rosemead, CA 91770

Summary

The Balsam Meadow Project is located approximately 55 miles northeast of Fresno, California, in the Sierra Nevada (Fig. 8). Major components of this project include an underground powerhouse for a 200 MW unit and a high pressure power tunnel/shaft. Extensive site exploration included over 12,400 ft of core drilling with 5 deep drill holes in the powerhouse area. Core from these holes indicated excellent quality quartz diorite with average joint spacing on the order of 10 to 15 ft and core lengths up to 200 ft without joints.

During tunnel construction, a series of anomalous, open rock joints was encountered at depths greater than 1,000 ft below the surface in the lower portion of the Project. These features indicate that the lower tunnel complex including the powerhouse chamber is constructed in a part of the rock mass that contains an extensive and interconnected fracture system that is, under the present ground water regime, highly water bearing. Water from this joint system, with initial flow exceeding 1,200 gpm, was first encountered in the access tunnel near its intersection with the power chamber. As the water-bearing joints were intersected at lower elevations in the tailrace tunnel, the joints in the access and construction tunnels, at higher elevation, ceased flowing almost immediately, indicating the presence of a highly interconnected joint system.

In the lower power tunnel, indications of the open and interconnected nature of this type of fracture system were demonstrated by several observations: (1) the presence of several open joints, with apertures of 1 to 5 inches; (2) apertures with mineral coatings, thus indicating the presence of "ancient" ground water percolation; (3) observation of joints exhibiting aspiration and expelling of air resulting from distant subsurface blasting; and (4) rapid travel of dye injected into subhorizontal joints between a point in the lower power tunnel and a location 1,400 ft distant in the tailrace tunnel, at a rate of approximately 5.5 ft/min.

Geochemical interpretations, combined with the existing conventional data on the geology and hydrogeology of the area, resulted in a conceptual model of the hydrogeologic conditions controlling seepage to the tunnels. A sump system successfully contained and controlled the ground water inflow. From observations of water recovery in joints and piezometers made during test periods when the sump pumps were shut down, it is also evident that in the absence of the sump, leakage would enter the power chamber and pressure would increase behind the concrete walls.

Project Description

Basic layout of the Balsam Meadow Hydroelectric Project is sketched in Figure 9. The project diverts water from an existing tunnel, constructed in the 1920s as a reservoir overflow, into the Balsam Meadow forebay by way of a new 5,900 ft long diversion tunnel. The forebay is formed by a 1,300 ft long, 122 ft high zoned rockfill dam with a concrete face. The reservoir has approximately 2,000 acre-ft of usable storage.

From the forebay, the water enters the 2,800 ft long, 18 ft diameter, upper power tunnel, then drops 1,000 ft in a 13.5 ft diameter vertical power shaft to a 1,300 ft long lower power tunnel. From there the water enters the powerhouse, passes through the 200 MW turbine, and finally discharges into the existing Shaver Lake via the approximately 7,500 ft long, 18 ft diameter tailrace tunnel. The tunnels were constructed by conventional drill and blast method and were designed to be unlined, excluding a 330 ft steel lined section leading to the powerhouse chamber. A 13.5 ft diameter raise bore drilling machine was employed to construct the 1,000 ft vertical power shaft.

Site Investigations

Surface Mapping

The Balsam Meadow project is located between Huntington and Shaver Lakes in the central Sierra Nevada. Surface elevations in the project area range between 5,500 and 7,000 ft.

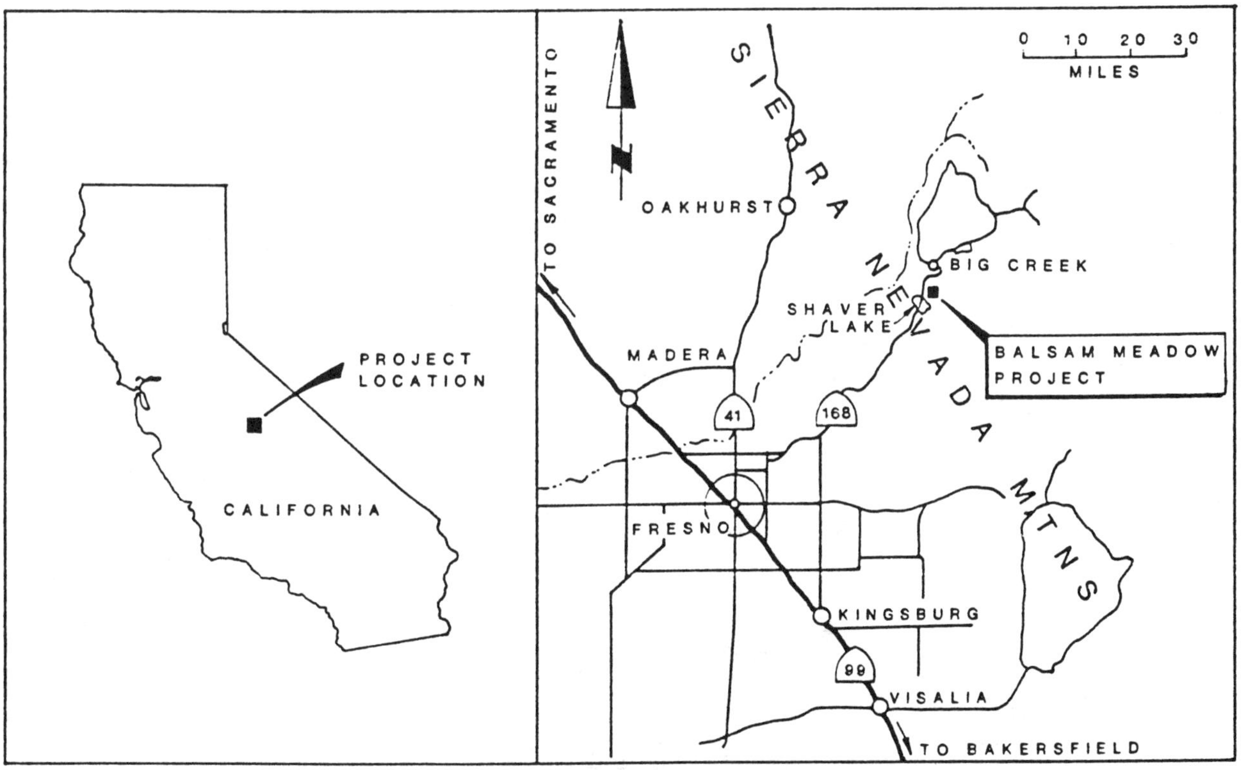

Figure 8. Balsam Meadow project

The region is underlain primarily by Mesozoic intrusive rocks of the composite Sierra Nevada batholith. The project is situated entirely within a single intrusive body previously mapped as the Dinkey Creek pluton by Bateman and Wones (1972) and Lockwood and Bateman (1976). Within the project area, the pluton consists of relatively coarse-grained, quartz diorite containing numerous small inclusions of diorite.

The rock mass contains joints spaced at wide, irregular intervals. Joint orientations vary from place to place, but tend to form a rectilinear pattern consisting of three sets of joints with each set oriented nearly 90° to the other two. The most prominent joint set trends northeast and is nearly vertical. Less prominent sets include a steeply inclined set trending northwestward and a gently inclined or subhorizontal set.

The exposed quartz diorite of the Dinkey Creek pluton's present surface expression was created by deep erosion during the Cenozoic facilitated by near-surface weathering. The presence of large natural exposures of hard, unweathered rock in many places within the project area, and the common occurrence of unweathered rock beneath weathered material observed in road cuts, suggests that the base of weathering zone is shallow in most places.

Case histories from tunneling projects in the Sierra batholith indicate that photolineations are useful for the evaluation of major discontinuities. The lineations are usually controlled by some geologic feature such as a sheared or deeply weathered zone or regional joint patterns. Photolineation maps produced for the project show a relatively consistent pattern. The most prominent orientation is N60^{o}E and corresponds to the principal joint set in the project region. No surface faults were mapped in the project area, therefore, the lineations are interpreted as surface manifestations of the joint systems.

Subsurface Investigations

The subsurface exploration program was designed to verify rock conditions at the dam, portals, powerhouse, shafts, and along zones of potential weakness discovered in the geologic mapping. Over 12,400 ft of core was drilled. Five deep core holes, each exceeding 1,100 ft in lengths, were drilled in the powerhouse area (Fig. 10).

Three core holes, each greater than 350 ft long, were drilled along the access and tailrace tunnels. Core from these holes indicated excellent quality rock with average joint spacing on the order of 10 to 15 ft and core lengths up to 200 ft without

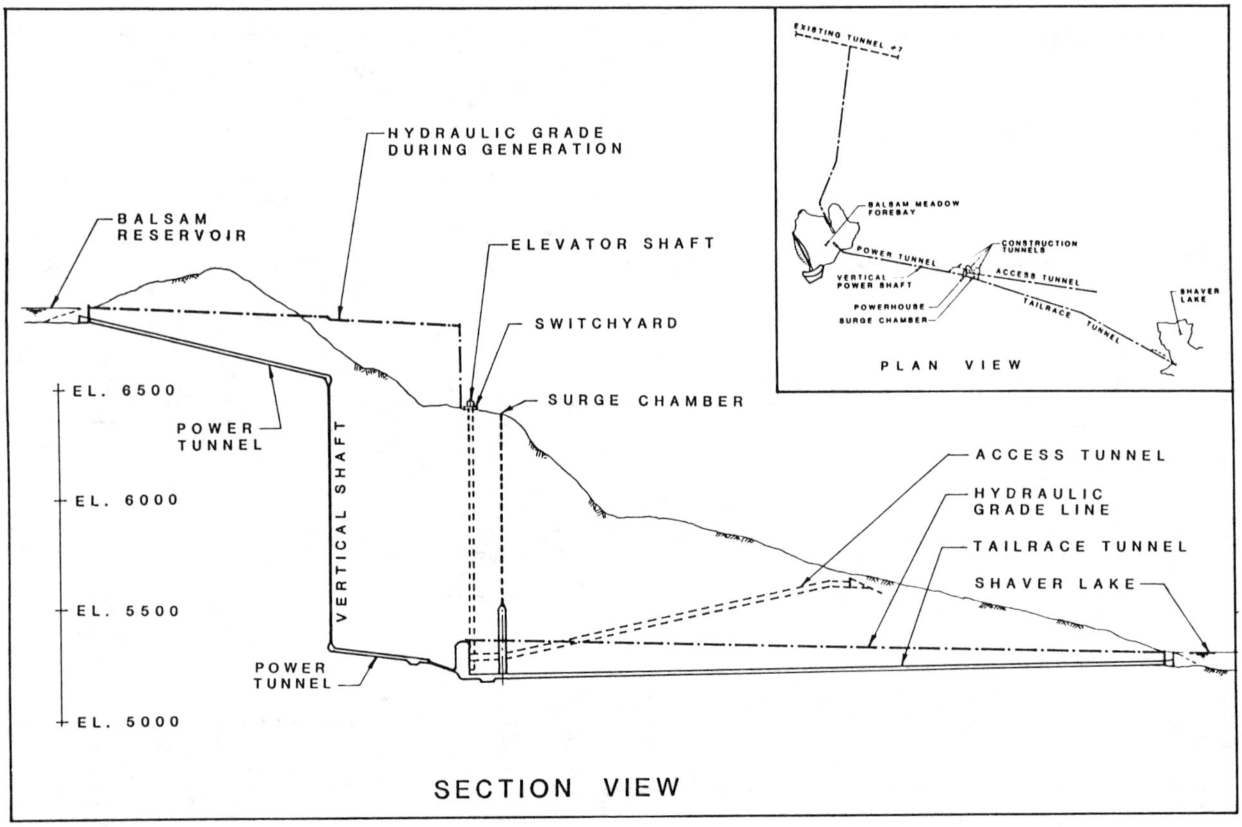

Figure 9. Plan and section of Balsam Meadow project

joints. A few open joints were encountered, however, allowing either total drill water loss or high water loss during water pressure tests. Although these joints were some tens of feet apart, their presence indicated the possibility of water flow into underground openings. The most open condition encountered was a near vertical joint at the site of the surge chamber (core hole SC-1).

Rock tests on core samples indicated an average uniaxial compressive strength of 145 MPa (21,000 psi) and tensile strength of 12 MPa (1,720 psi). The elastic modulus was on the order of 62 GPa (9.0 x 10^6 psi).

In-situ stress measurements were performed in the vertical core hole, PH-1, drilled into the powerhouse area. Based on the stress measurements, the position and stability of the powerhouse chamber was evaluated and the need for steel liner within the lower power tunnel was determined.

Ground Water

The regional hydrologic system in the central Sierra Nevada reveals surface features such as lakes, streams, meadows, canyons, and large gullies which are recognized long-term sources of water infiltration (ground water recharge) derived from snow melt. Physiographic features such as these, which would be considered as likely sources of ground water inflow toward the Project area, are few. However, the majority of the natural water courses and impoundments are crossed by linear, structurally controlled surface expressions, representing near-vertical fracture zones (photolineations). These zones of relatively high permeability provide the likely conduits or preferred paths responsible for the lateral transmission and downward percolation of meteoric waters entering the bedrock mass. Circulating drill-hole water was lost partially or entirely on several occasions in a number of deep holes indicating the presence of open joints.

Construction

Water Inflow

In March 1984, the first significant ground water inflow noted during Project construction was encountered in the access tunnel at Sta. 19+30. The water emanated from northeast-trending joints and reached a maximum flow of 150 gpm. Approximately 13 acre-ft of water were drained from the rock mass in 30 days. These joints, believed to be

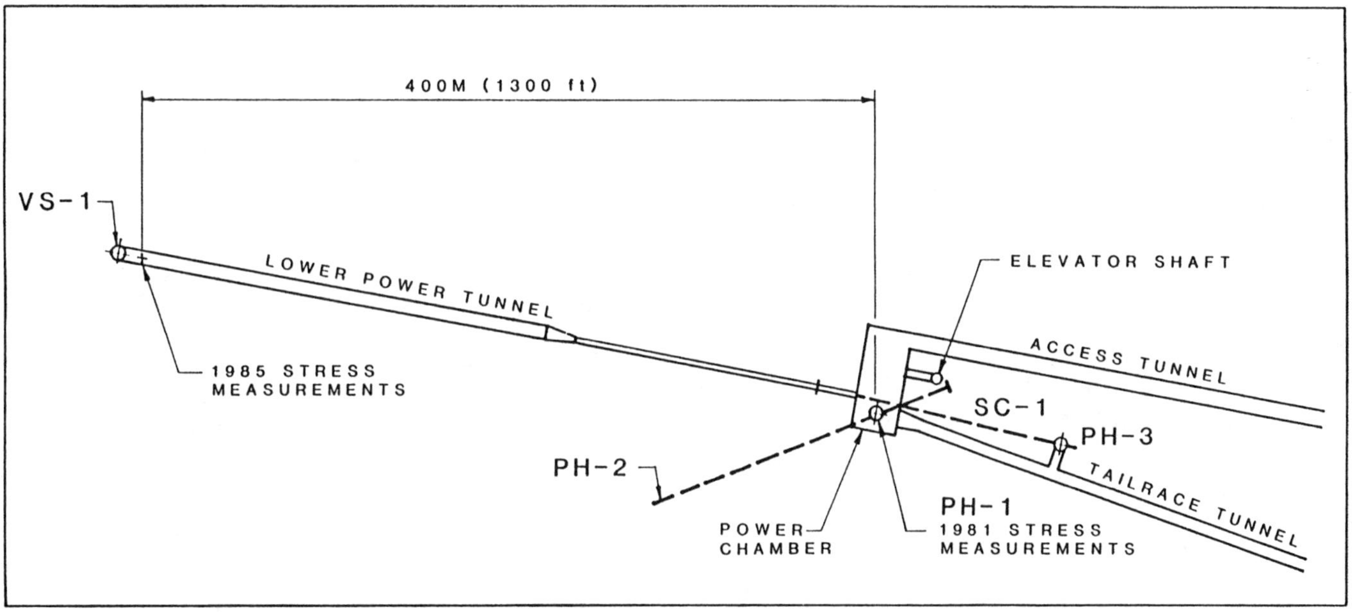

Figure 10. Subsurface Investigations. PH-1, SC-1 and VS-1 are vertical core holes at the powerhouse, surge chamber, and vertical powershaft locations. PH-2 and 3 are angle core holes over the powerhouse along alignment shown by dashed line.

related to surface photolineations, continued to yield about 6 gpm in December 1987.

In May 1984, a major inflow was tapped in the access tunnel near Sta. 42+53 (Fig. 11). A series of near-vertical, N60°E joints and subhorizontal joints was encountered with inflows of up to 1,700 gpm. As the tunneling advanced northerly, additional water was encountered in the construction tunnels along the N60°E-trending joints. Water was further encountered at the intersection of CT 4 and the tailrace tunnel, a location which coincided with the lowest elevation of the underground atmospheric facilities. Within a short time (few days), all major underground water-producing joints ceased to flow in deference to flow concentrated into this "low point."

As the various tunnels were advanced, it became evident that the water-bearing N60oE vertical joint system encountered in the power chamber area, was persistent, predictable and interconnected by subhorizontal joints from which water inflows were also observed. It was anticipated that other unseen subhorizontal joints might occur at lower depths.

Geology of the Lower Power Tunnel

Numerous prominent joints were encountered during the excavation of the lower power tunnel (LPT). From its intersection with CT 5 northward along the LPT, the major joints are near-vertical about N60°E. These consist of 5 to 10 closely spaced joints with decomposed granitic-type filling, which were wet when exposed. One of these joints was approximately 2 ft wide with filling of layered sand, silt, and clay (Fig. 12). We believe these layered granular infillings were water-transported and deposited in much the same manner as are sediments in a small stream channel. Joint zones within decomposed granitic rock occur over an interval of 200 ft northerly from the CT 5 intersection. The next 250 ft section revealed fewer individual joints. Where present, the joints appeared generally tight. Where open, they had minor mineral fillings, showed slight hydrothermal alteration, and had sparse wet spots.

The first extremely open joint zone, without individual joints containing filling material, was encountered 450 ft north of the CT 5 intersection. This zone was about 20 ft wide and included at least 30 individual joints striking roughly N80°E with steep dips in both NW and SE directions (Fig. 13). Many of the joints were open, with apertures up to 5 inches in width. Coatings of calcite crystals occurred along most of the open joints.

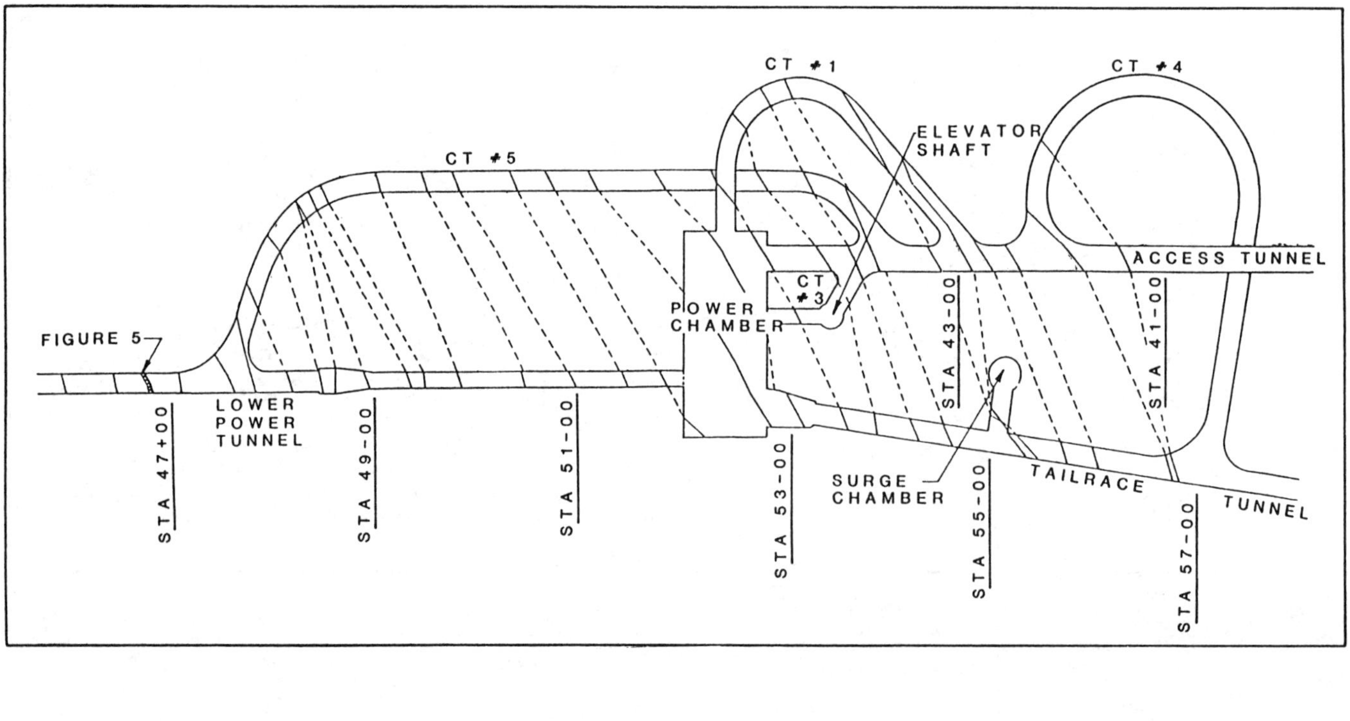

Figure 11 Plan view of lower portion of project: acess, tailrace, lower power, construction tunnels and powerhouse. Dashed lines indicate near vertical fracture system.

Along the remaining 350 feet interval, several additional open joints, with individual joint-apertures from 1 to 5 inches wide, were encountered in zones of up to six joints per zone. The open joints are oriented N70°E to E-W, 80°NW to 80°SE.

Open Fracture System

The appearance of the wall rock exposed during the construction of the lower power tunnel described above with its inherent integrity so manifestly disrupted by the numerous open joints, suggests the surrounding rock had undergone intense localized dilatent deformation. This occurrence, causing disruption of the rock mass at such extreme depth and presumably under very high original confining pressure, appears to be singularly unique in this district of the Sierra Nevada and was indeed unanticipated (Fig. 14).

During construction, several unusual physical phenomena occurred in association with tunnel excavation through the rock containing the open fracture system, which demonstrated the interconnected nature of the joint network. Examples were: (1) when an open joint was encountered during the drilling of tunnel-excavation shot holes, the holes aspirated as though a vacuum had been tapped, (2) a core hole drilled into a heading of the lower power tunnel was used in hydrofracture testing of the local rock, crossed an open joint which expelled air into the drill hole, (3) during blasting in the tailrace tunnel, shock waves, from detonations there, caused compressed air and associated dust to expel from an open joint in the lower power tunnel, some 2,850 ft distant, (4) in verifying the interconnected nature of the open joints, uranine dye was introduced into an open joint in order to trace its travel path and velocity. Water containing the dye was injected at a rate of about 100 gpm for several hours into an open joint crossing the invert of the lower power tunnel at Sta. 41+40 about 250 ft from the lower elbow.

The joint network readily accepted the quantity injected and the dye was observed 4 1/2 hours later flowing from a piezometer hole located in the tailrace tunnel some 1,400 ft horizontally distant, and 100 ft lower in elevation, than the point of injection. These data yield a minimum travel rate of 330 ft/hr, 5.5 ft/min., thus indicating extraordinarily open intercommunication within the joint systems.

Hydrofracturing Stress Measurements

The open joints encountered in the lower power tunnel would normally not be expected to occur in a stress regime, such as that calculated from in-situ stress measurements made during the geologic exploration. Accordingly, additional stress measurements were later performed to verify the conclusions reached in the original exploratory-phase tests to determine if a different stress field might exist in the rock mass near the lower elbow of the power shaft.

Figure 12. Lower power tunnel. Layered sand, silt and clay in a filled joint open 2 feet. Over 1200 ft. of cover at this location. Position shown in Fig 9.

Figure 13. Lower power tunnel. Fracture zone 20 ft wide composed of at least 30 joints with apertures up to 5 inches.

These tests, described by Hamilton and Korbin (1987), demonstrated the in-situ minimum rock stress to be considerably lower than that concluded from the earlier tests - in the order of 2.07 MPa (300 psi), rather that 4.83 MPa (700 psi). The operating hydraulic pressure would be in the order of 4.14 MPa (600 psi) - a pressure which, based upon these findings, would severely tax the strength of the rock.

Tunnel Lining

Because of: (1) the presence of unpredicted open joints, (2) low minimum in-situ rock stress measured locally, and (3) high design hydraulic operating pressures, it was concluded that to assure integrity of the rock mass and prevent outflow of water under high pressure into rock joints surrounding the lower power tunnel, significant protective measures would be required. Accordingly, a wide variety of standard and innovate grout applications and lining alternatives were studied. It was concluded that the installation of steel lining for the entire length of the 1,300 ft long power tunnel and the lower 400 ft of the vertical power shaft, would most economically provide the desired safeguards.

The portion of the tailrace tunnel between the draft tube and the CT 4 was within the major water-producing fracture system. Concrete lining was prescribed for this reach, including the orifice tunnel and surge chamber, to prevent water under operating pressure from entering the open joint system transecting the area.

Figure 14. Lower power tunnel. Open joint with aperture of 5 inches. White band is an aplite vein with apparent offset. Over 1400 feet of cover in this location.

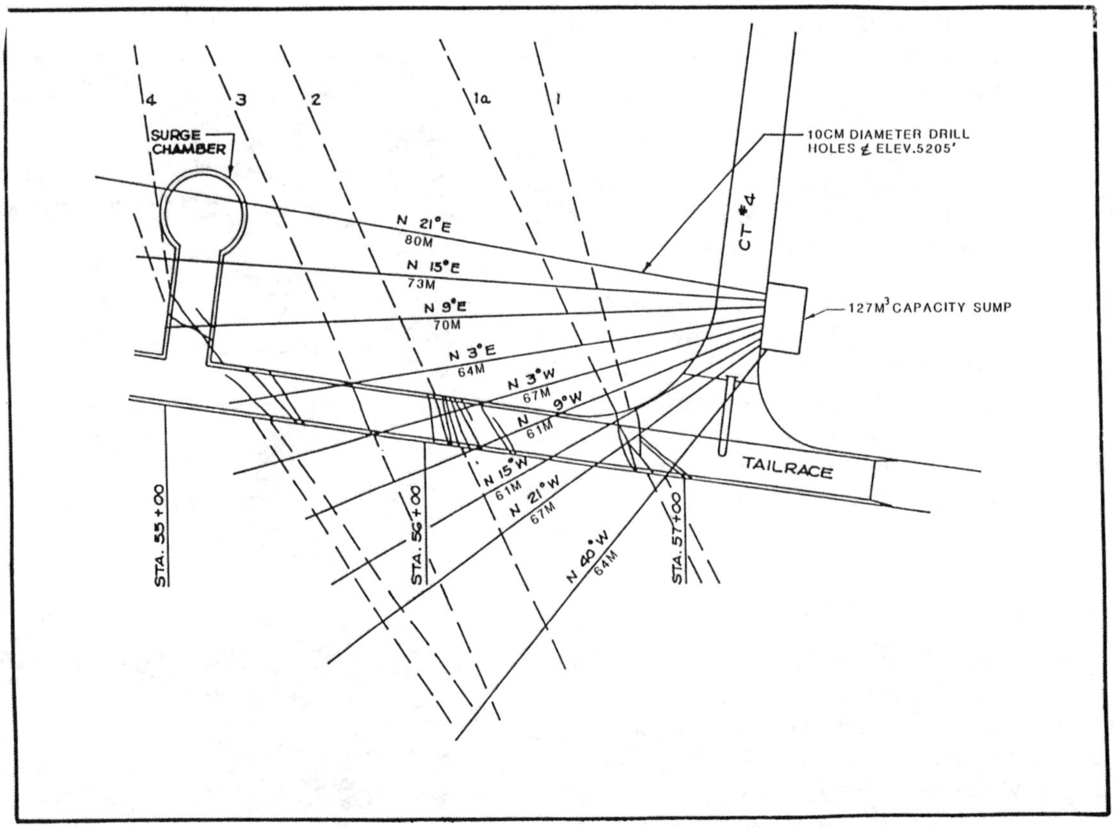

Figure 15. Water Control Plan. CT#4 sump with horizontal drain holes. Dashed lines indicate near vertical fracture system.

Water Control

Hydraulic Study and Model

A hydrogeologic study was made to estimate: (1) the rate and duration of ground water inflow during underground construction, and (2) the impact of tunnel pressurization under operational conditions on water flow to and from the tunnel complex.

This investigation relied heavily on chemical and isotopic interpretations to determine the sources of ground water entering the tunnels. The geochemical interpretations, combined with the existing conventional data on the geology and hydrogeology of the area, resulted in a conceptual model of the hydrogeologic conditions controlling seepage to the tunnels. The data and interpretations of the study are consistent with a model representing a ground water flow system contributing seepage to the tunnel complex from two immediate source zones derived from a common area of origin, as discussed below.

The hydrogeologic data indicate that an extensive recharge system exists over a broad region to the north and northeast of the Project area. Oxygen isotope ratios suggests the source(s) of infiltration water encountered in the tunnels originates from elevations above 7,000 ft. Surface water chemistry in the Project area is similar to that for average snow melt for the Sierra Nevada. However, the chemistry of the lower tunnel inflow samples was distinctly different from that of average snow melt.

Chemical analyses indicated there are two different sources providing seepage water into the tunnel complex. The first water (termed "A" water) is contained in the joint zones reflected on the surface by the photolineations. This water was encountered in the access tunnel near Sta. 20+00 and again in the tailrace tunnel near Sta. 80+00. It shows a short residency time within the rock as indicated by its low salt content and low temperature of 40°F. Low tritium content implies that there is no rapid recharge to the tunnels from surface streams or through fracture zones connecting directly to the ground surface. The low tritium content also indicates the "A" water has an average age of greater than 40 years.

The second water (termed "B" water) is that encountered flowing from the N60°E joint complex near the power chamber. This water is characterized by a high salt content and an elevated temperature of 64°F indicating prolonged water-to-rock contact or residency. The tritium content of the "B" water is also low. The concept of long residence time required to achieve both chloride leaching and temperature increase, was supported by radio-carbon dating, indicating an age greater than 3,200 years. This latter water is identified as typical of the deep groundwaters encountered during tunnel excavation near Sta. 42+00 in the access tunnel and later in the tailrace tunnel between Sta. 55+00 and Sta. 57+00. Tunneling operations activated this groundwater system allowing several thousand acre-feet of water to drain from subsurface storage since June 1984. The flow rate demonstrated a steady decline from a maximum of approximately 1,700 gpm to an excess of 800 gpm as of December 1989. There appears to be neither mixing nor physical connection between these two waters within the Project area. Results of several samples obtained during the three-year monitoring period revealed no chemistry or temperature changes of the two waters.

Water Control Plan

The conceptual hydrogeologic model indicated that the inflow through the interconnected joint system drains to the lowest atmospheric level in the underground system. Accordingly, a 34,000-gallon capacity sump was excavated 22 ft long by 12 ft wide by 17 ft deep at the lowest elevation of the underground system in CT 4, near its intersection with the tailrace tunnel. One-half meter above the sump invert, nine 10-cm diameter horizontal drain holes were drilled in a fan-shape pattern into the rock mass to intersect the water-bearing joints. As shown on Figure 15, the holes ranged between 200 and 260 ft in length. After completion of sump construction, shut-off valves were installed in the drain pipes for flow control and were importantly used for both rock-joint dewatering and pressure testing during the tailrace tunnel lining installation and subsequent grouting operations.

It is obvious that the CT 4 sump system has successfully contained and controlled the groundwater inflow. From observations of water recovery in joints and piezometers made during test periods when the sump pumps were shut down, it is also evident that in the absence of the CT 4 sump, leakage would enter the power chamber and pressure would increase behind the concrete walls.

References

Bateman, P. C. and Wones, D. R., 1972, Geologic Map of Huntington Lake Quadrangle, Central Sierra Nevada, California: U. S. Geol. Survey, Map GQ-987.

Hamilton, P. and Korbin, G., 1987, Importance of multiple stress measurements in design, Balsam Meadow Hydroelectric Project: RETC Proceedings, AIME/ASCE, NY, vol. 2, p. 850-866.

Lockwood, J. and Bateman, P. C., 1976, Geologic Map of Shaver Lake Quadrangle, Central Sierra Nevada, California: U. S. Geol. Survey, Map GQ-1271.